System Identification

IEEE Press
445 Hoes Lane
Piscataway, NJ 08854

IEEE Press Editorial Board
Lajos Hanzo, *Editor in Chief*

R. Abhari	M. El-Hawary	O. P. Malik
J. Anderson	B-M. Haemmerli	S. Nahavandi
G. W. Arnold	M. Lanzerotti	T. Samad
F. Canavero	D. Jacobson	G. Zobrist

Kenneth Moore, *Director of IEEE Book and Information Services (BIS)*

Technical Reviewers

Keith Godfrey
University of Warwick
Coventry, UK

Jérôme Antoni
Vibrations and Acoustic Laboratory
INSA of Lyon
Villeurbanne Cedex, France

System Identification
A Frequency Domain Approach

Second Edition

Rik Pintelon
Johan Schoukens

MATLAB® examples

◆IEEE
IEEE PRESS

WILEY

A JOHN WILEY & SONS, INC., PUBLICATION

Copyright © 2012 by the Institute of Electrical and Electronics Engineers. All rights reserved.

Published by John Wiley & Sons, Inc., Hoboken, New Jersey.

Published simultaneously in Canada.

MATLAB and Simulink are registered trademarks of The MathWorks, Inc. See www.mathworks.com/trademarks for a list of additional trademarks. **The MathWorks Publisher Logo identifies books that contain MATLAB® content. Used with permission. The MathWorks does not warrant the accuracy of the text or exercises in this book or in the software downloadable from** http://www.wiley.com/WileyCDA/WileyTitle/productCd-047064477X.html **and** http://www.mathworks.com/matlabcentral/fileexchange/?term=authorid%3A80973. **The book's or downloadable software's use or discussion of MATLAB® software or related products does not constitute endorsement or sponsorship by The MathWorks of a particular use of the MATLAB® software or related products.**

For MATLAB® and Simulink® product information, or information on other related products, please contact:

The MathWorks, Inc.
3 Apple Hill Drive
Natick, MA 01760-2098 USA
Tel: 508-647-7000
Fax: 508-647-7001
E-mail: info@mathworks.com
Web: www.mathworks.com

No part of this publication may be reproduced, stored in a retrieval system, or transmitted in any form or by any means, electronic, mechanical, photocopying, recording, scanning, or otherwise, except as permitted under Section 107 or 108 of the 1976 United States Copyright Act, without either the prior written permission of the Publisher, or authorization through payment of the appropriate per-copy fee to the Copyright Clearance Center, Inc., 222 Rosewood Drive, Danvers, MA 01923, (978) 750-8400, fax (978) 750-4470, or on the web at www.copyright.com. Requests to the Publisher for permission should be addressed to the Permissions Department, John Wiley & Sons, Inc., 111 River Street, Hoboken, NJ 07030, (201) 748-6011, fax (201) 748-6008, or online at http://www.wiley.com/go/permission.

Limit of Liability/Disclaimer of Warranty: While the publisher and author have used their best efforts in preparing this book, they make no representations or warranties with respect to the accuracy or completeness of the contents of this book and specifically disclaim any implied warranties of merchantability or fitness for a particular purpose. No warranty may be created or extended by sales representatives or written sales materials. The advice and strategies contained herein may not be suitable for your situation. You should consult with a professional where appropriate. Neither the publisher nor author shall be liable for any loss of profit or any other commercial damages, including but not limited to special, incidental, consequential, or other damages.

For general information on our other products and services or for technical support, please contact our Customer Care Department within the United States at (800) 762-2974, outside the United States at (317) 572-3993 or fax (317) 572-4002.

Wiley also publishes its books in a variety of electronic formats. Some content that appears in print may not be available in electronic formats. For more information about Wiley products, visit our web site at www.wiley.com.

Library of Congress Cataloging-in-Publication Data:

Pintelon, R. (Rik)
  System identification : a frequency domain approach / Rik Pintelon, Johan Schoukens. — 2nd ed.
    p. cm.
  ISBN 978-0-470-64037-1 (cloth)
1. System identification. I. Schoukens, J. (Johan) II. Title.
  QA402.P56 2012
  003'.1—dc23   2011044212

To my grandchildren, Lamine, Aminata, Alia-Rik, Malick, Khadidiatou, Babacar, Ibrahima, Khadidiatou and those to be born

Rik

To Annick, Maarten, Sanne, and Ine

Johan

Contents

Preface to the First Edition xxv

Preface to the Second Edition xxix

Acknowledgments xxxiii

List of Operators and Notational Conventions xxxv

List of Symbols xxxix

List of Abbreviations xliii

CHAPTER 1 An Introduction to Identification 1
- 1.1 What Is Identification? 1
- 1.2 Identification: A Simple Example 2
  - 1.2.1 Estimation of the Value of a Resistor 2
  - 1.2.2 Simplified Analysis of the Estimators 6
  - 1.2.3 Interpretation of the Estimators: A Cost Function–Based Approach 11
- 1.3 Description of the Stochastic Behavior of Estimators 12
  - 1.3.1 Location Properties: Unbiased and Consistent Estimates 13
  - 1.3.2 Dispersion Properties: Efficient Estimators 14
- 1.4 Basic Steps in the Identification Process 17
  - 1.4.1 Collect Information about the System 17
  - 1.4.2 Select a Model Structure to Represent the System 17
  - 1.4.3 Match the Selected Model Structure to the Measurements 19
  - 1.4.4 Validate the Selected Model 19
  - 1.4.5 Conclusion 19
- 1.5 A Statistical Approach to the Estimation Problem 19
  - 1.5.1 Least Squares Estimation 20

1.5.2 Weighted Least Squares Estimation 22
1.5.3 The Maximum Likelihood Estimator 23
1.5.4 The Bayes Estimator 25
1.5.5 Instrumental Variables 27
1.6 Exercises 29

CHAPTER 2 Measurement of Frequency Response Functions – Standard Solutions 33

2.1 Introduction 33
2.2 An Introduction to the Discrete Fourier Transform 34
    2.2.1 The Sampling Process 34
    2.2.2 The Discrete Fourier Transform (DFT-FFT) 35
    2.2.3 DFT Properties of Periodic Signals 40
    2.2.4 DFT of Burst Signals 42
    2.2.5 Conclusion 43
2.3 Spectral Representations of Periodic Signals 43
2.4 Analysis of FRF Measurements Using Periodic Excitations 44
    2.4.1 Measurement Setup 44
    2.4.2 Error Analysis 45
2.5 Reducing FRF Measurement Errors for Periodic Excitations 49
    2.5.1 Basic Principles 49
    2.5.2 Processing Repeated Measurements 51
    2.5.3 Improved Averaging Methods for Nonsynchronized Measurements 52
    2.5.4 Coherence 53
2.6 FRF Measurements Using Random Excitations 54
    2.6.1 Basic Principles 54
    2.6.2 Reducing the Noise Influence 54
    2.6.3 Leakage Errors 59
    2.6.4 Indirect FRF Measurements 61
    2.6.5 Improved FRF Measurements Using Overlapping Segments 62
2.7 FRF Measurements of Multiple-Input, Multiple-Output Systems 64
    2.7.1 One Experiment 64
    2.7.2 Multiple Experiments 65
    2.7.3 Discussion 67
2.8 Guidelines for FRF Measurements 68
    2.8.1 Guideline 1: Use Periodic Excitations 68
    2.8.2 Guideline 2: Select the Best FRF Estimator 68
    2.8.3 Guideline 3: Pretreatment of Data 69
2.9 Conclusion 69
2.10 Exercises 69
2.11 Appendixes 70
    Appendix 2.A Radius of a Circular Confidence Region of the FRF 70
    Appendix 2.B Asymptotic Behavior of Averaging Techniques 71
    Appendix 2.C Covariance of the FRM Measurement 72

CHAPTER 3 Frequency Response Function Measurements in the Presence of Nonlinear Distortions 73

3.1 Introduction 73
3.2 Intuitive Understanding of the Behavior of Nonlinear Systems 74
3.3 A Formal Framework to Describe Nonlinear Distortions 75

- 3.3.1 Class of Excitation Signals 76
- 3.3.2 Selection of a Model Structure for the Nonlinear System 77
- 3.4 Study of the Properties of FRF Measurements in the Presence of Nonlinear Distortions 78
  - 3.4.1 Study of the Expected Value of the FRF for a Constant Number of Harmonics 80
  - 3.4.2 Asymptotic Behavior of the FRF if the Number of Harmonics Tends to Infinity 81
  - 3.4.3 Further Comments on the Best Linear Approximation 83
  - 3.4.4 Further Comments on the Output Stochastic Nonlinear Contributions 86
- 3.5 Extension to Discrete-Time Modeling 88
  - 3.5.1 Periodic Signals 88
  - 3.5.2 Random Signals 89
- 3.6 Experimental Illustration 90
  - 3.6.1 Visualization of the Nonlinearity Using Stepped Sine Measurements 90
  - 3.6.2 Measurement of the Best Linear Approximation 91
- 3.7 Multivariable Systems 92
- 3.8 Best Linear Approximation of a System Operating in Closed Loop 93
- 3.9 Conclusion 95
- 3.10 Exercises 95
- 3.11 Appendixes 96
  - Appendix 3.A Bias and Stochastic Contributions of the Nonlinear Distortions 96
  - Appendix 3.B Study of the Moments of the Stochastic Nonlinear Contributions 97
  - Appendix 3.C Mixing Property of the Stochastic Nonlinear Contributions 101
  - Appendix 3.D Structure of the Indecomposable Sets 104
  - Appendix 3.E Distribution of the Stochastic Nonlinearities 105
  - Appendix 3.F Extension to Random Amplitudes and Nonuniform Phases 109
  - Appendix 3.G Response of a Nonlinear System to a Gaussian Excitation 109
  - Appendix 3.H Proof of Theorem 3.12 111
  - Appendix 3.I Proof of Theorem 3.15 113
  - Appendix 3.J Proof of Theorem 3.16 113
  - Appendix 3.K Proof of Theorem 3.17 113
  - Appendix 3.L Proof of Theorem 3.18 114
  - Appendix 3.M Proof of Theorem 3.21 117
  - Appendix 3.N Covariance of the Multivariate BLA 117
  - Appendix 3.O Proof of Theorem 3.22 118

CHAPTER 4 Detection, Quantification, and Qualification of Nonlinear Distortions in FRF Measurements 119

- 4.1 Introduction 119
- 4.2 The Riemann Equivalence Class of Excitation Signals 120
  - 4.2.1 Definition of the Excitation Signals 121
  - 4.2.2 Definition of the Riemann Equivalences 122

4.2.3   Invariance of the Best Linear Approximation and the Variance of the Stochastic Nonlinear Distortions 124
4.2.4   Experimental Illustration 126
4.3   Detection of Nonlinear Distortions Using Random Phase Multisines 130
4.3.1   The Robust Method 130
4.3.2   The Fast Method 135
4.3.3   Discussion 139
4.3.4   Experimental Illustration 141
4.4   Guidelines for Measuring the Best Linear Approximation 146
4.5   Appendixes 147
Appendix 4.A   Proof of the Nonlinearity Detection Properties of Random Phase Multisines 147
Appendix 4.B   Influence Generator Noise on the FRF Estimate 147
Appendix 4.C   Rationale for (4-50) and (4-52) 148

CHAPTER 5   Design of Excitation Signals 151

5.1   Introduction 151
5.2   General Remarks on Excitation Signals for Nonparametric Frequency Response Measurements 152
5.2.1   Quantifying the Quality of an Excitation Signal 153
5.2.2   Stepped Sine versus Broadband Excitations 154
5.3   Study of Broadband Excitation Signals 155
5.3.1   General Purpose Signals 155
5.3.2   Optimized Test Signals 162
5.3.3   Advanced Test Signals 165
5.4   Optimization of Excitation Signals for Parametric Measurements 167
5.4.1   Introduction 167
5.4.2   Optimization of the Power Spectrum of a Signal 168
5.5   Experiment Design for Control 173
5.6   Appendix 174
Appendix 5.A   Minimizing the Crest Factor of a Multisine 174

CHAPTER 6   Models of Linear Time-Invariant Systems 177

6.1   Introduction 177
6.2   Plant Models 182
6.3   Relation Between the Input-Output DFT Spectra 184
6.3.1   Models for Periodic Signals 185
6.3.2   Models for Arbitrary Signals 185
6.3.3   Models for Records with Missing Data 188
6.3.4   Models for Concatenated Data Sets 189
6.4   Models for Damped (Complex) Exponentials 190
6.5   Identifiability 191
6.5.1   Models for Periodic Signals 191
6.5.2   Models for Arbitrary Signals 192
6.5.3   Models for Records with Missing Data 192
6.5.4   Model for Concatenated Data Sets 193
6.6   Multivariable Systems 193
6.7   Noise Models 195
6.7.1   Introduction 195

Contents   xi

    6.7.2 Nonparametric Noise Model 195
    6.7.3 Parametric Noise Model 195
 6.8 Nonlinear Systems 202
 6.9 Exercises 203
 6.10 Appendixes 204
    Appendix 6.A Stability and Minimum Phase Regions 204
    Appendix 6.B Relation between DFT Spectra and Transfer Function for Arbitrary Signals 206
    Appendix 6.C Parameterizations of the Extended Transfer Function Model 209
    Appendix 6.D Convergence Rate of the Equivalent Initial Conditions 209
    Appendix 6.E Some Integral Expressions 210
    Appendix 6.F Convergence Rate of the Residual Alias Errors 212
    Appendix 6.G Relation between DFT Spectra and Transfer Function for Arbitrary Signals with Missing Data 216
    Appendix 6.H Relationship between DFT Spectra of Concatenated Data Sets and Transfer Function 217
    Appendix 6.I Free Decay Response of a Finite-Dimensional System 217
    Appendix 6.J Relation between the Free Decay Parameters and the Partial Fraction Expansion 218
    Appendix 6.K Some Properties of Polynomials 218
    Appendix 6.L Proof of the Identifiability of Transfer Function Model (6-32) (Theorem 6.9) 219
    Appendix 6.M Proof of the Identifiability of Transfer Function Model (6-34) 220
    Appendix 6.N Rank of the Residue Matrices of Multivariable Transfer Function Models 221
    Appendix 6.O Band-Limited Observation of Continuous-Time Noise (Theorem 6.14) 221
    Appendix 6.P Correlation Noise Transient with Noise Input (Theorem 6.15) 222
    Appendix 6.Q Correlation Noise Alias Error with Noise Input (Theorem 6.16) 224
    Appendix 6.R Correlation Plant Transient and Plant Alias Error with Plant Input (Theorem 6.17) 224

CHAPTER 7 Measurement of Frequency Response Functions – The Local Polynomial Approach 225
 7.1 Introduction 225
 7.2 Arbitrary Excitations 227
    7.2.1 Problem Statement and Assumptions 227
    7.2.2 The Local Polynomial Method 228
    7.2.3 The Spectral Analysis Method 233
    7.2.4 Confidence Regions of the FRM 236
    7.2.5 Comparison of the Methods 237
    7.2.6 Bias-Variance Trade-Off 239
    7.2.7 The Noisy Input, Noisy Output Case 240
    7.2.8 Nonlinear Systems 241
    7.2.9 Concatenating Data Records 243
    7.2.10 Experimental Illustration 246
 7.3 Periodic Excitations 248
    7.3.1 Introduction 248

7.3.2 Suppression of the Noise Transient (Leakage) Errors in Periodic Signals 250
7.3.3 The Robust Method for Measuring the Frequency Response Matrix 252
7.3.4 The Fast Method for Measuring the Frequency Response Matrix 253
7.3.5 Nonlinear Systems 254
7.3.6 The Robust Method for Measuring the Best Linear Approximation 254
7.3.7 The Fast Method for Measuring the Best Linear Approximation 258
7.3.8 Non-Steady State Conditions 260
7.3.9 Experimental Illustration 261
7.4 Comparison Periodic – Random Excitations 263
    7.4.1 Discussion 263
    7.4.2 Experimental Illustration 264
7.5 Guidelines for Advanced FRF Measurements 268
7.6 Appendixes 269
    Appendix 7.A Proof of Equation (7-10) 269
    Appendix 7.B Proof of Equation (7-13) 269
    Appendix 7.C System Leakage Contribution to the Bias on the FRM Estimates 270
    Appendix 7.D Proof of Equations (7-17) and (7-18) 271
    Appendix 7.E Proof of Equation (7-19) 272
    Appendix 7.F Proof of Equation (7-50) 273
    Appendix 7.G Proof of Equation (7-55) 273
    Appendix 7.H Proof of Equation (7-74) 274
    Appendix 7.I Proof of Equation (7-76) 275
    Appendix 7.J Zero Pattern of the Inverse of a Matrix 276
    Appendix 7.K Covariance Matrix of (7-86) 276
    Appendix 7.L Expected Value of the Sample Noise and Total Covariances (7-93) 277
    Appendix 7.M The Generator Noise Does Not Contribute to the Covariance of the FRM 277
    Appendix 7.N Bias Robust FRM Estimate under Transient Conditions 278

CHAPTER 8   An Intuitive Introduction to Frequency Domain Identification 279

8.1 Intuitive approach 279
8.2 The Errors-in-Variables Formulation 280
8.3 Generating Starting Values 282
8.4 Comparison with the "Classical" Time Domain Identification Framework 283
8.5 Extensions of the Model: Dealing with Unknown Delays and Transients 284

CHAPTER 9   Estimation with Known Noise Model 285

9.1 Introduction 285
9.2 Frequency Domain Data 286
9.3 Plant Model 288

Contents

- 9.4 Estimation Algorithms 289
- 9.5 Quick Tools to Analyze Estimators 291
- 9.6 Assumptions 293
  - 9.6.1 Stochastic Convergence 293
  - 9.6.2 Stochastic Convergence Rate 295
  - 9.6.3 Systematic and Stochastic Errors 295
  - 9.6.4 Asymptotic Normality 295
  - 9.6.5 Deterministic Convergence 296
  - 9.6.6 Consistency 297
  - 9.6.7 Asymptotic Bias 297
  - 9.6.8 Asymptotic Efficiency 297
- 9.7 Asymptotic Properties 298
- 9.8 Linear Least Squares 301
  - 9.8.1 Introduction 301
  - 9.8.2 Linear Least Squares 301
  - 9.8.3 Iterative Weighted Linear Least Squares 303
  - 9.8.4 A Simple Example 304
- 9.9 Nonlinear Least Squares 305
  - 9.9.1 Output Error 305
  - 9.9.2 Logarithmic Least Squares 308
  - 9.9.3 A Simple Example—Continued 310
- 9.10 Total Least Squares 310
  - 9.10.1 Introduction 310
  - 9.10.2 Total Least Squares 312
  - 9.10.3 Generalized Total Least Squares 313
- 9.11 Maximum Likelihood 314
  - 9.11.1 The Maximum Likelihood Solution 314
  - 9.11.2 Discussion 316
  - 9.11.3 Asymptotic Properties 317
  - 9.11.4 Calculation of Uncertainty Bounds 318
- 9.12 Approximate Maximum Likelihood 319
  - 9.12.1 Introduction 319
  - 9.12.2 Iterative Quadratic Maximum Likelihood 319
  - 9.12.3 Bootstrapped Total Least Squares 320
  - 9.12.4 Weighted (Total) Least Squares 321
- 9.13 Instrumental Variables 323
- 9.14 Subspace Algorithms 324
  - 9.14.1 Model Equations 324
  - 9.14.2 Subspace Identification Algorithms 325
  - 9.14.3 Stochastic Properties 329
- 9.15 Illustration and Overview of the Properties 330
  - 9.15.1 Simulation Example 1 330
  - 9.15.2 Simulation Example 2 332
  - 9.15.3 Real Measurement Examples 334
  - 9.15.4 Overview of the Properties 339
- 9.16 High-Order Systems 341
  - 9.16.1 Scalar Orthogonal Polynomials 341
  - 9.16.2 Vector Orthogonal Polynomials 342
  - 9.16.3 Application to the Estimators 343
  - 9.16.4 Notes 344
- 9.17 Systems with Time Delay 344
- 9.18 Identification in Feedback 345
- 9.19 Modeling in the Presence of Nonlinear Distortions 346

9.20 Missing Data 346
9.21 Multivariable Systems 348
9.22 Transfer Function Models with Complex Coefficients 349
9.23 Exercises 350
9.24 Appendixes 351
- Appendix 9.A A Second-Order Simulation Example 351
- Appendix 9.B Signal-to-Noise Ratio of DFT Spectra Measurements of Random Excitations 351
- Appendix 9.C Signal-to-Noise Ratio of DFT Spectra Measurements of Periodic Excitations 352
- Appendix 9.D Asymptotic Behavior Cost Function for a Time Domain Experiment 353
- Appendix 9.E Asymptotic Properties of Frequency Domain Estimators with Deterministic Weighting (Theorem 9.21) 354
- Appendix 9.F Asymptotic Properties of Frequency Domain Estimators with Stochastic Weighting (Corollary 9.22) 358
- Appendix 9.G Expected Value of an Analytic Function 358
- Appendix 9.H Total Least Squares Solution – Equivalences (Lemma 9.23) 359
- Appendix 9.I Expected Value Total Least Squares Cost Function 361
- Appendix 9.J Explicit Form of the Total Least Squares Cost Function 361
- Appendix 9.K Rank of the Column Covariance Matrix 362
- Appendix 9.L Calculation of the Gaussian Maximum Likelihood Estimate 363
- Appendix 9.M Number of Free Parameters in an Errors-in-Variables Problem 366
- Appendix 9.N Uncertainty of the BTLS Estimator in the Absence of Model Errors 367
- Appendix 9.O Asymptotic Properties of the Instrumental Variables Method 367
- Appendix 9.P Equivalences between Range Spaces 367
- Appendix 9.Q Estimation of the Range Space 368
- Appendix 9.R Subspace Algorithm for Discrete-Time Systems (Algorithm 9.24) 370
- Appendix 9.S Subspace Algorithm for Continuous-Time Systems (Algorithm 9.25) 373
- Appendix 9.T Sensitivity Estimates to Noise Model Errors 377
- Appendix 9.U IWLS Solution in Case of Vector Orthogonal Polynomials 379
- Appendix 9.V Asymptotic Properties in the Presence of Nonlinear Distortions 379
- Appendix 9.W Consistency of the Missing Data Problem 380
- Appendix 9.X Normal Equation for Complex Parameters and Analytic Residuals 381
- Appendix 9.Y Total Least Squares for Complex Parameters 382

CHAPTER 10 Estimation with Unknown Noise Model – Standard Solutions 383

10.1 Introduction 383
10.2 Discussion of the Disturbing Noise Assumptions 385
- 10.2.1 Assuming Independent Normally Distributed Noise for Time Domain Experiments 385

Contents

- 10.2.2 Considering Successive Periods as Independent Realizations 386
- 10.3 Properties of the ML Estimator Using a Sample Covariance Matrix 386
  - 10.3.1 The Sample Maximum Likelihood Estimator: Definition of the Cost Function 386
  - 10.3.2 Properties of the Sample Maximum Likelihood Estimator 387
  - 10.3.3 Discussion 389
  - 10.3.4 Estimation of Covariance Matrix of the Model Parameters 389
  - 10.3.5 Properties of the Cost Function in Its Global Minimum 389
- 10.4 Properties of the GTLS Estimator Using a Sample Covariance Matrix 390
- 10.5 Properties of the BTLS Estimator Using a Sample Covariance Matrix 392
- 10.6 Properties of the SUB Estimator Using a Sample Covariance Matrix 395
- 10.7 Identification in the Presence of Nonlinear Distortions 396
- 10.8 Illustration and Overview of the Properties 398
  - 10.8.1 Real Measurement Example 398
  - 10.8.2 Overview of the Properties 399
- 10.9 Identification of Parametric Noise Models 401
  - 10.9.1 Generalized Output Error Stochastic Framework 402
  - 10.9.2 A Frequency Domain Solution 403
  - 10.9.3 Asymptotic Properties of the Gaussian Maximum Likelihood Estimator 406
  - 10.9.4 Discussion 409
  - 10.9.5 Experimental Illustration 410
- 10.10 Identification in Feedback 411
- 10.11 Appendixes 413
  - Appendix 10.A Expected Value and Variance of the Inverse of Chi-Square Random Variable 413
  - Appendix 10.B First and Second Moments of the Ratio of the True and the Sample Variance of the Equation Error 413
  - Appendix 10.C Calculation of Some First- and Second-Order Moments 414
  - Appendix 10.D Proof of Theorem 10.3 415
  - Appendix 10.E Approximation of the Derivative of the Cost Function 416
  - Appendix 10.F Loss in Efficiency of the Sample Estimator 417
  - Appendix 10.G Mean and Variance of the Sample Cost in Its Global Minimum 418
  - Appendix 10.H Asymptotic Properties of the SGTLS Estimator (Theorem 10.6) 420
  - Appendix 10.I Relationship between the GTLS and the SGTLS Estimates (Theorem 10.7) 421
  - Appendix 10.J Asymptotic Properties of SBTLS Estimator (Theorem 10.8) 421
  - Appendix 10.K Relationship between the BTLS and the SBTLS Estimates (Theorem 10.9) 422
  - Appendix 10.L Asymptotic Properties of SSUB Algorithms (Theorem 10.10) 422
  - Appendix 10.M Best Linear Approximation of a Cascade of Nonlinear Systems 424
  - Appendix 10.N Sum of Analytic Function Values over a Uniform Grid of the Unit Circle 424

Appendix 10.O Gaussian Log-likelihood Function (Theorem 10.18) 426
Appendix 10.P Proof of Corollary 10.19 427
Appendix 10.Q Proof of Corollary 10.20 427
Appendix 10.R Proof of Theorem 10.25 427

CHAPTER 11 Model Selection and Validation 431

11.1 Introduction 431
11.2 Assessing the Model Quality: Quantifying the Stochastic Errors 432
    11.2.1 Uncertainty Bounds on the Calculated Transfer Functions 433
    11.2.2 Uncertainty Bounds on the Residuals 433
    11.2.3 Uncertainty Bounds on the Poles/Zeros 435
11.3 Avoiding Overmodeling 437
    11.3.1 Introduction: Impact of an Increasing Number of Parameters on the Uncertainty 437
    11.3.2 Balancing the Model Complexity versus the Model Variability 438
11.4 Detection of Undermodeling 441
    11.4.1 Undermodeling: A Good Idea? 441
    11.4.2 Detecting Model Errors 442
    11.4.3 Qualifying and Quantifying the Model Errors 444
    11.4.4 Illustration on a Mechanical System 448
11.5 Model Selection 449
    11.5.1 Model Structure Selection Based on Preliminary Data Processing: Initial Guess 450
    11.5.2 "Postidentification" Model Structure Updating 451
11.6 Guidelines for the User 452
11.7 Exercises 453
11.8 Appendixes 453
    Appendix 11.A Proof of Equation (11-3) 453
    Appendix 11.B Proof of Equation (11-4) 454
    Appendix 11.C Properties of the Global Minimum of the Maximum Likelihood Cost Function (Theorem 11.2) 455
    Appendix 11.D Calculation of Improved Uncertainty Bounds for the Estimated Poles and Zeros 455
    Appendix 11.E Sample Correlation at Lags Different from Zero (Proof of Theorem 11.5) 457
    Appendix 11.F Sample Correlation at Lag Zero (Proof of Theorem 11.5) 459
    Appendix 11.G Variance of the Sample Correlation (Proof of Theorem 11.5) 459
    Appendix 11.H Study of the Sample Correlation at Lag One (Proof of Theorem 11.7) 460
    Appendix 11.I Expected Value Sample Correlation 461
    Appendix 11.J Standard Deviation Sample Correlation 461

CHAPTER 12 Estimation with Unknown Noise Model – The Local Polynomial Approach 463

12.1 Introduction 463
12.2 Generalized Sample Mean and Sample Covariance 464
    12.2.1 Arbitrary Excitations 465
    12.2.2 Periodic Excitations 468

Contents   xvii

- 12.2.3 Choice Frequency Width of the Local Polynomial Approach 469
- 12.2.4 Overview of the Properties 470
- 12.3 Sample Maximum Likelihood Estimator 470
  - 12.3.1 Sample Maximum Likelihood Cost Function 470
  - 12.3.2 Asymptotic Properties 472
  - 12.3.3 Computational Issues 474
  - 12.3.4 Calculation of the Asymptotic Covariance Matrix 475
  - 12.3.5 Generation of Starting Values 476
  - 12.3.6 Model Selection and Validation 477
- 12.4 Identification in the Presence of Nonlinear Distortions 479
- 12.5 Experimental Illustration 480
- 12.6 Guidelines for Parametric Transfer Function Modeling 483
- 12.7 Appendices 484
  - Appendix 12.A Proof of Lemma 12.1 484
  - Appendix 12.B Proof of Lemma 12.2 485
  - Appendix 12.C Proof of Lemma 12.3 486
  - Appendix 12.D Proof of Lemma 12.4 486
  - Appendix 12.E Proof of Equation (12-7) 487
  - Appendix 12.F Proof of Lemma 12.5 487
  - Appendix 12.G Proof of Theorem 12.7 488
  - Appendix 12.H Proof of Theorem 12.8 489
  - Appendix 12.I Proof of the Pseudo-Jacobian (12-25) 491
  - Appendix 12.J Proof of (12-30) 492
  - Appendix 12.K Proof of Theorem 12.9 493
  - Appendix 12.L Properties Generalized Sample Means and Sample Covariances in the Presence of Nonlinear Distortions 494
  - Appendix 12.M Covariance Model Parameters and Variance SML Cost Function in the Presence of Nonlinear Distortions 495
  - Appendix 12.N Linear Plant and Nonlinear Actuator or Controller 496

CHAPTER 13 Basic Choices in System Identification 497

- 13.1 Introduction 497
- 13.2 Intersample Assumptions: Facts 498
  - 13.2.1 Formal Description of the Zero-Order-Hold and Band-Limited Assumptions 498
  - 13.2.2 Relation between the Intersample Behavior and the Model 500
  - 13.2.3 Mixing the Intersample Behavior and the Model 502
  - 13.2.4 Experimental Illustration 504
- 13.3 The Intersample Assumption: Appreciation of the Facts 509
  - 13.3.1 Intended Use of the Model 509
  - 13.3.2 Impact of the Intersample Assumption on the Setup 511
  - 13.3.3 Impact of the Intersample Behavior Assumption on the Identification Methods 512
- 13.4 Nonparametric Noise Models: Facts 513
- 13.5 Nonparametric Noise Models: Detailed Discussion and Appreciation of the Facts 513
  - 13.5.1 The Quality of the Noise Model 513
  - 13.5.2 Improved/Simplified Model Validation 514
  - 13.5.3 Simplified Model Selection/Minimization Cost Function 514
  - 13.5.4 Errors-in-Variables Identification and Identification in Feedback 515
  - 13.5.5 Increased Uncertainty of the Plant Model 515
  - 13.5.6 Not Suitable for Output Data Only 516

- 13.6 Periodic Excitations: Facts 516
- 13.7 Periodic Excitations: Detailed Discussion and Appreciation of the Facts 516
  - 13.7.1 Data Reduction Linked to an Improved Signal-to-Noise Ratio of the Raw Data 516
  - 13.7.2 Elimination of Nonexcited Frequencies 517
  - 13.7.3 Improved Frequency Response Function Measurements 518
  - 13.7.4 Detection, Qualification, and Quantification of Nonlinear Distortions 519
  - 13.7.5 Detection and Removal of Trends 520
  - 13.7.6 Reduced Frequency Resolution 520
  - 13.7.7 Increased Uncertainty if the Nonlinear Distortions Dominate over the Noise 520
- 13.8 Periodic versus Random Excitations: User Aspects 520
  - 13.8.1 Design Aspects: Required User Interaction 520
- 13.9 Time and Frequency Domain Identification 522
- 13.10 Time and Frequency Domain Identification: Equivalences 522
  - 13.10.1 Initial Conditions: Transient versus Leakage Errors 522
  - 13.10.2 Windowing in the Frequency Domain, (Noncausal) Filtering in the Time Domain 523
  - 13.10.3 Cost Function Interpretation 524
- 13.11 Time and Frequency Domain Identification: Differences 525
  - 13.11.1 Choice of the Model 525
  - 13.11.2 Unstable Plants 525
  - 13.11.3 Noise Models: Parametric or Nonparametric 526
  - 13.11.4 Extended Frequency Range: Combination of Different Experiments 526
  - 13.11.5 The Errors-in-Variables Problem 527
- 13.12 Imposing Constraints on the Identified Model 528
- 13.13 Conclusions 529
- 13.14 Exercises 530

CHAPTER 14 Guidelines for the User 531

- 14.1 Introduction 531
- 14.2 Selection of an Identification Scheme 531
  - 14.2.1 Questions – Proposed Solutions 531
- 14.3 Identification Step-by-Step 533
  - 14.3.1 Check and Selection of the Experimental Setup 533
  - 14.3.2 Design of an Experiment 534
  - 14.3.3 Choice Noise Model 536
  - 14.3.4 Preprocessing 536
  - 14.3.5 Identification 539
- 14.4 Validation 542
- 14.5 Conclusion 543
- 14.6 Appendixes 544
  - Appendix 14.A Independent Experiments 544
  - Appendix 14.B Relationship between Averaged DFT Spectra and Transfer Function for Arbitrary Excitations 544

CHAPTER 15 Some Linear Algebra Fundamentals 545

- 15.1 Notations and Definitions 545

Contents

- 15.2 Operators and Functions 546
- 15.3 Norms 547
- 15.4 Decompositions 548
  - 15.4.1 Singular Value Decomposition 548
  - 15.4.2 Generalized Singular Value Decomposition 549
  - 15.4.3 The QR Factorization 550
  - 15.4.4 Square Root of a Positive (Semi-)Definite Matrix 550
- 15.5 Moore-Penrose Pseudoinverse 550
- 15.6 Idempotent Matrices 551
- 15.7 Kronecker Algebra 552
- 15.8 Isomorphism between Complex and Real Matrices 553
- 15.9 Derivatives 554
  - 15.9.1 Derivatives of Functions and Vectors w.r.t. a Vector 554
  - 15.9.2 Derivative of a Function w.r.t. a Matrix 555
- 15.10 Inner Product 556
- 15.11 Gram-Schmidt Orthogonalization 558
- 15.12 Calculating the Roots of Polynomials 560
  - 15.12.1 Scalar Orthogonal Polynomials 560
  - 15.12.2 Vector Orthogonal Polynomials 561
- 15.13 Sensitivity of the Least Squares Solution 562
- 15.14 Exercises 563
- 15.15 Appendix 565
  - Appendix 15.A Calculation of the Roots of a Polynomial 565

CHAPTER 16 Some Probability and Stochastic Convergence Fundamentals 567

- 16.1 Notations and Definitions 567
- 16.2 The Covariance Matrix of a Function of a Random Variable 571
- 16.3 Sample Variables 572
- 16.4 Mixing Random Variables 573
  - 16.4.1 Definition 573
  - 16.4.2 Properties 574
- 16.5 Preliminary Example 576
- 16.6 Definitions of Stochastic Limits 578
- 16.7 Interrelations between Stochastic Limits 579
- 16.8 Properties of Stochastic Limits 582
- 16.9 Laws of Large Numbers 583
- 16.10 Central Limit Theorems 585
- 16.11 Properties of Estimators 586
- 16.12 Cramér-Rao Lower Bound 588
- 16.13 How to Prove Asymptotic Properties of Estimators? 591
  - 16.13.1 Convergence—Consistency 591
  - 16.13.2 Convergence Rate 592
  - 16.13.3 Asymptotic Bias 594
  - 16.13.4 Asymptotic Normality 595
  - 16.13.5 Asymptotic Efficiency 595
- 16.14 Pitfalls 595
- 16.15 Preliminary Example—Continued 596
  - 16.15.1 Consistency 597
  - 16.15.2 Convergence Rate 598
  - 16.15.3 Asymptotic Normality 598

16.15.4 Asymptotic Efficiency 599
16.15.5 Asymptotic Bias 600
16.15.6 Robustness 600
16.16 Properties of the Noise after a Discrete Fourier Transform 601
16.17 Exercises 605
16.18 Appendixes 606
  Appendix 16.A Indecomposable Sets 606
  Appendix 16.B Proof of Lemma 16.5 608
  Appendix 16.C Proof of Lemma 16.8 608
  Appendix 16.D Almost Sure Convergence Implies Convergence in Probability 610
  Appendix 16.E Convergence in Mean Square Implies Convergence in Probability 610
  Appendix 16.F The Borel-Cantelli Lemma 610
  Appendix 16.G Proof of the (Strong) Law of Large Numbers for Mixing Sequences 611
  Appendix 16.H Proof of the Central Limit Theorem for Mixing Sequences 613
  Appendix 16.I Generalized Cauchy-Schwarz Inequality for Random Vectors 614
  Appendix 16.J Proof of the Generalized Cramér-Rao Inequality (Theorem 16.18) 614
  Appendix 16.K Proof of Lemma 16.23 615
  Appendix 16.L Proof of Lemma 16.24 616
  Appendix 16.M Proof of Lemma 16.26 617
  Appendix 16.N Proof of Lemma 16.27 618
  Appendix 16.O Proof of Theorem 16.28 619
  Appendix 16.P Proof of Theorem 16.29 621
  Appendix 16.Q Proof of Corollary 16.30 622
  Appendix 16.R Proof of Lemma 16.31 622
  Appendix 16.S Proof of Theorem 16.32 623
  Appendix 16.T Proof of Theorem 16.33 624

CHAPTER 17 Properties of Least Squares Estimators with Deterministic Weighting 627

17.1 Introduction 627
17.2 Strong Convergence 628
  17.2.1 Strong Convergence of the Cost Function 629
  17.2.2 Strong Convergence of the Minimizer 630
17.3 Strong Consistency 631
17.4 Convergence Rate 632
  17.4.1 Convergence of the Derivatives of the Cost Function 633
  17.4.2 Convergence Rate of $\hat{\theta}(z)$ to $\tilde{\theta}(z_0)$ 633
  17.4.3 Convergence Rate of $\tilde{\theta}(z_0)$ to θ_* 634
17.5 Asymptotic Bias 634
17.6 Asymptotic Normality 636
17.7 Asymptotic Efficiency 637
17.8 Overview of the Asymptotic Properties 637
17.9 Exercises 638
17.10 Appendixes 639
  Appendix 17.A Proof of the Strong Convergence of the Cost Function (Lemma 17.3) 639

Contents   xxi

    Appendix 17.B Proof of the Strong Convergence of the Minimizer
           (Theorem 17.6) 640
    Appendix 17.C Lemmas 641
    Appendix 17.D Proof of the Convergence Rate of the Minimizer
          (Theorem 17.19) 646
    Appendix 17.E Proof of the Improved Convergence Rate of the Minimizer
          (Theorem 17.21) 647
    Appendix 17.F Equivalence between the Truncated and the Original
          Minimizer (Lemma 17.27) 648
    Appendix 17.G Proof of the Asymptotic Bias on the Truncated Minimizer
          (Theorem 17.28) 648
    Appendix 17.H Cumulants of the Partial Sum of a Mixing Sequence 649
    Appendix 17.I Proof of the Asymptotic Distribution of the Minimizer
          (Theorem 17.29) 649
    Appendix 17.J Proof of the Existence and the Convergence of the
          Covariance Matrix of the Truncated Minimizer
          (Theorem 17.30) 650

CHAPTER 18 Properties of Least Squares Estimators with Stochastic Weighting 651

  18.1 Introduction—Notational Conventions 651
  18.2 Strong Convergence 652
    18.2.1 Strong Convergence of the Cost Function 653
    18.2.2 Strong Convergence of the Minimizer 653
  18.3 Strong Consistency 654
  18.4 Convergence Rate 654
    18.4.1 Convergence of the Derivatives of the Cost Function 655
    18.4.2 Convergence Rate of $\hat{\theta}(z)$ to $\tilde{\theta}(z_0)$ 656
  18.5 Asymptotic Bias 657
  18.6 Asymptotic Normality 658
  18.7 Overview of the Asymptotic Properties 659
  18.8 Exercises 660
  18.9 Appendixes 661
    Appendix 18.A Proof of the Strong Convergence of the Cost Function
          (Lemma 18.4) 661
    Appendix 18.B Proof of the Convergence Rate of the Minimizer
          (Theorem 18.16) 661
    Appendix 18.C Proof of the Asymptotic Bias of the Truncated Minimizer
          (Theorem 18.22) 662
    Appendix 18.D Proof of the Asymptotic Normality of the Minimizer
          (Theorem 18.25) 663

CHAPTER 19 Identification of Semilinear Models 665

  19.1 The Semilinear Model 665
    19.1.1 Signal Model 666
    19.1.2 Transfer Function Model 666
  19.2 The Markov Estimator 666
    19.2.1 Real Case 666
    19.2.2 Complex Case 668
  19.3 Cramér-Rao Lower Bound 668
    19.3.1 Real Case 668

		19.3.2	Complex Case 669
	19.4	Properties of the Markov Estimator 670
		19.4.1	Consistency 671
		19.4.2	Strong Convergence 672
		19.4.3	Convergence Rate 672
		19.4.4	Asymptotic Normality 673
		19.4.5	Asymptotic Efficiency 674
		19.4.6	Robustness 674
		19.4.7	Practical Calculation of Uncertainty Bounds 674
	19.5	Residuals of the Model Equation 675
		19.5.1	Real Case 675
		19.5.2	Complex Case 677
	19.6	Mean and Variance of the Cost Function 678
		19.6.1	Real Case 678
		19.6.2	Complex Case 679
	19.7	Model Selection and Model Validation 680
		19.7.1	Real Case 680
		19.7.2	Complex Case 682
	19.8	Exercises 683
	19.9	Appendixes 684
		Appendix 19.A Constrained Minimization (19-6) 684
		Appendix 19.B Proof of the Cramér-Rao Lower Bound for Semilinear Models 685
		Appendix 19.C Markov Estimates of the Observations for Signal Models and Transfer Function Models with Known Input 687
		Appendix 19.D Proof of the Convergence Rate of the Markov Estimates for Large Signal-to-Noise Ratios and Small Model Errors (Theorem 19.2) 688
		Appendix 19.E Proof of the Asymptotic Distribution of the Markov Estimates without Model Errors 690
		Appendix 19.F Proof of the Asymptotic Efficiency of the Markov Estimates (Theorem 19.4) 691
		Appendix 19.G Proof of the Convergence Rate of the Residuals (Lemma 19.8) 691
		Appendix 19.H Properties of the Projection Matrix in Lemma 19.9 691
		Appendix 19.I Proof of the Improved Convergence Rate of the Residuals (Lemma 19.9) 692
		Appendix 19.J Proof of the Properties of the Sample Correlation of the Residuals (Theorem 19.10) 692
		Appendix 19.K Proof of the Convergence Rate of the Minimum of the Cost Function (Lemma 19.11) 694
		Appendix 19.L Proof of the Properties of the Cost Function (Theorem 19.12) 697
		Appendix 19.M Model Selection Criteria 697
		Appendix 19.N Proof of the Modified AIC and MDL Criteria (Theorem 19.15) 697

CHAPTER 20 Identification of Invariants of (Over)Parameterized Models 699

	20.1	Introduction 699
	20.2	(Over)Parameterized Models and Their Invariants 700
	20.3	Cramér-Rao Lower Bound for Invariants of (Over)Parameterized Models 702

20.4 Estimates of Invariants of (Over)Parameterized Models – Finite Sample Results 703
    20.4.1 The Estimators 703
    20.4.2 Main Result 705
20.5 A Simple Numerical Example 706
20.6 Exercises 708
20.7 Appendixes 708
    Appendix 20.A Proof of Theorem 20.8 (Cramér-Rao Bound of (Over)Parameterized Models) 708
    Appendix 20.B Proof of Theorem 20.15 (Jacobian Matrix of (Over)Parameterized Models) 709
    Appendix 20.C Proof of Theorem 20.16 709

References 711

Subject Index 729

Author Index 739

About the Authors 743

Preface to the First Edition

Identification is a powerful technique for building accurate models of complex systems from noisy data. It consists of three basic steps, which are interrelated: (1) the design of an experiment; (2) the construction of a model, black box or from physical laws; and (3) the estimation of the model parameters from the measurements. The art of modeling lies in proper use of the skills and specialized knowledge of experts in the field of study, who decide what approximations can be made, suggest how to manipulate the system, reveal the important aspects, and so on. Consequently, modeling should preferably be executed by these experts themselves. Naturally, they require relevant tools for extracting information of interest. However, most experts will not be familiar with identification theory and will struggle in each new situation with the same difficulties while developing their own identification techniques, losing time over problems already solved in the literature of identification.

This book presents a thorough description of methods to model linear dynamic time-invariant systems by their transfer function. The relations between the transfer function and the physical parameters of the system are very dependent upon the specific problem. Because transfer function models are generally valid, we have restricted the scope of the book to these alone, so as to develop and study general purpose identification techniques. This should not be unnecessarily restricting for readers who are more interested in the physical parameters of a system: the transfer function still contains all the information that is available in the measurements, and it can be considered to be an intermediate model between the measurements and the physical parameters. Also, the transfer function model is very suitable for those readers looking for a black box description of the input-output relations of a system. And, of course, the model is directly applicable to predict the output of the system.

In this book, we use mainly frequency domain representations of the data. This opens many possibilities to identify continuous-time (Laplace-domain) or discrete-time (z-domain) models, if necessary extended with an arbitrary and unknown delay. Although we advocate using periodic excitations, we also extend the methods and models to deal with arbitrary excitations. The "classical" time-domain identification methods that are specifically directed toward these signals are briefly covered and encapsulated in the identification framework that we offer to the reader.

This book provides answers to questions at different levels, such as: What is identification and why do I need it? How to measure the frequency response function of a linear dynamic system? How to identify a dynamic system? All these are very basic questions, directly focused on the interests of the practitioner. Especially for these readers, we have added guidelines to many chapters for the user, giving explicit and clear advice on what are good choices in order to attain a sound solution. Another important part of the material is intended for readers who want to study identification techniques at a more profound level. Questions on how to analyze and prove the properties of an identification scheme are addressed in this part. This study is not restricted to the identification of linear dynamic systems; it is valid for a very wide class of weighted, nonlinear least squares estimators. As such, this book provides a great deal of information for readers who want to set up their own identification scheme to solve their specific problem.

The structure of the book can be split into four parts: (1) collection of raw data or nonparametric identification; (2) parametric identification; (3) comparison with existing frameworks, and guidelines; (4) profound development of theoretical tools.

In the *first part*, after the introductory chapter on identification, we discuss the collection of the raw data: How to measure a frequency response function of a system. What is the impact of nonlinear distortions? How to recognize, qualify, and quantify nonlinear distortions. How to select the excitation signals in order to get the best measurements. This nonparametric approach to identification is discussed in detail in Chapters 2 to 5, and 7.[1]

In the *second part*, we focus on the identification of parametric models. Signal and system models are presented, using a frequency and a time domain representation. The equivalence and impact of leakage effects and initial conditions are shown. Nonparametric and parametric noise models are introduced. The estimation of the parameters in these models is studied in detail. Weighted (nonlinear) least squares methods, maximum likelihood, and subspace methods are discussed and analyzed. First, we assume that the disturbing noise model is known; next, the methods are extended to the more realistic situation of unknown noise models that have to be extracted from the data, together with the system model. Special attention is paid to the numerical conditioning of the sets of equations to be solved. Taking some precautions, very high order systems, with 100 poles and zeros or even more, can be identified. Finally, validation tools to verify the quality of the models are explained. The presence of unmodeled dynamics or nonlinear distortions is detected, and simple rules to guide even the inexperienced user to a good solution are given. This material is presented in Chapters 6, and 8 to 12[1].

The *third part* begins with an extensive comparison of what is classically called time and frequency domain identification. It is shown that, basically, both approaches are equivalent, but some questions are more naturally answered in one domain instead of the other. The most important question is nonparametric versus parametric noise models. Next, we provide the practitioner with detailed guidelines to help avoid pitfalls from the very beginning of the process (collecting the raw data), over the selection of appropriate identification methods until the model validation. This part covers Chapters 13 and 14.

The *last part* of the book is intended for readers who want to acquire a thorough understanding of the material or those who want to develop their own identification scheme. Not only do we give an introduction to the stochastic concepts we use, but we also show, in a structured approach, how to prove the properties of an estimator. This avoids the need for each freshman in this field to find out, time and again, the basic steps to solve such a problem. Starting from this background, a general but detailed framework is set up to analyze the properties of nonlinear least squares estimators with deterministic and stochastic weighting.

---

1. Chapters 4, 7, and 12 were not in the first edition.

For the special and quite important class of semilinear models, it is possible to make this analysis in much more detail. This material is covered in Chapters 15 to 20.

It is possible to extract a number of undergraduate courses from this book. In most of the chapters that can be used in these courses, we added exercises that introduce the students to the typical problems that appear when applying the methods to solve practical problems.

A first, quite general undergraduate course subject is the measurement of frequency response functions (FRF) of dynamic systems, as discussed in Chapters 2 to 5. Chapter 7[1] deals with advanced techniques for measuring FRFs and is more suited for a graduate course on the topic. To understand Chapter 7 one should first master Chapters 1 to 6.

Another possibility is a first introduction to the identification of linear dynamic systems. Such an undergraduate course should include Chapter 1 and some selected parts of Chapters 6, and 8 to 11. Chapter 12[1] generalizes the results of Chapters 10 and 11 to non-steady state conditions and arbitrary excitations and is more suited for a graduate course on linear system identification. To understand Chapter 12 one should first read Chapters 7, 10, and 11.

A final course, at the graduate level, is an advanced course on identification based on the methods that are explained in Chapters 17 to 20. This gives an excellent introduction for students who want to develop their own algorithms.

A MATLAB® toolbox, which includes most of the techniques developed in this book, is available. It can be used with a graphical user interface, avoiding most problems and difficult questions for the inexperienced user. At the basic level, this toolbox produces almost autonomously a good model. At the intermediate or advanced level, the user obtains access to some of the parameters in order to optimize the operation of the toolbox to solve dedicated modeling problem. Finally, for those who want to use it as a research tool, there is also a command level that gives full access to all the parameters that can be set to optimize and influence the behavior of the algorithms. More information on this package can be obtained by sending an E-mail to one of the authors: rik.pintelon@vub.ac.be or johan.schoukens@vub.ac.be

Rik Pintelon
Department of Electrical Engineering
Vrije Universiteit Brussel
BELGIUM

Johan Schoukens
Department of Electrical Engineering
Vrije Universiteit Brussel
BELGIUM

---

1. Chapters 4, 7, and 12 were not in the first edition.

Preface to the Second Edition

During the 10 years since the first edition appeared, frequency domain system identification has evolved considerably. In the second edition we have added new material that reflects our personal view on this development. The book has been updated and new sections and chapters have been added. These mainly deal with arbitrary excitations; periodic excitations under non-steady state conditions; discrete-time and continuous-time parametric noise modeling in the frequency domain; the detection, quantification, and qualification of nonlinear distortions; the best linear approximation of nonlinear systems operating in feedback; and multi-input, multi-output systems. Finally, a large number of new experiments have been included throughout the chapters. In the sequel, we explain these extensions in more detail.

In the first edition the emphasis was strongly put on the use of periodic excitations because, at that time, it was the only way to obtain nonparametric noise models in a pre-processing step, which considerably simplifies the system identification task. Although very successful, this approach has a number of shortcomings: (i) it does not account for the noise leakage that increases the variability of the frequency response function (FRF) estimate and introduces a correlation among consecutive signal periods; (ii) it is sensitive to plant transients that introduce a bias in the FRF and the nonparametric noise models; and (iii) it cannot handle arbitrary excitations. Solutions for these problems are presented in Chapters 7 (nonparametric models) and 12 (parametric transfer function models) of the second edition. These new methods have lead to new insights and, hence, also to new guidelines for the user (see Chapter 14).

The first edition studied the properties of the best linear approximation of a nonlinear plant operating in open loop for the class of Gaussian excitation signals. It was unclear whether these results could be extended to nonlinear systems operating in closed loop configuration and to non-Gaussian inputs. These issues are handled in Section 3.5.2 (non-Gaussian excitations) and Sections 3.8, 7.2.8, and 7.3.5 (nonlinear plants operating in closed loop) of the second edition.

In the first edition odd-odd (only every second odd harmonic is excited) random phase multisines were proposed to detect, qualify, and quantify the level of the nonlinear distortions in FRF measurements. The level of the nonlinear distortions at the non-excited harmonics (= detection lines) was used for quantifying roughly the level of the nonlinear distortions at the nearest excited harmonics. No theoretical justification was given for this extrapolation. In Chapter 4 of the second edition it is proven that the extrapolation is asymptotically (for the number of excited harmonics going to infinity) exact if the detection lines (non-excited) harmonics are randomly distributed over the frequency band of interest. It results in the so-called full or odd random phase multisines with random harmonic grid (see Section 4.2).

The first edition handled the frequency domain identification of parametric discrete-time noise models under the restriction that the DFT frequencies cover the unit circle uniformly. In the second edition (see Section 10.9) this is generalized to continuous-time and discrete-time noise models, identified on (a) part(s) of the imaginary axis or unit circle, respectively. The link with the classical time domain prediction error framework is also discussed in detail.

The first edition was mostly devoted to single-input, single-output systems. In the second edition a full extension to the multivariable case is made for the design of periodic excitations (see Sections 2.7 and 3.7), the nonparametric frequency response matrix (FRM) measurement using periodic and random excitations (see Section 2.7 and Chapter 7), the detection and quantification of nonlinear distortions in FRM measurements using periodic excitations (see Sections 3.7, and 7.3.5 to 7.3.7), and the parametric transfer function modeling (see Chapter 12).

In the first edition the experiments were mainly concentrated in one chapter. This chapter has been deleted and replaced in the second edition by new experiments (see Chapters 4, 7, 10, 12, and 13) that use the new insights and the newly developed identification methods.

To guide the reader through this book a number of "lecture maps" for the following topics are provided: introduction to identification; nonparametric FRF measurements; identification of linear dynamic systems; measurement and modeling of multiple-input, multiple-output systems; measurement and modeling of nonlinear systems; and analysis of the stochastic properties of estimators. These selected topics can be used as undergraduate (u) and/or graduate (g) courses.

Introduction to Identification$^{(u)}$
Chapters 1 and 2; Chapter 5, Section 5.1–5.3; and Chapter 8.

Nonparametric FRF Measurements$^{(u, g)}$
Part I$^{(u)}$ — Basics: Chapter 1, Section 1.3; Chapter 2; and Chapter 5, Sections 5.1–5.3.
Part II$^{(u, g)}$ — Influence of Nonlinear Distortions: Chapters 3 and 4.
Part III$^{(g)}$ — Advanced Methods for Arbitrary and Periodic Excitations under Non-Steady State Conditions: Chapter 6, Sections 6.1–6.3, and 6.6; and Chapter 7.

Identification of Linear Dynamic Systems$^{(u, g)}$
Part I$^{(u)}$ — Basics of Frequency Domain System Identification: Chapter 1; Chapter 6, Sections 6.1–6.3, and 6.5; and Chapters 8, 9, 11, and 20.
Part II$^{(u, g)}$ — Estimation with Unknown Noise Model: Chapter 6, Sections 6.7 and 6.8; and Chapters 10 and 13.
Part III$^{(g)}$ — Advanced Methods for Arbitrary and Periodic Excitations under Non-Steady State Conditions: Chapters 7, 12, and 14.

Measurement and Modeling of Multiple-Input, Multiple-Output Systems$^{(g)}$
Chapter 2, Section 2.7; Chapter 3, Section 3.7; Chapter 6, Section 6.6; Chapter 7; Chapter 9, Section 9.21; and Chapter 12.

Measurement and Modeling of Nonlinear Systems$^{(u, g)}$
Chapters 3 and 4; Chapter 6, Section 6.8; Chapter 9, Section 9.19; Chapter 10, Section 10.7; and Chapter 12, Section 12.4.

Analysis of the Stochastic Properties of Estimators$^{(g)}$
Chapters 16–19.

Finally, software support for identifying multivariable systems is freely available at the website

http://booksupport.wiley.com

via MATLAB® m-files (design of multi-input periodic excitations, nonparametric frequency response matrix measurements using periodic and random excitations, detection and quantification of nonlinear distortions in FRM measurements, parametric transfer function modeling using nonparametric noise models, and simultaneous parametric identification of noise and plant models).

Acknowledgments

This book is the culmination of 30 years of research in the identification group of the Department of Fundamental Electricity and Instrumentation (ELEC) of the Vrije Universiteit Brussel (VUB). It is the result of close and harmonious cooperation between many of the workers in the department, and we would like to thank all of them for exchanges of ideas and important discussions that have taken place over the gestation period of this, the end result. We are greatly indebted to the former heads of the department ELEC, Jean Renneboog and Alain Barel, to the R&D Department of the VUB, to the FWO-Vlaanderen (the Fund for Scientific Research–Flanders), to the Flemish government (Concerted Action–GOA, and Methusalem research programs), and to the federal government (Interuniversity Attraction Poles–IAP research programs). Without their sustained support, this work would never have seen the light of day.

We are grateful to many colleagues and coworkers in the field who imparted new ideas and methods to us, took time for discussions, and were prepared to listen to some of our less conventional ideas that did not fit in the mainstream of the "classical" identification. Their critical remarks and constructive suggestions contributed significantly to our view on the field.

Last, but not least, we want to thank Yves Rolain for contributing to our work as a friend, a colleague, and a coauthor of many of our papers. Without his sustained support, we would never have been able to access the advanced measurement equipment we used for the many experiments that are reported in this book.

Rik Pintelon

Johan Schoukens

List of Operators and Notational Conventions

$\mathbb{A}$	outline uppercase font denotes a set, for example, $\mathbb{N}$, $\mathbb{Z}$, $\mathbb{Q}$, $\mathbb{R}$ and $\mathbb{C}$ are, respectively, the natural, the integer, the rational, the real, and the complex numbers
$\otimes$	the Kronecker matrix product
$*$	convolution operator
Re( )	real part of
Im( )	imaginary part of
$\arg\min_x f(x)$	the minimizing argument of $f(x)$
$O(x)$	an arbitrary function with the property $\lim_{x \to 0} |O(x)/x| < \infty$
$o(x)$	an arbitrary function with the property $\lim_{x \to 0} |o(x)/x| = 0$
$\hat{\theta}$	estimated value of θ
$\bar{x}$	complex conjugate of x
subscript 0	true value
subscript Re	$A_{\text{Re}} = \begin{bmatrix} \text{Re}(A) & -\text{Im}(A) \\ \text{Im}(A) & \text{Re}(A) \end{bmatrix}$
subscript re	$A_{\text{re}} = \begin{bmatrix} \text{Re}(A) \\ \text{Im}(A) \end{bmatrix}$
subscript u	with respect to the input of the system
subscript y	with respect to the output of the system
subscript $*$	limiting estimate
superscript T	matrix transpose

superscript $-T$	transpose of the inverse matrix
superscript H	Hermitian transpose: complex conjugate transpose of a matrix
superscript $-H$	Hermitian transpose of the inverse matrix
superscript $+$	Moore-Penrose pseudoinverse
superscript $\perp$	orthogonal complement of a subspace or a matrix
$\angle x$	phase (argument) of the complex number x
$X_{[i]}(s)$	ith entry of the vector function $X(s)$
$A_{[i,j]}(s)$	i,jth entry of the matrix function $A(s)$
$A_{[:,j]}$	jth column of A
$A_{[i,:]}$	ith row of A
$A_{[n:m,p:q]}$	rows n to m and columns p to q of A
$X^{[k]}(s)$	kth realization of a random process $X(s)$
$\lambda(A)$	eigenvalue of a square matrix A
$\sigma(A)$	singular value of an $n \times m$ matrix A
$\kappa(A) = (\max_i \sigma_i(A))/(\min_i \sigma_i(A))$	condition number of an $n \times m$ matrix A
$|x| = \sqrt{(\text{Re}(x))^2 + (\text{Im}(x))^2}$	magnitude of a complex number x
$\|A\|_1 = \max_{1 \leq j \leq m} \sum_{i=1}^{n} |A_{[i,j]}|$	1-norm of an $n \times m$ matrix A
$\|A\|_2 = \max_{1 \leq i \leq m} \sigma_i(A)$	2-norm of an $n \times m$ $(n \geq m)$ matrix A
$\|X\|_2 = \sqrt{X^H X}$	2-norm of the column vector X
$\|A\|_\infty = \max_{1 \leq i \leq n} \sum_{j=1}^{m} |A_{[i,j]}|$	$\times$-norm of an $n \times m$ matrix A
$\|A\|_F = \sqrt{\text{tr}(A^H A)}$	Frobenius norm of an $n \times m$ matrix A
$\text{diag}(A_1, A_2, ..., A_K)$	block diagonal matrix with blocks A_k, $k = 1, 2, ..., K$
$\text{herm}(A) = (A + A^H)/2$	Hermitian symmetric part of an $n \times m$ matrix A
$\text{null}(A)$	null space of the $n \times m$ matrix A, linear subspace of $\mathbb{C}^m$ defined by $Ax = 0$
$\text{range}(A)$	range of the $n \times m$ matrix A, linear subspace of $\mathbb{C}^n$ that is reachable by making linear combinations of the columns of A ($\text{range}(A) = (\text{null}(A^T))^\perp$)
$\text{rank}(A)$	rank of the $n \times m$ matrix A, maximum number of linear independent rows (columns) of A
$\text{std}(x) = (\mathbb{E}\{|x - \mathbb{E}\{x\}|^2\})^{1/2}$	standard deviation of x

List of Operators and Notational Conventions

span$\{a_1, a_2, ..., a_m\}$	the span of the vectors $a_1, a_2, ..., a_m$ is the linear subspace obtained by making all possible linear combinations of $a_1, a_2, ..., a_m$
tr$(A) = \sum_{i=1}^{n} A_{[i,i]}$	trace of an $n \times n$ matrix A
var$(x) = \mathbb{E}\{|x - \mathbb{E}\{x\}|^2\}$	variance of x
vec(A)	a column vector formed by stacking the columns of the matrix A on top of each other
a.s.lim	almost sure limit, limit with probability one
l.i.m.	limit in mean square
plim	limit in probability
Lim	limit in distribution
$\mathbb{E}\{\ \}$	mathematical expectation
Prob( )	probability
$b_X = X - \mathbb{E}\{X\}$	bias of the estimate X
Cov$(X, Y) = \mathbb{E}\{(X - \mathbb{E}\{X\})(Y - \mathbb{E}\{X\})^H\}$	cross-covariance matrix of X and Y
covar$(x, y) = \mathbb{E}\{(x - \mathbb{E}\{x\})(y - \mathbb{E}\{y\})\}$	covariance of x and y
cum( )	cumulant
$C_X = $ Cov$(X) = $ Cov(X, X)	covariance matrix of X
$\hat{C}_X = \frac{1}{M-1} \sum_{m=1}^{M} (X^{[m]} - \hat{X})(X^{[m]} - \hat{X})^H$	sample covariance matrix of M realizations of X
$C_{XY} = $ Cov(X, Y)	cross-covariance matrix of X and Y
$\hat{C}_{XY} = \frac{1}{M-1} \sum_{m=1}^{M} (X^{[m]} - \hat{X})(Y^{[m]} - \hat{Y})^H$	sample cross-covariance matrix of M realizations of X and Y
$CR(X)$	Cramér-Rao lower bound on X
DFT$(x(t))$	discrete Fourier transform of the samples $x(t)$, $t = 0, 1, ..., N-1$
$Fi(X)$	Fisher information matrix with respect to the parameters X
I_m	$m \times m$ identity matrix
q	backward shift operator: $qu(kT_s) = u((k-1)T_s)$
MSE$(X) = \mathbb{E}\{(X - X_0)(X - X_0)^H\}$	mean square error of the estimate X
$R_{xx}(\tau) = \mathbb{E}\{x(t)x^H(t-\tau)\}$	autocorrelation of $x(t)$
$R_{xy}(\tau) = \mathbb{E}\{x(t)y^H(t-\tau)\}$	cross-correlation of $x(t)$ and $y(t)$
$S_{xx}(j\omega)$	Fourier transform of $R_{xx}(\tau)$ (autopower spectrum of $x(t)$)
$S_{xy}(j\omega)$	Fourier transform of $R_{xy}(\tau)$ (cross-power spectrum of $x(t)$ and $y(t)$)
$\hat{S}_{XX}(k) = \frac{1}{M} \sum_{m=1}^{M} X^{[m]}(k) X^{[m]H}(k)$	Sample autopower spectrum of $x(t)$

$\hat{S}_{XY}(k) = \frac{1}{M}\sum_{m=1}^{M} X^{[m]}(k) Y^{[m]H}(k)$	Sample cross-power spectrum of $x(t)$ and $y(t)$
$\hat{X} = \frac{1}{M}\sum_{m=1}^{M} X^{[m]}$	sample mean of M realizations (experiments) of X
$\mu_x = \mathbb{E}\{x\}$	mean value of x
$\sigma_x^2 = \text{var}(x)$	variance of the x
$\hat{\sigma}_x^2 = \frac{1}{M-1}\sum_{m=1}^{M} |x^{[m]} - \hat{x}|^2$	sample variance of M realizations of x
$\sigma_{xy}^2 = \text{covar}(x, y)$	covariance of x and y
$\hat{\sigma}_{xy}^2 = \frac{1}{M-1}\sum_{m=1}^{M} (x^{[m]} - \hat{x})\overline{(y^{[m]} - \hat{y})}$	sample covariance of M realizations of x and y

List of Symbols

$A(\Omega, \theta) = \sum_{r=0}^{n_p} a_r p_r(\Omega)$ — denominator polynomial plant model expanded in the polynomial basis $p_r(\Omega)$

$A(\Omega, \theta) = \sum_{r=0}^{n_a} a_r \Omega^r$ — denominator polynomial plant model

$B(\Omega, \theta) = \sum_{r=0}^{n_q} b_r q_r(\Omega)$ — numerator plant model expanded in the polynomial basis $q_r(\Omega)$

$B(\Omega, \theta) = \sum_{r=0}^{n_b} b_r \Omega^r$ — numerator polynomial plant model

$C(z^{-1}, \theta) = \sum_{r=0}^{n_c} c_r z^{-r}$ — numerator polynomial noise model

$Cr(u)$ — crest factor of the signal $u(t)$: see (5-3)

$D(z^{-1}, \theta) = \sum_{k=0}^{n_d} d_k z^{-k}$ — denominator polynomial noise model

$e(t)$ — white noise at time t

$E(k)$ — discrete Fourier transform of the samples $e(tT_s)$, $t = 0, 1, \ldots, N-1$

f — frequency

F — number of frequency domain data samples

f_s — sampling frequency

$G(j\omega)$ — frequency response function

$G_{\text{BLA}}(j\omega)$ — best linear approximation of a nonlinear plant

$G(\Omega, \theta) = B(\Omega, \theta)/A(\Omega, \theta)$ — parametric plant model

$H(z^{-1}, \theta) = C(z^{-1}, \theta)/D(z^{-1}, \theta)$ — parametric noise model

$I(\Omega, \theta) = \sum_{r=0}^{n_i} i_r \Omega^r$ — polynomial of the initial and the final conditions of the plant model $B(\Omega, \theta)/A(\Omega, \theta)$

j — $j^2 = -1$

xxxix

Symbol	Description
$J(z^{-1}, \theta) = \sum_{r=0}^{n_d} j_r z^{-r}$	polynomial of the initial and the final conditions of the noise model $C(z^{-1}, \theta)/D(z^{-1}, \theta)$
M	number of (repeated) experiments
N	number of time domain data samples
n_a, n_b, n_c, n_d, n_i and n_j	order of the polynomials $A(\Omega, \theta)$, $B(\Omega, \theta)$, $C(z^{-1}, \theta)$, $D(z^{-1}, \theta)$, $I(\Omega, \theta)$, and $J(z^{-1}, \theta)$
n_θ	dimension of the parameter vector θ
$n_u(t), n_y(t)$	disturbing time domain noise on the input $u(t)$ and output $y(t)$ signals, respectively
$N_U(k), N_Y(k)$	discrete Fourier transform of the samples $n_u(tT_s)$ and $n_y(tT_s)$, $k = 0, 1, \ldots, N-1$, respectively
s	Laplace transform variable
s_k	Laplace transform variable evaluated along the imaginary axis at DFT frequency k: $s_k = j\omega_k$
t	continuous or discrete-time variable
$Tf(u)$	time factor of the signal $u(t)$: see (5-7)
$T_G(\Omega, \theta) = I(\Omega, \theta)/A(\Omega, \theta)$	parametric transient model of the plant $B(\Omega, \theta)/A(\Omega, \theta)$
$T_H(\Omega, \theta) = J(\Omega, \theta)/D(\Omega, \theta)$	parametric transient model of the noise $C(\Omega, \theta)/D(\Omega, \theta)$
T_s	sampling period
$u(t), y(t)$	input and output time signals
$U(e^{j\omega T_s}), Y(e^{j\omega T_s})$	Fourier transform of $u(tT_s)$ and $y(tT_s)$
$U(k), Y(k)$	discrete Fourier transform of the samples $u(tT_s)$ and $y(tT_s)$, $t = 0, 1, \ldots, N-1$
$U(j\omega), Y(j\omega)$	Fourier transform of $u(t)$ and $y(t)$
$U(s), Y(s)$	one-sided Laplace transform of $u(t)$ and $y(t)$
$U(z), Y(z)$	one-sided Z-transform of $u(tT_s)$, $y(tT_s)$
U_k, Y_k	Fourier coefficients of the periodic signals $u(t)$, $y(t)$
$V_*(\theta)$	asymptotic ($F \to \infty$) cost function
$V_F(\theta, z)$	cost function based on F measurements
$V_F'(\theta, z)$	derivative cost function w.r.t. θ (dimension $1 \times n_\theta$)
$V_F''(\theta, z)$	second-order derivative (Hessian) cost function w.r.t. θ (dimension $n_\theta \times n_\theta$)
$Z(k) = [Y(k) U(k)]^T$	data vector containing the measured input and output (DFT) spectra at (DFT) frequency k

List of Symbols

$Z = [Z^T(1)Z^T(2)\ldots Z^T(F)]^T$	data vector containing the measured input and output DFT spectra (dimension $2F$)
z	Z-transform variable
z_k	Z-transform variable evaluated along the unit circle at DFT frequency k: $z_k = e^{j\omega_k T_s} = e^{j2\pi k/N}$
$\delta(x)$ with $x \in \mathbb{R}$	Dirac function: $\delta(0) = \infty$ and $\delta(x) = 0$ for $x \neq 0$
$\delta(k)$ with $k \in \mathbb{Z}$	Kronecker delta: $\delta(0) = 1$ and $\delta(k) = 0$ for $k \neq 0$
$\varepsilon(\theta, Z)$	column vector of the (weighted) model residuals (dimension F)
θ	column vector of the model parameters
$\tilde{\theta}(Z_0)$	minimizing argument of the cost function $V_F(\theta)$
$\hat{\theta}(Z)$	estimated model parameters, minimizing argument of the cost function $V_F(\theta, Z)$
$\underline{\hat{\theta}}(Z)$	truncated estimator
$\sigma_U^2(k) = \text{var}(U(k))$	variance of the measured input DFT spectrum
$\sigma_Y^2(k) = \text{var}(Y(k))$	variance of the measured output DFT spectrum
$\sigma_{YU}^2(k) = \text{covar}(Y(k), U(k))$	covariance of the measured output and input DFT spectra
τ	time delay (normalized with the sampling period for discrete-time systems)
$J(\theta, Z) = \partial\varepsilon(\theta, Z)/\partial\theta$	gradient of residuals $\varepsilon(\theta, Z)$ w.r.t. the parameters θ (dimension $F \times n_\theta$)
$\omega = 2\pi f$	angular frequency
Ω	generalized transform variable: Laplace domain $\Omega = s$, Z-domain $\Omega = z^{-1}$, Richards domain $\Omega = \tanh(\tau_R s)$, and diffusion phenomena $\Omega = \sqrt{s}$
Ω_k	generalized transform variable evaluated at DFT frequency k: Laplace domain $\Omega_k = j\omega_k$, Z-domain $\Omega_k = e^{-j\omega_k T_s}$, Richards domain $\Omega_k = \tanh(\tau_R j\omega_k)$, and diffusion phenomena $\Omega_k = \sqrt{j\omega_k}$, with $\omega_k = 2\pi k/N$

List of Abbreviations

ARMA	AutoRegressive Moving Average
ARMAX	AutoRegressive Moving Average with eXternal input
ARX	AutoRegressive with eXternal input
BJ	Box-Jenkins (model structure)
BLA	Best Linear Approximation
BTLS	Bootstrapped Total Least Squares
CRB	Cramér-Rao Bound for biased estimators
DFT	Discrete Fourier Transform
DUT	Device Under Test
EV	Errors-in-Variables
FFT	Fast Fourier Transform
FRF	Frequency Response Function
FRM	Frequency Response Matrix
GSVD	Generalized Singular Value Decomposition
GTLS	Generalized Total Least Squares
iid	independent identically distributed
IV	Instrumental Variables
IWLS	Iterative Weighted Linear Least Squares
IQML	Iterative Quadratic Maximum Likelihood
LPM	Local Polynomial Method
LS	Least Squares
MIMO	Multiple-Input, Multiple-Output
ML	Maximum Likelihood
NLS	Nonlinear Least Squares

NLS-FRF	Nonlinear Least Squares based on FRF measurements
NLS-IO	Nonlinear Least Squares based on Input-Output measurements
OE	Output Error (model structure)
pdf	probability density function
PE	Prediction Error
rms	root mean square value
SBTLS	Sample BTLS
SGTLS	Sample GTLS
SISO	Single Input, Single Output
SML	Sample ML
SNR	Signal-to-Noise Ratio
SSUB	Sample SUB
SUB	Subspace
SVD	Singular Value Decomposition
TLS	Total Least Squares
UCRB	Cramér-Rao Bound for Unbiased estimators
w.p.1	with probability one
WGTLS	Weighted Generalized Total Least Squares
WLS	Weighted Least Squares

1

An Introduction to Identification

Abstract: In this chapter a brief, intuitive introduction to identification theory is given. By means of a simple example, the reader is made aware of a number of pitfalls associated with a model built from noisy measurements. Starting from this example, the advantages of an identification approach for measuring and modeling are shown, and finally a family of estimators is introduced. A comprehensive introduction to identification can be found in, among others, Beck and Arnold (1977), Goodwin and Payne (1977), Norton (1986), Sörenson (1980), and Kendall and Stuart (1979). Basic concepts of statistics such as the expected value, covariance matrix, and probability density function are assumed to be known.

1.1 WHAT IS IDENTIFICATION?

From birth onwards, we interact with our environment. Intuitively, we learn to control our actions by predicting their effect. These predictions are based on an inborn model fitted to reality, using past experiences. Starting from very simple actions (if I push a ball, it rolls), we soon are able to deal with much more complicated challenges (walking, running, biking, playing Ping-Pong). Finally, this process culminates in the design of complicated systems such as radios, airplanes, and mobile phones. We even build models to get a better understanding of our universe: What does the life cycle of the sun look like? Can we predict the weather of this afternoon, tomorrow, next week, next month? From these examples it is seen that we never deal with the whole of nature at once: we focus on the aspects we are interested in and do not try to describe all of reality using one coherent model. The job is split up, and efforts are concentrated on just one part of reality at a time. This part is called the system, the rest of nature being referred to as the environment of the system. Interactions between the system and its environment are described by input and output ports. For a very long time in the history of mankind the models were qualitative, and even today we describe most real-life situations using this "simple" approach. For example, a ball will roll downhill; temperature will rise if the heat has been switched on; it seems it will rain because the sky looks very dark. In the last centuries, this qualitative approach was complemented with quantitative models based on advanced mathematics, and, until the last decade, this seemed to be the most successful approach in many fields of science. Most physical laws are quantitative models describing some part of our impression of reality. It soon became clear, however, that it can be very difficult to match a mathematical model to the available observations and experiences. Consequently, qualitative logical methods typified by fuzzy modeling became more popular, once more. In this book we deal with the mathematical, quantitative modeling approach. Fitting these models to our observations creates new problems. We look at the world through "dirty" glasses: when we measure a length, the weight of a mass, the current or voltage, and so on, we always make errors because the instruments we use are not perfect. Also, the models are imperfect; reality is far more complex than the rules we apply. Many systems are not deterministic. They also show a stochastic behavior that makes it impossible to predict exactly their output. Noise in a radio receiver, Brownian motion of small particles, and variation of the wind speed in a thunderstorm are illustrations of this nature. Usually we split the model into a deterministic part and a stochastic part. The deterministic aspects are captured by the mathematical system model, while the stochastic behavior is modeled as a noise distortion. The aim of identification theory is to provide a systematic approach to fit the mathematical model, as well as possible, to the deterministic part, eliminating the noise distortions as much as possible.

Later in this book the meaning of terms such as "system" and "goodness of fit" will be precisely defined. Before formalizing the discussion, we want to motivate the reader by analyzing a very simple example, illustrating many of the aspects and problems that appear in identification theory.

1.2 IDENTIFICATION: A SIMPLE EXAMPLE

1.2.1 Estimation of the Value of a Resistor

Two groups of students have to measure a resistance. Their measurement setup is shown in Figure 1-1. They pass a constant but unknown current through the resistor. The

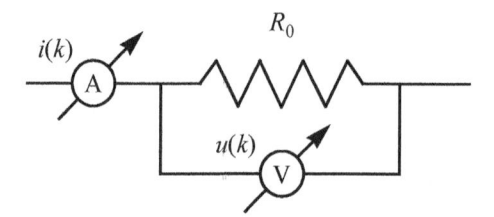

Figure 1-1. Measurement of a resistor using an ammeter (A) and a voltmeter (V).

voltage u_0 across the resistor and the current i_0 through it are measured using a voltmeter and an ampere meter. The input impedance of the voltmeter is very large compared with the unknown resistor so that all the measured current is assumed to pass through the resistor. A set of voltage and current measurements, respectively, $u(k)$, $i(k)$ with $k = 1, 2, ..., N$ is made. The measurement results of each group are shown in Figure 1-2. Because the measurements are very noisy, the groups decide to average their results. Following a lengthy discussion, three estimators for the resistance are proposed:

$$\hat{R}_{\text{SA}}(N) = \frac{1}{N}\sum_{k=1}^{N} \frac{u(k)}{i(k)} \qquad (1\text{-}1)$$

$$\hat{R}_{\text{LS}}(N) = \frac{\frac{1}{N}\sum_{k=1}^{N} u(k)i(k)}{\frac{1}{N}\sum_{k=1}^{N} i^2(k)} \qquad (1\text{-}2)$$

Section 1.2 ■ Identification: A Simple Example

$$\hat{R}_{\text{EV}}(N) = \frac{\frac{1}{N}\sum_{k=1}^{N} u(k)}{\frac{1}{N}\sum_{k=1}^{N} i(k)} \tag{1-3}$$

The index N indicates that the estimate is based on N observations. Note that the three estimators result in the same estimate on noiseless data. Both groups process their measurements, and their results are given in Figure 1-3. From this figure a number of interesting observations can be made:

- All estimators have large variations for small values of N and seem to converge to an asymptotic value for large values of N, except $\hat{R}_{\text{SA}}(N)$ of group A. This corresponds to the intuitively expected behavior: if a large number of data points are processed we should be able to eliminate the noise influence by the averaging effect.

- The asymptotic values of the estimators depend on the kind of averaging technique that is used. This shows that there is a serious problem: at least two out of the three methods converge to a wrong value. It is not even certain that any one of the estimators is doing well. This is quite catastrophic: even an infinite amount of measurements does not guarantee that the exact value is found.

- The $\hat{R}_{\text{SA}}(N)$ of group A behaves very strangely. Instead of converging to a fixed value, it jumps irregularly up and down before convergence is reached.

These observations prove very clearly that a good theory is needed to explain and understand the behavior of candidate estimators. This will allow us to make a sound selection out of many possibilities and to indicate in advance, before running expensive experiments, whether the selected method is prone to serious shortcomings.

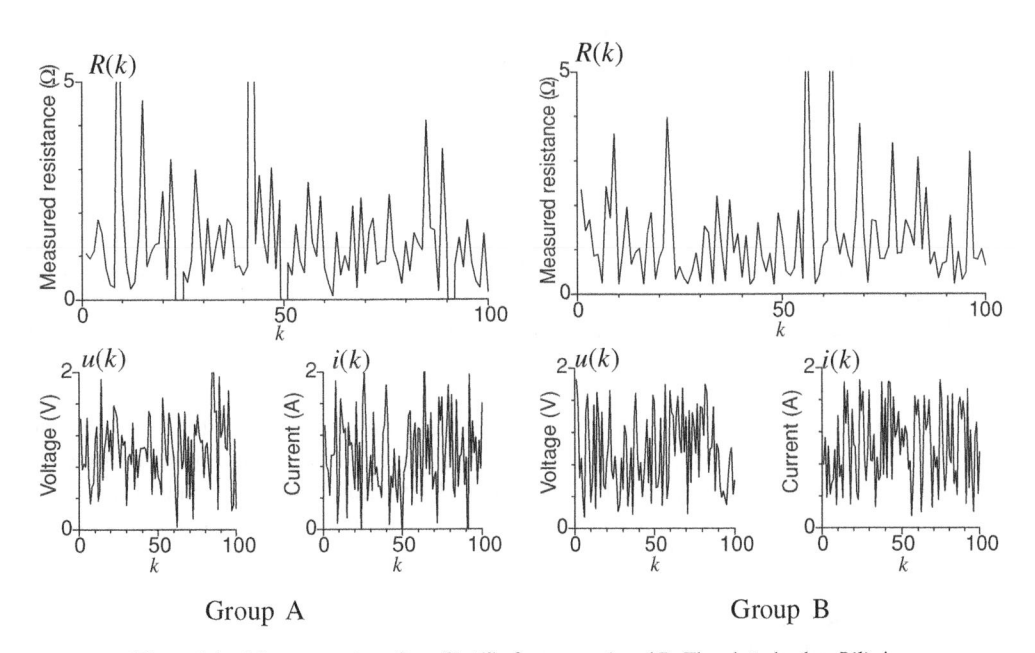

Figure 1-2. Measurement results $u(k)$, $i(k)$ for groups A and B. The plotted value $R(k)$ is obtained by direct division of the voltage by the current: $R(k) = u(k)/i(k)$.

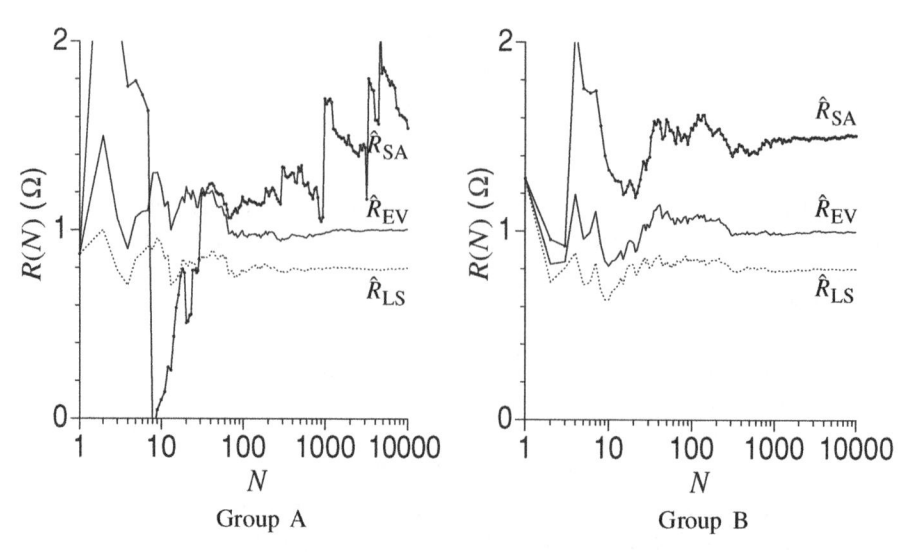

Figure 1-3. Estimated resistance values $\hat{R}(N)$ for both groups as a function of the number of processed data N.

In order to get a better understanding of their results, the students repeat their experiments many times and look to the histogram of $\hat{R}(N)$ for $N = 10, 100$, and 1000. Normalizing these histograms gives an estimate of the pdf (probability density function) of $\hat{R}(N)$ as shown in Figure 1-4. Again, the students can learn a lot from these figures:

- For small values of N the estimates are widely scattered. As the number of processed measurements increases, the pdf becomes more concentrated.
- The estimates $\hat{R}_{LS}(N)$ are less scattered than $\hat{R}_{EV}(N)$, while for $\hat{R}_{SA}(N)$ the odd behavior in the results of group A appears again. The distribution of this estimate does not contract for growing values of N for group A, while it does for group B.
- Again it is clearly visible that the distributions are concentrated around different values.

At this point in the exercise, the students still cannot decide which estimator is the best. Moreover, there seems to be a serious problem with the measurements of group A because $\hat{R}_{SA}(N)$ behaves very oddly. First they decide to focus on the scattering of the different estimators, trying to get more insight into the dependence on N. In order to quantify the scattering of the estimates, their standard deviation is calculated and plotted as a function of N in Figure 1-5.

- The standard deviation of $\hat{R}(N)$ decreases monotonically with N except for the pathological case, $\hat{R}_{SA}(N)$, of group A. Moreover, it can be concluded by comparing with the broken line that the standard deviation is proportional to $1/\sqrt{N}$. This is in agreement with the rule of thumb that states that the uncertainty on an averaged quantity obtained from independent measurements decreases as $1/\sqrt{N}$.

- The uncertainty in this experiment depends on the estimator. Moreover, the proportionality to $1/\sqrt{N}$ is obtained only for sufficiently large values of N for $\hat{R}_{LS}(N)$ and $\hat{R}_{EV}(N)$.

Section 1.2 ■ Identification: A Simple Example

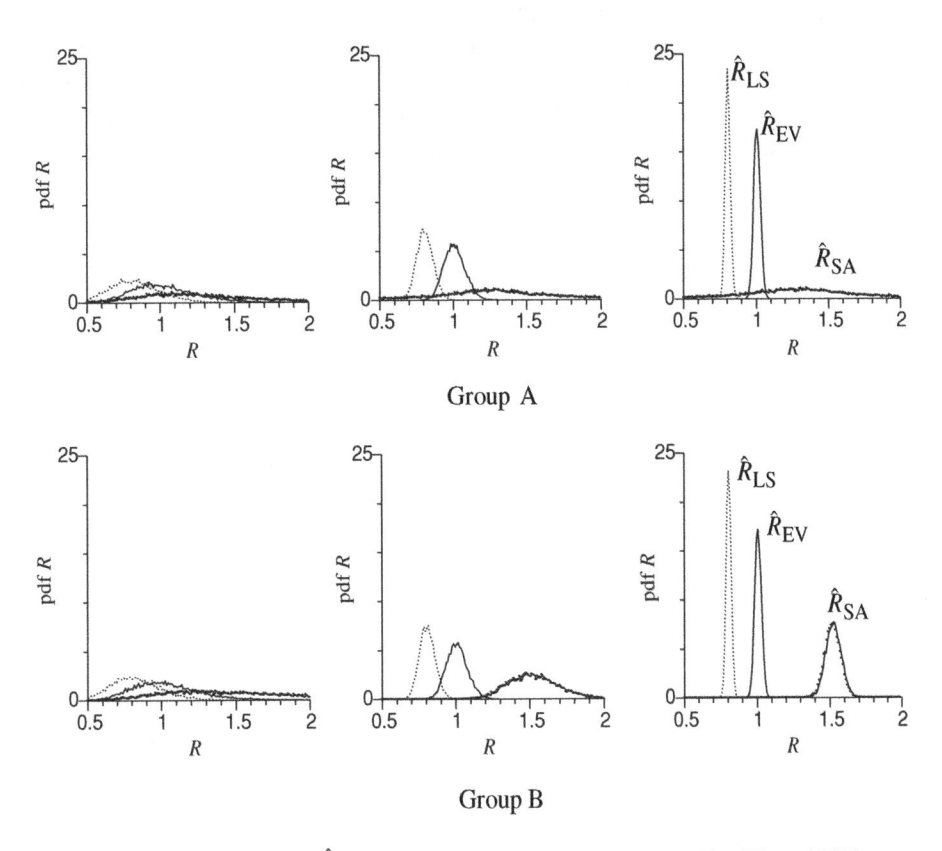

Figure 1-4. Observed pdf of $\hat{R}(N)$ for both groups, from left to right $N = 10$, 100, and 1000.

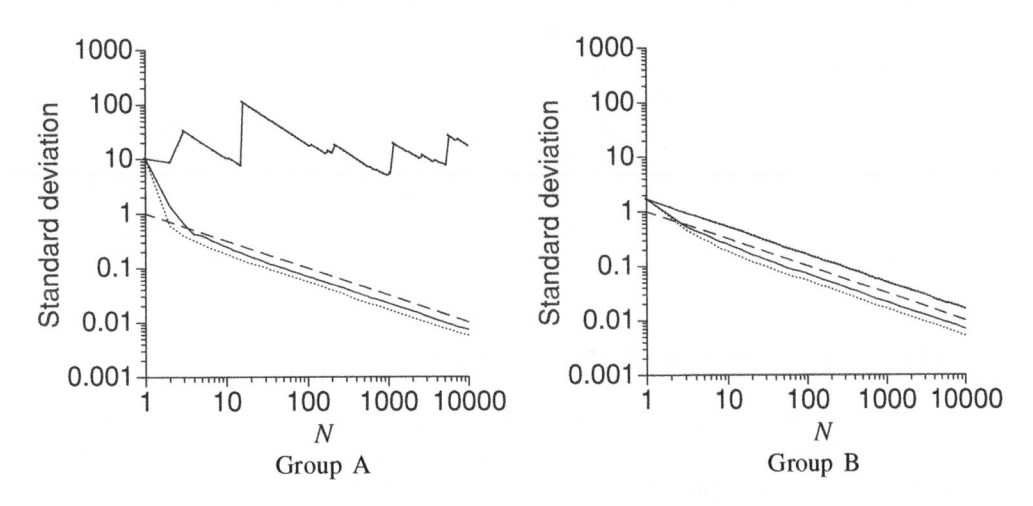

Figure 1-5. Standard deviation of $\hat{R}(N)$ for the different estimators and comparison with $1/\sqrt{N}$; full dotted line: $\hat{R}_{SA}(N)$; dotted line: $\hat{R}_{LS}(N)$, full line: $\hat{R}_{EV}(N)$, dashed line $1/\sqrt{N}$.

Because both groups of students use the same programs to process their measurements, they conclude that the strange behavior of $\hat{R}_{SA}(N)$ in group A should be due to a difference in the raw data. For that reason they take a closer look at the time records given in Figure 1-2. Here it can be seen that the measurements of group A are a bit more scattered than those of group B. Moreover, group A measures some negative values for the current while group B does not. In order to get a better understanding, they make a histogram of the raw current data as shown in Figure 1-6.

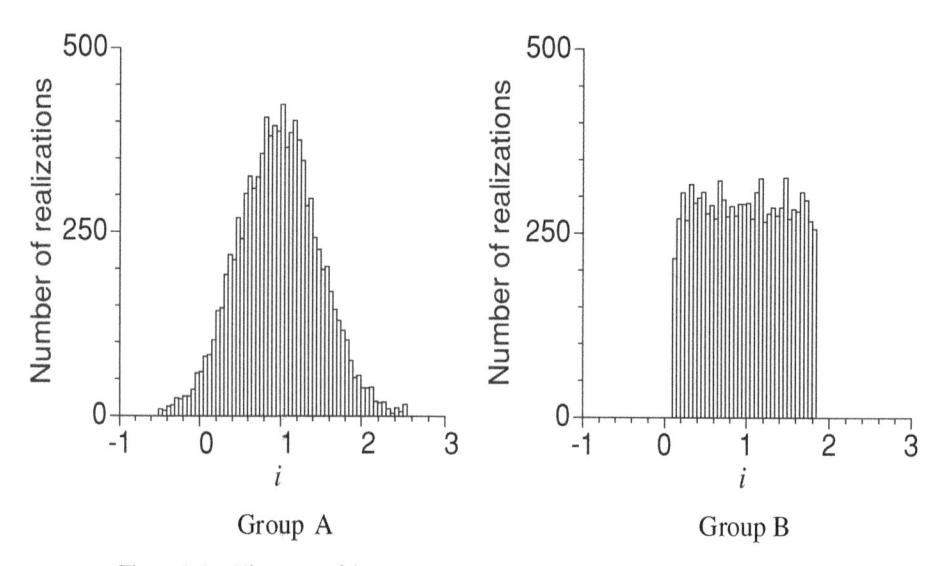

Figure 1-6. Histogram of the current measurements.

These histograms clarify the strange behavior of $\hat{R}_{SA}$ of group A. The noise on the measurements of group A looks completely different from that of group B. Because of the noise on the current measurements, there is a significant risk of getting current values that are very close to zero for group A, whereas this is not so for group B. These small current measurements blow up the estimate $\hat{R}(k) = u(k)/i(k)$ for some k, so that the running average $\hat{R}_{SA}$ cannot converge, or more precisely, the expected value $\mathbb{E}\{u(k)/i(k)\}$ does not exist. This will be discussed in more detail later in this chapter. This example shows very clearly that there is a strong need for methods that can generate and select between different estimators. Before setting up a general framework, the resistance problem is further elaborated.

It is also remarkable to note that, although the noise on the measurements is completely differently distributed, the distribution of the estimated resistance values $\hat{R}_{LS}$ and $\hat{R}_{EV}$ seems to be the same in Figure 1-4 for both groups.

1.2.2 Simplified Analysis of the Estimators

With knowledge obtained from the previous series of experiments, the students eliminate $\hat{R}_{SA}$, but they are still not able to decide whether $\hat{R}_{LS}$ or $\hat{R}_{EV}$ is the best. More advanced analysis techniques are needed to solve this problem. As the estimates are based on a combination of a finite number of noisy measurements, there are bound to be stochastic variables. Therefore, an analysis of the stochastic behavior is needed to select between both estimators. This is done by calculating the limiting values and making series expansions of the estimators. In order to keep the example simple, we will use some of the limit concepts quite

Section 1.2 ■ Identification: A Simple Example

loosely. Precise definitions are postponed to Section 16.6. Three observed problems are analyzed in the following:

- Why do the asymptotic values depend on the estimator?
- Can we explain the behavior of the variance?
- Why does the $\hat{R}_{SA}$ estimator behave strangely for group A?

To do this it is necessary to specify the stochastic framework: how are the measurements disturbed with the noise (multiplicative, additive), and how is the noise distributed? For simplicity, we assume that the current and voltage measurements are disturbed by additive zero mean, independently and identically distributed noise, formally formulated as:

$$i(k) = i_0 + n_i(k) \qquad u(k) = u_0 + n_u(k) \qquad (1\text{-}4)$$

where i_0 and u_0 are the exact but unknown values of the current and the voltage, $n_i(k)$ and $n_u(k)$, are the noise on the measurements.

Assumption 1.1 (Disturbing Noise): $n_i(k)$ and $n_u(k)$ are mutually independent, zero mean, independent and identically distributed (iid) random variables with a symmetric distribution and with variance σ_u^2 and σ_i^2.

1.2.2.1 Asymptotic Value of the Estimators. In this section the limiting value of the estimates for $N \to \infty$ is calculated. The calculations are based on the observation that the sample mean of iid random variables $x(k)$, $k = 1, \ldots, N$ converges to its expected value (see Section 16.9), $\mathbb{E}\{x\}$

$$\lim_{N \to \infty} \frac{1}{N} \sum_{k=1}^{N} x(k) = \mathbb{E}\{x\} \qquad (1\text{-}5)$$

Moreover, if $x(k)$ and $y(k)$ obey Assumption 1.1, then

$$\lim_{N \to \infty} \frac{1}{N} \sum_{k=1}^{N} x(k)y(k) = 0 \qquad (1\text{-}6)$$

Because we are dealing here with stochastic variables, the meaning of this statement should be defined more precisely, but in this section we will just use this formal notation and make the calculations straightforwardly (see Section 16.6 for a formal definition).

The first estimator we analyze is $\hat{R}_{LS}(N)$. Taking the limit of (1-2) gives

$$\begin{aligned}
\lim_{N \to \infty} \hat{R}_{LS}(N) &= \lim_{N \to \infty} \frac{\sum_{k=1}^{N} u(k)i(k)}{\sum_{k=1}^{N} i^2(k)} \\
&= \frac{\lim_{N \to \infty} \sum_{k=1}^{N} (u_0 + n_u(k))(i_0 + n_i(k))}{\lim_{N \to \infty} \sum_{k=1}^{N} (i_0 + n_i(k))^2}
\end{aligned} \qquad (1\text{-}7)$$

Or, after dividing the numerator and denominator by N,

$$\lim_{N \to \infty} \hat{R}_{LS}(N) = \frac{\lim_{N \to \infty}\left[u_0 i_0 + \frac{u_0}{N}\sum_{k=1}^{N} n_i(k) + \frac{i_0}{N}\sum_{k=1}^{N} n_u(k) + \frac{1}{N}\sum_{k=1}^{N} n_u(k)n_i(k) \right]}{\lim_{N \to \infty}\left[i_0^2 + \frac{1}{N}\sum_{k=1}^{N} n_i^2(k) + \frac{2i_0}{N}\sum_{k=1}^{N} n_i(k) \right]}$$

Because n_i and n_u are zero mean iid, it follows from (1-5) and (1-6) that

$$\lim_{N \to \infty} \frac{1}{N}\sum_{k=1}^{N} n_u(k) = 0, \quad \lim_{N \to \infty} \frac{1}{N}\sum_{k=1}^{N} n_i(k) = 0, \text{ and } \lim_{N \to \infty} \frac{1}{N}\sum_{k=1}^{N} n_u(k)n_i(k) = 0$$

However, the sum of the squared current noise distributions does not converge to zero but converges to a constant value different from zero

$$\lim_{N \to \infty} \frac{1}{N}\sum_{k=1}^{N} n_i^2(k) = \sigma_i^2$$

so that the asymptotic value becomes:

$$\lim_{N \to \infty} \hat{R}_{LS}(N) = \frac{u_0 i_0}{i_0^2 + \sigma_i^2} = R_0 \frac{1}{1 + \sigma_i^2/i_0^2} \tag{1-8}$$

This simple analysis gives insight into the behavior of the $\hat{R}_{LS}(N)$ estimator. Asymptotically, this estimator underestimates the value of the resistance due to quadratic noise contributions in the denominator. Although the noise disappears in the averaging process of the numerator, it contributes systematically in the denominator. This results in a systematic error (called bias) that depends on the signal-to-noise ratio (SNR) of the current measurements: i_0/σ_i.

The analysis of the second estimator $\hat{R}_{EV}(N)$ is completely similar. Using (1-3), we get

$$\lim_{N \to \infty} \hat{R}_{EV}(N) = \lim_{N \to \infty} \frac{\frac{1}{N}\sum_{k=1}^{N} u(k)}{\frac{1}{N}\sum_{k=1}^{N} i(k)} = \frac{\lim_{N \to \infty} \frac{1}{N}\sum_{k=1}^{N}(u_0 + n_u(k))}{\lim_{N \to \infty} \frac{1}{N}\sum_{k=1}^{N}(i_0 + n_i(k))} \tag{1-9}$$

or

$$\lim_{N \to \infty} \hat{R}_{EV}(N) = \frac{u_0 + \lim_{N \to \infty} \frac{1}{N}\sum_{k=1}^{N} n_u(k)}{i_0 + \lim_{N \to \infty} \frac{1}{N}\sum_{k=1}^{N} n_i(k)} = \frac{u_0}{i_0} = R_0 \tag{1-10}$$

so that we can conclude now that $\hat{R}_{EV}(N)$ converges to the true value and should be preferred over $\hat{R}_{LS}(N)$. These conclusions are also confirmed by the students' results in Figure 1-3, where it is seen that the asymptotic value of $\hat{R}_{LS}(N)$ is much smaller than that of $\hat{R}_{EV}(N)$.

1.2.2.2 Strange Behavior of the "Simple Approach".
Finally, we have to analyze $\hat{R}_{SA}(N)$ in order to understand its strange behavior. Can't we repeat the previous analysis here? Consider

$$\hat{R}_{SA}(N) = \frac{1}{N}\sum_{k=0}^{N}\frac{u(k)}{i(k)} = \frac{1}{N}\sum_{k=0}^{N}\frac{u_0 + n_u(k)}{i_0 + n_i(k)} \qquad (1\text{-}11)$$

A major difference from the previous estimators is the order of summing and dividing: here the measurements are first divided and then summed together, whereas for the other estimators we first summed the measurements together before making the division. In other words, for $\hat{R}_{LS}(N)$ and $\hat{R}_{EV}(N)$ we first applied an averaging process (summing over the measurements) before making the division. This makes an important difference.

$$\hat{R}_{SA}(N) = \frac{1}{N}\frac{u_0}{i_0}\sum_{k=0}^{N}\frac{1 + n_u(k)/u_0}{1 + n_i(k)/i_0} \qquad (1\text{-}12)$$

In order to process $\hat{R}_{SA}(N)$ along the same lines as the other estimators, we should get rid of the division, for example, by making a Taylor series expansion:

$$\frac{1}{1+x} = \sum_{l=0}^{\infty}(-1)^l x^l \quad \text{for } |x| < 1 \qquad (1\text{-}13)$$

with $x = n_i(k)/i_0$. Because the terms $n_i^{2l+1}(k)$ and $n_u^l(k)n_i^l(k)$ disappear in the averaging process (the pdfs are symmetrical), the limiting value becomes

$$\lim_{N \to \infty}\hat{R}_{SA}(N) = R_0\left(1 + \frac{1}{N}\sum_{k=1}^{N}(n_i(k)/i_0)^2 + \frac{1}{N}\sum_{k=1}^{N}(n_i(k)/i_0)^4 + \ldots\right) \qquad (1\text{-}14)$$

with $|n_i(k)/i_0| < 1$. If we neglect all terms of order 4 or more, the final result becomes

$$\lim_{N \to \infty}\hat{R}_{SA}(N) = R_0(1 + \sigma_i^2/i_0^2) \qquad (1\text{-}15)$$

if $|n_i(k)/i_0| < 1, \forall k$.

From this analysis we can draw two important conclusions:

- The asymptotic value exists only if the following condition on the measurements is met: the series expansion must exist otherwise (1-15) is NOT valid. The measurements of group A violate the condition that is given in (1-14), while those of group B obey it (see Figure 1-6). A more detailed analysis shows that this condition is too rigorous. In practice, it is enough that the expected value $\mathbb{E}\{\hat{R}_{SA}(N)\}$ exists (see Chapter 17). Because this value depends on the pdf of the noise, a more detailed analysis of the measurement noise would be required. For some noise distributions the expected value exists even if the Taylor expansion does not!

- If the asymptotic value exists, (1-15) shows that it will be too large. This is also seen in the results of group B in Figure 1-3. We know already that $\hat{R}_{EV}(N)$ converges to the exact value, and $\hat{R}_{SA}(N)$ is clearly significantly larger.

1.2.2.3 Variance Analysis. In order to get a better understanding of the sensitivity of the different estimators to the measurement noise, the students make a variance analysis using first-order Taylor series approximations.

Again they begin with the $\hat{R}_{LS}(N)$. Starting from (1-7) and neglecting all second-order contributions such as $n_u(k)n_i(k)$ or $n_i^2(k)$, it is found that

$$\hat{R}_{LS}(N) \approx R_0\left(1 + \frac{1}{N}\sum_{k=1}^{N}(n_u(k)/u_0 - n_i(k)/i_0)\right) = R_0 + \Delta R \qquad (1\text{-}16)$$

The approximated variance $\text{var}(\hat{R}_{LS}(N))$ is (using Assumption 1.1)

$$\text{var}(\hat{R}_{LS}(N)) = \mathbb{E}\{(\Delta R)^2\} = \frac{R_0^2}{N}\left(\frac{\sigma_u^2}{u_0^2} + \frac{\sigma_i^2}{i_0^2}\right) \qquad (1\text{-}17)$$

with $\mathbb{E}\{x\}$ the expected value of x. Note that during the calculation of the variance, the shift of the mean value of $\hat{R}_{LS}(N)$ is not considered because it is a second-order contribution.

For the other two estimators, exactly the same results are found:

$$\text{var}(\hat{R}_{EV}(N)) = \text{var}(\hat{R}_{SA}(N)) = \frac{R_0^2}{N}\left(\frac{\sigma_u^2}{u_0^2} + \frac{\sigma_i^2}{i_0^2}\right) \qquad (1\text{-}18)$$

The result $\text{var}(\hat{R}_{SA}(N))$ is valid only if the expected values exist.

Again, a number of interesting conclusions can be drawn from this result

- The standard deviation is proportional to $1/\sqrt{N}$, as was found before in Figure 1-5.
- Although it is possible to reduce the variance by averaging over repeated measurements, this is no excuse for sloppy experiments because the uncertainty is inversely proportional to the SNR of the measurements. Increasing the SNR requires many more measurements in order to get the same final uncertainty on the estimates.
- The variances of the three estimators should be the same. This seems to conflict with the results of Figure 1-5. However, the theoretical expressions are based on first-order approximations. If the SNR drops to values that are too small, the second-order moments are no longer negligible. In order to check this, the students set up a simulation and tune the noise parameters so that they get the same behavior as they observed in their measurements. These values are: $i_0 = 1$ A, $u_0 = 1$ V, $\sigma_i = 1$ A, $\sigma_u = 1$ V. The noise of group A is normally distributed and uniformly distributed for group B. Next they vary the standard deviations and plot the results in Figure 1-7 for $\hat{R}_{EV}(N)$ and $\hat{R}_{LS}(N)$. Here it is clear that for higher SNR the uncertainties coincide, whereas they differ significantly for the lower SNR. To give closed form mathematical expressions for this behavior, it is not enough any more to specify the first- and second-order moments of the noise (mean, variance); the higher order moments or the pdf of the noise are also required (see Section 16.15).

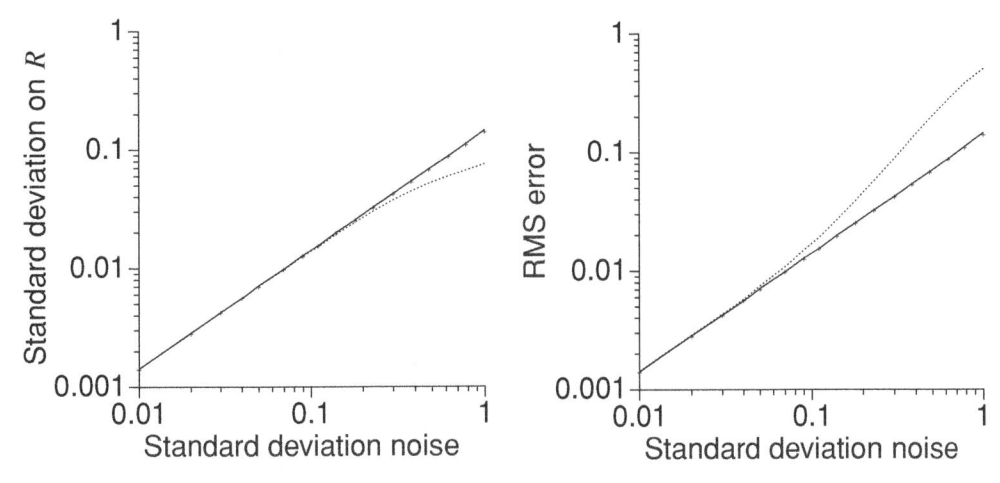

Figure 1-7. Evolution of the standard deviation and the rms error on the estimated resistance value as a function of the standard deviation of the noise ($\sigma_u = \sigma_i$). Solid lines: $\hat{R}_{EV}(N)$, dotted lines: $\hat{R}_{LS}(N)$, and '+' the theoretical value σ_R.

- Although $\hat{R}_{LS}(N)$ has a smaller variance than $\hat{R}_{EV}(N)$ for low SNR, its total root mean square (rms) error (difference with respect to the true value) is significantly larger because of its systematic error. The following is quite a typical observation: many estimators reduce the stochastic error at the cost of systematic errors. For the $\hat{R}_{EV}$ the rms error is completely due to the variability of the estimator because the rms error coincides completely with the theoretical curve of the standard deviation.

1.2.3 Interpretation of the Estimators: A Cost Function–Based Approach

The previous section showed that there is not just one single estimator for each problem. Moreover, the properties of the estimators can vary quite a lot. This raises two questions: how can we generate good estimators and how can we evaluate their properties? The answers are given in this and the following sections. In order to recognize good estimators it is necessary to specify what a good estimator is. This is done in the next section. First we will deal with the question of how estimators are generated. Again, there exist different approaches. A first group of methods starts from a deterministic approach. A typical example is the observation that the noiseless data should obey some model equations. The system parameters are then extracted by intelligent manipulation of these equations, usually inspired by numerical or algebraic techniques. Next, the same procedure is used on noisy data. The major disadvantage of this approach is that it does not guarantee at all that the resulting estimator has good noise behavior. The estimates can be extremely sensitive to disturbing noise. The alternative is to embed the problem in a stochastic framework. A typical question to be answered is: where does the disturbing noise sneak into my problem and how does it behave? To answer this question, it is necessary to make a careful analysis of the measurement setup. Next, the best parameters are selected using statistical considerations. In most cases these methods lead to a cost function interpretation and the estimates are found as the arguments that minimize the cost function. The estimates of the previous section can be found as the minimizers of the following cost functions:

$\hat{R}_{SA}(N)$: Consider the successive resistance estimates $R(k) = u(k)/i(k)$. The overall estimate after N measurements is then the argument minimizing the following cost function:

$$\hat{R}_{SA}(N) = \arg\min_R V_{SA}(R, N) \text{ with } V_{SA}(R, N) = \frac{1}{2}\sum_{k=1}^N (R(k) - R)^2 \qquad (1\text{-}19)$$

This is the simplest approach ("SA" stands for simple approach) of the estimation problem. As seen before, it has very poor properties.

$\hat{R}_{LS}(N)$: A second possibility is to minimize the equation errors in the model equation $u(k) - Ri(k) = e(k, R)$ in least squares (LS) sense. For noiseless measurements $e(k, R_0) = 0$, with R_0 the true resistance value,

$$\hat{R}_{LS}(N) = \arg\min_R V_{LS}(R, N) \text{ with } V_{LS}(R, N) = \frac{1}{2}\sum_{k=1}^N e^2(k, R) \qquad (1\text{-}20)$$

$\hat{R}_{EV}(N)$: The basic idea of the last approach is to express that the current as well as the voltage measurements are disturbed by noise. This is called the errors-in-variables (EV) approach. The idea is to estimate the exact current and voltage (i_0, u_0), parameterized as (i_p, u_p), keeping in mind the model equation $u_0 = Ri_0$.

$$\hat{R}_{EV}(N) = \arg\min_{R, i_p, u_p} V_{EV}(R, i_p, u_p, N) \text{ subject to } u_p = Ri_p$$
$$V_{EV}(R, i_p, u_p, N) = \frac{1}{2}\sum_{k=1}^N (u(k) - u_p)^2 + \frac{1}{2}\sum_{k=1}^N (i(k) - i_p)^2 \qquad (1\text{-}21)$$

This wide variety of possible solutions and motivations illustrates the need for a more systematic approach. In this book we put the emphasis on a stochastic embedding approach, selecting a cost function on the basis of a noise analysis of the general measurement setup that is used.

All the cost functions presented here are of the "least squares" type. Again, there exist many other possibilities, for example, the sum of the absolute values. There are two reasons for choosing a quadratic cost: first, it is easier to minimize than other functions, and second, we will show that normally distributed disturbing noise leads to a quadratic criterion. This does not imply that it is the best choice from all points of view. If it is known that some outliers in the measurements can appear (due to exceptionally large errors, a temporary sensor failure, a transmission error, etc.), it may be better to select a least absolute values cost function (sum of the absolute values) because these outliers are strongly emphasized in a least squares concept (Huber, 1981; Van den Bos, 1985). Sometimes a mixed criterion is used; for example, the small errors are quadratically weighted while the large errors only appear linear in the cost to reduce the impact of outliers (Ljung, 1995).

1.3 DESCRIPTION OF THE STOCHASTIC BEHAVIOR OF ESTIMATORS

Because the estimates are obtained as a function of a finite number of noisy measurements, they are stochastic variables as well. Their pdf is needed in order to characterize them completely. However, in practice it is usually very hard to derive it, so that the behavior of the estimates is described by a few numbers only, such as their mean value (as a description of the location) and the covariance matrix (to describe the dispersion). Both aspects are discussed in the following. A detailed discussion is given in Chapter 16.

1.3.1 Location Properties: Unbiased and Consistent Estimates

The choice for the mean value is not obvious at all from a theoretical point of view. Other location parameters such as the median or the mode (Stuart and Ord, 1987) could also be used, but the latter are much more difficult to analyze in most cases. As it can be shown that many estimates are asymptotically normally distributed under weak conditions, this choice is not so important because, in the normal case, these location parameters coincide. It seems very natural to require that the mean value equals the true value, but it turns out to be impractical. What are the true parameters of a system? We can speak about true parameters only if an exact model exists. It is clear that this is a purely imaginary situation because in practice we always stumble on model errors so that only excitation-dependent approximations can be made. For theoretical reasons, it still makes sense to consider the concept of "true parameters," but it is clear at this point that we have to generalize to more realistic situations. One possible generalization is to consider the estimator evaluated in the noiseless situation as the "best" approximation. These parameters are then used as a reference value to compare the results obtained from noisy measurements. The goal is then to remove the influence of the disturbing noise so that the estimator converges to this reference value.

Definition 1.2 (Unbiasedness): An estimator $\hat{\theta}$ of the parameters θ_0 is unbiased if $\mathbb{E}\{\hat{\theta}\} = \theta_0$ for all true parameters θ_0. Otherwise, it is a biased estimator.

If the expected value equals the true value only for an infinite number of measurements, then the estimator is called asymptotically unbiased. In practice, it turns out that (asymptotic) unbiasedness is a hard requirement to deal with.

Example 1.3 (Unbiased and Biased Estimators): At the end of their experiments, the students want to estimate the value of the voltage over the resistor. Starting from the measurements (1-4), they first carry out a noise analysis of their measurements by calculating the sample mean value and the sample variance:

$$\hat{u}(N) = \frac{1}{N}\sum_{k=1}^{N} u(k) \quad \text{and} \quad \hat{\sigma}_u^2(N) = \frac{1}{N}\sum_{k=1}^{N} (u(k) - \hat{u}(N))^2 \qquad (1\text{-}22)$$

Applying the previous definition, it is readily seen that

$$\mathbb{E}\{\hat{u}(N)\} = \frac{1}{N}\sum_{k=1}^{N} \mathbb{E}\{u(k)\} = \frac{1}{N}\sum_{k=1}^{N} u_0 = u_0 \qquad (1\text{-}23)$$

because the noise is zero mean, so that their voltage estimate is unbiased. The same can be done for the variance estimate:

$$\mathbb{E}\{\hat{\sigma}_u^2(N)\} = \frac{N-1}{N}\sigma_u^2 \qquad (1\text{-}24)$$

This estimator shows a systematic error of σ_u^2/N and is thus biased. However, as $N \to \infty$ the bias disappears, and following the definitions it is asymptotically unbiased. It is clear that a better estimate would be $\sum_{k=1}^{N}(u(k) - \hat{u}(N))^2/(N-1)$, which is the expression that is found in the handbooks on statistics. □

For many estimators, it is very difficult or even impossible to find the expected value analytically. Sometimes it does not even exist, as is the case for $\hat{R}_{SA}(N)$ of group A. Moreover, unbiased estimators can still have a bad distribution; for example, the pdf of the estimator is symmetrically distributed around its mean value, with a minimum at the mean value. Consequently, a more handy tool (e.g., consistency) is needed.

Definition 1.4 (Consistency): An estimator $\hat{\theta}(N)$ of the parameters θ_0 is weakly consistent if it converges in probability to θ_0: $\plim_{N \to \infty} \hat{\theta}(N) = \theta_0$ and strongly consistent if it converges with probability one (almost surely) to θ_0: $\operatorname*{a.s.lim}_{N \to \infty} \hat{\theta}(N) = \theta_0$.

The precise explanation of these probability limits is given in Section 16.6. Loosely explained, it means that the pdf of $\hat{\theta}(N)$ contracts around the true value θ_0, or $\lim_{N \to \infty} \operatorname{Prob}(|\hat{\theta}(N) - \theta_0| > \delta > 0) = 0$. The major advantage of the consistency concept is purely mathematical: it is much easier to prove consistency than unbiasedness using probabilistic theories starting from the cost function interpretation. A general outline of how to prove consistency is given in Section 17.3. Another nice property of the plim is that it can be interchanged with a continuous function: $\plim f(a) = f(\plim(a))$ if both limits exist (see Section 16.8). In fact, it was this property that we applied during the calculations of the limit values of $\hat{R}_{LS}$ and $\hat{R}_{EV}$, for example,

$$\plim_{N \to \infty} \hat{R}_{EV}(N) = \plim_{N \to \infty} \frac{\frac{1}{N}\sum_{k=1}^{N} u(k)}{\frac{1}{N}\sum_{k=1}^{N} i(k)} = \frac{\plim_{N \to \infty} \frac{1}{N}\sum_{k=1}^{N} u(k)}{\plim_{N \to \infty} \frac{1}{N}\sum_{k=1}^{N} i(k)} = \frac{u_0}{i_0} = R_0 \quad (1\text{-}25)$$

Consequently, $\hat{R}_{EV}(N)$ is a weakly consistent estimator. Calculating the expected value is much more involved in this case due to the division. Therefore, consistency is better suited than (asymptotic) unbiasedness to study it.

1.3.2 Dispersion Properties: Efficient Estimators

In this book the covariance matrix is used to measure the dispersion of an estimator, that is, to ascertain how much the actual estimator is scattered around its limiting value. Again, this choice, among other possibilities (for example, percentiles), is highly motivated from a mathematical point of view. Within the stochastic framework used, it will be quite easy to calculate the covariance matrix, whereas it is much more involved to obtain the other measures. For normal distributions, all dispersion measures are obtainable from the covariance matrix so that for most estimators this choice is not too restrictive because their distribution converges to a normal one.

As users, we are highly interested in estimators with minimal errors. However, because we can collect only a finite number of noisy measurements, it is clear that there are limits on the accuracy and precision we can reach. This is precisely quantified in the Cramér-Rao inequality. This inequality provides a lower bound on the covariance matrix of a(n) (un)biased estimator starting from the likelihood function. First we introduce the likelihood function, then we present the Cramér-Rao lower bound.

Consider the measurements $z \in \mathbb{R}^N$ obtained from a system described by a hypothetical, exact model that is parameterized in θ. These measurements are disturbed by noise and are, hence, stochastic variables that are characterized by a probability density function $f(z|\theta_0)$ that depends on the exact model parameters θ_0 with $\int_{z \in \mathbb{R}^N} f(z|\theta_0)dz = 1$. Next we can interpret this relation conversely, namely, how likely is it that a specific set of measurements $z = z_m$ are generated by a system with parameters θ? In other words, we now consider a given set of measurements and view the model parameters as the free variables:

$$L(z_m|\theta) = f(z = z_m|\theta) \tag{1-26}$$

with θ the free variables. $L(z_m|\theta)$ is called the likelihood function. In many calculations the log likelihood function $l(z|\theta) = \ln(L(z|\theta))$ is used. In (1-26) we used z_m to indicate explicitly that we use the numerical values of the measurements that were obtained from the experiments. From here on, we just use z as a symbol because it will be clear from the context what interpretation should be given to z. The reader should be aware that $L(z|\theta)$ is not a probability density function with respect to θ because $\int_\theta L(z|\theta)d\theta \neq 1$. Notice the subtle difference in terminology; that is, probability is replaced by likeliness.

The Cramér-Rao lower bound gives a lower limit on the covariance matrix of parameters. Under quite general conditions, this limit is universal and independent of the selected estimator: no estimator that violates this bound can be found. It is given by (see Section 16.12)

$$CR(\theta_0) = \left(I_{n_\theta} + \frac{\partial b_\theta}{\partial \theta}\right)^T Fi^{-1}(\theta_0)\left(I_{n_\theta} + \frac{\partial b_\theta}{\partial \theta}\right)$$

$$Fi(\theta_0) = \mathbb{E}\left\{\left(\frac{\partial l(z|\theta)}{\partial \theta}\right)^T\left(\frac{\partial l(z|\theta)}{\partial \theta}\right)\right\} = -\mathbb{E}\left\{\frac{\partial^2 l(z|\theta)}{\partial \theta^2}\right\} \tag{1-27}$$

The derivatives are calculated in $\theta = \theta_0$, and $b_\theta = \mathbb{E}\{\hat{\theta}\} - \theta_0$ is the bias on the estimator. Note that for biased estimators ($\partial b_\theta/\partial \theta \neq 0$) the lower bound (1-27) can be zero: $CR(\theta_0) = 0$ (see Example 16.20 on page 590). For unbiased estimators (1-27) reduces to $CR(\theta_0) = Fi^{-1}(\theta_0)$.

$Fi(\theta)$ is called the Fisher information matrix; it is a measure of the information in an experiment: the larger the matrix, the more information there is. In (1-27) it is assumed that the first and second derivatives of the log likelihood function exist with respect to θ.

Example 1.5 (Influence of the Number of Parameters on the Cramér-Rao Lower Bound): A group of students want to determine the flow of tap water by measuring the height $h_0(t)$ of the water in a measuring jug as a function of time t. However, their work is not precise and in the end they are not sure about the exact starting time of their experiment. They include it in the model as an additional parameter: $h_0(t) = a(t - t_{\text{start}}) = at + b$, and $\theta = [a, b]^T$. Assume that the noise $n_h(k)$ on the height measurements is iid zero mean normally distributed $N(0, \sigma^2)$, and the noise on the time instances is negligible $h(k) = at_k + b + n_h(k)$; then the following stochastic model can be used:

$$\text{Prob}(h(k), t_k) = \text{Prob}(h(k) - (at_k + b)) = \text{Prob}(n_h(k))$$

where $\text{Prob}(h(k), t_k)$ is the probability of making the measurements $h(k)$ at t_k. The likelihood function for the set of measurements $h = \{(h(1), t_1), ..., (h(N), t_N)\}$ is

$$L(h|a, b) = \frac{1}{(2\pi\sigma^2)^{N/2}} e^{-\frac{1}{2\sigma^2}\sum_{k=1}^{N}(h(k) - at_k - b)^2} \qquad (1\text{-}28)$$

and the log likelihood function becomes

$$l(h|a, b) = -\frac{N}{2}\log(2\pi\sigma^2) - \frac{1}{2\sigma^2}\sum_{k=1}^{N}(h(k) - at_k - b)^2 \qquad (1\text{-}29)$$

The Fisher information matrix and the Cramér-Rao lower bound are found using (1-27):

$$Fi(a, b) = \frac{N}{\sigma^2}\begin{bmatrix} s^2 & \mu \\ \mu & 1 \end{bmatrix} \rightarrow CR(a, b) = Fi^{-1}(a, b) = \frac{\sigma^2}{N(s^2 - \mu^2)}\begin{bmatrix} 1 & -\mu \\ -\mu & s^2 \end{bmatrix} \qquad (1\text{-}30)$$

with $\mu = \sum_{k=1}^{N} t_k/N$ and $s^2 = \sum_{k=1}^{N} t_k^2/N$. These expressions are very informative. First, we can note that the attainable uncertainty is proportional to the standard deviation of the noise. This means that inaccurate measurements result in poor estimates, or identification is no excuse for sloppy measurements. The uncertainty decreases as $\sqrt{N}$, which can be used as a rule of thumb whenever independent measurements are processed. Finally, it can also be noted that the uncertainty depends on the actual time instances used in the experiment. In other words, by making a proper design of the experiment, it is possible to influence the uncertainty on the estimates. This idea will be exploited fully in Chapter 5. Another question we can now answer is what price is paid to include the additional model parameter b to account for the unknown starting time. By comparing $Fi^{-1}(a, b)$ with $Fi^{-1}(a)$ (assuming that b is known), it is found that

$$\sigma_a^2(a, b) = \frac{\sigma^2}{N(s^2 - \mu^2)} \geq \frac{\sigma^2}{Ns^2} = \sigma_a^2(a) \qquad (1\text{-}31)$$

where $\sigma_a^2(a, b)$ is the lower bound on the variance of a if both parameters are estimated, else $\sigma_a^2(a)$ is the lower bound if only a is estimated. This shows that adding additional parameters to a model increases the minimum attainable uncertainty on it. Of course, these parameters may be needed to remove systematic errors so that a balance between stochastic errors and systematic errors is achieved. This is further elaborated in Chapter 11. □

The Cramér-Rao lower bound is a conservative estimate of the smallest possible covariance matrix that is not always attainable (the values may be too small). Tighter bounds exist (Abel, 1993), but these are more involved to calculate. Consequently, the Cramér-Rao bound is the criterion most used to verify the efficiency of an estimator.

Definition 1.6 (Efficiency): An unbiased estimator is called efficient if its covariance matrix is smaller than that of any other unbiased estimator.

An unbiased estimator that reaches the Cramér-Rao lower bound is also an efficient estimator. For biased estimators, a generalized expression should be used (see Section 16.12).

1.4 BASIC STEPS IN THE IDENTIFICATION PROCESS

Each identification session consists of a series of basic steps. Some of them may be hidden or selected without the user being aware of his/her choice. Clearly, this can result in poor or suboptimal results. In each session the following actions should be taken:

- Collect information about the system.
- Select a model structure to represent the system.
- Choose the model parameters to fit as well as possible the model to the measurements: selection of a "goodness of fit" criterion.
- Validate the selected model.

Each of these points is discussed in more detail below.

1.4.1 Collect Information about the System

If we want to build a model for a system, we should get information about it. This can be done by just watching the natural fluctuations (e.g., vibration analysis of a bridge that is excited by normal traffic), but most often it is more efficient to set up dedicated experiments that actively excite the system (e.g., controlled excitation of a mechanical structure using a shaker). In the latter case, the user has to select an excitation that optimizes his/her own goal (for example, minimum cost, minimum time, or minimum power consumption for a given measurement accuracy) within the operator constraints (e.g., the excitation should remain below a maximum allowable level). The quality of the final result can depend heavily on the choices that are made. Later, we will thoroughly discuss the selection of the excitation signals.

1.4.2 Select a Model Structure to Represent the System

A choice should be made within all the possible mathematical models that can be used to represent the system. Again, a wide variety of possibilities exist, such as

- Parametric versus nonparametric models
  In a parametric model, the system is described using a limited number of characteristic quantities called the parameters of the model, whereas in a nonparametric model the system is characterized by measurements of a system function at a large number of points. Examples of parametric models are the transfer function of a filter described by its poles and zeros and the motion equations of a piston. An example of a nonparametric model is the description of a filter by its impulse response at a large number of points.
  Usually it is simpler to create a nonparametric model than a parametric one because the modeler needs less knowledge about the system itself in the former case. However, physical insight and concentration of information are more substantial for parametric models than for nonparametric ones. We will concentrate on transfer function models (parametric models), but the problem of frequency response function measurements (nonparametric model) will also be elaborated.
- White box models versus black box models
  In the construction of a model, physical laws whose availability and applicability depend on the insight and skills of the experimenter can be used (Kirchhoff's laws,

Newton's laws, etc.). Specialized knowledge related to different scientific fields may be brought into this phase of the identification process. The modeling of a loudspeaker, for example, requires extensive understanding of mechanical, electrical, and acoustical phenomena. The result may be a physical model, based on comprehensive knowledge of the internal functioning of the system. Such a model is called a white box model.

Another approach is to extract a black box model from the data. Instead of making a detailed study and developing a model based upon physical insight and knowledge, a mathematical model is proposed that allows sufficient description of any observed input and output measurements. This reduces the modeling effort significantly. For example, instead of modeling the loudspeaker using physical laws, an input-output relation, taking the form of a high-order transfer function, can be proposed.

The choice between the different methods depends on the aim of the study: the white box approach is better for gaining insight into the working principles of a system, but a black box model may be sufficient if the model will be used only for prediction of the output.

Although, as a rule of thumb, it is advisable to include as much prior knowledge as possible during the modeling process, it is not always easy to do. If we know, for example, that a system is stable, it is not simple to express this information if the polynomial coefficients are used as parameters.

- Linear models versus nonlinear models
  In real life, almost every system is nonlinear. Because the theory of nonlinear systems is very involved, these are mostly approximated by linear models, assuming that in the operation region the behavior can be linearized. This kind of approximation makes it possible to use simple models without jeopardizing properties that are of importance to the modeler. This choice depends strongly on the intended use of the model. For example, a nonlinear model is needed to describe the distortion of an amplifier, but a linear model will be sufficient to represent its transfer characteristics if the linear behavior is dominant and is the only interest.

- Linear-in-the-parameters versus nonlinear-in-the-parameters
  A model is called linear-in-the-parameters if there exists a linear relation between these parameters and the error that is minimized. This does not imply that the system itself is linear. For example, $\varepsilon = y - (a_1 u + a_2 u^2)$ is linear in the parameters a_1 and a_2 but describes a nonlinear system. On the other hand,

$$\varepsilon(j\omega) = Y(j\omega) - \frac{a_0 + a_1 j\omega}{b_0 + b_1 j\omega} U(j\omega)$$

describes a linear system but the model is nonlinear in the b_1 and b_2 parameters. Linearity in the parameters is a very important aspect of models because it has a strong impact on the complexity of the estimators if a (weighted) least squares cost function is used. In that case, the problem can be solved analytically for models that are linear in the parameters so that an iterative optimization problem is avoided. This is illustrated in Section 1.5.1.

1.4.3 Match the Selected Model Structure to the Measurements

Once a model structure is chosen (e.g., a parametric transfer function model), it should be matched as well as possible with the available information about the system. Mostly, this is done by minimizing a criterion that measures a goodness of the fit. The choice of this criterion is extremely important because it determines the stochastic properties of the final estimator. As seen from the resistance example, many choices are possible and each of them can lead to a different estimator with its own properties. Usually, the cost function defines a distance between the experimental data and the model. The cost function can be chosen on an ad hoc basis using intuitive insight, but there also exists a more systematic approach based on stochastic arguments as explained in Section 1.5. Simple tests on the cost function exist (necessary conditions) to check even before deriving the estimator whether it can be consistent (see Chapter 9, Section 9.5).

1.4.4 Validate the Selected Model

Finally, the validity of the selected model should be tested: does this model describe the available data properly or are there still indications that some of the data are not well modeled, indicating remaining model errors? In practice, the best model (meaning the one with the smallest errors) is not always preferred. Often a simpler model that describes the system within user-specified error bounds is preferred. Tools will be provided that guide the user through this process by separating the remaining errors into different classes, for example, unmodeled linear dynamics and nonlinear distortions. From this information, further improvements of the model can be proposed, if necessary.

During the validation tests it is always important to keep the application in mind. The model should be tested under the same conditions as will be used later. Extrapolation should be avoided as much as possible. The application also determines what properties are critical.

1.4.5 Conclusion

This brief overview of the identification process shows that it is a complex task with a number of interacting choices. It is important to pay attention to all aspects of this procedure, from the experiment design to the model validation, in order to get the best results. The reader should be aware that, besides this list of actions, other aspects are also important. A short inspection of the measurement setup can reveal important shortcomings that can jeopardize a lot of information. Good understanding of the intended applications helps to set up good experiments, and is very important to make the proper simplifications during the model-building process. Many times, choices are made that are not based on complicated theories but are dictated by the practical circumstances. In these cases, a good theoretical understanding of the applied methods will help users to be aware of the sensitive aspects of their techniques. This will enable them to put all their effort on the most critical decisions. Moreover, they will become aware of the weak points of the final model.

1.5 A STATISTICAL APPROACH TO THE ESTIMATION PROBLEM

In the previous sections it was shown that an intuitive approach to a parameter estimation problem can cause serious errors without even being noticed. To avoid severe mistakes, a theoretical framework is needed. Here, a statistical development of the parameter estimation theory is made. Four related estimators are studied: the least squares (LS) estimator, weighted least squares (WLS) estimator, maximum likelihood (ML) estimator, and, finally, the Bayes estimator. It should be clear that, as mentioned before, it is still possible to use other estimators, such as the least absolute values. However, a comprehensive overview of all possible techniques is beyond the scope of this book.

To use the Bayes estimator, the a priori probability density function (pdf) of the unknown parameters and the pdf of the noise on the measurements are required. Although it seems, at first, quite strange that the parameters have a pdf, we will illustrate in the next section that we use this concept regularly in daily life. The ML estimator requires only knowledge of the pdf of the noise on the measurements, and the WLS estimator can be applied optimally if the covariance matrix of the noise is known. Even if this information is lacking, the LS method is usable. Each of these estimators will be explained in more detail and illustrated in the following sections.

1.5.1 Least Squares Estimation

One of the simplest estimation techniques is the least squares estimator. In this case, the match between the model and the measurements is quantified by a least squares cost function. As this is an arbitrary choice, initially, it is clear that the result is not necessarily optimal. By choosing other cost functions such as the sum of the least absolute values, it is possible to find other estimators, with different properties, that perform better in specific situations. Some of these are studied explicitly in the literature. In this book we concentrate on least squares, a choice strongly motivated by numerical aspects: minimizing a least squares cost function is usually less involved than the alternative cost functions. Later on, this choice will also be shown to be motivated from the stochastic point of view. Normally distributed noise leads, naturally, to least squares estimation. As seen in the resistance example, even within the class of least squares estimators, there are different possibilities resulting in completely different estimators. A full treatment of the problem is beyond the scope of this book, hence, we focus only on the aspects that are of direct importance to our major goal.

Consider a multiple-input, single-output system modeled by $y_0(k) = g(u_0(k), \theta_0)$ with k the measurement index, $y(k) \in \mathbb{R}$, $u_0(k) \in \mathbb{R}^{1 \times n_u}$, and $\theta_0 \in \mathbb{R}^{n_\theta}$ the true parameter vector. The aim is to estimate the parameters from noisy observations at the output of the system: $y(k) = y_0(k) + n_y(k)$. This is done by minimizing the sum of the squared errors $e(k, \theta) = y(k) - y(k, \theta)$, with $y(k, \theta)$ the modeled output:

$$\hat{\theta}_{\mathrm{NLS}}(N) = \arg\min_{\theta} V_{\mathrm{NLS}}(\theta, N), \text{ with } V_{\mathrm{NLS}}(\theta, N) = \frac{1}{2}\sum_{k=1}^{N} e^2(k, \theta) \qquad (1\text{-}32)$$

In general, the analytical solution of the nonlinear least squares problem (1-32) is not known, so numerical methods must be used. A number of techniques are described in the literature (Fletcher, 1991), and many are found in commercially available mathematical packages. They vary from very simple techniques such as simplex methods that require no derivatives at all, through gradient or steepest descent methods (based on first-order derivatives), to Newton methods that make use of second-order derivatives. The optimal choice strongly depends on the specific problem. However, the Gauss-Newton method is very well suited to deal with the least squares minimization problem because it makes explicit use of the structure of the cost function. The second derivatives of the cost function (the Hessian matrix) are approximated in this method by the first-order derivatives of $e(\theta)$. Define the Jacobian matrix $J(\theta) \in \mathbb{R}^{N \times n_\theta}$: $J(\theta) = \partial e(\theta)/\partial \theta$ and consider the Hessian matrix:

Section 1.5 ■ A Statistical Approach to the Estimation Problem

$$\frac{\partial^2 V_{\text{NLS}}(\theta, N)}{\partial \theta^2} = J^T(\theta)J(\theta) - \sum_{k=1}^{N} e(k, \theta)\frac{\partial^2 g(u_0(k), \theta)}{\partial \theta^2} \tag{1-33}$$

If the second term in (1-33) is small (for example, $\|e(\theta)\|_2$ is "small") with respect to the first one, then $J^T(\theta)J(\theta)$ will be a good approximation for the second-order derivatives of the cost function. The numerical solution is then found by applying the following iterative process:

$$\theta^{(i+1)} = \theta^{(i)} + \Delta\theta^{(i+1)} \text{ with } J^T(\theta^{(i)})J(\theta^{(i)})\Delta\theta^{(i+1)} = -J^T(\theta^{(i)})e(\theta^{(i)}) \tag{1-34}$$

Equation (1-34) reveals two important advantages. First, only the gradient needs to be calculated, and not the Hessian, thus reducing the calculation time. Moreover, very often, the condition number of the Hessian matrix is the square of that of the Jacobian. This leads us to the second advantage: using, for example, singular value decomposition (SVD) or QR decomposition techniques, (1-34) can be solved without forming the product $J^T(\theta^{(i)})J(\theta^{(i)})$ so that more complex problems can be solved, because the numerical errors are significantly reduced (see Exercise 1.12). If (1-34) converges to the global minimum of (1-32), then $\hat{\theta}_{\text{NLS}}(N) = \theta^{(\infty)}$.

Because there are no explicit expressions available for the estimator as a function of the measurements, it is not straightforward to study its properties. For this reason, special theories are developed to analyze the properties of the estimator by analyzing the cost function. These techniques are covered in detail in Section 19.4. Under quite general assumptions on the noise (for example, iid noise with finite second- and fourth-order moments), some regularity conditions on the model $g(u_0(k), \theta)$, and the excitation (choice of $u_0(k)$), consistency of the least squares estimator is proved. Also, an approximate expression for the covariance matrix $\text{Cov}(\hat{\theta}_{\text{NLS}}(N))$ is available:

$$\text{Cov}(\hat{\theta}_{\text{NLS}}(N)) \approx (J^T(\theta)J(\theta))^{-1}J^T(\theta)\text{Cov}(n_y)J(\theta)(J^T(\theta)J(\theta))^{-1}\Big|_{\theta = \hat{\theta}_{\text{LS}}(N)} \tag{1-35}$$

with $\text{Cov}(n_y) = \mathbb{E}\{n_y n_y^T\}$. Note that this approximation is still a stochastic variable because it depends on $\hat{\theta}_{\text{NLS}}(N)$, while the exact expression should be in θ_0. If the model is linear-in-the-parameters, $y_0 = K(u_0)\theta_0$, and $e(\theta) = y - K(u_0)\theta$, then (1-32) reduces to a linear least squares cost function, and explicit expressions are available for the estimator (note that $K = -\partial e(\theta)/\partial\theta = -J(\theta)$ is parameter independent in this case). In order to keep the expressions compact, we do not include the arguments of K in the following:

$$\hat{\theta}_{\text{LS}}(N) = (K^T K)^{-1} K^T y \tag{1-36}$$

The covariance matrix still equals (1-35) with $J(\hat{\theta}_{\text{LS}}(N))$ replaced by $-K$, but now it is an exact expression and no longer an approximation. Moreover, it is possible to prove that the estimator is unbiased for zero mean noise:

$$\mathbb{E}\{\hat{\theta}_{\text{LS}}(N)\} = (K^T K)^{-1} K^T \mathbb{E}\{y\} = (K^T K)^{-1} K^T y_0 = (K^T K)^{-1} K^T K \theta_0 = \theta_0 \tag{1-37}$$

This result is valid only if K is not disturbed by noise. If the inputs u are also disturbed by noise, it is no longer possible to bring $(K^TK)^{-1}K^T$ outside the expectation. In this case, additional quadratic noise contributions appear in K^TK so that $\hat{\theta}_{LS}(N)$ underestimates the true values. This was visible in the estimation of the resistance ($K_{[k]} = i(k)$, $y(k) = u(k)$, $\theta = R$) where (1-8) shows the impact of the quadratic contributions of the input noise.

Example 1.7 (Weighing a Loaf of Bread): John is asked to estimate the weight of a loaf of bread from N noisy measurements $y(k) = \theta_0 + n_y(k)$ with θ_0 the true but unknown weight, $y(k)$ the weight measurement, and $n_y(k)$ the measurement noise. From a prior analysis, making repeated measurements, it turns out that $n_y(k)$ is zero mean iid with variance σ_y^2. The model becomes $y = K\theta + n_y$ with $K = (1, 1, \ldots, 1)^T$. Using (1-36), the estimate is

$$\hat{\theta}_{LS}(N) = (K^TK)^{-1}K^Ty = \frac{1}{N}\sum_{k=1}^{N} y(k) \tag{1-38}$$

with variance

$$\text{var}(\hat{\theta}_{LS}(N)) = (K^TK)^{-1}K^T(\sigma_y^2 I_N)K(K^TK)^{-1} = \sigma_y^2/N \tag{1-39}$$

This example shows that it is much easier to get the solution when it is possible to formulate the problem under the standard conditions. □

This short analysis shows that the least squares estimator is applicable to a very wide range of problems. No prior information is required to use it, which explains its success. However, its specific properties depend on the actual situation. General statements can be made only if some noise characteristics are known. In that case it is also possible to improve the quality of the estimates by using this knowledge in the estimator. If, for example, the covariance matrix of the noise is known, a weighted least squares can be used.

1.5.2 Weighted Least Squares Estimation

In (1-32) all measurements are equally weighted. In many problems it is desirable to put more emphasis on one measurement with respect to the other. This can be done to make the difference between measurements and model smaller in some regions, but it can also be motivated by stochastic arguments. If the covariance matrix of the noise is known, then it seems logical to suppress measurements with high uncertainty and to emphasize those with low uncertainty. In practice, it is not always clear what weighting should be used. If it is, for example, known that model errors are present, then the user may prefer to put in a dedicated weighting in order to keep the model errors small in some specific operation regions instead of using the weighting dictated by the covariance matrix.

In general, the weighted nonlinear least squares estimate $\hat{\theta}_{WNLS}(N)$ is

$$\hat{\theta}_{WNLS}(N) = \arg\min_{\theta} V_{WNLS}(\theta, N) \text{ with } V_{WNLS}(\theta, N) = \frac{1}{2}e^T(\theta)We(\theta) \tag{1-40}$$

where $W \in \mathbb{R}^{N \times N}$ is a symmetric positive definite weighting matrix (the asymmetric part does not contribute to a quadratic form). The evaluation of this cost function requires $O(N^2)$ operations, which are very time consuming. Consequently, (block) diagonal weighting matrices are preferred in many problems, reducing the number of operations to $O(N)$. All the remarks on the numerical aspects of the least squares estimator are also valid for the weighted least squares. This can be understood easily by applying the following transformation: $\varepsilon(\theta) = Se(\theta)$ with $S^T S = W$ so that $V_{\text{WNLS}}(\theta, N) = \varepsilon^T(\theta)\varepsilon(\theta)/2$, which is a least squares estimator in the transformed variables. This also leads to the following Gauss-Newton algorithm to minimize the cost function

$$\theta^{(i+1)} = \theta^{(i)} + \Delta\theta^{(i+1)} \text{ with } J^T(\theta^{(i)})WJ(\theta^{(i)})\Delta\theta^{(i+1)} = -J^T(\theta^{(i)})We(\theta^{(i)}) \quad (1\text{-}41)$$

Equation (1-35) is generalized to (noticing that $W^T = W$)

$$\text{Cov}(\hat{\theta}_{\text{WNLS}}(N)) \approx (J^T(\theta)WJ(\theta))^{-1}J^T(\theta)WC_{n_y}WJ(\theta)(J^T(\theta)WJ(\theta))^{-1}\Big|_{\theta = \hat{\theta}_{\text{WNLS}}(N)} \quad (1\text{-}42)$$

with $C_{n_y} = \text{Cov}(n_y)$. By choosing $W = C_{n_y}^{-1}$, the expression simplifies to

$$\text{Cov}(\hat{\theta}_{\text{WNLS}}(N)) \approx [J^T(\hat{\theta}_{\text{WNLS}}(N))C_{n_y}^{-1}J(\hat{\theta}_{\text{WNLS}}(N))]^{-1} \quad (1\text{-}43)$$

In Exercise 1.16 it is shown that among all possible positive definite choices for W, the best one is $W = C_{n_y}^{-1}$ because this minimizes the covariance matrix. The results for models that are linear-in-the-parameters are immediately found, analogous to the least squares estimator. Also, in this case, the weighted least squares is unbiased under the same conditions as the least squares estimator.

1.5.3 The Maximum Likelihood Estimator

Using the covariance matrix of the noise as the weighting matrix allows prior knowledge about the noise on the measurements. However, a full stochastic characterization requires the pdf of the noise distortions. If this knowledge is available, it may be possible to get better results than those attained with a weighted least squares. Maximum likelihood estimation offers a theoretical framework to incorporate the knowledge about the distribution in the estimator. The pdf f_{n_y} of the noise also determines the conditional pdf $f(y|\theta_0)$ of the measurements, given the hypothetical exact model, $y_0 = G(u_0, \theta_0)$, that describes the system and the inputs that excite the system. Assuming, again, an additive noise model $y = y_0 + n_y$, with $y, y_0, n_y \in \mathbb{R}^N$, the likelihood function becomes:

$$f(y|\theta_0, u_0) = f_{n_y}(y - G(u_0, \theta_0)) \quad (1\text{-}44)$$

The maximum likelihood procedure consists of two steps. First the numerical values y_m of the actual measurements are plugged into (1-44) for the variables y, and next the model parameters θ_0 are considered as the free variables. This results in the so-called likelihood function. The maximum likelihood estimate is then found as the maximizer of the likelihood function

$$\hat{\theta}_{\text{ML}}(N) = \arg\max_{\theta} f(y_m|\theta, u_0) \quad (1\text{-}45)$$

From now on, we will no longer explicitly indicate the numerical values y_m but just use the symbol y for the measured values.

Example 1.8 (Weighing a Loaf of Bread—Continued): Consider Example 1.7 again, but assume that more information about the noise is available. This time John knows that the distribution f_y of n_y is normal with zero mean and standard deviation σ_y. With this information he can build an ML estimator:

$$f(y|\theta) = \frac{1}{\sqrt{2\pi\sigma_y^2}} e^{-\frac{(y-\theta)^2}{2\sigma_y^2}} \tag{1-46}$$

and the estimated weight becomes $\hat{\theta}_{ML} = y$. It is therefore not possible to give a better estimate than the measured value itself. If John makes repeated independent measurements $y(1), \ldots, y(N)$, the likelihood function is

$$f(y|\theta) = \frac{1}{(2\pi\sigma_y^2)^{N/2}} e^{-\frac{1}{2\sigma_y^2}\sum_{k=1}^{N}(y(k)-\theta)^2} \tag{1-47}$$

Because $(2\pi\sigma_y^2)^{-N/2}$ is parameter independent, the ML estimate is given by the minimizer of $\sum_{k=1}^{N}(y(k)-\theta)^2/(2\sigma_y^2)$ and becomes

$$\hat{\theta}_{ML}(N) = \frac{1}{N}\sum_{k=1}^{N} y(k) \tag{1-48}$$

This is nothing other than the sample mean of the measurements. It is again easy to check that this estimate is unbiased. Note that in this case the ML estimator and the (weighted) least squares estimator are the same. This is the case only for normally distributed errors. □

The unbiased behavior may not be generalized because the MLE can also be biased. For example, the sample mean and sample variance are shown to be the ML estimates for the mean and the variance of measurements that are identically independent and normally distributed:

$$\hat{\mu}_{ML} = \frac{1}{N}\sum_{k=1}^{N} y(k), \quad \hat{\sigma}_{ML}^2 = \frac{1}{N}\sum_{k=1}^{N}(y(k)-\hat{\mu}_{ML})^2.$$

Although the first estimate is unbiased, the second one can be shown to be prone to a bias of σ^2/N that asymptotically disappears in N:

$$\mathbb{E}\{\hat{\sigma}_{ML}\} = \frac{N-1}{N}\sigma^2.$$

This shows that there is a clear need to understand the properties of the ML estimator better. In the literature, a series of important properties is tabled assuming well-defined experimental conditions. Each time these conditions are met, the user knows in advance, before passing through the complete development process, what the properties of the estimator would be. On the other hand, if the conditions are not met, nothing is guaranteed anymore and a dedicated analysis is, again, required. In this introductory chapter we just make a loose statement of the properties; a very precise description can be found in the literature (Goodwin and Payne, 1977; Caines, 1988).

Properties of the ML Estimator

- *Principle of invariance:* if $\hat{\theta}_{ML}$ is an ML estimator of $\theta \in \mathbb{R}^{n_\theta}$, then $\hat{\theta}_g = g(\hat{\theta}_{ML})$ is an ML estimator of $g(\theta)$ where g is a function, $\hat{\theta}_g \in \mathbb{R}^{n_g}$, and $n_g \leq n_\theta$, with n_θ a finite number.

- *Consistency:* if $\hat{\theta}_{ML}(N)$ is an ML estimator based on N iid random variables, with n_θ independent of N, then $\hat{\theta}_{ML}(N)$ converges to θ_0 almost surely: $\text{a.s.}\lim_{N \to \infty} \hat{\theta}_{ML}(N) = \theta_0$.
  If n_θ depends on N, the property is no longer valid, and the consistency should be checked again. See, for example, the errors-in-variables estimator in the previous section where not only is the resistance value estimated, but also the currents $i(1), \ldots, i(N)$ and voltages $u(1), \ldots, u(N)$. In this case $n_\theta = N + 1$, e.g., the N current values and the unknown resistance value, and the voltage is calculated from the estimated current and resistance value.

- *Asymptotic normality:* if $\hat{\theta}_{ML}(N)$ is an ML estimator based on N iid random variables, with n_θ independent of N, then $\hat{\theta}_{ML}(N)$ converges in law to a normal random variable.
  The importance of this property is that it not only allows one to calculate uncertainty bounds on the estimates but also guarantees that most of the probability mass gets more and more unimodally concentrated around its limiting value.

- *Asymptotic efficiency:* if $\hat{\theta}_{ML}(N)$ is an ML estimator based on N iid random variables, with n_θ independent of N, then $\hat{\theta}_{ML}(N)$ is asymptotically efficient ($\text{Cov}(\hat{\theta}_{ML}(N))$ reaches asymptotically the Cramér-Rao lower bound).

1.5.4 The Bayes Estimator

As described before, the Bayes estimator requires the most prior information before it is applicable, namely the pdf of the noise on the measurements and the pdf of the unknown parameters. The kernel of the Bayes estimator is the conditional pdf of the unknown parameters θ with respect to the measurements y: $f(\theta|u, y)$. This pdf contains complete information about the parameters θ, given a set of measurements y. This makes it possible for the experimenter to determine the best estimate of θ for the given situation. To select this best value, it is necessary to lay down an objective criterion, for example, the minimization of a risk function $C(\theta|\theta_0)$ that describes the cost of selecting the parameters θ if θ_0 are the true but unknown parameters. The estimated parameters $\hat{\theta}$ are found as the minimizers of the risk function weighted with the probability $f(\theta|u, y)$:

$$\hat{\theta}(N) = \arg\min_{\theta_0} \int_{\theta \in \mathbb{D}} C(\theta|\theta_0) f(\theta|u, y) d\theta \qquad (1\text{-}49)$$

For some specific choices of $C(\theta|\theta_0)$, the solution of (1-49) is well known; for example, $C(\theta|\theta_0) = |\theta - \theta_0|^2$ leads to the mean value, and $C(\theta|\theta_0) = |\theta - \theta_0|$ results in the median, which is less sensitive to outliers because these contribute less to the second criterion than to the first (Eykhoff, 1974).

Another objective criterion is to choose the estimate as

$$\hat{\theta}_{Bayes}(N) = \arg\max_{\theta} f(\theta|u, y) \qquad (1\text{-}50)$$

The first and second examples are "minimum risk" estimators, and the last is the Bayes estimator. In practice, it is very difficult to select the best out of these. In the next section, we study the Bayes estimator in more detail. To search for the maximizer of (1-50) the Bayes rule is applied:

$$f(\theta|u, y) = \frac{f(y|\theta, u)f(\theta)}{f(y)} \tag{1-51}$$

In order to maximize the right-hand side of this equation it is sufficient to maximize its numerator, because the denominator is independent of the parameters θ, so that the solution is given by looking for the maximum of $f(y|\theta, u)f(\theta)$. This simple analysis shows that a lot of a priori information is required to use the Bayes estimator: $f(y|\theta, u)$ (also appearing in the ML estimator) and $f(\theta)$. In many problems the parameter distribution $f(\theta)$ is unavailable, and this is one of the main reasons why the Bayes estimator is rarely used in practice (Norton, 1986).

Example 1.9 (Use of the Bayes Estimator in Daily Life): We commonly use some important principles of the Bayes estimator without being aware of it. This is illustrated in the following story: Joan was walking at night in Belgium and suddenly saw a large animal in the far distance. She decided that it was either a horse or an elephant Prob(observation|elephant) = Prob(observation|horse). However, the probability of seeing an elephant in Belgium is much lower than that of seeing a horse: Prob(elephant in Belgium) « Prob(horse in Belgium) so that from the Bayes principle Joan concludes she was seeing a horse. If she was on safari in Kenya instead of Belgium, the conclusion would be opposite, because Prob(elephant in Kenya) » Prob(horse in Kenya).

Joan continued her walk. When she came closer she saw that the animal had big feet, a small tail, and also a long trunk so that she had to review her previous conclusion on the basis of all this additional information: there was an elephant walking on the street. When she passed the corner, she saw that a circus had arrived in town. □

From the previous example it is clear that in a Bayes estimator the prior knowledge of the pdf of the estimated parameters is very important. It also illustrates that it balances our prior knowledge with the measurement information. This is more quantitatively illustrated in the next example.

Example 1.10 (Weighing a Loaf of Bread—Continued): Consider again Example 1.8 but assume this time that the baker told John that the bread normally weighs about $w = 800$ g. However, the weight can vary around this mean value as a result of humidity, the temperature of the oven, and so on, in a normal way with a standard deviation σ_w. With all this information, John knows enough to build a Bayes estimator. Using normal distributions and noticing that $f(y|\theta) = f_y(n_y) = f_y(y-\theta)$, the Bayes estimator is found by maximizing

$$f(y|\theta)f(\theta) = \frac{1}{\sqrt{2\pi\sigma_y^2}} e^{-\frac{(y-\theta)^2}{2\sigma_y^2}} \frac{1}{\sqrt{2\pi\sigma_w^2}} e^{-\frac{(\theta-w)^2}{2\sigma_w^2}} \tag{1-52}$$

and the estimated weight becomes

$$\hat{\theta}_{\text{Bayes}} = \frac{y/\sigma_y^2 + w/\sigma_w^2}{1/\sigma_y^2 + 1/\sigma_w^2} \qquad (1\text{-}53)$$

In this result, two parts can be distinguished: y, the information derived from the measurement, and w, the a priori information from the baker. If the quality of the prior information is high compared with that of the measurements ($\sigma_w \ll \sigma_y$), the estimate is determined mainly by the prior information. If the quality of the prior information is very low compared with the measurements ($\sigma_w \gg \sigma_y$), the estimate is determined mainly by the information from the measurements.

After making several independent measurements $y(1), \ldots, y(N)$ the Bayes estimator becomes

$$\hat{\theta}_{\text{Bayes}}(N) = \frac{\sum_{k=1}^{N} y(k)/\sigma_y^2 + w/\sigma_w^2}{N/\sigma_y^2 + 1/\sigma_w^2} \qquad (1\text{-}54)$$

The previous conclusions remain valid. However, when the number of measurements increases, the first term dominates the second one such that the impact of the prior information is reduced (Sörenson, 1980). Finally, when N becomes infinite, the estimate is completely determined by the measurements. □

Conclusion. From these examples it is seen that a Bayes estimator combines prior knowledge of the parameters with information from measurements. When the number of measurements is increased, the measurement information becomes more important and the influence of the prior information decreases. If there is no information about the distribution of the parameters, the Bayes estimator reduces to the ML estimator. If the noise is normally distributed, the ML estimator reduces to the weighted least squares. If the noise is white, the weighted least squares boils down to the least squares estimator.

1.5.5 Instrumental Variables

In this section we will discuss a final parameter estimation method that is very suitable when both the input and the output are disturbed by noise. Although it does not belong directly to the previous family of estimators, we include it in this chapter for use later, to interpret one of the proposed identification schemes. In the resistance estimation examples, it was shown that the least squares method $\hat{R}_{\text{LS}}(N)$ is biased because of the quadratic noise contributions appearing in the denominator:

$$\hat{R}_{\text{LS}}(N) = \frac{\frac{1}{N}\sum_{k=1}^{N} u(k)i(k)}{\frac{1}{N}\sum_{k=1}^{N} i^2(k)}, \text{ with } \lim_{N \to \infty} \hat{R}_{\text{LS}}(N) = R_0 \frac{1}{1 + \sigma_i^2/i_0^2} \qquad (1\text{-}55)$$

This systematic error can be removed by replacing $i(k)$ in the numerator and denominator by $i(k-1)$ so that the new estimate becomes:

$$\hat{R}_{\text{IV}}(N) = \frac{\frac{1}{N}\sum_{k=1}^{N} u(k)i(k-1)}{\frac{1}{N}\sum_{k=1}^{N} i(k)i(k-1)} \tag{1-56}$$

Making the same analysis as in Section 1.2.2.1, it is seen that all quadratic noise contributions are eliminated by this choice, so that

$$\lim_{N \to \infty} \hat{R}_{\text{IV}}(N) = R_0 \tag{1-57}$$

The idea used to generate (1-56) can be generalized as follows. Consider the linear-in-the-parameters model structure $y_0 = K(u_0)\theta_0$ in Section 1.5.1, and replace K^T in (1-36) by G^T, to get

$$\hat{\theta}_{\text{IV}}(N) = (G^T K(u))^{-1} G^T y \tag{1-58}$$

The choice of G, a matrix of the same size as $K(u)$, will be defined later. $\hat{\theta}_{\text{IV}}(N)$ is the instrumental variables estimate. Consistency is proved by considering the plim for $N \to \infty$ (Norton, 1986). For simplicity, we assume all the plim exists, namely

$$\begin{aligned}
\text{plim}\, \hat{\theta}_{\text{IV}} &= \text{plim}\, \{(G^T K(u))^{-1} G^T y\} \\
&= (\text{plim}\, \{G^T K(u_0 + n_u)\})^{-1} (\text{plim}\{G^T y_0 + G^T n_y\}) \\
&= (\text{plim}\, \{G^T K(u_0 + n_u)/N\})^{-1} (\text{plim}\, \{G^T K(u_0)/N\}\, \theta_0 + \text{plim}\, \{G^T n_y/N\})
\end{aligned}$$

If

$$\text{plim}\, \{G^T K(u_0 + n_u)/N\} = \text{plim}\, \{G^T K(u_0)/N\} \quad \text{and} \quad \text{plim}\, \{G^T n_y/N\} = 0 \tag{1-59}$$

then

$$\lim_{N \to \infty} \text{plim}\, \hat{\theta}_{\text{IV}}(N) = \theta_0 \tag{1-60}$$

Equation (1-59) defines the necessary conditions for G to get a consistent estimate. Loosely stated, G should not be correlated with the noise on $K(u_0 + n_u)$ and the output noise n_y. The variables used for building the entries of G are called the instrumental variables.

If the covariance for $C_{n_y} = \sigma^2 I_N$, then an approximate expression for the covariance matrix of the estimates is (Norton, 1986):

$$\text{Cov}(\hat{\theta}_{\text{IV}}(N)) \approx \sigma^2 R_{GK}^{-1} R_{GG} R_{GK}^{-T} \quad \text{with} \quad R_{GK} = G^T K(u)/N \text{ and } R_{GG} = G^T G/N \tag{1-61}$$

This reveals another condition on the choice of the instrumental variables G: although they should be "uncorrelated" with the noise on the output observation n_y, they should be correlated maximally with K, otherwise R_{GK} tends to zero and $\text{Cov}(\hat{\theta}_{\text{IV}}(N))$ would become very large. In the case of the resistance estimate, the instrumental variables are the shifted input. Because we used a constant current, no problem arises. In practice, this technique can be generalized to varying inputs under the condition that the power spectrum of the noise is much wider than the power spectrum of the input. In the following exercises the instrumental variables method is applied to the resistance example.

1.6 EXERCISES

1.1. Set up a simulation to measure the value of the resistance using

$$i(k) = i_0 + n_i(k) \qquad u(k) = u_0 + n_u(k) \qquad (1\text{-}62)$$

Use for n_i and n_u zero mean iid noise with standard deviation σ_i and σ_u. Consider uniformly and normally distributed noise and use $i_0 = 1$ A, $u_0 = 1$ V, $\sigma_i = 0.5(1)$ A, and $\sigma_u = 0.5(1)$ V. Plot $R(k) = u(k)/i(k)$ for $k = 1, ..., 100$.

1.2. Apply the estimators $\hat{R}_{LS}, \hat{R}_{EV}, \hat{R}_{SA}$ from (1-1) to (1-3) to the results of the simulator in Exercise 1.1 and plot the results as a number of the processed measurements N.

1.3. Measure the histogram for the three estimators of Exercise 1.2 for $N = 10, 100, 1000$ and plot the approximated pdf.

1.4. Use the simulator of Exercise 1.1 to estimate the variance of the three estimators of Exercise 1.2 as a function of N and plot the results on a log-log scale. Check the $1/\sqrt{N}$ rule of thumb. Vary N between 1 and 10^4.

1.5. Derive the variance expressions $\text{var}(\hat{R}_{LS}(N))$, $\text{var}(\hat{R}_{EV}(N))$, and $\text{var}(\hat{R}_{SA}(N))$ under Assumption 1.1 using linear approximations as illustrated in (1-16) and (1-17).

1.6. Use the simulator of Exercise 1.1 to estimate the variance of the three estimators of Exercise 1.2 for $N = 100$ as a function of the SNR of the current and the voltage measurements. Compare the results with the theoretical level (see (1-17) and (1-18)) and discuss the results.

1.7. The bias compensated least squares solution of the resistor problem is given by

$$\hat{R}_{BC}(N) = \frac{\frac{1}{N}\sum_{k=1}^{N} u(k)i(k)}{\frac{1}{N}\sum_{k=1}^{N} i^2(k) - \sigma_i^2} \qquad (1\text{-}63)$$

Use the simulator of Exercise 1.1 to estimate the variance and the mean square error of $\hat{R}_{LS}, \hat{R}_{EV}$, and $\hat{R}_{BC}$ from (1-2), (1-3), and (1-63) with $N = 100$. Vary the noise-to-signal ratio between 0.01 and 1, and plot the results on a log-log scale. Compare the mean square error with the variance. What do you conclude?

1.8. Derive the estimators $\hat{R}_{LS}(N)$, $\hat{R}_{EV}(N)$, and $\hat{R}_{SA}(N)$ by minimizing the cost functions (1-19), (1-20), and (1-21).

1.9. Reformulate the cost functions (1-19), (1-20), and (1-21) for the case that the current is varying from measurement to measurement (the current is no longer a DC source), and derive the new expressions of the estimators. Show that the errors-in-variables estimator $\hat{R}_{EV}(N)$ minimizes the following cost function w.r.t. R

$$\sum_{k=1}^{N} \frac{(u(k) - Ri(k))^2}{\sigma_u^2(k) + R^2\sigma_i^2(k)} \qquad (1\text{-}64)$$

with $\sigma_i^2(k)$ and $\sigma_u^2(k)$ the variances of, respectively, the current and voltage measurements (hint: eliminate $i_p(k)$ and $u_p(k) = Ri_p(k)$ in (1-21) via $\partial V_{EV}/\partial i_p(k) = 0$, $k = 1, 2, ..., N$).

1.10. Consider a signal

$$y_0(k) = \sin(2\pi f k T_s + \varphi) \qquad (1\text{-}65)$$

and its measurement

$$y(k) = y_0(k) + n_y(k), \text{ for } k = 1, \ldots, 1024 \qquad (1\text{-}66)$$

where $n_y(k)$ is iid normally distributed noise with zero mean and variance σ_y^2. Calculate the Cramér-Rao for the estimates (f, φ). What is the best choice for T_s if we want to estimate the frequency with minimum variance?

1.11. Consider a polynomial model:

$$y_0(k) = \sum_{p=1}^{P} a_p u^p(k) \qquad (1\text{-}67)$$

that is identified from a set of measurements $y(k) = y_0(k) + n_y(k)$, with $u(k) = [-N:N]/N$ and $n_y(k)$ zero mean iid distributed noise with variance σ_y^2. Set up the least squares estimator for this problem, and observe the condition number for growing values of P (put $N = 1000$). What is the maximum order that can be reliably identified?

1.12. Consider the least squares solution $\hat{\theta}_{LS}(N) = (J^T J)^{-1} J^T y$ of the overdetermined set $J\theta = y$ (as they appear in (1-36)). Show that this solution can be calculated using the SVD method of Section 15.5 on matrix algebra without forming the product $J^T J$ as $\hat{\theta}_{LS}(N) = J^+ y$, with $J^+ = V\Sigma^+ U^T$.

1.13. Apply the method of Exercise 1.12. to the polynomial problem of Exercise 1.11, and find the maximum order that can be identified reliably.

1.14. The polynomial identification problem is an ill-posed problem because of the poor numerical conditioning of the normal equations. Using the SVD method, it is already possible to solve higher order problems, but even then the numerical conditioning decreases quickly. A much better solution is to change the model representation and to use orthogonal polynomials $T_p(u)$ such that

$$y_0(k) = \sum_{p=1}^{P} a_p u^p(k) = \sum_{p=1}^{P} t_p T_p(u(k)) \qquad (1\text{-}68)$$

where $T_p(u) = \sum_{k=1}^{p} a_{pk} u^k$ is a polynomial of degree p. The coefficients a_{pk} are set s.t.

$$\sum_{k=1}^{N} T_r(u(k)) T_s(u(k)) = \delta(r-s)$$

Note that the actual form of $T_p(u)$ (the choice of a_{pk}) depends on the set of input values $u(k)$ that appears in the problem. Reformulate the polynomial identification problem using the orthogonal basis and discuss the condition number of the new estimator.

Remarks:

For the given set of input values, the orthogonal polynomials $T_p(u)$ are given by the following recurrence relation (Ralston and Rabinowitz, 1984):

$$\frac{1}{\alpha_{j+1}} T_{j+1}(u) = \frac{u}{\alpha_j} T_j(u) - \frac{\beta_j}{\alpha_{j-1}} T_{j-1}(u)$$

$$\beta_j = \frac{j^2[(2N+1)^2 - j^2]}{4(4j^2 - 1)}, \qquad \alpha_j = \frac{(2j)!}{(j!)^2 (2N)^j} \qquad (1\text{-}69)$$

with $T_0(u) = 1$ and $T_{-1}(u) = 0$ for $j = 0, 1, \ldots$. When using orthogonal polynomials the reader should take care not to use the explicit polynomial expressions, but only the values of the orthogonal polynomials. Otherwise the numerical stability is not guaranteed. As a result, it is also not possible to calculate the coefficients a_p of the original solution; only the value of the solution can be calculated (see Ralston and Rabinowitz, 1984).

1.15. Prove expression (1-42) for the covariance matrix of a weighted least squares for models that are linear-in-the-parameters.

1.16. Show that the covariance matrix of the weighted least squares estimator becomes minimal for $W = C_{n_y}^{-1}$ (hint: use the Schwarz inequality $B^T B \geq (B^T A)(A^T A)^{-1}(A^T B)$, see Eykhoff, 1974, p. 525, and put $C_{n_y}^{-1} = C^T C$, $B = CJ$, and $A = C^{-T} WJ$).

1.17. Consider the linear-in-the-parameters model $y_0 = K(u_0)\theta_0$ and calculate the variance of the modeled output $\hat{y} = K(u_0)\hat{\theta}$ starting from the covariance matrix C_θ given in (1-43).

1.18. Show that the variance on the output of the polynomial model in Exercise 1.11. is independent of the model representation $y_0(k) = \sum_{p=1}^{P} a_p u^p(k)$ or $y_0(k) = \sum_{p=1}^{P} t_p T_p(u(k))$. Check this by a simulation using the estimators of Exercises 1.11. and 1.14. for a polynomial of degree 5 (so that the numerical conditioning of the problem remains acceptable for the direct estimation).

1.19. Consider the system $y_0 = au$. Construct the least squares and the weighted least squares estimator for a starting from the measurements $y(k) = au(k) + n_y(k)$ with $\mathbb{E}\{n_y(k)\} = 0$ and $\sigma_{n_y}^2(k) = u(k)$. Compare the bias and the variance of both estimators for $u(k) = 1, 2, \ldots, 10$. Verify your results by means of a simulation.

1.20. Construct $\hat{R}_{IV}(N)$ for the resistance example of Section 1.2.1 using (1-58). Use the time-shifted current as an instrumental variable. Study the behavior of the estimator (mean value and variance) as a function of the shift by means of a simulation.

1.21. Study the behavior of $\hat{R}_{IV}(N)$ (mean value and variance) of the previous exercise for the situation where $i_0(k)$ is generated as low-pass filtered noise (bandwidth of the filter at $f_s/50$) as a function of the applied delay by means of a simulation.

2

Measurement of Frequency Response Functions – Standard Solutions

Abstract: Frequency response function (FRF) measurements are an interesting intermediate step in the identification process. The complexity of the modeling problem is visualized before starting parametric modeling; the quality of the measurements is assessed in an early phase. In this chapter a number of basic FRF measurement methods are discussed. The reader is referred to Chapter 7 for the advanced techniques. An analysis of the bias and efficiency of the FRF measurements is made, and their dependence on the experimental conditions and on the excitation signal is analyzed. Averaging techniques are proposed to improve the quality of the FRF measurement. Guidelines given at the critical steps of the FRF measurement process enable the less experienced user to start modeling from good raw data.

2.1 INTRODUCTION

Consider the linear dynamic system $G(j\omega)$ between the input $u(t)$ and the output $y(t)$ as shown in Figure 2-1. The aim of this book is to build a parametric model for this system, identifying, for example, a transfer function $G(j\omega, \theta)$. Such a model is called parametric because it employs a finite-dimensional parameter vector. Parametric modeling requires a series of user decisions (e.g., selection of the order of numerator and denominator of $G(s, \theta)$; thus, it is strongly advised to get a good initial idea about the system under test. Step or impulse response measurements provide this information. Also, frequency response function (FRF) measurements are very valuable. An FRF consists of transfer function measurements $G(j\omega_k)$ at a discrete set of frequencies ω_k, $k = 1, ..., F$. All these models are called nonparametric because the information is not condensed into a small set of parameters. In this chapter we focus, exclusively, on FRF measurements. A series of basic questions is addressed:

- How are the bias and efficiency of the FRF measurements influenced by the experimental conditions?
- How should the excitation signal be chosen?
- Can we improve the quality of the FRF using averaging methods?
- Can we quantify the quality of the FRF measurements?

- What is the impact of nonlinear distortions on the measured FRF and how can we detect their existence?

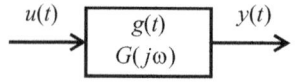

Figure 2-1. Block diagram of the system.

All these aspects are discussed here or, for the last question, in the next chapter. Starting with a straightforward solution, the more advanced techniques are introduced step by step, showing each time what additional problems are addressed by these more advanced techniques. As FRF measurement techniques rely heavily on the transformation of sampled signals from the time to the frequency domain, we will spend some time on the most important aspects of the discrete Fourier transform.

2.2 AN INTRODUCTION TO THE DISCRETE FOURIER TRANSFORM

In most situations, real-life systems are naturally continuous in time. However, most signal processing is now done on digital computers that operate on discrete-time signals. In practice, the continuous-time signals are discretized (sampled) and quantized (digitized) so that the signal can finally be stored in the memory of a digital computer. Next, the spectrum of these signals is needed in order to calculate the FRF of the system. This is done using the discrete Fourier transform (DFT), usually calculated with the fast Fourier transform (FFT) algorithm. Each of these steps creates errors, and it is important for a user to understand their behavior to minimize the impact of the errors on his/her results. In this section only a brief introduction is given. For an extended overview, the reader is referred to Brigham's (1974) book. First we discuss, briefly, the sampling process, next we show how to "measure" the Fourier spectrum of a signal, and finally we focus on the spectral properties of periodic excitations and how to exploit them to minimize measurement errors.

2.2.1 The Sampling Process

The continuous-time signal is sampled at an equidistant time grid and is represented by the equivalent discrete-time sequence $u_d(n) = u(nT_s)$. In the time domain, the sampling process can be formulated as a multiplication with a periodically repeated Dirac impulse (Brigham, 1974):

$$\tilde{u}_d(t) = u(t)\delta_{T_s}(t) \text{ with } \delta_{T_s}(t) = \sum_{n=-\infty}^{\infty}\delta(t-nT_s) \qquad (2\text{-}1)$$

Note that in this framework the discrete-time signal $u_d(n)$ is formally represented by a continuous-time signal $\tilde{u}_d(t)$ that carries all its power at the discrete-time instances nT_s. Define the spectrum of the discrete-time signal as

$$U_d(e^{j2\pi fT_s}) = \sum_{n=-\infty}^{\infty} u_d(n)e^{-j2\pi fnT_s} \qquad (2\text{-}2)$$

Then the following relation exists:

$$U_d(e^{j2\pi fT_s}) = \tilde{U}_d(j2\pi f) = F\{\tilde{u}_d(t)\} = \int_{-\infty}^{+\infty}\tilde{u}_d(t)e^{-j2\pi ft}dt \qquad (2\text{-}3)$$

Section 2.2 ■ An Introduction to the Discrete Fourier Transform

The spectrum $U_d(e^{j\omega T_s})$ is linked to $U(j\omega)$ by noticing that a multiplication in the time domain, $u(t)\delta_{T_s}(t)$, corresponds to the convolution of the spectra in the frequency domain, $U(j2\pi f) * (f_s \delta_{f_s}(f))$, with $f_s \delta_{f_s}(f)$ the spectrum of $\delta_{T_s}(t)$, and $\delta_{f_s}(f)$ a periodically repeated Dirac impulse with period $f_s = 1/T_s$

$$\delta_{f_s}(f) = \sum_{k=-\infty}^{+\infty} \delta(f - kf_s) \tag{2-4}$$

Using (2-4), we get

$$U_d(e^{j2\pi fT_s}) = U(j2\pi f) * (f_s \delta_{f_s}(f)) = \frac{1}{T_s}\sum_{k=-\infty}^{+\infty} U(j2\pi(f - kf_s)) \tag{2-5}$$

The convolution of the spectra is illustrated in Figure 2-2. It shows that the sampling process results in a repeated spectrum in the frequency domain with period f_s. If the bandwidth f_B of the sampled signal is larger than half the sampling frequency, the shifted spectra overlap and information is lost. Therefore, it is important to restrict the bandwidth below half the sampling frequency $f_B < f_s/2$ in order to avoid errors. This error is called the aliasing error and the condition on the sample frequency is known as Shannon's sampling theorem. In practice, it is often necessary to add anti-alias filters to eliminate the high-frequency spectral content of the signal.

2.2.2 The Discrete Fourier Transform (DFT-FFT)

Three basic steps have to be taken to measure the spectrum of a continuous-time signal:

- Discretization in time: sample the continuous-time signal at an equidistant time grid.
- Restrict the length of the data record: our computers can deal with only a finite number of data. Thus, the length of the record is restricted to N samples, excluding the rest. This is called windowing.
- Discretization in frequency: the finite length discrete-time signal still has a continuous frequency spectrum. The value of this spectrum will be calculated only at an equidistant set of frequencies.

The impact of all these steps is illustrated in more detail in the following, in a simple example. The continuous-time signal $u(t) = \cos(2\pi f_0 t)$, with $f_0 = 5.5$ Hz is sampled at $f_s = 64$ Hz during 1 second. From these measurements we will calculate the discrete Fourier transform step by step.

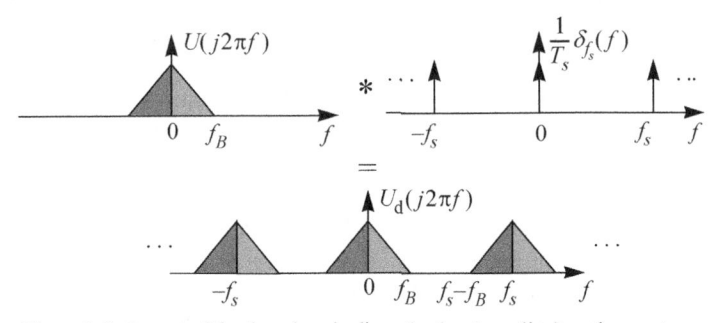

Figure 2-2. Impact of the time domain discretization (sampling) on the spectrum.

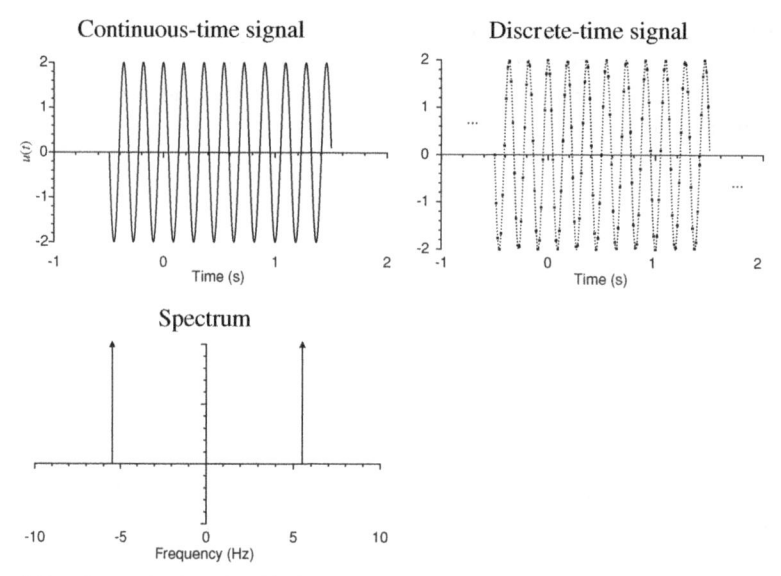

Figure 2-3. The time signal before and after sampling together with the spectrum in the frequency band [−10 Hz, 10 Hz].

2.2.2.1 Discretization in Time. The sampling process has already been discussed in the previous section. Figure 2-3 shows the signal together with its spectrum before and after sampling. In order to keep enough detail in the figures shown, a zoom is made in the frequency band [−10 Hz, 10 Hz]. The periodic repetitions of the spectrum of the discrete-time sequence are not shown. Note that if no aliasing appears, the spectra of the continuous-time and the discrete-time signal are equal to each other within a scale factor.

Mathematical description:

$$\text{time domain:} \quad \tilde{u}_d(t) = \sum_{n=-\infty}^{+\infty} u(t)\delta(t - nT_s)$$

$$\text{frequency domain:} \quad U_d(e^{j2\pi fT_s}) = T_s^{-1}\sum_{k=-\infty}^{+\infty} U(j2\pi(f - kf_s)) \quad (2\text{-}6)$$

2.2.2.2 Windowing. The sampled signal still has an infinite length $(-\infty, \infty.)$ Because the computer can process only a finite number of samples, we have to restrict the measurement length. We consider only samples that appear in the measurement window:

$$w(t) = 1 \text{ if } 0 \leq t < T \quad \text{and} \quad w(t) = 0 \text{ elsewhere} \quad (2\text{-}7)$$

This rectangular window, together with its spectrum (the phase is omitted), is shown in Figure 2-4. This window is called a rectangular window and its major characteristic is its width T. Its spectrum $W(j2\pi f)$ is a sinc-like signal, see (2-8), with zero crossings at the multiples of $1/T$. In this example, $T = 1$ s. This window is multiplied with the sampled signal to obtain a new signal that is different from zero in only a finite number of samples.

The spectra have to be convoluted in the frequency domain. Remembering that a convolution with a Dirac impulse is nothing other than a shift of the origin to the position of the impulse, the result of Figure 2-5 is found. The broken lines in the spectra indicate the position of the original frequency components. As can be seen, the restriction of the signal to a finite interval in the time domain smears the power in the frequency domain over the neighboring frequencies. This phenomenon is called leakage.

Section 2.2 ■ An Introduction to the Discrete Fourier Transform

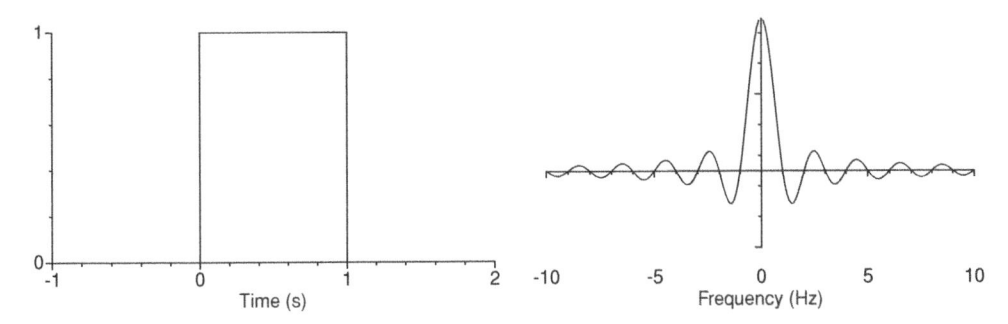

Figure 2-4. Rectangular window and its spectrum (the phase is omitted).

Mathematical description:

$$\begin{aligned} \text{time domain:} &\quad w(t)\tilde{u}_d(t) \\ \text{frequency domain:} &\quad W(j2\pi f) * U_d(e^{j2\pi fT_s}) \end{aligned} \quad (2\text{-}8)$$

with $W(j\omega) = Te^{-j\omega T/2}\text{sinc}(\omega T/2)$ and $\text{sinc}(x) = \sin(x)/x$.

2.2.2.3 Discretization in Frequency. As can be seen in Figure 2-5, the spectrum of the sampled and windowed signal is still a continuous frequency signal. Because the spectrum can be calculated in only a finite number of frequencies, the frequencies considered should also be restricted to a discrete grid. An equidistant grid with spacing $1/T$ is selected. Hence, the spectrum is calculated only at the frequencies $f_k = k/T$ Hz. This can be considered as frequency sampling or discretization in frequency. The resulting sampled spectrum, shown in Figure 2-6, is quite disappointing. Although the shape of the original spectrum (Figure 2-3) can still be recognized, it seems that all detailed information about it has definitely been lost. The basic reason for this problem is that the original frequency (5.5 Hz) does not correspond to one of the sampled frequencies in the DFT (multiples of $1/T = 1$ Hz).

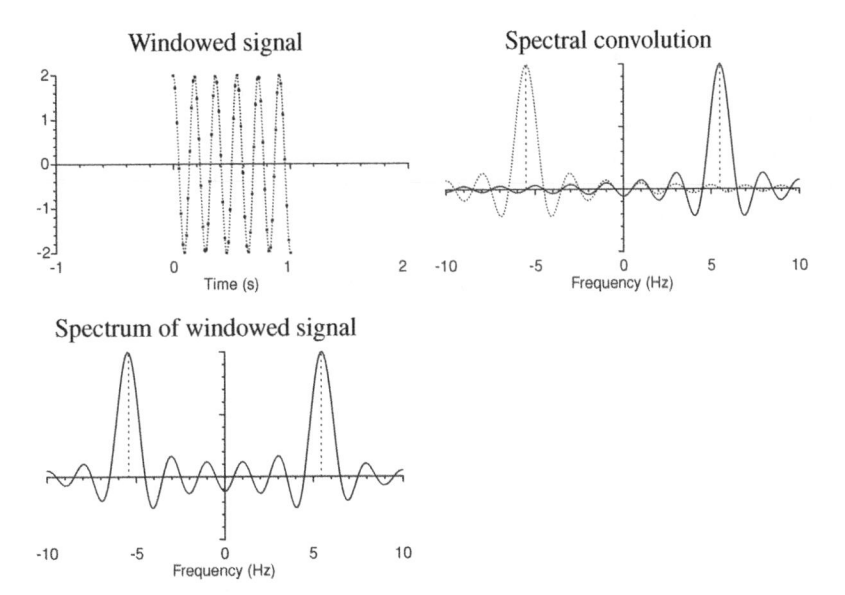

Figure 2-5. Spectrum of the sampled signal after applying a rectangular window.

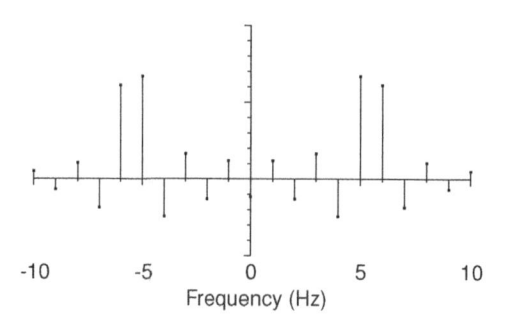

Figure 2-6. DFT result.

This can also be seen in the time domain representation of the DFT result. Sampling in the frequency domain at multiples of $1/T$ is described as a multiplication with a Dirac train (see Section 2.2.1) so that in the time domain a convolution should be made with a Dirac train $T\delta_T(t)$. This results in a periodic repetition with period T of the sampled and windowed signal as shown in Figure 2-7. However, T is not a multiple of the signal period, resulting in a discontinuity that appears at the borders of the window as seen in Figure 2-7 ($T = 1$ s in this case).

Mathematical description:

$$\text{time domain:} \quad (w(t)\tilde{u}_d(t))*(T\delta_T(t))$$
$$\text{frequency domain:} \quad (W(j2\pi f)*U_d(e^{j2\pi fT_s}))\delta_{1/T}(f) \quad (2\text{-}9)$$

From (2-9) it follows that the relationship between the time domain samples $u_d(n) = u(nT_s)$ (amplitudes of the Dirac impulses of the time domain signal in (2-9)) and the frequency domain samples $U_{\text{DFT}}(k)$ (amplitudes of the Dirac impulses of the spectrum in (2-9)), is given by

$$U_{\text{DFT}}(k) = \sum_{n=0}^{N-1} u(nT_s)e^{-j2\pi nk/N}, \quad k = 0, 1, ..., N-1 \quad (2\text{-}10)$$

Equation (2-10) is called the discrete Fourier transform (DFT) of the samples $u(nT_s)$, $n = 0, 1, ..., N-1$.

If an integer number of periods is measured, the DFT will give an exact copy of the discrete spectrum of the periodic signal. This is illustrated in Figure 2-8, showing the spectra after windowing and after discretization for $u(t) = \cos(2\pi f_0 t)$, $f_0 = 5$ Hz, $T = 1$ s. This time, no leakage is observed. The basic reason for this remarkable difference is that the continuous-time spectrum equals zero at the frequencies where the spectrum is sampled because the window length is an exact multiple of the period length. Also, the time domain interpretation in Figure 2-9 illustrates the result: this time, the periodic repetition coincides with the period of the signal (no discontinuities appear at the multiples of T).

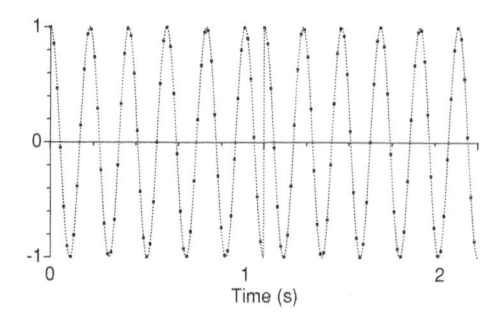

Figure 2-7. Interpretation of the DFT result in the time domain.

Section 2.2 ■ An Introduction to the Discrete Fourier Transform

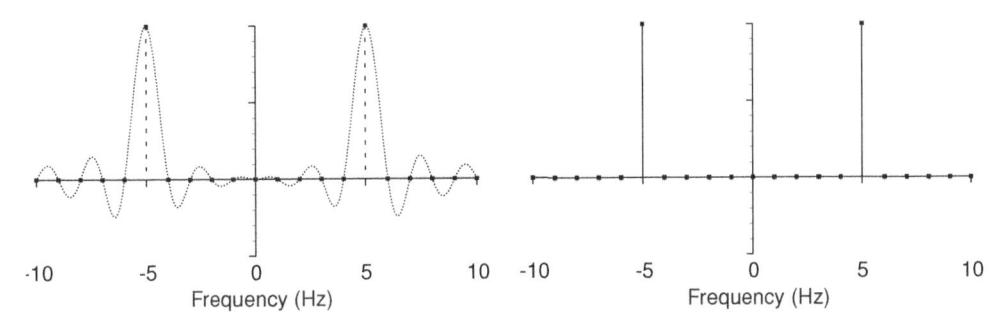

Figure 2-8. DFT spectrum for a periodic signal when an integer number of periods is measured.

At a glance, this seems to be a theoretical result without practical value. The probability of getting an exact match between the signal and the window length is in general, indeed, zero. However, in many FRF measurements, the user masters the generator and the acquisition. In these experimental setups both systems are driven by mother clocks that are synchronized with each other. It is therefore possible for the user to create this ideal match, which eliminates the leakage effect completely. We strongly advise realization of such a setup whenever possible. If for some reason it is impossible to get synchronized measurements, there exist other less attractive alternatives based on windows other than the rectangular window. An extended discussion of the window properties can be found in Harris (1978). In Section 2.2.3 we will briefly touch on this topic.

2.2.2.4 The DFT Expressions. For the samples $u(nT_s)$, $n = 0, 2, ..., N-1$, the DFT relations between the time and frequency domain sequences are

$$U_{\text{DFT}}(k) = \sum_{n=0}^{N-1} u(nT_s) e^{-j2\pi nk/N} \quad \text{and} \quad u(nT_s) = \frac{1}{N} \sum_{k=0}^{N-1} U_{\text{DFT}}(k) e^{j2\pi kn/N}$$

(see (2-10)). In this book the scaling factor $1/N$ is symmetrically distributed over both transforms using $1/\sqrt{N}$, and the notation $U_{\text{DFT}}(k)$ will be replaced by $U(k)$ in order not to overload the equations. This gives

$$U(k) = \frac{1}{\sqrt{N}} \sum_{n=0}^{N-1} u(nT_s) e^{-j2\pi nk/N} \quad \text{and} \quad u(nT_s) = \frac{1}{\sqrt{N}} \sum_{k=0}^{N-1} U(k) e^{j2\pi kn/N} \quad (2\text{-}11)$$

The straightforward evaluation of (2-11) requires $O(N^2)$ operations. However, if N is a power of two, a very efficient implementation known as the FFT (fast Fourier transform) is available: it calculates the transforms in $O(N\log_2 N)$ operations (Brigham, 1974). If N is not

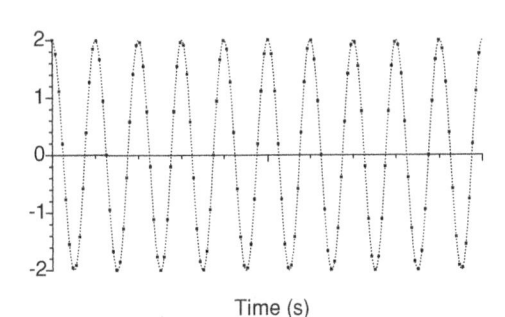

Figure 2-9. Interpretation of the DFT result in the time domain when an integer number of periods is measured.

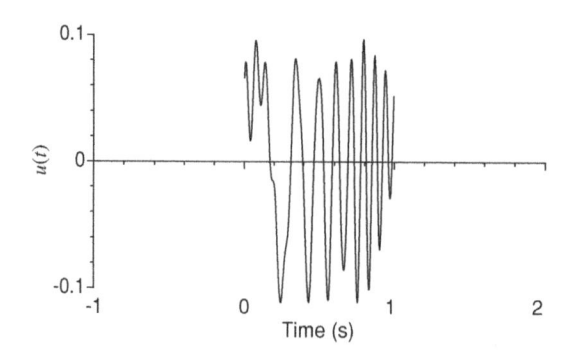

Figure 2-10. Example of a periodic excitation consisting of the sum of 15 sines with equal amplitude and frequencies kf_0, $k = 1, 2, ..., 15$.

a power of two, there still exist fast implementations such as the chirp-z transform (Rabiner and Gold, 1975). The FFT algorithm is available in many numerical packages.

2.2.3 DFT Properties of Periodic Signals

2.2.3.1 Integer Number of Periods Measured. Consider the periodic signal $u(t) = \sum_{k=1}^{15} \cos(2\pi k f_0 t + \phi_k)$ in Figure 2-10. Using the same sample frequency, this signal is measured over 1 period and over 10 periods. For both data records the DFT is calculated and the first 150 lines of the DFT spectrum are plotted in Figure 2-11. In both cases an exact recovery of the signal spectrum is made because each time an integer number of periods is measured. However, by measuring 10 times longer ($10N$ data points), the spectral resolution is increased from $1/T = f_s/N$ to $1/(10T) = f_s/(10N)$. Whereas in the former time the spectral lines appear at harmonics $k = 1, 2, ..., 15$, they are placed in the latter time at $k = 10, 20, ..., 150$. The gaps between these spectral lines can be used later on to extract noise information because the noise is nonperiodic and excites all spectral lines.

2.2.3.2 No Integer Number of Periods Measured. From Section 2.2.2 it is known that leakage errors appear if no integer number of periods is measured. A sound solution for this problem is to change the setup and measure an integer number. If this is impossible we can try to minimize the impact of the leakage on the measurement. A classical technique is to apply a window other than the rectangular one. A concise review of windows and their properties is given by Harris (1978). Here, we present only one of the possibilities: the Hanning or cosine window

$$w(t) = 1 - \cos(2\pi t/T) \text{ if } 0 \leq t < T \text{ and } w(t) = 0 \text{ elsewhere} \qquad (2\text{-}12)$$

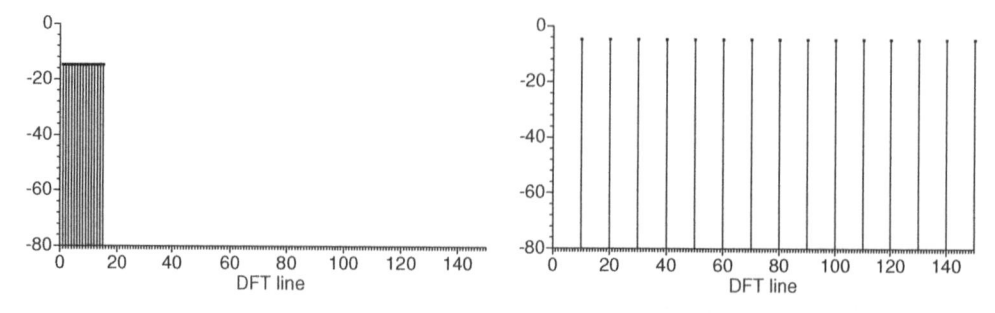

Figure 2-11. DFT spectrum (amplitude in dB) of a periodic signal with 15 components. On the left, 1 period in the window; on the right, 10 periods in the window.

Section 2.2 ■ An Introduction to the Discrete Fourier Transform

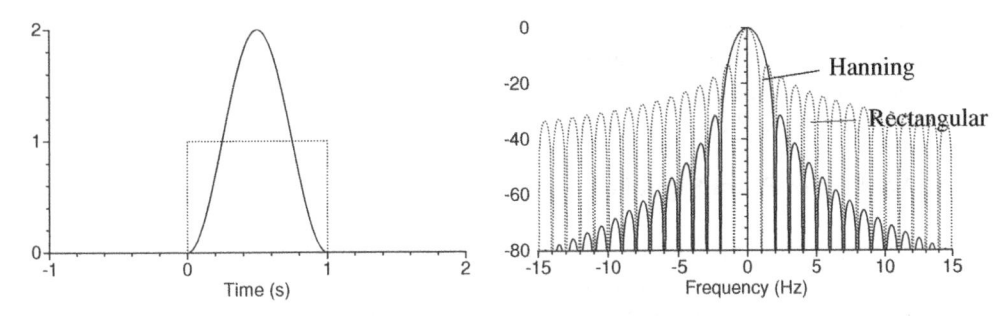

Figure 2-12. Comparison of the rectangular window with the Hanning window in the time domain (left) and the frequency domain (right, amplitude spectrum in dB).

The aim of all the alternative windows is to taper the signal at the beginning and at the end of the window in order to decrease the discontinuities of the periodically reconstructed signal because they are the basic source of the leakage errors. In Figure 2-12 the rectangular (2-7) window and the Hanning (2-12) window are shown. By applying such an alternative window, we do not eliminate the leakage effect but only reshape its impact. Windowing in the time domain is equivalent to a convolution with its spectrum in the frequency domain. The spectrum of the Hanning window decreases much faster than that of a rectangular window, keeping the leakage effect more localized. On the other hand, the main lobe of the Hanning window (first lobe around zero) is two times wider than that of the rectangular window; hence, for components that are close to each other (less than four DFT bins) the interference will increase. This is a typical effect of these windows: they minimize the far leakage effects (far from the position of the original frequency) at a cost of a loss in resolution. The choice of the window also affects the noise sensitivity, the maximum error on the amplitude of the spectral components, etc. We refer the interested reader to Harris (1978) for more information.

To illustrate the effect of the window on the spectrum, we considered 10.5 periods of the periodic signal and calculated the DFT, first with a rectangular window and second with the Hanning window (Figure 2-13). The separation between the components becomes much more visible for the Hanning than for the rectangular window. The interference is reduced from −30 dB (3%) to less than −60 dB (0.1%).

Conclusion. The best solution is to measure an integer number of periods. If this is impossible, the leakage interference between the different spectral components can be reduced by measuring enough periods and using, for example, a Hanning window. For M measured periods, the leakage errors are an $O(M^{-1})$ effect for the rectangular window and an

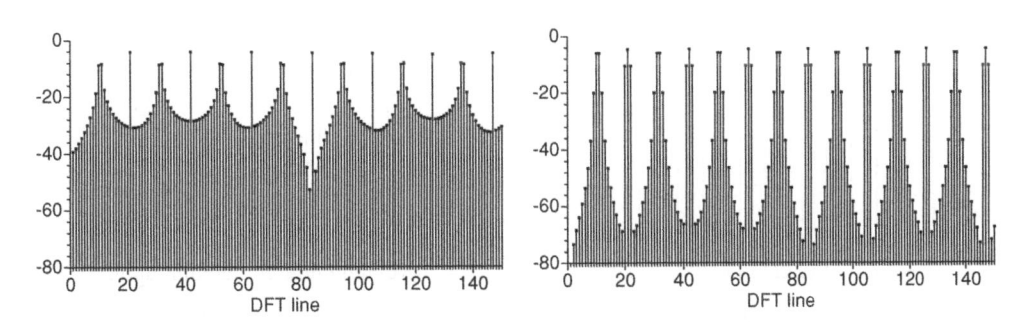

Figure 2-13. Impact of the rectangular (left) and Hanning window (right) on the spectrum for 10.5 measured periods.

$O(M^{-2})$ for the Hanning window. Notice that if at least three or more integer periods are measured, the Hanning window also allows perfect recovery of the original spectral lines. This is a very specific property of the Hanning window that is due to the fact that its zeros coincide with those of the rectangular window except for the main lobe (Figure 2-12).

2.2.4 DFT of Burst Signals

The study of the DFT properties showed that no leakage errors occur if periodic signals are analyzed and an integer number of periods is measured. There is an important exception to this general rule: using a DFT, it is possible to sample the continuous spectrum of a burst signal.

Definition 2.1 (Burst Signal): $u(t)$ is a burst signal if $u(t) = 0 \ \forall t \notin [0, T_B]$.

Remark. A time-limited signal cannot be band-limited ($|U(j2\pi f)| = 0$ if $|f| > f_B$); thus the time discretization of such a signal always creates aliasing errors. In practice, most burst signals are low-pass filtered signals, which minimize these aliasing effects if a reasonable design is made. In Figure 2-14, an example of such a signal is given. This is an exponentially damped signal that is not exactly zero at the end of the window. So the "burst" condition is not exactly met, but again the errors are negligible for a good design.

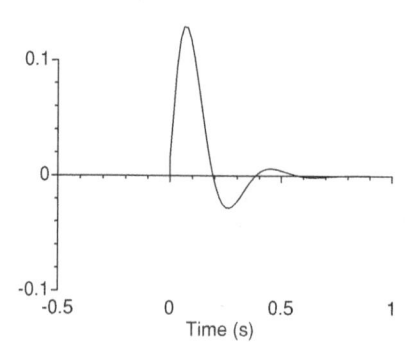

Figure 2-14. Burst signal.

In Section 2.2.2, it was shown that the DFT eventually makes a periodic reconstruction of the original sequence. Because this sequence is zero outside the window ($T > T_B$) this reconstruction does not create discontinuities at the borders and hence the calculated spectrum is a perfect copy of the original one at the DFT lines. This is illustrated in Figure 2-15, where the DFT spectrum of the burst signal in Figure 2-14 is shown. In the first case, the window length was 1 s, resulting in a frequency resolution of 1 Hz, and in the second case, the win-

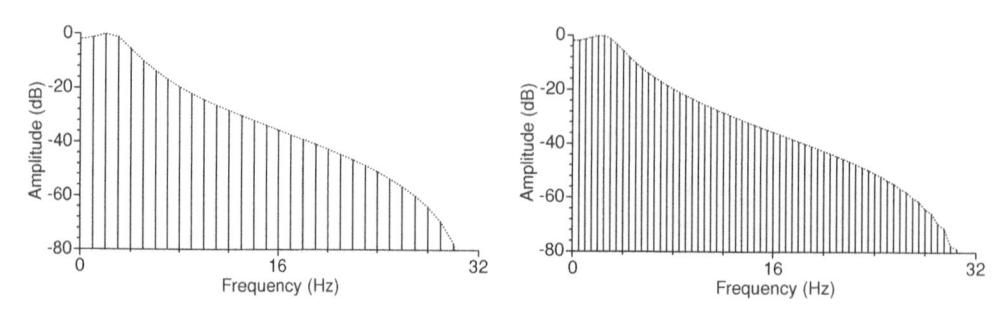

Figure 2-15. DFT spectrum of a burst signal. Left: window length 1 s (64 points); right: window length 2 s (128 points). The dotted line is the original continuous spectrum.

dow length was 2 s (this can be done by zero appending: N zeros are appended to extend the record length to $2N$), resulting in a frequency resolution of 0.5 Hz.

2.2.5 Conclusion

It is possible to calculate the spectra of sampled signals using the DFT (FFT), but two errors can occur. The first is the aliasing error: the power at higher frequencies is mirrored at the lower frequencies. To avoid this, the sampling frequency should be set high enough ($f_s > 2f_B$). The second error is leakage: the spectrum of the signal is smeared out due to the finite length of the measurements. In two special, but in practice very important, situations it can be completely avoided. For example, the spectrum of periodic signals measured over an integer number of periods is perfectly calculated by the DFT. It is an exact copy of the spectrum of the continuous-time signals, at least up to half the sample frequency for band-limited signals. We strongly advise the readers to get as close as possible to this ideal situation whenever they have enough freedom during the experiment design. If it is not possible to realize the previous conditions, errors will appear, but it is still possible to reshape these errors to minimize their effect on the results.

2.3 SPECTRAL REPRESENTATIONS OF PERIODIC SIGNALS

In this book we will use three different spectral representations of a periodic signal: the Fourier series, the Fourier transform, and the discrete Fourier transform. Because these all describe the same signal, it is clear that there are close connections between them. Consider a periodic signal, described by its Fourier series representation:

$$u(t) = \sum_{k=1}^{F} |A_k| \cos(k\omega_0 t + \angle A_k) = \sum_{k=-F, k \neq 0}^{F} (A_k/2) e^{jk\omega_0 t} \quad (2\text{-}13)$$

with $A_k = |A_k| e^{j\angle A_k}$.

- The Fourier coefficient at line k is then

$$U_k = A_k/2 \quad (2\text{-}14)$$

- The Fourier transform is $F\{u(t)\} = U(j\omega) = \int_{-\infty}^{+\infty} u(t) e^{-j\omega t} dt$, with

$$U(j\omega) = \sum_{k=-F, k \neq 0}^{F} (A_k/2) \delta(f - kf_0) \quad (2\text{-}15)$$

The Dirac impulses account for the convergence problems of the Fourier integral on periodic signals.

- The discrete Fourier transform of one period ($f_s = 1/T_s = Nf_0$) of $u_d(n) = u(nT_s)$ is given by

$$U(k) = \frac{1}{\sqrt{N}} \sum_{n=0}^{N-1} u_d(n) e^{-j2\pi kn/N} = \sqrt{N} A_k/2 \quad (2\text{-}16)$$

The difference in notation between the Fourier transform $U(j\omega)$, the discrete Fourier transform $U(k)$, and the Fourier coefficient U_k is indicated by the argument ($j\omega$, k, or subscript k).

2.4 ANALYSIS OF FRF MEASUREMENTS USING PERIODIC EXCITATIONS

In this section we study the principal techniques to measure the FRF of a linear system. During the first part of this analysis we assume that the plant is periodically excited and that an integer number of periods of the steady-state response is measured. The aim of the study is to understand the impact of the disturbing noise on the measured transfer function. Next, we will also consider the use of arbitrary excitations.

2.4.1 Measurement Setup

The typical measurement setup for an FRF measurement is given in Figure 2-16. The generator signal (e.g., a ZOH-reconstructed signal) is applied to the plant (e.g., a mechanical system) using an actuator (e.g., an electromechanical shaker). The input $u_1(t)$ and output $y_1(t)$ (e.g., the applied force and resulting acceleration) are passed through the anti-alias filter before sampling, resulting in $u_{AA}(t)$ and $y_{AA}(t)$. For simplicity we assume that the anti-aliasing filters are perfect, leading to the following assumption:

Assumption 2.2 (Band-Limited Measurements): $u_{AA}(t)$, $y_{AA}(t)$ are band-limited copies of $u_1(t)$, $y_1(t)$ obeying the Shannon theorem: e.g., $U_{AA}(j\omega) = U_1(j\omega)$ for $|\omega| < \omega_s/2$, and $U_{AA}(j\omega) = 0$ for $|\omega| \geq \omega_s/2$.

These time domain signals are finally transformed to the frequency domain using the discrete Fourier transform (DFT), implemented as an FFT (fast Fourier transform). In this section we assume that an integer number of periods is measured so that no leakage errors appear. The FRF at frequency f_k is eventually given by

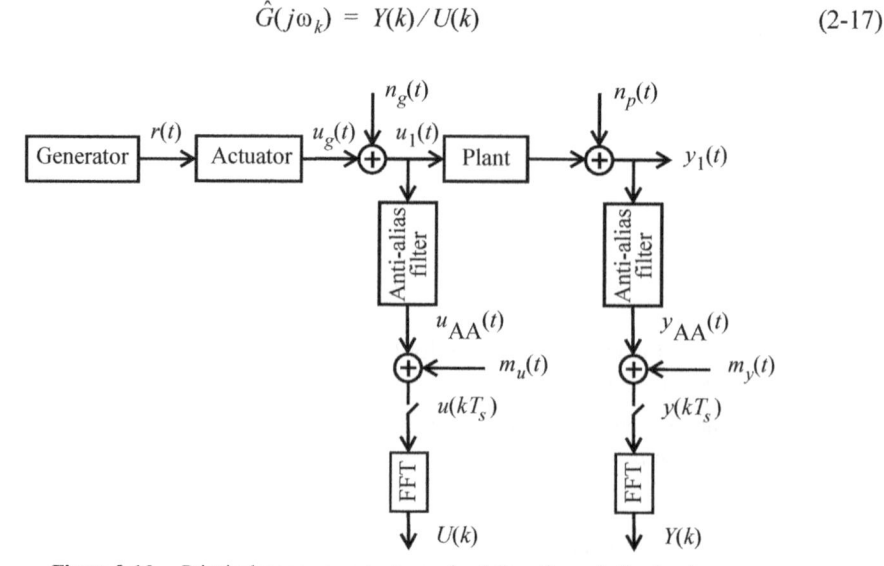

$$\hat{G}(j\omega_k) = Y(k)/U(k) \qquad (2\text{-}17)$$

Figure 2-16. Principal measurement setup and notations for periodic signals.

Section 2.4 ■ Analysis of FRF Measurements Using Periodic Excitations

with $f_k = k/T$, and $T = NT_s$ the length of the measured record. This process is disturbed at different points with noise as shown in Figure 2-16. Generator noise $n_g(t)$ distorts the actual, applied excitation; $m_u(t)$ models the measurement noise (e.g., amplifier noise, quantization noise) on the measured input; $m_y(t)$ stands for the output measurement noise; and the process noise (generated by the plant itself) is given by $n_p(t)$. Notice that although the generator noise $n_g(t)$ acts as a proper excitation signal, it is considered in the periodic setup as a noise source because it is a nonperiodic signal. Later in the chapter, the consequences of this decision will be analyzed in detail. After the DFT we find, at frequency f_k, that

$$Y(k) = Y_0(k) + N_Y(k)$$
$$U(k) = U_0(k) + N_U(k) \tag{2-18}$$

where $N_U(k)$ and $N_Y(k)$ are the contributions of the noise to the measured Fourier coefficients. The impact of the DFT on the noise is intensively studied. Under very mild conditions on the time domain noise, it is shown that (see Section 16.16) these noise contributions are circular complex normally distributed. For our purpose the most important properties of such a distribution are repeated in the following assumption:

Assumption 2.3 (Disturbing Noise): The input $N_U(k)$ and output $N_Y(k)$ errors satisfy

$$\mathbb{E}\{N_U^l(k)\} = 0, \quad \mathbb{E}\{N_Y^l(k)\} = 0, \quad l = 1, 2, \ldots$$
$$\mathbb{E}\{|N_U(k)|^2\} = \sigma_U^2(k), \quad \mathbb{E}\{|N_Y(k)|^2\} = \sigma_Y^2(k) \tag{2-19}$$
$$\mathbb{E}\{N_Y(k)\bar{N}_U(k)\} = \sigma_{YU}^2(k) = \bar{\sigma}_{UY}^2(k), \quad \mathbb{E}\{N_Y(k)N_U(k)\} = 0$$

for $k = 1, 2, \ldots, F$.

At a glance it can be surprising that a squared variable has a zero mean ($\mathbb{E}\{x^2\} = 0$), but the reader should keep in mind that we deal here with complex variables (see also Exercise 16.8). Using these properties, it is easy to carry out a simplified calculation of $\mathbb{E}\{\hat{G}(j\omega_k)\}$ and $\sigma_{\hat{G}}^2(k) = \text{var}(\hat{G}(j\omega_k))$.

2.4.2 Error Analysis

In this section we calculate the bias (systematic error) and the variability (variance) of the measured FRF. In order to address the essential aspects carefully, the analysis is simplified significantly using a Taylor series, assumed to converge. At the end of the section more precise results are included.

Consider the measured FRF $\hat{G}(j\omega_k)$:

$$\hat{G}(j\omega_k) = \frac{Y_0(k) + N_Y(k)}{U_0(k) + N_U(k)} = G_0(j\omega_k)\frac{1 + N_Y(k)/Y_0(k)}{1 + N_U(k)/U_0(k)} \tag{2-20}$$

The Taylor series expansion of $\hat{G}(j\omega_k)$ is

$$\hat{G}(j\omega_k) = G_0(j\omega_k)\left(1 + \frac{N_Y(k)}{Y_0(k)}\right)\left(1 - \frac{N_U(k)}{U_0(k)} + \left(\frac{N_U(k)}{U_0(k)}\right)^2\right) + \text{higher order terms} \tag{2-21}$$

In order to calculate the mean value and the variance of $\hat{G}(j\omega_k)$ it is necessary to make an assumption on the relation between the noise $N_U(k), N_Y(k)$ and the undisturbed signals $U_0(k), Y_0(k)$:

Assumption 2.4 (Disturbing Noise—Continued): The disturbing noise $N_U(k)$, $N_Y(k)$ is independent of the undisturbed signals $U_0(k), Y_0(k)$.

In many cases this is not a difficult assumption. However, in some applications such as measurements in feedback, this assumption is not met if arbitrary excitations are used, leading to systematic errors.

2.4.2.1 Bias Error on the FRF. Under Assumptions 2.3 and 2.4 it follows directly from (2-21) that $\mathbb{E}\{\hat{G}(j\omega_k)\} = G_0(j\omega_k)$. This result can be extended easily to the higher order terms of the Taylor expansion. It shows that if the Taylor series converges, the expected value equals the exact value. However, it is well known that the Taylor series of $1/(1+x)$ converges only if $|x| < 1$, or in this case $|N_U(k)/U_0(k)| < 1$. For normally distributed noise this condition is always violated by a fraction of the realizations. For high SNR ($\sigma_U(k) < |U_0(k)|$) the previous result will be a very good approximation, but for low SNR a significant bias pops up. If $U_0(k)$ is fixed and the noise is normally distributed, an exact calculation of the expected value can be obtained without using the Taylor series approximation (Guillaume et al., 1992b). For uncorrelated input-output noise ($\sigma_{YU}(k) = 0$) the relative bias $b(k)$ is

$$b(k) = \frac{\mathbb{E}\{\hat{G}(j\omega_k)\}}{G_0(j\omega_k)} - 1 = -\exp(-|U_0(k)|^2/\sigma_U^2(k)) \quad (2\text{-}22)$$

This shows that, even for a moderate SNR, small bias errors exist, for example, for an SNR of 6 dB ($[|U_0(k)|/\sigma_U(k) = 2]$), $|b(k)| = 0.018$, but the reader should be aware that significant outliers on $\hat{G}(j\omega_k)$ can appear. For an SNR of 10 dB, $|b(k)| = 5\times10^{-5}$.

If the input noise and output noise are linearly correlated, as in the case of feedback, a more complicated expression is found (see Appendix 9.G or Pintelon and Schoukens, 2001):

$$b(k) = -\exp(-|U_0(k)|^2/\sigma_U^2(k))\left(1 - \rho(k)\frac{U_0(k)/\sigma_U(k)}{Y_0(k)/\sigma_Y(k)}\right) \quad (2\text{-}23)$$

with $\rho(k) = \sigma_{YU}^2(k)/(\sigma_U(k)\sigma_Y(k))$. Also, in this case the maximal relative bias (2-23) is quite small. It is smaller than 1×10^{-4} if the worst case input and output signal-to-noise ratios $|U_0(k)|/\sigma_U(k)$, $|Y_0(k)|/\sigma_Y(k)$ are larger than 10 dB.

This good behavior is due to the use of periodic excitations. If $U_0(k)$ is also a stochastic variable, as is the case for random excitations, the analysis is much more involved. It turns out that in this case the FRF methods are much more sensitive to the noise, leading rapidly to large systematic errors. This discussion is postponed to Section 2.6 but, just as an illustration, it can be mentioned that the bias errors in this case grow to more than 20% for an SNR of 6 dB.

2.4.2.2 Variance Analysis of the FRF. Under Assumption 2.4 and restricting the Taylor expansion in (2-21) to the first-order terms, gives

Section 2.4 ■ Analysis of FRF Measurements Using Periodic Excitations

$$\hat{G}(j\omega_k) \approx G_0(j\omega_k)\left(1 + \frac{N_Y(k)}{Y_0(k)} - \frac{N_U(k)}{U_0(k)}\right) = G_0(j\omega_k) + N_G(k) \qquad (2\text{-}24)$$

where $N_G(k) = G_0(j\omega_k)(N_Y(k)/Y_0(k) - N_U(k)/U_0(k))$. Because $\mathbb{E}\{N_G(k)\} = 0$ the variance $\sigma_{\hat{G}}^2(k) = \text{var}(\hat{G}(j\omega_k))$ is given by

$$\sigma_{\hat{G}}^2(k) \approx \sigma_G^2(k) = \mathbb{E}\{|N_G(k)|^2\} = |G_0(j\omega_k)|^2 \left(\frac{\sigma_Y^2(k)}{|Y_0(k)|^2} + \frac{\sigma_U^2(k)}{|U_0(k)|^2} - 2\text{Re}\left(\frac{\sigma_{YU}^2(k)}{Y_0(k)\bar{U}_0(k)}\right)\right) \qquad (2\text{-}25)$$

The variance is inversely proportional to the square of the signal-to-noise ratio (SNR) of the measurements. This result facilitates the excitation design and answers the question of how the power spectrum of the excitation signal should be chosen to cause a small uncertainty.

Remark. In the previous calculations, an approximate expression for the variance $\sigma_G^2(k)$ is obtained. A detailed analysis (Broersen, 1995) shows, however, that this variance does not exist because of the presence of outliers that appear when the denominator comes very close to zero. This risk disappears for improving input SNR. The variance $\sigma_G^2(k)$ (2-25) can then be interpreted as that of the limiting distribution for the input SNR $\rightarrow \infty$. Guillaume et al. (1992b) showed that the problem can be removed by eliminating the measurements with a "too small" denominator so that no outliers appear anymore.

For the special case that the generator noise dominates ($m_u(t) = 0$, $m_y(t) = 0$ and $n_p(t) = 0$), the following relations exist: $n_y(t) = g_0(t)*n_u(t)$, with $g_0(t)$ the plant impulse response, so that

$$\sigma_Y^2(k) \approx |G_0(j\omega_k)|^2 \sigma_U^2(k) \quad \text{and} \quad \sigma_{YU}^2(k) = \mathbb{E}\{N_Y(k)\bar{N}_U(k)\} \approx G_0(j\omega_k)\sigma_U^2(k)$$

The approximations are due to the leakage effect that appears when random signals are Fourier transformed with the DFT. Substituting these results into (2-25) gives $\sigma_G^2(k) \approx 0$, which implies that the generator noise does not contribute to the uncertainty on the FRF measurements. It also does not contribute to better knowledge of the system because the $n_g(t)$ contributions disappear in the periodic averaging process. This means that some information is lost because $n_g(t)$ can also be considered as an excitation signal.

A number of possibilities are available to reduce the variance (2-25). The most simple solution is to inject more power into the system, increasing $|U_0(k)|$ and $|Y_0(k)|$. In Chapter 5, methods are proposed to maximize this power, while the peak value of the excitations remains below a user-specified level, so that nonlinear operation of the plant is avoided. A second possibility is to measure the FRF frequency by frequency, making stepped sine measurements that concentrate all power at one frequency at a time, so that the SNR is maximized. The disadvantage of this method is that it can become extremely slow because at each frequency point sufficient waiting time should be added until all transients due to the frequency change have disappeared. The alternative is to use well-designed broadband excitations in combination with good averaging methods. This solution depends, again, strongly on the periodic or random nature of the excitation signal, leading to completely different methods.

2.4.2.3 Confidence Regions of the FRF. To construct confidence regions with a given confidence level, one needs the probability density function (pdf) of the measured FRF $\hat{G}(j\omega_k)$ (2-20). Assuming that the input and output errors are circular complex normally distributed ($N_U(k)$, $N_Y(k)$ are normally distributed and Assumption 2.3 is satisfied), the pdf of $\hat{G}(j\omega_k)$ (2-20) is given by (for simplicity of notation we drop the frequency argument)

$$f_{\hat{G}}(\hat{G}) = |G_0|^{-2} f_z(\hat{G}/G_0) \qquad (2\text{-}26)$$

where $f_z(z)$ is the pdf of the ratio $z = \hat{G}/G_0$

$$f_z(z) = \frac{1}{\pi a(z)} \left(\frac{\sigma_u^2 \sigma_y^2 (1-|\rho|^2)}{a(z)} + \left| \frac{b(z)}{a(z)} + 1 \right|^2 \right) e^{-|z-1|^2/a(z)} \qquad (2\text{-}27)$$

with

$$\begin{aligned} a(z) &= \sigma_y^2 + |z|^2 \sigma_u^2 - 2\sigma_y \sigma_u \operatorname{Re}(\rho \bar{z}) \\ b(z) &= (z-1)(\bar{\rho}\sigma_y - \bar{z}\sigma_u)\sigma_u \end{aligned} \qquad (2\text{-}28)$$

σ_u^2, σ_y^2, σ_{yu}^2 the (co)variances of the normalized errors $u = N_U/U_0$ and $y = N_Y/Y_0$, and $\rho = \sigma_{yu}^2/(\sigma_y \sigma_u)$ the corresponding correlation coefficient (proof: see Pintelon et al., 2003). Using (2-27), a circular $100 \times p\%$ ($0 < p < 1$) confidence region with center $\mathbb{E}\{G\}$ and radius $R = r|G_0|$ can be constructed as

$$\operatorname{Prob}(|\hat{G} - \mathbb{E}\{\hat{G}\}| \leq R) = \operatorname{Prob}(|z - \mathbb{E}\{z\}| \leq r) = p \qquad (2\text{-}29)$$

where the probability Prob( ) is calculated via numerical integration (see Pintelon et al., 2003 for the details).

If the input signal-to-noise ratio $1/\sigma_u = |U_0|/\sigma_U$ is larger than 20 dB, then the circular $100 \times p\%$ confidence region (2-29) can be constructed via a circular complex Gaussian approximation of (2-26)

$$f_{\hat{G}}(\hat{G}) \approx \frac{1}{|G_0|^2 \pi \sigma_G^2} e^{-|\hat{G} - \mathbb{E}\{\hat{G}\}|^2/\sigma_G^2} \qquad (2\text{-}30)$$

where $\mathbb{E}\{\hat{G}\}$ and σ_G^2 are defined in, respectively, (2-22) or (2-23) and (2-25). Using (2-30), the radius R of the circular $100 \times p\%$ confidence region (2-29) equals

$$R = \sqrt{-\ln(1-p)} \, \sigma_G \qquad (2\text{-}31)$$

(proof: see Appendix 2.A).

Remarks

(i) Although the variance of the FRF measurement (2-20) does not exist in case of input noise ($\sigma_U \neq 0$), the $100 \times p\%$ uncertainty bound defined in (2-29) and (2-31) is asymptotically (for $|U_0|/\sigma_U \to \infty$) exact.

(ii) For real normal random variables ellipsoids are the most compact $100 \times p \%$ confidence regions (Stuart and Ord, 1987). For a circular complex normal random variable the ellipse reduces to a circle.

(iii) The 95% ($p = 0.95$) confidence bound on the FRF measurement is a circle with radius $\sqrt{3}\sigma_G$, which should not be confused with the 95% confidence interval $\pm 2\sigma_G$, valid for real normal random variables.

2.5 REDUCING FRF MEASUREMENT ERRORS FOR PERIODIC EXCITATIONS

This section shows how to reduce the bias and the variance of FRF measurements using well-designed averaging techniques. Because the solutions strongly depend on the periodic or random behavior of the excitation, the discussion is split into two parts. In the first part we deal with periodic signals because they lead to the best solutions, while the algorithms are very simple. In the next section, random excitations are considered because they are still very popular, even if they lead to inferior results compared with periodic excitations.

All FRF averaging techniques start from M input-output data blocks $u^{[l]}(t)$, $y^{[l]}(t)$, $l = 1, 2, ..., M$. To study the stochastic behavior of theses averaging methods we need an assumption concerning the way the data blocks $u^{[l]}(t)$, $y^{[l]}(t)$, $l = 1, 2, ..., M$ are collected.

Assumption 2.5 (Measurement Data Blocks): The M input-output data blocks $u^{[l]}(t)$, $y^{[l]}(t)$, $l = 1, 2, ..., M$ stem either (i) from M independent (possibly repeated) experiments where the disturbing noise $n_u^{[l]}(t)$, $n_y^{[l]}(t)$ has finite Pth order moments and is independent over l or (ii) from a single experiment where the disturbing noise $n_u(t)$, $n_y(t)$ can be written as filtered white noise with finite Pth order moments.

Intuitively, Assumption 2.5(ii) boils down to saying that the correlation length of the noise should be much smaller than the total measurement time.

2.5.1 Basic Principles

In this section we assume again explicitly that the excitation signal $u_0(t)$ is periodic with period T, such that the sampled signal $u_0(nT_s) = u_0((n + N_p)T_s)$. Notice that this also imposes a constraint on the sampling period because the signal period should be a multiple of the sampling period $T = N_p T_s$. For notational simplicity, we drop the sampling period T_s in the argument of the signals; for example, $x(nT_s)$ is denoted as $x(n)$. When periodic excitations are applied, it is possible to collect M successive periods (with length N_p) and to average the measurements in the time domain over these repeated periods, exemplified by the output measurement (Figure 2-17):

$$\hat{y}(n) = \frac{1}{M}\sum_{l=0}^{M-1} y(n + lN_p) = \frac{1}{M}\sum_{l=1}^{M} y^{[l]}(n) \text{ with } y^{[l]}(n) = y(n + (l-1)N_p) \quad (2\text{-}32)$$

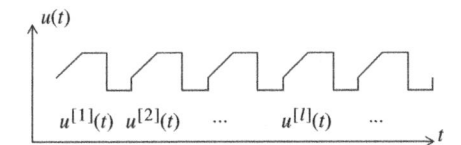

Figure 2-17. Processing periodic excitations.

and the DFT is $\hat{Y}(k) = \text{DFT}(\hat{y}(n))$. The FRF estimate is

$$\hat{G}_{\text{ML}}(j\omega_k) = \frac{\hat{Y}(k)}{\hat{U}(k)} \qquad (2\text{-}33)$$

$\hat{G}_{\text{ML}}$ is the maximum likelihood solution for Gaussian disturbances if the repeated measurements $u^{[l]}, y^{[l]}$ can be considered to be independent over l. It is clear that due to the averaging process, the noise is reduced as $1/\sqrt{M}$ under Assumptions 2.4 and 2.5 ($P = 2$), so that, asymptotically,

$$\underset{M \to \infty}{\text{a.s.lim}} \hat{G}_{\text{ML}}(j\omega_k) = \frac{\underset{M \to \infty}{\text{a.s.lim}} \hat{Y}(k)}{\underset{M \to \infty}{\text{a.s.lim}} \hat{U}(k)} = \frac{Y_0(k)}{U_0(k)} = G_0(j\omega_k)$$

$$\hat{G}_{\text{ML}}(j\omega_k) = G_0(j\omega_k) + O_p(M^{-1/2}) \qquad (2\text{-}34)$$

in the absence of other systematic error sources typified by instrumentation errors (proof: see Appendix 2.B). Moreover, under Assumptions 2.4 and 2.5 (i, $P = 2 + \varepsilon$) or (ii, $P = \infty$), the FRF estimate $\hat{G}_{\text{ML}}(j\omega_k)$ (2-33) is asymptotically normally distributed (see Appendix 2.B). Many dynamic signal analyzers offer this averaging option; for example, $M = 128$ averages are made over $N_p = 2048$ points. Because this improves the results at a very low computational cost, it is strongly advised to make full use of this option. In practice, M is determined by the maximum measurement time T and the minimum required frequency resolution f_0: $M = Tf_0$.

Although the computational effort is minimized by first averaging the measurements in the time domain before calculating the DFTs, it also makes sense to calculate the spectrum of each individual subrecord and perform the averaging in the frequency domain. In the latter case it is also possible to estimate the noise (co)variance. Because the DFT is a linear operator, the order of the operations does not influence the result. Consider the DFTs of the subrecords

$$U^{[l]}(k) = \text{DFT}(u^{[l]}(n)), \qquad Y^{[l]}(k) = \text{DFT}(y^{[l]}(n)) \qquad (2\text{-}35)$$

and calculate the sample mean

$$\hat{U}(k) = \frac{1}{M}\sum_{l=1}^{M} U^{[l]}(k), \quad \hat{Y}(k) = \frac{1}{M}\sum_{l=1}^{M} Y^{[l]}(k), \text{ with } \hat{G}_{\text{ML}}(j\omega_k) = \frac{\hat{Y}(k)}{\hat{U}(k)} \qquad (2\text{-}36)$$

and the sample (co)variances

$$\hat{\sigma}_U^2(k) = \frac{1}{M-1}\sum_{l=1}^{M}|U^{[l]}(k) - \hat{U}(k)|^2, \quad \hat{\sigma}_Y^2(k) = \frac{1}{M-1}\sum_{l=1}^{M}|Y^{[l]}(k) - \hat{Y}(k)|^2$$

$$\hat{\sigma}_{YU}^2(k) = \frac{1}{M-1}\sum_{l=1}^{M}(Y^{[l]}(k) - \hat{Y}(k))\overline{(U^{[l]}(k) - \hat{U}(k))} \qquad (2\text{-}37)$$

Section 2.5 ■ Reducing FRF Measurement Errors for Periodic Excitations

These are unbiased estimates of the true (co)variances. Under Assumptions 2.4 and 2.5 (i, $P = 2$), the asymptotic variance of $\hat{G}_{\text{ML}}(j\omega_k)$ (2-33) is given by (2-25) (see Appendix 2.B). Using (2-37), it can be approximated as

$$\hat{\sigma}_{\hat{G}}^2(k) \approx \frac{|\hat{G}_{\text{ML}}(j\omega_k)|^2}{M}(\hat{\sigma}_Y^2(k)/|\hat{Y}(k)|^2 + \hat{\sigma}_U^2(k)/|\hat{U}(k)|^2 - 2\text{Re}(\hat{\sigma}_{YU}^2(k)/(\hat{Y}(k)\overline{\hat{U}(k)}))) \quad (2\text{-}38)$$

The additional division by M is due to the averaging effect that reduces the noise variance by a factor M if the noise can be considered to be uncorrelated from one subrecord to the other. Note that the circular $100 \times p\%$ confidence region on the FRF estimate (2-29), with R defined in (2-31), needs the value of the variance $\sigma_{\hat{G}}^2(k)$ (2-25), which is unknown. In the sequel of this section we construct a circular $100 \times p\%$ confidence region on the FRF estimate (2-36) using the sample variance $\hat{\sigma}_{\hat{G}}^2(k)$ (2-38).

Assuming that $N_G(k)$ (2-24) is circular complex normally distributed, and that the sample variance $\hat{\sigma}_{\hat{G}}^2(k)$ (2-38) is based on M independent observations, then

$$|\hat{G}_{\text{ML}}(j\omega_k) - G_0(j\omega_k)|^2 / \hat{\sigma}_{\hat{G}}^2(k) \quad (2\text{-}39)$$

is the ratio of two independent chi-squared distributed random variables with 2 and $2M - 2$ degrees of freedom, respectively (the sample mean and sample variance of Gaussian random variables are independently distributed; Stuart and Ord, 1987). Hence, the ratio (2-39) is $F(2, 2M - 2)$-distributed, and a $100 \times p\%$ confidence region for $\hat{G}_{\text{ML}}(j\omega_k)$ can be constructed as a circle with center $\hat{G}_{\text{ML}}(j\omega_k)$ and radius $\hat{\sigma}_{\hat{G}}(k)\sqrt{F_p(2, 2M - 2)}$,

$$\text{Prob}(|\hat{G}_{\text{ML}} - G_0| \leq \hat{\sigma}_{\hat{G}}\sqrt{F_p(2, 2M - 2)}) = p \quad (2\text{-}40)$$

where $F_p(2, 2M - 2)$ is the $100 \times p\%$ percentile of an $F(2, 2M - 2)$-distributed random variable. If M is sufficiently large ($M \geq 20$), then $F_p(2, 2M - 2) \approx -\ln(1 - p)$, and the radius of the circular confidence bound (2-40) reduces to (2-31).

2.5.2 Processing Repeated Measurements

Many instruments do not have enough memory to store long data records. Instead, they make repeated synchronized (starting each time at the same point in the period) measurements of the periodic signal by using a good trigger. In practice, a slight variation appears from measurement to measurement, resulting in time jitter. Consider, for simplicity, noiseless measurements. Then

$$u^{[l]}(nT_s) = u_0(nT_s - \tau^{[l]}) \quad (2\text{-}41)$$

with $\tau^{[l]}$ the variation with respect to the perfect starting point of the measurement. The expected value becomes

$$\mu_u(nT_s) = \mathbb{E}\{u^{[l]}(nT_s)\} = \int_{-\infty}^{+\infty} u_0(nT_s - \tau)f_\tau(\tau)d\tau \quad (2\text{-}42)$$

with $f_\tau(\tau)$ the probability density function of the jitter, and its spectrum is

$$M_u(e^{j\omega T_s}) = U_0(e^{j\omega T_s})F_\tau(j\omega) \tag{2-43}$$

with $F_\tau(j\omega) = F\{f_\tau(\tau)\}$ the characteristic function of $f_\tau(\tau)$. This shows that the jitter acts as a linear filter on the data (Souders et al., 1990). It creates no systematic errors if the jitter is the same for the input and the output error. However, the uncertainty on the FRF measurement increases, especially at the higher frequencies because $F_\tau(j\omega)$ has a low-pass behavior. For example, for normally distributed jitter $N(0, \alpha^2 T_s^2)$,

$$F_\tau(j\omega) = e^{-(\omega^2\alpha^2 T_s^2)/2} = e^{-(\alpha\omega/\omega_s)^2 2\pi^2} \tag{2-44}$$

For jitter with a standard deviation of one sample, a loss of 11 dB appears at $f_s/4$ and 43 dB at $f_s/2$. This clearly shows that it is extremely important to pay sufficient attention to the quality of the triggering if full band measurements are made.

2.5.3 Improved Averaging Methods for Nonsynchronized Measurements

Sometimes it is impossible to get a proper trigger signal that guarantees good synchronization of the measurements. A prime possibility to solve this problem is to perform a post synchronization, estimating each time the delay with respect to the reference record (for example, the first one) and adding a corresponding phase shift $e^{j\omega\tau^{[l]}}$ to the measurements. An alternative is to calculate the FRF of each individual measurement (the division $Y(k)/U(k)$ eliminates the varying delay). As explained in Section 2.4.2.1, this can create bias errors if the simple arithmetic mean is used to average the individual FRF measurements. In Guillaume et al. (1992b) nonlinear averaging methods have been developed that are more robust on this aspect, without increasing the variance significantly. The most robust method turned out to be

$$\begin{aligned}|\hat{G}_{H_{\log}}(j\omega_k)| &= \exp\left(\frac{1}{M}\sum_{l=1}^{M}\text{Re}\left(\log\frac{Y^{[l]}(k)}{U^{[l]}(k)}\right)\right) \\ \angle\hat{G}_{H_{\log}}(j\omega_k) &= \angle\hat{S}_{YU}(k) \quad \text{with} \quad \hat{S}_{YU}(k) = \frac{1}{M}\sum_{l=1}^{M}Y^{[l]}(k)\overline{U^{[l]}(k)}\end{aligned} \tag{2-45}$$

The split between amplitude and phase is made to avoid the phase wrapping problems of the complex logarithm. For circular complex normally distributed errors it is shown under Assumptions 2.4 and 2.5(i) that the relative amplitude error $|\hat{G}_{H_{\log}}(j\omega_k)|/|G_0(j\omega_k)| - 1$ converges for $M \to \infty$ to

$$\exp\left(\frac{1}{2}\text{Ei}\left(-\frac{|U_0(k)|^2}{\sigma_U^2(k)}\right) - \frac{1}{2}\text{Ei}\left(-\frac{|Y_0(k)|^2}{\sigma_Y^2(k)}\right)\right) - 1 \tag{2-46}$$

with Ei(.) the exponential integral functions (Gradshteyn and Ryzhik, 1980). This result is also valid for correlated input-output noise (still assuming that $U_0(k)$ is independent of the disturbing noise). This results in very small bias errors, even for poor SNR, as given in Figure 2-18. A comparison with other classical methods that were originally developed for random excitations is given in Section 2.6.

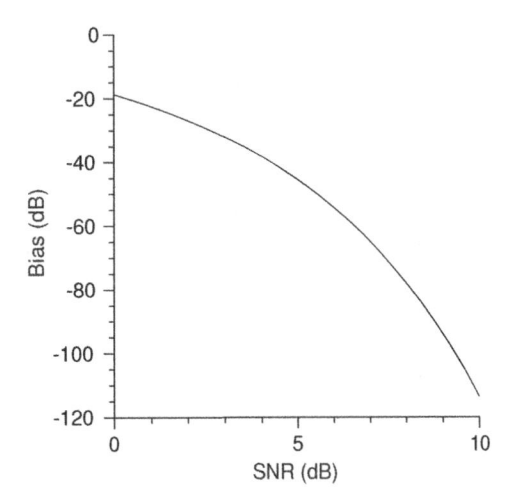

Figure 2-18. Maximum relative bias of $|\hat{G}_{H_{\log}}(j\omega_k)|$ for a given worst case SNR (on input or output).

Remarks

(i) The relative amplitude error (2-46) is also valid in the presence of correlated noise because the log operator in (2-45) separates both noise sources.

(ii) The phase estimate is unbiased if the noise is uncorrelated $\sigma_{YU}(k) = 0$.

2.5.4 Coherence

A measure often used to quantify the quality of the obtained FRF is the coherence $\gamma^2(\omega)$ defined as

$$\gamma^2(\omega) = \frac{|S_{yu}(j\omega)|^2}{S_{uu}(j\omega)S_{yy}(j\omega)} \qquad (2\text{-}47)$$

It measures how much of the output power is coherent (linearly related) with the input power (Bendat and Piersol, 1980; Cadzow and Solomon, 1987). It is shown to be captured between 0 and 1:

$$0 \leq \gamma^2(\omega) \leq 1 \qquad (2\text{-}48)$$

If $\gamma(\omega)$ is smaller than 1 it indicates the presence of

- Extraneous noise in the measurements
- Leakage errors of the DFT
- A nonlinear distortion (only for random excitations)
- Other inputs besides $u(t)$ contributing to the output

For periodic signals (2-47) becomes

$$\hat{\gamma}^2(\omega_k) = \frac{\left|\frac{1}{M}\sum_{l=1}^{M} Y^{[l]}(k)\overline{U}^{[l]}(k)\right|^2}{\left(\frac{1}{M}\sum_{l=1}^{M}|U^{[l]}(k)|^2\right)\left(\frac{1}{M}\sum_{l=1}^{M}|Y^{[l]}(k)|^2\right)} = \frac{\left|1 + \frac{\hat{\sigma}_{YU}^2(k)}{Y_0(k)\overline{U}_0(k)}\right|^2}{\left(1 + \frac{\hat{\sigma}_U^2(k)}{|U_0(k)|^2}\right)\left(1 + \frac{\hat{\sigma}_Y^2(k)}{|Y_0(k)|^2}\right)} \quad (2\text{-}49)$$

where the exact (co)variances are replaced by sample (co)variances. Notice that $\gamma^2(\omega_k) = 1$ when there is only generator noise and the leakage errors are neglected. Sometimes coherence is used to detect nonlinear distortions, although its value is unity for periodic excitations in the absence of noise ($\sigma_U^2(k) = 0$, $\sigma_Y^2(k) = 0$ and $\sigma_{YU}^2(k) = 0$), independent of the presence of nonlinearities (McCormack et al., 1994b). Hence, better alternatives, given in Chapter 3, are sought for the detection of nonlinear distortions.

The variance on the measured FRF can be estimated directly from the coherence by

$$\hat{\sigma}_{\hat{G}}^2(k) \approx |\hat{G}(j\omega_k)|^2 \frac{1 - \gamma^2(\omega_k)}{\gamma^2(\omega_k)} \quad (2\text{-}50)$$

This follows directly from substitution of (2-49) into (2-50), assuming that

$$\frac{\hat{\sigma}_U^2(k)\hat{\sigma}_Y^2(k)}{|U_0(k)|^2|Y_0(k)|^2} \ll \frac{\hat{\sigma}_U^2(k)}{|U_0(k)|^2} + \frac{\hat{\sigma}_Y^2(k)}{|Y_0(k)|^2} \quad \text{and} \quad \left|\frac{\hat{\sigma}_{YU}^2(k)}{Y_0(k)\overline{U}_0(k)}\right| \ll 1$$

This estimate will be very useful in the case of random excitations, where it is impossible to estimate $\hat{\sigma}_U^2(k)$, $\hat{\sigma}_Y^2(k)$, and $\hat{\sigma}_{YU}^2(k)$ directly from the data.

2.6 FRF MEASUREMENTS USING RANDOM EXCITATIONS

In this section we focus on methods that are also applicable to random excitations. The major difference compared with periodic excitations is the variation of the excitation from one realization (subrecord) to the other. This requires other methods to get acceptable results. A comprehensive overview of dedicated FRF measurement techniques for random signals is given in the book of Bendat and Piersol (1980). In this section we give a brief introduction and an alternative to improve the classical methods.

2.6.1 Basic Principles

Consider a linear system driven with random excitations, so that $u_0(t)$ is no longer periodic. Under these conditions the analysis of the previous section is no longer valid. For example, it is no longer possible to consider a fixed value $U_0(k)$ in the Taylor expansion as was done in Section 2.5. A more detailed analysis is needed because the excitation signal varies from one realization to the other. These aspects will be tackled first and dedicated solutions to deal with random excitations are proposed in Section 2.6.2. Also, leakage errors appear (see Section 2.2.2). In general, the spectrum of random signals does not even exist (Bendat and Piersol, 1980; Papoulis, 1981) so that again a detailed analysis is required to understand exactly what is going on.

2.6.2 Reducing the Noise Influence

When measuring the FRF using random excitations, the same approach could be made as for periodic data. The full record is again split into M subrecords with input and output

Section 2.6 ■ FRF Measurements Using Random Excitations

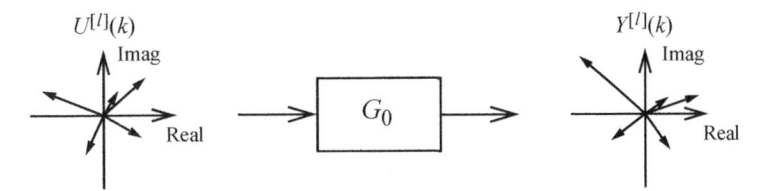

Figure 2-19. Successive realizations of $U^{[l]}(k)$ and $Y^{[l]}(k)$.

DFT spectra $U^{[l]}(k)$, $Y^{[l]}(k)$ for block l. Eventually, the FRF for block l is then $Y^{[l]}(k)/U^{[l]}(k)$. Broersen (1995) showed that this direct calculation has an infinite variance. From (2-20) it is also seen that bias errors are created because $\mathbb{E}\{1/(1 + N_U(k)/U_0(k))\} \neq 1$. This bias is mainly induced by the nonlinear behavior of the division. The bias will be small only if $|N_U(k)/U_0(k)| \ll 1$. It is, therefore, necessary to reduce the noise by averaging before making the division. However, because $\mathbb{E}\{U^{[l]}(k)\} = 0$, it is clear that this cannot be done straightforwardly. The reason for this problem is that the vector $U^{[l]}(k)$ has a random phase, uniformly distributed between $[0, 2\pi)$ so that its averaged value is zero (see Figure 2-19). A possibility to avoid this problem is to eliminate the phase of $U^{[l]}(k)$ by multiplying it with its complex conjugate, to get vectors with a fixed phase as shown in Figure 2-20. It is also possible to average before making the division:

$$\hat{G}(j\omega_k) = \frac{\hat{S}_{YU}(k)}{\hat{S}_{UU}(k)} = \frac{\frac{1}{M}\sum_{l=1}^{M} Y^{[l]}(k)\overline{U}^{[l]}(k)}{\frac{1}{M}\sum_{l=1}^{M} |U^{[l]}(k)|^2} \qquad (2\text{-}51)$$

Readers who are familiar with this field will observe that this expression is nothing other than the discrete implementation of the Wiener-Hopf equation (see Bendat and Piersol, 1980, Eq. (4.7), and Eykhoff, 1974, Eq. (8.10)), relating the cross-power with the autopower spectrum: $S_{yu}(j\omega) = G(j\omega)S_{uu}(j\omega)$. The asymptotic properties can be obtained easily by splitting the measurements into the undisturbed parts $U_0(k)$, $Y_0(k)$ (neglecting the leakage effects) and the distortions $N_U(k)$, $N_Y(k)$. Under Assumptions 2.4 and 2.5 ($P = 4$), the systematic errors and the variability can be calculated.

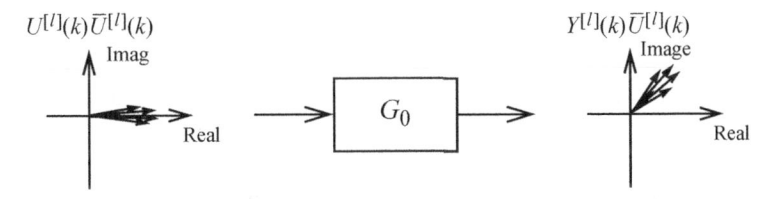

Figure 2-20. Successive realizations of $U^{[l]}(k)\overline{U}^{[l]}(k)$ and $Y^{[l]}(k)\overline{U}^{[l]}(k)$.

2.6.2.1 Systematic Errors.
Under Assumptions 2.4 and 2.5 ($P = 4$), the estimate (2-51) converges to

$$\underset{M \to \infty}{\text{a.s.lim}}\, \hat{G}(j\omega_k) = \frac{\underset{M \to \infty}{\text{a.s.lim}}\, \frac{1}{M}\sum_{l=1}^{M} Y^{[l]}(k)\overline{U}^{[l]}(k)}{\underset{M \to \infty}{\text{a.s.lim}}\, \frac{1}{M}\sum_{l=1}^{M} |U^{[l]}(k)|^2} = \frac{\mathbb{E}\{Y_0(k)\overline{U}_0(k)\} + \sigma_{YU}^2(k)}{\mathbb{E}\{|U_0(k)|^2\} + \sigma_U^2(k)} \qquad (2\text{-}52)$$

at the rate $O_p(M^{-1/2})$ (see Appendix 2.B). Moreover, under Assumption 2.4 and Assumption 2.5(i, $P = 4 + \varepsilon$) or (ii, $P = \infty$), the FRF estimate $\hat{G}(j\omega_k)$ (2-51) is asymptotically normally distributed (see Appendix 2.B). Neglecting the leakage effects, (2-52) becomes

$$\underset{M \to \infty}{\text{a.s.lim}} \hat{G}(j\omega_k) \approx G_0(j\omega_k) \frac{1 + \sigma_{YU}^2(k)/\mathbb{E}\{Y_0(k)\overline{U}_0(k)\}}{1 + \sigma_U^2(k)/\mathbb{E}\{|U_0(k)|^2\}} \quad (2\text{-}53)$$

Notice that for random signals $\mathbb{E}\{|U_0(k)|^2\}$ cannot be replaced by $|U_0(k)|^2$ because $U_0(k)$ varies from one realization to the other. Equation (2-53) shows that there is a systematic error that did not appear in the previous approach. This is the price to be paid for using random instead of periodic excitations. If the input signal can be measured free of noise, $\sigma_U(k) = 0$, the bias disappears. The method (2-51) is sometimes called the H_1 method. If the SNR at the output is much higher than that at the input, then it is better to use the following alternative:

$$\hat{G}(j\omega_k) = \frac{\hat{S}_{YY}(k)}{\hat{S}_{UY}(k)} = \frac{\frac{1}{M}\sum_{l=1}^{M}|Y^{[l]}(k)|^2}{\frac{1}{M}\sum_{l=1}^{M}U^{[l]}(k)\overline{Y}^{[l]}(k)} \quad (2\text{-}54)$$

which is called the H_2 method. The H_2 method (2-54) under the same noise assumptions has the same asymptotic ($M \to \infty$) properties as the H_1 method (2-51). Neglecting the leakage effects, the asymptotic value of (2-54) is

$$\underset{M \to \infty}{\text{a.s.lim}} \hat{G}(j\omega_k) \approx G_0(j\omega_k) \frac{1 + \sigma_Y^2(k)/\mathbb{E}\{|Y_0(k)|^2\}}{1 + \sigma_{UY}^2(k)/\mathbb{E}\{Y_0(k)\overline{U}_0(k)\}} \quad (2\text{-}55)$$

For uncorrelated noise, $\sigma_{UY}^2(k) = 0$, (2-53) and (2-55) reduce to, respectively,

$$|G_0(j\omega_k)|/|1 + \sigma_U^2(k)/\mathbb{E}\{|U_0(k)|^2\}| \quad \text{and} \quad |G_0(j\omega_k)||1 + \sigma_Y^2(k)/\mathbb{E}\{|Y_0(k)|^2\}|$$

Hence,

$$\left| \underset{M \to \infty}{\text{a.s.lim}} \hat{G}_{H_1}(j\omega_k) \right| \leq |G_0(j\omega_k)| \leq \left| \underset{M \to \infty}{\text{a.s.lim}} \hat{G}_{H_2}(j\omega_k) \right| \quad (2\text{-}56)$$

where $\hat{G}_{H_1}(j\omega_k)$ and $\hat{G}_{H_2}(j\omega_k)$ are given by, respectively, (2-51) and (2-54). This result cannot be generalized to the case of correlated noise.

2.6.2.2 Variance. An approximate expression for the variance of $\hat{G}(j\omega_k)$ (valid for the H_1 and H_2) is found by considering only the linear noise contributions to (2-51):

$$\hat{G}(j\omega_k) = G_0(j\omega_k)\frac{1 + N_1(k)}{1 + N_2(k)} \approx G_0(j\omega_k)(N_1(k) - N_2(k)) \quad (2\text{-}57)$$

with

$$N_1(k) = \frac{\sum_{l=1}^{M} N_Y^{[l]}(k)\overline{U}_0^{[l]}(k) + Y_0^{[l]}(k)\overline{N}_U^{[l]}(k)}{\sum_{l=1}^{M} Y_0^{[l]}(k)\overline{U}_0^{[l]}(k)}, \quad N_2(k) = \frac{\sum_{l=1}^{M} N_U^{[l]}(k)\overline{U}_0^{[l]}(k) + U_0^{[l]}(k)\overline{N}_U^{[l]}(k)}{\sum_{l=1}^{M} |\overline{U}_0^{[l]}(k)|^2}$$

Section 2.6 ■ FRF Measurements Using Random Excitations

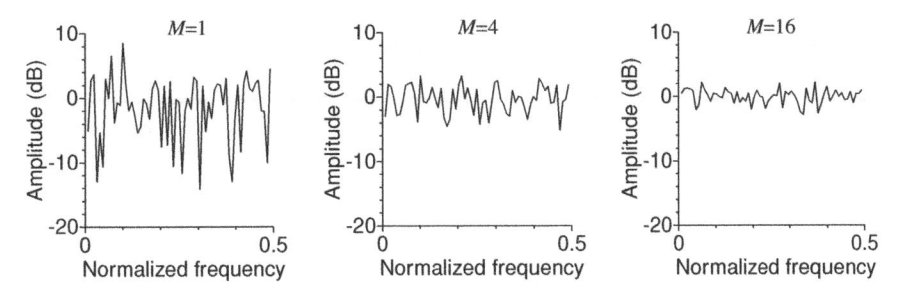

Figure 2-21. Realized power spectrum $S_{UU}^{(M)}(j\omega_k)$ for a white noise sequence ($M = 1, 4, 16$).
Note that for a periodic signal a flat line at 0 dB would be found.

Next, the variance of (2-57) is obtained assuming that the M data blocks (subrecords) are independent, Assumption 2.5(i), and by taking the expected value $\mathbb{E}\{|G_0(j\omega_k)(N_1(k) - N_2(k))|^2\}$ with respect to the noise and not to the random excitation signal. This means that we calculate the variance that would be obtained if the experiment was repeated with the same noise realizations for the excitation signal. Neglecting the leakage errors,

$$G_0(j\omega_k) \approx \frac{Y_0^{[l]}(k)}{U_0^{[l]}(k)} \approx \frac{\sum_{l=1}^{M} Y_0^{[l]}(k)\overline{U}_0^{[l]}(k)}{\sum_{l=1}^{M} |U_0^{[l]}(k)|^2} \approx \frac{\sum_{l=1}^{M} |Y_0^{[l]}(k)|^2}{\sum_{l=1}^{M} \overline{Y}_0^{[l]}(k)U_0^{[l]}(k)}$$

we find

$$\sigma_{\hat{G}}^2(k) = |G_0(j\omega_k)|^2 \left[\frac{\sigma_Y^2(k)}{\sum_{l=1}^{M}|Y_0^{[l]}(k)|^2} + \frac{\sigma_U^2(k)}{\sum_{l=1}^{M}|U_0^{[l]}(k)|^2} - 2\text{Re}\left(\frac{\sigma_{YU}^2(k)}{\sum_{l=1}^{M} Y_0^{[l]}(k)\overline{U}_0^{[l]}(k)}\right) \right] \quad (2\text{-}58)$$

If the number of blocks $M \to \infty$ and we assume that the random excitation is stationary, the variance becomes

$$\underset{M \to \infty}{\text{a.s.lim}} M\sigma_{\hat{G}}^2(k) = |G_0(j\omega_k)|^2 \left(\frac{\sigma_Y^2(k)}{S_{Y_0Y_0}(j\omega_k)} + \frac{\sigma_U^2(k)}{S_{U_0U_0}(j\omega_k)} - 2\text{Re}\left(\frac{\sigma_{YU}^2(k)}{S_{Y_0U_0}(j\omega_k)}\right) \right) \quad (2\text{-}59)$$

so that for M sufficiently large, the following approximate expression can be used:

$$\sigma_{\hat{G}}^2(k) \approx \frac{|G_0(j\omega_k)|^2}{M} \left(\frac{\sigma_Y^2(k)}{S_{Y_0Y_0}(j\omega_k)} + \frac{\sigma_U^2(k)}{S_{U_0U_0}(j\omega_k)} - 2\text{Re}\left(\frac{\sigma_{YU}^2(k)}{S_{Y_0U_0}(j\omega_k)}\right) \right) \quad (2\text{-}60)$$

This expression is similar to (2-38) and shows that the uncertainty $\sigma_{\hat{G}}(k)$ decreases as $O(M^{-1/2})$. However, for small M,

$$\hat{S}_{U_0U_0}(k) = \frac{1}{M}\sum_{l=1}^{M}|U_0^{[l]}(k)|^2 \quad (2\text{-}61)$$

which can be significantly different from $S_{u_0u_0}(j\omega_k)$; thus (2-58) should be used. At some frequencies large drops in the realized power spectrum can appear, jeopardizing the FRF measurement completely. Therefore, it is strongly advised to average over a number of blocks to avoid these dips. In Figure 2-21, the realized power spectrum $\hat{S}_{U_0U_0}(k)$, after processing M

TABLE 2-1 Study of the Stochastic Behavior of the Averaged Spectrum of a Random Signal

N	Ratio 95% Upper/95% Lower Bound (dB)	Ratio 1/95% Lower Bound (dB)
1	22	13
2	14	7.5
4	9	4.7
8	6.2	3.0
16	4.3	2.1
32	3.1	1.4
64	2.1	1.0
128	1.5	0.7
256	1.1	0.5

blocks of a white noise excitation, is shown in dB (S_{xx} in dB is given by $10\log_{10}S_{xx}$). It is clearly seen that, compared with the limit value $\hat{S}_{U_0U_0}(k)$ (a constant value of 0 dB) for $M \to \infty$, a significant loss can occur. The normalized power spectrum $2M\hat{S}_{U_0U_0}(k)/S_{u_0u_0}(j\omega_k)$ is χ^2 distributed, having $2M$ degrees of freedom because it consists of the sum of $2M$ squared, independent, zero mean, normally distributed variables with equal variance (the real and imaginary part). In Table 2-1 the 95% uncertainty regions of the amplitude spectrum are described by their upper and lower bounds. The ratio of the lower bound to the rms value is also tabulated to illustrate the loss in SNR of the weakest components because of the stochastic nature of the excitations.

In Figure 2-22 the loss in the SNR for random signals when compared with a deterministic signal with flat amplitude spectrum is shown as a function of the number of processed blocks M. It shows that for small M the SNR increases very rapidly because dips in the averaged input power spectrum disappear. It also shows that four experiments are needed to guarantee that 95% of the measurement points have an SNR corresponding to that of a well-designed, deterministic excitation after only one period (SNR normalized at 0 dB). This is one of the reasons why we strongly advocate the use of periodic excitations.

The coherence $\gamma^2(\omega_k)$, as given in (2-47), can be used again to give an overall impression of the quality of the measurement. In practice, the variance on the FRF is estimated from the coherence using (2-50).

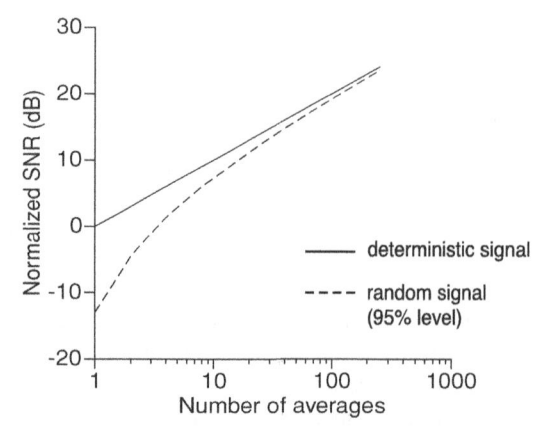

Figure 2-22. Loss in SNR for random excitations as a function of the number of processed blocks M.

2.6.3 Leakage Errors

In the previous section we assumed that it was possible to pass easily from a continuous-time signal $u(t)$ to its Fourier transform $U(j\omega)$. In practice, the DFT of random signals suffers from leakage errors (see Section 2.2.2). So even for undisturbed signals ($n_u(t) = 0$, $n_y(t) = 0$, and $n_g(t) = 0$ in Figure 2-16) the FRF measurement is incorrect, that is, $G_0(j\omega_k) \neq Y_0(k)/U_0(k)$, where $U_0(k)$, $Y_0(k)$ denote the DFT of u_0, y_0. Ljung (1999) shows for discrete-time systems that the error on the FRF disappears as $O(N^{-1/2})$. This result can also be extended to FRF measurements of continuous-time systems. This is formulated precisely in the following theorem:

Theorem 2.6 (Leakage Errors on FRF Measurements of Continuous-Time Systems): Consider the signals $y(t)$ and $u(t)$ obeying Assumption 2.2 and related by the strictly stable system $G_0(j\omega) = F\{g_0(t)\}$ ($y(t) = g_0(t) * u(t)$). Let

$$U(k) = \frac{1}{\sqrt{N}} \sum_{n=0}^{N-1} u(nT_s) e^{-j2\pi kn/N}, \quad Y(k) = \frac{1}{\sqrt{N}} \sum_{n=0}^{N-1} y(nT_s) e^{-j2\pi kn/N} \quad (2\text{-}62)$$

be the DFT spectra of the sampled signals $u(nT_s)$ and $y(nT_s)$. If $u(nT_s)$ is uniformly bounded, filtered white noise, then

$$Y(k) = G_0(j\omega_k)U(k) + T_G(j\omega_k) \quad (2\text{-}63)$$

with $T_G(j\omega_k) = O(N^{-1/2})$ uniformly over the frequency k.

Proof. See Section 6.5.2. □

Formula (2-63) shows that the leakage errors can be interpreted as a transient effect. This is illustrated in Figure 2-23.

Remarks

(i) The DFT for random signals is defined with a scaling factor $1/\sqrt{N}$ so that the DTF spectrum behaves as $O(N^0)$.
(ii) If the excitation signal is a periodic signal and the number of data points is increased by repeating this signal (so that no additional frequencies are excited), the previous result can be formulated more strongly as $|T(j\omega_k)| \leq O(N^{-1})$.
(iii) This theorem shows that the leakage error decreases with an increasing number of data, but it does not guarantee that the errors are small for finite N.

Example 2.7 (Leakage Errors on FRF Measurement): To illustrate the impact of the leakage effect, a simulation is made on a second-order discrete-time system with a narrow resonance peak of 30 dB. The system is driven with white normally distributed noise, without disturbing noise. The record is split into $M = 100$ subrecords of length 256 data points each, and the DFT spectrum of each windowed record (2-8) is calculated for a rectangular (2-7) and a Hanning (2-12) window. Next, the FRF is estimated using (2-51) and the results are shown in Figure 2-24 for the rectangular and the Hanning windows. The errors can become very large, especially around the resonance frequency, where fast variations of the FRF occur. Re-

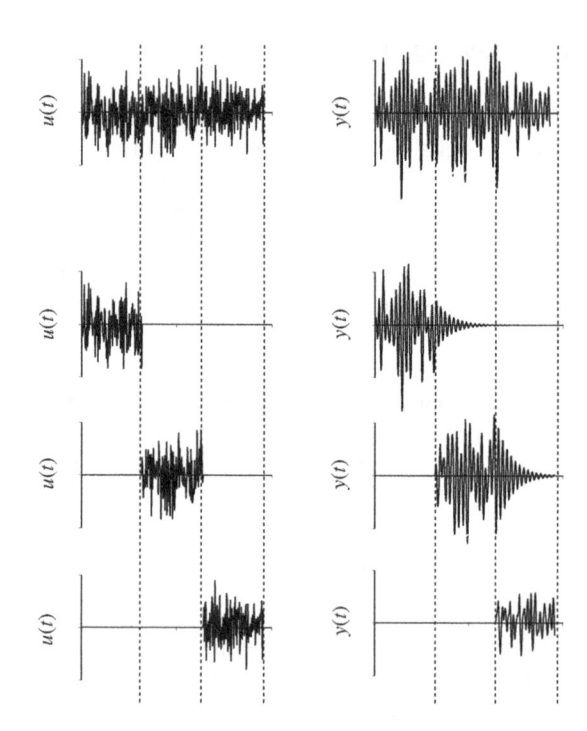

Figure 2-23. Interpretation of the leakage error as a transient effect.

placing the rectangular window with a Hanning window reduces the errors significantly at most frequencies, but the problem at the resonance persists. Note also that these results are obtained after 100 averages. So the systematic errors dominate in these results, which shows that leakage not only increases random errors but also creates a bias. These errors are proportional to the second derivative $d^2 G_0(j\omega)/d\omega^2$ (Bendat and Piersol, 1980). In Figure 2-25 the coherence calculated with (2-49) is shown. Although it is poor everywhere for the rectangular window, it is quite good for the Hanning window except around the resonance frequency. □

Conclusion: Leakage can jeopardize the quality of the FRF measurements significantly. Averaging reduces the random appearance, resulting in smoother measurements, but cannot eliminate the systematic errors. Using other windows makes it possible to reshape the leakage errors, but they remain large in the frequency bands with fast variations of the FRF. Often these bands carry most information (e.g., the resonance frequency). To avoid leakage, the

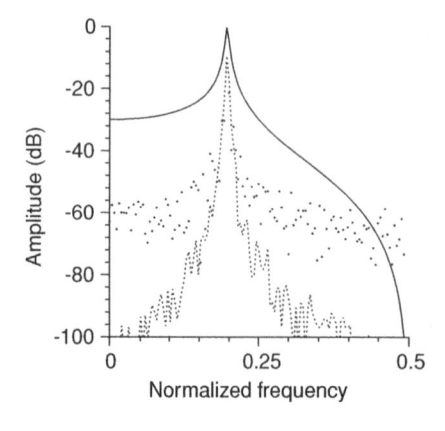

Figure 2-24. Illustration of the leakage effect. ——: $G_0(e^{-j\omega T_s})$, +: complex errors with a rectangular window, ----: complex errors with a Hanning window.

Section 2.6 ■ FRF Measurements Using Random Excitations 61

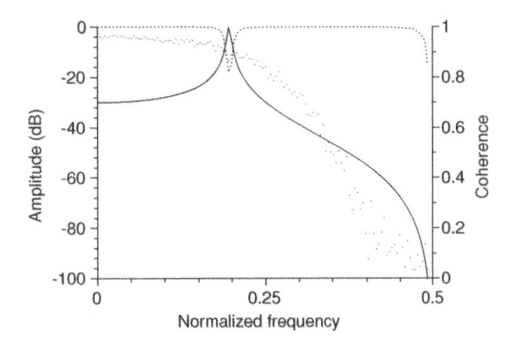

Figure 2-25. Coherence of the measurements in Figure 2-24. ——: $G_0(e^{-j\omega T_s})$, +: coherence with a rectangular window, ---- coherence with a Hanning window.

best solution is use of periodic excitations and measurements of an integer number of periods. Alternative methods that exploit the particular nature of leakage errors are given in Chapter 4. A last possibility is to use burst random excitations as explained in Section 2.2.4.

2.6.4 Indirect FRF Measurements

Consider the measurement setup of Figure 2-26, where the reference signal $r(t)$ is random. Due to the input measurement noise $m_u(t)$ (open and closed loop) and the process noise $n_p(t)$ (closed loop only), the FRF estimate (2-51) is inconsistent (see (2-53) with $\sigma_U^2 \neq 0$ and $\sigma_{YU}^2 \neq 0$). The bias in (2-53) is avoided via the indirect FRF estimate

$$\hat{G}(j\omega_k) = \frac{\hat{S}_{YR}(k)}{\hat{S}_{UR}(k)} = \frac{\frac{1}{M}\sum_{m=1}^{M} Y^{[m]}(k)\overline{R}^{[m]}(k)}{\frac{1}{M}\sum_{m=1}^{M} U^{[m]}(k)\overline{R}^{[m]}(k)} \qquad (2\text{-}64)$$

where $R^{[m]}(k)$ is the DFT spectrum of the mth subrecord of the known reference signal (Wellstead, 1977 and 1981). Since all the noise sources in Figure 2-26 are independent of the reference signal $r(t)$, we find, assuming that the leakage errors can be neglected,

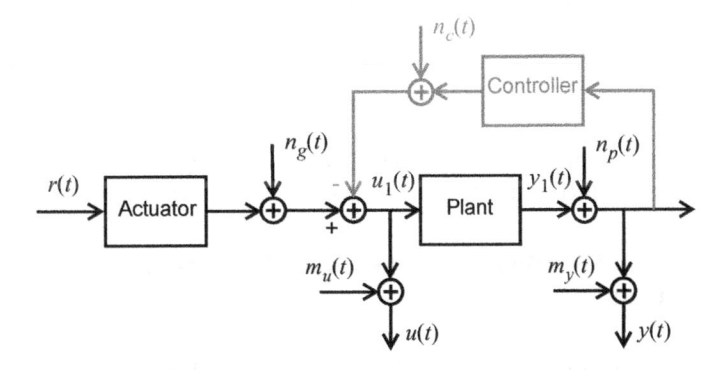

Figure 2-26. FRF measurement of a plant operating in open (black lines only) or closed (black and gray lines) loop. $u_1(t)$ and $y_1(t)$ are the true input-output signals and $n_g(t)$, $n_c(t)$, $n_p(t)$, and $m_u(t)$, $m_y(t)$ are, respectively, the generator noise, the controller noise, the process noise, and the input-output measurement noise.

$$\underset{M \to \infty}{\text{a.s.}\lim} \hat{G}(j\omega_k) = \frac{\underset{M \to \infty}{\text{a.s.}\lim} \hat{S}_{YR}(k)}{\underset{M \to \infty}{\text{a.s.}\lim} \hat{S}_{UR}(k)} = \frac{\dfrac{G_0(j\omega_k)G_{\text{act}}(j\omega_k)}{1 + G_0(j\omega_k)M_0(j\omega_k)}S_{RR}(k)}{\dfrac{G_{\text{act}}(j\omega_k)}{1 + G_0(j\omega_k)M_0(j\omega_k)}S_{RR}(k)} = G_0(j\omega_k) \quad (2\text{-}65)$$

where $G_{\text{act}}(j\omega)$ and $M_0(j\omega)$ are, respectively, the true actuator and controller frequency response functions. The drawback of the indirect method (2-64) is that the generator $n_g(t)$ (open and closed loop) and the controller $n_c(t)$ (closed loop only) noise parts of the true plant input $u_1(t)$ are considered as noise. Indeed, $n_g(t)$ and $n_c(t)$ are not correlated with the reference signal $r(t)$ and, hence, vanish asymptotically ($M \to \infty$) in the cross-power spectra $\hat{S}_{YR}(k)$ and $\hat{S}_{UR}(k)$.

If $G_{\text{act}}(j\omega) = 1$ and if the process noise is the only disturbance in Figure 2-26 ($n_g = 0$, $n_c = 0$, $m_u = 0$, and $m_y = 0$), then the controller transfer function $M_0(j\omega)$ is exactly known (since $r(t)$ is known and since $u(t)$ and $y(t)$ are observed without measurement errors, the input and output of the feedback branch in Figure 2-26 are exactly known). Under these conditions the indirect estimate (2-64) can be written as

$$\hat{G}(j\omega_k) = \frac{\hat{G}_{\text{cl}}(j\omega_k)}{1 - M_0(j\omega_k)\hat{G}_{\text{cl}}(j\omega_k)} \quad (2\text{-}66)$$

where $\hat{G}_{\text{cl}}(j\omega_k) = \hat{S}_{YR}(k)/\hat{S}_{RR}(k)$ is an estimate of the closed loop transfer function. The finite sample properties (bias and variance) of the indirect estimate (2-66) have been studied in detail in Heath (2001) and Welsh and Goodwin (2002).

Remarks

(i) The indirect method (2-64) can be interpreted as an instrumental variables method where the reference signal plays the role of instrumental variable (Norton, 1986).

(ii) Dividing the numerator and denominator of $\hat{G}(j\omega_k)$ in (2-64) by $\hat{S}_{RR}(k)$ shows that the indirect FRF estimate is the ratio of the FRF from reference to output $\hat{G}_{ry}(j\omega_k)$ to the FRF from reference to input $\hat{G}_{ru}(j\omega_k)$

$$\hat{G}(j\omega_k) = \hat{G}_{ry}(j\omega_k)/\hat{G}_{ru}(j\omega_k) \quad (2\text{-}67)$$

(iii) For periodic reference signals $r(t)$ we have that $R^{[m]}(k) = R(k)$ and, hence, the indirect FRF estimate (2-64) reduces to the direct FRF estimate (2-33).

2.6.5 Improved FRF Measurements Using Overlapping Segments

The leakage errors in the direct (2-51) and indirect (2-64) FRF estimates are reduced by multiplying the input-output signals with a time domain window $w(t)$, $t = 0, 1, ..., N-1$, (e.g., a Hanning window) before calculating the input-output DFT spectra (see Example 2.7). They can be reduced further by using overlapping subrecords (see Figure 2-27), resulting in the so-called weighted overlapped segment averaging (WOSA) introduced by Welsh (1967)

Section 2.6 ■ FRF Measurements Using Random Excitations

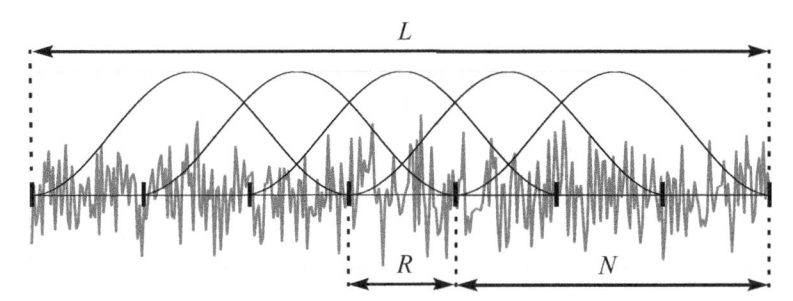

Figure 2-27. Principle of the weighted overlapped segment averaging: the signal (gray line) of length L is divided into M weighted segments (black lines) of length N with an overlap of $N-R$ samples. The figure shows an overlap of 2/3.

for estimating noise power spectra. The autopower and cross-power spectra in (2-51) and (2-64) are then replaced by

$$\hat{S}_{X_W Z_W}(k) = \frac{1}{M}\sum_{m=0}^{M-1} X_W^{[m]}(k)\overline{Z_W^{[m]}}(k) \qquad (2\text{-}68)$$

where

$$X_W^{[m]}(k) = \frac{1}{\sqrt{\sum_{t=0}^{N-1} w^2(t)}} \sum_{t=0}^{N-1} w(t)x(t+mR)e^{-j2\pi kt/N} \qquad (2\text{-}69)$$

with $X, Z \in \{U, Y, R\}$; $x, z \in \{u, y, r\}$; $w(t)$, $t = 0, 1, ..., N-1$, the time domain window; N the length of the subrecord (segment); M the number of subrecords (segments); and $N-R$ the number of common samples in two consecutive overlapping subrecords (segments). The total length of the signals equals $L = RM + N - R = R(M-1) + N$ samples, and the fraction of overlap between adjacent segments is $1 - R/N$. The scaling factor in the DFT (2-69) is such that white noise $e(t)$ with variance σ^2 results in complex noise $E_W(k)$ with the same variance σ^2. Note that the computational effort of the WOSA estimate (2-68) increases with the fraction of overlap.

The properties of the WOSA estimator have been studied in Carter and Nuttall (1980) and Nuttall and Carter (1982) within the context of spectrum estimation ((2-68) with $Z = X$ and $z = x$). They reported that a Hanning window (2-12) combined with 1/2 overlap is a good compromise between leakage error suppression and computational effort. These settings are often the default choice in digital spectrum analyzers.

Assuming that the subrecord (segment) length N is larger than the system (plant) time constant, it has been shown in Antoni and Schoukens (2007 and 2009) that the mean square error of the leakage errors on the direct FRF estimate

$$\hat{G}(j\omega_k) = \hat{S}_{Y_W U_W}(k)/\hat{S}_{U_W U_W}(k) \qquad (2\text{-}70)$$

are minimized by the half-sine window, $w(t) = \sin(\pi t/N)$, $t = 0, 1, ..., N-1$, combined with 2/3 overlap ($R = N/3$ in (2-69)). Compared with the estimate without overlap ($R = N$ in (2-69)), the 2/3 overlap reduces the variance of the leakage errors and the noise by a factor of 3 and 1.7, respectively, for the half-sine window. For the Hanning window

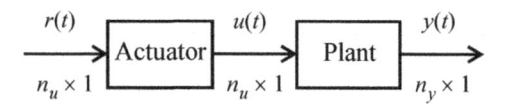

Figure 2-28. Linear time-invariant plant with n_u inputs and n_y outputs, excited by a periodic $n_u \times 1$ signal $r(t)$ via a (non)linear actuator.

(2-12) these factors are, respectively, 3.8 and 2. Compared with the half-sine window, the bias due to the leakage errors is 1.09 times larger for the Hanning window.

2.7 FRF MEASUREMENTS OF MULTIPLE-INPUT, MULTIPLE-OUTPUT SYSTEMS

The results that are presented in this chapter are also valid for multiple-input, multiple-output (MIMO) systems. Excitation signals that are suitable for single input, single output (SISO) systems also form a good basis to start MIMO measurements. However, additional precautions have to be taken because the FRF of a MIMO system is described by an $n_y \times n_u$ matrix at each frequency:

$$G(j\omega_k) \in \mathbb{C}^{n_y \times n_u} \tag{2-71}$$

with n_u and n_y the numbers of inputs and outputs of the system. Indeed, since the input-output relationship

$$Y(k) = G(j\omega_k)U(k) \tag{2-72}$$

has n_y equations with $n_y n_u$ unknowns, the frequency response matrix (FRM) $G(j\omega_k)$ cannot be identified from one experiment (2-72) unless constraints are imposed on either the spectral content of the input signal (see Section 2.7.1) or the smoothness of the FRM (see Chapter 7). Another possibility consists in performing n_u experiments with n_u different excitations signals. This approach is discussed in detail in Section 2.7.2.

2.7.1 One Experiment

It is possible to calculate the FRM $G(j\omega_k)$ via (2-72) if an excited frequency only appears at one input. For example, if the input signals are chosen such that each input contains F excited frequencies at an interleaved frequency grid (Figwer and Niederlinski, 1995; Verbeeck et al., 1999b)

$$\begin{cases} U_{[p]}(n_u(k-1)+p) \neq 0 & k = 1, 2, \ldots, F; p = 1, 2, \ldots, n_u \\ U_{[p]}(n_u(k-1)+r) = 0 & r = 1, 2, \ldots, n_u; r \neq p \end{cases} \tag{2-73}$$

one obtains the so-called zippered multisines (see Figure 2-29, left column, for the case $n_u = 3$). Entry $[q, p]$ of the FRM is then found as

$$G_{[q,p]}(j\omega_{n_u(k-1)+p}) = \frac{Y_{[q]}(n_u(k-1)+p)}{U_{[p]}(n_u(k-1)+p)} \tag{2-74}$$

Section 2.7 ■ FRF Measurements of Multiple-Input, Multiple-Output Systems

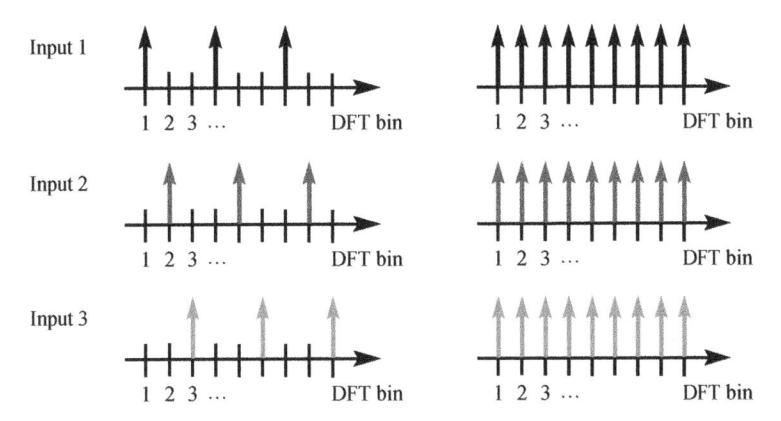

Figure 2-29. Input DFT spectrum of an $n_u = 3$ input system (black: input 1; dark gray: input 2; light gray: input 3). Left: zippered multisine; and right: Hadamard and (random) orthogonal multisines.

for $k = 1, 2, ..., F$, $p = 1, 2, ..., n_u$, and $q = 1, 2, ..., n_y$. It shows that periodic signals with a zippered spectrum (2-73) are uncorrelated for a finite number of samples N. The zippered multisine solution requires n_u different generators to create the signals. Note that it is also possible to design ternary signals (take only the values $\{-a, 0, a\}$) with a zippered spectrum (Tan et al., 2009).

The zippered input spectrum approach is useful in control applications where a model is needed from the generator signal (or output of the controller) $r(t)$ (see Figure 2-28) to the output of the plant (Rivera et al., 2007 and 2009). Due to the (non)linear interaction between the actuators and the system (Verbeeck et al., 1999b), condition (2-73) is, in general, not fulfilled at the input $u(t)$ of the multivariable system (see Figure 2-28). In that case, the zippered input spectrum approach cannot be used to measure the FRM of the system itself from one single experiment. For the same reason it can also not be used for direct FRM estimates under closed loop conditions.

2.7.2 Multiple Experiments

There are two kinds of approaches to perform the n_u experiments: either the inputs are excited one after the other, or all inputs are excited simultaneously. The main advantage of the first approach is that it requires only one generator. The disadvantages of the first w.r.t. the second approach are: (i) for the same frequency resolution and input rms value, the signal-to-noise ratio is $\sqrt{n_u}$ times smaller, or for the same signal-to-noise ratio and input rms value, the measurement time is n_u times longer; and (ii) the experiments do not mimic the operational conditions, which might be a problem if the system behaves nonlinearly. Therefore, in this section we do not consider the first measurement approach.

In the noiseless case the relation between the input and output DFT spectra of the n_u experiments is

$$\mathbf{Y}_0(k) = G_0(j\omega_k)\mathbf{U}_0(k) \tag{2-75}$$

with $\mathbf{U}_0(k) \in \mathbb{C}^{n_u \times n_u}$, $\mathbf{Y}_0(k) \in \mathbb{C}^{n_y \times n_u}$, and where the entry $\mathbf{X}_{[p,q]}(k)$ corresponds to the pth input-output signal of the qth experiment. It is clear that solving (2-75) for $G_0(j\omega_k)$ puts a strong condition on the excitation design: the matrix $\mathbf{U}_0(k)$ should be regular, otherwise $G_0(j\omega_k)$ is not identifiable. In the noisy case the FRM estimate is then obtained as

$$\hat{G}(j\omega_k) = \mathbf{Y}(k)\mathbf{U}^{-1}(k) \qquad (2\text{-}76)$$

The covariance of the FRM estimate (2-76) is related to the input-output noise covariances as

$$\text{Cov}(\text{vec}(\hat{G}(j\omega_k))) \approx \overline{(\mathbf{U}_0(k)\mathbf{U}_0^H(k))^{-1}} \otimes (V(k)C_Z(k)V^H(k))$$

$$V(k) = \begin{bmatrix} I_{n_y} & -G_0(j\omega_k) \end{bmatrix} \qquad (2\text{-}77)$$

$$C_Z(k) = \begin{bmatrix} C_Y(k) & C_{YU}(k) \\ C_{YU}^H(k) & C_U(k) \end{bmatrix}$$

with $\otimes$ the Kronecker product (see Section 15.7), and $C_U(k)$, $C_Y(k)$, and $C_{YU}(k)$ the input-output noise covariance matrices of one experiment (one column of $\mathbf{Y}(k)$ and $\mathbf{U}(k)$). The proof of (2-77) is given in Appendix 2.C. From (2-77) it follows that the sensitivity of $\hat{G}(j\omega_k)$ (2-76) to input-output measurement errors depends strongly on the condition number of $\mathbf{U}_0(k)$ ($\text{cond}(\mathbf{U}_0(k)\mathbf{U}_0^H(k)) = (\text{cond}(\mathbf{U}_0(k)))^2$), and a careful design is necessary in order to avoid deterioration of the results. Sometimes the number of experiments is even higher than the number of inputs. In that case $\mathbf{U}^{-1}(k)$ in (2-76) is replaced by the Moore-Penrose pseudoinverse $\mathbf{U}^+(k)$ (see Section 15.5).

Two solutions guaranteeing a good condition number for $\mathbf{U}_0(k)$ are available. In the first solution the circular shifted versions of the entries of one $n_u \times 1$ zippered multisine signal $r(t)$ (2-73) are used as inputs for the n_u experiments (Verbeeck et al., 1999b). The drawback of this approach is the poor frequency resolution. The second solution starts from one scalar multisine $r_{\text{SISO}}(t)$ that excites all frequencies in the band of interest (see Figure 2-29, right column). Next, the n_u linearly independent reference signals are obtained by multiplying the spectrum $R_{\text{SISO}}(k)$ with an arbitrary orthogonal $n_u \times n_u$ matrix T ($T^{-1} = T^H$, with T^H the complex conjugate transpose of T)

$$\mathbf{R}(k) = R_{\text{SISO}}(k)T \qquad (2\text{-}78)$$

(the pth column of $\mathbf{R}(k)$ represents the DFT spectrum of the $n_u \times 1$ reference signal of the pth experiment). With the choice (2-78), the reference signals of the different experiments are orthogonal to each other for a finite value of the number of samples N. In addition, in case of an ideal actuator ($u(t) = r(t)$ in Figure 2-28), (2-78) minimizes the determinant of the covariance matrix of the FRM estimate (Guillaume et al., 1996b; Guillaume, 1998; Dobrowiecki et al., 2006), i.e. it is D-optimal.

If the number of inputs is a power of 2, then one can choose T in (2-78) to be equal to the Hadamard matrix of order $n_u = 2^m$

$$T = \frac{1}{\sqrt{n_u}} H_{2^m} \text{ with } H_{2^m} = H_2 \otimes H_{2^{m-1}}, \text{ and } H_2 = \begin{bmatrix} 1 & 1 \\ 1 & -1 \end{bmatrix}, \qquad (2\text{-}79)$$

leading to the so-called Hadamard multisines (Guillaume, 1998). Note that (2-79) can be generated using only one generator and a set of inverters to perform the multiplication by -1 where needed.

For any value of n_u, one can choose T in (2-78) to be equal to the $n_u \times n_u$ DFT matrix

$$T_{[p,q]} = n_u^{-1/2} e^{j2\pi(p-1)(q-1)/n_u} \text{ with } p, q = 1, 2, \ldots, n_u \qquad (2\text{-}80)$$

leading to the so-called orthogonal multisines (Dobrowiecki et al., 2006). Contrary to the Hadamard solution, n_u generators are needed here to create the signals, but n_u is not confined to a power of two.

In many applications, the power spectra of the operational perturbations differ over the inputs. To cope with this requirement, the matrix in (2-78) is multiplied by a frequency-dependent diagonal $n_u \times n_u$ matrix $D_A(k)$ that defines the shape of the input amplitude spectra

$$\mathbf{R}(k) = R_{\text{SISO}}(k) D_A(k) T \qquad (2\text{-}81)$$

where $D_{A[p,q]}(k) = A_p(k)\delta(p-q)$, with $\delta(p-q)$ the Kronecker delta and $A_p(k) \geq 0$. Independent of the choice of the orthogonal matrix T, (2-81) always requires n_u different generators to create the signals.

Note that $\mathbf{R}(k)$ in (2-78) and (2-81) always can be written as

$$\mathbf{R}(k) = D_{|R|}(k) T_{\angle R}(k) \qquad (2\text{-}82)$$

where $D_{|R|}(k)$ is a diagonal matrix with pth diagonal element equal to the 2-norm of the pth row of $\mathbf{R}(k)$, and with $T_{\angle R}(k)$ an orthogonal matrix such that $T_{\angle R[p,q]}(k) = e^{j\angle \mathbf{R}_{[p,q]}(k)}$. A numerical stable inverse of (2-82) is then obtained as $\mathbf{R}^{-1}(k) = T_{\angle R}^H(k) D_{|R|}^{-1}(k)$.

2.7.3 Discussion

If the zippered (single experiment) and the orthogonal multisines (multiple experiments) are designed with the same number of samples N per period (see Figure 2-29), then the frequency resolution and the measurement time of the multiple experiments with the orthogonal multisines (2-78) or (2-81) are n_u times larger than those of the single experiment with the zippered multisine. Conversely, for the same frequency resolution, the measurement times of both experiments are the same if we do not consider the time to wait for steady state. The zippered multisine experiment has only one transient time, while the orthogonal multisines have n_u transient times. The multiple experiment is, however, robust to (non)linear harmonic interference introduced by the actuator and, therefore, is the preferred solution for modeling from noisy input-output observations. If the spectral purity of the inputs can be guaranteed (e.g., in open loop control applications where the actuator is a part of the model), then the single experiment with zippered multisines is the prime choice (simpler experiment and calculations).

The proposed periodic excitations (zippered, orthogonal, and Hadamard multisines) minimize the correlation among the different input signals. Otherwise, the input power spectrum matrix can be almost singular, which results in unreliable FRM estimates. However, the FRM of a highly interactive process can be ill-conditioned by the nature of the system itself. Accurate identification of the low gain directions of the FRM then requires high amplitude correlated inputs, and this is in conflict with the previous requirement for uncorrelated excitations. The reader is referred to Zhu and Stec (2006) and Rivera et al. (2009) for more information about this challenging problem.

68   Chapter 2 ■ Measurement of Frequency Response Functions – Standard Solutions

2.8 GUIDELINES FOR FRF MEASUREMENTS

The aim of this section is to condense the information from the previous sections to a short list of guidelines. Following these guidelines does not always guarantee good measurements but at least ensures avoidance of a number of common mistakes.

2.8.1 Guideline 1: Use Periodic Excitations

If one can impose the excitation, then periodic signals are preferred over random signals because the former lead to consistent estimates, even in feedback (see Section 2.5.1), and allow estimation simultaneously with the (co)variances of the input-output noise. The following are recommended in order of importance: (i) measure multiple periods in one record; (ii) select a good synchronization; (iii) collect a number of single measurements. The design of periodic and random excitations is discussed in detail in Chapter 5.

2.8.2 Guideline 2: Select the Best FRF Estimator

2.8.2.1 Periodic Excitations. Use $\hat{G}_{\mathrm{ML}}(j\omega_k)$ if multiple periods are measured or if repeated measurements with good synchronization are made (Section 2.5.2); otherwise, in case of poor or no synchronization, select $\hat{G}_{H_{\log}}(j\omega_k)$ (2-45), $\hat{G}_{H_1}(j\omega_k)$ (2-51), or $\hat{G}_{H_2}(j\omega_k)$ (2-54) depending on the SNR of the measurements using Figure 2-30. Use a rectangular window in the DFT.

If it is impossible to measure an integer number of periods precisely (even after selecting a smaller number of samples), a Hanning window can be used to reduce the errors from $O(N^{-1})$ to $O(N^{-2})$ at the excited frequency lines (see Section 2.2.3) if at least four periods are captured.

2.8.2.2 Random Excitations. Select a Hanning window (2-12) in the DFT to reduce the leakage errors. Use $\hat{G}_{H_1}(j\omega_k)$ (2-51) if the input SNR is best and $\hat{G}_{H_2}(j\omega_k)$ (2-54) if the output SNR output is best to estimate the FRF. Keep in mind that the measurements are biased if both input and output are prone to noise distortions, or if the system operates in closed loop. The reader is referred to Chapter 7 for more advanced leakage suppression techniques.

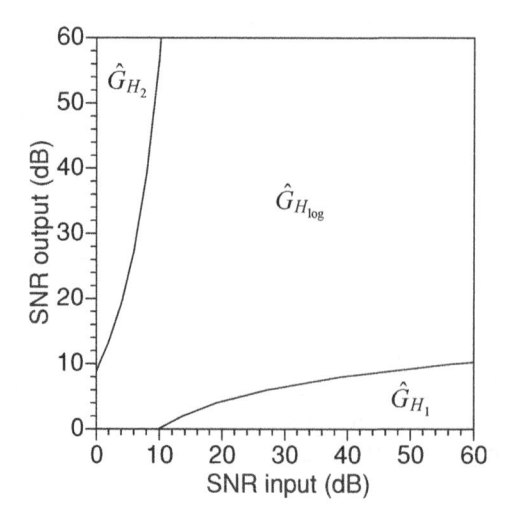

Figure 2-30. Selection between $\hat{G}_{H_{\log}}(j\omega_k)$, $\hat{G}_{H_1}(j\omega_k)$, and $\hat{G}_{H_2}(j\omega_k)$ as a function of the SNR in case the repeated measurements are not well synchronized.

Section 2.10 ■ Exercises

If computation time is not an issue, then the WOSA estimate (2-70) combined with a half-sine window and 2/3 overlap is the prime choice because it minimizes the mean square error of the leakage. If a known reference signal is available, then the bias in the direct estimates (2-51) and (2-70) due to input noise and/or a feedback loop is avoided by the indirect estimate (2-64) and its WOSA equivalent.

2.8.3 Guideline 3: Pretreatment of Data

Before processing the data, we strongly advise effecting a visual inspection for anomalies such as (periodic) spikes, outliers, overload, drift, and offset. Some of these problems can also be detected automatically. A slow drift can be removed by use of polynomial regression (McCormack et al., 1994a; Peirlinckx et al., 1996). Outliers can be detected using the periodic nature of the excitation by observing the variations from one period to the next. A sound solution is to perform a new experiment. If this is not possible, a simple alternative is to replace the erroneous data by the equivalent value of the neighboring periods.

2.9 CONCLUSION

FRF measurements give a great deal of information about the device or plant under test. Very often the FRF is easily accessible and it is strongly advised to take this intermediate step in the identification process. It provides not only much qualitative information about the complexity of the problem but also quantitative information about the plant and the measurement quality. This can be used to set up a measurement-driven weighting function for the identification step and also gives very valuable information for the model validation. The user has significant influence on the measurement quality by generating a good excitation and selecting the proper algorithms to process the raw measurement data. For these reasons, we strongly encourage the reader to take the time to understand the basic principles of FRF measurements. Good nonparametric measurements will simplify the task of building parametric models significantly.

2.10 EXERCISES

Remark. In these exercises (and also in the next chapters) we will use the MATLAB® notation. MATLAB is a high-performance language for technical computing developed by Mathworks Inc. More information can be found at http://www.mathworks.com/.

2.1. Calculate the signal $u_0(t) = \sum_{l=1}^{255} A_l \sin(2\pi f_0 l t T_s + \phi_l)$, $t = 0, ..., N-1$ with $f_0 = 1$, $T_s = 1/1024$, and ϕ_l independently and uniformly distributed in, $[0, 2\pi[$.
Calculate $U_0(k) = \text{DFT}(u_0(t))$ using the MATLAB FFT instruction for $N = 1024, 1500, 4096, 5000$ and plot the amplitude spectrum in dB $U_{dB}(k) = 20\log_{10}|U_0(k)|$. Use for the first time a rectangular window and for the second time a Hanning window, and discuss your results. (What is the impact of leakage? What happens if a Hanning window is applied on a record consisting of an integer number of periods?)
Note that this routine also works for $N \neq 2^n$ but that it becomes significant more slowly.

2.2. This exercise shows how a very fast calculation of a periodic signal is achieved, starting from its spectrum.
Define U = ZEROS(1024, 1) %; this is a 1024×1 vector with all entries zero.
Set U(2:256) = exp($2\pi j$rand(255, 1)).
u = 2*REAL(IFFT(U))
Compare the computational effort of this approach with that of Exercise 2.1 Give an explanation of the algorithm.

Remark: In MATLAB $\underset{\sim}{U}(1)$ contains the DC component and $\underset{\sim}{U}(k)$ the Fourier coefficient of the harmonic $(k-1)f_0$. The underscore $\sim$ indicates that we refer to a spectrum in the MATLAB notation: $\underset{\sim}{U}(k) = U(k+1)$.

2.3. Calculate one period of $u_0(t)$ with random phased components (with unit amplitude) at the frequencies lf_0, $l = 1, 2, ..., 255$, $f_0 = 10$ Hz putting $N_p = 512$ points in one period. Use the method of Exercise 2.2
What is the sample period T_s that is needed to generate this signal?
Set up the signals $u_{0_l} \in \mathbb{R}^{lN_p \times 1}$ containing l successive periods using the REPMAT instruction for $l = 1, 2, 4$ and study the relation between the fundamental frequency, N_p, and the line number of the spectral components of the repeated signal.

2.4. Define the discrete-time system $G_0(z^{-1})$: [b,a] = CHEBY1(2, 10, 0.5) (this is a second-order system with resonance frequency at $0.25 f_s$).
Plot the amplitude of the transfer function of this system in dB (use the function FREQZ). Consider the signals u_{0_l} of Exercise 2.3 and calculate the responses $y_{0_l}(t) = g_0(t)*u_{0_l}(t)$ using the filter operation $y_{0_l} = $ FILTER(b, a, u_{0_l}) of MATLAB.
Estimate the FRF of the system at the excited frequency lines as $\hat{G}_l(z_r^{-1}) = \underset{\sim}{Y}_{0_l}(k_r)/\underset{\sim}{U}_{0_l}(k_r)$. The indices k_r should be properly chosen to select only the lines where the system is excited. Compare the measured FRF with the exact one and discuss the result. What is the origin of the errors?

2.5. Repeat the previous exercise for $l = 2$ but eliminate the first period in u_{0_2}, y_{0_2} before calculating the DFT spectra. Explain why the errors disappeared.

2.6. Generate an iid random signal with zero mean $u_0(t)$, $t = 1, ..., 512M$. Normalize the rms value of this signal to 1. Calculate $y_0 = g_0(t)*u_0(t)$ (Exercise 2.4) and estimate $\hat{G}_{H_1}$ (2-51) for $M = 1, 4, 16, 64$. Discuss the results. Repeat the exercise but this time eliminate the transient effects using the technique of Exercise 2.5.

2.7. Generate an iid random signal with zero mean $u_0(t)$, $t = 1, ..., 512M$. Normalize the rms value of this signal to 1. Calculate $y(t) = g_0(t)*u_0(t) + n_y(t)$ (Exercise 2.4) with $n_y(t)$ iid normally distributed noise with zero mean and $\sigma_y = 0.1$. Estimate $\hat{G}_{H_1}$ (2-51) for $M = 1, 4, 16, 64$. Discuss the results.

2.8. Generate an iid random signal with zero mean $u_0(t)$, $t = 1, ..., 512M$. Normalize the rms value of this signal to 1. Calculate $y_0 = g_0(t)*u_0(t)$ (Exercise 2.4). Generate $u(t) = u_0(t) + n_u(t)$, and $y(t) = y_0(t) + n_y(t)$, with $n_u(t), n_y(t)$ iid normally distributed noise with zero mean and $\sigma_u = 0.5$, $\sigma_y = 0.1$. Estimate $\hat{G}_{H_1}(j\omega_k)$ (2-51) for $M = 1, 4, 16, 64$. Discuss the results. Can you suggest a better method?

2.9. Estimate the variance of $\hat{G}_{H_1}(j\omega_k)$ (2-51) for the setup of Exercise 2.7 using the coherence $\gamma^2(\omega_k)$. Put $M = 16$. Repeat the simulation 50 times and calculate $\sigma_G^2(k)$ from the repeated estimates. Compare both results.

2.10. Consider the signal $u_{0_{16}}$ of Exercise 2.4 and calculate the output for the input $u(t) = u_{0_{16}}(t) + n_g(t)$ where $n_g(t)$ is zero mean iid generator noise with $\sigma_{n_g} = 0.1$. Calculate the system output $y(t) = g(t)*u(t)$ and skip the first period to avoid transients. Calculate the FRF $\hat{G}_{ML}(j\omega_k)$ (2-33) and estimate the variance of the FRF for this setup using the coherence. Explain why the impact of generator noise on the variance is so small.

2.11 APPENDIXES

Appendix 2.A Radius of a Circular Confidence Region of the FRF

Consider a circular complex normally distributed random variable v with mean μ_v and variance σ_v^2. The $p \times 100\%$ confidence circle with center μ_v and radius $r\sigma_v$ is defined as $\text{Prob}(|v - \mu_v| \leq r\sigma_v) = p$. Using polar coordinates $r_1 e^{j\theta_1} = (v - \mu_v)/\sigma_v$, it can be written as

$$\int_0^{2\pi}\int_0^r f_v(\sigma_v r_1 e^{j\theta_1} + \mu_v)\sigma_v^2 r_1 dr_1 d\theta_1 = p \qquad (2\text{-}83)$$

where $f_v(v) = e^{-|v-\mu_v|^2/\sigma_v^2}/(\pi\sigma_v^2)$ is the probability density function of v (see (16-14), page 569). Elaborating (2-83) gives

$$2\int_0^r e^{-r_1^2} r_1 dr_1 = p \quad \Rightarrow \quad r = \sqrt{-\ln(1-p)} \qquad (2\text{-}84)$$

which proves (2-31). □

Appendix 2.B Asymptotic Behavior of Averaging Techniques

The proofs of the asymptotic ($M \to \infty$) properties of the averaging techniques (2-33), (2-51), and (2-54) follow the lines of Sections 16.13 (general theory) and 16.15 (application on the measurement of a resistance). To understand these proofs fully we advise reading Sections 16.13 and 16.15 first.

We will prove the results for the ML estimator (2-33); the proofs for the H_1 (2-51) and H_2 (2-54) methods follow exactly the same lines. The only difference is that the H_1 and H_2 methods require the existence of the fourth-order moments of the disturbing noise instead of the second-order moments for the ML method (2-33). This is due to the squaring operation of the noise in (2-51) and (2-54). We split the proof in two parts: (i) the data blocks (subrecords) are independent, Assumption 2.5(i), and (ii) the data blocks are correlated, Assumption 2.5(ii).

2.B.1 Independent Data Blocks.
In (2-33) sums of the form

$$S(M)/M = M^{-1}\sum_{l=1}^M N^{[l]}(k) \qquad (2\text{-}85)$$

occur with $N^{[l]}(k)$ the DFT of $n_u^{[l]}(t)$ or $n_y^{[l]}(t)$, $t = 0, 1, \ldots, N_p - 1$. Under Assumption 2.5(i, $P = 2$) the noise $N^{[l]}(k)$, $l = 1, 2, \ldots, M$, is independent over l and has finite second-order moments. Hence, $S(M)/M$ converges with probability one (w.p. 1) at the rate $O_p(M^{-1/2})$ to its expected value (see Section 16.9, version 2 of the law of large numbers). The expected value of $S(M)$ is zero because

$$\mathbb{E}\{N^{[l]}(k)\} = N_p^{-1/2}\sum_{t=0}^{N_p-1}\mu_n^{[l]} e^{-j2\pi kt/N_p} = 0 \text{ for } k \neq 0$$

where $\mu_n^{[l]} = \mathbb{E}\{n^{[l]}(t)\}$ Using the results of Sections 16.13.1 and 16.13.2 it follows directly that the estimate $\hat{G}_{\text{ML}}(j\omega_k)$ (2-33) converges w.p. 1 (almost surely) at the rate $O_p(M^{-1/2})$ to $G_0(j\omega_k)$.

Under Assumption 2.5(i, $P = 2 + \varepsilon$), the noise $N^{[l]}(k)$ is independent over l and has finite moments of order $2 + \varepsilon$. Hence, $S(M)/\sqrt{M}$ is asymptotically normally distributed (see Section 16.10, version 2 of the central limit theorem). Using the results of Section 16.13.4 it follows directly that $\hat{G}_{\text{ML}}(j\omega_k)$ is asymptotically normally distributed and that its variance is asymptotically given by

$$\sigma_{\hat{G}}^2(k) = \frac{|G_0(j\omega_k)|^2}{M}\left(\frac{\sigma_Y^2(k)}{|Y_0(k)|^2} + \frac{\sigma_U^2(k)}{|U_0(k)|^2} - 2\text{Re}\left(\frac{\sigma_{YU}^2(k)}{Y_0(k)\overline{U}_0(k)}\right)\right)$$

where $\sigma_U^2(k)$, $\sigma_Y^2(k)$, and $\sigma_{YU}^2(k)$ are the noise (co)variances of one data block (subrecord).

2.B.2 Correlated Data Blocks. The proof follows the same lines of the previous section. The only difference is that other versions of the strong law of large numbers and the central limit theorem are used. The sum (2-85) can be written as

$$S(M)/M = \mathrm{DFT}(s(M)/M) \text{ where } s(M)/M = M^{-1}\sum_{l=1}^{M} n^{[l]}(t) \tag{2-86}$$

with $n^{[l]}(t) = n_u(t)$ or $n_y(t)$. Under Assumption 2.5(ii, P) the disturbing noise $n(t)$ in (2-86) can be written as filtered white noise $e(t)$ with finite moments of order P, so that $n(t)$ is mixing over t of order P (see Example 16.6). Hence, the subrecord $n^{[l]}(t) = n(t + lN_p)$ is mixing over l of order P. We conclude that under Assumption 2.5(ii, $P = 2$), the sum $s(M)/M$ converges w.p. 1 at the rate $O_p(M^{-1/2})$ to its expected value (see Section 16.9, version 3 of the law of large numbers), and that under Assumption 2.5(ii, $P = \infty$), $s(M)/\sqrt{M}$ is asymptotically normally distributed (see Section 16.10, version 3 of the central limit theorem). This is also valid for $S(M)/M$, with $\mathbb{E}\{S(M)\} = 0$, because the number of elements N_p in the DFT sum does not increase with M. Using the results of Sections 16.13.1, 16.13.2, and 16.13.4, it follows directly that the estimate $\hat{G}_{\mathrm{ML}}(j\omega_k)$ (2-33) converges w.p. 1 (almost surely) at the rate $O_p(M^{-1/2})$ to $G_0(j\omega_k)$ and that $\hat{G}_{\mathrm{ML}}(j\omega_k)$ is asymptotically normally distributed.

Appendix 2.C Covariance of the FRM Measurement

For ease of notation we drop the frequency arguments in this appendix. Using $\mathbf{X} = \mathbf{X}_0 + \mathbf{N_X}$, with $\mathbf{X}_0$ the true value, $\mathbf{N_X}$ the noise contribution, and $X = Y$ or U, the FRM estimate (2-76) can be approximated via a first order Taylor series as

$$\hat{G} \approx G_0 + N_G \text{ with } N_G = \mathbf{N_Y}\mathbf{U}_0^{-1} - G_0\mathbf{N_U}\mathbf{U}_0^{-1} \tag{2-87}$$

Rewriting N_G as

$$N_G = V\mathbf{N_Z}\mathbf{U}_0^{-1} \text{ with } V = \begin{bmatrix} I_{n_y} & -G_0 \end{bmatrix} \text{ and } \mathbf{N_Z} = \begin{bmatrix} \mathbf{N}_Y^T & \mathbf{N}_U^T \end{bmatrix}^T \tag{2-88}$$

and applying the vec( ) operator to (2-87) we find, using $\mathrm{vec}(ABC) = (C^T \otimes A)\mathrm{vec}(B)$,

$$\begin{aligned}\mathrm{Cov}(\mathrm{vec}(\hat{G})) &\approx (\mathbf{U}_0^{-T} \otimes V)\mathrm{Cov}(\mathrm{vec}(\mathbf{N_Z}))(\overline{\mathbf{U}_0^{-1}} \otimes V^H) \\ &\approx (\mathbf{U}_0^{-T} \otimes V)(I_{n_u} \otimes C_Z)(\overline{\mathbf{U}_0^{-1}} \otimes V^H) \\ &\approx (\overline{\mathbf{U}_0\mathbf{U}_0^H})^{-1} \otimes (VC_Z V^H)\end{aligned} \tag{2-89}$$

The second equality uses the fact that the columns of $\mathbf{N_Z}$ are independently distributed (each column represents an independent experiment) and have the same covariance matrix $C_Z = \mathrm{Cov}(\mathbf{N}_{Z[:,p]})$, $p = 1, 2, \ldots, n_u$. The last equality applies twice $(A \otimes B)(C \otimes D) = (AC) \otimes (BD)$.

3

Frequency Response Function Measurements in the Presence of Nonlinear Distortions

Abstract: In this book we deal with the measurement and identification of linear dynamic systems. However, in reality the linearity assumption is only approximately valid. Many systems that are assumed to be linear are disturbed by nonlinear distortions. The aim of this chapter is not to show how nonlinear systems should be modeled, because this problem is beyond the scope of this book. The goal is to provide the reader with an insight into the impact of nonlinear distortions on FRF measurements. Finally, it will be shown how we can still use the linear framework under these conditions.

3.1 INTRODUCTION

The aim of this chapter is not to model nonlinear systems, because this problem is beyond the scope of this book. The goal is to provide the reader with insight into the behavior of nonlinear distortions and their impact on frequency response function (FRF) measurements. This allows not only a better understanding of the error mechanism but also knowledge that can be used during the design of the experiment in order to get the best results under the imposed operational conditions. To do so, the user should clearly specify the goal of his/her measurements. In order to formalize this discussion, we use the general structure given in Figure 3-1. The measured output $y(t)$ consists of a linear $y_L(t)$ and a nonlinear $y_{NL}(t)$ contribution. For simplicity we assume that the linear contribution dominates the nonlinear one for sufficiently small inputs:

$$\lim_{u_{rms} \to 0} \frac{(y_{NL})_{rms}}{(y_L)_{rms}} = 0 \qquad (3\text{-}1)$$

Under this assumption we have two basic options: (i) The goal of the measurement is to get the FRF of the underlying linear system, minimizing the impact of the NLS on the measurements. If (3-1) is not valid, the theory that is developed in this chapter is still applicable, but it is no longer possible to define an underlying linear system. (ii) Trying to find the best linear approximation to the global system, including the NLS. The first option is the best choice if

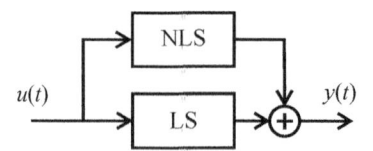

Figure 3-1. General setup of the nonlinear distortion.

some underlying linear physical model exists and the user wants to identify it as well as possible. In that case, the rms value of the excitation should be chosen as small as possible. The second choice is preferred if the model will be used to describe the relation between input and output using a linear model. Then, the nonlinearity will be linearized around the operation point of the test. This is the topic of the present chapter and of Chapter 4.

The chapter is structured along the following lines: first we give a simple introduction to the behavior of nonlinear systems; next we develop the theory for continuous-time systems operating in open loop and excited by Gaussian-like signals (random phase multisines, periodic noise, and Gaussian noise); and finally the results are extended to discrete-time systems, non-Gaussian excitations, multivariable systems, and systems operating in feedback. Detection techniques for nonlinear distortions, and optimal measurement of the best linear approximation are handled in Chapter 4.

3.2 INTUITIVE UNDERSTANDING OF THE BEHAVIOR OF NONLINEAR SYSTEMS

Consider the static nonlinear system $y = u + u^2 + u^3$ excited with a sine wave $u(t) = A\sin(2\pi f_0 t)$. The response of this system is split into its linear, quadratic, and cubic contributions. The corresponding amplitude spectra are given in Figure 3-2. It shows that nonlinear systems create additional harmonics. On the one hand, this allows the detection of nonlinear contributions, but it also shows that the FRF measurements are disturbed. The cubic subsystem also puts power at the original frequency f_0 that cannot be separated from the linear contributions using only a single sine measurement. More advanced methods that are beyond the scope of this book are needed to solve this problem (e.g., Bendat, 1998). In general, for a multiharmonic periodic signal, the frequencies of quadratic terms are found by looking for all combinations $f_i + f_j$ over the positive and negative frequencies of the signal. For the cubic terms triple sums $f_i + f_j + f_k$ should be considered, and in general n frequencies should be combined for a nonlinearity of degree n. This shows that for periodic signals having only odd frequency components (at $f_0, 3f_0, 5f_0, \ldots$), the even nonlinearities do not disturb the FRF measurements (the sum of two odd frequencies is always even), but it is impossible to avoid disturbances from the odd nonlinearities (e.g., $f_0 + f_0 - f_0 = f_0$).

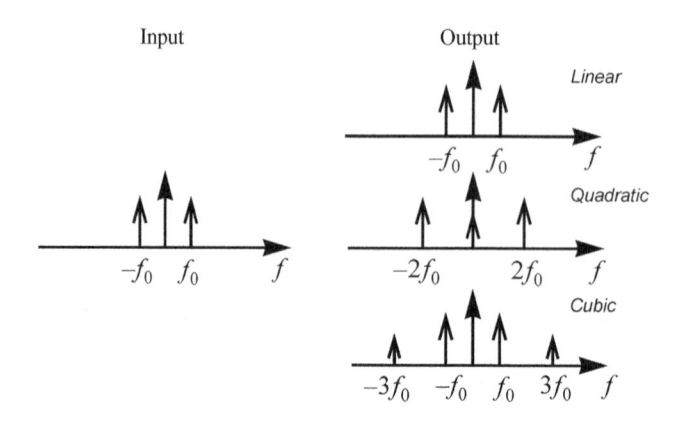

Figure 3-2. Impact of linear, quadratic, and cubic systems on the spectrum of a sine.

These results can be generalized using Volterra systems. A concise introduction to this technique is given in the book of Schetzen (1980). The basic idea is to extend the linear model to a nonlinear one using multidimensional convolutions, for example,

$$y(t) = \int_{-\infty}^{+\infty} g_1(\tau)u(t-\tau)d\tau + \int_{-\infty}^{+\infty}\int_{-\infty}^{+\infty} g_2(\tau_1,\tau_2)u(t-\tau_1)u(t-\tau_2)d\tau_1 d\tau_2 + \cdots \quad (3\text{-}2)$$

For static nonlinear systems this relation simplifies to a Taylor expansion:

$$y(t) = g_1 u(t) + g_2 u^2(t) + \cdots \quad (3\text{-}3)$$

The autocorrelation $R_{yu}(\tau)$ no longer depends on the second-order moments of u only but also on the higher order ones. Consequently, the nonlinear distortions of the FRF measurement also depend on the amplitude distribution of the excitation, for example, normally, uniformly, or binary distributed excitations. If the aim is to get the best linear approximation, it is important to use the same kind of excitations (power spectrum and amplitude distribution) as will be applied later on to the system, otherwise the linear approximation can become invalid.

For periodic excitations with $F = N/2 - 1$ harmonics at frequencies kf_s/N, $k = 1, \ldots, F$, relation (3-2) simplifies to a sum over all possible frequency combinations adding to the output Fourier coefficient Y_k at frequency kf_s/N (Chua and Ng, 1979):

$$Y_k = \sum_{\alpha=1}^{\infty} Y_k^\alpha \quad (3\text{-}4)$$

with Y_k^α the contribution of degree α

$$Y_k^\alpha = \sum_{k_1, k_2, \ldots k_{\alpha-1} = -N/2+1}^{N/2-1} G_{L_k, k_1, k_2, \ldots, k_{\alpha-1}}^\alpha U_{k_1} U_{k_2} \cdots U_{k_{\alpha-1}} U_{L_k} \quad (3\text{-}5)$$

$$L_k = k - \sum_{i=1}^{\alpha-1} k_i$$

and U_r the input Fourier coefficient at frequency rf_s/N (see Section 2.3 for the relationship between the Fourier coefficient and the DFT spectrum of a periodic signal). $G_{L_k, k_1, k_2, \ldots, k_{\alpha-1}}^\alpha$ is the symmetrized frequency domain representation of the Volterra kernel of degree α (Schetzen, 1980) so that the order of the frequencies $L_k, k_1, k_2, \ldots, k_{\alpha-1}$ has no importance

$$G_{k_1, k_2, \ldots, k_\alpha}^\alpha = \int_{-\infty}^{+\infty} \cdots \int_{-\infty}^{+\infty} g_\alpha(\tau_1, \ldots, \tau_\alpha) e^{-j2\pi f_0(k_1\tau_1 + \ldots k_\alpha\tau_\alpha)} d\tau_1 \ldots d\tau_\alpha \quad (3\text{-}6)$$

The convergence of this sum is later guaranteed in Definition 3.5.

3.3 A FORMAL FRAMEWORK TO DESCRIBE NONLINEAR DISTORTIONS

Describing nonlinear systems is a tedious job because it is necessary to guarantee convergence of the Volterra series (3-4). Moreover, the limiting value also depends on the amplitude distribution of the excitation. A normally distributed excitation can result in a different limiting value than a uniform distribution, even if the power spectra of both excitation signals are the same. For these reasons it is necessary to state, precisely, the validity of these theories. This depends on the class of excitation signals and the class of nonlinear distortions that will be considered.

3.3.1 Class of Excitation Signals

As mentioned before, FRF measurements in the presence of nonlinear distortions depend on the class of excitation signals. We focus on random multisines. These are periodic random excitations with a user-defined amplitude spectrum. When an integer number of periods is measured, the amplitude spectrum is perfectly realized, which is not the case for a random excitation (see also Chapter 5 on excitation signals). All the results can be generalized easily to (periodic) random signals (random amplitude and random phase), at a price of taking an additional expectation with respect to the amplitudes in the expressions, as is commented on after Theorem 3.7. This generalizes the results to the wider class of normally distributed random excitations. However, from the experimental point of view, we have a strong preference to use periodic excitations with well-controlled amplitude spectra, as explained in the previous chapter.

Definition 3.1 (Random Phase Multisine): $u(t)$ is a random phase multisine if

$$u(t) = \sum_{k=-N/2+1}^{N/2-1} U_k e^{j2\pi f_s kt/N} \tag{3-7}$$

with $U_k = \bar{U}_{-k} = |U_k|e^{j\varphi_k}$, f_s the clock frequency of the arbitrary waveform generator, $F = N/2 - 1$ the number of frequency components, $N \in \mathbb{N}$ the number of samples in one signal period, and the phases φ_k a realization of an independent distributed random process on $[0, 2\pi)$ such that $\mathbb{E}\{e^{j\varphi_k}\} = 0$.

Remarks

(i) A possible choice for φ_k could be to select it as a uniformly distributed noise sequence, but other choices will also do. For example, φ_k can also be chosen to have a discrete distribution.

(ii) If the amplitude spectrum $|U_k|$ is random, then (3-7) equals periodic noise.

(iii) For simplicity, U_0 is set to zero, considering the DC component as the operating point of the system. Also, the output bias of the nonlinear system depends nonlinearly on the input. Consequently, linear models cannot describe the variations of the output bias as a function of the input. The DC information of the input and the output will not be used during the linear identification process.

(iv) It is strongly advised to use FFT techniques to calculate multisine signals, otherwise the computation time becomes very long (see Exercises 2.1 and 2.2).

We will study the asymptotic behavior of the nonlinear distortions for multisines with a growing number of harmonics. In order to keep excitations with a finite power for $N \to \infty$, the signals are scaled with $1/\sqrt{N}$. This leads finally to the class of normalized random multisines $\mathbb{E}_N$ and the class of periodic noise excitations $\mathbb{P}_N$ that we will use in this study.

Definition 3.2 (Normalized Random Phase Multisine): The class of normalized random multisines $\mathbb{E}_N$ is given by the set of random multisines $u_N(t)$ (3-7) having a normalized amplitude spectrum: $|U_k| = \hat{U}(kf_s/N)/\sqrt{N}$. The deterministic amplitudes $\hat{U}(kf_s/N) \in \mathbb{R}^+$ are uniformly bounded, $\hat{U}(f) \leq M_U$, where the function $\hat{U}(f)$ has a finite number of discontinuities on the interval $[0, f_s/2]$. The phases φ_k are the realization of an independent (over k) random process satisfying $\mathbb{E}\{e^{j\varphi_k}\} = 0$. The DC component of the $u_N(t)$ is set to zero, $U_0 = 0$, and the clock frequency f_s is independent of N.

Definition 3.3 (Normalized Periodic Noise): The class of normalized periodic noise excitations $\mathbb{P}_N$ is given by the set of random multisines $u_N(t)$ (3-7) having a normalized random amplitude spectrum: $|U_k| = \hat{U}(kf_s/N)/\sqrt{N}$. The amplitudes $\hat{U}(kf_s/N) \in \mathbb{R}^+$ and the phases φ_k are the realization of independent (jointly, and over k) random processes satisfying the following conditions: $\hat{U}(kf_s/N)$ has uniformly bounded moments of any order $\mathbb{E}\{\hat{U}^\alpha(f)\} \leq M_U^\alpha$, the function $\mathbb{E}\{\hat{U}^2(f)\}$ has a finite number of discontinuities on the interval $[0, f_s/2]$, and $\mathbb{E}\{e^{j\varphi_k}\} = 0$. The DC component of the $u_N(t)$ is set to zero, $U_0 = 0$, and the frequency f_s is independent of N.

In the sequel of the book, a more general signal will be used. Because it is closely related to the concept of normalized multisines, we prefer to define it here. The ideas developed in this chapter can even be applied to this class of excitation signals, if some of the assumptions are modified (e.g., the convergence assumption in Definition 3.5). However, the reader should be aware that the limiting value of the measured FRF can depend on the specific signal in this generalized case.

Definition 3.4 (Normalized Periodic Signals): The class of normalized periodic signals is given by the set of periodic signals $u_N(t)$ (3-7) that have a normalized amplitude or power spectrum. For signals with a deterministic amplitude spectrum, we have $|U_k| = O(N^{-1/2})$. For signals with a random amplitude spectrum, the expected value $\mathbb{E}\{|U_k|^2\}$ is normalized: $\mathbb{E}\{|U_k|^2\} = O(1/N)$. For deterministic signals the peak value $(\max_t |u(t)| \leq C < \infty$ for any t, including $t = \infty$) should be bounded.

3.3.2 Selection of a Model Structure for the Nonlinear System

In this section we set up a mathematical description for the nonlinear distortions. Although we are not interested, at all, in extracting these models from the measurement, a formal description is needed in order to characterize and quantify the impact of the nonlinear distortions. One of the most general descriptions for nonlinear systems is the Volterra models (3-2) splitting the relation between input and output in different contributions of increasing degree of nonlinearity (Schetzen, 1980).

Convergence aspects are a central issue when dealing with these models. Uniform convergence requires that there exists an upper bound on the output error (= system output − model output) amplitude that is independent of the input and decreases to zero if the number of terms n_α in $\sum_{\alpha=1}^{n_\alpha} Y_k^\alpha$ goes to infinity. It can be shown only for a very restricted set of systems, e.g., the underlying nonlinear function is analytic for all considered inputs. The class of allowable systems is considerably extended if the uniform convergence is replaced by mean square convergence. In that case it is no longer necessary that the output converges everywhere in the domain of interest. Only the power (or root mean square value) of the error signal should converge to zero for a specified class of excitations. Thus, at a discrete set of isolated points the model does not necessarily converge (similar to the convergence of a Fourier series to a discontinuous function). Under mean square convergence relays, quantizers and other discontinuous nonlinear systems can be included in the model set. The reader should be aware that this set of systems is not complete; for example, bifurcations can still not be modeled within this concept. These ideas are very similar to the idea of Wiener series as explained by Schetzen (1980). Because the FRF measurements can be considered as the minimizers of a weighted least squares cost function, it is clear that the input-output relationship of the nonlinear distortions is approximated in least square sense. This motivates the following assumption:

Definition 3.5 (Class of Nonlinear Systems): $\mathbb{S}$ is the set of nonlinear systems such that for random multisines $u_N \in \mathbb{E}_N$ (see Definition 3.2) or periodic noise $u_N \in \mathbb{P}_N$ (see Definition 3.3)

$$\sum_{\alpha=1}^{\infty} M_{G^\alpha} M_U^\alpha \leq C_1 < \infty \tag{3-8}$$

with $M_{G^\alpha} = \max \left| G^\alpha_{L_k, k_1, k_2, \ldots, k_{\alpha-1}} \right|$ and where M_U^α is defined in Definition 3.2 or Definition 3.3.

Under condition (3-8) there exists a uniformly bounded Volterra series whose output converges in mean square sense to the output of the nonlinear distortion for $u_N \in \mathbb{E}_N$. The FRF measurement $G(j\omega_k)$ at frequency f_k for nonlinear systems belonging to the set $\mathbb{S}$ excited with $u_N \in \mathbb{E}_N$ or $u_N \in \mathbb{P}_N$ is the sum of the nonlinear contributions of degree α, $G^\alpha(j\omega_k)$ (see (3-5)):

$$\begin{aligned} G(j\omega_k) &= \frac{Y_k}{U_k} = \sum_{\alpha=1}^{\infty} G^\alpha(j\omega_k) \\ G^\alpha(j\omega_k) &= \frac{Y_k^\alpha}{U_k} = \sum_{k_1, \ldots, k_{\alpha-1} = -N/2+1}^{N/2-1} G^\alpha_{L_k, k_1, k_2, \ldots, k_{\alpha-1}} \frac{U_{k_1} U_{k_2} \ldots U_{k_{\alpha-1}} U_{L_k}}{U_k} \end{aligned} \tag{3-9}$$

3.4 STUDY OF THE PROPERTIES OF FRF MEASUREMENTS IN THE PRESENCE OF NONLINEAR DISTORTIONS

In this section, profound insight is given into the impact of the nonlinear distortions on the FRF measurements for normalized random multisine excitations. It is shown in Appendix 3.A that the contributions to the FRF can be partitioned into two sets, the first consisting of contributions that do not depend on the random phases of the excitation and the second containing the contributions that depend on the random phases:

(i) Systematic contributions $G_B(j\omega_k)$: There exists a linear dynamic system $G_{\text{BLA}}(j\omega_k)$ to which the expected value of the FRF estimate converges under weak conditions. It differs from the underlying linear system $G_0(j\omega_k)$ by the systematic contributions $G_B(j\omega_k)$ of the nonlinear distortions. We will show that for the class of normally distributed signals (including random multisines and noise excitations) this linear system is the best linear approximation (BLA) to the nonlinear system. The contributions of $G_{\text{BLA}}(j\omega_k)$ do not depend upon the random phases of the input.

(ii) Stochastic contributions $G_S(j\omega_k)$: Even for a very large number of frequencies and in the absence of disturbing noise, the FRF measurement is not smooth as a function of the frequency. It is scattered around its expected value, and these deviations do not converge to zero. They are called the stochastic nonlinear distortions. The contributions to $G_S(j\omega_k)$ depend on the random phases of the input.

These concepts are formalized below. For a system belonging to the set $\mathbb{S}$ and a normalized random multisine excitation $u_N \in \mathbb{E}_N$ (or normalized periodic noise $u_N \in \mathbb{P}_N$), the measured FRF consists of three parts:

$$G(j\omega_k) = G_{\text{BLA}}(j\omega_k) + G_S(j\omega_k) + N_G(k) \tag{3-10}$$

with $G_{\text{BLA}}(j\omega_k)$ the best linear approximation (BLA), $G_S(j\omega_k)$ the stochastic nonlinear contributions, and $N_G(k)$ the errors due to the output noise.

The best linear approximation $G_{\text{BLA}}(j\omega_k)$ consists of two parts:

$$G_{\text{BLA}}(j\omega_k) = G_0(j\omega_k) + G_B(j\omega_k) \tag{3-11}$$

with $G_0(j\omega_k)$ the underlying linear system and $G_B(j\omega_k)$ the bias or systematic errors due to the nonlinear distortions.

$G_S(j\omega_k)$ is called a stochastic contribution because it behaves as uncorrelated (over the frequencies) noise, although the reader should be aware that it is not a random signal once the excitation signal is fixed. Because of this noisy behavior, the presence of nonlinear distortions is often not recognized.

$N_G(k)$ describes the impact of the disturbing noise on the FRF measurement. For simplicity, we assume that the input measurements are noise free (dominating output noise), resulting in a noise distortion $N_G(k)$ having the following properties:

Assumption 3.6 (Measurement Noise): The noise $N_G(k)$ on the FRF measurement has the following properties.

(i) $\mathbb{E}\{N_G(k)\} = 0$

(ii) $\mathbb{E}\{N_G(k)\bar{N}_G(l)\} = \sigma_G^2(k)\delta_{kl}$ and $\mathbb{E}\{|N_G(j\omega_k)|^2\} = \sigma_G^2(k)$

(iii) $\mathbb{E}\{N_G(l)|N_G(k)|^2\} = 0$ for $k, l \neq 0$

(iv) $\mathbb{E}\{(|N_G(k)|^2 - \sigma_G^2(k))(|N_G(l)|^2 - \sigma_G^2(l))\} = \begin{cases} 0 & k \neq l \\ O(N^0) & k = l \end{cases}$

The different contributions to the FRF are studied in more detail in the following for two situations. In the first case we look for the average value if the experiment is repeated for a constant number of harmonics in the excitation. The second case deals with the asymptotic behavior if the number of harmonics $N \to \infty$.

3.4.1 Study of the Expected Value of the FRF for a Constant Number of Harmonics

What happens if the FRF measurement is averaged for different realizations of a normalized random multisine excitation, keeping its amplitude spectrum constant? Or more formally: what is the expected value $\mathbb{E}\{G(j\omega_k)\}$ for N fixed? Thereto the mathematical expectation $\mathbb{E}\{G^\alpha(j\omega_k)\}$ is calculated with respect to the phases. This means that the measured frequency response function of the system is averaged over different realizations of the random multisine excitation, keeping the frequency grid and the amplitude of the Fourier coefficients U_k of the excitation signal $u(t)$ constant.

Theorem 3.7 (Response Nonlinear System): For a system belonging to the set $\mathbb{S}$ (see Definition 3.5), excited with independent realizations of a normalized random multisine $u_N \in \mathbb{E}_N$ (see Definition 3.2) or normalized periodic noise $u_N \in \mathbb{P}_N$ (see Definition 3.3), we have:

1. The expected value of $G(j\omega_k)$ is given by

$$\mathbb{E}\{G(j\omega_k)\} = G_{\text{BLA}}(j\omega_k) = G_0(j\omega_k) + G_B(j\omega_k) \tag{3-12}$$

with

$$G_B(j\omega_k) = \begin{cases} \sum_{\alpha=2}^{\infty} G_B^{2\alpha-1}(j\omega_k) & \text{uniform continuous phase distribution} \\ \sum_{\alpha=2}^{\infty} G_B^{2\alpha-1}(j\omega_k) + O(N^{-1}) & \text{otherwise} \end{cases}$$

$$G_B^{2\alpha-1}(j\omega_k) = \mathbb{E}\{G^{2\alpha-1}(j\omega_k)\}.$$

2. The expected value of $G^\alpha(j\omega_k)$ is given by

$$\mathbb{E}\{G^{2\alpha-1}(j\omega_k)\} = c_\alpha \sum_{\substack{s_1,\ldots,s_{\alpha-1}=1}}^{N/2-1} G^{2\alpha-1}_{k,-s_1,s_1,\ldots,-s_{\alpha-1},s_{\alpha-1}} \mathbb{E}_{\text{amp}}\{|U_{s_1}|^2\cdots|U_{s_{\alpha-1}}|^2\}$$

$$+ O_\alpha(N^{-1}) \tag{3-13}$$

$$\mathbb{E}\{G^{2\alpha}(j\omega_k)\} = \begin{cases} 0 & \text{uniform continuous phase distribution} \\ O_\alpha(N^{-3/2}) & \text{otherwise} \end{cases}$$

with $c_\alpha = 2^{\alpha-1}(2\alpha-1)!!$, $\sum_{\alpha=2}^{\infty} O_\alpha(N^{-\beta}) = O(N^{-\beta})$, and $\mathbb{E}_{\text{amp}}\{.\}$ the expected value with respect to the random amplitudes of the periodic noise.

Proof. See Appendix 3.A. □

Remarks

(i) Note that from (3-13) it follows that $\mathbb{E}\{G^{2\alpha-1}(j\omega_k)\} = O(N^0)$ because in the sum $N^{\alpha-1}$ terms of $O(1/N^{\alpha-1})$ are added together ($|U_{s_1}| = O(N^{-1/2})$; see Definitions 3.2 and 3.3).

(ii) The best linear approximation depends on the number of frequencies N that are used in the random phase multisine and the periodic noise (see (3-12) and (3-13), and Evans and Rees, 2000). Therefore, it would be better to denote it as $G_{\text{BLA},N}$. However, later on it will be shown that the limit for $N \to \infty$ exists: $\lim_{N \to \infty} G_{\text{BLA},N}(j\omega) = G_{\text{BLA}}(j\omega)$. For that reason we preferred not to overload the notation, leaving out the dependence on N.

(iii) Instead of $G_B^\alpha(j\omega_k)$ being considered as the expected value (see (3-12)), it can be interpreted as that part of the transfer function contribution of degree α that is independent of the random phase of the random multisine excitation. All the components that still depend on the random phase have a zero mean value because $\mathbb{E}\{e^{j\varphi_k}\} = 0$ and as such do not contribute to the bias term. Consequently, $G_B^\alpha(j\omega_k)$ is independent of the random phases of the excitation; in the contributing terms the random phases of the excitation cancel each other, resulting in a systematic contribution of the nonlinear distortion to the FRF. $G_S^\alpha(j\omega_k)$ depends on the random phases of the excitation so that it is a random component, modeling the stochastic contribution of the nonlinear distortion of degree α to the FRF.

(iv) A typical example of a discrete phase distribution is $\varphi_k \in \{0, \pi\}$. For discrete phase distributions, the even degree terms also have a bias contribution that disappears as an $O(N^{-1})$.

An important conclusion of this section is that only the odd terms $G^{2\alpha-1}(j\omega_k)$ contribute to the best linear approximation; it does (asymptotically) not depend on the even nonlinear distortions. This result will be used later on to formulate optimized measurement strategies. The theorem also gives a possibility to measure $G_{\text{BLA}}(j\omega_k)$. It can be obtained by averaging over a sufficient number of experiments with different realizations of the random multisine so that the stochastic nonlinear contributions are averaged to zero.

3.4.2 Asymptotic Behavior of the FRF if the Number of Harmonics Tends to Infinity

From the previous section we know that, besides the disturbing noise $N_G(k)$, the measured FRF consists of two remaining components: a deterministic one $G_{\text{BLA}}(j\omega_k)$ and a stochastic one $G_S(j\omega_k)$. A first possibility to measure $G_{\text{BLA}}(j\omega_k)$ is to average over a large number of experiments so that the contribution $G_S(j\omega_k)$ is averaged to zero for a fixed number of frequency components N in the random multisine (periodic noise). Because in each realization we should calculate and load each time a new random multisine (periodic noise sequence) in the generator memory and wait until the transients in the measured signals disappear, it is tempting to stick to one experiment, but using a very dense ($N \to \infty$) multisine (periodic noise). One might hope that the resulting measurement of the FRF would become smooth because the stochastic nonlinear contributions would average to zero. It turns out that this is not the case. Neither of the contributions ($G_{\text{BLA}}(j\omega_k)$ and $G_S(j\omega_k)$) decreases if the number of frequencies N of the excitation increases; the FRF does not become smooth for $N \to \infty$. Also the bias contribution $G_B(j\omega_k)$ does not decrease when N increases because it is an $O(N^0)$. This is formalized in the next theorem.

Theorem 3.8 (Asymptotic Behavior of the Systematic and Stochastic Nonlinearities): Consider a system belonging to the system set $\mathbb{S}$ (see Definition 3.5), excited with a random multisine $u_N \in \mathbb{E}_N$ (see Definition 3.2) or periodic noise $u_N \in \mathbb{P}_N$ (see Definition 3.3). The systematic $G_B(j\omega_k)$ and stochastic $G_S(j\omega_k)$ contributions to the transfer function

$G(j\omega_k) = G_{BLA}(j\omega_k) + G_S(j\omega_k)$, with $G_{BLA}(j\omega_k) = G_0(j\omega_k) + G_B(j\omega_k)$, do not decrease to zero as $N \to \infty$: $G_B(j\omega_k)$ is an $O(N^0)$ and $G_S(j\omega_k)$ is an $O_{m.s.}(N^0)$.

Proof. See remarks in Section 3.4.1 on $G_B(j\omega_k)$ and Appendix 3.B on $G_S(j\omega_k)$. □

The stochastic behavior of $G_S(j\omega_k)$ can be further characterized, showing that its second-order properties are completely similar to those of the noise $N_G(k)$. This explains why it is difficult to distinguish between noise and nonlinear distortions. It is also the reason why nonlinear distortions are often not recognized.

Theorem 3.9 (Properties of the Stochastic Nonlinearities $G_S(j\omega_k)$): For a system belonging to the system set $\mathbb{S}$ (see Definition 3.5), excited with a random multisine $u_N \in \mathbb{E}_N$ (see Definition 3.2) or periodic noise $u_N \in \mathbb{P}_N$ (see Definition 3.3), the following properties are valid:

(i) $\mathbb{E}\{G_S(j\omega_k)\} = 0$

(ii) $\mathbb{E}\{G_S(j\omega_l)\overline{G}_S(j\omega_k)\} = O(N^{-1})$ if $k \neq l$ and $\mathbb{E}\{|G_S(j\omega_k)|^2\} \equiv \sigma^2_{G_S}(k) = O(N^0)$

(iii) $\mathbb{E}\{G_S(j\omega_l)|G_S(j\omega_k)|^2\} = O(N^{-1})$ for $k \neq l$

(iv) $\mathbb{E}\{(|G_S(j\omega_k)|^2 - \sigma^2_{G_S}(k))(|G_S(j\omega_l)|^2 - \sigma^2_{G_S}(l))\} = \begin{cases} O(N^{-1}) & k \neq l \\ O(N^0) & k = l \end{cases}$

Proof. See Appendix 3.B. □

Remark. These observations are in agreement with the classical result showing that the output of a nonlinear system can be split into two parts (Bendat, 1998; Forssell and Ljung, 2000c): a first part that is linearly related to the input (in our case leading to $G_{BLA}(j\omega_k)$) and a second part that is uncorrelated with the input (leading to $G_S(j\omega_k)$). Theorem 3.9 tells more about the second and higher order properties of the uncorrelated part.

In the previous theorem, the moments of the nonlinear contributions up to the fourth order were studied. In general, it is even possible to tell more about these nonlinear distortions. In the next theorem, it is shown that they are mixing (see also Section 16.4). Loosely speaking, this means that the dependence of the nonlinear contributions decreases fast enough to zero if the frequency distance between the contributions increases.

Theorem 3.10 (Mixing Property of the Stochastic Nonlinearities $G_S(j\omega_k)$): The nonlinear contributions for a system belonging to the system set $\mathbb{S}$ (see Definition 3.5), excited with a random multisine $u_N \in \mathbb{E}_N$ (see Definition 3.2) or periodic noise $u_N \in \mathbb{P}_N$ (see Definition 3.3), are mixing of order infinity.

Proof. See Appendix 3.C. □

Theorem 3.11 (Asymptotic Distribution of the Stochastic Nonlinearities $G_S(j\omega_k)$): For a system belonging to the system set $\mathbb{S}$ (see Definition 3.5), excited with a random multisine $u_N \in \mathbb{E}_N$ (see Definition 3.2) or periodic noise $u_N \in \mathbb{P}_N$ (see Definition 3.3), the stochastic nonlinearities are asymptotically ($N \to \infty$) circular complex normally distributed (convergence in law at the rate $O(N^{-1})$).

Proof. See Appendix 3.E. □

3.4.3 Further Comments on the Best Linear Approximation

In this section a physical interpretation is given for the best linear approximation (BLA) $G_{\text{BLA}}(j\omega)$. First, it will be shown that normally distributed random excitations and random multisines (see Definition 3.2) result in the same BLA if both excitations are generated from the same power spectral density ($S_{uu}(j\omega) = \mathbb{E}\{\hat{U}^2(f)\}/f_s$). Next, it will be shown that $G_{\text{BLA}}(j\omega)$ corresponds to the best linear approximation, in least squares sense, of the nonlinear system; finally it will be shown that asymptotically $G_{\text{BLA}}(j\omega)$ is smooth. In Section 4.2 the asymptotic ($N \to \infty$) equivalence is extended to a broader class of signals.

3.4.3.1 Connecting the Random Multisine to Normally Distributed Noise. If the system is excited with Gaussian noise, the limit value of the estimated FRF (after averaging over an infinite number of blocks and neglecting leakage effects, see Chapter 2) is given by

$$G_{\text{BLA}}(j\omega) = S_{yu}(j\omega)/S_{uu}(j\omega) \qquad (3\text{-}14)$$

which is the direct method for measuring the BLA. Splitting $Y(j\omega)$ into its contribution of degree α results in $Y(j\omega) = \sum_{\alpha=1}^{\infty} Y^\alpha(j\omega)$ and shows that the nonlinear contribution of degree α to $G_{\text{BLA}}(j\omega)$ should be calculated as: $G^\alpha(j\omega) = S_{yu}^\alpha(j\omega)/S_{uu}(j\omega)$. To interpret $S_{yu}^\alpha(j\omega)$, higher order spectra can be used (Bendat and Piersol, 1980; Bendat, 1998; Billings, 1980; Brillinger, 1981; Mendel, 1991; Nikias and Mendel, 1993; Nikias and Petropulu, 1993). Because these higher order spectra depend not only on the power spectrum of the excitation noise but also on their higher order moments, it is clear that the value of $G^\alpha(j\omega)$ also depends on its pdf. In the case of zero mean normally distributed noise, the higher order spectra can be calculated easily and the contribution of degree α is given by

$$G^{2\alpha-1}(j\omega) = c_\alpha \int_0^\infty \cdots \int_0^\infty G^{2\alpha-1}_{f,f_1,-f_1,\ldots,-f_{\alpha-1}} S_{uu}(j\omega_1)\ldots S_{uu}(j\omega_{\alpha-1}) df_1 \ldots df_{\alpha-1} \qquad (3\text{-}15)$$

$$G^{2\alpha}(j\omega) = 0$$

with $c_\alpha = 2^{\alpha-1}(2\alpha-1)!!$ (see Appendix 3.G). This result allows us a better understanding of the BLA as it was obtained for the random multisine: (3-15) is similar to (3-13). Integrals have to be considered over the continuous power spectrum of the noise instead of sums over discrete spectral components of the periodic signal. In Section 3.4.3.3 a formal statement is given on the asymptotic ($N \to \infty$) equivalence of $G_{\text{BLA}}(j\omega)$ for the three considered classes of excitations: random multisines, periodic noise, and normally distributed noise.

3.4.3.2 Interpretation of the Best Linear Approximation. When a nonlinear system is approximated using a linear system, it is important to be sure that the best approximation is made. This is actually the case for $G_{\text{BLA}}(j\omega)$. This follows directly from the fact that (3-14) is shown to give the best linear approximation in least square sense (see Eykhoff, 1974; and Bendat and Piersol, 1980 for the continuous-time case; and see Enqvist, 2005 for the discrete-time case). The estimated impulse response (and the corresponding FRF) minimizes the mean square value of $e(t) = y(t) - g(t) * u(t)$ over the measurement interval. For periodic excitations, the direct method (3-14) boils down to $G(j\omega_k) = S_{yu}(j\omega_k)/S_{uu}(j\omega_k) = Y_k/U_k$, which is exactly the starting expression used in (3-9). So the best linear approximation is also the best linear approximation for the class of random multisine excitations. The reader should be aware that this approximation is a function of the power spectrum (rms value and coloring) of the excitation.

3.4.3.3 Asymptotic Equivalences.

The following theorem states that the asymptotic best linear approximation $G_{\text{BLA}}(j\omega)$ is the same for random phase multisines (Definition 3.2), periodic noise (Definition 3.3), and Gaussian noise with the same (power) spectra. Hence, $G_{\text{BLA}}(j\omega)$ can be used to predict the response of the nonlinear system to any signal belonging to these three classes. Note, however, that the prediction error is bounded below by the stochastic nonlinear contributions $y_s(t) = \lim_{N \to \infty} y_{s,N}(t)$ (the notation $y_{s,N}$ is used here to indicate explicitly the dependence on the number of components). If this error is too large for a particular application, then the only way to improve the prediction quality is to model also the nonlinear behavior of the system.

The advantage of using random phase multisines over periodic noise to measure $G_{\text{BLA}}(j\omega)$ is that additional averages over the random amplitudes are avoided. The advantage of using periodic noise over Gaussian noise to measure $G_{\text{BLA}}(j\omega)$ is that the leakage errors are avoided.

Assuming that FRF measurements with M different excitations signals are made, the asymptotic best linear approximation $G_{\text{BLA}}(j\omega_k)$ can be estimated as

$$\hat{G}_{\text{BLA}}(j\omega_k) = \frac{1}{M} \sum_{m=1}^{M} Y^{[m]}(k)/U^{[m]}(k) \qquad (3\text{-}16)$$

for random phase multisines (see (3-14)) and as

$$\hat{G}_{\text{BLA}}(j\omega_k) = \sum_{m=1}^{M} Y^{[m]}(k)\overline{U}^{[m]}(k) / \sum_{m=1}^{M} |U^{[m]}(k)|^2 \qquad (3\text{-}17)$$

for periodic and Gaussian noise, where $U^{[m]}(k)$ and $Y^{[m]}(k)$ are the input and output DFT spectra of the mth FRF measurement.

Theorem 3.12 (Asymptotic Best Linear Approximation): Consider the following three classes of excitation signals: (i) random phase multisines (see Definition 3.2) with $\hat{U}^2(f) \equiv S_{\hat{U}\hat{U}}(f)$, (ii) periodic noise (see Definition 3.3) with $\mathbb{E}\{\hat{U}^2(f)\} \equiv S_{\hat{U}\hat{U}}(f)$, and (iii) Gaussian noise with power spectrum $S_{uu}(j\omega) \equiv S_{\hat{U}\hat{U}}(f)/f_s$ for $|f| < f_s/2$ and zero elsewhere. For these three classes of excitation signals, the best linear approximations $G_{\text{BLA},N}(j\omega)$ (H_1-FRF measurement) of a nonlinear system belonging to the class $\mathbb{S}$ (see Definition 3.5) converge (measurement time and $N \to \infty$) at the rate $O(N^{-1})$ to the same limit value $G_{\text{BLA}}(j\omega)$. If the joint second-order derivatives of the odd degree kernels $G^{2\alpha-1}_{f,f_1,-f_1,\ldots,f_{\alpha-1},-f_{\alpha-1}}$, $\alpha = 1, 2, \ldots, \infty$, w.r.t. $f, f_1, \ldots, f_{\alpha-1}$, are bounded for $f, f_1, \ldots, f_{\alpha-1} \in [0, f_s/2]$, then $G_{\text{BLA}}(j\omega)$ is given by

$$G_{\text{BLA}}(j\omega) = G_0(j\omega) + G_B(j\omega) = G_0(j\omega) + \sum_{\alpha=2}^{\infty} C_1^{\alpha}(j\omega)$$
$$C_1^{\alpha} = c_{\alpha} \int_0^{f_s/2} \cdots \int_0^{f_s/2} G^{2\alpha-1}_{f,f_1,-f_1,\ldots,f_{\alpha-1}} S_{uu}(j\omega_1)\ldots S_{uu}(j\omega_{\alpha-1}) df_1 \ldots df_{\alpha-1} \qquad (3\text{-}18)$$

with $c_{\alpha} = 2^{\alpha-1}(2\alpha-1)!!$.

Proof. See Appendix 3.H. □

Section 3.4 ■ Study of the Properties of FRF Measurements in the Presence of Nonlinear Distortions   85

$$u(t) \rightarrow \boxed{\text{Linear system } R(j\omega)} \xrightarrow{w(t)} \boxed{\text{Static nonlinear system}} \xrightarrow{z(t)} \boxed{\text{Linear system } S(j\omega)} \xrightarrow{y(t)}$$

$$G_{\text{BLA}}(j\omega) \div R(j\omega)S(j\omega)$$

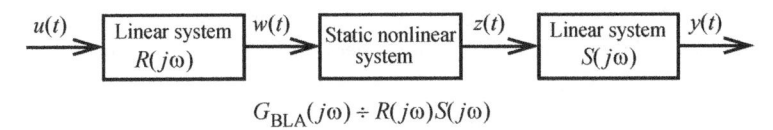

Figure 3-3. Nonlinear Wiener-Hammerstein system and its best linear approximation.

From Theorem 3.12 it follows that the asymptotic best linear approximation $G_{\text{BLA}}(j\omega)$ depends only on the second-order moments $S_{\hat{U}\hat{U}}(f)$ of the input spectrum. Note also that (3-15), with $S_{uu}(j\omega) = S_{\hat{U}\hat{U}}(f)/f_s$ for $|f| < f_s/2$ and zero elsewhere, reduces to C_1^α in (3-18).

3.4.3.4 Smoothness. Additional assumptions are required to guarantee the smoothness of $G_{\text{BLA}}(j\omega)$. This restricts further the class of allowable nonlinear systems.

Assumption 3.13 (Even and/or Odd Degree): For any $\omega \in [0, \omega_s/2]$, the even and/or odd degree Volterra kernels $G^\alpha_{f,f_1,f_2,\ldots,f_{\alpha-1}}$, $\alpha = 1, 2, \ldots$, are continuous functions of ω with continuous Pth order derivative w.r.t. ω.

For example, systems consisting of the cascade and parallel connection of linear systems and multipliers result in rational Volterra kernels for which Assumption 3.13 is satisfied (Schetzen, 1980).

Assumption 3.14: The series $\sum_{\alpha=2}^Q C_1^\alpha(j\omega)$, with C_1^α defined in (3-18), and its derivatives of order $1, 2, \ldots, P$ w.r.t. ω converge ($Q \to \infty$) uniformly in $\omega \in [0, \omega_s/2]$ to their limit sum.

Note that Assumptions 3.13 and 3.14 do not exclude the possible nonuniform (point wise or mean square) convergence of the output of the Volterra series model $y_Q(t)$ to $y(t)$.

Theorem 3.15 (Smoothness Best Linear Approximation): Under the conditions of Theorem 3.12 and Assumptions 3.13(odd degree) and 3.14, the asymptotic best linear approximation $G_{\text{BLA}}(j\omega)$ is a continuous function of $\omega \in [0, \omega_s/2]$ with continuous Pth order derivative.

Proof. See Appendix 3.I.   □

From this theorem it follows that $G_{\text{BLA}}(j\omega)$ and its higher order derivatives w.r.t. ω are continuous functions of ω. This explains why $G_{\text{BLA}}(s)$ can be approximated very well by a rational function of s of sufficiently high order.

3.4.3.5 Special Case: Wiener-Hammerstein Systems. In the case of a Wiener-Hammerstein system, consisting of a linear system with transfer function $R(j\omega)$, followed by a static nonlinearity, $v(t) = \sum_{k=0}^\infty a_k u^k(t)$, with $a_k \in \mathbb{R}$, and a second linear system $S(j\omega)$ (see Figure 3-3), the previous expressions can be simplified further. The Volterra kernel of degree α at frequency k is (Schetzen, 1980)

$$\begin{aligned} G^{2\alpha-1}_{k_1,\ldots,k_{2\alpha-1}} &= a_{2\alpha-1} S(j\omega_k) R(j\omega_{k_1})\ldots R(j\omega_{k_{2\alpha-1}}) \\ G^{2\alpha}_{k_1,\ldots,k_{2\alpha}} &= 0 \end{aligned} \qquad (3\text{-}19)$$

with $\omega_k = \sum_{i=1}^{\beta} \omega_{k_i}$, $\beta = 2\alpha - 1$ or $\beta = 2\alpha$, and $a_{2\alpha-1}$ a constant independent of the frequencies and the input signal. Using Theorems 3.7 and 3.12, we find

$$G_0(j\omega_k) = a_1 R(j\omega_k)S(j\omega_k)$$
$$G_\beta^{2\alpha-1}(j\omega_k) = a_{2\alpha-1} D_\alpha R(j\omega_k)S(j\omega_k) + O_\alpha(N^{-1})$$
$$D_\alpha = c_\alpha \int_0^{f_s/2} \cdots \int_0^{f_s/2} |R(f_1)|^2 \cdots |R(f_{\alpha-1})|^2 S_{uu}(j\omega_1) \cdots S_{uu}(j\omega_{\alpha-1}) df_1 \cdots df_{\alpha-1} \quad (3\text{-}20)$$
$$G_\beta^{2\alpha}(j\omega_k) = 0$$

with $c_\alpha = 2^{\alpha-1}(2\alpha-1)!!$, $R(f) = R(j\omega)$, and $\sum_{\alpha=1}^{\infty} O_\alpha(N^{-1}) = O(N^{-1})$. Hence, the asymptotic ($N \to \infty$) best linear approximation is given by

$$G_{\text{BLA}}(j\omega) = C(U, R)R(j\omega)S(j\omega) \quad (3\text{-}21)$$

with $C(U, R) = \sum_{\alpha=1}^{\infty} a_{2\alpha-1} D_\alpha$. As a result, for Wiener-Hammerstein systems, the asymptotic best linear approximation $G_{\text{BLA}}(j\omega)$ equals the underlying linear system within a real frequency-independent scale factor $C(U, R)$ that depends on the excitation signal and the system $R(j\omega)$. Similar results were also reported (Billings and Fakhour, 1982; Nikias and Petropulu, 1993) for special classes of excitation signals such as white zero mean Gaussian noise.

Remark. Sometimes the structure in Figure 3-3 is called the "general model" (e.g., Billings and Fakhour, 1982).

3.4.4 Further Comments on the Output Stochastic Nonlinear Contributions

Also for the output stochastic nonlinear contributions, the smoothness and the equivalence results can be obtained. In Section 4.2 these asymptotic ($N \to \infty$) results are extended to a broader class of excitation signals.

Using (3-9) and (3-10) with $N_G(k) = 0$ (no disturbing noise), the relation between the input and output Fourier coefficients at frequency $f_k = kf_s/N$ can be written as

$$Y_k = G_{\text{BLA}}(j\omega_k)U_k + Y_{Sk} \quad (3\text{-}22)$$

with $Y_{Sk} = G_S(j\omega_k)U_k$ the stochastic nonlinear contributions observed at the output of the system. Because $U_k = O(N^{-1/2})$ (see Definitions 3.2 and 3.3) and $|G_S(j\omega_k)| = O(N^0)$ (see Theorem 3.9) it follows that $Y_{Sk} = O(N^{-1/2})$. The following two theorems study the (asymptotic $N \to \infty$) behavior of $Y_S(k) = \sqrt{N}Y_{Sk}$ for random phase multisines, periodic noise, and Gaussian noise excitations.

Theorem 3.16 (Properties of the Stochastic Nonlinear Contributions $Y_S(k)$): For a system belonging to the system set $\mathbb{S}$ (see Definition 3.5), excited with a random multisine $u_N \in \mathbb{E}_N$ (see Definition 3.2) or periodic noise $u_N \in \mathbb{P}_N$ (see Definition 3.3), the stochastic nonlinear contribution $Y_S(k) = G_S(j\omega_k)U(k)$ in the output DFT spectrum

$$Y(k) = G_{\text{BLA}}(j\omega_k)U(k) + Y_S(k) \quad (3\text{-}23)$$

Section 3.4 ■ Study of the Properties of FRF Measurements in the Presence of Nonlinear Distortions

has the following (asymptotic $N \to \infty$) properties:

(i) $Y_S(k)$ has zero mean and is uncorrelated with – but not independent of – $U(k)$
(ii) $Y_S(k)$ has the same properties of $G_S(j\omega_k)$ in Theorem 3.9
(iii) $Y_S(k)$ is mixing over k of order infinity
(iv) $Y_S(k)$ is asymptotically ($N \to \infty$) circular complex normally distributed (convergence in law at the rate $O(N^{-1})$)

Proof. See Appendix 3.J. □

Since $U(k)$ is a random variable one could think that the asymptotic normality of $Y_S(k) = G_S(j\omega_k)U(k)$ (property iv of Theorem 3.16) is in contradiction with the asymptotic normality of $G_S(j\omega_k)$ (Theorem 3.11) and vice versa ($G_S(j\omega_k) = Y_S(k)/U(k)$). This is not the case because the random variable $G_S(j\omega_k)$ ($Y_S(k)$) is correlated with $U(k)$. Note also that $Y_S(k)e^{-j\angle U(k)}$ has exactly the same asymptotic ($N \to \infty$) properties as $G_S(j\omega_k)$ (Theorems 3.9, 3.10, and 3.11) because the phase of $Y_S(k)e^{-j\angle U(k)}$ is equal to that of $G_S(j\omega_k)$. The (asymptotic $N \to \infty$) equivalence of var($Y_S(k)$) for random phase multisines, periodic noise, and Gaussian noise excitations is established in the next theorem. To denote the dependence of the results on the number of frequencies N a subscript N is added.

Theorem 3.17 (Asymptotic Variance of the Stochastic Nonlinear Contributions $Y_S(k)$): Consider the following three classes of excitation signals: (i) random phase multisines (see Definition 3.2) with $\hat{U}^2(f) \equiv S_{\hat{U}\hat{U}}(f)$, (ii) periodic noise (see Definition 3.3) with $\mathbb{E}\{\hat{U}^2(f)\} \equiv S_{\hat{U}\hat{U}}(f)$, and (iii) Gaussian noise with power spectrum $S_{uu}(j\omega) \equiv S_{\hat{U}\hat{U}}(f)/f_s$ for $|f| < f_s/2$ and zero elsewhere. For these three classes of excitation signals, the variances var($Y_S(k)$) of the stochastic nonlinear distortions $Y_S(k) = \sqrt{N}Y_{Sk,N}$ of a nonlinear system belonging to the class $\mathbb{S}$ (see Definition 3.5) converge (measurement time and $N \to \infty$) at the rate $O(N^{-1})$ to the same limit value $\sigma_{Y_S}^2(f)$.

Proof. See Appendix 3.K. □

Theorem 3.18 (Asymptotic Variance of the Best Linear Approximation): The asymptotic variance ($N, M \to \infty$) of the FRF estimate $\hat{G}_{BLA}(j\omega_k)$ (3-16) and (3-17) due to the stochastic nonlinear distortions is given by

$$\lim_{N, M \to \infty} M\text{var}(\hat{G}_{BLA}(j\omega_k)) = \frac{\sigma_{Y_S}^2(f)}{f_s S_{uu}(j\omega)} \text{ with } \sigma_{Y_S}^2(f) = \lim_{N \to \infty} \text{var}(\sqrt{N}Y_{Sk,N}) \quad (3\text{-}24)$$

where $S_{\hat{U}\hat{U}}(f) = f_s S_{uu}(j\omega)$. The convergence rate in N of the variance of the BLA is an $O(N^{-1})$. For random phase multisines (3-24) is valid for any finite value $M \geq 1$.

Proof. See Appendix 3.L. □

Although result (3-24) is straightforward for random phase multisines, it is not for periodic and Gaussian noise excitations because $Y_S(k)$ and $U(k)$ in (3-17) are not independently distributed. The practical consequence of Theorem 3.18 is that no distinction should be made between the noise and the stochastic nonlinear distortions for generating uncertainty bounds on the nonparametric estimates (3-16) and (3-17) of the BLA. Hence, the $100 \times p\%$ confidence region (2-40), where $\hat{\sigma}_{\hat{G}}$ is the sample standard deviation of the BLA over the M independent realizations, remains valid.

From Theorems 3.17 and 3.18 it follows that the variance (3-24) of the FRF measurement (3-16) and (3-17) depends only on the second-order moments $S_{\tilde{U}\tilde{U}}(f)$ of the input spectrum. Hence, it is the same for random phase multisines, periodic noise, and Gaussian noise excitations. To establish the smoothness of $\sigma_{Y_S}^2(f)$ in Theorems 3.17 and 3.18 we need the following assumption:

Assumption 3.19: Define $C^{\alpha,\beta}(j\omega) = \lim_{N \to \infty} C_N^{\alpha,\beta}$ where $C_N^{\alpha,\beta}$ is given in (3-98). The series $\sum_{\alpha=2}^{Q_\alpha} \sum_{\beta}^{Q_\beta} C^{\alpha,\beta}(j\omega)$ and its derivatives of order 1, 2, ..., P converge ($Q_\alpha, Q_\beta \to \infty$) uniformly to their limit sum.

Theorem 3.20 (Smoothness Asymptotic Variance $\sigma_{Y_S}^2(f)$): Under the conditions of Theorem 3.17 and Assumptions 3.13(even and odd degree) and 3.19 the asymptotic variance $\sigma_{Y_S}^2(f)$ is a continuous function of $f \in [0, f_s/2]$ with continuous Pth order derivative.

Proof. Follow the same lines of Appendix 3.I. □

Theorem 3.20 explains why $Y_{Sk,N}$ can be modeled very well as filtered (band-limited) white noise.

3.5 EXTENSION TO DISCRETE-TIME MODELING

3.5.1 Periodic Signals

The results of Sections 3.4.1 to 3.4.4 were obtained for continuous-time systems. In this section we will show that these can be extended to discrete-time models. Some precautions should be taken because, for the discrete-time domain, the frequency axis is finite: $\omega \in [-\pi, \pi)$. In the nonlinear operations, higher frequencies can be created (e.g., $k\omega$), but these are folded back to the previous interval by the modulo operation: $\omega_{folded} = [(\omega + \pi) \bmod 2\pi] - \pi$, so that new frequency combinations appear that were not present in the previous sections. We show subsequently that the folding operation does not change the nature of these components (systematic or stochastic contributions). To do so, we consider the unfolded frequency $\tilde{\omega}$, as it results from the frequency combinations in the nonlinear system. In the next theorem we show that for a nonlinear system, excited by a band-limited random multisine excitation ($U(j\omega) = 0$ for $|\omega| > \omega_s/2$), its output components at frequencies $|\tilde{\omega}| > \omega_s/2$ can only be stochastic contributions. This means that they cannot be combined with any component of the random multisine excitation to result in a phase-independent combination.

Remark. To formalize this result in a theorem, we have to consider discrete-time random multisines. These are obtained directly from Definition 3.2 by replacing t by the discrete-time variable k, with $k = 0, 2, ..., N-1$. The frequencies of a discrete-time random multisine are restricted to the grid $2\pi/N$ in order to get periodic discrete-time signals (see Oppenheim et al., 1997: not all frequencies result in a periodic signal in the discrete-time domain!).

Theorem 3.21 (Stochastic Behavior of the Out-of-Band Components): For a (discrete-time) system belonging to the system set $\mathbb{S}$ (see Definition 3.5), excited with independent realizations of a (discrete-time) normalized random multisine $u_N \in \mathbb{E}_N$ (see Definition 3.2 and the previous note) or (discrete-time) normalized periodic noise $u_N \in \mathbb{P}_N$

Section 3.5 ■ Extension to Discrete-Time Modeling

(see Definition 3.3 and the previous note), with maximum angular frequency $\omega_{max} = l_{max}\omega_s/N$ ($l_{max} = N/2$), we have for $\underset{\sim}{\omega}_L = L\omega_s/N$, $|L| > l_{max}$:

$$\mathbb{E}\{Y_L^\alpha e^{-j\angle U_l}\} = 0 \qquad (3\text{-}25)$$

Proof. See Appendix 3.M. □

Note that this theorem is valid for continuous-time and discrete-time systems (using $\underset{\sim}{\omega}$). A direct result of this theorem is that all results of the previous sections can also be applied to discrete-time systems. Because none of the "out-of-band" components can create systematic contributions, the folding process does not change the nature of the output contributions of a nonlinear system, and, hence, the previous proofs remain valid.

3.5.2 Random Signals

Consider the class of discrete-time filtered white noise excitations $u(t) = L(q)e(t)$, with q the backward shift operator ($qx(t) = x(t-1)$), $e(t)$ discrete-time white noise with mean value μ and variance σ^2, and $L(z^{-1})$ a stable filter. The best linear approximation (BLA) of a nonlinear system belonging to the system set $\mathbb{S}$ (see Definition 3.5) can be defined as

$$G_{\text{BLA}}(q) = \arg\min_G \mathbb{E}\{(y_1(t) - G(q)u_1(t))^2\} \qquad (3\text{-}26)$$

with $y_1(t) = y(t) - \mathbb{E}\{y(t)\}$ and $u_1(t) = u(t) - \mathbb{E}\{u(t)\}$. The BLA (3-26) equals the linear time invariant second order equivalent (LTI-SOE) defined in Enqvist and Ljung (2005) and Enqvist (2005), except that $G(q)$ is not restricted to the class of stable and causal models. In Enqvist and Ljung (2005) and Enqvist (2005) it is shown that the solution of the unconstrained minimization problem (3-26) is given by

$$G_{\text{BLA}}(e^{-j\omega T_s}) = \frac{F\{\mathbb{E}\{y_1(t)u_1(t-\tau)\}\}}{F\{\mathbb{E}\{u_1(t)u_1(t-\tau)\}\}} = \frac{S_{y_1u_1}(j\omega)}{S_{u_1u_1}(j\omega)} \qquad (3\text{-}27)$$

which coincides with (3-14). The BLA or LTI-SOE (3-26) has the following properties:

1. For normally distributed noise $e(t)$, the BLA (3-26) coincides with the BLA (3-12) obtained using random phase multisines with $\hat{U}^2(f) \equiv |L(e^{-j\omega T_s})|^2 \sigma^2$ (see Definition 3.2 where $N \to \infty$). In addition: (i) the BLA of a static nonlinear system is static; (ii) the BLA of a nonlinear finite impulse response system has a finite impulse response; (iii) the BLA is independent of the phase of $L(e^{-j\omega T_s})$; and (iv) the BLA has small sensitivity to small nonlinearities (proof: see Sections 3.4.1–3.4.4 and Enqvist, 2005).

2. For non-Gaussian noise $e(t)$, the output residuals $y_s(t) = y_1(t) - G_{\text{BLA}}(q)u_1(t)$, with $G_{\text{BLA}}(q)$ the minimizer of (3-26), are still uncorrelated with $u_1(t)$. However, (3-26) has some counterintuitive properties: (i) the BLA of a static nonlinear system is not necessarily static; (ii) the BLA of a nonlinear finite impulse response system does not necessarily have a finite impulse response; (iii) the BLA can depend on the phase of $L(e^{-j\omega T_s})$; and (iv) the BLA can have a high sensitivity to small nonlinearities (proof: see Enqvist, 2005).

3.6 EXPERIMENTAL ILLUSTRATION

A nonlinear mechanical resonating system (mass, viscous damping, nonlinear spring) is simulated with an electrical circuit. The displacement $y(t)$ (output) is related to the force $u(t)$ (input) by the following nonlinear, second-order differential equation:

$$m\frac{d^2y(t)}{dt^2} + d\frac{dy(t)}{dt} + k(y(t))y(t) = u(t) \tag{3-28}$$

The nonlinear spring is described by a static but position-dependent stiffness

$$k(y) = a + by^2 \tag{3-29}$$

For small excitations, the spring becomes almost linear so that the underlying linear system consists of a second-order resonance system. A series of experimental results on this system are shown. First, the nonlinear behavior will be illustrated using stepped sine measurements. Next, the split of the transfer function into the underlying linear system $G_0(j\omega_k)$, the best linear approximation $G_{BLA}(j\omega_k)$, the stochastic nonlinear distortions $G_S(j\omega_k)$, and the noise contributions $N_G(k)$ are shown.

3.6.1 Visualization of the Nonlinearity Using Stepped Sine Measurements

To visualize the nonlinear behavior of the system, a stepped sine measurement is made (Figure 3-4). The frequency of the sine is first stepped upward until the maximum frequency is reached and then stepped down again. At each frequency a measurement is made over an integer number of periods. During the experiment we took care to have a continuous excitation signal; no discontinuities appeared at the frequency-changing instants. The nonlinear behavior of the system is clearly visible. The measured transfer function depends, strongly, on the amplitude of the sine excitation. Moreover, the measurements also show that the actual output of the system depends on the past inputs: the up-path differs from the down-path for large excitations. Such behavior cannot be described using Volterra-based descriptions. Nevertheless, we will still apply the previously developed theory to this system. This can be done because the bifurcation appears only for large excitations, injecting a lot of power close to the resonance frequency of the system. If we use normalized random multisines, only a fraction of the power is injected in this band so that the bifurcation problem does not disturb the measurements anymore.

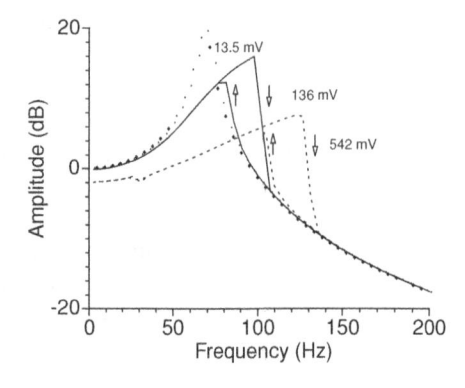

Figure 3-4. Stepped sine measurement at different amplitudes (rms values given). An up and down sweep is made. For the 13.5 mV measurement: black boxes up sweep, white boxes down sweep. For the others: ↓ up sweep, ↑ down sweep.

Section 3.6 ■ Experimental Illustration

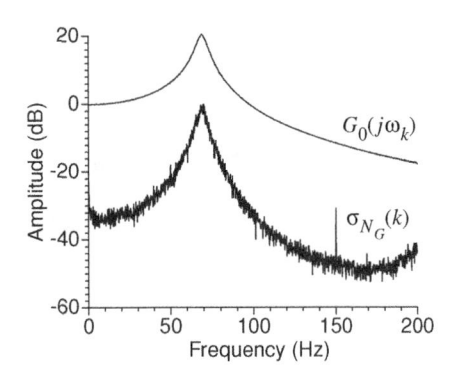

Figure 3-5. Measurement of the underlying linear system $G_0(j\omega_k)$ and its standard deviation.

3.6.2 Measurement of the Best Linear Approximation

In a second step, the underlying linear system is measured using a normalized random multisine ($f_k = (2k+1)f_0$, $k = 0, 1, \ldots, 1340$ and $f_0 \approx 0.0745$ Hz) with a small amplitude (rms value of 34.2 mV). The standard deviation $\sigma_{N_G}(k)$ is calculated from 10 consecutive periods. The results are shown in Figure 3-5.

The impact of the nonlinearity is made visible by increasing the excitation level of the normalized random multisine to an rms value of 127 mV. The measurement was repeated for 10 different realizations of the excitation signal so that $\sigma_S(k)$ could also be measured. The measurement results are shown in Figure 3-6. On the left side, the best linear approximation is compared with the underlying linear system. A number of observations can be made: the resonance frequency is shifted to the right, the peak value is decreased, and the measurement becomes more noisy.

The shift to the right of the resonance frequency is due to the nonlinear behavior of the hardening spring. For larger excursions, the average stiffness increases and so also does the resonance frequency. Note that if the $G_0(j\omega_k)$ measurement were not available, there would be no indication at all that this system is strongly nonlinear. This shows, clearly, why we need dedicated tools to detect the presence of nonlinear distortions. The difference between $G_{BLA}(j\omega_k)$ and $G_0(j\omega_k)$ is due to the systematic contributions $G_B(j\omega_k)$.

The increased noise level can be understood only from the previous, explained theory; it is due to the stochastic contributions $G_S(j\omega_k)$. Changing the excitation level did not change the disturbing noise, but $G_S(j\omega_k)$ became much larger. This is visualized on the left side of the figure. The standard deviation $\sigma_{G_S}(k)$ is obtained by measuring the FRF from 10 realizations of the normalized random multisine. For the small excitation level, it is completely dominated by the measurement noise $\sigma_{N_G}(k)$, whereas for the large excitation, $\sigma_{G_S}(k)$

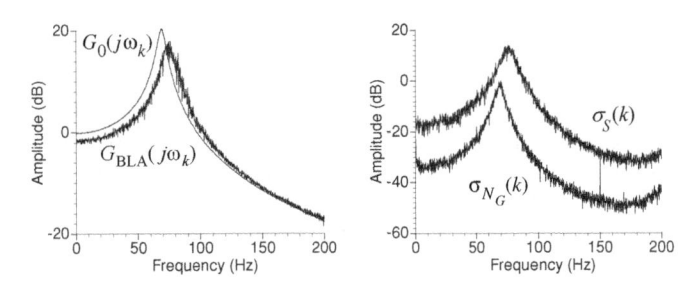

Figure 3-6. Comparison of the measured best linear approximation $G_{BLA}(j\omega_k)$ obtained from 10 realizations and the underlying linear system $G_0(j\omega_k)$.

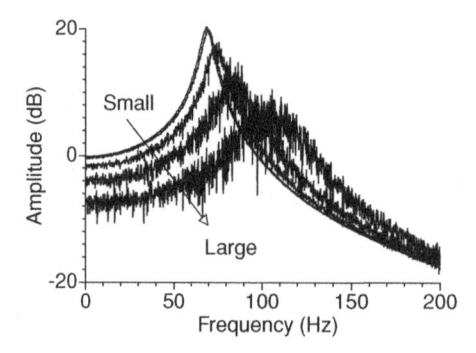

Figure 3-7. Evolution of the best linear approximation for growing excitation levels: rms values of 34 mV, 54 mV, 127 mV, 253 mV, and 507 mV.

dominates. This is also illustrated in Figure 3-7, where the evolution of the measured FRF is shown as a function of the excitation level. As can be seen, the stochastic contributions grow with the level while the measurement conditions (and, hence, the disturbing noise) remain the same. Again, it is very difficult to understand this result without the previously gained insight into the behavior of nonlinear systems. This also suggests a first test to detect the presence of nonlinear distortions. The standard deviation calculated from a set of consecutive periods (without changing the excitation signal) should be the same as that calculated from repeated measurements, using different realizations of the excitation signal.

3.7 MULTIVARIABLE SYSTEMS

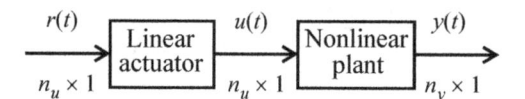

Figure 3-8. Nonlinear plant with n_u inputs and n_y outputs, excited with $n_u \times 1$ random phase multisines $r(t)$ via a linear actuator.

In Dobrowiecki and Schoukens (2007a, b) it has been shown that the results proven for the single-input, single-output case are also valid for multiple-input, multiple-output nonlinear systems. To measure the $n_y \times n_u$ BLA, one should ensure the randomness of the $n_y \times 1$ error term $Y_S(k)$ in (3-23). Therefore, the nonlinear multivariate system is excited by n_u different random phase multisines $r_{[p]}(t)$, $p = 1, 2, ..., n_u$, (see Definition 3.1) with user-defined amplitudes $|R_{[p]}(k)|$ and where the phases $\angle R_{[p]}(k)$ are independently chosen over the frequency k and the input p. Hence, the n_u reference signals $r(t)$ defined by (2-81) are replaced by

$$\mathbf{R}(k) = D_R(k)T \qquad (3\text{-}30)$$

where $D_R(k) \in \mathbb{C}^{n_u \times n_u}$ is a diagonal matrix with entries $D_{R[p,q]}(k) = R_{[p]}(k)\delta(p-q)$, and with T the $n_u \times n_u$ orthogonal matrix (2-80). Since the stochastic nonlinear distortions $Y_S(k)$ of the n_u experiments with the *random orthogonal multisines* (3-30) are not independent of each other, their covariance matrix cannot be estimated. However, their contribution to the BLA can still be measured (Dobrowiecki and Schoukens, 2007a and b; Wernholt and Gunnarsson, 2008). This dependency problem is solved by adding the same random phase $\phi_e(k)$ to each input, such that $\mathbb{E}\{e^{j\phi_e(k)}\} = 0$, where $\phi_e(k)$ is randomly chosen over the frequency k and over the experiment e. Hence, (3-30) is replaced by

$$\mathbf{R}(k) = D_R(k) T D_\phi(k) \tag{3-31}$$

where $D_{\phi\,[p,\,e]}(k) = e^{j\phi_e(k)}\delta(p-e)$. Compared with (3-30), the *full random orthogonal multisines* (3-31) allows us to estimate the covariance matrix of the stochastic nonlinear distortions $Y_S(k)$ (see Section 7.3.6). Both solutions require n_u generators. Similarly to (2-81), $\mathbf{R}(k)$ (3-31) can be also be written under the form (2-82), which allows us to calculate the inverse $\mathbf{R}^{-1}(k)$ in a numerically stable way.

Finally, using the random orthogonal (3-30) or full random orthogonal (3-31) multisines an estimate of the $n_y \times n_u$ BLA $G_{\mathrm{BLA}}(j\omega_k)$ is obtained as

$$\hat{G}_{\mathrm{BLA}}(j\omega_k) = \mathbf{Y}(k)\mathbf{U}^{-1}(k) \tag{3-32}$$

where the pth column of $\mathbf{U}(k)$ and $\mathbf{Y}(k)$ corresponds to pth multisine experiment (pth column of $\mathbf{R}(k)$). Averaging (3-32) over M independent random phase realizations of the (full) random orthogonal multisines gives an improved estimate

$$\hat{G}_{\mathrm{BLA}}(j\omega_k) = \frac{1}{M} \sum_{m=1}^{M} \mathbf{Y}^{[m]}(k)(\mathbf{U}^{[m]}(k))^{-1} \tag{3-33}$$

that converges for $M \to \infty$ to $G_{\mathrm{BLA}}(j\omega_k)$. For full random orthogonal multisines the covariance $\hat{G}_{\mathrm{BLA}}(j\omega_k)$ (3-33) is obtained as follows

$$\mathrm{Cov}(\mathrm{vec}(\hat{G}_{\mathrm{BLA}}(j\omega_k))) = \frac{1}{M} \overline{(G_{\mathrm{act}}(j\omega_k) D_{|R|^2}(k) G_{\mathrm{act}}^H(j\omega_k))}^{-1} \otimes C_{Y_S}(k) \tag{3-34}$$

with $G_{\mathrm{act}}(j\omega_k)$ the $n_u \times n_u$ actuator FRM (see Figure 3-8), $D_{|R|^2}(k)$ a diagonal matrix containing the squared 2-norm of each row of $\mathbf{R}^{[m]}(k)$, and where $C_{Y_S}(k) = \mathrm{Cov}(Y_S(k))$ is the covariance of the stochastic nonlinear contributions of one experiment (proof: see Appendix 3.N).

3.8 BEST LINEAR APPROXIMATION OF A SYSTEM OPERATING IN CLOSED LOOP

Consider a nonlinear plant operating within a linear feedback loop that is excited with a random phase multisine $r(t)$ via a linear actuator (see Figure 3-9). Due to the feedback loop the plant input $u(t)$ depends on the output of the nonlinear plant and, therefore, $u(t)$ is no longer a random phase multisine (3-7). Hence, the properties proven for the best linear approximation (BLA) obtained via the direct methods (3-9) and (3-14) are no longer valid in closed loop. To handle this problem we redefine the BLA via the indirect method for measuring FRFs (see Section 2.6.4)

$$G_{\mathrm{BLA}}(j\omega) = \frac{S_{yr}(j\omega)}{S_{ur}(j\omega)} = \frac{S_{yr}(j\omega)/S_{rr}(j\omega)}{S_{ur}(j\omega)/S_{rr}(j\omega)} = \frac{G_{ry}(j\omega)}{G_{ru}(j\omega)} \tag{3-35}$$

where the expected values in the cross-power spectra are taken w.r.t. the random realization of the reference signal $r(t)$. The last equality in (3-35) shows that the indirect BLA can be

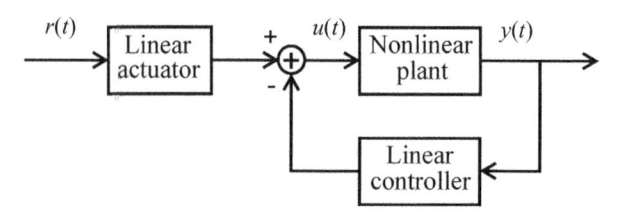

Figure 3-9. Nonlinear plant operating within a linear feedback loop. The plant is excited with a random phase multisine $r(t)$ via a linear actuator.

written as the ratio of the BLA $G_{ry}(j\omega)$ from reference to output to the BLA $G_{ru}(j\omega)$ from reference to input. In the open loop case the indirect method (3-35) reduces to the direct method (3-14) because $S_{yr}(j\omega) = S_{yu}(j\omega)/\overline{G}_{act}(j\omega)$ and $S_{ur}(j\omega) = S_{uu}(j\omega)/\overline{G}_{act}(j\omega)$, with $G_{act}(j\omega)$ the actuator FRF. Using definition (3-35) it is shown in the following theorem that the BLA and the stochastic nonlinear contributions of a nonlinear system operating in closed loop have similar properties as in the open loop case.

Theorem 3.22 (BLA of a Nonlinear System Operating in Closed Loop): Consider a nonlinear system operating within a linear feedback loop (see Figure 3-9) that is excited with a random phase multisine $r_N \in \mathbb{E}_N$ (see Definition 3.2) or periodic noise $r_N \in \mathbb{P}_N$ (see Definition 3.3). Assume furthermore that the nonlinear plant and the closed loop system belong to the set $\mathbb{S}$ (Definition 3.5), and that the BLA is calculated using the indirect method (3-35). The input-output DFT spectra of the nonlinear plant are then related as

$$Y(k) = G_{BLA}(j\omega_k)U(k) + Y_S(k) \tag{3-36}$$

where $G_{BLA}(j\omega_k)$ and $Y_S(k)$ have the following properties:

(i) $G_{BLA}(j\omega_k)$ does not depend on the even degree nonlinearities

(ii) Under the assumptions of Theorem 3.15 the asymptotic $(N \to \infty)$ BLA $G_{BLA}(j\omega)$ is a smooth function of the frequency with continuous P th order derivative

(iii) $Y_S(k)$ is uncorrelated with – but not independent of – the reference signal $R(k)$ and satisfies properties (ii-iv) of Theorem 3.16

(iv) Under the assumptions of Theorem 3.20 the asymptotic $(N \to \infty)$ variance of $Y_S(k)$ is a smooth function of the frequency with continuous P th order derivative

Proof. See Appendix 3.O. □

The main differences with the open loop case are that the stochastic nonlinear contributions $Y_S(k)$ are uncorrelated with the reference signal and not the plant input, and that the plant input depends on $Y_S(k)$ via the feedback law $U(k) = G_{act}(j\omega_k)R(k) - M_0(j\omega_k)Y(k)$, where $G_{act}(j\omega)$ and $M_0(j\omega)$ are, respectively, the actuator and controller frequency response functions. Contrary to the linear case (see Section 2.6.4), the indirect method (3-35) for measuring the BLA of a nonlinear system operating in feedback does not reduce to the direct method (3-14) for the class of random phase multisines.

3.9 CONCLUSION

FRF measurements give a great deal of information about the device or plant under test. Very often it is easily accessible and it is strongly advised to take this intermediate step in the identification process. It provides a lot of qualitative information about the complexity of the problem, as well as quantitative information about the plant and the measurement quality. This can be used to set up a measurement-driven weighting function for the identification step and also provides very valuable information for the model validation. The user has a large impact on the measurement quality by generating a good excitation and selecting the proper algorithms to process the raw measurement data. For these reasons, we strongly encourage the reader to take the time to understand the basic principles of FRF measurements. Good, nonparametric measurements will significantly simplify the task of building parametric models.

3.10 EXERCISES

3.1. Generate a random multisine $u_0(k)$ (see Exercise 2.2), with N_p points in one period, that excites the frequency lines, $4k+1$ for $k = 1, ..., \text{fix}(N_p/12)$ with equal power. Normalize the rms value of $u(t)$ to 1. Calculate

$$y_0(t) = u_0(t) + 0.1 u_0^2(t) + 0.01 u_0^3(t) \tag{3-37}$$

Calculate the output spectrum and discuss it. Observe the spectral behavior inside and outside the excited frequency band. Does the behavior depend on the value of N_p?

3.2. Repeat the previous exercise for $u_{\text{rms}} = 10$ and $u_{\text{rms}} = 100$. Discuss the behavior of the even and odd spectral lines.

3.3. Measure the FRF for $u_{\text{rms}} = 1, 10, 100$. Consider, for each situation, 100 realizations of the random multisine. Study the mean value and the standard deviation of the FRF. Extract $G_B(j\omega_k)$, $G_S(j\omega_k)$, $\sigma_{G_S}^2(k)$ and discuss your results.

3.4. Repeat Exercise 3.3 but, this time, use a random multisine that excites all spectral lines between 1 and $\text{fix}(N_p/12)$. Compare both results and explain the different behavior.

3.5. Repeat Exercise 3.3 but use a zero mean random noise excitation that has approximately the same power spectrum as the excitation in Exercise 3.4.

3.6. Construct a discrete-time Wiener-Hammerstein system $y_0 = WH(u_0)$ (see Figure 3-3) with static nonlinearity: $z = x + 0.1x^2 + 0.01x^3$. Measure $G_{\text{BLA}}(j\omega_k)$ (make a motivated choice for the power spectrum of the excitation signal) for $u_{\text{rms}} = 1, 10, 100$. Scale the gain of the first system so that the power of the contribution of degree 3 generates 1% of the linear output power for the first input amplitude. Discuss the results.

3.7. Consider the Wiener-Hammerstein system of Exercise 3.6 and add white, zero mean disturbing noise to the output.

$$y_0 = WH(u_0) \quad \text{and} \quad y(t) = y_0(t) + n_y(t) \tag{3-38}$$

Measure $G_{\text{BLA}}(j\omega_k)$ again (consider 100 realizations) and calculate $\sigma_{N_G}^2(k)$ and $\sigma_{G_S}^2(k)$. Use repeated periods to separate the measurement noise $n_y(t)$ from the nonlinear distortions.

3.11 APPENDIXES

Appendix 3.A Bias and Stochastic Contributions of the Nonlinear Distortions

In this appendix we assume a deterministic amplitude and a uniform continuous phase distribution. The random amplitude, the discrete phase, and the nonuniform continuous phase distributions are commented on in Appendix 3.F.

In order not to overload the notations, the following simplifications are made in this appendix: $G(j\omega_k)$, $G_{BLA}(j\omega_k)$, and $G_S(j\omega_k)$ are denoted as $G(k)$, $G_{BLA}(k)$, and $G_S(k)$, respectively.

Consider the contribution of degree α to the FRF:

$$G^\alpha(k) = \sum_{k_1, \ldots, k_{\alpha-1} = -N/2+1}^{N/2-1} G^\alpha_{L_k, k_1, k_2, \ldots, k_{\alpha-1}} \frac{U_{k_1} U_{k_2} \ldots U_{k_{\alpha-1}} U_{L_k}}{U_k}$$

$$= \sum_{k_1, \ldots, k_{\alpha-1} = -N/2+1}^{N/2-1} |G^\alpha_{L_k, k_1, k_2, \ldots, k_{\alpha-1}}| \frac{|U_{k_1}||U_{k_2}|\ldots|U_{k_{\alpha-1}}||U_{L_k}|}{|U_k|} e^{j\phi(k_1, k_2, \ldots, k_{\alpha-1}, L_k)}$$

(3-39)

with $k = L_k + \sum_{i=1}^{\alpha-1} k_i$, $\phi(k_1, \ldots, L_k) = \sum_{i=1}^{\alpha-1} \varphi_{k_i} + \varphi_{L_k} + \varphi_G - \varphi_k$, $\varphi_{k_i} = \angle U_{k_i}$ and $\varphi_G = \angle G^\alpha_{L_k, k_1, k_2, \ldots, k_{\alpha-1}}$. Define the disjoint sets

$$\mathbb{K}_{Bk,\alpha} = \{(k_1, k_2, \ldots, k_{\alpha-1}) | \phi(k_1, k_2, \ldots, k_{\alpha-1}, L_k) \text{ is independent of } \phi\}$$

$$\mathbb{K}_{Sk,\alpha} = \{(k_1, k_2, \ldots, k_{\alpha-1}) | \phi(k_1, k_2, \ldots, k_{\alpha-1}, L_k) \text{ depends on } \phi\}$$

(3-40)

with $\phi = \{\varphi_1, \ldots, \varphi_N\}$. The set $\mathbb{K}_{Bk,\alpha}$ corresponds to the situation where all frequencies but one (equal to k) can be grouped in pairs $(-l, l)$ so that their phases cancel. This results, by definition, in contributions to $G^\alpha_B(k)$, while this is not the case for the set $\mathbb{K}_{Sk,\alpha}$ (the phases cannot cancel) so that, by definition, these contribute to $G^\alpha_S(k)$. Eq. (3-39) becomes

$$G^\alpha(k) = G^\alpha_S(k) + G^\alpha_B(k)$$

$$G^\alpha_S(k) = \sum_{K \in \mathbb{K}_{Sk,\alpha}} G^\alpha_{L_k, k_1, k_2, \ldots, k_{\alpha-1}} U_{k_1} U_{k_2} \ldots U_{k_{\alpha-1}} U_{L_k} / U_k$$

$$G^\alpha_B(k) = \sum_{K \in \mathbb{K}_{Bk,\alpha}} G^\alpha_{L_k, k_1, k_2, \ldots, k_{\alpha-1}} U_{k_1} U_{k_2} \ldots U_{k_{\alpha-1}} U_{L_k} / U_k$$

(3-41)

with $K = (k_1, k_2, \ldots, k_{\alpha-1})$, and where $\sum_{K \in \mathbb{K}_{Sk,\alpha}}$ and $\sum_{K \in \mathbb{K}_{Bk,\alpha}}$ denote the sum over all combinations belonging to the sets $\mathbb{K}_{Sk,\alpha}$ and $\mathbb{K}_{Bk,\alpha}$, respectively.

In the second part of this appendix, we prove (3-13). From the definition of $\mathbb{K}_{Bk,\alpha}$, it follows that the only contributions different from zero are those with α odd. For that reason we focus from now forward on $G^{2\alpha-1}_B(k)$. The factor c_α in (3-13) is due to the fact that each of the terms in this sum appears multiple times in the original expression (3-41) or (3-9) where the frequency indices run from $-N/2+1$ to $N/2-1$. The number of appearances when starting the sums at zero will be different, and c_α compensated for that. The number of contributions to the sum (3-41) for a given frequency combination $\in \mathbb{K}_{Bk,\alpha}$ depends on the fact that some of the paired frequencies are equal to each other or not. If some of the paired frequencies are equal to each other or equal to k, there remain less degrees of freedom (because not all paired frequency values can be freely chosen), and, hence, they contribute to the final result only as an $O_\alpha(N^{-p})$, $p \geq 1$ (see also the following appendices) with $\sum_{\alpha=2}^{\infty} O_\alpha(N^{-p}) = O(N^{-p})$ (the Volterra series converges). Hence, we can focus completely on the situation where all paired frequencies are different from each other and from k. Each such frequency combination appears $(2\alpha - 1)!$ times in (3-41) for $G_B^{2\alpha-1}(k)$, keeping in mind the symmetrical Volterra kernels. In (3-13) each contributing combination appears only $(\alpha - 1)!$ times. Hence, the following correction term is needed

$$\frac{(2\alpha - 1)!}{(\alpha - 1)!} = 2^{\alpha - 1}(2\alpha - 1)!! \qquad (3\text{-}42)$$

Appendix 3.B Study of the Moments of the Stochastic Nonlinear Contributions

In this appendix we assume a deterministic amplitude and a uniform continuous phase distribution. The random amplitude, the discrete phase, and the nonuniform continuous phase distributions are commented on in Appendix 3.F.

In order not to overload the notations, the following simplifications are made in this appendix: $G(j\omega_k)$, $G_{\text{BLA}}(j\omega_k)$, and $G_S(j\omega_k)$ are denoted as $G(k)$, $G_{\text{BLA}}(k)$, and $G_S(k)$, respectively.

In this appendix, the moments of the stochastic nonlinear contributions $G_S(k)$ are calculated for nonlinear systems belonging to the set $\mathbb{S}$ (Definition 3.5), assuming a normalized random multisine excitation (Definition 3.2). From (3-41) it follows that the stochastic nonlinear contributions to the measurement, at frequency k, are given by multidimensional sums with $(k_1, k_2, ..., k_{\alpha-1}) \in \mathbb{K}_{Sk,\alpha}$, for which it is not possible to partition all the frequencies but one in pairs $(-l, l)$. As a consequence, these terms have a random phase such that $\mathbb{E}\{e^{j\varphi_k}\} = 0$. It follows directly that $\mathbb{E}\{G_S^\alpha(k)\} = 0$, and, hence, $\mathbb{E}\{G_S(k)\} = \mathbb{E}\{\sum_{\alpha=2}^{\infty} G_S^\alpha(k)\} = 0$. The study of the higher order moments is much more complicated. The basic idea is first to prove that

$$\left| \mathbb{E}\{G_S^{r_1}(k_1) G_S^{r_2}(k_2) ... G_S^{r_n}(k_n)\} \right| \leq O(N^{-p}) M_U^{-n} \prod_{i=1}^{n} M_{G^{r_i}} M_U^{r_i} \qquad (3\text{-}43)$$

for arbitrary n, where p depends on the actual situation. $G_S^{r_i}(k_i)$ stands for the stochastic nonlinear contribution of degree r_i at frequency k_i. Next, using (3-43) and Definition 3.5, we find

$$\left| \mathbb{E}\{G_S(k_1) G_S(k_2) ... G_S(k_n)\} \right| \leq \sum_{r_1=2}^{\infty} \cdots \sum_{r_n=2}^{\infty} \left| \mathbb{E}\{G_S^{r_1}(k_1) G_S^{r_2}(k_2) ... G_S^{r_n}(k_n)\} \right|$$

$$\leq O(N^{-p}) M_U^{-n} \prod_{i=1}^{n} \sum_{r_i=2}^{\infty} M_{G^{r_i}} M_U^{r_i} \qquad (3\text{-}44)$$

$$\leq O(N^{-p}) M_U^{-n} C_1^n$$

so that $\mathbb{E}\{G_S(k_1) G_S(k_2) ... G_S(k_n)\}$ converges to zero, at least, as an $O(N^{-p})$.

Lemma 3.23 (number of nonzero contributions): Consider a system belonging to the set $\mathbb{S}$, excited with a random multisine $u_N \in \mathbb{E}_N$. Under Definitions 3.2 and 3.5, the expected value $\mathbb{E}\{G_S^{r_1}(k_1)G_S^{r_2}(k_2)...G_S^{r_n}(k_n)\}$ is bounded by

$$\left|\mathbb{E}\{G_S^{r_1}(k_1)G_S^{r_2}(k_2)...G_S^{r_n}(k_n)\}\right| \le O(N^{-\nu})M_U^{-n}\prod_{i=1}^{n}M_{G^{r_i}}M_U^{r_i} \qquad (3\text{-}45)$$

with $\nu = \text{int}((n-2m+1)/2)$ and where m is the number of pairs $(k_i, k_j = -k_i)$ that can be formed in the set $\{k_1, k_2, ..., k_n\}$.

Note: If the number of unpaired frequencies k_i is odd, then $\nu = 1$, while $\nu = 0$ if it is even.

Proof. The basic idea is to count the number of nonzero contributions in $\mathbb{E}\{G_S^{r_1}(k_1)G_S^{r_2}(k_2)...G_S^{r_n}(k_n)\}$ as a function of N. Note that each of the terms in the product $G_S^{r_i}(k_i)$ consists of a multiple sum over the frequencies; see (3-41). The terms in the product $G_S^{r_1}(k_1)G_S^{r_2}(k_2)...G_S^{r_n}(k_n)$ that have a nonzero expected value are those where all phases of the participating frequency components cancel each other. This means that we have to look for frequency pairs $(l, -l)$ having a zero phase contribution.

Consider the frequencies that contribute to $G_S^{r_i}(k_i)$, $i = 1, ..., n$:

$$\begin{array}{ll} -k_1 & (l_1(k_1), l_2(k_1), ..., l_{r_1-1}(k_1), l_{r_1}(k_1)) \\ -k_2 & (l_1(k_2), l_2(k_2), ..., l_{r_2-1}(k_2), l_{r_2}(k_2)) \\ & \cdots \\ -k_n & (l_1(k_n), l_2(k_n), ..., l_{r_n-1}(k_n), l_{r_n}(k_n)) \end{array} \qquad (3\text{-}46)$$

with

$$l_{r_i}(k_i) = k_i - \sum_{p=1}^{r_i-1} l_p(k_i), \quad i = 1, ..., n \qquad (3\text{-}47)$$

The frequency $-k_i$ (called denominator frequencies) comes from the denominator in (3-41), where the minus sign accounts for the negative phase contribution of the denominator term; $(l_1(k_i), l_2(k_i), ..., l_{r_i}(k_i))$ are the frequencies in the numerator of (3-41) (called numerator frequencies) and their sum should be equal to k_i in order to get a contribution at frequency k_i. The total number of numerator frequencies participating in the sums is $F_a = \sum_{i=1}^{n} r_i$. Equation (3-47) imposes n constraints, so that the total number of degrees of freedom at this moment is $F_a - n$.

The only nonzero contributions to the expected value (3-43) are those where all frequencies (numerator frequencies and unpaired denominator frequencies) can be grouped in pairs $(-l, l)$ such that their phase contributions are canceled. This pairing process will be imposed step by step (first on the denominator frequencies k_i, next on the remaining numerator frequencies $l_j(k_h)$), and the additional constraints on the free frequencies in (3-46) will be checked.

3.B.1 Denominator Frequencies

(i) First pair the denominator frequencies $(-k_i = k_j)$. Assume there are m such pairs.

(ii) All remaining $n-2m$ unpaired denominator frequencies k_i should be paired with one of the numerator frequencies $l_j(k_h)$, $h = 1, ..., n$ and $j = 1, ..., r_h$. Because the denominator frequencies have fixed values (no summing over k_i), this fixes $n-2m$ numerator frequencies.

(iii) Eventually, after pairing all denominator frequencies, the number of free frequencies is $F_a - n - (n-2m) = F_a - 2n + 2m$. Note that the worst case situation appears when m is maximized because this leaves the maximum number of numerator frequencies free.

3.B.2 Numerator Frequencies. Next the remaining numerator frequencies should be paired. These can be partitioned in two groups: the free numerator frequencies ($F_a - 2n + 2m$) and the (n) dependent frequencies $l_{r_i}(k_i)$. We impose pairs only on the free frequencies, assuming that the dependent frequencies are then automatically paired. This is again the worst case situation (the largest number of free frequencies), since in the other case additional constraints would be imposed. Note also that pairing is a worst case phase canceling process: grouping four or more frequencies together is a stronger restriction than making pairs of two frequencies. Two situations will be considered: n is even or n is odd.

(i) n is even (F_a is even, otherwise there would always remain an unpaired frequency and these terms have zero mean): all free frequencies can be paired, resulting in $(F_a - 2n + 2m)/2$ pairs where the frequency can be freely chosen. So the maximum number of zero phase terms in $G_S^{r_1}(k_1) G_S^{r_2}(k_2) ... G_S^{r_n}(k_n)$ is an $O(N^{v_0})$ with $v_0 = (F_a - 2n + 2m)/2$. From Definitions 3.2 and 3.5 and (3-9), it follows that each term in the sum of $G_S^{r_i}(k_i)$ is an $O(N^{v_i}) M_{G^{r_i}} M_U^{r_i - 1}$, with $v_i = (1 - r_i)/2$, and, hence, the expected value is bounded by

$$\left| \mathbb{E}\{G_S^{r_1}(k_1) G_S^{r_2}(k_2) ... G_S^{r_n}(k_n)\} \right| \leq O(N^{-v}) M_U^{-n} \prod_{i=1}^{n} M_{G^{r_i}} M_U^{r_i} \quad (3\text{-}48)$$

with $v = -v_0 - \sum_{i=1}^{n} v_i = (n - 2m)/2$ for n even.

(ii) n is odd (F_a is odd, otherwise there would always remain an unpaired frequency). In this case, not all the free numerator frequencies ($F_a - 2n + 2m$) can be paired since they are odd in number. So $(F_a - 2n + 2m - 1)/2$ pairs of free frequencies can be formed, and there remains one unpaired free frequency that should be combined with one dependent frequency. Again we can assume that the other $n-1$ (an even number) dependent frequencies are then automatically paired (worst case). So the question is whether the last pairing step (the independent frequency equals minus the dependent frequency) creates a new constraint. To answer this question, it is important to note that not all but one numerator frequencies in a row of (3-46) can be paired to each other, because this would be a systematic contribution (see Appendix 3.A). As a consequence, the dependent frequency $l_{r_i}(k_i)$ (3-46) cannot be paired with another frequency in its own row. This would either impose a new constraint in this row (put $l_{r_i}(k_i) = -l_p(k_i)$ for a given p in (3-47)) or create a systematic contribution. So the last pair (independent frequency, dependent frequency) should be formed over two different rows (connected to k_i, k_j, $i \neq j$). Because the constraints (3-47) are active only row by row (they combined frequencies of the same row), this creates an additional constraint, and, hence, the frequency of the last pair is fixed by this constraint. So the number of free pairs is an $O(N^{v_0})$ with $v_0 = (F_a - 2n + 2m - 1)/2$. Because each contribution in the sum of $G_S^{r_i}(k_i)$ is an $O(N^{v_i})$, with $v_i = (1 - r_i)/2$, it is clear that the expected value is bounded by

$$\left| \mathbb{E}\{G_S^{r_1}(k_1)G_S^{r_2}(k_2)...G_S^{r_n}(k_n)\} \right| \leq O(N^{-v})M_U^{-n}\prod_{i=1}^{n} M_{G^{r_i}}M_U^{r_i} \qquad (3\text{-}49)$$

with $v = -v_0 - \sum_{i=1}^{n} v_i = (n - 2m + 1)/2$ for n odd.

The bound in the results, (3-48) and (3-49), can be written as $O(N^{-\text{int}((n-2m+1)/2)})$, which proves the lemma. □

Theorem 3.24 (Moments Stochastic Nonlinear Contributions): Consider a system belonging to the set $\mathbb{S}$ (see Definition 3.5), excited with a random multisine $u_N \in \mathbb{E}_N$ (see Definition 3.2). The expected value $\mathbb{E}\{G_S(k_1)G_S(k_2)...G_S(k_n)\}$ is bounded by $\left| \mathbb{E}\{G_S(k_1)G_S(k_2)...G_S(k_n)\} \right| \leq O(N^{-v})M_U^{-n}C_1^n$, with $v = \text{int}((n - 2m + 1)/2)$.

Proof. The proof follows directly from Lemma 3.23, by the fact that v is independent of r_i, $i = 1, 2, ..., n$. Hence, Lemma 3.23 can be directly generalized to Theorem 3.24. □

Theorem 3.25 (Properties Stochastic Nonlinear Contributions): Consider a system belonging to the set $\mathbb{S}$ (Definition 3.5), excited with a random multisine $u_N \in \mathbb{E}_N$ (Definition 3.2). The stochastic nonlinear contributions $G_S(k)$ have the following properties:

1. $\mathbb{E}\{G_S(k)\overline{G}_S(l)\} = O(N^{-1})$ for $k \neq l$
2. $\mathbb{E}\{|G_S(k)|^2\} \equiv \sigma_S^2(k) = O(N^0)$
3. $\mathbb{E}\{G_S(k)|G_S(l)|^2\} = O(N^{-1})$
4. $\mathbb{E}\{(|G_S(k)|^2 - \sigma_S^2(k))(|G_S(l)|^2 - \sigma_S^2(l))\} = O(N^{-1})$ for $k \neq l$
5. $\mathbb{E}\{G_S(k)\overline{G}_S(k+m)\overline{G}_S(l)\overline{G}_S(l+m)\} = O(N^{-2})$ for $k \neq l, -k \neq l + m, -l \neq k + m$, $m \neq 0$ (all frequencies differ from each other)
6. $\mathbb{E}\{|G_S(k)|^2|G_S(k+m)|^2\} = O(N^0)$ for $m \neq 0$

Proof. The proof consists of a straightforward application of Theorem 3.24. Note that v is maximal if the number of paired numerator frequencies is maximized.

1. $\mathbb{E}\{G_S(k)\overline{G}_S(l)\} = G_S(k)G_S(-l) = O(N^{-1})$ for $k \neq l$.
   $m = 0$, $n = 2$ hence $v = \text{int}((n - 2m + 1)/2) = \text{int}((2 - 0 + 1)/2) = 1$.
2. $\mathbb{E}\{|G_S(k)|^2\} = \mathbb{E}\{G_S(k)G_S(-k)\} = O(N^0)$.
   $m = 1$, $n = 2$ hence $v = \text{int}((n - 2m + 1)/2) = \text{int}((2 - 2 + 1)/2) = 0$.
3. $\mathbb{E}\{G_S(k)|G_S(l)|^2\} = \mathbb{E}\{G_S(k)G_S(l)G_S(-l)\} = O(N^{-1})$.
   $m = 1$, $n = 3$ hence $v = \text{int}((n - 2m + 1)/2) = \text{int}((3 - 2 + 1)/2) = 1$.
4. $\mathbb{E}\{(|G_S(k)|^2 - \sigma_S^2(k))(|G_S(l)|^2 - \sigma_S^2(l))\} = O(N^{-1})$ for $k \neq l$.

Here, some precautions have to be taken. In order to simplify the proof, the expected value is rewritten as

$$\mathbb{E}\{(|G_S(k)|^2 - \sigma_S^2(k))(|G_S(l)|^2 - \sigma_S^2(l))\} = \mathbb{E}\{|G_S(k)|^2|G_S(l)|^2\} - \sigma_S^2(k)\sigma_S^2(l) \qquad (3\text{-}50)$$

We study the first term in the right-hand side of (3-50)

$$\mathbb{E}\{|G_S(k)|^2|G_S(l)|^2\} = \mathbb{E}\{G_S(k)G_S(-k)G_S(l)G_S(-l)\}$$

Here, two disjoint situations can be considered. In the first situation (A), all denominator frequencies are paired $(k, -k)$ and $(l, -l)$ so that $v = \text{int}((4-4+1)/2) = 0$, while in the second situation (B), at least one of the pairs $(k, -k)$ or $(l, -l)$ is not created in the pairing process so that $m \leq 1$, and $v = 1$. The expected value can be split over these two types of contributions.

$$\mathbb{E}\{|G_S(k)|^2|G_S(l)|^2\} = \mathbb{E}\{|G_S(k)|^2|G_S(l)|^2\}_A + \mathbb{E}\{|G_S(k)|^2|G_S(l)|^2\}_B \tag{3-51}$$

(i) First, situation (A) is studied. Again two possibilities exist: (1) some pairs link the k-lines to the l-lines; (2) no such links appear.

First we deal with possibility (1): From claims 2 and 3 in Appendix 3.E, it follows that such combinations are an $O(N^{-1})$, so these terms do not act as the dominating contributions. Possibility (2): Here the k-lines are not lined to the l-lines. Because the combinations no longer depend on the phase (sum of the phases is zero), they are deterministic contributions and, hence,

$$\begin{aligned}\mathbb{E}\{|G_S(k)|^2|G_S(l)|^2\}_A &= \mathbb{E}\{|G_S(k)|^2\}\mathbb{E}\{|G_S(k)|^2\} + O(N^{-1}) \\ &= \sigma_S^2(k)\sigma_S^2(l) + O(N^{-1})\end{aligned} \tag{3-52}$$

Clearly (3-52) cancels the second term in (3-50).

(ii) In set (B), we have that $v = 1$, and, hence, it has again an $O(N^{-1})$ contribution to (3-50).

We conclude that (3-50) is an $O(N^{-1})$.

5. The proofs of 5 and 6 are completely similar to any one of the previously studied situations.

Appendix 3.C Mixing Property of the Stochastic Nonlinear Contributions

In this appendix we assume a deterministic amplitude and a uniform continuous phase distribution. The random amplitude, the discrete phase, and the nonuniform continuous phase distributions are commented on in Appendix 3.F.

In this appendix, the proof of Theorem 3.10 is given: Consider a system belonging to the set $\mathbb{S}$ (Definition 3.5), excited with a random multisine $u_N \in \mathbb{E}_N$ (Definition 3.2). The (stochastic) nonlinear contributions $G_S(k)$ are mixing of arbitrary order n.

Proof. We prove the mixing property for the nonlinear contributions $G_B(k) + G_S(k)$. Because $G_B(k)$ is deterministic, the mixing property of $G_S(k)$ follows immediately. We show that $G^{r_1}(k_1)G^{r_2}(k_2)...G^{r_n}(k_n)$ are mixing, for an arbitrary n. The theorem follows then from Definition 3.5 and the linearity property of mixing variables (Lemma 16.4). $G^{r_1}(k_1)G^{r_2}(k_2)...G^{r_n}(k_n)$ is mixing if

$$\max_{k_n} \sum_{k_1, k_2, ..., k_{n-1} = -N/2+1}^{N/2-1} |\text{cum}(G^{r_1}(k_1), G^{r_2}(k_2), ..., G^{r_{n-1}}(k_{n-1}), G^{r_n}(k_n))| \leq c_n < \infty \tag{3-53}$$

for any N, infinity included, with c_n independent of N. Using Lemma 16.4 and Definition 3.5, it turns out that it is sufficient to prove that

$$\max_{k_n} \sum_{k_1,\ldots,k_{n-1}=-N/2+1}^{N/2-1} \sum_{l_i(k_i)=-N/2+1}^{N/2-1} \left| \text{cum}(U_{k_1}^{-1} \prod_{i=1}^{r_1} U_{l_i(k_1)}, \ldots, U_{k_n}^{-1} \prod_{i=1}^{r_n} U_{l_i(k_n)}) \right| \leq c_n < \infty \quad (3\text{-}54)$$

for any N, infinity included, with c_n independent of N. In this expression $\sum_{l_i(k_i)=-N/2+1}^{N/2-1}$ stands for the sum over all numerator frequencies $l_1(k_1)$, $l_2(k_1)$, ..., $l_{r_1}(k_1)$, $l_1(k_2)$, ..., $l_{r_2}(k_2)$, ..., $l_{r_n}(k_n)$ (see Appendix 3.B) appearing in $G^{r_1}(k_1)G^{r_2}(k_2)\ldots G^{r_n}(k_n)$. To calculate the cumulant we have to set up a table with all participating input Fourier coefficients (characterized by their frequency) and consider next all indecomposable sets in this table (see Appendix 16.A). The table is given by (see also 3-46)

$$\begin{aligned} -k_1 & \quad (l_1(k_1), l_2(k_1), \ldots, l_{r_1-1}(k_1), l_{r_1}(k_1)) \\ -k_2 & \quad (l_1(k_2), l_2(k_2), \ldots, l_{r_2-1}(k_2), l_{r_2}(k_2)) \\ & \quad \ldots \\ -k_n & \quad (l_1(k_n), l_2(k_n), \ldots, l_{r_n-1}(k_n), l_{r_n}(k_n)) \end{aligned} \quad (3\text{-}55)$$

with

$$k_i - \sum_{p=1}^{r_i-1} l_p(k_i) - l_{r_i}(k_i) = 0, \; i = 1, \ldots, n \quad (3\text{-}56)$$

All frequencies but one (k_n) appear as a summation index in (3-54). We will count again the number of nonzero cumulants over the indecomposable partitions that appear in the sum. To do so, we have to determine the maximum number of degrees of freedom, taking into account all restrictions that will appear. The following constraints will be considered:

(i) The $\text{cum}(U_{j_1}, U_{j_2}, \ldots, U_{j_s})$ is different from zero only if $|j_1| = |j_2| = \ldots = |j_s|$ and the terms are paired. Hence, only cumulants over sets with an even number of elements can be different from zero. The sum of the frequencies in such a set is zero.

(ii) All the row constraints (3-56) should be respected.

(iii) All frequencies are different from zero $j_i \neq 0$.

(iv) Only indecomposable partitions are considered.

The constraint (3-56) can also be written as

$$A_1 J_a = 0 \quad (3\text{-}57)$$

where J_a is a vector containing all frequencies that participate in (3-55). The entries of A_1 are 1, −1, or 0 depending on how the corresponding frequency in J_a contributes to the corresponding row. Note that the "indecomposability" property is completely preserved in A_1.

Partitioning. Consider an indecomposable partition of (3-55) and select the partitions that have nonzero cumulants. On each subset of such partition we can associate one frequency (see condition 1 above). All these frequencies are put in the vector J_p, and we replace the set of equations (3-57) by

$$AJ_p = 0 \qquad (3\text{-}58)$$

Some of the subsets will combine only frequencies belonging to one row. Because the sum over all these frequencies in such a subset is zero (see condition 1 above), their entry in A is zero. So only subsets that combine frequencies from different rows can have an entry in A that is different from zero. If such a subset (over different rows) has zero entries in A, it can be split in smaller subsets with nonzero entries (the partition remains indecomposable). This is a worst case situation because a smaller number of frequencies are linked to each other, and, hence, a larger number of free frequencies remains. For example,

$$\begin{bmatrix} l & -l \\ -l & l \end{bmatrix} \rightarrow \begin{bmatrix} l_1 \\ -l_1 \end{bmatrix} \begin{bmatrix} l_2 \\ -l_2 \end{bmatrix} \qquad (3\text{-}59)$$

With these replacements, the structure of A and A_1 is the same with respect to the indecomposable partitions: A is indecomposable $\Leftrightarrow$ A_1 is indecomposable. So from now on we focus completely on A.

Note that the entries corresponding to a given frequency in J_p appear at most in one column in A.

A can also have subsets with an odd number of entries (e.g., three). However, because each subset covers an even number of frequencies, such a subset corresponds to a subset in A_1 with an even number of entries, e.g.,

$$\begin{bmatrix} 1 \\ 1 \\ -2 \end{bmatrix} \text{ in } A \quad \leftrightarrow \quad \text{corresponds, for example, to } \begin{bmatrix} 1 \\ 1 \\ -1 & -1 \end{bmatrix} \text{ in } A_1 \qquad (3\text{-}60)$$

Such a set can always be broken into

$$\begin{bmatrix} 1 \\ -1 \end{bmatrix}\begin{bmatrix} 1 \\ -1 \end{bmatrix} \qquad (3\text{-}61)$$

without changing the indecomposable structure. Again, this is a worst case situation. So, we should consider only subsets with an even number of entries in A.

Indecomposable Partitions. Only the indecomposable partitions are considered. It is possible to select a submatrix in A, A_{ind}, that is indecomposable. After rearranging the order of the columns, A can be written as $A = [A_{\text{ind}} \; A_{\text{rest}}]$.

No Zero Frequencies. None of the frequencies in J_p (3-58) may be equal to zero. So every row in A should contain at least two entries that are different from zero, otherwise (3-58) forces at least one frequency to be zero. Hence, it is possible to form a matrix $\tilde{A}$ by extending A_{ind} with additional columns of A_{rest}, such that each row of $\tilde{A}$ contains at least two nonzero entries.

Structure of $\tilde{A}$. We study the structure of $\tilde{A}$ in more detail in Appendix 3.D, where it is shown that $\tilde{A}$ can always be rearranged (some columns might be shifted back to A_{rest}) to a matrix with $\sum_{k=1}^{k_{\max}} 2k N_{2k} \geq 2n$ entries grouped in $\sum_{k=1}^{k_{\max}} N_{2k}$ columns, and $\text{rank}(\tilde{A}) = \sum_{k=1}^{k_{\max}} N_{2k} - 1$ (N_{2k} is the number of subsets in $\tilde{A}$ with $2k$ elements). Because $\text{rank}(A) \geq \text{rank}(\tilde{A})$ it follows that at most one frequency can be freely chosen.

Number of Degrees of Freedom. J_p contains at most $F_a + n - 2$ free frequencies, with $F_a = \sum_{i=1}^{n} r_i$, because at least one frequency is paired with $-k_n$. Each entry of $\tilde{A}$ corresponds to at least one free frequency in (3-55), so $\sum_{k=1}^{k_{\max}} 2k N_{2k} \geq 2n$ frequencies of (3-47) are used in $\tilde{A}$ while at most one is free (see above). The maximum number of degrees of freedom (worst case) appears when all remaining free frequencies (in A_{rest}) $F_a + n - 2 - \sum_{k=1}^{k_{\max}} 2k N_{2k} \leq F_a + n - 2 - 2n$ are grouped in pairs. So the free number of frequencies (including the free one of $\tilde{A}$) is given by

$$F_{\text{free}} \leq (F_a - n - 2)/2 + 1 \leq (F_a - n)/2 \qquad (3\text{-}62)$$

Each of these frequencies can be freely chosen out of the $2(N/2 - 1)$ input frequencies. The number of degrees of freedom is thus an $O(N^{(F_a - n)/2})$.

Mixing. Because $U_l = O(N^{-1/2})$, each cumulant in the sum (3-54) is an $O(N^{(n - F_a)/2})$. Each cumulant in (3-54) is calculated as the sum over all indecomposable partitions of table (3-55), which reduces the number of free frequencies in the sums (3-54) to $(F_a - n)/2$ (see (3-62)). Hence, (3-54) is an $O(N^{(F_a - n)/2}) O(N^{(n - F_a)/2}) = O(N^0)$, which proves the theorem.

Appendix 3.D Structure of the Indecomposable Sets

In this appendix, we assume a deterministic amplitude and a uniform continuous phase distribution. The random amplitude, the discrete phase, and the nonuniform continuous phase distributions are commented on in Appendix 3.F.

The matrix $\tilde{A}$ contains an indecomposable set extended with additional columns such that each row contains at least two nonzero entries. These additional columns might create additional links between the rows so that it might be possible to "break" larger subsets to smaller ones, the smallest ones corresponding to pairs, while the number of degrees of freedom is not decreased (so the worst case is maintained). The breaking process can be continued until all subsets are reduced to pairs, or the remaining subset is an "essential" set $\mathbb{S}_e$ with $2k$ ($k \geq 2$) entries in $\tilde{A}$ that cannot be broken without losing the indecomposability of $\tilde{A}$. This leads to the following definition.

Definition 3.26: The subset $\mathbb{S}_e$ with $2k$ ($k \geq 2$) entries in $\tilde{A}$ is an essential subset if it is possible to define a partition on the entries of $\tilde{A}$, $\{\mathbb{S}_e, \{\tilde{A}_1\}, \ldots, \{\tilde{A}_{2k}\}\}$, where each of the subsets $\{\tilde{A}_i\}$ is indecomposable and linked to only one element of $\mathbb{S}_e$.

Lemma 3.27: Consider a subset $\mathbb{S}_i$ with $2k$ ($k \geq 2$) entries in $\tilde{A}$. Either it is possible to brake it into two subsets $\mathbb{S}_{i1}$ and $\mathbb{S}_{i2}$, without losing the indecomposability of $\tilde{A}$, or $\mathbb{S}_i$ is an essential subset.

Proof. The lemma follows directly from the definition. If $\mathbb{S}_i$ is not an essential subset, there is a partitioning in $\tilde{A}$, where at least one of the subsets is linked to two elements of $\mathbb{S}_i$. Hence, $\mathbb{S}_i$ can be broken into two parts, each containing one of these elements, without losing the indecomposability of the partition. □

After repetitively applying Lemma 3.27, the matrix $\tilde{A}$ is partitioned in pairs and essential subsets. Consider, for example, a situation with one essential subset:

$$\begin{bmatrix} x & x & & & & \\ x & & x & & & \\ x & & & & & \\ x & & & & & \\ & x & \{\tilde{A}_1\} & & & \\ & & \cdots & & & \\ & & & x & \{\tilde{A}_2\} & \\ & & & & \cdots & \end{bmatrix} \quad (3\text{-}63)$$

with x a nonzero entry in $\tilde{A}$, and $\{\tilde{A}_i\}$ indecomposable sets consisting of pairs. Hence, their structure can always be written as

$$\begin{bmatrix} x & & & & & \\ x & x & & & & y \\ & x & x & \cdots & & \\ & & x & & & \\ & & & \cdots & & \\ & & & & x & \\ & & & & x & x \end{bmatrix} \quad (3\text{-}64)$$

The entry y in the last column can appear at any of the rows but the last one. It is clear that the rank of this square matrix is the number of columns -1 because the sum of all entries in one column is zero. So only one frequency is free. This is the frequency that is linked to the essential set, so that no free frequency remains. This idea can be further extended to situations with multiple essential sets or no essential set (where one of the pairs can be considered as a special case of essential set). The conclusion is that the rank of $\tilde{A}$ is the number of columns -1.

Note. During the breaking process, additional dependent columns might appear. These are shifted back from $\tilde{A}$ to A_{rest}.

Lemma 3.28: The matrix $\tilde{A}$ can be reduced using the breaking and column-removing process to a matrix with $\text{rank}(\tilde{A}) = \sum_{k=1}^{k_{\max}} N_{2k} - 1$, with N_{2k} the number of sets with $2k$ entries.

Appendix 3.E   Distribution of the Stochastic Nonlinearities

In this appendix, we assume a deterministic amplitude and a uniform continuous phase distribution. The random amplitude, the discrete phase, and the nonuniform continuous phase distributions are commented on in Appendix 3.F.

In this appendix, the proof of Theorem 3.11 is given: for a system belonging to the system set $\mathbb{S}$, excited with a random multisine $u_N \in \mathbb{E}_N$, the stochastic nonlinearities are circular normally distributed. The frequency index k is sometimes omitted for notational simplicity.

The proof consists of the following two main steps.

- The Volterra series can be written as the sum of contributions up to degree M $(G_S)^+$ plus a rest term, $G_{\tilde{S}}$, which is an $O(\varepsilon)$.
- Each of the M terms is normally distributed, and their variances are an $O(\varepsilon^0)$. Also, the variance of G_S^+ is an $O(\varepsilon^0)$, while the variance of the rest term is an $O(\varepsilon^2)$. So, G_S converges in distribution to G_S^+, which is a finite sum of circular normally distributed variables. So, G_S^+ is also circular normally distributed.

We prove these results now step by step.

(i) $G_S(k) = G_S^+(k) + G_{\tilde{S}}(k)$, with $\sigma_{G^-}^2 = O(\varepsilon^2)$, $\sigma_{G^+}^2 = O(\varepsilon^0)$, and ε arbitrary small.

Proof. $G_S(k) = \sum_{\alpha=2}^{\infty} G_S^\alpha(k)$, with $\sum_{\alpha=1}^{\infty} M_{G^\alpha} M_U^\alpha \leq C_1 < \infty$ (Definition 3.5). So,

$$\forall \varepsilon, \exists M \text{ s.t. } \sum_{\alpha=M+1}^{\infty} M_{G^\alpha} M_U^\alpha < \varepsilon \Rightarrow |G_{\tilde{S}}(k)| = \left| \sum_{\alpha=M+1}^{\infty} G_S^\alpha(k) \right| = O(\varepsilon)$$

The variance of $G_{\tilde{S}}(k)$ can be bounded above by

$$\sigma_{G^-}^2(k) = \mathbb{E}\{G_{\tilde{S}} \overline{G_{\tilde{S}}}\} = \sum_{\alpha_1, \alpha_2 = M+1}^{\infty} \mathbb{E}\{G_S^{\alpha_1} \overline{G_S^{\alpha_2}}\} \leq \sum_{\alpha_1, \alpha_2 = M+1}^{\infty} \left| \mathbb{E}\{G_S^{\alpha_1} \overline{G_S^{\alpha_2}}\} \right| \quad (3\text{-}65)$$

From Lemma 3.23 ($n = 2$, $m = 1 \to \nu = 0$), it follows that

$$\left| \mathbb{E}\{G_S^{\alpha_1} \overline{G_S^{\alpha_2}}\} \right| \leq O(N^0) M_U^{-2} M_U^{\alpha_1 + \alpha_2} M_{G^{\alpha_1}} M_{G^{\alpha_2}} \quad (3\text{-}66)$$

Combining (3-65) and (3-66) gives

$$\sigma_{G^-}^2(k) \leq \frac{O(N^0)}{M_U^2} \left(\sum_{\alpha_1 = M+1}^{\infty} M_{G^{\alpha_1}} M_U^{\alpha_1} \right) \left(\sum_{\alpha_2 = M+1}^{\infty} M_{G^{\alpha_2}} M_U^{\alpha_2} \right) \leq O(\varepsilon^2) \quad (3\text{-}67)$$

Similarly, it is shown that $\sigma_{G^+}^2(k) = O(\varepsilon^0)$.

(ii) Study of the odd moments $\mathbb{E}\{(G_S^\alpha(k))^{2p+1}\}$

Using Lemma 3.23, with $n = 2p+1$, $m = p$, it follows that $\nu = 1$. Hence,

$$\mathbb{E}\{(G_S^\alpha(k))^{2p+1}\} = O(N^{-1}) \quad (3\text{-}68)$$

(iii) Study of the even moments $\mathbb{E}\{|G_S^\alpha(k)|^{2p}\}$

We use again the notation of Appendix 3.C. In (3-55) we put $k_{2i-1} = k$ ($i = 1, \ldots, p$, and $k_{2i} = -k$, and define the set of equations

$$K = BJ_p, \text{ with } K = (k, -k, \ldots, k, -k)^T \quad (3\text{-}69)$$

From Appendix 3.C, we know that the worst case (maximum number of combinations) is given if the denominator frequencies are paired with each other, because this leaves the largest number of frequencies free. Hence, the numerator frequencies should be partitioned s.t. the phases cancel each other. Just as in Appendix 3.C, the subsets can each time be restricted to depend on only one frequency (otherwise they can be broken into smaller subsets without changing their contribution). Next we prove a number of additional properties on the grouping process.

Claim 1: *Partitions that contain subsets linking more than two rows in (3-58) give only $O(N^{-v})$, $v \geq 1$ contributions.*

Proof. Consider the set of equations (3-58). Each row in B has more than one entry different from zero, because otherwise it would be a systematic contribution instead of a stochastic one (all frequencies but one are paired). So there is a submatrix $\tilde{B}$ in B, after rearranging the columns, that contains at least $4p$ entries. Using the definitions of Appendix 3.C, the number of entries in $\tilde{B}$ is $\sum_{k=1}^{k_{max}} 2kN_{2k} \geq 4p$, and the number of columns (set frequencies) is $\sum_{k=1}^{k_{max}} N_{2k}$, where, for the same reason as explained in (3-60), only subsets with an even number of entries are considered. So after pairing, the total number of independent frequencies is

$$\frac{(F_a - 2p)_1 - (\sum_{k=1}^{k_{max}} 2kN_{2k})_2}{2} + (\sum_{k=1}^{k_{max}} N_{2k})_3 = \frac{F_a - 2p}{2} - \sum_{k=1}^{k_{max}} (k-1) N_{2k} \qquad (3\text{-}70)$$

with $(\ )_1$ the total number of independent frequencies after imposing the row constraints (3-47), $(\ )_2$ the number of entries used in $\tilde{B}$, and $(\ )_3$ the number of set frequencies in $\tilde{B}$. Each of these combinations is an $O(N^{(2p-F_a)/2})$. If $\exists\, k > 1$ s.t. $N_{2k} \neq 0$, then the second part in (3-70) is negative, and consequently the claim is proved. $\square$

Conclusion. Only pairs should be considered.

Claim 2: *Partitions, using pairs as subsets, that link more than two rows $(k, -k)$ in (3-58) give only $O(N^{-v})$, $v \geq 1$ contributions.*

Proof. For such a partition, keeping in mind that each row should contain at least two entries different from zero, B should contain at least the following submatrix $\tilde{B}$:

$$\begin{array}{c} k \\ -k \\ k \\ -k \end{array} \begin{bmatrix} 0 & x & 0 & x & 0 \\ x & x & 0 & x & 0 \\ x & 0 & x & 0 & x \\ 0 & 0 & x & 0 & x \end{bmatrix} \qquad (3\text{-}71)$$

where $x = \pm 1$. It is clear that $\tilde{B}$, consisting of q columns, has rank 3 and uses $2q$ entries. Assuming that the row conditions for the corresponding lines are automatically met, we get that the number of free frequencies in $\tilde{B}$ is $q - 3$. The remaining $2p - 4$ row conditions should still be met, so that there are $2p - 4$ dependent variables. Hence, the number of free pairs is

$$(F_a - (2p - 4) - 2q)/2 + q - 3 = (F_a - 2p)/2 - 1 \qquad (3\text{-}72)$$

Because $U_l = O(N^{-1/2})$, each term in the sums of $\mathbb{E}\{|G_S^\alpha(k)|^{2p}\}$ is an $O(N^{-(F_a-n)/2})$. The number of free summation variables in $\mathbb{E}\{|G_S^\alpha(k)|^{2p}\}$ is given by (3-72). Hence,

$$\mathbb{E}\{|G_S^\alpha(k)|^{2p}\} = O(N^{(F_a-2p)/2-1})O(N^{-(F_a-n)/2}) = O(N^{-1})$$

since $n = 2p$. □

Claim 3: *Partitions that link pairs of rows (k, l), $l \neq -k$ in (3-58) give only $O(N^{-\nu})$, $\nu \geq 1$ contributions.*

Proof. For such a partition, keeping in mind that each row should contain at least two entries different from zero, B should contain at least the following submatrix $\tilde{B}$:

$$k \begin{bmatrix} 1 & 1 \\ -1 & -1 \end{bmatrix} \quad \text{or} \quad k \begin{bmatrix} 1 & -1 \\ -1 & 1 \end{bmatrix} \quad \text{or similar} \tag{3-73}$$

Because the rank of $\tilde{B}$ is 1, and the rank of the augmented matrix

$$\begin{bmatrix} k & \tilde{B} \\ l & \end{bmatrix} \tag{3-74}$$

is 2, this set has no solution. Hence, at least an additional link with another row is needed to increase the rank of $\tilde{B}$ to 2. Claim 3 then follows from the previous Claim 2. Note that pairing (k, k) is a special case of this claim. □

Claim 4: *Partitions that contain rows that are not linked to another row do not exist.*

Proof. Because each row corresponds to a stochastic contribution, it is clear that not all the frequencies in one row can be paired within this row.

From Claims 1 to 4, it follows that the only contributions of $O(N^0)$ to $\mathbb{E}\{|G_S^\alpha(k)|^{2p}\}$ are those where the partitions link all the rows per two with the denominator frequencies of the form $(k, -k)$. $\mathbb{E}\{|G_S^\alpha(k)|^{2p}\}$ is given, within an $O(N^{-1})$, by the sum of all these contributions

$$\mathbb{E}\{|G_S^\alpha(k)|^{2p}\} = \sum_{\text{all distinct combinations of pairs}} \mathbb{E}\{G_S^\alpha(k)G_S^\alpha(-k)\} \ldots \mathbb{E}\{G_S^\alpha(k)G_S^\alpha(-k)\} \tag{3-75}$$

In this expression "all distinct combinations of pairs" indicates all permutations that can be formed over the rows (3-69) such that distinct products of pairs $(k, -k)$ are formed. For example, if we have four rows $(1, 2, 3, 4)$ with frequencies $k, -k, k, -k$, we should consider $(1, 2)(3, 4)$; $(1, 4)(2, 3)$. The combination $(1, 3)(2, 4)$ forms pairs (k, k) and does not contribute. From Picinbono (1993, p. 112, Eq. (4.95)) it follows that this corresponds to the moments of a circular, normal distribution. As convergence in the moments implies convergence in distribution (see Lemma 16.11), it follows that $G_S^\alpha(k)$ (3-41) is asymptotically $(N \to \infty)$ normally distributed.

The moments are the coefficients in the Taylor series expansion of the characteristic function $\phi(t)$ (Stuart and Ord, 1987). Since the convergence rate of the odd (3-68) even (3-75) moments is an $O(N^{-1})$, it follows that $\phi(t)$ corresponding to $G_S^\alpha(k)$ (3-41) equals that of a normal random variable within an $O(N^{-1})$, uniformly in t. Because the characteristic function is related to the probability density function by the Fourier integral, it follows that the distribution function $F_N(y)$ of $G_S^\alpha(k)$ equals that of a normal random variable within an $O(N^{-1})$, uniformly in y. □

Appendix 3.F Extension to Random Amplitudes and Nonuniform Phases

Because the random amplitude has uniformly bounded moments of any order and is independently distributed of the phase, we can calculate the expected value w.r.t. the phase, independently of the amplitude distribution. Hence, all previous proofs in Appendix 3.A to 3.E remain valid for random amplitudes.

The basic reason that a discrete phase or nonuniform continuous distribution needs special attention is that $\mathbb{E}\{U_k^2\}$ can be different from zero, e.g., $\varphi_k \in \{0, \pi\}$. However, a careful check shows that all previous proofs in Appendix 3.A to 3.E remain valid if (l, l) is also considered as a paired frequency. Notice that for such a pair the sum of the frequencies is no longer zero (no major impact on the proofs). A second difference is the fact that such a pair is represented by one element in the A and B matrices, but notice that there are still two frequencies linked to this single element. The $\varphi_k \in \{0, \pi\}$ distribution is a worst case. Discrete distributions with more elements link more frequencies to generate a nonzero expected value.

The major difference is the expected value $\mathbb{E}\{G^\alpha\}$. Additional $O(N^{-1})$ terms appear, also for the even nonlinearities.

A typical odd degree bias contribution for a discrete phase distribution $\varphi \in \{0, \pi\}$ would be $-k \quad l_1, l_1, ..., l_e, l_e, m_1, -m_1, ..., m_o, -m_o, k$. It is important to notice that $\sum_{i=1}^{e} l_i = 0$ in order to meet the frequency constraint and, hence, an additional frequency constraint becomes active. Using arguments similar to those in the previous appendices, the sum of all these contributions is an $O(N^{-1})$.

An example of an even degree systematic for a nonuniform phase contribution is $-k \quad l_1, l_1, l_1, l_2, l_2, k$ with $3l_1 + 2l_2$. Note that in this case at least three frequencies are linked in one "pair" so that an $O(N^{-3/2})$ results.

Appendix 3.G Response of a Nonlinear System to a Gaussian Excitation

For noise excitations, the FRF is measured using the H_1 method (2-51), and its limit value is given by

$$G^\alpha(j\omega) = \frac{\mathbb{E}\{Y^\alpha(j\omega)\overline{U}(j\omega)\}}{\mathbb{E}\{U(j\omega)\overline{U}(j\omega)\}} = \frac{S_{yu}(j\omega)}{S_{uu}(j\omega)} \quad (3\text{-}76)$$

The cross-spectrum $S_{yu}(j\omega)$ is the Fourier transform of the cross-correlation $R_{yu}(\tau)$ between the input and the output and depends on higher order spectra. In the case of zero mean normal distributed noise, these higher order spectra can easily be calculated. Consider the contribution of degree α:

110 Chapter 3 ■ Frequency Response Function Measurements in the Presence of Nonlinear Distortions

$$y^\alpha(t) = \int_{-\infty}^{+\infty} \cdots \int_{-\infty}^{+\infty} g_\alpha(\tau_0, ..., \tau_{\alpha-1}) u(t-\tau_0)...u(t-\tau_{\alpha-1}) d\tau_1...d\tau_\alpha$$

$$R_{y^\alpha u}(\tau_0) = \mathbb{E}\{y^\alpha(t)u(t-\tau_0)\} \quad (3\text{-}77)$$

$$= \int_{-\infty}^{+\infty} \cdots \int_{-\infty}^{+\infty} g_\alpha(\tau_1, ..., \tau_\alpha) \mathbb{E}\{u(t-\tau_0)u(t-\tau_1)...u(t-\tau_\alpha)\} d\tau_1...d\tau_\alpha$$

For zero mean jointly normally distributed noise, the higher order moments are given by (Schetzen, 1980, p. 218):

$$\mathbb{E}\{u_1 u_2...u_M\} = \begin{cases} 0 & \text{if M is odd} \\ \Sigma\Pi \mathbb{E}\{u_i u_j\} & \text{if M is even} \end{cases} \quad (3\text{-}78)$$

The $\Sigma\Pi$ stands for the summation over all distinct ways of partitioning the M random variables into products of averages of pairs. It is shown that there are $(M-1)!!$ such combination for M even (Schetzen, 1980) and zero if M is odd. Hence, $R_{y^{2\alpha}u}(\tau_0) = 0$ and

$$G_B^{2\alpha}(j\omega) = \frac{S_{y^{2\alpha}u}(j\omega)}{S_{uu}(j\omega)} = \frac{F\{R_{y^{2\alpha}u}(\tau_0)\}}{S_{uu}(j\omega)} = 0$$

From here on, it is assumed that α is odd so that an even number of input terms appear.

Using (3-78), the expected value in (3-77) becomes

$$\mathbb{E}\{u(t-\tau_0)u(t-\tau_1)...u(t-\tau_\alpha)\} = \Sigma\Pi R_{uu}(\tau_i - \tau_j) \quad (3\text{-}79)$$

Using the relationship between the autocorrelation and the power spectrum of the input,

$$R_{uu}(\tau) = \int_{-\infty}^{+\infty} S_{uu}(j\omega) e^{j\omega\tau} df \quad (3\text{-}80)$$

Eq. (3-77) can be rewritten as

$$R_{y^{2\alpha-1}u}(\tau_0) =$$

$$\int_{-\infty}^{+\infty} \cdots \int_{-\infty}^{+\infty} g_{2\alpha-1}(\tau_1, ..., \tau_{2\alpha-1}) \Sigma\Pi S_{uu}(j\omega_r) e^{j\omega_r(\tau_i - \tau_j)} d\tau_1...d\tau_{2\alpha-1} df_1...df_\alpha \quad (3\text{-}81)$$

In order to calculate this expression, the contribution of one term of $\Sigma\Pi$ is analyzed in detail for the partition $(\tau_0, \tau_1), (\tau_2, \tau_3), ..., (\tau_{2\alpha-2}, \tau_{2\alpha-1})$:

$$\int_{-\infty}^{+\infty} \cdots \int_{-\infty}^{+\infty} g_{2\alpha-1}(\tau_1, ..., \tau_{2\alpha-1}) \prod_{r=1}^{\alpha} S_{uu}(j\omega_r) e^{j\omega_r(\tau_{2r-2} - \tau_{2r-1})} d\tau_1...d\tau_{2\alpha-1} df_1...df_\alpha$$

Define

$$G_{f_1,-f_2,f_2,...,-f_\alpha f_\alpha}^{2\alpha-1} = \int_{-\infty}^{+\infty} \cdots \int_{-\infty}^{+\infty} g_{2\alpha-1}(\tau_1, ..., \tau_{2\alpha-1}) e^{-j\omega_1 \tau_1} \prod_{r=2}^{\alpha} e^{j\omega_r(\tau_{2r-2} - \tau_{2r-1})} d\tau_1...d\tau_{2\alpha-1}$$

Section 3.11 ■ Appendixes

Because $G^{2\alpha-1}_{f_1,-f_2,f_2,...,-f_\alpha,f_\alpha}$ is a symmetrical kernel, it does not depend on the order of its arguments. So, all possible terms in the partitioning give the same result, thus (3-81) becomes

$$R_{y^{2\alpha-1}u}(\tau_0) = (2\alpha-1)!!\int_{-\infty}^{\infty}...\int_{-\infty}^{\infty} G^{2\alpha-1}_{f_1,-f_2,f_2,...,-f_\alpha,f_\alpha} \prod_{r=2}^{\alpha} S_{uu}(j\omega_r)S_{uu}(j\omega_1)e^{j\omega_1\tau_0}df_1...df_\alpha \quad (3\text{-}82)$$

Note that the power spectrum of $R_{y^{2\alpha-1}u}(\tau_0)$ is given by

$$S_{y^{2\alpha-1}u}(j\omega) = \int_{-\infty}^{\infty} R_{y^{2\alpha-1}u}(\tau_0)e^{-j\omega\tau_0}d\tau_0 \quad (3\text{-}83)$$

Applying (3-83) to (3-82) results in

$$S_{y^{2\alpha-1}u}(j\omega) = S_{uu}(j\omega)(2\alpha-1)!!\int_{-\infty}^{\infty}...\int_{-\infty}^{\infty} G^{2\alpha-1}_{f,-f_2,f_2,...,-f_\alpha,f_\alpha}S_{uu}(j\omega_2)...S_{uu}(j\omega_\alpha)df_2...df_\alpha$$

Dividing $S_{y^{2\alpha-1}u}(j\omega)$ by $S_{uu}(j\omega)$ gives $G^{2\alpha-1}_B(j\omega)$:

$$G^{2\alpha-1}_B(j\omega) = \frac{S_{y^{2\alpha-1}u}(j\omega)}{S_{uu}(j\omega)}$$

$$= (2\alpha-1)!!\int_{-\infty}^{\infty}...\int_{-\infty}^{\infty} G^{2\alpha-1}_{f,-f_2,f_2,...,-f_\alpha,f_\alpha}S_{uu}(j\omega_2)...S_{uu}(j\omega_\alpha)df_2...df_\alpha$$

$$= (2\alpha-1)!!2^{\alpha-1}\int_{0}^{\infty}...\int_{0}^{\infty} G^{2\alpha-1}_{f,-f_2,f_2,...,-f_\alpha,f_\alpha}S_{uu}(j\omega_2)...S_{uu}(j\omega_\alpha)df_2...df_\alpha$$

where in the last step the double-sided spectra are replaced by single-sided spectra.

Appendix 3.H Proof of Theorem 3.12

Note: In this appendix, we denote explicitly the dependence of the results on the number of frequencies $F = N/2 - 1$ by adding a subscript N.

We elaborate the first term in the right-hand side of (3-13):

$$c_\alpha \sum_{k_1,...,k_{\alpha-1}=1}^{N/2-1} G^{2\alpha-1}_{k,-k_1,k_1,...,-k_{\alpha-1},k_{\alpha-1}} \mathbb{E}\{|U_{k_1}|^2...|U_{k_{\alpha-1}}|^2\} \quad (3\text{-}84)$$

Splitting the sums in (3-84) as $\sum_{k_i} = \sum_{\text{all } k_i \text{ different}} + \sum_{\text{not all } k_i \text{ different}}$ and using

$$\mathbb{E}\{|U_{k_1}|^2...|U_{k_{\alpha-1}}|^2\} = \prod_{i=1}^{\alpha-1}\mathbb{E}\{|U_{k_i}|^2\} \text{ for all } k_i\text{'s different} \quad (3\text{-}85)$$

makes it possible to rewrite (3-84) as

$$c_\alpha \left(\sum_{\text{all } k_i \text{ different}} G^{2\alpha-1}_{k,-k_1,k_1,\ldots,-k_{\alpha-1},k_{\alpha-1}} \prod_{i=1}^{\alpha-1} |U_{k_i}|^2 \right.$$

$$\left. + \sum_{\text{not all } k_i \text{ different}} G^{2\alpha-1}_{k,-k_1,k_1,\ldots,-k_{\alpha-1},k_{\alpha-1}} \mathbb{E}\{|U_{k_1}|^2 \cdots |U_{k_{\alpha-1}}|^2\} \right) \tag{3-86}$$

Adding and subtracting $\sum_{\text{not all } k_i \text{ different}}$ in the first summation of (3-86) gives, using $|U_{k_i}|^2 = S_{\hat{U}\hat{U}}(f_{k_i})/N$,

$$C^\alpha_{1N} + C^\alpha_{2N} \tag{3-87}$$

where

$$C^\alpha_{1N} = \frac{c_\alpha}{N^{\alpha-1}} \sum_{k_1,\ldots,k_{\alpha-1}=1}^{N/2-1} G^{2\alpha-1}_{k,-k_1,k_1,\ldots,-k_{\alpha-1},k_{\alpha-1}} \prod_{i=1}^{\alpha-1} S_{\hat{U}\hat{U}}(f_{k_i})$$

$$C^\alpha_{2N} = c_\alpha \sum_{\text{not all } k_i \text{ different}} G^{2\alpha-1}_{k,-k_1,k_1,\ldots,-k_{\alpha-1},k_{\alpha-1}} \Delta_{\alpha-1} \tag{3-88}$$

$$\Delta_{\alpha-1} = \mathbb{E}\{|U_{k_1}|^2 \cdots |U_{k_{\alpha-1}}|^2\} - \prod_{i=1}^{\alpha-1} |U_{k_i}|^2$$

Because $S_{\hat{U}\hat{U}}(f_{k_i})$ and $|G^{2\alpha-1}_{k,-k_1,k_1,\ldots,-k_{\alpha-1},k_{\alpha-1}}|$ are uniformly bounded (see Definitions 3.2 to 3.4), $|U_{k_i}|^2 = O(N^{-1})$ (see Definitions 3.2 and 3.3), and the sum $\sum_{\text{not all } k_i \text{ different}}$ contains at most $\alpha-2$ independent k_i's, we find

$$|C^\alpha_{1N}| \le \frac{c_\alpha}{N^{\alpha-1}} O(N^{\alpha-1}) = O(N^0) \quad \text{and} \quad |C^\alpha_{2N}| \le \frac{c_\alpha}{N^{\alpha-1}} O(N^{\alpha-2}) = O(N^{-1}) \tag{3-89}$$

Collecting (3-12), (3-13), (3-87), and (3-89), we get

$$G_{B,N}(s_k) = \sum_{\alpha=2}^\infty G^{2\alpha-1}_{B,N}(s_k) = \sum_{\alpha=2}^\infty C^\alpha_{1N} + O(N^{-1}) \tag{3-90}$$

Because $S_{\hat{U}\hat{U}}(f_{k_i})$ is by assumption the same for the three classes of excitation signals, it follows from (3-88) and (3-90) that for these three classes $G_{B,N}(s_k)$ converges ($N \to \infty$) at the rate $O(N^{-1})$ to the same limit value $G_B(s_k)$. Under some additional assumptions on the odd degree kernels it is possible to calculate an explicit expression for $G_B(s_k)$.

Because the joint second-order derivatives of

$$G^{2\alpha-1}_{k,-k_1,k_1,\ldots,-k_{\alpha-1},k_{\alpha-1}} = G^{2\alpha-1}_{f_k,-f_{k_1},f_{k_1},\ldots,-f_{k_{\alpha-1}},f_{k_{\alpha-1}}}$$

w.r.t. $f_{k_1}, f_{k_2}, \ldots f_{k_{\alpha-1}}$ and f_k are bounded for $f_{k_1}, f_{k_2}, \ldots, f_{k_{\alpha-1}}, f_k \in [0, f_s/2]$, the Riemann sum

$$C^\alpha_{1N} = \frac{c_\alpha}{N^{\alpha-1}} \sum_{k_1,\ldots,k_{\alpha-1}=1}^{N/2-1} G^{2\alpha-1}_{k,-k_1,k_1,\ldots,-k_{\alpha-1},k_{\alpha-1}} \prod_{i=1}^{\alpha-1} S_{\hat{U}\hat{U}}(f_{k_i}) \tag{3-91}$$

Section 3.11 ■ Appendixes

where $f_{k_i} - f_{k_{i-1}} = f_s/N$, converges to C_1^α

$$C_1^\alpha = \frac{c_\alpha}{f_s^{\alpha-1}} \int_0^{f_s/2} \cdots \int_0^{f_s/2} G_{f_k, -f_{k_1}, f_{k_1}, \ldots, f_{k_{\alpha-1}}}^{2\alpha-1} S_{\hat{U}\hat{U}}(f_1) \ldots S_{\hat{U}\hat{U}}(f_{\alpha-1}) df_1 \ldots df_{\alpha-1}$$

at the rate $O(N^{-2})$ (Ralston and Rabinowitz, 1984; midpoint rule (4.10-10)). Together with (3-90) and $S_{\hat{U}\hat{U}}(f) = S_{uu}(j\omega)f_s$ it shows that $\lim_{N \to \infty} G_{\text{BLA},N}(j\omega) = G_{\text{BLA}}(j\omega)$, with $G_{\text{BLA}}(j\omega)$ defined in (3-18).

Appendix 3.I   Proof of Theorem 3.15

The sum of a uniformly convergent series of continuous functions is continuous (see Kaplan, 1993, Theorem 31). Hence, under Assumptions 3.13 and 3.14, the sum $G_B(j\omega) = \sum_{\alpha=2}^\infty C_1^\alpha(j\omega)$, and its derivatives of order $1, 2, \ldots, P$ w.r.t. ω, are continuous functions of $\omega \in [0, \omega_s/2]$.

Appendix 3.J   Proof of Theorem 3.16

From Theorem 3.8 it follows that the input-output DFT spectra are related to the best linear approximation $G_{\text{BLA}}(s)$ by (3-23) where $Y_S(k) = G_S(j\omega_k)U(k)$ with $G_S(j\omega_k)$ the stochastic contributions to the transfer function. The correlation between $Y_S(k) = G_S(j\omega_k)U(k)$ and $U(k)$ can be elaborated as

$$\mathbb{E}\{Y_S(k)\overline{U}(k)\} = \mathbb{E}\{G_S(j\omega_k)|U(k)|^2\} = \sum_{\alpha=2}^\infty \mathbb{E}\{G_S^\alpha(j\omega_k)|U(k)|^2\} = 0 \quad (3-92)$$

with $G_S^\alpha(j\omega_k)$ defined in (3-41), and where the last equality uses the property that $|U(k)|$ is independently distributed of the phases of the Fourier coefficients (see Definitions 3.2 and 3.3), and the fact that $\mathbb{E}\{G_S^\alpha(j\omega_k)\} = 0$ is due to the random phase behavior of each term in the sum that defines $G_S^\alpha(j\omega_k)$.

Equations (3-93) and (3-94) of Appendix 3.K give an explicit expression for the stochastic nonlinear contribution $Y_S(k) = \sqrt{N}Y_{Sk}$. Since the sums in $\sqrt{N}Y_{Sk}^\alpha$ (3-93) are subject to the same frequency and phase constraints as those of $G_S^\alpha(j\omega_k)$ in (3-39), the proofs in Appendices 3.A-3.E can be redone for $\sqrt{N}Y_{Sk}^\alpha$, showing that the results remain valid.

Appendix 3.K   Proof of Theorem 3.17

Note: In this appendix, we denote explicitly the dependence of the results on the number of frequencies $F = N/2 - 1$ by adding a subscript N.

From (3-9), (3-10), and (3-22), it follows that the stochastic nonlinear contributions $Y_{Sk,N}$ are given by

$$\sqrt{N}Y_{Sk,N} = \sum_{\alpha=2}^\infty \sqrt{N}Y_{Sk,N}^\alpha$$

$$\sqrt{N}Y_{Sk,N}^\alpha = \sqrt{N} \sum_{k_1, \ldots, k_{\alpha-1} = -N/2+1}^{N/2-1} G_{k_1, \ldots, k_\alpha}^\alpha U_{k_1} \ldots U_{k_\alpha} \quad (3-93)$$

with constraints

$$k = \sum_{i=1}^{\alpha} k_i, \; \sum_{i=1}^{\alpha} \varphi_{k_i} \neq \varphi_k, \; k \neq 0, \text{ and } k_i \neq 0 \text{ for } i = 1, \ldots, \alpha \tag{3-94}$$

and where $G_{k_1, \ldots, k_\alpha}^{\alpha} = G_{f_{k_1}, \ldots, f_{k_\alpha}}^{\alpha}$ with $f_{k_i} = k_i f_s/N$. The variance of $\sqrt{N} Y_{Sk, N}$ equals

$$\text{var}(\sqrt{N} Y_{Sk, N}) = N \sum_{\alpha, \beta = 2}^{\infty} \mathbb{E}\{Y_{Sk, N}^{\alpha} \bar{Y}_{Sk, N}^{\beta}\} = \sum_{\alpha, \beta = 2}^{\infty} C_N^{\alpha, \beta} \tag{3-95}$$

with

$$C_N^{\alpha, \beta} = N \sum_{\substack{k_1, \ldots, k_{\alpha-1} = -N/2+1 \\ l_1, \ldots, l_{\beta-1} = -N/2+1}}^{N/2-1} G_{k_1, \ldots, k_\alpha}^{\alpha} \bar{G}_{l_1, \ldots, l_\beta}^{\beta} \mathbb{E}\{U_{k_1} \ldots U_{k_\alpha} \bar{U}_{l_1} \ldots \bar{U}_{l_\beta}\} \tag{3-96}$$

Because $U_k = N^{-1/2} \hat{U}(f_k) e^{j\varphi_k}$ with $f_k = kf_s/N$, $\mathbb{E}\{e^{j\varphi_k}\} = 0$ and φ_k independent of $\hat{U}(f_k)$, it follows that

$$\mathbb{E}\{U_{k_1} \ldots U_{k_\alpha} \bar{U}_{l_1} \ldots \bar{U}_{l_\beta}\} \neq 0 \Leftrightarrow \sum_{i=1}^{\alpha} \varphi_{k_i} = \sum_{i=1}^{\beta} \varphi_{l_i} \tag{3-97}$$

Taking into account the constraints (3-94), the phase condition in (3-97) can be met only if the frequencies are paired as $(m_j, -m_j)$ with $m_j \in \{k_1, \ldots, k_\alpha, -l_1, \ldots, -l_\beta\}$ and where not all m_j should be different. The maximum number of terms in the sums (3-96) is obtained by maximizing the number of independent m_j's (number of independent pairs). Because $|U_k|^2 = S_{\hat{U}\hat{U}}(f_k)/N$ and the maximum number of independent pairs equals $\gamma = (\alpha + \beta)/2 - 1$, (3-96) can be written as

$$C_N^{\alpha, \beta} = N^{-\gamma} \sum_{m_1, \ldots, m_\gamma = -N/2+1}^{N/2-1} \left(\sum_{k_i, l_i} G_{k_1, \ldots, k_\alpha}^{\alpha} \bar{G}_{l_1, \ldots, l_\beta}^{\beta} \right) S_{\hat{U}\hat{U}}(f_{m_1}) \ldots S_{\hat{U}\hat{U}}(f_{m_\gamma}) + O(N^{-1}) \tag{3-98}$$

where the sum $\sum_{k_i, l_i}$ extends over the choices of $m_j \in \{k_1, \ldots, k_\alpha, -l_1, \ldots, -l_\beta\}$ resulting in γ independent pairs $(m_j, -m_j)$, and where the second term stems from the nonzero contributions in (3-96) containing at most $\gamma - 1$ independent m_j's. Note that the first term in (3-98) is an $O(N^0)$ and that it is the same for random phase multisines, periodic noise, and Gaussian noise with the same (power) spectra $S_{\hat{U}\hat{U}}(f)$. Collecting (3-95) and (3-98) gives

$$\text{var}(\sqrt{N} Y_{Sk, N}) = \sigma_{Y_S, N}^2(k) + O(N^{-1}) \tag{3-99}$$

where $\sigma_{Y_S, N}^2(k) = O(N^0)$ is the same for the three classes of excitation signals. Taking the limit $N \to \infty$ of (3-99) proves the theorem with $\sigma_{Y_S}^2(f) = \lim_{N \to \infty} \sigma_{Y_S, N}^2(k)$.

Appendix 3.L Proof of Theorem 3.18

In this section we study the variability of the BLA measurement (3-16) and (3-17) due to the stochastic nonlinear distortions only.

Section 3.11 ■ Appendixes

3.L.1 Random Phase Multisines. Calculating the variance of (3-16) taking into account (3-23) and the following properties (i) $Y_S^{[m]}(k)$ is independently distributed over the realization m, (ii) $\text{var}(Y_S^{[m]}(k))$ is independent of m, (iii) $Y_S^{[m]}(k)$ is uncorrelated with $U^{[m]}(k)$, and (iv) $|U^{[m]}(k)|$ is deterministic and independent of m, gives

$$\text{var}(\hat{G}_{\text{BLA}}(j\omega_k)) = \frac{1}{M^2}\sum_{m=1}^{M} \mathbb{E}\left\{\frac{|Y_S^{[m]}(k)|^2}{|U^{[m]}(k)|^2}\right\} = \frac{\text{var}(Y_S^{[m]}(k))}{M|U^{[m]}(k)|^2} \quad (3\text{-}100)$$

Combining (3-99) with (3-100) using $\sqrt{N}Y_{Sk,N} = Y_S^{[m]}(k)$ and $S_{\hat{U}\hat{U}}(f_k) = |U^{[m]}(k)|^2$ shows that $M\text{var}(\hat{G}_{\text{BLA}}(j\omega_k))$ converges at the rate $O(N^{-1})$ to $\sigma_{Y_S}^2(f)/S_{\hat{U}\hat{U}}(f)$.

3.L.2 Periodic Noise. Using (3-23), the estimated BLA (3-17) can be written as

$$\hat{G}_{\text{BLA}}(j\omega_k) = G_{\text{BLA}}(j\omega_k) + \hat{S}_{Y_S U}(k)/\hat{S}_{UU}(k) \quad (3\text{-}101)$$

For M sufficiently large, $1/\hat{S}_{UU}(k)$ can be approximated as

$$\frac{1}{\hat{S}_{UU}(k)} = \frac{1}{S_{\hat{U}\hat{U}}(f_k) + O(M^{-1/2})} = \frac{1}{S_{\hat{U}\hat{U}}(f_k)(1 + O(M^{-1/2}))} \approx S_{\hat{U}\hat{U}}^{-1}(f_k)(1 - O(M^{-1/2})) \quad (3\text{-}102)$$

(convergence in mean square sense at the rate $O(M^{-1/2})$). Combining (3-101) and (3-102) gives

$$\hat{G}_{\text{BLA}}(j\omega_k) - G_{\text{BLA}}(j\omega_k) = \hat{S}_{Y_S U}(k)/S_{\hat{U}\hat{U}}(f_k)(1 - O(M^{-1/2})) \quad (3\text{-}103)$$

Using (3-103) and the following properties (i) $Y_S^{[m]}(k)$ is independently distributed over the realization m, and (ii) $\mathbb{E}\{|Y_S^{[m]}(k)|^2|U^{[m]}(k)|^2\}$ is independent of m, we conclude that the asymptotic ($M \to \infty$) variance of the BLA equals

$$\lim_{M \to \infty} M\,\text{var}(\hat{G}_{\text{BLA}}(j\omega_k)) = \mathbb{E}\{|Y_S^{[m]}(k)|^2|U^{[m]}(k)|^2\}/S_{\hat{U}\hat{U}}^2(f_k) \quad (3\text{-}104)$$

In the sequel of this section we will show that

$$\mathbb{E}\{|Y_S^{[m]}(k)|^2|U^{[m]}(k)|^2\} = \mathbb{E}\{|Y_S^{[m]}(k)|^2\}\mathbb{E}\{|U^{[m]}(k)|^2\} + O(N^{-1}) \quad (3\text{-}105)$$

which proves (3-24).

To simplify the notations we omit now the realization superscript $[m]$ in (3-105). From (3-4) and $Y_k = Y(k)/\sqrt{N}$, it follows that during the calculation of the numerator of (3-104), mixed terms of nonlinear degrees α_1 and α_2 appear, viz.

$$\mathbb{E}\{|Y_S(k)|^2|U(k)|^2\} = \sum_{\alpha_1,\alpha_2=2}^{\infty} \mathbb{E}\{Y_S^{\alpha_1}(k)\bar{Y}_S^{\alpha_2}(k)|U(k)|^2\} \quad (3\text{-}106)$$

Using $G_S(k) = Y_S(k)/U(k)$, $U_k = U(k)/\sqrt{N}$, $G_S(k) = \sum_{\alpha=1}^{\infty} G_S^{\alpha}(k)$ with $G_S^{\alpha}(k)$ defined in (3-41), one term $\mathbb{E}\{Y_S^{\alpha_1}(k)\bar{Y}_S^{\alpha_2}(k)|U(k)|^2\}$ in (3-106) can be explicitly spelled out as

$$\mathbb{E}\{Y_S^{\alpha_1}(k)\bar{Y}_S^{\alpha_2}(k)|U(k)|^2\} = N^{-\frac{(\alpha_1+\alpha_2)}{2}+1} \sum_{\substack{K \in \mathbb{K}_{Sk,\alpha_1} \\ I \in \mathbb{K}_{Sk,\alpha_2}}} G_{k_1,k_2,\ldots,k_{\alpha_1}}^{\alpha_1} G_{i_1,i_2,\ldots,i_{\alpha_2}}^{\alpha_2} \quad (3\text{-}107)$$

$$\times \mathbb{E}\{U(k_1)U(k_2)\ldots U(k_{\alpha_1})\bar{U}(i_1)\bar{U}(i_2)\ldots \bar{U}(i_{\alpha_2})|U(k)|^2\}$$

with $K = (k_1, k_2, \ldots, k_{\alpha_1-1})$, $I = (i_1, i_2, \ldots, i_{\alpha_2-1})$, and where $\mathbb{K}_{Sk,\alpha}$ is defined in (3-40). For periodic noise the expected value in (3-107) can be split as the product of the expected value over the amplitudes and that over the phases (the amplitudes and phases are – by construction – independently distributed). Hence, only those terms where the phases pair will contribute, which requires $\alpha_1 + \alpha_2$ to be even.

Two possible situations can be considered during the pairing of the phases: either all frequencies i_r of the pairs $U(i_r)\bar{U}(i_r)$ are different from each other or not.

1. All paired frequencies are different. Since $U(k)\bar{U}(k)$ is already a pair, the remaining $\alpha_1 + \alpha_2$ indexes in (3-107) should be paired. Notice that the indexes $k_1, k_2, \ldots, k_{\alpha_1}$ and $i_1, i_2, \ldots, i_{\alpha_2}$ in (3-107) have, respectively, $\alpha_1 - 1$ and $\alpha_2 - 1$ degrees of freedom, because the frequency constraints $\sum_{r=1}^{\alpha_1} k_r = k$ and $\sum_{r=1}^{\alpha_2} i_r = k$ should be fulfilled. Hence, after pairing, the total number of degrees of freedom is $(\alpha_1 + \alpha_2)/2 - 1$ and, under these conditions, the expected value (3-107) is an $O(N^0)$. Since the amplitudes are – by construction – independently distributed over the frequency, the expected value (3-107) is proportional to $\mathbb{E}\{|U(k)|^2\} = S_{\hat{U}\hat{U}}(f_k)$.

2. Some paired frequencies are equal. When the frequencies of at least two pairs are equal to each other ($i_{r_1} = i_{r_2}$), then the number of degrees of freedom after pairing drops with at least 1. As a result, these terms are at most an $O(N^{-1})$ and can be neglected compared with the contributions of situation 1. Remark that in this case higher order $\mathbb{E}\{|U(k)|^{2q}\}$ with $q > 1$ appear, but from Definition 3.3 it follows that these are finite.

We conclude that

$$\mathbb{E}\{Y_S^{\alpha_1}(k)\bar{Y}_S^{\alpha_2}(k)|U(k)|^2\} = S_{\hat{U}\hat{U}}(f_k)\mathbb{E}\{Y_S^{\alpha_1}(k)\bar{Y}_S^{\alpha_2}(k)\} + O(N^{-1}) \quad (3\text{-}108)$$

Combining (3-106) and (3-108) proves (3-105).

Remark We implicitly assumed that the phases of the periodic noise have a uniform continuous distribution. Following the lines of Appendix 3.F, the results can easily be extended to non-uniform continuous and discrete phase distributions.

3.L.3 Gaussian Noise. For arbitrary excitations the relationship between the input-output DFT spectra is given by

$$Y(k) = G_{\text{BLA}}(j\omega_k)U(k) + T_G(j\omega_k) + Y_S(k) \quad (3\text{-}109)$$

where $T_G(j\omega_k) = O(N^{-1/2})$ represents the leakage (transient) error (see Section 6.3.2). Since $U(k)$ and $Y_S(k)$ are both an $O(N^0)$, it follows from (3-109) that $T_G(j\omega_k)$ can asymptotically

($N \to \infty$) be neglected in the variance analysis of the BLA. For filtered white Gaussian noise excitations $u(t)$, the input DFT spectrum $U(k)$ can be written as

$$U(k) = L(z_k^{-1})E(k) + T_L(z_k^{-1}) \qquad (3\text{-}110)$$

where $E(k) = O(N^0)$ is circular complex normally distributed and with $T_L(z_k^{-1}) = O(N^{-1/2})$ the leakage (transient) error (see Section 6.3.2). Hence, $T_L(z_k^{-1})$ can asymptotically ($N \to \infty$) be neglected in (3-110). It shows that filtered white Gaussian noise has asymptotically ($N \to \infty$) the same properties as periodic noise. A detailed analysis shows that T_G and T_L have an $O(N^{-1})$ contribution to $\text{var}(\hat{G}_{\text{BLA}}(j\omega_k))$. This is left as an exercise for the reader.

Appendix 3.M Proof of Theorem 3.21

Consider the contributions to Y_L^α, $|L| > l_{\max}$ (see (3-5)). These are of the form

$$G^\alpha_{L, k_1, k_2, \ldots, k_{\alpha-1}} U_{k_1} U_{k_2} \ldots U_{k_{\alpha-1}} U_{k_\alpha} \text{ with } k_\alpha = L - \sum_{i=1}^{\alpha-1} k_i \qquad (3\text{-}111)$$

To get systematic contributions, α should be odd because even nonlinearities cannot create systematic contributions. Assume that $\exists\, l$ s.t. the phase of $U_{-l}(U_{k_1} U_{k_2} \ldots U_{k_{\alpha-1}} U_{k_\alpha})$ is zero (these combinations create the systematic contributions). We will check whether such combinations can exist.

The only possibility to get zero phase is that U_{-l} is paired with one of the components U_{k_i}. In that case there exists a k_i s.t. the phase $U_{-l}U_{k_i}$ is zero. After rearranging the order of the components, we can put U_{k_i} in the last place. Also, the components that pair are put together, and eventually the contributions can be written as

$$U_{r_1} U_{r_2} \ldots U_{r_{2\beta}} (U_{s_1} U_{-s_1}) \ldots (U_{s_p} U_{s_{-p}})(U_{k_\alpha} U_{-l}) \qquad (3\text{-}112)$$

with U_{r_i} the unpaired components. Now there are two possibilities:

(i) There are no unpaired components left, $\beta = 0$. In that case, the combination in (3-111) contributes to the frequency $L = l = k_\alpha$, which is by definition in the excitation band (k_α is an excitation frequency). This violates that $|L| > l_{\max}$.

(ii) There are unpaired components ($\beta \neq 0$). In that case not all frequencies in (3-111) are paired, and, hence, the phase is not zero. So, this situation cannot also result in systematic contributions.

This proves the theorem. □

Appendix 3.N Covariance of the Multivariate BLA

The input-output DFT spectra in (3-33) are related by

$$\mathbf{Y}^{[m]}(k) = G_{\text{BLA}}(j\omega_k)\mathbf{U}^{[m]}(k) + \mathbf{Y}^{[m]}_S(k) \qquad (3\text{-}113)$$

where the columns of $\mathbf{Y}^{[m]}_S(k)$ are uncorrelated for full random orthogonal multisines. Hence,

$$\text{Cov}(\text{vec}(\mathbf{Y}_S^{[m]}(k))) = I_{n_u} \otimes C_{Y_S}(k) \tag{3-114}$$

with $C_{Y_S}(k) = \text{Cov}(Y_S(k))$ the covariance of the stochastic nonlinear contributions of one experiment. Using (3-31) and $\mathbf{U}^{[m]}(k) = G_{\text{act}}(j\omega_k)\mathbf{R}^{[m]}(k)$ (see Figure 3-8), we find that

$$\mathbf{U}^{[m]}(k)\mathbf{U}^{[m]H}(k) = G_{\text{act}}(j\omega_k)D_{|R|^2}(k)G_{\text{act}}^H(j\omega_k) \tag{3-115}$$

where $D_{|R|^2}(k)$ is a diagonal matrix containing the squared 2-norm of each row of $\mathbf{R}^{[m]}(k)$, is independent of the realization number m. Applying (2-77) to (3-33) taking into account (3-114) and (3-115), proves (3-34).

Appendix 3.O Proof of Theorem 3.22

The input-output DFT spectra $Z(k) = [Y(k) \; U(k)]^T$ of the nonlinear plant operating in closed loop (see Figure 3-9) are related to the DFT spectrum $R(k)$ of the reference signal as

$$Z(k) = G_{rz}(j\omega_k)R(k) + \tilde{Z}_S(k) \tag{3-116}$$

with $G_{rz}(j\omega_k)$ the BLA from the reference signal $r(t)$ to $z(t) = [y(t) \; u(t)]^T$, $\tilde{Z}_S(k)$ the DFT transform of the stochastic nonlinear distortions $\tilde{z}_s(t)$, and where $\tilde{Z}_S(k)$ is uncorrelated with $R(k)$ (proof: the system from $r(t)$ to $z(t)$ belongs to the set $\mathbb{S}$ and operates in open loop so that the results of Dobrowiecki and Schoukens, 2007b are valid). The $(n_y + n_u) \times n_u$ matrix $G_{rz}(j\omega_k)$ and the $(n_y + n_u) \times 1$ vector $\tilde{Z}_S(k)$ can be split in the first n_y output rows and the last n_u input rows giving

$$G_{rz}(j\omega_k) = \begin{bmatrix} G_{ry}(j\omega_k) \\ G_{ru}(j\omega_k) \end{bmatrix} \text{ and } \tilde{Z}_S(k) = \begin{bmatrix} \tilde{Y}_S(k) \\ \tilde{U}_S(k) \end{bmatrix} \tag{3-117}$$

Using (3-116), (3-117), and the definition (3-35) of the BLA of a nonlinear plant operating in feedback, the difference $Y_S(k)$ between the actual output of the nonlinear plant and the output of the BLA is found to be equal to

$$Y_S(k) = Y(k) - G_{\text{BLA}}(j\omega_k)U(k) = \tilde{Y}_S(k) - G_{\text{BLA}}(j\omega_k)\tilde{U}_S(k) \tag{3-118}$$

Since $\tilde{Z}_S(k)$ is uncorrelated with $R(k)$, it follows immediately from (3-118) that $Y_S(k)$ is also uncorrelated with $R(k)$

$$\mathbb{E}\{Y_S(k)\bar{R}(k)\} = [1 \; -G_{\text{BLA}}(j\omega_k)] \, \mathbb{E}\{\tilde{Z}_S(k)\bar{R}(k)\} = 0 \tag{3-119}$$

Since $\tilde{Z}_S(k)$ satisfies properties (ii-iv) of Theorem 3.16 and Theorem 3.20 (multivariate open loop case) this is also valid for $Y_S(k)$ (3-118), which proves properties (iii-iv) of Theorem 3.22.

The indirect BLA (3-35) is the ratio of two BLAs (open loop case), each satisfying Theorems 3.7 and 3.15. Hence, these properties are also valid for (3-35), which proves properties (i-ii) of Theorem 3.22. □

4

Detection, Quantification, and Qualification of Nonlinear Distortions in FRF Measurements

Abstract: Full characterization of frequency response function (FRF) measurements on nonlinear systems requires the simultaneous quantification of the best linear approximation (BLA), its noise variance, and the variance of the stochastic nonlinear distortions. Two measurement methods satisfying these requirements are presented in this chapter. They are based on specially designed random phase multisine excitations. In the first part of this chapter the invariance of the BLA and the variance of the stochastic nonlinear distortions within this special class of excitation signals is shown. The second part describes the algorithmic details of the two measurement methods. The theory and algorithms are illustrated on real measurement examples.

4.1 INTRODUCTION

The literature describes a series of methods, different from those presented in the sequel of this chapter. Here, we will touch only on a few of them; an extended list of references is available in Natke et al. (1988) and Vanhounacker et al. (2002). Also Haber (1985) gives a brief review of nonlinearity tests.

- A simple method is to *scale the input* $u(t) \rightarrow \alpha u(t)$ and verify whether the output also scales with α, after taking care for the offsets. In practice, this method is less appealing. Two separate measurements are needed, and in many applications it is not simple to impose a scaled input due to the nonlinear load of the generator with the input impedance of the tested system. This problem is not disposed only in the special case where a discrete-time model is built between a signal in a computer memory and the output of the physical system (see Section 13.2). In this special situation the user has full control over the excitation signal. Moreover, the small nonlinearities have to be detected by taking the difference between two large, measured signals, making the method extremely sensitive to all possible measurement errors due to this indirect nature.

- Another popular test is to check the *coherence*. This method does not allow separation of noise disturbances from nonlinearity problems and it fails completely for periodic excitations. Extending the test to higher order spectra by probing directly for higher order correlations that are typical for nonlinear systems may eliminate these drawbacks, but these methods are very time consuming, especially for random excitations.

- Also, *Hilbert transform* tests have been proposed (Tomlinson, 1987). Actually, these methods do not, directly, detect the nonlinear behavior itself. The method checks for a noncausality in the impulse response of the linear approximation (FRF) that might be induced by the nonlinear behavior, although there is no guarantee at all that there is a one-to-one relation between both effects. The method imposes significant constraints (e.g., only working on lowly damped systems) and a series of correction terms should be added because an FRF measurement can be made only in a restricted frequency band, while the theory requires data from DC to ∞.

- Finally, the *sine test* is the simplest test for characterizing directly the nonlinear behavior by verifying the presence of higher harmonics in the output spectrum. However, this approach has a number of serious drawbacks. It is not only very slow (see Section 5.2.2), but can also be insensitive to the nonlinear behavior. Indeed, if the nonlinear system consists of a nonlinearity followed by a linear dynamic lowpass system, then the higher harmonics can be attenuated below the noise level.

In this chapter we will present nonlinearity tests based on random phase multisine excitations that do not suffer from the aforementioned problems. The possibility to detect nonlinear distortions with these signals will be embedded by a careful selection of their amplitude spectrum, only a selected set of harmonics will be excited. This idea has already been suggested by Evans et al. (1994) and McCormack et al. (1994b).

The ideal FRF-measurement method should provide the measured FRF, and at the same time the presence of nonlinear distortions should be detected, qualified (even or odd distortions), and quantified (the level of the distortions). Because the prime interest in these measurements is the FRF, it is unacceptable that most of the time would be spent on the detection of the nonlinear distortion at the cost of a reduced quality of the FRF measurement. This excludes most existing methods that require a series of dedicated measurements to make the nonlinearity test. In general, it is impossible to realize this ideal; however, when random phase multisine excitations are applied, we can come close to it.

This chapter is organized as follows. First, we define an equivalence class of excitation signal consisting of Gaussian noise and random phase multisines with randomly selected non-excited harmonics. Within this equivalence class it is shown that the best linear approximation (BLA) and its variance are the same. Next, two practical methods for measuring the BLA, its noise variance, and the variance of the nonlinear distortions using random phase multisines are proposed. The first method – called the "robust" method - applies directly the definition (3-12) of the BLA, while the second method – called the "fast" method – exploits explicitly the information at the non-excited harmonics. Contrary to the "robust" method, the "fast" method can also distinguish between odd and even nonlinear contributions. Finally, some guidelines for measuring the BLA and its uncertainty are given.

4.2 THE RIEMANN EQUIVALENCE CLASS OF EXCITATION SIGNALS

In Sections 3.4.3 and 3.4.4 the asymptotic equivalence of the best linear approximation and its variance has been established for Gaussian noise, periodic Gaussian noise, and random

4.2.1 Definition of the Excitation Signals

4.2.1.1 Random Phase Multisines. Consider the zero mean random phase multisine (3-7)

$$u(t) = \sum_{k=-N/2+1, k \neq 0}^{N/2-1} U_k e^{j2\pi f_s kt/N} \tag{4-1}$$

where the Fourier coefficients U_k are either zero (the harmonic is not excited) or satisfy $|U_k| = \hat{U}(kf_s/N)/\sqrt{N}$, with $S_{\hat{U}\hat{U}}(f) = \hat{U}^2(f)$ a uniformly bounded positive function ($0 \leq S_{\hat{U}\hat{U}}(f) \leq M_U < \infty$) with a finite number of discontinuities on the interval $[0, f_s/2]$ (see Definition 3.2 on page 77). According to the choice of the excited harmonics we distinguish the following cases:

- *Full random phase multisines*: All harmonics in the frequency band of interest are excited, for example, $k = 1, 2, 3, ..., F$ with $F = O(N) < N/2$ in (4-1).
- *Odd random phase multisines*: All odd harmonics and none of the even harmonics are excited, for example, $2k - 1$ with $k = 1, 2, 3, ..., F$ and $2F - 1 = O(N) < N/2$ in (4-1).
- *Random phase multisines with random harmonic grid*: Within each group of N_{sub} successive harmonics a number of randomly selected harmonics is not excited. The probability that a harmonic is excited is called p, for example, $p = 3/5$ if two out of five consecutive harmonics ($N_{\text{sub}} = 5$) are not excited. To guarantee a uniform coverage of the frequency grid in (4-1), we take a small number of successive harmonics and eliminate randomly one excited harmonic, for example, $N_{\text{sub}} = 2, 3$, or 4, and, respectively, $p = 1/2, 2/3$, and $3/4$. The random harmonic grid can be applied to the full and odd random phase multisines.

These multisines have the following nonlinearity detection properties:

(i) *Full random phase multisines*: The sum of the odd and even nonlinear distortions appear in the output spectrum at the excited frequencies. In general no distinction can be made between odd and even in-band (= the excited frequency band) distortions.

(ii) *Full random phase multisines with random harmonic grid*: The sum of the odd and even nonlinear distortions appear in the output spectrum at the excited frequencies and the non-excited in-band harmonics. The non-excited in-band harmonics serve as nonlinearity detection lines where the level of $Y_S(k)$ can be retrieved. In general, no distinction can be made between odd and even in-band distortions.

(iii) *Odd random phase multisines*: The even and odd nonlinear distortions appear in the output spectrum at, respectively, the even and odd harmonics. Hence, the excited odd harmonics are only disturbed by the odd nonlinear distortions, while the even in-band harmonics detect the even nonlinear distortions.

(iv) *Odd random phase multisines with random harmonic grid*: The even and odd nonlinear distortions appear in the output spectrum at, respectively, the even and odd harmonics. The even and odd non-excited in-band harmonics detect the presence of, respectively, the even and odd nonlinear distortions. The odd excited harmonics are only disturbed by the odd nonlinear distortions.

(proof: see Appendix 4.A). In practice, the nonlinear distortions should be detected in the presence of noise. Therefore, a number of consecutive periods of the steady state response are measured and the level of the non-excited in-band harmonics is compared with the noise sample standard deviation (see Section 4.3.2 for more details).

4.2.1.2 Periodic Noise. Consider the periodic signal (4-1) where the amplitudes $|U_k|$ of the Fourier coefficients are either zero, or the realization of an independent (of the phases $\angle U_k$, and over k) random process, with $S_{\hat{U}\hat{U}}(f) = \mathbb{E}\{\hat{U}^2(f)\}$ a uniformly bounded function with a finite number of discontinuities on the interval $[0, f_s/2]$ (see Definition 3.2 on page 77). According to the choice of the excited harmonics we can also distinguish between *full periodic noise, odd periodic noise, full periodic noise with random harmonic grid,* and *odd periodic noise with random harmonic grid* (follow the same lines of Section 4.2.1.1). Notice that the only difference between periodic noise and a random phase multisine is the random nature of amplitude spectrum of periodic noise.

4.2.1.3 Gaussian Noise. Consider stationary Gaussian noise excitations $u(t)$ with a power spectrum (power spectral density) $S_{uu}(j\omega)$, that is piecewise continuous with finite number of discontinuities and with $S_{uu}(j\omega) = 0$ for $|\omega| \geq \omega_s/2$. Using a rectangular window, the squared amplitude of the DFT spectrum $U(k)$ (2-11) of N samples $u(t)$, $t = 0, 1, ..., N-1$, is related to $S_{uu}(j\omega_k)$ as

$$\frac{1}{N}\mathbb{E}\{|U(k)|^2\} = S_{uu}(j\omega_k)\frac{f_s}{N} + O(N^{-2}) \tag{4-2}$$

where the expectation is taken over the random realizations of the Gaussian excitation $u(t)$ (the leakage error in $\mathbb{E}\{|U(k)|^2\}$ is an $O(N^{-1})$ for rectangular windows, see Brillinger, 1981).

4.2.2 Definition of the Riemann Equivalences

In this section we define the Riemann equivalence class of excitation signals that collects all signals that are (asymptotically) normally distributed and have asymptotically ($N \rightarrow \infty$) the same power in each finite frequency interval. Comparison of the power in a finite frequency interval is needed because the discrete power spectrum of a periodic signal can never be equal to the continuous power spectrum of a random signal. This leads to the definition of Riemann equivalent power spectra of signals (periodic and/or random).

Definition 4.1 (Riemann Equivalent Power Spectra): The power spectra (power spectral density) $S_{y_1 y_1}(j\omega)$ and $S_{y_2 y_2}(j\omega)$ of two signals (periodic and/or random) $y_1(t)$ and $y_2(t)$ are Riemann equivalent if

$$\int_{\omega_1}^{\omega_2} S_{y_1 y_1}(j\omega)df = \int_{\omega_1}^{\omega_2} S_{y_2 y_2}(j\omega)df + O(N^{-1}) \tag{4-3}$$

for any $0 < \omega_1 < \omega_2 < \omega_s/2$. If $y_i(t)$ is a periodic signal then $S_{y_i y_i}(j\omega)$ is a sum of Dirac functions and the integral in (4-3) is replaced by the Riemann sum

Section 4.2 ■ The Riemann Equivalence Class of Excitation Signals

$$\int_{\omega_1}^{\omega_2} S_{y_i y_i}(j\omega) df = \frac{1}{N} \sum_{k=k_1}^{k_2} |Y_i(k)|^2 \quad (4\text{-}4)$$

with $k_1 = \lceil Nf_1/f_s \rceil$ and $k_2 = \lfloor Nf_2/f_s \rfloor$, and where $\lceil x \rceil$ ($\lfloor x \rfloor$) is the smallest (largest) integer larger (smaller) than or equal to x.

From this definition, it follows that Riemann equivalent power spectra asymptotically ($N \to \infty$) have the same power (continuous or discrete) in each finite frequency band. Hence, signals with different periodicity or even and odd signals (e.g., only harmonics $2k$ or $2k+1$ are excited in (4-1)) can have Riemann equivalent power spectra. Using Definition 4.1 we can define now the class of asymptotically ($N \to \infty$) normally distributed signals with Riemann equivalent power spectra.

Definition 4.2 (Riemann Equivalence Class of Asymptotically Normally Distributed Excitation Signals): $\mathbb{N}_{S_{uu}}$ is the class of asymptotically ($N \to \infty$) normally distributed signals $u(t)$ with Riemann equivalent power spectrum (see Definition 4.1) $S_{uu}(j\omega)$, which is piecewise continuous with a finite number of discontinuities.

Examples of signals belonging to the Riemann equivalence class $\mathbb{N}_{S_{uu}}$ (see Definition 4.2) are:

- Random phase multisines: full and odd with or without random harmonic grid (see Section 4.2.1.1).
- Periodic noise: full and odd with or without random harmonic grid (see Section 4.2.1.2).
- Gaussian noise (see Section 4.2.1.3).

From (4-2)–(4-4) it is possible to define for these periodic signals a relationship between the probability $p(k)$ that a frequency line k is excited, the expected value of the magnitude squared of their DFT spectra $U(k)$, and the power spectrum $S_{uu}(j\omega)$ of the Gaussian noise as

$$\frac{1}{N} \mathbb{E}\{|U(k)|^2\} = \frac{1}{N} p(k) S_{\hat{U}\hat{U}}(k\frac{f_s}{N}) = S_{uu}(j\omega_k)\frac{f_s}{N} + O(N^{-2}) \quad (4\text{-}5)$$

where $S_{\hat{U}\hat{U}}(f)$ represents the power spectrum (mean square value) of the excited harmonics. Note that the expected value in (4-5) is taken w.r.t. the random amplitude and/or phase spectrum of the non-zero harmonics and the probability that a frequency line is excited. From (4-5) it follows that the rms value of the non-zero amplitudes of the periodic noise and random phase multisines are given by

$$S_{\hat{U}\hat{U}}^{1/2}(k\frac{f_s}{N}) = \sqrt{f_s \frac{S_{uu}(j\omega_k)}{p(k)}} + O(N^{-1}) \quad (4\text{-}6)$$

For periodic signals that do not use all harmonics, for example, odd random phase multisines, an equivalent probability can be defined. Since an odd multisine excites only one harmonic out of two, we can use the factor $p(k) = 0.5$ in (4-5) as a measure for the "fill-factor" of the multisine (no longer a probability interpretation), such that we get the correct harmonic amplitudes in (4-6). If the excited harmonics are logarithmically distributed over the frequency band of interest, then the "fill-factor" $p(k)$ decreases as an $O(k^{-1})$.

4.2.3 Invariance of the Best Linear Approximation and the Variance of the Stochastic Nonlinear Distortions

Using the Riemann equivalence class of asymptotically normally distributed signals we can formulate precisely the asymptotic invariance claims for the best linear approximation and the variance of the stochastic nonlinear distortions.

Theorem 4.3 (Asymptotic Invariance BLA and Variance of the Stochastic Nonlinear Distortions): Consider a nonlinear system belonging to the set $\mathbb{S}$ (see Definition 3.5). For all excitations $u(t)$ belonging to the equivalence class $\mathbb{N}_{S_{uu}}$ (see Definition 4.2) the best linear approximations $G_{\text{BLA},N}(j\omega_k)$ are asymptotically ($N \to \infty$) the same, and the power spectra $S_{y_s y_s}(j\omega)$ of the stochastic nonlinear distortions $y_s(t)$ are Riemann equivalent (see Definition 4.1) with

$$G_{\text{BLA},N}(j\omega_k) = G_{\text{BLA}}(j\omega_k) + O(N^{-1})$$
$$\text{var}(Y_{S,N}(k)) = \sigma_{Y_S}^2(k) + O(N^{-1}) \tag{4-7}$$

where $G_{\text{BLA}}(j\omega)$ is defined in (3-18), $\sigma_{Y_S}^2(k) = S_{y_s y_s}(j\omega_k) f_s$, and where subscript N indicates the dependency on the number of excited frequencies $F = O(N)$ in the signal $u(t) \in \mathbb{N}_{S_{uu}}$.

Proof. Follow exactly the same lines of the proofs of Theorems 3.12 (Appendix 3.H) and 3.17 (Appendix 3.K) where $S_{\hat{U}\hat{U}}(f_k)$ is replaced by $p(k) S_{\hat{U}\hat{U}}(f_k) = S_{uu}(j\omega_k) f_s$. □

Remark An important consequence of Theorem 4.3 is that for full (odd) random harmonic grid signals (random phase multisines and periodic noise) the level of the stochastic nonlinear distortions at the excited (odd) harmonics can be estimated as the signal level at the nearest non-excited (odd) harmonic. This property is used in the "fast" method for detecting the nonlinear distortions (see Section 4.3.2).

In the sequel of this section the Riemann equivalence of the power spectra $S_{y_s y_s}(j\omega)$ in Theorem 4.3 is discussed in detail for Gaussian noise, periodic noise, and random phase multisines belonging to the equivalence class $\mathbb{N}_{S_{uu}}$. Table 4-1 gives an overview of the discussion.

4.2.3.1 Gaussian Noise. Gaussian noise $u(t)$ with power spectrum (power spectral density) $S_{uu}(j\omega)$ is used as reference excitation signal. The corresponding power spectrum $S_{y_s y_s}(j\omega)$ of the stochastic nonlinear distortions $y_s(t)$ serves as a reference for the other excitation signals. It can be split in even ($S_{y_s y_s, \text{even}}(j\omega)$) and odd ($S_{y_s y_s, \text{odd}}(j\omega)$) contributions originating from, respectively, the even (e.g., $y = x^2$) and odd (e.g., $y = x^3$) nonlinearities

$$S_{y_s y_s}(j\omega) = S_{y_s y_s, \text{even}}(j\omega) + S_{y_s y_s, \text{odd}}(j\omega) \tag{4-8}$$

Similarly to (4-2), the DFT spectrum $Y_S(k)$ of $y_s(t)$ is related to $S_{y_s y_s}(j\omega)$ as

TABLE 4-1 Comparison of the Riemann Equivalent Gaussian Signals. $S_{\hat{U}\hat{U}}(f_k)$: Mean Square Value of the Excited Harmonics of the Periodic Noise and Random Phase Multisine, and $\text{var}(Y_S(k))$ the Variance of the Corresponding Stochastic Nonlinear Distortions. $S_{uu}(j\omega_k)$: Power Spectrum of the Gaussian Noise Excitation, and $S_{y_s y_s}(j\omega_k)$ the Power Spectrum of the Corresponding Stochastic Nonlinear Distortions that Can Be Split in Even ($S_{y_s y_s,\text{even}}(j\omega_k)$) and Odd ($S_{y_s y_s,\text{odd}}(j\omega_k)$) Contributions.

Excited harmonics	$S_{\hat{U}\hat{U}}(f_k)$	$\text{var}(Y_S(k))$
full	$S_{uu}(j\omega_k) f_s$	$S_{y_s y_s}(j\omega_k) f_s$
odd	$2 S_{uu}(j\omega_k) f_s$	k even: $2 S_{y_s y_s,\text{even}}(j\omega_k) f_s$ k odd: $2 S_{y_s y_s,\text{odd}}(j\omega_k) f_s$
full random with prob. $p(k)$	$S_{uu}(j\omega_k) f_s / p(k)$	$S_{y_s y_s}(j\omega_k) f_s$
odd random with prob. $p(k)$	$2 S_{uu}(j\omega_k) f_s / p(k)$	k even: $2 S_{y_s y_s,\text{even}}(j\omega_k) f_s$ k odd: $2 S_{y_s y_s,\text{odd}}(j\omega_k) f_s$

$$\frac{1}{N} \mathbb{E}\{|Y_S(k)|^2\} = S_{y_s y_s}(j\omega_k) \frac{f_s}{N} + O(N^{-2}) \tag{4-9}$$

where the expected value is taken over the random realizations of the Gaussian noise $u(t)$.

4.2.3.2 Full Random Phase Multisines and Full Periodic Noise.

Since all harmonics in the frequency band of interest are excited, (4-5) is valid with $p(k) = 1$. Hence, for full multisine and full periodic noise excitations the power spectra $S_{y_s y_s}^{\text{full}}(j\omega)$ of the stochastic nonlinear distortions $y_s^{\text{full}}(t)$ are Riemann equivalent to $S_{y_s y_s}(j\omega)$. Moreover, $Y_S^{\text{full}}(k)$, the DFT spectrum of $y_s^{\text{full}}(t)$, satisfies (4-9).

4.2.3.3 Odd Random Phase Multisines and Odd Periodic Noise.

Since all odd harmonics in the frequency band of interest are excited, (4-5) is valid with "fill factor" $p(k) = 0.5$. Hence, the resulting power spectrum $S_{y_s y_s}^{\text{odd}}(j\omega)$ of the stochastic nonlinear distortions $y_s^{\text{odd}}(t)$ is still Riemann equivalent to $S_{y_s y_s}(j\omega)$. However, since for odd periodic excitations the even nonlinear distortions are only present at the even output harmonics, whereas the odd nonlinearities are only active at the odd output harmonics, a "fill factor" of 0.5 must be introduced in the relationship between $Y_S^{\text{odd}}(k)$, the DFT spectrum of $y_s^{\text{odd}}(t)$, and $S_{y_s y_s}(j\omega_k)$, giving

$$\frac{0.5}{N} \mathbb{E}\{|Y_S^{\text{odd}}(2k+1)|^2\} = S_{y_s y_s,\text{odd}}(j\omega_{2k+1}) \frac{f_s}{N} + O(N^{-2})$$
$$\frac{0.5}{N} \mathbb{E}\{|Y_S^{\text{odd}}(2k)|^2\} = S_{y_s y_s,\text{even}}(j\omega_{2k}) \frac{f_s}{N} + O(N^{-2}) \tag{4-10}$$

where the expected value is taken over the random realizations of the odd periodic excitation.

4.2.3.4 Random Harmonic Grid Multisines and Periodic Noise.

Multisines and periodic noise excitations that do not excite all the harmonics should carefully be designed. When randomly selected harmonics are not excited, then it is necessary to choose the rms value of the excited harmonics according to (4-6), where $p(k)$ is the probability that harmonic k is excited. The power spectrum $S_{y_s y_s}^{\text{harm}}(j\omega)$ of the stochastic nonlinear distortions $y_s^{\text{harm}}(t)$ is Riemann equivalent to $S_{y_s y_s}(j\omega)$, and the DFT spectrum $Y_S^{\text{harm}}(k)$ of $y_s^{\text{harm}}(t)$ satisfies (4-9). For odd random harmonic signals we should also take into account the "fill factor" of 0.5 giving the same result (4-10), where Y_S^{odd} is replaced by $Y_S^{\text{odd, harm}}$.

4.2.4 Experimental Illustration

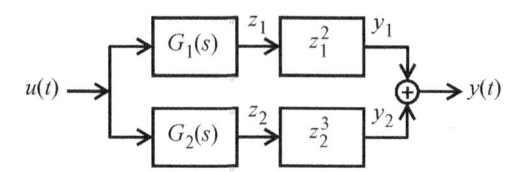

Figure 4-1. Block schematic of the parallel Wiener system.

The goal of the measurement example is to illustrate Theorem 4.3 for the following signals belonging to the equivalence class: Gaussian noise, full and odd random phase multisines, and full and odd random phase multisines with a random harmonic grid. John Lataire (Department ELEC of the Vrije Universiteit Brussel) has provided us with the experimental data (Schoukens et al., 2009).

4.2.4.1 Measurement Setup. A nonlinear electronic test system has been designed to illustrate all aspects of Theorem 4.3. It consists of the parallel connection of two Wiener systems (see Figure 4-1). The upper branch in Figure 4-1 is the cascade of a lowpass filter $G_1(s)$ with cutoff frequency of 1 kHz followed by an even static nonlinearity $y_1 = z_1^2$; while the lower branch consists of a highpass filter with cutoff frequency of 2 kHz followed by an odd static nonlinearity $y_2 = z_2^3$. The gains of both branches are tuned to get about the same power in the even and odd nonlinear distortions at the output. In the crossover of both frequency bands the even and odd nonlinearities are equally important.

The excitation signals $u(t)$ are generated with an arbitrary waveform generator (HPE1445A) in ZOH mode without reconstruction filter, while the input-output signals are measured with two data acquisition cards (HPE1430A) that have a high linearity and have anti-alias filters on board. The sampling frequency of the setup is 39.0625 kHz, and $N = 3256$ data points per period are measured.

The full random phase multisines excite harmonics $k = 1, 2, \ldots, 333$, with equal amplitudes $|U_k|$, generator clock frequency $f_s = 39.0625$ kHz, and $N = 3256$ in (4-1). Hence, the input signal covers the frequency band [12 Hz, 4 kHz], while output signal energy can be expected till 12 kHz ($= 3 \times 4$ kHz). The latter is well below the Nyquist frequency $f_s/2 = 19.5$ kHz of the acquisition. For the odd random phase multisines we set the amplitudes of the even harmonics in the full multisine to zero. To generate the random harmonic grid with $100 \times p\%$ excited frequencies, $100 \times (1-p)\%$ of the F excited harmonics are eliminated (full: $F = 333$; odd: $F = 166$). Experiments with $M = 200$ different random realizations of the harmonic grid and/or the phases are performed, and each time two consecutive periods of the steady state response are measured. The best linear approximation (BLA) and its sample variance are calculated at the excited harmonics k_{exc}

$$\hat{G}_{\text{BLA}}(j\omega_{k_{\text{exc}}}) = \frac{1}{M(k_{\text{exc}})} \sum_{m \in \mathbb{M}(k_{\text{exc}})} \frac{Y^{[m]}(k_{\text{exc}})}{U^{[m]}(k_{\text{exc}})}$$

$$\hat{\sigma}_{\hat{G}}^2(k_{\text{exc}}) = \frac{1}{M(k_{\text{exc}}) - 1} \sum_{m \in \mathbb{M}(k_{\text{exc}})} \left| \frac{Y^{[m]}(k_{\text{exc}})}{U^{[m]}(k_{\text{exc}})} - \hat{G}_{\text{BLA}}(j\omega_{k_{\text{exc}}}) \right|^2$$

(4-11)

with $X^{[m]}(k)$, $X = Y, U$, the mean of the DFT spectra of the two periods, $\mathbb{M}(k)$ the set of realization indices m where harmonic k is excited, and $M(k) \leq M$ the number of elements in the set $\mathbb{M}(k)$; while the mean output power is calculated at the non-excited harmonics $k_{\text{non-exc}}$

$$|Y(k_{\text{non-exc}})|^2 = \frac{1}{\overline{M}(k_{\text{non-exc}})} \sum_{m \in \overline{\mathbb{M}}(k_{\text{non-exc}})} |Y^{[m]}(k_{\text{non-exc}})|^2 \qquad (4\text{-}12)$$

with $\overline{\mathbb{M}}(k)$ the set of realization indices m where harmonic k is not excited, and $\overline{M}(k) \leq M$ the number of elements in the set $\overline{\mathbb{M}}(k)$. Using (4-11) and (4-12) an estimate of the variance of the stochastic nonlinear distortions at the excited k_{exc} and non-excited $k_{\text{non-exc}}$ harmonics is obtained as

$$\begin{aligned}\text{var}(Y_S(k_{\text{non-exc}})) &\approx \hat{\sigma}_{Y_s}^2(k_{\text{non-exc}}) = |Y(k_{\text{non-exc}})|^2 \\ \text{var}(Y_S(k_{\text{exc}})) &\approx \hat{\sigma}_{Y_s}^2(k_{\text{exc}}) = \hat{\sigma}_S^2(k_{\text{exc}})|U(k_{\text{exc}})|^2\end{aligned} \qquad (4\text{-}13)$$

Finally, the mean output power of the BLA at the excited frequencies is given by

$$|Y_{\text{BLA}}(k_{\text{exc}})|^2 = |\hat{G}_{\text{BLA}}(j\omega_{k_{\text{exc}}})|^2 \frac{1}{M(k_{\text{exc}})} \sum_{m \in \mathbb{M}(k_{\text{exc}})} |U^{[m]}(k_{\text{exc}})|^2 \qquad (4\text{-}14)$$

with $\hat{G}_{\text{BLA}}(j\omega_{k_{\text{exc}}})$, $M(k_{\text{exc}})$, and $\mathbb{M}(k_{\text{exc}})$ defined in (4-11).

The Gaussian noise experiments are performed as follows. First, the DFT of $3N$ white Gaussian noise samples is calculated, and the spectrum above 4 kHz is set to zero. The inverse DFT of the windowed spectrum gives periodic band-limited white Gaussian noise of length $3N$. Next, the periodic Gaussian noise is applied to the nonlinear test system and $2N$ samples of the input-output signals are measured (same experiment time as for the multisine excitations). This is repeated for $M = 200$ independent realizations. Further, the BLA and the variance of the stochastic nonlinear distortions are estimated as

$$\begin{aligned}\hat{G}_{\text{BLA}}(j\omega_k) &= \frac{\hat{S}_{YU}(k)}{\hat{S}_{UU}(k)} \\ \text{var}(Y_S(k)) &\approx \hat{\sigma}_{Y_s}^2(k) = \frac{M}{M-1}\left(\hat{S}_{YY}(k) - \frac{|\hat{S}_{YU}(k)|^2}{\hat{S}_{UU}(k)}\right)\end{aligned} \qquad (4\text{-}15)$$

with $\hat{S}_{YU}(k) = M^{-1}\sum_{m=1}^{M} Y^{[m]}(k)\overline{U}^{[m]}(k)$ the cross-power spectrum based on the M input-output DFT spectra (rectangular window) of the $2N$ samples (Brillinger, 1981). Finally, the mean output power of the BLA is given by

$$\hat{S}_{Y_{\text{BLA}}Y_{\text{BLA}}}(k) = |\hat{G}_{\text{BLA}}(j\omega_{k_{\text{exc}}})|^2 \hat{S}_{UU}(k) \qquad (4\text{-}16)$$

Note that – by construction – the frequency resolution of the BLA measurement with the Gaussian noise excitation is twice that of the periodic measurements.

4.2.4.2 Results Full Multisines. Figure 4-2 shows the mean output power spectrum (4-12) and the variance of the stochastic nonlinear distortions (4-13) for the full random phase multisines with random harmonic grid p = 0.1, 0.4, and 1. As p decreases the amplitudes of the excited output harmonics ('+') increase as $1/\sqrt{p}$ (see (4-5)). It can be seen that the variance of the stochastic nonlinear distortions var($Y_S(k)$) is independent of the value of p. Moreover, as predicted by Theorem 4.3 (see Table 4-1), var($Y_S(k)$) at the excited harmonics coincides with var($Y_S(k)$) at the nearest non-excited harmonic for all p-values. Note also that the noise level (extracted from the M residuals over the two signal periods) is well below the level of the nonlinear distortions, which justifies formulas (4-13) for calculating var($Y_S(k)$).

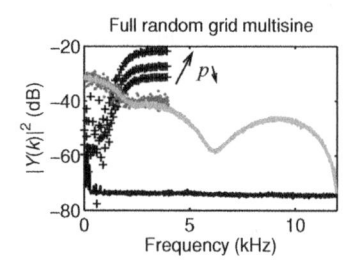

Figure 4-2. Output power spectrum of the parallel Wiener system excited by full random phase multisines with a random harmonic grid ($100 \times p$ = 10%, 40%, 100%). For all p-values, '+': $|Y_{BLA}(k)|^2$ (4-14) at the excited harmonics (only 1 out of 8 are shown); black lines: noise variance; light gray lines: var($Y_S(k)$) (4-13) at the non-excited harmonics; and dark gray dots: var($Y_S(k)$) (4-13) at the excited harmonics.

4.2.4.3 Comparison Gaussian Noise and Full Multisines. Figure 4-3 compares the mean output power spectrum (4-12) and the variance of the stochastic nonlinear distortions (4-13) of the full random phase multisine with random harmonic grid (p = 0.4) to that of the Gaussian noise excitation (see (4-12) and (4-15)). Since the multisine excites only $100 \times p$ = 40% of its harmonics, the amplitudes of the multisine harmonics (gray '+') are about 4 dB larger than the noise power spectrum (black '+'), which is in agreement with (4-5). As predicted by Theorem 4.3 (see Table 4-1), the variances of the stochastic nonlinear distortions are the same for both excitation signals (var($Y_S^{full}(k)$) = var($Y_S^{noise}(k)$)).

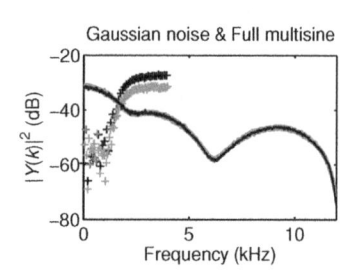

Figure 4-3. Output power spectrum of the parallel Wiener system excited by Gaussian noise (gray) and full random phase multisines (black) with a random harmonic grid ($100 \times p = 40\%$). '+': $|Y_{BLA}(k)|^2$ (4-14) and $\hat{S}_{Y_{BLA}Y_{BLA}}(k)$ (4-16) at the excited harmonics (only 1 out of 8 are shown); black line: var($Y_S^{full}(k)$) (4-13) of the full multisine; and gray line (coincides with the black line): var($Y_S^{noise}(k)$) (4-15) of the Gaussian noise.

4.2.4.4 Comparison Odd and Full Multisines. The left plot of Figure 4-4 compares the mean output power spectrum (4-12) and the variance of the stochastic nonlinear distortions (4-13) of the odd and full random phase multisine with random harmonic grid (p = 0.8). Since the odd multisine contains two times less harmonics than the full multisine, the odd harmonic amplitudes (gray '+') are 3 dB larger than those of the full multisine (black '+'). While for the odd multisines the odd (light gray line) and even (dark gray line) nonlinear distortions appear at, respectively, the odd and even harmonics only; all harmonics of the full multisine are disturbed by the sum of the odd and even nonlinear distortions (black line). As predicted by Theorem 4.3 (see Table 4-1), the mean of the odd and even nonlinear distortions of the odd multisine (see Figure 4-4, right plot, light gray) equals the total distortion of the full multisine (see Figure 4-4, right plot, dark gray): var($Y_S^{full}(2k+1)$) ≈ (var($Y_S^{odd}(2k+1)$) + var($Y_S^{odd}(2k)$))/2.

From the left plot of Figure 4-4 it can also be seen that the even nonlinear distortions are dominant below 2 kHz and almost drop to the noise level above 4 kHz, whereas the odd nonlinearities are present in the whole frequency band [0 Hz, 11 kHz]. It illustrates that measurements with odd random phase multisines with a random harmonic grid fully characterize a dynamic system from a single experiment: not only is the BLA obtained but also the nature (even and/or odd) and the level of the nonlinear distortions.

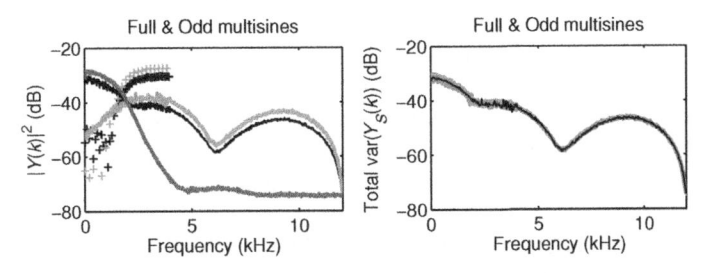

Figure 4-4. Output power spectra of the parallel Wiener system excited by full (black) and odd (gray) random phase multisine with a random harmonic grid ($100 \times p = 80\%$). '+' left figure: $|Y_{\text{BLA}}(k)|^2$ (4-14) at the excited harmonics (only 1 out of 8 are shown); black lines left and right figures: $\text{var}(Y_S^{\text{full}}(k))$ (4-13) full multisine; light and dark gray lines left figure: $\text{var}(Y_S^{\text{odd}}(2k+1))$ (4-13) and $\text{var}(Y_S^{\text{odd}}(2k))$ (4-13) odd multisine; and light gray line right figure: $(\text{var}(Y_S^{\text{odd}}(2k+1)) + \text{var}(Y_S^{\text{odd}}(2k)))/2$.

4.2.4.5 Best Linear Approximation. Figure 4-5 compares the best linear approximation (full lines) and its variance (dots) of the Gaussian noise excitation to that of the full (left plot) and odd (right plot) random phase multisines with random harmonic grid ($p = 0.1$, 0.4, 1). In agreement with Theorem 4.3, the best linear approximation (BLA) is the same for all the excitation signals. Because the BLA of even nonlinearities is (asymptotically for $N \to \infty$) zero for all excitations $u(t)$ belonging to the equivalence class $\mathbb{N}_{S_{uu}}$ (see Theorem 3.7 and (3-15)), the BLA of the nonlinear test system in Figure 4-1 is proportional to $G_2(j\omega_k)$ (see Section 3.4.3.5). It explains why the BLA in Figure 4-5 drops to zero at the lower frequencies.

From the left plot of Figure 4-5 it can be seen that the variance of the Gaussian noise BLA coincides with that of the full random phase multisine with $p = 1$. For decreasing values of p, the variance of the BLA decreases proportionally to p, which is not in contradiction with Theorem 4.3. Indeed, since the variances of the output stochastic nonlinear distortions are the same for the Gaussian noise and the full multisines (all p-values), and since the power of the excited harmonics of the full multisine increases with p (see (4-5)), it follows from (2-38) that the variance of the full multisine BLA is proportional to p.

Below 2.5 kHz, the variance of the odd random phase multisine BLAs is for all p-values smaller than that of the Gaussian noise (see Figure 4-5, right plot). This is explained by

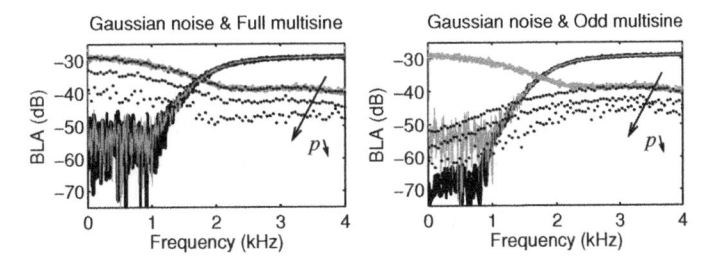

Figure 4-5. Best linear approximation (BLA) measured using Gaussian noise (left and right figure: gray), and full (left figure: black) and odd (right figure: black) random phase multisines with random harmonic grid ($100 \times p = 10\%, 40\%, 100\%$) - parallel Wiener system. Solid lines: BLA, and '•': variance BLA.

(i) for odd multisines, the even nonlinearities are only present at the even output harmonics and, hence, do not disturb the BLA measurements; and (ii) the even distortions are dominant in the band [0 kHz, 2.5 kHz]. Since the odd nonlinearities are dominant above 2.5 kHz, the variance of the Gaussian BLA coincides with that of the odd random phase multisine with $p = 1$ (see Table 4-1: the signal-to-distortion ratios are the same).

Note that the frequency resolution of the Gaussian BLA is $2/p$ ($4/p$) times larger than that of the full (odd) multisine BLA. Hence, the information content of the Gaussian BLA measurement is twice that of the full multisine BLA for all p-values (local averaging of the Gaussian BLA over $2/p$ neighboring frequencies reduces the variance by $2/p$). Similarly, if the odd distortions are dominant, then the information content of the Gaussian BLA measurement is four times that of the odd multisine BLA for all p-values.

4.3 DETECTION OF NONLINEAR DISTORTIONS USING RANDOM PHASE MULTISINES

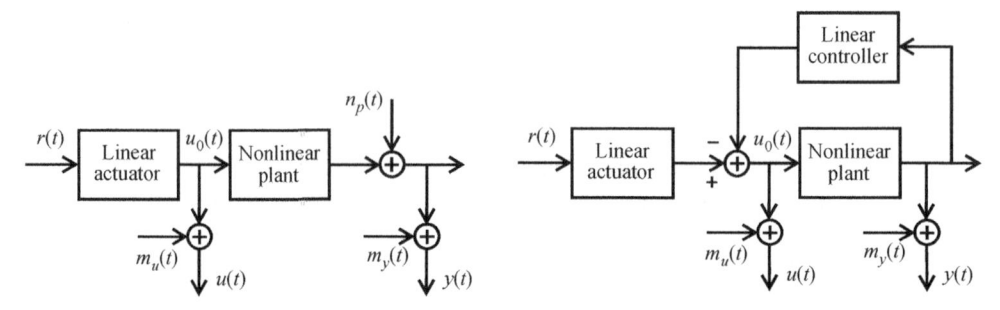

Figure 4-6. Setup for measuring the best linear approximation of a nonlinear system operating in open (left) or closed (right) loop. The nonlinear system is excited via a linear actuator by the reference signal $r(t)$. $u(t)$ and $y(t)$ are the noisy input-output observations and $n_p(t)$, $m_u(t)$, and $m_y(t)$ are, respectively, the process noise and the input-output measurement errors.

In this section we present two methods for fully characterizing FRF measurements of nonlinear dynamic systems operating in open or closed loop (see Figure 4-6). Beside the best linear approximation (BLA), the noise variance and the variance of the stochastic nonlinear distortions are also quantified. The first method – called the "robust" method – detects the nonlinear behavior via averaging of the FRF over multiple experiments with full or odd random phase multisines (direct application of definition (3-12)), while the second method – called the "fast" method – quantifies the nonlinear behavior via the non-excited output harmonics of one experiment with a full or odd random phase multisine with random harmonic grid (application of Theorem 4.3).

4.3.1 The Robust Method

4.3.1.1 Basic Idea. The robust method starts from multiple experiments with full or odd random phase multisines. For each experiment a number of consecutive periods of the steady state response are measured, and the FRF corresponding to each period is calculated. Averaging of the FRFs over the consecutive periods quantifies the noise level (the stochastic nonlinear distortions have the same periodicity as the multisine excitations). Averaging of these mean FRFs over the multiple experiments quantifies the sum of the remaining noise level and the level of the stochastic nonlinear distortions (the stochastic nonlinear distortions dependent on the random phase realization of the multisine excitations). Finally, the difference between the total distortion level (averaging over the experiments) and the noise level

Section 4.3 ■ Detection of Nonlinear Distortions Using Random Phase Multisines   131

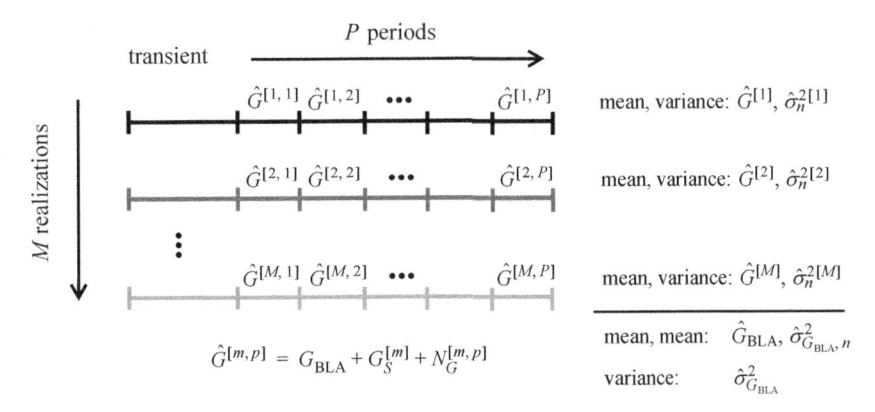

Figure 4-7. The robust procedure for measuring the BLA: P consecutive periods of the steady state response to M independent realizations of full or odd random phase multisine excitations are measured. $\hat{G}^{[m,p]}$ is the FRF estimate of the pth period of the mth experiment, which depends on the BLA G_{BLA}, the stochastic nonlinear distortions $G_S^{[m]}$, and the noise $N_G^{[m,p]}$. $\hat{G}^{[m]}$, $\hat{\sigma}_n^{2[m]}$ are, respectively, the sample mean and sample noise variance over the periods of the mth experiment. $\hat{G}_{\text{BLA}}$, $\hat{\sigma}_{G_{\text{BLA}}}^2$ are, respectively, the sample mean and sample total variance over the M experiments. Finally, $\hat{\sigma}_{G_{\text{BLA}},n}^2$ is the mean sample noise variance over the M realizations.

(averaging over the periods) is an estimate of the stochastic nonlinear distortions. The whole procedure is summarized in Figure 4-7 and discussed in detail in the next section.

4.3.1.2 Measurement Procedure. First, we explain in detail the procedure assuming that the system operates in open loop and that the input is known exactly (see Figure 4-6, left plot with $m_u(t) = 0$). Next, we generalize the method to the noisy input – noisy output case (see Figure 4-6, left plot). For the latter we must make the distinction between two situations: either the reference signal is available or it is unknown. Finally, we handle systems operating in closed loop (see Figure 4-6, right plot).

Known Input–Open Loop. If the input is known, then the FRF of each period can be calculated from the steady state response without introducing a bias error

$$\hat{G}^{[m,p]}(j\omega_k) = \frac{Y^{[m,p]}(k)}{U_0^{[m]}(k)} = G_{\text{BLA}}(j\omega_k) + G_S^{[m]}(k) + N_G^{[m,p]} \tag{4-17}$$

with $G_S^{[m]}(k) = Y_S^{[m]}(k)/U_0^{[m]}(k)$, $N_G^{[m,p]} = N_Y^{[m,p]}/U_0^{[m]}(k)$, $m = 1, 2, ..., M$, $p = 1, 2, ..., P$, (see Figure 4-7). Note that $Y_S^{[m]}(k)$ does not depend on the period index p because $y_s^{[m]}(t)$ has the same periodicity as the multisine excitation $u_0^{[m]}(t)$. Calculating the sample mean and the sample variance of the FRF estimates (4-17) over the P periods, gives

$$\hat{G}^{[m]}(j\omega_k) = \frac{1}{P}\sum_{p=1}^{P}\hat{G}^{[m,p]}(j\omega_k)$$
$$\hat{\sigma}_n^{2[m]}(k) = \frac{1}{P(P-1)}\sum_{p=1}^{P}\left|\hat{G}^{[m,p]}(j\omega_k) - \hat{G}^{[m]}(j\omega_k)\right|^2 \tag{4-18}$$

where $\hat{\sigma}_n^{2[m]}$ is the sample noise variance of the sample mean $\hat{G}^{[m]}(j\omega_k)$, which explains the extra factor P. Finally, we calculate over the M realizations the sample mean and the sample variance of the FRF estimates (4-18), and the mean of the sample noise variances (4-18)

$$\hat{G}_{\text{BLA}}(j\omega_k) = \frac{1}{M}\sum_{m=1}^{M}\hat{G}^{[m]}(j\omega_k)$$

$$\hat{\sigma}^2_{\hat{G}_{\text{BLA}}}(k) = \frac{1}{M(M-1)}\sum_{m=1}^{M}\left|\hat{G}^{[m]}(j\omega_k) - \hat{G}_{\text{BLA}}(j\omega_k)\right|^2 \quad (4\text{-}19)$$

$$\hat{\sigma}^2_{\hat{G}_{\text{BLA}},n}(k) = \frac{1}{M^2}\sum_{m=1}^{M}\hat{\sigma}^{2[m]}_n(k)$$

where $\hat{\sigma}^2_{\hat{G}_{\text{BLA}}}(k)$ and $\hat{\sigma}^2_{\hat{G}_{\text{BLA}},n}(k)$ are, respectively, the sample total variance and the sample noise variance of the BLA estimate $\hat{G}_{\text{BLA}}(j\omega_k)$, which explains the extra factor M.

The expected values of the sample total and noise variances (4-19) are given by

$$\mathbb{E}\{\hat{\sigma}^2_{\hat{G}_{\text{BLA}}}(k)\} = \frac{\sigma^2_{Y_S}(k)}{M|U_0(k)|^2} + \frac{\sigma^2_{Y,n}(k)}{MP|U_0(k)|^2}$$

$$\mathbb{E}\{\hat{\sigma}^2_{\hat{G}_{\text{BLA}},n}(k)\} = \frac{\sigma^2_{Y,n}(k)}{MP|U_0(k)|^2} \quad (4\text{-}20)$$

with $\sigma^2_{Y,n}(k) = \text{var}(N_Y^{[m,p]})$ the output noise variance of one period, $\sigma^2_{Y_S}(k) = \text{var}(Y_S^{[m]}(k))$ the variance of the output stochastic nonlinear distortions of one experiment, and where $|U_0(k)|^2$ is independent of the random phase realization. From (4-19) it follows that an estimate of the variance of the stochastic nonlinear distortions $G_S^{[m]}(k)$ on the BLA w.r.t. one multisine experiment, $\sigma^2_S(k) = \text{var}(G_S^{[m]}(k)) = \sigma^2_{Y_S}(k)/|U_0(k)|^2$, can be readily obtained as

$$\hat{\sigma}^2_S(k) = \begin{cases} M(\hat{\sigma}^2_{\hat{G}_{\text{BLA}}}(k) - \hat{\sigma}^2_{\hat{G}_{\text{BLA}},n}(k)) & \hat{\sigma}^2_{\hat{G}_{\text{BLA}}}(k) > \hat{\sigma}^2_{\hat{G}_{\text{BLA}},n}(k) \\ 0 & \hat{\sigma}^2_{\hat{G}_{\text{BLA}}}(k) \leq \hat{\sigma}^2_{\hat{G}_{\text{BLA}},n}(k) \end{cases} \quad (4\text{-}21)$$

An estimate of $\sigma^2_{Y_S}(k) = \text{var}(Y_S^{[m]}(k))$ is then given by $\hat{\sigma}^2_{Y_S}(k) = \hat{\sigma}^2_S(k)|U_0(k)|^2$.

Noisy Input–Unknown Reference–Open Loop. To avoid a bias in the FRF estimate

$$\hat{G}^{[m,p]}(j\omega_k) = Y^{[m,p]}(k)/U^{[m,p]}(k)$$

the noisy input-output DFT spectra are first averaged over the P periods before calculating the ratio. This gives the following input-output sample means and input-output sample noise (co)variances

$$\hat{X}^{[m]}(k) = \frac{1}{P}\sum_{p=1}^{P}\hat{X}^{[m,p]}(k)$$

$$\hat{\sigma}^{2[m]}_{XZ,n}(k) = \frac{1}{P(P-1)}\sum_{p=1}^{P}(\hat{X}^{[m,p]}(k) - \hat{X}^{[m]}(k))\overline{(\hat{Z}^{[m,p]}(k) - \hat{Z}^{[m]}(k))} \quad (4\text{-}22)$$

with $X, Z = Y$ and/or U. The FRF and its noise variance are then obtained as

Section 4.3 ■ Detection of Nonlinear Distortions Using Random Phase Multisines

$$\hat{G}^{[m]}(j\omega_k) = \frac{\hat{Y}^{[m]}(k)}{\hat{U}^{[m]}(k)}$$

$$\hat{\sigma}_n^{2[m]}(k) = |\hat{G}^{[m]}(j\omega_k)|^2 \left(\frac{\hat{\sigma}_{Y,n}^{2[m]}(k)}{|\hat{Y}^{[m]}(k)|^2} + \frac{\hat{\sigma}_{U,n}^{2[m]}(k)}{|\hat{U}^{[m]}(k)|^2} - 2\text{Re}\left(\frac{\hat{\sigma}_{YU,n}^{2[m]}(k)}{\hat{Y}^{[m]}(k)\overline{\hat{U}^{[m]}(k)}} \right) \right) \quad (4\text{-}23)$$

(see (2-25)). The rest of the procedure follows exactly the same lines as in the known input case (see (4-19)–(4-21) where $N_Y(k)$ is replaced by $N_Y(k) - G_{\text{BLA}}(j\omega_k)N_U(k)$.

Noisy Input–Known Reference–Open Loop. To reduce the bias in the FRF estimate (4-23) due to the input noise even more, the mean input-output DFT spectra should be averaged over the M realizations before calculating the ratio. However, straightforward averaging is impossible because of the independent random phase realizations over the experiments. Therefore, before averaging, the phases of the mean input-output DFT spectra $\hat{U}^{[m]}(k)$ and $\hat{Y}^{[m]}(k)$ must be turned back by the phase of a known reference signal $R^{[m]}(k)$ (typically the signal $r(t)$ stored in the arbitrary waveform generator; see Figure 4-6, left plot) for each realization, giving

$$\hat{X}_R^{[m]}(k) = \hat{X}^{[m]}(k)/e^{j\angle R^{[m]}(k)} \quad (4\text{-}24)$$

with $X = U, Y$ ($|R^{[m]}(k)|$ is independent of m). The sample means, and sample noise and total (co)variances of the projected input-output DFT spectra (4-24) equal

$$\hat{X}_R(k) = \frac{1}{M} \sum_{m=1}^{M} \hat{X}_R^{[m]}(k)$$

$$\hat{\sigma}_{X_R Z_R}^2(k) = \frac{1}{M(M-1)} \sum_{m=1}^{M} (\hat{X}_R^{[m]}(k) - \hat{X}_R(k))(\overline{\hat{Z}_R^{[m]}(k) - \hat{Z}_R(k)}) \quad (4\text{-}25)$$

$$\hat{\sigma}_{X_R Z_R, n}^2(k) = \frac{1}{M^2} \sum_{m=1}^{M} \hat{\sigma}_{XZ,n}^{2[m]}(k)$$

with $X, Z = Y$ and/or U, and where the last equality uses $\hat{\sigma}_{X_R Z_R, n}^{2[m]}(k) = \hat{\sigma}_{XZ,n}^{2[m]}(k)$. Using (4-25), the BLA and its noise and total variances are then estimated as

$$\hat{G}_{\text{BLA}}(j\omega_k) = \hat{Y}_R(k)/\hat{U}_R(k)$$

$$\hat{\sigma}_{\hat{G}_{\text{BLA}}}^2(k) = |\hat{G}_{\text{BLA}}(j\omega_k)|^2 \left(\frac{\hat{\sigma}_{Y_R}^2(k)}{|\hat{Y}_R(k)|^2} + \frac{\hat{\sigma}_{U_R}^2(k)}{|\hat{U}_R(k)|^2} - 2\text{Re}\left(\frac{\hat{\sigma}_{Y_R U_R}^2(k)}{\hat{Y}_R(k)\overline{\hat{U}_R(k)}} \right) \right) \quad (4\text{-}26)$$

$$\hat{\sigma}_{\hat{G}_{\text{BLA}}, n}^2(k) = |\hat{G}_{\text{BLA}}(j\omega_k)|^2 \left(\frac{\hat{\sigma}_{Y_R, n}^2(k)}{|\hat{Y}_R(k)|^2} + \frac{\hat{\sigma}_{U_R, n}^2(k)}{|\hat{U}_R(k)|^2} - 2\text{Re}\left(\frac{\hat{\sigma}_{Y_R U_R, n}^2(k)}{\hat{Y}_R(k)\overline{\hat{U}_R(k)}} \right) \right)$$

Finally, the variance of the stochastic nonlinear distortions $\sigma_S^2(k)$ ($\sigma_{Y_S}^2(k)$) is derived from (4-26) as in (4-21).

Closed Loop. If the system operates in closed loop (see Figure 4-6, right plot), then the best linear approximation is defined via the indirect method (3-35). This is exactly what is calculated in (4-24)–(4-26) because

$$\hat{G}_{\mathrm{BLA}}(j\omega_k) = \frac{\hat{S}_{\hat{Y}R}(k)}{\hat{S}_{\hat{U}R}(k)} = \frac{\sum_{m=1}^{M} \hat{Y}^{[m]}(k)\overline{R^{[m]}(k)}}{\sum_{m=1}^{M} \hat{U}^{[m]}(k)\overline{R^{[m]}(k)}} = \frac{\hat{Y}_R(k)}{\hat{U}_R(k)} \quad (4\text{-}27)$$

where the last equality uses the fact that $|R^{[m]}(k)|$ is independent of m. Hence, we can follow the same procedure as for the noisy input–known reference case to estimate the BLA, its noise and total variance, and the variance of the stochastic nonlinear distortions $\sigma_S^2(k)$. The variance of the output stochastic nonlinear distortions obtained from $\hat{\sigma}_S^2(k)$ as $\hat{\sigma}_{Y_S}^2(k) = \hat{\sigma}_S^2(k)|\hat{U}_R(k)|^2$ is exactly the variance of $Y_S(k)$ in Theorem 3.22. Using (3-118) it can also be calculated from the input-output sample noise and sample total (co)variances (4-25) as

$$\hat{\sigma}_{Y_S}^2(k) = \hat{\sigma}_{\hat{Y}_S}^2(k) + |\hat{G}_{\mathrm{BLA}}(j\omega_k)|^2 \hat{\sigma}_{\hat{U}_S}^2(k) - 2\mathrm{Re}(\hat{\sigma}_{\hat{Y}_S \hat{U}_S}^2(k)\overline{\hat{G}_{\mathrm{BLA}}(j\omega_k)}) \quad (4\text{-}28)$$

with $\hat{\sigma}_{\hat{X}_S \hat{Z}_S}^2(k) = M(\hat{\sigma}_{\hat{X}_R \hat{Z}_R}^2(k) - \hat{\sigma}_{\hat{X}_R \hat{Z}_R, n}^2(k))$, $X, Z = Y$ and/or U, an estimate of the (co)variances of the input-output stochastic nonlinear distortions.

Remarks

- If the nonlinear system is excited by a non-ideal actuator, then, due to the interaction between the actuator and the nonlinear system, the input is also disturbed by stochastic nonlinear distortions. In that case, the averaging over the M realizations in (4-25) also reduces the bias of the estimated BLA due to the input stochastic nonlinear distortions. The same is valid for the averaging over M in (4-27).

- The optimal choice of the number of periods P and the number of independent random phase realizations (experiments) M in the robust method (see Figure 4-7) for a given measurement time $T = M \times P \times NT_s$ depends on the ultimate goal of the identification experiment. If the aim is to minimize the total variance of the best linear approximation, while maintaining the ability to distinguish the noise from the nonlinear distortions, then one should maximize the averaging of the nonlinear distortions. Hence, $P = 2$ and $M = T/(P \times NT_s)$ is the best choice (see (4-20)). If the objective is to maximize the nonlinear detection sensitivity, then the noise should be suppressed while keeping the level of the nonlinear distortions. In this case $M = 2$ and $P = T/(M \times NT_s)$ is the optimal choice (see (4-20)).

- Whether the full or the odd random phase multisines are optimal for minimizing the total variance of the BLA measurement depends on the nature of the nonlinear distortions. If the odd nonlinear distortions are dominant $(S_{y_s y_s, \mathrm{odd}}(j\omega_k)) \gg S_{y_s y_s, \mathrm{even}}(j\omega_k))$, then the signal-to-distortion ratios of the full and odd random phase multisines are the same (see Table 4-1), and the information content of the full multisine BLA measurement is twice that of the odd multisine (the frequency resolution of the full multisine is twice as large, and averaging the full multisine BLA over two neighboring frequencies reduces its variance by a factor 2). However, if the even nonlinear distortions are dominant $(S_{y_s y_s, \mathrm{even}}(j\omega_k) \gg S_{y_s y_s, \mathrm{odd}}(j\omega_k))$, then the signal-to-distortion power ratio of the odd multisine is much larger than that of the full multisine (see Table 4-1). Hence, the information content of the odd multisine BLA measurement is much larger than that of the full multisine (infinitely larger if there are no odd distortions). If the even and odd nonlinear distortions are of the same order of magnitude $(S_{y_s y_s, \mathrm{odd}}(j\omega_k) \sim S_{y_s y_s, \mathrm{even}}(j\omega_k))$, then the information content of both full and odd multisine BLA measurements is the same.

Section 4.3 ■ Detection of Nonlinear Distortions Using Random Phase Multisines 135

4.3.2 The Fast Method

4.3.2.1 Basic Idea. The fast method starts from one experiment with a full or odd random phase multisine with random harmonic grid. The random harmonic grid is generated by grouping the excited (odd) harmonics in N_{sub} consecutive harmonics, and eliminating exactly $100 \times (1-p)\%$ randomly selected (odd) harmonics in each group (see Figure 4-8, black arrows). A number of consecutive periods of the steady state response are measured, and the input-output DFT spectra of each period are calculated. The sample mean and sample standard deviation of these input-output DFT spectra over the periods quantify, respectively, the signal and the noise levels (see Figure 4-8, black arrows and thin black horizontal line). Nonlinear distortions are present if the signal level is above the noise level at the non-excited harmonics (see Figure 4-8, left plot, gray arrows at harmonics 2–4, 6, 10–14, and 16; and Figure 4-8, right plot, gray arrows at harmonics 3, 5, and 7). Finally, using Theorem 4.3, the level of the nonlinear distortions at the excited (odd) harmonics is estimated as the signal level of the nearest non-excited (odd) harmonic (see Figure 4-8, dark gray arrows).

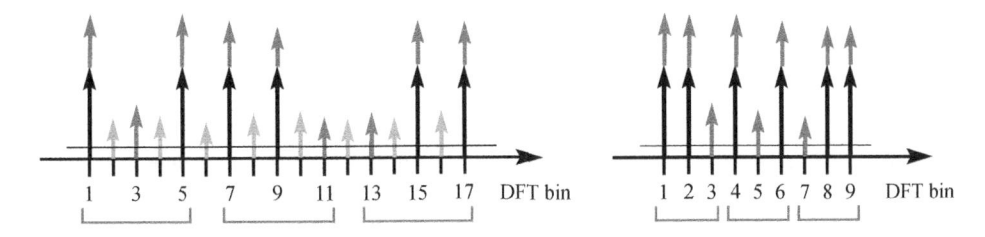

Figure 4-8. DFT spectrum of the steady state response of a nonlinear system to a random phase multisine with random harmonic grid ($N_{\text{sub}} = 3, p = 2/3$). Left: odd multisine; and right: full multisine. Black arrows: the linear contribution; light gray arrows: the even nonlinear distortions; dark gray arrows: the odd (left) or odd plus even (right) nonlinear contributions; and thin black horizontal lines: noise level.

4.3.2.2 Measurement Procedure. First, we explain in detail the procedure for nonlinear system operating in open loop (see Figure 4-6, left plot). Next, we extend the method to the noisy input data. Finally, we discuss the closed loop case (see Figure 4-6, right plot).

Known Input–Open Loop. Since the nonlinear system operates in open loop, the output DFT spectrum of each period of the steady state response to a full or odd random phase multisine with random harmonic grid is given by

$$Y^{[p]}(k) = Y_0(k) + N_Y^{[p]}(k) + Y_S(k) \tag{4-29}$$

with $p = 1, 2, \ldots, P$. The sample mean and sample noise covariance of (4-29) over the P periods equal

$$\hat{Y}(k) = \frac{1}{P} \sum_{p=1}^{P} Y^{[p]}(k)$$

$$\hat{\sigma}_{\hat{Y}, n}^2(k) = \frac{1}{P(P-1)} \sum_{p=1}^{P} |Y^{[p]}(k) - \hat{Y}(k)|^2 \tag{4-30}$$

with $\hat{\sigma}_{\hat{Y}, n}^2(k)$ the sample noise variance of the sample mean $\hat{Y}(k)$, which explains the extra factor P. The BLA and its noise variance are readily obtained from (4-30) as

$$\hat{G}_{\text{BLA}}(j\omega_{k_{\text{exc}}}) = \frac{\hat{Y}(k_{\text{exc}})}{U_0(k_{\text{exc}})} \text{ and } \hat{\sigma}^2_{\hat{G}_{\text{BLA}},n}(k_{\text{exc}}) = \frac{\hat{\sigma}^2_{\hat{Y},n}(k_{\text{exc}})}{|U_0(k_{\text{exc}})|^2} \qquad (4\text{-}31)$$

where k_{exc} is an excited (odd) harmonic. Note that the expected value of the sample variance (4-31) is given by

$$\mathbb{E}\{\hat{\sigma}^2_{\hat{G}_{\text{BLA}},n}(k_{\text{exc}})\} = \frac{\sigma^2_{Y,n}(k_{\text{exc}})}{P|U_0(k_{\text{exc}})|^2} \qquad (4\text{-}32)$$

with $\sigma^2_{Y,n}(k) = \text{var}(N_Y^{[p]}(k))$ the output noise variance of one period. In the sequel we analyze (4-30) at the non-excited harmonics to detect (and classify) the nonlinear distortions and to estimate the level of the stochastic nonlinear distortions on the BLA estimate (4-31).

Assuming that no signal energy is present at the non-excited output harmonic $k_{\text{non-exc}}$ ($Y_0(k_{\text{non-exc}}) = 0$), the ratio $|\hat{Y}(k_{\text{non-exc}})|^2 / \hat{\sigma}^2_{\hat{Y},n}(k_{\text{non-exc}})$ is $F(2, 2P-2)$- distributed (Stuart and Ord, 1987) and the following null hypothesis test can be constructed. If

$$|\hat{Y}(k_{\text{non-exc}})|^2 / \hat{\sigma}^2_{\hat{Y},n}(k_{\text{non-exc}}) \leq F_p(2, 2P-2) \qquad (4\text{-}33)$$

with $F_p(2, 2P-2)$ the $100 \times p\%$ percentile of an $F(2, 2P-2)$- distributed random variable, then the null hypothesis $Y_0(k_{\text{non-exc}}) = 0$ is accepted, otherwise it is rejected. Performing the test (4-33) at all non-excited harmonics allows one to detect, quantify, and classify (odd multisines only) the nonlinear distortions.

The total variance (sum of the variance of the stochastic nonlinear distortions and the noise variance) at an excited harmonic k_{exc} is estimated as the magnitude squared of the level of the nearest non-excited (odd) harmonic

$$\hat{\sigma}^2_{\hat{Y}}(k_{\text{exc}}) = |\hat{Y}(k_{\text{non-exc}})|^2 \qquad (4\text{-}34)$$

Using Theorem 4.3, the expected value of (4-34) is found to be equal to

$$\mathbb{E}\{\hat{\sigma}^2_{\hat{Y}}(k_{\text{exc}})\} = \sigma^2_{Y,n}(k_{\text{exc}})/P + \sigma^2_{Y_S}(k_{\text{exc}}) + O(N^{-1}) \qquad (4\text{-}35)$$

with $\sigma^2_{Y_S}(k) = \text{var}(Y_S(k))$ the variance of the output stochastic nonlinear distortions. Dividing (4-34) by the magnitude squared of the input DFT spectrum gives an estimate of the total variance of the BLA

$$\hat{\sigma}^2_{\hat{G}_{\text{BLA}}}(k_{\text{exc}}) = \hat{\sigma}^2_{\hat{Y}}(k_{\text{exc}})/|U_0(k_{\text{exc}})|^2 \qquad (4\text{-}36)$$

Finally, the variance of the stochastic nonlinear distortions on the BLA is estimated as

$$\hat{\sigma}^2_{\hat{S}}(k_{\text{exc}}) = \begin{cases} \hat{\sigma}^2_{\hat{G}_{\text{BLA}}}(k_{\text{exc}}) - \hat{\sigma}^2_{\hat{G}_{\text{BLA}},n}(k_{\text{exc}}) & \hat{\sigma}^2_{\hat{G}_{\text{BLA}}}(k_{\text{exc}}) > \hat{\sigma}^2_{\hat{G}_{\text{BLA}},n}(k_{\text{exc}}) \\ 0 & \hat{\sigma}^2_{\hat{G}_{\text{BLA}}}(k_{\text{exc}}) \leq \hat{\sigma}^2_{\hat{G}_{\text{BLA}},n}(k_{\text{exc}}) \end{cases} \qquad (4\text{-}37)$$

(proof: collect (4-31), (4-32), (4-35), and (4-36)).

Section 4.3 ■ Detection of Nonlinear Distortions Using Random Phase Multisines 137

Noisy Input–Open Loop. If the input is noisy, then $U^{[p]}(k) = U_0(k) + N_U^{[p]}(k)$ is added to (4-29), and (4-30) is replaced by

$$\hat{X}(k) = \frac{1}{P} \sum_{p=1}^{P} \hat{X}^{[p]}(k)$$

$$\hat{\sigma}_{\hat{X}\hat{Z},n}^2(k) = \frac{1}{P(P-1)} \sum_{p=1}^{P} (\hat{X}^{[p]}(k) - \hat{X}(k))(\overline{\hat{Z}^{[p]}(k) - \hat{Z}(k)})$$

(4-38)

with $X, Z = Y$ and/or U. The BLA and its noise variance are calculated using (4-38) at the excited harmonics k_{exc}

$$\hat{G}_{\text{BLA}}(j\omega_{k_{\text{exc}}}) = \frac{\hat{Y}(k_{\text{exc}})}{\hat{U}(k_{\text{exc}})}$$

$$\hat{\sigma}_{\hat{G}_{\text{BLA}},n}^2(k_{\text{exc}}) = |\hat{G}_{\text{BLA}}(j\omega_{k_{\text{exc}}})|^2 \left(\frac{\hat{\sigma}_{\hat{Y},n}^2(k_{\text{exc}})}{|\hat{Y}(k_{\text{exc}})|^2} + \frac{\hat{\sigma}_{\hat{U},n}^2(k_{\text{exc}})}{|\hat{U}(k_{\text{exc}})|^2} - 2\operatorname{Re}(\frac{\hat{\sigma}_{\hat{Y}\hat{U},n}^2(k_{\text{exc}})}{\hat{Y}(k_{\text{exc}})\overline{\hat{U}(k_{\text{exc}})}}) \right)$$

(4-39)

Note that the noise variance of the BLA depends on both the input $N_U^{[p]}(k)$ and output $N_Y^{[p]}(k)$ noise, while the total variance of the BLA calculated as in (4-34)–(4-36) is independent of the input noise. Hence, (4-37), where $\hat{\sigma}_{G_{\text{BLA}}}^2(k_{\text{exc}})$ is calculated as in (4-36), is inappropriate for estimating the level of the stochastic nonlinear distortions because it contains a bias term depending on the input noise. In the sequel we handle this problem.

To account for the input noise and the possible correlation between the input and output noise, we replace the output spectrum at the non-excited harmonics in (4-34) by the corrected output $\hat{Y}_c(k_{\text{non-exc}})$ defined as

$$\hat{Y}_c(k_{\text{non-exc}}) = \hat{Y}(k_{\text{non-exc}}) - \hat{G}_{\text{BLA}}(j\omega_{k_{\text{non-exc}}})\hat{U}(k_{\text{non-exc}})$$

(4-40)

where $\hat{G}_{\text{BLA}}(j\omega_{k_{\text{non-exc}}})$ is obtained via linear interpolation of the BLA (4-39) at the closest excited (odd) harmonics $k_{1\text{exc}} < k_{\text{non-exc}} < k_{2\text{exc}}$

$$\hat{G}_{\text{BLA}}(j\omega_{k_{\text{non-exc}}}) = \frac{(k_{2\text{exc}} - k_{\text{non-exc}})\hat{G}_{\text{BLA}}(j\omega_{k_{1\text{exc}}}) + (k_{\text{non-exc}} - k_{1\text{exc}})\hat{G}_{\text{BLA}}(j\omega_{k_{2\text{exc}}})}{k_{2\text{exc}} - k_{1\text{exc}}}$$

(4-41)

Using Theorem 4.3, $G_{\text{BLA}}(j\omega_{k_{\text{non-exc}}}) = G_{\text{BLA}}(j\omega_{k_{\text{exc}}}) + O(N^{-1})$, and neglecting the noise on $\hat{G}_{\text{BLA}}(j\omega_{k_{\text{non-exc}}})$ as a second order effect, the expected value of the total variance estimate at an excited harmonic

$$\hat{\sigma}_{\hat{Y}_c}^2(k_{\text{exc}}) = |\hat{Y}_c(k_{\text{non-exc}})|^2$$

(4-42)

is given by

$$\mathbb{E}\{\hat{\sigma}^2_{\hat{Y}_c}(k_{\text{exc}})\} \approx |U_0(k_{\text{exc}})|^2 \sigma^2_{G_{\text{BLA}},n}(k_{\text{exc}}) + \sigma^2_{Y_S}(k_{\text{exc}}) + O(N^{-1})$$

$$\sigma^2_{G_{\text{BLA}},n}(k_{\text{exc}}) = \frac{|G_{\text{BLA}}(j\omega_{k_{\text{exc}}})|^2}{P} \left(\frac{\sigma^2_{Y,n}(k_{\text{exc}})}{|Y_0(k_{\text{exc}})|^2} + \frac{\sigma^2_{U,n}(k_{\text{exc}})}{|U_0(k_{\text{exc}})|^2} - 2\operatorname{Re}\left(\frac{\sigma^2_{YU,n}(k_{\text{exc}})}{Y_0(k_{\text{exc}})\overline{U}_0(k_{\text{exc}})} \right) \right) \quad (4\text{-}43)$$

Hence, (4-37), where $\hat{\sigma}^2_{G_{\text{BLA}},n}(k_{\text{exc}})$ is given by (4-39) and where the total variance $\hat{\sigma}^2_{G_{\text{BLA}}}(k_{\text{exc}})$ is calculated as

$$\hat{\sigma}^2_{G_{\text{BLA}}}(k_{\text{exc}}) = \hat{\sigma}^2_{\hat{Y}_c}(k_{\text{exc}}) / |\hat{U}(k_{\text{exc}})|^2, \quad (4\text{-}44)$$

correctly eliminates the influence of the input noise in the estimated variance of the stochastic nonlinear distortions (proof: collect (4-39) and (4-42)–(4-44) using $\mathbb{E}\{\hat{\sigma}^2_{\hat{X}\hat{Z},n}(k)\} = \sigma^2_{YU,n}(k)/P$).

Closed Loop. If the nonlinear system operates in closed loop (see Figure 4-6, right plot), then both the input and output DFT spectra are disturbed by the stochastic nonlinear distortions, and (4-29) is replaced by

$$\begin{aligned} Y^{[p]}(k) &= Y_0(k) + N_Y^{[p]}(k) + \tilde{Y}_S(k) \\ U^{[p]}(k) &= U_0(k) + N_U^{[p]}(k) + \tilde{U}_S(k) \end{aligned} \quad (4\text{-}45)$$

The difficulty here is that the signal level at the non-excited output harmonics $\tilde{Y}_S(k_{\text{non-exc}})$ is no longer a measure of the stochastic nonlinear distortions $Y_S(k)$ (difference between the actual output of the nonlinear system and the output of the BLA). This problem is solved via the results of Theorem 3.22 on page 94 and Appendix 3.O on page 118 where it is shown that

$$Y_0(k) = G_{\text{BLA}}(j\omega_k)U_0(k) \text{ and } Y_S(k) = \tilde{Y}_S(k) - G_{\text{BLA}}(j\omega_k)\tilde{U}_S(k) \quad (4\text{-}46)$$

Comparing (4-40) to (4-46) it follows that the corrected output DFT spectrum $\hat{Y}_c(k_{\text{non-exc}})$ (4-40) compensates properly the output nonlinear distortions for the presence of the input nonlinear distortions. We conclude that the fast procedure for noisy input – noisy output signals correctly handles nonlinear systems operating in closed loop.

Remarks

- The fast method can be refined by replacing the nearest neighbor estimates (4-34) and (4-42) by a linear interpolation.
- The total variance estimates of the BLA in (4-36) and (4-44) are based on one output residual (see (4-34) and (4-42)). The variability of these estimates can be reduced at the cost of introducing an $O(N^{-1})$ bias by averaging (4-34) and (4-42) over neighboring non-excited (odd) harmonics.
- The value of N_{sub} (number of consecutive excited (odd) harmonics) for generating the random harmonic grid is chosen small (typically 2, 3, and 4) in order to guarantee a uniform frequency resolution of the BLA measurement and the nonlinearity detection. Via the fraction p of the randomly selected excited harmonics in N_{sub}, a trade-off is made between the frequency resolution of the BLA measurement

(pf_s/N) and the nonlinearity detection $((1-p)f_s/N)$. Typical choices are $(N_{sub}, p) = (2, 0.5)$, $(3, 2/3)$, and $(4, 3/4)$, which corresponds to leaving out 1 in 2, 3, or 4 consecutive excited frequencies.

- The optimal choice of the number of periods P for fixed measurement time depends on the goal of the identification experiment. If the purpose is to maximize the sensitivity of the nonlinearity detection, then P should be chosen as large as possible. However, if the aim is to maximize the information content (Fisher information matrix) of the BLA measurement (minimize the total variance of the parametric BLA modeling), while maintaining the ability to distinguish the noise from the distortions, then $P = 2$ is optimal.

- For nonlinear systems operating in closed loop (see Figure 4-6), the FRF calculated at the non-excited harmonics equals minus one over the feedback dynamics. Indeed, assuming that no noise is present, the input-output DFT spectra are related as

$$U(k) = G_{act}(j\omega_k)R(k) - M_0(j\omega_k)Y(k) \quad (4\text{-}47)$$

with $G_{act}(j\omega)$ and $M_0(j\omega)$, respectively, the linear actuator and controller (feedback) dynamics. Finally, evaluating (4-47) at the non-excited harmonics

$$U(k_{non\text{-}exc}) = -M_0(j\omega_{k_{non\text{-}exc}})Y(k_{non\text{-}exc}) \quad (4\text{-}48)$$

proves the statement.

4.3.3 Discussion

4.3.3.1 Nonlinear Actuator. If the actuator behaves nonlinearly and the plant is linear, then the actuator distortions act as generator noise: the estimated BLA is equal to the linear plant dynamics and its total variance only depends on the input-output noise (co)variances (proof: see Appendix 4.B).

If both the actuator and the plant are nonlinear, then the correction of the output DFT spectrum (4-40) in the fast method is a first order compensation for the spectral impurity of the plant input (it is exact for nonlinear plants operating in feedback and linear actuators, or for linear plants and nonlinear actuators). If the signal-to-distortion ratio at the output of the actuator is sufficiently large (e.g., > 20 dB), then the total variance of the BLA (4-44) predicted by the fast method coincides with that (4-26) of the robust method. Note that the robust method does not need the assumption of a sufficiently large actuator signal-to-distortion ratio.

4.3.3.2 Classification in Even and Odd Distortions. Using odd random phase multisines with a random harmonic grid, the FRF measurement of a nonlinear system operating in open loop and driven by a linear actuator can be fully characterized (see Section 4.2): besides estimates of the BLA, its noise variance, and its total variance, the level of the odd and even nonlinear distortions is also quantified. The fast method discussed in Section 4.3.2 extracts this information from one single experiment. If the actuator behaves nonlinearly, then the fast method still estimates the correct BLA and its noise and total variances, provided the actuator signal-to-distortion ratio is sufficiently large. However, the estimated level of the even and odd nonlinear distortions might be biased. This is illustrated in the following example.

Consider a Hammerstein system (see Figure 3-3 on page 85 with $R(j\omega) = 1$) with static nonlinearity $z(t) = \alpha u^2(t) + \beta u^3(t)$. Since $u^2(t)$ and $u^3(t)$ combine, respectively, two and three frequencies of $u(t)$ (see Section 3.2), the dominant stochastic nonlinear contributions at the even detection lines $2m$ in $z(t)$ are of the form

$$\begin{aligned}\alpha u^2(t): \quad & \alpha U(2k_1+1)U(2k_2+1) & 2k_1+2k_2+2 = 2m \\ \beta u^3(t): \quad & \beta U(2k_3+1)U(2k_4+1)U(2l) & 2k_3+2k_4+2l+2 = 2m\end{aligned} \qquad (4\text{-}49)$$

with $U(2k_i+1)$, $i = 1, \ldots, 4$, the excited odd harmonics and $U(2l)$ an even distortion line. If $|\beta U(2l)|$ is not much smaller than $|\alpha|$, then the third degree contribution in (4-49) will bias the estimated level of the even nonlinear distortions. As a rule of thumb, this bias can be neglected if the ratio of the power spectra (power spectral densities) of the odd to even nonlinear distortions is much smaller than the ratio of the power spectra (power spectral densities) of the odd excited harmonics to the even distortion lines

$$\frac{S_{y_s y_s, \text{odd}}(j\omega)}{S_{y_s y_s, \text{even}}(j\omega)} \ll \frac{S_{uu, \text{odd}}(j\omega)}{S_{uu, \text{even}}(j\omega)} \Rightarrow \text{no bias on even distortions} \qquad (4\text{-}50)$$

(see Appendix 4.C for a rationale). The relationship between the DFT spectra and the power spectral densities is given in, respectively, (4-5) and (4-10).

Similarly, the dominant stochastic nonlinear contributions at the odd detection lines $2m+1$ in $z(t)$ are of the type

$$\begin{aligned}\alpha u^2(t): \quad & \alpha U(2k_1+1)U(2l) & 2k_1+2l+1 = 2m+1 \\ \beta u^3(t): \quad & \beta U(2k_2+1)U(2k_3+1)U(2k_4+1) & 2k_2+2k_3+2k_4+3 = 2m+1\end{aligned} \qquad (4\text{-}51)$$

with $U(2k_i+1)$, $i = 1, \ldots, 4$, the excited odd harmonics and $U(2l)$ an even distortion line. If $|\alpha U(2l)|$ is not much smaller than $|\beta U(2k_2+1)U(2k_3+1)|$, then the even nonlinear contribution in (4-51) will bias the estimated level of the odd nonlinear distortions. As a rule of thumb, this bias can be neglected if

$$\frac{S_{y_s y_s, \text{even}}(j\omega)}{S_{y_s y_s, \text{odd}}(j\omega)} \ll \frac{S_{uu, \text{odd}}(j\omega)}{S_{uu, \text{even}}(j\omega)} \Rightarrow \text{no bias on odd distortions} \qquad (4\text{-}52)$$

is satisfied (see Appendix 4.C for a rationale).

Conclusion. The bias on the estimated levels of the even and odd nonlinear distortions can be neglected if the even and odd distortions are of the same order of magnitude ($S_{y_s y_s, \text{even}}(j\omega) \sim S_{y_s y_s, \text{odd}}(j\omega)$), and if the input signal-to-even-distortion ratio is sufficiently large ($S_{uu, \text{odd}}(j\omega)/S_{uu, \text{even}}(j\omega) \gg 1$). If, for example, the even distortions are dominant ($S_{y_s y_s, \text{even}}(j\omega) \gg S_{y_s y_s, \text{odd}}(j\omega)$), then (4-50) is automatically satisfied and the level of the even nonlinear distortions is correctly estimated. However, unbiased estimation of the level of the odd distortions puts a severe constraint on the input signal-to-even-distortion ratio (see (4-52)). If the odd distortions are dominant, then the opposite is true: the level of the odd nonlinear distortions is correctly estimated, while unbiased estimation of the level of the even distortions puts a strong condition on the input signal-to-even-distortion ratio (see (4-50) with $S_{y_s y_s, \text{odd}}(j\omega) \gg S_{y_s y_s, \text{even}}(j\omega)$). These results are also valid for nonlinear systems operating in feedback. Note that conditions (4-50) and (4-52) can easily be checked a posteriori.

Figure 4-9. Basic block diagram for measuring the open loop gain of an operational amplifier (opamp). The × 1 voltage buffers prevent loading of the circuit and the opamp, and the resistor values are matched $R_1 \approx R_3 \approx 300\ \Omega$, and $R_2 \approx R_4 \approx 12\ k\Omega$. The circuit is excited by the voltage source $v_g(t)$ with output impedance $R_g = 50\ \Omega$, and the input $u(t)$ and output $y(t)$ voltages of the opamp are measured.

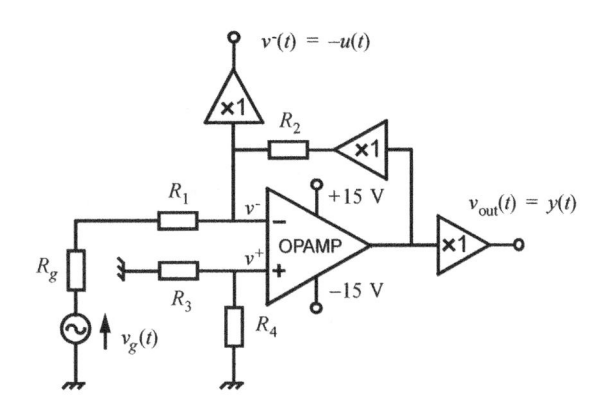

4.3.4 Experimental Illustration

The goal of the measurement example is threefold: (i) experimental illustration of the fast method (Section 4.3.2) on a nonlinear system operating in feedback; (ii) comparison of the fast (Section 4.3.2) and the robust (Section 4.3.1) methods; and (iii) experimental verification of (4-48).

4.3.4.1 Measurement Setup. Figure 4-9 shows the basic setup for measuring the open loop gain $A(j\omega) = V_{\text{out}}(j\omega)/(V^+(j\omega) - V^-(j\omega))$ of an operation amplifier (opamp). Because of its very high gain, measuring the opamp in open loop would immediately drive the output into saturation. Therefore, to limit the output voltage, a feedback resistor R_2 connecting the output of the opamp to its negative input and a resistor R_1 in series with the voltage source $v_g(t)$ are added. To prevent loading of the output of the opamp by the resistor R_2, a voltage buffer (gain 1, very high input impedance, and 50 Ω output impedance) is put in series to R_2. The negative input v^- and output v_{out} of the opamp are buffered (× 1 voltage buffers with very high input impedance and 50 Ω output impedance) before being sent to the acquisition units (HP E1430A). The systematic errors introduced by the dynamics of the voltage buffers and the acquisition units are eliminated via a relative calibration (see Section 13.2.2.2 on page 501 and Pintelon et al., 2004b for the details).

The odd random phase multisines with random harmonic grid $v_g(t)$ are generated by an arbitrary waveform generator (HP E1445A) at the sampling frequency $f_s = 10\ \text{MHz}/2^4 = 625\ \text{kHz}$. The generator signal is lowpass filtered (cut off frequency of 250 kHz) before being applied to the circuit. The acquisition and generator units are synchronized, and their sampling frequencies are derived from the same mother clock. Of each signal, $P = 5$ consecutive periods of the steady state response with $N = 2^{16} = 64 \times 1024$ points per period are measured at the sampling rate $f_s = 625\ \text{kHz}$.

The frequencies of the odd random phase multisines (4-1) are logarithmically distributed between $f_{\min} = f_s/N \approx 9.5\ \text{Hz}$ and $f_{\max} = 9999 f_s/N \approx 95\ \text{kHz}$, and the amplitude spectrum is chosen to be flat. Of each group of three consecutive harmonics ($N_{\text{sub}} = 3$), exactly one odd excited harmonic is randomly eliminated ($p = 2/3$). The resulting odd random phase multisines with random harmonic grid contain $F = 299$ odd excited harmonics with equal amplitudes. These amplitudes are chosen such that the rms value of the negative input $v^-(t)$ of the opamp equals 6.1 mV.

While the fast method starts from one experiment with an odd random phase multisine, the robust method requires multiple experiments. Therefore, $M = 25$ experiments are performed with $M = 25$ independent random phase realizations of odd random phase multisines with the same random harmonic grid and the same amplitude spectrum.

Figure 4-10. Measured input (left) and output (right) Fourier coefficients of the operational amplifier excited by an odd random phase multisine with a logarithmic random harmonic grid (not all excited and non-excited frequencies are shown). '+': excited odd harmonics, 'o': non-excited odd harmonics, 'gray ∗': non-excited even harmonics, and black line: noise variance (all harmonics).

4.3.4.2 Fast Method. The fast method (Section 4.3.2) starts from $P = 5$ consecutive periods of the steady state response to one particular random phase realization of the odd random phase multisines with random harmonic grid. Figure 4-10 shows the mean value of the Fourier coefficients of the input $u(t) = -v^-(t)$ and output $y(t) = v_{out}(t)$ signals over the $P = 5$ periods of the first experiment. The '+' indicates the excited odd harmonics, the 'o' indicates the odd non-excited harmonics, the 'gray ∗' indicates the even non-excited harmonics, and the black line indicates the noise standard deviation of the mean value of the Fourier coefficients (excited and non-excited harmonics). The following two observations can be made:

1. The odd (o) and even (gray ∗) non-excited harmonics are well above the noise level (black line), except the even input harmonics below 200 Hz.

2. The odd (o) nonlinear distortions are significantly larger than the even nonlinear distortions (gray ∗), indicating a dominant odd nonlinear behavior.

Moreover, from the output Fourier coefficients (Figure 4-10, right plot) one would wrongly conclude that the nonlinear distortions (o and gray ∗) of the opamp are about 50 dB (o) to 70 dB (gray ∗) below the linear contributions. This is due to the linearizing effect of the feedback loop in the test circuit (resistor R_2 in Figure 4-9).

Linear compensation (4-40) of the output Fourier coefficients for the parasitic power at the non-excited input harmonics gives the corrected output Fourier coefficients shown in Figure 4-11. The horizontal and oblique black lines indicate the noise standard deviation of, respectively, the excited and non-excited harmonics. In addition to the previous observations the following can be seen:

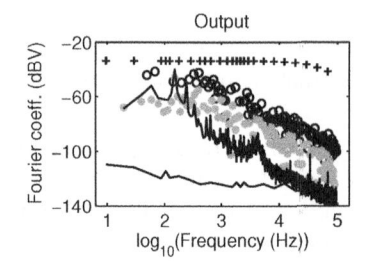

Figure 4-11. Corrected output Fourier coefficients (4-40) – operational amplifier (not all harmonics are shown). '+': excited odd harmonics, 'o': odd non-excited harmonics, 'gray ∗': even non-excited harmonics, bottom black line: noise variance excited harmonics, and top black line: noise variance non-excited harmonics.

Section 4.3 ■ Detection of Nonlinear Distortions Using Random Phase Multisines

1. Below 10 kHz the noise level of the non-excited output harmonics is much larger after correction (compare the oblique and the horizontal black lines). This is due to the input noise and the high gain of the operational amplifier.

2. The nonlinear distortions in the corrected output spectrum are much larger than in the original output spectrum (compare Figures 4-10 and 4-11). Hence, the linear correction (4-40) opens the feedback loop.

3. The nonlinear distortions are quite large in the band [10 Hz, 300 Hz] and decrease with increasing frequency. In the band [10 Hz, 200 Hz], the even (gray $*$) nonlinear distortions are below the noise level.

Although the level of the nonlinear distortions in the compensated output spectrum (Figure 4-11) quantifies correctly the level of the nonlinear distortions on the BLA measurement, the relationship between the true even and odd nonlinear behavior of the opamp and the level of the even and odd non-excited harmonics in the corrected output spectrum is valid only if inequalities (4-50) and (4-52) are satisfied.

To verify whether these inequalities are satisfied or not, the mean power spectra over the $M = 25$ experiments of the input signal $u(t)$ and the output stochastic nonlinear distortions $y_s(t)$ (level non-excited harmonics in the corrected output spectrum) are calculated at the even and odd harmonics. This is done via mean square averaging of the amplitudes of the Fourier coefficients taking into account their fill factor. Figure 4-12 shows the power ratios $S_{uu,\,odd}/S_{uu,\,even}$ (black line), $S_{y_s y_s,\,odd}/S_{y_s y_s,\,even}$ (light gray line), and $S_{y_s y_s,\,even}/S_{y_s y_s,\,odd}$ (dark gray line). The following conclusions can be drawn:

1. Over the whole frequency band $S_{y_s y_s,\,even}/S_{y_s y_s,\,odd}$ (dark gray) is much smaller than $S_{uu,\,odd}/S_{uu,\,even}$ (black) so that inequality (4-52) is fulfilled. Hence, the level of the odd non-excited harmonics in the corrected output spectrum (see Figure 4-11, 'o') is a correct indication of the level of the odd nonlinear distortions.

2. Above 1 kHz, the power ratio $S_{y_s y_s,\,odd}/S_{y_s y_s,\,even}$ (light gray) is much smaller than $S_{uu,\,odd}/S_{uu,\,even}$ (black) so that inequality (4-50) is satisfied. Hence, the level of the even non-excited harmonics in the corrected output spectrum (see Figure 4-11, 'gray $*$') is a correct indication of the level of the even nonlinear distortions. This is no longer true in the band [200 Hz, 1 kHz]. Below 200 Hz, the even output harmonics are at the noise level (see Figure 4-11).

Figure 4-12. Comparison between the input signal-to-even-distortion power ratio $S_{uu,\,odd}/S_{uu,\,even}$ (black), and the power ratio of the even to the odd $S_{y_s y_s,\,even}/S_{y_s y_s,\,odd}$ (dark gray) and the odd to the even $S_{y_s y_s,\,odd}/S_{y_s y_s,\,even}$ (light gray) distortions.

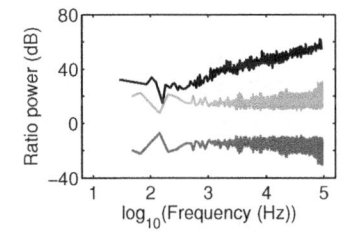

Finally, the best linear approximation (4-39), its noise variance (4-39), and its total variance (4-44) are calculated from the input and corrected output Fourier coefficients. The results are shown in Figure 4-13: the bold black line represents the best linear approximation of the open loop gain, the thin black line the noise variance, and the gray line the total variance. The following observations can be made:

144 Chapter 4 ■ Detection, Quantification, and Qualification of Nonlinear Distortions in FRF Measurements

1. The total variance (gray line) is well above the noise variance (thin black line), indicating that the nonlinear distortions are dominant. Hence, the non-smooth frequency behavior of the amplitude and phase characteristics of the open loop gain is solely due to the nonlinear distortions.

2. The nonlinear distortion (4-37) to open loop gain ratio $\hat{\sigma}_S(k_{exc})/|\hat{G}_{BLA}(j\omega_{k_{exc}})|$ is maximal at the low frequencies and decreases with increasing frequency. It indicates that the operational amplifier behaves linearly for small gains and nonlinearly for large gains.

Note also that no information about the nonlinear distortions is available at the first two odd excited harmonics 1 and 3. This is due to the fact that harmonic 5 is the first non-excited odd harmonic, and that the distortion levels at the excited odd harmonics are obtained by (linear) interpolation – and not extrapolation – of the levels at the odd non-excited harmonics.

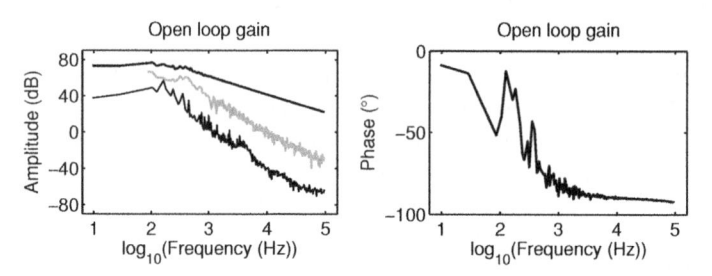

Figure 4-13. Measured open loop gain of the operational amplifier – fast method. Bold black lines: amplitude (left) and phase (right), thin black line: noise variance, and gray line: total variance (noise + nonlinear distortions).

4.3.4.3 Comparison with the Robust Method. The robust method (Section 4.3.1) starts from the $M = 25$ experiments with different random phase realizations of the odd random phase multisines with fixed random harmonic grid and amplitude spectrum. The results are shown in Figure 4-14 (solid lines): it can be seen that the total variance (gray line) is everywhere much larger (20 dB and more) than the noise variance (thin black line). It indicates that the nonlinear distortions are the dominant error source in the open loop gain measurement.

Comparing Figure 4-14 to Figure 4-13, it can be concluded that both figures have the same *qualitative* information content and lead to the same conclusions. However, the noise and total variances in Figure 4-14 are about 14 dB lower and less noisy than in Figure 4-13. These two phenomena are due to the additional averaging over the $M = 25$ random phase realizations of the odd multisines in the robust method. Furthermore, in Figure 4-14, the total variance is available at all odd excited harmonics.

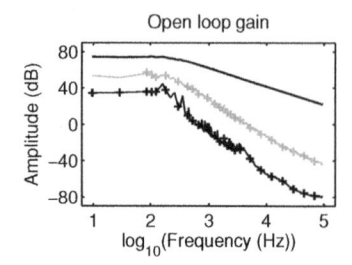

Figure 4-14. Comparison between the robust (solid lines) and fast ('+') estimates of the open loop gain. Black solid line: open loop gain (both estimates coincide), thin solid line and black '+': noise variance, and gray line and gray '+': total variance (noise + nonlinear distortion).

To verify that the fast and robust method are *quantitatively* equivalent, the fast method is applied to each of the $M = 25$ experiments. Next, the mean values of the open loop gain, its noise variance, and its total variance are calculated. Finally, the mean noise and total variances are divided by M. Figure 4-14 shows the results ('+'): it can be seen that the fast estimates of the noise ('black +') and total ('gray +') variances coincide with those of the robust estimates (black and gray solid lines).

4.3.4.4 FRF at the Non-excited Frequencies. For each of the $M = 25$ experiments, the BLA is calculated at all non-excited frequencies

$$\hat{G}_{\text{BLA}}^{[m]}(j\omega_{k_{\text{non-exc}}}) = \hat{Y}^{[m]}(k_{\text{non-exc}})/\hat{U}^{[m]}(k_{\text{non-exc}}) \qquad (4\text{-}53)$$

where $\hat{X}^{[m]}(k)$, $X = Y, U$, is defined in (4-22), and with $m = 1, 2, \ldots, M$. The mean and variance (4-19) of (4-53) over the $M = 25$ experiments are shown in Figure 4-15 (horizontal black and gray lines). The following observations can be made:

1. The variance of the BLA at the non-excited odd harmonics is significantly smaller than that at the non-excited even harmonics. The explanation follows immediately from Figure 4-10: the input-output odd-distortion-to-noise levels are about 20 dB larger than the input-output even-distortion-to-noise levels.

2. Calculating the mean value over the frequency of $\hat{G}_{\text{BLA}}(j\omega_{k_{\text{non-exc}}})$ gives 30.5 dB. This should be compared to the theoretical expected value

$$-1/M_0(j\omega) = -(R_1 + R_g + R_2)/(R_1 + R_g) \approx 31.0 \text{ dB}$$

(see (4-48) and Figure 4-16, where R_2 includes the 50 Ω output impedance of the voltage buffer).

The second observation is consistent with the results of Theorem 3.22 on page 94. It leads to the equivalent block schematic of an operational amplifier (opamp) shown in Figure 4-16, where the output of the opamp is modeled as the sum of two voltage sources $-Av(t)$ and $v_s(t)$. The latter is uncorrelated with – but not independent of – the generator $v_g(t)$. The part of $-Av(t)$ that is correlated with $v_g(t)$ contains energy at the excited harmonics only, while $v_s(t)$ contains energy at all harmonics.

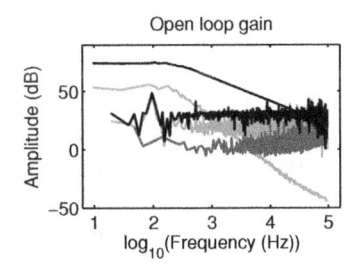

Figure 4-15. Frequency response function evaluated at the excited (top black line) and non-excited (horizontal black line) harmonics and the corresponding total variances (top light gray line: excited harmonics; horizontal dark gray line: non-excited odd harmonics; and horizontal light gray line: non-excited even harmonics).

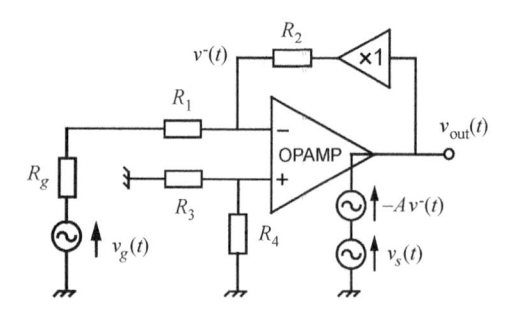

Figure 4-16. Block schematic of the best linear approximation of the operational amplifier (opamp) excited by a random phase multisine $v_g(t)$: the voltage source $v_s(t)$ of the stochastic nonlinear distortions is uncorrelated with – but not independent of – the generator $v_g(t)$.

4.4 GUIDELINES FOR MEASURING THE BEST LINEAR APPROXIMATION

A first important rule is that the excitations used for the identification experiment should mimic as closely as possible the operational perturbations (same rms value, same power spectrum, same probability density function). If the operational perturbations are normally distributed, then random phase multisine with the appropriate rms value and power spectrum are well suited. Further, we distinguish two different cases: either nothing is known about the (possible) nonlinear behavior of the system, or it is known beforehand that the nonlinear distortions are the dominant error source in the FRF measurement. Each case leads to different recommendations.

Case 1: No Prior Knowledge about the Error Sources. If it is unknown whether nonlinear distortions are present/important or not, then one should perform experiments with random phase multisines because of their ability to distinguish noise from nonlinear distortions. The optimal experiment and the optimal type of random phase multisine depend on the ultimate goal of the FRF measurement.

- *Maximize the Sensitivity of the Nonlinear Detection.* In this case, the noise averaging should be maximized for a given measurement time T and a given frequency resolution f_0 of the FRF measurement. Hence, the optimal choice is measuring a maximal number of consecutive periods $P = pTf_0$ ($P = pTf_0/2$) of the steady state response to a full (odd) random phase multisine with a random harmonic grid of $100 \times p\%$ excited (odd) harmonics. The measurements are processed with the fast method (Section 4.3.2). An odd random phase multisine is used if classification of the nonlinear distortions in even and odd contributions is needed.

- *Minimize the Variance of the BLA Measurement.* In this case, the averaging of both the noise and the stochastic nonlinear distortions should be maximized for a given measurement time T and a given frequency resolution f_0 of the FRF measurement, while keeping the ability to quantify the noise and the nonlinear distortions. The optimal choice is measuring $P = 2$ consecutive periods of the steady state response to a maximal number $M = Tf_0/P$ ($M = TF_0/(2P)$) of different random phase realizations of full (odd) random phase multisines. The measurements are processed with the robust method (Section 4.3.1). The choice between full or odd random phase multisines depends on the type of the nonlinearities:

    (a) Full multisines are optimal if the odd distortions are dominant.
    (b) Odd multisines are optimal if the even distortions are dominant.

If the odd and even distortions are of the same order of magnitude, then the information content of the full and odd multisine BLA measurements is the same.

Case 2: Nonlinear Distortions Are Dominant. If it is known beforehand that the nonlinear distortions are the dominant error source in the FRF measurement, then the averaging of the stochastic nonlinear distortions should be maximized for a given measurement time T and a given frequency resolution f_0 of the FRF measurement. The robust method with $P = 1$ and $M = Tf_0$ ($M = Tf_0/2$) different experiments with full (odd) random phase multisines is then optimal. The choice between between full and odd multisines is the same as in Case 1 – minimize the variance of the BLA measurement.

4.5 APPENDIXES

Appendix 4.A Proof of the Nonlinearity Detection Properties of Random Phase Multisines

From the Volterra theory (see (3-4) and (3-5)) it follows that even and odd nonlinearities combine, respectively, an even and odd number of input frequencies. For baseband multisines (the first excited frequency is f_s/N), this leads to the following conclusions:

1. If all input harmonics are excited (full random phase multisine), then – in general – all output harmonics are disturbed by both the even and odd nonlinear distortions. Hence, at the non-excited output harmonics we detect the sum of the odd and even nonlinear contributions.

2. If the even input harmonics are not excited (odd random phase multisine), then the even and odd nonlinear distortions appear at, respectively, the even and odd output harmonics (the sum of, respectively, an even and an odd number of odd harmonics results in, respectively, an even and odd harmonic). Hence, at the even output harmonics and the non-excited odd output harmonics we detect, respectively, the even and odd nonlinear distortions; and the excited odd output harmonics are only disturbed by the odd nonlinear distortions.

For bandpass multisines (only the frequencies in the band $[k_{min}, k_{max}]f_s/N$ are excited with $k_{min} > 1$ and $k_{max} < N/2$), the in-band output harmonics might only be disturbed by the odd nonlinearties. For example, if $k_{max} < 2k_{min}$ ($k_{max} < 1.5k_{min}$) then the second (fourth) degree nonlinear distortions fall outside the excited frequency band.

Appendix 4.B Influence Generator Noise on the FRF Estimate

We consider here the open and closed loop setup of Figure 4-6 where the actuator is nonlinear and the plant linear. The feedback dynamics remain linear.

4.B.1 The Best Linear Approximation. Applying the definition of the BLA (3-14) for nonlinear systems operating in open loop gives

$$G_{BLA}(j\omega) = \frac{S_{yu}(j\omega)}{S_{uu}(j\omega)} = \frac{G_0(j\omega)S_{uu}(j\omega)}{S_{uu}(j\omega)} = G_0(j\omega) \qquad (4\text{-}54)$$

where the second equality uses the property that the plant has linear dynamics $G_0(j\omega)$. Similarly, we find for the BLA (3-35) of nonlinear systems operating in closed loop

$$G_{\text{BLA}}(j\omega) = \frac{S_{yr}(j\omega)}{S_{ur}(j\omega)} = \frac{G_0(j\omega)S_{ur}(j\omega)}{S_{ur}(j\omega)} = G_0(j\omega) \qquad (4\text{-}55)$$

4.B.2 The Total Variance of the Best Linear Approximation. We assume here that the plant operates in open loop. The proof for the closed loop case follows exactly the same lines. The DFT spectrum of the output of the nonlinear actuator can be written as

$$U_0(k) = G_{\text{BLA, act}}(j\omega_k)R(k) + N_S(k) = U_1(k) + N_S(k) \qquad (4\text{-}56)$$

with $G_{\text{BLA, act}}(j\omega_k)$ the BLA of the nonlinear actuator and $N_S(k)$ the corresponding stochastic nonlinear distortions. To simplify the notations we will calculate the variance of the input-output DFT spectra over the random realizations of the excitation $r(t)$, assuming that the input-output measurement noise and the process noise sources are zero. Under these assumptions the input-output errors are given by

$$N_U(k) = N_S(k) \text{ and } N_Y(k) = G_0(j\omega_k)N_S(k) \qquad (4\text{-}57)$$

with corresponding (co)variances

$$\sigma_U^2(k) = \text{var}(N_S(k)), \quad \sigma_Y^2(k) = |G_0(j\omega_k)|^2 \text{var}(N_S(k)), \text{ and}$$

$$\sigma_{YU}^2(k) = G_0(j\omega_k)\text{var}(N_S(k)) \qquad (4\text{-}58)$$

Evaluating the variance (2-25) on page 47, where the noise (co)variances are replaced by (4-58), and where $U_0(k)$ and $Y_0(k)$ are replaced by, respectively, $U_1(k)$ defined in (4-56) and $Y_1(k) = G_0(j\omega_k)U_1(k)$, gives

$$\sigma_G^2(k) = |G_0(j\omega_k)|^2 \left(\frac{|G_0(j\omega_k)|^2}{|Y_1(k)|^2} + \frac{1}{|U_1(k)|^2} - 2\text{Re}(\frac{G_0(j\omega_k)}{Y_1(k)\overline{U_1(k)}}) \right) \text{var}(N_S(k)) = 0 \qquad (4\text{-}59)$$

It shows that the stochastic nonlinear distortions $N_S(k)$ of the nonlinear actuator do not influence the total variance (4-26) and (4-44) of the BLA estimate.

Appendix 4.C Rationale for (4-50) and (4-52)

The following two assumptions are made for deriving (4-50) and (4-52): (i) the phases of the plant input multisine are independently distributed, and (ii) the plant input signal-to-even-distortion ratio is frequency independent. Assumption (i) is only approximately valid for nonlinear systems operating in open loop and driven by a nonlinear actuator, and for nonlinear systems operating in feedback (and driven by a linear actuator). In general, assumption (ii) will be an approximation too. Therefore, the reasoning in this appendix is a rationale and not a strict proof of (4-50) and (4-52).

Section 4.5 ■ Appendixes

4.C.1 Rationale for (4-50). The bias on the estimated level of the even distortions can be neglected if the sum of all stochastic nonlinear contributions of the form (4-49) satisfy

$$\left|\sum_{k_3, k_4} \beta U(2k_3 + 1)U(2k_4 + 1)U(2l)\right|^2 \ll \left|\sum_{k_1} \alpha U(2k_1 + 1)U(2k_2 + 1)\right|^2 \quad (4\text{-}60)$$

where l depends on k_3 and k_4, and k_2 depends on k_1 (see (4-49)), and where

$$\sum_{k_1} \alpha U(2k_1 + 1)U(2k_2 + 1) = Y_{S,\text{even}}(2m) \quad (4\text{-}61)$$

We elaborate now the left-hand side of (4-60), assuming that the input signal-to-even-distortion ratio $|U(2l+1)|/|U(2l)|$ is independent of l, and that the phases of the multisine are independently distributed. The factor $U(2l)$ in the left-hand side of (4-60) can be rewritten as

$$U(2l) = |U(2l+1)|\frac{U(2l)}{|U(2l+1)|} = |U(2l+1)|e^{j\angle U(2l)}\frac{|U(2l)|}{|U(2l+1)|} \quad (4\text{-}62)$$

Replacing $U(2l)$ in the left-hand side of (4-60) by (4-62), taking into account that $|U(2l+1)|/|U(2l)|$ is independent of l, gives

$$\left|\sum_{k_3, k_4} \beta U(2k_3 + 1)U(2k_4 + 1)U(2l)\right|^2 = \frac{|U(2m)|^2}{|U(2m+1)|^2} \times \\ \left|\sum_{k_3, k_4} \beta U(2k_3 + 1)U(2k_4 + 1)|U(2l+1)|e^{j\angle U(2l)}\right|^2 \quad (4\text{-}63)$$

Since the phases of $U(2l+1)$ and $U(2l)$ are independently distributed, we can approximate the sum in the right-hand side of (4-63) as

$$\sum_{k_3, k_4} \beta U(2k_3 + 1)U(2k_4 + 1)|U(2l+1)|e^{j\angle U(2l)} \approx Y_{S,\text{odd}}(2m+1) \quad (4\text{-}64)$$

Collecting (4-60), (4-61), (4-63), and (4-64) and multiplying the result with $|U(2m+1)|^2$ we find

$$|U(2m)|^2|Y_{S,\text{odd}}(2m+1)|^2 \ll |U(2m+1)|^2|Y_{S,\text{even}}(2m)|^2 \quad (4\text{-}65)$$

Taking the expected value of (4-65) and dividing by $\mathbb{E}\{|U(2m)|^2\}$ gives

$$\text{var}(Y_{S,\text{odd}}(2m+1)) \ll \frac{\mathbb{E}\{|U(2m+1)|^2\}}{\mathbb{E}\{|U(2m)|^2\}}\text{var}(Y_{S,\text{even}}(2m)) \quad (4\text{-}66)$$

Combining (4-2) and (4-10) with (4-66) finally proves (4-50).

4.C.2 Rationale for (4-52). The bias on the estimated level of the odd distortions can be neglected if the sum of all stochastic nonlinear contributions of the form (4-51) satisfy

$$\left|\sum_{k_1} \alpha U(2k_1 + 1)U(2l)\right|^2 \ll \left|\sum_{k_2, k_3} \beta U(2k_2 + 1)U(2k_3 + 1)U(2k_4 + 1)\right|^2 \quad (4\text{-}67)$$

where the sum in the right-hand side is equal to $Y_{S,\text{odd}}(2m+1)$. We now replace $U(2l)$ in the left-hand side of (4-67) by (4-62), where $U(2l+1)$ is replaced by $U(2l-1)$, and take into account that

$$\sum_{k_1} \alpha U(2k_1+1)|U(2l-1)|e^{j\angle U(2l)} \approx Y_{S,\text{even}}(2m) \qquad (4\text{-}68)$$

Following the same lines of Section 4.C.1, we get

$$\text{var}(Y_{S,\text{even}}(2m)) \ll \frac{\mathbb{E}\{|U(2m+1)|^2\}}{\mathbb{E}\{|U(2m)|^2\}} \text{var}(Y_{S,\text{odd}}(2m+1)) \qquad (4\text{-}69)$$

Combining (4-2) and (4-10) with (4-69) finally proves (4-52).

5

Design of Excitation Signals

Abstract: Good experiments are the best guarantee of building good models. The selection of good excitation signals is an important step in the design of the experiment. In this chapter we explain how to get such signals. In the first part, three classes of excitations are considered:

(i) General purpose signals that can be applied without any optimization. The only parameters to be selected are the bandwidth of the excitation signal and the frequency resolution of the measurement.
(ii) Optimized test signals: these facilitate excitation of the system with a user-specified power spectrum, for example, a semilogarithmic distributed spectrum.
(iii) Dedicated test signals: these are signals with optimized characteristics for special situations; for example, the signal and its derivative do not exceed a user-specified value.

The second part of this chapter deals briefly with:

(i) The design of optimal power spectra so that the available power is used at the frequencies where it contributes most to the knowledge of the system.
(ii) Experiment design for control.

5.1 INTRODUCTION

In most system-analysis applications, the dynamic behavior of the system is derived from measurements of the input and output signals. In some situations the input signal is imposed by the environment and it is impossible to excite the device under test with an arbitrarily chosen input (for example, in biological systems, where the choice of excitation is very limited). In other situations, only binary signals may be applicable. However, in a wide variety of cases, the only restriction on input signals is that of a limitation in the permitted amplitude range.

A very common method used in transfer function measurements, until the end of the 1960s, was that of the combination of a slowly swept sine with a tracking filter. Since the development of advanced digital signal processing algorithms, and especially since the efficient implementation of the discrete Fourier transform (DFT) with the fast Fourier transform (FFT), it became possible to use more complex input signals. Instead of exciting the unknown system frequency by frequency, sophisticated waveforms with a broadband spectrum are generated, enabling collection of all the required spectral information from a single measurement. This can result in a considerable reduction of the measurement time but also in an undesired loss of accuracy if no precautions are taken. We will analyze the trade-off between accuracy and measurement time, but before starting we must choose between a nonparametric and a parametric modeling approach. In the nonparametric representation, the system is characterized by measurements of the frequency response at a large number of frequencies, whereas in a parametric model, the system is described by a mathematical transfer function model with a limited number of parameters. It is precisely these parameters that have to be estimated in the parametric modeling approach. The optimum spectrum of the excitation in the parametric case will be different from that in the nonparametric case: this is principally because the parametric model combines the information available from all frequencies in only a few parameters. In a direct nonparametric frequency response measurement, there is no relation between the measurements at the various frequencies and, therefore, the excitation should be designed to achieve a predefined accuracy in the frequency bands of interest: for example, maximizing the absolute or relative accuracy of the measurements. In a parametric approach, the energy will be concentrated at the frequencies where it contributes most to the knowledge about the model parameters.

To design an optimized excitation signal, it is necessary to specify the final goal. For the nonparametric case, we will look for signals that *maximize the minimum accuracy obtained in a fixed measurement time for a specified maximum peak value of the excitation:*

$$\min(\max_{k \in \mathbb{F}} \sigma_G^2(k)) \quad \text{with} \quad \max_t |u(t)| \leq u_{\max} \tag{5-1}$$

where $\mathbb{F}$ is the set of frequencies at which the frequency response is measured. In the parametric case, the determinant of the information matrix will be maximized, as discussed later.

We first focus our attention on the design of excitation signals for non-parametric measurements. The parametric modeling approach will be studied in the second part.

5.2 GENERAL REMARKS ON EXCITATION SIGNALS FOR NONPARAMETRIC FREQUENCY RESPONSE MEASUREMENTS

In this section the nonparametric frequency response function (FRF) measurement problem is studied. It should be clear to the reader that signals that provide good FRF measurements are also very well suited for use in a parametric identification step, which gives this section more general value.

Before starting a detailed comparison of some candidate excitation signals, we first introduce two quality measures for excitation signals. In general, these measures depend on the actual measured FRF and on the properties of the disturbing noise (e.g., its power spectrum). However, in order to simplify the discussion, we assume that we deal with flat systems (the amplitude of the FRF is a constant) in the presence of white noise. If necessary, we will indicate how the conclusions should be modified to the general situation of arbitrary systems with colored noise distortions.

5.2.1 Quantifying the Quality of an Excitation Signal

In Chapter 2 it was shown that the uncertainty on the FRF at ω_k after M averages is

$$\sigma_G^2(k) \approx \frac{|G_0(j\omega_k)|^2}{M}\left(\frac{\sigma_Y^2(k)}{S_{Y_0Y_0}(k)} + \frac{\sigma_U^2(k)}{S_{U_0U_0}(k)} - 2\operatorname{Re}\left(\frac{\sigma_{YU}^2(k)}{S_{Y_0U_0}(k)}\right)\right) \tag{5-2}$$

The uncertainty is inversely proportional to the total power of the excitation signal and also to the shape of its power spectrum. In order to have a constant variance $\sigma_G^2(k)$ at all frequencies, the power distribution should be proportional to the impact of the disturbing noise. This leads to the definition of two characteristics for excitation signals: the crest factor and the time factor.

Definition 5.1 (Crest Factor): The crest factor $Cr(u)$ of a signal $u(t)$ is given by the ratio of the peak value u_{peak} of the signal to its rms value u_{rmse} in the frequency band of interest

$$Cr(u) = \frac{u_{\text{peak}}}{u_{\text{rmse}}} = \frac{\max_{t \in [0,T]} |u(t)|}{u_{\text{rms}}\sqrt{P_{\text{int}}/P_{\text{tot}}}} \text{ with } u_{\text{rms}}^2 = \frac{1}{T}\int_0^T u^2(t)dt \tag{5-3}$$

with T the measurement time, u_{rms} the rms value of the signal, P_{tot} the total power of the signal, and P_{int} the power in the frequency band of interest.

The crest factor gives an idea of the compactness of the signal. Signals with an impulsive behavior (having a large crest factor) inject much less power into the system than signals having the same peak value and a small crest factor. The effective rms value u_{rmse} is used to express that only the power in the frequency band of interest contributes to the knowledge of the system.

The time factor of an excitation signal also accounts for the power distribution of the signal over the frequencies. If this is unequally distributed with respect to the noise, some FRF points will be poorly measured. We will require that the worst measurements still reach a minimum quality. For the sake of general conclusions, we make the following simplifying assumptions: $|G_0(j\omega_k)|^2$, $\sigma_U^2(k)$, $\sigma_Y^2(k)$, $\sigma_{YU}^2(k)$ are constant. Expression (5-1) reduces to

$$\sigma_G^2(k) \sim \frac{1}{M|U(k)|^2} \tag{5-4}$$

and the number of averages to reach a specified variance is proportional to $M \sim 1/|U(k)|^2$. The total measurement time T is proportional to the required number of averages M. Also notice that $Cr^2(u) = u_{\text{peak}}^2/u_{\text{rmse}}^2$ and $u_{\text{rmse}}^2 = 2FU_{\text{rmse}}^2$ with F the number of frequencies in the frequency band of interest and, $U_{\text{rmse}}^2 = \sum_{k=1}^F |U(k)|^2/F$. Then

$$T \sim \max_k \frac{1}{|U(k)|^2} \sim \max_k \frac{U_{\text{rmse}}^2}{|U(k)|^2 U_{\text{rmse}}^2} \sim \max_k \frac{Cr^2(u)}{\frac{|U(k)|^2 u_{\text{peak}}^2}{U_{\text{rmse}}^2} F} \tag{5-5}$$

and the required measurement time per frequency line for a specified peak value u_{peak} becomes proportional to

$$\frac{T}{F} \equiv Tf(u) \sim \max_{k} \frac{Cr^2(u)U_{\text{rmse}}^2}{|U(k)|^2} \tag{5-6}$$

The proportionality factor is fixed by normalizing $Tf(u) \equiv 1$ for a sine wave. Thus, the time factor $Tf(u)$ indicates the required measurement time per frequency point that is needed to guarantee a minimum SNR on the FRF measurement, and this is compared with a stepped sine excitation.

Definition 5.2 (Time Factor): The time factor $Tf(u)$ of a signal $u(t)$ is given by

$$Tf(u) = \max_{k \in \mathbb{F}} 0.5 Cr^2(u) U_{\text{rmse}}^2 / |U(k)|^2 \tag{5-7}$$

This result can be generalized for situations with frequency-dependent noise levels and varying transfer functions. However, it is still impossible to make general comparisons on the excitation signals. The ability of the excitation signals to deal with these situations depends on their flexibility to impose a user-specified power spectrum.

5.2.2 Stepped Sine versus Broadband Excitations

In this chapter we use the time factor of the sine as a reference to qualify the broadband excitations. However, the reader should be aware that this quality measure deals only with the SNR properties of the signal. In practice, other aspects also influence the total measurement time. To make this clear, the measurement time of a stepped sine experiment (consisting of a series of single sine measurements at the desired frequencies) is compared with the measurement time with a broadband excitation having the same time factor. Two extreme situations are considered, assuming a very good SNR the first time and a very poor SNR the second time. Finally, the intermediate situation is analyzed.

5.2.2.1 Very Good SNR. For the stepped sine, the measurement time is determined by two elements. At least one period of the sine should be measured and, after each frequency step, a waiting time $T_w(k)$ should be included, allowing the transients (of the plant and the measurement system) to disappear. For highly damped systems, these transients are short, but they are very long for lightly damped systems, as they appear, for example, in many mechanical applications. For simplicity, we assume that $T_w(k)$ is a frequency-independent constant. Under these conditions, the total measurement time is $T_{ss} = \sum_{k=1}^{F} 1/f_k + FT_w$ for the stepped sine and $T_{bs} = 1/\Delta f + T_w$ for the broadband measurement, where Δf is the frequency resolution (one period of the broadband measurement equals $1/\Delta f$). If $f_k = kf_0$ and $\Delta f = f_0$, these expressions become

$$T_{ss} = \frac{1}{f_0}\sum_{k=1}^{F} 1/k + FT_w \approx \frac{1}{f_0}(0.58 + \ln F) + FT_w \quad \text{and} \quad T_{bs} = \frac{1}{f_0} + T_w$$

This shows that a significant gain in measurement time is obtained using the broadband excitation.

5.2.2.2 Very Poor SNR. When the SNR is poor, the measurement time needed to get an acceptable uncertainty is much larger than the waiting time T_w, and it is proportional to $\max_k |U(k)|^{-2}$ (if we assume for simplicity a constant noise level on the measurements). A broadband signal distributes the power over F frequencies, while a stepped sine measurement keeps all power focused on one line at each partial measurement: $|U_{ss}(k)|^2 = F|U_{bs}(k)|^2$. Hence, to get the same SNR, the measurement time at one frequency will be F times smaller for the single sine measurement compared with the broadband excitation measurement. However, for a single sine measurement, F measurements should be made, one after another, while all information is captured at once for the broadband measurement, so that, eventually, the total measurement time is the same.

5.2.2.3 Intermediate Situation: Balancing the Transient Errors versus the Noise Errors. In general, the user faces a tricky situation with measurements of medium quality (for example, an SNR of 40 dB). In that case, (5-8) gives a rough rule of thumb for estimating the required waiting time so that the equivalent output noise errors equal the transient errors (Schoukens et al., 2000):

$$T_w = \frac{\tau}{2}\ln(\frac{\tau}{2T}\text{SNR}^2) \text{ with SNR}^2 = \frac{S_{yy}(j\omega_k)}{\sigma_Y^2(k) + \sigma_U^2(k)|G(\Omega_k)|^2 - 2\text{Re}(\sigma_{YU}^2(k)\overline{G}(\Omega_k))} \quad (5\text{-}8)$$

with τ the dominating time constant of the system in the considered frequency band, T the length of the time record, and SNR expressed as the ratio of the output power $S_{yy}(j\omega_k)$ to the equivalent output noise. For example, for $T = 10$ s, $\tau = 1$ s, and an SNR of 40 dB (SNR = 100), the waiting time becomes at least 3 s after each frequency change.

Conclusion. The total measurement time required for step sine measurements will always be larger than that of broadband measurements, provided that we can design the latter excitations with a time factor close to 1. As the damping of a system decreases (time constants increase), the SNR where the stepped sine becomes competitive increases. For most practical situations, the broadband measurement results in a significantly reduced measurement time. For this reason, we focus completely on broadband excitations.

5.3 STUDY OF BROADBAND EXCITATION SIGNALS

The excitation signals are split into three classes: general purpose signals (no optimization involved), optimized test signals (passing through a fully automatic optimization procedure), and, finally, advanced test signals that have some very dedicated properties to deal with specific situations, for example, optimizing not only the signal but also its first and second derivative (such as displacement, velocity, and acceleration).

5.3.1 General Purpose Signals

In this section we study and compare the properties of some general purpose excitation signals. This means that no special optimization is performed to deal with specific situations. These signals should be able to excite the system with an almost flat power spectrum in a user-specified frequency band. From the previous section, we know already that an optimum signal should have a low crest and time factor. Besides these two conditions, it is also important that no leakage appears during the analysis of the measurements, as explained in Section 2.2.3. Therefore, we are strongly in favor of periodic excitations. Leakage errors cannot be avoided if aperiodic signals are used, and it will be necessary to average over a large number of measurements, even if a nonuniform time window is used. This considerably increases the measurement time required for a specified accuracy. Bursts, or time-limited signals, are exceptions to this rule: the continuous spectra of these signals are correctly sampled with the

DFT if the amplitude spectrum is sufficiently band-limited for the aliasing effect to be neglected (see Section 2.2.4). Six general purpose signals are considered: swept sine, also called periodic chirp; multisine excitation; maximum length binary sequences; white noise; burst white noise; and pulse testing. At the end of this section, the signals are compared with each other in an example.

5.3.1.1 Swept Sine

Definition 5.3 (Swept Sine): A swept sine (also called periodic chirp) is a sine sweep test, where the frequency is swept up and/or down in one measurement period, and this is repeated in such a way that a periodic signal is created (Brown et al., 1977).

$$u(t) = A\sin((at+b)t) \qquad 0 \le t < T_0 \qquad (5\text{-}9)$$

with T_0 the period, $a = \pi(k_2 - k_1)f_0^2$, $b = 2\pi k_1 f_0$, $f_0 = 1/T_0$, $k_2 > k_1 \in \mathbb{N}$, and $k_1 f_0$, $k_2 f_0$ the lowest and the highest frequency, respectively.

Properties

- Periodic signal with period $T_0 = 1/f_0$ → no leakage.
- Frequency resolution $1/T_0$.
- Most of the power is equally distributed in the user-selected frequency band $[k_1, k_2]f_0$ with $k_2 > k_1 \in \mathbb{N}$.
- Crest factor typically 1.45, time factor typically between 1.5 and 4.

Discussion. A swept sine has a low crest factor (comparable to the crest factor of a sine wave) but the amplitude spectrum is not actually flat (see Figure 5-5). This introduces frequency components with a lower SNR, resulting in a longer measurement time for a given accuracy. Although a swept sine can create band spectra, it is not possible to generate a signal with an arbitrary amplitude spectrum. A second drawback is that not only are the frequency lines of interest excited, but also a number of other spectral lines appear. This is unimportant with linear systems, but it can be very disturbing in systems with nonlinear behavior.

5.3.1.2 Schroeder Multisine

Definition 5.4 (Schroeder Multisine): A Schroeder multisine is a sum of harmonically related sine waves

$$u(t) = \sum_{k=1}^{F} A\cos(2\pi f_k t + \phi_k) \qquad (5\text{-}10)$$

with Schroeder phases $\phi_k = -k(k-1)\pi/F$ and $f_k = l_k f_0$ with $l_k \in \mathbb{N}$.

Properties

- Periodic signal with period $T_0 = 1/f_0$ → no leakage.
- Frequency resolution $1/T_0$.
- All the power at the user-selected frequencies that can be chosen without restriction on the discrete grid kf_0.
- Crest factor typically 1.7, time factor typically 1.5.

Discussion. For the general purpose signal we selected a flat amplitude spectrum $A_k = A$ for the harmonic components of the multisine. However, in general, the user can make an arbitrary choice.

For simplicity, we also used the Schroeder phases (Schroeder, 1970). Although these are not optimal, they give good results for flat amplitude spectra of multisines where a successive set of frequencies is excited. Smaller crest factors can be found by optimizing the phases by using a numerical optimization routine. Dedicated methods are discussed in the next section on optimized test signals, reducing the crest factor from, typically, 1.7 to about 1.4.

Remark. It is strongly advised to use FFT techniques to calculate multisine signals, otherwise the computation time becomes very long (see Exercises 2.1 and 2.2).

5.3.1.3 Pseudorandom Binary Sequence

Definition 5.5 (Pseudorandom Binary Sequence): A pseudorandom binary sequence (PRBS) is a deterministic, periodic sequence of length N that switches between one level (e.g., +1) and another level (e.g., −1). The switches can occur only on a discrete-time grid at multiples of the clock period T_c ($k_l T_c$, $k_l \in \mathbb{N}$) and are chosen such that the autocorrelation is as given in Figure 5-1 (Godfrey, 1993a, 1993b).

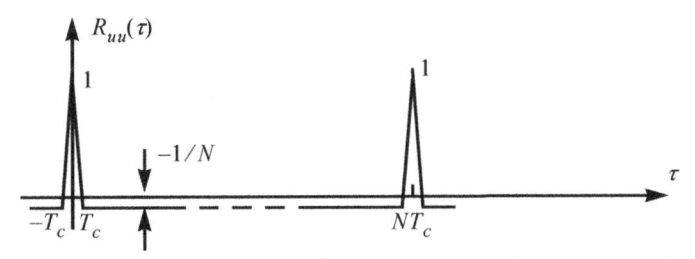

Figure 5-1. Autocorrelation function of a PRBS of length N, switching between ±1.

Properties

- Periodic signal with period $T = NT_c$ → no leakage.
- Frequency resolution $1/T$.
- Most of the power below $0.4 f_c = 0.4/T_c$ (see Figure 5-3). Optimal choice of the clock frequency $f_c = 2.5 f_{\max}$, with $f_{\max}$ the maximum frequency of interest.
- Crest factor is 1 if all power is considered, time factor typically 1.5.

Discussion. The PRBS has a spectrum whose components decrease in inverse proportion to the frequency. The amplitude $A(k)$ of the Fourier coefficient U_k of a PRBS is given by

$$A(0) = \frac{a}{N} \quad \text{and} \quad A(k) = a\frac{\sqrt{N+1}}{N}\text{sinc}(k\pi/N) \text{ for } k = 1, 2, ..., N-1 \quad (5\text{-}11)$$

with $\text{sinc}(x) = \sin(x)/x$, $2a$ the peak-to-peak amplitude of the sequence, and $f_k = k(f_c/N)$.

It is not possible to find a binary sequence for every arbitrary length N. However, there are a number of possibilities to generate these sequences, hence, there is still much freedom in choosing N.

First possibility: Use a quadratic residue code method (Godfrey, 1993b). This method generates a PRBS with length $N = 4k - 1$ where N should also be a prime number (e.g., $N = 3, 7, 11, 19, 23, 31, \ldots$). The signals can be generated by the following MATLAB® code:

$$x = -\text{ones}(N, 1); x(\text{mod}([1:N].\wedge 2, N) + 1) = 1; x(1) = 1$$

These sequences can easily be generated, nowadays, using arbitrary waveform generators.

Second possibility: For a long time (in the 1960s and 1970s), it was technically not possible to generate the previous sequences and for that reason other PRBS signals such as the maximum length binary sequences (MLBS) were preferred. These can be generated with the setup shown in Figure 5-2 (Godfrey, 1969, 1980, 1993b; Eykhoff, 1974; Norton, 1986). From all possible binary sequences that can be generated with a fixed register length, the MLBS has the longest period and the shortest correlation length. This means that the spectrum is as flat as possible. The feedback choice determines whether a sequence with the maximum period

$$T_{max} = (2^R - 1)T_c \tag{5-12}$$

is generated. Here, R is the register length and T_c is the clock period.

Because the length N (in clock cycles) does not equal 2^n samples, it is not possible to apply a straightforward FFT analysis. Instead, the chirp-z transform can be used, which permits efficient calculation of the DFT for an arbitrary number of data points (Rabiner and Gold, 1975; Oppenheim and Schafer, 1975). Most numerical packages can calculate the DFT for arbitrary lengths.

In Figure 5-3, details of the first lobe of the amplitude spectrum are given for an MLBS generated with lengths $N = 15, 31$, and 63. The amplitude of the individual components decreases with increasing length. The crest factor varies as a function of the spectral band ($0 < f \leq f_{max}$) in use, decreasing to 1 as the bandwidth increases to infinity. However, the time factor has a different behavior, as seen in Figure 5-3(b): it decreases for low frequencies but increases to infinity if f_{max} approaches f_c, as the amplitudes decrease to zero at this frequency. The time factor is less than 1.5 if the upper limit of the frequency band is taken between 0.2 and 0.6 of the PRBS generator clock frequency. The optimal value of the upper frequency limit is around $0.4f_c$, resulting in a time factor of 1.1 corresponding to a clock frequency f_c equal to 2.5 times the maximum frequency of interest.

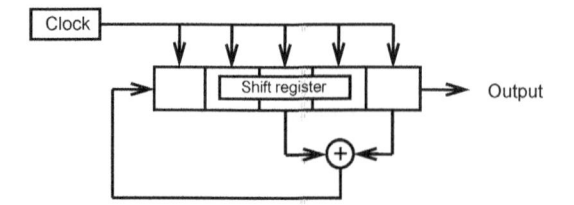

Figure 5-2. Generation of a maximum length binary signal with a shift register (can be initialized with an arbitrary nonzero code).

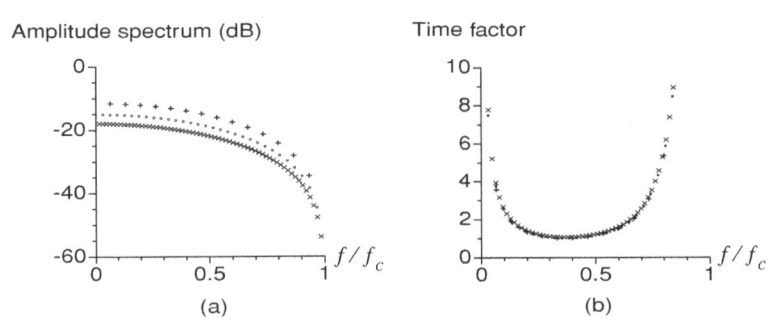

Section 5.3 ■ Study of Broadband Excitation Signals

Figure 5-3. (a) Part of the amplitude spectrum and (b) the time factor of an MLBS as function of the bandwidth used ($0 \rightarrow f_{\max}$), lengths $N = 15, 31,$ and 63.

5.3.1.4 Random Noise

Definition 5.6 (Random Noise): Random noise is a noise sequence whose power spectrum can be influenced by digital filters (Brown et al., 1977; Van Brussel, 1975).

Properties

- Random excitation → leakage problems.
- Equivalent frequency resolution $1/T$.
- Shaping of the power spectrum using a digital filter.
- Crest factor, typically 2–3, and time factor 4.5.

Discussion. In practice, the extreme values of the random signal are clipped (for example, outside the 2 sigma interval) to avoid excessive peak values. The major disadvantages of random excitations are the leakage problems and the drops in the amplitude spectrum if only one realization is processed. In Section 2.6.2, we explain how to deal with these problems.

To maximize the power injected into the system, it is advantageous to use binary noise. This is done by retaining only the sign of the original noise signal (Schoukens et al., 1995). In order to maintain the binary nature, all prefiltering should be done before the sign operation. Because the sign operation is a nonlinear operation, it distorts the power spectrum. Consequently, it is impossible to keep full control over the power spectrum and the crest factor at the same time. This is illustrated in Figure 5-4. A white noise sequence is filtered, and then only the sign is retained so that a binary sequence is generated. The actual, realized spectrum is compared with the desired power spectrum. As can be seen, the power spectrum is only partly under control. Most of the power is injected in the frequency band of interest, but there is still a large fraction generated outside this band.

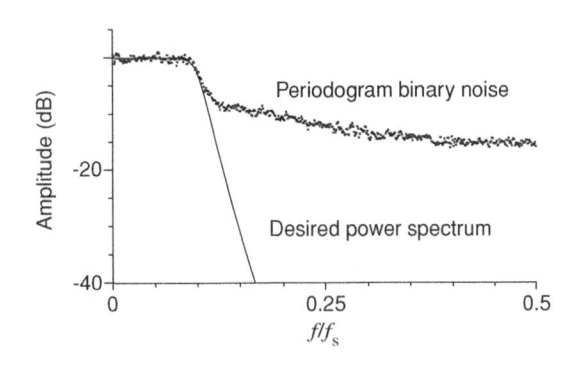

Figure 5-4. Comparison of the spectrum of a filtered noise sequence before and after the sign function.

5.3.1.5 Random Burst

Definition 5.7 (Random Burst): A random burst is a noise sequence that is imposed on the system during the first part of the measurement sequence, and a zero input is applied for the rest of the measurement period (Herlufsen, 1984).

$$u(t) = w(t)r(t)$$

$$w(t) = \begin{cases} 1 & 0 \leq t < T_1 \\ 0 & T_1 \leq t < T \end{cases}$$

with $r(t)$ a random variable and $w(t)$ a window function.

Properties

- Random excitation, no leakage if the system response becomes negligible before the end of the measurement window (T).
- Equivalent frequency resolution $1/T$.
- Shaping of the power spectrum using a digital filter.
- Crest factor, typically, $3(T/T_1)^{1/2}$, minimum time factor $\geq 4.5 T/T_1$.

Discussion. The crest factor of a random burst sequence is equal to that of the random sequence multiplied by $\sqrt{T/T_1}$. For systems with low damping factors, the relative width T_1/T of the burst must be very small, resulting in a high crest factor. The biggest advantage of using a random burst is that there are no leakage errors (a uniform window should be used to calculate the DFT). The power spectrum of a random burst is a random variable, as it is for a periodic noise sequence, and so the same restrictions are valid as those mentioned for periodic noise.

5.3.1.6 Pulse-Impact Testing

Definition 5.8 (Pulse): The impulse response is measured directly in the time domain by exciting the plant with a short pulse (Halvorsen and Brown, 1977). For example, for a single pulse,

$$u(t) = \begin{cases} A & 0 \leq t < T_1 \\ 0 & T_1 < t \leq T \end{cases}$$

with T_1 the pulse width and T the measurement period.

Properties

- Deterministic excitation, no leakage if the system response becomes negligible before the end of the measurement window (T).
- Equivalent frequency resolution $1/T$.
- Shaping of the power spectrum by modifying the pulse shape.
- Optimal choice $T_1 = 1/(2.5 f_{max})$.
- Minimum crest factor $\sqrt{T/T_1}$, minimum time factor is T/T_1.

Discussion. The autocorrelation of the impulse response is the same as that of the MLBS, so their amplitude spectra are the same. To get the same input energy, the amplitude must be increased by a factor of $\sqrt{T/T_1}$. The minimum time factor is reached for the same upper frequency limit as for the MLBS. More sophisticated impulse generation techniques are given by Halvorsen and Brown (1977), but the general characteristics remain the same. In mechanical testing, the impulse (or hammer) excitation is still popular because it can be applied very simply: no shakers or other expensive equipment are needed to create the input.

5.3.1.7 Example: Comparison of the General Purpose Excitations. In order to get a better understanding of the behavior of the general purpose signals, they are compared with each other in this section. The aim is to excite a frequency band between 1 and 42 Hz, using signals with a length of 256 samples and a sampling frequency of 256 Hz. The resulting signals and their amplitude spectrum are shown in Figure 5-5.

For the MLBS, a clock frequency of 128 Hz was used in order to get better coverage of the frequency band ($N = 127$). The peak value of every signal was scaled to one. The random excitations consist of filtered Gaussian noise (Butterworth filter of order 7 with a cutoff fre-

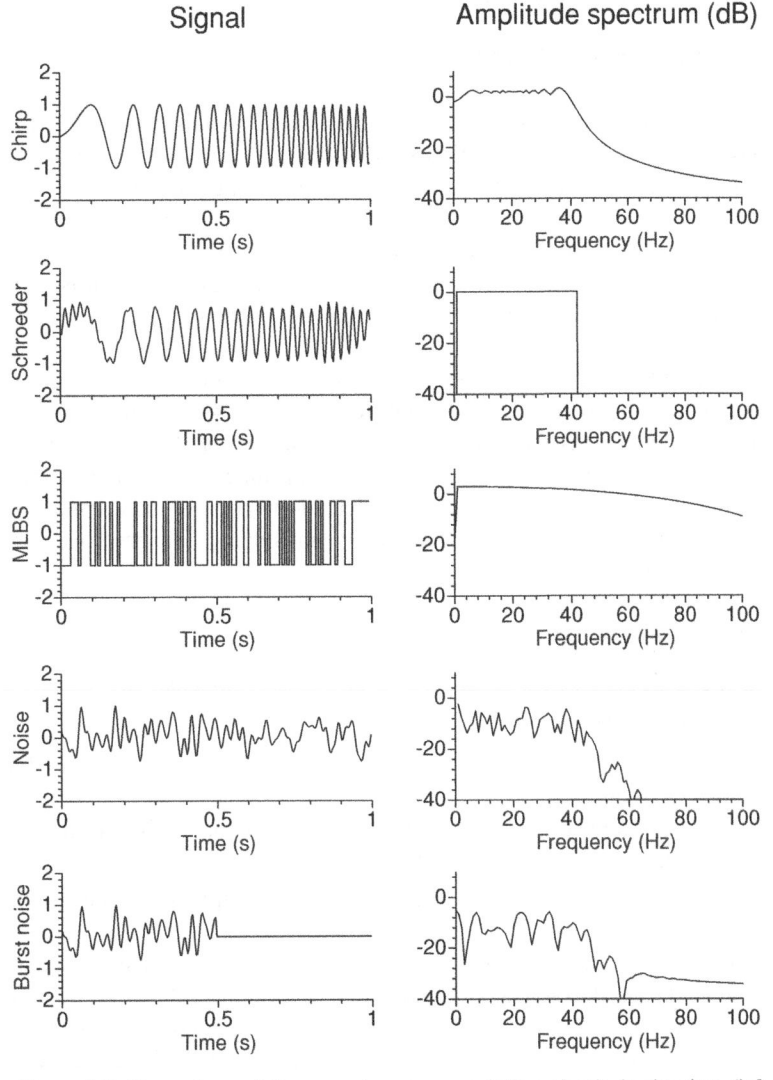

Figure 5-5. Comparison of the general purpose excitation signals in the time (left column) and frequency (right column) domains.

quency of 42 Hz). The figure is very informative. The multisine is the only signal that exclusively excites the frequency band of interest. All the other signals also excite outside this band. The first three signals inject considerably more power into the system than the noise excitations. After normalization, the power in the frequency band of interest is 1 for the MLBS, 0.81 for the periodic chirp, 0.60 for the multisine, and 0.08 and 0.05 for the random and burst random signals, respectively. The worst measurements will appear at the lines with the smallest amplitude spectrum. This was 0.009 and 0.004 for the random and burst random, 0.55 for the chirp, 1.037 for the multisine, and 1.18 for the MLBS. The amplitude of the chirp drops only at a few border lines of the frequency band of interest; it is slightly above the multisine on most other lines. From this we can conclude that the chirp, multisine, and MLBS have about the same quality and the selection should be based on personal preference, technical possibilities, and second rank arguments that are important for specific situations (e.g., no power outside the band). The random excitations have inferior properties compared with the first three deterministic excitations. They are prone to leakage and inject significantly less power into the system, resulting in a poor SNR.

5.3.2 Optimized Test Signals

Whereas in the previous section we considered signals that could be applied directly, we consider in this section excitation signals where an iterative algorithm is needed to optimize their design. Because of the continuously increasing computer power, this is not a real drawback. The design time runs from a few seconds for simple signals to a few minutes for complex signals with a few hundred frequency components.

Two classes of signals are considered. First the design of multisine excitations with minimized crest factor is discussed, then optimized binary sequences are designed.

5.3.2.1 Optimized Multisines. These are classical multisines where the user chooses the excited frequencies on the equidistant frequency grid kf_0 and also selects the desired amplitude spectrum. This is the signal preferred by the authors because it gives maximal flexibility combined with a minimum measurement time and a maximum quality of the measurements. Moreover, by making a dedicated selection of the components of the excitation signal, it is even possible to detect, qualify, and quantify the presence of nonlinear distortions (see Chapter 3).

Properties

- Periodic signal with period $T_0 = 1/f_0$ → no leakage.
- Frequency resolution $1/T_0$.
- All the power at the user-selected frequencies that can be freely chosen on the discrete grid kf_0.
- The amplitude of the harmonic components can be freely chosen and is exactly realized, no out-of-band power appears.
- Crest factor from 1.4 to 2, depending on the complexity of the amplitude spectrum.

Discussion. Instead of using explicit phase relations for the multisine, a numerical search method is used to select optimal phases that minimize the crest factor. In the literature, many crest factor minimization methods have been presented. In (5-10) the explicit expressions of the Schroeder phases are given, allowing a direct calculation of the phases. For multisines with a sparse spectrum, where the frequency lines are few and far apart, or for multisines with an amplitude spectrum that is not flat, the Schroeder phases give no better results

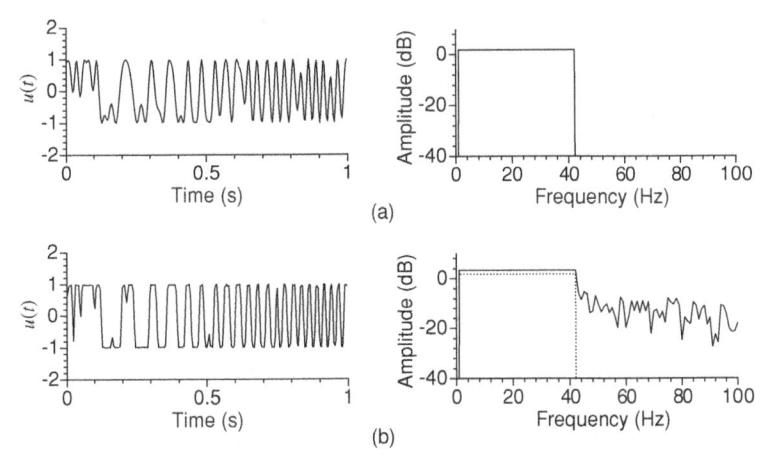

Section 5.3 ■ Study of Broadband Excitation Signals

Figure 5-6. Example of the general purpose multisine after optimizing phases. (a) Without snow, (b) with snow: ⋯ (reference signal without snow as in a), —— with snow.

than those obtained with a random phase selection, uniformly distributed in $[0, 2\pi)$. In these situations, more sophisticated methods are needed and no explicit formulas are available. Two algorithms are proposed. The first one is a clipping procedure that cuts the largest peaks of the signal. The second one is based on the successive minimization of a series of $l_{2p}(\phi)$ norms w.r.t. for increasing p

$$l_{2p}(\phi) = \|u(t, \phi)\|_{2p} = \left(\frac{1}{T_0}\int_0^{T_0} u^{2p}(t, \phi)dt\right)^{\frac{1}{2p}}$$
$$u(t, \phi) = \sum_{k=1}^{F} A_k \cos(2\pi f_k t + \phi_k)$$
(5-13)

with $\phi^T = [\phi_1 \phi_2 \ldots \phi_F]$ the phases of the multisine $u(t)$, T_0 the period of the multisine, and $p = 2, 4, 8, 16, \ldots$. Compared with the first one, it gives smaller crest factors but needs a larger memory, especially for multisines with a large number of components. With increasing computing power, the last method becomes more and more attractive. Both algorithms are discussed in Appendix 5.A.

Note: Both algorithms can be generalized easily to generate a signal with a power spectrum $S_{uu}(j\omega) + S_{aa}(j\omega)$, with $S_{uu}(j\omega)$ the desired power spectrum and $S_{aa}(j\omega)$ additional power that is added by the algorithm at other frequencies such that the crest factor of the signal decreases further (e.g., by adding additional harmonics to a sine wave, a block-like signal results, pushing the crest factor well below $\sqrt{2}$) (Guillaume et al., 1991). This is called snowing. During the calculation of the crest factor, the additional power is not considered when calculating u_{rsme}.

Example 5.9 (Flat (Snow) Multisine): The signal of the previous section is also optimized with the l_{2p} algorithm, resulting in a crest factor of 1.42 (compared with 1.67 for the Schroeder multisine). It is shown in Figure 5-6(a). Next, snowing was allowed on the lines 43–255, pushing down the crest factor to 1.19. This made it possible to get 40% more power in the frequency band of interest, compared with the original signal, which had no snow. Compared with the PRBS, 19% more power is injected in the frequency band of interest. About 5% of the totally available power is "wasted" at the snow lines. □

Example 5.10 (Quasi-Logarithmic Excitation): The advantage of the iterative algorithms becomes most obvious when dealing with more complex power spectra. In this example, a quasi-logarithmic multisine is generated, depositing the power at an almost logarithmic frequency grid ($N_{\log} = 4096$, $f_{\max} = 0.4 N_{\log}$, $f_{k+1}/f_k \approx 1.05$). Each time, the frequencies are shifted to the nearest harmonic line. After optimization, the crest factor is 2.0 (Schroeder phases: 3.3) so that almost three times more power can be injected for the same peak value of the excitation. In this example the crest factor is reduced using the successive minimization algorithm (5-13). The alternative is to use the clipping algorithm (Van der Ouderaa and Renneboog, 1988), but the first algorithm gives better results in a shorter time, at a cost of needing a larger memory. The signal is shown in Figure 5-7. □

5.3.2.2 Discrete Interval Binary Sequence (DIBS). The second class of optimized excitation signals are the discrete interval binary sequences. These are periodic binary sequences, where the sign can change only at an equidistant discrete set of points in time (Van den Bos, 1974; Paehlike and Rake, 1979; Van den Bos and Krol, 1979). The amplitude spectrum of the sequence can be optimized by choosing a good switching sequence so that the energy is concentrated within the frequency band of interest.

Properties

- Periodic signal with arbitrary period length $T_0 = 1/f_0$ → no leakage.
- Frequency resolution $1/T_0$.
- The power is concentrated at the user-selected frequencies that can be freely chosen on the discrete grid kf_0, but the other frequencies are also excited.
- The amplitude of the harmonic components can be freely chosen but is only approximately realized.
- The crest factor depends on the complexity of the signal but is usually rather small.

Discussion. The generation of a DIBS is based on an iterative algorithm proposed by Van den Bos and Krol (1979). The procedure is begun a number of times from different starting values, and the best signal is retained. With a DIBS, it is possible to concentrate the energy in a discrete set of spectral lines. The crest factor is greater than one because not all of the power is concentrated at the frequencies of interest; but even then, most of the energy can be confined to the frequency band required, which is not possible with the MLBS. Paehlike and Rake (1979) have presented an iterative scheme for putting more of the energy into the

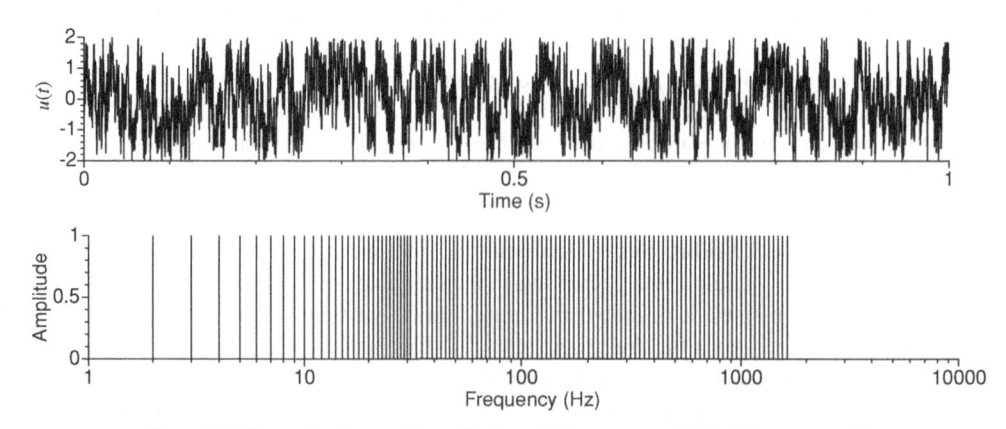

Figure 5-7. Example of a quasi-logarithmic multisine on an equidistant frequency grid.

weakest spectral lines, thus improving the SNR and decreasing the time factor. Compared with the PRBS, the DIBS can be generated for any sequence length with an arbitrary power spectrum.

Example 5.11 (Low-Pass Spectrum): The general purpose signal of the previous section was also recalculated using this method and is compared with the results of the MLBS in Figure 5-8. The crest factor of the DIBS signal is 1.36, compared with 1.5 for the MLBS. □

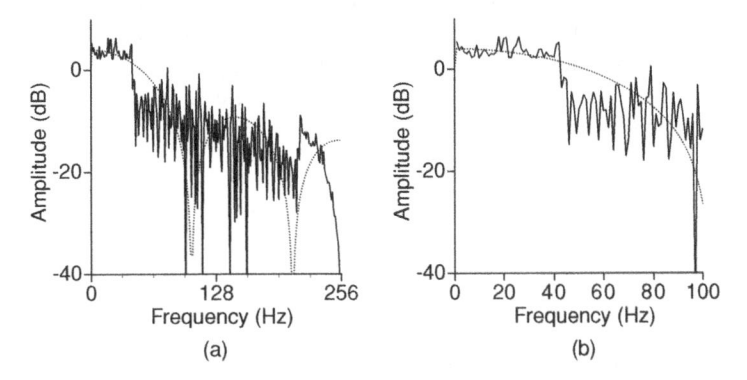

Figure 5-8. Comparison of the spectrum of a DIBS (f_c = 256 Hz, N = 256) and an PRBS (N = 103, f_c = 103 Hz) to generate a flat spectrum in a band 1–42 Hz. (a) Global view, (b) zoom on the frequency band of interest, —— DIBS, ---- PRBS.

Example 5.12 (Bandpass Spectrum): Figure 5-9 illustrates the possibility of creating a bandpass spectrum using a DIBS (crest factor 1.29). Note that this is not possible at all with an MLBS. □

5.3.3 Advanced Test Signals

In this section we discuss some excitation signals with very specialized properties, for example, signals where the crest factor of the first or second derivative is also minimized. These should be used only in critical conditions, where the special shape of the excitation gives a significant advantage. Even for these signals, the additional design time is quite restricted (from a few seconds to a few minutes), but their proper design and application requires a good user's insight into the properties of these signals and their application.

5.3.3.1 Crest Factor Minimization of Linearly Related Multiple Multisines. In some problems, it is not sufficient to keep the crest factor of the excitation low; the system output should also have a small crest factor. In other applications, the signal and its first or second derivative should be small. For example, in mechanical systems the acceleration should be restricted in order to avoid excessive forces, while excessive displacements are

Figure 5-9. Spectrum of a DIBS (f_c = 256 Hz, N = 256) designed to generate a flat spectrum in the band 40–60 Hz.

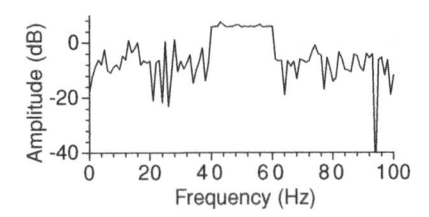

TABLE 5-1 Crest Factor Minimization of $u(t)$ and $d^2u(t)/dt^2$

	Crest Factor Input	Crest Factor Output
Input min. (5-13)	1.39	2.85
Input/output min. (5-14)	1.61	1.63

avoided to keep the stroke of the shaker small and to maintain a linear behavior of the system. Again, it would be useful if the crest factor of both signals is minimized at the same time.

The l_{2p} crest factor minimization algorithm of the previous section makes it possible to optimize multiple multisines linked by linear systems, e.g., $Y(j\omega) = G(j\omega)U(j\omega)$. Criterion (5-13) is generalized to

$$\left\| \frac{u(t,\phi)}{u_{\text{rms}}}, \frac{y(t,\phi)}{y_{\text{rms}}} \right\|_{2p} = \left(\frac{1}{T_0} \int_0^{T_0} \left(\frac{u^{2p}(t,\phi)}{u_{\text{rms}}^{2p}} + \frac{y^{2p}(t,\phi)}{y_{\text{rms}}^{2p}} \right) dt \right)^{\frac{1}{2p}}$$

$$u(t,\phi) = \sum_{k=1}^{F} A_k \cos(2\pi f_k t + \phi_k) \quad (5\text{-}14)$$

$$y(t,\phi) = \sum_{k=1}^{F} A_k |G_0(\Omega_k)| \cos(2\pi f_k t + \phi_k + \angle G_0(\Omega_k))$$

with $\phi^T = [\phi_1 \phi_2 \dots \phi_F]$, $\Omega_k = j\omega_k$ for continuous-time systems and $\Omega_k = e^{-j\omega_k T_s}$ for discrete-time systems. In Guillaume et al. (1991), it is shown that the minimum of (5-14), with respect to ϕ, for p growing to infinity results in two multisines with equal and minimum crest factors. Sometimes, it is more advantageous to minimize the scaled peak values of both multisines, allowing optimal use of the full scale of the measurement equipment. This is done by minimizing

$$\left\| u(t,\phi), \frac{y(t,\phi)}{S} \right\|_{2p} \quad (5\text{-}15)$$

with S a scaling factor. When S is chosen as the ratio of the rms values, signals with equal crest factors are obtained. Clearly, when S is chosen too large (or too small), the problem reduces to the minimization of $\|u(t,\phi)\|_{2p}$ or $\|y(t,\phi)\|_{2p}$.

Example 5.13 (Crest Factor Minimization of Linearly Related Multisines): A multisine $u(t)$ with $F = 512$ consecutive components is designed to have minimum crest, together with its second derivative $d^2u(t)/dt^2$. Table 5-1 gives the crest factors that are obtained using l_{2p} and the resulting output signals $d^2u(t)/dt^2$ are shown in Figure 5-10. As can be seen, the crest factor is reduced to 60% of its original value. In the case of a mechanical system this allows a significant reduction of the forces and, hence, the dimensions and the cost of the shaker used to generate the signals. □

5.3.3.2 Multilevel Excitation Signals. The DIBS (see Section 5.3.2) is a binary sequence that excites two levels only. In some applications, ternary signals can be used (e.g., levels −1, 0, 1), allowing greater flexibility during the design. In general, this leads to the following results:

The total power of the signal decreases: since the signal is set equal to zero at some points (instead of −1 or 1), it is clear that less power is available in the design.

The out-of-band power is reduced: the greater flexibility due to the additional level gives better control over the power spectrum. This makes it possible to reduce the out-of-band power.

The lowest in-band level is about the same: although less power is generated, the lowest amplitude at a frequency line of interest remains almost the same. This guarantees that the minimum uncertainty of the measurement will be the same for binary and three-level signals. However, by using the ternary signal, less power is wasted.

The design of multilevel signals is extensively discussed by McCormack et al. (1995) and Barker and Zhuang (1997).

5.3.3.3 Harmonic Suppression. In Section 4.2 it was shown that periodic signals with an odd (spectral lines $2k + 1$ present) spectrum with randomly selected non-excited odd harmonics make it possible to eliminate the even nonlinearities and detect the presence of odd nonlinearities. Such signals can easily be obtained from multisines where the amplitudes of the corresponding lines are put to zero. It is also possible to create such signals from binary (Tan et al., 2005) and ternary (Tan et al., 2009) sequences. The inversely repeated sequence $[u(t), -u(t)]$ has no even components in its spectrum. Using multilevel designs (Barker and Zhuang, 1997), it is also possible to suppress the second and third harmonics of a set of specified primes. Finally, it is also possible to design sparse harmonic multisines that facilitate a direct probing of the second and third degree Volterra kernels (Evans, 1998; Boyd et al., 1983) with a minimum interference.

5.4 OPTIMIZATION OF EXCITATION SIGNALS FOR PARAMETRIC MEASUREMENTS

5.4.1 Introduction

Here, the parametric measurement problem is studied. We will concentrate on the parameters θ of the mathematical model $G(\Omega_k, \theta)$, with $\Omega_k = j\omega_k$ for continuous-time systems and $\Omega_k = e^{-j\omega_k T_s}$ for discrete-time systems, which describes the measured transfer function $G_0(\Omega_k)$. To fit the model $G(\Omega_k, \theta)$ on the measurements $G(\Omega_k)$, a cost function $V(\theta, Z)$, with Z a vector containing the measured input-output DFT spectra, which is an indication of the quality of the fit, is minimized. As explained in Chapter 1, a simple and very

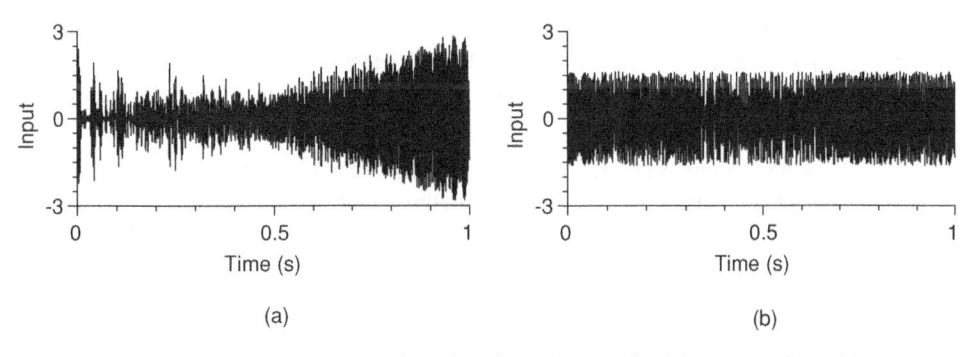

Figure 5-10. The second order derivative of a multisine with minimum crest factor (a) and input-output optimized crest factor (b).

popular choice for $V(\theta, Z)$ is the least squares cost function in which the squared differences between the model and the measurements are summed together. Another possibility is to embed the choice of the cost function in a statistical framework, as done for the maximum likelihood (ML) estimator, resulting in a weighted least squares estimator if the disturbing noise is normally distributed (Chapter 1). The quality of the estimates strongly depends on the excitation signals applied during the experiment. As in the nonparametric case, the excitation signal will be optimized in two steps, the first being the selection of an optimized power spectrum followed by a crest factor minimization of the involved signals in the second step.

To optimize the input spectrum, we need a scalar criterion that is sensitive to the accuracy of all the parameters of the system. The determinant of the covariance matrix, which is equal to the volume of the uncertainty ellipsoid, is such a criterion.

A range of criteria, other than the determinant, can be found in the literature, optimization of the trace being the most popular. For the sake of brevity, we limit ourselves in this text to examining the minimization of the determinant of the covariance matrix. For more information on other criteria, the reader is referred to other publications (Federov, 1972; Goodwin and Payne, 1977; Zarrop, 1979; Walter and Pronzato, 1997).

For computational simplicity, the covariance matrix is approximated by the Cramér-Rao lower bound (inverse information matrix) because the latter is easier to calculate (Chapter 1). General expressions of the information matrix can be obtained (without specifying an estimator) and the problem of minimizing the determinant of the covariance matrix is replaced by maximizing the determinant of the information matrix. This approximation is valid if the covariance matrix of the actual estimator approximates the Cramér-Rao lower bound sufficiently close for the considered experiments.

5.4.2 Optimization of the Power Spectrum of a Signal

5.4.2.1 Preliminary Aspects. The information matrix is the kernel of optimizing algorithms. It is a real symmetric and semipositive definite $n_\theta \times n_\theta$ matrix, where n_θ is the number of unknown model parameters. Each optimal design in the frequency domain can be reduced to a design consisting of a discrete set of $n_\theta(n_\theta + 1)/2 + 1$ frequencies (Federov, 1972; Goodwin and Payne, 1977), which corresponds to the number of free parameters in a symmetric $n_\theta \times n_\theta$ matrix + 1. The minimum number of frequencies required to avoid a nonsingular information matrix is $\text{int}(n_\theta/2)$ (with $\text{int}(x)$ the integer part of x). When using classical optimizing algorithms, the computer time needed to search for an extreme value depends strongly on the number of frequencies. From a modeling point of view, however, the minimum number is undesirable, because if an estimate of n_θ parameters is made using $\text{int}(n_\theta/2)$ frequencies, there is no possibility of detecting model errors. A second drawback of working with the minimum number of frequencies is that it is more difficult to compress the signals in the time domain.

Most algorithms presented in the literature searched for optimal designs with the minimum number of frequencies. We present a method for designing optimal power spectra based on a discrete frequency grid: this is not in itself a restriction because we look for periodic signals that have discrete spectra. This will lead to a significant reduction of the computation time. The method can be applied in the Laplace domain (continuous-time systems) as well as in the z-domain (discrete-time systems). In order to stress this equivalence, we use Ω as the frequency variable in the following interchangeable manner: $\Omega = j\omega$ (Laplace), or $\Omega = e^{-j\omega T_s}$ (z-domain). The following function is used in the optimization algorithm.

Definition 5.14 (Dispersion Function): The dispersion function $v(\chi, \Omega_k)$ for a given input power spectrum $\chi(\Omega) = (|U(1)|^2 ... |U(F)|^2)$, with $\sum_{k=1}^{F} |U(k)|^2 = \wp$ is

$$v(\chi, \Omega_k) = \text{trace}([Fi(\chi)]^{-1} fi(\Omega_k))$$

with $Fi(\chi)$ the information matrix resulting from the design $\chi(\Omega)$, $fi(\Omega_k)$ the information matrix corresponding to a single frequency input with a normalized power spectrum $|U(k)|^2 = \wp$, and Ω_k the frequency.

The dispersion function has the following properties:

- The dispersion function can be related to the input and output noise on the measurements (Schoukens and Pintelon, 1991) as

$$v(\chi, \Omega_k) = \frac{2\sigma_G^2(\Omega_k, \theta)\wp}{\sigma_U^2(k)|G(\Omega_k)|^2 + \sigma_Y^2(k) - 2\text{Re}(\sigma_{YU}^2(k)\overline{G}(\Omega_k))} \quad (5\text{-}16)$$

with $\sigma_G^2(\Omega_k, \theta)$ the uncertainty on the transfer function using the Cramér-Rao lower bound as covariance matrix for the model parameters. The dispersion can be interpreted as the ratio of the variance of the system frequency response, calculated with the estimated parameters, to the noise power of the measurements referred to the output of the system at the frequency Ω_k.

- The dispersion function is a normalized quantity (Goodwin and Payne, 1977):

$$\sum_{k=1}^{F} v(\chi, \Omega_k) \frac{|U(k)|^2}{\wp} = n_\theta \quad (5\text{-}17)$$

- The maximum of the dispersion function $v(\chi, \Omega_k)$ over the frequency grid is larger than or equal to the number of parameters n_θ (Goodwin and Payne, 1977).

These three properties will be used in the algorithm for designing an optimized excitation signal.

5.4.2.2 An Efficient Algorithm for Maximizing the Information Matrix. Although the optimal input may be found analytically for simple situations, in general, no closed form solution can be found. Therefore, an iterative design is required. Most algorithms carry out a search in the continuous frequency space to find the frequency with the maximum dispersion and then add extra energy at this frequency. The resulting spectrum is normalized, and the procedure is repeated until the variations are negligible. More sophisticated algorithms combine this procedure with a mechanism that removes components from the spectrum (Federov, 1972; Zarrop, 1979). The search for a maximum is very time consuming, and the final spectrum is difficult to generate because the optimal frequencies are not harmonically related. For these reasons, it is better to reduce the frequency space to a discrete set of frequencies in the analysis; the implications of this restriction for the attainable accuracy are studied in more detail by Van den Eijnde and Schoukens (1991), and it turns out that there is no significant loss in attainable accuracy if the discrete set of frequencies is sufficiently dense.

In general, any discrete set of frequencies can be used, but if only periodic signals are retained, it is obvious that the selected frequencies should be harmonically related. For the initial design, the simplest first choice is that of equally spaced spectral lines within the frequency band of interest, with the total fixed input power uniformly distributed over the F frequencies in this set. The resulting spectrum constitutes the initial design χ_0. The response dispersion function $v(\chi_0, \Omega_k)$ is computed for every spectral line Ω_k in the set, and the available power is redistributed over all spectral lines proportionally to the corresponding values of the dispersion function. The optimal input is found by repeating this procedure; the iteration can be stopped when the variation of the determinant of the information matrix is small. This method was described in the late 1970s (see Walter and Pronzato, 1997, pp. 305–306, and the references therein). If we express this approach in mathematical terms, we end up with an algorithm with the following consecutive steps:

Algorithm 5.15 (Optimization Power Spectrum)

1. Initiation:

   Select a set $\mathbb{F}$ of F frequencies $\Omega_1, ..., \Omega_F$ within the frequency band of interest: $\mathbb{F} = \{\Omega_1, ..., \Omega_F\}$. Distribute the input power equally over these F frequencies. This constitutes the initial design χ_0.

2. Iteration:

   2a. Set $i = i + 1$ and compute the response dispersion function $v(\chi_i, \Omega_k)$ for $k = 1, ..., F$.

   2b. Compose a new design in the following way:

   $$\chi_{i+1}(\Omega_k) = \chi_i(\Omega_k) v(\chi_i, \Omega_k) / n_\theta \text{ for } k = 1, ..., F \tag{5-18}$$

   2c. If $\max(v(\chi_i, \Omega_k) - n_\theta) < \varepsilon$ with ε sufficiently small and $\Omega_k \in \mathbb{F}$, then the optimum design is found; otherwise go to step 2a.

Proof. See Van den Eijnde and Schoukens (1991) and Delbaen (1990). □

It has been shown (Walter and Pronzato, 1997; Delbaen, 1990) that each run of this algorithm yields a superior input design and that consecutive designs converge monotonously to a design with the optimum dispersion function and, hence, the minimum determinant of the Cramér-Rao bound (= D-optimality).

5.4.2.3 Importance of Crest Factor Minimization. In a second step, after the selection of the power spectrum, the crest factor of the corresponding multisine(s) should be minimized. To compare different excitations, it is necessary to scale the determinant of the Cramér-Rao lower bound and the dispersion function with the optimized crest factor so that all signals are compared for the same peak value.

$$\det(CR_{\text{scaled}}(\theta)) = \det(CR(\theta)) Cr^{2n_\theta}(u) \tag{5-19}$$

5.4.2.4 Practical Implementation. It is obvious that the calculation of the optimum amplitude spectrum is possible only if enough knowledge of the system is available. In most situations, a two-step procedure is required, restricting the applicability of these methods significantly. In the first step, the unknown parameters are estimated using a multisine with a flat amplitude spectrum; in the second step, these estimated values are used to optimize the amplitude spectrum. The covariance matrix of the estimated, unknown model parameters should be close enough to the Cramér-Rao lower bound.

Section 5.4 ■ Optimization of Excitation Signals for Parametric Measurements 171

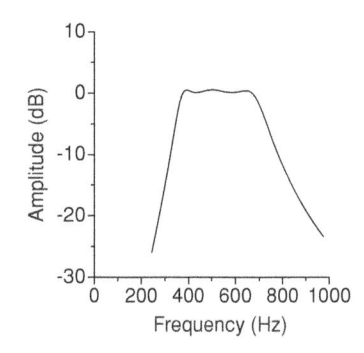

Figure 5-11. Amplitude transfer characteristic of the studied system.

5.4.2.5 Example: An Experimental Verification. The power spectrum optimization for a parametric measurement is illustrated in the following example:

$$G_0(s) = \frac{b_2 s^2 + b_3 s^3 + b_4 s^4}{a_0 + a_1 s + \ldots + a_6 s^6} \quad (5\text{-}20)$$

The coefficients are given in Table 5-2 and the corresponding amplitude characteristic is given in Figure 5-11. The system is excited with a multisine at the frequencies $f_k = kf_0$, with $k = 25, 26, \ldots, 100$ and $f_0 = 50/2048$ MHz. The rms value of the multisine is set equal to $1/\sqrt{2}$. Two multisines are considered, the first one having a flat amplitude spectrum and the second one being optimized on the basis of the procedure described before. The evolution of the power spectrum optimization process is given in Figure 5-12. The optimization is stopped before the final convergence is reached (after three iterations) to avoid signals with a sparse spectrum. These are very difficult to compress and have a large crest factor. From (5-19) it is seen that this would jeopardize the accuracy gain that is obtained with the design of an optimal spectrum. In this example the determinant of the corresponding Cramér-Rao lower bound was reduced with a factor 43 after three iterations.

The crest factors or peak values of the multisine at the input and output are minimized using the l_{2p} algorithm (5-15) and the results are given in Table 5-3. Three situations are considered:

- Minimization of the crest factor of the input signal
- Simultaneous minimization of the crest factors of the input and output
- Simultaneous minimization of the peak values of the input and output

For our purpose, the last possibility is the most interesting because it will determine the settings of the full scale of the measurement instruments. In Table 5-3, it is seen that the peak values of the multisine, with the optimized power spectrum, are equal to those of the multisine with flat power spectrum (optimization c). So the settings of the measurement instru-

TABLE 5-2 Coefficients of the Transfer Function of the Sixth-Order Continuous-Time Bandpass Filter

		b_2	b_3	b_4		
		8.973e-10	5.5155e-12	3.2010e-17		

a_0	a_1	a_2	a_3	a_4	a_5	a_6
1	2.5017e-4	3.5869e-7	5.5550e-11	3.36031e-14	2.5351e-18	1.0131e-21

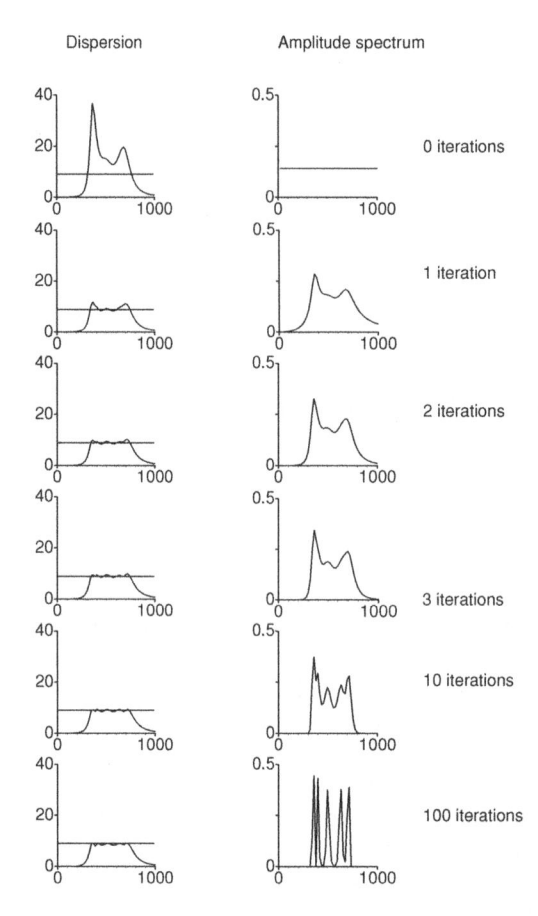

Figure 5-12. Evolution of the power spectrum optimization process.

ments can remain the same for both excitations, and, consequently, the noise on both measurements will be equal. However, the uncertainty on the estimated parameters will be smaller in the second case because the determinant $\det(Fi(\theta))$ is much smaller than in the first case, resulting in a smaller uncertainty on the calculated transfer characteristics.

From experimental tests, it turned out that these signals can be generated in practice; small disturbances at the amplitudes or the phases in the generator (and reconstruction filter) do not result in an excessive growth of the crest factor. In Figure 5-13(b) measurements of the calculated multisines are given. They were generated with a 12-bit arbitrary waveform generator with 2048 points in one period (sampling frequency 20 kHz). The generator was followed by a reconstruction filter (a Cauer filter with a cutoff frequency of 2 kHz). No phase or amplitude compensation was made for the distortion introduced by the reconstruction filter. If this amplitude/phase distortion becomes disturbing, it is always possible to give a pre-compensation to the amplitudes/phases of the multisine. The measurements were made with an 8-bit digitizer (full scale ± 1 V) at 512 points with a sampling frequency of 5 kHz.

Figure 5-13(a) compares the uncertainty $\sigma_G(\Omega_k, \theta)$ on the estimated transfer function model in case a multisine with a flat and an optimized amplitude spectrum is used. These results were experimentally verified using the setup described before. Sixty measurements were made to measure the standard deviation of the FRF measurement. The results are shown in Figure 5-13(b). It is obvious that this result is relevant only if the model errors of the parametric model in the identification step are smaller than the identification uncertainty due to the noise.

Section 5.5 ■ Optimization of Excitation Signals for Parametric Measurements

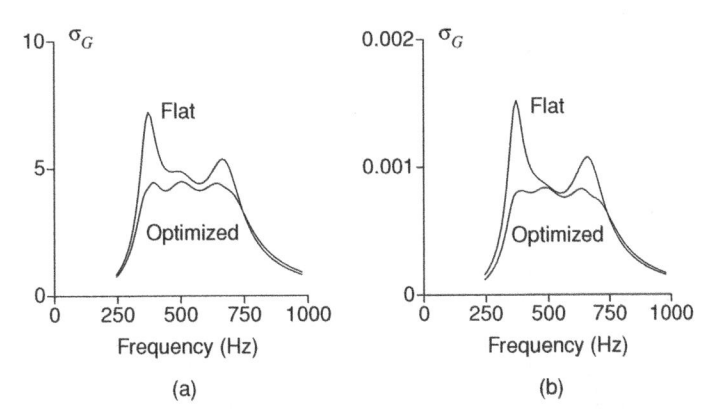

Figure 5-13. Comparison of the model uncertainty with the flat and the optimized power spectrum, (a) theoretical (scaled) results, (b) experimental results.

5.5 EXPERIMENT DESIGN FOR CONTROL

Although experiment design has a long history (see, for example, Mehra, 1974; Goodwin and Payne, 1977), it was only in the late 1990s that the field obtained a renewed interest (Antoulas and Anderson, 1999; Gevers, 2005). Most approaches (see, for example, Gevers and Ljung, 1986; Forssell and Ljung, 2000b) focus on designing a power constrained input signal (input spectrum) that minimizes some control-oriented criterion (e.g., model mismatch between the nominal and actual closed loop system). In practice it boils down to finding the constrained input signal that minimizes a measure of a control-oriented model uncertainty set. A dual point of view for the optimal experiment design problem has been formulated. It looks for the least costly identification experiment that gives an uncertainty set lying within a given maximum allowable control-oriented model uncertainty measure (see, for example, Bombois et al., 2006; Barenthin and Hjalmarsson, 2008). The identification cost can be measured as the input energy, or the experiment time, or the performance degradation during experimentation due to the additional excitation signal, or a combination of these.

As is the case for the power spectrum optimized signals (see Section 5.4), the optimal control-oriented experiment can only be calculated if the true system is known. Hence, an iterative procedure is needed. The reader is referred to the literature for the algorithmic details (see, for example, Bombois et al., 2006; Barenthin et al., 2008).

TABLE 5-3 Crest Factor or Peak Value Minimization of Two Multisines Related by (5-20)

	Input		Output	
	Crest Factor	Peak Value	Crest Factor	Peak Value
Flat input power spectrum				
a	1.459	1.031	2.749	1.418
b	1.667	1.170	1.667	0.862
c	1.509	1.067	2.065	1.067
Optimized input power spectrum				
c	1.459	1.031	1.860	1.200
b	1.582	1.118	1.582	1.026
c	1.508	1.066	1.643	1.066

a: minimization of the crest factor of the input
b: simultaneous minimization of the crest factors of the input and output
c: simultaneous minimization of the peak values of the input and output

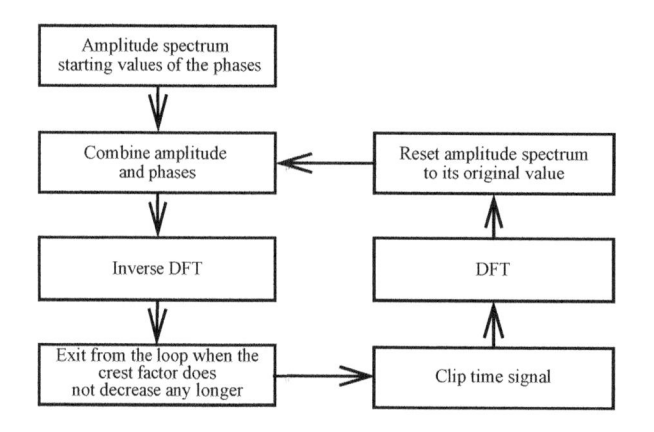

Figure 5-14. Minimization of the crest factor of a multisine: clipping algorithm.

5.6 APPENDIX

Appendix 5.A   Minimizing the Crest Factor of a Multisine

5.A.1 Clipping Algorithm. In Van der Ouderaa et al. (1988a, 1988b) an iterative method has been developed to optimize the phases. The method is very close to an algorithm presented by Van den Bos (1987). The basic idea behind this method is a clipping procedure, which is illustrated in Figure 5-14. For a given amplitude spectrum, a time signal with a minimum peak value has to be found. The iteration procedure is started from the specified amplitude spectrum, and arbitrary phases are taken as starting values. Using the inverse Fourier transform, the signal is calculated at a set of discrete equidistant times. A new time signal is then generated by clipping off all the values larger than a given maximum, and the new modified spectrum and phases are calculated using the FFT. These new phases are retained as a first approximation to the solution, but the modified amplitude spectrum is rejected in favor of the original one. This procedure is repeated until no further significant reduction of the crest factor is obtained. During the iteration process, the clipping level is changed from a low value in the beginning (e.g., 0.7 u_{max}) to almost no clipping (e.g., 0.999 u_{max}) at the end of the process, for strongly compressed signals. In general, the algorithm needs a few hundred iterations to obtain useful signals (for example, a flat multisine with a crest factor of 1.5), but in order to obtain near-optimal crest factors (of 1.4) a few hundred thousand iteration steps are more likely to be required. This algorithm is called the clipping algorithm.

5.A.2 Infinity Norm Algorithm. In Guillaume et al. (1991) an algorithm has been developed based on the minimization of the l_{2p} norm

$$l_{2p}(\phi) = \|u(t, \phi)\|_{2p} = \left(\frac{1}{T_0}\int_0^{T_0} u^{2p}(t, \phi)dt\right)^{\frac{1}{2p}}$$

$$u(t, \phi) = \sum_{k=1}^{F} A_k \cos(2\pi f_k t + \phi_k)$$

(5-21)

with T_0 the period of the multisine and $p = 2, 4, 8, 16, \ldots$. It is shown that the $l_{2p}(\phi)$ norm is equal to

$$l_{2p}(\phi) = \left(\frac{1}{N}\sum_{t=0}^{N-1} u^{2p}(tT_s, \phi)\right)^{\frac{1}{2p}} \text{ if } N \geq 2pf_{max}T_0 + 1$$

(5-22)

with $f_{\max}$ the maximum frequency occurring in the multisine and N the number of samples in one period. Condition $N \geq 2pf_{\max}T_0 + 1$ in (5-22) expresses that no alias contribution may appear on the DC component.

The l_{2p} norm is minimized with respect to the phases using a Marquardt algorithm for values of p that are gradually increased during the iteration process. This defines a descent algorithm that converges to a local minimum. From our experiences, it turned out that the results of this algorithm were better than those obtained with the previous method. In practice, conditions (5-22) may be violated as long as a sufficiently large number of points is considered (e.g., $N \geq 16f_{\max}T_0 + 1$), leading to a significant reduction of the calculation time.

6

Models of Linear Time-Invariant Systems

Abstract: This chapter presents the nonparametric and parametric system (signal) and noise models used throughout this book. The models are described in the frequency domain and cover linear time-invariant discrete-time systems (z-domain), continuous-time systems (s-domain), diffusion phenomena ($\sqrt{s}$-domain), commensurate microwave systems ($\tanh(\tau_R s)$), and damped (complex) exponentials. The classical transfer function models describing the relationship between the DFT spectra of the input and output signals are valid for periodic and time-limited signals only. These models are extended to arbitrary excitations for discrete-time and continuous-time systems. Extended transfer function models are also derived in case samples are missing at the input and/or output signals. The identifiability issues of the different models are discussed and generalizations to the multivariable case are given. The basic concepts of linear system theory are assumed to be known. Textbooks on the topic are by Oppenheim et al. (1997), Kailath (1980), and Kwakernaak and Sivan (1991).

6.1 INTRODUCTION

Although most real-life processes are nonlinear and time variant, they can often be approximated very well by linear time-invariant systems. Linear time-invariant continuous-time systems are described by differential equations (finite dimensional or lumped systems) or partial differential equations (infinite dimensional or distributed systems) with constant coefficients. The transfer function between the input $u(t)$ and the output $y(t)$ of the process is calculated assuming that the initial conditions are zero.

Example 6.1 (Lumped Continuous-Time System): Consider the LC resonator of Figure 6-1.

The input of the system is the voltage source $u(t)$ and the output is the voltage $y(t)$ across the capacitor. Both are related by a second-order differential equation,

$$LC\frac{d^2y(t)}{dt^2} + y(t) = u(t) \qquad (6\text{-}1)$$

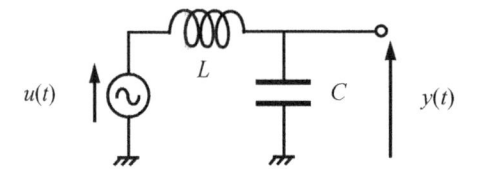

Figure 6-1. *LC* series resonator.

Taking the Laplace transform of (6-1), assuming that the initial conditions are zero ($y(0) = 0$ and $y'(0) = 0$), gives the transfer function

$$G(s) = \frac{Y(s)}{U(s)} = \frac{1}{1 + LCs^2} \qquad (6\text{-}2)$$

Note that $G(s)$ has one complex conjugate pole pair $s = \pm j/\sqrt{LC}$ on the imaginary axis. □

Example 6.2 (Distributed Continuous-Time System): Consider the clamped beam of Figure 6-2.

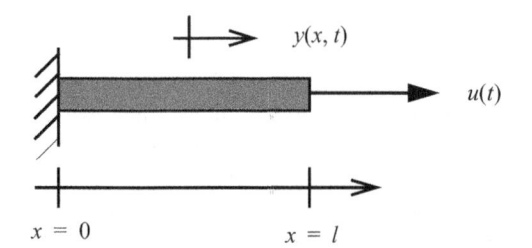

Figure 6-2. Longitudinal vibrations of a clamped beam.

The input of the system is the force per unit area $u(t)$ and the output is the longitudinal displacement $y(x, t)$. Both are related by a second-order partial differential equation,

$$\frac{\partial^2 y(x, t)}{\partial t^2} = \frac{E}{\rho} \frac{\partial^2 y(x, t)}{\partial x^2} \qquad (6\text{-}3)$$

with boundary conditions $y(0, t) = 0$ and $\partial y(x, t)/\partial x|_{x=l} = u(t)/E$. E, ρ are, respectively, the elasticity modulus and the density of the beam. The transfer function between the force per unit area $u(t)$ and the longitudinal displacement at the end of the beam $y(l, t)$ is calculated, assuming zero initial conditions $y(x, 0) = 0$, $\partial y(x, t)/\partial t|_{t=0} = 0$. We find

$$G(s) = \frac{Y(l, s)}{U(s)} = \frac{l}{E} \frac{\tanh(\tau s)}{\tau s} \qquad (6\text{-}4)$$

with $\tau = \sqrt{\rho l^2/E}$. Note that $G(s)$ has an infinite number of complex conjugate pole pairs $s = \pm j(2k+1)\pi/(2\tau)$, $k \in \mathbb{N}$ on the imaginary axis (see Exercise 6.1). According to the Mittag-Leffler theorem (Henrici, 1974), (6-4) can be expanded in an infinite series of partial fractions (see Exercise 6.2)

$$G(s) = \frac{l}{E} \sum_{k=0}^{\infty} \frac{2}{(\tau s)^2 + (\pi(2k+1)/2)^2} \qquad (6\text{-}5)$$

Because the active frequency range of $|2/((\tau s)^2 + (\pi(2k+1)/2)^2)|_{s=j\omega}$ is limited, it follows from (6-5) that, within a given frequency band, (6-4) can be approximated very well by a rational transfer function of finite order in s. □

The conclusions of Example 6.2 are valid for most physical infinite-dimensional processes: their irrational transfer functions have an infinite (countable) number of poles (those at infinity included) and can be approximated well in a limited frequency band by a rational form of finite order in s. The advantage of using a rational approximation is that the form of the model is robust w.r.t. (small) changes in the geometry and/or the boundary conditions. This is not the case for the irrational transfer function models, because they must be recalculated for each particular geometry and boundary condition. The disadvantage of the rational approximation is that the model contains too many parameters; for example, the irrational transfer function (6-4) has two independent parameters while a rational approximation of order two uses five independent parameters.

The irrational transfer functions of systems where diffusion phenomena such as mass or heat transfer are important are very often a function of $\sqrt{s}$. For such systems it might be a good idea to use a rational approximation in $\sqrt{s}$ instead of s (see, for example, Pintelon et al., 2005). Examples of such systems are electrochemical processes where the charge transport, controlled by diffusion, is modeled by an impedance (Warburg impedance) that is proportional to $\sqrt{s}$ (Wang, 1987). Practical applications can be found in visco-elasticity (Baker et al., 1996), mechanics (Sakakibara, 1997), electrochemistry (Durbha et al., 1999), heat conduction (Battaglia et al., 2001), and control (Moreau et al., 2002).

The irrational transfer functions of lossless commensurate microwave devices are a rational function of the Richards variable $S = \tanh(\tau_R s)$ (Rizzi, 1988). For real (lossy) microwave devices it might be a good idea to use rational approximations in $\tanh(\tau_R s)$ instead of s.

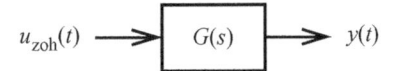

When a lumped continuous-time system is excited by a piecewise constant signal, then there exists a discrete-time model that, exactly, describes the input-output behavior of system at the sampling instances (see Example 6.3). This result is used in control applications where the input of the system (plant) is the piecewise constant output of a digital controller.

Example 6.3 (Discrete-Time System): Consider a lumped continuous-time system (see Figure 6-3) excited by a piecewise constant excitation signal

$$u_{\text{zoh}}(t) = \sum_{r=0}^{\infty} u(r) \text{zoh}(t - rT_s) \tag{6-6}$$

with $\text{zoh}(t) = 1$ for $t \in [0, T_s)$ and $\text{zoh}(t) = 0$ elsewhere. The Laplace transform of the output $y(t)$ equals

$$Y(s) = \frac{G(s)}{s}(1 - z^{-1})U(z)\bigg|_{z = e^{sT_s}} \tag{6-7}$$

with $U(z)$ the Z-transform of the samples $u(k)$. Applying the residue formula (Selby, 1973)

Figure 6-3. Lumped continuous-time system excited by a piecewise constant signal.

$u_{\text{zoh}}(t) \rightarrow \boxed{G(s)} \rightarrow y(t)$

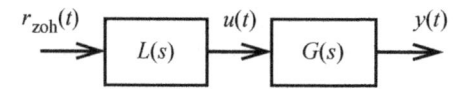

Figure 6-4. Cascade of continuous-time systems excited by a piecewise constant signal.

$$Z\{Y(s)\} = \sum_{\text{poles } Y(s)} \text{Res}(\frac{z}{z - e^{sT_s}} Y(s)) \qquad (6\text{-}8)$$

to (6-7), we find the Z-transform of the sampled output $y(kT_s)$

$$Y(z) = (1 - z^{-1})U(z)Z\{G(s)/s\} \qquad (6\text{-}9)$$

It follows that there exists a discrete-time model with transfer function

$$G_{\text{ZOH}}(z^{-1}) = Y(z)/U(z) = (1 - z^{-1})Z\{G(s)/s\} \qquad (6\text{-}10)$$

that exactly describes the input-output behavior of the continuous-time model at the sampling times $t = kT_s$. Formula (6-10) is known as the step-invariant transformation. Result (6-10) can be generalized to the cascade of two systems (see Figure 6-4). However, in this case the discrete-time model relating the sampled input $u(t)$ to the sampled output $y(t)$ of the plant $G(s)$

$$G_d(z^{-1}) = Y(z)/U(z) = \frac{Y(z)/R(Z)}{U(z)/R(Z)} = \frac{Z\{L(s)G(s)/s\}}{Z\{L(s)/s\}} \qquad (6\text{-}11)$$

depends on the characteristics of the preceding system $L(s)$ (see Exercise 6.4). □

The results of Example 6.3 can be generalized to a certain class of nonlinear continuous-time systems. If a continuous-time Volterra system is excited by a piecewise constant signal, then there exists a discrete-time Volterra model that, exactly, describes the input-output behavior of the system at the sampling instances (see Example 6.4).

Example 6.4 (Nonlinear Discrete-Time System): The output $y(t)$ of a time-invariant continuous-time Volterra system can be written as

$$\begin{aligned} y(t) &= \sum_{\alpha = 1}^{\infty} y_\alpha(t) \\ y_\alpha(t) &= \int_0^\infty \int_0^\infty \cdots \int_0^\infty g_\alpha(\tau_1, \tau_2, \ldots, \tau_\alpha) u(t - \tau_1) u(t - \tau_2) \ldots u(t - \tau_\alpha) d\tau_1 d\tau_2 \ldots d\tau_\alpha \end{aligned} \qquad (6\text{-}12)$$

with $u(t)$ the input, $y_\alpha(t)$ the nonlinear contribution of degree α, and $g_\alpha(\tau_1, \ldots, \tau_\alpha)$ the multidimensional impulse response of degree α (Schetzen, 1980). Note that $y_\alpha(t)$ is written as a multidimensional convolution of $g_\alpha(\tau_1, \ldots, \tau_\alpha)$ with the input. The contribution of degree α in (6-12) can always be written as

$$y_\alpha(t) = \sum_{n_1, \ldots, n_\alpha = 1}^{\infty} \int_{(n_1 - 1)T_s}^{n_1 T_s} \cdots \int_{(n_\alpha - 1)T_s}^{n_\alpha T_s} g_\alpha(\tau_1, \ldots, \tau_\alpha) u(t - \tau_1) \ldots u(t - \tau_\alpha) d\tau_1 \ldots d\tau_\alpha \qquad (6\text{-}13)$$

Section 6.1 ■ Introduction

Evaluating (6-13) at $t = kT_s$ for piecewise constant inputs $u_{zoh}(t)$ (6-6), taking into account that $u_{zoh}(kT_s - \tau) = u(k-n)$ for $\tau \in ((n-1)T_s, nT_s]$, gives

$$y(kT_s) = \sum_{\alpha=1}^{\infty} y_\alpha(kT_s)$$

$$y_\alpha(kT_s) = \sum_{n_1, n_2, \ldots, n_\alpha = 1}^{\infty} g_{\alpha zoh}(n_1, n_2, \ldots, n_\alpha) u(k-n_1) u(k-n_2) \ldots u(k-n_\alpha)$$

(6-14)

where $g_{\alpha zoh}(n_1, n_2, \ldots, n_\alpha)$ is defined as

$$g_{\alpha zoh}(n_1, \ldots, n_\alpha) = \int_{(n_1-1)T_s}^{n_1 T_s} \ldots \int_{(n_\alpha-1)T_s}^{n_\alpha T_s} g_\alpha(\tau_1, \ldots, \tau_\alpha) d\tau_1 \ldots d\tau_\alpha \quad (6\text{-}15)$$

Equation (6-14) is a shift-invariant discrete-time Volterra model (Brillinger, 1981) that exactly describes the input-output behavior of the time-invariant continuous-time Volterra system (6-12) at the sampling times $t = kT_s$.

Note that the Z-transform of the linear contribution in (6-14),

$$y_1(kT_s) = \sum_{n_1=1}^{\infty} g_{1zoh}(n_1) u(k-n_1) \text{ with } g_{1zoh}(n_1) = \int_{(n_1-1)T_s}^{n_1 T_s} g_1(\tau_1) d\tau_1$$

is exactly (6-9) and (6-10). □

We conclude from Examples 6.1 to 6.3 that rational transfer function models of some generalized frequency variable are appropriate for describing a broad class of (in)finite-dimensional linear time-invariant systems. The stable and minimum phase regions of the poles and zeros in the s-, z- and $\sqrt{s}$- domains are shown in Figure 6-5 (proof: see Appendix 6.A). It follows that the impulse response of a stable s- or z- domain system decreases exponentially to zero while that of a stable $\sqrt{s}$- domain system decreases algebraically to zero as an $O(t^{-3/2})$. For unstable systems (s-, z- and $\sqrt{s}$- domains) the impulse response growths exponentially (proof: see Appendix 6.A). In what follows, we discuss several possible parameterizations of transfer function models and establish the relationship with the discrete Fourier transforms (DFTs) of the input and output signals.

For lumped continuous-time and discrete-time systems the transfer function models, and their relationship with the input-output DFT spectra, are obtained by taking, respectively, the Laplace transform of the following differential equation:

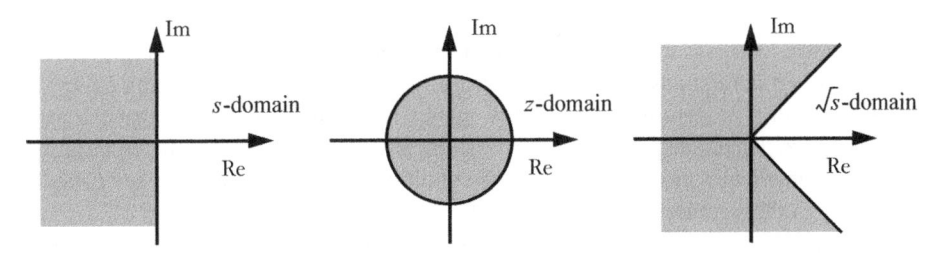

Figure 6-5. Gray area: stable and minimum phase regions of, respectively, the poles and zeros. s-domain: $\text{Re}(s) < 0$, z-domain: $|z| < 1$, and $\sqrt{s}$-domain: $|\angle\sqrt{s}| > \pi/4$.

$$\sum_{n=0}^{n_a} a_n y^{(n)}(t) = \sum_{m=0}^{n_b} b_m u^{(m)}(t) \qquad (6\text{-}16)$$

and the Z-transform of the following difference equation:

$$\sum_{n=0}^{n_a} a_n y(t-n) = \sum_{m=0}^{n_b} b_m u(t-m) \qquad (6\text{-}17)$$

If the system is proper, $n_a \geq n_b$, then (6-16) and (6-17) can be written under their state space representation form as, respectively,

$$\begin{aligned} \frac{dx(t)}{dt} &= Ax(t) + Bu(t) \\ y(t) &= Cx(t) + Du(t) \end{aligned} \qquad (6\text{-}18)$$

and

$$\begin{aligned} x(t+1) &= Ax(t) + Bu(t) \\ y(t) &= Cx(t) + Du(t) \end{aligned} \qquad (6\text{-}19)$$

where $x(t) \in \mathbb{R}^{n_a}$ is the state vector (Kailath, 1980). The parameters $A \in \mathbb{R}^{n_a \times n_a}$, $B \in \mathbb{R}^{n_a \times 1}$, $C \in \mathbb{R}^{1 \times n_a}$, and $D \in \mathbb{R}$ of the state space equations (6-18) and (6-19) can easily be related to the a_n and b_m coefficients of (6-16) and (6-17) (see Exercise 6.6).

6.2 PLANT MODELS

The parametric model that will be used mostly throughout this book is a *rational form*

$$G(\Omega, \theta) = \frac{B(\Omega, \theta)}{A(\Omega, \theta)} = \frac{\sum_{r=0}^{n_b} b_r \Omega^r}{\sum_{r=0}^{n_a} a_r \Omega^r} \qquad (6\text{-}20)$$

with $\Omega = s$ for lumped continuous-time systems, $\Omega = z^{-1}$ for discrete-time systems, $\Omega = \sqrt{s}$ for diffusion phenomena, $\Omega = \tanh(\tau_R s)$ for commensurate microwave devices, and $\theta \in \mathbb{R}^{n_\theta}$ the vector of the plant model parameters

$$\theta^T = [a_0 a_1 \ldots a_{n_a} b_0 b_1 \ldots b_{n_b}] \qquad (6\text{-}21)$$

The reason for this is that it is very easy to get good starting values for (6-20) (see Chapter 9). For lumped continuous-time and discrete-time systems, (6-20) is obtained by taking the Laplace and Z-transform of (6-17) and (6-16), respectively, assuming that the initial conditions are zero. For large order systems (typically $n_a, n_b > 30$) parameterization (6-20) becomes numerically unstable (leads to ill-conditioned normal equations, see Chapter 9), thus requiring other representations to be used.

In modal analysis (Ewins, 1991) and nuclear magnetic resonance modeling (see Section 6.4) a *partial fraction expansion* of (6-20) is often used. Assuming that $G(\Omega, \theta)$ has simple poles, it has the form (Henrici, 1974)

Section 6.2 ■ Plant Models

$$G(\Omega, \theta) = \sum_{\substack{r=-p \\ r \neq 0}}^{p} \frac{L_r}{\Omega - \lambda_r} + \sum_{r=1}^{q} \frac{S_r}{\Omega - \sigma_r} \qquad (6\text{-}22)$$

for strictly proper ($n_b < n_a$) continuous-time models ($\Omega = s$, $\sqrt{s}$ or $\tanh(\tau_R s)$) and

$$G(z^{-1}, \theta) = \sum_{\substack{r=-p \\ r \neq 0}}^{p} \frac{L_r z^{-1}}{1 - \lambda_r z^{-1}} + \sum_{r=1}^{q} \frac{S_r z^{-1}}{1 - \sigma_r z^{-1}} \qquad (6\text{-}23)$$

for proper discrete-time models ($n_b \leq n_a$) with $b_0 = 0$ (see Exercise 6.5). In both cases we have $L_{-r} = \bar{L}_r$, $\lambda_{-r} = \bar{\lambda}_r$ and $S_r, \sigma_r \in \mathbb{R}$ with $2p + q = n_a$ so that

$$\theta^T = [\sigma_1 \ldots \sigma_q \text{Re}(\lambda_1)\text{Im}(\lambda_1)\ldots\text{Re}(\lambda_p)\text{Im}(\lambda_p) S_1 \ldots S_q \text{Re}(L_1)\text{Im}(L_1)\ldots\text{Re}(L_p)\text{Im}(L_p)] \qquad (6\text{-}24)$$

Because parameterizations (6-22) and (6-23) are numerically more stable than (6-20) (except in the case of poles of multiplicity larger than one), one could think of using these models to identify high-order systems (typically $n_a, n_b > 30$). In practice, these representations are not really helpful because the starting values, generated by using parameterization (6-20), are of insufficient quality for higher order systems, resulting in poor transfer function estimates (6-22) and (6-23) (one gets stuck in a local minimum). The disadvantage of parameterizations (6-22) and (6-23) is that they do not allow the choice of the order n_b of the numerator polynomial of $G(\Omega, \theta)$. The advantage is that they can deal very easily with constraints on the residues and the poles (see Section 5.4).

An alternative solution for high-order systems is to *factorize* transfer function (6-20) in its poles and zeros. Assuming that $G(\Omega, \theta)$ has simple poles and zeros, we get

$$G(\Omega, \theta) = K \frac{\prod_{r=1}^{n_b} (\Omega - \zeta_r)}{\prod_{r=1}^{n_a} (\Omega - \lambda_r)}$$

However, this representation suffers from the same problems as (6-22) and (6-23): (i) starting values should be generated via (6-20), and (ii) it leads to ill-conditioned normal equations if the true plant model contains multiple poles and/or zeros. Note that the latter is not the case for parameterization (6-20).

To handle high-order systems (typical $n_a, n_b > 30$) the numerator and denominator polynomials of the transfer function (6-20) are expanded in *scalar or vector orthogonal polynomials* (see Section 15.11 and Exercise 1.14)

$$G(\Omega, \theta) = \frac{B(\Omega, \theta)}{A(\Omega, \theta)} = \frac{\sum_{r=0}^{n_q} b_r q_r(\Omega)}{\sum_{r=0}^{n_p} a_r p_r(\Omega)} \qquad (6\text{-}25)$$

For *scalar orthogonal polynomials* we have $n_p = n_a$, $n_q = n_b$ and $p_r(\Omega)$, $q_r(\Omega)$ are polynomials of order r; for *vector orthogonal polynomials* $b_r = a_r$, $n_q = n_p = n_a + n_b + 1$ and $p_r(\Omega)$, $q_r(\Omega)$ are polynomials of increasing order with $p_{n_p}(\Omega)$, $q_{n_q}(\Omega)$ polynomials of order n_a, n_b, respectively. These are chosen such that they maximize the numerical stability of the model (minimize the condition number of the normal equations, see Chapter 9).

The *state space representation* form of a proper ($n_b \leq n_a$) transfer function (6-20) is

$$G(s, \theta) = C(sI_{n_a} - A)^{-1} B + D \tag{6-26}$$

for lumped continuous-time systems and

$$G(z^{-1}, \theta) = z^{-1} C(I_{n_a} - z^{-1} A)^{-1} B + D \tag{6-27}$$

for discrete-time systems. Equations (6-26) and (6-27) are obtained by taking the Laplace and Z-transform of (6-19) and (6-18), respectively, assuming that the initial conditions are zero. In both cases we have $A \in \mathbb{R}^{n_a \times n_a}$, $B \in \mathbb{R}^{n_a \times 1}$, $C \in \mathbb{R}^{1 \times n_a}$, and $D \in \mathbb{R}$, so that

$$\theta^T = [\text{vec}^T(A) \ B^T \ C \ D] \tag{6-28}$$

The disadvantages of the state space representation are that it does not exist for improper systems ($n_b > n_a$) and that it does not allow one to choose the order n_b of the numerator polynomial of $G(s, \theta)$. The advantage is that it allows straightforward extension to multivariable systems (see Section 6.6).

A *time delay* can be added to transfer function models (6-20), (6-22), (6-23), (6-25), (6-26), and (6-27). For example, for continuous-time models ($\Omega = s$, $\sqrt{s}$ or $\tanh(\tau_R s)$) (6-20) becomes

$$G(\Omega, \theta) = e^{-\tau s} \frac{B(\Omega, \theta)}{A(\Omega, \theta)} = e^{-\tau s} \frac{\sum_{r=0}^{n_b} b_r \Omega^r}{\sum_{r=0}^{n_a} a_r \Omega^r} \tag{6-29}$$

and for discrete-time models

$$G(z^{-1}, \theta) = z^{-\tau/T_s} \frac{B(z^{-1}, \theta)}{A(z^{-1}, \theta)} = z^{-\tau/T_s} \frac{\sum_{r=0}^{n_b} b_r z^{-r}}{\sum_{r=0}^{n_a} a_r z^{-r}} \tag{6-30}$$

where $\tau \in \mathbb{R}$ is an arbitrary time delay (not necessarily an integer multiple of the sampling period T_s). Then the vector of the model parameters θ also contains the delay τ.

6.3 RELATION BETWEEN THE INPUT-OUTPUT DFT SPECTRA

In this section we establish the relationship between the DFTs of the sampled input and output signals of a linear dynamic system

Section 6.3 ■ Relation Between the Input-Output DFT Spectra

$$U(k) = \frac{1}{\sqrt{N}} \sum_{t=0}^{N-1} u(tT_s) z_k^{-t}, \quad Y(k) = \frac{1}{\sqrt{N}} \sum_{t=0}^{N-1} y(tT_s) z_k^{-t} \text{ with } z_k = e^{j2\pi k/N} \quad (6\text{-}31)$$

and the transfer function models $G(\Omega, \theta)$ of Section 6.2. We start with periodic excitation signals, proceed with arbitrary signals, and finally handle the case where data samples are missing at the input and/or output signals. For the continuous-time systems ($\Omega = s$, $\sqrt{s}$ or $\tanh(\tau_R s)$) we will assume that the excitation is band-limited.

6.3.1 Models for Periodic Signals

Assume that we apply a periodic signal $u(t)$ with harmonically related frequencies hf_0, $h \in \mathbb{H} \subset \mathbb{N}$, and period $T_0 = 1/f_0$ to the system and that we observe the steady-state response during an integer number of periods $NT_s = MT_0$ with $M \in \mathbb{N}$. If the excitation is band-limited (continuous-time systems) or piecewise constant (discrete-time systems), then the ratio of the output to the input DFT spectra at the excited frequency lines $k = Mh$, $h \in \mathbb{H}$, gives the true transfer function

$$Y(k) = G(\Omega_k, \theta) U(k) \quad (6\text{-}32)$$

where $\Omega_k = s_k$, z_k^{-1}, $\sqrt{s_k}$ or $\tanh(\tau_R s_k)$ with $s_k = j\omega_k$ and $z_k = e^{j\omega_k T_s}$, and where $G(\Omega, \theta)$ can take any parameterization of Section 6.2 (Brigham, 1974; Oppenheim et al., 1997). For single sine excitations (6-32) is valid at arbitrary (not related to a DFT grid) frequencies.

6.3.2 Models for Arbitrary Signals

6.3.2.1 Introduction. Spectral leakage occurs in the calculation of the DFT of nonperiodic signals and of periodic signals observed at a noninteger number of periods (see Section 2.2.3 and Brigham, 1974). For these signals, relationship (6-32) is no longer valid and should, therefore, be generalized. We will show that the DFT spectra $Y(k)$, $U(k)$ satisfy an extended transfer function model that includes the beginning and end effects of the data record (see Figure 2-23 on page 60). The relationship is exact, without any approximation for discrete-time systems, and is approximately valid within some spectral alias errors for (lumped) continuous-time systems.

6.3.2.2 The Extended Transfer Function Model. The DFT spectra $Y(k)$, $U(k)$ of the observed samples $y(t)$, $u(t)$, $t = 0, T_s, ..., (N-1)T_s$ satisfy

$$A(\Omega_k, \theta) Y(k) = B(\Omega_k, \theta) U(k) + I(\Omega_k, \theta) + \Delta(\Omega_k) \text{ with } \Delta(z_k^{-1}) = 0 \quad (6\text{-}33)$$

($\Omega = s$, $\sqrt{s}$ or z^{-1}), and where the polynomial $I(\Omega, \theta) = \sum_{r=0}^{n_i} i_r \Omega^r$ with $n_i = \max(n_a, n_b) - 1$ is independently parameterized of the plant model parameters (6-21) (proof: see Appendix 6.B). The coefficients i_r are a linear function of the difference between the initial and final conditions of the system and, therefore, will be called the equivalent initial conditions. The term $\Delta(\Omega_k)$ in (6-33), with $\Omega \neq z^{-1}$, represents the residual alias error and is present even if the signals have been low-pass filtered before sampling. Dividing (6-33) by $A(\Omega_k, \theta)$ gives the extended transfer function models

$$Y(k) = G(\Omega_k, \theta)U(k) + T_G(\Omega_k, \theta) + \delta(\Omega_k) \text{ with } \delta(z_k^{-1}) = 0 \qquad (6\text{-}34)$$

($\Omega = s, \sqrt{s}$ or z^{-1}), and where $G(\Omega, \theta)$ and $T_G(\Omega, \theta)$ can take any parameterization of Section 6.2. $T_G(\Omega, \theta)$ is called the plant transient term.

For the *rational form*, representation $G(\Omega, \theta)$ is as in (6-20) and

$$T_G(\Omega, \theta) = \frac{I(\Omega, \theta)}{A(\Omega, \theta)} = \frac{\sum_{r=0}^{n_i} i_r \Omega^r}{\sum_{r=0}^{n_a} a_r \Omega^r} \qquad (6\text{-}35)$$

where $i_0 i_1 \ldots i_{n_i}$ is added to θ (6-21). For the *partial fraction expansion*, $G(\Omega, \theta)$ is as in (6-22), (6-23), and

$$T_G(\Omega, \theta) = \sum_{\substack{r=-p \\ r \neq 0}}^{p} \frac{l_r}{\Omega - \lambda_r} + \sum_{r=1}^{q} \frac{s_r}{\Omega - \sigma_r} \text{ with } \Omega = s, \sqrt{s} \qquad (6\text{-}36)$$

$$T_G(z^{-1}, \theta) = \sum_{\substack{r=-p \\ r \neq 0}}^{p} \frac{l_r}{1 - \lambda_r z^{-1}} + \sum_{r=1}^{q} \frac{s_r}{1 - \sigma_r z^{-1}} \qquad (6\text{-}37)$$

where $s_1 \ldots s_q \text{Re}(l_1)\text{Im}(l_1)\ldots \text{Re}(l_p)\text{Im}(l_p)$ is added to θ (6-24). For the *orthogonal polynomials*, $G(\Omega, \theta)$ is as in (6-25) and

$$T_G(\Omega, \theta) = \frac{\sum_{r=0}^{n_r} i_r r_r(\Omega)}{\sum_{r=0}^{n_p} a_r p_r(\Omega)} \qquad (6\text{-}38)$$

For *scalar orthogonal polynomials*, $n_p = n_a$, $n_q = n_b$, $n_r = n_i$, and $p_r(\Omega)$, $q_r(\Omega)$, $r_r(\Omega)$ are polynomials of order r; for *vector orthogonal polynomials*, $a_r = b_r = i_r$, $n_p = n_q = n_r = n_a + n_b + n_i + 2$ and $p_r(\Omega)$, $q_r(\Omega)$, $r_r(\Omega)$ are polynomials of increasing order with $p_{n_p}(\Omega)$, $q_{n_q}(\Omega)$, $r_{n_r}(\Omega)$ polynomials of order n_a, n_b, n_i, respectively. These are chosen such that they maximize the numerical stability of the model (minimize the condition number of the normal equations, see Chapter 9). Finally, for the *state space representation*, $G(\Omega, \theta)$ is as in (6-26), (6-27), and

$$T_G(s, \theta) = C(sI_{n_a} - A)^{-1} x_I \qquad (6\text{-}39)$$

$$T_G(z^{-1}, \theta) = C(I_{n_a} - z^{-1} A)^{-1} x_I \qquad (6\text{-}40)$$

where $x_I \in \mathbb{R}^{n_a}$ is added to θ (6-28) (proof: see Appendix 6.C).

The convergence rate to zero of the transient term $T_G(\Omega_k, \theta)$ and the alias term $\delta(\Omega_k)$ in the extended transfer function model (6-34) is established in the following two lemmas.

Section 6.3 ■ Relation Between the Input-Output DFT Spectra

Lemma 6.5 (Convergence Rate $T_G(\Omega_k, \theta)$): Consider bounded excitations $u(t)$ (z-domain), bounded excitations $u(t)$ with finite left $(\max(n_a, n_b) - 1)$th order derivative (s-domain), or bounded excitations $u(t)$ with finite left $(\max(n_a, n_b) - 1)$th fractional order derivative ($\sqrt{s}$-domain). For these inputs applied to stable plants or unstable plants captured within a stabilizing feedback loop, the transient term $T_G(\Omega_k, \theta)$ tends to zero as $O(N^{-1/2})$. For bounded random excitations $T_G(z_k^{-1}, \theta)$ is an $O_{\text{a.s.}}(N^{-1/2})$.

Proof. See Appendix 6.D. □

Lemma 6.6 (Convergence Rate $\delta(\Omega_k)$): Consider band-limited periodic signals, $U(j\omega) = 0$ for $|\omega| > \omega_B$, and band-limited random signals, $S_{uu}(j\omega) = 0$ for $|\omega| > \omega_B$, with $\omega_B < \omega_s/2$. Assume furthermore that these signals have finite nonzero power

$$\frac{1}{NT_s} \int_{-NT_s/2}^{+NT_s/2} \mathbb{E}\{x^2(t)\} dt = O(N^0) > 0 \tag{6-41}$$

for any N, ∞ included. The residual alias error $\delta(\Omega_k)$ tends to zero as $O(N^{-1/2})$ for band-limited periodic excitations, and $O_{\text{m.s.}}(N^{-1/2})$ for band-limited random excitations with differentiable power spectrum $S_{uu}(j\omega)$ ($dS_{uu}(j\omega)/d\omega < \infty$ for $|\omega| \leq \omega_B$).

Proof. See Appendix 6.F. □

Using Lemmas 6.5 and 6.6, we can calculate how fast the extended transfer function model (6-34) tends to the transfer function model (6-32) as $N \to \infty$.

Theorem 6.7 (Convergence Rate Extended Transfer Function Models): Under the assumptions of Lemma 6.5, the convergence rates of discrete-time model (6-34) to (6-32) are $O(N^{-1/2})$ for normalized periodic signals (see Definition 3.4, $F = O(N)$), $O(N^{-1})$ at the excited DFT frequencies for periodic signals with a fixed number of frequencies ($F = O(N^0)$), and $O_{\text{a.s.}}(N^{-1/2})$ for random excitations with differentiable power spectrum. Under the assumptions of Lemmas 6.5 and 6.6, the convergence rates of continuous-time model (6-34) to (6-32) are $O(N^{-1/2})$ for normalized periodic signals (see Definition 3.4, $F = O(N)$), $O(N^{-1})$ at the excited DFT frequencies for periodic signals with a fixed number of frequencies ($F = O(N^0)$), and $O_p(N^{-1/2})$ for random excitations with differentiable power spectrum and $O(N^{-1/2})$.

Proof. It follows directly from Lemmas 6.5 and 6.6 and the fact that the DFT spectrum of band-limited signals with finite nonzero power is $O(N^0)$ for random signals, $O(N^0)$ for normalized periodic signals, and $O(N^{1/2})$ at the excited DFT frequencies for periodic signals with a fixed number of frequencies. □

6.3.2.3 Discussion. The extended transfer model (6-34) shows that the leakage errors on the input and output DFT spectra can be modeled by a rational function and are, in fact, an initial condition (transient) problem. This is illustrated in Figure 2-23 on page 60. The difference from time domain identification is that the equivalent initial conditions take into account the beginning as well as the end effects of the finite data record. In the time domain the initial conditions remain the same as the number of data N increases, whereas in the frequency domain they vary with N (not only due to the scaling factor $N^{-1/2}$ but also due to the varying final conditions of the experiment). Asymptotically ($N \to \infty$), the extended transfer function model (6-34) reduces to (6-32) (Theorem 6.7).

Theorem 6.7 shows that the classical transfer function model $Y(k) = G(s_k, \theta)U(k)$ contains no asymptotic ($N \to \infty$) approximation errors in the complete frequency band from DC to Nyquist for band-limited input signals with finite nonzero power.

The transient term $T_G(\Omega, \theta)$ is zero if the initial and final conditions of the experiment are the same (see Appendix 6.B, (6-93) and (6-100)). This is the case for periodic signals observed during an integer number of periods and for time-limited signals. For the band-limited versions of these signals the alias term $\delta(\Omega_k)$ is also zero.

From Lemmas 6.5 and 6.6 it follows that the transient term $T_G(\Omega_k, \theta)$ and the alias error $\delta(\Omega_k)$ ($\Omega \neq z^{-1}$) tend to zero at the same rate. Hence, $\delta(\Omega_k)$ cannot be neglected w.r.t. $T_G(\Omega_k, \theta)$, even for "large" values of N. However, practice has shown that the alias error $\delta(\Omega_k)$ can be approximated well by a polynomial (Pintelon and Schoukens, 1997b). Therefore, to reduce $\delta(\Omega_k)$ in (6-34), the order of the polynomial $I(\Omega, \theta)$ ($\Omega \neq z^{-1}$) is increased: $n_i \geq \max(n_a, n_b) - 1$.

6.3.3 Models for Records with Missing Data

6.3.3.1 Introduction. Because of temporary sensor failure and/or data transmission errors, it may happen that data samples are missing in the measured signals. The best thing to do then is to throw away the data set and to repeat the experiment. This is not always possible because, for example, the experiment is too expensive, or some of the data are collected in an irregular way using laboratory analysis. Sometimes the output is sampled at a lower rate than the input, which is a periodic missing output data problem (Goodwin and Adams, 1994; Albertos et al., 1999). Treating the missing data as unknown parameters, a generalized version of the extended transfer model (6-34) is constructed. It can handle missing input and/or output data and does not assume any particular missing data pattern.

6.3.3.2 The Extended Transfer Function Model. For simplicity of notation we will assume that M_u consecutive input samples starting at $t = K_u T_s$ and M_y consecutive output samples starting at $t = K_y T_s$ are missing. The sets $\mathbb{M}_u$ and $\mathbb{M}_y$ describing the time instances of the missing input and output samples are then

$$\mathbb{M}_x = \{K_x, K_x + 1, \ldots, K_x + M_x - 1\} \tag{6-42}$$

where $x = u, y$. Define $x^m(tT_s)$, $t = 0, 1, \ldots, N-1$ as the data set where the missing samples are replaced by zeros

$$x^m(tT_s) = \begin{cases} 0 & t \in \mathbb{M}_x \\ x(tT_s) & \text{elsewhere} \end{cases} \tag{6-43}$$

and $X^m(k)$ as the corresponding DFT spectrum ($X = U, Y$ and $x = u, y$). The DFT spectra $Y^m(k)$, $U^m(k)$ of the observed samples (missing data sets) $y^m(tT_s)$, $u^m(tT_s)$, $t = 0, 1, \ldots, N-1$ satisfy

$$\begin{aligned} A(\Omega_k, \theta) Y^m(k) = B(\Omega_k, \theta) U^m(k) + I(\Omega_k, \theta) + \\ z_k^{-K_u} B(\Omega_k, \theta) I_u(z_k^{-1}, \psi) - z_k^{-K_y} A(\Omega_k, \theta) I_y(z_k^{-1}, \psi) + \Delta(\Omega_k) \end{aligned} \tag{6-44}$$

with $\Delta(z_k^{-1}) = 0$ ($\Omega = s, \sqrt{s}$ or z^{-1}), and where the polynomials $I_x(z^{-1}, \psi) = N^{-1/2}\sum_{t=0}^{M_x - 1} x(K_x + t)z^{-t}$, $x = u, y$, contain the missing data and ψ is the parameter vector of the missing samples

$$\psi^T = [u(K_u T_s) \ldots u((K_u + M_u - 1)T_s)\, y(K_y T_s) \ldots y((K_y + M_y - 1)T_s)] \qquad (6\text{-}45)$$

(proof: see Appendix 6.G). Note that model (6-44) is bilinear in the parameters θ, ψ. Dividing (6-44) by $A(\Omega_k, \theta)$ gives the extended transfer function model

$$\begin{aligned} Y^m(k) = \; & G(\Omega_k, \theta) U^m(k) + T_G(\Omega_k, \theta) + \\ & z_k^{-K_u} G(\Omega_k, \theta) I_u(z_k^{-1}, \psi) - z_k^{-K_y} I_y(z_k^{-1}, \psi) + \delta(\Omega_k) \end{aligned} \qquad (6\text{-}46)$$

with $\delta(z_k^{-1}) = 0$, and where $G(\Omega, \theta)$ and $T_G(\Omega, \theta)$ can take any parameterization of Sections 6.2 and 6.3.2. The alias error $\delta(\Omega_k)$, $\Omega \neq z^{-1}$, has the same properties as in Section 6.3.2. Hence, its influence can be captured by choosing the order of the polynomial $I(\Omega_k, \theta) \geq \max(n_a, n_b) - 1$. The generalization of (6-46) to the case where data are missing at more than one place is straightforward (see Exercise 6.7).

6.3.4 Models for Concatenated Data Sets

6.3.4.1 Introduction. Sometimes several data sets of experiments on a plant operating under exactly the same conditions are available (e.g., the historical data sets in the process or power plant industry). Due to temporary sensor failure and/or transmission errors it may also happen that a large number of consecutive input and output samples are missing at several time instances in one experiment. If the number of consecutive missing samples is larger than the dominant time constant of the plant, then the approach of Section 6.3.3 no longer makes sense. A first solution consists of handling the complete input-output data sets as independent experiments. A better solution consists of expressing that the different data sets have been collected under exactly the same operational conditions. This is done by concatenating the different data sets. It is shown that the concatenated data sets satisfy a generalized version of the extended transfer function model (6-34).

6.3.4.2 The Extended Transfer Function Model. Consider, for example, two input-output data sets $u_i(tT_s), y_i(tT_s)$, with $i = 1, 2$ and $t = 0, 1, \ldots, N_i - 1$, of the same system gathered under the same operational conditions. Define $x^c(tT_s)$, with $t = 0, 1, \ldots, N-1$ and $N = N_1 + N_2$, as the concatenated data set

$$x^c(tT_s) = \begin{cases} x_1(tT_s) & t = 0, 1, \ldots, N_1 - 1 \\ x_2(tT_s) & t = N_1, N_1 + 1, \ldots, N - 1 \end{cases} \qquad (6\text{-}47)$$

and $X^c(k)$ as the corresponding DFT spectrum ($X = U, Y$ and $x = u, y$). The DFT spectra $Y^c(k)$, $U^c(k)$ of the concatenated data sets $y^c(tT_s)$, $u^c(tT_s)$, $t = 0, 1, \ldots, N-1$ satisfy

$$Y^c(\Omega_k) = G(\Omega_k, \theta) U^c(k) + T_G(\Omega_k, \theta) + z_k^{-N_1} T_G^c(\Omega_k, \theta) + \delta(\Omega_k) \qquad (6\text{-}48)$$

with $\delta(z_k^{-1}) = 0$, and where $G(\Omega, \theta)$, $T_G(\Omega, \theta)$, and $T_G^c(\Omega_k, \theta) = I^c(\Omega_k, \theta)/A(\Omega_k, \theta)$ can take any parameterization of Sections 6.2 and 6.3.2 (proof: see Appendix 6.H). $I^c(\Omega_k, \theta)$ is a polynomial of order $\max(n_a, n_b) - 1$. Since the alias error $\delta(\Omega_k)$, $\Omega \neq z^{-1}$, has the same properties as in Section 6.3.2, its contribution is captured by the polynomials $I(\Omega_k, \theta)$ and $I^c(\Omega_k, \theta)$ where the order is chosen $\geq \max(n_a, n_b) - 1$. The generalization of (6-48) to the case where more than two data sets are concatenated is straightforward (see Exercise 6.8).

6.4 MODELS FOR DAMPED (COMPLEX) EXPONENTIALS

In some applications an impulse excitation is applied to the system and only the free decay response is observed, which consists of the sum of (complex) exponentially damped cosines. For real strictly proper lumped continuous-time systems ($\theta \in \mathbb{R}^{n_\theta}$ in (6-35)) with simple complex conjugate pole pairs, it has the form

$$y(t) = 2\sum_{r=1}^{n_a} a_r e^{-d_r(t+\tau)} \cos(\omega_r(t+\tau) + \phi_r) \tag{6-49}$$

with $a_r \in \mathbb{R}^+$ the amplitude, $d_r \in \mathbb{R}^+$ the decay, $\omega_r \in \mathbb{R}^+$ the angular frequency, and $\phi_r \in \mathbb{R}$ the phase of the rth exponentially damped cosine. τ is the (known) delay between the beginning of the free decay experiment and the start of the observations. In modal analysis (6-49) is parameterized in the resonant angular frequency $\omega_0 = \sqrt{d^2 + \omega^2}$ and the damping coefficient $\zeta = d/\omega_0$, while in circuit theory the resonant angular frequency ω_0 and the quality factor $Q = 1/(2\zeta)$ are used. For complex strictly proper lumped continuous-time systems ($\theta \in \mathbb{C}^{n_\theta}$ in (6-35)) with simple complex poles the response is

$$y(t) = \sum_{r=1}^{n_a} a_r e^{j\phi_r} e^{(-d_r + j\omega_r)(t+\tau)} \tag{6-50}$$

Examples of (6-49) and (6-50) are, respectively, impact testing in modal analysis (Ewins, 1991) and nuclear magnetic resonance (NMR) measurements (Kumaresan et al., 1990). In the first application the mechanical structure is excited with an impulse, and the free decay response of the structure, for example, the displacement or the acceleration, is measured at a given location. In the second application the free decay responses of a magnetic field in two orthogonal directions are combined into one complex signal.

The DFT spectrum $Y(k)$ of the free decay response $y(t)$ of a strictly proper lumped continuous-time system or a proper discrete-time system with $b_0 = 0$ satisfies

$$Y(k) = T_G(z_k^{-1}, \theta) \tag{6-51}$$

where $T_G(z^{-1}, \theta)$ is the rational function (6-35) with $n_i = n_a - 1$ (proof: see Appendix 6.I). $T_G(z^{-1}, \theta)$ can also be parameterized as in (6-37), (6-38), and (6-40). The parameters of the free decay responses (6-49) and (6-50) can easily be related to the parameters of the partial fraction expansion (6-37) with $q = 0$ ($\lambda_{-r} \neq \bar{\lambda}_r$ and $l_{-r} \neq \bar{l}_r$ for complex systems). In both cases we have

$$a_r e^{j\phi_r} = \frac{l_r \sqrt{N}}{\lambda_r^{\tau/T_s}(1 - \lambda_r^N)}, \quad -d_r + j\omega_r = \frac{1}{T_s}\ln(\lambda_r) \tag{6-52}$$

(proof: see Appendix 6.J).

In NMR measurements the response is typically of the form (6-50) where each term corresponds to the response of a particular chemical substance in a (human) tissue. The amplitude a_k is a measure of the concentration of the substance. Often it is known that a particular substance with known frequency f_r is present in the tissue. Sometimes the chemical structure of the substance imposes the ratio of some amplitudes. All this prior information results in parameter constraints that can easily be taken into account in the partial fraction expansion (6-37). This is not the case for the other parameterizations, which explains why representation (6-37) is popular in NMR modeling. Parameterization (6-35) is appropriate for obtaining starting values for (6-37).

6.5 IDENTIFIABILITY

Loosely speaking, a parametric model $M(\theta, Z)$ is identifiable when the parameters θ can be estimated uniquely using the data Z. It requires that the data are informative enough to distinguish between different models (= condition on the experiment) and that different parameter values give different models (= condition on the model structure). More formally, the identifiability concept can be defined as follows.

Definition 6.8 (Identifiability): A model $M(\theta, Z)$, with θ the model parameters and Z the data, is identifiable at θ_1 if for any θ in a (possibly small) neighborhood of θ_1, $M(\theta, Z) = M(\theta_1, Z)$ implies that $\theta = \theta_1$.

Note that Definition 6.8 gives a definition of *local identifiability*. If the implication in Definition 6.8 is valid for almost all θ and θ_1 values, then one has *global identifiability* (see Ljung, 1999 for a detailed discussion of this issue). In this section we give necessary conditions for the identifiability of the transfer function models of Section 6.3. These conditions can be split into constraints on the parameters θ (identifiable parameterization) and constraints on the input signal (persistent excitation).

6.5.1 Models for Periodic Signals

The identifiability of transfer function model (6-32) depends on the particular parameterization of $G(\Omega, \theta)$. The *rational forms* (6-20) and (6-25) are not identifiable because replacing θ by $\lambda\theta$, with $\lambda \in \mathbb{R}_0$, results in the same input-output description: $G(\Omega, \lambda\theta) = G(\Omega, \theta)$. This parameter ambiguity is removed by constraining the model parameters, for example, $\theta_{[1]} = 1$ or $\|\theta\|_2 = 1$. For transfer functions with a time delay (6-29) and (6-30), the parameter ambiguity is removed by constraining the numerator and denominator coefficients, but not the delay. The *partial fraction expansions* (6-22) and (6-23) contain no parameter ambiguities and, hence, are identifiable. Replacing (A, B, C, D) by $(TAT^{-1}, TB, CT^{-1}D)$ in the *state space representations* (6-26) and (6-27) with $T \in \mathbb{R}^{n_a \times n_a}$, a regular matrix ($\det(T) \neq 0$), leaves $G(\Omega, \theta)$ unchanged. This parameter ambiguity is removed by imposing n_a^2 constraints on θ, which leads to the so-called identifiable state space representations (Van Overbeek and Ljung, 1982). Besides possible constraints on θ, the identifiability of transfer function model (6-32) also puts conditions on the DFT spectrum $U(k)$ of the input signal.

Theorem 6.9 (Identifiability Transfer Function Model (6-32)): Transfer function model (6-32), parameterized as in (6-20) and (6-25) with, for example, constraint $a_{n_a} = 1$, is identifiable if and only if

1. The polynomials $A(\Omega, \theta)$ and $B(\Omega, \theta)$ have no common roots.

2. The input DFT spectrum $U(k)$ is different from zero for at least $(n_a + n_b + 1)/2$ different DFT frequencies, where DC ($k = 0$) and Nyquist ($N/2$) each count for $1/2$.

Proof. See Appendix 6.L. □

With appropriate additional assumptions on $G(\Omega, \theta)$, Theorem 6.9 also applies for the other parameterizations. For example, the partial fraction expansions (6-22) and (6-23) assume that $G(\Omega, \theta)$ has simple poles. The condition on $U(k)$ is fulfilled, for example, if $u(t)$ consists of the sum of at least $(n_a + n_b + 1)/2$ sine waves. Note that for complex systems, $\theta \in \mathbb{C}^{n_\theta}$, $(n_a + n_b + 1)$ frequencies are required.

6.5.2 Models for Arbitrary Signals

The identifiability of transfer function model (6-34) depends on the particular parameterization of $G(\Omega, \theta)$ and $T_G(\Omega, \theta)$. The *partial fraction expansions* (6-22), (6-23) and (6-36), (6-37) are identifiable, while the same parameter ambiguities occur as in the periodic case (see Section 6.5.1) for the *rational forms* (6-20), (6-25) and (6-35), (6-38) ($G(\Omega, \lambda\theta) = G(\Omega, \theta)$, $T_G(\Omega, \lambda\theta) = T_G(\Omega, \theta)$) and the *state space representations* (6-26), (6-27) and (6-39), (6-40) (replacing (A, B, C, D) by $(TAT^{-1}, TB, CT^{-1}, D)$ leaves $G(\Omega, \theta)$ and $T_G(\Omega, \theta)$ unchanged). Compared with the periodic case, the identifiability of transfer function model (6-34) requires additional conditions on the DFT spectrum, $U(k)$, of the input signal. Necessary conditions for the identifiability of transfer function model (6-34) are

1. The polynomials $A(\Omega, \theta)$, $B(\Omega, \theta)$, and $I(\Omega, \theta)$ have no common roots.
2. The input DFT spectrum $U(k)$ is different from zero for at least $(n_b + n_i + 2)/2$ different DFT frequencies, where DC ($k = 0$) and Nyquist ($N/2$) each count for $1/2$.
3. $U(k)$ cannot be written as a rational form in Ω_k of order n_i over n_b or less.

(Proof: See Appendix 6.M). □

Note that condition 1 does not exclude $A(\Omega, \theta)$ and $B(\Omega, \theta)$ for having common roots and/or $B(\Omega, \theta)$ and $I(\Omega, \theta)$ for having common roots (see Exercise 5.9). If condition 3 is not fulfilled, then the terms $G(\Omega_k, \theta)U(k)$ and $T_G(\Omega_k, \theta)$ are indistinguishable. This is, for example, the case when the DFT spectrum $U(k)$ is a constant ($u(t)$ is an impulse (Dirac)).

6.5.3 Models for Records with Missing Data

The identifiability of transfer function model (6-44) depends on the particular parameterization of $G(\Omega, \theta)$ and $T_G(\Omega, \theta)$, the missing data pattern, and the input DFT spectrum $U^m(k)$. The same parameter constraints should be applied on θ as in Section 6.5.2. A similar analysis, as in Section 6.5.2, gives the following necessary conditions on $U^m(k)$ and the missing data pattern:

1. The polynomials $A(\Omega, \theta)$, $B(\Omega, \theta)$, and $I(\Omega, \theta)$ have no common roots.
2. The input DFT spectrum $U^m(k)$ is different from zero for at least $(n_b + n_i + 2)/2$ different DFT frequencies, where DC ($k = 0$) and Nyquist ($N/2$) each count for $1/2$.

Section 6.6 ■ Multivariable Systems

3. It is not possible to write $U^m(k)$ as a rational form in Ω_k of order n_i over n_b or less.

4. For discrete-time systems, it is not possible to write $U^m(k) + z_k^{-K_u}I_u(z_k^{-1}, \psi)$ as a rational form in z_k^{-1} of order n_i over n_b or less.

5. For discrete-time systems it is not possible to write $z_k^{-(K_y - K_u)}I_y(z_k^{-1}, \psi)/I_u(z_k^{-1}, \psi)$ as a rational form in z_k^{-1} of order n_b over n_a or larger.

Condition 5 constrains the missing data pattern. For example, discrete-time systems are not identifiable (the condition number of (6-44) is infinitely large) if the input and output samples are missing at the same place, $K_u = K_y$ and the number of consecutive missing samples is larger than or equal to the system order, $M_u, M_y \geq \max(n_a, n_b)$. The missing input and output samples ψ cannot be estimated and the plant model parameters θ should be estimated from the two sets of complete input-output data. Everything happens as if two experiments with full data are available (see Section 6.3.4). Continuous-time systems are still identifiable if $K_u = K_y$ and $M_u = M_y \geq \max(n_a, n_b)$; however, the condition number of model (6-44) increases quickly with the number of consecutive missing samples. For too large an $M_u = M_y$, (6-44) is no longer identifiable within a given finite arithmetic precision (Pintelon and Schoukens, 1999b). The identifiability conditions can easily be extended to the case where data are missing at more than one place.

6.5.4 Model for Concatenated Data Sets

The identifiability of transfer function model (6-48) depends on the particular parameterization of $G(\Omega, \theta)$, $T_G(\Omega, \theta)$, and $T_G^c(\Omega, \theta)$, and the input DFT spectrum $U^c(k)$. The same parameter constraints should be applied on θ as in Section 6.5.2. A similar analysis, as in Section 6.5.2, shows that the necessary conditions for the identifiability of transfer function model (6-48) are conditions 1–3 of Section 6.5.2, where the polynomial $I_c(\Omega_k, \theta)$ is added to condition 1, and where $U(k)$ is replaced by $U^c(k)$.

6.6 MULTIVARIABLE SYSTEMS

The n_y outputs and the n_u inputs of a multivariable system are related by an $n_y \times n_u$ transfer function matrix $G(\Omega, \theta)$, where each entry $G_{[i,j]}(\Omega, \theta)$, with $i = 1, 2, ..., n_y$ and $j = 1, 2, ..., n_u$, is a rational function of Ω ($\Omega = s$, $\sqrt{s}$, $\tanh(\tau_R s)$ or z^{-1}, see Section 6.2). If no relationships exist between the coefficients of the different transfer functions $G_{[i,j]}(\Omega, \theta)$, then the multivariable system is the parallel connection of separate multiple-input, single-output (MISO) systems. Often, the transfer functions $G_{[i,j]}(\Omega, \theta)$ have the same denominator, for example, in modal analysis (Ewins, 1991) and the two port description of LC, LR, and RC circuits (Balabanian and Bickart, 1969). This leads to the *common denominator* model

$$G(\Omega, \theta) = \frac{B(\Omega, \theta)}{A(\Omega, \theta)} = \frac{\sum_{r=0}^{n_b} B_r \Omega^r}{\sum_{r=0}^{n_a} a_r \Omega^r} \tag{6-53}$$

where $A(\Omega, \theta)$, with $a_r \in \mathbb{R}$, is the common denominator polynomial and $B(\Omega, \theta)$, with $B_r \in \mathbb{R}^{n_y \times n_u}$, a polynomial matrix.

A natural generalization of the scalar transfer function (6-20) is the so-called matrix-fraction descriptions (Kailath, 1980). Writing the transfer function matrix as a *left matrix fraction* gives

$$G(\Omega, \theta) = A^{-1}(\Omega, \theta)B(\Omega, \theta) = \left(\sum_{r=0}^{n_a} A_r \Omega^r\right)^{-1} \left(\sum_{r=0}^{n_b} B_r \Omega^r\right) \tag{6-54}$$

where $A(\Omega, \theta)$, with $A_r \in \mathbb{R}^{n_y \times n_y}$, and $B(\Omega, \theta)$, with $B_r \in \mathbb{R}^{n_y \times n_u}$, are polynomial matrices. Writing the transfer function matrix as a *right matrix fraction* gives

$$G(\Omega, \theta) = B(\Omega, \theta)A^{-1}(\Omega, \theta) \tag{6-55}$$

where $A(\Omega, \theta)$ and $B(\Omega, \theta)$ are, respectively, $n_u \times n_u$ and $n_y \times n_u$ polynomial matrices.

The *partial fraction expansion* of $G(\Omega, \theta)$ has the same form (6-22), (6-23) where each residue matrix $L_r, S_r \in \mathbb{R}^{n_y \times n_u}$ may have a different rank. Sometimes the rank is known beforehand and this should be taken into account in the parameterization. For example, in modal analysis the residue matrices have rank one (Heylen et al., 1997) and are written as $L_r = v_r w_r^T$ with $v_r \in \mathbb{R}^{n_y}$ and $w_r \in \mathbb{R}^{n_u}$ the modal vectors.

The *state space representation* has the same form (6-26), (6-27) with $A \in \mathbb{R}^{n_a \times n_a}$, $B \in \mathbb{R}^{n_a \times n_u}$, $C \in \mathbb{R}^{n_y \times n_a}$, and $D \in \mathbb{R}^{n_y \times n_u}$.

Remarks

(i) The relation to the input and output DFT spectra and the identifiability issues of the multivariable parametric models are similar to the single input, single output case. For example, the left matrix fraction description (6-54) is made identifiable with the parameter constraint $A_{n_a} = I_{n_a}$.

(ii) The common denominator (6-53) and the left matrix fraction (6-54) descriptions allow straightforward generalization of the scalar relationship (6-33) between the numerator and denominator polynomials of the transfer function model and the input and output DFT spectra. This is important for generating starting values (see Chapter 9). Formula (6-33) is then valid with $A(\Omega, \theta)$, $B(\Omega, \theta)$ as defined in (6-53) and (6-54) and $I(\Omega, \theta) = \sum_{r=0}^{n_b} I_r \Omega^r$, $I_r \in \mathbb{R}^{n_y}$, a polynomial vector. This is not the case for the right matrix fraction description (6-55), which can, however, be used if the identification starts from the measured frequency response matrix $G(\Omega_k)$

$$G(\Omega_k) = B(\Omega_k, \theta)A^{-1}(\Omega_k, \theta) \Rightarrow G(\Omega_k)A(\Omega_k, \theta) = B(\Omega_k, \theta)$$

(iii) The residue matrices $L_r, S_r \in \mathbb{R}^{n_y \times n_u}$ of the partial fraction expansions (6-22) and (6-23) of the left and right matrix fractions (6-54) and (6-55) and the state space models (6-26) and (6-27) have rank one (proof: see Appendix 6.N). This is not the case for the common denominator model (6-53) which, in general, has full rank residue matrices. In those applications where it is known beforehand that the residue matrices have rank one (e.g., in modal analysis), the common denominator model uses too many parameters and, hence, the left and right matrix fraction descriptions and the state space model are preferred. However, the common denominator model should be used in all applications where the rank of the residue matrices is unknown (e.g., the best linear approximation of a multivariable nonlinear system).

6.7 NOISE MODELS

6.7.1 Introduction

In practice, disturbing noise sources occur everywhere in the measurement setup (see Figure 2-16). The DFT spectra $U(k)$ and $Y(k)$ of the observed input $u(t)$ and output $y(t)$ signals are noisy replicas of the true (unknown) DFT spectra $U_0(k)$ and $Y_0(k)$

$$Y(k) = Y_0(k) + N_Y(k)$$
$$U(k) = U_0(k) + N_U(k)$$
(6-56)

where $N_U(k) = \text{DFT}(n_u(t))$ and $N_Y(k) = \text{DFT}(n_y(t))$ are functions of the measurement noise, the process noise, and possibly the generator noise (see Section 2.4). In order to put a quality label (uncertainty bounds) on the measured frequency response function (see Chapters 2 and 4) and the estimated transfer function model (see Chapters 9, 10, and 12), we need a model for the disturbing errors $N_U(k)$ and $N_Y(k)$.

6.7.2 Nonparametric Noise Model

As a nonparametric noise model, we will take the (co)variances of the discrete Fourier transform of the input and output errors

$$\sigma_U^2(k) = \text{var}(N_U(k)), \quad \sigma_Y^2(k) = \text{var}(N_Y(k)), \quad \sigma_{YU}^2(k) = \text{covar}(N_Y(k), N_U(k)) \quad (6\text{-}57)$$

at the DFT frequencies k of interest. It can be obtained via a noise analysis without excitation signal ($r(t) = 0$ in Figure 2-16 on page 44) or via independent, repeated experiments with the same excitation signal $r(t)$. The last approach is strongly recommended because it reduces the total measurement time (the frequency response function and the noise model are measured at the same time), and because the noise model is measured at nominal operating conditions. In practice, the independent, repeated experiments are obtained from consecutive periods of the (steady state) response of the system to a periodic excitation $r(t)$ (see Chapters 10 and 12). For arbitrary (random) excitations $r(t)$, a nonparametric noise model for the measured input-output signals $u(t)$ and $y(t)$ can still be obtained if $r(t)$ in Figure 2-16 is known (see Chapter 12).

6.7.3 Parametric Noise Model

6.7.3.1 Introduction. Most stochastic processes in engineering applications have an intrinsic continuous-time nature. Think, for example, of the thermal noise generated by resistors (Pyati, 1992) or the flicker and generation-recombination noise generated by semiconductor devices (Lowen and Teich, 1990). Despite this fact, the impact of noise in system identification has mostly been modeled as discrete-time filtered white noise (Åström, 1970; Hannan and Deistler, 1988; Söderström and Stoica, 1989; Middleton and Goodwin, 1990; and Ljung, 1999). The two main reasons for this are: (i) the success of digital control and the related discrete-time modeling, and (ii) the mathematical difficulty of handling stochastic differential equations. In various other disciplines such as operational modal analysis (Cauberghe et al., 2003; Pintelon et al., 2006b), signal processing (Fan et al., 1999), astrophysics (Phadke and Wu, 1974), and econometrics (Bergström, 1990), continuous-time modeling is of considerable importance. For example, in operational modal analysis of civil engineering structures (bridges, buildings, ...) the system is excited by an unobserved input (turbulent wind flow, traffic, ...), and the observed response (acceleration) carries information about the system poles (resonance frequencies and damping ratios). Knowledge of the latter is important for security/maintenance reasons. In the sequel of this section we handle both the discrete-time and continuous-time cases.

6.7.3.2 Discrete-Time Models. In control applications the input is assumed to be known, $n_u(t) = 0$, and the disturbance $n_y(t)$ is modeled at the sampling instances as filtered white noise $e(t)$

$$n_y(t) = H(q, \theta)e(t) \tag{6-58}$$

with $q = z^{-1}$ the backward shift operator, $e(t)$ a stationary white noise sequence with zero mean and variance σ^2, and

$$H(z^{-1}, \theta) = \frac{C(z^{-1}, \theta)}{D(z^{-1}, \theta)} = \frac{\sum_{r=0}^{n_c} c_r z^{-r}}{\sum_{r=0}^{n_d} d_r z^{-r}} \tag{6-59}$$

The unknown parameters are $c_0, c_1, ..., c_{n_c}$, $d_0, d_1, ..., d_{n_d}$, and σ. Model (6-58) contains two parameter ambiguities: replacing c_r, d_r and σ by $\lambda_1^{-1}\lambda_2 c_r$, $\lambda_2 d_r$ and $\lambda_1 \sigma$, with $\lambda_1, \lambda_2 \neq 0$, leaves (6-58) unchanged ($e(t)$ is multiplied with the same factor as σ). These parameter ambiguities are removed by adding two constraints on the numerator and denominator coefficients of (6-59). In most cases, the choice $d_0 = c_0 = 1$ is made (monic transfer function).

Under the following assumption the relationship between the discrete-time model (6-58) and the true underlying continuous-time stochastic process can be established.

Assumption 6.10 (Wiener Process or Zero-Order-Hold Noise): The unobserved driving noise source $e_c(t)$ in Figure 6-6 is either a Wiener stochastic process (= process with continuous-time white Gaussian noise increments, also called Brownian motion) or is piecewise constant in between white noise samples (zero-order-hold).

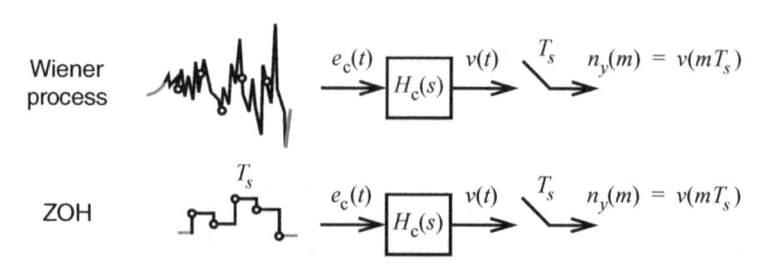

Figure 6-6. Two continuous-time noise-generating mechanisms $e_c(t)$ (Wiener process and piecewise constant white noise) within a zero-order-hold acquisition setup (no anti-alias filter), leading to discrete-time modeling. The "o" indicate $e_c(mT_s)$.

Section 6.7 ■ Noise Models

Theorem 6.11 (Discrete-Time Noise Model): Under Assumption 6.10 the noise $n_y(t)$ at the sampling instances is described exactly by a discrete-time model (6-58). The poles of the discrete-time model $H(z^{-1})$ (6-59) are related to those of the continuous-time transfer function $H_c(s)$ by the impulse invariant transformation $z = e^{sT_s}$.

Proof. Under the zero-order-hold input assumption on $e_c(t)$, $H(z^{-1})$ (6-59) is related to $H_c(s)$ via the step-invariant transformation (6-10). For Wiener processes $e_c(t)$, the theorem is proven in Åström (1970) and Jazwinski (1970). □

One can wonder whether Wiener stochastic processes, whose variance increases linearly in time, and zero-order-hold white noise processes are realistic descriptions of the true noise-generating mechanism. This is rarely the case in practice. However, by increasing the orders n_c and n_d of the polynomials in (6-59) and/or the sampling frequency $f_s = 1/T_s$, the approximation errors in (6-58) can be made sufficiently small. In general, (6-58) will be only approximately true and any physical interpretation of the results should be done with care.

6.7.3.3 Continuous-Time Models. Since continuous-time white noise has infinite variance (power), it is rather difficult to link it to a physical noise-generating mechanism. Therefore, the concept of continuous-time band-limited white noise has been introduced in Åström (1970).

Definition 6.12 (Continuous-Time Band-Limited White Gaussian Noise): A continuous-time stochastic process $e_c(t)$ is band-limited white Gaussian noise if $e_c(t)$ is normally distributed and if its power spectral density (Fourier transform of its autocorrelation function) $S_{e_c e_c}(j\omega)$ satisfies

$$S_{e_c e_c}(j\omega) = \begin{cases} \sigma_{e_c}^2/(2f_B) & |f| \leq f_B \\ 0 & |f| > f_B \end{cases} \quad (6\text{-}60)$$

where $\sigma_{e_c}^2 = \text{var}(e_c(t))$, and with f_B the bandwidth of the power spectral density.

The actual continuous-time noise process $\eta(t)$ is then modeled as a filtered version of the band-limited white noise source $e_c(t)$

$$\eta(t) = H(p, \theta)e_c(t) \quad (6\text{-}61)$$

with $p = d/dt$ the derivative operator, and

$$H(s, \theta) = \frac{C(s, \theta)}{D(s, \theta)} = \frac{\sum_{r=0}^{n_c} c_r s^r}{\sum_{r=0}^{n_d} d_r s^r} \quad (6\text{-}62)$$

The unknown parameters are $c_0, c_1, ..., c_{n_c}, d_0, d_1, ..., d_{n_d}$, and σ_{e_c}. Similarly to (6-58), model (6-62) contains two parameter ambiguities that are removed by adding two constraints on the numerator and denominator coefficients of (6-62).

Under the following relaxed assumption on the driving noise source $e_c(t)$, an important property of the band-limited observation $v(t)$ of the noise process $\eta(t)$ (6-61) can be shown.

Assumption 6.13 (Generalized Band-Limited White Gaussian Noise): The power spectral density of the unobserved driving noise source $e_c(t)$ in Figure 6-7 satisfies

$$S_{e_c e_c}(j\omega) = \begin{cases} \sigma_{e_c}^2/(2f_B) & |f| \leq f_B \\ O(f^{-(1+\varepsilon)}) & |f| > f_B \end{cases} \qquad (6\text{-}63)$$

with $\varepsilon > 0$.

Theorem 6.14 (Band-Limited Observation of Continuous-Time Noise): Under Assumption 6.13, with $f_B \geq f_s/2$, the band-limited observation $v(t)$ of $\eta(t)$ (see Figure 6-7) can be written as

$$v(t) = H(p, \theta)\varepsilon_c(t) \qquad (6\text{-}64)$$

where $\varepsilon_c(t)$ is band-limited white Gaussian noise with $f_B = f_s/2$ in (6-60). At the sampling instances $t = mT_s$, $e(m) = \varepsilon_c(mT_s)$ is independent, normally distributed, discrete-time white noise.

Proof. See Appendix 6.O. □

Remarks

(i) The condition $S_{e_c e_c}(j\omega) = O(f^{-(1+\varepsilon)})$ for $|f| > f_B$ in Assumption 6.13 is the weakest decay giving a finite variance

$$\text{var}(e_c(t)) = \int_{-\infty}^{+\infty} S_{e_c e_c}(j\omega) df$$

An example of band-limited white noise satisfying (6-63) is the thermal noise generated by resistors in electrical circuits.

(ii) In practice, a perfect anti-alias filter does not exist and, consequently, $\varepsilon_c(t)$ in (6-64) is only approximately band-limited white noise. However, the only requirement to be satisfied by the anti-alias filter is that the attenuation in the stop band ($f > f_s/2$) is sufficiently large (attenuations of more than 100 dB are easily realizable), even at the price of an increased passband ripple. Indeed, since the phase of the anti-alias filter does not influence the power spectral density $S_{\varepsilon_c \varepsilon_c}(j\omega) = |AA(j\omega)|^2 S_{e_c e_c}(j\omega)$, it is sufficient to compensate the passband ripple of $|AA(j\omega)|$ via an absolute amplitude calibration.

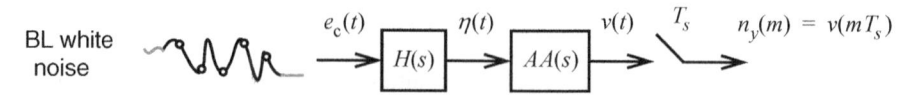

Figure 6-7. Continuous-time band-limited white noise generating mechanisms $e_c(t)$ within a band-limited acquisition setup ($AA(s)$ is the anti-alias filter), leading to continuous-time modeling. The "o" indicate $e_c(mT_s)$.

Section 6.7 ■ Noise Models

(iii) Replacing $p = d/dt$ by the fractional derivative $d^{1/2}/dt^{1/2}$ (see Appendix 6.B, Section 6.B.3) in (6-61), and s by $\sqrt{s}$ in (6-62) shows that the results are also valid for diffusion phenomena. An example is the $1/f$ noise generated by semiconductor devices: $S_{\eta\eta}(j\omega)$ in Fig. 6-7 is then proportional to $1/|f|$, which implies that the continuous-time transfer function H is proportional to $1/\sqrt{s}$.

(iv) It is tempting to think that the results of Section 6.7.3.2 are valid, irrespective of the intersample behavior of the noise-generating mechanism. The following reasoning shows that this is wrong. Consider a filtered continuous-time band-limited white noise process with an autocorrelation function $R(\tau)$. Its power spectral density $S(j\omega) = F\{R(\tau)\}$ is a rational function of $j\omega$ for $|f| \le f_B$. Consider now a discrete-time stochastic process with an autocorrelation function $R_d(n) = R(nT_s)$. At the sampling instants, this discrete-time process has the same first and second order moments as the continuous-time process. The power spectrum $S_d(j\omega)$ of the discrete-time process is related to the power spectral density $S(j\omega)$ as

$$S_d(j\omega) = \sum_{n=-\infty}^{+\infty} R_d(n) e^{-jn2\pi fT_s} = \frac{1}{T_s} \sum_{n=-\infty}^{+\infty} S(j(\omega - n\omega_s))$$

It clearly shows that, in general, $S_d(j\omega)$ is not a rational spectrum in $\exp(-j\omega T_s)$. Hence, a "high" order rational function $H_d(z^{-1})$ is needed to model $S_d(j\omega)$ accurately, and physical interpretation of $H_d(z^{-1})$ is dangerous, especially at low sampling rates.

6.7.3.4 Relation Between the Input-Output DFT Spectra. Taking the DFT of (6-58) and (6-64), it follows that the DFT spectrum $N_Y(k)$ of the observed noise samples $n_y(m) = v(mT_s)$ is related to the DFT spectrum $E(k)$ of the unobserved driving noise source samples $e(m) = e_c(mT_s)$ or $\varepsilon_c(mT_s)$ as

$$N_Y(k) = H(\Omega_k, \theta)E(k) + T_H(\Omega_k, \theta) + \delta(\Omega_k) \qquad (6\text{-}65)$$

with $\delta(z_k^{-1}) = 0$ ($\Omega = s$, $\sqrt{s}$ or z^{-1}), and where $E(k)$ is (asymptotically for $N \to \infty$) circular complex normally distributed (proof: apply (6-34) to (6-58) and (6-64), and apply Lemma 16.24 to $e(m)$). $T_H(z^{-1}, \theta)$ is the noise transient term,

$$T_H(\Omega, \theta) = \frac{J(\Omega, \theta)}{D(\Omega, \theta)} = \frac{\sum_{r=0}^{n_j} j_r \Omega^r}{\sum_{r=0}^{n_d} d_r \Omega^r} \qquad (6\text{-}66)$$

with $n_j = (\max(n_c, n_d) - 1)$, and where the coefficients j_r are a function of the difference between the initial and final conditions of the noise process. The convergence rate to zero of T_H and δ in (6-65) as $N \to \infty$ is given by Lemmas 6.5 and 6.6, respectively. Because the sum $T_H(\Omega, \theta) + \delta(\Omega)$ decreases to zero as $O_p(N^{-1/2})$ (see Theorem 6.7), and $H(\Omega_k, \theta)E(k)$ is an $O_p(N^0)$ (see Section 16.16), (6-65) is often approximated by $N_Y(k) = H(\Omega_k, \theta)E(k)$.

Since the numerator coefficients of the transient term $T_H(\Omega)$ only depend on a finite number $n_j = \max(n_c, n_d) - 1$ of initial and final conditions, one can expect that $T_H(\Omega_k)$ is only weakly correlated with the input DFT spectrum $E(k)$. This result is confirmed in the following theorem.

Theorem 6.15 (Correlation Noise Transient with Noise Input): Under the assumptions of Theorems 6.11 and 6.14, the correlation of the transient term $T_H(\Omega_k)$ with the unobserved input DFT spectrum $E(k)$ converges to zero at the rate $(N \to \infty)$

$$\mathbb{E}\{T_H(\Omega_k)\bar{E}(k)\} = \begin{cases} O(1/N) & \Omega = z^{-1} \\ O(\ln(N)/N) & \Omega = s, \sqrt{s} \end{cases} \quad (6\text{-}67)$$

Proof. See Appendix 6.P. □

If we replace the ideal anti-alias filter in Theorem 6.14 with a real lowpass filter such that the power spectral density $S_{\varepsilon_c \varepsilon_c}(j\omega)$ of $\varepsilon_c(t)$ in (6-64) is a rational function of $j\omega$, then $O(\ln(N)/N)$ in (6-67) reduces to an $O(1/N)$ (see Appendix 6.P). Under the same conditions, the residual alias error $\delta(s_k)$ in (6-65) is also weakly correlated with $E(k)$.

Theorem 6.16 (Correlation Noise Alias Error with Noise Input): Consider the assumptions of Theorem 6.14, where the ideal anti-alias filter is replaced by a real lowpass filter such that the power spectral density $S_{\varepsilon_c \varepsilon_c}(j\omega)$ of $\varepsilon_c(t)$ in (6-64) is a rational function of $j\omega$. The correlation of the alias error $\delta(s_k)$ with the unobserved input DFT spectrum $E(k)$ decreases to zero at the rate $(N \to \infty)$

$$\mathbb{E}\{\delta(s_k)\bar{E}(k)\} = O(1/N) \quad (6\text{-}68)$$

Proof. See Appendix 6.Q. □

6.7.3.5 Model Structures. The parametric noise model (6-65) can be combined with any plant model of Section 6.3. For example, combining (6-34) and (6-65) within a generalized output error framework ((6-56) with $N_U(k) = 0$), gives

$$Y(k) = G(\Omega_k, \theta)U(k) + T_G(\Omega_k, \theta) + H(\Omega_k, \theta)E(k) + T_H(\Omega_k, \theta) + \delta(\Omega_k) \quad (6\text{-}69)$$

with $\delta(z^{-1}) = 0$ and where $c_1, \ldots, c_{n_c}$, $d_1, \ldots, d_{n_d}$ and possibly $j_0, \ldots, j_{n_j}$ are added to the parameter vector θ. Model (6-69) with $\Omega = z^{-1}$ represents the classical time domain model structures, for example, ARX (AutoRegressive with eXogenous input) for $C(z^{-1}, \theta) = 1$ and $D(z^{-1}, \theta) = A(z^{-1}, \theta)$,

$$\text{ARX:} \quad Y(k) = \frac{B(z_k^{-1}, \theta)}{A(z_k^{-1}, \theta)}U(k) + \frac{1}{A(z_k^{-1}, \theta)}E(k) + \frac{K(z_k^{-1}, \theta)}{A(z_k^{-1}, \theta)} \quad (6\text{-}70)$$

with $K(z^{-1}, \theta) = I(z_k^{-1}, \theta) + J(z_k^{-1}, \theta)$ and $n_k = \max(n_a, n_b) - 1$; ARMAX (AutoregRessive Moving Average with eXogenous input) for $D(z^{-1}, \theta) = A(z^{-1}, \theta)$,

$$\text{ARMAX:} \quad Y(k) = \frac{B(z_k^{-1}, \theta)}{A(z_k^{-1}, \theta)}U(k) + \frac{C(z_k^{-1}, \theta)}{A(z_k^{-1}, \theta)}E(k) + \frac{K(z_k^{-1}, \theta)}{A(z_k^{-1}, \theta)} \quad (6\text{-}71)$$

with $K(z^{-1}, \theta) = I(z_k^{-1}, \theta) + J(z_k^{-1}, \theta)$ and $n_k = \max(n_a, n_b, n_c) - 1$; ARMA (AutoRegressive Moving Average) for $G(z^{-1}, \theta) = 0$ and $T(z^{-1}, \theta) = 0$,

Section 6.7 ■ Noise Models

$$\text{ARMA:} \quad Y(k) = \frac{C(z_k^{-1}, \theta)}{D(z_k^{-1}, \theta)} E(k) + \frac{J(z_k^{-1}, \theta)}{D(z_k^{-1}, \theta)} \tag{6-72}$$

OE (Output Error) for $H(z^{-1}, \theta) = 1$ and $T_H(z^{-1}, \theta) = 0$,

$$\text{OE:} \quad Y(k) = \frac{B(z_k^{-1}, \theta)}{A(z_k^{-1}, \theta)} U(k) + \frac{I(z_k^{-1}, \theta)}{A(z_k^{-1}, \theta)} + E(k) \tag{6-73}$$

and BJ (Box-Jenkins)

$$\text{BJ:} \quad Y(k) = \frac{B(z_k^{-1}, \theta)}{A(z_k^{-1}, \theta)} U(k) + \frac{I(z_k^{-1}, \theta)}{A(z_k^{-1}, \theta)} + \frac{C(z_k^{-1}, \theta)}{D(z_k^{-1}, \theta)} E(k) + \frac{J(z_k^{-1}, \theta)}{D(z_k^{-1}, \theta)} \tag{6-74}$$

when the plant $G(z^{-1}, \theta)$ and the noise $H(z^{-1}, \theta)$ models have no common parameters (Ljung, 1999). Replacing z^{-1} by s or $\sqrt{s}$ in (6-70) to (6-74) gives the continuous-time counterparts C-ARX, C-ARMAX, C-ARMA, C-OE and C-BJ.

Combining a continuous-time plant model ((6-34) with $\Omega = s$ or $\sqrt{s}$) with a discrete-time noise model ((6-65) with $\Omega = z^{-1}$) in a generalized output error framework ((6-56) with $N_U(k) = 0$), gives a hybrid Box-Jenkins model:

$$Y(k) = G(\Omega_k, \theta) U(k) + T_G(\Omega_k, \theta) + H(z_k^{-1}, \theta) E(k) + T_H(z_k^{-1}, \theta) + \delta(\Omega_k) \tag{6-75}$$

with $\Omega = s$ or $\sqrt{s}$, and where $\delta(\Omega_k)$ is independent of the noise dynamics (Pintelon et al., 2000; Young et al., 2006; and Young et al., 2008).

The plant transient $T_G(\Omega_k, \theta)$ and the noise transient $T_H(\Omega_k, \theta)$ terms in (6-69) are not always distinguishable (separately identifiable), for example,

$$T_G(\Omega, \theta) + T_H(\Omega, \theta) = \frac{I(\Omega, \theta) + J(\Omega, \theta)}{A(\Omega, \theta)} \tag{6-76}$$

for (C-)ARX and (C-)ARMAX models and only the sum $i_r + j_r$ of the coefficients can be identified. Therefore, we replaced $I(\Omega, \theta) + J(\Omega, \theta)$ by $K(\Omega, \theta)$ in (6-70) and (6-71). For (C-)BJ models we have

$$T_G(\Omega, \theta) + T_H(\Omega, \theta) = \frac{I(\Omega, \theta) D(\Omega, \theta) + A(\Omega, \theta) J(\Omega, \theta)}{A(\Omega, \theta) D(\Omega, \theta)} \tag{6-77}$$

where $T_G(\Omega_k, \theta)$ and $T_H(\Omega_k, \theta)$ are distinguishable (i_r and j_r are identifiable) if $A(\Omega, \theta)$ and $D(\Omega, \theta)$ have no common roots and if $n_b \leq n_a$ and $n_c \leq n_d$ (see Exercise 6.12). If $A(\Omega, \theta)$ and $D(\Omega, \theta)$ have common roots then the parameterization should be adapted accordingly (see Exercise 5.12). Although the transient terms $T_H(\Omega, \theta)$ and $T_G(\Omega, \theta)$ are often neglected, they are important in model validation tests (see Section 13.10.1) and in lowly damped systems (e.g., vibrating mechanical structures).

Under the assumptions of Theorems 6.15 and 6.16 the noise transient $T_H(\Omega_k, \theta)$ and the noise alias error $\delta(\Omega_k)$ in (6-69) are weakly correlated with $E(k)$. A similar result is true for the plant transient term $T_G(\Omega_k, \theta)$ and the plant alias errors if the input $u(t)$ can be written as filtered (band-limited) white noise.

Theorem 6.17 (Correlation Plant Transient and Plant Alias Error with Plant Input): If the input power spectrum (power spectral density) $S_{uu}(j\omega)$ is a rational function of $\Omega = j\omega$, $\sqrt{j\omega}$ or $e^{-j\omega T_s}$, then, as $N \to \infty$, the correlation of the plant transient $T_G(\Omega_k, \theta)$ and the plant alias error $\delta(s_k)$ with the input DFT spectrum $U(k)$ decrease to zero as

$$\mathbb{E}\{T_G(\Omega_k)\bar{U}(k)\} = O(1/N) \qquad \Omega = s, \sqrt{s}, z^{-1}$$
$$\mathbb{E}\{\delta(s_k)\bar{U}(k)\} = O(1/N) \tag{6-78}$$

Proof. See Appendix 6.R.

6.8 NONLINEAR SYSTEMS

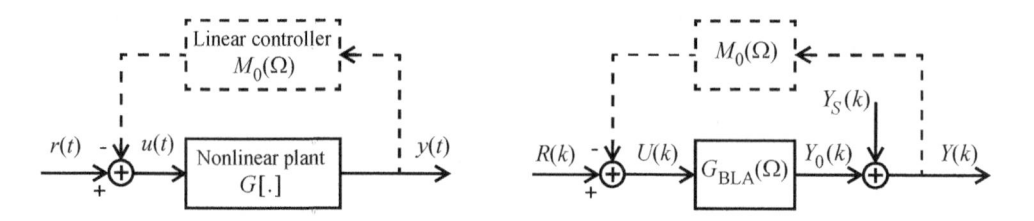

Figure 6-8. Nonlinear system $y(t) = G[u(t)]$ operating in open (solid lines only) or closed (solid and dashed lines) loop. Left: time domain representation; right: equivalent frequency domain representation using the best linear approximation (BLA).

In this section the response of the nonlinear system $y(t) = G[u(t)]$ (see Figure 6-8, left block diagram) is studied for random phase multisine (Definition 3.2) and periodic noise (Definition 3.3) excitations $u(t)$. These are periodic signals with a deterministic (random multisine) or random (periodic noise) amplitude spectrum and a random phase spectrum. The class of nonlinear distortions considered is restricted to the nonlinear systems that can be approximated arbitrarily well in least squares sense by a Volterra series on a given input domain (see Definition 3.5). It makes it possible to describe strongly nonlinear phenomena such as saturation (e.g., amplifiers) and discontinuities (e.g., relays, quantizers).

From Theorems 3.16 and 3.22 it follows that the input-output DFT spectra are related to the best linear approximation (BLA) $G_{BLA}(\Omega)$ as

$$Y(k) = G_{BLA}(\Omega_k)U(k) + Y_S(k) \tag{6-79}$$

where $G_{BLA}(\Omega_k)$ consists of the sum of the true underlying linear system $G_0(\Omega_k)$ and the bias term $G_B(\Omega_k)$, which depends on the odd nonlinear distortions and the power spectrum of the input signal (Theorem 3.7). The stochastic nonlinear contribution $Y_S(k)$ has the following properties: it is uncorrelated with – but not independent of – the input $U(k)$ (open loop setup, Figure 6-8, left block diagram) or the reference signal $R(k)$ (closed loop setup, Figure 6-8, left block diagram); its variance is a smooth function of the frequency, it is mixing over k of order infinity; and it is is asymptotically ($N \to \infty$) circular complex normally distributed (see Theorem 3.16). These observations motivate the right block diagram of Figure 6-8.

6.9 EXERCISES

6.1. Show that the transfer function between the force per unit area and the longitudinal displacement of the clamped beam is given by (6-4). Calculate the poles of (6-4) (hint: values of s such that $\cosh(\tau s) = 0$).

6.2. Calculate the partial fraction expansion (6-5) of transfer function (6-4) (hint: note that $G(\infty) = 0$ and calculate $\sum_{k=-\infty}^{\infty} R_{2k+1}/(s-s_{2k+1})$ with R_k the residue of the pole s_k).

6.3. Consider the charging of a capacitor (see Figure 6-9). Show that the transfer function between the piecewise constant voltage source $u(t)$ and the sampled voltage $y(t)$, $t = kT_s$, across the capacitor is given by

$$G_{\text{ZOH}}(z^{-1}) = (1 - e^{-T_s/(RC)})/(z - e^{-T_s/(RC)}) \tag{6-80}$$

(hint: first show that $G(s) = 1/(1 + RCs)$)..

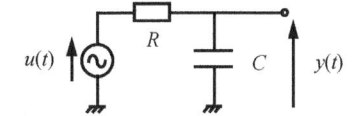

Figure 6-9. Charging of a capacitor with a voltage $u(t)$.

6.4. Consider the cascade of two continuous-time systems shown in Figure 6-4 and show that discrete-time model (6-11) describes the input-output behavior of the continuous-time model exactly at the sampling times $t = kT_s$ (hint: apply (6-10) on the transfer functions from $r_{\text{zoh}}(t)$ to $u(t)$ and from $r_{\text{zoh}}(t)$ to $y(t)$).

6.5. Show that the partial fraction expansion of a proper ($n_b \leq n_a$) discrete-time system $G(z^{-1}, \theta)$ with $b_0 = 0$ is given by (6-23) (hint: multiply the numerator and denominator polynomial of $G(z^{-1}, \theta)$ with z^{n_a} and calculate the partial fraction expansion in z).

6.6. Show that a state space representation of difference equation (6-17) is given by (6-19) with

$$A = \begin{bmatrix} -\frac{a_1}{a_0} & -\frac{a_2}{a_0} & \cdots & -\frac{a_{n_a-1}}{a_0} & -\frac{a_{n_a}}{a_0} \\ 1 & 0 & \cdots & 0 & 0 \\ 0 & 1 & \cdots & \cdots & \cdots \\ \cdots & \cdots & \cdots & 0 & \cdots \\ 0 & \cdots & 0 & 1 & 0 \end{bmatrix}, B = \begin{bmatrix} \frac{1}{a_0} \\ 0 \\ 0 \\ \cdots \\ 0 \end{bmatrix},$$

$$C = \begin{bmatrix} b_1 - a_1\frac{b_0}{a_0} & \cdots & b_{n_b} - a_{n_b}\frac{b_0}{a_0}, & -a_{n_b+1}\frac{b_0}{a_0} & \cdots & -a_{n_a}\frac{b_0}{a_0} \end{bmatrix} \text{ and } D = \frac{b_0}{a_0}$$

(hint: eliminate the state vector in (6-19)).

6.7. Assume that $M_{u[i]}$ input samples are missing at time instants $t = K_{u[i]}T_s$, $i = 1, 2, \ldots, M_{u[i]}$, and $M_{y[j]}$ output samples are missing at time instants $t = K_{y[j]}T_s$, $j = 1, 2, \ldots, M_{y[j]}$. Show that the extended transfer function model for discrete-time systems is given by

$$Y^m(k) = G(z_k^{-1}, \theta)U^m(k) + T_G(z_k^{-1}, \theta) +$$

$$G(z_k^{-1}, \theta)\sum_{i=1}^{M_{u[i]}} z_k^{-K_{u[i]}} I_{u[i]}(z_k^{-1}, \psi) - \sum_{j=1}^{M_{y[j]}} z_k^{-K_{y[j]}} I_{y[j]}(z_k^{-1}, \psi)$$

where $I_{x[i]}(z^{-1}, \psi) = N^{-1/2}\sum_{n=0}^{M_{x[i]}-1} x(K_{x[i]} + n)z^{-n}$, $x = u, y$ (hint: follow the lines of Appendix 6.G).

6.8. Assume that M_c input-output data sets $u_i(tT_s), y_i(tT_s)$, $t = 0, 1, ..., N_i - 1$, are available. Show that the DFT spectra of the concatenated input-output signals satisfy

$$Y^c(k) = G(\Omega_k, \theta)U^c(k) + T_G(\Omega_k, \theta) + \sum_{i=1}^{M_c-1} z_k^{-\tau_i} T_{G[i]}^c(\Omega_k, \theta) + \delta(\Omega_k) \quad (6\text{-}81)$$

with $\delta(z_k^{-1}) = 0$, $\tau_i = \sum_{r=1}^{i} N_r$, and where $T_{G[i]}^c(\Omega_k, \theta) = I_{[i]}^c(\Omega_k, \theta)/A(\Omega_k, \theta)$ with $n_{c[i]} = n_i = \max(n_a, n_b) - 1$ (hint: follow the lines of Appendix 6.H).

6.9. Show relation (6-52) for the real case (6-49) (hint: use $\cos(x) = (e^{jx} + e^{-jx})/2$ and follow the lines of Appendix 6.J).

6.10. Show that a_0 in $a_0 y(t) + y'(t) = a_0 u(t) + u'(t)$, with $x'(t)$ the derivative of $x(t)$ w.r.t. t, is identifiable if and only if $y(0-) \neq u(0-)$. Note that $y(t)$ can never be made different from $u(t)$ if $y(0-) = u(0-)$. Only internal action in the system can make $y(0-) \neq u(0-)$.

6.11. Show that the functions $f_r(\Omega_k) = \Omega_k^r U(k)$, $r = 0, 1, ..., n_b$, and $k = 0, 1, ..., N/2$ are linearly independent if and only if $U(k) \neq 0$ for at least $(n_b + 1)/2$ DFT frequencies where DC and Nyquist each count for $1/2$ (hint: study $\sum_{r=0}^{n_b} \beta_r f_r(\Omega_k) = 0$ at the DFT frequencies where $U(k) \neq 0$).

6.12. Consider model (6-69) where $n_b \leq n_a$, $n_c \leq n_d$, $A(z^{-1}, \theta)$ and $D(z^{-1}, \theta)$ have no common roots, and $G(z^{-1}, \theta)$, $H(z^{-1}, \theta)$ have respective minimal orders n_b over n_a and n_c over n_d. Show that $T_G(z^{-1}, \theta)$ and $T_H(z^{-1}, \theta)$ are identifiable (hint: suppose that $A(z^{-1}, \theta)$ and $D(z^{-1}, \theta)$ are given and write $T_G(z_k^{-1}, \theta) + T_H(z_k^{-1}, \theta)$ as $\sum_{r=0}^{n_i} i_r f_r(z_k^{-1}) + \sum_{r=0}^{n_j} j_r g_r(z_k^{-1})$ with $f_r(z_k^{-1}) = z_k^{-r}/A(z_k^{-1}, \theta)$, $g_r(z_k^{-1}) = z_k^{-r}/D(z_k^{-1}, \theta)$; next show, following the lines of Appendix 6.K, that $f_r(z_k^{-1})$, $g_r(z_k^{-1})$ are independent functions).

6.13. Consider model (6-69) where $n_b \leq n_a$, $n_c \leq n_d$, $A = \tilde{A}F$, $D = \tilde{D}F$, $\tilde{A}$ and $\tilde{D}$ have no common roots, and G, H have respective minimal orders n_b over n_a and n_c over n_d. Show that

$$G = \frac{B}{\tilde{A}F}, \quad H = \frac{C}{\tilde{D}F} \text{ and } T_G + T_H = \frac{1}{F}\left(I_2 + \frac{\tilde{I}}{\tilde{A}} + \frac{\tilde{C}}{\tilde{D}}\right),$$

where I_2, $\tilde{I}$, and $\tilde{C}$ are polynomials in z^{-1} of respective orders $n_f - 1$, $n_a - n_f - 1$, and $n_d - n_f - 1$, is an identifiable parameterization.

6.10 APPENDIXES

Appendix 6.A Stability and Minimum Phase Regions

To determine the stability regions of the poles we expand the rational form $G(\Omega, \theta)$ in partial fractions. We find

$$G(s, \theta) = \sum_r \frac{L_r}{s - \lambda_r}, \quad G(z^{-1}, \theta) = \sum_r \frac{L_r}{z - \lambda_r} \text{ and } G(\sqrt{s}, \theta) = \sum_r \frac{L_r}{\sqrt{s} - \lambda_r} \quad (6\text{-}82)$$

(see (6-22) and (6-23)). Calculating the impulse responses of $G(\Omega, \theta)$ in (6-82) gives

$$g(t) = L^{-1}\{G(s, \theta)\} = \sum_r L_r e^{\lambda_r t} \qquad \text{(a)}$$

$$g(n) = Z^{-1}\{G(z^{-1}, \theta)\} = \sum_r L_r \lambda_r^{(n-1)}, \text{ for } n > 0 \qquad \text{(b)} \qquad (6\text{-}83)$$

$$g(t) = L^{-1}\{G(\sqrt{s}, \theta)\} = \sum_r L_r \left(\frac{1}{\sqrt{\pi t}} + \lambda_r e^{\lambda_r^2 t} \text{erfc}(-\lambda_r \sqrt{t})\right) \qquad \text{(c)}$$

with $L^{-1}\{\ \}$ the inverse Laplace transform, $Z^{-1}\{\ \}$ the inverse Z-transform, and erfc( ) the complementary error function (Selby, 1973; Spiegel, 1965). The asymptotic ($t \to \infty$) behavior of (6-83c) is studied in Lemma 6.18. It follows that the impulse responses are asymptotically zero (poles are stable) if and only if $\text{Re}(\lambda_r) < 0$ in the s-domain (6-83a), $|\lambda_r| < 1$ in the z-domain (6-83b), and $|\angle \lambda_r| > \pi/4$ in the $\sqrt{s}$-domain (6-83c). The convergence rate to zero of the impulse response is exponentially for s- and z-domain systems, while it is algebraically as an $O(t^{-3/2})$ for $\sqrt{s}$-domain systems (proof: see Lemma 6.18). For unstable s-, z-, and $\sqrt{s}$-domain systems the impulse response growths exponentially. By definition, the minimum phase region of the zeros equals the stable region of the poles.

Lemma 6.18 (Asymptotic Behavior Impulse Response (6-83c)): For $t \to \infty$ the function

$$f(t) = \frac{1}{\sqrt{\pi t}} + \lambda e^{\lambda^2 t} \text{erfc}(-\lambda \sqrt{t}) \qquad (6\text{-}84)$$

can be written as

$$f(t) = \begin{cases} 2\lambda e^{\lambda^2 t} + O(t^{-3/2}) & |\angle \lambda| \leq \pi/4 \\ \dfrac{1}{2\lambda^2 \sqrt{\pi} t^{3/2}} + O(t^{-5/2}) & |\angle \lambda| > \pi/4 \end{cases} \qquad (6\text{-}85)$$

It shows that $f(t)$ (i) grows exponentially for $|\angle \lambda| < \pi/4$ ($\Rightarrow \text{Re}(\lambda^2) > 0$), (ii) decreases algebraically to zero as $O(t^{-3/2})$ for $|\angle \lambda| > \pi/4$, and (iii) converges to a periodic solution for $|\angle \lambda| = \pi/4$ ($\Rightarrow \text{Re}(\lambda^2) = 0$) as $O(t^{-3/2})$.

Proof. Using $\text{erfc}(z) = 1 - \text{erf}(z)$ and $\text{erf}(-z) = -\text{erf}(z)$, with $\text{erf}(z)$ the error function, (6-84) can be rewritten as

$$f(t) = \frac{1}{\sqrt{\pi t}} + \lambda e^{\lambda^2 t}(1 + \text{erf}(\lambda \sqrt{t})) = \frac{1}{\sqrt{\pi t}} + \lambda e^{\lambda^2 t}(2 - \text{erfc}(\lambda \sqrt{t})) \qquad (6\text{-}86)$$

Applying the following asymptotic expansions ($z \to \infty$) of $\text{erfc}(z)$

$$\begin{aligned}
\text{erfc}(-z) &= -\frac{e^{-z^2}}{z\sqrt{\pi}}\left(1 + \sum_{m=1}^{\infty}(-1)^m \frac{(2m-1)!!}{(2z^2)^m}\right) & |\angle z| &> \pi/4 \\
\text{erfc}(z) &= \frac{e^{-z^2}}{z\sqrt{\pi}}\left(1 + \sum_{m=1}^{\infty}(-1)^m \frac{(2m-1)!!}{(2z^2)^m}\right) & |\angle z| &< 3\pi/4
\end{aligned} \qquad (6\text{-}87)$$

(see Abramowitz and Stegun, 1970) to (6-84) and (6-86) gives (6-85). With the property $\text{erf}(\bar{z}) = \overline{\text{erf}(z)}$ the results for the complex conjugate root $\bar{\lambda}$ can easily be derived from (6-84) to (6-87). □

Appendix 6.B   Relation between DFT Spectra and Transfer Function for Arbitrary Signals

First, the result is proved for discrete-time systems, and next, for (lumped) continuous-time systems. In both cases we assume that the input and output samples are known (no measurement and no process noise) exactly for $t = 0, 1, ..., N-1$ and are unknown elsewhere.

6.B.1 Discrete-Time Systems (z-domain).   The discrete input and output samples satisfy difference equation (6-17) for any t. Taking the one-sided Z-transform of both sides of (6-17) using

$$\sum_{t=0}^{\infty} x(t-n)z^{-t} = z^{-n}(X(z) + X_1(z))$$

where $X(z) = \sum_{t=0}^{\infty} x(t)z^{-t}$ ($X = U, Y$ and $x = u, y$) is the one-sided Z-transform of $x(t)$ and $X_1(z) = \sum_{t=1}^{n} x(-t)z^t$, gives

$$A(z^{-1})Y(z) = B(z^{-1})U(z) + I_1(z^{-1}) \qquad (6\text{-}88)$$

$A(z^{-1})$ and $B(z^{-1})$ are, respectively, the denominator and numerator polynomials of the plant transfer function (6-20) and $I_1(z^{-1})$ stands for the influence of the initial conditions of the experiment (past samples of $u(t)$ and $y(t)$)

$$I_1(z^{-1}) = \sum_{m=1}^{n_b} \sum_{t=1}^{m} b_m u(-t) z^{t-m} - \sum_{n=1}^{n_a} \sum_{t=1}^{n} a_n y(-t) z^{t-n} \qquad (6\text{-}89)$$

Model (6-88) cannot be evaluated on the unit circle because the input and output samples for $t \geq N$ are unknown. These samples must, hence, be eliminated. We solve, thereto, difference equation (6-17) for $t = N, N+1, ..., \infty$. Multiplying both sides of (6-17) with z^{-t} and making the summation over $t = N, N+1, ..., \infty$ using

$$\sum_{t=N}^{\infty} x(t-n)z^{-t} = z^{-n}(\tilde{X}(z) + z^{-N}X_2(z))$$

where $\tilde{X}(z) = \sum_{t=N}^{\infty} x(t)z^{-t}$ and $X_2(z) = \sum_{t=1}^{n} x(N-t)z^t$ ($X = U, Y$ and $x = u, y$), gives

$$A(z^{-1})\tilde{Y}(z) = B(z^{-1})\tilde{U}(z) + z^{-N}I_2(z^{-1}) \qquad (6\text{-}90)$$

$I_2(z^{-1})$ stands for the influence of the final conditions (samples of $u(t)$ and $y(t)$ at the end of the experiment)

$$I_2(z^{-1}) = \sum_{m=1}^{n_b} \sum_{t=1}^{m} b_m u(N-t) z^{t-m} - \sum_{n=1}^{n_a} \sum_{t=1}^{n} a_n y(N-t) z^{t-n} \qquad (6\text{-}91)$$

Subtracting (6-90) from (6-88) gives

Section 6.10 ■ Appendixes

$$A(z^{-1})Y_N(z) = B(z^{-1})U_N(z) + I_1(z^{-1}) - z^{-N}I_2(z^{-1}) \tag{6-92}$$

where $X_N(z) = X(z) - \tilde{X}(z) = \sum_{t=0}^{N-1} x(t)z^{-t}$ ($X = U, Y$ and $x = u, y$). Evaluation of (6-92) on the unit circle at the DFT frequencies $z_k = \exp(j2\pi k/N)$, taking into account that $z_k^N = 1$, $Y_N(z_k) = N^{1/2}Y(k)$ and $U_N(z_k) = N^{1/2}U(k)$, finally gives

$$A(z_k^{-1})Y(k) = B(z_k^{-1})U(k) + I(z_k^{-1}) \tag{6-93}$$

where $I(z^{-1}) = N^{-1/2}(I_1(z^{-1}) - I_2(z^{-1}))$ is a polynomial of order $n_i = \max(n_a, n_b) - 1$. The polynomial $I(z^{-1})$ can be parameterized independently of the numerator and denominator coefficients of $G(z^{-1})$ (6-20) because its coefficients i_r depend, for a given plant model, linearly on $\max(n_a, n_b)$ independent initial conditions.

6.B.2 Lumped Continuous-Time Systems (s-domain). The proof follows the same lines as in the previous section. We assume that the excitation $u(t)$ is band-limited. The input and output continuous-time signals satisfy differential equation (6-16). Taking the one-sided Laplace transform of (6-16) gives

$$A(s)Y(s) = B(s)U(s) + I_1(s) \tag{6-94}$$

where $U(s)$ and $Y(s)$ are the one-sided Laplace transforms of $u(t)$ and $y(t)$, respectively. $A(s)$, $B(s)$ are, respectively, the numerator and denominator polynomials of the plant transfer function (6-20) and $I_1(s)$ represents the influence of the initial conditions (value and derivatives of $u(t)$ and $y(t)$ at $t = 0-$)

$$I_1(s) = \sum_{n=1}^{n_a} \sum_{r=0}^{n-1} a_n s^{n-r-1} y^{(r)}(0-) - \sum_{m=1}^{n_b} \sum_{r=0}^{m-1} b_m s^{m-r-1} u^{(r)}(0-) \tag{6-95}$$

The integrals appearing in the model (6-94) cannot be evaluated because the input and output signals are unknown for $t \geq NT_s$. The differential equation (6-16) is, therefore, solved for $t \geq NT_s$ using the one-sided Laplace transform. Multiplying both sides of (6-16) by e^{-st} and integrating from $t = NT_s$ to $t = \infty$ gives

$$A(s)\tilde{Y}(s) = B\tilde{U}(s) + e^{-NT_s s} I_2(s) \tag{6-96}$$

where $\tilde{X}(s) = \int_{NT_s}^{\infty} e^{-st} x(t) dt$ ($X = U, Y$ and $x = u, y$). $I_2(s)$ stands for the influence of the final conditions (value and derivatives of $u(t)$ and $y(t)$ at $t = NT_s-$)

$$I_2(s) = \sum_{n=1}^{n_a} \sum_{r=0}^{n-1} a_n s^{n-r-1} y^{(r)}(NT_s-) - \sum_{m=1}^{n_b} \sum_{r=0}^{m-1} b_m s^{m-r-1} u^{(r)}(NT_s-) \tag{6-97}$$

Subtracting (6-96) from (6-94) gives

$$A(s)Y_N(s) = B(s)U_N(s) + I_1(s) - e^{-NT_s s} I_2(s) \tag{6-98}$$

where $X_N(s) = X(s) - \tilde{X}(s) = \int_0^{NT_s} e^{-st}x(t)dt$ ($X = U, Y$ and $x = u, y$). Evaluating (6-98) along the $j\omega$-axis at the DFT frequencies $s_k = j2\pi f_s k/N$ using the relationship between the DFT and the Fourier integral (Brigham, 1974)

$$X(k) = \frac{1}{\sqrt{N}}\sum_{t=0}^{N-1} x(tT_s)z_k^{-t} = \frac{1}{T_s\sqrt{N}}\sum_{n=-\infty}^{n=+\infty} X_N(s_k - nj\omega_s) \qquad (6\text{-}99)$$

and taking into account that $e^{-NT_s s_k} = 1$, results in

$$A(s_k)Y(k) = B(s_k)U(k) + I(s_k) + \Delta(s_k) \qquad (6\text{-}100)$$

$I(s) = N^{-1/2}(I_1(s) - I_2(s))/T_s$ is a polynomial of order $\max(n_a, n_b) - 1$, and $\Delta(s_k)$ is the residual spectral alias error

$$\Delta(s_k) = \frac{1}{T_s\sqrt{N}}\sum_{\substack{n=-\infty \\ n \neq 0}}^{n=+\infty} [B(s_k)U_N(s_k - nj\omega_s) - A(s_k)Y_N(s_k - nj\omega_s)] \qquad (6\text{-}101)$$

Note that the spectral alias error is due to the piecewise constant approximation of the Fourier integrals $U_N(j\omega)$ and $Y_N(j\omega)$ by the discrete Fourier transforms $U(k)$ and $Y(k)$ (see (6-99)): it is present even if the signals $u(t)$ and $y(t)$ passed through a low-pass filter before sampling. For the same reason as in the previous section, the polynomial $I(s)$ is parameterized independent of the numerator and denominator coefficients of $G(s)$ (6-20).

6.B.3 Diffusion Phenomena ($\sqrt{s}$-domain). The proof follows the same lines as in Section 6.B.2, but now we start from the following fractional differential equation

$$\sum_{n=0}^{n_a} a_n y^{(n/2)}(t) = \sum_{m=0}^{n_b} b_m u^{(m/2)}(t) \qquad (6\text{-}102)$$

where $x^{(i+1/2)}(t)$, with $i \in \mathbb{N}$, is the Riemann-Liouville fractional derivative of order $i + 1/2$

$$\frac{d^{i+1/2}}{dt^{i+1/2}} = \frac{d^i}{dt^i}\left(\frac{d^{1/2}}{dt^{1/2}}\right) \text{ with } \frac{d^{1/2}x(t)}{dt^{1/2}} = \frac{1}{\sqrt{\pi}}\frac{d}{dt}\int_0^t \frac{x(\tau)}{\sqrt{t-\tau}}d\tau \qquad (6\text{-}103)$$

(see Oldham and Spanier, 1974). Taking the one-sided Laplace transform of (6-103) gives

$$A(\sqrt{s})Y(s) = B(\sqrt{s})U(s) + I_1(s) \qquad (6\text{-}104)$$

(see Oldham and Spanier, 1974) with $A(\sqrt{s})$ and $B(\sqrt{s})$, respectively, the numerator and denominator polynomials of the plant transfer function. $I_1(s)$ represents the influence of the initial conditions (value and fractional derivatives of $u(t)$ and $y(t)$ at $t = 0$)

$$I_1(s) = \frac{1}{\sqrt{N}} \left(\sum_{n=0}^{n_a} a_n \sum_{r=0}^{\lceil n/2 \rceil - 1} s^r(y^{(n/2-r-1)}(0) - y^{(n/2-r-1)}(NT_s)) \right.$$
$$\left. - \sum_{m=0}^{n_b} b_m \sum_{r=0}^{\lceil m/2 \rceil - 1} s^r(u^{(m/2-r-1)}(0) - u^{(m/2-r-1)}(NT_s)) \right)$$
(6-105)

with $\lceil x \rceil$ the smallest integer value larger than or equal to x. Although $I_1(s)$ is a polynomial function of s, it can be written as a polynomial in $\sqrt{s}$ of order $n_i = \max(n_a, n_b) - 1$. The remainder of the proof follows exactly the same lines of Section 6.B.2.

Appendix 6.C Parameterizations of the Extended Transfer Function Model

The partial fraction expansions (6-36) and (6-37) follow directly from the fact that $T_G(\Omega, \theta)$ has the same poles as $G(\Omega, \theta)$. The particular form (6-37) is obtained by rewriting $T_G(z^{-1}, \theta)$ as $z(z^{-1} T_G(z^{-1}, \theta))$, where the partial fraction expansion of $z^{-1} T_G(z^{-1}, \theta)$ has the form (6-23), because the orders of the numerator and the denominator of $z^{-1} T_G(z^{-1}, \theta)$ are equal (see Exercise 6.5).

The state space equations of a proper ($n_b \leq n_a$) discrete-time system are given by (6-19). Following the lines of Appendix 6.B, we solve (6-19) for $t = 0, 1, \ldots, \infty$ and for $t = N, N+1, \ldots, \infty$ using the one-sided Z-transform. Using the same notations as in Appendix 6.B, we find

$$Y(z) = G(z^{-1}) U(z) + C(I_{n_a} - z^{-1} A)^{-1} x(0) \tag{6-106}$$

$$\tilde{Y}(z) = G(z^{-1}) \tilde{U}(z) + z^{-N} C(I_{n_a} - z^{-1} A)^{-1} x(N) \tag{6-107}$$

where $G(z^{-1})$ is given by (6-27). Subtracting (6-107) from (6-106) and evaluating the result at the DFT frequencies $z = z_k$ gives (6-40) with $x_I = N^{-1/2}(x(0) - x(N))$.

The state space equations of a proper ($n_b \leq n_a$) lumped continuous-time system are given by (6-18). Following the lines of Appendix 6.B, we solve (6-18) for $t \in [0, \infty]$ and for $t \in [NT_s, \infty]$ using the one-sided Laplace transform. Using the same notations as in Appendix 6.B, we find

$$Y(s) = G(s) U(s) + C(sI_{n_a} - A)^{-1} x(0-) \tag{6-108}$$

$$\tilde{Y}(s) = G(s) \tilde{U}(s) + e^{-NT_s s} C(sI_{n_a} - A)^{-1} x(NT_s-) \tag{6-109}$$

where $G(s)$ is given by (6-26). Subtracting (6-109) from (6-108) and evaluating the result at the DFT frequencies $s = s_k$ gives (6-39) with $x_I = N^{-1/2}(x(0-) - x(NT_s-))/T_s$. □

Appendix 6.D Convergence Rate of the Equivalent Initial Conditions

We prove the result for a discrete-time system ($\Omega = z^{-1}$). The proof for continuous-time systems ($\Omega = s, \sqrt{s}$) follows the same lines. Using (6-89), (6-91), and (6-93) of Appendix 6.B, we find the following relationship between the coefficients of the polynomial $I(z^{-1}, \theta) = \sum_{r=0}^{n_i} i_r z^{-r}$ in the plant model (6-33) and the initial and final conditions of the experiment:

$$I(z^{-1}, \theta) = N^{-1/2}\left(\sum_{m=1}^{n_b}\sum_{t=1}^{m} b_m \Delta_N u(t) z^{t-m} - \sum_{n=1}^{n_a}\sum_{t=1}^{n} a_n \Delta_N y(t) z^{t-n}\right) \tag{6-110}$$

where $\Delta_N x(t) = x(-t) - x(N-t)$ with $x = u, y$. It shows that the coefficients i_r, $r = 0, 1, \ldots, n_i$, of $I(z^{-1}, \theta)$ depend linearly on $2n_a$ output and $2n_b$ input samples (finite number independent of N). A bounded input applied to a stable linear system results in a bounded output; see Kailath (1980). The same is true for unstable plants captured in a stabilizing feedback loop. Therefore, it follows from (6-110) that i_r in (6-35) is an $O(N^{-1/2})$.

For bounded random inputs we still have $|i_r| \leq C/\sqrt{N}$ with probability one, so that $i_r = O_{a.s.}(N^{-1/2})$. Because the residues of the poles of a rational function are proportional to its numerator coefficients, the same conclusions hold for the residues l_r and s_r in (6-37). Similar reasoning proves the results for x_I in (6-40) (see Appendix 6.C for explicit expressions of x_I). □

Appendix 6.E   Some Integral Expressions

6.E.1 Definite Integrals Involving sin(x)/x Functions.   Using

$$\cos(ax)\sin(bx) = 0.5[\sin((a+b)x) - \sin((a-b)x)] \tag{6-111}$$

$$\int_{-\infty}^{+\infty} \frac{\sin(cx)}{x} dx = \begin{cases} \pi & c > 0 \\ -\pi & c < 0 \end{cases} \tag{6-112}$$

and $0 < a < b$ we find

$$\int_{-\infty}^{+\infty} \frac{\cos(ax)\sin(bx)}{\pi x} dx = \int_{-\infty}^{+\infty} \frac{\sin((a+b)x) - \sin((a-b)x)}{2\pi x} dx = \frac{1}{2} - \left(-\frac{1}{2}\right) = 1 \tag{6-113}$$

Note that (6-113) is zero if $a > b$. Because $\sin(ax)\sin(bx)/(\pi x)$ is a uniformly bounded odd function of x in $[-\infty, +\infty]$, we have

$$\int_{-\infty}^{+\infty} \frac{\sin(ax)\sin(bx)}{\pi x} dx = 0 \tag{6-114}$$

for any value of a and b.

6.E.2 Convergence Rate of Integrals Involving sin(x)/x Functions.   In this section we study the convergence rate to zero of

Section 6.10 ■ Appendixes

$$\int_N^{+\infty} f_1(x)\,dx, \quad \int_N^{+\infty} f_2(x)\,dx, \quad \int_{-\infty}^{-N} f_1(x)\,dx \text{ and } \int_{-\infty}^{-N} f_2(x)\,dx \tag{6-115}$$

as $N \to \infty$, where

$$\begin{aligned} f_1(x) &= \frac{\cos(ax)\sin(bx)}{\pi x} = \frac{\sin((a+b)x) - \sin((a-b)x)}{2\pi x} \\ f_2(x) &= \frac{\sin(ax)\sin(bx)}{\pi x} = \frac{\cos((a-b)x) - \cos((a+b)x)}{2\pi x} \end{aligned} \tag{6-116}$$

are uniformly bounded functions of x. Because $f_1(x)$ and $f_2(x)$ are, respectively, even and odd functions of x, it follows from (6-115) and (6-116) that it is sufficient to analyze the convergence rate of

$$\int_N^{+\infty} \frac{\sin(cx)}{x}\,dx \tag{6-117}$$

with $c > 0$. The basic idea is to write the integral (6-117) as an infinite sum of integrals over one period $2\pi/c$ of the $\sin(cx)$ function

$$\int_N^{+\infty} \frac{\sin(cx)}{x}\,dx = \int_N^{2k_1\pi/c} \frac{\sin(cx)}{x}\,dx + \sum_{k=k_1}^{+\infty} \int_{2k\pi/c}^{2(k+1)\pi/c} \frac{\sin(cx)}{x}\,dx \tag{6-118}$$

with $k_1 = \mathrm{int}(Nc/(2\pi)) + 1$ and $\mathrm{int}(x)$ the integer part of the real number x. Each integral in the infinite sum can be bounded above by

$$\begin{aligned} \int_{2k\pi/c}^{2(k+1)\pi/c} \frac{\sin(cx)}{x}\,dx &= \int_{2k\pi/c}^{(2k+1)\pi/c} \frac{\sin(cx)}{x}\,dx + \int_{(2k+1)\pi/c}^{2(k+1)\pi/c} \frac{\sin(cx)}{x}\,dx \\ &\leq \int_{2k\pi/c}^{(2k+1)\pi/c} \frac{\sin(cx)}{2k\pi/c}\,dx + \int_{(2k+1)\pi/c}^{2(k+1)\pi/c} \frac{\sin(cx)}{2(k+1)\pi/c}\,dx \end{aligned} \tag{6-119}$$

because $\sin(cx) \geq 0$ for any x in $[2k\pi/c, (2k+1)\pi/c]$ and $\sin(cx) \leq 0$ for any x in $[(2k+1)\pi/c, 2(k+1)\pi/c]$. Working out the integrals in (6-119) gives

$$\int_{2k\pi/c}^{2(k+1)\pi/c} \frac{\sin(cx)}{x}\,dx \leq \frac{1}{\pi k} - \frac{1}{\pi(k+1)} = \frac{1}{\pi k(k+1)} \tag{6-120}$$

Hence, the second term in the right-hand side of (6-118) can be bounded above by

$$\sum_{k=k_1}^{+\infty} \int_{2k\pi/c}^{2(k+1)\pi/c} \frac{\sin(cx)}{x} dx \le \sum_{k=k_1}^{+\infty} \frac{1}{\pi k(k+1)} \le C_1 \int_{k_1}^{+\infty} \frac{dx}{x^2} = \frac{C_1}{k_1} = O(N^{-1}) \qquad (6\text{-}121)$$

with C_1 a constant independent of N. The second inequality is due to the Cauchy integral test (Gradshteyn and Ryzhik, 1980). The first term in the right-hand side of (6-118) can be bounded above by

$$\left| \int_N^{2k_1\pi/c} \frac{\sin(cx)}{x} dx \right| \le \frac{2k_1\pi/c - N}{N} \le \frac{2\pi}{cN} = O(N^{-1}) \qquad (6\text{-}122)$$

Collecting (6-121) and (6-122) proves that the integral (6-118) is an $O(N^{-1})$.

Appendix 6.F Convergence Rate of the Residual Alias Errors

We proof the result lumped continuous-time system (s-domain). The proof for diffusion phenomena ($\sqrt{s}$-domain) follows the same lines. Because the output Fourier spectrum $Y(j\omega)$ is related to the input Fourier spectrum $U(j\omega)$ by $Y(j\omega) = G(j\omega)U(j\omega)$ with $G(s)$ stable, the output signal has exactly the same spectral properties as the input signal, for example, band-limited, discrete Fourier spectrum. Similarly, because $S_{yy}(j\omega) = |G(j\omega)|^2 S_{uu}(j\omega)$ with $G(s)$ stable, the output power spectrum $S_{yy}(j\omega)$ has the same spectral properties as the input power spectrum $S_{uu}(j\omega)$, for example, band-limited, differentiable power spectrum. Therefore, to study $\delta(s_k) = \Delta(s_k)/A(s_k)$ (see (6-101)) it is sufficient to study the spectral content of a band-limited signal $x(t)$, observed during a time NT_s. The errors in the DFT spectra giving the term $\delta(s_k)$ are in fact, leakage errors that can be interpreted as alias errors. Indeed, due to the multiplication of $x(t)$ with a rectangular window $w_N(t)$, sharp transitions occur at the edges of $x_N(t) = x(t)w_N(t)$. These sharp transitions have a high frequency content. For ease of notation, we will take the time origin in the middle of the observation window $w_N(t)$: $w_N(t) = 1$ for $t \in [-N/2, N/2)T_s$ and zero elsewhere. First, we prove the result for normalized periodic signals (see Definition 3.4), and next, for random signals.

6.F.1 Periodic Signals.
A normalized periodic signal has the form

$$x(t) = \sum_{k=1}^{F} \frac{A_k}{\sqrt{N}} \sin(\omega_k t + \phi_k) \qquad (6\text{-}123)$$

where $A_k > 0$ and where F increases with N, $F = O(N)$ (see Definition 3.4). By assumption, the signal $x(t)$ is band-limited so that $\max_k f_k \le f_B < f_s/2$. The outline of the proof is as follows. First, we calculate the high frequency content $x_a(t)$ of the observed signal $x_N(t) = w_N(t)x(t)$. Next, the energy of $x_a(t)$ is compared with that of $x_N(t)$. Finally, via Parseval's theorem, we draw conclusions concerning the Fourier spectra $X_N(j\omega)$ and $X_a(j\omega)$ of $x_N(t)$ and $x_a(t)$, respectively.

The high frequency content $x_a(t)$ is found by multiplying $X_N(j\omega)$ with a window $P(f)$ that excludes all frequencies in the band $[-f_s/2, f_s/2]$, and by taking afterward the inverse Fourier transform. We find

Section 6.10 ■ Appendixes

$$X_a(j2\pi f) = X_N(j2\pi f)P(f) \Rightarrow x_a(t) = x_N(t)*p(t) \quad (6\text{-}124)$$

with $p(t)$ the inverse Fourier transform of $P(f)$ and $*$ the convolution product. The window $P(f)$ can be written as $P(f) = 1 - B(f)$, where $B(f) = 1$ for $|f| \leq f_s/2$ and zero elsewhere, so that

$$p(t) = \delta(t) - \sin(\omega_s t/2)/(\pi t) \quad (6\text{-}125)$$

with $\sin(\omega_s t/2)/(\pi t)$ the inverse Fourier transform of $B(f)$. Using (6-125), we get the following expression for $x_a(t)$:

$$\begin{aligned} x_a(t) &= x_N(t) - x_N(t)*(\sin(\omega_s t/2)/(\pi t)) \\ &= x_N(t) - \int_{t-NT_s/2}^{t+NT_s/2} x(t-\tau)\frac{\sin(\omega_s \tau/2)}{\pi \tau} d\tau \end{aligned} \quad (6\text{-}126)$$

Putting (6-123) in (6-126) gives, using $\sin(a-b) = \sin(a)\cos(b) - \cos(a)\sin(b)$,

$$x_a(t) = x_N(t) - x_1(t) + x_2(t) \quad (6\text{-}127)$$

with

$$\begin{aligned} x_1(t) &= \sum_{k=1}^{F} \frac{A_k}{\sqrt{N}}\sin(\omega_k t + \phi_k) \int_{t-NT_s/2}^{t+NT_s/2} f_1(\tau)d\tau, \quad f_1(\tau) = \frac{\cos(\omega_k \tau)\sin(\omega_s \tau/2)}{\pi \tau} \\ x_2(t) &= \sum_{k=1}^{F} \frac{A_k}{\sqrt{N}}\cos(\omega_k t + \phi_k) \int_{t-NT_s/2}^{t+NT_s/2} f_2(\tau)d\tau, \quad f_2(\tau) = \frac{\sin(\omega_k \tau)\sin(\omega_s \tau/2)}{\pi \tau} \end{aligned} \quad (6\text{-}128)$$

We now study (6-127) for the four following cases: (i) $t \in (-NT_s/2, NT_s/2)$, (ii) $t = -NT_s/2$ and $t = NT_s/2$, (iii) $t > NT_s/2$, and (iv) $t < -NT_s/2$.

If $t \in (-NT_s/2, NT_s/2)$, then we can split up the integrals in (6-128) as

$$\int_{t-NT_s/2}^{t+NT_s/2} f_i(\tau)d\tau = \int_{-\infty}^{+\infty} f_i(\tau)d\tau - \int_{-\infty}^{t-NT_s/2} f_i(\tau)d\tau - \int_{t+NT_s/2}^{+\infty} f_i(\tau)d\tau \quad (6\text{-}129)$$

for $i = 1, 2$. From Appendix 6.E it follows that

$$\int_{-\infty}^{+\infty} f_1(\tau)d\tau = 1 \qquad \int_{-\infty}^{+\infty} f_2(\tau)d\tau = 0 \quad (6\text{-}130)$$

and that for $N \to \infty$ and t fixed, or $t = \alpha NT_s/2$ with $\alpha \in (-1, 1)$,

$$\int_{-\infty}^{t-NT_s/2} f_i(\tau)d\tau = O(\frac{1}{t-NT_s/2}) \qquad \int_{t+NT_s/2}^{+\infty} f_i(\tau)d\tau = O(\frac{1}{t+NT_s/2}) \qquad (6\text{-}131)$$

The first integral in (6-130) is valid only if $x(t)$ (6-123) is band-limited, $\omega_k < \omega_s/2$ for $k = 1, 2, ..., F$, while the second integral follows from the fact that $f_2(\tau)$ is an odd function of τ. Collecting (6-127) to (6-131) gives, using $x(t) = O(N^0)$,

$$x_a(t) = O(\frac{1}{t+NT_s/2}) + O(\frac{1}{t-NT_s/2}) \text{ for } t = \alpha NT_s/2 \text{ with } \alpha \in (-1, 1) \qquad (6\text{-}132)$$

For $t = -NT_s/2$ and $NT_s/2$ the integrals in (6-128) are finite for any N, ∞ included. The same is true for $t = -NT_s/2 \pm \Delta t$ and $NT_s/2 \pm \Delta t$, with Δt independent of N. Together with $x(t) = O(N^0)$, it follows that

$$x_a(t) = O(N^0) \text{ for } t = -NT_s/2 \pm \Delta t \text{ and } NT_s/2 \pm \Delta t \qquad (6\text{-}133)$$

with Δt independent of N.

If $t > \alpha NT_s/2$ with $\alpha > 1$, then we can split up the integrals in (6-128) as

$$\int_{t-NT_s/2}^{t+NT_s/2} f_i(\tau)d\tau = \int_{t-NT_s/2}^{+\infty} f_i(\tau)d\tau - \int_{t+NT_s/2}^{+\infty} f_i(\tau)d\tau \qquad (6\text{-}134)$$

where, according to Appendix 6.E,

$$\int_{t-NT_s/2}^{+\infty} f_i(\tau)d\tau = O(\frac{1}{t-NT_s/2}) \qquad \int_{t+NT_s/2}^{+\infty} f_i(\tau)d\tau = O(\frac{1}{t+NT_s/2}) \qquad (6\text{-}135)$$

Collecting (6-127), (6-128), (6-134), and (6-135) gives, using $x(t) = O(N^0)$ and $x_N(t) = 0$ for $t > NT_s/2$,

$$x_a(t) = O(\frac{1}{t-NT_s/2}) + O(\frac{1}{t+NT_s/2}) \text{ for } t > \alpha NT_s/2 \text{ with } \alpha > 1 \qquad (6\text{-}136)$$

Following the same lines, we find for $t < -NT_s/2$,

$$x_a(t) = O(\frac{1}{t-NT_s/2}) + O(\frac{1}{t+NT_s/2}) \text{ for } t < -\alpha NT_s/2 \text{ with } \alpha > 1 \qquad (6\text{-}137)$$

Section 6.10 ■ Appendixes   215

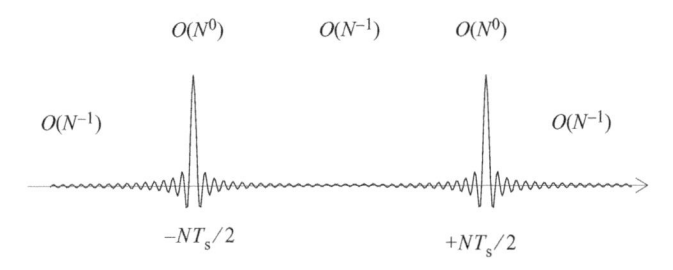

$O(N^{-1})$ $O(N^0)$ $O(N^{-1})$ $O(N^0)$ $O(N^{-1})$

$-NT_s/2$ $+NT_s/2$

Figure 6-10. Visualization of the high frequency content $x_a(t)$ of the observed signal $x_N(t)$.

From (6-132), (6-133), (6-136), and (6-137) we conclude that $x_a(t)$ tends to zero everywhere as $O(N^{-1})$, except in a close, N-independent, neighborhood of $t = -NT_s/2$ and $t = NT_s/2$ where it behaves as an $O(N^0)$. A graphical representation of $x_a(t)$ is shown in Figure 6-10. The ringing at the edges of the observation window are known as the Gibbs phenomenon. Note that the difference $x_N(t) - x_a(t)$ is band-limited.

Using (6-132), (6-133), (6-136), and (6-137), we can calculate the energy of $x_a(t)$. We find

$$\int_{-\infty}^{+\infty} x_a^2(t)\,dt = O(N^0) \tag{6-138}$$

From (6-123) it follows that

$$\int_{-\infty}^{+\infty} x_N^2(t)\,dt = \int_{-NT_s/2}^{+NT_s/2} x^2(t)\,dt = O(N) \tag{6-139}$$

Applying Parceval's theorem to (6-138) and (6-139), we get

$$\int_{-\infty}^{+\infty} x_a^2(t)\,dt = \int_{-\infty}^{+\infty} |X_a(j2\pi f)|^2\,df = 2\int_{f_s/2}^{+\infty} |X_a(j2\pi f)|^2\,df = O(N^0) \tag{6-140}$$

$$\int_{-f_s/2}^{+f_s/2} |X_N(j2\pi f)|^2\,df = \int_{-\infty}^{+\infty} |X_N(j2\pi f)|^2\,df - \int_{-\infty}^{+\infty} |X_a(j2\pi f)|^2\,df$$

$$= \int_{-\infty}^{+\infty} x_N^2(t)\,dt - \int_{-\infty}^{+\infty} x_a^2(t)\,dt \tag{6-141}$$

$$= O(N)$$

It follows that the ratio of the energy above Nyquist ($|f| > f_s/2$) to the energy below Nyquist ($|f| < f_s/2$) of $x_N(t)$ is an $O(N^{-1})$. By construction, the energy of the normalized periodic signal $x(t)$ is continuously spread over the $F = O(N)$ frequencies f_k (see Definition 3.4), so that the DFT spectrum $X(k)$ of $x_N(t)$ is an $O(N^0)$. As the energy of the pulse-like signal $x_a(t)$ is also continuously spread over the frequency, it follows directly that $\delta(s_k) = O(N^{-1/2})$.

Note that formulas (6-138) to (6-141) are also valid for periodic signals with a fixed number of frequencies $F = O(N^0)$ and fixed amplitudes $A_k/\sqrt{N} = O(N^0)$ in (6-123). The difference with the normalized periodic signals is that the signal energy is concentrated at a fixed number of frequencies f_k. Hence, at the excited frequencies f_k, $X(k)$ is an $O(N^{1/2})$, while $\delta(s_k)$ is still an $O(N^{-1/2})$. It shows that the relative convergence rate of $\delta(s_k)$ is an $O(N^{-1})$ at f_k.

6.F.2 Random Signals. The autocorrelation function $R_N(\tau, t)$ of the observed random signal $x_N(t)$ is related to the autocorrelation $R(\tau)$ of the complete signal $x(t)$ by

$$R_N(\tau, t) = \mathbb{E}\{x_N(t)x_N(t+\tau)\} = w_N(t)w_N(t+\tau)R(\tau) \tag{6-142}$$

Taking the Fourier transform of (6-142) w.r.t. τ gives the power spectrum

$$\begin{aligned} S_N(j\omega, t) &= w_N(t)[S(j\omega)*(W_N(j\omega)e^{j\omega t})] \\ &= w_N(t)\int_{-\infty}^{+\infty} S(j2\pi g)e^{j2\pi(f-g)t}W_N(j2\pi(f-g))dg \\ &= w_N(t)\int_{-f_B}^{+f_B} S(j2\pi g)e^{j2\pi(f-g)t}W_N(j2\pi(f-g))dg \end{aligned} \tag{6-143}$$

with $W_N(j\omega) = 2\omega^{-1}\sin(\omega NT_s/2)$ the spectrum of the window $w_N(t)$, and where the last equality is due to the fact that $x(t)$ is band-limited, $S(j\omega) = 0$ for $f > f_B$. Because $S(j2\pi g)/(f-g)$ is finite for any $f > f_s/2$ and $|g| \le f_B$, we can apply partial integration to (6-143). We find for $f > f_s/2$

$$\begin{aligned} S_N(j\omega, t) = &\frac{w_N(t)}{N}\left[\frac{S(j2\pi g)e^{j2\pi(f-g)t}}{f-g}\frac{\cos(\pi NT_s(f-g))}{\pi^2 T_s}\right]_{-f_B}^{+f_B} \\ &- \frac{w_N(t)}{N}\int_{-f_B}^{+f_B}\frac{\cos(\pi(f-g)NT_s)}{\pi^2 T_s}\frac{d}{dg}\left(\frac{S(j2\pi g)e^{j2\pi(f-g)t}}{f-g}\right)dg \end{aligned} \tag{6-144}$$

Clearly, the first term in the right-hand side of (6-144) is an $O(N^{-1})$. Because $S(j2\pi g)$ is differentiable for $|g| \le f_B$ and $f-g \ne 0$ for any $|f| > f_s/2$ and $|g| \le f_B$, the integral in (6-144) is finite for any N, ∞ included. Hence, the second term in the right-hand side of (6-144) is also an $O(N^{-1})$, so that $S_N(j\omega, t) = O(N^{-1})$ for $f > f_s/2$. This establishes the mean square convergence (see Chapter 16) of the signal energy above $f_s/2$ to zero (the power spectrum is a second-order moment). As $S_N(j\omega, t) = O(N^0)$ for $|f| \le f_B$ and the DFT spectrum $X(k)$ is an $O(N^0)$ for stationary random signals, we have $\delta(s_k) = O_{\text{m.s.}}(N^{-1/2})$.

Appendix 6.G Relation between DFT Spectra and Transfer Function for Arbitrary Signals with Missing Data

The DFT spectrum $X(k) = \text{DFT}(x(t))$ of the complete set (no missing data) can be split into the contributions of the known and the unknown samples

Section 6.10 ■ Appendixes

$$X(k) = X^m(k) + z_k^{-K_x} I_x(z_k^{-1}) \tag{6-145}$$

where $X^m(k) = \text{DFT}(x^m(t))$, $I_x(z_k^{-1}) = N^{-1/2} \sum_{t=0}^{M_x - 1} x(K_x + t) z^{-t}$, and $x^m(t)$ is defined in (6-43). Applying (6-145), with $X = U, Y$ and $x = u, y$, to (6-33) gives (6-44). □

Appendix 6.H Relationship between DFT Spectra of Concatenated Data Sets and Transfer Function

The proof will be given for the discrete-time case ($\Omega = z^{-1}$). That of the continuous-time case ($\Omega = s, \sqrt{s}$) is similar and the analogy with the discrete-time case is the same as in Appendix 6.B.

The Z-transforms $X_i(z^{-1})$, $i = 1, 2$ and $X = U, Y$, of the input-output signals of the two data sets $x_i(tT_s)$, $i = 1, 2$ and $t = 0, 1, ..., N_i - 1$, with $x = u, y$, satisfy, respectively, (see Appendix 6.B, (6-92))

$$A(z^{-1}) Y_1(z^{-1}) = B(z^{-1}) U_1(z^{-1}) + I_1(z^{-1}) - z^{-N_1} I_2(z^{-1}) \tag{6-146}$$

$$A(z^{-1}) Y_2(z^{-1}) = B(z^{-1}) U_2(z^{-1}) + I_3(z^{-1}) - z^{-N_2} I_4(z^{-1}) \tag{6-147}$$

where $I_i(z^{-1})$, $i = 1, 2, 3, 4$, are polynomials of order $n_i = \max(n_a, n_b) - 1$ that represent the initial or final state of the system at $t = 0$ (I_1, I_3), $t = N_1$ (I_2) and $t = N_2$ (I_4). Multiplying (6-147) by z^{-N_1} and adding the result to (6-146) gives

$$A(z^{-1}) Y^c(z^{-1}) = B(z^{-1}) U^c(z^{-1}) + I_1(z^{-1}) - z^{-N} I_4(z^{-1}) + z^{-N_1}(I_3(z^{-1}) - I_2(z^{-1})) \tag{6-148}$$

where $X^c(z^{-1}) = X_1(z^{-1}) + z^{-N_1} X_2(z^{-1})$, $X = U, Y$, is the Z-transform of $x^c(tT_s)$ defined in (6-47), with $x = u, y$. Evaluating (6-148) along the unit circle at the DFT frequencies $z_k = \exp(2\pi jk/N)$, taking into account that $z_k^{-N} = 1$, gives, after division by $\sqrt{N}$,

$$A(z_k^{-1}) Y^c(k) = B(z_k^{-1}) U^c(k) + I(z_k^{-1}) + z_k^{-N_1} I^c(z_k^{-1}) \tag{6-149}$$

with $X^c(k) = X^c(z_k^{-1}) / \sqrt{N}$, $X = U, Y$, the scaled DFT spectra of the concatenated data sets $x^c(nT_s)$, $n = 0, 1, ..., N$, $x = u, y$, and where $I(z_k^{-1}) = (I_1(z_k^{-1}) - I_4(z_k^{-1}))/\sqrt{N}$ and $I^c(z_k^{-1}) = (I_3(z_k^{-1}) - I_2(z_k^{-1}))/\sqrt{N}$ are polynomials in z_k^{-1} of order n_i. Equation (6-149) results in parametric model (6-48) with $\theta = [a_0 a_1 ... a_{n_a} b_0 b_1 ... b_{n_b} i_0 i_1 ... i_{n_i} i_{c0} i_{c1} ... i_{cn_i}]^T$. □

Appendix 6.I Free Decay Response of a Finite-Dimensional System

For discrete-time systems, (6-51) follows directly from (6-33) with $U(k) = 0$. For strictly proper lumped continuous-time systems we use (6-98) with $U_N(s) = 0$

$$Y_N(s) = \frac{I_1(s)}{A(s)} - e^{-NT_s s} \frac{I_2(s)}{A(s)} \tag{6-150}$$

and where the polynomials $I_1(s)$, $I_2(s)$ have order $n_a - 1$. Taking the one-sided Z-transform of the sampled inverse Laplace transform of (6-150) gives

$$Y_N(z) = T_1(z^{-1}) - z^{-N} T_2(z^{-1}) \tag{6-151}$$

with $Y_N(z) = \sum_{t=0}^{N-1} y(tT_s) z^{-t}$ and $T_1(z^{-1})$, $T_2(z^{-1})$ rational forms in z^{-1} of order $(n_a - 1)$ over n_a. The poles z_p of $T_1(z^{-1})$, $T_2(z^{-1})$ are related to the roots s_p of $A(s)$ by the impulse invariant transformation, $z_p = \exp(s_p T_s)$. Evaluating (6-151) at the DFT frequencies $z_k = e^{j2\pi k/N}$ with $z_k^{-N} = 1$ gives (6-51), after division by $\sqrt{N}$. □

Appendix 6.J Relation between the Free Decay Parameters and the Partial Fraction Expansion

We will prove (6-52) for the complex case (6-50). Using $\sum_{t=0}^{N-1} x^t = (1-x^N)/(1-x)$ and $z_k^N = 1$, the DFT transform $Y(k) = N^{-1/2} \sum_{t=0}^{N-1} y(tT_s) z_k^{-t}$ of $y(t)$, (6-50) becomes

$$Y(k) = \frac{1}{\sqrt{N}} \sum_{r=1}^{n_a} a_r e^{j\phi_r} \lambda_r^{\tau/T_s} \frac{1 - \lambda_r^N}{1 - \lambda_r z_k^{-1}} \tag{6-152}$$

where $\lambda_r = e^{(-d_r + j\omega_r) T_s}$. Comparing (6-152) with (6-37) where, $q = 0$, $\lambda_{-r} \neq \bar{\lambda}_r$, and $l_{-r} \neq \bar{l}_r$, gives (6-52). The proof of the real case follows the same lines (see Exercise 6.9). □

Appendix 6.K Some Properties of Polynomials

Lemma 6.19: Consider the polynomial $P(\Omega)$,

$$P(\Omega) = \left(\sum_{r=0}^{n_b} \beta_r \Omega^r\right)\left(\sum_{r=0}^{n_a} a_r \Omega^r\right) + \left(\sum_{r=0}^{n_a-1} \alpha_r \Omega^r\right)\left(\sum_{r=0}^{n_b} b_r \Omega^r\right) \tag{6-153}$$

with a_r, b_r fixed coefficients and α_r, β_r free parameters, and suppose that $P(\Omega) = 0$ must be fulfilled for any Ω. All the parameters α_r and β_r are zero if and only if the polynomials $A(\Omega, \theta)$ and $B(\Omega, \theta)$ have no common roots.

Proof. If the parameters α_r and β_r are not all zero, then we can rewrite $P(\Omega) = 0$ as

$$\frac{B(\Omega, \theta)}{A(\Omega, \theta)} = -\frac{\sum_{r=0}^{n_b} \beta_r \Omega^r}{\sum_{r=0}^{n_a-1} \alpha_r \Omega^r} \tag{6-154}$$

(if all α_r are zero in $P(\Omega) = 0$ then all β_r are zero and vice versa so that at least one α_r and one β_r should be different from zero).

If the polynomials $A(\Omega, \theta)$ and $B(\Omega, \theta)$ have *no common roots*, then $B(\Omega, \theta)/A(\Omega, \theta)$ has minimal order n_b over n_a. Equation (6-154) implies that $B(\Omega, \theta)/A(\Omega, \theta)$ can be written as a rational form of order n_b over $n_a - 1$, which is impossible. Hence, $P(\Omega) = 0$ can be true for any Ω only if the parameters α_r, β_r are all zero.

If the polynomials $A(\Omega, \theta)$ and $B(\Omega, \theta)$ have *common roots*, then the minimal order of $B(\Omega, \theta)/A(\Omega, \theta)$ is less than n_b over n_a, and (6-154) is fulfilled with α_r and β_r not all zero. □

Lemma 6.20: Consider the polynomial $P(\Omega)$ in (6-153) and suppose that $P(\Omega) = 0$ for at least $(n_a + n_b + 1)/2$ DFT frequencies Ω_k where DC ($k = 0$) and Nyquist ($k = N/2$) each count for $1/2$. The free parameters α_r, β_r are zero if and only if the polynomials $A(\Omega, \theta)$ and $B(\Omega, \theta)$ have no common roots.

Proof. The polynomial equation $P(\Omega) = 0$ is fulfilled for any Ω if and only if the coefficients of all the powers of Ω are zero. We will show that this is also true if $P(\Omega) = 0$ for at least $(n_a + n_b + 1)/2$ DFT frequencies. Applying Lemma 6.19 proves the lemma.

Evaluating $P(\Omega) = \sum_{r=0}^{n_p} p_r \Omega^r$, with $n_p = n_a + n_b$, at F DFT frequencies $k_j \in \{1, 2, ..., N/2 - 1\}$, $j = 1, 2, ..., F$, gives

$$V(\Omega_{k_1}, \Omega_{k_2}, ..., \Omega_{k_F})p = 0 \quad (6\text{-}155)$$

with $p^T = [p_0 p_1 ... p_{n_p}]$ and where the matrix $V(\Omega_{k_1}, \Omega_{k_2}, ..., \Omega_{k_F}) \in \mathbb{C}^{F \times (n_p + 1)}$ has a Vandermonde structure

$$V(\Omega_{k_1}, \Omega_{k_2}, ..., \Omega_{k_F}) = \begin{bmatrix} 1 & \Omega_{k_1} & \Omega_{k_1}^2 & ... & \Omega_{k_1}^{n_p} \\ 1 & \Omega_{k_2} & \Omega_{k_2}^2 & ... & \Omega_{k_2}^{n_p} \\ ... & ... & ... & ... & ... \\ 1 & \Omega_{k_F} & \Omega_{k_F}^2 & ... & \Omega_{k_F}^{n_p} \end{bmatrix} \quad (6\text{-}156)$$

The Vandermonde matrix (6-156) is of full rank if and only if the F DFT frequencies Ω_{k_j} are all different (see Golub and Van Loan, 1996 and Exercise 15.6). Adding the F complex conjugate DFT frequencies to (6-155) gives

$$V(\Omega_{k_1}, ..., \Omega_{k_F}, \Omega_{-k_1}, ..., \Omega_{-k_F})p = 0 \quad (6\text{-}157)$$

where $V(\Omega_{k_1}, ..., \Omega_{k_F}, \Omega_{-k_1}, ..., \Omega_{-k_F}) \in \mathbb{C}^{2F \times (n_p + 1)}$ is of full rank ($\Omega_{k_j} \neq \Omega_{-k_j}$). Hence from (6-157), it follows that $p = 0$ if and only if $F \geq (n_p + 1)/2$. The same reasoning holds if DC ($k = 0$) and Nyquist ($k = N/2$) are added to the frequencies. However, since Ω_0, $\Omega_{N/2}$ are real numbers, they increase the rank of $V(\Omega_{k_1}, ..., \Omega_{k_F}, \Omega_{-k_1}, ..., \Omega_{-k_F})$ by one instead of two as for each complex Ω_k. □

Appendix 6.L Proof of the Identifiability of Transfer Function Model (6-32) (Theorem 6.9)

Choosing $a_{n_a} = 1$, transfer function model (6-32) can be written as

$$\Omega_k^{n_a} Y(k) = \sum_{r=0}^{n_b} b_r f_r(\Omega_k) - \sum_{r=0}^{n_a - 1} a_r g_r(\Omega_k) \quad (6\text{-}158)$$

with $f_r(\Omega_k) = \Omega_k^r U(k)$ and $g_r(\Omega_k) = \Omega_k^r Y(k)$. The coefficients $a_0, a_1, \ldots, a_{n_a-1}, b_0, \ldots, b_{n_b}$ in (6-158) are identifiable if and only if the functions $f_r(\Omega_k)$, $r = 0, 1, \ldots, n_b$ and $g_r(\Omega_k)$, $r = 0, 1, \ldots, n_a - 1$ are linearly independent. This is the case if and only if

$$\sum_{r=0}^{n_b} \beta_r f_r(\Omega_k) + \sum_{r=0}^{n_a-1} \alpha_r g_r(\Omega_k) = 0 \qquad (6\text{-}159)$$

$k = 0, 1, \ldots, N/2$, implies that all parameters α_r, β_r are zero. Multiplying (6-159) by $A(\Omega_k, \theta)$ and using $Y(k) = G(\Omega_k, \theta)U(k)$ gives $P_1(\Omega_k)U(k) = 0$, $k = 0, 1, \ldots, N/2$, with

$$P_1(\Omega) = \left(\sum_{r=0}^{n_b} \beta_r \Omega^r\right)\left(\sum_{r=0}^{n_a} a_r \Omega^r\right) + \left(\sum_{r=0}^{n_a-1} \alpha_r \Omega^r\right)\left(\sum_{r=0}^{n_b} b_r \Omega^r\right) \qquad (6\text{-}160)$$

At the DFT frequencies where $U(k) \neq 0$, $P_1(\Omega_k)U(k) = 0$ is equivalent to $P_1(\Omega_k) = 0$. The free parameters α_r and β_r in $P_1(\Omega) = 0$ are zero if and only if $A(\Omega, \theta)$ and $B(\Omega, \theta)$ have no common roots and $U(k) \neq 0$ for at least $(n_a + n_b + 1)/2$ different DFT frequencies (proof: see Appendix 6.K). □

Appendix 6.M Proof of the Identifiability of Transfer Function Model (6-34)

Choosing $a_{n_a} = 1$, transfer function model (6-33) can be written as

$$\Omega_k^{n_a} Y(k) = \sum_{r=0}^{n_b} b_r f_r(\Omega_k) - \sum_{r=0}^{n_a-1} a_r g_r(\Omega_k) + \sum_{r=0}^{n_i} i_r \Omega_k^r + \Delta(\Omega_k) \qquad (6\text{-}161)$$

with $f_r(\Omega_k) = \Omega_k^r U(k)$, $g_r(\Omega_k) = \Omega_k^r Y(k)$ and with $\Delta(z_k^{-1}) = 0$ and $\Delta(\Omega_k)$ given by (6-101). A necessary condition for the identifiability of the coefficients $a_0, \ldots, a_{n_a-1}, b_0, \ldots, b_{n_b}, i_0, \ldots, i_{n_i}$ in (6-161) is that $f_r(\Omega_k)$ and Ω_k^r are linearly independent. If $f_r(\Omega_k)$ and Ω_k^r are linearly dependent, then there exist coefficients β_r, γ_r, not all zero, such that

$$\sum_{r=0}^{n_b} \beta_r f_r(\Omega_k) + \sum_{r=0}^{n_i} \gamma_r \Omega_k^r = 0 \qquad (6\text{-}162)$$

As the functions $f_r(\Omega_k)$ and Ω_k^r are themselves linearly independent (see Exercise 6.11) not all β_r and not all γ_r are zero. From (6-162) it follows, then, that

$$U(k) = -\left(\sum_{r=0}^{n_i} \gamma_r \Omega_k^r\right) / \left(\sum_{r=0}^{n_b} \beta_r \Omega_k^r\right) \qquad (6\text{-}163)$$

$k = 0, 1, \ldots, N/2$. We conclude that the functions $f_r(\Omega_k)$ and Ω_k^r are linearly dependent if and only if $U(k)$ can be written as a rational form of order n_i over n_b or less, otherwise they are linearly independent. If $U(k) \neq 0$ for less than $(n_b + n_i + 2)/2$ different DFT frequencies, then $U(k)$ can always be written as in the form (6-163) (a rational function of order n_i over n_b fits exactly $(n_b + n_i + 1)/2$ arbitrary complex numbers).

Transfer function model (6-34) can be written as

$$Y(k) = G(\Omega_k, \theta)U(k) + T_G(\Omega_k, \theta) + \delta(\Omega_k) \qquad (6\text{-}164)$$

Section 6.10 ■ Appendixes

with $\delta(z_k^{-1}) = 0$ and

$$\delta(s_k) = \frac{1}{T_s\sqrt{N}} \sum_{\substack{n=-\infty \\ n \neq 0}}^{n=+\infty} [G(s_k, \theta)U_N(s_k - nj\omega_s) - Y_N(s_k - nj\omega_s)] \qquad (6\text{-}165)$$

(see (6-101)). If the polynomials $A(\Omega, \theta)$, $B(\Omega, \theta)$, and $I(\Omega, \theta)$ have common roots, then the rational functions $G(\Omega, \theta)$ and $T_G(\Omega, \theta)$ can be simplified, leaving (6-164) unchanged. Clearly, the roots that have been removed are not identifiable. □

Appendix 6.N Rank of the Residue Matrices of Multivariable Transfer Function Models

The transfer function model of the left and right matrix fraction descriptions (6-54) and (6-55), and the state space models (6-26) and (6-27), depends on the inverse of a polynomial matrix. Let $A(s)$ be an $n \times n$ polynomial matrix with simple zeros (roots of $\det(A(s)) = 0$). The residue matrix L_0 of a simple pole s_0 of $A^{-1}(s)$ ($\det(A(s_0)) = 0$) is given by

$$L_0 = \lim_{s \to s_0} (s - s_0)A^{-1}(s) = \lim_{s \to s_0} (s - s_0)\frac{\text{adj}(A(s))}{\det(A(s))} = \text{adj}(A(s_0))\lim_{s \to s_0}\frac{(s - s_0)}{\det(A(s))} \qquad (6\text{-}166)$$

with $\text{adj}(A(s))$ the adjoint matrix of $A(s)$ (Gantmacher, 1990), and where the last limit in (6-166) is different from zero. Since $\text{adj}(A(s)) = \det(A(s))A^{-1}(s)$, $\det(\alpha A(s)) = \alpha^n\det(A(s))$, and $\det(A^{-1}(s)) = 1/\det(A(s))$, it follows that $\det(\text{adj}(A(s))) = (\det(A(s)))^{n-1}$. It shows that s_0 is a root of multiplicity $n-1$ of $\det(\text{adj}(A(s)))$ and, therefore, the rank of $\text{adj}(A(s_0))$ cannot be larger than 1 ($n-1$ eigenvalues of $\text{adj}(A(s_0))$ are zero). Finally, $\text{rank}(L_0) = \text{rank}(\text{adj}(A(s_0)))$, which concludes the proof. □

Appendix 6.O Band-Limited Observation of Continuous-Time Noise (Theorem 6.14)

The power spectral density of the stochastic process $v(t)$ at the output of the anti-alias filter $AA(j\omega)$ can be written as

$$S_{vv}(j\omega) = |H(j\omega)|^2|AA(j\omega)|^2 S_{e_c e_c}(j\omega) = |H(j\omega)|^2 S_{\varepsilon_c \varepsilon_c}(j\omega) \qquad (6\text{-}167)$$

Since $|AA(j\omega)| = 1$ for $|f| \leq f_s/2$ and zero elsewhere, $S_{\varepsilon_c \varepsilon_c}(j\omega) = |AA(j\omega)|^2 S_{e_c e_c}(j\omega)$ is constant for $|f| \leq f_s/2$ and zero elsewhere. Because filtered Gaussian noise remains normally distributed, it shows that $\varepsilon_c(t)$ is a band-limited white Gaussian noise process (6-60) with $f_B = f_s/2$. Taking the inverse Fourier transform of $S_{\varepsilon_c \varepsilon_c}(j\omega)$ gives the autocorrelation function $R_{\varepsilon_c \varepsilon_c}(\tau)$ of $\varepsilon_c(t)$

$$R_{\varepsilon_c \varepsilon_c}(\tau) = \int_{-f_s/2}^{+f_s/2} \frac{\sigma^2}{f_s} df = \sigma^2 \text{sinc}(\pi f_s \tau) \qquad (6\text{-}168)$$

with $\text{sinc}(x) = \sin(x)/x$ and where $\sigma^2 = \text{var}(\varepsilon_c(t))$. Since $R_{\varepsilon_c \varepsilon_c}(mT_s) = 0$ for $m \neq 0$, it follows that $e(m) = \varepsilon_c(mT_s)$ is uncorrelated over m. Finally, using the normality of $\varepsilon_c(t)$, it proves that $e(m)$ is independent, normally distributed, discrete-time white noise with variance σ^2. □

Appendix 6.P Correlation Noise Transient with Noise Input (Theorem 6.15)

Since the proof does not use the stability of the noise transfer function $H(\Omega)$ ($T_H(\Omega)$ and $H(\Omega)$ have the same poles), the result (6-67) is also valid for unstable systems captured in a stabilizing feedback loop.

6.P.1 Discrete-Time Systems (z-domain).

For discrete-time systems $H(z^{-1})$ of order n_c over n_d, the coefficients of the numerator polynomial of the rational form $T_H(\Omega)$ depend on $(e(N-r) - e(-r))/\sqrt{N}$, $r = 1, 2, ..., n_c$ and $(n_y(N-r) - n_y(-r))/\sqrt{N}$, $r = 1, 2, ..., n_d$ (see Appendix 6.B, Section 6.B.1). Hence, in $\mathbb{E}\{T_H(z_k^{-1})\bar{E}(k)\}$ terms of the form

$$\frac{1}{\sqrt{N}}\mathbb{E}\{(e(N-r) - e(-r))\bar{E}(k)\} = \frac{1}{N}\sum_{t=0}^{N-1} \mathbb{E}\{(e(N-r) - e(-r))e(t)\}e^{j2\pi kt/N}$$

$$\frac{1}{\sqrt{N}}\mathbb{E}\{(n_y(N-r) - n_y(-r))\bar{E}(k)\} = \frac{1}{N}\sum_{t=0}^{N-1} \mathbb{E}\{(n_y(N-r) - n_y(-r))e(t)\}e^{j2\pi kt/N}$$
(6-169)

should be calculated. Since $e(t)$ is discrete-time white noise it follows immediately that

$$\sum_{t=0}^{N-1} \mathbb{E}\{(e(N-r) - e(-r))e(t)\}e^{j2\pi kt/N} = \sigma^2 e^{j2\pi k(N-r)/N} = O(N^0) \quad (6-170)$$

In the sequel we study the terms depending on the cross-correlation between $n_y(t)$ and $e(t)$. The cross-power spectrum $S_{n_y e}(j\omega)$ between $n_y(t)$ and $e(t)$ is given by $S_{n_y e}(j\omega) = \sigma^2 H(e^{-j\omega T_s})$, with $\sigma^2 = \text{var}(e(t))$. As a consequence, the cross-correlation $R_{n_y e}(\tau)$, which is the inverse Fourier transform of $S_{n_y e}(j\omega)$, consists of the sum of damped (complex) exponentials: the poles of $H(z^{-1})$ inside the unit circle define $R_{n_y e}(\tau)$ for $\tau \geq 0$, and those outside the unit circle define $R_{n_y e}(\tau)$ for $\tau < 0$ (Kwakernaak and Sivan, 1991). Hence, $R_{n_y e}(\tau)$ can be bounded above as $|R_{n_y e}(\tau)| \leq \beta e^{-\alpha \tau}$, with $\alpha, \beta > 0$, and

$$\left|\sum_{t=0}^{N-1} \mathbb{E}\{n_y(n)e(n-t)\}e^{j2\pi kt/N}\right| \leq \beta \sum_{t=0}^{N-1} e^{-\alpha t} = O(N^0) \quad (6-171)$$

From (6-170) and (6-171) we conclude that (6-169) is an $O(N^{-1})$, which proves the first equality of (6-67).

6.P.2 Lumped Continuous-Time Systems (s-domain).

For continuous-time systems $H(s)$ of order n_c over n_d, the coefficients of the numerator polynomial of the rational form $T_H(s)$ depend on $(\varepsilon_c^{(r)}(NT_s) - \varepsilon_c^{(r)}(0))/\sqrt{N}$, $r = 1, 2, ..., n_c$ and $(v^{(r)}(NT_s) - v^{(r)}(0))/\sqrt{N}$, $r = 1, 2, ..., n_d$ (apply the results of Appendix 6.B, Section 6.B.2, to (6-64)). Hence, in $\mathbb{E}\{T_H(s_k)\bar{E}(k)\}$ terms of the form

Section 6.10 ■ Appendixes

$$\frac{1}{\sqrt{N}}\mathbb{E}\{(\varepsilon_c^{(r)}(NT_s) - \varepsilon_c^{(r)}(0))\bar{E}(k)\} = \frac{1}{N}\sum_{n=0}^{N-1}\mathbb{E}\{(\varepsilon_c^{(r)}(NT_s) - \varepsilon_c^{(r)}(0))e(n)\}z_k^n$$
$$\frac{1}{\sqrt{N}}\mathbb{E}\{(v^{(r)}(NT_s) - v^{(r)}(0))\bar{E}(k)\} = \frac{1}{N}\sum_{n=0}^{N-1}\mathbb{E}\{(v^{(r)}(NT_s) - v^{(r)}(0))e(n)\}z_k^n$$
(6-172)

with $e(n) = \varepsilon_c(nT_s)$ and $z_k = e^{j2\pi k/N}$, should be calculated. We analyze the cross-correlation $R_{v^{(r)}\varepsilon_c}(\tau)$ between the r th derivative of $v(t)$ and the input $\varepsilon_c(t)$ (the analysis of $R_{\varepsilon_c^{(r)}\varepsilon_c}(\tau)$ follows exactly the same lines). It is calculated via the inverse Fourier transform of the cross-power spectral density $S_{v^{(r)}\varepsilon_c}(j\omega)$

$$S_{v^{(r)}\varepsilon_c}(j\omega) = (j\omega)^r H(j\omega) S_{\varepsilon_c\varepsilon_c}(j\omega) \qquad (6-173)$$

with $S_{\varepsilon_c\varepsilon_c}(j\omega)$ the power spectral density of the band-limited white noise source $\varepsilon_c(t)$ with variance σ^2 (see Definition 6.12 with $f_B = f_s/2$). We find

$$R_{v^{(r)}\varepsilon_c}(\tau) = F^{-1}\{(j\omega)^r H(j\omega)\} * F^{-1}\{S_{\varepsilon_c\varepsilon_c}(j\omega)\}$$
$$= \int_{-\infty}^{+\infty} R(t)\sigma^2 \text{sinc}(\pi f_s(\tau - t))dt \qquad (6-174)$$

with $F^{-1}\{.\}$ the inverse Fourier transform, $\text{sinc}(x) = \sin(x)/x$, and where $R(t) = F^{-1}\{(j\omega)^r G(j\omega)\}$ is a sum of damped (complex) exponentials: the poles of $s^r G(s)$ in the left half plane define $R(t)$ for $t \geq 0$, and those in the right half plane define $R(t)$ for $t < 0$ (Papoulis, 1981). Using (6-174) we get for a general term of (6-172)

$$\frac{1}{N}\sum_{n=0}^{N-1}\mathbb{E}\{v^{(r)}(NT_s)e(n)\}e^{j2\pi kn/N} = \frac{1}{N}\int_{-\infty}^{+\infty} R(t)\sigma^2 S(N-n, t)dt \qquad (6-175)$$

where

$$S(N-n, t) = \sum_{n=0}^{N-1}\text{sinc}(\pi f_s((N-n)T_s - t))e^{j2\pi kn/N} = O(\log(N)) \qquad (6-176)$$

Hence, (6-175) can be bounded above by

$$\left|\frac{1}{N}\int_{-\infty}^{+\infty} R(t)\sigma^2 S(N-n, t)dt\right| \leq O(\log(N)/N)\int_{-\infty}^{+\infty}|R(t)|dt = O(\log(N)/N)O(1) \qquad (6-177)$$

where the last equality uses the fact that $R(t)$ is a sum of damped (complex) exponentials. Collecting (6-172), (6-175), and (6-177) proves the second equality of (6-67) for $\Omega = s$.

Remark If the perfect anti-alias filter is replaced by a real lowpass filter such that $S_{\varepsilon_c\varepsilon_c}(j\omega)$ is a rational function of $j\omega$ then $R_{v^{(r)}\varepsilon_c}(\tau)$ (6-174) is a sum of damped (complex) exponentials, and

$$\sum_{n=0}^{N-1}\mathbb{E}\{v^{(r)}(NT_s)e(n)\}e^{j2\pi kn/N} = \sum_{n=0}^{N-1}R_{v^{(r)}\varepsilon_c}((N-n)T_s)e^{j2\pi kn/N} = O(N^0) \qquad (6-178)$$

(proof: apply $\sum_{n=0}^{N-1} x^n = (1-x^N)/(1-x)$ to the sum of complex exponentials). Combining (6-172) with (6-178) shows that $\mathbb{E}\{T_H(s_k)\bar{E}(k)\} = O(1/N)$.

6.P.3 Diffusion Phenomena ($\sqrt{s}$- domain). The only difference with the lumped continuous-time case is that $R(t)$ in (6-177) is a sum of $O(t^{-3/2})$ terms instead of a sum of damped (complex) exponentials (see Appendix 6.A, Lemma 6.18). Results (6-177) and (6-178) remain valid because $\int_1^{+\infty} O(t^{-3/2}) dt = O(1)$.

Appendix 6.Q Correlation Noise Alias Error with Noise Input (Theorem 6.16)

In this appendix we study the asymptotic ($N \to \infty$) behavior of $\mathbb{E}\{\delta(s_k)\bar{E}(k)\}$, with $\delta(s_k) = (\Delta(s_k)/A(s_k))$, and where $\Delta(s_k)$ is defined in (6-101). In $\mathbb{E}\{\Delta(s_k)\bar{E}(k)\}$, terms of the form

$$\frac{1}{\sqrt{N}} \mathbb{E}\{E_N(s_k - jm\omega_s)\bar{E}(k)\} = \frac{1}{N} \int_0^{NT_s} e^{-(s_k - jm\omega_s)t} \sum_{n=0}^{N-1} R_{\varepsilon_c \varepsilon_c}(t - nT_s) z_k^n dt$$

$$\frac{1}{\sqrt{N}} \mathbb{E}\{V_N(s_k - jm\omega_s)\bar{E}(k)\} = \frac{1}{N} \int_0^{NT_s} e^{-(s_k - jm\omega_s)t} \sum_{n=0}^{N-1} R_{v\varepsilon_c}(t - nT_s) z_k^n dt$$

(6-179)

with $m \neq 0$, $e(n) = \varepsilon_c(nT_s)$, $s_k = j\omega_k$ and $z_k = e^{j2\pi k/N}$, should be calculated. Since $S_{v\varepsilon_c}(j\omega) = H(j\omega) S_{\varepsilon_c \varepsilon_c}(j\omega)$, where $S_{\varepsilon_c \varepsilon_c}(j\omega)$ is by assumption a rational function of $j\omega$, $R_{v\varepsilon_c} = F^{-1}\{S_{v\varepsilon_c}(j\omega)\}$ is a sum of damped (complex) exponentials $\alpha e^{\beta|t|}$ with $\text{Re}(\beta) < 0$. It can be verified that the sum $\sum_{n=0}^{N-1} R_{v\varepsilon_c}(t - nT_s) z_k^n$ consists of terms of the from

$$e^{\beta t}, e^{\pm \beta(t - MT_s)}, e^{\beta(NT_s - t)}$$

(6-180)

with $(M-1)T_s < t \leq MT_s$, and where $M \in \mathbb{N}$ depends on t (proof: apply $\sum_{n=0}^{N-1} x^n = (1-x^N)/(1-x)$ to the sum of complex exponentials). Since

$$\int_0^{NT_s} e^{-(s_k - jm\omega_s)t} g(t) dt = O(N^0) \text{ for } g(t) = e^{\beta t}, e^{\pm \beta(t - MT_s)}, e^{\beta(NT_s - t)} \text{ and } m \neq 0 \quad (6\text{-}181)$$

it follows that the integrals in (6-179) are an $O(N^0)$, which proves the theorem. □

Appendix 6.R Correlation Plant Transient and Plant Alias Error with Plant Input (Theorem 6.17)

Since $S_{uu}(j\omega)$ is, by assumption, a rational function of Ω_k, all cross-power spectra, for example, $S_{yu}(j\omega) = G(\Omega) S_{uu}(j\omega)$, are also a rational functions of Ω. Hence, the corresponding cross-correlation functions, for example, $R_{yu}(\tau) = F^{-1}\{S_{yu}(j\omega)\}$, are a sum of damped (complex) exponentials (s-, z- domains), or a sum of $O(t^{-3/2})$ terms, which is the key property needed for proving Theorems 6.15 and 6.16 (see Appendices 6.P and 6.Q). □

7

Measurement of Frequency Response Functions – The Local Polynomial Approach

Abstract: The classical methods for measuring the frequency response function (see Chapter 2) suffer from a number of shortcomings: the bias and/or variance caused by the residual plant and/or noise leakage errors, the frequency resolution is much smaller than the inverse of the experiment duration, and separate data records (random excitations) or transient data (periodic excitations) cannot be handled. This chapter describes advanced methods for measuring the frequency response function of multivariable n_u input, n_y output systems using arbitrary and periodic excitations. They reduce the aforementioned problems by exploiting explicitly the smooth behavior of the frequency response function and the leakage (transient) errors as a function of the frequency. The key idea is to approximate locally the frequency response function and/or the transient errors by a low degree polynomial.

7.1 INTRODUCTION

The major problem in estimating the nonparametric frequency response function (FRF) and noise covariance matrix using *arbitrary* excitations is the suppression of the system and noise leakage errors. These errors are introduced when transforming a finite number N of time domain samples to the frequency domain via the discrete Fourier transform (DFT). Spectral analysis methods handle this problem via time domain windowing (see Section 2.6). To reduce the noise on the estimates, the record of N samples is divided in M subrecords of length N/M, which decreases the frequency resolution from f_s/N to Mf_s/N, and the results are averaged over the M subrecords. Hence, choosing M is making a trade-off between, on the one hand, the leakage elimination and the frequency resolution (the larger M, the larger the leakage errors and the smaller the frequency resolution), and, on the other hand, the noise suppression (the variance of the estimates decreases by M). Another shortcoming of the classical spectral analysis method is that it cannot handle the concatenation of separate data records.

In Section 7.2 of this chapter we present a method for estimating the frequency response matrix (FRM) and its covariance at full frequency resolution f_s/N from an experiment with *random* excitations. The basic assumption made is that the plant and noise transfer functions (and hence also the plant and noise transient terms) are smooth functions of the frequency that can locally be approximated by a low degree polynomial. This so-called local polynomial method suppresses much better the leakage (transient) errors, while maintaining a useful noise averaging effect that is at least as good as that of the spectral analysis methods. Hence, even if the noise errors dominate the leakage errors, the local polynomial method is preferred because the FRF is estimated at the full frequency resolution of the experiment (= the inverse of the experiment duration). Moreover, it can also handle concatenated data records (Section 7.2.9). The major disadvantage of stationary random excitations is that no distinction can be made between noise and nonlinear distortions.

Via steady state experiments with *periodic* inputs one can easily separate the noise from the nonlinear distortions in FRM estimates (see Chapter 4). Moreover, uncorrelated input signals can be constructed for finite samples (see Sections 2.7 and 3.7). The major disadvantages are the reduced frequency resolution (much smaller than the inverse of the total experiment duration), and the fact that no transient data can be handled. Section 7.3 of this chapter presents two methods for increasing the frequency resolution (or decreasing the measurement time) of FRM estimates using periodic excitations, while keeping the ability to separate the noise contribution from the nonlinear distortion. The first (= "robust") method requires multiple (at least n_u) experiments with uncorrelated sets of inputs, and puts no condition on the FRM; while the second (= "fast") method starts from one single experiment, but assumes that the FRM can locally be approximated by a polynomial.

Due to the non-periodic nature of noise, the steady state response of a dynamic system to a *periodic* input is still subject to noise transients (noise leakage errors). For lightly damped systems these noise transients (significantly) increase the variance of FRF estimates. Section 7.3 of this chapter presents a technique for suppressing the noise transients in FRM estimates using periodic excitations. It is based on a local polynomial approximation of the noise leakage error. Irrespective of the number of inputs n_u and outputs n_y, it is shown in Section 7.3.2 that two periods of the steady state response are enough to suppress the noise transients and to estimate the input-output noise covariance matrix. Since no distinction can be made between the system and noise transients, the presented method is also applicable to the first two periods of the transient response of the system to a periodic input (see Section 7.3.8). For lightly damped systems this results in a significant reduction of the experimental time.

To separate the noise from the nonlinear distortions in the FRM estimate of a nonlinear system, the "robust" method needs a number of FRM estimates with independent random realizations of Gaussian-like input signals (see Chapter 4). Irrespective of the number of inputs n_u and outputs n_y, it is shown in Section 7.3.6 that two independent realizations are enough. For lightly damped systems this results again in a significant reduction of the experimental time.

The "fast" method estimates the FRM, the noise level, and the level of the nonlinear distortions from two consecutive periods of the transient response to one set of uncorrelated periodic input signals that – for all inputs – excites all harmonics in the frequency band of interest (see Section 7.3.7). Hence, irrespective of the number of inputs n_u and outputs n_y, the loss in frequency resolution (or increase in measurement time) of the "fast" method w.r.t. stationary random excitations is a factor 2. Compared with the "zippered multisine" approach (see Section 2.7.1), the "fast" method does not suffer from the (non)linear spectral distortions introduced by an actuator or a feedback loop, and has a larger frequency resolution (at least a factor n_u). Similarly to the random input case (see Section 7.2.2), the "fast" method for periodic signals requires a local polynomial approximation of the FRM. This is the price to be paid for relaxing the experimental conditions w.r.t. the "zippered multisine" approach (all

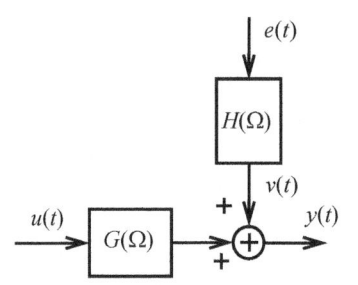

Figure 7-1. Linear dynamic system with known input $u(t)$ and noisy output $y(t)$. The output noise $v(t)$ is written as filtered (band-limited) white noise.

instead of 1 out of n_u harmonics are excited) and the "robust" method (1 instead of n_u experiments).

7.2 ARBITRARY EXCITATIONS

7.2.1 Problem Statement and Assumptions

Consider the linear dynamic multivariable system of Figure 7-1 with n_u inputs and n_y outputs. The arbitrary (random) excitation $u(t)$ is assumed to be known and the output $y(t)$ is disturbed by filtered (band-limited) white noise $v(t)$ (= generalized output error framework, see Section 6.7.3 on page 195). The input-output discrete Fourier transform (DFT) spectra $U(k)$, $Y(k)$ of N samples of the input-output signals $u(t)$, $y(t)$ are related as

$$Y(k) = G(\Omega_k)U(k) + T_G(\Omega_k) + H(\Omega_k)E(k) + T_H(\Omega_k) \tag{7-1}$$

where $G(\Omega)$ and $H(\Omega)$ are, respectively, the $n_y \times n_u$ system and $n_y \times n_y$ noise rational transfer function matrices, with $T_G(\Omega)$ and $T_H(\Omega)$ rational $n_y \times 1$ vector functions that have the same poles as, respectively, $G(\Omega)$ and $H(\Omega)$ (see (6-69) on page 200). $T_G(\Omega)$ and $T_H(\Omega)$ are, respectively, the system and noise transient (leakage) terms. The goal is to obtain a nonparmetric estimate of the noise covariance matrix

$$C_V(k) = \text{Cov}(V(k)) \text{ with } V(k) = H(\Omega_k)E(k) \tag{7-2}$$

and the frequency response matrix (FRM) $G(\Omega_k)$ starting from the measured input-output DFT spectra $U(k)$, $Y(k)$ in the frequency band of interest. The two difficulties in obtaining accurate estimates of $C_V(k)$ are the presence of the exogenous part $G(\Omega_k)U(k)$ and the system and noise leakage terms $T_G(\Omega_k)$ and $T_H(\Omega_k)$ in (7-1). While the dominant term $G(\Omega_k)U(k)$ can easily be eliminated via a nonparametric estimate of the FRM, the major problem is the suppression of the leakage terms. The latter is also valid for the FRM estimate.

Assuming that the unobserved (band-limited) white noise source $e(t)$ is independent and identically distributed (iid) at the sampling instances, the DFT spectrum $E(k)$ of $e(t)$, $k = 1, 2, ..., N/2 - 1$ (DC and Nyquist are excluded), has the following properties (see Section 16.16 on page 601):

(i) $E(k)$ is uncorrelated (over k), circular complex distributed ($\mathbb{E}\{E^2(k)\} = 0$) for any N if $e(t)$ has finite second order moments.

(ii) $E(k)$ is asymptotically (for $N \rightarrow \infty$) independent (over k) circular complex normally distributed if $e(t)$ has existing moments of any order.

(iii) $E(k)$ is independent (over k), circular complex normally distributed for any N if $e(t)$ is Gaussian.

While the first property is sufficient to estimate the noise covariance matrix (see Section 7.2.2.2), the second property is needed to calculate uncertainty bounds on the measured frequency response function with a given confidence level (see Section 7.2.4). The last property is needed if the estimated noise covariance matrix is used as weighting for the parametric identification of the system transfer function (see Chapter 12).

Although the local polynomial approach presented in Section 7.2.2.2 can be applied to both random and deterministic (periodic and non-periodic) input signals, its bias and covariance matrix will be studied assuming that the input is a stationary stochastic process that can be modeled as filtered white iid noise with existing moments of any order. Hence, similarly to the disturbing noise, the input DFT spectrum is asymptotically (for $N \rightarrow \infty$) independent (over k) circular complex normally distributed. In addition, it is assumed that the input $u(t)$ is independent of the disturbing noise $v(t)$, which implies that the system operates in open loop. If a known reference signal is available, then these two assumptions can be relaxed, and the noisy input case and systems operating in feedback can also be handled (see Section 7.2.7).

7.2.2 The Local Polynomial Method

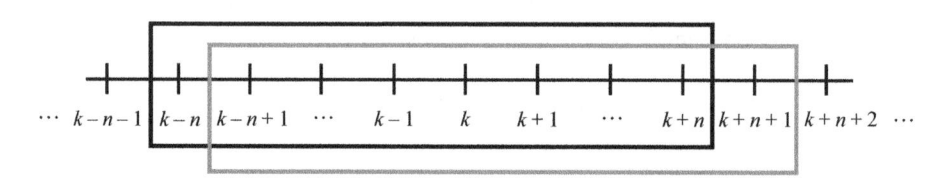

Figure 7-2. The $2n+1$ DFT frequencies used by the local polynomial method at frequency k (black rectangle) and frequency $k+1$ (gray rectangle).

7.2.2.1 The Basic Idea. Since $G(\Omega)$ and $T(\Omega) = T_G(\Omega) + T_H(\Omega)$ are all smooth functions of the frequency, they can be approximated in the band $[k-n, k+n]$ (see Figure 7-2, black rectangle) by a polynomial. This is done via a Taylor series expansion of $G(\Omega_{k \pm r})$ and $T(\Omega_{k \pm r})$ at Ω_k for $r = 0, 1, ..., n$. The coefficients of these expansions can be estimated from the data via a linear least squares fit, provided that the exogeneous term $G(\Omega_k)U(k)$ in (7-1) can be distinguished from the transient term $T(\Omega_k)$. Since $T(\Omega)$ and $G(\Omega)$ are smooth functions of the frequency it requires that the input $U(k)$ varies sufficiently "wild" over the frequency. This condition is fulfilled for filtered white noise and random phase multisine excitations. The result is an estimate of the frequency response matrix $\hat{G}(\Omega_k)$ and the transient term $\hat{T}(\Omega_k)$ at DFT frequency k. From the residuals of the linear least squares fit we also obtain an estimate $\hat{C}_V(k)$ of the noise covariance matrix $C_V(k) = \text{Cov}(H(\Omega_k)E(k))$. Next, the DFT frequencies are shifted over one DFT bin and the same procedure is repeated in the band $[k-n+1, k+n+1]$ (see Figure 7-2, gray rectangle), giving an estimate of the frequency response matrix $\hat{G}(\Omega_{k+1})$, the transient term $\hat{T}(\Omega_{k+1})$, and the noise covariance matrix $\hat{C}_V(k+1)$ at DFT frequency $k+1$. Proceeding in this way estimates of the frequency response matrix, the transient term, and the noise covariance matrix are obtained at all frequencies in the frequency band of interest. Since the consecutive frequency bands used for the linear least squares fits are overlapping (see Figure 7-2, black and gray rectangles), the estimates at consecutive DFT frequencies are correlated. From Figure 7-2 it can easily be seen that the correlation length of the estimates equals $\pm 2n$.

7.2.2.2 The Basic Algorithm.

First the output DFT spectrum at DFT line $k+r$ is rewritten as

$$Y(k+r) = G(\Omega_{k+r})U(k+r) + T(\Omega_{k+r}) + V(k+r) \qquad (7\text{-}3)$$

where $V(k)$ is defined in (7-2) and $T(\Omega) = T_G(\Omega) + T_H(\Omega)$. For continuous-time systems, where $n_i \geq \max(n_a, n_b) - 1$ and $n_j \geq \max(n_c, n_d) - 1$, $T(\Omega)$ represents the sum of the transient (leakage) term and the residual alias term. This is permitted since the transient and the residual alias terms have exactly the same properties (see Lemmas 6.5 and 6.6 and Theorems 6.15 and 6.16). Next, since $G(\Omega)$ and $T(\Omega)$ are, respectively, rational matrix and rational vector functions, they have continuous derivatives up to any order. Hence, $G(\Omega_{k+r})$ and $T(\Omega_{k+r})$ can be expanded at Ω_k as

$$\begin{aligned}
G(\Omega_{k+r}) &= G(\Omega_k) + \sum_{s=1}^{R} g_s(k) r^s + O((r/N)^{R+1}) \\
T(\Omega_{k+r}) &= T(\Omega_k) + \sum_{s=1}^{R} t_s(k) r^s + N^{-1/2} O((r/N)^{R+1})
\end{aligned} \qquad (7\text{-}4)$$

(proof: apply Taylor's formula with remainder, see Theorem 41 of Kaplan, 1993), where $O(N^{-(R+1)})$ in the remainders stems from $f_{k+r} - f_k = rf_s/N$, and where the additional factor $N^{-1/2}$ in the remainder of $T(\Omega_{k+r})$ is due to the fact that $T_G(\Omega)$ and $T_H(\Omega)$ are both an $O(N^{-1/2})$ (see Section 6.3.2 on page 185 and Section 6.7.3 on page 195). One could think of using different orders for the Taylor series expansion of the system transfer function $G(\Omega)$ and the leakage term $T(\Omega)$. However, it turns out that if an Rth order polynomial approximation is suitable for $G(\Omega)$, it is also appropriate for $T(\Omega)$. The basic reason for this is that $G(\Omega)$ and $T_G(\Omega)$ have exactly the same poles, and that the system leakage term $T_G(\Omega)$ mostly dominates over the noise leakage term $T_H(\Omega)$. The remainders in the polynomial expansions (7-4) of $G(\Omega_{k+r})$ and $T(\Omega_{k+r})$ are, respectively, the system interpolation error and the sum of the residual system and noise leakage errors.

Assuming further that the remainders in (7-4) can be neglected, (7-3) becomes

$$\begin{aligned}
Y(k+r) &= (G(\Omega_k) + \sum_{s=1}^{R} g_s(k) r^s) U(k+r) + T(\Omega_k) + \sum_{s=1}^{R} t_s(k) r^s + V(k+r) \\
&= \Theta K(k+r) + V(k+r)
\end{aligned} \qquad (7\text{-}5)$$

where Θ is the $n_y \times (R+1)(n_u+1)$ matrix of the unknown complex parameters

$$\Theta = \begin{bmatrix} G(\Omega_k) & g_1(k) & g_2(k) & \ldots & g_R(k) & T(\Omega_k) & t_1(k) & t_2(k) & \ldots & t_R(k) \end{bmatrix} \qquad (7\text{-}6)$$

and $K(k+r)$ is the $(R+1)(n_u+1) \times 1$ input data vector

$$K(k+r) = \begin{bmatrix} K_1(r) \otimes U(k+r) \\ K_1(r) \end{bmatrix} \text{ with } K_1(r) = \begin{bmatrix} 1 \\ r \\ \ldots \\ r^R \end{bmatrix} \qquad (7\text{-}7)$$

with $\otimes$ the Kronecker product ($A \otimes B$ is the matrix with block elements $A_{[i,j]}B$, see Section 15.7 on page 552). Collecting (7-5) for $r = -n, -n+1, \ldots, 0, \ldots, n$ finally gives

$$Y_n = \Theta K_n + V_n \tag{7-8}$$

where Y_n, K_n, and V_n are, respectively, $n_y \times (2n+1)$, $(R+1)(n_u+1) \times (2n+1)$, and $n_y \times (2n+1)$ matrices

$$X_n = [\ X(k-n)\ X(k-n+1)\ \ldots\ X(k)\ \ldots\ X(k+n)\ ] \tag{7-9}$$

with $X = Y$, K, and V.

Eq. (7-8) is an overdetermined set of equations in the unknown Θ that can be solved in least squares sense as

$$\hat{\Theta} = Y_n K_n^H (K_n K_n^H)^{-1} \tag{7-10}$$

if $2n + 1 \geq (R+1)(n_u+1)$, where x^H is the Hermitian (complex conjugate) transpose of x (proof: see Appendix 7.A). The residual of the least squares fit,

$$\hat{V}_n = Y_n - \hat{\Theta} K_n = Y_n P_n, \text{ where } P_n = I_{2n+1} - K_n^H (K_n K_n^H)^{-1} K_n \tag{7-11}$$

is an idempotent matrix (see Section 15.6), is related to the noise V_n as

$$\hat{V}_n = V_n P_n \tag{7-12}$$

Hence, an estimate of the noise covariance matrix (7-2) is given by

$$\hat{C}_V(k) = \frac{1}{q} \hat{V}_n \hat{V}_n^H \text{ with } q = 2n + 1 - (R+1)(n_u+1) \tag{7-13}$$

and where $q = \text{rank}(P_n)$ is the number of the degrees of freedom (*dof*) of the residual $\hat{V}_n$ (proof: see Appendix 7.B).

Since (7-13) implicitly assumes that the true noise covariance matrix $C_V(k+r)$ is equal to $C_V(k)$ for $r = -n, -n+1, \ldots, 0, \ldots, n$, the value of n should be chosen as small as possible. The minimal values of n are such that the number of degrees of freedom $dof_{\text{arb}} = q$ in the residuals (7-13) satisfies

$$q = dof_{\text{arb}} \geq n_y + m \tag{7-14}$$

where

$$\begin{array}{lll} m = 0 & \text{rank}(\hat{C}_V(k)) = \text{rank}(C_V(k)) & \\ m = 1 & \mathbb{E}\{\hat{C}_V^{-1}(k)\} \text{ exists} & (7\text{-}15) \\ m = 2 & \mathbb{E}\{\hat{C}_V^{-2}(k)\} \text{ exists} & \end{array}$$

Section 7.2 ■ Arbitrary Excitations

(see Mahata et al., 2006). The value $m = 0$ is sufficient for calculating uncertainty bounds on the FRF, while values $m > 0$ are needed when $\hat{C}_V(k)$ is used as weighting for the parametric identification of the system transfer function (see Chapter 12). Note that $\text{rank}(\hat{C}_V(k)) < n_y$ for $m = -n_y + 1, -n_y + 2, \ldots, -1$.

Using (7-10), the local polynomial estimate of the frequency response function is obtained as

$$\hat{G}(\Omega_k) = \hat{\Theta} \begin{bmatrix} I_{n_u} \\ 0 \end{bmatrix} = \hat{\Theta}_{[:,1:n_u]} \quad (7\text{-}16)$$

where $X_{[:,1:n_u]}$ selects the first n_u columns of the matrix X.

Since DFT lines $k + r$, $r = -n, -n + 1, \ldots, n$, are used for estimating the noise covariance matrix (7-13) and the FRM (7-16), it follows that $\hat{C}_V(k)$ and $\hat{G}(\Omega_k)$ are correlated with, respectively, $\hat{C}_V(k + r)$ and $\hat{G}(\Omega_{k+r})$, $r = -2n, -2n + 1, \ldots, 2n$. It explains why more parameters Θ ($n_y \times (R+1)(n_u+1) \times F$: see (7-6) with $k = 1, 2, \ldots, F$) can be identified than the number of measurements ($n_y \times F$: see (7-1) with $k = 1, 2, \ldots, F$).

7.2.2.3 Bias Error. The bias error of the noise covariance estimate (7-13) consists of four contributions: the zero order interpolation error $O_{\text{int}H}$ of the noise covariance matrix over $2n + 1$ frequencies, the polynomial interpolation error $O_{\text{int}G}$ of the frequency response matrix over $2n + 1$ frequencies, the residual noise leakage error $O_{\text{leak}H}$, and the residual leakage error O_{leak} that depends on both the system and noise dynamics:

$$\mathbb{E}\{\hat{C}_V(k)\} = C_V(k) + C_V^{(1)}(k)O_{\text{int}H}(n/N) + C_V^{(1)}(k)O_{\text{leak}H}((n/N)^{(R+2)}) + \ldots \\ G^{(R+1)}(\Omega_k)O_{\text{int}G}((n/N)^{2(R+1)})(G^{(R+1)}(\Omega_k))^H + O_{\text{leak}}((n/N)^{(2R+3)}) \quad (7\text{-}17)$$

(proof: see Appendix 7.D). In (7-17) $X^{(m)}$ is the mth order derivative of X w.r.t. the frequency. Note that $C_V^{(1)}(k) = 0$ for white noise ($H^{(1)}(\Omega) = 0$) or at a transmission zero of the noise ($H(\Omega_k) = 0$), because $C_V(k) = H(\Omega_k)\text{Cov}(E(k))H^H(\Omega_k)$. For white noise excitations, and n sufficiently large, it can be shown that $O_{\text{int}H}(n/N)$ in (7-17) is replaced by $O_{\text{int}H}((n/N)^2)$ (see Pintelon et al., 2010a).

The bias error of the frequency response matrix (FRM) estimate (7-16) consists of two contributions: the polynomial interpolation error $O_{\text{int}G}$ of the FRM over $2n + 1$ frequencies, and the residual system leakage errors $O_{\text{leak}G}$ in (7-4)

$$\mathbb{E}\{\hat{G}(\Omega_k)\} = G(\Omega_k) + G^{(R+1)}(\Omega_k)O_{\text{int}G}((n/N)^{(R+1)}) + O_{\text{leak}G}((n/N)^{(R+2)}) \quad (7\text{-}18)$$

(proof: see Appendix 7.D). Note that the noise leakage errors do not affect the expected value of the FRM estimate.

From (7-17) it can be seen that the optimal choice of the polynomial degree R is case dependent: the noise interpolation error $O_{\text{int}H}$ increases with R (n is proportional to R, see (7-13) and (7-14)), while the other error terms decrease for increasing values of R. Extensive simulations show that a second order polynomial approximation $R = 2$ is often a good compromise. However, for lowly damped systems larger values ($R = 3, 4$) are sometimes needed, especially around the resonance frequency.

7.2.2.4 Covariance Frequency Response Matrix.
The covariance matrix of the FRF estimate (7-16) is given by

$$\mathrm{Cov}(\mathrm{vec}(\hat{G}(\Omega_k))) = \overline{\mathbb{E}\{S^H S\}} \otimes C_V(k) + O_{\mathrm{int}H}(n^0/N) \qquad (7\text{-}19)$$

with

$$S = K_n^H (K_n K_n^H)^{-1} \begin{bmatrix} I_{n_u} \\ 0 \end{bmatrix}, \quad \mathbb{E}\{S^H S\} = O(N^0/n), \quad C_V(k) = O(N^0) \qquad (7\text{-}20)$$

where $\mathrm{vec}(X)$ puts the columns of matrix X on top of each other, and where the expected value is taken w.r.t. the random input (proof: see Appendix 7.E). From (7-19) it can be seen that increasing the frequency width n of the local polynomial estimate (7-10) will decrease the covariance of the FRF at the rate $O(n^{-1})$ as long as the noise interpolation error contribution $O_{\mathrm{int}H}$ can be neglected. Once $O_{\mathrm{int}H}$ becomes dominant, it makes no sense to increase n further.

In practice, the expected value $\mathbb{E}\{S^H S\}$ and the true noise covariance matrix $C_V(k)$ are unknown and both are replaced by, respectively, $S^H S$ and $\hat{C}_V(k)$ (7-13), giving

$$\mathrm{Cov}(\mathrm{vec}(\hat{G}(\Omega_k))) \approx \overline{S^H S} \otimes \hat{C}_V(k) \qquad (7\text{-}21)$$

7.2.2.5 Numerically Stable Implementation.
Using the singular value decomposition (SVD)

$$K_n^H = U_K \Sigma_K V_K^H \qquad (7\text{-}22)$$

the matrix inversion in the calculation of the projection matrix P_n (7-11) can be avoided

$$P_n = I_{2n+1} - U_K U_K^H = U_K^\perp (U_K^\perp)^H \qquad (7\text{-}23)$$

where $U_K^\perp$ is the orthogonal complement of U_K ($U_K^H U_K^\perp = 0$). As a result the residual $\hat{V}_n$ (7-12) can be calculated even if the input DFT spectrum $U(k+r)$ is zero for $r = -n, -n+1, \ldots, 0, \ldots, n$. Hence, the local polynomial estimate of the noise covariance matrix (7-13) is robust to lack of excitation in the frequency band of interest. The SVD (7-22) also allows to calculate $\hat{\Theta}$ (7-10) in a numerically reliable way as

$$\hat{\Theta} = Y_n U_K \Sigma_K^{-1} V_K^H \qquad (7\text{-}24)$$

giving a numerically stable estimate $\hat{G}(\Omega_k)$ (7-16).

The numerical stability of the calculations can be improved further by scaling the rows of K_n before calculating the SVD

$$K_n \rightarrow D_{\mathrm{scale}}^{-1} K_n \qquad (7\text{-}25)$$

where D_{scale} is a diagonal matrix with entry $D_{\mathrm{scale}[i,i]} = \|K_{n[i,:]}\|_2$ if $\|K_{n[i,:]}\|_2 \neq 0$ and otherwise $D_{\mathrm{scale}[i,i]} = 1$ ($X_{[i,:]}$ denotes the i th row of X). Note that after scaling (7-25), formulas (7-11) and (7-23) remain valid with $(D_{\text{scale}}^{-1} K_n)^H = U_K \Sigma_K V_K^H$, while (7-16) is modified as

$$\hat{G}(\Omega_k) = (Y_n U_K \Sigma_K^{-1} V_K^H D_{\text{scale}}^{-1}) \begin{bmatrix} I_{n_u} \\ 0 \end{bmatrix} \qquad (7\text{-}26)$$

The covariance expression (7-21) is calculated in a numerically reliable way via the SVD of K_n^H (7-22):

$$S^H S = \begin{bmatrix} I_{n_u} \\ 0 \end{bmatrix}^H (K_n K_n^H)^{-1} \begin{bmatrix} I_{n_u} \\ 0 \end{bmatrix} = S_1^H S_1 \text{ with } S_1 = \Sigma_K^{-1} V_K^H \begin{bmatrix} I_{n_u} \\ 0 \end{bmatrix} \qquad (7\text{-}27)$$

If the rows of K_n have been scaled as in (7-25), then S_1 in (7-27) is replaced by

$$S_1 \rightarrow \Sigma_K^{-1} V_K^H D_{\text{scale}}^{-1} \begin{bmatrix} I_{n_u} \\ 0 \end{bmatrix} \qquad (7\text{-}28)$$

7.2.2.6 Border Effects. Model equation (7-8) assumes that n frequencies are available at the left- and right-hand sides of the DFT frequency k, which is not the case at the left and right borders of the frequency band. At those borders, $r = -n, -n+1, ..., 0, ..., n$ in (7-8) is replaced by

$$r = -n+p, -n+p+1, ..., 0, ..., n+p \qquad (7\text{-}29)$$

with $p = n, n-1, ..., 1$ for the first n DFT frequencies (left border), and $p = -1, -2, ..., -n$ for the last n DFT frequencies (right border). All formulas remain valid for (7-29), except the bias error terms (7-17) and (7-18), where n/N should be replaced by $(n+|p|)/N$, $p = \pm 1, \pm 2, ..., \pm n$.

Since K_n has a Vandermonde type of matrix structure in the powers of r (Golub and Van Loan, 1996), and since at the first n and last n DFT frequencies the r-values (7-29) are in absolute value larger than those at the other DFT frequencies $k+r$, $r = -n, -n+1, ..., 0, ..., n$, it follows that the condition number of K_n is larger at the borders. Hence, it can be concluded from (7-27) that the variance of the estimated FRF (7-19) increases at the borders. Simulations indicate that the order of magnitude of the increase in variance is $((n+|p|)/n)^{(R+1)}$, $p = \pm 1, \pm 2, ..., \pm n$, with a maximum of $2^{(R+1)}$ for the first and the last DFT frequencies.

7.2.3 The Spectral Analysis Method

7.2.3.1 The Basic Algorithm. The spectral analysis method (also called the H_1 estimator) calculates the frequency response matrix (FRM) and the output noise covariance matrix via the cross- and autopower spectra of the input-output signals

$$\begin{aligned} G(\Omega_k) &= S_{yu}(j\omega_k) S_{uu}^{-1}(j\omega_k) \\ C_V(k) &= S_{yy}(j\omega_k) - S_{yu}(j\omega_k) S_{uu}^{-1}(j\omega_k) S_{yu}^H(j\omega_k) \end{aligned} \qquad (7\text{-}30)$$

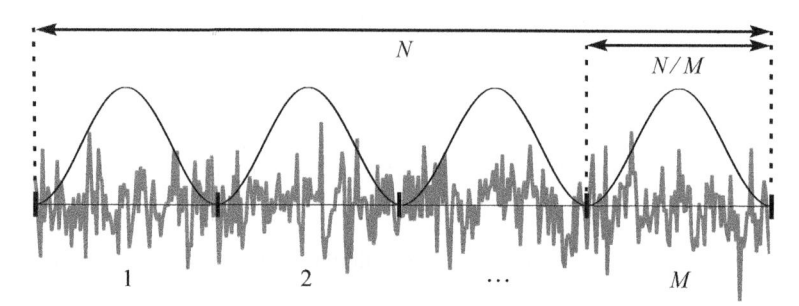

Figure 7-3. Principle of the spectral analysis method: the signal (gray line) of length N is divided in M weighted segments (black lines) of length N/M.

(see Bendat and Piersol, 1980 and Brillinger, 1981). In practice, the true cross- and auto-power spectra are unknown, and are replaced by sample estimates obtained by splitting the input-output records into M non-overlapping segments of length N/M (see Figure 7-3). To suppress the influence of the system and noise leakage terms in the DFT spectra, a window $w(t)$ is applied to the time signals, for example,

$$\hat{S}_{Y_W U_W}(k) = \frac{1}{M}\sum_{m=1}^{M} Y_W^{[m]}(k)(U_W^{[m]}(k))^H \qquad (7\text{-}31)$$

and similarly for $\hat{S}_{Y_W Y_W}(k)$ and $\hat{S}_{U_W U_W}(k)$, where $X_W^{[m]}(k)$, with $X = Y, U$, is the DFT

$$X_W^{[m]}(k) = \frac{1}{\sqrt{\sum_{t=0}^{N/M-1}|w(t)|^2}}\sum_{t=0}^{N/M-1} w(t)x(t+(m-1)N/M)e^{-j2\pi kt/N} \qquad (7\text{-}32)$$

of the mth weighted segment. The scaling in (7-32) is such that the transformation preserves the rms value of the signal. The FRM and the noise covariance matrix are then estimated as

$$\begin{aligned}\hat{G}^{\text{win}}(\Omega_k) &= \hat{S}_{Y_W U_W}(k)\hat{S}_{U_W U_W}^{-1}(k) \\ \hat{C}_V^{\text{win}}(k) &= \frac{M}{q}(\hat{S}_{Y_W Y_W}(k) - \hat{S}_{Y_W U_W}(k)\hat{S}_{U_W U_W}^{-1}(k)\hat{S}_{Y_W U_W}^H(k))\end{aligned} \qquad (7\text{-}33)$$

with $q = M - n_u$ the number of the degrees of freedom in the residual

$$\hat{S}_{Y_W Y_W}(k) - \hat{S}_{Y_W U_W}(k)\hat{S}_{U_W U_W}^{-1}(k)\hat{S}_{Y_W U_W}^H(k) \qquad (7\text{-}34)$$

(see Brillinger, 1981). Note that the variance of the FRM estimate (7-33) can be decreased further by using overlapping segments (see Section 2.6.5 on page 62).

The properties of the spectral analysis estimator are studied for the following windows

$$\begin{aligned}\text{rectangular (win = rect, } W = \text{)}: &\quad w(t) = 1 \\ \text{diff (win = diff, } W = D\text{)}: &\quad w(t) = 1 - e^{j2\pi tM/N} \\ \text{half-sine (win = sine, } W = S\text{)}: &\quad w(t) = \sin(\pi tM/N) \\ \text{Hanning (win = hann, } W = H\text{)}: &\quad w(t) = 0.5(1 - \cos(2\pi tM/N))\end{aligned} \qquad (7\text{-}35)$$

In practice the diff window is calculated as the difference of the DFT spectrum

$$X_D(k) = (X(k+1) - X(k))/\sqrt{2} \tag{7-36}$$

(see Schoukens et al., 2006a). Formula (7-36) shows that the diff estimates of the noise covariance matrix and FRF (7-33) are in fact estimates of the true values at half the DFT frequency: $C_V(k + 1/2)$ and $G(\Omega_{k+1/2})$. The Hanning window can be interpreted as a scaled double difference of the DFT spectrum

$$X_H(k) = (2X(k) - X(k-1) - X(k+1))/\sqrt{6} \tag{7-37}$$

(see Harris, 1978), and the half-sine window is related to the diff window by

$$X_S(k) = -X_D(k+1/2)/j \tag{7-38}$$

(see Antoni and Schoukens, 2007). In Antoni and Schoukens (2007) it has been shown that the half-sine window is optimal for suppressing the stochastic and systematic leakage errors in FRF measurements. The same is true for the diff window, as shown by (7-38).

To minimize the leakage errors in the FRM and noise covariance matrix estimates (7-33), the number of blocks M should be chosen as small as possible. The minimal values of M are such that the number of degrees of freedom q in the residuals (7-34) satisfies (7-14) where $m = 0$, 1, or 2 according to the required property of the estimate $\hat{C}_V(k)$ (see (7-15)). M is bounded above by the requirement that N/M should be larger than the dominant time constant of the system.

Since the rectangular, diff, and Hanning windows combine, respectively, zero, two, and three DFT frequencies, the correlation length over the frequency of the estimated FRM and noise covariance matrix (7-33) equals, respectively, zero, ±1, and ±2 DFT frequencies.

7.2.3.2 Bias Error. The expected value of the noise covariance matrix $\hat{C}_V^{\text{win}}(k)$ in (7-33) is given by

$$\begin{aligned}
\mathbb{E}\{\hat{C}_V^{\text{rect}}(k)\} &= C_V(k) + O_{\text{leak}}(M/N) \\
\mathbb{E}\{\hat{C}_V^{\text{diff, sine}}(k)\} &= C_V(k_1) + \frac{1}{8}C_V^{(2)}(k_1)O_{\text{int}H}((M/N)^2) + C_V^{(1)}(k_1)O_{\text{leak}H}((M/N)^2) + \ldots \\
&\quad \frac{1}{8}G^{(1)}(\Omega_{k_1})O_{\text{int}G}((M/N)^2)(G^{(1)}(\Omega_{k_1}))^H + O_{\text{leak}}((M/N)^3) \\
\mathbb{E}\{\hat{C}_V^{\text{hann}}(k)\} &= C_V(k) + \frac{1}{6}C_V^{(2)}(k)O_{\text{int}H}((M/N)^2) + C_V^{(1)}(k)O_{\text{leak}H}((M/N)^3) + \ldots \\
&\quad \frac{1}{6}G^{(1)}(\Omega_k)O_{\text{int}G}((M/N)^2)(G^{(1)}(\Omega_k))^H + G^{(1)}(\Omega_k)O_{\text{leak}G}((M/N)^4) + \ldots \\
&\quad O_{\text{leak}}((M/N)^5)
\end{aligned} \tag{7-39}$$

where $k_1 = k + 1/2$ or $k_1 = k$ for, respectively, the diff and half-sine window, and where O_{leak} depends on both the system and the noise dynamics (proof: see Pintelon et al., 2010a). It can be seen that the bias $\mathbb{E}\{\hat{C}_V^{\text{win}}(k)\} - C_V(k)$ is only due to leakage O_{leak} for the rectangular window, while for the diff, half-sine, and Hanning windows the bias error consists of an interpolation error $O_{\text{int}H}$ and $O_{\text{int}G}$ of, respectively, $H(\Omega)$ and $G(\Omega)$ over two (diff and half-sine) or three (Hanning) neighboring DFT frequencies, and a leakage error O_{leak}.

The expected value of the FRM $\hat{G}^{\text{win}}(\Omega_k)$ in (7-33) equals

$$\mathbb{E}\{\hat{G}^{\text{rect}}(\Omega_k)\} = G(\Omega_k) + O_{\text{leak}G}(M/N)$$

$$\mathbb{E}\{\hat{G}^{\text{diff, sine}}(\Omega_k)\} = G(\Omega_{k_1}) + \frac{1}{4}G^{(1)}(\Omega_{k_1})O_{\text{int}G}((M/N)^2) + O_{\text{leak}G}((M/N)^2) \qquad (7\text{-}40)$$

$$\mathbb{E}\{\hat{G}^{\text{hann}}(\Omega_k)\} = G(\Omega_k) + \frac{1}{3}G^{(1)}(\Omega_{k_1})O_{\text{int}G}((M/N)^2) + O_{\text{leak}G}((M/N)^3)$$

where k_1 is defined in (7-39) (proof: see Pintelon et al., 2010a). It can be seen that the bias $\mathbb{E}\{\hat{G}^{\text{win}}(\Omega_k)\} - G(\Omega_k)$ only depends on the system leakage error $O_{\text{leak}G}$ for the rectangular window, while for the diff, half-sine, and Hanning windows the bias error has two contributions, an interpolation error $O_{\text{int}G}$ of the FRM over two (diff and half-sine) or three (Hanning) neighboring DFT frequencies, and a system leakage error $O_{\text{leak}G}$. Note also that the noise leakage errors do not affect the expected value of the FRM estimates.

From (7-39) and (7-40) it can be seen that the system and noise interpolation errors of the Hanning window are 4/3 times larger (1.25 dB for $\hat{C}_V$ in (7-39), and 2.5 dB for $\hat{G}$ in (7-40)) than those of the diff and half-sine windows. This is due to the fact that the Hanning window combines three DFT frequencies, while the diff and half-sine windows only combine two DFT frequencies (see (7-36), (7-37), and (7-38)). The system and noise leakage errors of the Hanning window are, however, much smaller.

7.2.3.3 Covariance Frequency Response Matrix. The covariance matrix of the FRM estimate (7-33) is given by

$$\text{Cov}(\text{vec}(\hat{G}^{\text{win}}(\Omega_k))) = \frac{1}{M}\mathbb{E}\{\hat{S}^{-1}_{U_W U_W}(k)\} \otimes C_V(k) + O^{\text{win}}_{\text{int}H}((M/N)^2) \qquad (7\text{-}41)$$

where the expected value is taken w.r.t. the random input, and where $O^{\text{win}}_{\text{int}H}((M/N)^2)$ is zero for the rectangular window $O^{\text{rect}}_{\text{int}H} = 0$ (proof: see Pintelon et al., 2010a). In practice the expected value $\mathbb{E}\{\hat{S}^{-1}_{U_W U_W}(k)\}$ and the true noise covariance matrix $C_V(k)$ are unknown and are replaced by, respectively, $\hat{S}^{-1}_{U_W U_W}(k)$ and $\hat{C}^{\text{win}}_V(k)$ (7-33), giving

$$\text{Cov}(\text{vec}(\hat{G}^{\text{win}}(\Omega_k))) \approx \frac{1}{M}\hat{S}^{-1}_{U_W U_W}(k) \otimes \hat{C}^{\text{win}}_V(k) \qquad (7\text{-}42)$$

Since $\hat{C}^{\text{win}}_V(k)$ in (7-33) estimates the covariance matrix of the residual of model equation (7-1), the covariance matrices (7-42) of the FRM estimates correctly account for the variability due to the leakage terms (see Pintelon et al., 2010a).

7.2.4 Confidence Regions of the FRM

For fixed inputs, the FRM estimates $\hat{G}(\Omega_k)$ (7-16) and (7-33) are asymptotically ($N \to \infty$) circular complex normally distributed with mean value $G(\Omega_k)$ and covariance matrix $\text{Cov}(\text{vec}(\hat{G}(\Omega_k)))$ (7-21) and (7-42). Hence, the most compact $100 \times p\%$ confidence bounds are ellipsoids (Stuart and Ord, 1987). For example, the $100 \times p\%$ confidence ellipsoid of $\text{vec}(\hat{G}(\Omega_k))$ is defined by those values of $X \in \mathbb{C}^{n_y n_u}$ such that

Section 7.2 ■ Arbitrary Excitations

$$(X - \text{vec}(\hat{G}(\Omega_k)))^H \text{Cov}^{-1}(\text{vec}(\hat{G}(\Omega_k)))(X - \text{vec}(\hat{G}(\Omega_k))) \leq \frac{1}{2}\chi_p^2(2n_y n_u) \quad (7\text{-}43)$$

with $\chi_p^2(2n_y n_u)$ the $100 \times p\%$ percentile of a χ^2- distributed random variable with $2n_y n_u$ degrees of freedom. For entry $[i,j]$ of the frequency response matrix $\hat{G}(\Omega_k)$, (7-43) reduces to a circle with center $\hat{G}_{[i,j]}(\Omega_k)$ and radius $(-\ln(1-p))^{1/2}\text{std}(\hat{G}_{[i,j]}(\Omega_k))$ (see Section 2.4.2.3 on page 48).

If the true covariance matrix in (7-43) is replaced by an asymptotically ($N \rightarrow \infty$) complex Wishart distributed estimate $\hat{C}_{\text{vec}\hat{G}}(k)$ (7-21) or (7-42) with dof degrees of freedom, then a $100 \times p\%$ confidence bound is constructed via the Hotelling's T^2- statistic (Giri, 1965)

$$(X - \text{vec}(\hat{G}(\Omega_k)))^H \hat{C}_{\text{vec}\hat{G}}^{-1}(k)(X - \text{vec}(\hat{G}(\Omega_k))) \leq \frac{n_1 dof}{n_2} F_p(n_1, n_2) \quad (7\text{-}44)$$

with $n_1 = 2n_u n_y$, $n_2 = 2dof - 2n_u n_y + 2$, and $F_p(n_1, n_2)$ the $100 \times p\%$ percentile of a $F(n_1, n_2)$-distributed random variable (proof: use (16-32b) with $n = n_y n_u$ and $R = dof + 1$). For entry $[i,j]$ of the frequency response matrix $\hat{G}(\Omega_k)$, (7-44) reduces to the F-test (2-40) on page 51 where $M = dof + 1$.

7.2.5 Comparison of the Methods

We first compare the local polynomial approach with the spectral analysis method using non-overlapping segments. Next, we discuss the benefits of using overlapping segments in FRF measurements.

7.2.5.1 Bias and Covariance of the Estimates. From (7-17), (7-18), (7-39), and (7-40) it can be concluded that the spectral analysis method with the rectangular window has the largest leakage bias error in the noise covariance matrix $\hat{C}_V(k)$ and in the frequency response matrix $\hat{G}(\Omega_k)$ estimates. Compared with the spectral analysis method with the diff, half-sine, and Hanning windows, the local polynomial approach has, for N sufficiently large and $R \geq 2$, (i) the smallest system interpolation error for $\hat{C}_V(k)$ (compare (7-17) and (7-39)) and $\hat{G}(\Omega_k)$ (compare (7-18) and (7-40)), and (ii) the smallest noise and/or system leakage error for $\hat{C}_V(k)$ and $\hat{G}(\Omega_k)$. Moreover, the frequency resolution of the local polynomial estimates is M times larger than that of the spectral analysis methods, and the noise covariance matrix estimates $\hat{C}_V(k)$ are robust to lack of excitation in the frequency band of interest, which is not the case for the spectral analysis methods. The noise interpolation error $O_{\text{int}H}$ for the local polynomial estimate $\hat{C}_V(k)$, however, decreases slower to zero as $N \rightarrow \infty$.

Although an analytic comparison of the bias (7-18) and (7-40) and the covariance matrices (7-21) and (7-42) of the FRM estimates (7-16) with $R \geq 2$ and (7-33) is not feasible, a rationale will be given for the following ranking

$$\begin{aligned} \text{bias}(\hat{G}(\Omega_k)) &< \text{bias}(\hat{G}^{\text{win}}(\Omega_k)) & \text{(a)} \\ \text{bias}(\hat{C}_V(\Omega_k)) &< \text{bias}(\hat{C}_V^{\text{win}}(\Omega_k)) & \text{(b)} \\ \text{Cov}(\text{vec}(\hat{G}(\Omega_k))) &< \text{Cov}(\text{vec}(\hat{G}^{\text{win}}(\Omega_k))) & \text{(c)} \end{aligned} \quad (7\text{-}45)$$

with $\text{bias}(\hat{x}) = \mathbb{E}\{\hat{x}\} - x$ the bias of x. Since no exact proof is given, the ranking should be interpreted as a general trend, without guarantee that no counter examples can be found for particular systems and values of N. The inequality (7-45a) for the bias of $\hat{G}$ follows immediately from (7-18) and (7-40) by comparing the powers of N in $O_{\text{int}G}$ and $O_{\text{leak}G}$. Since the spectral estimates (7-33) of the noise covariance matrix include the variability of the (residual) leakage errors (see Pintelon et al., 2010a), and since in practice $O_{\text{int}H}$ in (7-17) can be neglected w.r.t. the other terms, we conclude from (7-17) and (7-39) that $\mathbb{E}\{\hat{C}_V(k)\} < \mathbb{E}\{\hat{C}_V^{\text{win}}(k)\}$. Hence, $\text{bias}(\hat{C}_V(\Omega_k)) < \text{bias}(\hat{C}_V^{\text{win}}(\Omega_k))$, which proves the inequality (7-45b). Combining (7-45b) with (7-21) and (7-42) already justifies (7-45c) at those frequencies where the system and/or noise leakage errors are important. Simulations reveal that $\text{Cov}(\text{vec}(\hat{G}(\Omega_k)))$ can be larger than $\text{Cov}(\text{vec}(\hat{G}^{\text{win}}(\Omega_k)))$ for the same value of m (or q) in (7-14), in the frequency bands where the interpolation/leakage errors can be neglected (see Pintelon et al., 2010a). However, the comparison is not fair because the frequency resolution of the local polynomial method is M times larger than that of the spectral analysis methods. Therefore, the local polynomial FRF estimate $\hat{G}(\Omega_k)$ should first be averaged over M neighboring frequencies before calculating the covariance matrix. In order to avoid the resulting first order interpolation bias error, the averaging of the FRF is done by increasing m (n) in (7-14) with M ($m \to m + M$). Proceeding in this way, n increases and the covariance matrix of the local polynomial FRF estimates can be made smaller than that of the spectral analysis methods as long as $O_{\text{int}H}$ in (7-19) can be neglected w.r.t. the first term. Since, by assumption $O_{\text{int}}, O_{\text{leak}} \ll C_V$, the covariance of the FRF (7-19) can be reduced without detectable increase of the bias (7-18). This is not in contradiction with the results of Schoukens et al. (2006b) where, similarly to the spectral analysis methods, the local polynomial FRF estimate (called the Taylor method) was averaged over M subrecords of length N/M at the frequency resolution Mf_s/N, while here, the averaging is performed at the maximal frequency resolution f_s/N by increasing m (n) in (7-14).

Comparing the powers of (M/N) in (7-39), it can easily be deduced from (7-39) and (7-42) that

$$\begin{aligned}
&\text{Cov}(\text{vec}(\hat{G}^{\text{diff}})) = \text{Cov}(\text{vec}(\hat{G}^{\text{sine}})) < \text{Cov}(\text{vec}(\hat{G}^{\text{rect}}))\\
&\text{Cov}(\text{vec}(\hat{G}^{\text{hann}})) < \text{Cov}(\text{vec}(\hat{G}^{\text{rect}}))\\
&\text{Cov}(\text{vec}(\hat{G}^{\text{hann}})) < \text{Cov}(\text{vec}(\hat{G}^{\text{diff}})) \quad \text{if } O_{\text{int}} < O_{\text{leak}}\\
&\text{Cov}(\text{vec}(\hat{G}^{\text{diff}})) \leq \text{Cov}(\text{vec}(\hat{G}^{\text{hann}})) \leq \frac{4}{3}\text{Cov}(\text{vec}(\hat{G}^{\text{diff}})) \quad \text{if } O_{\text{int}} \geq O_{\text{leak}}\\
&\text{Cov}(\text{vec}(\hat{G}^{\text{hann}})) \approx \text{Cov}(\text{vec}(\hat{G}^{\text{diff}})) \approx \text{Cov}(\text{vec}(\hat{G}^{\text{rect}})) \quad \text{if } O_{\text{int}}, O_{\text{leak}} \ll C_V
\end{aligned} \quad (7\text{-}46)$$

This ranking, where 4/3 is replaced by 16/9 in the last upper bound, is also valid for the mean square error (mse = Cov + bias bias^T), because (7-46) also applies for the bias (7-40) on the FRM estimates. We conclude that, among the spectral analysis estimates, the diff and half-sine windows perform somewhat better (smaller mean square error) than the Hanning window when the interpolation errors dominate, while the Hanning window outperforms (smallest mean square error) in those frequency bands where the leakage errors are dominant. Simulations show that the latter is typically the case in the neighborhood of a transmission zero of the system or noise transfer functions, even if the measurement time is much larger than the dominant system time constant. It refines the conclusions of Schoukens et al. (2006a) and Antoni and Schoukens (2007) that the diff/half-sine windows are optimal for FRM measurements. Note that the upper bound in the last inequality of (7-46) is reached if the interpolation errors are the largest terms in the noise covariance matrix (7-39). This is consistent with the single-input, single-output results of Schoukens et al. (2006a).

7.2.5.2 Spectral Analysis with Overlapping Segments.
The benefit of using overlapping segments (see Section 2.6.5 on page 62) is the variance reduction of the FRM estimate (7-30). This reduction is maximal for 100% of overlap (Antoni and Schoukens, 2007). Compared with the non-overlapping case a maximal noise variance decrease of a factor 1.49 (1.2 dB), 1.70 (2.3 dB) and 2.08 (3 dB) is obtained for, respectively, the rectangular, half-sine/diff, and Hanning windows (Antoni and Schoukens, 2007). The maximal reduction of the sum of the system interpolation variance and system leakage variance is a factor 3.54 (5.5 dB) and 3.86 (5.9 dB) for, respectively, the half-sine/diff and Hanning windows (Antoni and Schoukens, 2007). However, the convergence rate as a function of M/N of the bias (7-40) and covariance (7-39), (7-42) of the FRM estimates (7-30) are qualitatively the same for the spectral analysis methods with and without overlapping segments. Hence, the following conclusions can be drawn:

1. The ranking (7-46) for the covariance matrix and the mean square error remains valid for the spectral analysis with overlapping segments, except that the last equality must be replaced by

$$\text{Cov}(\text{vec}(\hat{G}^{\text{hann}})) < \text{Cov}(\text{vec}(\hat{G}^{\text{diff}})) < \text{Cov}(\text{vec}(\hat{G}^{\text{rect}})) \quad \text{if} \quad O_{\text{int}}, O_{\text{leak}} \ll C_V$$

2. In those frequency bands where the system interpolation and system leakage errors are dominant, ranking (7-45) still applies.

3. Compared with the non-overlapping case, m (n) in (7-14) should be increased in order to satisfy (7-45) in those frequency bands where the interpolation and leakage errors can be neglected ($O_{\text{int}}, O_{\text{leak}} \ll C_V$).

7.2.5.3 Correlation of the Estimates over the Frequency.
The correlation length over the frequency of the noise covariance matrix and FRF estimates equals $\pm 2n$ DFT frequencies at the resolution f_s/N for the local polynomial method; while for the spectral analysis method with the rectangular, diff, and Hanning windows it is, respectively, zero, ± 1, and ± 2 at the resolution Mf_s/N. This correlation length should be taken into account when constructing confidence bounds of the estimated FRM over the frequency. The influence of the correlation over the frequency of the nonparameteric estimates in parametric modeling of the system transfer function is discussed in Chapter 12.

7.2.6 Bias-Variance Trade-Off

Comparing (7-18) and (7-19) it follows that the mean square error (mse) of the local polynomial FRF estimate (mse = Cov + bias bias^H) might be decreased by increasing the bandwidth n of the local polynomial approximation (Cov = $O(n^{-1}) + O_{\text{int}H}(n^0)$ and $b = O_{\text{int}G}(n^{R+1}) + O_{\text{leak}G}(n^{R+2})$). It is clear that increasing n (m) is beneficial in those frequency bands where the interpolation/leakage errors can be neglected ($O_{\text{int}}, O_{\text{leak}} \ll C_V$). However, in all other cases a bias-variance trade-off must be made. Following the lines of Fan and Gijbels (1995) a frequency-dependent data-driven bandwidth n can be selected giving an optimal bias-variance trade-off. One could, for example, choose that value of n (m) in (7-14) that minimizes Akaike's information criterion (AIC) of the local least squares approximation (7-10) (see Schoukens and Pintelon, 2010a for the details).

Another way to reduce the mean square error of the local polynomial FRF estimate consists of exploiting the relationships between the polynomial coefficients (7-6) in neighboring frequency windows. For a fixed value of the bandwidth n, the resulting constrained

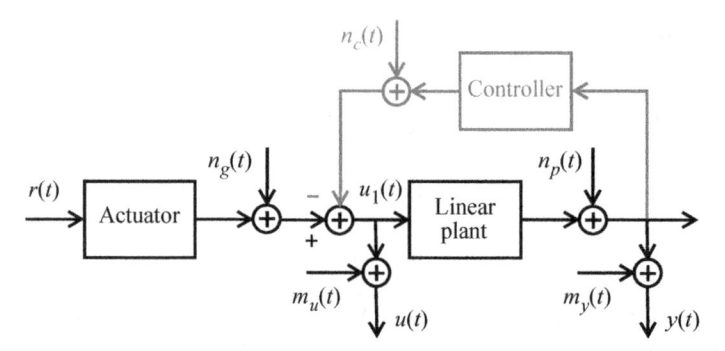

Figure 7-4. Measurement of the frequency response function matrix of a linear multivariable system operating in open (black lines only) or closed loop (black and gray lines). The linear system is excited by the known reference signal $r(t)$ via an actuator. $u(t)$ and $y(t)$ are the noisy input-output observations and $n_c(t)$, $n_g(t)$, $n_p(t)$, $m_u(t)$, and $m_y(t)$ represent, respectively, the controller, generator, and process noise, and the input-output measurement errors. The actuator or the controller may behave nonlinearly. If the controller behaves nonlinearly, then $n_g(t) = 0$, $n_p(t) = 0$, and $n_c(t) = 0$.

local polynomial method makes a trade-off between the variance reduction and the bias increase induced by the constraints (see Gevers et al., 2011 for the details).

7.2.7 The Noisy Input, Noisy Output Case

7.2.7.1 Linear Actuator and Linear Controller. Consider first the closed loop set up of Figure 7-4 with a linear actuator and a linear feedback loop. Note that feedback is inherently present in any experimental setup where the system loads the actuator. For example, a voltage generator with non-zero output impedance that is driving a circuit with finite input impedance. As a consequence the actual input of the system depends on both the actuator and the system dynamics. Hence, the actual input ($u_1(t)$ in Figure 7-4) is mostly unknown and should be measured. Moreover, due to the feedback loop, $u_1(t)$ is correlated with the process noise $n_p(t)$ (see Figure 7-4). These are the two major difficulties of identification in closed loop.

If the input observations are noisy and/or the input depends on the process noise due to a feedback loop, then the local polynomial and spectral analysis estimates of the FRM and the noise covariance matrix are biased. To avoid this bias, the $n_u \times 1$ reference signal $r(t)$ in Figure 7-4 – typically the signal stored in the arbitrary waveform generator – should be known, and model (7-1) is replaced by

$$Z(k) = G_{rz}(\Omega_k)R(k) + V_Z(k) + T_{rz}(\Omega_k) \qquad (7\text{-}47)$$

with

$$Z(k) = \begin{bmatrix} Y(k) \\ U(k) \end{bmatrix}, \; G_{rz}(\Omega_k) = \begin{bmatrix} G_{ry}(\Omega_k) \\ G_{ry}(\Omega_k) \end{bmatrix} \text{ and } V_Z(k) = H_{rz}(\Omega_k)E_Z(k) = \begin{bmatrix} V_Y(k) \\ V_U(k) \end{bmatrix} \qquad (7\text{-}48)$$

and where $V_Z(k)$ models that part of the input-output DFT spectra that does not depend on the reference signal. From the FRM estimate $\hat{G}_{rz}(\Omega_k)$ we obtain a consistent estimate of the FRM $G(\Omega_k)$

Section 7.2 ■ Arbitrary Excitations

$$\hat{G}(\Omega_k) = \hat{G}_{ry}(\Omega_k)\hat{G}_{ru}^{-1}(\Omega_k) \qquad (7\text{-}49)$$

Eq. (7-49), where the spectral analysis method (7-33) is used for estimating the FRM $G_{rz}(\Omega_k)$, is the multivariable version of the indirect FRF estimate (see Section 2.6.4 on page 61). The drawback of the indirect method (7-49) is that the part of the true system input $u_1(t)$ (see Figure 7-4) depending on the process $n_p(t)$, controller $n_c(t)$, and generator $n_g(t)$ noise sources is considered as a disturbance and is suppressed in the FRM estimate. The covariance matrix of $\hat{G}$ (7-49) is related to that of $\hat{G}_{rz}$ by

$$\text{Cov}(\text{vec}(\hat{G})) \approx (\hat{G}_{ru}^{-T} \otimes [I_{n_y} - \hat{G}])\text{Cov}(\text{vec}(\hat{G}_{rz}))(\hat{G}_{ru}^{-T} \otimes [I_{n_y} - \hat{G}])^H \qquad (7\text{-}50)$$

(proof: see Appendix 7.F) where $\text{Cov}(\text{vec}(\hat{G}_{rz}))$ depends on the estimated noise covariance matrix $\hat{C}_{V_Z}(k)$ as given by (7-21) for the local polynomial estimate, and by (7-42) for the spectral analysis methods.

7.2.7.2 Nonlinear Actuator and Linear Controller. If the actuator behaves nonlinearly, then we can replace its block diagram in Figure 7-4 by the best linear approximation (BLA) of the actuator plus an additive output source representing the stochastic nonlinear distortions $r_s(t)$ (proof: combine Theorem 3.16 on page 86 with (6-69) on page 200). These stochastic nonlinear distortions $r_s(t)$ are uncorrelated with – but not independent of – the reference signal $r(t)$. Eq. (7-47) becomes

$$Z(k) = G_{rz}(\Omega_k)R(k) + V_Z(k) + Z_S(k) + T_{rz}(\Omega_k) \qquad (7\text{-}51)$$

where $Z_S(k)$ depends on $R_S(k)$, and the plant and controller dynamics, and where $G_{rz}(\Omega_k)$ depends on the BLA of the actuator, and the plant and controller dynamics. Since the plant and the controller are linear, $G_{rz}(\Omega_k)$ and $Z_S(k)$ still satisfy

$$G(\Omega_k) = G_{ry}(\Omega_k)G_{ru}^{-1}(\Omega_k) \text{ and } Y_S(k) = G(\Omega_k)U_S(k) \qquad (7\text{-}52)$$

with $G_{ry}(\Omega_k)$ and $G_{ru}(\Omega_k)$ defined in (7-48), $Y_S(k)$ the first n_y rows of $Z_S(k)$, and $U_S(k)$ the last n_u rows of $Z_S(k)$. $Z_S(k)$ is uncorrelated with $R(k)$ and, therefore, it follows from (7-51) and (7-52) that the conclusions of Section 7.2.7.1 are also valid here.

7.2.7.3 Linear Actuator and Nonlinear Controller. If the controller behaves nonlinearly, then we assume that no generator, process, or controller noise sources are present. The controller block diagram in Figure 7-4 is replaced by the best linear approximation (BLA) of the controller plus an additive output source representing the stochastic nonlinear distortions $v_s(t)$ (proof: combine Theorem 3.22 on page 94, where the role of the plant and the controller are interchanged, with (6-69) on page 200). Eqs. (7-51) and (7-52), where $Z_S(k)$ depends on $V_S(k)$ and is correlated with – but not independent of – the reference signal $R(k)$, remain valid. Hence, the results of Section 7.2.7.1 also apply here.

7.2.8 Nonlinear Systems

The setup for measuring the best linear approximation (BLA) of a nonlinear system is shown in Figure 7-5. The main difference with the setup for measuring the FRM of a linear

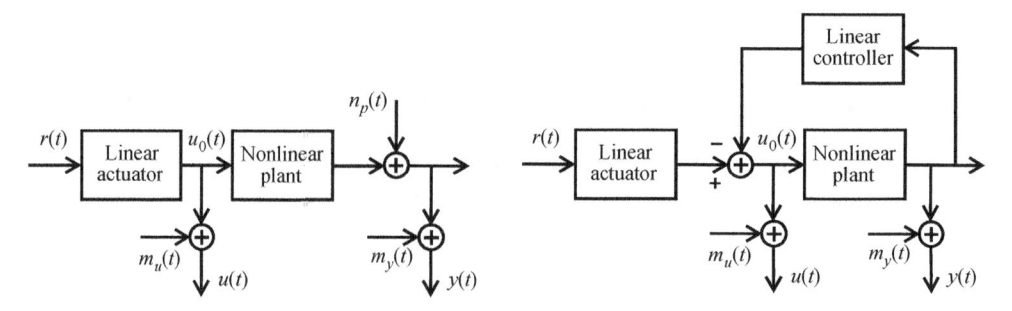

Figure 7-5. Setup for measuring the best linear approximation of a multivariable nonlinear system (n_u inputs, n_y outputs) operating in open (left) or closed (right) loop. The nonlinear system is excited via a linear actuator by the known $n_u \times 1$ reference signal $r(t)$. $u(t)$ and $y(t)$ are the noisy input-output observations and $n_p(t)$, $m_u(t)$, and $m_y(t)$ are, respectively, the process noise, and the input-output measurement errors.

system (see Figure 7-4) is that both the actuator and the feedback controller should be linear, and that the reference signal $r(t)$ is assumed to be normally distributed. The input-output DFT spectra of the measured input-output signals are then related by

$$U(k) = U_0(k) + V_U(k) + T_U(\Omega_k)$$
$$Y(k) = G_{\text{BLA}}(\Omega_k)U_0(k) + V_Y(k) + Y_S(k) + T_Y(\Omega_k) \qquad (7\text{-}53)$$

where the input-output errors $V_U(k) = H_U(\Omega_k)E_U(k)$ and $V_Y(k) = H_Y(\Omega_k)E_Y(k)$ depend on the noise sources in Figure 7-5 (proof: combine Theorems 3.16 and 3.22 with (6-69)). $U_0(k)$ depends on the reference signal $R(k)$ (open and closed loop) and the stochastic nonlinear distortion $Y_S(k)$ (closed loop only). The zero mean stochastic nonlinear distortion $Y_S(k)$ is uncorrelated with – but not independent of – the input $U_0(k)$ (open loop) or the reference signal $R(k)$ (closed loop); and the input-output transient terms $T_U(\Omega_k)$ and $T_Y(\Omega_k)$ comprise the plant and noise leakage errors (discrete-time and continuous-time) and the residual alias errors (continuous-time only). Comparing (7-53) to (7-47), it can be seen that the indirect method of Section 7.2.7 estimates the BLA and its covariance matrix. The latter depends on both the input-output noise $V_U(k)$, $V_Y(k)$ and the output stochastic nonlinear distortion $Y_S(k)$.

If the linearity assumption on the actuator is not satisfied, then the BLA of the nonlinear plant also depends on the actuator characteristics. However, in those frequency bands where the signal-to-distortion ratio is larger than 10 dB, the BLA of the cascade of the nonlinear actuator and the nonlinear plant is in very good approximation equal to the product of the BLA of the actuator and the BLA of the plant (Dobrowiecki and Schoukens, 2006). Equation (7-53) is then replaced by

$$U(k) = U_0(k) + V_U(k) + U_S(k) + T_U(\Omega_k)$$
$$Y(k) = G_{\text{BLA}}(\Omega_k)U_0(k) + V_Y(k) + Y_S(k) + T_Y(\Omega_k) \qquad (7\text{-}54)$$

where the input-output stochastic nonlinear distortions $U_S(k)$ and $Y_S(k)$ are correlated via the nonlinear distortions of the actuator. Comparing (7-54) to (7-47) it follows that the indirect method explained in Section 7.2.7 automatically accounts for the input stochastic nonlinear distortion $U_S(k)$ and for the correlation between $U_S(k)$ and $Y_S(k)$. The nonlinear

Section 7.2 ■ Arbitrary Excitations

contribution of the plant is eventually found as $Y_S(k) - G_{BLA}(\Omega_k)U_S(k)$ (proof: see Appendix 3.O).

It can be concluded that using stationary random excitations it is impossible to distinguish the stochastic nonlinear distortions from the measurement and/or process noise in Figure 7-5. At the price of a loss in frequency resolution of at least a factor 2, it is possible to detect the nonlinear behavior and to estimate an upper bound on the noise level via an experiment with nonstationary random signals (Zhang et al., 2010).

We analyze now the bias error of the BLA (7-16) and the covariance (7-13) estimate assuming that the nonlinear plant operates in open loop and that the input is known exactly. Extension to noisy inputs and/or nonlinear systems operating in feedback and/or nonlinear actuators is straightforward via the indirect method of Section 7.2.7. Since $Y_S(k)$ is uncorrelated with the input, the expected value (7-18), where $G(\Omega_k)$ is replaced by $G_{BLA}(\Omega_k)$, remains valid. However, an additonal term appears in the expected value (7-17) because $Y_S(k)$ is only asymptotically ($N \to \infty$) uncorrelated over the frequency. We find

$$\mathbb{E}\{\hat{C}_V\} = C_V(k) + C_{Y_S}(k) + (C_V^{(1)}(k) + C_{Y_S}^{(1)}(k))O_{\text{int}H}(n/N) + O(4n^2/(qN)) \quad (7\text{-}55)$$

with $C_{Y_S}(k) = \text{Cov}(Y_S(k))$, q defined in (7-13), and where the bias term $O(4n^2/(qN))$ stems from the correlation of $Y_S(k)$ over the frequency (proof: see Appendix 7.G).

7.2.9 Concatenating Data Records

7.2.9.1 The Basic Algorithm. Given M_c data records originating from experiments on the same plant under the same operational conditions, we will estimate the frequency response matrix (FRM) and its covariance at the full frequency resolution of the concatenated data records. We first assume that only the output is disturbed by noise (output error stochastic framework), and next handle the noisy input, noisy output case (errors-in-variables stochastic framework).

Combining the extended transfer function model for concatenated data records (6-48) with the output noise model (6-65) gives

$$Y^c(k) = G(\Omega_k)U^c(k) + T(\Omega_k) + \sum_{i=1}^{M_c-1} z_k^{-\tau_i} T_i^c(\Omega_k) + H(\Omega_k)E(k) \quad (7\text{-}56)$$

with $\tau_i = \sum_{r=1}^{i} N_r$, N_i the length (in samples) of the ith record, and where $T = T_G + T_H$ and $T_i^c = T_{iG}^c + T_{iH}^c$ represent the sum of the plant and noise transient (leakage) contributions. For continuous-time systems ($\Omega = s, \sqrt{s}$) the contribution of the residual alias error $\delta(\Omega_k)$ is included in the transient terms T and T_i^c by increasing the order of the numerator polynomials. Note that (7-56) can be rewritten as

$$Y^c(k) = \tilde{G}(\Omega_k)\tilde{U}^c(k) + T(\Omega_k) + H(\Omega_k)E(k) \quad (7\text{-}57)$$

where $\tilde{G}(\Omega_k)$ and $\tilde{U}^c(k)$ are the extended FRM and the extended input DFT spectrum

$$\begin{aligned}\tilde{G}(\Omega_k) &= \begin{bmatrix} G(\Omega_k) & T_1^c(\Omega_k) & \ldots & T_{M_c-1}^c(\Omega_k) \end{bmatrix} \\ \tilde{U}^c(k) &= \begin{bmatrix} (U^c(k))^T & z_k^{-N_1} & \ldots & z_k^{-N_{M_c-1}} \end{bmatrix}^T\end{aligned} \quad (7\text{-}58)$$

Comparing (7-57) to (7-1) it can be concluded that the local polynomial method of Section 7.2.2 is directly applicable to concatenated data records. One only needs to extend the known $n_u \times 1$ input $u(t)$ by $M_c - 1$ signals $\delta(t - \sum_{r=1}^{i} N_r)$, $i = 1, 2, ..., M_c - 1$, with $\delta(t) = 0$ for $t \neq 0$ and $\delta(0) = 1$.

To handle the noisy input, noisy output case (errors-in-variables stochastic framework) it is sufficient to add the $M_c - 1$ signals $\delta(t - \sum_{r=1}^{i} N_r)$, $i = 1, 2, ..., M_c - 1$, to the known reference $r(t)$ in the indirect method of Section 7.2.7.

7.2.9.2 Bias Error.
The bias error of the local polynomial estimates is analyzed in the special case that the M_c data records have equal length N_0. We compare the bias of the concatenated data records method with that of the non-concatenated approach where the estimates of each data record separately are averaged. For both methods the same degree R of the local polynomial approximation and the same number of degrees of freedom $dof_{\text{arb}} = q$ (7-13) of the residuals of the least squares fit. The latter implies that the difference between the number of complex equations and the number of complex parameters in (7-8) remains constant in the analysis.

From (7-18) it follows that the ratio $R_{\text{bias}, G}$ of the Frobenius norm of the bias error on the mean FRM estimate $\hat{G}_{\text{mean}}(\Omega_k)$ over the non-concatenated data sets to that of the FRM estimate $\hat{G}_{\text{concat}}(\Omega_k)$ using the concatenated data records is given by

$$R_{\text{bias}, G} = O(\xi^{R+1}(M_c)) \tag{7-59}$$

where

$$\xi(M_c) = M_c \frac{q - 1 + (R+1)(n_u + 1)}{q - 1 + (R+1)(n_u + M_c)} \text{ with } \xi(\infty) = \frac{q - 1 + (R+1)(n_u + 1)}{R + 1} \tag{7-60}$$

It can easily be verified that $R_{\text{bias}, G} > 1$ for any value of q, R, and n_u if $M_c \geq 2$, which shows that the bias error on the FRM estimate is reduced by concatenating data records. The bias reduction (7-59) of $\hat{G}_{\text{concat}}(\Omega_k)$ w.r.t. $\hat{G}_{\text{mean}}(\Omega_k)$ is maximal as the number of records M_c tends to infinity

$$R_{\text{bias}, G}^{\max} = \lim_{M_c \to \infty} R_{\text{bias}, G} = O(\xi^{R+1}(\infty)) \tag{7-61}$$

For example, for $q = 6$ (a typical value needed in single-input, single-output parametric transfer function modeling, see Chapter 12), $R = 2$, and $n_u = 1$, $R_{\text{bias}, G}$ (7-59) becomes

$$R_{\text{bias}, G} = O\left(\left(\frac{11 M_c}{8 + 3 M_c}\right)^3\right)$$

and the corresponding maximal bias reduction equals $R_{\text{bias}, G}^{\max} = 33.9$ dB. (see Figure 7-6).

Following the same lines, bounds on the ratio R_{bias, C_V} of the Frobenius norm of the bias errors on the noise covariance estimates (7-17) are found

$$O(\xi) \leq R_{\text{bias}, C_V} \leq O(\xi^{2(R+1)}) \tag{7-62}$$

where ξ is defined in (7-60). The lower and upper bounds in (7-62) are reached when, respectively, the noise and the plant interpolation errors in (7-17) are dominant. Using the same numerical example as for the FRM estimate, the maximal bias error reduction on the noise covariance estimate $R_{\text{bias}, C_V}^{\max}$ varies between 5.6 dB and 33.9 dB.

7.2.9.3 Covariance Frequency Response matrix.

Under the same conditions as for the bias error (see Section 7.2.9.2), we compare the covariance of the frequency response matrix (FRM) estimate $\hat{G}_{\text{concat}}(\Omega_k)$ of the concatenated data sets with the covariance of the mean FRM estimate $\hat{G}_{\text{mean}}(\Omega_k)$ over the non-concatenated records. Since q (7-13) is constant in the analysis, the difference between the number of columns and rows of K_n in (7-8) is independent of the number of data records M_c. Hence, $\overline{\mathbb{E}\{S^H S\}}$ in (7-19) will be more or less the same for $\hat{G}_{\text{mean}}(\Omega_k)$ and $\hat{G}_{\text{concat}}(\Omega_k)$, and it is sufficient to compare the expected value of the noise covariance estimate $\hat{C}_V(k)$ of both approaches. Two limiting cases are considered: (i) the term in (7-17) depending on the plant interpolation error $O_{\text{int}G}$ is dominant, and (ii) the noise contribution $C_V(k)$ in (7-17) is dominant.

If the plant interpolation error is dominant in (7-17), then it follows from (7-19) that the ratio $R_{\text{cov}, G}$ of the Frobenius norm of the covariance of $\hat{G}_{\text{mean}}(\Omega_k)$ to that of $\hat{G}_{\text{concat}}(\Omega_k)$ is given by $R_{\text{cov}, G} = O(\xi^{2(R+1)}(M_c))/M_c$ where ξ is defined in (7-60). The additional factor $1/M_c$ quantifies the covariance reduction of $\hat{G}_{\text{mean}}(\Omega_k)$ due to the averaging over the M_c data records. As the number of records M_c tends to infinity, the ratio $R_{\text{cov}, G}$ tends to $O(\xi^{2(R+1)}(\infty))/M_c$. Hence, for M_c sufficiently large, the covariance of $\hat{G}_{\text{mean}}(\Omega_k)$ becomes smaller than that of $\hat{G}_{\text{concat}}(\Omega_k)$. However, since the frequency resolution of $\hat{G}_{\text{concat}}(\Omega_k)$ is M_c times larger than that of $\hat{G}_{\text{mean}}(\Omega_k)$, the information content of $\hat{G}_{\text{concat}}(\Omega_k)$ is still $O(\xi^{2(R+1)}(\infty))$ times larger than that of $\hat{G}_{\text{mean}}(\Omega_k)$.

If the noise contribution $C_V(k)$ is dominant in (7-17), then it follows from (7-19) that the ratio equals $R_{\text{cov}, G} = M_c^{-1}$. Since the frequency resolution of $\hat{G}_{\text{concat}}(\Omega_k)$ is M_c times larger than that of $\hat{G}_{\text{mean}}(\Omega_k)$, it follows that both estimates have the same information content.

We conclude that the ratio $R_{\text{cov}, G}$ is bounded by

$$\frac{1}{M_c} \leq R_{\text{cov}, G} \leq \frac{1}{M_c} O(\xi^{2(R+1)}(M_c)) \qquad (7\text{-}63)$$

where ξ is defined in (7-60). Figure 7-6 shows the bounds (7-63) for the same numerical example as the bias error in Section 7.2.3.2. The upper bound (7-63) becomes maximal in Figure 7-6 for $M_c = 13$: $R_{\text{cov}, G}^{\max} = 17.9$ dB.

7.2.9.4 Conclusion.

Concatenating data records reduces significantly the bias of the FRM and noise covariance estimates. In those frequency bands where the plant interpolation/leakage errors are important, there is also a significant reduction in the covariance of the FRM estimates. Taking into account the increased frequency resolution, the concatenation does not increase the noise sensitivity of the FRM estimate in those frequency bands where the noise error is dominant.

Figure 7-6. Comparison between the frequency response function (FRF) estimates using the non-concatenated ($\hat{G}_{\text{mean}}$) and the concatenated ($\hat{G}_{\text{concat}}$) data records for the case $q = 6$, $R = 2$, and $n_u = 1$. Black line: the bias ratio bias($\hat{G}_{\text{mean}}$)/bias($\hat{G}_{\text{concat}}$), and gray lines: upper and lower bound of the variance ratio var($\hat{G}_{\text{mean}}$)/var($\hat{G}_{\text{concat}}$).

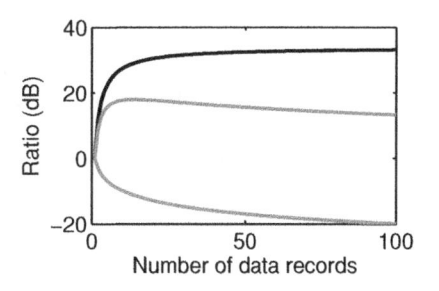

7.2.10 Experimental Illustration

The goal of the measurement example is threefold: (i) comparison of the local polynomial estimates (Section 7.2.2) with the classical spectral analysis method (Section 7.2.3); (ii) comparison of the direct (Section 7.2.2) and indirect (Section 7.2.7) local polynomial estimates on noisy input-output data; and (iii) experimental performance analysis of the concatenation of data records (Section 7.2.9).

7.2.10.1 Measurement Setup. A steel beam ($\rho = 7800$ kg/m^3) of length 61 cm, height 2.47 cm, and width 4.93 mm is hung by two nylon threads perpendicular to the excitation direction. It is excited in its transverse direction by a mini-shaker (B&K 4810) at 10 cm from the end of the beam. The force (input) and acceleration (output) at the excitation point are measured with an impedance head (B&K 8001). These signals are amplified (charge amplifier B&K 2635) and buffered ($Z_{in} > 5$ MΩ, $Z_{out} = 50$ Ω) before being applied to the acquisition channels (HPE 1430A, $Z_{in} = 50$ Ω) of the VXI measurement device. The excitations are generated by an arbitrary waveform generator (HPE 1445A, $Z_{out} = 50$ Ω) at a sampling frequency $f_s = 10$ MHz/$2^9 \approx 19.53$ kHz. The output of the arbitrary waveform generator is amplified before being applied to the mini-shaker. In order to reduce the effect of the inductive impedance of the shaker on the amplifier unit, an 18 Ω/5 W resistance is put in series with its input.

An experiment is performed with a random binary sequence covering the band from DC to 6 kHz ($u_{rms} = 241$ mV). $N = 50 \times 1024$ samples of the reference signal (signal stored in the arbitrary waveform generator), the force (input), and the acceleration (output) are collected. Based on these signals the indirect (7-49) and direct (7-26), (7-33) local polynomial and spectral analysis (diff window) estimates of the frequency response function are calculated.

7.2.10.2 Comparison Local Polynomial and Spectral Analysis Estimates. The indirect FRF estimates (7-49) are calculated for the local polynomial method (7-26) with $R = 2$ and n in (7-13) chosen such that $dof_{arb} = q = 13$, and the spectral analysis method (7-33) using the diff window where $M = 13$ in (7-31). With this choice of the parameters, the useful noise averaging in the FRF estimate of the spectral analysis method over the $M = 13$ blocks is equivalent to that of the local polynomial method over the $2n + 1$ neighboring frequencies ($q = 13$). Figure 7-7 shows the results: compared with the local polynomial estimates (black lines), the spectral analysis method (gray lines) has a much lower frequency resolution and a larger variance. The latter is due to the superior leakage (transient) suppression of the local polynomial method.

7.2.10.3 Comparison Errors-in-Variables and Output Error Estimates. To illustrate the importance of the noise on the input data, we compare in Figure 7-8 the indirect local polynomial estimate using the reference signal and the input-output data (= errors-in-variables estimate, see Section 7.2.7), with the direct local polynomial estimate using the input-output data only (= output error estimate, see Section 7.2.2). Note the large bias and large uncertainty of the output error (direct) estimate at the resonance frequencies. This is explained by the fact that, at the resonance frequencies, the amplitude of the input spectrums drops to zero, resulting in a poor input signal-to-noise ratio. This invalidates the hypothesis of a high input signal-to-noise ratio made by the output error approach.

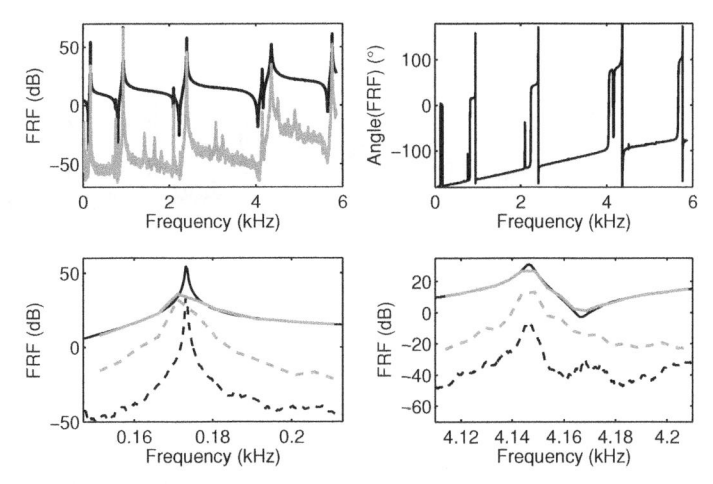

Figure 7-7. Measured force-to-acceleration frequency response function (FRF) – flexural vibrations of a steel beam. Top row: local polynomial estimate (black lines) and its standard deviation (gray line). Bottom row: zoom around the first (left figure) and fifth (right figure) resonance of the local polynomial (black) and the spectral analysis (gray) estimates and their corresponding standard deviation (dashed lines).

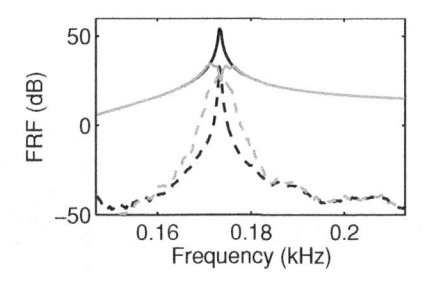

Figure 7-8. Comparison between the local polynomial estimates of the frequency response function (FRF) using an errors-in-variables (black) and the output error (gray) approach – zoom around the first resonance. Solid lines: FRF; dashed lines: standard deviation FRF.

7.2.10.4 Concatenation of Data Records.

To illustrate the concatenation of data sets, the reference signal and the input-output data are divided in four blocks of equal length $N/4$. Next, the concatenation of the blocks in the reversed order $(4, 3, 2, 1)$ is processed using the indirect local polynomial method of Section 7.2.7, where three signals $\delta(t - iN/4)$, $i = 1, 2, 3$ and $t = 0, 1, ..., N-1$ with $\delta(t) = 0$ for $t \neq 0$ and $\delta(0) = 1$, are added to the known reference signal (see Section 7.2.9). The result is the frequency response function estimate $\hat{G}_{\text{concat}}(\Omega_k)$. Further, the following indirect local polynomial estimates are calculated: the mean $\hat{G}_{\text{mean}}(\Omega_k)$ of the FRFs of each block of length $N/4$ separately, and the FRF $\hat{G}_{\text{full}}(\Omega_k)$ using the full data record.

Figure 7-9 compares the mean FRF over the four blocks ($\hat{G}_{\text{mean}}(\Omega_k)$) with the FRF using the concatenated data blocks ($\hat{G}_{\text{concat}}(\Omega_k)$). It follows that both estimates coincide: 49.9% of the differences between the FRFs (top left figure black line) fall within the 50% confidence bound of $\hat{G}_{\text{concat}}(\Omega_k)$ (top left figure, gray line). In the frequency bands where the plant interpolation errors are dominant, the variance of $\hat{G}_{\text{mean}}(\Omega_k)$ is about 23.1 dB larger than the variance of $\hat{G}_{\text{concat}}(\Omega_k)$; while in the bands where the noise error is dominant the $\text{var}(\hat{G}_{\text{mean}}(\Omega_k))$ is about -4.5 dB smaller than $\text{var}(\hat{G}_{\text{concat}}(\Omega_k))$. These values are in good agreement with the theoretical upper (19.5 dB) and lower (-6 dB) bounds in (7-63).

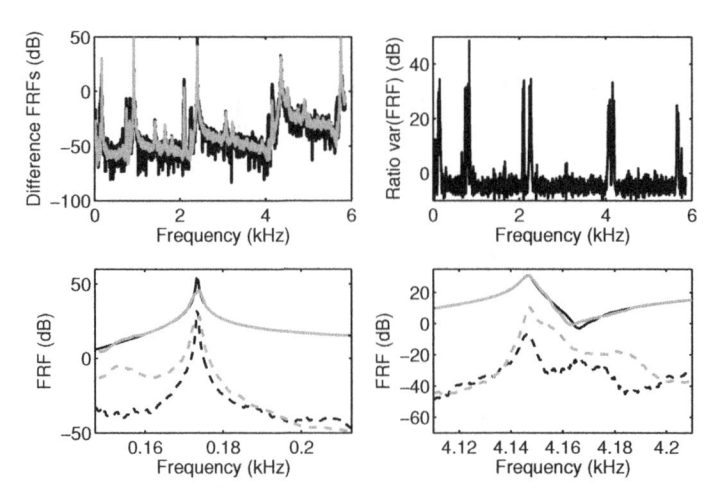

Figure 7-9. Comparison between the frequency response function (FRF) estimate using the non-concatenated ($\hat{G}_{\text{mean}}$) and concatenated ($\hat{G}_{\text{concat}}$) records – flexural vibrations of a steel beam. Top left figure: the difference $|\hat{G}_{\text{concat}} - \hat{G}_{\text{mean}}|$ (black) and the 50% confidence bound of $\hat{G}_{\text{concat}}$ (gray). Top right figure: the variance ratio $\text{var}(\hat{G}_{\text{mean}})/\text{var}(\hat{G}_{\text{concat}})$. Bottom row: zoom around the first (left figure) and fifth (right figure) resonance of $\hat{G}_{\text{concat}}$ (black) and $\hat{G}_{\text{mean}}$ (gray) and the corresponding standard deviations (dashed lines).

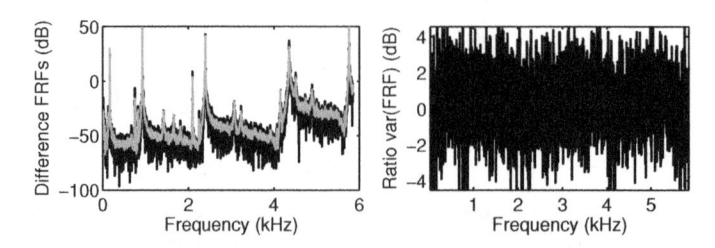

Figure 7-10. Comparison between the indirect local polynomial estimates of the frequency response function (FRF) using the full ($\hat{G}_{\text{full}}$) and the concatenated ($\hat{G}_{\text{concat}}$) data sets – flexural vibrations of a steel beam. Left figure: the difference $|\hat{G}_{\text{concat}} - \hat{G}_{\text{full}}|$ (black) and the 50% confidence bound of $\hat{G}_{\text{concat}}$ (gray). Right figure: the variance ratio $\text{var}(\hat{G}_{\text{full}})/\text{var}(\hat{G}_{\text{concat}})$.

Figure 7-10 compares the FRF estimates using the full data record ($\hat{G}_{\text{full}}(\Omega_k)$) and the concatenated data records ($\hat{G}_{\text{concat}}(\Omega_k)$). It follows that both estimates coincide and have the same variance: 61.5% of the differences between the FRFs (left figure, black line) fall within the 50% confidence bound of the FRF estimate (left figure, gray line), and the mean ratio of the variances equals 1.05 (right figure).

7.3 PERIODIC EXCITATIONS

7.3.1 Introduction

Consider a linear multivariable n_u- input, n_y- output system operating in open or closed loop (see Figure 7-4). The input-output DFT spectra of the measured steady state response of that system to a periodic reference signal $r(t)$ are related as

Section 7.3 ■ Periodic Excitations

$$\begin{aligned} U(k) &= U_0(k) + V_U(k) + T_{H_U}(\Omega_k) \\ Y(k) &= G(\Omega_k)U_0(k) + V_Y(k) + T_{H_Y}(\Omega_k) \end{aligned} \qquad (7\text{-}64)$$

with $T_{H_U}(\Omega_k)$ and $T_{H_Y}(\Omega_k)$ the $n_u \times 1$ input and the $n_y \times 1$ output noise transient (leakage) errors, and where $V_U(k) = H_U(\Omega_k)E_U(k)$ and $V_Y(k) = H_Y(\Omega_k)E_Y(k)$ depend on the noise sources in Figure 7-4 (proof: use (6-69) on page 200 with $T_G = 0$). The input-output noise transient errors increase the variability of the classical frequency response matrix estimators (see Sections 2.4 and 2.7), and introduce a correlation between the DFT spectra of consecutive periods. The goal is to obtain, at the excited frequencies, a noise leakage-free FRM estimate and a non-parametric estimate of the input-output noise covariances

$$C_U(k) = \mathrm{Cov}(V_U(k)), \ C_Y(k) = \mathrm{Cov}(V_Y(k)), \text{ and } C_{YU}(k) = \mathbb{E}\{V_Y(k)V_U^H(k)\} \qquad (7\text{-}65)$$

starting from the measured input-output DFT spectra $U(k)$, $Y(k)$. We hereby assume that the DFT spectra $E_U(k)$ and $E_Y(k)$ satisfy properties (i–iii) on page 227.

First an algorithm for suppressing nonparametrically the noise transient (leakage) errors in periodic signals is presented in Section 7.3.2. It is the first step of the robust method for measuring the frequency response matrix of linear systems (Sections 7.3.3), and the robust and fast methods for measuring the best linear approximation (BLA) of nonlinear systems (Sections 7.3.6 and 7.3.7). Next, the methodology is extended to non-steady state conditions (Section 7.3.8). Finally, the theory is illustrated on real measurement examples in Section 7.3.9.

To guide the reader through the different algorithms, we give in Table 7-1 an overview of the experimental conditions (required number of experiments, periods P, and random phase realizations M) and the approximations made (smooth behavior of the FRF/BLA and/ or the transient errors) when measuring the frequency response function (FRF) of a linear time-invariant system, or the best linear approximation (BLA) of a nonlinear PISPO system. For a given measurement time, a maximal frequency resolution is obtained by selecting the minimal values of M and P.

TABLE 7-1 Overview of the Experimental Conditions and the Approximations Made When Measuring an n_u Input, n_y Output System

Algorithm	Frequency Response Function (FRF)	Best Linear Approximation (BLA)
Robust	• n_u experiments with multisines of Sections 2.7.2 or 3.7 • $P \geq 2$ periods • local polynomial approximation of the transient (leakage) error	• $M \times n_u$ experiments with full random orthogonal multisines (3-31) • $P \geq 2$ periods and $M \geq 2$ random phase realizations • local polynomial approximation of the transient (leakage) error
Fast	• 1 experiment with uncorrelated inputs • $P \geq 1$ periods • local polynomial approximation of the FRF[a] and the transient (leakage) error	• 1 experiment with uncorrelated multisines • $P \geq 2$ periods • local polynomial approximation of the BLA and the transient (leakage) error

a. Not for single input systems ($n_u = 1$).

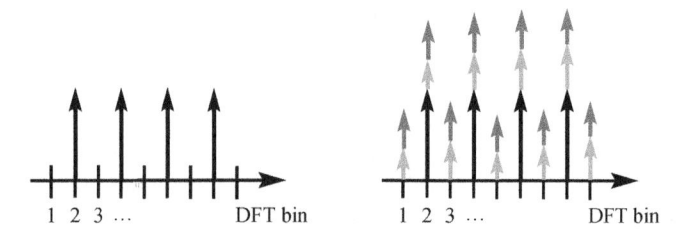

Figure 7-11. DFT spectrum of $P = 2$ consecutive periods of a noiseless (left) and a noisy (right) periodic signal. The light gray arrows represent the noise (random function of the frequency), the dark gray arrows the noise transient (smooth function of the frequency), and the black arrows the periodic signal.

7.3.2 Suppression of the Noise Transient (Leakage) Errors in Periodic Signals

7.3.2.1 Basic Idea. The method starts from the input-output DFT spectra of P periods of the steady state response to a broadband periodic excitation signal

$$Z(k) = \frac{1}{\sqrt{PN}} \sum_{t=0}^{PN-1} z(t) e^{-j2\pi k t/(PN)} \tag{7-66}$$

with N the number of samples per period, and where $z(t)$ is a $(n_y + n_u) \times 1$ vector of the outputs and inputs stacked on top of each other

$$z(t) = \begin{bmatrix} y(t) \\ u(t) \end{bmatrix} \tag{7-67}$$

Since the DFT spectrum $Z_0(k)$ of the noiseless signal $z_0(t)$ is exactly zero at the DFT frequencies $kP + r$, for $k = 0, 1, ..., N/2 - 1$ and $r = 1, 2, ..., P - 1$ (see Figure 7-11, left plot for the case $P = 2$), it can only contain signal energy at DFT frequencies kP (see Figure 7-11, left plot). Therefore, $2n$ non-excited DFT lines $Z(kP + r)$ – the n first lines to the left and the n first lines to the right of $Z(kP)$ (see Figure 7-11, right plot, odd DFT frequencies) – are used for estimating the input-output noise covariance matrices

$$C_Z^{\text{noise}}(kP) = \text{Cov}(V_Z(kP)) \text{ with } V_Z(kP) = \begin{bmatrix} V_Y(kP) \\ V_U(kP) \end{bmatrix} \tag{7-68}$$

(see Figure 7-11, right plot: light gray arrows), and the input-output noise transient terms

$$T_{H_Z}(\Omega_{kP}) = \begin{bmatrix} T_{H_Y}(\Omega_{kP}) \\ T_{H_U}(kP) \end{bmatrix} \tag{7-69}$$

(see Figure 7-11, right plot: dark gray arrows) in (7-64). This is done via a local polynomial least squares approximation of degree $R \geq 1$ of the noise transient terms at the non-excited frequencies

$$T_{H_Z}(\Omega_{kP+r}) = T_{H_Z}(\Omega_{kP}) + \sum_{s=1}^{R} t_s(k)r^s + (PN)^{-1/2}O((r/(PN))^{R+1}) \quad (7\text{-}70)$$

Removing the estimated noise transient term $\hat{T}_{H_Z}(\Omega_{kP})$ from the signal line $Z(kP)$ (see Figure 7-11, right plot: dark gray arrows at the even DFT frequencies) finally gives the sample mean of the input-output DFT spectra. The whole procedure is a simplified version of the local polynomial method for arbitrary excitations (see Section 7.2.2), and is summarized in the next section.

7.3.2.2 Algorithm. The noise transient $\hat{T}_{H_Z}(\Omega_{kP})$ and the noise covariance $\hat{C}_Z^{\text{noise}}(kP)$ estimates are calculated from the $(n_y + n_u) \times (R+1)$ local polynomial least squares approximation $\hat{\Theta}$ of degree R of the noise transient as

$$\hat{T}_{H_Z}(\Omega_{kP}) = \hat{\Theta}_{[:,1]}$$

$$\hat{C}_Z^{\text{noise}}(kP) = \frac{1}{q^{\text{noise}}}(Z_n - \hat{\Theta}K_n)(Z_n - \hat{\Theta}K_n)^H \quad (7\text{-}71)$$

$$\hat{\Theta} = Z_n K_n^H (K_n K_n^H)^{-1}$$

with $q^{\text{noise}} = 2n - (R+1)$ the number of degrees of freedom of the least squares residuals, and $x_{[:,1]}$ the first column of x. The $(n_y + n_u) \times 2n$ matrix Z_n and the $(R+1) \times 2n$ matrix K_n have the following form

$$X_n = \begin{bmatrix} X(kP - r_n) & \ldots & X(kP - r_1) & X(kP + r_1) & \ldots & X(kP + r_n) \end{bmatrix} \quad (7\text{-}72)$$

with $X = Z, K$; $Z(kP \pm r_i)$ the input-output DFT spectra at the non-excited DFT lines; and $K(kP \pm r_i) = [1 \ (\pm r_i) \ \ldots \ (\pm r_i)^R]^T$ with R the degree of the polynomial approximation. The numbers r_i, $i = 1, 2, \ldots, n$, are the n first elements of the set $\mathbb{N} \setminus \{kP | k \in \mathbb{N}\}$ (e.g., the gray arrows at the odd lines in the right plot of Figure 7-11). Combining (7-68) and (7-69) with (7-71) finally gives the local polynomial estimates of the input-output transient terms and noise covariances.

Subtracting the transient term $\hat{T}_{H_Z}(\Omega_{kP})$ (7-71) from the excited DFT lines $Z(kP)$ defines the sample mean of the input-output spectra over the P periods

$$\hat{Z}(kP) = Z(kP) - \hat{T}_{H_Z}(\Omega_{kP}) \quad (7\text{-}73)$$

The sample covariance $\hat{C}_{\hat{Z}}^{\text{noise}}(kP)$ of the sample mean $\hat{Z}(kP)$ (7-73) is related to the sample noise covariance $\hat{C}_Z^{\text{noise}}(kP)$ (7-71) as

$$\hat{C}_{\hat{Z}}^{\text{noise}}(kP) = \mu_{\text{poly}} \hat{C}_Z^{\text{noise}}(kP)$$

$$\mu_{\text{poly}} = 1 + \|\Sigma_K^{-1} V_{K[1,:]}^H\|_2^2 \quad (7\text{-}74)$$

with $\|x\|_2$ the 2-norm of x, $x_{[1,:]}$ the first row of x, and $U_K \Sigma_K V_K^H$ the singular value decomposition of K_n^H (proof: see Appendix 7.H). The factor μ_{poly} quantifies the increase in noise variance of the sample mean $\hat{Z}(kP)$ (7-73) w.r.t. the uncorrected DFT spectrum $Z(kP)$. It is induced by the estimate $\hat{T}_{H_Z}(\Omega_{kP})$ of the transient term in (7-73), and is typically 1 dB.

The degrees of freedom (dof^{noise}) of the sample noise covariance matrices (7-71) and (7-74) are set by q^{noise} in (7-71)

$$dof^{\text{noise}} = q^{\text{noise}} = 2n - (R+1) \qquad (7\text{-}75)$$

To ensure that $\hat{C}_Z^{\text{noise}}(kP)$ has the same rank as the true noise covariance $C_Z^{\text{noise}}(kP)$, one needs to fulfil the condition $dof^{\text{noise}} \geq n_y + n_u$. As a consequence the frequency width $[kP - r_n, kP + r_n]$ of the local polynomial least squares approximation is larger for multivariable than for single-input, single-output systems. However, in any case $P = 2$ periods remain sufficient for the noise analysis.

Remarks

(i) The two main differences between (7-71) and the local polynomial method for arbitrary excitations (7-10) are: (i) only the transient term is estimated in (7-71), while (7-10) estimates both the FRM and the transient, and (ii) only non-excited neighboring frequencies are used in (7-71), while (7-10) uses all neighboring frequencies.

(ii) Following the same lines of Section 7.2.2.5, the least squares estimate (7-71) is calculated in a numerically stable way via the singular value decomposition of K_n^H.

7.3.2.3 Bias Error. The expected values of the sample mean (7-73) and the sample noise covariance matrix (7-71) equal

$$\begin{aligned} \mathbb{E}\{\hat{Z}(kP)\} &= Z_0(kP) \\ \mathbb{E}\{\hat{C}_Z^{\text{noise}}(kP)\} &= C_Z^{\text{noise}}(kP) + O_{\text{int}H}((r_n/(PN))^2) + O_{\text{leak}H}((r_n/(PN))^{R+2}) \end{aligned} \qquad (7\text{-}76)$$

with $Z_0(kP)$ the true DFT spectrum, r_n defined in (7-72), $C_Z^{\text{noise}}(kP)$ the true noise covariance (7-68), and where $O_{\text{int}H}$ and $O_{\text{leak}H}$ are the bias contributions of the noise interpolation error and the residual noise leakage error, respectively (proof: see Appendix 7.I). The interpolation errors stem from the noise coloring over the $2n$ neighboring non-excited DFT lines in the linear least squares estimate (7-71). Notice that the noise interpolation error in (7-17) is larger than in (7-76). This is due to the coloring of the input power spectrum over the neighboring $2n + 1$ excited frequencies.

7.3.3 The Robust Method for Measuring the Frequency Response Matrix

7.3.3.1 Basic Algorithm. The robust method for measuring the frequency response matrix (FRM) starts from $P \geq 2$ consecutive periods of the steady state response to n_u linearly independent $n_u \times 1$ broadband reference signals $r^{[e]}(t)$, $e = 1, 2, ..., n_u$, for example, the Hadamard or orthogonal multisines of Section 2.7.2 on page 65, or the (full) random orthogonal multisines of Section 3.7 on page 92. Next, the noise transient suppression algorithm of Section 7.3.2 is applied to each experiment separately, giving the following sample means and sample covariances of the input-output DFT spectra

$$\hat{Z}^{[e]}(kP), \hat{C}_{\hat{Z}}^{\text{noise}[e]}(kP) \text{ for } e = 1, 2, ..., n_u \qquad (7\text{-}77)$$

Section 7.3 ■ Periodic Excitations

$$\hat{\mathbf{Z}}(kP) = [\hat{Z}^{[1]}(kP) \; \hat{Z}^{[2]}(kP) \; \ldots \; \hat{Z}^{[n_u]}(kP)] \tag{7-78}$$

(see (7-73) and (7-74)). Finally, the FRM is calculated from (7-78) as

$$\hat{G}(j\omega_k) = \hat{\mathbf{Y}}(kP)\hat{\mathbf{U}}^{-1}(kP) \tag{7-79}$$

with $\hat{\mathbf{Y}}(kP)$ and $\hat{\mathbf{U}}(kP)$, respectively, the first n_y and last n_u rows of $\hat{\mathbf{Z}}(kP)$, and where $\omega_k = 2\pi k f_s / N$.

7.3.3.2 Bias Error.
The expected value of sample noise covariance in (7-77) equals

$$\mathbb{E}\{\hat{C}_{\hat{Z}}^{\text{noise}[e]}(kP)\} = \mu_{\text{poly}} C_Z^{\text{noise}}(kP) + O_{\text{int}H}((r_n/(PN))^2) \tag{7-80}$$

with μ_{poly} defined in (7-74), r_n defined in (7-72), and where $C_Z^{\text{noise}}(kP)$ is the true noise covariance (7-68) of one multisine experiment (proof: combine (7-74) and (7-76), and neglect $O_{\text{leak}H}$ w.r.t. $O_{\text{int}H}$).

Since $\hat{T}_{H_Z}(\Omega_{kP})$ is an unbiased estimate of $T_{H_Z}(\Omega_k)$ ((7-71) with $\mathbb{E}\{T_{H_Z}(\Omega_k)\} = 0$ and $\mathbb{E}\{V_Z(k)\} = 0$), it follows that the FRM estimate (7-79) is unbiased in the absence of input noise. Although the FRM estimate (7-79) is biased in the presence of input noise, the bias can be neglected if the input signal-to-noise ratio is larger than 10 dB (see Section 2.4.2.1 on page 46).

7.3.3.3 Covariance of the Frequency Response Matrix.
The covariance of the FRM estimate (7-79) is calculated as in (2-77) on page 66, where $G_0(j\omega_k)$, $\mathbf{U}_0(k)$ and $C_Z(k)$ are replaced by, respectively, $\hat{G}(j\omega_k)$, $\hat{\mathbf{U}}(kP)$ and the mean over the n_u experiments of the sample noise covariance estimate (7-77)

$$\hat{C}_{\hat{Z}}^{\text{noise}}(kP) = \frac{1}{n_u} \sum_{e=1}^{n_u} \hat{C}_{\hat{Z}}^{\text{noise}[e]}(kP)$$

7.3.4 The Fast Method for Measuring the Frequency Response Matrix

The fast method for measuring the frequency response matrix (FRM) starts from $P = 1$ period of the steady state response to a known $n_u \times 1$ broadband reference signal $r(t)$ with a random phase DFT spectrum. Examples of such signals are one column of the Hadamard or orthogonal multisines of Section 2.7.2 on page 65, where $r_{\text{SISO}}(t)$ is a random phase multisine, or one column of the (full) random orthogonal multisines of Section 3.7 on page 92. Next, the local polynomial method for arbitrary excitations of Section 7.2.7 is applied to the known $n_u \times 1$ input $r(t)$, noisy $(n_y + n_u) \times 1$ output data $z(t) = [y^T(t) \; u^T(t)]^T$, giving the FRM estimate (7-49) and its covariance (7-50). Note that the random phase condition of the reference signal is sufficient to separate in (7-47) the exogenous term $G_{rz}(\Omega_k)R(k)$ (random function over the frequency) from the transient term $T_{rz}(\Omega_k)$ (smooth function of the frequency).

For random phase multisines the bias expressions (7-17) and (7-18) are valid with $O_{\text{leak}G} = 0$. The main differences w.r.t. the random excitations are that (i) the amplitude spectrum of the multisine is deterministic instead of random, and (ii) the transient (leakage) is only due to the noise dynamics.

7.3.5 Nonlinear Systems

Figure 7-5 shows the setup for measuring the best linear approximation (BLA) of a nonlinear system operating in open or closed loop. The input-output DFT spectra of the steady state response to a periodic reference signal $r(t)$ are then related as

$$U(k) = U_0(k) + U_S(k) + V_U(k) + T_{H_U}(\Omega_k)$$
$$Y(k) = G_{\text{BLA}}(\Omega_k)U_0(k) + Y_S(k) + V_Y(k) + T_{H_Y}(\Omega_k) \quad (7\text{-}81)$$

with $G_{\text{BLA}}(\Omega_k)$ the best linear approximation, $U_S(k)$ and $Y_S(k)$ the input-output stochastic nonlinear distortions, $T_{H_U}(\Omega_k)$ and $T_{H_Y}(\Omega_k)$ the input-output noise transient (leakage) errors, $V_U(k) = H_U(\Omega_k)E_U(k)$ and $V_Y(k) = H_Y(\Omega_k)E_Y(k)$ the input-output errors depending on the noise sources in Figure 7-5, and where $U_0(k)$ is the DFT spectrum of the periodic part of the plant input that is correlated with the reference signal

$$U_0(k) = (I_{n_u} + G_{\text{cont}}(\Omega_k)G_{\text{BLA}}(\Omega_k))^{-1}G_{\text{act}}(\Omega_k)R(k) \quad (7\text{-}82)$$

(proof: combine Theorems 3.16 and 3.22 with (6-69)). $G_{\text{act}}(\Omega)$ and $G_{\text{cont}}(\Omega)$ are, respectively, the actuator and controller transfer functions, and $U_S(k)$ and $Y_S(k)$ are correlated via the feedback loop and/or the nonlinear distortions of the actuator (see Section 7.2.8).

The robust and fast methods explained in, respectively, Sections 7.3.6 and 7.3.7 are indirect methods: as in (7-49), the BLA is obtained via the FRMs from reference to output and from reference to input. Hence, they can handle the open as well as the closed loop cases, and automatically account for the input stochastic nonlinear distortions $U_S(k)$ and for the correlation between $U_S(k)$ and $Y_S(k)$.

7.3.6 The Robust Method for Measuring the Best Linear Approximation

7.3.6.1 Basic Algorithm. The robust method for measuring the best linear approximation (BLA) requires that the n_u steady state experiments are performed with the full random orthogonal multisines (3-31) on page 93, and that the reference signal $r(t)$ is known. Since the stochastic nonlinear distortions $u_s(t)$ and $y_s(t)$ have the same periodicity as the reference signal, no information about $u_s(t)$ and $y_s(t)$ can be gathered by comparing consecutive periods of these experiments. Therefore, the experiments with the full random orthogonal multisines are repeated for $M \geq 2$ independent random phase realizations ($\angle R_{[p]}(k)$ as well as $\phi_e(k)$ in (3-31), for $p, e = 1, 2, ..., n_u$ and $k = 1, 2, ..., F$). Since $y_s(t)$ depends on the particular random phase realization of $r(t)$, comparing the FRM estimates over the M different full random phase experiments allows us to estimate the BLA and the covariance of the stochastic nonlinear distortions as explained in detail in the sequel of this section.

Applying the robust method for measuring the FRM (see Section 7.3.3) to each realization of the full random orthogonal multisines gives the following sample means and sample noise covariances of the input-output DFT spectra

$$\hat{Z}^{[m,e]}(kP), \hat{C}_{\hat{Z}}^{\text{noise}[m,e]}(kP) \text{ for } e = 1, 2, ..., n_u \text{ and } m = 1, 2, ..., M \quad (7\text{-}83)$$

$$\hat{Z}^{[m]}(kP) = [\hat{Z}^{[m,1]}(kP) \; \hat{Z}^{[m,2]}(kP) \; ... \; \hat{Z}^{[m,n_u]}(kP)] \quad (7\text{-}84)$$

Section 7.3 ■ Periodic Excitations

Averaging of $\hat{C}_{\hat{Z}}^{\text{noise}[m,\,e]}(kP)$ over the M realizations gives an improved estimate of the input-output noise covariances

$$\hat{C}_{\hat{Z}}^{\text{noise}[e]}(kP) = \frac{1}{M}\sum_{m=1}^{M} \hat{C}_{\hat{Z}}^{\text{noise}[m,\,e]}(kP) \qquad (7\text{-}85)$$

Straightforward averaging of the input-output DFT spectra $\hat{\mathbf{Z}}^{[m]}(kP)$ over the realizations m is not possible because of the random choice of the phases of the reference signal (3-31) over m. To allow for averaging over m, the input-output DFT spectra $\hat{\mathbf{Z}}^{[m]}(kP)$ must first be referred to the reference signal. This is done as follows.

First, note that for each random phase realization m, the full random orthogonal multisines (3-31) can be written under the form (2-82) on page 67, where $D_{|R|}(kP)$ is a diagonal matrix containing the – by construction – realization independent input amplitude spectra, and where $T_{\angle R}^{[m]}(kP)$ is an orthogonal matrix ($T_{\angle R}^{[m]-1} = T_{\angle R}^{[m]H}$) containing the random realization of the input phases. Next, the input-output DFT spectra $\hat{\mathbf{Z}}^{[m]}(kP)$ are right divided by $T_{\angle R}^{[m]}(kP)$ giving

$$\hat{\mathbf{Z}}_R^{[m]}(kP) = \hat{\mathbf{Z}}^{[m]}(kP) T_{\angle R}^{[m]H}(kP) \qquad (7\text{-}86)$$

Notice that the columns of $\hat{\mathbf{Z}}_R^{[m]}(kP)$ are uncorrelated because the columns of $\hat{\mathbf{Z}}^{[m]}(kP)$ are – by construction – independently distributed, and because $\text{vec}(\hat{\mathbf{Z}}_R^{[m]}(kP))$ and $\text{vec}(\hat{\mathbf{Z}}^{[m]}(kP))$ have the same covariance (proof: see Appendix 7.K). Further, the sample mean of the DFT spectra is obtained by averaging (7-86) over the M realizations, and the total sample covariance of each column of $\hat{\mathbf{Z}}_R^{[m]}(kP)$ is calculated by an averaging over the M realizations and $2n_E + 1$ neighboring excited frequencies. The latter is necessary for getting a covariance estimate with sufficient degrees of freedom. The resulting algorithm becomes

$$\hat{\mathbf{Z}}_R(kP) = \frac{1}{M}\sum_{m=1}^{M} \hat{\mathbf{Z}}_R^{[m]}(kP) \qquad (7\text{-}87)$$

$$\hat{C}_{\hat{Z}_R}^{[e]}(kP) = \frac{1}{M}\sum_{m=1}^{M} \sum_{i=-n_E}^{n_E} \frac{r_Z^{[m,\,e]}(k_i) r_Z^{[m,\,e]H}(k_i)}{(M-1)(2n_E+1)} \qquad (7\text{-}88)$$

$$r_Z^{[m,\,e]}(k_i) = \hat{\mathbf{Z}}_{R[:,\,e]}^{[m]}(k_i P) - \hat{\mathbf{Z}}_{R[:,\,e]}(k_i P)$$

where $X_{[:,\,e]}$ denotes column e of X, $\hat{C}_{\hat{Z}_R}^{[e]}(kP)$ is the sample total covariance of the sample mean $\hat{\mathbf{Z}}_R(kP)$ (7-87), and where the integer numbers k_i, with $k_0 = k$, indicate the first n_E excited harmonics left ($i = -n_E, ..., -1$) or right ($i = 1, ..., n_E$) from k. Note that the total covariance $\hat{C}_{\hat{Z}_R}^{[e]}(kP)$ (7-88) includes the influence of the noise and the stochastic nonlinear distortions. The noise covariance of each column of (7-87) is given by (7-85) divided by M

$$\hat{C}_{\hat{Z}_R}^{\text{noise}[e]}(kP) = \frac{1}{M}\hat{C}_{\hat{Z}}^{\text{noise}[e]}(kP) \qquad (7\text{-}89)$$

Finally, the best linear approximation is calculated from (7-87) as

$$\hat{G}_{\text{BLA}}(j\omega_k) = \hat{\mathbf{Y}}_R(kP)\hat{\mathbf{U}}_R^{-1}(kP) \qquad (7\text{-}90)$$

with $\hat{\mathbf{Y}}_R(kP)$ and $\hat{\mathbf{U}}_R(kP)$, respectively, the first n_y and last n_u rows of $\hat{\mathbf{Z}}_R(kP)$ (7-87), and where $\omega_k = 2\pi k f_s/N$.

The degrees of freedom of the sample noise (7-89) and total (7-88) covariances are given by, respectively,

$$dof_{\text{robust}}^{\text{noise}} = M(2n-(R+1)) \text{ and } dof_{\text{robust}} = (M-1)(2n_E+1) \qquad (7\text{-}91)$$

It follows that via an appropriate choice of n and n_E, measuring $P = 2$ periods of $M = 2$ realizations is enough for estimating accurately the noise and total covariances. However, the minimal values of n and n_E increase with the number of inputs n_u and outputs n_y (e.g., for the sample noise covariance (7-89) the condition $dof_{\text{robust}}^{\text{noise}} \geq n_y + n_u$ must be fulfilled).

Remarks

(i) The major difference with the robust method of Section 4.3.1 on page 130 is that $M = 2$ independent random phase realizations are sufficient for estimating the total covariance. This result is achieved by averaging the sample covariance over neighboring excited frequencies. The latter, however, introduces a bias in the estimate (see Section 7.3.6.2), that can be important for highly correlated input-output errors (see Section 7.3.6.5).

(ii) The estimated BLA (7-90) and the sample noise (7-89) and total (7-88) covariances are correlated over the frequency. The correlation length is $\pm 2n/P$ at the frequency resolution f_s/N for the BLA (7-90) and the sample noise covariance (7-89), while it is $\pm 2n_E$ for the sample total covariance (7-88).

(iii) The sample noise (7-89) and total (7-88) covariance estimates of the n_u experiments are used as weighting in parametric transfer function modeling (see Chapter 12). To calculate uncertainty bounds on the nonparametric FRM estimates the sample covariances are averaged over the n_u experiments,

$$\hat{C}_{\hat{Z}_R}^{\text{noise}}(kP) = \frac{1}{n_u}\sum_{e=1}^{n_u}\hat{C}_{\hat{Z}_R}^{\text{noise}[e]}(kP) \text{ and } \hat{C}_{\hat{Z}_R}(kP) = \frac{1}{n_u}\sum_{e=1}^{n_u}\hat{C}_{\hat{Z}_R}^{[e]}(kP), \quad (7\text{-}92)$$

which increases the degrees of freedom (7-91) with n_u.

7.3.6.2 Bias Error. Multiplying the sample noise (7-89) and total (7-88) covariances by M, and taking the expected value gives

$$\begin{aligned} M\,\mathbb{E}\{\hat{C}_{\hat{Z}_R}^{\text{noise}[e]}(kP)\} &= \mu_{\text{poly}}C_Z^{\text{noise}}(kP) + O_{\text{int}H}((r_n/(PN))^2) \\ M\,\mathbb{E}\{\hat{C}_{\hat{Z}_R}^{[e]}(kP)\} &= \mu_{\text{poly}}C_Z^{\text{noise}}(kP) + PC_{Z_S}(kP) + O_{\text{int}H}((r_{n_E}/N)^\alpha) \end{aligned} \qquad (7\text{-}93)$$

with μ_{poly} defined in (7-74), $C_Z^{\text{noise}}(kP)$ the true noise covariance (7-68) of one multisine experiment, $C_{Z_S}(kP)$ the true covariance of the stochastic nonlinear distortions $Z_S = [Y_S^T \; U_S^T]^T$ of one period of one multisine experiment, r_n defined in (7-72), $r_{n_E} = k_{n_E} - k$, and where $\alpha = 1, 2$ for, respectively, non-uniformly and uniformly distributed excited harmonics (proof: see Appendix 7.L). The interpolation bias $O_{\text{int}H}$ in the sample total covariance originates from the averaging over the $2n_E + 1$ neighboring excited frequencies in (7-88).

Section 7.3 ■ Periodic Excitations 257

Since the estimated input-output DFT spectra are unbiased $\mathbb{E}\{\hat{\mathbf{Z}}_R(kP)\} = \mathbf{Z}_0(kP)$, the estimated BLA (7-90) is unbiased in the absence of input errors (noise and/or nonlinear distortions). In the presence of input errors the BLA (7-90) is biased. However, if the input signal-to-noise and signal-to-distortion ratios are larger than 10 dB, then the bias can be neglected (see Section 2.4.2.1 on page 46).

7.3.6.3 Covariance of the Stochastic Nonlinear Distortions.

From (7-93) it can be concluded that the difference between the total (7-88) and the noise (7-89) sample covariances is a measure of the covariance of the stochastic nonlinear distortions $C_{Z_S}(kP)$, where $Z_S = [Y_S^T \ U_S^T]^T$. Hence, an estimate $\hat{C}_{Z_S}(kP)$ of $C_{Z_S}(kP)$ can be obtained as the mean value over the n_u experiments of these differences

$$\hat{C}_{Z_S}(kP) = \frac{M}{Pn_u}\sum_{e=1}^{n_u}(\hat{C}_{\hat{Z}_R}^{[e]}(kP) - \hat{C}_{\hat{Z}_R}^{\text{noise}[e]}(kP)) \qquad (7\text{-}94)$$

7.3.6.4 Covariance of the Best Linear Approximation.

The noise and total covariances of the BLA estimate (7-90) are calculated as in (2-77) on page 66, where $G_0(j\omega_k)$ and $\mathbf{U}_0(k)$ and are replaced by, respectively, $\hat{G}_{\text{BLA}}(j\omega_k)$ and $\hat{\mathbf{U}}_R(kP)$ (see (7-90)), and where $C_Z(k)$ is replaced by, respectively, $\hat{C}_{\hat{Z}_R}^{\text{noise}}(kP)$ and $\hat{C}_{\hat{Z}_R}(kP)$ (7-92).

To estimate the level of the stochastic nonlinear distortions on the BLA estimate w.r.t. one multisine experiment we replace $C_Z(k)$ in (2-77) by $\hat{C}_{Z_S}(kP)$ (7-94).

7.3.6.5 Special Case: Dominant Generator Errors.

The contribution of the generator errors $n_g(t)$ to the input-output errors $V_U(k)$ and $V_Y(k)$ in (7-64) and (7-81) are related by the true system transfer function (see Figure 7-4). Therefore, the generator error contributions to the input-output covariance matrices are cancelled in the covariance expression of the FRM (proof: see Appendix 7.M). Hence, if the generator error in Figure 7-4 is the dominant noise source, then the slightest error made in the input-output covariance estimates causes a large error in the covariance estimate of the FRM (the difference of large terms should almost be zero). This high sensitivity to small errors on the input-output covariance estimates is avoided by performing the sample mean and sample covariance calculations on the FRM instead of on the input-output DFT spectra. The bias introduced by the divisions in (7-79) and (7-90) can be neglected if the input signal-to-noise and signal-to-distortion ratios are larger than 10 dB (see Section 2.4.2.1 on page 46).

In modal analysis experiments, the nonlinear distortions generated by the shakers act as dominant generator errors that contribute to the input-output total covariances but not to the input-output noise covariances. Indeed, the nonlinear distortions are random (over the random phase realization of $r(t)$) periodic signals that are uncorrelated with the reference signal (see Section 3.8 on page 93). Since the robust method of Section 7.3.6 introduces small interpolation errors in the total covariance estimates (the $O_{\text{int}H}$ term in (7-93)), the sample mean and sample covariance calculations (7-87) and (7-88) are replaced by, respectively,

$$\hat{G}(j\omega_k) = \frac{1}{M}\sum_{m=1}^{M}\hat{G}^{[m]}(j\omega_k) \text{ with } \hat{G}^{[m]}(j\omega_k) = \hat{\mathbf{Y}}^{[m]}(kP)(\hat{\mathbf{U}}^{[m]}(kP))^{-1} \qquad (7\text{-}95)$$

$$\hat{C}_{\text{vec}G}(k) = \frac{1}{M}\sum_{m=1}^{M}\sum_{i=-n_E}^{n_E}\frac{r_{\text{vec}G}^{[m]}(k_i)r_{\text{vec}G}^{[m]H}(k_i)}{(M-1)(2n_E+1)} \qquad (7\text{-}96)$$

$$r_{\text{vec}G}^{[m]}(k) = \text{vec}(G^{[m]}(j\omega_k)) - \text{vec}(\hat{G}(j\omega_k))$$

where vec(X) puts the columns of the matrix X on top of each other, and with $\hat{\mathbf{Y}}^{[m]}(kP)$ and $\hat{\mathbf{U}}^{[m]}(kP)$, respectively, the first n_y and last n_u rows of $\hat{\mathbf{Z}}^{[m]}(kP)$ (7-84). Although the input-output noise covariances are also subject to small interpolation errors (the $O_{\text{int}H}$ term in (7-93)), they are still calculated via the procedure of Section 7.3.3, because the generator noise is mostly not dominant in modal analysis experiments.

7.3.7 The Fast Method for Measuring the Best Linear Approximation

7.3.7.1 Basic Algorithm. The fast method for measuring the best linear approximation (BLA) requires one experiment with a set of known uncorrelated random phase multisines $r(t)$ (e.g., one column of the (full) random orthogonal multisines $\mathbf{R}(k)$ in (3-30) or (3-31)). Starting from $P \geq 2$ periods of the steady state response to $r(t)$, the noise transient (leakage) errors in the input-output signals (7-81) are suppressed first via the procedure of Section 7.3.2, giving

$$\hat{Z}(kP) = G_{rz}(\Omega_{kP})R(kP) + Z_S(kP) + V_{\hat{Z}}(kP)$$
$$V_{\hat{Z}}(kP) = V_Z(kP) + T_{H_Z}(\Omega_{kP}) - \hat{T}_{H_Z}(\Omega_{kP}) \quad (7\text{-}97)$$

and the noise covariance estimate $\hat{C}_{\hat{Z}}^{\text{noise}}(kP)$ (7-74), where $G_{rz}(\Omega)$, $V_Z(kP)$, $T_{H_Z}(\Omega)$, $\hat{T}_{H_Z}(\Omega_{kP})$ and $\hat{Z}(kP)$ are defined in, respectively, (7-48), (7-68), (7-69), (7-71), and (7-73), and with $Z_S = [Y_S^T \ U_S^T]^T$. Next, a modified version of the local polynomial method for arbitrary excitations of Section 7.2.7 where no transient term estimated is applied to (7-97). The result is an estimate of the BLA $G_{\text{BLA}}(j\omega_k)$ and the total covariance $\text{Cov}(\hat{Z}(kP))$ of the input-output DFT spectra. The latter can be written as the sum of the covariance of the stochastic nonlinear distortions and the noise covariance

$$\text{Cov}(\hat{Z}(kP)) = \text{Cov}(Z_S(kP)) + \text{Cov}(V_{\hat{Z}}(kP)) \quad (7\text{-}98)$$

where a sample estimate of $\text{Cov}(V_{\hat{Z}}(kP))$ is given by (7-74). The modified local polynomial method is explained in the sequel of this section.

Since the transient term has already been suppressed in (7-97), we only need to estimate the FRM $G_{rz}(\Omega_{kP})$. Applying the local polynomial method of Section 7.2.7 without transient term to $2n_E + 1$ neighboring excited frequencies $[(k-r_{n_E})P, (k-r_{n_E}+1)P, \ldots, kP, \ldots, (k+r_{n_E})P]$ gives the following algorithm

$$\hat{G}_{rz}(\Omega_{kP}) = \hat{\Psi}_{[:,1:n_u]}$$
$$\hat{C}_{\hat{Z}}(kP) = \frac{1}{q}(\hat{Z}_{n_E} - \hat{\Psi} L_{n_E})(\hat{Z}_{n_E} - \hat{\Psi} L_{n_E})^H \quad (7\text{-}99)$$
$$\hat{\Psi} = \hat{Z}_{n_E} L_{n_E}^H (L_{n_E} L_{n_E}^H)^{-1}$$

with $q = 2n_E + 1 - (R+1)n_u$ the number of degrees of freedom of the least squares residuals, and $x_{[:,1:n_u]}$ the first n_u columns of x. The $(n_y + n_u) \times (2n_E + 1)$ matrix $\hat{Z}_{n_E}$ and the $(R+1)n_u \times (2n_E + 1)$ matrix L_{n_E} have the following form

Section 7.3 ■ Periodic Excitations

$$X_{n_E} = \begin{bmatrix} X((k-r_{n_E})P) & \ldots & X(kP) & \ldots & X((k+r_{n_E})P) \end{bmatrix} \quad (7\text{-}100)$$

with $X = Z, L$; $Z((k \pm r_i)P)$ the input-output DFT spectra at the excited DFT lines; and

$$L((k \pm r_i)P) = \begin{bmatrix} 1 \\ (\pm r_i) \\ \ldots \\ (\pm r_i)^R \end{bmatrix} \otimes R((k \pm r_i)P) \quad (7\text{-}101)$$

$\hat{C}_{\tilde{Z}}(kP)$ in (7-99) is the sample total covariance of the input-output DFT spectra that includes the influence of the noise and the stochastic nonlinear distortions. Combining (7-49) and (7-99) gives an estimate of the BLA of the nonlinear system

$$\hat{G}_{\text{BLA}}(j\omega_k) = \hat{G}_{ry}(\Omega_{kP})\hat{G}_{ru}^{-1}(\Omega_{kP}) \quad (7\text{-}102)$$

with $\omega_k = 2\pi k/N$, and where $\hat{G}_{ry}(\Omega_{kP})$ and $\hat{G}_{ru}(\Omega_{kP})$ are, respectively, the first n_y and last n_u rows of $\hat{G}_{rz}(\Omega_{kP})$.

The degrees of freedom of the sample noise (7-74) and total (7-99) covariances are given by, respectively,

$$dof_{\text{fast}}^{\text{noise}} = 2n - (R+1) \text{ and } dof_{\text{fast}} = 2n_E + 1 - (R+1)n_u \quad (7\text{-}103)$$

It follows that via an appropriate choice of n and n_E, measuring $P = 2$ consecutive periods of one multisine experiment is enough for estimating accurately the noise and total covariances. However, the minimal values of n and n_E increase with the number of inputs n_u and outputs n_y (e.g., for the sample noise covariance (7-74) the condition $dof_{\text{fast}}^{\text{noise}} \geq n_y + n_u$ must be fulfilled).

Remarks

(i) The major difference with the fast method of Section 4.3.2 on page 135 is that the information about the nonlinear distortions is obtained via a local polynomial approximation of the BLA over the excited frequency, while in Section 4.3.2 the distortion information is extracted from the non-excited multisine frequencies.

(ii) Following the same lines of Section 7.2.2.5, the least squares estimate (7-99) is calculated in a numerically stable way via the singular value decomposition of $L_{n_E}^H$.

(iii) The estimated BLA (7-102) and the sample noise (7-74) and total (7-99) covariances are correlated over the frequency. The correlation length is $\pm 2n/P$ at the frequency resolution f_s/N for the sample noise covariance (7-74), while it is $\pm 2n_E$ for the BLA (7-102) and the sample total covariance (7-99).

7.3.7.2 Bias Error. The bias error of the sample noise covariance (7-74) is given in (7-76). Taking the expected value of the sample total covariance $\hat{C}_{\tilde{Z}}(kP)$ (7-99) gives

$$\mathbb{E}\{\hat{C}_{\tilde{Z}}(kP)\} = \mu_{\text{poly}} C_{\tilde{Z}}^{\text{noise}}(kP) + PC_{Z_S}(kP) + O_{\text{int}H}(r_{n_E}/N) + O(4n_E^2/(dof_{\text{fast}}N)) \quad (7\text{-}104)$$

with $C_{Z_S}(kP)$ the true covariance of the stochastic nonlinear distortions $Z_S = [Y_S^T \; U_S^T]^T$ of one period, and where r_{n_E} and dof_{fast} are defined in (7-99) and (7-103), respectively (proof: use (7-55), where $C_V(k)$ is replaced by $\text{Cov}(\hat{Z}(kP))$ (7-98)).

If the input signal-to-noise and signal-to-distortion ratios are larger than 10 dB, then the bias on the BLA estimate is mainly due to the plant interpolation error

$$\mathbb{E}\{\hat{G}_{\text{BLA}}(j\omega_k)\} = G_{\text{BLA}}(j\omega_k) + O_{\text{int}G}((r_{n_E}/N)^{(R+1)}) \qquad (7\text{-}105)$$

with r_{n_E} defined in (7-99) (proof: use (7-18) with $O_{\text{leak}G} = 0$).

7.3.7.3 Covariance of the Stochastic Nonlinear Distortions. From the expected values (7-76) and (7-104) it can be seen that the difference between the sample total covariance (7-99) and the sample noise covariance (7-74) is an estimate of the covariance of the stochastic nonlinear distortions $Z_S = [Y_S^T \; U_S^T]^T$

$$\hat{C}_{Z_S}(kP) = \frac{1}{P}(\hat{C}_{\hat{Z}}(kP) - \hat{C}_{\hat{Z}}^{\text{noise}}(kP)) \qquad (7\text{-}106)$$

7.3.7.4 Covariance of the Best Linear Approximation. The noise and total covariances of the BLA estimate (7-102) are calculated as in (7-21) and (7-50), where K_n is replaced by L_{n_E} (see (7-99) and (7-101)), $\hat{G}$ by $\hat{G}_{\text{BLA}}$ (7-102), and $\hat{C}_V$ by, respectively, $\hat{C}_{\hat{Z}}^{\text{noise}}(kP)$ (7-74) and $\hat{C}_{\hat{Z}}(kP)$ (7-99).

To estimate the level of the stochastic nonlinear distortions on BLA estimate, we replace $\hat{C}_V$ in (7-21) by $\hat{C}_{Z_S}(kP)$ (7-106).

7.3.8 Non-Steady State Conditions

In the previous sections the analysis was done under steady state operation of the dynamic systems in Figure 7-4. If the plant is measured under transient conditions, then (7-64) is replaced by

$$\begin{aligned} U(k) &= U_0(k) + V_U(k) + T_U(\Omega_k) \\ Y(k) &= G(\Omega_k)U_0(k) + V_Y(k) + T_Y(\Omega_k) \end{aligned} \qquad (7\text{-}107)$$

with $U_0(k)$ the DFT spectrum of the periodic part of $u(t)$, and where $T_U(\Omega_k)$ and $T_Y(\Omega_k)$ are, respectively, $n_u \times 1$ and $n_y \times 1$ rational vector functions depending on the actuator, the controller (closed loop only), the plant, and the noise dynamics, and on the difference between the initial and final conditions of the experiment (proof: use (6-69) on page 200). Following the same lines, it can be seen that (7-81), where T_{H_U} and T_{H_Y} are replaced by, respectively, T_U and T_Y, is also valid under transient conditions. Since no distinction can be made between the noise and the plant transients (leakage) errors, the robust method for linear systems (Section 7.3.3) and the robust and fast methods for nonlinear systems (Sections 7.3.6 and 7.3.7) can handle the first two periods of the transient response to a periodic input. Similarly, one period of the transient response is sufficient for the fast method of Section 7.3.4.

The major difference between the plant and noise transient (leakage) errors is that the plant leakage errors introduce a bias in the estimated frequency response matrix (or best linear approximation). For the fast methods (Sections 7.3.4 and 7.3.7) the plant leakage bias $O_{\text{leak}G}$ can be neglected w.r.t. the plant interpolation bias $O_{\text{int}G}$ (see (7-18)). For the robust methods (Sections 7.3.3 and 7.3.6) the plant leakage bias is of the order

Section 7.3 ■ Periodic Excitations

$$O_{\text{leak}G}(\sqrt{N}\left(\frac{r_n}{PN}\right)^{R+2}) \tag{7-108}$$

(proof: see Appendix 7.N).

7.3.9 Experimental Illustration

The goal of the measurement example is twofold: (i) comparison of the robust (Section 7.3.6) and the fast (Section 7.3.7) estimates of the best linear approximation (BLA) of a weakly nonlinear system; and (ii) comparison of the robust local polynomial method (Section 7.3.6) with the classical robust method that neglects the noise transients (Section 4.3.1).

7.3.9.1 Measurement Setup. A plexiglass beam (density 1200 kg/m³, height 2.00 cm, and width 1.03 cm) under free-free boundary conditions is excited in its longitudinal direction by a periodic force applied at one of its ends (see Pintelon et al., 2004 for a detailed description of the experimental setup). The force (excitation $u(t)$) and acceleration (response $y(t)$) are measured at the excitation point (see Section 7.2.10.1 for a detailed description of the analog signal conditioning: generation, amplification, impedance matching). The generator and acquisition units operate at the same sampling frequency $f_s = 10 \text{ MHz}/2^{10} \approx 9.77$ kHz.

Experiments are performed with odd random phase multisine excitations $r(t)$ consisting of the sum of $F = 1137$ harmonically related frequencies $(2k+1)f_s/N$ ($k = 0, 1, \ldots F-1$, and $N = 8192$ samples per period) in the band [100 Hz, 4 kHz], with equal harmonic amplitudes. These signals are applied to the plexiglass beam via a mini-shaker. The rms value of the resulting force signal $u(t)$ equals 124 mV. $P = 2$ consecutive periods of the steady state response $u(t)$ and $y(t)$ to $M = 3$ different random phase realizations of the multisine excitation $r(t)$ are measured. Based on these signals the best linear approximation is estimated using the robust method of Section 7.3.6 (first two experiments) and the fast method of Section 7.3.7 (third experiment).

7.3.9.2 Comparison Robust and Fast Estimates. Using $u(t)$ and $y(t)$ of the first two experiments, the robust estimate (7-95) of the best linear approximation (BLA) are calculated with $R = 2$ (second order local polynomial approximation of the noise transients) and n, n_E chosen such that $dof_{\text{robust}}^{\text{noise}} = dof_{\text{robust}} = 7$ in (7-91). The fast estimate (7-102) of the BLA is calculated using $r(t)$, $u(t)$, and $y(t)$ of the third experiment with $R = 4$ (fourth order local polynomial approximation of the BLA and the noise transient) and n, n_E chosen such that $dof_{\text{fast}}^{\text{noise}} = dof_{\text{fast}} = 8$ in (7-103). The degrees R of the local polynomial approximations are chosen such that the bias error of the respective BLA estimates are below the noise level. Since the fast method also approximates the BLA, a higher degree is needed than for the robust method. The results are shown in Figure 7-12: it can be seen that the total variance (gray lines) is significantly larger than the noise variance (black thin lines), indicating that the nonlinear distortions are the dominant error source in the BLA estimate. The level of the stochastic nonlinear distortions is estimated from this difference as in (7-94) and (7-106) for, respectively, the robust and fast methods.

Figure 7-13 compares the robust and the fast BLA estimates. From the top left plot it can be concluded that the BLAs coincide within their total uncertainty (61.6% of the residuals lie outside the 50% confidence bound). From the right column of Figure 7-13 it follows that the mean variance ratios $\text{var}(\hat{G}_{\text{BLA}}^{\text{robust}})/\text{var}(\hat{G}_{\text{BLA}}^{\text{fast}})$ of the noise, the total errors, and the

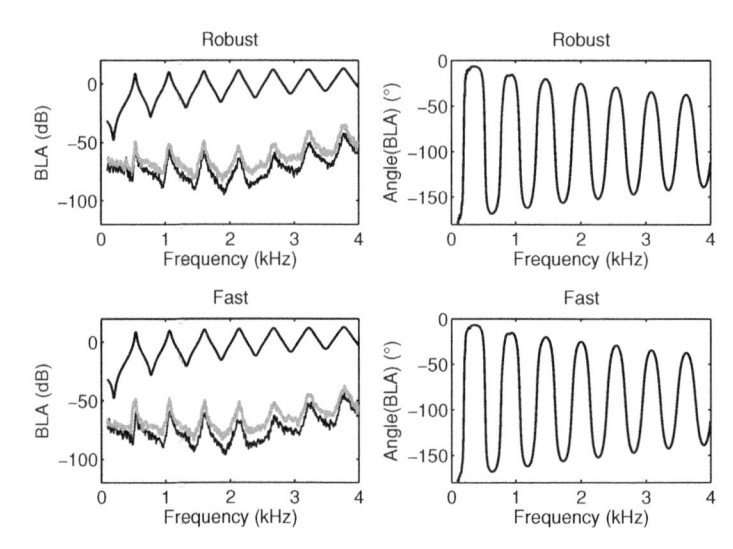

Figure 7-12. Robust (top row) and fast (bottom row) estimates of the force-to-acceleration best linear approximation (BLA) – longitudinal vibrations of a plexiglass beam. Left column: magnitude BLA (bold black lines), noise variance BLA (thin black lines), and total variance BLA (gray lines). Right column: phase BLA.

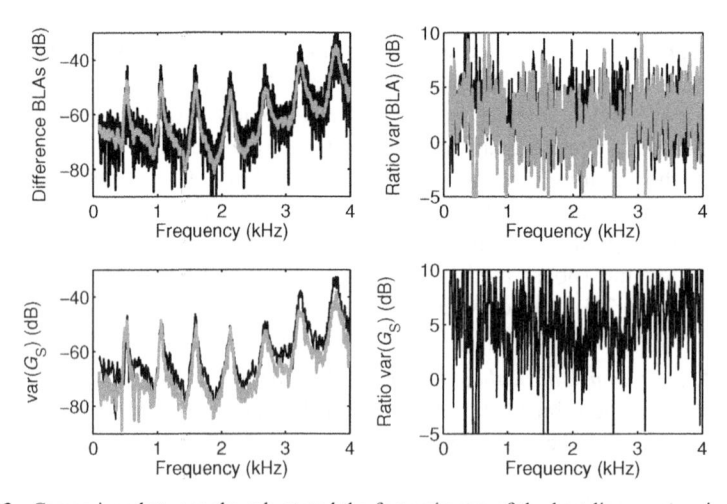

Figure 7-13. Comparison between the robust and the fast estimates of the best linear approximation (BLA) – longitudinal vibrations of a plexiglass beam. Top left: the difference $|\hat{G}_{BLA}^{robust} - \hat{G}_{BLA}^{fast}|$ (black) and its 50% confidence bound (gray). Top right: the noise (black) and total (gray) variance ratios $\text{var}(\hat{G}_{BLA}^{robust})/\text{var}(\hat{G}_{BLA}^{fast})$. Bottom left: variance of the stochastic nonlinear distortions G_S on the robust (black) and fast (gray) BLA estimates. Bottom right: variance ratio $\text{var}(G_S^{robust})/\text{var}(G_S^{fast})$.

stochastic nonlinear distortions are, respectively, 3.1 dB, 2.3 dB, and 5.9 dB. This can be explained by the averaging effect over $2n_E + 1$ neighboring frequencies in the fast BLA estimate (7-99) and (7-102). However, the smaller variances of the fast estimate do not necessarily imply that the information content of the fast BLA estimate is larger than that of the robust BLA estimate. Indeed, the robust BLA estimates are uncorrelated over the frequency, while the correlation length of the fast estimates equals $\pm n_E = \pm 12$ excited frequencies. Also, increasing n_E in (7-99) decreases the variance of the fast BLA estimate (7-102), but increases the correlation length over the frequency.

Section 7.4 ■ Comparison Periodic – Random Excitations

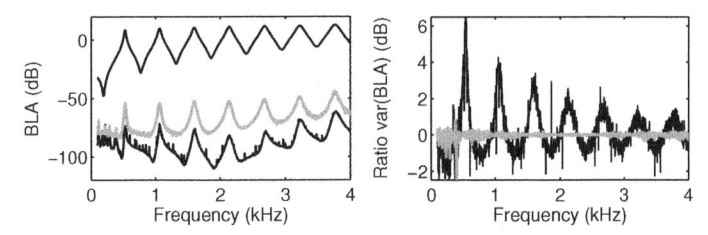

Figure 7-14. Comparison between the local polynomial and the classical robust estimates – longitudinal vibrations of plexiglass beam. Left: robust local polynomial estimate of the BLA (black bold line), its noise variance (black line), and its total variance (gray line). Right: noise (gray) and total (black) variance ratios of the classical robust estimate without transient suppression to the robust local polynomial estimates.

7.3.9.3 Comparison with the Classical Robust Estimates. The previous experiment is repeated measuring $P = 10$ consecutive periods of the steady state response for $M = 25$ different random phase realizations. Figure 7-14 compares the robust local polynomial approach that suppresses the noise transients (Section 7.3.6 with $R = 2$ and $dof_{\text{robust}} = dof_{\text{robust}}^{\text{noise}} = 24$) with the classical robust method that neglects the noise transients (Section 4.3.1 on page 130). The gain in noise variance of the local polynomial method ranges from 6 dB at the resonance with the lowest damping to 1 dB at the resonance with the highest damping. In between the resonances the variance of the local polynomial method is about 1 dB larger than that of the classical method (variance ratio of -1 dB in Figure 7-14, right plot). This observation is consistent with the increase in noise uncertainty due to the transient suppression, which is quantified by μ_{poly} in (7-74). It nicely illustrates the importance of the noise transient (leakage) errors in lightly damped systems. Since the nonlinear distortions are much larger than the noise errors (see Figure 7-14, left plot), and since the system operates in steady state, the total variances of both approaches are equals (0 dB ratio in Figure 7-14, right plot). Note that the noise and total variances in Figure 7-14 (left) are, respectively, $(10 \times 25)/(2 \times 2) = 21$ dB and $25/2 = 11$ dB smaller than in Figure 7-12 (top left).

7.4 COMPARISON PERIODIC – RANDOM EXCITATIONS

7.4.1 Discussion

Since the structure of the covariance of the robust FRM estimate (2-77) on page 66 is totally different from the covariance of the fast FRM estimate (7-21), it is very difficult (if not impossible) to compare them. The structure of the covariance matrix of the fast estimate is, however, exactly the same as that for random excitations (see Section 7.3.7.4), if the same order R and frequency width $2n + 1$ ($n = n_E$) are used for the local polynomial approximations. Therefore, we compare in this section the fast method of Section 7.3.7 applied to the first $P = 2$ periods (N samples per period) of the transient response to a (full) random orthogonal multisine $r(t)$, with the local polynomial method of Section 7.2.2 applied to the first $2N$ samples of the response to stationary Gaussian inputs. The experiment time, the excitation rms value, and the coloring of the input power spectrum are in both cases the same.

- The frequency resolution of the frequency response matrix (FRM) measurement using the random phase multisines $\hat{G}_{\text{fast}}$ is 2 times smaller (2 times less frequencies) than that of the Gaussian excitation $\hat{G}_{\text{arb}}$. However, the random phase multisine excitations allow us to separate the noise from the nonlinear distortions, which is not the case for the stationary Gaussian inputs.

- Since the signal spectra are averaged over two periods in the fast method, the signal-to-noise ratios of the input-output DFT spectra and the FRM measurement using the random phase multisines are $\sqrt{2}$ times larger than those using the Gaussian excitation. However, since the arbitrary excitation has 2 times more frequencies, the noise sensitivity of both FRM estimates is exactly the same (e.g., averaging $\hat{G}_{arb}$ over two neighboring frequencies increases the signal-to-noise ratio by $\sqrt{2}$).

- Because the rms values of the random phase multisine and the Gaussian noise excitations are the same, the signal-to-distortion ratios of the input-output DFT spectra and the FRM measurement for both excitation signals are the same (see Section 4.2 on page 120 and Schoukens et al., 2009). However, the arbitrary excitation has 2 times more frequencies and, therefore, the sensitivity of $\hat{G}_{arb}$ to the nonlinear distortions is $\sqrt{2}$ times smaller than that of $\hat{G}_{fast}$ (e.g., averaging $\hat{G}_{arb}$ over two neighboring frequencies increases the signal-to-distortion ratio by $\sqrt{2}$).

- In both cases, a local polynomial approximation of degree R of the leakage (transient) errors and the FRM is made. The bias error on the FRM estimate introduced by the local polynomial approximation of the FRM is 2^{R+1} times larger for the random phase multisines. This is due to the loss of a factor 2 in frequency resolution w.r.t. the Gaussian noise excitations: compare (7-18), where N is replaced by $2N$, with (7-105).

Summarized, the fast method using $P = 2$ periods of the transient response to a (full) random orthogonal multisine allows us to quantify the noise level and the level of the nonlinear distortions on the FRM estimates. The drawbacks w.r.t. stationary random noise are a reduced frequency resolution (factor 2), a larger bias error on the FRM estimates (factor 2^{R+1}), and an increased sensitivity w.r.t. the nonlinear distortions (factor $\sqrt{2}$). The noise sensitivity of both approaches is exactly the same.

7.4.2 Experimental Illustration

The goal of the measurement example is threefold: (i) illustration of the local polynomial approach on a multivariable system; (ii) illustration of the importance of system and/or noise transient (leakage) errors in frequency response function (FRM) estimates; and (iii) comparison of the fast method using periodic excitations (Section 7.3.7) with the local polynomial method using arbitrary excitations (Section 7.2.7). The latter will be referred to as the "arb" estimate.

7.4.2.1 Measurement Setup. As test case we take an aluminum tooling plate (PE 200) of size 30.4 cm x 61.8 cm x 6.7 mm (see Pintelon et al., 2011b for a detailed description of the experimental setup). The plate is excited under free-free boundary conditions by two mini-shakers spaced 25.5 cm apart, and the forces (2×1 excitation $u(t)$) and accelerations (2×1 response $y(t)$) are measured at the two excitation points (see Section 7.2.10.1 for a detailed description of the analog signal conditioning: generation, amplification, impedance matching). The two generators and two acquisition units are synchronized and operate at the sampling frequency $f_s = 10 \text{ MHz}/2^{12} \approx 2.44 \text{ kHz}$.

In a first experiment, one column of the full random orthogonal multisine (3-31) is applied to the $n_u = 2$ input, $n_y = 2$ output system. The multisines (3-31) have a flat amplitude spectrum ($D_{|R|}(k)$ in (2-82) on page 67 is independent of k) with equal rms values for all inputs, and contain $F = 12887$ harmonically related frequencies that are uniformly distributed in the band [120 Hz, 600 Hz] (kf_s/N with $k = 3221, 3222, \ldots, 16107$ and

Section 7.4 ■ Comparison Periodic – Random Excitations

$N = 64 \times 1024$ points per period). The frequency resolution of the corresponding FRM measurement is 37.3 mHz. We measure $P = 2$ periods ($2N$ samples) of the force (= input) and acceleration (= output) signals after a waiting time of $N/64 = 1024$ samples. These signals are used for calculating the fast estimate (Section 7.3.7) of the best linear approximation (BLA).

Instead of Gaussian noise, we use a full random orthogonal multisine (3-31) with period length $2N$ samples in the second experiment. The reason for this is that the amplitude spectrum of the multisine is deterministic while that of the Gaussian noise is random (Rayleigh) distributed. The random phase multisines have the same rms value as in the first experiment, and contain $F = 25772$ harmonically related frequencies in the band [120 Hz, 600 Hz] (kf_s/N with $k = 6442, 6443, ..., 32213$ and $N = 128 \times 1024$ points per period) with flat amplitude spectra. The frequency resolution of the corresponding FRM measurement is 18.6 mHz. Only $P = 1$ period ($2N$ samples) of the force and acceleration signals are measured after a waiting time of $N/32 = 2048$ samples. These signals are used for calculating the local polynomial estimate (Section 7.2.7) of the best linear approximation (BLA). The latter is called the "arb" estimate in the sequel of Section 7.4.2.

7.4.2.2 The Fast Estimate of the BLA. The fast estimate (7-102) of the BLA is calculated using $r(t)$, $u(t)$, and $y(t)$ of the first experiment with $R = 4$ (fourth order local polynomial approximation of the BLA and the transient) and n, n_E chosen such that $dof_{\text{fast}}^{\text{noise}} = dof_{\text{fast}} = 9$ in (7-103). Figure 7-15 shows the results. It can be seen that $G_{[1,1]}$ contains less resonances than $G_{[2,2]}$, which is a nice illustration that the multivariable approach reduces the risk of missing important resonances. The total variance (dark gray) is only significantly larger than the noise variance (light gray) around the resonance frequencies (see also Figure 7-17). Note also the presence of the third mains harmonic (150 Hz) in the noise variance of $G_{[2,1]}$ and $G_{[2,2]}$.

To illustrate the importance of the system and/or noise transient in the time signals, the fast estimates are repeated without transient suppression. For this purpose we replace the sample mean (7-97) and the corresponding sample noise covariance (7-74) by

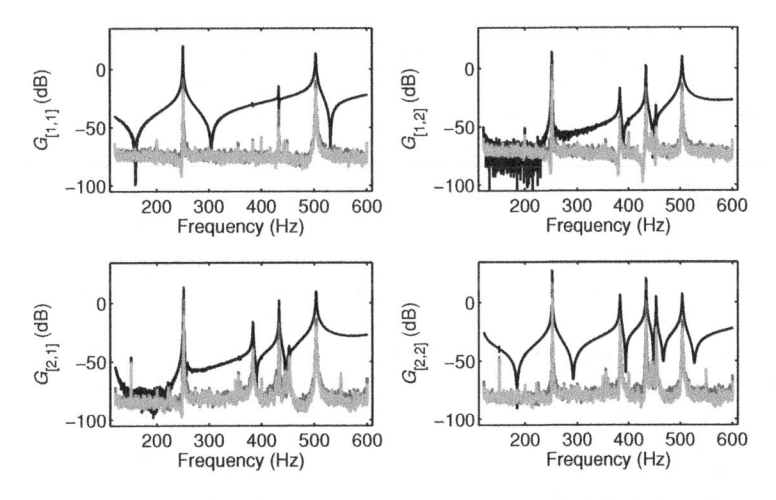

Figure 7-15. Fast estimate of the best linear approximation (black), its noise variance (light gray), and its total variance (dark gray) – aluminum plate.

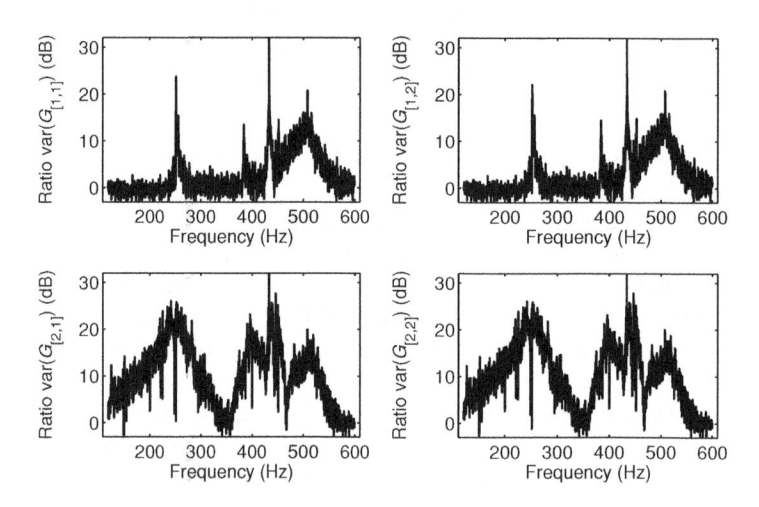

Figure 7-16. Ratio of the total variances of the fast best linear approximation estimates without and with transient suppression – aluminum plate.

$$\hat{Z}(kP) = Z(kP) \text{ and } \hat{C}_{\hat{Z}}^{\text{noise}}(kP) = Z_n Z_n^H / (2n+1)$$

where Z_n is defined in (7-72). Figure 7-16 shows the ratio of the total variance of the BLA estimate without transient suppression to the variance of the BLA with transient suppression. At the resonances the uncertainty of the estimates without transient suppression is about 10 to 30 dB larger. In those frequency bands where the transient (leakage) errors can be neglected, the noise uncertainty of the BLA estimate with transient suppression is about 1 dB larger than that without transient suppression. This observation is consistent with the increase in uncertainty quantified by μ_{poly} in (7-74).

7.4.2.3 Comparison Fast and Arb Estimates. The arb estimate (7-49) of the BLA is calculated using $r(t)$, $u(t)$, and $y(t)$ of the second experiment with $R = 4$ (fourth order local polynomial approximation of the BLA and the transient) and n chosen such that $dof_{\text{arb}} = q$ in (7-13) equals 10. Figures 7-17 to 7-19 compare the fast with the arb estimates.

The arb estimates of the BLA have twice the frequency resolution of the fast method, however, they gives no indication about the possible nonlinear behavior of the system (Figure 7-17). In the frequency bands where the noise is dominant, the mean ratio of the total variance of the fast estimate to that of the arb estimate is -2.8 dB (see Figure 7-18), which is in good agreement with the theoretical expected value of -3 dB (see Section 7.4.1). In the neighborhood of those resonance peaks where the total variance is significantly larger than the noise variance, the mean total variance ratio is about 0.15 dB, which is close to the theoretical predicted value of 0 dB (see Section 7.4.1).

Finally, it can be concluded from Figure 7-19 that both BLA estimates agree fairly well but not exactly. Indeed, although about 54% of the residuals lie outside the 50% confidence bound, some small systematic deviations between both estimates can be observed in the band [480 Hz, 500 Hz] for $G_{[2,1]}$ and $G_{[2,2]}$ and below 220 Hz for $G_{[2,2]}$, and around some of the sharp resonances of some of the entries of G.

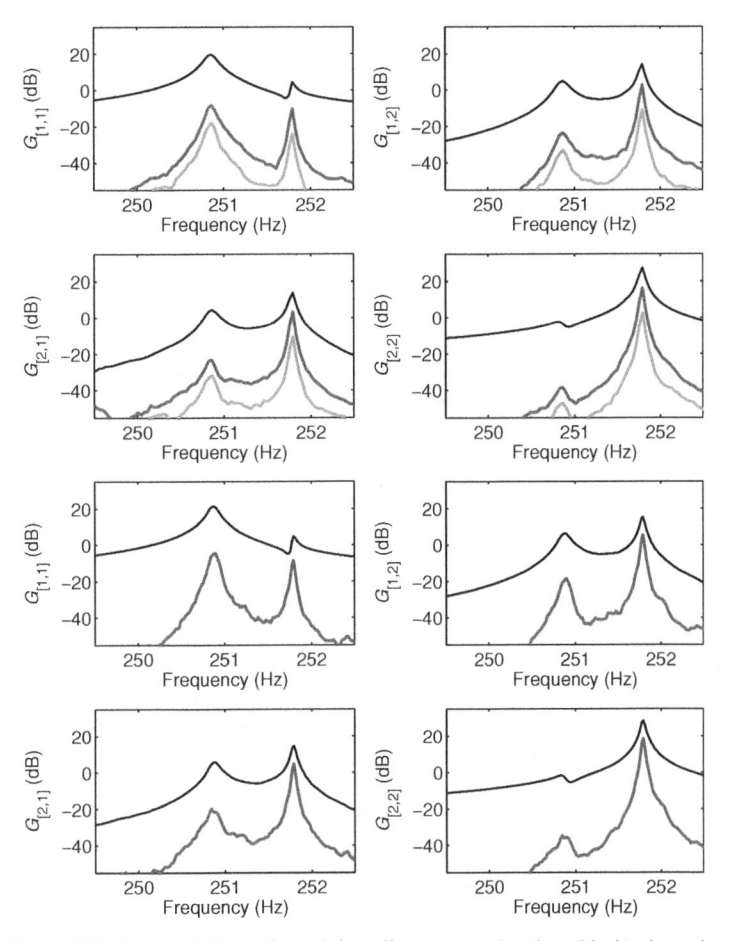

Figure 7-17. Zoom of the estimated best linear approximation (black), its noise variance (light gray), and its total variance (dark gray) – aluminum plate. Top rows: fast estimate, and bottom rows: arb estimates.

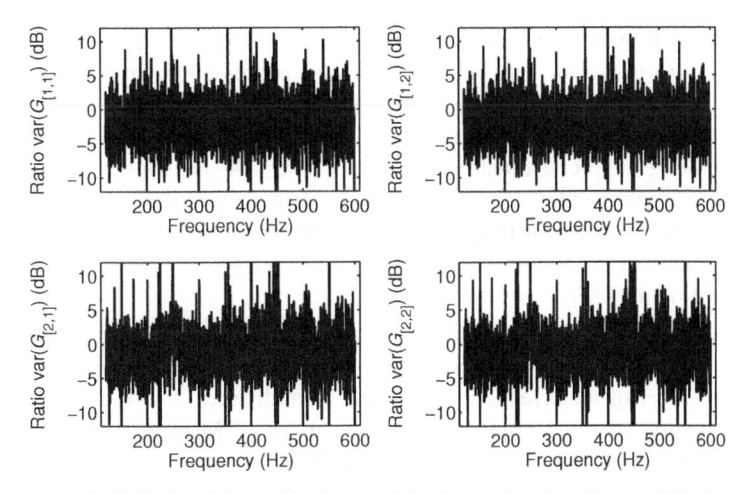

Figure 7-18. Ratio of the total variances of the fast to the arb estimates of the best linear approximation – aluminum plate.

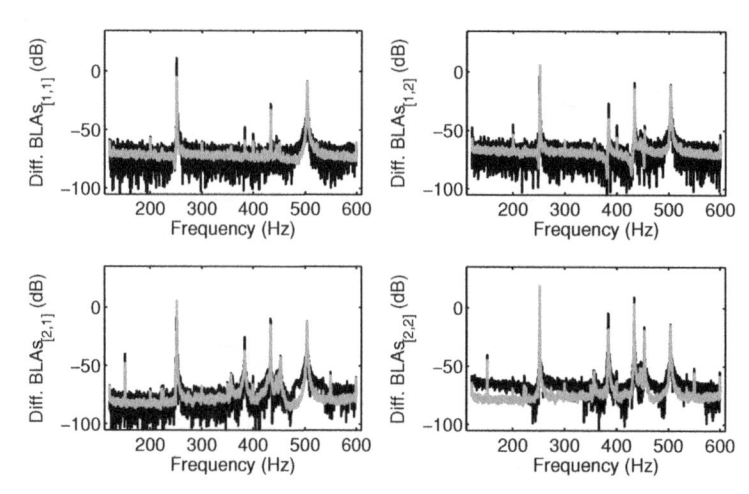

Figure 7-19. Differences between the estimated best linear approximations $|\hat{G}_{\text{fast}} - \hat{G}_{\text{arb}}|$ (black) and their 50% confidence bounds (gray) – aluminum plate.

7.5 GUIDELINES FOR ADVANCED FRF MEASUREMENTS

- *Guideline 1: Use Uncorrelated Inputs.* It is strongly recommended to minimize the correlation among the different excitation signals of the multivariable system. Otherwise, the input power spectrum matrix can be close to singular, resulting in poor frequency response matrix (FRM) estimates (7-16) and (7-79). However, sometimes the FRM of a highly interactive system can be ill-conditioned by the nature of the system itself (Zhu and Stec, 2006; and Rivera et al., 2009). Accurate identification of the low gain directions of the FRM then requires highly correlated inputs, which is in contradiction with the previous requirement of uncorrelated inputs. A two step procedure can solve this problem. First, an estimate of the FRM is obtained with uncorrelated inputs. Next, the system is excited in its low gain direction with a highly correlated input designed using the low gain direction estimated in the first step. Iterative refinement is possible.

- *Guideline 2: Keep Track of the Reference Signal.* The local polynomial method assumes that the input is known exactly. If the input is disturbed by noise and/or the system operates in feedback, then the direct local polynomial estimate of the FRM (7-16) is biased. This bias is avoided by adding a known reference signal (typically the signal stored in the arbitrary waveform generator) to the problem: the FRM is estimated from the known reference to the inputs and outputs of the plant, and the resulting indirect estimate of the plant FRM (7-49) is consistent. Hence, it is strongly recommended to keep the known reference signal together with the measured input-output signals. The only drawback of the indirect method is that the part of the plant input that is not correlated with the reference signal does not contribute to the FRM estimate (7-49).

- *Guideline 3: Use Periodic Excitation Signals.* The prime choice is the fast method (Section 7.3.4) applied to the first two periods of the transient response of the plant to a set of uncorrelated random phase multisines. For a given experiment time, it maximizes the frequency resolution of the FRM measurement, while keeping the ability to distinguish the noise from the nonlinear distortions. The major drawback of the fast method compared with the robust method (Section 7.3.3) is the plant interpolation bias error in the FRM estimate.

Section 7.6 ■ Appendixes

Using stationary random excitations it is not possible to detect the presence of nonlinear distortions in the FRM measurement. However, if frequency resolution is the major concern, then random excitations or one period of the transient response to uncorrelated random phase multisines are a good choice. The latter has the advantage of having a deterministic amplitude spectrum. Compared with the fast method applied to two periods, the sensitivity to nonlinear distortions of these solutions is $\sqrt{2}$ smaller.

Conclusion. For lowly damped systems the local polynomial methods result in a significant reduction of the measurement time or, for a given experiment time, in a significant increase of the frequency resolution. In addition these methods also reduce the required skills of the users because there is no need anymore to select the steady state part of the measurements (periodic excitations), or to decide upon the length of subrecords for calculating the cross-power spectra (arbitrary excitations).

7.6 APPENDIXES

Appendix 7.A Proof of Equation (7-10)

Applying $\text{vec}(ABC) = (C^T \otimes A)\text{vec}(B)$ (see Section 15.7 on page 552) to (7-8), where $\text{vec}(x)$ stacks the columns of x on top of each other (see (15-9)), we get

$$y_n = k_n \theta + v_n \qquad (7\text{-}109)$$

with $y_n = \text{vec}(Y_n)$, $k_n = K_n^T \otimes I_{n_y}$, $\theta = \text{vec}(\Theta)$, and $v_n = \text{vec}(V_n)$. The least squares solution of (7-109) is given by

$$\hat{\theta} = (k_n^H k_n)^{-1} k_n^H y_n \qquad (7\text{-}110)$$

Using $(A \otimes B)^H = A^H \otimes B^H$, $(A \otimes B)^{-1} = A^{-1} \otimes B^{-1}$, and $(A \otimes B)(C \otimes D) = (AC) \otimes (BD)$ (see Section 15.7), (7-110) can be written as

$$\text{vec}(\hat{\Theta}) = (\overline{(K_n K_n^H)^{-1} K_n} \otimes I_{n_y})\text{vec}(Y_n) \qquad (7\text{-}111)$$

with $\bar{x}$ the complex conjugate of x. Applying $(C^T \otimes A)\text{vec}(B) = \text{vec}(ABC)$ to (7-111) gives $\text{vec}(\hat{\Theta}) = \text{vec}(Y_n K_n^H (K_n K_n^H)^{-1})$, which is exactly (7-10) where all columns are stacked on top of each other. □

Appendix 7.B Proof of Equation (7-13)

Using $P_n P_n^H = P_n$, entry $[i,j]$ of $\hat{C}_V(k)$ (7-13) can be written as

$$(\hat{C}_V(k))_{[i,j]} = \frac{1}{q}(V_n P_n V_n^H)_{[i,j]} = \frac{1}{q}V_{n[i,:]} P_n V_{n[j,:]}^H = \frac{1}{q}\text{trace}(P_n V_{n[j,:]}^H V_{n[i,:]}) \qquad (7\text{-}112)$$

where $X_{[i,:]}$ denotes the ith row of X and with $\text{trace}(X) = \sum_i X_{[i,i]}$. Since $V(k)$ is uncorrelated over the frequency k (see Section 7.2.1), the expected value of (7-112) equals

$$\mathbb{E}\{(\hat{C}_V(k))_{[i,j]}\} = \frac{1}{q}\text{trace}(P_n E\{V_{n[j,:]}^H V_{n[i,:]}\}) = \frac{1}{q}\text{trace}(P_n C_n) \quad (7\text{-}113)$$

where

$$C_n = \text{diag}((C_V(k-n))_{[i,j]}, \ldots, (C_V(k))_{[i,j]}, \ldots, (C_V(k+n))_{[i,j]}) \quad (7\text{-}114)$$

The noise covariance matrix is a smooth function of the frequency and, therefore,

$$C_V(k+r) = C_V(k) + C_V^{(1)}(k) O_{\text{int}H}(r/N) \quad (7\text{-}115)$$

with $C_V^{(1)}(k)$ the derivative of $C_V(k)$ w.r.t. the frequency (proof: apply the mean value theorem with $f_{k+r} - f_k = rf_s/N$, see Theorem 41 of Kaplan, 1993). Collecting (7-113), (7-114), and (7-115) gives

$$\mathbb{E}\{(\hat{C}_V(k))_{[i,j]}\} = \frac{1}{q}(C_V(k))_{[i,j]}\text{trace}(P_n) + (C_V^{(1)}(k))_{[i,j]} O_{\text{int}H}(n/N) \quad (7\text{-}116)$$

Since P_n (7-11) and $K_n^H(K_n K_n^H)^{-1} K_n$ are idempotent matrices it follows that

$$\text{trace}(P_n) = \text{rank}(P_n) = 2n + 1 - \text{rank}(K_n) = 2n + 1 - (R+1)(n_u + 1) = q \quad (7\text{-}117)$$

(see Section 15.6 on page 551), which concludes the proof. □

Appendix 7.C System Leakage Contribution to the Bias on the FRM Estimates

In this appendix we study the contribution of the system leakage error to the bias on the FRM estimates. It will be shown that for the polynomial approach

$$\mathbb{E}\{O_{\text{leak}}((n/N)^{(R+3/2)} K_n^H (K_n K_n^H)^{-1})\} = O_{\text{leak}G}((n/N)^{(R+2)}) \quad (7\text{-}118)$$

where O_{leak} stems from (7-122), and where $O_{\text{leak}G}$ is the term in (7-124).

The matrix K_n (see (7-7) and (7-9)) depends the input DFT spectrum $U(k+r)$, $r = -n, \ldots, 0, \ldots, n$, and, hence, is correlated with the system leakage term $T_G(\Omega_k)$. To calculate the expected value (7-118), a first order Taylor series approximation of $(K_n K_n^H)^{-1}$ is made

$$(K_n K_n^H)^{-1} = (\mathbb{E}\{K_n K_n^H\} + \Delta)^{-1} \approx (\mathbb{E}\{K_n K_n^H\})^{-1} - (\mathbb{E}\{K_n K_n^H\})^{-1}\Delta(\mathbb{E}\{K_n K_n^H\})^{-1} \quad (7\text{-}119)$$

where Δ is a linear combination of U, $\bar{U}$, UU^H. Using (7-119), the fact that K_n is an affine function of U, and the fact that O_{leak} in (7-122) is an affine function of $T_G^{(R+1)}$,

$$O_{\text{leak}}((n/N)^{(R+3/2)}) = O((n/N)^{(R+1)})(T_G^{(R+1)} + T_H^{(R+1)})$$

(proof: combine (7-3), (7-4), and (7-8)), it can be seen that (7-118) is a linear combination of terms of the form

$$\mathbb{E}\{T_{G[i]}^{(R+1)}\overline{U}_{[j]}\}, \quad \mathbb{E}\{T_{G[i]}^{(R+1)}\overline{U}_{[j]}U_{[k]}\}, \quad \mathbb{E}\{T_{G[i]}^{(R+1)}\overline{U}_{[j]}\overline{U}_{[k]}\},$$
$$\text{and } \mathbb{E}\{T_{G[i]}^{(R+1)}\overline{U}_{[j]}U_{[k]}\overline{U}_{[l]}\} \tag{7-120}$$

Since $T_G^{(R+1)}$ and U are zero mean, jointly normally distributed random variables, the third order moments in (7-120) are zero, and the fourth order moment is given by

$$\mathbb{E}\{T_{G[i]}^{(R+1)}\overline{U}_{[j]}U_{[k]}\overline{U}_{[l]}\} = \mathbb{E}\{T_{G[i]}^{(R+1)}\overline{U}_{[j]}\}\mathbb{E}\{U_{[k]}\overline{U}_{[l]}\} + \mathbb{E}\{T_{G[i]}^{(R+1)}U_{[k]}\}\mathbb{E}\{\overline{U}_{[j]}\overline{U}_{[l]}\}$$
$$+ \mathbb{E}\{T_{G[i]}^{(R+1)}\overline{U}_{[l]}\}\mathbb{E}\{U_{[k]}\overline{U}_{[j]}\} \tag{7-121}$$

(see Picinbono, 1993). Combining (6-67) and (6-68) on page 200 (for continuous-time systems T_G with $n_i \geq \max(n_a, n_b) - 1$ represents the sum of the transient and the residual alias terms) with (7-120) and (7-121) proves (7-118). Using the property that an even order moment of Gaussian random variables can always be written as the sum of products of second order moments (Stuart and Ord, 1987), it can easily be verified that (7-118) remains valid if higher order terms are included in the Taylor series expansion (7-119).

Appendix 7.D Proof of Equations (7-17) and (7-18)

7.D.1 *Proof of equation (7-18).* Collecting (7-3) and (7-4) for $r = -n, -n+1, \ldots, 0, \ldots, n$, and taking into account that the remainder of the FRM in (7-4) depends on the $(R+1)$th derivative of the FRF $G^{(R+1)}(\Omega_k)$, the matrix Y_n can be written as

$$Y_n = \Theta K_n + V_n + G^{(R+1)}(\Omega_k)O_{\text{int}G}((n/N)^{(R+1)}) + O_{\text{leak}}((n/N)^{(R+3/2)}) \tag{7-122}$$

with Θ the true parameter value, and where $O_{\text{int}G}$ stems from the remainder of the polynomial approximation of the FRF, and O_{leak} from the remainder of the leakage term. Note that O_{leak} consists of two contributions: the noise leakage term $T_H(\Omega_k)$ and the system leakage term $T_G(\Omega_k)$. The former is correlated with the noise V_n, while the latter is correlated with the input and hence with K_n (see Theorems 6.15 and 6.16: for continuous-time systems T_G with $n_i \geq \max(n_a, n_b) - 1$ and T_H with $n_j \geq \max(n_c, n_d) - 1$ represent the sum of the transient and the residual alias terms).

Since $T_G(\Omega)$ is a rational form whose numerator coefficients (i) depend linearly on the difference between the initial and final conditions of the experiment, and (ii) decrease as an $O(N^{-1/2})$, it can be shown that under the excitation assumption of Section 7.2.1

$$\mathbb{E}\{T_G(\Omega_k)U^H(k)\} = O(N^{-1}) \Rightarrow \mathbb{E}\{O_{\text{leak}}((n/N)^{(R+3/2)})K_n^H\} = O_{\text{leak}G}((n/N)^{(R+2)}) \tag{7-123}$$

(proof: see Theorems 6.15 and 6.16, taking into account that for continuous-time systems T_G with $n_i \geq \max(n_a, n_b) - 1$ represents the sum of the transient and the residual alias terms). Taking the expected value of (7-10), using (7-122) and (7-123), gives

$$\mathbb{E}\{\hat{\Theta}\} = \Theta + G^{(R+1)}(\Omega_k)O_{\text{int}G}((n/N)^{(R+1)}) + O_{\text{leak}G}((n/N)^{(R+2)}) \tag{7-124}$$

(proof for the leakage term: see Appendix 7.C). Combining (7-16) and (7-124) proves (7-18).

7.D.2 Proof of equation (7-17). Using (7-122), the residual of the least squares fit $\hat{V}_n$ (7-11) can be written as

$$\hat{V}_n = Y_n P_n = V_n P_n + G^{(R+1)}(\Omega_k) O_{\text{int}G}((n/N)^{(R+1)}) + O_{\text{leak}}((n/N)^{(R+3/2)}) \quad (7\text{-}125)$$

where $O_{\text{int}G}$ is independent of the noise V_n, and where O_{leak} is correlated with V_n through the noise leakage part $T_H(\Omega_k)$ of $T(\Omega_k)$.

Since $T_H(\Omega_k)$ is zero for white noise ($H(\Omega_k) = 1$ in (7-2)), its correlation with $E(k)$ (see Theorems 6.15 and 6.16, taking into account that for continuous-time systems T_H with $n_j \geq \max(n_c, n_d) - 1$ represents the sum of the transient and the residual alias terms) can be written as

$$\mathbb{E}\{T_H(\Omega_k) E^H(k)\} = O(H^{(1)}(\Omega_k) N^{-1}) \quad (7\text{-}126)$$

with $H^{(1)}(\Omega)$ the derivative of $H(\Omega)$ w.r.t. the frequency. From (7-126) and the definition of $V(k)$ (7-2), it immediately follows that

$$\mathbb{E}\{O_{\text{leak}}((n/N)^{(R+3/2)}) V_n^H\} = O_{\text{leak}H}(H^{(1)}(\Omega_k)(n/N)^{(R+2)}) \quad (7\text{-}127)$$

Using (7-125) and (7-127), one finds for the expected value of $\hat{C}_V(k)$ (7-13)

$$\mathbb{E}\{\hat{C}_V(k)\} = \frac{1}{q}\mathbb{E}\{V_n P_n V_n^H\} + G^{(R+1)}(\Omega_k) O_{\text{int}G}((n/N)^{2(R+1)})(G^{(R+1)}(\Omega_k))^H + \ldots \\ O_{\text{leak}H}(H^{(1)}(\Omega_k)(n/N)^{(R+2)}) + O_{\text{leak}}((n/N)^{(2R+3)}) \quad (7\text{-}128)$$

where the first term is elaborated in Appendix 7.B. Note that at a zero of the noise model, $H(\Omega_k) = 0$, the correlation (7-127) can be neglected because $V_n \approx 0$ for N sufficiently large ($V(k) = 0$ and $V(k+r) \approx 0$, $r = -n, \ldots, -1, 1, \ldots, n$). Hence, since

$$C_V^{(1)}(k) = 2\text{herm}(H^{(1)}(\Omega_k) \text{Cov}(E(k)) H^H(\Omega_k)) \quad (7\text{-}129)$$

with $\text{herm}(X) = (X + X^H)/2$, the leakage term $O_{\text{leak}H}$ in (7-128) can be replaced by

$$C_V^{(1)}(k) O_{\text{leak}H}((n/N)^{(R+2)}) \quad (7\text{-}130)$$

Combining (7-116), (7-128), and (7-130) proves (7-17). □

Appendix 7.E Proof of Equation (7-19)

Combining (7-10) and (7-16) shows that (for notational simplicity we omit the frequency argument)

$$\hat{G} = G + V_n S \quad (7\text{-}131)$$

where S is defined in (7-20). Using $\text{vec}(ABC) = (C^T \otimes A)\text{vec}(B)$ (see Section 15.7) we find $\text{vec}(\hat{G} - G) = (S^T \otimes I_{n_y})\text{vec}(V_n)$. Combining this with $(A \otimes B)^H = A^H \otimes B^H$ gives

$$\text{Cov}(\text{vec}(\hat{G})) = \mathbb{E}\{(S^T \otimes I_{n_y})\text{Cov}(\text{vec}(V_n))(\bar{S} \otimes I_{n_y})\} \tag{7-132}$$

where $\text{Cov}(\text{vec}(V_n)) = I_{2n+1} \otimes C_V(k) + O_{\text{int}H}(n/N)$ (see (7-114) and (7-115)), and where the expected value is taken w.r.t. the random input. Applying twice $(A \otimes B)(C \otimes D) = (AC) \otimes (BD)$ to the right-hand side of the covariance expression (7-132), taking into account that $\mathbb{E}\{S^H S\} = O(N^0/n)$, proves (7-19). □

Appendix 7.F   Proof of Equation (7-50)

Defining $\Delta\hat{X} = (\hat{X} - X)$, the noise on the FRF estimate $\hat{G} = \hat{G}_{ry}\hat{G}_{ru}^{-1}$ (7-49) can be obtained as a function of the noise on $\hat{G}_{rz}$ via a first order Taylor series approximation

$$\begin{aligned}
\hat{G} &= \hat{G}_{ry}\hat{G}_{ru}^{-1} \\
&= (G_{ry} + \Delta\hat{G}_{ry})(G_{ru} + \Delta\hat{G}_{ru})^{-1} \\
&= (G_{ry} + \Delta\hat{G}_{ry})(I_{n_u} + G_{ru}^{-1}\Delta\hat{G}_{ru})^{-1}G_{ru}^{-1} \\
&\approx (G_{ry} + \Delta\hat{G}_{ry})(I_{n_u} - G_{ru}^{-1}\Delta\hat{G}_{RU})G_{ru}^{-1} \\
&\approx G + (\Delta\hat{G}_{ry} - G\Delta\hat{G}_{ru})G_{ru}^{-1} \\
&\approx G + [I_{n_y} \ -G]\Delta\hat{G}_{rz}G_{ru}^{-1}
\end{aligned} \tag{7-133}$$

Applying $\text{vec}(ABC) = (C^T \otimes A)\text{vec}(B)$ to (7-133) gives

$$\text{vec}(\Delta\hat{G}) \approx (G_{ru}^{-T} \otimes [I_{n_y} \ -G])\text{vec}(\Delta\hat{G}_{ru}) \tag{7-134}$$

Calculating the covariance matrix of (7-134), and replacing the true unknown values by the estimates gives (7-50), which concludes the proof. □

Appendix 7.G   Proof of Equation (7-55)

The proof follows the same lines of Appendixes 7.B and 7.D, where the output noise $V(k)$ is replaced by the sum of the output noise and the stochastic nonlinear distortions $V(k) + Y_S(k)$. Since $V(k)$ is independently distributed of $Y_S(k)$ we can analyze the contribution of $Y_S(k)$ separately. The fact that $V(k)$ is uncorrelated over the frequency, while $Y_S(k)$ is only asymptotically ($N \to \infty$) uncorrelated over the frequency

$$\mathbb{E}\{Y_S(k)Y_S^H(l)\} = O(N^{-1}) \text{ for } k \neq l \tag{7-135}$$

(see Theorem 3.16) is the key technical difference that matters in the proofs of Appendixes 7.B and 7.D. Due to the correlation, the matrix C_n (7-114) is no longer diagonal. While the main diagonal has a similar contribution as in (7-116), where C_V is replaced by $C_{Y_S} = \text{Cov}(Y_S)$, the $2n(2n+1)$ non-diagonal elements decrease as an $O(N^{-1})$, giving

$$\mathbb{E}\{\hat{C}_V(k)\} = C_{Y_S}(k) + C_{Y_S}^{(1)}(k)O_{\text{int}H}(n/N) + \frac{1}{q}O(4n^2/N) \qquad (7\text{-}136)$$

Combining (7-136) with (7-116) proves (7-55).

Appendix 7.H Proof of Equation (7-74)

Using the singular value decomposition $K_n^H = U_K \Sigma_K V_K^H$, the estimated noise transient term (7-71) can be written as

$$\hat{T}_{H_Z}(\Omega_{kP}) = T_{H_Z}(\Omega_{kP}) + V_n b_1 \qquad (7\text{-}137)$$

where V_n has the structure (7-72) with $V = V_Z$ and V_Z defined in (7-68), and with

$$b_1 = (K_n^H(K_n K_n^H)^{-1})_{[:,1]} = U_K \Sigma_K V_{K[1,:]}^H \qquad (7\text{-}138)$$

Combining (7-73) with (7-137), taking into account that at the non-excited DFT frequencies $Z(kP + r_i) = V_Z(kP + r_i) + T_{H_Z}(\Omega_{kP+r_i})$, gives

$$\hat{Z}(kP) = Z_0(kP) + V_Z(kP) + V_n b_1 \qquad (7\text{-}139)$$

Since V_n is independent of $V_Z(kP)$ (see (7-72)), the covariance of (7-139) is given by

$$\text{Cov}(\hat{Z}(kP)) = C_Z^{\text{noise}}(kP) + \text{Cov}(V_n b_1) \qquad (7\text{-}140)$$

Entry $[i,j]$ of the second term in (7-140) can be elaborated as

$$(\text{Cov}(V_n b_1))_{[i,j]} = \mathbb{E}\{V_{n[i,:]} b_1 b_1^H V_{n[j,:]}^H\} = \text{trace}(b_1 b_1^H \mathbb{E}\{V_{n[j,:]}^H V_{n[i,:]}\}) \qquad (7\text{-}141)$$

Using a first order Taylor series expansion with remainder

$$C_Z^{\text{noise}}(kP \pm r_i) = C_Z^{\text{noise}}(kP) \pm C_Z^{\text{noise}(1)}(k_1) r_i / (PN) \qquad (7\text{-}142)$$

with $k_1 \in [kP, kP \pm r_i]$, and $C_Z^{\text{noise}(1)}$ the first order derivative of C_Z^{noise} w.r.t. its argument, it can be verified that

$$\mathbb{E}\{V_{n[j,:]}^H V_{n[i,:]}\} = ((C_Z^{\text{noise}}(kP))_{[i,j]} + O(r_n/(PN))) I_{2n} \qquad (7\text{-}143)$$

where I_{2n} is the $2n \times 2n$ identity matrix. Collecting (7-141) to (7-143) gives

$$(\text{Cov}(V_n b_1))_{[i,j]} = (C_Z^{\text{noise}}(kP))_{[i,j]} \|b_1\|_2^2 + O(r_n/(PN)) \qquad (7\text{-}144)$$

Combining (7-140) and (7-144), taking into account that $U_K^H U_K = I_{2n}$, finally gives

Section 7.6 ■ Appendixes   275

$$\text{Cov}(\hat{Z}(kP)) = \mu_{\text{poly}} C_Z^{\text{noise}}(kP) + O(r_n/(PN)) \quad (7\text{-}145)$$

It shows that (7-74) is the asymptotic $(PN \to \infty)$ covariance of (7-73). □

Appendix 7.I   Proof of Equation (7-76)

The noise interpolation error is analyzed assuming that the residual leakage error originating from the local polynomial approximation is zero. Expanding the noise covariance matrix in Taylor series with remainder

$$C_Z^{\text{noise}}(kP + r_i) = C_Z^{\text{noise}}(kP) + C_Z^{\text{noise}(1)}(kP)\frac{r_i}{PN} + \frac{1}{2}C_Z^{\text{noise}(2)}(kP)\left(\frac{r_i}{PN}\right)^2 + O\left(\left(\frac{r_i}{PN}\right)^3\right) \quad (7\text{-}146)$$

and combining this result with (7-113) and (7-114), where $C_V(k) = C_Z^{\text{noise}}(kP)$ (compare (7-71) to (7-13)), gives the following bias expression

$$\mathbb{E}\{(\hat{C}_Z^{\text{noise}}(kP))_{[i,j]}\} - (C_Z^{\text{noise}}(kP))_{[i,j]} = \frac{\text{trace}(P_n D_n)}{q^{\text{noise}}} + C_Z^{\text{noise}(2)}(kP) O_{\text{int}}\left(\left(\frac{r_i}{PN}\right)^2\right) \quad (7\text{-}147)$$

where $P_n = I_{2n} - K_n^H(K_n K_n^H)^{-1} K_n$, with K_n defined in (7-72), and $D_n = \text{diag}(-r_n, \ldots, -r_1, r_1, \ldots, r_n)$. Since $\text{trace}(D_n) = 0$, the first term in (7-147) can be elaborated as

$$\text{trace}(P_n D_n) = \text{trace}(D_n) - \text{trace}((K_n K_n^H)^{-1} K_n D_n K_n^H) = -\text{trace}(M_1^{-1} M_2) \quad (7\text{-}148)$$

where

$$\begin{aligned} M_1 &= K_n K_n^H = \sum_{i=-n}^{n} K(kP + r_i) K^H(kP + r_i) \\ M_2 &= K_n D_n K_n^H = \sum_{i=-n}^{n} r_i K(kP + r_i) K^H(kP + r_i) \end{aligned} \quad (7\text{-}149)$$

The $(R+1) \times (R+1)$ matrix $K(kP + r_i) K^H(kP + r_i)$, with $K(kP + r_i) = [1 \ r_i \ \ldots \ r_i^R]^T$, has a Hankel structure

$$K(kP + r_i) K^H(kP + r_i) = \begin{bmatrix} 1 & r_i & r_i^2 & \ldots & r_i^R \\ r_i & r_i^2 & r_i^3 & \ldots & r_i^{R+1} \\ \ldots & \ldots & \ldots & \ldots & \ldots \\ r_i^R & r_i^{R+1} & r_i^{R+2} & \ldots & r_i^{2R} \end{bmatrix} \quad (7\text{-}150)$$

Substituting (7-150) in (7-149) shows that all entries of M_1 with even row and column indexes are zero $M_{1[2i, 2j]} = 0$; while all entries of M_2 with odd row and column indexes are zero $M_{2[2i+1, 2j+1]} = 0$. Since M_1^{-1} has the same zero pattern as M_1 (proof: see Appendix 7.J), the matrix product $M_1^{-1} M_2$ has zero entries on its main diagonal. Hence, $\text{trace}(M_1^{-1} M_2) = 0$, which shows that the noise interpolation bias (7-147) is an $O_{\text{int}}((r_i/(PN))^2)$. □

Appendix 7.J Zero Pattern of the Inverse of a Matrix

In this appendix we show that $M_1^{-1} = \text{adj}(M_1)/\text{det}(M_1)$ has the same zero pattern as M_1 ($M_{1[2i,2j]} = 0$). For example, for $R = 5$, M_1 has the following form

$$\begin{bmatrix} x & 0 & x & 0 & x & 0 \\ 0 & x & 0 & x & 0 & x \\ x & 0 & x & 0 & x & 0 \\ 0 & x & 0 & x & 0 & x \\ x & 0 & x & 0 & x & 0 \\ 0 & x & 0 & x & 0 & x \end{bmatrix} \tag{7-151}$$

where x denotes a non-zero entry. The adjoint matrix $\text{adj}(M_1)$ is obtained by replacing each entry by its co-factor and next taking the matrix transpose. The co-factor of a zero-entry of M_1 is obtained by deleting the corresponding row and column, calculating the determinant, and multiplying the result by -1. For example, for the zero entry [1, 4] in (7-151)

$$\begin{bmatrix} \cancel{x} & \cancel{0} & \cancel{x} & \cancel{0} & \cancel{x} & \cancel{0} \\ 0 & x & 0 & \vert & 0 & x \\ x & 0 & x & \vert & x & 0 \\ 0 & x & 0 & \vert & 0 & x \\ x & 0 & x & \vert & x & 0 \\ 0 & x & 0 & \vert & 0 & x \end{bmatrix} \tag{7-152}$$

This reduces the number of non-zero entries for R odd (or even) from $(R+1)/2$ (or $R/2+1$) to $(R-1)/2$ (or $R/2$) rows with the same zero pattern (in (7-151): rows 2, 4 and 6 with column 4 deleted). It shows that these $(R+1)/2$ (or $R/2+1$) rows are linearly dependent and, hence, the co-factor corresponding to a zero entry is zero. Since the zero pattern of M_1 is symmetric, it proves that M_1^{-1} has the same zero pattern as M_1. □

Appendix 7.K Covariance Matrix of (7-86)

Using $\text{vec}(ABC) = (C^T \otimes A)\text{vec}(B)$ and removing the frequency arguments for notational simplicity, we find for the covariance of (7-86)

$$\begin{aligned} \text{Cov}(\text{vec}(\hat{\mathbf{Z}}_R^{[m]})) &= (\overline{T_{\angle R}^{[m]}} \otimes I_{n_z})\text{Cov}(\text{vec}(\hat{\mathbf{Z}}^{[m]}))(T_{\angle R}^{[m]T} \otimes I_{n_z}) \\ &= (\overline{T_{\angle R}^{[m]}} \otimes I_{n_z})(I_{n_u} \otimes C_Z)(T_{\angle R}^{[m]T} \otimes I_{n_z}) \end{aligned} \tag{7-153}$$

with $n_z = n_y + n_u$, and where the second equality in (7-153) uses the following properties of the n_u experiments with the full random orthogonal multisines (3-31) of Section 3.7

1. The noise is independent and identically distributed over the n_u experiments and the M random phase realizations.

2. By construction of the full random orthogonal multisines, the stochastic nonlinear distortions are uncorrelated over the n_u experiments and independent over the M random phase realizations.

3. Since the power spectrum of the reference signal $\mathbb{E}\{\hat{\mathbf{R}}_{[:,e]}^{[m]} \hat{\mathbf{R}}_{[:,e]}^{[m]H}\}$, where $X_{[:,e]}$ denotes column e of X, is independent of the experiment e and the realization m, the covariance of the stochastic nonlinear distortions $\mathbb{E}\{\hat{\mathbf{Y}}_{S[:,e]}^{[m]} \hat{\mathbf{Y}}_{S[:,e]}^{[m]H}\}$, is also independent of e and m.

Elaborating (7-153) using $(A \otimes B)(C \otimes D) = AC \otimes BD$, finally gives

$$\text{Cov}(\text{vec}(\hat{\mathbf{Z}}_R^{[m]})) = \overline{T_{\angle R}^{[m]} T_{\angle R}^{[m]H}} \otimes C_Z = I_{n_u} \otimes C_Z = \text{Cov}(\text{vec}(\hat{\mathbf{Z}}^{[m]})) \quad (7\text{-}154)$$

The second equality in (7-154) uses the orthogonality of the matrix $T_{\angle R}^{[m]}$. Note that we also have proven that the columns of $\hat{\mathbf{Z}}_R^{[m]}$ are uncorrelated and have the same covariance. □

Appendix 7.L   Expected Value of the Sample Noise and Total Covariances (7-93)

The expected value of the sample noise covariance follows immediately from (7-74) and (7-76) where O_{leak} has been neglected w.r.t. O_{int}.

Let $c(k_i)$ be one entry of the matrix $r_Z^{[m,e]}(k_i) r_Z^{[m,e]H}(k_i)$ in (7-88). Using a second order Taylor series with remainder of $c(k_i)$

$$c(k_i) = c(k) + c^{(1)}(k)\frac{k_i - k}{N} + \frac{1}{2}c^{(2)}(l_i)\left(\frac{k_i - k}{N}\right)^2 \quad (7\text{-}155)$$

with $x^{(n)}$ the nth order derivative of x, $l_i \in [k, k_i]$, it can be found that

$$\sum_{i=-n_E}^{n_E} \frac{c(k_i)}{2n_E + 1} - c(k) = \frac{c^{(1)}(k)}{N(2n_E + 1)} \sum_{i=-n_E}^{n_E} (k_i - k) + O\left(\left(\frac{n_E}{N}\right)^2\right) \quad (7\text{-}156)$$

with $n_E = k_{n_E} - k$. For uniformly distributed excited harmonics we have that

$$\sum_{i=-n_E}^{n_E} (k_i - k) = 0$$

($k_{-i} = -k_i$) and, hence the bias in (7-156) is an $O((n_E/N)^2)$; otherwise it is an $O(n_E/N)$. The multiplying factor μ_{poly} quantifies the increase in noise variance due to the leakage removal in the sample mean $\hat{Z}^{[m,e]}(kP)$ for each realization m, and each experiment e (see (7-73) and (7-74)). □

Appendix 7.M   The Generator Noise Does Not Contribute to the Covariance of the FRM

Assuming that only generator errors $n_g(t)$ are present in the open loop setup of Figure 7-4, the input $V_U(k)$ and output $V_Y(k)$ errors are related by $V_Y(k) = G_0(j\omega_k)V_U(k)$, with G_0 the true FRM. The covariance matrix C_Z of $Z = [Y^T\ U^T]^T$ can then be written as

$$C_Z = \begin{bmatrix} G_0 C_U G_0^H & G_0 C_U \\ C_U G_0^H & C_U \end{bmatrix} = \begin{bmatrix} G_0 \\ I_{n_u} \end{bmatrix} C_U \begin{bmatrix} G_0^H & I_{n_u} \end{bmatrix} \qquad (7\text{-}157)$$

Substituting (7-157) in the expression of the covariance matrix of the FRM (2-77) on page 66, taking into account that $[G_0^H \ I_{n_u}]V^H(k) = 0$, with $V(k)$ defined in (2-77), shows that $\text{Cov}(\text{vec}(\hat{G})) = 0$.

Appendix 7.N Bias Robust FRM Estimate under Transient Conditions

To simplify the notations the proof is given for single-input, single output systems with exactly known steady state input, and exactly known transient output. Extension to multivariable systems with noisy non-steady state inputs and outputs is straightforward. From Section 7.3.3 it follows that the FRF estimate (7-79) can be written as

$$\hat{G}(j\omega_k) = \frac{\hat{Y}(kP)}{\hat{U}(kP)} = G(j\omega_k) + \frac{T_G(\Omega_{kP}) - \hat{T}_G(\Omega_{kP})}{U_0(kP)} \qquad (7\text{-}158)$$

Taking into account that $|U_0(kP)|$ is deterministic, the bias of (7-158) equals

$$\mathbb{E}\{\hat{G}(j\omega_k)\} - G(j\omega_k) = \frac{\mathbb{E}\{(T_G(\Omega_{kP}) - \hat{T}_G(\Omega_{kP}))\overline{U_0(kP)}\}}{|U_0(kP)|^2} \qquad (7\text{-}159)$$

where the expected value is taken w.r.t. the random phases of the input. For multisines with a fixed rms value and a number of harmonics increasing as an $O(N)$, the input DFT spectrum satisfies $|U_0(kP)|^2 = O(P)$. Combining this result with (7-159) and

$$T_G(\Omega_{kP}) - \hat{T}_G(\Omega_{kP}) = O((r_n/(PN))^{R+3/2})$$

(use (7-70)), proves that (7-108) is an upper bound for the bias. □

8

An Intuitive Introduction to Frequency Domain Identification

Abstract: In the next three chapters a detailed study of frequency domain identification schemes will be made. A wide class of methods is discussed, and it will be shown how the properties of the estimators are set by the choice of their cost function. Those readers who just want to solve their modeling problem, without passing through all these underlying theories, might still profit from a basic understanding of the methods they will use. For that reason we decided to provide, in this chapter, an intuitive insight into the frequency domain identification problem. First, a straightforward approach will be discussed, starting from the measured FRF of the systems transfer function; next a more general formulation will be made, based on the errors-in-variables concept, leading to a very robust identification method. Finally, it is discussed, briefly, how the general method can be applied to specific situations: no input noise present; the FRF is measured, and so on.

8.1 INTUITIVE APPROACH

The basic aim of this book is to measure and model the transfer function $G_0(\Omega)$ of a plant, starting from noisy input and output measurements (see Figure 8-1). An intuitive approach is to extract, first, a measured FRF $G(\Omega_k)$, $k = 1, ..., F$ of the systems' transfer function at a set of well-chosen frequencies (see Chapter 2 for a detailed discussion). Next, these measurements are approximated by a parametric model $G(\Omega_k, \theta)$ that explains the measurements as much as possible. As explained in Chapter 1, the quality of the match between measurements and model is measured by the cost function. The parameters are then tuned to minimize the cost function so that a best match is obtained. There is no unique choice for the cost function, and because each cost leads to an estimator, it is possible to find different estimators for the same problem. An intuitive choice of the cost function is

$$V_F(\theta, Z) = \frac{1}{F} \sum_{k=1}^{F} \frac{|G(\Omega_k) - G(\Omega_k, \theta)|^2}{\sigma_G^2(k)} \qquad (8\text{-}1)$$

The weighted least squares distance between the measurement and the model is minimized. Measurements with a small uncertainty ($\sigma_G^2(k)$ is small) are more important than those with a

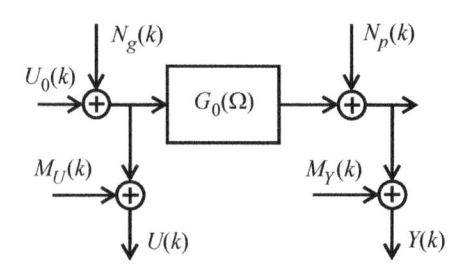

Figure 8-1. Frequency domain representation of the measurement process. N_g, N_p, M_U, and M_Y are, respectively, the generator noise, the process noise, and the input/output measurement errors. Note that the system can be captured in a feedback loop.

large uncertainty ($\sigma_G^2(k)$ is large). Although this method works amazingly well in many cases, it suffers from a major drawback. It is not always that easy to get a good measurement of G_0 due to the presence of the noise $M_U(k)$ on the input. If the classical correlation methods (H_1 method) are used, a bias appears (see Chapter 2). The measured transfer function converges for an increasing number of averages to:

$$\lim_{M \to \infty} G(\Omega_k) = \frac{S_{YU}(\Omega_k)}{S_{UU}(\Omega_k)} = G_0(\Omega_k) \frac{1}{1 + S_{M_U M_U}(\Omega_k)/S_{UU}(\Omega_k)} \qquad (8\text{-}2)$$

It is then easy to show that the parametric approximation will also be biased. When periodic excitations are used, alternative methods based on the direct division of output and input spectrum are available: $G(\Omega_k) = Y(k)/U(k)$. Although this method is less sensitive to the bias problem (see Chapter 2) for sufficiently large SNR at the input (better than 6 dB), it can be shown that in general its variance $\sigma_G^2(k)$ does not exist. (Guillaume et al., 1996a; Broersen, 1995). Especially for a low SNR at the input, large spikes frequently appear in the estimate, thus the variance estimate does not converge anymore. For larger SNRs the risk of encountering this problem becomes negligible in practice. However, this puts the user in a situation where he has to decide himself whether the method is applicable or not. For that reason a more robust alternative is formulated in the next section. Although it looks, at a glance, more complicated, it turns out that the computational complexity is not higher than that of the intuitive approach if periodic excitations are used. The major advantage for the less experienced user is that a check is no longer needed to verify whether the operational conditions on the intuitive technique are met or not. The algorithm can be automated fully and included in a general purpose package for public usage by laymen in the identification domain.

8.2 THE ERRORS-IN-VARIABLES FORMULATION

The intuitive methods of the previous section run into problems due to the presence of a division $Y(k)/U(k)$ that is a highly nonlinear operator. The denominator can become almost zero (the noise cancels the input) at some frequencies and this creates outliers. The errors-in-variables approach avoids a direct division of both measured spectra. Instead, the input and output spectra are considered as unknown parameters, connected by the parametric transfer function model:

$$\begin{aligned} Y(k) &= Y_0(k) + N_Y(k) \\ U(k) &= U_0(k) + N_U(k) \end{aligned} \qquad (8\text{-}3)$$

Section 8.2 ■ The Errors-in-Variables Formulation

with $Y_0(k) = G(\Omega_k, \theta)U_0(k)$ and where $N_Y(k)$ and $N_U(k)$ include the generator noise, the process noise, and the measurement noise. Because the exact Fourier coefficients $Y_0(k)$ and $U_0(k)$ are unknown, they are replaced by the parameters $Y_p(k)$ and $U_p(k)$, which are estimated by minimizing the distance between the measurements and the parameters ($|U(k) - U_p(k)|$, $|Y(k) - Y_p(k)|$), leading to a new constraint optimization problem. If the input and output measurements are uncorrelated with each other, the following least squares cost function can be used:

$$V_F(\theta, Z) = \frac{1}{F} \sum_{k=1}^{F} \frac{|U(k) - U_p(k)|^2}{\sigma_U^2(k)} + \frac{|Y(k) - Y_p(k)|^2}{\sigma_Y^2(k)}$$

$$= \frac{1}{F} \sum_{k=1}^{F} \begin{pmatrix} Y(k) - Y_p(k) \\ U(k) - U_p(k) \end{pmatrix}^H \begin{bmatrix} \sigma_Y^2(k) & 0 \\ 0 & \sigma_U^2(k) \end{bmatrix}^{-1} \begin{pmatrix} Y(k) - Y_p(k) \\ U(k) - U_p(k) \end{pmatrix}$$

(8-4)

to be minimized under the constraints

$$Y_p(k) = G(\Omega_k, \theta)U_p(k) \qquad k = 1, 2, ..., F \qquad (8\text{-}5)$$

In reality, the noise $N_U(k)$ and $N_Y(k)$ is correlated ($\sigma_{YU}^2(k) \neq 0$), and the full weighted least squares cost function should be considered:

$$V_F(\theta, Z) = \frac{1}{F} \sum_{k=1}^{F} \begin{pmatrix} Y(k) - Y_p(k) \\ U(k) - U_p(k) \end{pmatrix}^H \begin{bmatrix} \sigma_Y^2(k) & \sigma_{YU}^2(k) \\ \sigma_{UY}^2(k) & \sigma_U^2(k) \end{bmatrix}^{-1} \begin{pmatrix} Y(k) - Y_p(k) \\ U(k) - U_p(k) \end{pmatrix}$$

(8-6)

where $\sigma_{UY}(k) = \bar{\sigma}_{YU}(k)$. This cost function should be minimized with respect to the model parameters θ and also to the Fourier coefficients $U_p(k)$, $Y_p(k)$, $k = 1, ..., F$. As F can be very large, this appears to be a very hard task. However, this cost function can be simplified further. It is possible to eliminate $U_p(k)$, $Y_p(k)$ explicitly from the problem, simplifying the cost function to:

$$V_F(\theta, Z) = \frac{1}{F} \sum_{k=1}^{F} \frac{|Y(k) - G(\Omega_k, \theta)U(k)|^2}{\sigma_Y^2(k) + \sigma_U^2(k)|G(\Omega_k, \theta)|^2 - 2\text{Re}(\sigma_{YU}^2(k)\bar{G}(\Omega_k, \theta))}$$

(8-7)

Some of the advantages and properties of this formulation are discussed below.

8.2.1.1 Robustness with Respect to Bad Measurements.
Compared with (8-1), division of the measured Fourier coefficients is no longer needed. The cost function does not degenerate, even if the measured input equaled zero at some frequencies. The user should not bother anymore with the selection of an appropriate method to measure the FRF.

8.2.1.2 Symmetric Formulation.
By replacing in the cost function $G(\Omega_k, \theta) = B(\Omega_k, \theta)/A(\Omega_k, \theta)$ and multiplying the numerator and denominator with $|A(\Omega_k, \theta)|^2$, a complete symmetric formulation is found. The input and output have exactly the same role in the problem:

$$V_F(\theta, Z) = \frac{1}{F} \sum_{k=1}^{F} \frac{|A(\Omega_k, \theta)Y(k) - B(\Omega_k, \theta)U(k)|^2}{\sigma_Y^2(k)|A(\Omega_k, \theta)|^2 + \sigma_U^2(k)|B(\Omega_k, \theta)|^2 - 2\text{Re}(\sigma_{YU}^2(k)A(\Omega_k, \theta)\bar{B}(\Omega_k, \theta))} \quad (8\text{-}8)$$

8.2.1.3 Measuring the Noise Model.
The cost function depends on the exact values $\sigma_U^2(k)$, $\sigma_Y^2(k)$, and $\sigma_{YU}^2(k)$. In practice, these should be obtained from measured data. Section 2.5.1 shows how the sample covariance matrix can easily be extracted from repeated measurements and, later on, it will be shown that it is sufficient to use only four or six repetitions to guarantee that the properties of the estimator are not lost. This means, again, that a fully identifiability procedure can be set up. If the user can apply a periodic excitation, all other information can be extracted automatically without worrying about determining the noise model. If interested, however, the user can use this information to evaluate the quality of the experiments before starting the actual identification. For example, by examining the measured FRF together with its uncertainty, the complexity of the problem and the quality of the measurements can be revealed.

8.2.1.4 Dealing with Exactly Known Inputs.
In some applications (e.g., control problems) a model that links the output of the process directly to the digital controller output is built. In these cases the input signal is exactly known because it is stored in the memory of a computer. The errors-in-variables approach is automatically adapted to this situation by putting $\sigma_U^2(k)$ and $\sigma_{YU}^2(k)$ equal to zero.

8.2.1.5 Starting from Measured FRF.
Sometimes the user has only the measured FRF available. In that case it is still possible to use the previous approach by putting $Y(k) = G(\Omega_k)$, and $\sigma_Y^2(k) = \sigma_G^2(k)$, the input is set to $U(k) = 1$, with $\sigma_U^2(k) = 0$. The variance $\sigma_G^2(k)$ can be obtained directly from the coherence as explained in Section 2.5.4.

8.2.1.6 Properties.
The properties of the estimator are studied in detail in the next chapter, and it is shown that under weak conditions the estimates converge (for an increasing number of data points) to the parameters θ_* that would be found in the noiseless case. The uncertainty on the estimates approaches the smallest possible level for estimates without systematic errors. The covariance matrix $\text{Cov}(\hat{\theta})$ can be calculated at the end of the identification process. Starting from $\text{Cov}(\hat{\theta})$, it is easy to generate uncertainty bounds on other θ-dependent quantities; for example, for the FRF of the transfer function we get that

$$\text{var}(G(\Omega, \hat{\theta})) \approx \left(\frac{\partial G(\Omega, \theta)}{\partial \theta}\right) \text{Cov}(\hat{\theta}) \left(\frac{\partial G(\Omega, \theta)}{\partial \theta}\right)^H \bigg|_{\theta = \mathbb{E}\{\hat{\theta}\}} \quad (8\text{-}9)$$

(see Section 16.2). In practice the derivatives are evaluated in the estimated value $\hat{\theta}$. Also, the uncertainty bounds on the poles and zeros (see Section 11.2.3) or on the residuals (difference between measured and modeled FRF) (Section 11.2.2) can be obtained.

8.3 GENERATING STARTING VALUES

The cost function (8-8) is highly nonlinear in the parameters θ because they appear in the numerator and denominator. As a result, the minimization of the cost becomes quite difficult. It is possible to solve the problem analytically only in extremely simple cases. In all other situations a numerical search procedure is needed. The convergence of these methods depends strongly on the generation of good starting values for the model parameters θ. In general, it is impossible to get this information from physical principles because the link between the coefficients of the transfer function and the underlying physical systems is very nonlinear, especially for higher order systems. Moreover, often the user does not want to make the effort to collect all the required knowledge because the goal of the experiment is to generate a black box model that describes the input-output behavior. For that reason we need self-starting algorithms that generate the starting values from the measured data and not from unavailable prior knowledge.

A possibility to make the optimization self-starting is to change the cost function in the first step so that its global minimum can be calculated directly. There are a number of possibilities to reach this goal. The simplest solution is just to remove the denominator in (8-8) so that the problem becomes linear-in-the-parameters and the minimum is found by solving a linear set of equations (see Section 9.8.2). The disadvantage of this straightforward approach is that the solution becomes extremely noise sensitive. For that reason attempts were made to make a parameter-independent reconstruction of the denominator of (8-8) using measurement information only (Section 9.12.4). This results in significantly improved starting values. A second possibility to generate starting values is to continue with the nonlinear cost function but to modify it such that the global minimum can easily be found using advanced, but widely available, numerical techniques such as singular value decomposition. This leads to the generalized, total least squares type of solutions (see Section 9.10.3) that minimize a cost function of the form

$$V_F(\theta, Z) = \frac{\sum_{k=1}^{F} |A(\Omega_k, \theta)Y(k) - B(\Omega_k, \theta)U(k)|^2}{\sum_{k=1}^{F} \sigma_Y^2(k)|A(\Omega_k, \theta)|^2 + \sigma_U^2(k)|B(\Omega_k, \theta)|^2 - 2\mathrm{Re}(\sigma_{YU}^2(k)A(\Omega_k, \theta)\bar{B}(\Omega_k, \theta))} \quad (8\text{-}10)$$

Although the efficiency of this method is lower than that of the original MLE, it provides good candidate starting values. Again, it is possible to improve the quality by adding a nonparametric frequency weighting as explained in the next chapter. A third possibility to get starting values is to use subspace methods (see Section 9.14) that are based on state space models. Compared with the previous algorithms, this method is less flexible because it is not possible to choose the number of poles different from the numbered zeros; but despite this disadvantage, good quality starting values are generated. A major advantage of subspace methods is that they are very well suited to multiinput, multioutput (MIMO) problems.

As a general procedure, we advise the reader to combine these techniques by calculating two or three candidate starting values and to select, out of these, the solution that results in the smallest MLE cost (8-8).

8.4 COMPARISON WITH THE "CLASSICAL" TIME DOMAIN IDENTIFICATION FRAMEWORK

Identification has a long tradition. Over the years, the attention shifted almost completely to the use of discrete-time models that were identified starting from arbitrary (no periodicity required) excitations. The major difference with the preceding approach is that a parametric noise model is used (Ljung, 1999; see also Section 10.9). Ljung (1999) gives a frequency domain interpretation of the cost function that is minimized with these techniques. By neglecting the leakage effects, the following equivalent frequency domain representation of the time domain cost is found:

$$V_N(\theta, Z) \approx \frac{1}{N} \sum_{k=0}^{N-1} \frac{|Y(k) - G(z_k^{-1}, \theta) U(k)|^2}{|H(z_k^{-1}, \theta)|^2} \tag{8-11}$$

where $|H(z^{-1}, \theta)|^2$ is a parametric model for the power spectrum of the process noise. These methods work well if the measurement noise $(M_U(k), M_Y(k))$ is negligible, otherwise the results will be prone to systematic errors. The major advantage of this approach is that no periodic signals are needed. Its major disadvantage is the need to estimate an additional model $H(z^{-1}, \theta)$. A more detailed discussion is given in Section 10.9.

8.5 EXTENSIONS OF THE MODEL: DEALING WITH UNKNOWN DELAYS AND TRANSIENTS

In the previous sections, the simplest model was used. The results can be generalized to systems with an unknown delay τ. To do so, the model is extended to $G(\Omega, \theta)e^{-\tau s}$ for continuous-time systems or to $z^{-\tau/T_s} G(z^{-1}, \theta)$ for discrete-time systems (see Section 6.2). The reader has to realize that the corresponding optimization problem is much more difficult to solve because it is very sensitive to local minima. Consequently, a good starting value of the delay is needed.

Another generalization is the extension of the model to include transients (before the system reaches its steady-state behavior) or to cover, also, the situation with arbitrary (nonperiodic) excitations. Again, this is simply solved by adding an additional rational term to the model (Section 6.3.2):

$$G(\Omega, \theta) + \frac{I(\Omega, \theta)}{A(\Omega, \theta)} \tag{8-12}$$

Because the additional rational term has the same denominator, the complexity of the numerical optimization process is almost not affected by this generalization. A similar extension can be used to process experiments with missing data (Section 6.3.3).

9

Estimation with Known Noise Model

Abstract: This chapter gives an overview of frequency domain identification methods for single input, single output systems. Estimators such as the (weighted) linear least squares, the weighted nonlinear least squares, the maximum likelihood, the (weighted) total least squares, the instrumental variables, and the subspace algorithms are discussed in detail. The interrelations between the different approaches are highlighted through a study of the (equivalent) cost functions. Special attention is also paid to global minimizers that try to approximate the maximum likelihood estimator. The properties of the different approaches are illustrated by means of an "on-line" simulation example. The chapter ends with an overview of the properties of the estimators and a brief discussion of the particularities of estimating high-order systems, systems with time delay, systems in feedback, systems with missing data, multivariable systems, and transfer function models with complex coefficients.

9.1 INTRODUCTION

In this chapter we handle the identification of the plant model assuming that the noise model is known exactly. We give an overview of frequency domain identification methods for single input, single output systems (Sections 9.8 to 9.15). Afterward, the particularities of high-order systems (Section 9.16), systems with time delay (Section 9.17), systems in feedback (Section 9.18), the missing data problem (Section 9.20), and multivariable systems (Section 9.21) are discussed. A second-order system $G(s, \theta) = 1/(1 + s + s^2)$ is used as an "on-line" illustration through Sections 9.8 to 9.14. Figure 9-1 shows the true transfer function and the simulated noisy frequency response data (see Appendix 9.A for more information concerning the generation of the simulation data). Readers who want only a quick taste of the basics of frequency domain estimation (and accept the claimed properties as they are) may skip the last paragraph of Section 9.4 and Sections 9.5 to 9.7 but should still go through Sections 9.2 and 9.3 before tackling the description of the estimators (Sections 9.8 to 9.15).

Before starting with the overview, we discuss the type of data (experiments) we can handle (Section 9.2), introduce some notations for the parametric plant models (Section 9.3), and present the general form of the identification algorithms (Section 9.4). Section 9.5, quick tools to analyze estimators, is intended for readers who are not interested in the technical details of the proofs but still want to get some insight into the derivation of some basic properties. Combined with Section 9.7, which discusses the general asymptotic properties of esti-

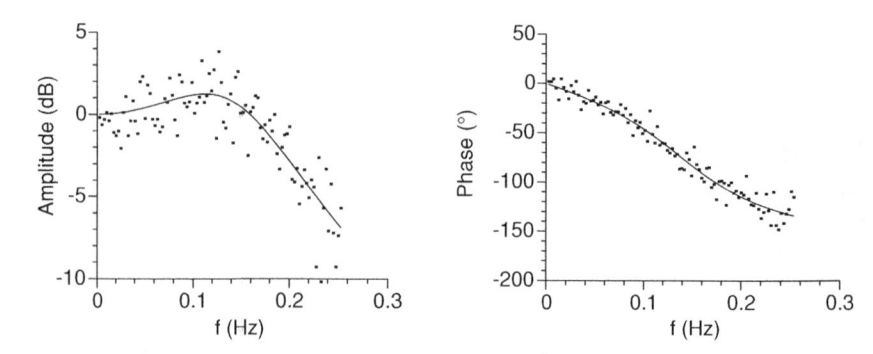

Figure 9-1. Second-order example $G(s, \theta) = 1/(1 + s + s^2)$: true transfer function (solid line) and simulated noisy data (dots).

mators minimizing a cost function that is quadratic-in-the-measurements, it will allow them to easily verify and understand the properties of the different estimators described in Sections 9.8 to 9.14. Those who are interested in the technical details will find a comprehensive list of the basic assumptions needed to prove the asymptotic properties of the estimators (Section 9.6). The proofs of the theorems are given in the Appendix and rely completely on the results of Chapters 17, 18, and 19. The reader is referred to these chapters for more background information concerning the way properties are proved.

9.2 FREQUENCY DOMAIN DATA

The identification starts from measured input-output discrete Fourier transform (DFT) spectra $U(k)$, $Y(k)$,

$$Y(k) = Y_0(k) + N_Y(k)$$
$$U(k) = U_0(k) + N_U(k) \qquad (9\text{-}1)$$

with $U_0(k)$, $Y_0(k)$ the true unknown values, or from a measured frequency response function $G(\Omega_k)$,

$$G(\Omega_k) = G_0(\Omega_k) + N_G(k) \qquad (9\text{-}2)$$

with $G_0(\Omega_k)$ the true unknown value, at a set of F frequencies Ω_k, $k = 1, 2, ..., F$, which may be a (sub)set of the DFT frequencies. Note that (9-2) is a special case of (9-1) with $Y(k) = G(\Omega_k)$ and $U(k) = 1$. The $2F$ complex-valued vector Z contains the measured input-output (DFT) spectra

$$Z^T = [Z^T(1)\ Z^T(2)\ ...\ Z^T(F)] \text{ with } Z^T(k) = [Y(k)\ U(k)] \qquad (9\text{-}3)$$

where $k = 1, 2, ..., F$. It is related to the true values by $Z = Z_0 + N_Z$, where the disturbing noise N_Z has zero mean and is independent of Z_0.

The frequency domain data (9-1), (9-2) can be obtained via time domain or frequency domain experiments. In a *time domain experiment* a broadband random or normalized periodic (see Definition 3.4) excitation is applied to the plant and N samples of the input and output signals are measured (see Figure 9-2). For the periodic signals the steady-state re-

Section 9.2 ■ Frequency Domain Data

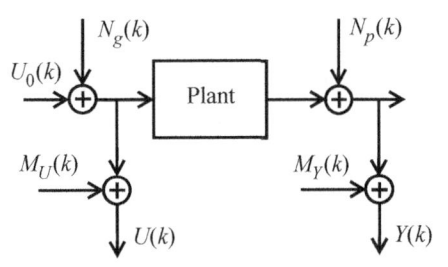

Figure 9-3. Frequency domain experiment: a single sine excitation $u_g(t) = A\sin(2\pi f_k t + \phi)$ is applied to the plant and the input-output spectra of the steady-state response are measured at frequency f_k. This experiment is repeated at F different frequencies. $N_g(k)$ is the generator noise, $M_U(k)$ and $M_Y(k)$ are the input and output measurement errors, and $N_p(k)$ is the process noise.

sponse is observed over an integer number of periods. These N input-output samples are transformed to the frequency domain using the discrete Fourier transform. $F \leq N/2 + 1$ DFT frequencies of the input and output DFT spectra are used for the identification. For arbitrary signals $u_g(t)$ the generator noise $n_g(t)$ is a part of the excitation, $u_0(t) = u_g(t) + n_g(t)$, so that the frequency domain errors $N_U(k)$ and $N_Y(k)$ in (9-1) are related to the disturbing noise sources in Figure 9-2 as

$$N_Y(k) = \text{DFT}(m_y(t) + n_p(t))$$
$$N_U(k) = \text{DFT}(m_u(t))$$
(9-4)

For periodic signals the generator noise $n_g(t)$ is a disturbing noise source, $u_0(t) = u_g(t)$, which causes a correlation between the input and output errors. Indeed, the frequency domain errors $N_U(k)$ and $N_Y(k)$ in (9-1) are then related to the disturbing noise sources in Figure 9-2 as

$$N_Y(k) = \text{DFT}(n_g(t) * g_0(t) + n_p(t) + m_y(t))$$
$$N_U(k) = \text{DFT}(n_g(t) + m_u(t))$$
(9-5)

with * the convolution operator and $g_0(t)$ the impulse response of the plant. In a *frequency domain experiment*, a single sine excitation is applied to the plant and the input-output spectra of the steady-state response are measured at the excited frequency. This experiment is re-

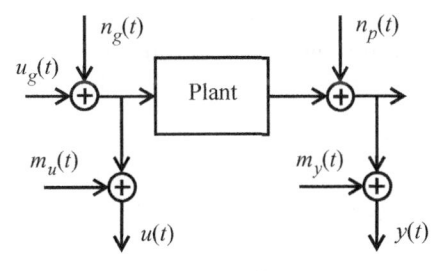

Figure 9-2. Time domain experiment: a broadband excitation $u_g(t)$ is applied to the plant. The DFT spectra of N observed input-output samples are calculated. $F = O(N)$ DFT frequencies of the input-output DFT spectra are retained. $n_g(t)$ is the generator noise, $m_u(t)$ and $m_y(t)$ are the input and output measurement errors, and $n_p(t)$ is the process noise.

peated at F different frequencies. For example, high-frequency network analyzers (microwave measurements) and impedance analyzers follow this measurement procedure. Also, most dynamic signal analyzers have such a measurement mode. The frequency domain errors $N_U(k)$ and $N_Y(k)$ in (9-1) are related to the noise sources in Figure 9-3 as

$$N_Y(k) = N_g(k)G_0(\Omega_k) + M_Y(k) + N_P(k)$$
$$N_U(k) = N_g(k) + M_U(k)$$
(9-6)

with $G_0(\Omega_k)$ the plant transfer function.

Due to the imperfections of the measurement devices, it is recommended not to use measurements at DC and in the neighborhood of the Nyquist frequency. Indeed, acquisition units mostly introduce DC offset errors and anti-alias protection is mostly guaranteed only up to about 80% of the Nyquist frequency. The measurements can also be the result of a linearization of a nonlinear system at an operating point. This will introduce DC values in the input and output signals that are not compatible with the linear model and, hence, should be removed.

An important question asked when (re)designing an experiment is: "What will happen with the estimates (uncertainty, bias, ...) if one gathered, for example, four times more data?" Ideally, we would like to answer this question for each finite value of F. Except for the (weighted) linear least squares, this is possible only for "sufficiently large" values of F. To analyze the stochastic properties of the estimators for F "sufficiently large" we will make a mental experiment where the number of frequencies F tends to infinity. For a frequency domain experiment this implies that the number of single sine measurements F tends to infinity, while for a time domain experiment this implies that the number of measured time domain samples N tends to infinity. Note that we do not consider time domain experiments ($N \to \infty$) with periodic signals containing a fixed number (independent of N) of frequencies F. Indeed, for such experiments the signal-to-noise ratio tends to infinity as $N \to \infty$ at the excited DFT frequencies (see Appendix 9.C), and, hence, all the estimators considered in this chapter would be consistent in a trivial manner. For random and normalized multisine ($F = O(N)$, see Definition 3.2) excitations the signal-to-noise ratio per spectral line remains an $O(N^0)$ (see Appendices 9.B and 9.C) so that consistency is a nontrivial issue.

9.3 PLANT MODEL

Unless mentioned otherwise, we will assume in this chapter that the parameterization of the plant model is identifiable (see Definition 6.8). It implies that the parameter vector θ contains only the free parameters of the model, for example, all the numerator and denominator coefficients of the rational form $G(\Omega, \theta) = B(\Omega, \theta)/A(\Omega, \theta)$ except $a_0 = 1$. Note, however, that from a numerical point of view it is often better to use the full overparameterized form in combination with dedicated numerical methods. Chapter 20 discusses this issue in detail.

For any parameterization of Sections 6.2 and 6.3 (rational form, partial fraction expansion, and state space representation) we can use the *output error*, which is the difference between the observed output $Y(k)$ and the modeled output $Y(\Omega_k, \theta)$. From transfer function models (6-32), and (6-34) we get

$$Y(\Omega_k, \theta) = G(\Omega_k, \theta)U(k)$$
(9-7)

for periodic signals ($\Omega = z^{-1}$, s, $\sqrt{s}$ or $\tanh(\tau_R s)$) and

Section 9.4 ■ Estimation Algorithms

$$Y(\Omega_k, \theta) = G(\Omega_k, \theta)U(k) + T_G(\Omega_k, \theta) \tag{9-8}$$

for arbitrary excitations ($\Omega = z^{-1}$, s, or $\sqrt{s}$).

For the *rational forms,* (6-20), (6-25), (6-35), and (6-38), it is convenient also to introduce the *equation error* $e(\Omega_k, \theta, Z(k))$, which is the difference between the left- and right-hand sides of transfer function models (6-32) and (6-34), after multiplication by $A(\Omega_k, \theta)$. We get

$$e(\Omega_k, \theta, Z(k)) = A(\Omega_k, \theta)Y(k) - B(\Omega_k, \theta)U(k) \tag{9-9}$$

for periodic signals ($\Omega = z^{-1}$, s, $\sqrt{s}$ or $\tanh(\tau_R s)$) and

$$e(\Omega_k, \theta, Z(k)) = A(\Omega_k, \theta)Y(k) - B(\Omega_k, \theta)U(k) - I(\Omega_k, \theta) \tag{9-10}$$

for arbitrary excitations ($\Omega = z^{-1}$, s, or $\sqrt{s}$). The equation error $e(\Omega_k, \theta, Z(k))$ is not exactly zero because the observations $Y(k)$ and $U(k)$ are disturbed by noise and θ does not equal the true value θ_0 (if it exists).

Note that (9-8) and (9-10) are valid only at a (sub)set of the DFT frequencies but (9-7) and (9-9) are also valid at arbitrary (not related to a DFT grid) frequencies. For concatenated data sets, (9-8) and (9-10) are extended with terms of the form $z_k^{-N_i} T^c_{G[i]}(\Omega_k, \theta)$ and $z_k^{-N_i} I^c_{[i]}(\Omega_k, \theta)$, respectively (see (6-48) and Exercise 6.8).

9.4 ESTIMATION ALGORITHMS

Most algorithms discussed in this chapter minimize (in each step) a "quadratic-like" cost function $V(\theta, Z)$

$$V(\theta, Z) = \varepsilon^H(\theta, Z)\varepsilon(\theta, Z) = \sum_{k=1}^{F} |\varepsilon(\Omega_k, \theta, Z(k))|^2 \tag{9-11}$$

where $\varepsilon(\theta, Z) \in \mathbb{C}^F$ is some kind of measure of the difference between the measurements and the model. The residual $\varepsilon(\theta, Z) \in \mathbb{C}^F$ is a (non)linear vector function of the model parameters θ and the measurements Z. Note that $\varepsilon_{[k]}(\theta, Z) = \varepsilon(\Omega_k, \theta, Z(k))$ depends only on the measurements at frequency Ω_k.

A first important subclass of (9-11) consists of the cost functions $V(\theta, Z)$, which are *quadratic-in-the-measurements* Z. For these cost functions the residual $\varepsilon(\theta, Z)$ is linear in Z and can be written as

$$\varepsilon(\theta, Z) = \varepsilon(\theta, Z_0) + \Delta(\theta, N_Z) \tag{9-12}$$

with $\Delta_{[k]}(\theta, N_Z) = \Delta(\Omega_k, \theta, N_Z(k))$ and $\Delta(\Omega_k, \theta, 0) = 0$. Hence, (9-11) becomes

$$\begin{aligned} V(\theta, Z) &= V(\theta, Z_0) + v(\theta, N_Z) + 2\mathrm{Re}(\varepsilon^H(\theta, Z_0)\Delta(\theta, N_Z)) \\ v(\theta, N_Z) &= \Delta^H(\theta, N_Z)\Delta(\theta, N_Z) \end{aligned} \tag{9-13}$$

where $v(\theta, N_Z)$ represents that part of the cost function depending on N_Z only.

A second important subclass of (9-11) consists of the cost functions $V(\theta, Z) = f(\theta, \eta(Z), Z)$, which depend on an initial guess $\eta(Z)$ of the model parameters,

$$V(\theta, Z) = \varepsilon^H(\theta, Z)\varepsilon(\theta, Z) = \sum_{k=1}^{F} |\varepsilon(\Omega_k, \theta, \eta(Z), Z(k))|^2 \qquad (9\text{-}14)$$

and which are quadratic-in-the-measurements Z when $\eta(Z)$ in (9-14) is replaced by a non-random vector η.

Often, a Newton-Gauss type of algorithm is used to find the minimizer $\hat{\theta}(Z)$ of (9-11). Rewriting (9-11) as $V(\theta, Z) = \varepsilon_{re}^T(\theta, Z)\varepsilon_{re}(\theta, Z)$, where $(\ )_{re}$ stacks the real and imaginary parts on top of each other (see Section 15.8),

$$\varepsilon_{re}(\theta, Z) = \begin{bmatrix} \text{Re}(\varepsilon(\theta, Z)) \\ \text{Im}(\varepsilon(\theta, Z)) \end{bmatrix} \qquad (9\text{-}15)$$

the ith iteration step of this algorithm is given by (see also Section 1.5.1)

$$J_{re}^T(\theta^{(i-1)}, Z) J_{re}(\theta^{(i-1)}, Z)\Delta\theta^{(i)} = -J_{re}^T(\theta^{(i-1)}, Z)\varepsilon_{re}(\theta^{(i-1)}, Z) \qquad (9\text{-}16)$$

with $\Delta\theta^{(i)} = \theta^{(i)} - \theta^{(i-1)}$ and $J(\theta, Z) = \partial\varepsilon(\theta, Z)/\partial\theta$ the Jacobian of the vector $\varepsilon(\theta, Z)$. Using complex numbers, (9-16) can be written as

$$\text{Re}(J^H(\theta^{(i-1)}, Z)J(\theta^{(i-1)}, Z))\Delta\theta^{(i)} = -\text{Re}(J^H(\theta^{(i-1)}, Z)\varepsilon(\theta^{(i-1)}, Z)) \qquad (9\text{-}17)$$

If the algorithm converges to the global minimum, then $\hat{\theta}(Z) = \theta^{(\infty)}$. When identifying continuous-time systems in the s- and $\sqrt{s}$- domains, it is indispensable to scale the frequency axis (and, hence, also the parameters) to guarantee the numerical stability of the normal equations (9-16). Without scaling, identification in the s- and $\sqrt{s}$- domains is often impossible with the available computing precision, even for modest orders of the transfer function. Although the scale factor that minimizes the condition number of $J_{re}(\theta^{(i-1)}, Z)$ is plant and model dependent, a good compromise is to use the median of the set of angular frequencies in the frequency band of interest: $\omega_{scale} = \text{median}\{\omega_1, \omega_2, ..., \omega_F\}$ (see Pintelon and Kollár, 2005). For example, the term $a_m s^m$ becomes $a_m \omega_{scale}^m (s/\omega_{scale})^m$ after scaling and $a_m \omega_{scale}^m$ is estimated. For any domain (s, $\sqrt{s}$, z), the numerical stability of the calculations is improved significantly by solving the overdetermined set of equations

$$J_{re}(\theta^{(i-1)}, Z)\Delta\theta^{(i)} = -\varepsilon_{re}(\theta^{(i-1)}, Z) \qquad (9\text{-}18)$$

instead of (9-16), for example, using the singular value decomposition or a QR factorization (see Section 15.13). Finally, the condition number of the Jacobian matrix $J_{re}(\theta^{(i-1)}, Z)$ in (9-18) is reduced by scaling each column by its 2-norm: $J_{re}(\theta^{(i-1)}, Z) \to J_T = J_{re}(\theta^{(i-1)}, Z)T^{-1}$ and $\Delta\theta^{(i)} \to \Delta\theta_T = T\Delta\theta^{(i)}$ where

$$T = \text{diag}(\|J_{re[:,1]}(\theta^{(i-1)}, Z)\|_2, \|J_{re[:,2]}(\theta^{(i-1)}, Z)\|_2, ..., \|J_{re[:,n_\theta]}(\theta^{(i-1)}, Z)\|_2) \qquad (9\text{-}19)$$

(Van Huffel and Vandewalle, 1991). The transformed equation $J_T \Delta\theta_T = -\varepsilon_{re}(\theta^{(i-1)}, Z)$ is then solved via an SVD or QR decomposition of J_T, and the actual parameter variation is found as $\Delta\theta^{(i)} = T^{-1}\Delta\theta_T$. Note that the convergence region of the Newton-Gauss algorithm can be enlarged by using a Levenberg-Marquardt version of (9-16) and (9-18) (see Fletcher, 1991 and Section 9.L.4 of Appendix 9.L).

To study the asymptotic behavior of the identification algorithms, it is convenient to scale the cost function with the number of frequencies, $V_F(\theta, Z) = V(\theta, Z)/F$, $v_F(\theta, N_Z) = v(\theta, N_Z)/F$, and $f_F(\theta, \eta(Z), Z) = f(\theta, \eta(Z), Z)/F$. The expected values of the cost function $V_F(\theta) = \mathbb{E}\{V_F(\theta, Z)\}$ and its minimizer $\tilde{\theta}(Z_0)$ play an important role in the convergence analysis of the estimate $\hat{\theta}(Z)$. All the asymptotic properties ($F \to \infty$) of the estimate $\hat{\theta}(Z)$ will be formulated w.r.t. the minimizer $\tilde{\theta}(Z_0)$ of the expected value of the cost function. The conditions under which $\hat{\theta}(Z)$ converges to $\tilde{\theta}(Z_0)$ will be studied. This is a stochastic convergence problem that mainly depends on the disturbing noise properties. When model errors are present, $\tilde{\theta}(Z_0)$ will vary as the number of frequencies F increases. We may wonder then whether $\tilde{\theta}(Z_0)$ converges to some limit value $\theta_* = \lim_{F \to \infty} \tilde{\theta}(Z_0)$ which is the minimizer of the limit cost function $V_*(\theta) = \lim_{F \to \infty} V_F(\theta)$. This is a deterministic convergence problem that depends on the way data (frequencies) are added in the time or frequency domain experiment. The notations introduced are summarized in Table 9-1.

TABLE 9-1 Overview of Notations Frequently Used: $\eta(Z)$ Is an (Initial) Estimate of the Model Parameters and η_* Is Its Limit Value

Cost function	$V_F(\theta, Z)$, $f_F(\theta, \eta(Z), Z)$	$V_F(\theta) = \mathbb{E}\{V_F(\theta, Z)\}$, $V_F(\theta) = \mathbb{E}\{f_F(\theta, \eta_*, Z)\}$	$V_*(\theta) = \lim_{F \to \infty} V_F(\theta)$
Minimizer	$\hat{\theta}(Z)$	$\tilde{\theta}(Z_0)$	θ_*

9.5 QUICK TOOLS TO ANALYZE ESTIMATORS

The minimum we can expect from a "sound" estimator is that in the noiseless case we get the true answer (correctness property). In the noisy case we should get asymptotically ($F \to \infty$) the true answer (consistency property) and hopefully a "small" uncertainty (efficiency property). We may also wonder whether the estimates depend on the particular parameter constraint chosen ($a_0 = 1$, or $\|\theta\|_2^2 = 1$, or …), how fast the estimates converge, and what happens with the estimates if the true model does not belong to the considered model set. All these questions are thoroughly studied in this chapter.

Some of the previously raised questions can easily be analyzed using the following quick tools. The first step in the analysis consists of calculating the (equivalent) cost function $V(\theta, Z)$ of the identification method. Next we verify the following:

1. *(Asymptotic) correctness:* assuming that the true model belongs to the model set, the identification algorithm is (asymptotically) correct if it produces the true model for an (in)finite amount of noiseless ($N_Z = 0$) data. This is true if $V_F(\theta, Z_0)$ ($\lim_{F \to \infty} V_F(\theta, Z_0)$) is minimal in the true model parameters θ_0. All the identification algorithms of this chapter are correct for transfer function models (9-7) with $\Omega = z^{-1}$, s, $\sqrt{s}$, or $\tanh(\tau_R s)$ and (9-8) with $\Omega = z^{-1}$, where $G(\Omega, \theta)$ and $T(\Omega, \theta)$ can take any parameterization of Sections 6.2 and 6.3. They are asymptotically correct for continuous-time models using arbitrary excitations, model (9-8) with $\Omega = s$.

2. *Consistency:* the (equivalent) cost function minimized by most identification methods in this chapter is a quadratic function of the measurements Z. The expected value of such cost functions can be written as

$$V_F(\theta) = \mathbb{E}\{V_F(\theta, Z)\} = \mathbb{E}\{V_F(\theta, Z_0)\} + \mathbb{E}\{v_F(\theta, N_Z)\} \quad (9\text{-}20)$$

(see (9-13), Z_0 and N_Z are independent). A necessary condition for consistency is that the limit of the expected value of the cost function $V_*(\theta) = \lim_{F \to \infty} V_F(\theta)$ is minimal in θ_0 (Theorem 17.15). It follows from (9-20) that this condition is satisfied if $\mathbb{E}\{v_F(\theta, N_Z)\}$ is a θ-independent constant. Hence, for correct methods we have $\tilde{\theta}(Z_0) = \theta_0$, while for asymptotically correct methods $\theta_* = \theta_0$. For cost functions of the form (9-14), we replace $\eta(Z)$ by its limit value η_* before taking the expected value of the cost function. The same analysis is then performed on $V_F(\theta) = \mathbb{E}\{f_F(\theta, \eta_*, Z)\}$.

3. *Convergence to the noiseless solution:* if model errors exist, for example, because of a wrong choice of the order of the numerator and/or denominator polynomials, or because a true linear lumped model simply does not exist, then $\hat{\theta}(Z)$ converges to $\tilde{\theta}(Z_0) \neq \theta_0$. Under some conditions, the value $\tilde{\theta}(Z_0)$ is independent of the noise level of the measurements. To verify this, we replace C_{N_Z} by $\upsilon^2 C_{N_Z}$ in the cost function (9-20), with υ a real number. If this transforms $\mathbb{E}\{V_F(\theta, Z_0)\}$ into $f(\upsilon^2)\mathbb{E}\{V_F(\theta, Z_0)\}$ and if $\mathbb{E}\{v_F(\theta, N_Z)\}$ is a θ-independent constant then $\tilde{\theta}(Z_0)$, and, hence, also $\theta_* = \lim_{F \to \infty} \tilde{\theta}(Z_0)$ (if it exists), is independent of the noise level υ. This is true for any υ and, hence, also for $\upsilon \to 0$, which defines, asymptotically, the noiseless solution. Note, however, that the noiseless solution $\tilde{\theta}(Z_0)$ defined in this way may still depend on the noise coloring and the noise covariance matrix C_{N_Z}, for example, the ratio of the output variance $\sigma_Y^2(k)$ to the input variance $\sigma_U^2(k)$ (see Section 9.11). For cost functions of the form (9-14), the analysis is performed on $V_F(\theta) = \mathbb{E}\{f_F(\theta, \eta_*, Z)\}$ and the same conclusions hold if η_*, the limit value of $\eta(Z)$, is independent of υ.

4. *Dependence on the parameter constraint:* from a numerical point of view it is also handy that the estimate of the plant transfer function $G(\Omega_k, \hat{\theta}(Z))$ is independent of the particular parameter constraint chosen, for example, $a_i = 1$, or $b_j = 1$, or $\|\theta\|_2^2 = 1$ ... Indeed, if we fix a zero coefficient to one, then the normal equations (9-16) become ill conditioned. To avoid this problem, it is better to use the constraint $\|\theta\|_2^2 = 1$ (see also Chapter 20). The estimated plant model $G(\Omega, \hat{\theta}(Z))$ is independent of the parameter constraint chosen if, for any $\lambda \neq 0$, $V_F(\lambda \theta, Z) = V_F(\theta, Z)$, with θ the full overparameterized form (proof: see Chapter 20).

5. *Numerical reliability of the normal equations:* the Hessian of the expected value of the cost function has full rank in the true parameter values: $\text{rank}(V_F''(\theta_0)) = \dim(\theta) =$ number of free model parameters (' is the derivative w.r.t. θ). If the Hessian is not of full rank, then the cost function cannot be approximated by a quadratic function in the neighborhood of the solution θ_0. This is problematic for most of the nonlinear minimization algorithms.

6. *Influence of the noise level and the model errors:* to study the influence of small measurement errors, we replace N_Z by υN_Z and C_{N_Z} by $\upsilon^2 C_{N_Z}$ and analyze the expression for $\upsilon \to 0$. Model errors are present if $e(\tilde{\theta}(Z_0), Z_0) \neq 0$. To study the influence of small model errors we replace $e(\tilde{\theta}(Z_0), Z_0)$ by $\mu e(\tilde{\theta}(Z_0), Z_0)$ and analyze the expressions for $\mu \to 0$.

9.6 ASSUMPTIONS

In this section we give an overview of all the assumptions required to analyze the asymptotic ($F \to \infty$) behavior of the estimate $\hat{\theta}(Z)$. They are grouped per property in increasing order of complexity: stochastic convergence, stochastic convergence rate, systematic and stochastic errors, consistency, asymptotic bias, asymptotic normality, and asymptotic efficiency. Hereby, we make the distinction between a time and a frequency domain experiment because the signal and disturbing noise properties are easiest to describe in the respective domains. It allows the reader to verify, easily, what kind of assumptions are required for a particular property and experiment in each theorem of this chapter. For theoretical convenience, the disturbing noise in the time domain experiment is modeled as discrete-time filtered white noise. Physical interpretation of these noise models should, however, be done with care (see Section 6.7.3 for more details).

The cost function (9-11) and its higher order derivatives w.r.t. θ may not exist for some values of the model parameters θ. To avoid the resulting technical difficulties in the proof of the theorems, a regular set $\mathbb{P}_r$ of θ-values is constructed where $V_F(\theta, Z)$ and its higher order derivatives exist and are finite. By construction, we make this set closed and bounded compact. The minimizer of (9-11) is then defined as

$$\hat{\theta}(Z) = \arg \min_{\theta \in \mathbb{P}_r} V_F(\theta, Z) \tag{9-21}$$

(for the maximum likelihood estimation of ARMAX models the compactness assumption of the parameter space can be avoided, see Hannan and Deistler (1988)).

The properties of (9-21) will be studied using the results of Chapter 17 for Sections 9.8.2, 9.9, and 9.10; of Chapter 18 for Sections 9.8.3, 9.12, and 9.14; and of Chapter 19 for Section 9.11. The reader is referred to these chapters for detailed background information concerning the proof of the theorems. There she or he will also find answers to questions such as "Why do we need a particular assumption and what is it used for?" and "What is the main philosophy behind the proof of a particular property?" Other basic questions such as "Which statistical tools are available?" and "How should they be used?" are tackled in Chapter 16.

9.6.1 Stochastic Convergence

To show the convergence ($F \to \infty$) of the estimator $\hat{\theta}(Z)$ (9-21) to $\tilde{\theta}(Z_0)$ we need conditions on the excitation signal, the disturbing noise, and the cost function. For example, the persistence of excitation Assumption 9.7 requires that the excitation signal satisfies, at least, the identifiability conditions of Section 6.5. Note that we do not require the existence of a true model.

Assumption 9.1 (Excitation Signal—Time Domain Experiment): The excitation $u(t)$ either is a normalized periodic signal (see Definitions 3.2, 3.3, and 3.4) or can be written at the sampling instances as filtered white noise $u(t) = H_u(q)e_u(t)$, where $H_u(z^{-1})$ is a stable rational filter. $e_u(t)$ is independently distributed and has stationary first- and second-order moments and uniformly bounded fourth-order moments. For periodic excitations the input-output signals of the steady-state response are observed over an integer number of periods. N samples of the input and output signals are transformed to the frequency domain using the DFT. $F \leq N/2 + 1$ DFT frequencies of the input-output DFT spectra are used for the identification. The number of selected frequencies F is proportional to N: $F = O(N)$.

In classical time domain system identification the excitation signal $u(t)$ should be *quasi-stationary* (Ljung, 1999), which means that

$$\mathbb{E}\{u(t)\} = \mu_u(t) \qquad |\mu_u(t)| \le c_1 < \infty$$
$$\mathbb{E}\{u(t)u(r)\} = R_{uu}(t,r) \qquad |R_{uu}(t,r)| \le c_2 < \infty \qquad (9\text{-}22)$$
$$R_{uu}(\tau) = \lim_{N \to \infty} \frac{1}{N} \sum_{t=1}^{N} R_{uu}(t, t-\tau)$$

should be satisfied for any t, r, and τ, with c_1, c_2 constants independent of t, r. The class of excitation signals defined by Assumption 9.1 forms a subset of the class of quasi-stationary signals (9-22) and, hence, is less general (see Exercise 9.1). This restriction is the price to pay to allow noncausal filtering (removal of DFT frequencies) of the input and output DFT spectra. Note that Assumption 9.1 is easily met if the excitation stems from an arbitrary waveform generator.

Assumption 9.2 (Excitation Signal—Frequency Domain Experiment): The plant is measured in steady state with a single sine excitation. This experiment is repeated at F different frequencies $f_{\min} \le f_k \le f_{\max}$, $k = 1, 2, ..., F$, with $f_{\min}$ and ($f_{\max} < \infty$), respectively, the minimum and maximum excitation frequencies.

Assumption 9.3 (Disturbing Noise—Time Domain Experiment): At the sampling instances the disturbing time domain noise sources $n_y(t)$, $n_u(t)$ are jointly correlated filtered white noise sequences

$$\begin{bmatrix} n_y(t) \\ n_u(t) \end{bmatrix} = \begin{bmatrix} H_{11}(q) & H_{12}(q) \\ H_{21}(q) & H_{22}(q) \end{bmatrix} \begin{bmatrix} e_1(t) \\ e_2(t) \end{bmatrix} \text{ or } n_z(t) = H(q)e(t) \qquad (9\text{-}23)$$

with $n_z^T(t) = [n_y(t)\ n_u(t)]$, $e^T(t) = [e_1(t)\ e_2(t)]$ and where $H(z^{-1})$ is a stable filter. $e(t)$ is independently distributed (over t and over its entries) with continuous probability density function, has stationary first- and second-order moments, uniformly bounded fourth-order moments, and is independent of the true (unknown) excitation $u_0(t)$. The frequency domain errors $N_Y(k)$, $N_U(k)$ are related to the time domain errors $n_y(t)$, $n_u(t)$ by the discrete Fourier transform: $N_Y(k) = \text{DFT}(n_y(t))$ and $N_U(k) = \text{DFT}(n_u(t))$.

Assumption 9.4 (Disturbing Noise—Frequency Domain Experiment): The frequency domain errors $N_Y(k)$, $N_U(k)$ are independent (over k), jointly correlated, zero mean random variables with uniformly bounded absolute moments of order four. $N_Y(k)$, $N_U(k)$ are independent of the true (unknown) excitation $U_0(k)$.

Assumption 9.5 (Frequency Domain Errors): The (co)variances $\sigma_Y^2(k) = \text{var}(N_Y(k))$, $\sigma_U^2(k) = \text{var}(N_U(k))$, and $\sigma_{YU}^2(k) = \text{covar}(N_Y(k), N_U(k))$ of the frequency domain errors $N_Y(k)$, $N_U(k)$ are known.

Assumption 9.6 (Continuity Cost Function): The cost function $V_F(\theta, Z)$ is a continuous function of θ in the compact set $\mathbb{P}_r$.

Assumption 9.7 (Persistence of Excitation): There exists an F_0 such that for any $F \geq F_0$, ∞ included, the expected value of the cost function $V_F(\theta) = \mathbb{E}\{V_F(\theta, Z)\}$ has a unique global minimum $\tilde{\theta}(Z_0)$, which is an interior point of $\mathbb{P}_r$.

If $V_F(\theta)$ is not convex, then in the presence of model errors $V_F(\theta)$ can have more than one global minimum. An example of this can be found in Kabaila (1983). To handle these cases we restrict the compact set $\mathbb{P}_r$ in Assumption 9.7 such that $V_F(\theta)$ contains a unique global minimum in $\mathbb{P}_r$.

9.6.2 Stochastic Convergence Rate

When designing a new time or frequency domain experiment based on the results of a previous experiment, one must choose the number of frequencies F. To make a motivated choice it is important to know how fast the difference between the estimate and its limit value $\hat{\theta}(Z) - \tilde{\theta}(Z_0)$ converges to zero as $F \to \infty$. To establish the convergence rate of $\hat{\theta}(Z)$ to $\tilde{\theta}(Z_0)$, we need suitable assumptions concerning the first- and second-order derivatives of the cost function w.r.t. θ. We also need a persistence-of-excitation condition that is stronger than Assumption 9.7. In addition to Assumptions 9.1 to 9.6, we require:

Assumption 9.8 (Continuity First- and Second-Order Derivatives Cost Function): The cost function $V_F(\theta, Z)$ has continuous first- and second-order derivatives w.r.t. θ in $\mathbb{P}_r$ for any value of F, ∞ included.

Assumption 9.9 (Persistence of Excitation): There exists an F_0 such that for any $F \geq F_0$, ∞ included, the Hessian of the expected value of the cost function is regular at the unique global minimizer $\tilde{\theta}(Z_0)$, which is an interior point of $\mathbb{P}_r$: $c_1 I_{n_\theta} \leq V_F''(\tilde{\theta}(Z_0)) \leq c_2 I_{n_\theta}$ where $0 < c_1 \leq c_2 < \infty$ and c_1, c_2 are F-independent constants.

9.6.3 Systematic and Stochastic Errors

A more profound analysis makes it possible to distinguish between the asymptotic behavior of the stochastic and the systematic deviations in the residual $\hat{\theta}(Z) - \tilde{\theta}(Z_0)$. In addition to Assumptions 9.1 to 9.6, 9.8, and 9.9, we require:

Assumption 9.10 (Continuity Third-Order Derivative Cost Function): The cost function has continuous third-order derivatives w.r.t. θ in $\mathbb{P}_r$ for any value of F, ∞ included.

9.6.4 Asymptotic Normality

To calculate uncertainty regions with a given confidence level, we need the probability density function of the estimate $\hat{\theta}(Z)$. A good approximation can be found if the asymptotic distribution function of $\hat{\theta}(Z)$ is known. Whereas the consistency and convergence rate analysis of $\hat{\theta}(Z)$ requires finite moments of order 4, the convergence and the convergence rate analysis of the distribution function of $\hat{\theta}(Z)$ needs the existence of the moments of any order for a time domain experiment and of order 6 for a frequency domain experiment. In addition to Assumptions 9.1 to 9.6 and 9.8 to 9.10, we require:

Assumption 9.11 (Excitation Signal—Time Domain Experiment): The excitation signals $u(t)$ in Assumption 9.1 have finite moments of any order. The excitation noise $e_u(t)$ is independent and identically distributed.

Assumption 9.12 (Disturbing Noise—Time Domain Experiment): The disturbing noise $e(t)$ in Assumption 9.3 is independent and identically distributed with finite moments of any order.

For a frequency domain experiment, these conditions can be relaxed because the successive frequency measurements are independent (see Assumptions 9.2 and 9.4), whereas they are correlated for a time domain experiment (see Assumption 9.3).

Assumption 9.13 (Disturbing Noise—Frequency Domain Experiment): (a) For the asymptotic normality: the disturbing noise N_Z satisfies $\sum_{k=1}^{F} \text{Cov}(N_Z(k)) = O(F)$ and has uniformly bounded absolute moments of order $4 + \varepsilon$ with $\varepsilon > 0$, for example, $\mathbb{E}|N_Y(k)|^{4+\varepsilon} \leq c_1 < \infty$ with c_1 independent of F. (b) For the convergence rate: in addition, the disturbing noise N_z has uniformly bounded absolute moments of order six, for example, $\mathbb{E}|N_U(k)|^6 \leq c_2 < \infty$ with c_2 independent of F.

9.6.5 Deterministic Convergence

To study the deterministic convergence and the convergence rate of $\tilde{\theta}(Z_0)$ to θ_*, we must define the strategy of adding new frequencies to the data. We need this information because the model errors depend on the power spectrum of the excitation. In addition to Assumptions 9.1 to 9.6, 9.8, and 9.9, we require:

Assumption 9.14 (Strategy of Adding Frequencies): As $F \to \infty$ the frequencies f_k cover the frequency interval $[f_{\min}, f_{\max}]$ with a density function $n(f)$ defined as

$$n(f) = \lim_{\Delta f \to 0} \lim_{F \to \infty} \frac{N_F(f + \Delta f) - N_F(f)}{F \Delta f} \qquad (9\text{-}24)$$

where $N_F(f)$ is the number of frequencies in the interval $[0, f]$ when the total number of frequencies is F. The density $n(f)$ is continuous with bounded second-order derivative w.r.t. f in $[f_{\min}, f_{\max}]$ except at a finite number of frequencies.

Special cases are a uniform ($n(f)$ independent of f) or a logarithmic ($n(f)$ is proportional to f^{-1}) distribution of the number frequencies in $[f_{\min}, f_{\max}]$.

Assumption 9.15 (Constraint on the Residual): The second-order derivatives w.r.t. the frequency f of the residual $\mathbb{E}\{|\varepsilon(\Omega(f), \theta, Z(f))|^2\}$ and its first- and second-order derivatives w.r.t. θ, are bounded in the frequency band $[f_{\min}, f_{\max}]$, except at a finite number of frequencies ($\Omega(f) = j2\pi f$, $e^{j2\pi f T_s}$, $\sqrt{j2\pi f}$ or $\tanh(\tau_R j2\pi f)$).

Assumption 9.15 puts some conditions on the limit power spectrum $|U_0(f)|^2$ or $S_{uu}(j\omega)$ of the periodic or random excitation; it should be a continuous function of f with bounded second-order derivative.

9.6.6 Consistency

Contrary to the stochastic convergence, consistency can be shown only if a true linear model exists and if it belongs to the considered model set. It also imposes some conditions on the expected value of the cost function, which should be verified for each estimator. To study, under these conditions, the stochastic convergence, the stochastic convergence rate, the improved stochastic convergence rate, and the asymptotic normality, we require, in addition to the assumptions of Sections 9.6.1 to 9.6.4, the following:

Assumption 9.16 (Existence of a True Linear Plant Model): There is an identifiable parameterization $\theta_0 \in \mathbb{P}_r$ such that $G(\Omega_k, \theta_0) U_0(k)$, $G(z_k^{-1}, \theta_0) U_0(k) + T_G(z_k^{-1}, \theta_0)$, or $G(s_k, \theta_0) U_0(k) + T_G(s_k, \theta_0) + \delta(s_k)$ with $G(s, \theta_0)$ stable represents the true output $Y_0(k)$.

Assumption 9.17 (Consistency Condition on the Cost Function): The expected value of the cost function $V_F(\theta) = \mathbb{E}\{V_F(\theta, Z)\}$, or its limit value $V_*(\theta) = \lim_{F \to \infty} V_F(\theta)$, is minimal in the true model parameters θ_0.

9.6.7 Asymptotic Bias

Speaking about systematic or bias errors makes sense only if a true model exists and if it belongs to the considered model set. Studying the bias is possible only if the expected value of the estimate $\hat{\theta}(Z)$ exists. To ensure the existence of the expected value, we remove "large," "highly improbable" values of $\hat{\theta}(Z)$. This results in the truncated estimate $\underline{\hat{\theta}}(Z)$, which is defined as

$$\underline{\hat{\theta}}(Z) = \begin{cases} \hat{\theta}(Z) & \|\hat{\theta}(Z) - \tilde{\theta}(Z_0)\|_2 \leq L \\ 0 & \|\hat{\theta}(Z) - \tilde{\theta}(Z_0)\|_2 > L \end{cases} \quad (9\text{-}25)$$

where L is an (arbitrarily) large number ($0 < L < \infty$) independent of F. Lemma 17.27 guarantees that there exists an F_0 such that for any $F \geq F_0$ $\underline{\hat{\theta}}(Z) = \hat{\theta}(Z)$ with probability one (in probability). We require that Assumptions 9.1 to 9.6, 9.8 to 9.10, 9.16, and 9.17 are valid.

9.6.8 Asymptotic Efficiency

A basic step in the analysis of the asymptotic efficiency of the estimate $\hat{\theta}(Z)$ is the calculation of the Fisher information matrix. It inherently assumes the existence of a true model and knowledge of the probability density function of the disturbing noise in the frequency domain. Therefore, in addition to Assumptions 9.1 to 9.6, 9.8 to 9.13, 9.16, and 9.17, we require:

Assumption 9.18 (Circular Complex Frequency Domain Errors): The frequency domain errors $N_Y(k)$, $N_U(k)$ are independent (over k), jointly correlated, zero mean, circular complex distributed random variables.

Assumption 9.19 (pdf Frequency Domain Errors): The observations Z_0 are deterministic and the frequency domain errors $N_Y(k)$, $N_U(k)$ are normally distributed random variables.

Assumption 9.20 (Efficiency Condition Frequency Domain Errors): The number of noncoherent noise sources equals 1. This is true if and only if one of the three following conditions is fulfilled for $k = 1, 2, \ldots, F$: (i) no input noise $\sigma_U^2(k) = 0$, (ii) no output noise $\sigma_Y^2(k) = 0$, or (iii) totally correlated input-output errors $|\sigma_{YU}^2(k)|/(\sigma_Y(k)\sigma_U(k)) = 1$.

For example, Assumption 9.20 is fulfilled in feedback when only process noise is present (no measurement errors and no controller noise, see Section 9.18 and Exercise 9.2).

9.7 ASYMPTOTIC PROPERTIES

In this section we give an overview and an elaborated discussion of the asymptotic properties of the minimizer $\hat{\theta}(Z)$ of cost functions $V_F(\theta, Z)$ which are quadratic-in-the-measurements Z. The overview starts with general estimators, proceeds with consistent estimators, and ends with the maximum likelihood estimator. Afterward, the results are generalized to cost functions of the form (9-14) that are nonquadratic in Z. In a first reading, one may skip Theorems 9.21 and 9.28 and go directly to the discussion of the properties.

Theorem 9.21 (Asymptotic Properties $\hat{\theta}(Z)$): Consider models (9-7) and (9-8) with any identifiable parameterization of Sections 6.2 and 6.3. Let $\hat{\theta}(Z)$ be the minimizer of a cost function $V_F(\theta, Z)$ of the form (9-11) that is quadratic-in-the-measurements Z. Under the assumptions of Section 9.6, the minimizer $\hat{\theta}(Z)$ has the following asymptotic $(F \to \infty)$ properties,

1. *Stochastic convergence:* $\hat{\theta}(Z)$ converges strongly to $\tilde{\theta}(Z_0)$, the minimizer of $V_F(\theta) = \mathbb{E}\{V_F(\theta, Z)\}$ (assumptions Section 9.6.1).

2. *Stochastic convergence rate:* $\hat{\theta}(Z)$ converges in probability at the rate $O_p(F^{-1/2})$ to $\tilde{\theta}(Z_0)$ (assumptions Section 9.6.2).

3. *Systematic and stochastic errors:* $\hat{\theta}(Z)$ converges in probability to $\tilde{\theta}(Z_0)$ with

$$\hat{\theta}(Z) = \tilde{\theta}(Z_0) + \delta_\theta(Z) + b_\theta(Z)$$
$$\delta_\theta(Z) = -V_F''^{-1}(\tilde{\theta}(Z_0))V_F'^T(\tilde{\theta}(Z_0), Z) \quad (9\text{-}26)$$

where $\delta_\theta(Z) = O_p(F^{-1/2})$, with $\mathbb{E}\{\delta_\theta(Z)\} = 0$, is the dominating stochastic error and where $b_\theta(Z) = O_p(F^{-1})$ contains the contribution of the systematic errors (assumptions Section 9.6.3).

4. *Asymptotic normality:* $\sqrt{F}(\hat{\theta}(Z) - \tilde{\theta}(Z_0))$ converges in law at the rate $O(F^{-1/2})$ to a Gaussian random variable with zero mean and covariance matrix $\text{Cov}(\sqrt{F}\delta_\theta(Z))$

$$\text{Cov}(\sqrt{F}\delta_\theta(Z)) = V_F''^{-1}(\tilde{\theta}(Z_0))Q_F(\tilde{\theta}(Z_0))V_F''^{-1}(\tilde{\theta}(Z_0))$$
$$Q_F(\tilde{\theta}(Z_0)) = F\mathbb{E}\{V_F'^T(\tilde{\theta}(Z_0), Z)V_F'(\tilde{\theta}(Z_0), Z)\} \quad (9\text{-}27)$$

(assumptions Section 9.6.4).

5. *Deterministic convergence:* $\tilde{\theta}(Z_0)$ converges to θ_*, the minimizer of

$$V_*(\theta) = \int_{f_{\min}}^{f_{\max}} \mathbb{E}\{|\varepsilon(\Omega(f), \theta, Z(f))|^2\}n(f)df \quad (9\text{-}28)$$

Section 9.7 ■ Asymptotic Properties   299

with $\Omega(f) = j2\pi f$, $e^{j2\pi fT_s}$, $\sqrt{j2\pi f}$, or $\tanh(\tau_R j2\pi f)$. The convergence rate is an $O(F^{-2})$ for frequency domain experiments and $O(F^{-1})$ for time domain experiments (assumptions Section 9.6.5).

If in addition $V_F(\theta, Z)$ satisfies the consistency conditions then,

6. *Consistency:* $\hat{\theta}(Z)$ is strongly (weakly for model (9-8) with $\Omega = s$) consistent; replace in properties 1 to 4 $\tilde{\theta}(Z_0)$ ($\lim_{F \to \infty} \tilde{\theta}(Z_0) = \theta_*$ for model (9-8) with $\Omega = s$) by θ_0 (assumptions Section 9.6.6).

7. *Asymptotic bias:* The asymptotic bias $b_\theta = \mathbb{E}\{b_\theta(Z)\}$, and its derivative w.r.t. θ_0, $\partial b_\theta / \partial \theta_0$, of $\hat{\theta}(Z)$ are an $O(F^{-1})$ ($O(F^{-1/2})$ for model (9-8) with $\Omega = s$ and random excitation) for all $\theta_0 \in \mathbb{P}_r$ (assumptions Section 9.6.7).

If in addition $V_F(\theta, Z)$ is the maximum likelihood cost function then,

8. *Asymptotic efficiency:* The Gaussian maximum likelihood estimate $\hat{\theta}_{ML}(Z)$ is asymptotically efficient: $\text{Cov}(\delta_\theta(Z)) = Fi^{-1}(\theta_0)$ with $Fi(\theta_0) = FV_F''(\theta_0)$ the Fisher information matrix. Moreover, we have

$$\lim_{F \to \infty} (\text{Cov}(\sqrt{F}\hat{\theta}(Z)) - \text{Cov}(\sqrt{F}\delta_\theta(Z))) = 0 \qquad (9\text{-}29)$$

(assumptions Section 9.6.8).

Proof. See Appendix 9.E.   □

Corollary 9.22 (Asymptotic Properties $\hat{\theta}(Z)$—continued): Let $\hat{\theta}(Z)$ be the minimizer of a cost function $V_F(\theta, Z) = f_F(\theta, \eta(Z), Z)$ of the form (9-14) where $f_F(\theta, \eta, Z)$ is quadratic-in-the-measurements Z. Assume that the cost function $f_F(\theta, \eta, Z)$ and its third-order derivatives w.r.t. $x = [\theta^T \, \eta^T]^T$ are continuous and that $f_F(\theta, \eta, Z)$ fulfills the assumptions of Section 9.6. Define, furthermore, $g_F(\theta, \eta(Z), Z) = V_F'^T(\theta, Z)$ and $g_F(\theta, \eta) = \mathbb{E}\{g_F(\theta, \eta, Z)\}$. If Theorem 9.21 is valid for the (initial) estimate $\eta(Z)$, then the minimizer $\hat{\theta}(Z)$ has the asymptotic properties of Theorem 9.21 with the following three modifications:

(i) To calculate $V_F(\theta)$ and $V_*(\theta)$ we first replace $\eta(Z)$ by its limit value η_* before taking the expected value, which gives

$$V_F(\theta) = \frac{1}{F} \sum_{k=1}^{F} \mathbb{E}\{|\varepsilon(\Omega_k, \theta, Z(k), \eta_*)|^2\}$$
$$V_*(\theta) = \int_{f_{\min}}^{f_{\max}} \mathbb{E}\{|\varepsilon(\Omega(f), \theta, Z(f), \eta_*)|^2\} n(f) df \qquad (9\text{-}30)$$

(ii) $\mathbb{E}\{\delta_\theta(Z)\}$ is not necessarily zero or may not even exist.

(iii) $\delta_\theta(Z)$ in the expression of the covariance matrix (9-27) is replaced by $d_\theta(Z)$

$$d_\theta(Z) = -V_F''^{-1}(\tilde{\theta}(Z_0)) d_F(Z)$$
$$d_F(Z) = g_F(\tilde{\theta}(Z_0), \eta_*, Z) + \frac{\partial g_F(\tilde{\theta}(z_0), \eta)}{\partial \eta_*} \delta_\eta(Z) \qquad (9\text{-}31)$$

where $\delta_\eta(Z)$ is given by (9-26), and with $d_\theta(Z) = O_p(F^{-1/2})$, $\mathbb{E}\{d_\theta(Z)\} = 0$.

Proof. See Appendix 9.F.   □

We are ready now to answer the question we posed in Section 9.2: "What will happen with one's estimates (uncertainty, bias ...) if one gathered, for example, four times more data?" Property 1 ensures that $\hat{\theta}(Z)$ is likely to be closer to the minimizer $\tilde{\theta}(Z_0)$ of the expected value of the cost function. Property 2 tells us that $\hat{\theta}(Z)$ is likely to be two times closer to $\tilde{\theta}(Z_0)$. From property 3 it follows that the systematic and stochastic errors in the residual $\hat{\theta}(Z) - \tilde{\theta}(Z_0)$ are likely to decrease with a factor of 4 and 2, respectively. Finally, property 4 ensures that the distribution function of $\hat{\theta}(Z)$ is likely to be two times closer to a normal distribution. Similar results are obtained when no model errors are present $\tilde{\theta}(Z_0) = \theta_0$.

Expression (9-27) allows a theoretical calculation of the covariance matrix of the estimates in the presence of model errors. It requires, however, knowledge of the fourth-order moments of the noise and of the minimizer $\tilde{\theta}(Z_0)$ of the expected value of the cost function. Although $\tilde{\theta}(Z_0)$ can be approximated by the actual estimate $\hat{\theta}(Z)$, the fourth-order moments of the noise are mostly unknown. For the maximum likelihood estimator, the covariance expression (9-27) can be significantly simplified (only second-order moments of the noise are required) and a good approximation of the covariance matrix results as a by-product of the nonlinear minimization scheme (9-16) (see Section 9.11). Property 4 then makes it possible to calculate uncertainty regions around $\hat{\theta}(Z)$ that contain $\tilde{\theta}(Z_0)$ with some user-defined probability level. The same can be done for any model-related quantity (see Sections 16.2 and 19.4.7).

If model errors exist, then $\hat{\theta}(Z)$ converges to a value $\tilde{\theta}(Z_0) \neq \theta_0$ ($\theta_* \neq \theta_0$) that still depends on F. Property 5 guarantees that $\tilde{\theta}(Z_0)$ converges at the rate $O(F^{-1})$ or faster to its limit value θ_*, while according to property 2 the stochastic convergence rate of $\hat{\theta}(Z)$ to $\tilde{\theta}(Z_0)$ is an $O_p(F^{-1/2})$. Therefore, $\tilde{\theta}(Z_0)$ can be replaced everywhere by θ_* in properties 1 to 4. In case of model errors, we may also wonder whether $\hat{\theta}(Z)$ still converges to the same solution if the same experiment is repeated with a higher signal-to-noise ratio (lower noise levels). To verify this, we apply quick analysis tool number 3 (see Section 9.5) to the cost function. If so, then $\tilde{\theta}(Z_0)$ (θ_*) can be interpreted as the solution of the noiseless problem.

Property 6 guarantees that the estimate $\hat{\theta}(Z)$ converges to the true model parameters θ_0 for cost functions satisfying the consistency conditions 9.16 and 9.17. This does, however, not imply that the (equivalent) initial conditions in model (9-8) are consistently estimated. Indeed, the part of θ_0 corresponding to the (equivalent) initial conditions decreases to zero as $F^{-1/2}$ (use Lemma 6.5 taking into account that $F = O(N)$ for a time domain experiment), while the difference $\hat{\theta}(Z) - \theta_0$ is an $O_p(F^{-1/2})$. Hence, the relative difference $|(\hat{\theta}_{[i]}(Z) - \theta_{0[i]})/\theta_{0[i]}|$ between the estimated and the true initial conditions does not decrease to zero as the number of frequencies F increases to infinity, which shows that the initial conditions are not consistently estimated. This result can easily be understood in the time domain. The (equivalent) initial conditions (transient term $T_G(\Omega, \theta)$ in (9-8)) correspond to an exponentially decaying transient in the time domain. Observing the input and output signals during a longer period does not give more information about the transient, hence, it cannot be estimated consistently. For the same reason, properties 7 and 8 do not imply that the estimated equivalent initial conditions are asymptotically efficient and have an $O(F^{-1})$ bias. Although they cannot be estimated consistently, we still include the initial conditions in model (9-8) because it turns out that they improve the finite sample behavior (F is not "large") of the estimated plant model $G(\Omega, \hat{\theta}(Z))$. Note also that the influence of the transient term $T_G(\Omega, \theta)$ to the cost function $V_F(\theta, Z)$ is an $O_p(F^{-1})$ (see Appendix 9.D).

The asymptotic efficiency of the maximum likelihood estimator (property 8 of Theorem 9.21) has been shown under some restrictive noise assumptions (see Assumption 9.20); for example, the input must be known exactly. In general, the maximum likelihood solution is not asymptotically efficient. This is not in contradiction with the general properties of maximum likelihood estimators (Section 1.5.3) because the number of estimated parameters in the errors-in-variables problem increases with F (see Section 9.11).

For deterministic vectors $\eta(Z) = \eta_*$ the term $d_N(Z)$ in (9-31) reduces to $g_F(\tilde{\theta}(z_0), \eta_*, Z)$. Therefore, modification number iii of Corollary 9.22 shows that in general the stochastic vector will increase the uncertainty of the estimates. If, however, $\partial g_F(\tilde{\theta}(z_0), \eta)/\partial \eta_* = o(F^0)$, then there is no asymptotic increase in uncertainty (see, for example, Section 9.12.3).

9.8 LINEAR LEAST SQUARES

9.8.1 Introduction

A reasonable measure of goodness of fit is to compare the observed output $Y(k)$ with the modeled output $Y(k, \theta)$ (9-7) or (9-8), where $G(\Omega, \theta)$ (and $T(\Omega, \theta)$) can take any parameterization of Section 6.2 (and Section 6.3). The plant model parameters are then obtained by minimizing the sum of the squared residuals

$$V_{\text{NLS}}(\theta, Z) = \sum_{k=1}^{F} |Y(k) - Y(k, \theta)|^2 \qquad (9\text{-}32)$$

w.r.t. to θ. Because $Y(k, \theta)$ is a nonlinear function of θ, the cost function (9-32) is a nonquadratic function of θ. All the estimation methods presented in this section try to minimize (9-32) by (successive) linear least squares approximation(s). The key idea is to make (9-32) quadratic in θ by parameterizing $G(\Omega, \theta)$ (and $T(\Omega, \theta)$) as a rational form $B(\Omega, \theta)/A(\Omega, \theta)$ (and $I(\Omega, \theta)/A(\Omega, \theta)$) and by multiplying each residual $Y(k) - Y(k, \theta)$ in the cost function (9-32) by $A(\Omega_k, \theta)$.

9.8.2 Linear Least Squares

Multiplying each residual $Y(k) - Y(k, \theta)$ in the cost function (9-32) by $A(\Omega_k, \theta)$ gives the linear least squares (LS) cost function

$$V_{\text{LS}}(\theta, Z) = \sum_{k=1}^{F} |e(\Omega_k, \theta, Z(k))|^2 \qquad (9\text{-}33)$$

with $e(\Omega_k, \theta, Z(k))$ the equation error (9-9) or (9-10). The linear least squares (LS) estimate $\hat{\theta}_{\text{LS}}(Z)$ is found by minimizing (9-33) w.r.t. θ using the constraint $a_i = 1$ or $b_i = 1$. In Levy (1959) the linear least squares approach was applied for the first time to identify continuous-time models starting from transfer function measurements ((9-33) with equation error (9-9), $\Omega = s$, $Y(k) = G(s_k)$, and $U(k) = 1$). The linearization of the output error $Y(k) - Y(\Omega_k, \theta)$ has two major drawbacks when identifying continuous-time models ($\Omega = s$, $\sqrt{s}$, and $\tanh(\tau_R s)$): the overemphasizing of high-frequency errors in (9-33) and the large dynamic range of the numbers in the normal equation (9-16). Indeed, $e(\Omega_k, \theta, Z(k))$ is a polynomial in Ω_k and, hence, the contribution of the disturbing noise at frequency Ω_k to the cost function increases with $|\Omega_k|^{2\max(n_a, n_b)}$. This may result in poor low-frequency fits (see Fig. 9-4) and ill-conditioned normal equations for identification problems with a large dynamic frequency range. Similar problems occur for discrete-time models ($\Omega = z^{-1}$) when identified on a "small" part of the unit circle.

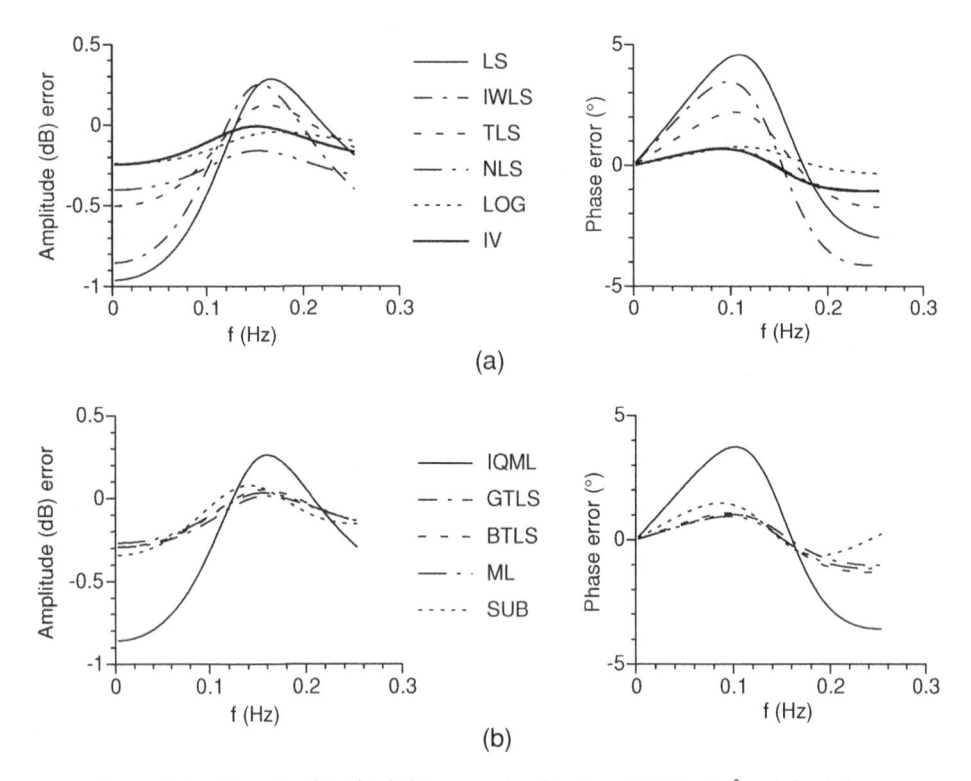

Figure 9-4. Second-order simulation example $G(s,\theta) = 1/(1+s+s^2)$ defined in Appendix 9.A (see also Figure 9-1). Left: difference between the estimated amplitude in dB and the true amplitude in dB, and right: phase error in degrees. (a) Estimators requiring no noise information, (b) estimators requiring the noise covariance.

Because $V_{LS}(\theta, Z)$ is quadratic-in-the-measurements Z, the asymptotic properties proved in Theorem 9.21, with $V_F(\theta, Z) = V_{LS}(\theta, Z)/F$, are valid for $\hat{\theta}_{LS}(Z)$. To reveal the major properties of $\hat{\theta}_{LS}(Z)$ we use the quick analysis tools of Section 9.5. Taking the expected value of (9-33) gives (9-20) with

$$\mathbb{E}\{v_F(\theta, N_Z)\} = \frac{1}{F}\sum_{k=1}^{F} \sigma_e^2(\Omega_k, \theta) \qquad (9\text{-}34)$$

(see Exercise 9.3). $\sigma_e^2(\Omega_k, \theta) = \text{var}(e(\Omega_k, \theta, N_Z(k)))$ is the variance of the equation error where the measurements Z have been replaced by the noise on the measurements N_Z

$$\sigma_e^2(\Omega_k, \theta) = \sigma_Y^2(k)|A(\Omega_k, \theta)|^2 + \sigma_U^2(k)|B(\Omega_k, \theta)|^2 - 2\text{Re}(\sigma_{YU}^2(k)A(\Omega_k, \theta)\bar{B}(\Omega_k, \theta)) \qquad (9\text{-}35)$$

Applying quick tool 2 (see Section 9.5) to (9-34) shows that the linear least squares estimate $\hat{\theta}_{LS}(Z)$ is, in general, inconsistent because (9-35) is, in general, θ- dependent. It is consistent if $\sigma_e^2(\Omega_k, \theta)$ is independent of θ, for example, no input noise ($\sigma_U^2(k) = 0$, $\sigma_{YU}^2(k) = 0$) and a polynomial plant model ($A(\Omega, \theta) = 1$). Replacing C_{N_Z} by $v^2 C_{N_Z}$ in the expected value of the cost function gives, taking into account (9-34),

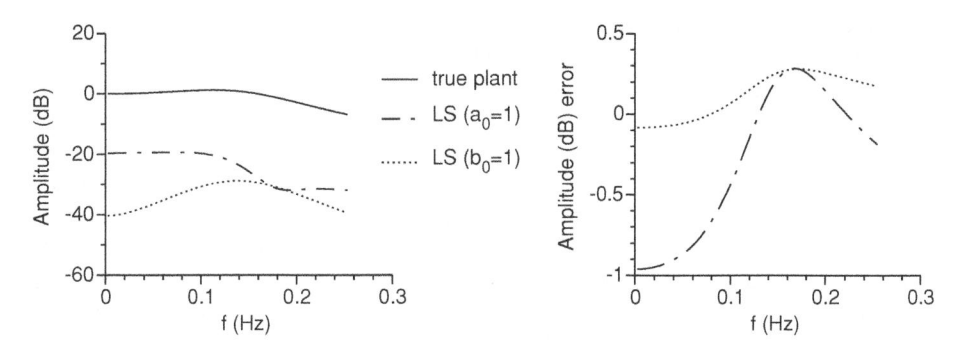

Figure 9-5. Second-order simulation example $G(s, \theta) = 1/(1+s+s^2)$ defined in Appendix 9.A (see also Figure 9-1). Comparison of the linear least squares estimates using the constraint $a_0 = 1$ and the linear least squares estimates using the constraint $b_0 = 1$. Left figure, true plant model (solid line) and magnitude of the complex error between the estimated and the true plant model. Right figure, difference between the estimated amplitude in dB and the true amplitude in dB.

$$V_F(\theta) = \mathbb{E}\{V_F(\theta, Z_0)\} + \upsilon^2 \mathbb{E}\{v_F(\theta, N_Z)\} \quad (9\text{-}36)$$

It shows that, in general, $\tilde{\theta}_{LS}(Z_0)$ and θ_{*LS} depend on the disturbing noise level υ and, hence, cannot be considered as the noiseless solutions (see Section 9.5, quick tool 3). They are the noiseless solutions if $\sigma_e^2(\Omega_k, \theta)$ is independent of θ. From (9-33) it follows directly that $V_{LS}(\lambda \theta, Z) = \lambda^2 V_{LS}(\theta, Z)$ so that $\hat{\theta}_{LS}(Z)$ depends on the particular constraint chosen, for example, $a_i = 1$ or $b_i = 1$ (see Section 9.5, quick tool 4). This is illustrated in Figure 9-5. Note that on the average the estimate with $a_0 = 1$ is too small (underbiased), while the estimate with $b_0 = 1$ is too large (overbiased). This is in agreement with the results of De Moor et al. (1994). See Table 9-5 for an overview of the properties of the LS estimator.

9.8.3 Iterative Weighted Linear Least Squares

To overcome the lack of sensitivity to low-frequency errors of the linear least squares estimator, the equation error $e(\Omega_k, \theta, Z(k))$ in (9-33) is divided by an initial guess of the denominator polynomial $A(\Omega_k, \theta^{(0)})$. The obtained weighted linear least squares estimate $\theta^{(1)}$ can be used to calculate a (hopefully) better estimate of the denominator polynomial $A(\Omega_k, \theta^{(1)})$, resulting in a (hopefully) better estimate $\theta^{(2)}$, and so on .... The ith step of the iterative procedure consists of minimizing

$$V_{IWLS}^{(i)}(\theta^{(i)}, Z) = \sum_{k=1}^{F} \frac{|e(\Omega_k, \theta^{(i)}, Z(k))|^2}{|A(\Omega_k, \theta^{(i-1)})|^2} \quad (9\text{-}37)$$

with $e(\Omega_k, \theta, Z(k))$ the equation error (9-9) or (9-10), w.r.t. $\theta^{(i)}$ using the constraint $a_j = 1$ or $b_j = 1$. In most cases the linear least squares estimate is used as starting value $\theta^{(0)} = \hat{\theta}_{LS}(Z)$, and when convergent the iterative weighted linear least squares (IWLS) estimate is $\hat{\theta}_{IWLS}(Z) = \theta^{(\infty)}$. In Sanathanan and Koerner (1963) this iterative procedure was applied for the first time to identify continuous-time models starting from transfer function measurements ((9-37) with equation error (9-9), $\Omega = s$, $Y(k) = G(s_k)$ and $U(k) = 1$). From Figure 9-4 it can be seen that the low-frequency errors of the IWLS fit are indeed smaller than those of the LS fit. When convergent ($\theta^{(i)} = \theta^{(i-1)}$ for i sufficiently large) the IWLS cost (9-37) tends to the nonlinear least squares cost (9-32). Although this property is very appealing, it does not guarantee that the global minima of both cost functions are the same. Therefore, one requires that the derivatives of these cost functions w.r.t. θ are asymptotically ($i \to \infty$) the same. In general, this is not true and, hence, $\hat{\theta}_{\text{IWLS}}(Z) \neq \hat{\theta}_{\text{NLS}}(Z)$. However, as the elementwise difference between the Jacobians is proportional to the equation error $e(\Omega_k, \theta^{(i-1)}, Z(k))$ (see Exercise 9.4), both estimates will coincide ($\hat{\theta}_{\text{IWLS}}(Z) \approx \hat{\theta}_{\text{NLS}}(Z)$) for "sufficiently high" signal-to-noise ratios and "sufficiently small" modeling errors, otherwise the difference may be large. This is illustrated by the "high noise" simulation example of Figure 9-4 (compare IWLS with NLS), and the "low noise" simulation example of Figure 9-8 (compare IWLS to NLS).

Analysis of the statistical properties of the estimate $\theta^{(\infty)}$ is in general impossible. It is, however, feasible to analyze the properties of the first step of the iterative procedure (9-37). If the initial guess $\theta^{(0)}$ is deterministic and independent of the number of frequencies F, then Theorem 9.21 is valid and $\hat{\theta}_{\text{IWLS}}(Z) = \theta^{(1)}$ has asymptotic ($F \to \infty$) properties similar to those of $\hat{\theta}_{\text{LS}}(Z)$ (see Section 9.8.2). If the choice $\theta^{(0)} = \hat{\theta}_{\text{LS}}(Z)$ is made, then the cost function (9-37) is no longer a quadratic function of the measurements Z. Indeed, $\hat{\theta}_{\text{LS}}(Z)$ depends on Z and appears in the denominator of (9-37). Although this complicates the analysis, it turns out that Theorem 9.21 is still valid for $\hat{\theta}_{\text{IWLS}}(Z) = \theta^{(1)}$ with three minor modifications (see Corollary 9.22). Hence, $\hat{\theta}_{\text{IWLS}}(Z) = \theta^{(1)}$ has the same asymptotic ($F \to \infty$) properties as $\hat{\theta}_{\text{LS}}(Z)$. We conclude that in general the estimate $\hat{\theta}_{\text{IWLS}}(Z)$ is inconsistent, depends on the particular constraint chosen, and does not converge to a noiseless solution.

Many modifications of and extensions to the original method of Sanathanan and Koerner (1963) have been published. Almost all of them fit within the following (iterative) weighted least squares framework:

$$\sum_{k=1}^{F} W^2(\Omega_k, \theta^{(i-1)}) |e(\Omega_k, \theta^{(i)}, Z(k))|^2 \tag{9-38}$$

where $W(\Omega_k, \theta^{(i-1)})$ is a well-chosen real weighting function (see Pintelon et al., 1994 for an overview). One particular weighting is interesting, namely

$$W(\Omega_k, \theta^{(i-1)}) = \frac{1}{|A(\Omega_k, \theta^{(i-1)})|^r} \text{ with } r \in [0, \infty) \tag{9-39}$$

Two special cases of (9-39) are the linear least squares method for $r = 0$ and the iterative weighted linear least squares method (9-37) for $r = 1$. Powers r, different from one, may result in smaller output errors $Y(k) - Y(k, \theta)$; for example, if the iterative scheme (9-37) does not converge, then relaxation ($r < 1$) is helpful. In 't Mannetje (1973), the relaxation idea was applied for the first time to identify continuous-time models starting from transfer function measurements ((9-37) with equation error (9-9), $\Omega = s$, $Y(k) = G(s_k)$, and $U(k) = 1$). The asymptotic ($F \to \infty$) properties of the minimizer of (9-38) are similar to those of $\hat{\theta}_{\text{IWLS}}(Z)$ (9-37), so that in general the minimizers of (9-38) and (9-32) are different. See Table 9-5 for an overview of the properties of the IWLS estimator.

9.8.4 A Simple Example

Consider the identification of an integrator $G(s, \theta) = b_0/(a_1 s)$, starting from frequency response data $G(s_k) = G_0(s_k) + N_G(k)$, perturbed with independent (over the frequency), zero mean, circular complex noise $N_G(k)$ with variance $\text{var}(N_G(k)) = \sigma^2$ and finite

Section 9.9 ■ Nonlinear Least Squares

fourth-order moments. The iterative weighted linear least squares estimate (9-38) is calculated using the weight (9-39) and the constraint $b_0 = 1$

$$(\hat{a}_1)_{\text{IWLS}} = \frac{\sum_{k=1}^{F} |s_k|^{-2r} \operatorname{Re}(s_k G(s_k))}{\sum_{k=1}^{F} |s_k|^{2(1-r)} |G(s_k)|^2} \qquad (9\text{-}40)$$

Applying the strong law of large numbers (see Section 16.9, version 2) to the numerator and denominator of (9-40) and the interchangeability property of the almost sure limit and a continuous function (see Section 16.8, property 1), we find

$$\operatorname*{a.s.lim}_{F \to \infty} (\hat{a}_1)_{\text{IWLS}} = (a_1)_0 \frac{1}{1 + \lim_{F \to \infty} \frac{\sum_{k=1}^{F} |s_k|^{-2r} \sigma^2 / |G_0(s_k)|^2}{\sum_{k=1}^{F} |s_k|^{-2r}}} \qquad (9\text{-}41)$$

with $(a_1)_0$ the true value. As predicted by the theory (apply quick tool number 2), it clearly follows from (9-41) that $(\hat{a}_1)_{\text{IWLS}}$ and, hence, also $G(s_k, \hat{\theta}_{\text{IWLS}}(Z))$ are inconsistent estimates. Taking, for example, $F = 100$ angular frequencies equally spaced between 0.1 and 2 and $\sigma^2 = 0.5$; the right-hand side of (9-41) is then equal to $0.587(a_1)_0$ and $0.916(a_1)_0$ for, respectively, $r = 0$ (LS solution (9-33)) and $r = 1$ (IWLS solution (9-37)). It shows that weighting the linear least squares residual with an initial guess of the denominator polynomial indeed improves the estimates. In this numerical example, values of $r > 1$ give even better results compared with $r = 1$.

Making the same calculations for the IWLS estimate with constraint $a_1 = 1$, we get

$$(\hat{b}_0)_{\text{IWLS}} = \frac{\sum_{k=1}^{F} |s_k|^{-2r} \operatorname{Re}(s_k G(s_k))}{\sum_{k=1}^{F} |s_k|^{-2r}}$$

$$\operatorname*{a.s.lim}_{F \to \infty} (\hat{b}_0)_{\text{IWLS}} = \frac{\lim_{F \to \infty} \frac{1}{F} \sum_{k=1}^{F} |s_k|^{-2r} \operatorname{Re}(s_k G_0(s_k))}{\lim_{F \to \infty} \frac{1}{F} \sum_{k=1}^{F} |s_k|^{-2r}} = (b_0)_0 \qquad (9\text{-}42)$$

with $(b_0)_0$ the true value. As predicted by the theory (apply quick tool number 2) $(\hat{b}_0)_{\text{IWLS}}$ and, hence, also $G(s_k, \hat{\theta}_{\text{IWLS}}(Z))$ are consistent estimates. It illustrates nicely the dependence of $G(s_k, \hat{\theta}_{\text{IWLS}}(Z))$ on the parameter constraint used (quick tool number 4).

Putting $r = 0$ in (9-40) to (9-42) shows that the same conclusions hold for the least squares estimate $(\hat{a}_0)_{\text{LS}}$ and $(\hat{b}_0)_{\text{LS}}$.

9.9 NONLINEAR LEAST SQUARES

9.9.1 Output Error

The nonlinear least squares (NLS) estimator $\hat{\theta}_{\text{NLS}}(Z)$ minimizes the sum of the squared residuals between the observed output $Y(k)$ and the modeled output $Y(k, \theta)$ (9-7) or (9-8), where $G(\Omega, \theta)$ (and $T_G(\Omega, \theta)$) can take any parameterization of Section 6.2 (and Section 6.3)

$$V_{\mathrm{NLS}}(\theta, Z) = \sum_{k=1}^{F} |Y(k) - Y(k, \theta)|^2 \qquad (9\text{-}43)$$

The Newton-Gauss minimization scheme (9-18) is used to calculate $\hat{\theta}_{\mathrm{NLS}}(Z)$, and as with most nonlinear minimization problems, the method may converge to a local minimum of (9-43) ($\theta^{(\infty)} \neq \hat{\theta}_{\mathrm{NLS}}(Z)$). Therefore, it is important to have starting values of "sufficiently high" quality. The (iterative) weighted linear least squares solution (9-37) can be used for this purpose. In Van den Enden et al. (1977) and Van den Enden and Leenknegt (1986) this scheme was used for the first time to identify, respectively, continuous-time and discrete-time models starting from transfer function measurements ((9-43) with output model (9-7), $\Omega = s$ or z^{-1}, $Y(k) = G(s_k)$, and $U(k) = 1$).

Because $V_{\mathrm{NLS}}(\theta, Z)$ is quadratic-in-the-measurements Z, the asymptotic properties proved in Theorem 9.21, with $V_F(\theta, Z) = V_{\mathrm{NLS}}(\theta, Z)/F$, are valid for $\hat{\theta}_{\mathrm{NLS}}(Z)$. We use the quick analysis tools of Section 9.5 to reveal the major properties of $\hat{\theta}_{\mathrm{NLS}}(Z)$. Taking the expected value of (9-43) gives (9-20) with

$$\mathbb{E}\{v_F(\theta, N_Z)\} = \frac{1}{F} \sum_{k=1}^{F} \sigma_Y^2(\Omega_k, \theta) \qquad (9\text{-}44)$$

(see Exercise 9.5). $\sigma_Y^2(\Omega_k, \theta)$ is the variance of the output error where the measurements Z have been replaced by the noise on the measurements N_Z

$$\sigma_Y^2(\Omega_k, \theta) = \sigma_Y^2(k) + \sigma_U^2(k)|G(\Omega_k, \theta)|^2 - 2\mathrm{Re}(\sigma_{YU}^2(k)\overline{G}(\Omega_k, \theta)) \qquad (9\text{-}45)$$

Applying quick tool 2 (see Section 9.5) to (9-44) shows that in general the nonlinear least squares estimate $\hat{\theta}_{\mathrm{NLS}}(Z)$ is inconsistent. It is consistent if $\sigma_Y^2(\Omega_k, \theta)$ is independent of θ, which is the case for transfer function measurements (9-2) ($Y(k) = G(\Omega_k)$, $U(k) = 1$, $\sigma_U^2(k) = 0$, and $\sigma_{YU}^2(k) = 0$) or input-output measurements (9-1) with exactly known input ($\sigma_U^2(k) = 0$, $\sigma_{YU}^2(k) = 0$). Replacing C_{N_Z} by $v^2 C_{N_Z}$ in the expected value of the cost function gives, taking into account (9-44),

$$V_F(\theta) = \mathbb{E}\{V_F(\theta, Z_0)\} + v^2 \mathbb{E}\{v_F(\theta, N_Z)\} \qquad (9\text{-}46)$$

It shows that in general $\tilde{\theta}_{\mathrm{NLS}}(Z_0)$ and $\theta_{*\mathrm{NLS}}$ depend on the disturbing noise level v and, hence, cannot be considered as the noiseless solutions (see Section 9.5, quick tool 3). They are the noiseless solutions for transfer function measurements and input-output measurements with exactly known input, because $\sigma_Y^2(\Omega_k, \theta)$ and, hence, also $\mathbb{E}\{v_F(\theta, N_Z)\}$ are then independent of θ. From (9-43) it follows immediately that $V_{\mathrm{NLS}}(\lambda\theta, Z) = V_{\mathrm{NLS}}(\theta, Z)$ so that $\hat{\theta}_{\mathrm{NLS}}(Z)$ is independent of the particular parameter constraint $a_i = 1$, $b_i = 1$ or $\|\theta\|_2^2 = 1$ chosen (see Section 9.5, quick tool 4).

We conclude from the previous discussion that the NLS estimator is inconsistent for noisy input-output measurements, while it is consistent for transfer function measurements. This suggests that for transfer function model (9-7) the bias in the estimates could be removed if the input-output measurements (9-1) are transformed into a transfer function measurement (9-2) with $G(k) = Y(k)/U(k)$. The nonlinear least squares estimate then minimizes

$$V_{\mathrm{NLS}}(\theta) = \sum_{k=1}^{F} |Y(k)/U(k) - G(\Omega_k, \theta)|^2 \qquad (9\text{-}47)$$

w.r.t. θ. From a theoretical point of view the minimizer of (9-47) is inconsistent because the mean value of the noise on $Y(k)/U(k)$ is not zero,

$$Y(k)/U(k) = G_0(\Omega_k) + N_G(k)$$
$$N_G(k) = G_0(k)\left(\frac{1 + N_Y(k)/Y_0(k)}{1 + N_U(k)/U_0(k)} - 1\right) \quad (9\text{-}48)$$

with $\mathbb{E}\{N_G(k)\} \neq 0$. Moreover, the moments of order 2 and higher of $N_G(k)$ do not exist (Guillaume et al., 1996a). We first study the bias term as a function of the signal-to-noise ratios and next tackle the nonexistence of the higher order moments.

For zero mean, circular complex distributed errors $N_Y(k)$, $N_U(k)$ (Assumption 9.18) with even probability density function the bias $\mathbb{E}\{N_G(k)\}$ is a function of the fourth-order moments of the noise (see Appendix 9.G). Assume now that the input-output errors are linearly correlated,

$$N_Y(k) = N(k) + \rho(k)\frac{\sigma_U(k)\sigma_Y(k)}{\sigma_V^2(k)}N_V(k)$$
$$N_U(k) = M(k) + N_V(k) \quad (9\text{-}49)$$

where $N(k)$, $M(k)$, and $N_V(k)$ are mutually independent random variables, and with $\rho(k) = \sigma_{YU}^2(k)/(\sigma_U(k)\sigma_Y(k))$ the correlation coefficient. Note that a correlation of the form (9-49) occurs, for example, in linear feedback systems (see Section 9.18). If $N_Y(k)$, $N_U(k)$ are, in addition, circular complex normally distributed (Assumptions 9.18 and 9.19), then an analytic expression can be found for the relative bias $b(k) = \mathbb{E}\{N_G(k)\}/G_0(\Omega_k)$ (see Appendix 9.G)

$$b(k) = -\exp(-|U_0(k)|^2/\sigma_U^2(k))\left(1 - \rho(k)\frac{U_0(k)/\sigma_U(k)}{Y_0(k)/\sigma_Y(k)}\right) \text{ for } k \neq 0, N/2 \quad (9\text{-}50)$$

For uncorrelated input-output errors, $\rho(k) = 0$, (9-50) reduces to a real number

$$b(k) = -\exp(-|U_0(k)|^2/\sigma_U^2(k)) \quad (9\text{-}51)$$

and, hence, the bias does not affect the phase. From (9-50) it follows that the relative bias $|b(k)|$ is maximal for totally correlated input-output errors, $|\rho(k)| = 1$ and $\angle \rho(k) = \pi + \angle G_0(\Omega_k)$,

$$\max_{\rho(k)}|b(k)| = \exp(-|U_0(k)|^2/\sigma_U^2(k))\left(1 + \frac{|U_0(k)|/\sigma_U(k)}{|Y_0(k)|/\sigma_Y(k)}\right) \quad (9\text{-}52)$$

The relative bias $|b(k)|$ (9-51) is smaller than 5×10^{-5} for signal-to-noise ratios $|U_0(k)|/\sigma_U(k)$ larger than 10 dB, and the maximal relative bias (9-52) is smaller than 1×10^{-4} if the worst case input and output signal-to-noise ratios $|U_0(k)|/\sigma_U(k)$, $|Y_0(k)|/\sigma_Y(k)$ are larger than 10 dB.

To ensure the existence of the higher order moments of $N_G(k)$ we exclude large, highly improbable values of $G(\Omega_k) = Y(k)/U(k)$. Define the truncated ratio $\underline{G}(\Omega_k)$ as

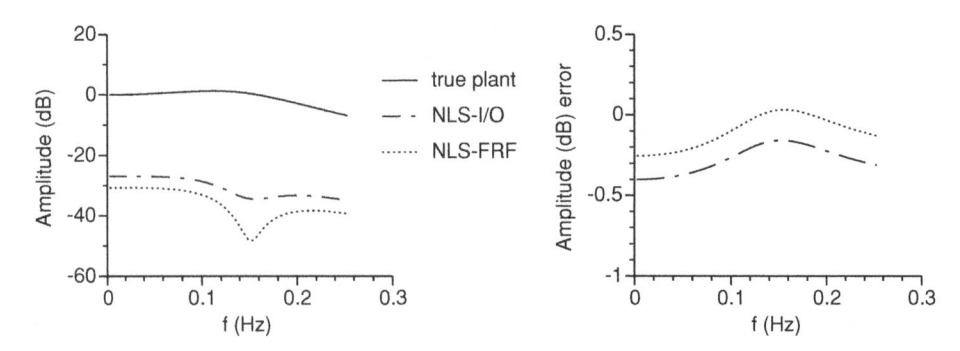

Figure 9-6. Second-order simulation example $G(s, \theta) = 1/(1 + s + s^2)$ defined in Appendix 9.A (see also Figure 9-1). Comparison of the nonlinear least squares estimates using the input-output spectra $Y(k)$, $U(k)$ (NLS-I/O) and the nonlinear least squares estimates using the frequency response function $G(k) = Y(k)/U(k)$ (NLS-FRF). Left, true plant model (solid line) and magnitude of the complex error between the estimated and the true plant model. Right, difference between the estimated amplitude in dB and the true amplitude in dB.

$$\underline{G}(\Omega_k) = \begin{cases} Y(k)/U(k) & |U(k)/U_0(k)| \geq L \\ 0 & |U(k)/U_0(k)| < L \end{cases} \quad (9\text{-}53)$$

with L an arbitrarily small number. Note that this is exactly what we do in practice: if the ratio $Y(k)/U(k)$ is unacceptably large, then we reject it. For input signal-to-noise ratios larger than 10 dB and $L = 1 \times 10^{-3}$ the change in bias of $\underline{G}(\Omega_k)$ w.r.t. $G(\Omega_k)$ is negligible and the variance of the truncated estimate is in good approximation given by the variance obtained via linearization (see (2-25) and Guillaume et al., 1996a)

$$\sigma_G^2(k) = |G_0(\Omega_k)|^2 [\sigma_Y^2(k)/|Y_0(k)|^2 + \sigma_U^2(k)/|U_0(k)|^2 - 2\text{Re}(\sigma_{YU}^2(k)/(Y_0(k)\overline{U}_0(k)))] \quad (9\text{-}54)$$

Hence, from a practical point of view, we may say that $N_G(k)$ has zero mean with existing higher order moments and that Assumption 9.4 is valid for $N_G(k)$ if $N_U(k)$, $N_Y(k)$ satisfy Assumptions 9.18 and 9.19. Because the cost function (9-47) is quadratic-in-the-measurements $Y(k)/U(k)$, we conclude that Theorem 9.21 is "practically valid" for the estimate $\hat{\theta}_{\text{NLS}}(Z)$ if the worst case input and output signal-to-noise ratio is at least 10 dB. Figure 9-6 shows that the errors of the NLS-I/O estimate (9-43) based on the input-output spectra are larger than those of the NLS-FRF estimate based on the frequency response function (9-47). As predicted by the theory, the NLS-I/O estimate is biased while the NLS-FRF estimate is "practically" consistent (compare NLS-FRF of Figure 9-6 to ML of Figure 9-4). See Table 9-5 for an overview of the properties of the NLS-FRF and NLS-IO estimators.

9.9.2 Logarithmic Least Squares

For frequency response functions with a large dynamic range, the nonlinear least squares estimator (9-43) of rational transfer function model (9-7) parameterized in powers of Ω_k (see (6-20)) may become ill conditioned. The dynamic range of the frequency response function can be limited by taking the natural logarithm of the model equation

Section 9.9 ■ Nonlinear Least Squares

$Y(k) = G(\Omega_k, \theta)U(k)$ giving $\ln(Y(k)/U(k)) = \ln(G(\Omega_k, \theta))$. The logarithmic least squares (LOG) estimator then minimizes

$$V_{\text{LOG}}(\theta, Z) = \sum_{k=1}^{F} |\ln(Y(k)/U(k)) - \ln(G(\Omega_k, \theta))|^2 \qquad (9\text{-}55)$$

w.r.t. θ. Besides its improved numerical stability (Sidman et al., 1991), the logarithmic least squares estimate $\hat{\theta}_{\text{LOG}}(Z)$ is particularly robust with respect to outliers in the measurements (Guillaume et al., 1995). Good starting values for the LOG estimator are the LS (9-33) and the IWLS (9-37) estimates.

From a theoretical point of view the logarithmic least squares estimator is inconsistent because the noise on $\ln(Y(k)/U(k))$ has no zero mean,

$$\begin{aligned}\ln(Y(k)/U(k)) &= \ln(G_0(\Omega_k)) + N(k) \\ N(k) &= \ln(1 + N_Y(k)/Y_0(k)) - \ln(1 + N_U(k)/U_0(k))\end{aligned} \qquad (9\text{-}56)$$

with $\mathbb{E}\{N(k)\} \neq 0$. The higher order moments of $N(k)$, however, do exist (Guillaume et al., 1996a). For zero mean, circular complex distributed errors $N_Y(k)$, $N_U(k)$ (Assumption 9.18) with even probability density function, the bias $b(k) = \mathbb{E}\{N(k)\}$ is a function of the fourth-order moments of the noise (see Appendix 9.G). If the errors are, in addition, normally distributed (Assumption 9.19), then an analytic expression can be found for $b(k)$ (see Appendix 9.G)

$$b(k) = \frac{1}{2}\text{Ei}(-\frac{|U_0(k)|^2}{\sigma_U^2(k)}) - \frac{1}{2}\text{Ei}(-\frac{|Y_0(k)|^2}{\sigma_Y^2(k)}) \text{ for } k \neq 0, N/2 \qquad (9\text{-}57)$$

with Ei(.) the exponential integral function (Gradshteyn and Ryzhik, 1980). Note that this expression is also valid for correlated input-output errors. It follows that the maximum bias error $|b(k)|$ is smaller than 2×10^{-6} for signal-to-noise-ratios $|Y_0(k)|/\sigma_Y(k)$ and $|U_0(k)|/\sigma_U(k)$ larger than 10 dB (see also Figure 2-18 on page 53). Hence, from a practical point of view, we may say that $N(k)$ has zero mean and that Assumption 9.4 is valid for $N(k)$ if $N_Y(k)$ and $N_U(k)$ satisfy Assumptions 9.18 and 9.19. Because the cost function (9-55) is quadratic-in-the-measurements $\ln(Y(k)/U(k))$, we conclude that Theorem 9.21, with $V_F(\theta, Z) = V_{\text{LOG}}(\theta, Z)/F$, is "practically valid" for the logarithmic LS estimate $\hat{\theta}_{\text{LOG}}(Z)$ if the worst case signal-to-noise ratio is at least 10 dB. The expected value of (9-55) then equals (9-20) with

$$\mathbb{E}\{v_F(\theta, N_Z)\} = \frac{1}{F}\sum_{k=1}^{F}\mathbb{E}\{|\ln(1 + N_Y(k)/Y_0(k)) - \ln(1 + N_U(k)/U_0(k))|^2\} \qquad (9\text{-}58)$$

Because $\mathbb{E}\{v_F(\theta, N_Z)\}$ is independent of θ, $\hat{\theta}_{\text{LOG}}(Z)$ is "practically consistent" and $\tilde{\theta}_{\text{LOG}}(Z_0)$, $\theta_{*\text{LOG}}$ are "practically" the noiseless solutions in case model errors are present (apply quick tools number 2 and 3 of Section 9.5). From (9-55) it follows that $V_{\text{LOG}}(\lambda\theta, Z) = V_{\text{LOG}}(\theta, Z)$, and, hence, $\hat{\theta}_{\text{LOG}}(Z)$ is independent of the particular, chosen parameter constraint $a_i = 1$, $b_i = 1$, or $\|\theta\|_2^2 = 1$ (see Section 9.5, quick tool 4). See Table 9-5 for an overview of the properties of the LOG estimator.

9.9.3 A Simple Example—Continued

We use the example of Section 9.8.4 to calculate the nonlinear least squares estimate (9-43) of the integrator model $G(s, \theta) = b_0/(a_1 s)$, using the constraint $b_0 = 1$. Making similar calculations as in Section 9.8.4, we get

$$(\hat{a}_1)_{\text{NLS}} = \frac{\sum_{k=1}^{F} |s_k|^{-2}}{\sum_{k=1}^{F} \text{Re}(G(s_k)/\tilde{s}_k)}$$

$$\underset{F \to \infty}{\text{a.s.lim}} (\hat{a}_1)_{\text{NLS}} = \frac{\underset{F \to \infty}{\lim} \frac{1}{F} \sum_{k=1}^{F} |s_k|^{-2}}{\underset{F \to \infty}{\lim} \frac{1}{F} \sum_{k=1}^{F} \text{Re}(G_0(s_k)/\tilde{s}_k)} = (a_1)_0$$

which shows that the nonlinear least squares estimator $(\hat{a}_1)_{\text{NLS}}$ and, hence, $G(s_k, \hat{\theta}_{\text{NLS}}(Z))$ are, indeed, consistent for transfer function measurements. It is easy to verify that the NLS estimate using the constraint $a_0 = 1$ equals $(\hat{b}_1)_{\text{NLS}} = 1/(\hat{a}_1)_{\text{NLS}}$. Hence, $(\hat{b}_1)_{\text{NLS}}$ and $G(s_k, \hat{\theta}_{\text{NLS}}(Z))$ are consistent estimates, which illustrates the independence of $G(s_k, \hat{\theta}_{\text{NLS}}(Z))$ on the parameter constraint used (quick tool number 4).

9.10 TOTAL LEAST SQUARES

9.10.1 Introduction

The total least squares (TLS) approach requires a model equation that is linear in the model parameters θ. Transfer function models (9-7) and (9-8), where $G(\Omega, \theta)$ and $T_G(\Omega, \theta)$ are parameterized as rational forms (6-20), (6-25) and (6-35), (6-38), can be made linear in θ by multiplication with the denominator polynomial $A(\Omega_k, \theta)$. This is not the case for the other parameterizations and, therefore, the TLS estimators can only be applied to rational forms without delay. Hence, the linear set of equations that needs to be solved in total least square sense is

$$e(\Omega_k, \theta, Z(k)) \approx 0 \quad k = 1, 2, \ldots, F \quad (9\text{-}59)$$

with $e(\Omega_k, \theta, Z(k))$ the equation error (9-9) or (9-10). They can be written as

$$J(Z)\theta \approx 0 \text{ or } J_{\text{re}}(Z)\theta \approx 0 \quad (9\text{-}60)$$

with $J(Z) = \partial e(\theta, Z)/\partial \theta$ the Jacobian of the vector $e(\theta, Z)$ ($e_{[k]}(\theta, Z) = e(\Omega_k, \theta, Z(k))$), and where $(\ )_{\text{re}}$ stacks the real and imaginary parts on top of each other

$$J_{\text{re}}(Z) = \begin{bmatrix} \text{Re}(J(Z)) \\ \text{Im}(J(Z)) \end{bmatrix} \quad (9\text{-}61)$$

(see Section 15.8). Operation (9-61) is necessary to ensure that the solution θ is real. A left and a right weighting can be applied to (9-60)

Section 9.10 ■ Total Least Squares

$$(WJ(Z)C^{-1})(C\theta) \approx 0 \quad \text{or} \quad (W_{\text{Re}} J_{\text{re}}(Z) C^{-1})(C\theta) \approx 0 \qquad (9\text{-}62)$$

where $W \in \mathbb{C}^{F \times F}$ and $C \in \mathbb{R}^{n_\theta \times n_\theta}$ are regular matrices and where $(WJ(Z))_{\text{re}} = W_{\text{Re}} J_{\text{re}}(Z)$ with

$$W_{\text{Re}} = \begin{bmatrix} \text{Re}(W) & -\text{Im}(W) \\ \text{Im}(W) & \text{Re}(W) \end{bmatrix} \qquad (9\text{-}63)$$

(see Lemma 15.4). A diagonal left weighting matrix W influences each row of $J(Z)$ separately and makes it possible to introduce a frequency-dependent weighting of the residuals $e(\Omega_k, \theta, Z(k))$. The right weighting matrix C influences each row of $J(Z)$ in exactly the same way and, hence, will not introduce a frequency-dependent weighting of the residuals $e(\Omega_k, \theta, Z(k))$. It can be used to influence the noise characteristics of $J(Z)$ (see Section 9.10.3).

The total least squares solution of the weighted problem (9-62) tries to find a modified matrix $\tilde{J}_{\text{re}}$, which is as close as possible to $J_{\text{re}}(Z)$ (in Frobenius norm, see Section 15.3), and a vector θ satisfying $\tilde{J}_{\text{re}} \theta = 0$. The unknown parameters in the total least squares problem are, hence, the matrix $\tilde{J}_{\text{re}}$ ($2Fn_\theta$ real parameters) and the model parameters θ (n_θ real numbers). These parameters are related to each other by the model equation $\tilde{J}_{\text{re}} \theta = 0$ ($2F$ real equations), so that the total number of free parameters equals $(2F+1)n_\theta - 2F$. This should be compared with the measured matrix $J_{\text{re}}(Z)$ ($2Fn_\theta$ real numbers), which gives a redundancy of $2F - n_\theta$. It shows that increasing F will (most probably) give more information about θ, but not about $\tilde{J}_{\text{re}}$. Indeed, no additional information can be accumulated about $2Fn_\theta$ real parameters starting from $2F$ real measurements.

The matrix $\tilde{J}_{\text{re}}$ and the vector θ are the solution of

$$\arg\min_{\tilde{J}_{\text{re}}, \theta} \left\| W_{\text{Re}} (J_{\text{re}}(Z) - \tilde{J}_{\text{re}}) C^{-1} \right\|_F^2 \text{ subject to } \tilde{J}_{\text{re}} \theta = 0 \text{ and } \|C\theta\|_2^2 = 1 \qquad (9\text{-}64)$$

(Van Huffel and Vandewalle, 1991). After elimination of $\tilde{J}_{\text{re}}$ in (9-64), we get the following equivalences.

Lemma 9.23 (Total Least Squares Solution—Equivalences): The total least squares problem (9-64) is equivalent to

1. $\arg\min_{\theta} \|WJ(Z)\theta\|_2^2 / \|C\theta\|_2^2$
2. $\arg\min_{\theta} \|WJ(Z)\theta\|_2^2$ subject to $\|C\theta\|_2^2 = 1$
3. finding the eigenvector θ corresponding to the smallest generalized eigenvalue λ of the generalized eigenvalue problem $(W_{\text{Re}} J_{\text{re}}(Z))^T (W_{\text{Re}} J_{\text{re}}(Z)) \theta = \lambda C^T C \theta$

Proof. See Appendix 9.H. □

Although we have assumed, during the proof, that the matrices W and C are nonsingular, it follows from Lemma 9.23 that the TLS solution remains well defined for singular weighting matrices W and C. In these cases, we take Lemma 9.23 as a definition of the total least squares solution. Equivalences 1 and 2 of Lemma 9.23 (nonlinear minimization of a cost function) are used to analyze the asymptotic properties ($F \to \infty$) of the TLS solution, while equivalence 3 is used to calculate the solution. The generalized eigenvalue problem (equivalence 3 of Lemma 9.23) can be calculated in a numerically stable way, even when C is singular, through the generalized singular value decomposition (GSVD) of the matrix pair $(W_{\text{Re}} J_{\text{re}}(Z), C)$ (see Section 15.4.2). The TLS solution is then the generalized right singular vector corresponding to the smallest generalized singular value of $(W_{\text{Re}} J_{\text{re}}(Z), C)$. When $C = I_{n_\theta}$ then the generalized eigenvalue problem reduces to an ordinary eigenvalue problem, which is solved in a numerically stable way through the singular value decomposition (SVD) of the matrix $W_{\text{Re}} J_{\text{re}}(Z)$ (see Section 15.4.2). The TLS solution is then the right singular vector corresponding to the smallest singular value of $W_{\text{Re}} J_{\text{re}}(Z)$.

9.10.2 Total Least Squares

Putting $W = I_F$ and $C = I_{n_\theta}$ in (9-64) gives the total least squares estimate $\hat{\theta}_{\text{TLS}}(Z)$. According to equivalence 2 of Lemma 9.23, $\hat{\theta}_{\text{TLS}}(Z)$ is the minimizer of

$$V_{\text{TLS}}(\theta, Z) = \sum_{k=1}^{F} |e(\Omega_k, \theta, Z(k))|^2 \text{ subject to } \|\theta\|_2^2 = 1 \quad (9\text{-}65)$$

with $e(\Omega_k, \theta, Z(k))$ the equation error (9-9) or (9-10) (proof: see Appendix 9.J). It shows that the total least squares solution (9-65) is nothing other than the linear least squares solution (9-33) with parameter constraint $\|\theta\|_2^2 = 1$. Hence, $\hat{\theta}_{\text{TLS}}(Z)$ has the same asymptotic properties ($F \to \infty$) as $\hat{\theta}_{\text{LS}}(Z)$: in general, $\hat{\theta}_{\text{TLS}}(Z)$ is inconsistent and $\tilde{\theta}_{\text{TLS}}(Z_0)$, $\theta_{*\text{TLS}}$ depend on the signal-to-noise ratio. To reveal when $\hat{\theta}_{\text{TLS}}(Z)$ is consistent, we use equivalence 1 of Lemma 9.23

$$V_{\text{TLS}}(\theta, Z) = \|J(Z)\theta\|_2^2 / \|\theta\|_2^2 = \sum_{k=1}^{F} |e(\Omega_k, \theta, Z(k))|^2 / \|\theta\|_2^2 \quad (9\text{-}66)$$

Taking the expected value of (9-66) gives (9-20) with $V_F(\theta) = \mathbb{E}\{V_{\text{TLS}}(\theta, Z)\}/F$, and

$$\mathbb{E}\{v_F(\theta, N_Z)\} = \frac{1}{F} \theta^T C_J \theta / \|\theta\|_2^2 = \frac{1}{F} \sum_{k=1}^{F} \sigma_e^2(\Omega_k, \theta) / \|\theta\|_2^2 \quad (9\text{-}67)$$

where $\sigma_e^2(\Omega_k, \theta)$ is defined in (9-35) and where

$$C_J = \mathbb{E}\{j_{\text{re}}^T(N_Z) j_{\text{re}}(N_Z)\} = \mathbb{E}\{\text{Re}(j^H(N_Z) j(N_Z))\} \text{ with } j(N_Z) = J(Z) - J(Z_0) \quad (9\text{-}68)$$

is the column covariance matrix of $j_{\text{re}}(N_Z)$ (see Appendix 9.I). Note that $j(N_Z) \neq J(N_Z)$ for model (9-10). Applying quick analysis tool number 2 (see Section 9.5) to (9-67) shows that the total least squares estimator $\hat{\theta}_{\text{TLS}}(Z)$ is consistent if C_J is proportional to I_{n_θ}: $C_J = \sigma^2 I_{n_\theta}$.

Like the LS estimate, the total least squares solution can be improved by adding an appropriate frequency-dependent weighting. The TLS version of (9-38) is found by making the choice $C = I_{n_\theta}$ and

$$W = \text{diag}(W(\Omega_1, \theta^{(i-1)}), W(\Omega_2, \theta^{(i-1)}), \ldots, W(\Omega_F, \theta^{(i-1)})) \quad (9\text{-}69)$$

with $W(\Omega_k, \theta^{(i-1)}) \in \mathbb{R}$, in (9-64) (proof: see Appendix 9.J). The weighted total least squares solution is calculated as the right singular vector corresponding to the smallest singular value of $W_{\text{Re}} J_{\text{re}}(Z)$.

9.10.3 Generalized Total Least Squares

The total least squares estimator (9-65) is inconsistent because the column covariance matrix C_J (9-68) is different from $\sigma^2 I_{n_\theta}$ (see Section 9.10.2). Taking as right weighting C a square root of C_J

$$C = C_J^{1/2} \text{ such that } C^T C = C_J \tag{9-70}$$

(see Section 15.4.4), then the column covariance matrix of $j_{re}(N_Z) C^{-1}$, with $j(N_Z) = J(Z) - J(Z_0)$, becomes

$$\mathbb{E}\{C^{-T} j_{re}^T(N_Z) j_{re}(N_Z) C^{-1}\} = C^{-T} C_J C^{-1} = I_{n_\theta} \tag{9-71}$$

It shows that the total least squares estimator can be made consistent by an appropriate choice of the right weighting matrix C. Note that the calculation of C requires knowledge of the noise (co)variances (Assumption 9.5).

Putting $W = I_F$ and $C = C_J^{1/2}$ in (9-64) gives the generalized total least squares (GTLS) estimate $\hat{\theta}_{GTLS}(Z)$. According to equivalence 1 of Lemma 9.23, $\hat{\theta}_{GTLS}(Z)$ is the minimizer of

$$V_{GTLS}(\theta, Z) = \frac{\sum_{k=1}^{F} |e(\Omega_k, \theta, Z(k))|^2}{\sum_{k=1}^{F} \sigma_e^2(\Omega_k, \theta)} \tag{9-72}$$

with $\sigma_e^2(\Omega_k, \theta) = \mathrm{var}(e(\Omega_k, \theta, N_Z(k)))$ (see (9-35)) and $e(\Omega_k, \theta, Z(k))$ the equation error (9-9) or (9-10) (proof: see Appendix 9.J). In Swevers et al. (1992) the generalized total least squares method was applied for the first time to identify discrete-time models from noisy input-output measurements ((9-72) with equation error (9-9) and $\Omega = z^{-1}$). Due to the equal weighting of the residuals $e(\Omega_k, \theta, Z(k))$ over all frequencies in (9-72), the GTLS estimate suffers from the same problem as the LS and TLS estimates: it overemphasizes the high-frequency errors. Although this effect is not apparent in the second-order simulation example (see Figure 9-4), it is visible on more complex systems (see Figure 9-8).

Because $V_{GTLS}(\theta, Z)$ is quadratic-in-the-measurements Z, Theorem 9.21, with $V_F(\theta, Z) = V_{GTLS}(\theta, Z)$, is valid for $\hat{\theta}_{GTLS}(Z)$. Due to the denominator in (9-72), the expression for the limit cost $V_*(\theta)$ in property 5 is somewhat more complicated (see Exercise 9.6). Taking the expected value of (9-72) gives (9-20) with

$$\mathbb{E}\{v_F(\theta, Z)\} = 1 \tag{9-73}$$

As $\mathbb{E}\{v_F(\theta, Z)\}$ is independent of θ, the generalized total least squares estimate $\hat{\theta}_{GTLS}(Z)$ is consistent, and $\tilde{\theta}_{GTLS}(Z_0)$, θ_{*GTLS} are the noiseless solutions when model errors are present (apply quick analysis tools number 2 and 3 of Section 9.5). From (9-72), it follows that $V_{GTLS}(\lambda \theta, Z) = V_{GTLS}(\theta, Z)$ so that $\hat{\theta}_{GTLS}(Z)$ is independent of the particular, chosen constraint $a_i = 1$, $b_i = 1$, or $\|\theta\|_2^2 = 1$ (quick tool number 4). See Table 9-5 for an overview of the properties of the GTLS estimator.

To deemphasize the high frequency errors in (9-72), a left weighting matrix W should be added, and at the same time, to keep the consistency, the right weighting C should be adapted. For example, the choice,

$$W = \text{diag}(W(\Omega_1), W(\Omega_2), \ldots, W(\Omega_F)) \text{ with } W(\Omega_k) \in \mathbb{R} \quad (9\text{-}74)$$

$$C = C_{WJ}^{1/2} \text{ such that } C^T C = C_{WJ} = \mathbb{E}\{\text{Re}((Wj(N_Z))^H (Wj(N_Z)))\} \quad (9\text{-}75)$$

in (9-64), with C_{WJ} the column covariance matrix of $W_{\text{Re}} j_{\text{re}}(N_Z)$ (see (9-68)), gives the following weighted generalized total least squares cost function:

$$V_{\text{WGTLS}}(\theta, Z) = \frac{\sum_{k=1}^{F} W^2(\Omega_k) |e(\Omega_k, \theta, Z(k))|^2}{\sum_{k=1}^{F} W^2(\Omega_k) \sigma_e^2(\Omega_k, \theta)} \quad (9\text{-}76)$$

(see Appendix 9.J). Although the weight $W(\Omega_k)$ does not affect the consistency of the weighted generalized total least squares estimate $\hat{\theta}_{\text{WGTLS}}(Z)$, it can seriously influence its uncertainty. A motivated choice will be presented in Section 9.12.3. Apart from this effect, $\hat{\theta}_{\text{WGTLS}}(Z)$ has the same asymptotic properties as $\hat{\theta}_{\text{GTLS}}(Z)$. The estimate $\hat{\theta}_{\text{WGTLS}}(Z)$ is calculated as the generalized right singular vector corresponding to the smallest generalized singular value of the matrix pair $(W_{\text{Re}} J_{\text{re}}(Z), C_{WJ}^{1/2})$. Note that the column covariance matrix C_{WJ} in (9-75) is singular under Assumption 9.20(i) or 9.20(ii) (see Appendix 9.K).

9.11 MAXIMUM LIKELIHOOD

9.11.1 The Maximum Likelihood Solution

To construct the maximum likelihood solution, starting from the frequency domain data (9-1) or (9-2), we need the probability density function (pdf) of the frequency domain errors $N_Z(k) = [N_Y(k) \ N_U(k)]^T$, $k = 1, 2, \ldots, F$. For a frequency domain experiment $N_Z(k)$ is independent over k (Assumption 9.4), while for a time domain experiment $N_Z(k)$ is asymptotically ($F \to \infty$) independent over k and circular complex normally distributed (see Sections 9.6.1 and 16.16). Therefore, it is reasonable to construct the maximum likelihood (ML) solution under the assumption that $N_Z(k)$ is independent (over k) circular complex normally distributed with known covariance matrix (Assumptions 9.5, 9.18, and 9.19). We also assume that the true excitation $U_0(k)$ and, hence, also the true response $Y_0(k)$ are deterministic (Assumption 9.19).

Because the true input $U_0(k)$ and output $Y_0(k)$ DFT spectra in (9-1) are unknown, they should be estimated and parameterized as $U_p(k)$, $Y_p(k)$. The unknown parameters in the errors-in-variables approach are, hence, the unknown input $U_p(k)$ and output $Y_p(k)$ DFT spectra ($4F$ real numbers) and the model parameters θ (n_θ real numbers). These parameters are related to each other by the model equations

$$e(\Omega_k, \theta, Z_p(k)) = 0 \quad k = 1, 2, \ldots, F \quad (9\text{-}77)$$

with $e(\Omega_k, \theta, Z(k))$ the equation error (9-9) or (9-10) ($2F$ real equations); thus, the total number of free parameters equals $2F + n_\theta$. This should be compared with the number of measured input $U(k)$ and output $Y(k)$ spectra ($4F$ real numbers), which gives a redundancy of $2F - n_\theta$. It shows that increasing F will (most probably) give more information about θ but not about $U_p(k)$ and $Y_p(k)$. Indeed, four new real parameters are added for each frequency.

Under Assumptions 9.5, 9.18, and 9.19 the negative log-likelihood function is

Section 9.11 ■ Maximum Likelihood

$$-\ln f_{N_Z}(Z, Z_p, \theta) = (Z - Z_p)^H C_{N_Z}^+ (Z - Z_p) + c$$
$$C_{N_Z} = \text{diag}(\text{Cov}(N_Z(1)), \text{Cov}(N_Z(2)), \ldots, \text{Cov}(N_Z(F)))$$
(9-78)

with + the Moore-Penrose pseudoinverse and c a constant, independent of Z_p and θ (see Appendix 9.L). (9-78) should be minimized w.r.t. Z_p and θ subject to the constraints (9-77). This constrained minimization problem can be solved using Lagrange multipliers $\lambda \in \mathbb{C}^F$

$$(Z - Z_p)^H C_{N_Z}^+ (Z - Z_p) + \text{Re}(\lambda^H e(\theta, Z_p))$$
(9-79)

Elimination of Z_p in (9-79) gives the maximum likelihood cost function

$$V_{\text{ML}}(\theta, Z) = \sum_{k=1}^{F} \frac{|e(\Omega_k, \theta, Z(k))|^2}{\sigma_e^2(\Omega_k, \theta)}$$
(9-80)

(see Appendix 9.L) with $e(\Omega_k, \theta, Z(k))$ the equation error (9-9) or (9-10) and $\sigma_e^2(\Omega_k, \theta)$ the variance of the equation error where the measurements Z have been replaced by the noise on the measurements N_Z; see (9-35). If DC (Ω_0) and Nyquist ($\Omega_{N/2}$) are present in the data, then

$$\frac{1}{2} \frac{|e(\Omega_0, \theta, Z(0))|^2}{\sigma_e^2(\Omega_0, \theta)} + \frac{1}{2} \frac{|e(\Omega_{N/2}, \theta, Z(N/2))|^2}{\sigma_e^2(\Omega_{N/2}, \theta)}$$
(9-81)

should be added to the cost function (9-80) (see Appendix 9.L). Dividing the numerator and denominator of each term in the sum (9-80) by $|A(\Omega_k, \theta)|^2$ gives

$$V_{\text{ML}}(\theta, Z) = \sum_{k=1}^{F} \frac{|Y(k) - Y(\Omega_k, \theta)|^2}{\sigma_Y^2(\Omega_k, \theta)}$$
(9-82)

with $\sigma_Y^2(\Omega_k, \theta)$ the variance of the output error, where the measurements Z have been replaced by the noise on the measurements N_Z; see (9-45). Under this form, it is suitable for any parameterization of the transfer function model (see Sections 6.2 and 6.3). Cost functions (9-80) and (9-82) can also be written as

$$V_{\text{ML}}(\theta, Z) = \sum_{k=1}^{F} |\varepsilon(\Omega_k, \theta, Z(k))|^2$$
(9-83)

where $\varepsilon(\Omega_k, \theta, Z(k))$ are the respective weighted residuals,

$$\varepsilon(\Omega_k, \theta, Z(k)) = e(\Omega_k, \theta, Z(k))/\sigma_e(\Omega_k, \theta)$$
(9-84)

$$\varepsilon(\Omega_k, \theta, Z(k)) = (Y(k) - Y(\Omega_k, \theta))/\sigma_Y(\Omega_k, \theta)$$
(9-85)

with $\text{var}(\varepsilon(\Omega_k, \theta, N_Z(k))) = 1$. The maximum likelihood estimate $\hat{\theta}_{\text{ML}}(Z)$ is the minimizer of (9-83) (see Appendix 9.L, Section 9.L.4 for the numerical implementation).

Using $\hat{\theta}_{\mathrm{ML}}(Z)$, the maximum likelihood estimates $\hat{U}_{\mathrm{ML}}(k)$ and $\hat{Y}_{\mathrm{ML}}(k)$ of the input and output DFT spectra can be calculated, namely

$$\hat{Y}_{\mathrm{ML}}(k) = Y(k) - (\sigma_Y^2(k)\bar{A}(\Omega_k, \hat{\theta}) - \sigma_{YU}^2(k)\bar{B}(\Omega_k, \hat{\theta}))\frac{e(\Omega_k, \hat{\theta}, Z(k))}{\sigma_e^2(\Omega_k, \hat{\theta})}$$

$$\hat{U}_{\mathrm{ML}}(k) = U(k) - (\bar{\sigma}_{YU}^2(k)\bar{A}(\Omega_k, \hat{\theta}) - \sigma_U^2(k)\bar{B}(\Omega_k, \hat{\theta}))\frac{e(\Omega_k, \hat{\theta}, Z(k))}{\sigma_e^2(\Omega_k, \hat{\theta})}$$

(9-86)

with $\hat{\theta} = \hat{\theta}_{\mathrm{ML}}(Z)$ (see Appendix 9.L). If the input is known ($\sigma_U^2(k) = 0$ and $\sigma_{YU}^2(k) = 0$), then (9-86) reduces to

$$\hat{Y}_{\mathrm{ML}}(k) = G(\Omega_k, \hat{\theta}_{\mathrm{ML}}(Z))U_0(k) \ (+ T_G(\Omega_k, \hat{\theta}_{\mathrm{ML}}(Z)))$$
$$\hat{U}_{\mathrm{ML}}(k) = U_0(k)$$

(9-87)

and the ML estimate $\hat{Y}_{\mathrm{ML}}(k)$ is nothing other than the output predicted by the model.

9.11.2 Discussion

The maximum likelihood solution (9-83) weights the equation or output error at each frequency Ω_k with its measurement uncertainty, so that frequency bands with high-quality measurements ($\sigma_Y^2(k)$ and $\sigma_U^2(k)$ are "small") contribute more to the ML cost than frequency bands with poor-quality measurements ($\sigma_Y^2(k)$ and $\sigma_U^2(k)$ are "large"). Hence, in a natural way, the ML cost gives much confidence to accurate measurements while it rejects noisy measurements. Inspection of the variance of the output error (9-45) leads to the following observations:

(i) In the uncorrelated case ($\sigma_{YU}^2(k) = 0$) the relative importance of the input disturbance w.r.t. the output disturbance is given by the model-dependent ratio

$$\frac{|G(\Omega_k, \theta)|^2 \sigma_U^2(k)}{\sigma_Y^2(k)}$$

(9-88)

(ii) The significance of the correlation between the input and output disturbances is assessed by the model-dependent ratio

$$\rho(k) = \frac{-2\mathrm{Re}(\sigma_{YU}^2(k)\bar{G}(\Omega_k, \theta))}{\sigma_Y^2(k) + |G(\Omega_k, \theta)|^2 \sigma_U^2(k)}$$

(9-89)

(iii) If the measurement errors $M_Y(k)$ and $M_U(k)$ in Figure 9-3 are uncorrelated, then the sign of $\rho(k)$ in (9-89) determines the behavior of the generator noise $N_g(k)$. If $\rho(k) < 0$, then the variance $\sigma_Y^2(\Omega_k, \theta)$ (see (9-45)) is decreased w.r.t. the uncorrelated case ($\sigma_{YU}^2(k) = 0$), which means that $N_g(k)$ contributes constructively to the excitation signal $U_0(k) + N_g(k)$ at frequency Ω_k. If $\rho(k) > 0$, then the variance $\sigma_Y^2(\Omega_k, \theta)$ is increased w.r.t. the uncorrelated case, which means that $N_g(k)$ acts as a disturbing noise source at frequency Ω_k.

If the Assumptions 9.5, 9.18, and 9.19 made to construct (9-83) are not fulfilled, for example, the errors $N_Z(k)$ are not normally distributed, then (9-83) is no longer the maximum likelihood solution of the problem. The same is true if the excitation is not deterministic. If

Section 9.11 ■ Maximum Likelihood

the errors $N_Z(k)$ are non-Gaussian, independent (over k), circular complex distributed random variables, then (9-83) is a Markov estimator (see Section 19.2.2), for which all the results of Chapter 19 apply. If the errors are not circular complex distributed, $\mathbb{E}\{N_Z(k)N_Z^T(k)\} \neq 0$, then $\mathrm{Cov}(N_Z(k))$ does not contain all the information included in $\mathrm{Cov}((N_Z(k))_{\mathrm{re}})$, and (9-83) is no longer the Markov solution of the problem (see Exercise 9.7, and Section 19.2). In that case (9-83) is just a weighted nonlinear least squares solution. With some misuse of terminology $\hat{\theta}_{\mathrm{ML}}(Z)$ will, independent of the true noise properties, denote the minimizer of (9-83).

9.11.3 Asymptotic Properties

The general maximum likelihood properties listed in Section 1.5.3 are not valid for the maximum likelihood solution (9-83) of the errors-in-variables problem. Indeed, they have been shown under the assumption that the number of estimated parameters does not increase with the amount of data, while the number of free parameters in the errors-in-variables problem is $2F + n_\theta$ and increases with the number of frequencies F. Therefore, even under the ideal Assumptions 9.5, 9.18, and 9.19, the consistency, asymptotic normality, and asymptotic efficiency still have to be proved, and it is not self-evident at all that the ML solution (9-83) will have nice asymptotic ($F \to \infty$) properties. We will first study the properties of $\hat{\theta}_{\mathrm{ML}}(Z)$ under less restrictive noise assumptions than those made to construct the ML solution.

Because $V_{\mathrm{ML}}(\theta, Z)$ is quadratic-in-the-measurements Z, Theorem 9.21, with $V_F(\theta, Z) = V_{\mathrm{ML}}(\theta, Z)/F$, is valid for $\hat{\theta}_{\mathrm{ML}}(Z)$. Taking the expected value of (9-83) gives (9-20) with

$$\mathbb{E}\{v_F(\theta, Z)\} = 1 \qquad (9-90)$$

It shows that $\hat{\theta}_{\mathrm{ML}}(Z)$ is consistent and, if there are model errors, that $\tilde{\theta}_{\mathrm{ML}}(Z_0)$, $\theta_{*\mathrm{ML}}$ are the noiseless solutions (apply quick analysis tools number 2 and 3 of Section 9.5). The noiseless solutions are obtained by decreasing the input and output noise levels simultaneously to zero while maintaining the ratios $\sigma_Y^2(k)/\sigma_U^2(k)$ and $\sigma_{YU}^2(k)/\sigma_U^2(k)$ constant (see quick tool number 3). Changing the ratios $\sigma_Y^2(k)/\sigma_U^2(k)$ and $\sigma_{YU}^2(k)/\sigma_U^2(k)$ introduces a frequency-dependent modification of $\sigma_e^2(\Omega_k, \theta)$ or $\sigma_Y^2(\Omega_k, \theta)$ in the cost function (9-83) and, hence, changes the noiseless solutions. We also have $V_{\mathrm{ML}}(\lambda\theta, Z) = V_{\mathrm{ML}}(\theta, Z)$ (see (9-83)) so that $\hat{\theta}_{\mathrm{ML}}(Z)$ is independent of the particular constraint chosen, for example, $a_i = 1$, $b_i = 1$, or $\|\theta\|_2^2 = 1$ (quick tool number 4). We conclude that $\hat{\theta}_{\mathrm{ML}}(Z)$ is, in general, consistent and asymptotically normally distributed. From property 8 of Theorem 9.21, it follows that $\hat{\theta}_{\mathrm{ML}}(Z)$ is, in general, inefficient. It is asymptotically efficient only if the input-output disturbances stem from one noncoherent noise source (see Assumption 9.20).

It can be seen from (9-86) that the estimates $\hat{U}_{\mathrm{ML}}(k)$ and $\hat{Y}_{\mathrm{ML}}(k)$ of the input and output DFT spectra are in general inconsistent, even if $\hat{\theta}_{\mathrm{ML}}(Z)$ is consistent. This can easily be understood as follows: making more measurements (increasing F) will not increase the knowledge of the input and output DFT spectra at one particular frequency (no noise averaging effect occurs). Because they are inconsistent, it makes no sense to calculate, for example, an "improved" frequency response function estimate using $\hat{U}_{\mathrm{ML}}(k)$ and $\hat{Y}_{\mathrm{ML}}(k)$. If the input is known and $\hat{\theta}_{\mathrm{ML}}(Z)$ is consistent, then $\hat{Y}_{\mathrm{ML}}(k)$ is consistent (see (9-87)). Similarly, if the output is known and $\hat{\theta}_{\mathrm{ML}}(Z)$ is consistent, then $\hat{U}_{\mathrm{ML}}(k)$ is consistent.

As the properties of $\hat{\theta}_{\mathrm{ML}}(Z)$ are also valid under the more restrictive Assumptions 9.5, 9.18, and 9.19, it follows from Theorem 9.21 that the maximum likelihood estimator ((9-83) with Assumptions 9.5, 9.18, and 9.19) is consistent and asymptotically normally distributed but that it is not asymptotically efficient (note the difference from the general maximum likelihood properties of Section 1.5.3). An inefficiency term is present; it tends to zero as the noise level υ tends to zero

$$\text{Cov}(\delta_\theta(Z)) = Fi^{-1}(\theta_0)(I_{n_\theta} + O(\upsilon))$$
$$Fi(\theta_0) = 2\text{Re}\left(\left(\frac{\partial \varepsilon(\theta, Z_0)}{\partial \theta_0}\right)^H \left(\frac{\partial \varepsilon(\theta, Z_0)}{\partial \theta_0}\right)\right) \tag{9-91}$$

with $Fi(\theta_0)$ the Fisher information matrix (see Appendix 9.E —asymptotic efficiency). For errors N_Z with an even pdf, the deviation in (9-91) is an $O(\upsilon^2)$. An explicit expression for the inefficiency term can be found in Pintelon and Hong (2007), where it is shown that the deviation from the Cramér-Rao lower bound is important only if both the input and output SNRs are low (< 3 dB) in certain frequency bands, and if at those frequencies the input-output errors are weakly correlated. Hence, in practice the inefficiency term will be neglected when calculating the covariance matrix of the estimates (see Section 9.11.4). The ML estimator is asymptotically efficient if only one noncoherent disturbing noise source is present (Theorem 9.21). This corresponds to the case where the total number of estimated parameters does not increase with F (see Appendix 9.M), thus the general maximum likelihood properties of Section 1.5.3 are valid. Note that the consistency and asymptotic normality properties of the ML estimator ((9-83) with Assumptions 9.5, 9.18, and 9.19) have been shown in Theorem 9.21 under much less restrictive noise assumptions than those made to construct the ML solution. The errors $N_Z(k)$ may be non-Gaussian, correlated over the frequencies k, and non-circular complex $\mathbb{E}\{N_Z(k)N_Z^T(k)\} \neq 0$. It shows the *robustness* of the consistency and asymptotic normality properties of the ML estimator w.r.t. Assumptions 9.5, 9.18, and 9.19. See Table 9-5 for an overview of the properties of the ML estimator.

9.11.4 Calculation of Uncertainty Bounds

According to property 3 of Theorem 9.21, the covariance matrix of the truncated estimator $\hat{\theta}_{\text{ML}}(Z)$ (see (9-25)) is asymptotically ($F \to \infty$) given by expression (9-27)

$$\text{Cov}(\hat{\theta}_{\text{ML}}(Z)) = \text{Cov}(\delta_\theta(Z))(I_{n_\theta} + O(F^{-1/2})) \tag{9-92}$$

(see Theorem 17.30). Expression (9-27) for $\text{Cov}(\delta_\theta(Z))$ is not really tractable because it requires, for example, the third- and fourth-order moments of the noise, which are mostly unknown. An approximation for "small" model errors ($\mu \to 0$) and "large" signal-to-noise ratios ($\upsilon \to 0$) can be calculated. Applying quick tool number 6 of Section 9.5 to (9-27) yields

$$\text{Cov}(\delta_\theta(Z)) = C_\theta(I_{n_\theta} + O(\upsilon) + O(\mu) + O(\mu^2 \upsilon^{-2})\lambda(Z_0))$$
$$C_\theta = \left[\mathbb{E}\left\{2\text{Re}\left(\frac{\partial \varepsilon(\theta, Z_0)}{\partial \tilde{\theta}_{\text{ML}}(Z_0)}\right)^H \left(\frac{\partial \varepsilon(\theta, Z_0)}{\partial \tilde{\theta}_{\text{ML}}(Z_0)}\right)\right\}\right]^{-1} = \upsilon^2 O(F^{-1}) \tag{9-93}$$

where $\lambda(Z_0) = 1$ for random Z_0 and $\lambda(Z_0) = 0$ for deterministic Z_0 (see Exercise 19.10). If model errors are present ($\mu \neq 0$), then the uncertainty of the estimated model parameters

(9-93) does not decrease to zero for random excitations ($\lambda(Z_0) = 1$) as the noise level υ tends to zero. To calculate (9-93) we need the true observations Z_0 and the minimizer $\tilde{\theta}_{\mathrm{ML}}(Z_0)$ of the expected value of the cost function, which are not available. An approximation is calculated by replacing Z_0 by Z and $\tilde{\theta}_{\mathrm{ML}}(Z_0)$ by $\hat{\theta}_{\mathrm{ML}}(Z)$, giving

$$\mathrm{Cov}(\hat{\theta}_{\mathrm{ML}}(Z)) \approx \left[2\mathrm{Re}\left(\left(\frac{\partial \varepsilon(\theta, Z)}{\partial \hat{\theta}_{\mathrm{ML}}(Z)}\right)^H \left(\frac{\partial \varepsilon(\theta, Z)}{\partial \hat{\theta}_{\mathrm{ML}}(Z)}\right)\right)\right]^{-1} \qquad (9\text{-}94)$$

Note that the expression between brackets in (9-94) equals, within a factor of 2, the matrix of the normal equation in the last Newton-Gauss step (9-17). Together with property 4 of Theorem 9.21 and the results of Section 16.2, (9-94) allows the calculation of uncertainty regions with a given confidence level for any model-related quantity (see also Section 19.4.7).

9.12 APPROXIMATE MAXIMUM LIKELIHOOD

9.12.1 Introduction

Compared with the maximum likelihood solution, the iterative weighted linear least squares (IWLS) and weighted generalized total least squares (WGTLS) estimators have a big advantage as global minimizers. Their noise sensitivity can, however, be poor. The basic idea of this section is to construct estimators that combine the global minimization properties of the IWLS and WGTLS estimators with the good statistical properties of the ML estimator. The key to the solution of this problem is an appropriate choice of the frequency-dependent weighting. Comparing the IWLS and WGTLS cost functions (9-38) and (9-76) with the maximum likelihood solution (9-80) suggests that the "optimal" weighting is $W(\Omega_k) = \sigma_e^{-1}(\Omega_k, \theta)$. Because θ is unknown, it should be reconstructed iteratively as

$$W(\Omega_k, \theta^{(i-1)}) = \sigma_e^{-1}(\Omega_k, \theta^{(i-1)}) \qquad (9\text{-}95)$$

The weighting (9-95) can even be relaxed as in (9-39)

$$W(\Omega_k, \theta^{(i-1)}) = \sigma_e^{-r}(\Omega_k, \theta^{(i-1)}) \text{ with } r \in [0, 1] \qquad (9\text{-}96)$$

Special cases are no weighting, $r = 0$, and "full" weighting, $r = 1$.

Just as in Sections 9.8 and 9.10, the estimators of this section require that the plant transfer function $G(\Omega, \theta)$ and the transient term $T_G(\Omega, \theta)$ are parameterized as rational forms $B(\Omega, \theta)/A(\Omega, \theta)$ (see (6-20), (6-25)) and $I(\Omega, \theta)/A(\Omega, \theta)$ (see (6-35), (6-38)), respectively.

9.12.2 Iterative Quadratic Maximum Likelihood

Making the choice (9-96) in the IWLS cost function (9-38) gives the iterative quadratic maximum likelihood method,

$$V_{\mathrm{IQML}}(\theta^{(i)}, Z) = \sum_{k=1}^{F} \frac{|e(\Omega_k, \theta^{(i)}, Z(k))|^2}{\sigma_e^{2r}(\Omega_k, \theta^{(i-1)})} \qquad (9\text{-}97)$$

with $e(\Omega_k, \theta, Z(k))$ the equation error (9-9) or (9-10). If convergent ($\theta^{(i)} = \theta^{(i-1)}$ for i sufficiently large), the "full" IQML cost ((9-97) with $r = 1$) tends to the ML cost (9-80). This does not, however, imply that $\hat{\theta}_{\text{IQML}}(Z) = \hat{\theta}_{\text{ML}}(Z)$. Indeed, therefore, one needs that the derivatives of both cost functions w.r.t. θ are the same. This is not the case here so that $\hat{\theta}_{\text{IQML}}(Z) \neq \hat{\theta}_{\text{ML}}(Z)$. However, because the elementwise difference between both Jacobians is proportional to the residual $\varepsilon(\Omega_k, \theta^{(i-1)}, Z(k))$ (9-84) (see Exercise 9.8), both estimates will coincide ($\hat{\theta}_{\text{IQML}}(Z) \approx \hat{\theta}_{\text{ML}}(Z)$) for "sufficiently high" signal-to-noise ratios and "sufficiently small" modeling errors; otherwise the difference may be large. This is illustrated by the "high noise" simulation example of Figure 9-4 (compare IQML and ML) and the "low noise" simulation example of Figure 9-8 (compare IQML and ML). We conclude that the IQML estimator (9-97) is related to the ML solution (9-80) as the IWLS estimator (9-37) to the nonlinear least squares solution (9-43).

Because (9-97) is a special case of (9-38), the estimate $\hat{\theta}_{\text{IQML}}(Z)$ has the same asymptotic ($F \to \infty$) properties as $\hat{\theta}_{\text{IWLS}}(Z)$ (see Section 9.8.3): $\hat{\theta}_{\text{IQML}}(Z)$ is inconsistent, depends on the particular constraint chosen, and does not converge to a noiseless solution. See Table 9-5 for an overview of the properties of the IQML estimator.

9.12.3 Bootstrapped Total Least Squares

Making the choice (9-96) in the WGTLS estimator (9-76) gives the bootstrapped total least squares (BTLS) method

$$V_{\text{BTLS}}(\theta^{(i)}, Z) = \frac{\sum_{k=1}^{F} \frac{|e(\Omega_k, \theta^{(i)}, Z(k))|^2}{\sigma_e^{2r}(\Omega_k, \theta^{(i-1)})}}{\sum_{k=1}^{F} \frac{\sigma_e^2(\Omega_k, \theta^{(i)})}{\sigma_e^{2r}(\Omega_k, \theta^{(i-1)})}} \tag{9-98}$$

with $\sigma_e^2(\Omega_k, \theta) = \text{var}(e(\Omega_k, \theta, N_Z(k)))$ (see (9-35)) and $e(\Omega_k, \theta, Z(k))$ the equation error (9-9) or (9-10). Relaxation of the weighting ($r < 1$) may be necessary if a lowly damped pole and zero are very close (relative to the spacing of the frequency grid) to each other. If convergent ($\theta^{(i)} = \theta^{(i-1)}$ for i sufficiently large), the "full" BTLS cost ((9-98) with $r = 1$) tends to the ML cost (9-80). The Jacobians of both estimators are, however, different, even for $i \to \infty$, and therefore $\hat{\theta}_{\text{BTLS}}(Z) \neq \hat{\theta}_{\text{ML}}(Z)$. Likewise, for IQML (see Section 9.12.2), the elementwise difference between both Jacobians is proportional to ML residual $\varepsilon(\Omega_k, \theta^{(i-1)}, Z(k))$ (9-84). In practice, it turns out that the difference is small for large signal-to-noise ratios such that the bootstrapped total least squares estimate $\hat{\theta}_{\text{BTLS}}(Z)$ is mostly (very) close to the maximum likelihood estimate $\hat{\theta}_{\text{ML}}(Z)$ (see Figure 9-4 and Section 9.15). The estimate $\hat{\theta}_{\text{BTLS}}(Z)$ is calculated numerically in exactly the same way as the weighted generalized total least squares in Section 9.10.3.

The asymptotic ($F \to \infty$) properties of the first step of the iterative procedure (9-98) can be analyzed using Theorem 9.21 and Corollary 9.22. If the initial guess $\theta^{(0)}$ is deterministic, then Theorem 9.21 is valid and the bootstrapped total least squares estimate $\hat{\theta}_{\text{BTLS}}(Z) = \theta^{(1)}$ has the same properties as $\hat{\theta}_{\text{WGTLS}}(Z)$ (see Section 9.10.3). If the choice $\theta^{(0)} = \hat{\theta}(Z)$ is made, then it is obvious that (9-98) is no longer a quadratic function of the measurements Z. Assuming that the initial guess $\hat{\theta}(Z)$ satisfies the properties of Theorem 9.21, for example, $\hat{\theta}(Z) = \hat{\theta}_{\text{LS}}(Z)$ or $\hat{\theta}(Z) = \hat{\theta}_{\text{GTLS}}(Z)$, then Theorem 9.21 is still valid for $\hat{\theta}_{\text{BTLS}}(Z) = \theta^{(1)}$ with three minor modifications (see Corollary 9.22). The first step of (9-98) can be written as

Section 9.12 ■ Approximate Maximum Likelihood

$$V_{\text{BTLS}}(\theta, Z) = f_F(\theta, \theta^{(0)}, Z) \tag{9-99}$$

Taking the expected value of the cost function (9-99), where $\theta^{(0)} = \hat{\theta}(Z)$ has been replaced by its limit ($F \to \infty$) value θ_* gives (9-20) with $V_F(\theta) = \mathbb{E}\{f_F(\theta, \theta_*, Z)\}$ and

$$V_F(\theta) = \frac{\sum_{k=1}^{F} \dfrac{\mathbb{E}\{|e(\Omega_k, \theta, Z_0(k))|^2\}}{\sigma_e^{2r}(\Omega_k, \theta_*)}}{\sum_{k=1}^{F} \dfrac{\sigma_e^2(\Omega_k, \theta)}{\sigma_e^{2r}(\Omega_k, \theta_*)}} + 1 \tag{9-100}$$

Hence, the bootstrapped total least squares estimate $\hat{\theta}_{\text{BTLS}}(Z) = \theta^{(1)}$ is consistent, even if $\theta^{(0)} = \hat{\theta}(Z)$ is inconsistent (apply quick analysis tool number 2 of Section 9.5). If the limit value θ_* does not depend on the noise level ν, then $\tilde{\theta}_{\text{BTLS}}(Z_0)$, $\theta_{*\text{BTLS}}$ are the noiseless solutions when there are model errors (quick analysis tool number 3). This is the case for $\theta^{(0)} = \hat{\theta}_{\text{GTLS}}(Z)$ but not for $\theta^{(0)} = \hat{\theta}_{\text{LS}}(Z)$. From (9-98) it follows that $V_{\text{BTLS}}(\lambda\theta, Z) = V_{\text{BTLS}}(\theta, Z)$ so that $\hat{\theta}_{\text{BTLS}}(Z)$ is independent of the particular, chosen constraint $a_i = 1$, $b_i = 1$, or $\|\theta\|_2^2 = 1$ (quick tool number 4). Because $\theta^{(1)} = \hat{\theta}_{\text{BTLS}}(Z)$ satisfies Theorem 9.21, the same reasoning can be applied to $\theta^{(2)}$, and so on, showing that the estimates obtained in the successive iteration steps have exactly the same properties as $\theta^{(1)}$. We conclude that the BTLS algorithm (9-98) generates consistent estimates in each iteration step. Hence, the iterative algorithm can be stopped at any iteration number (four iterations are usually sufficient). Further iteration (hopefully) decreases the uncertainty in the nonasymptotic case ($F \neq \infty$). In the absence of model errors, $\tilde{\theta}(Z_0) = \theta_0$ or $\theta_* = \theta_0$ for model (9-8) with $\Omega = s$, it follows from Corollary 9.22 that the asymptotic ($F \to \infty$) uncertainty of $\hat{\theta}_{\text{BTLS}}(Z) = \theta^{(1)}$ with $\theta^{(0)} = \hat{\theta}(Z)$ equals that of $\hat{\theta}_{\text{BTLS}}(Z) = \theta^{(1)}$ with $\theta^{(0)} = \theta_*$ (see Appendix 9.N). See Table 9-5 for an overview of the properties of the BTLS estimator.

9.12.4 Weighted (Total) Least Squares

The IQML and BTLS estimators need an initial guess of the model parameters to reconstruct the optimal ML weighting iteratively and, hence, are not self-starting. In this section, a noniterative approximation of the optimal ML weighting is given that does not require explicit knowledge of the model parameters θ.

The approximation is constructed as follows. Taking out the factor $|A(\Omega_k, \theta)||B(\Omega_k, \theta)|$ in the ML weighting (9-35) yields

$$\sigma_e^2(\Omega_k, \theta) = |A(\Omega_k, \theta)||B(\Omega_k, \theta)|(\sigma_Y^2(k)/|G(\Omega_k, \theta)| + \sigma_U^2(k)|G(\Omega_k, \theta)| \\ -2\text{Re}(\sigma_{YU}^2(k)\exp(-j\angle G(\Omega_k, \theta)))) \tag{9-101}$$

Replacing the unknown plant transfer function $G(\Omega_k, \theta)$ by the measured frequency response function $G(\Omega_k)$ or $Y(k)/U(k)$ and the factor $|A(\Omega_k, \theta)||B(\Omega_k, \theta)|$ by a θ-independent function $f(\Omega_k)$ in (9-101) gives the following approximation:

$$W^{-2}(\Omega_k) = f(\Omega_k)[\sigma_Y^2(k)/|G(\Omega_k)| + \sigma_U^2(k)|G(\Omega_k)| - 2\text{Re}(\sigma_{YU}^2(k)\exp(-j\angle G(\Omega_k)))] \tag{9-102}$$

The explicit form of the function $f(\Omega)$ depends on the particular domain Ω and is given below (see (9-104) and (9-105)). The reader is referred to Rolain and Pintelon (1999) for the rationale behind the construction of $f(\Omega)$. To avoid problems of division by zero in (9-102), regularization is applied in the frequency bands where $|G(\Omega_k)|$ is of the order of the magnitude of the noise standard deviation $\sigma_G(k)$:

$$W_{\text{reg}}^{-2}(\Omega_k) = \begin{cases} W^{-2}(\Omega_k) + \varepsilon W^{-2}(\Omega_{k+1}) & W^{-2}(\Omega_k) < \varepsilon W^{-2}(\Omega_{k+1}) \\ W^{-2}(\Omega_k) & \text{otherwise} \end{cases} \quad (9\text{-}103)$$

where ε is of the order of the numerical precision of the computer.

For *continuous-time systems*, $\Omega = s$, $\sqrt{s}$, or $\tanh(\tau_R s)$, the function $f(\Omega)$ has the form

$$f(\Omega_k) = \frac{1}{2}(g_{n_a}(\Omega_k)|G(\Omega_k)| + g_{n_b}(\Omega_k)/|G(\Omega_k)|)$$
$$g_n(\Omega) = (|\Omega|^{n+1} - 1)^2/(|\Omega| - 1)^2 \quad (9\text{-}104)$$

Recall that the frequency axis is scaled by $\omega_{\text{scale}} = \text{median}\{\omega_1, \omega_2, ..., \omega_F\}$ when identifying continuous-time systems (see Section 9.4), so that Ω in (9-104) represent the scaled frequency ($s \to s/\omega_{\text{scale}}$).

For *discrete-time systems*, $\Omega = z^{-1}$, the function $f(\Omega)$ has the form

$$f(z_k^{-1}) = (g^n(f_k, f_L) + g^n(f_k, f_U))^2$$
$$g(f_k, f) = |\cos(\omega_k T_s) - \cos(\omega T_s)| + (||\cos(\omega_L T_s)| - 0.5| + ||\cos(\omega_U T_s)| - 0.5|)/2 \quad (9\text{-}105)$$

with $n = \max(n_a, n_b) + 1$ and f_L, f_U the lower and upper frequencies of the "active" band of the plant. The active band $[f_L, f_U]$ is defined as the largest segment of continuous frequency points for which

$$h(k) > \frac{1}{F}\sum_{k=1}^{F} h(k)$$
$$h(k) = |G(z_k^{-1})|/\sigma_T + \sigma_T/|G(z_k^{-1})| \quad (9\text{-}106)$$

with σ_T^2 the mean (over the frequency) variance of the transfer function measurement $G(z_k^{-1})$ or $Y(k)/U(k)$,

$$\sigma_T^2 = \frac{1}{F}\sum_{k=1}^{F}|G(z_k^{-1})|^2\left(\frac{\sigma_Y^2(k)}{|Y(k)|^2} + \frac{\sigma_U^2(k)}{|U(k)|^2} - 2\text{Re}(\frac{\sigma_{YU}^2(k)}{Y(k)\overline{U}(k)})\right) \quad (9\text{-}107)$$

The noise influence on $h(k)$ in (9-106) is reduced by a running sum filter with a window length equal to 1% of the number of available frequency points.

The weighting (9-102) can be used to construct optimally weighted linear least squares (WLS) (9-38) or weighted generalized total least squares (WGTLS) (9-76) estimators. Because the weighting is a strong nonlinear function of the measurements Z, it is very difficult, if not impossible, to make precise statements about the asymptotic behavior of the WLS and WGTLS estimates obtained. They are inconsistent but (hopefully) lie within the attraction basin of the global minimum of the ML cost function. Although (9-102) may be a rough approximation due to the lack of knowledge about θ, a sensible improvement of the estimates w.r.t. to the unweighted case is obtained, even if the approximated and exact weight differ by as much as two orders of magnitudes. This low sensitivity is the key to the success of the proposed method. The power of the weighting is illustrated in Figure 9-9 on page 334 for a sixth-order discrete-time system.

9.13 INSTRUMENTAL VARIABLES

If two or more periods of the measured time signals are available, the measurements can be split up into two time records, each of them containing an integer number of signal periods. The DFT spectra calculated using the second time record can then be used as instrumental sequences for the linear least squares identification, based on the DFT spectra of the first time record (Van den Bos, 1991). The instrumental sequences obtained are almost ideal because they are strongly correlated with the true unknown DFT spectra and practically uncorrelated with the noise of the first time record (in case of colored noise a small but nonzero correlation may exist between the noise of the successive signal periods). The classical instrumental variable equations are asymmetric in the measurements and the instrumental sequences (see (1-58)). They can be made symmetric if the roles of the measurements and the instrumental sequences are interchanged and added to the original equations. Proceeding in this way, full use of the complete data set (measurements and instrumental sequences) is achieved. The equivalent cost function of the resulting enhanced instrumental variables estimator is

$$V_{IV}(\theta, Z) = \sum_{k=1}^{F} \text{Re}(e(\Omega_k, \theta, Z^{[1]}(k))\overline{e(\Omega_k, \theta, Z^{[2]}(k))}) \qquad (9\text{-}108)$$

where superscripts [1] and [2] indicate that the spectra are calculated using, respectively, the first and the second experiment (time record). Note that the cost function (9-108) can take negative values. Likewise, for the LS (9-33), TLS (9-65), and GTLS (9-72) cost functions, the high-frequency errors are overemphasized in (9-108).

Although the cost function $V_{IV}(\theta, Z)$ cannot be written under the quadratic form (9-11), Theorem 9.21, with $V_F(\theta, Z) = V_{IV}(\theta, Z)/F$, is still valid for $\hat{\theta}_{IV}(Z)$ (see Appendix 9.O). Assuming that the two experiments are independent, the expected value of (9-108) equals (9-20) with

$$\mathbb{E}\{v_F(\theta, N_Z)\} = 0 \qquad (9\text{-}109)$$

Applying quick analysis tools number 2 and 3 of Section 9.5 shows that $\hat{\theta}_{IV}(Z)$ is consistent and, when model errors are present $\tilde{\theta}_{IV}(Z_0)$, θ_{*IV} are the noiseless solutions. From (9-108) it follows that $V_{IV}(\lambda\theta, Z) = \lambda^2 V_{IV}(\theta, Z)$ so that $\hat{\theta}_{IV}(Z)$ depends on the particular constraint $a_i = 1$ or $b_i = 1$ chosen (apply quick tool number 4). See Table 9-5 for an overview of the properties of the IV estimator.

Note that the IV method lowers the bias of the corresponding LS estimates on the complete data set (DFT spectra of the first and second time records put together) at the price of a higher variance. The mean square error of the IV estimates tends asymptotically ($F \to \infty$) to zero, whereas that of the LS estimates tends asymptotically to the square of its bias. Hence, the IV method will perform better than the LS method for F sufficiently large. Compare, for example, the IV with the LS estimates in Figure 9-4.

9.14 SUBSPACE ALGORITHMS

9.14.1 Model Equations

Subspace identification methods estimate the state space representation of (9-7), namely

$$G(\xi, \theta) = C(\xi I_{n_a} - A)^{-1}B + D \tag{9-110}$$

where $\xi = z$ for discrete-time systems and $\xi = s$ for continuous-time systems. The identification procedure starts from a transformed version of the state space equations (6-18) and (6-19). These are constructed as follows. Assume that the input is periodic and that an integer number of periods of the steady-state response is observed. The discrete Fourier transform (DFT) of (6-18) and (6-19) then becomes

$$\begin{aligned} \xi_k X(k) &= AX(k) + BU(k) \\ Y(k) &= CX(k) + DU(k) \end{aligned} \tag{9-111}$$

with $X(k)$ the DFT of the state vector $x(t)$. By recursive use of the second and the first equation of (9-111) we find that

$$\begin{aligned} \xi_k^p Y(k) &= \xi_k^{p-1}(C\xi_k X(k) + D\xi_k U(k)) \\ &= \xi_k^{p-1}(CAX(k) + CBU(k) + D\xi_k U(k)) \\ &= \ldots \\ &= CA^p X(k) + (CA^{p-1}B + CA^{p-2}B\xi_k + \ldots + CB\xi_k^{p-1} + D\xi_k^p)U(k) \end{aligned} \tag{9-112}$$

Writing the last equation of (9-112) for $p = 0, 1, \ldots, r-1$ on top of each other gives

$$W_r(k)Y(k) = O_r X(k) + S_r W_r(k) U(k) \tag{9-113}$$

with

$$W_r(k) = \begin{bmatrix} 1 \\ \xi_k \\ \ldots \\ \xi_k^{r-1} \end{bmatrix}, O_r = \begin{bmatrix} C \\ CA \\ \ldots \\ CA^{r-1} \end{bmatrix} \text{ and } S_r = \begin{bmatrix} D & 0 & \ldots & 0 & 0 \\ CB & D & \ldots & 0 & 0 \\ \ldots & \ldots & \ldots & \ldots & \ldots \\ CA^{r-2}B & CA^{r-3}B & \ldots & CB & D \end{bmatrix} \tag{9-114}$$

Collecting (9-113) for $k = 1, 2, \ldots, F$ gives

$$\mathbf{Y} = O_r \mathbf{X} + S_r \mathbf{U} \tag{9-115}$$

with

$$\mathbf{Y} = \begin{bmatrix} W_r(1)Y(1) & W_r(2)Y(2) & \ldots & W_r(F)Y(F) \end{bmatrix}$$
$$\mathbf{U} = \begin{bmatrix} W_r(1)U(1) & W_r(2)U(2) & \ldots & W_r(F)U(F) \end{bmatrix} \quad (9\text{-}116)$$
$$\mathbf{X} = \begin{bmatrix} X(1) & X(2) & \ldots & X(F) \end{bmatrix}$$

The complex data matrices $\mathbf{Y}$ and $\mathbf{U}$ have r rows and F columns. $\mathbf{X}$ is a complex n_a by F matrix, and O_r and S_r are, respectively, real r by n_a and r by r matrices. Equation (9-115), with r larger than the model order n_a, is the basic model used in subspace identification.

The extended observability matrix O_r has the shift property

$$O_{r[1:r-1,\,:]} A = O_{r[2:r,\,:]} \quad (9\text{-}117)$$

which will be used in the identification procedure. O_r is not unique because it depends on the choice of the state variables. Indeed, replacing (A, B, C, D, X) by $(TAT^{-1}, TB, CT^{-1}, D, TX)$, with T an invertible matrix, in the state space equations (9-111) does not change the input-output transfer function (9-110) but does change O_r

$$O_r \rightarrow O_r T^{-1} \quad (9\text{-}118)$$

Note that $O_r \mathbf{X}$ and S_r in model equation (9-115) are invariant w.r.t. the invertible transformation T.

For identifiability purposes we will assume that the state space realization (9-111) is observable, $\text{rank}(O_r) = n_a$ for any $r \geq n_a$, and controllable,

$$\text{rank}([B \ AB \ \ldots \ A^{q-1}B]) = n_a \quad (9\text{-}119)$$

for any $q \geq n_a$.

For noisy input-output DFT spectra, $N_U(k) \neq 0$ and $N_Y(k) \neq 0$, model (9-115) becomes

$$\mathbf{Y} = O_r \mathbf{X} + S_r \mathbf{U} + \mathbf{N_Y} - S_r \mathbf{N_U} \quad (9\text{-}120)$$

where $\mathbf{N_Y}$ and $\mathbf{N_U}$ have the same structure as $\mathbf{Y}$ and $\mathbf{U}$ in (9-116).

9.14.2 Subspace Identification Algorithms

Subspace identification algorithms are basically a three-step procedure. First, an estimate $\hat{O}_r$ of the extended observability matrix is obtained using model (9-120). This is the most difficult step and consists mainly of eliminating the term depending on the input and reducing the noise influence. Next, $\hat{A}$ and $\hat{C}$ are found as the least squares solution of the overdetermined set of equations (9-117) and as the first row of $\hat{O}_r$ (see (9-114)), respectively. Finally, $\hat{B}$ and $\hat{D}$ are found as the linear least squares solution of

$$V_{\text{SUB}}(B, D, \hat{A}, \hat{C}, Z) = \sum_{k=1}^{F} W^2(\xi_k) \left| Y(k) - (\hat{C}(\xi_k I_{n_a} - \hat{A})^{-1} B + D) U(k) \right|^2 \quad (9\text{-}121)$$

where $W(\xi_k)$ is a well-chosen real weighting function.

We present two algorithms, one for discrete-time systems ($\xi = z$), based on McKelvey et al. (1996), and one for continuous-time system ($\xi = s$), based on Van Overschee and De Moor (1996a). The numerically efficient implementation of these algorithms is due to Verhaegen (1994).

Algorithm 9.24 (Subspace Algorithm for Discrete-Time Systems)

1. Estimate O_r given the data $Y(k)$, $U(k)$ and the noise (co)variances $\sigma_Y^2(k)$, $\sigma_U^2(k)$, $\sigma_{YU}^2(k)$:

    1a. Initialization:

    (i) If $\sigma_U^2(k) \neq 0$, replace $Y(k)$, $U(k)$, and $\sigma_Y^2(k)$ by, respectively, $Y(k)/U(k)$, 1, and $\sigma_G^2(k)$ (9-54).

    (ii) If the required transfer function model is improper, $n_a < n_b$, interchange the role of the input and output.

    (iii) Choose a value of $r > n_a$ and form the matrices

    $$Z = \begin{bmatrix} \text{Re}(\mathbf{U}) & \text{Im}(\mathbf{U}) \\ \text{Re}(\mathbf{Y}) & \text{Im}(\mathbf{Y}) \end{bmatrix} \text{ and } C_{\mathbf{Y}} = \text{Re}(\mathbf{C}\mathbf{C}^H)$$

    with $\mathbf{C} = [W_r(1)\sigma_Y(1) \; W_r(2)\sigma_Y(2) \; \ldots \; W_r(F)\sigma_Y(F)]$ and where $\mathbf{U}$ and $\mathbf{Y}$ are defined in (9-116), and $W_r(k)$ as in (9-114) with $\xi = z$.

    1b. Elimination of the input term in (9-120): calculate the QR factorization of Z^T, $Z^T = QR$, or $Z = R^T Q^T$,

    $$Z = \begin{bmatrix} R_{11}^T & 0 \\ R_{12}^T & R_{22}^T \end{bmatrix} \begin{bmatrix} Q_1^T \\ Q_2^T \end{bmatrix}$$

    where R_{ij} are r by r blocks of the upper triangular matrix R.

    1c. Reduction of the noise influence in (9-120): calculate the singular value decomposition of $C_{\mathbf{Y}}^{-1/2} R_{22}^T$,

    $$C_{\mathbf{Y}}^{-1/2} R_{22}^T = U \Sigma V^T$$

    where $C_{\mathbf{Y}}^{1/2}$ is a square root of $C_{\mathbf{Y}}$ (see Section 15.4.4), and estimate O_r as

Section 9.14 ■ Subspace Algorithms

$$\hat{O}_r = C_Y^{1/2} U_{[:, 1:n_a]}$$

2. Estimate A and C, given the estimate $\hat{O}_r$: solve the shift property (9-117) in least squares sense and select the first row of $\hat{O}_r$

$$\hat{A} = \hat{O}_{r[1:r-1,\,:]}^+ \hat{O}_{r[2:r,\,:]} \quad \text{and} \quad \hat{C} = \hat{O}_{r[1,\,:]}$$

with + the Moore-Penrose pseudoinverse (see Section 15.5).

3. Estimate B and D, given the estimates $\hat{A}$ and $\hat{C}$: minimize (9-121) w.r.t. B and D with $W(z_k) = 1/\sigma_Y(k)$.

Proof. See Appendix 9.R. □

One could use Algorithm 9.24 with $\xi = s$ for continuous-time systems. This works reasonably well for small values of r. However, for larger values, the matrix Z in Algorithm 9.24 becomes ill conditioned, resulting in poor estimates. This problem is solved by introducing two scalar orthogonal polynomial bases that orthogonalize, respectively, the first r rows of Z and the last r rows of Z. It can be shown that there are no other two scalar polynomial bases that result in a smaller condition number of Z (Rolain et al., 1995). The final algorithm is also a three-step procedure. First, a generalized extended observability matrix $\hat{O}_{r\perp}$ is estimated. This matrix has a generalized shift structure that is used to estimate A. Next, $\hat{A}$ and $\hat{C}$ are estimated using $\hat{O}_{r\perp}$. Finally, $\hat{B}$ and $\hat{D}$ are the linear least squares solution of (9-121).

Algorithm 9.25 (Subspace Algorithm for Continuous-Time Systems)

1. Estimate $O_{r\perp}$ given the data $Y(k)$, $U(k)$ and the noise (co)variances $\sigma_Y^2(k)$, $\sigma_U^2(k)$, $\sigma_{YU}^2(k)$:

   1a. Initialization:

   (i) If $\sigma_U^2(k) \neq 0$, replace $Y(k)$, $U(k)$, and $\sigma_Y^2(k)$ by, respectively, $Y(k)/U(k)$, 1, and $\sigma_G^2(k)$ (9-54).

   (ii) If the required transfer function model is improper $n_a < n_b$, interchange the role of the input and output.

   (iii) Choose a value of $r > n_a$ and normalize the frequencies s_k with $\omega_{\text{scale}} = \text{median}\{\omega_1, \omega_2, ..., \omega_F\}$ $(s_k \to s_k/\omega_{\text{scale}})$.

   1b. Orthogonalization of the output data: calculate the r by F matrix $\mathbf{Y}_\perp$ as follows: initialization:

   $$\mathbf{Y}_{\perp[1,\,:]} = \mathbf{Y}_{[1,\,:]}/\alpha_1 \quad \text{with} \quad \alpha_1 = \|\mathbf{Y}_{[1,\,:]}\|_2$$
   $$\mathbf{Y}_{\perp[2,\,:]} = \mathbf{Y}_{\perp[1,\,:]}D_s/\alpha_2 \quad \text{with} \quad \alpha_2 = \|\mathbf{Y}_{\perp[1,\,:]}D_s\|_2$$

   recursion: for $n = 3$ to r

   $$\mathbf{Y}_{\perp[n,\,:]} = (\mathbf{Y}_{\perp[n-1,\,:]}D_s + \alpha_{n-1}\mathbf{Y}_{\perp[n-2,\,:]})/\alpha_n \quad \text{with}$$
   $$\alpha_n = \|\mathbf{Y}_{\perp[n-1,\,:]}D_s + \alpha_{n-1}\mathbf{Y}_{\perp[n-2,\,:]}\|_2$$

where $Y_{[1,:]} = [Y(1)\ Y(2)\ ...\ Y(F)]$ and $D_s = \text{diag}(s_1, s_2, ..., s_F)$.

1c. Orthogonalization of the input data: perform the same calculation as in step 1b, but starting from $\mathbf{U}_{[1,:]} = [U(1)\ U(2)\ ...\ U(F)]$. The result is an r by F matrix $\mathbf{U}_\perp$ and numbers β_n, $n = 1, 2, ..., r$.

1d. Form the following matrices:

$$Z_\perp = \begin{bmatrix} \text{Re}(\mathbf{U}_\perp) & \text{Im}(\mathbf{U}_\perp) \\ \text{Re}(\mathbf{Y}_\perp) & \text{Im}(\mathbf{Y}_\perp) \end{bmatrix} \text{ and } C_{\mathbf{Y}_\perp} = \text{Re}(\mathbf{C}_\perp \mathbf{C}_\perp^H)$$

where $\mathbf{C}_\perp$ is calculated by starting from $\mathbf{C}_{[1,:]} = [\sigma_Y(1)\ \sigma_Y(2)\ ...\ \sigma_Y(F)]$, initialization:

$$\mathbf{C}_{\perp[1,:]} = \mathbf{C}_{[1,:]}/\alpha_1 \text{ and } \mathbf{C}_{\perp[2,:]} = \mathbf{C}_{\perp[1,:]} D_s/\alpha_2$$

recursion: for $n = 3$ to r

$$\mathbf{C}_{\perp[n,:]} = \mathbf{C}_{\perp[n-1,:]} D_s/\alpha_n + \alpha_{n-1}/\alpha_n \mathbf{C}_{\perp[n-2,:]}$$

1e. Elimination of the input term: calculate the QR factorization of $Z_\perp^T$, $Z_\perp^T = QR$ or,

$$Z_\perp = \begin{bmatrix} R_{11}^T & 0 \\ R_{12}^T & R_{22}^T \end{bmatrix} \begin{bmatrix} Q_1^T \\ Q_2^T \end{bmatrix}$$

where R_{ij} are r by r blocks of the upper triangular matrix R.

1f. Reduction of the noise influence: calculate the singular value decomposition of $C_{\mathbf{Y}_\perp}^{-1/2} R_{22}^T$,

$$C_{\mathbf{Y}_\perp}^{-1/2} R_{22}^T = U\Sigma V^T$$

where $C_{\mathbf{Y}_\perp}^{1/2}$ is a square root of $C_{\mathbf{Y}_\perp}$ (see Section 15.4.4), and estimate $O_{r\perp}$ as

$$\hat{O}_{r\perp} = C_{\mathbf{Y}_\perp}^{1/2} U_{[:,1:n_a]}$$

2. Estimate A and C given the estimate $\hat{O}_{r\perp}$: solve the generalized shift property in least squares sense and select the first row of $\hat{O}_{r\perp}$

$$\hat{A} = [D_1 \hat{O}_{r\perp[1:r-1,:]}]^+ [\hat{O}_{r\perp[2:r,:]} - b] \text{ and } \hat{C} = \alpha_1 \hat{O}_{r\perp[1,:]}$$

with $+$ the Moore-Penrose pseudoinverse (see Section 15.5) and

$$b = \begin{bmatrix} 0 \\ D_2 \hat{O}_{r\perp[1:r-2,:]} \end{bmatrix}, \begin{aligned} D_1 &= \text{diag}(1/\alpha_2, 1/\alpha_3, ..., 1/\alpha_r) \\ D_2 &= \text{diag}(\alpha_2/\alpha_3, \alpha_3/\alpha_4, ..., \alpha_{r-1}/\alpha_r) \end{aligned}$$

3. Estimate B and D, given the estimates $\hat{A}$ and $\hat{C}$: minimize (9-121) w.r.t. B and D with $W(s_k) = 1/\sigma_Y(k)$.
4. Denormalization of the estimates: multiply $\hat{A}$ and $\hat{C}$ by ω_{scale}.

Proof. See Appendix 9.S. □

Algorithm 9.25 differs in three ways from that described in Van Overschee and De Moor (1996a). First, the recursions in steps 1b and 1c are performed on rows with unit 2-norm. Next, the orthogonal projection is calculated via a QR factorization (see step 1e of Algorithm 9.25). Finally, one additional equation is used to estimate A (see step 2 of Algorithm 9.25). While the first two modifications improve the numerical stability of the algorithm, the third modification decreases the estimation error.

9.14.3 Stochastic Properties

The persistence-of-excitation condition is somewhat different for subspace algorithms compared with algorithms minimizing a cost function of the form (9-11). Therefore, we must add the following assumptions to Section 9.6.1.

Assumption 9.26 (Persistence of Excitation): There exists an F_0 such that for any $F \geq F_0$, ∞ included, $\mathrm{Re}(\mathbf{U}\mathbf{U}^H/F) \geq cI_r$ with $0 < c < \infty$ and c independent of F.

Assumption 9.27 (Identifiability Condition): There exists an F_0 such that for any $F \geq F_0$, ∞ included, $\mathrm{rank}(\mathbf{Y}_0^{\mathrm{re}}\Pi) \geq n_a$.

Note that under Assumptions 9.14 (distinct frequencies) and 9.16 (no model errors), Assumption 9.7 for (9-121) and Assumption 9.27 are fulfilled, if and only if (A, C) and (A, B) in (9-110) are, respectively, observable and controllable (see Appendix 9.R).

Theorem 9.28 (Asymptotic Properties $\hat{\theta}_{\mathrm{SUB}}(Z)$): Consider model (9-110), parameterized in its state space representation, and assume that the input-output data stem from the steady-state response of a system to a periodic excitation, observed during an integer number of periods (time or frequency domain experiment). The estimate $\hat{\theta}_{\mathrm{SUB}}(Z)$ obtained via Algorithm 9.24 or Algorithm 9.25 has the following asymptotic $(F \to \infty)$ properties:

1. *Stochastic convergence:* $\hat{\theta}_{\mathrm{SUB}}(Z)$ converges strongly to the noiseless solution $\theta_{*\mathrm{SUB}}$ (assumptions of Sections 9.6.1 and 9.6.5 and Assumptions 9.26 and 9.27).
2. *Stochastic convergence rate:* $\hat{\theta}_{\mathrm{SUB}}(Z)$ converges in probability at the rate $O_p(F^{-1/2})$ to $\theta_{*\mathrm{SUB}}$ (assumptions of Sections 9.6.2 and 9.6.5 and Assumptions 9.26 and 9.27).
3. *Consistency:* $\hat{\theta}_{\mathrm{SUB}}(Z)$ converges strongly to the true solution θ_0 (assumptions of Sections 9.6.5 and 9.6.6 and Assumptions 9.26 and 9.27).

Proof. See Appendix 9.R and Appendix 9.S. □

Although the subspace (SUB) estimates are strongly consistent $(F \to \infty)$ for any $r \geq n_a + 1$, with r independent of F, the finite sample properties of $\hat{\theta}_{\mathrm{SUB}}(Z)$ strongly depend on the choice of r. For example, values of r close to $n_a + 1$ usually result in poor estimates. An appropriate choice of r is therefore recommended. We propose to choose r such that

$$\sum_{k=1}^{F} \frac{|Y(k) - G(\Omega_k, \hat{\theta}_{\text{SUB}}(Z))U(k)|^2}{\sigma_Y^2(k)} \qquad (9\text{-}122)$$

is minimal. This optimization requires an exhaustive search for all $r \geq n_a + 1$ values (the cost function (9-122) is a craggy function of r, with many peaks and dips). In practice, we limit the search to the interval $[1.5n_a, 6n_a]$. However, sometimes it may be necessary to go beyond the upper limit $6n_a$ to find the optimum (see Section 9.15.3, modeling of a synchronous motor). It also turns out that the optimal value of r strongly depends on the plant and the noise characteristics.

The results of the SUB estimates (Algorithm 9.25 with $r = 5$) on the second-order simulation example are shown in Figure 9-4. Note that the SUB method estimates five free model parameters while the other methods estimate only three free model parameters. This is due to the fact that the subspace algorithms cannot impose the order of the numerator polynomial. See Table 9-5 for an overview of the properties of the SUB estimator.

9.15 ILLUSTRATION AND OVERVIEW OF THE PROPERTIES

The NLS-FRF (9-47), LOG (9-55), GTLS (9-72), ML (9-80), BTLS (9-98), IV (9-108), and SUB (see Section 9.14) estimators perform equally well on the second-order example (see Fig. 9-4 and Fig. 9-6). This is due to the very simple nature (low order, low amplitude dynamics, low frequency range, no model errors) of the simulation example. The differences are more apparent in the two simulation examples of this section. Two real measurement examples are also shown to illustrate an aspect that is not shown by the simulation examples: sensitivity to (small) model errors like unmodeled dynamics and nonlinearities. For all the simulation and real measurement examples, the optimal value of r in the SUB Algorithms 9.24 and 9.25 has been selected by an exhaustive search in the interval $[1.5n, 6n]$, except for the modeling of the electrical machine, where the search has been done in the interval $[1.5n, 18n]$.

9.15.1 Simulation Example 1

The simulated plant is a fifth-order continuous-time Butterworth filter with an extra transmission zero at $\omega = 3$ rad/s. The coefficients of the transfer function are given in Table 9-2 and the amplitude and phase characteristics are shown in Figure 9-7. A data set of $F = 100$ equally distributed frequencies is generated in the band [0.05 Hz, 5 Hz]

$$\begin{aligned} Y(k) &= G_0(s_k) + N_Y(k) \\ U(k) &= 1 + N_U(k) \end{aligned} \qquad (9\text{-}123)$$

with $N_Y(k)$ and $N_U(k)$, $k = 1, 2, ..., F$, independent, zero mean, circular complex Gaussian-distributed random variables with variance 2×10^{-6}. One hundred data sets of the type (9-123) are generated. For each set the LS, "full" IWLS, NLS, LOG, GTLS, ML, "full" IQML, and "full" BTLS estimates of model (6-20) with $n_a = 5$ and $n_b = 2$, and the SUB estimate (Algorithm 9.25 with $r = 20$) of model (6-26) with $n_a = 5$ are calculated. Note

TABLE 9-2 Coefficients of the Transfer Function of the Fifth-Order Butterworth Filter with a Transmission Zero

b_0	b_1	b_2			
1	0	1/9			

a_0	a_1	a_2	a_3	a_4	a_5
1	0.449941	0.101223	1.40740e-2	1.20939e-3	5.19623e-5

that the SUB estimate of model (6-26) is equivalent to that of model (6-20) with $n_a = n_b = 5$. All estimators use the constraint $b_1 = 0$ (the zero is forced to lie on the $j\omega$ axis) and $\|\theta\|_2^2 = 1$, except the LS, "full" IWLS, and SUB estimators. The LS and "full" IWLS use $b_0 = 1$ and $b_1 = 0$, and the SUB estimator uses no constraint at all. To perform a bias test, the normalized squared residuals of the mean parameter estimates are calculated for each set of 100 estimates of the model parameters,

$$b = (\langle \hat{\theta}(Z) \rangle - \theta_0)^T (\hat{C}_\theta / R)^+ (\langle \hat{\theta}(Z) \rangle - \theta_0) \qquad (9\text{-}124)$$

with θ_0 the true model parameters, $\langle \hat{\theta}(Z) \rangle$ and $\hat{C}_\theta$ the sample mean and sample covariance matrix of the data set,

$$\langle \hat{\theta}(Z) \rangle = \frac{1}{R} \sum_{r=1}^{R} \hat{\theta}^{[r]}(Z)$$
$$\hat{C}_\theta = \frac{1}{R-1} \sum_{r=1}^{R} (\hat{\theta}^{[r]}(Z) - \langle \hat{\theta}(Z) \rangle)(\hat{\theta}^{[r]}(Z) - \langle \hat{\theta}(Z) \rangle)^T \qquad (9\text{-}125)$$

and R the number of elements in the data set ($R = 100$). If $\hat{\theta}(Z)$ is an unbiased Gaussian estimate, then b is a Hotelling T^2- statistic that is

$$n_\theta \frac{(R-1)}{(R-n_\theta)} F(n_\theta, R - n_\theta) \qquad (9\text{-}126)$$

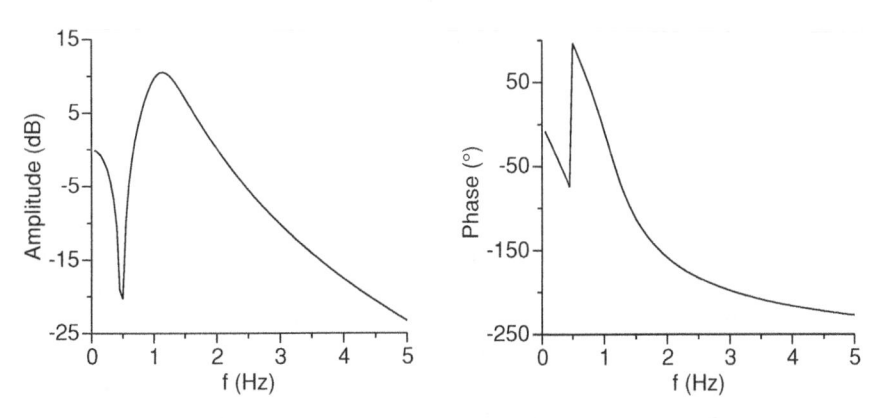

Figure 9-7. Fifth-order Butterworth filter with transmission zero (see Table 9-2): true transfer function.

TABLE 9-3 Bias Test on the Parameter Estimates: Unbiased if $b \leq 15.7$ ($b \leq 23.2$ for SUB)

Estimator	b (9-124)	Result Bias Test
LS	26000	biased
IWLS	3.9	unbiased
NLS	2.3	unbiased
LOG	5.8	unbiased
GTLS	6.4	unbiased
ML	1.7	unbiased
IQML	1.6	unbiased
BTLS	1.7	unbiased
SUB	12.5	unbiased

distributed with $n_\theta = 7$, the number of free model parameters, and $R = 100$ (see Section 16.3). Because $\hat{\theta}(Z)$ is asymptotically ($F \to \infty$) normally distributed (see Theorem 9.21, property 4), it is possible to perform a bias test on the estimates with a given confidence level. For example, the 95% percentile of (9-126) equals 15.7 for $n_\theta = 7$ (all estimates except SUB) and $R = 100$ and 23.2 for $n_\theta = 11$ (SUB) and $R = 100$. Hence, with 95% confidence, the estimates are unbiased if $b \leq 15.7$ ($b \leq 23.2$ for SUB), otherwise they are biased. According to Table 9-3, all the estimates, except the LS, are unbiased.

Using each set of 100 estimates of the model parameters, we can also calculate the relative mean square error of the transfer function estimate

$$\text{RMSE}(\underline{G}(s_k, \hat{\theta}(Z))) \approx \frac{1}{R} \sum_{r=1}^{R} |(G(s_k, \hat{\theta}^{[r]}(Z)) - G_0(s_k))/G_0(s_k)|^2 \qquad (9\text{-}127)$$

within an error of 1 dB and compare it with the Cramér-Rao lower bound on the relative transfer function error $(G(s_k, \hat{\theta}(Z)) - G_0(s_k))/G_0(s_k)$. The results are shown in Figure 9-8. It follows that BTLS has ML efficiency and that both estimators reach the Cramér-Rao lower bound. Both LS and GTLS estimators perform equally well; however, the mean square error (MSE) of the LS estimates is due to the bias (see Table 9-3) whereas that of the GTLS estimates is due to the variance (see Table 9-3). The bad performance of the LS and GTLS estimates is due to their inappropriate frequency weighting. The LOG, NLS, and SUB estimators deteriorate somewhat in efficiency w.r.t. the ML and BTLS estimates, but their efficiency is still much better than that of the GTLS method. Because of the high signal-to-noise ratio and the absence of model errors, the IWLS estimator performs as well as the NLS estimator, and the IQML and BTLS estimates coincide with the ML estimates.

9.15.2 Simulation Example 2

The goal of this simulation example is to compare different candidate starting value algorithms: LS (9-33), GTLS (9-72), WLS (9-38), and WGTLS (9-76) with weighting (9-102) and SUB (Algorithm 9.24 with $r = 32$). A sixth-order inverse Chebyshev discrete-time filter with a stopband attenuation of 40 dB and a cutoff frequency of 0.05 is selected as test example (see Figure 9-9a). The discrete-time system is excited at $F = 300$ equally spaced frequencies in the band $[0, 0.5]f_s$, with unit amplitude. Next, independent (over the frequency), zero mean, mutually uncorrelated ($\text{covar}(N_Y(k), N_U(k)) = 0$), circular complex uniformly distributed noise is added to both the input and output spectra with variances

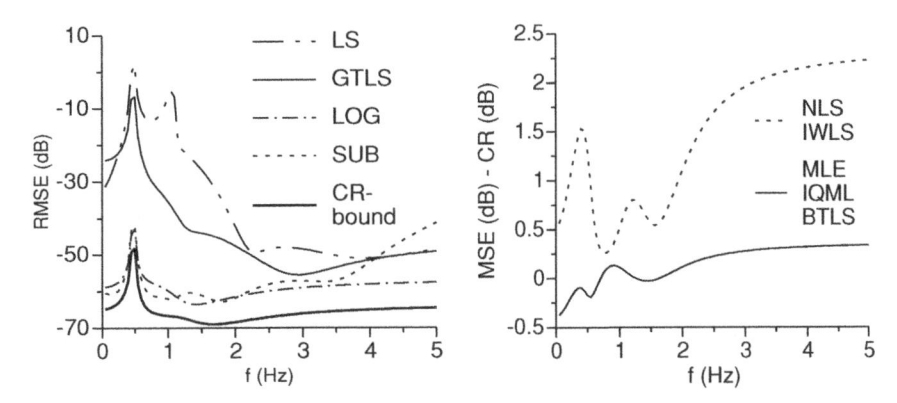

Section 9.15 ■ Illustration and Overview of the Properties

Figure 9-8. Fifth-order simulation example (see Figure 9-7 and Table 9-2): comparison of the (relative) mean square error (R)MSE of the transfer function estimate with the corresponding Cramér-Rao lower bound.

$$\text{var}(N_Y(k)) = 1 \times 10^{-6} + 9 \times 10^{-4} |G_0|^2, \quad \text{var}(N_U(k)) = 0.161 \quad (9\text{-}128)$$

The noisy frequency response function $G(z_k^{-1}) = Y(k)/U(k)$ is shown in Figure 9-9(b). The GTLS, WGTLS, and ML estimates are calculated using the constraint $\|\theta\|_2^2 = 1$, while the LS and WLS estimates use the constraint $a_0 = 1$. No constraint is used in the SUB estimate. Figure 9-9(c) and (d) show the estimated transfer functions in the band $[0, 0.25]f_s$, and Table 9-4 gives the value of the maximum likelihood cost function for the different solutions. Starting from the LS and GTLS solutions (see Figure 9-9(c)) the ML estimate gets stuck in a local minimum (see Figure 9-9(d) and Table 9-4). This is due to the fact that the LS and GTLS solutions place a transmission zero, completely out of the frequency band of interest. The two ML solutions are almost indistinguishable in the band $[0, 0.25]f_s$ but differ somewhat outside that band. Starting from the WLS, WGTLS, and SUB solutions (see Figure 9-9(c)), we find the global minimum of the ML cost function (see Figure 9-9(d) and Table 9-4). Although the WLS solution has a higher ML cost than the GTLS solution, it lies within the attraction basin of the global minimum of the ML estimator. This shows that it may be unsafe to select starting values based on the value of the ML cost.

TABLE 9-4 Maximum Likelihood (ML) Cost Function of the Starting Value Algorithms and the Corresponding ML Solution. Least squares: LS and ML (LS); generalized total least squares: GTLS and ML (GTLS); weighted least squares: WLS and ML (WLS); and weighted generalized total least squares: WGTLS and ML (WGTLS).

Estimator	LS	GTLS	WLS	WGTLS	SUB
ML cost function	2140	497	1770	354	333
Estimator (starting value)	ML (LS)	ML (GTLS)	ML (WLS)	ML (WGTLS)	ML (SUB)
ML cost function	436	431	317	317	317

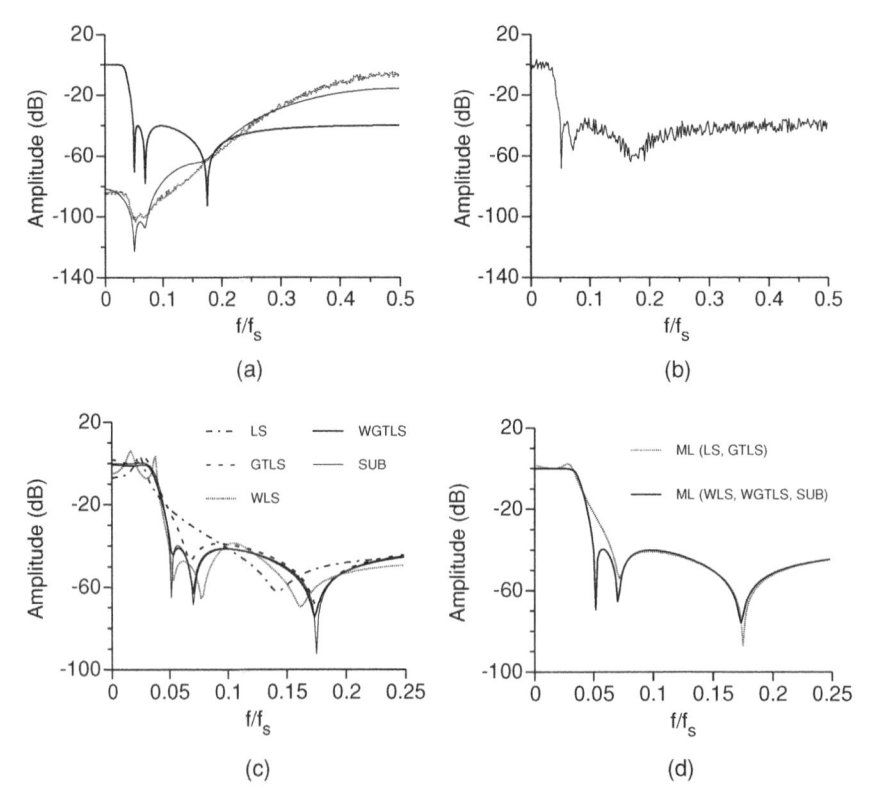

Figure 9-9. Second simulation example. (a) True frequency response function (bold line), maximum likelihood weighting (9-35) evaluated in θ_0 (solid line), and weighting (9-102) (dots); (b) noisy frequency response function; (c) LS, GTLS, WLS, WGTLS, and SUB solutions; (d) ML estimates starting from the solutions shown in (c).

9.15.3 Real Measurement Examples

Two measurement examples that illustrate the properties of the estimators particularly well are shown here. The norm constraint $\|\theta\|_2^2 = 1$ has been used in both examples for the NLS-FRF, LOG, GTLS, ML, IQML, and "full" BTLS estimators, and for the LS and IV estimators $b_0 = 1$ in the q-axis impedance model and $a_0 = 1$ in the flight flutter data model. No constraint is used in the SUB estimates. Because in both examples an improper model ($n_b > n_a$) is selected, the SUB estimates of model (6-26) are calculated using $1/G(s_k)$ instead of $G(s_k)$. The optimal value of r in the SUB Algorithm 9.25 is 61 and 35 for, respectively, the first and second examples. In the first measurement example the "full" IQML method was used, while in the second example it was necessary to relax the weighting of the IQML method ($r = 0.5$). For each measurement example, two sets of measured input and output spectra were available.

In the first measurement example (see Figures 9-10 and 9-11), the q-axis impedance of a 3.4 MW synchronous motor is modeled with a rational form in s of order $n_b = 4$ over $n_a = 3$. The measurements were carried out using a multisine excitation of 1000 A consisting of $F = 100$ frequencies logarithmically spaced in the band [12 mHz, 12 Hz]. The nonparametric noise model was obtained by analyzing $M = 30$ periods of the input and output signals. Note the particularly large dynamic range in both the amplitude and frequency band. All estimators use the averaged input-output spectra, $X(k) = M^{-1}\sum_{m=1}^{M} X^{[m]}(k)$ with $X = U$ and Y, except the IV estimator, which uses the two sets, $X_1(k) = 2M^{-1}\sum_{m=1}^{M/2} X^{[m]}(k)$ and

Section 9.15 ■ Illustration and Overview of the Properties

$X_2(k) = 2M^{-1}\sum_{m=M/2+1}^{M} X^{[m]}(k)$ with $X = U$ and Y. As expected, the LS, GTLS, and IV estimates are poor in the low-frequency range. The difference between the IQML (norm constraint), NLS-FRF, LOG, and ML estimates is almost indistinguishable. Referring to the large amplitude dynamics, the performance of the NLS-FRF estimator is remarkable. Figure 9-11 also shows the IQML solution using the constraint $b_0 = 1$. It illustrates, again, the influence of the parameter constraint on the estimates for cost functions that are not scale invariant.

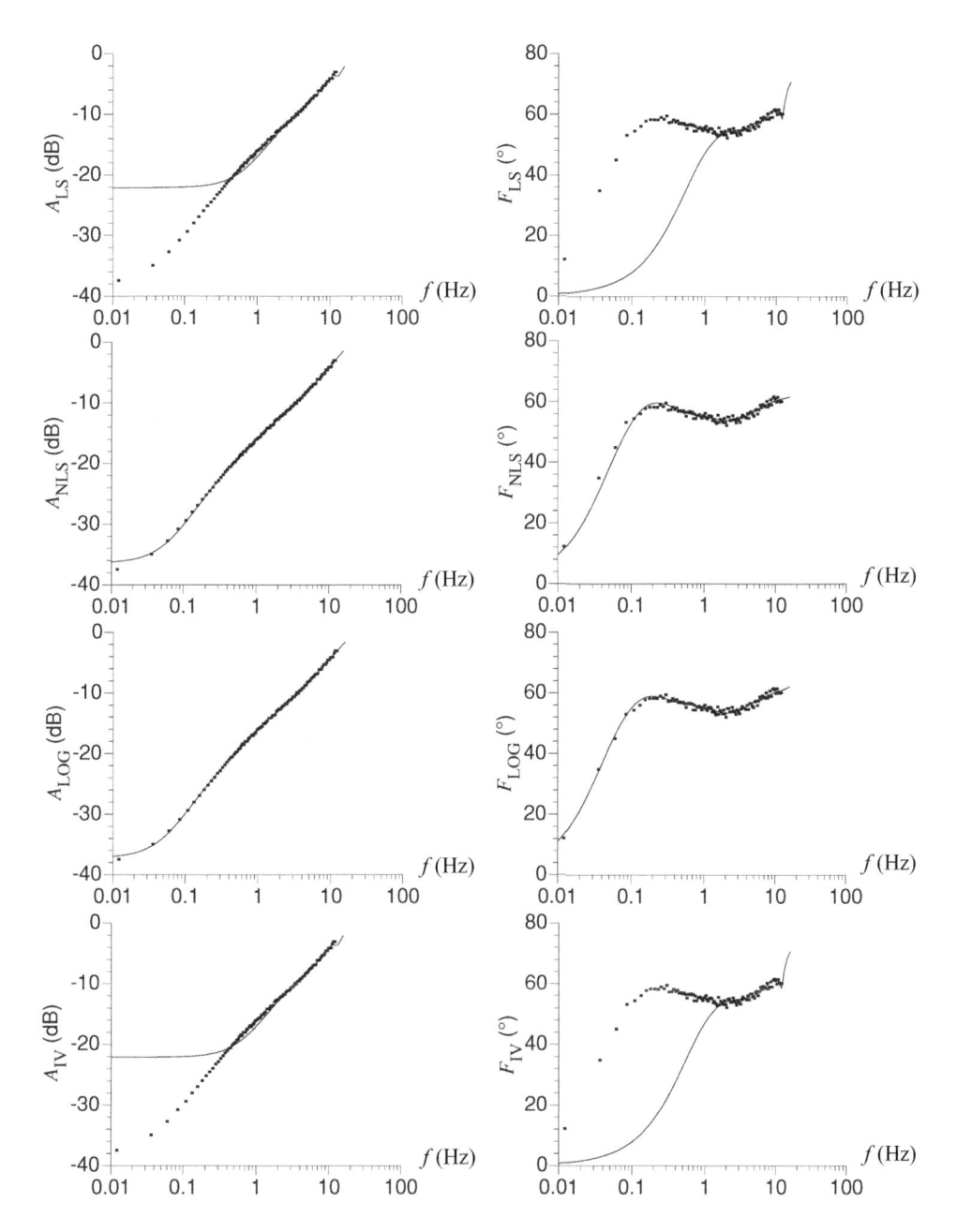

Figure 9-10. Comparison of the measurements (dots) and the estimates requiring no noise information (solid line) of the q-axis impedance of a synchronous machine (model $n_a = 3$, $n_b = 4$). Left: amplitude, and right: phase.

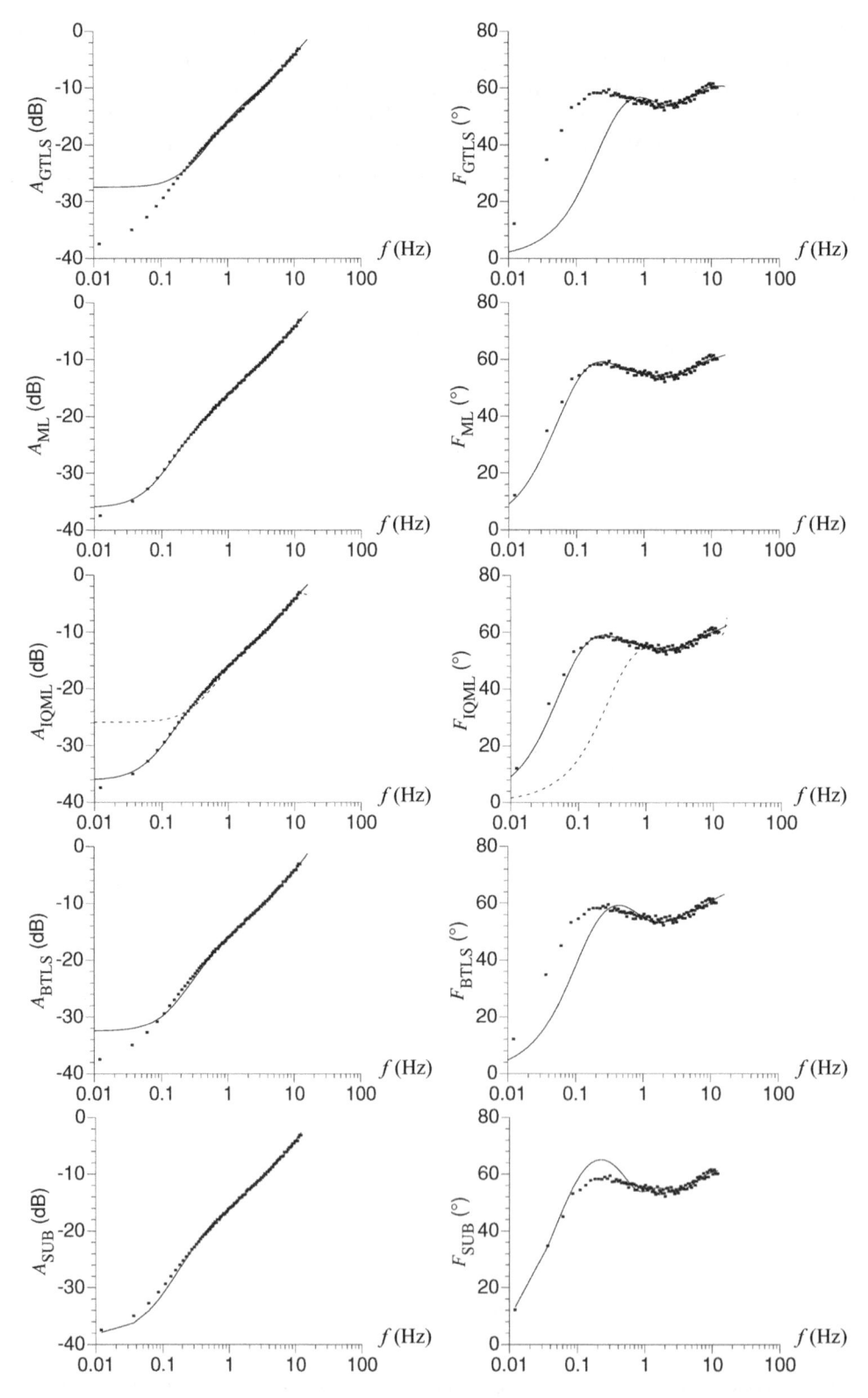

Figure 9-11. Comparison of the measurements (dots) and the estimates requiring noise information (solid line) of the q-axis impedance of a synchronous machine (model $n_a = 3$, $n_b = 4$). From left to right, amplitude and phase. For IQML, the estimates used the constraint $\|\theta\|_2^2 = 1$ (solid line) and $b_0 = 1$ (dashed line).

In the second measurement example (see Figures 9-12 and 9-13), the vibrations of the wings of an airplane are modeled with a rational form in s of order $n_b = 11$ over $n_a = 10$. LMS International (Belgium) have provided us with the experimental data. The measurements were carried out using a burst swept-sine excitation. Three sets of input-output signals of equal length are available. It is impossible to average the three measurements because they are not synchronized. 144 frequencies lie in the frequency band of interest [4 Hz, 11 Hz], giving three sets of 144 input-output DFT lines: $\{Y^{[m]}(k), U^{[m]}(k),$

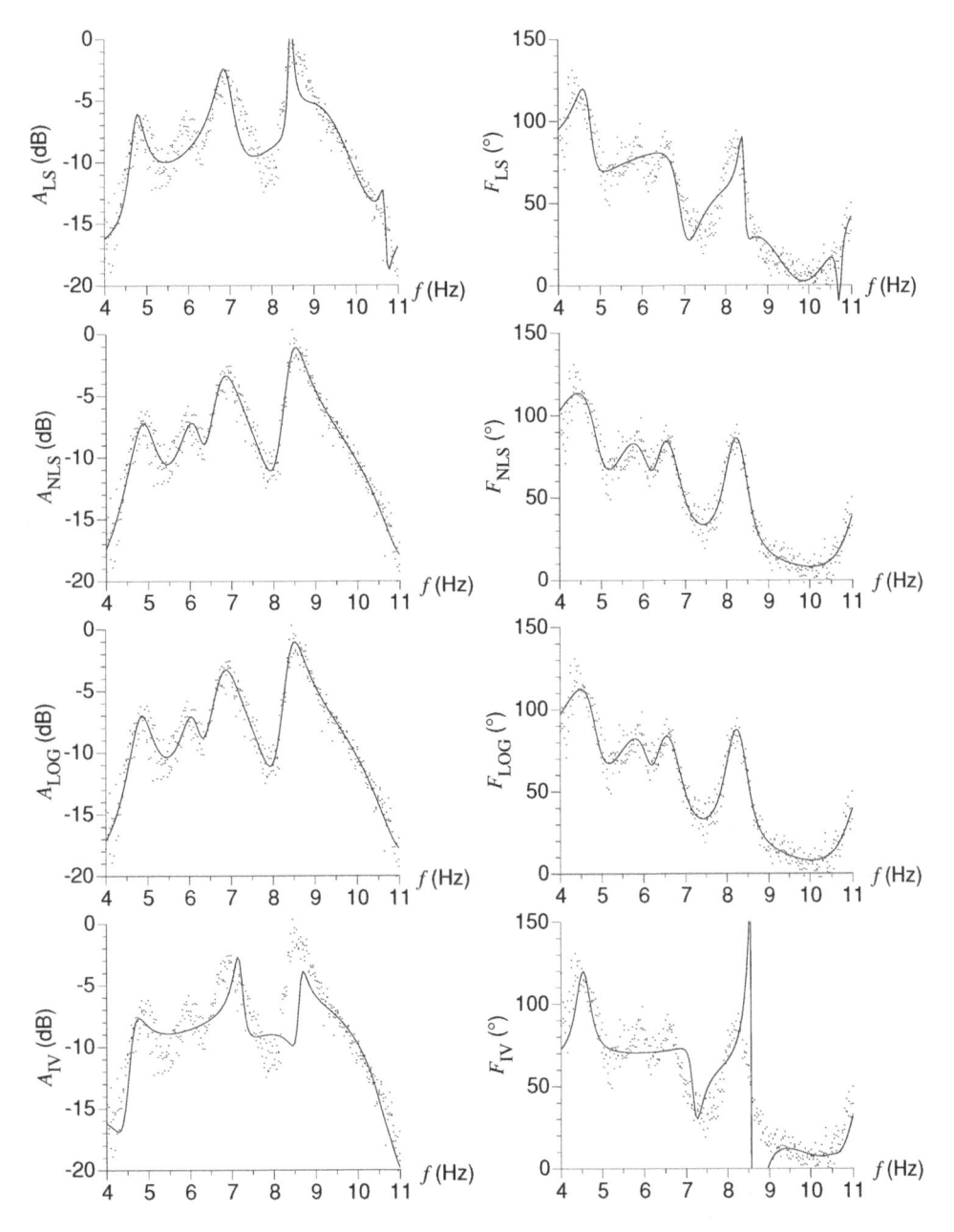

Figure 9-12. Comparison of the measurements (dots) and the estimates requiring no noise information (solid line) of the flight flutter data (model $n_a = 10$, $n_b = 11$). From left to right, amplitude and phase.

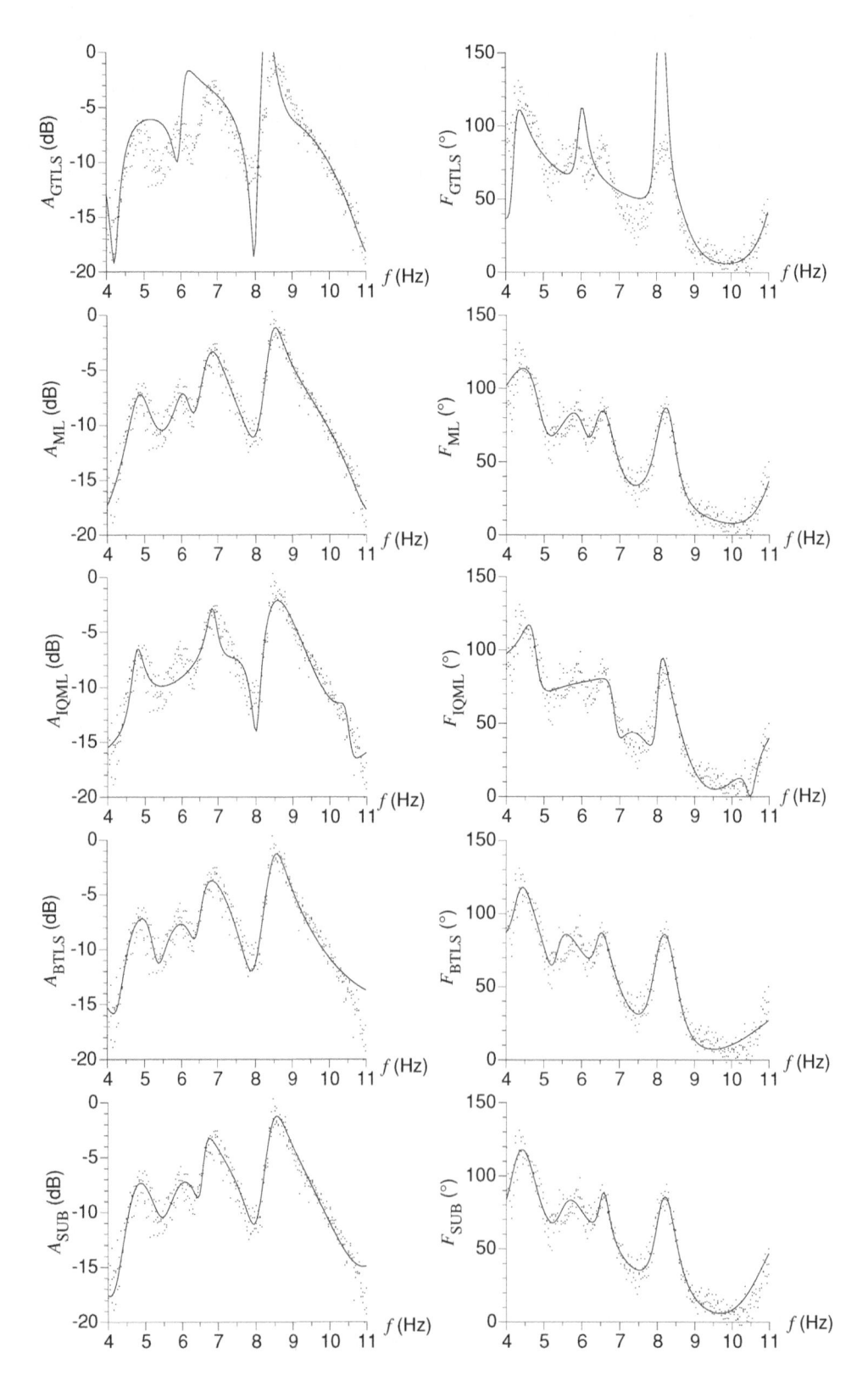

Figure 9-13. Comparison of the measurements (dots) and the estimates requiring noise information (solid line) of the flight flutter data (model $n_a = 10$, $n_b = 11$). From left to right, amplitude and phase.

$k = 1, 2, ..., 144\}$, $m = 1, 2, 3$. These $F = 3 \times 144$ input-output DFT lines are used for all the estimators except the IV and SUB estimators. The IV method uses one set as instrumental sequence, while the SUB algorithm uses the FRF measurement, averaged over the three sets $3^{-1}\sum_{m=1}^{3} Y^{[m]}(k)/U^{[m]}(k)$, $k = 1, 2, ..., 144$. The nonparametric noise model was obtained by analyzing the disturbing noise during the dead time in between consecutive bursts. Although the NLS, LOG, ML, and SUB estimates explain the measurements very well, a careful analysis of the ML cost reveals the presence of small plant model errors (a few tenths of a dB on the amplitude of the transfer function). These small modeling errors account for the better performance of the LS estimates w.r.t. the GTLS estimates. The poor quality of the LS and IV fits is due to the bad weighting of the residuals in their cost functions. Because of its more appropriate weighting of the residuals, the IQML estimator performs better than the GTLS in both measurement examples.

9.15.4 Overview of the Properties

Even if an identification method is based on sound theoretical principles, it can be put into practice only if the normal equations (9-16) or (9-18) are numerically stable and the corresponding cost function can easily be minimized. A global minimization property of the procedure or easy generation of reliable starting values is highly desirable. As constraint independence of the estimates allows the use of overparameterized models (see Chapter 20), it is important that the (equivalent) cost function of the identification method is scale invariant. Consistency and efficiency are important properties to assure that small stochastic deviations in the data do not result in, respectively, large systematic and large stochastic errors on the parameter estimates. Because in practice the true plant model does not often belong to the model set, it is desirable that the estimates are not sensitive to (small) plant modeling errors and that they converge to the noiseless solution. It is also important that the estimates are not sensitive to noise model errors, for example, incorrect noise (co)variances or noncircular complex noise. Table 9-5 gives an overview of these properties for some of the estimators discussed in the previous sections.

(i) If the noise is circular complex and if the worst case input and output signal-to-noise ratios are larger than 10 dB (see Section 9.9), then the nonlinear least squares estimator, based on frequency response function measurements (NLS-FRF), as well as the logarithmic least squares (LOG) estimator and the subspace algorithms (SUB) are "practically consistent," and when there are model errors they converge to the "practically noiseless solution." For circular complex noise $N_Z(k)$ with even pdf, the biases of the NLS-FRF and LOG estimates are a function of the fourth-order moments of the noise (co)covariances. However, if the noise is not circular complex, then the bias is a function of the second-order moments of the noise (co)variances (see Appendix 9.T). If the input is exactly known, then the NLS-FRF and SUB estimators are consistent or converge to the noiseless solution without any approximation.

(ii) The maximum likelihood (ML), generalized total least squares (GTLS), bootstrapped total least squares (BTLS), and subspace (SUB) estimates cannot be consistent if the wrong noise (co)variances are used. The resulting bias of the ML, GTLS, and BTLS estimates is proportional to the difference between the true and the actual noise (co)variances (see Appendix 9.T). To have a consistent ML estimate, it is sufficient that the actual noise covariance matrix $\hat{C}_{N_Z(k)}$ equals the true noise covariance matrix $C_{N_Z(k)} = \text{Cov}(N_Z(k))$ within a frequency-dependent scaling factor

TABLE 9-5 Overview of the Properties of Some Estimators in the General Case of Input-Output Errors

Estimator	Consistent	Conv. to Noiseless Solution	Efficiency	Prior Noise Knowledge	Global Minim. Procedure	Constraint Dependent	Sense. to Plant Model Errors	Sense. to Noise Model Errors
LS	No	No	Poor	No	Yes	Yes	Medium	—
IWLS	No	No	Poor	No	Yes	Yes	Medium	—
NLS-I/O	No	No	Medium	No	No	No	Very good	—
NLS-FRF	Yes[i]	Yes[i]	Medium	No	No	No	Very good	Very good[i]
LOG	Yes[i]	Yes	Good	No	No	No	Very good	Very good[i]
GTLS	Yes	Yes	Medium	Yes	Yes	No	Poor	Good[ii]
ML	Yes	Yes	Excellent	Yes	No	No	Very good	Good[ii]
IQML	No	No	see (iii)	Yes	Yes	Yes	Medium/Good[iii]	?
BTLS	Yes	Yes[iv]	Very good	Yes	Yes	No	Good	Good[ii]
IV	Yes	Yes	Poor	No	Yes	Yes	Medium	Very good[v]
SUB	Yes[i]	Yes	Very good	Yes	Yes	—	Good	Good[ii]

Note: the roman numbers in the table refer to the property number in the text.

Section 9.16 ■ High-Order Systems

$$\hat{C}_{N_Z(k)} = f(k) C_{N_Z(k)} \tag{9-129}$$

(see Appendix 9.T). Note that the consistency proofs of the ML, GTLS, BTLS, and SUB estimates do not require that the noise is circular complex (see Sections 9.10.3, 9.11, 9.12.3, and 9.14.3). Hence, the consistency property of the ML, GTLS, BTLS, and SUB estimates is robust w.r.t. to the circular complex noise assumption.

(iii) The efficiency of the iterative quadratic maximum likelihood (IQML) estimator strongly depends on the signal-to-noise ratio and on the presence of model errors. Its sensitivity to plant model errors is good if the parameter constraint $\|\theta\|_2^2 = 1$ is used.

(iv) BTLS converges to the noiseless solution if the limit value θ_* of the starting value $\theta^{(0)}$ is independent of the noise level υ.

(v) If the disturbing noise on the instrumental sequences is independent of the disturbing noise on the measurements, then the instrumental variables (IV) estimator is consistent and, when there are plant model errors, converges to the noiseless solution, irrespective of the true noise model. Otherwise, the estimate depends on the correlation between the disturbing noise on the instrumental sequences and the disturbing noise on the measurements.

9.16 HIGH-ORDER SYSTEMS

When identifying higher order transfer function models (9-7), (9-8) (typical $n_a, n_b > 30$) with the rational forms (6-20), (6-35), the condition number of the normal equations (9-18) can become so large that it is impossible to calculate a reliable solution within the available arithmetic precision. Therefore, to tackle high-order systems the numerator and denominator polynomials of the plant and transient models are expanded in scalar or vector orthogonal polynomials (6-25), (6-38), which are chosen such that they minimize (improve) the condition number. The polynomials are orthogonal w.r.t. some inner product defined by the cost function and, hence, dependent on the estimator used. The whole process will be explained for the IWLS estimator (9-38) in Sections 9.16.1 and 9.16.2 and afterward generalized to the other estimators in Section 9.16.3. In what follows, we assume that the parameter constraint $a_{n_p} = 1$ is used. In order to simplify the notations, we will limit the discussion to transfer function model (9-7). Generalization of the results to model (9-8) is straightforward.

9.16.1 Scalar Orthogonal Polynomials

The IWLS cost function (9-38) can be written as the sum of three terms

$$\sum_{k=1}^{F} W^2(\Omega_k, \theta^{(i-1)}) |Y(k)|^2 |A(\Omega_k, \theta^{(i)})|^2 + \sum_{k=1}^{F} W^2(\Omega_k, \theta^{(i-1)}) |U(k)|^2 |B(\Omega_k, \theta^{(i)})|^2 \\ -2\mathrm{Re}(\sum_{k=1}^{F} W^2(\Omega_k, \theta^{(i-1)}) Y(k) \bar{U}(k) A(\Omega_k, \theta^{(i)}) \bar{B}(\Omega_k, \theta^{(i)})) \tag{9-130}$$

Under the identifiability conditions of Theorem 6.9 the Jacobian matrix corresponding to the IWLS cost function (9-38),

$$J_{[k, r]}(\theta^{(i)}, Z) = W(\Omega_k, \theta^{(i-1)}) \partial e(\Omega_k, \theta, Z(k)) / \partial \theta_{[r]}^{(i)} \tag{9-131}$$

where $e(\Omega_k, \theta, Z(k))$ is given by (9-9) and $\theta^T = [a_0 a_1 \ldots a_{n_a-1} b_0 b_1 \ldots b_{n_b}]$ with $a_{n_a} = 1$, has full rank: $\text{rank}(J_{re}) = n_a + n_b + 1$. Hence, each of the first two terms in (9-130) defines an inner product of scalar polynomials

$$\langle x(\Omega), y(\Omega) \rangle_a = \text{Re}(\sum_{k=1}^{F} W^2(\Omega_k, \theta^{(i-1)}) |Y(k)|^2 x(\Omega) \bar{y}(\Omega))$$
$$\langle t(\Omega), z(\Omega) \rangle_b = \text{Re}(\sum_{k=1}^{F} W^2(\Omega_k, \theta^{(i-1)}) |U(k)|^2 t(\Omega) \bar{z}(\Omega))$$
(9-132)

with $x(\Omega)$, $y(\Omega)$ polynomials of order smaller than or equal to n_a and $t(\Omega)$, $z(\Omega)$ polynomials of order smaller than or equal to n_b (proof: see Lemma 15.6). Using definitions (9-132) and

$$A(\Omega, \theta) = \sum_{r=0}^{n_a} a_r p_r(\Omega), \quad B(\Omega, \theta) = \sum_{r=0}^{n_b} b_r q_r(\Omega) \tag{9-133}$$

the matrix $M = \text{Re}(J^H(\theta^{(i)}, Z) J(\theta^{(i)}, Z))$ of the normal equation (9-17) becomes

$$M_{[r+1, s+1]} = \langle p_s(\Omega), p_r(\Omega) \rangle_a \quad r, s = 0, \ldots, n_a$$
$$M_{[r+n_a+2, s+n_a+2]} = \langle q_s(\Omega), q_r(\Omega) \rangle_b \quad r, s = 0, \ldots, n_b$$
(9-134)

The polynomials $p_r(\Omega)$, $r = 0, 1, \ldots, n_a$, and $q_r(\Omega)$, $r = 0, 1, \ldots, n_b$, are calculated via a Gram-Schmidt orthogonalization (see Section 15.11) using inner products $\langle \, , \, \rangle_a$ and $\langle \, , \, \rangle_b$, respectively. Hence, $\langle p_s(\Omega), p_r(\Omega) \rangle_a = \delta(s-r)$, $\langle q_s(\Omega), q_r(\Omega) \rangle_b = \delta(s-r)$ and

$$\text{Re}(J^H(\theta^{(i)}, Z) J(\theta^{(i)}, Z)) = \begin{bmatrix} I_{n_a} & C_1 \\ C_1^T & I_{n_b+1} \end{bmatrix} \tag{9-135}$$

It can be shown that (9-135) is best conditioned: no other scalar polynomial bases for the numerator and denominator of the rational transfer function model can be found resulting in a better conditioned form $\text{Re}(J^H(\theta^{(i)}, Z) J(\theta^{(i)}, Z))$ (Forsythe and Strauss, 1955; Rolain et al., 1995). The IWLS solution is calculated by not using the special structure (9-135) but by solving the overdetermined set of equations (9-18). Proceeding in this way, the solution is insensitive to a loss of orthogonality among the computed basis polynomials. In Richardson and Formenti (1982), the scalar orthogonal polynomials were applied for the first time to improve the numerical conditioning of the linear least squares method ((9-130) with $W(\Omega_k, \theta^{(i-1)}) = 1$) in modal analysis problems ((9-33) with equation error (9-9) and $\Omega = s$).

9.16.2 Vector Orthogonal Polynomials

It is easy to verify that the IWLS cost function (9-38) can also be written as

$$\sum_{k=1}^{F} W^2(\Omega_k, \theta^{(i-1)}) \begin{bmatrix} A(\Omega_k, \theta^{(i)}) \\ B(\Omega_k, \theta^{(i)}) \end{bmatrix}^H \begin{bmatrix} |Y(k)|^2 & -\bar{Y}(k)U(k) \\ -Y(k)\bar{U}(k) & |U(k)|^2 \end{bmatrix} \begin{bmatrix} A(\Omega_k, \theta^{(i)}) \\ B(\Omega_k, \theta^{(i)}) \end{bmatrix} \tag{9-136}$$

Under the identifiability conditions of Theorem 6.9, the Jacobian matrix corresponding to the IWLS cost function (9-38) is

$$J_{[k, r]}(\theta^{(i)}, Z) = W(\Omega_k, \theta^{(i-1)}) \partial e(\Omega_k, \theta, Z(k)) / \partial \theta_{[r]}^{(i)} \tag{9-137}$$

where $e(\Omega_k, \theta, Z(k))$ is given by (9-9) and $\theta^T = [a_0 a_1 \ldots a_{n_a + n_b}]$ with $a_{n_a + n_b + 1} = 1$ has full rank: $\text{rank}(J_{\text{re}}) = n_a + n_b + 1$. Hence, the cost function (9-136) defines an inner product of vector polynomials

$$\langle x(\Omega), y(\Omega) \rangle = \text{Re}(\sum_{k=1}^{F} W^2(\Omega_k, \theta^{(i-1)}) y^H(\Omega_k) \begin{bmatrix} |Y(k)|^2 & -\bar{Y}(k)U(k) \\ -Y(k)\bar{U}(k) & |U(k)|^2 \end{bmatrix} x(\Omega)) \tag{9-138}$$

where $x(\Omega)$ and $y(\Omega)$ are 2 by 1 vector polynomials of order smaller than or equal to $n_a + n_b + 1$ (proof: see Lemma 15.7). The vector polynomials $x_r^T(\Omega) = [p_r(\Omega) \; q_r(\Omega)]$, $r = 0, 1, \ldots, n_a + n_b + 1$, are calculated via a Gram-Schmidt orthogonalization (see Section 15.11) using inner product (9-138). Hence we have

$$\langle x_s(\Omega), x_r(\Omega) \rangle = \delta(s - r) \tag{9-139}$$

$$\text{Re}(J^H(\theta^{(i)}, Z) J(\theta^{(i)}, Z)) = I_{n_a + n_b + 1} \tag{9-140}$$

Clearly, (9-140) has the smallest possible condition number: $\kappa(\text{Re}(J^H J)) = 1$. The IWLS solution is given by

$$G(\Omega, \hat{\theta}_{\text{IWLS}}(Z)) = q_{n_a + n_b + 1}(\Omega) / p_{n_a + n_b + 1}(\Omega) \tag{9-141}$$

(see Appendix 9.U). Note that solution (9-141) explicitly makes use of the orthogonality of the polynomial basis and, hence, is sensitive to a loss of orthogonality among the computed basis polynomials. A numerically stable and time-efficient implementation of the orthogonalization procedure can be found in Van Barel and Bultheel (1994) for discrete-time models ($\Omega = z^{-1}$). Following the same lines of Bultheel and Van Barel (1995), the time efficient implementation of the orthogonalization procedure has been extended in Bultheel et al. (2005) to multivariable continuous-time models ($\Omega = s$).

9.16.3 Application to the Estimators

Because the LS (9-33), IWLS (9-37), IQML (9-97), and WLS (9-102) estimators are special cases of the general IWLS estimator (9-38), the calculation of the orthogonal polynomials follows the same lines as in Sections 9.16.1 and 9.16.2. They are chosen such that they minimize the condition number of the normal equation (9-18).

For all the estimators whose (equivalent) cost function is a nonquadratic function of the model parameters, it is impossible to generate, in each iteration step of the Newton-Gauss procedure, a set of orthogonal polynomials that minimize the condition number of the normal equation (9-18). Indeed, the big difference between the IWLS solution and the nonlinear minimization scheme is that the former generates a solution in each iteration step, eventually based on an initial guess, and the latter generates an increment w.r.t. the initial guess. Because the initial guess and the increment should be calculated in the same polynomial basis, it is impossible to minimize the condition number of (9-18). However, it is still possible to make the solution well conditioned. This is done in the following way.

As already emphasized in Section 9.10, the solution of the total least squares estimators, GTLS (9-72), WGTLS (9-76), and BTLS (9-98), is not calculated via the nonlinear minimization scheme (9-18), but via the GSVD of the matrix pair $(W_{\text{Re}} J_{\text{re}}(Z), C)$. Compared with the IWLS cost function (9-38), $J(Z)$ is the Jacobian of the error vector $J(Z) = \partial e(\theta, Z)/\partial \theta^{(i)}$, W is a diagonal matrix with $W_{[k,k]} = W(\Omega_k, \theta^{(i-1)})$, and C is a square root of the column covariance matrix of $W_{\text{Re}} j_{\text{re}}(N_Z)$, with $j(N_Z) = J(Z) - J(Z_0)$. The orthogonal polynomial basis minimizing the condition number of the IWLS estimator is used for the corresponding TLS estimator with $W = I_F$ for GTLS, $W_{[k,k]} = W(\Omega_k, \theta^{(i-1)})$ for WGTLS, and $W_{[k,k]} = \sigma_e^{-1}(\Omega_k, \theta^{(i-1)})$ for BTLS. This choice minimizes the condition number of $W_{\text{Re}} J_{\text{re}}$.

For the NLS-IO (9-43), NLS-FRF (9-47), LOG (9-55), and ML (9-80) estimators we use the orthogonal basis of the starting value algorithm. This choice leads to well, but not best, conditioned normal equations.

9.16.4 Notes

The IWLS solution calculated in the vector orthogonal polynomial basis is given by the highest order vector polynomial (9-141): $a_0 = a_1 = \cdots = a_{n_a + n_b} = 0$ and $a_{n_a + n_b + 1} = 1$. If this solution is already of high quality, then the other estimators, (W)GTLS, BTLS, NLS-IO, NLS-FRF, LOG, and ML, calculated in the same basis, will only marginally perturb the solution: $|a_0|, |a_1|, \ldots, |a_{n_a + n_b}| \ll 1$ and $a_{n_a + n_b + 1} = 1$.

Because the inner products (9-132) and (9-138) depend on the measurements, the orthogonal basis depends on the disturbing noise. Therefore, the estimated numerator and denominator orthogonal polynomial coefficients of different experiments cannot be compared. Also, the properties of Theorem 9.21 cannot be applied to the estimated numerator and denominator coefficients. However, because the proof of Theorem 9.21 is independent of the parameterization, its properties are still valid for the invariants of the model, for example, the poles and the zeros.

Evaluating orthogonal polynomials at a particular frequency, or calculating the roots, should always be done through the recursion formula used to construct the orthogonal basis AND NOT via the expansion in powers of Ω, which is numerically ill conditioned for high orders (see Sections 15.11 and 15.12 and Exercise 1.14.). To preserve the numerical stability, the calculations for continuous-time systems ($\Omega = s$ or $\sqrt{s}$) should be performed using the normalized frequencies (see Section 9.4).

9.17 SYSTEMS WITH TIME DELAY

The main difficulty of estimating systems with an unknown time delay (plant model (6-29) or (6-30)) is that the corresponding NLS-IO (9-43), NLS-FRF (9-47), LOG (9-55), and ML (9-80) cost functions teem with local minima. A "sufficiently high" quality starting value for the delay is necessary to avoid the local minima. In time domain reflectometry, the time difference between the edges of the excitation pulse and the reflected (transmitted) pulse is a good initial guess of the delay (Pintelon and Van Biesen, 1990). This approach is no longer possible for overlapping pulses, periodic and random excitations. In these cases, a starting value can be obtained via the sample cross-correlation $\hat{R}_{yu}(\tau)$ between the output and the input signals,

Section 9.18 ■ Identification in Feedback

$$\hat{\tau} = \arg\max_{\tau} |\hat{R}_{yu}(\tau)| = \arg\max_{\tau} \left| \frac{1}{N-\tau} \sum_{t=\tau}^{N-1} \tilde{y}(t)\tilde{u}(t-\tau) \right| \qquad (9\text{-}142)$$

where $\tilde{x}(t) = x(t) - N^{-1}\sum_{t=0}^{N-1} x(t)$ with $x = y, u$, or via the mean slope of the unwrapped phase of the measured frequency response function

$$\hat{\tau} = -\frac{1}{k_2 - k_1} \sum_{k=k_1}^{k_2-1} \frac{\angle G(\Omega_{k+1}) - \angle G(\Omega_k)}{\omega_{k+1} - \omega_k} \qquad (9\text{-}143)$$

where $[\omega_{k_1}, \omega_{k_2}]$ defines the passband of the system. In both cases, the delay is an estimate of the sum of the true delay of the plant and the slope of the linearized phase of the rational part of the plant. Using the initial guess (9-142) or (9-143) as fixed value in the plant model (6-29) or (6-30), we can calculate starting values for the numerator and denominator coefficients in (6-29) or (6-30), for example, through the IWLS, WGTLS, or SUB estimates (see Sections 9.12.4 and 9.14).

9.18 IDENTIFICATION IN FEEDBACK

Figure 9-14 shows a block diagram of a basic linear feedback experiment. According to the nature of the reference signal $r(t)$, there is a subtle difference between what is considered as the true excitation of the plant and the disturbing noise. If the reference signal is *periodic,* then any deviation from the periodic behavior is considered as noise. The true input-output DFT spectra in (9-1) are then given by

$$\begin{aligned} Y_0(k) &= \frac{G_0(\Omega_k)}{1 + G_0(\Omega_k) M_0(\Omega_k)} R(k) \\ U_0(k) &= \frac{1}{1 + G_0(\Omega_k) M_0(\Omega_k)} R(k) \end{aligned} \qquad (9\text{-}144)$$

with $R(k)$ the DFT spectrum of the reference signal, where $G_0(\Omega)$, $M_0(\Omega)$ are the true plant and controller transfer functions, respectively. The frequency domain errors $N_U(k)$ and $N_Y(k)$ in (9-1) are related to the DFT spectra of the disturbing noise sources in Figure 9-14 as

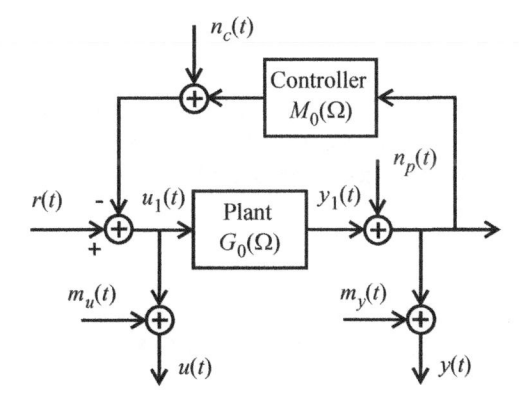

Figure 9-14. Feedback experiment with $r(t)$ the reference signal, $m_u(t)$, $m_y(t)$ the measurement noise sources, $n_p(t)$ the process noise, $n_c(t)$ the controller noise, and $u_1(t)$, $y_1(t)$ the input and output of the plant.

$$N_Y(k) = M_Y(k) + \frac{N_P(k) - G_0(\Omega_k)N_C(k)}{1 + G_0(\Omega_k)M_0(\Omega_k)}$$
$$N_U(k) = M_U(k) - \frac{N_C(k) + C_0(\Omega_k)N_P(k)}{1 + G_0(\Omega_k)M_0(\Omega_k)} \qquad (9\text{-}145)$$

Clearly, the disturbances $N_U(k)$ and $N_Y(k)$ are mutually correlated and are independent of the true input $U_0(k)$. Assumption 9.3 or 9.4 is fulfilled and, hence, Theorem 9.21 is valid for periodic excitations and systems in feedback. If the reference signal is *arbitrary*, then the controller noise $n_c(t)$ and the feedback part of the process noise $n_p(t)$ are indistinguishable from the contribution of the reference signal $r(t)$ to the excitation $u_1(t)$. Hence, the true input-output signals are $u_0(t) = u_1(t)$ and $y_0(t) = y_1(t)$. The technical difficulty arising, especially if the noise model is unknown (see Chapter 10), is that the true input signal $u_0(t)$ is correlated with the process noise $n_p(t)$ and, hence, also with the disturbing error at the output.

9.19 MODELING IN THE PRESENCE OF NONLINEAR DISTORTIONS

The goal of a linear identification experiment in the presence of nonlinear distortions can be the identification of the true underlying linear system, or the best linear approximation of the overall system, including the nonlinearities. The first case is useful for physical modeling and, if the system behaves linearly for small inputs, then crest factor optimized excitation signals are most suited for the identification experiment (see Chapter 5). The second case is useful if a linear input-output description is required for a certain class of excitation signals. In this section, we handle the second case. The validity (utility) of the linear model is application dependent and should be established in practice.

The identification starts from measured input-output DFT spectra of a time domain experiment with a random phase multisine (see Figure 9-15). Assuming that an integer number of periods of the steady-state response are observed, we have

$$Y(k) = G_{BLA}(s_k)U_0(k) + N_Y(k)$$
$$U(k) = U_0(k) + N_U(k) \qquad (9\text{-}146)$$

with $G_{BLA}(s)$ the best linear approximation (see Section 6.8), $N_U(k) = M_U(k)$, and $N_Y(k) = N_P(k) + Y_S(k) + M_Y(k)$. The properties of the stochastic nonlinear contributions $Y_S(k)$ are quite similar to those of the measurement and process noise in a time domain experiment (see Sections 6.8 and 9.6). Therefore, Theorem 9.21, where $G_0(s)$ is replaced by $G_{BLA}(s)$, remains valid (proof: see Appendix 9.V).

9.20 MISSING DATA

The form of the output $Y(\Omega, \theta)$, predicted by the model, basically changes if input and/or output samples are missing. Instead of (9-7) and (9-8), we get the following from transfer function model (6-46):

$$Y^m(\Omega_k, \Theta) = G(\Omega_k, \theta)U^m(k) + T_G(\Omega_k, \theta) + z_k^{-K_u}G(\Omega_k, \theta)I_u(z_k^{-1}, \psi) - z_k^{-K_y}I_y(z_k^{-1}, \psi) \quad (9\text{-}147)$$

Section 9.20 ■ Missing Data

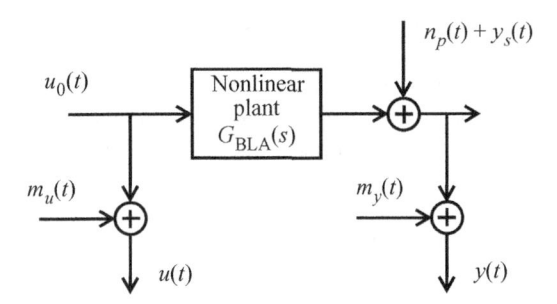

Figure 9-15. Time domain experiment: a random phase multisine is applied to a nonlinear plant $y(t) = G[u(t)]$. The DFT spectra of N observed input-output samples are calculated. $F = O(N)$ DFT frequencies of the input-output spectra are retained. $m_u(t)$ and $m_y(t)$ are the input and output measurement errors, $n_p(t)$ is the process noise, and $y_s(t)$ is the stochastic nonlinear contribution having the same periodicity as the excitation $u_0(t)$.

with $\Theta^T = [\theta^T \psi^T]$, ψ the vector containing the M_u missing input samples and M_y missing output samples, $Y^m(\Omega_k, \Theta)$ the output predicted by the model, $U^m(k)$ the DFT spectrum of the missing input data set, and $\Omega = z^{-1}$ or s (see Section 6.3.3). Inspired by the maximum likelihood solution (9-82), we can construct the following weighted nonlinear least squares (WNLS) estimator:

$$V_{\text{WNLS}}(\Theta, Z^m) = \sum_{k=0}^{N-1} \frac{|Y^m(k) - Y^m(\Omega_k, \Theta)|^2}{\sigma_Y^2(\Omega_k, \theta)} \quad (9\text{-}148)$$

with Z^m the missing data set and $Y^m(k)$ the DFT spectrum of the missing output data set. $\sigma_Y^2(\Omega_k, \theta)$ is the variance of the output error (9-45) calculated by using the (co)variances of the complete disturbing noise sequence (no missing samples), for example, $\sigma_U^2(k) = \text{var}(N_U(k))$ and $\sigma_U^2(k) \neq \text{var}(N_U^m(k))$. Minimizing (9-148) w.r.t. Θ gives the WNLS estimate $\hat{\Theta}_{\text{WNLS}}(Z^m)$ of the plant model parameters θ and the $M_u + M_y$ missing input and/or output samples ψ. To obtain starting values for the plant model parameters θ, the missing data are put equal to zero ($\psi = 0$ in (9-147)). If the number of consecutive missing samples is small, then better starting values for ψ can be obtained via linear interpolation of the known samples. This reduces the risk of being trapped in local minima (cost function (9-148) has more local minima than the problem without missing data).

The properties of $\hat{\theta}_{\text{WNLS}}(Z^m)$ can be studied, assuming that the fraction of the missing samples does not increase with the amount of data

$$\frac{M_u + M_y}{2N} = O(N^0) \quad (9\text{-}149)$$

To show the consistency of $\hat{\theta}_{\text{WNLS}}(Z^m)$, more restrictive assumptions are required than for the problem without missing data. In addition to the assumptions of Section 9.6.6 it is necessary that Assumptions 9.18, 9.19 and condition (9-149) are fulfilled (see Appendix 9.W). Note that the consistency proof relies entirely on the knowledge of the noise (co)variances. If the noise model is unknown and a parametric noise model is identified, then the estimates are no longer consistent. Hence, getting a consistent noise model is the key to the solution of the missing data problem.

More information about the missing output data problem in discrete-time modeling can be found in the literature on time series analysis (see, for example, Little and Rubin, 1987) and system identification (see, for example, Isaksson, 1993; Goodwin and Adams, 1994; Albertos et al., 1999). By considering the missing inputs as unknown parameters, the missing input data problem in discrete-time modeling can be solved by classical prediction error methods (Ljung, 1999).

9.21 MULTIVARIABLE SYSTEMS

Plant models (9-7) and (9-8) remain valid for multivariable systems. $Y(\Omega_k, \theta)$ is then the modeled n_y by 1 output vector, $G(\Omega_k, \theta)$ the n_y by n_u transfer function matrix, $U(k)$ the n_u by 1 vector of the input DFT spectra, and $T_G(\Omega_k, \theta)$ the n_y by 1 vector of the plant transients.

Following the lines of the scalar case, the multivariable versions of the IWLS (9-37), NLS-IO (9-43), NLS-FRF (9-47), LOG (9-55), and ML (9-80) estimators can be constructed for any of the multivariable parameterizations of $G(\Omega_k, \theta)$ and $T_G(\Omega_k, \theta)$ described in Section 6.6. For example, the IWLS estimator using a common denominator (Bayard, 1994a; Verboven et al., 2005), a left and right matrix fraction (de Callafon et al., 1996; Gaikwad and Rivera, 1997), or a state space (Bayard, 1994b) parameterization; the GTLS estimator using a common denominator (Verboven et al., 2004), or a left matrix fraction (Pintelon et al., 1998) parameterization; and the ML estimator using a common denominator (Guillaume et al., 1992a, 1996b; Peeters et al., 2000), a left matrix fraction (Pintelon et al., 1998), or a state space (Wills et al., 2009) parameterization. The numerical minimization of the cost function using the Newton-Gauss scheme (9-18) is somewhat more subtle for the multivariable estimators than for the scalar case (see Section 12.3.3 and Guillaume and Pintelon, 1996 for more details).

The IWLS (9-38), WGTLS (9-76), IQML (9-97), BTLS (9-98), and IV (9-108) estimators need a parameterization leading to an equation error that is linear in the model parameters. If the identification starts from measured input-output DFT spectra, then the common denominator model (6-53) and left matrix fraction description (6-54) are suitable. The equation errors (9-9) and (9-10) remain valid with $A(\Omega_k, \theta)$ the denominator polynomial (common denominator model (6-53)) or the n_y by n_y denominator matrix polynomial (left matrix fraction description (6-54)), $Y(k)$ the n_y by 1 vector of the output DFT spectra, $B(\Omega_k, \theta)$ the n_y by n_u numerator matrix polynomial, and $I(\Omega_k, \theta)$ the n_y by 1 vector of the plant equivalent initial conditions. If the identification starts from the measured frequency response matrix (see Section 2.7) then, besides the common denominator model and left matrix fraction description, we can also use the right matrix fraction description (6-55). The corresponding $n_y n_u \times 1$ equation error vectors are

$$e(\Omega_k, \theta, Z(k)) = \mathrm{vec}(A(\Omega_k, \theta)G(\Omega_k) - B(\Omega_k, \theta)) \qquad (9\text{-}150)$$

for the common denominator model and the left matrix fraction description, and

$$e(\Omega_k, \theta, Z(k)) = \mathrm{vec}(G(\Omega_k)A(\Omega_k, \theta) - B(\Omega_k, \theta)) \qquad (9\text{-}151)$$

for the right matrix fraction description, where $\mathrm{vec}(X)$ puts the columns of X on top of each other. Note that constructing an appropriate frequency weighting of the equation errors is somewhat more subtle for the multivariable WGTLS and BTLS estimators than for the scalar case (see Pintelon et al., 1998 for more details).

The subspace algorithms (see Section 9.14) require a multivariable version of model equation (9-120). It is easy to verify that (9-120) remains valid if **Y** and **U** in (9-116) are replaced by

$$\mathbf{Y} = \begin{bmatrix} W_r(1) \otimes Y(1) & W_r(2) \otimes Y(2) & \ldots & W_r(F) \otimes Y(F) \end{bmatrix}$$
$$\mathbf{U} = \begin{bmatrix} W_r(1) \otimes U(1) & W_r(2) \otimes U(2) & \ldots & W_r(F) \otimes U(F) \end{bmatrix}$$
(9-152)

with $\otimes$ the Kronecker product (see Section 15.7), and similarly for $\mathbf{N_Y}$ and $\mathbf{N_U}$. The multivariable versions of Algorithms 9.24 and 9.25 can be found in McKelvey et al. (1996) and Van Overschee and De Moor (1996a).

9.22 TRANSFER FUNCTION MODELS WITH COMPLEX COEFFICIENTS

Typical applications of transfer function modeling with complex coefficients can be found in nuclear magnetic resonance modeling (see Kumaresan et al., 1990 and Section 6.4) and the identification of rotor bearing systems (see Lee, 1993; Peeters et al., 2000). Because the cost functions of all the estimators for rational transfer function models have been developed without using the fact that θ is real, they remain valid for complex parameters θ. Also, the properties of the estimators remain the same. Indeed, to see this, it is sufficient to replace $\theta \in \mathbb{C}^{n_\theta}$ by $\theta_{re} \in \mathbb{R}^{2n_\theta}$ and to note that Theorem 9.21 is valid, independent of the particular parameterization chosen.

If $\theta \in \mathbb{C}^{n_\theta}$ is replaced by $\theta_{re} \in \mathbb{R}^{2n_\theta}$, then all the formulas for the real case apply to the complex case, except that the real part in the definition of the inner products (9-132), (9-138), and (9-249) should be removed. The modification of the inner product changes only the recursion formula used to calculate the orthogonal polynomial basis (see Section 15.11). For example, the normal equation (9-18) becomes

$$J_{re}(\theta_{re}^{(i-1)}, Z)\Delta\theta_{re}^{(i)} = -\varepsilon_{re}(\theta_{re}^{(i-1)}, Z) \quad (9\text{-}153)$$

with $J_{re}(\theta_{re}, Z) = \partial\varepsilon_{re}(\theta_{re}, Z)/\partial\theta_{re}$. If the weighted residual $\varepsilon(\theta, Z)$ is an analytic function of θ, then (9-153) is equivalent to

$$J(\theta^{(i-1)}, Z)\Delta\theta^{(i)} = -\varepsilon(\theta^{(i-1)}, Z) \quad (9\text{-}154)$$

with $J(\theta, Z) = \partial\varepsilon(\theta, Z)/\partial\theta$ (see Appendix 9.X). This is the case for IWLS (9-38), NLS-IO (9-43), NLS-FRF (9-47), LOG (9-55), IQML (9-97), and IV (9-108) estimators. This is not true for the ML estimator because $\sigma_e(\Omega, \theta)$ in (9-84) and $\sigma_Y(\Omega, \theta)$ in (9-85) are not analytic functions of θ. Hence, the ML normal equation (9-153) cannot be simplified to (9-154).

The solution of the WGTLS (9-76) and BTLS (9-98) estimators is calculated as the right generalized singular vector corresponding to the smallest generalized singular value of the matrix pair $(W_{Re}J_{re}(Z), C)$, with W the diagonal weighting matrix (9-74), $J(Z) = \partial e(\theta, Z)/\partial \theta$, and C a square root of the column covariance matrix of $W_{Re} j_{re}(N_Z)$, with $j(N_Z) = J(Z) - J(Z_0)$. Because $e(\theta, Z)$ is an analytic function of θ, the solution can also be calculated as the right generalized singular vector corresponding to the smallest generalized singular value of the matrix pair $(WJ(Z), C_c)$ with C_c a square root of the column covariance matrix of $Wj(N_Z)$ (see Appendix 9.Y).

9.23 EXERCISES

9.1. Show that the signals defined in Assumption 9.11 are quasi-stationary (9-22) (hint: assume that an integer number of periods is observed for periodic signals and use $u(t) = \text{IDFT}(U(k))$ with $U(N-k) = \bar{U}(k)$).

9.2. Consider the setup shown in Figure 9-14 with $r(t)$ a periodic signal and $m_u(t) = 0$, $m_y(t) = 0$ (no measurement errors). Show that Assumption 9.20(iii) is fulfilled (hint: use (9-145)).

9.3. Show that the contribution of the disturbing noise to the expected value of the linear least squares cost function is given by (9-34) (hint: use (9-12) with $\Delta(\Omega_k, \theta, N_Z(k)) = A(\Omega_k, \theta)N_Y(k) - B(\Omega_k, \theta)N_U(k)$).

9.4. Show that the difference between the Jacobians of the nonlinear least squares cost (9-32) and the iterative weighted least squares cost (9-37) is given by

$$(J_{\text{NLS}}(\theta^{(i-1)}, Z) - J_{\text{IWLS}}(\theta^{(i-1)}, Z))_{[k,l]} = -\frac{e(\Omega_k, \theta^{(i-1)}, Z(k)) \partial |A(\Omega_k, \theta^{(i-1)})|}{|A(\Omega_k, \theta^{(i-1)})|^2 \partial \theta^{(i-1)}_{[l]}}$$

(hint: compare the ith Newton-Gauss step (9-17) applied to the nonlinear least squares cost (9-32) with the ith normal equation of the IWLS cost function (9-37)).

9.5. Show that the contribution of the disturbing noise to the expected value of the nonlinear least squares cost function is given by (9-44) (hint: use (9-12) with $\Delta(\Omega_k, \theta, N_Z(k)) = N_Y(k) - G(\Omega_k, \theta)N_U(k)$).

9.6. Consider the weighted generalized total least squares estimator (9-76). Show that property 5 of Theorem 9.21 is still valid to

$$V_{*\text{WGTLS}}(\theta) = \frac{\int_{f_{\min}}^{f_{\max}} W^2(\Omega(f)) \, \mathbb{E}\{|e(\Omega(f), \theta, Z_0(f))|^2\} n(f) df}{\int_{f_{\min}}^{f_{\max}} W^2(\Omega(f)) \sigma_e^2(\Omega(f), \theta) n(f) df} + 1$$

(hint: divide the numerator and denominator of (9-76) by F and follow the lines of Appendix 9.E, Section 9.E.3).

9.7. Assume that the errors $N_Y(k)$, $N_U(k)$ on the measured input and output spectra $Y(k)$, $U(k)$ are independent (over k) random variables that are not circular complex distributed ($\mathbb{E}\{N_X^2(k)\} \neq 0$, $X = U, Y$). Show that the Markov estimator of model (6-32) minimizes

$$\frac{1}{2}\sum_{k=1}^{F} \frac{e_R^2(k, \theta)\sigma_I^2(k, \theta) + e_I^2(k, \theta)\sigma_R^2(k, \theta) - 2e_R(k, \theta)e_I(k, \theta)\sigma_{RI}^2(k, \theta)}{\sigma_R^2(k, \theta)\sigma_I^2(k, \theta) - \sigma_{RI}^4(k, \theta)}$$

with $e_R(k, \theta) = \text{Re}(e(\Omega_k, \theta, Z))$, $e_I(k, \theta) = \text{Im}(e(\Omega_k, \theta, Z))$, $\sigma_R^2(k, \theta) = \text{var}(e_R(k, \theta))$, $\sigma_I^2(k, \theta) = \text{var}(e_I(k, \theta))$, $\sigma_{RI}^2(k, \theta) = \text{covar}(e_R(k, \theta), e_I(k, \theta))$, and $e(\Omega_k, \theta, Z)$ given by (9-9) (hint: use (19-12) with $e_k(\theta, z_k) = (e(\Omega_k, \theta, Z))_{\text{re}}$ and $(\ )_{\text{re}}$ defined in (15-48)).

9.8. Show that the difference between the Jacobians of the ML cost (9-80) and the IQML cost (9-97) is given by

$$(J_{\text{ML}}(\theta^{(i-1)}, Z) - J_{\text{IQML}}(\theta^{(i-1)}, Z))_{[k,l]} = -\frac{e(\Omega_k, \theta^{(i-1)}, Z(k)) \partial \sigma_e(\Omega_k, \theta)}{\sigma_e^2(\Omega_k, \theta^{(i-1)})} \frac{\partial \sigma_e(\Omega_k, \theta)}{\partial \theta_{[l]}^{(i-1)}}$$

(hint: compare the ith Newton-Gauss step (9-17) applied to the ML cost (9-80) with the ith normal equation of the IQML cost function (9-97)).

9.9. Assume that the wrong noise (co)variances are used in the GTLS estimator (9-72). Show under Assumptions 9.16 and 9.17 that the bias $\tilde{\theta}(Z_0) - \theta_0$ is given by (9-278) with

$$V_F'(\theta_0) = 2 \frac{\sum_{k,l=1}^{F} |Y_0(k)|^2 |Y_0(l)|^2 \text{Re}\left(\frac{\partial \ln(G(\Omega_k, \theta))}{\partial \theta_0} W(k)\right)}{\sum_{k=1}^{F} |Y_0(k)|^2 (\hat{V}_U(k) + \hat{V}_Y(k))}$$

$$W(k) = V_U(k)(\hat{V}_U(l) + \hat{V}_Y(l)) - \hat{V}_U(k)(V_U(l) + V_Y(l))$$

where $V_U(k)$, $V_Y(k)$, $\hat{V}_U(k)$, and $\hat{V}_Y(k)$ are defined in Appendix 9.T. Show that the bias is not zero if the noise covariance matrix used, $\hat{C}_{N_Z(k)}$, satisfies (9-129). Note that the bias expression for the BTLS estimator (9-98) is similar to that of the GTLS estimator. (hint: follow the lines of Appendix 9.T; use $\hat{V}_Y(k) = f(k)V_Y(k)$, $\hat{V}_U(k) = f(k)V_U(k)$ to show that the bias is not zero under condition (9-129)).

9.24 APPENDIXES

Appendix 9.A A Second-Order Simulation Example

The second-order system $G(s, \theta) = 1/(1 + s + s^2)$ is excited at $F = 100$ frequencies, equally distributed in the band $[0.1, 10]/(2\pi)^2$ Hz. The true input $U_0(k) = 1$ and output $Y_0(k)$ spectra are disturbed by independent, zero mean, circular complex Gaussian noise with variance 0.04: $N_U(k), N_Y(k) \in N^c(0, 0.04)$ (see Section 16.1). Two sets of noisy simulated data $\{U^{[1]}(k), Y^{[2]}(k), k = 1, 2, ..., 100\}$ and $\{U^{[2]}(k), Y^{[2]}(k), k = 1, 2, ..., 100\}$ are generated. The noisy frequency response function $G(s_k)$, shown in Figure 9-1, is the ratio of the averaged output $Y(k) = (Y^{[1]}(k) + Y^{[2]}(k))/2$ to the averaged input $U(k) = (U^{[1]}(k) + U^{[2]}(k))/2$ spectra and is used as simulation data for the least squares (LS), iterative weighted least squares (IWLS), nonlinear least squares based on frequency response function (NLS-FRF), total least squares (TLS), logarithmic least squares (LOG), and subspace (SUB) estimators. The nonlinear least squares based on input-output data (NLS, NLS-IO), generalized total least squares (GTLS), maximum likelihood (ML), iterative quadratic maximum likelihood (IQML), and bootstrapped total least squares (BTLS) estimators use the averaged input $U(k)$ and output $Y(k)$ spectra as simulation data, and the instrumental variables (IV) method uses the two original noisy data sets separately. The constraint $\|\theta\|_2^2 = 1$ is used to calculate all the estimates except for the SUB algorithm, which uses no constraint, and for the LS, IWLS, IQML, and IV methods, which use $a_0 = 1$.

Appendix 9.B Signal-to-Noise Ratio of DFT Spectra Measurements of Random Excitations

Consider a random excitation $x(t)$, which is mixing of order 2 (the mixing condition limits the span of dependence of $x(t)$, see Section 16.4) and with $\text{var}(x(t)) > 0$ for any t, infinity included. The variance of its DFT spectrum $X(k) = N^{-1/2} \sum_{t=0}^{N-1} x(t) z_k^{-t}$ equals

$$\text{var}(X(k)) = \frac{1}{N} \sum_{t_1, t_2 = 0}^{N-1} \text{covar}(x(t_1), x(t_2)) z_k^{-t_1 + t_2} \qquad (9\text{-}155)$$

Because $x(t)$ is mixing of order 2, we have that (see (15-38) with $\text{cum}(x, y) = \text{covar}(x, y)$)

$$\sum_{t_1, t_2 = 0}^{N-1} |\text{covar}(x(t_1), x(t_2))| = O(N)$$

and, hence, (9-155) can be bounded above by

$$\text{var}(X(k)) \le \frac{1}{N} \sum_{t_1, t_2 = 0}^{N-1} |\text{covar}(x(t_1), x(t_2))| = O(N^0) \qquad (9\text{-}156)$$

The same reasoning holds for disturbing noise $v(t)$ satisfying the same conditions as $x(t)$, so that the signal-to-noise ratio $\text{std}(X(k))/\text{std}(V(k))$ is an $O(N^0)$.

Appendix 9.C Signal-to-Noise Ratio of DFT Spectra Measurements of Periodic Excitations

Consider a multisine $x(t)$ with finite power, $N^{-1} \sum_{t=0}^{N-1} x^2(tT_s) = O(N^0)$, and consisting of F harmonically related frequencies

$$x(t) = \sum_{r=1}^{F} A_r \sin(2\pi m_r f_0 t + \varphi_k) \qquad (9\text{-}157)$$

where $m_r \in \mathbb{N}$, $r = 1, 2, \ldots, F$ and $m_1 < m_2 < \cdots < m_F$. Assume that we observe the multisine during an integer number of periods, $NT_s/T_0 = Nf_0/f_s \in \mathbb{N}$, and that we respect the Nyquist condition $m_F f_0 < f_s/2$. Assume, furthermore, that the disturbing noise $v(t)$ is mixing of order 2 with $\text{var}(v(t)) > 0$ for any t, infinity included. Here, we handle two cases: F is independent of N, $F = O(N^0)$, and F increases with N, $F = O(N)$.

9.C.1 F Is Independent of N. Because F is independent of N and $x(t)$ has finite power, we have $A_r = O(N^0)$, $r = 1, 2, \ldots, F$. Using $\sin(x) = (e^{jx} - e^{-jx})/(2j)$, $\sum_{t=0}^{N-1} x^t = (1 - x^N)/(1-x)$, and $z_k^N = 1$, the DFT spectrum $X(k) = N^{-1/2} \sum_{t=0}^{N-1} x(tT_s) z_k^{-t}$ equals

$$X(k) = \begin{cases} \dfrac{\sqrt{N}}{2j} A_r e^{j\varphi_r} & k\dfrac{f_s}{N} = m_r f_0 \\ 0 & k\dfrac{f_s}{N} \ne m_r f_0 \end{cases} \qquad (9\text{-}158)$$

for $k = 0, 1, \ldots, N/2$ with $X(N-k) = \bar{X}(k)$ for $k = N/2 + 1, \ldots, N-1$. Because $A_r = O(N^0)$, we have $X(k) = O(N^{1/2})$ at the excited DFT frequencies. For disturbing noise $v(t)$, which is mixing of order 2, we have $\text{var}(V(k)) = O(N^0)$ (see Appendix 9.B) so that the signal-to-noise ratio $|X(k)|/\text{std}(V(k))$ of the multisine at the excited DFT frequencies increases as $O(N^{1/2})$.

Section 9.24 ■ Appendixes

9.C.2 F Increases as $O(N)$. Because $F = O(N)$ and $x(t)$ has finite power, we have $A_r = O(N^{-1/2})$, $r = 1, 2, ..., F$, and, hence, $|X(k)| = O(N^0)$ at the excited frequencies (see (9-158)). Combining this result with $\text{var}(V(k)) = O(N^0)$ (see Appendix 9.B) gives $|X(k)|/\text{std}(V(k)) = O(N^0)$.

Appendix 9.D Asymptotic Behavior Cost Function for a Time Domain Experiment

The cost functions that can handle time domain experiments can be written as

$$V_F(\theta, Z) = \frac{1}{F} \sum_{k \in \mathbb{F}} W^2(\Omega_k, \theta) |Y(k) - G(\Omega_k, \theta) U(k) - T_G(\Omega_k, \theta)|^2 \qquad (9\text{-}159)$$

with $\mathbb{F}$ a subset of the DFT frequencies $\{0, 1, ..., N-1\}$ and $W(\Omega, \theta)$ the absolute value of a rational function of Ω. We will show that

$$V_F(\theta, Z) = \frac{1}{F} \sum_{k \in \mathbb{F}} W^2(\Omega_k, \theta) |Y(k) - G(\Omega_k, \theta) U(k)|^2 + R(\theta) \qquad (9\text{-}160)$$

with $R(\theta) = O_p(F^{-1})$ uniformly in $\mathbb{P}_r$.

Elaborating (9-159) gives (9-160) with

$$\begin{aligned} R(\theta) &= \frac{1}{F} \sum_{k \in \mathbb{F}} W^2(\Omega_k, \theta) |T_G(\Omega_k, \theta)|^2 \\ &\quad - 2\text{Re}(\frac{1}{F} \sum_{k \in \mathbb{F}} W^2(\Omega_k, \theta)(Y(k) - G(\Omega_k, \theta) U(k)) \overline{T_G}(\Omega_k, \theta)) \end{aligned} \qquad (9\text{-}161)$$

The first term in (9-161) is an $O_p(F^{-1})$ because the numerator coefficients of $T_G(\Omega, \theta)$ tend to zero as $O_p(F^{-1/2})$ (Theorem 6.7). The second term in (9-161) can be written as the sum of two terms of the form

$$\text{Re}(\frac{1}{F^{3/2}} \sum_{k \in \mathbb{F}} X(k) F(\Omega_k, \theta)) \qquad (9\text{-}162)$$

with $X = Y$ or U and $F(\Omega, \theta)$ a noncausal, rational filter of finite order (independent of F). The additional factor $F^{-1/2}$ stems from the numerator coefficients of $T_G(\Omega, \theta)$. We now extend the sum in (9-162) to all the DFT frequencies

$$2\text{Re}(\frac{1}{F^{3/2}} \sum_{k \in \mathbb{F}} X(k) F(\Omega_k, \theta)) = \frac{1}{F^{3/2}} \sum_{k=0}^{N-1} X_1(k) F(\Omega_k, \theta) \qquad (9\text{-}163)$$

with

$$X_1(k) = \begin{cases} \bar{X}_1(N-k) = X(k) & k \in \mathbb{F} \\ 0 & \text{elsewhere} \end{cases} \qquad (9\text{-}164)$$

Using $X_f(k) = X_1(k) F(\Omega_k, \theta)$ and $F = O(N)$, (9-163) becomes

$$\frac{1}{F^{3/2}} \sum_{k=0}^{N-1} X_f(k) = \frac{\sqrt{N}}{F^{3/2}} x_f(0) = O(F^{-1}) x_f(0) \qquad (9\text{-}165)$$

where $x_f(t) = \text{IDFT}(X_f(k))$ is, within some transient effects, the response of $x_1(t)$ to the noncausal rational filter $F(\Omega, \theta)$. The original noisy signal $x(t)$ consists of the sum of a signal term $x_0(t)$ and a noise term $n_x(t)$ that satisfy, respectively, Assumptions 9.1 and 9.3. Therefore, the second-order moments of $x_0(t)$ and $n_x(t)$ are uniformly bounded. This is also valid for $x_1(t) = \text{IDFT}(X_1(k))$ because it is obtained by replacing the original DFT spectrum $X(k)$ by zeros at some DFT frequencies (see (9-164) with $\mathbb{F} \subset \{0, 1, ..., N-1\}$). Finally, the second-order moments of $x_f(t)$ are uniformly bounded because it is, within some transient effects, the response of $x_1(t)$ to the noncausal rational filter $F(\Omega, \theta)$. Hence, $x_f(0) = O_p(F^0)$ so that $R(\theta) = O_p(F^{-1})$ uniformly in θ_r, which concludes the proof. Note that using Lemma 16.23 exactly the same reasoning can be followed for the noise transient terms in $N_Y(k)$ and $N_U(k)$. □

Appendix 9.E Asymptotic Properties of Frequency Domain Estimators with Deterministic Weighting (Theorem 9.21)

If the frequency domain errors of the time and frequency domain experiments were mixing of order four (infinity), then Theorem 9.21 (except property 5) would follow immediately from the results of Chapters 17 and 19 (then the assumptions of Section 9.6 fulfill all the necessary conditions). For the frequency domain experiment the frequency domain errors are mixing of order four (Assumption 9.4) but not of order infinity (moments of order higher than $4 + \varepsilon$ do not necessarily exist, see Assumption 9.13). Hence, properties 1, 2, 3, 6, and 7 are valid but properties 4 and 8 still remain to be proved. Because after a DFT the noise is not mixing of order four (infinity) (see Section 16.16), all the properties of Theorem 9.21 remain to be proved for the time domain experiment. Fortunately, the resulting technical difficulties in the proofs can easily be solved using the results of Section 16.16. To understand fully the proofs of this appendix, we advise reading Chapter 17 first.

9.E.1 Stochastic Convergence (Properties 1, 2, 3, 6, and 7).

For the frequency domain experiment, properties 1, 2, 3, 6, and 7 follow directly from Theorems 17.6, 17.19, 17.21, 17.11, and 17.28, respectively. To show that the properties are valid for the time domain experiment, it is sufficient to show that the cost function and its higher order derivatives w.r.t. θ still converge strongly (weakly) and uniformly in $\mathbb{P}_r$ to their expected values when the mixing assumption of order four (Assumption 17.1 with $P = 4$) is replaced by Assumption 9.3. We will prove this for the cost function; the proof for its higher order derivatives w.r.t. θ follows exactly the same lines.

Because $\Delta(\Omega_k, \theta, N_Z(k))$ is linear in $N_Z(k)$ with $\Delta(\Omega_k, \theta, 0) = 0$ (see (9-12)) we have $\Delta(\Omega_k, \theta, N_Z(k)) = M_1(\Omega_k, \theta) N_Z(k)$ with $M_1(\Omega_k, \theta) \in \mathbb{C}^{1 \times 2}$. It facilitates rewriting (9-13) as

$$V_F(\theta, Z) = V_F(\theta, Z_0) + \frac{1}{F} \sum_{k=1}^{F} |M_1(\Omega_k, \theta) N_Z(k)|^2 \\ + 2\text{Re}(\frac{1}{F} \sum_{k=1}^{F} M_2(\Omega_k, \theta) N_Z(k)) \qquad (9\text{-}166)$$

Section 9.24 ■ Appendixes 355

with $M_2(\Omega_k, \theta) = \overline{\varepsilon(\Omega_k, \theta, Z_0(k))} M_1(\Omega_k, \theta)$. $V_F(\theta, Z_0)$, $M_1(\Omega_k, \theta)$, and $M_2(\Omega_k, \theta)$ are continuous in $\mathbb{P}_r$ (Assumption 9.6) and, therefore, also uniformly bounded in $\mathbb{P}_r$. If the input is random, then $V_F(\theta, Z_0)$ and $M_2(\Omega_k, \theta)$ have uniformly bounded second-order moments (see Assumption 9.1). Hence, under Assumptions 9.1, 9.3, and 9.6, the sums in (9-166),

$$V_F(\theta, Z_0), \frac{1}{F} \sum_{k=1}^{F} |M_1(\Omega_k, \theta) N_Z(k)|^2 \text{ and } 2\text{Re}(\frac{1}{F} \sum_{k=1}^{F} M_2(\Omega_k, \theta) N_Z(k)) \quad (9\text{-}167)$$

satisfy the conditions of Theorems 16.28 and 16.32 (strong laws of large numbers) and, therefore, converge uniformly w.p. 1 to their expected value at the rate $O_p(F^{-1/2})$ in the compact set $\mathbb{P}_r$. We conclude that $V_F(\theta, Z)$ converges uniformly w.p. 1 to its expected value $V_F(\theta)$ at the rate $O_p(F^{-1/2})$ in $\mathbb{P}_r$.

For the consistency and the bias (properties 6 and 7), we have to make a distinction between correct models ((9-7) with $\Omega = z^{-1}$, s, $\sqrt{s}$, or $\tanh(\tau_R s)$ and (9-8) with $\Omega = z^{-1}$) and asymptotically correct models ((9-8) with $\Omega = s, \sqrt{s}$). For the correct models we have $\tilde{\theta}(Z_0) = \theta_0$ and for the asymptotically correct model, $\theta_* = \lim_{F \to \infty} \tilde{\theta}(Z_0) = \theta_0$ (Assumption 9.17). Note that for the signals defined in the time domain experiment (Assumption 9.1), the model error $\delta(s_k)$ of (9-8) with $\Omega = s, \sqrt{s}$ converges weakly to zero at the rate $O_p(F^{-1/2})$ (Lemma 6.6, with $F = O(N)$ for a time domain experiment). Hence, $\hat{\theta}(Z)$ is weakly (and not strongly) consistent for model (9-8) with $\Omega = s, \sqrt{s}$. The convergence rate of the bias $b_\theta(Z)$ remains an $O(N^{-1})$ because $\mathbb{E}\{\delta(s_k) \bar{U}(k)\} = O(N^{-1})$ (see Theorem 6.17).

9.E.2 Asymptotic Normality (Properties 4 and 6). $F^{1/2}(\hat{\theta}(Z) - \tilde{\theta}(Z_0))$ is asymptotically normally distributed if and only if $\sqrt{F} \delta_\theta(Z)$, and, hence, the vector $F^{1/2} V_F'^T(\tilde{\theta}(Z_0), Z)$ is asymptotically normally distributed (see (9-26)). Taking the derivative of (9-11) w.r.t. θ gives for $F^{1/2} V_F'^T(\tilde{\theta}(Z_0), Z)$

$$\sqrt{F} V_F'^T(\tilde{\theta}(Z_0), Z) = \frac{1}{\sqrt{F}} \sum_{k=1}^{F} 2\text{Re}\left(\left(\frac{\partial \varepsilon(\Omega_k, \theta, Z(k))}{\partial \tilde{\theta}(Z_0)}\right)^H \varepsilon(\Omega_k, \tilde{\theta}(Z_0), Z(k)))\right)$$

$$= \frac{1}{\sqrt{F}} \sum_{k=1}^{F} x(k) \quad (9\text{-}168)$$

where $x(k)$ depends on zero, first, and second order powers of $N_Z(k)$. We will show here that (9-168) is asymptotically normally distributed for a frequency domain experiment (Assumption 9.13) and for a time domain experiment (Assumption 9.12).

Under Assumption 9.13 (frequency domain experiment) $x(k)$ is independently distributed (over the frequency k) with bounded absolute moments of order $2 + \varepsilon$ and 3 and with $\sum_{k=1}^{F} \text{var}(x(k)) = O(F)$. Hence, $F^{-1/2} \sum_{k=1}^{F} x(k)$ converges in law to a normal distribution at the rate $O(F^{-1/2})$ (see Section 16.10, version 2 of the central limit theorem).

Each entry of $x(k)$ in (9-168) can be written as the sum of terms of the form $X(k)\bar{V}(k)$ where $V(k)$ and $X(k)$ depend either on the disturbing noise or on the true input-output DFT spectra. We now study the sum

$$F^{-1/2} \sum_{k=1}^{F} X(k) \bar{V}(k) \quad (9\text{-}169)$$

under Assumptions 9.1, 9.11, and 9.12 for each combination of $X(k)$, $V(k)$ giving a random term $X(k)\bar{V}(k)$. If $V(k)$ and $X(k)$ depend on the DFT spectrum of one of the following signals, filtered iid noise, a normalized nonrandom periodic excitation (see Definition 3.4), a normalized random multisine (see Definition 3.2), or normalized periodic noise (see Definition 3.4), then (9-169) converges in law to a normal distribution at the rate $O(F^{-1/2})$ (proof: apply Theorems 16.29 and 16.33). We conclude that $F^{-1/2}\sum_{k=1}^{F} x(k)$ converges in law to a normal distribution at the rate $O(F^{-1/2})$.

9.E.3 Deterministic Convergence (Property 5). Theorem 17.24 is valid if Assumptions 17.4, 17.22, and 17.23 are fulfilled. We will show that the expected value of the cost function $V_F(\theta)$ converges uniformly in $\mathbb{P}_r$ to $V_*(\theta)$ at the rate $O(F^{-2})$. The proof for $V_F'(\theta)$ and $V_F''(\theta)$ follows the same lines. The expected value of the cost function equals (see (9-11))

$$V_F(\theta) = \frac{1}{F} \sum_{k=1}^{F} \mathbb{E}\{|\varepsilon(\Omega_k, \theta, Z(k))|^2\} \qquad (9\text{-}170)$$

Note that for a time domain experiment the influence of the transient term $T_G(\Omega_k, \theta)$ in the cost function $V_F(\theta, Z)$ can be neglected in the convergence rate analysis (see Appendix 9.D). Let $\mathbb{E}\{|\varepsilon(\Omega(f), \theta, Z(f))|^2\}$ be the limit value ($F \to \infty$) of $\mathbb{E}\{|\varepsilon(\Omega_k, \theta, Z(k))|^2\}$. We have

$$\mathbb{E}\{|\varepsilon(\Omega_k, \theta, Z(k))|^2\} = \mathbb{E}\{|\varepsilon(\Omega(f), \theta, Z(f))|^2\}|_{f=f_k} \qquad (9\text{-}171)$$

for a frequency domain experiment, whereas due to the leakage errors in the DFT spectra of the true input-output signals Z_0 and/or the disturbing noise $N_Z(k)$

$$\mathbb{E}\{|\varepsilon(\Omega_k, \theta, Z(k))|^2\} = \mathbb{E}\{|\varepsilon(\Omega(f), \theta, Z(f))|^2\}|_{f=f_k} + O(F^{-1}) \qquad (9\text{-}172)$$

for a time domain experiment (see Section 2.2 and Appendix 6.F with $F = O(N)$). Under Assumptions 9.14 and 9.15 the Riemann sum

$$\frac{1}{F} \sum_{k=1}^{F} \mathbb{E}\{|\varepsilon(\Omega(f), \theta, Z(f))|^2\}|_{f=f_k} \qquad (9\text{-}173)$$

converges to

$$V_*(\theta) = \int_{f_{\min}}^{f_{\max}} \mathbb{E}\{|\varepsilon(\Omega(f), \theta, Z(f))|^2\} n(f) df \qquad (9\text{-}174)$$

at the rate $O(F^{-2})$ (see Ralston and Rabinowitz, 1984; midpoint rule (4.10-10)). Hence, the convergence rate of $V_F(\theta)$ to $V_*(\theta)$ is an $O(F^{-2})$ for a frequency domain experiment and an $O(F^{-1})$ for a time domain experiment. Under Assumption 9.6 $\mathbb{E}\{|\varepsilon(\Omega_k, \theta, Z(k))|^2\}$ is a continuous function of $\theta \in \mathbb{P}_r$, and, hence, also uniformly bounded in $\mathbb{P}_r$. Therefore, the convergence of $V_F(\theta)$ to $V_*(\theta)$ is uniform in $\mathbb{P}_r$. Note that the integrand in (9-174) may be zero in some subintervals of $[f_{\min}, f_{\max}]$. In that case, the integral can be written as the sum of integrals. We conclude that Theorem 17.24 is valid with $K = 2$ for a frequency domain experiment and $K = 1$ for a time domain experiment.

9.E.4 Asymptotic Efficiency (Property 8). In the efficiency study, the covariance matrix of the limiting random variable $\delta_\theta(Z)$ or the truncated estimate $\hat{\underline{\theta}}(Z)$ is compared with the Cramér-Rao lower bound (16-88). As the bias b_θ and the derivative of the bias w.r.t. θ_0 of the truncated estimate $\hat{\underline{\theta}}(Z)$ tend to zero as $O(F^{-1})$ (property 7) and $\text{Cov}(\delta_\theta(Z))$ ($\text{Cov}(\hat{\underline{\theta}}(Z))$) tends to zero as $O(F^{-1})$ (Assumption 9.9), it is sufficient to compare $\text{Cov}(\delta_\theta(Z))$ or $\text{Cov}(\hat{\underline{\theta}}(Z))$ with $Fi^{-1}(\theta_0)$ in (16-88).

Under Assumptions 9.18 and 9.19 the Fisher information matrix of the model parameters is given by

$$Fi(\theta_0) = FV_F''(\theta_0) = 2\text{Re}\left(\left(\frac{\partial\varepsilon(\theta, Z_0)}{\partial\theta_0}\right)^H \left(\frac{\partial\varepsilon(\theta, Z_0)}{\partial\theta_0}\right)\right) \quad (9\text{-}175)$$

(see Section 19.3, formula (19-22)).

Under Assumption 9.18, (9-83) is a Markov estimator. For such an estimator the expression of the covariance matrix (9-27) can be elaborated. The cost function $V(\theta, Z)$ can be written as

$$V(\theta, Z) = \varepsilon^H(\theta, Z)\varepsilon(\theta, Z) = \frac{1}{2}(\sqrt{2}\varepsilon_{\text{re}}(\theta, Z))^T(\sqrt{2}\varepsilon_{\text{re}}(\theta, Z))$$

and similarly for $v(\theta, Z)$ in (9-13). Therefore, Theorem 19.3 is still valid when $\varepsilon(\theta, z)$ and $\Delta(\theta, n_z)$ are replaced by $\sqrt{2}\varepsilon_{\text{re}}(\theta, Z)$ and $\sqrt{2}\Delta_{\text{re}}(\theta, N_Z)$, respectively (compare (9-11) and (9-13) with (19-8)). Applying Lemmas 15.3 and 15.4 to expression (19-32) of Theorem 19.3, we get

$$\begin{aligned}\text{Cov}(\sqrt{F}\delta_\theta(Z)) &= V_F''^{-1}(\theta_0) + V_F''^{-1}(\theta_0)q_F(\theta_0)V_F''^{-1}(\theta_0) \\ q_F(\theta_0) &= F\, \mathbb{E}\{v_F'^T(\theta_0, N_Z)v_F'(\theta_0, N_Z)\} \\ &\quad + 2\text{Re}(2\text{herm}(\,\mathbb{E}\{\left(\frac{\partial\varepsilon(\theta, Z_0)}{\partial\theta_0}\right)^H\}\, \mathbb{E}\{\Delta(\theta_0, N_Z)v_F'(\theta_0, N_Z)\}))\end{aligned} \quad (9\text{-}176)$$

where $v_F(\theta, N_Z) = \Delta^H(\theta, N_Z)\Delta(\theta, N_Z)/F$ is defined in (9-13) and where the expected values are taken w.r.t. to the disturbing noise N_Z and the observations Z_0.

Under Assumptions 9.18 and 9.19, (9-83) is the maximum likelihood solution. Comparing (9-176) with (9-175) for deterministic Z_0 and Gaussian errors N_Z (Assumptions 9.18 and 9.19) shows that in general the maximum likelihood solution is asymptotically inefficient, $q_F(\theta_0) \neq 0$. Under Assumption 9.20 the rank of the $2F$ by $2F$ matrix C_{N_Z} equals F. Applying Theorem 19.4 with $r = 1$ and $t = 2$ shows, then, that $v_F(\theta, N_Z)$ is independent of θ, so that $q_F(\theta_0) = 0$. We conclude that the maximum likelihood estimate is asymptotically efficient under Assumption 9.20.

To analyze the influence of the noise level v on the inefficiency term in (9-176), we apply quick tool number 6 of Section 9.5. It follows that $V_F''^{-1}(\theta_0) = O(v^2)$ and $q_F(\theta_0) = O(v^{-1})$, which gives (9-91). If the pdf of N_Z is even, then the second term in the expression of $q_F(\theta_0)$ is zero, so that $q_F(\theta_0) = O(v^0)$. □

Appendix 9.F Asymptotic Properties of Frequency Domain Estimators with Stochastic Weighting (Corollary 9.22)

To understand fully the proof of this appendix, we advise reading Chapters 17 and 18 first. To prove the corollary, it is sufficient to verify that all the conditions of the theorems in Chapter 18 are fulfilled. The cost function $V(\theta, Z)/F$ (9-14), where the stochastic vector $\eta(Z)$ has been replaced by the deterministic vector η, is denoted by $f_F(\theta, \eta, Z)$. Clearly, $f_F(\theta, \eta, Z)$ satisfies the assumptions of Section 9.6. Therefore the cost function $f_F(\theta, \eta, Z)$ and its higher order derivatives w.r.t. θ converge w.p. 1 to their expected values (proof: follow the same lines as in Appendix 9.E). By assumption, the stochastic vector $\eta(Z)$ satisfies all the properties of Theorem 9.21, and the cost function $f_F(\theta, \eta, Z)$ has continuous third-order derivatives w.r.t. $x = [\theta^T \eta^T]^T$. We conclude that all the assumptions of Chapter 18 are fulfilled. From Theorems 18.5 and 18.6, it follows that $\tilde{\theta}(Z_0)$ and θ_* are the minimizers of, respectively, $V_F(\theta) = \mathbb{E}\{f_F(\theta, \eta_*, Z)\}$ and $V_*(\theta) = \lim_{F \to \infty} \mathbb{E}\{f_F(\theta, \eta_*, Z)\}$. The expected value of $\delta_\theta(Z)$ may not exist because the moments of $\eta(Z)$ in $V_F'(\tilde{\theta}(Z_0), Z) = f_F'(\theta, \eta(Z), Z)$ do not necessarily exist. Moreover, if it exists, we will have, in general, $\mathbb{E}\{\delta_\theta(Z)\} \neq 0$. Equation (9-31) follows from Theorem 18.25. Because $\eta(Z)$ satisfies, by assumption, Theorem 9.21, it follows directly that $\tilde{\eta}(Z) - \eta_*$ is given by $\delta_\eta(Z)$ (9-26) in Theorem 18.25.

Appendix 9.G Expected Value of an Analytic Function

Consider an analytic function $f(z)$ that has the property $f(0) = 0$. Its Taylor series expansion at the origin is then given by

$$f(z) = \sum_{r=1}^{\infty} \frac{f^{(r)}(0)}{r!} z^r \text{ for any } |z| < R \qquad (9\text{-}177)$$

with R the convergence radius. For zero mean circular complex errors z (see Assumption 9.18), we have $\mathbb{E}\{z\} = 0$ and $\mathbb{E}\{z^2\} = 0$. If, in addition, the errors have an even pdf, then $\mathbb{E}\{z^{2r+1}\} = 0$. Hence, for uniformly bounded random variables $|z| < R$ the expected value of (9-177) becomes

$$\mathbb{E}\{f(z)\} = \sum_{r=2}^{\infty} \frac{f^{(2r)}(0)}{(2r)!} \mathbb{E}\{z^{2r}\} = O(\mathbb{E}\{z^4\}) \qquad (9\text{-}178)$$

For circular complex normally distributed z we also have $\mathbb{E}\{z^r\} = 0$ (see Exercise 16.8) and, hence, $\mathbb{E}\{f(z)\} = 0$ if $R = \infty$.

The two functions of interest are $f(z) = 1/(1+z) - 1$ and $f(z) = \ln(1+z)$. Both functions have the property $f(0) = 0$ and the convergence radius of their Taylor series expansion at the origin is $R = 1$. Hence, for circular complex uniformly bounded noise $|z| < 1$ with even pdf we have

$$\mathbb{E}\{1/(1+z) - 1\} = O(\mathbb{E}\{z^4\}) \quad \text{and} \quad \mathbb{E}\{\ln(1+z)\} = O(\mathbb{E}\{z^4\}) \qquad (9\text{-}179)$$

For unbounded noise, the Taylor series expansion (9-177) diverges for all realizations $|z| > 1$ and (9-179) is no longer valid. However, for sufficiently large signal-to-noise ratios

$\mathbb{E}\{z^2\} \ll 1$, the probability to "hit" a value $|z| \geq 1$ is small and (9-178) is a very good approximation. For example, for Gaussian noise the right-hand sides of (9-179) would be zero, while the expected value is a very small number, given by

$$\mathbb{E}\{1/(1+z) - 1\} = -\exp(-1/\sigma_z^2) \text{ and } \mathbb{E}\{\ln(1+z)\} = -\frac{1}{2}\text{Ei}(-1/\sigma_z^2) \qquad (9\text{-}180)$$

with $\sigma_z^2 = \mathbb{E}\{|z|^2\}$ and Ei( ) the exponential integral function (Guillaume et al., 1992b). Applying (9-180) to (9-56) with $z = N_Y(k)/Y_0(k)$ and $z = N_U(k)/U_0(k)$ gives (9-57). Using (9-49), the expected value of $N_G(k)/G_0(\Omega_k)$ (9-48) can be written as

$$\mathbb{E}\{N_G(k)/G_0(\Omega_k)\} = \mathbb{E}\left\{\frac{1}{1+z} - 1\right\} + \rho(k)\frac{\sigma_U(k)\sigma_Y(k)}{\sigma_V^2(k)}\frac{U_0(k)}{Y_0(k)}\mathbb{E}\left\{\frac{v}{1+z}\right\} \qquad (9\text{-}181)$$

with $z = N_U(k)/U_0(k)$ and $v = N_V(k)/U_0(k)$. Using $z = m + v$ with $m = M(k)/U_0(k)$ we find

$$\frac{v}{1+z} = -\frac{1}{1+z} + \frac{1+v}{1+m+v} = \left(-\frac{1}{1+z} + 1\right) + \left(\frac{1}{1+m/(1+v)} - 1\right) \qquad (9\text{-}182)$$

The expected value of the second term in (9-182) is further elaborated. Because m and v are mutually independent, we have

$$\begin{aligned}\mathbb{E}\left\{\frac{1}{1+m/(1+v)} - 1\right\} &= \mathbb{E}\left\{\mathbb{E}\left\{\frac{1}{1+m/(1+v)} - 1\Big|v\right\}\right\} \\ &= \mathbb{E}\{-\exp(-|1+v|^2/\sigma_m^2)\} \\ &= -\exp(-1/\sigma_z^2)\sigma_m^2/\sigma_z^2\end{aligned} \qquad (9\text{-}183)$$

with $\sigma_z^2 = \sigma_m^2 + \sigma_v^2$. The second equality uses (9-180) and the third equality uses the circular complex normality of v. Collecting (9-180), (9-182), and (9-183) gives

$$\mathbb{E}\left\{\frac{v}{1+z}\right\} = \exp(-1/\sigma_z^2)\sigma_v^2/\sigma_z^2 \qquad (9\text{-}184)$$

Putting (9-184), with $\sigma_z^2 = \sigma_U^2(k)/|U_0(k)|^2$ and $\sigma_v^2 = \sigma_V^2(k)/|U_0(k)|^2$, into (9-181) gives (9-50). □

Appendix 9.H Total Least Squares Solution – Equivalences (Lemma 9.23)

To simplify the notations, we put $A = W_{\text{Re}}J_{\text{re}}(Z)C^{-1}$ and $x = C\theta$. This facilitates writing (9-62) and (9-64) as $Ax \approx 0$ and

$$\arg\min_{\tilde{A}, x} \|A - \tilde{A}\|_F^2 \text{ subject to } \tilde{A}x = 0 \text{ and } \|x\|_2^2 = 1, \tag{9-185}$$

respectively. Using the method of the Lagrange multipliers, the constrained minimization problem (9-185) can be reformulated as follows:

$$\arg\min_{\tilde{A}, x, \mu} \text{trace}((A - \tilde{A})(A - \tilde{A})^T) + \mu^T \tilde{A} x \text{ subject to } \|x\|_2^2 = 1 \tag{9-186}$$

where $\mu \in \mathbb{R}^{2F}$ is a Lagrange multiplier vector. Expressing the stationarity of the preceding cost function w.r.t. $\tilde{A}$ yields

$$-2(A - \tilde{A}) + \mu x^T = 0 \text{ or } 2(A - \tilde{A}) = \mu x^T \tag{9-187}$$

(use derivative rule (15-62) of Section 15.9.2 and $\mu^T \tilde{A} x = \text{trace}(\mu^T \tilde{A} x)$). Right multiplication of (9-187) by x, taking into account that $\tilde{A}x = 0$ (stationary cost function w.r.t. μ), gives $\mu = 2Ax/\|x\|_2^2$. Elimination of μ in (9-187) gives the following expression: $A - \tilde{A} = A x x^T / \|x\|_2^2$. Replacing $A - \tilde{A}$ in (9-186) by this expression and taking into account the constraint $\tilde{A} x = 0$ results in

$$\arg\min_x \|Ax\|_2^2 / \|x\|_2^2 \text{ subject to } \|x\|_2^2 = 1 \tag{9-188}$$

We will show that the constrained minimization problem (9-188) is equivalent to

1. $\arg\min_x \|Ax\|_2^2 / \|x\|_2^2$
2. $\arg\min_x \|Ax\|_2^2$ subject to $\|x\|_2^2 = 1$
3. Finding the eigenvector x corresponding to the smallest eigenvalue λ of the eigenvalue problem $A^T A x = \lambda x$.

In equivalent form number 1 the norm constraint is already included in the cost function, therefore the constraint $\|x\|_2^2 = 1$ in (9-188) can be removed. Equivalence 2 follows directly from equivalence 1. To prove equivalent form number 3, we reformulate equivalence 2 using a Lagrange multiplier λ

$$\arg\min_{x, \lambda} \|Ax\|_2^2 - \lambda(\|x\|_2^2 - 1) \tag{9-189}$$

Expressing the stationarity of the preceding cost function w.r.t. x yields

$$x^T A^T A - \lambda x^T = 0 \text{ or } A^T A x = \lambda x \tag{9-190}$$

subject to $\|x\|_2^2 = 1$ (stationarity cost function w.r.t. λ), which is an eigenvalue problem. Putting the solutions (x_k, λ_k), $k = 1, 2, ..., n_\theta$ of (9-190) in (9-189) taking into account the constraint $\|x\|_2^2 = 1$ gives

$$\arg\min_{x_k, k} \lambda_k \quad k = 1, 2, ..., n_\theta, \tag{9-191}$$

Section 9.24 ■ Appendixes

It shows that the eigenvector x_k corresponding to the smallest eigenvalue λ_k of A minimizes (9-189).

Substituting $A = W_{\text{Re}} J_{\text{re}}(Z) C^{-1}$ and $x = C\theta$ into equivalence 3 of (9-188) gives

$$C^{-T}(W_{\text{Re}} J_{\text{re}}(Z))^T (W_{\text{Re}} J_{\text{re}}(Z)) \theta = \lambda C \theta \qquad (9\text{-}192)$$

Left multiplication of (9-192) by C^T gives equivalence 3 of the lemma. Making the same substitution in equivalences 1 and 2 of (9-188), and taking into account that

$$\|Ax\|_2^2 = \|W_{\text{Re}} J_{\text{re}}(Z)\theta\|_2^2 = \|(WJ(Z)\theta)_{\text{re}}\|_2^2 = \|WJ(Z)\theta\|_2^2 \qquad (9\text{-}193)$$

(see Lemma 15.4), proves equivalences 1 and 2 of the lemma. □

Appendix 9.I   Expected Value Total Least Squares Cost Function

Because $\|j(N_Z)\theta\|_2^2$ is a real number, we have

$$\mathbb{E}\{\|j(N_Z)\theta\|_2^2\} = \text{Re}(\mathbb{E}\{\theta^T j^H(N_Z) j(N_Z) \theta\}) = \theta^T \mathbb{E}\{\text{Re}(j^H(N_Z) j(N_Z))\} \theta \qquad (9\text{-}194)$$

with $\text{Re}(j^H(N_Z) j(N_Z)) = j_{\text{re}}^T(N_Z) j_{\text{re}}(N_Z)$ (Lemma 15.4), which proves the first equality in (9-67). Using $j(N_Z)\theta = (J(Z) - J(Z_0))\theta = e(\theta, Z) - e(\theta, Z_0)$ we find

$$\mathbb{E}\{\|j(N_Z)\theta\|_2^2\} = \sum_{k=1}^{F} \sigma_e^2(\Omega_k, \theta) \qquad (9\text{-}195)$$

which proves the second equality in (9-67). □

Appendix 9.J   Explicit Form of the Total Least Squares Cost Function

9.J.1 Total Least Squares. Using (9-60), the cost function appearing in equivalence number 2 of Lemma 9.23, with $C = I_{n_\theta}$ and $W = I_F$, can be written as

$$\|J(Z)\theta\|_2^2 = \|e(\theta, Z)\|_2^2 = \sum_{k=1}^{F} |e(\Omega_k, \theta, Z(k))|^2 \qquad (9\text{-}196)$$

which is exactly (9-65). For the TLS with weight (9-69) we have

$$\|WJ(Z)\theta^{(i)}\|_2^2 = \|We(\theta^{(i)}, Z)\|_2 = \sum_{k=1}^{F} W^2(\Omega_k, \theta^{(i-1)}) |e(\Omega_k, \theta^{(i)}, Z(k))|^2 \qquad (9\text{-}197)$$

which is exactly (9-38).

9.J.2 Generalized Total Least Squares. According to equivalence number 1 of Lemma 9.23, with $W = I_F$, the GTLS cost function equals

$$V_{\text{GTLS}}(\theta, Z) = \|J(Z)\theta\|_2^2 / \|C\theta\|_2^2 \tag{9-198}$$

Using $C^T C = \mathbb{E}\{\text{Re}(j^H(N_Z)j(N_Z))\}$ and (9-195) we can rewrite $\|C\theta\|_2^2$ as

$$\|C\theta\|_2^2 = \mathbb{E}\{\theta^T \text{Re}(j^H(N_Z)j(N_Z))\theta\} = \mathbb{E}\{\|j(N_Z)\theta\|_2^2\} = \sum_{k=1}^{F} \sigma_e^2(\Omega_k, \theta) \tag{9-199}$$

Division of (9-196) by (9-199) gives (9-72).

For the WGTLS estimator with left and right weighting (9-74) and (9-75), we have

$$V_{\text{WGTLS}}(\theta, Z) = \|WJ(Z)\theta\|_2^2 / \|C\theta\|_2^2 \tag{9-200}$$

(see Lemma 9.23, equivalence number 1). Following the same lines as for (9-199), we find $\|C\theta\|_2^2 = \mathbb{E}\{\|Wj(N_Z)\theta\|_2^2\}$. Applying (9-197) to $\|WJ(Z)\theta\|_2^2$ and (9-195) to $\mathbb{E}\{\|Wj(N_Z)\theta\|_2^2\}$ gives (9-76) after division. □

Appendix 9.K Rank of the Column Covariance Matrix

We will show that the rank of the column covariance matrix C_{WJ} (9-75) is rank deficient under Assumption 9.20(i) or 9.20(ii). For the diagonal weighting (9-74), the kth row of $Wj(N_Z)$, with $j(N_Z) = J(Z) - J(Z_0)$, can be written as

$$(Wj(N_Z))_{[k, :]} = W(\Omega_k) N_Z^T(k) S^T(k) \tag{9-201}$$

with $N_Z^T(k) = [N_Y(k) \ N_U(k)]$,

$$S^T(k) = \begin{bmatrix} P_k^T(n_a) & 0 \\ 0 & -P_k^T(n_b) \end{bmatrix} \tag{9-202}$$

and $P_k^T(n) = \begin{bmatrix} 1 & \Omega_k & \ldots & \Omega_k^n \end{bmatrix}$. The column covariance matrix C_{WJ} (9-75) then becomes

$$C_{WJ} = \mathbb{E}\{\text{Re}((Wj(N_Z))^H (Wj(N_Z)))\} = \text{Re}(\sum_{k=1}^{F} W^2(\Omega_k) S(k) \text{Cov}(N_Z(k)) S^H(k)) \tag{9-203}$$

Under Assumption 9.20, $\text{Cov}(N_Z(k))$ has rank one for $k = 1, 2, \ldots, F$, so that

$$\text{Cov}(N_Z(k)) = c(k) c^H(k) \tag{9-204}$$

with $c(k) \in \mathbb{C}^2$ (see the SVD expansion (15-18)). Using (9-204), (9-203) can be written as $C_{WJ} = \text{Re}(B^H B) = B_{\text{re}}^T B_{\text{re}}$ where the kth row of B is given by

$$B_{[k, :]} = W(\Omega_k) c^H(k) S^H(k) = W(\Omega_k)[\bar{c}_{[1]}(k) P_k^H(n_a) \ -\bar{c}_{[2]}(k) P_k^H(n_b)] \tag{9-205}$$

and where the rank of B_{re} determines the rank of C_{WJ}. According to Assumption 9.20, we can distinguish three cases. Under Assumption 9.20(i), there is no input noise and

Section 9.24 ■ Appendixes   363

$c_{[1]}(k) \neq 0$, $c_{[2]}(k) = 0$ for any k, so that $\text{rank}(B_{\text{re}}) = n_a + 1$. Under Assumption 9.20(ii), there is no output noise, and $c_{[1]}(k) = 0$, $c_{[2]}(k) \neq 0$ for any k, so that $\text{rank}(B_{\text{re}}) = n_b + 1$. Under Assumption 9.20(iii), the input-output errors are totally correlated and $c_{[1]}(k) \neq 0$, $c_{[2]}(k) \neq 0$ for any k, so that in general B_{re} is of full rank. It is rank deficient only if, in addition, $c_{[1]}(k)/c_{[2]}(k)$ is real and independent of k. This is, for example, the case for totally correlated white noise errors $n_y(t)$, $n_u(t)$ ($\text{Cov}(N_Z(k))$ is then independent of k). □

Appendix 9.L   Calculation of the Gaussian Maximum Likelihood Estimate

9.L.1 Gaussian Log-Likelihood Function.   Under Assumptions 9.5, 9.18, and 9.19, the pdf of N_Z is given by

$$f_{N_Z}(N_Z) = \frac{1}{\pi^F \det(C_{N_Z})} \exp(-N_Z^H C_{N_Z}^{-1} N_Z) \tag{9-206}$$

(see (16-14)) with $C_{N_Z} = \text{Cov}(N_Z)$,

$$C_{N_Z} = \text{diag}(\text{Cov}(N_Z(1)), \text{Cov}(N_Z(2)), \ldots, \text{Cov}(N_Z(F)))$$

$$\text{Cov}(N_Z(k)) = \begin{bmatrix} \sigma_Y^2(k) & \sigma_{YU}^2(k) \\ \bar{\sigma}_{YU}^2(k) & \sigma_U^2(k) \end{bmatrix} \tag{9-207}$$

If C_{N_Z} is singular, then $C_{N_Z}^{-1}$ and $\det(C_{N_Z})$ are replaced by $C_{N_Z}^{+}$ and the product of the nonzero eigenvalues of C_{N_Z}, respectively. Replacing N_Z by $Z - Z_p$ in (9-206) and taking the negative of the natural logarithm gives the negative log-likelihood function

$$-\ln f_{N_Z}(Z, Z_p, \theta) = (Z - Z_p)^H C_{N_Z}^{+} (Z - Z_p) + c \tag{9-208}$$

with $c = F \ln(\pi) + \ln(\det(C_{N_Z}))$.

9.L.2 Elimination of the Unknown Input-Output DFT Spectra in the Cost Function.   Using Lemmas 15.3 and 15.4, Example 15.5 and Exercise 16.4, (9-79) can be written as

$$\frac{1}{2}(Z_{\text{re}} - Z_{\text{pre}})^T C_{N_{Z\text{re}}}^{+} (Z_{\text{re}} - Z_{\text{pre}}) + \lambda_{\text{re}}^T e_{\text{re}}(\theta, Z_p) \tag{9-209}$$

where $C_{N_{Z\text{re}}} = \text{Cov}(N_{Z\text{re}}) = 0.5(C_{N_Z})_{\text{Re}}$ with $C_{N_Z} = \text{Cov}(N_Z)$. Because $e_{\text{re}}(\theta, Z_p)$ is linear in Z_{pre}, minimization problem (9-209) is exactly equivalent to (19-6), and, hence, all the results of Section 19.2 are valid. Elimination of Z_{pre} in (9-209) gives (19-8)

$$V_{\text{ML}}(\theta, Z) = \frac{1}{2} e_{\text{re}}^T(\theta, Z) C_{e_{\text{re}}}^{-1}(\theta) e_{\text{re}}(\theta, Z) \tag{9-210}$$

with $C_{e_{\text{re}}}(\theta) = \text{Cov}(e_{\text{re}}(\theta, N_Z))$. Because the noise residual $e(\theta, N_Z)$ is linear in N_Z, it is complex circular

$$\mathbb{E}\{(e(\theta, N_Z) - \mathbb{E}\{e(\theta, N_Z)\})(e(\theta, N_Z) - \mathbb{E}\{e(\theta, N_Z)\})^T\} = 0 \qquad (9\text{-}211)$$

and $C_{e_{\text{re}}}(\theta) = 0.5(C_e(\theta))_{\text{Re}}$ (see Exercise 16.4). Note that $\mathbb{E}\{e(\theta, N_Z)\} \neq 0$ for model (9-10). Applying the result of 15.5 to (9-210) under Assumption 9.18 gives

$$\begin{aligned} V_{\text{ML}}(\theta, Z) &= e^H(\theta, Z) C_e^{-1}(\theta) e(\theta, Z) \\ C_e(\theta) &= \text{Cov}(e(\theta, N_Z)) = \text{diag}(\sigma_e^2(\Omega_1, \theta), \sigma_e^2(\Omega_2, \theta), \ldots, \sigma_e^2(\Omega_F, \theta)) \end{aligned} \qquad (9\text{-}212)$$

which is exactly (9-80).

In the derivation of (9-212), we implicitly assumed that no DC (Ω_0) and no Nyquist ($\Omega_{N/2}$) components were present in the data. Indeed, expressions (9-78) and (9-79) are valid only if all the elements of Z are complex, which is not the case for the DC and Nyquist components (real numbers). If DC and Nyquist are present, then under Assumption 9.18 the terms

$$\frac{1}{2} N_Z^T(0)(\text{Cov}(N_Z(0)))^{-1} N_Z^T(0) + \frac{1}{2} N_Z^T(N/2)(\text{Cov}(N_Z(N/2)))^{-1} N_Z^T(N/2) \\ + \lambda_0 e(\Omega_0, \theta, Z_p(0)) + \lambda_{N/2} e(\Omega_{N/2}, \theta, Z_p(N/2)) \qquad (9\text{-}213)$$

where $N_Z(0)$, $N_Z(N/2)$, λ_0, $\lambda_{N/2}$, $e(\Omega_0, \theta, Z_p(0))$, and $e(\Omega_{N/2}, \theta, Z_p(N/2))$ are real numbers, should be added to (9-79) and (9-209). Their contribution to (9-210) equals

$$\frac{1}{2} \frac{e^2(\Omega_0, \theta, Z(0))}{\sigma_e^2(\Omega_0, \theta)} + \frac{1}{2} \frac{e^2(\Omega_{N/2}, \theta, Z(N/2))}{\sigma_e^2(\Omega_{N/2}, \theta)} \qquad (9\text{-}214)$$

Because $e(\Omega_0, \theta, Z(0))$ and $e(\Omega_{N/2}, \theta, Z(N/2))$ are real numbers, the terms (9-214) remain unchanged in the transformation from (9-210) to (9-212).

9.L.3 Maximum Likelihood Estimate of the Input and Output DFT Spectra.

In the previous section, it was shown that all the results of Section 19.2 are valid for Z_{re}. Hence, the ML estimate $\hat{Z}_{\text{re}}$ of the true DFT spectra $Z_{0\text{re}}$ equals (19-11)

$$C_{N_{Z_{\text{re}}}} C_{N_{Z_{\text{re}}}}^+ \hat{Z}_{\text{re}} = C_{N_{Z_{\text{re}}}} C_{N_{Z_{\text{re}}}}^+ Z_{\text{re}} - C_{N_{Z_{\text{re}}}} M_1^T(\hat{\theta}) C_{e_{\text{re}}}^{-1}(\hat{\theta}) e_{\text{re}}(\hat{\theta}, Z) \qquad (9\text{-}215)$$

with $\hat{\theta} = \hat{\theta}_{\text{ML}}(Z)$ and

$$M_1(\theta) = \frac{\partial e_{\text{re}}(\theta, Z)}{\partial Z_{\text{re}}} = \begin{bmatrix} \dfrac{\partial \text{Re}(e(\theta, Z))}{\partial \text{Re}(Z)} & \dfrac{\partial \text{Re}(e(\theta, Z))}{\partial \text{Im}(Z)} \\ \dfrac{\partial \text{Im}(e(\theta, Z))}{\partial \text{Re}(Z)} & \dfrac{\partial \text{Im}(e(\theta, Z))}{\partial \text{Im}(Z)} \end{bmatrix} \qquad (9\text{-}216)$$

Because $e(\theta, Z)$ is an analytic function of Z, it satisfies the Cauchy-Riemann equations

Section 9.24 ■ Appendixes

$$\frac{\partial \text{Re}(e(\theta, Z))}{\partial \text{Re}(Z)} = \frac{\partial \text{Im}(e(\theta, Z))}{\partial \text{Im}(Z)}$$
$$\frac{\partial \text{Re}(e(\theta, Z))}{\partial \text{Im}(Z)} = -\frac{\partial \text{Im}(e(\theta, Z))}{\partial \text{Re}(Z)}$$
(9-217)

(Henrici, 1974). Applying (9-217) to (9-216) gives

$$M_1(\theta) = \begin{bmatrix} \dfrac{\partial \text{Re}(e(\theta,Z))}{\partial \text{Re}(Z)} & -\dfrac{\partial \text{Im}(e(\theta,Z))}{\partial \text{Re}(Z)} \\ \dfrac{\partial \text{Im}(e(\theta,Z))}{\partial \text{Re}(Z)} & \dfrac{\partial \text{Re}(e(\theta,Z))}{\partial \text{Re}(Z)} \end{bmatrix} = \left(\dfrac{\partial e(\theta,Z)}{\partial Z}\right)_{\text{Re}} \quad (9\text{-}218)$$

Using (9-218), $C_{e_{\text{re}}}(\theta) = 0.5(C_e(\theta))_{\text{Re}}$, $C_{N_{Z\text{re}}} = 0.5(C_{N_Z})_{\text{Re}}$, and Lemmas 15.3 and 15.4, (9-215) becomes

$$C_{N_Z}C_{N_Z}^+ \hat{Z} = C_{N_Z}C_{N_Z}^+ Z - C_{N_Z} M^H(\hat{\theta}) C_e^{-1}(\hat{\theta}) e(\hat{\theta}, Z)$$
$$M(\theta) = \frac{\partial e(\theta, Z)}{\partial Z} = \text{diag}([A(\Omega_1, \theta) \ -B(\Omega_1, \theta)], \ldots, [A(\Omega_F, \theta) \ -B(\Omega_F, \theta)])$$
(9-219)

Putting (9-207) and (9-212) in (9-219) assuming that C_{N_Z} is regular ($C_{N_Z}C_{N_Z}^+ = I_F$) gives (9-86). Note that the solution (9-86) remains well defined if C_{N_Z} is singular.

9.L.4 Minimization of the Maximum Likelihood Cost Function. The maximum likelihood cost function (9-83) is minimized using the Newton-Gauss algorithm (9-18). It requires the calculation of the Jacobian matrix $J(\theta, Z) = \partial \varepsilon(\theta, Z)/\partial \theta$, where $\varepsilon(\theta, Z)$ is given by (9-84) or (9-85). In this appendix an explicit expression of the Jacobian matrix $J(\theta, Z)$ is given for rational transfer function models (cost function (9-83) with weighted residual (9-84)). The calculation of $J(\theta, Z)$ for the other transfer function models (partial fraction expansion and state space representation) follows exactly the same lines (cost function (9-83) with weighted residual (9-85)).

Using (9-10), (9-35), and (9-84) we find for $k = 1, 2, \ldots, F$,

$$J_{[k, r+1]}(\theta, Z) = \frac{\partial \varepsilon(\Omega_k, \theta, Z(k))}{\partial a_r}$$
$$= \frac{\Omega_k^r Y(k)}{\sigma_e(\Omega_k, \theta)} - \frac{\varepsilon(\Omega_k, \theta, Z(k))}{\sigma_e^2(\Omega_k, \theta)} \text{Re}(\Omega_k^r[\sigma_Y^2(k)\bar{A}(\Omega_k, \theta) - \sigma_{YU}^2(k)\bar{B}(\Omega_k, \theta)])$$

with $r = 0, 1, \ldots, n_a$,

$$J_{[k, n_a + r + 2]}(\theta, Z) = \frac{\partial \varepsilon(\Omega_k, \theta, Z(k))}{\partial b_r}$$
$$= \frac{-\Omega_k^r U(k)}{\sigma_e(\Omega_k, \theta)} - \frac{\varepsilon(\Omega_k, \theta, Z(k))}{\sigma_e^2(\Omega_k, \theta)} \text{Re}(\Omega_k^r[\sigma_U^2(k)\bar{B}(\Omega_k, \theta) - \bar{\sigma}_{YU}^2(k)\bar{A}(\Omega_k, \theta)])$$

with $r = 0, 1, \ldots, n_b$, and

$$J_{[k, n_a + n_b + r + 3]}(\theta, Z) = \frac{\partial \varepsilon(\Omega_k, \theta, Z(k))}{\partial i_r} = \frac{-\Omega_k^r}{\sigma_e(\Omega_k, \theta)}$$

with $r = 0, 1, \ldots, n_i$. If a constraint of the form $\theta_{[j]} = 1$ is used, then the corresponding column in $J(\theta, Z)$ must be eliminated. If the constraint $\|\theta\|_2^2 = 1$ is used, then (9-18) is solved using the pseudoinverse (see Section 15.5) and $\theta^{(i)} = \theta^{(i-1)} + \Delta\theta^{(i)}$ is normalized ($\theta^{(i)} \to \theta^{(i)}/\|\theta^{(i)}\|_2$) before making a new iteration step ($i \to i + 1$).

The Levenberg-Marquardt version of (9-16) with constraint $\theta_{[j]} = 1$ is

$$(J_{\text{re}}^T(\theta^{(i-1)}, Z) J_{\text{re}}(\theta^{(i-1)}, Z) + \lambda^2 I_{n_\theta}) \Delta\theta^{(i)} = -J_{\text{re}}^T(\theta^{(i-1)}, Z) \varepsilon_{\text{re}}(\theta^{(i-1)}, Z) \quad (9\text{-}220)$$

(see Fletcher, 1991). The numerical stability of (9-220) is improved by solving the overdetermined set of equations

$$\begin{bmatrix} J_{\text{re}}(\theta^{(i-1)}, Z) \\ \lambda I_{n_\theta} \end{bmatrix} \Delta\theta^{(i)} = -\begin{bmatrix} \varepsilon_{\text{re}}(\theta^{(i-1)}, Z) \\ 0 \end{bmatrix} \quad (9\text{-}221)$$

using a QR factorization (see Section 15.4.3). If the constraint $\|\theta\|_2 = 1$ is used, then the Levenberg-Marquardt version of (9-18) is calculated as

$$\Delta\theta^{(i)} = -V \Lambda U^T \varepsilon_{\text{re}}(\theta^{(i-1)}, Z) \quad (9\text{-}222)$$

with

$$J_{\text{re}}(\theta^{(i-1)}, Z) = U \text{diag}(\sigma_1, \sigma_2, \ldots, \sigma_{n_\theta - 1}, 0) V^T$$

$$\Lambda = \text{diag}(\frac{\sigma_1}{\sigma_1^2 + \lambda^2}, \frac{\sigma_2}{\sigma_2^2 + \lambda^2}, \ldots, \frac{\sigma_{n_\theta - 1}}{\sigma_{n_\theta - 1}^2 + \lambda^2}, 0)$$

and $\sigma_1 \geq \sigma_2 \geq \ldots \geq \sigma_{n_\theta - 1}$. The initial value of λ in (9-221) and (9-222) is chosen proportional to the largest singular value of $J_{\text{re}}(\theta^{(0)}, Z)$, for example, $\lambda = \sigma_1/100$. If the iteration step $\theta^{(i)} = \theta^{(i-1)} + \Delta\theta^{(i)}$ is successful (the ML cost function decreases), then λ is decreased as $\lambda \to 0.4\lambda$; otherwise (the ML cost function increases) λ is increased as $\lambda \to 10\lambda$ and the iteration is restarted from $\theta^{(i-1)}$.

Appendix 9.M Number of Free Parameters in an Errors-in-Variables Problem

The total number of free parameters in an errors-in-variables problem (9-1) equals the sum of the number of free parameters in $\tilde{Z}_{0\text{re}}$ and the number of free model parameters in θ. According to Section 19.3, the number of free parameters in $\tilde{Z}_{0\text{re}}$ equals $\text{rank}(C_{N_{Z_{\text{re}}}}) - 2F$. Using $C_{N_{Z_{\text{re}}}} = 0.5(C_{N_Z})_{\text{Re}}$ and $\text{rank}((C_{N_Z})_{\text{Re}}) = 2\text{rank}(C_{N_Z})$ (Lemma 15.3) the total number of free parameters becomes

$$2\text{rank}(C_{N_Z}) - 2F + n_\theta \quad (9\text{-}223)$$

Under Assumption 9.20 we have $\text{rank}(C_{N_Z}) = F$ and (9-223) reduces to n_θ. We conclude that under Assumption 9.20 the total number of free parameters is independent of F. □

Appendix 9.N Uncertainty of the BTLS Estimator in the Absence of Model Errors

Cost function (9-100), where θ_* is replaced by η, equals $f_F(\theta, \eta, Z)$. The expected value of the derivative of $f_F(\theta, \eta, Z)$ w.r.t θ is denoted as $g_F(\theta, \eta) = \mathbb{E}\{f_F'(\theta, \eta, Z)\}$. In the absence of model errors, we have $\tilde{\theta}_{BTLS}(Z_0) = \theta_0$ ($\theta_{*BTLS} = \theta_0$ for model (9-8) with $\Omega = s$), so that $e(\Omega_k, \theta_0, Z_0(k)) = 0$ ($e(\Omega_k, \theta_0, Z_0(k)) = O(F^{-1/2})$, see Theorem 6.7), independently of η. Hence, from (9-100) and the definition of $g_F(\theta, \eta)$ it follows that $g_F(\theta_0, \eta) = 0$ ($g_F(\theta_0, \eta) = O(F^{-1})$) for any η, so that $\partial g_F(\theta_0, \eta)/\partial \eta = 0$ ($\partial g_F(\theta_0, \eta)/\partial \eta = O(F^{-1})$). We conclude that the second term in the right-hand side of (9-31) is zero (vanishes asymptotically w.r.t. the first term). It shows that the stochastic weighting does not increase the asymptotic uncertainty of the consistent BTLS estimate. □

Appendix 9.O Asymptotic Properties of the Instrumental Variables Method

The basic step in the proof of properties 1, 2, 3, 6, and 7 of Theorem 9.21 (see Appendix 9.E) is the strong convergence of the cost function $V_{IV}(\theta, Z)$ and its (higher order) derivatives w.r.t. θ to their expected values. It is easy to verify that the strong laws of large numbers used to prove the strong convergence can also be applied to cost functions that are bilinear in the measurements $Z^{[1]}$ and $Z^{[2]}$. To prove the asymptotic normality (properties 4 and 6) it is sufficient to note that the central limit theorems used in Appendix 9.E also apply to cost functions that are bilinear in the measurements $Z^{[1]}$ and $Z^{[2]}$. The deterministic convergence (property 5) follows the same lines of Appendix 9.E exactly. □

Appendix 9.P Equivalences between Range Spaces

In this appendix we show the following equivalences between the range spaces: (i) range($O_r \mathbf{X}$) = range(O_r), and (ii) range($O_r \mathbf{X}^{re} \Pi$) = range(O_r).

9.P.1 Proof of the First Equivalence. To prove the first equivalence, it is sufficient to show that $\mathbf{X}$ has full rank. The n_a by F matrix $\mathbf{X}$ in (9-116) is rank deficient if and only if there exists a (complex) row vector $C \neq 0$ such that $C\mathbf{X} = 0$. From (9-111) it follows that $X(k) = (\xi_k I_{n_a} - A)^{-1} B U(k)$. Assuming that $U(k) \neq 0$ it is possible to rewrite $C\mathbf{X} = 0$ as

$$C(\xi_k I_{n_a} - A)^{-1} B = 0 \text{ for } k = 1, 2, \dots, F \qquad (9\text{-}224)$$

Because, by assumption, at least $n_a + 1$ frequencies are distinct (Assumption 9.14) and the strictly proper system $G(\xi) = C(\xi I_{n_a} - A)^{-1} B$ can, at most, have $n_a - 1$ zeros, (9-224) can only be true if and only if $G(\xi) \equiv 0$. Assuming that the state space realization (9-111) is controllable, $G(\xi) \equiv 0$ can only be true if and only if $C = 0$ (Kailath, 1980). □

9.P.2 Proof of the Second Equivalence. The range space of $O_r \mathbf{X}^{re} \Pi$ equals the range space of O_r, unless rank cancellation occurs in $\mathbf{X}^{re} \Pi$. The rank cancellation does not occur if the intersection between the row spaces of $\mathbf{X}^{re}$ and $\mathbf{U}^{re}$ is empty. This is true if the $r + n_a$ by F matrix

$$\mathbf{Z} = \begin{bmatrix} \mathbf{U} \\ \mathbf{X} \end{bmatrix} \quad (9\text{-}225)$$

has rank $r + n_a$. $\mathbf{Z}$ is rank deficient if and only if there exists a row vector $L \neq 0$ such that $L\mathbf{Z} = 0$. Putting $L = [Dl_1 l_2 ... l_{r-1} C]$, and using $X(k) = (\xi_k I_{n_a} - A)^{-1} B U(k)$ (see (9-111)), $L\mathbf{Z} = 0$ can be written as

$$G(\xi_k) = 0 \text{ for } k = 1, 2, ..., F \quad (9\text{-}226)$$

with $G(\xi) = (D + C(\xi I_{n_a} - A)^{-1} B) + \sum_{m=1}^{r-1} l_m \xi^m$. Because, by assumption, at least $r + n_a + 1$ frequencies are distinct (Assumption 9.14) and $G(\xi)$ has, at most, $n_a + r - 1$ zeros, (9-226) can only be true if and only if $G(\xi) \equiv 0$. $G(\xi) \equiv 0$ is true if and only if $D = 0$, $l_m = 0$ for $m = 1, 2, ..., r-1$, and $C(\xi I_{n_a} - A)^{-1} B \equiv 0$. Assuming that the state space realization (9-111) is controllable, $C(\xi I_{n_a} - A)^{-1} B \equiv 0$ is true if and only if $C = 0$ (Kailath, 1980). □

Appendix 9.Q Estimation of the Range Space

The range space of a matrix equals the span of its left singular vectors corresponding to the nonzero singular values. In the first part of this appendix we study the estimation of the left singular vectors of a noisy r by F matrix A for $F \to \infty$. In the second part these results are applied to estimation of the range space of O_r.

9.Q.1 Asymptotic Properties of the Left Singular Vectors. Consider the real r by F matrix $A = A_0 + N$, where A_0 is the deterministic part and N is the zero mean noise contribution. The left singular vectors of A are equal to the eigenvectors $\hat{x}$ of AA^T (see Exercise 15.16)

$$(AA^T/F)\hat{x} = \hat{\lambda}\hat{x} \quad (9\text{-}227)$$

If NN^T/F and $A_0 N^T/F$ converge w.p. 1 to, respectively, C_N and 0, then AA^T/F converges w.p. 1 to

$$\mathbf{A_0 A_0^T} + C_N \text{ with } \mathbf{A_0 A_0^T} = \lim_{F \to \infty} A_0 A_0^T / F \quad (9\text{-}228)$$

Hence, (9-227) converges w.p. 1 to

$$(\mathbf{A_0 A_0^T} + C_N) x_* = \lambda_* x_* \quad (9\text{-}229)$$

where x_* and λ_* are the limit values of, respectively, $\hat{x}$ and $\hat{\lambda}$.

If $C_N = \sigma^2 I_r$ then (9-229) becomes $\mathbf{A_0 A_0^T} x_* = (\lambda_* - \sigma^2) x_*$. Clearly, $x_* = x_0$, with x_0 the left singular vector of $\mathbf{A_0}$. This proves the strong convergence of $\hat{x}$ to x_0.

If $C_N \neq \sigma^2 I_r$ then we form the matrix $B = C_N^{-1/2} A$, where $C_N^{1/2}$ is a square root of C_N, $C_N^{1/2} C_N^{T/2} = C_N$. Because $C_N^{-1/2}(NN^T/F) C_N^{-T/2}$ converges w.p. 1 to I_r, the equation $(BB^T/F)\hat{y} = \hat{\mu}\hat{y}$ converges w.p. 1 to

Section 9.24 ■ Appendixes

$$\mathbf{B}_0\mathbf{B}_0^T y_* = (\mu_* - 1)y_* \text{ with } \mathbf{B}_0 = C_N^{-1/2}\mathbf{A}_0 \tag{9-230}$$

Clearly $y_* = y_0$, with y_0 the left singular vector of $\mathbf{B}_0 = C_N^{-1/2}\mathbf{A}_0$. It follows that $y_0 = C_N^{-1/2}x_0$, which proves the strong convergence of $C_N^{1/2}\hat{y}$ to x_0.

Note that we have shown the strong convergence of the estimate $\hat{x}$ to the solution x_0 of the noiseless problem $\mathbf{A}_0$, without requiring the existence of a true model. If a true model exists and if it belongs to the considered model set, then x_0 equals the true value.

9.Q.2 Estimation of the Range Space of O_r. The results of the first part of this appendix are applicable to $\mathbf{Y}^{\text{re}}\Pi$ if $\mathbf{NN}^T/F$ and $\mathbf{X}^{\text{re}}\Pi\mathbf{N}^T/F$ converge w.p. 1 to, respectively, some C_N and 0. We will show that this is true under the assumptions of Section 9.6.1 and assuming that $N_U(k) = 0$.

Using (9-238) with $\mathbf{N}_U^{\text{re}} = 0$ we find

$$\mathbf{NN}^T = \mathbf{N}_Y^{\text{re}}\mathbf{N}_Y^{\text{re}\,T} - \mathbf{N}_Y^{\text{re}}\mathbf{U}^{\text{re}\,T}(\mathbf{U}^{\text{re}}\mathbf{U}^{\text{re}\,T})^{-1}\mathbf{U}^{\text{re}}\mathbf{N}_Y^{\text{re}\,T}$$
$$= \text{Re}(\mathbf{N}_Y\mathbf{N}_Y^H) - \text{Re}(\mathbf{N}_Y\mathbf{U}^H)(\text{Re}(\mathbf{UU}^H))^{-1}\text{Re}(\mathbf{UN}_Y^H) \tag{9-231}$$
$$\mathbf{X}^{\text{re}}\Pi\mathbf{N}^T = \text{Re}(\mathbf{XN}_Y^H) - \text{Re}(\mathbf{XU}^H)(\text{Re}(\mathbf{UU}^H))^{-1}\text{Re}(\mathbf{UN}_Y^H)$$

with

$$\mathbf{UU}^H = \sum_{k=1}^{F} |U_0(k)|^2 W_r(k)W_r^H(k)$$
$$\mathbf{XU}^H = \sum_{k=1}^{F} X(k)\bar{U}_0(k)W_r(k)W_r^H(k)$$

and

$$\mathbf{N}_Y\mathbf{N}_Y^H/F = \frac{1}{F}\sum_{k=1}^{F} |N_Y(k)|^2 W_r(k)W_r^H(k)$$
$$\mathbf{N}_Y\mathbf{U}^H/F = \frac{1}{F}\sum_{k=1}^{F} N_Y(k)\bar{U}_0(k)W_r(k)W_r^H(k) \tag{9-232}$$
$$\mathbf{XN}_Y^H/F = \frac{1}{F}\sum_{k=1}^{F} X(k)\bar{N}_Y(k)W_r(k)W_r^H(k)$$

By assumption $\text{Re}(\mathbf{UU}^H)/F > cI_r$, with $0 < c < \infty$ and c independent of F, for any F, ∞ included, and, hence, $(\text{Re}(\mathbf{UU}^H))^{-1} = O(F^{-1})$. For a frequency domain experiment, $N_Y(k)$ is independently distributed over k, while for a time domain experiment, $|N_Y(k)|^2$ and $N_Y(k)$ converge w.p. 1 to random variables that are mixing of order 2 (see Section 16.16). Hence, the sums in (9-232) converge w.p. 1 to their expected values (see Section 16.9, versions 2 and 3 of the law of large numbers)

$$\mathbb{E}\{\mathbf{N}_Y\mathbf{N}_Y^H/F\} = \frac{1}{F}\sum_{k=1}^{F} \sigma_Y^2(k)W_r(k)W_r^H(k)$$
$$\mathbb{E}\{\mathbf{N}_Y\mathbf{U}^H/F\} = 0 \tag{9-233}$$
$$\mathbb{E}\{\mathbf{XN}_Y^H/F\} = 0$$

Hence, $\mathbf{NN}^T/F$ converges w.p. 1 to

$$C_{\mathbf{Y}}/F = \mathbb{E}\{\operatorname{Re}(\mathbf{N_Y}\mathbf{N_Y^H})\}/F = \frac{1}{F}\operatorname{Re}(\sum_{k=1}^{F}\sigma_Y^2(k)W_r(k)W_r^H(k)) \qquad (9\text{-}234)$$

and $\mathbf{X}^{\mathrm{re}}\Pi\mathbf{N}^T/F$ converges w.p. 1 to 0.

We conclude that range($C_{\mathbf{Y}}^{1/2}U_{[:,1:n_a]}$) converges strongly to range($O_r\mathbf{X}^{\mathrm{re}}\Pi$), which is equal to range(O_r) (Appendix 9.P). Hence, we have established the strong consistency of $\hat{O}_r$. From versions 2 and 3 of the law of large numbers (see Section 16.9), it follows that $\mathbf{N}\mathbf{N}^T/F$ and $\mathbf{X}^{\mathrm{re}}\Pi\mathbf{N}^T/F$ converge in probability at the rate $O_p(F^{-1/2})$ to their limit value. Hence, this is also valid for $\mathbf{Y}^{\mathrm{re}}\Pi$ and $C_{\mathbf{Y}}^{1/2}U_{[:,1:n_a]}$ (9-242) so that $\hat{O}_r$ converges in probability at the rate $O_p(F^{-1/2})$ to O_r. In case of model errors Assumption 9.27 guarantees that the results of the first part of this appendix can be applied to equation (9-242) showing the strong convergence of $\hat{O}_r = \text{range}(C_{\mathbf{Y}}^{1/2}U_{[:,1:n_a]})$ to the solution O_{r*} of the noiseless problem ($\mathbf{N} = 0$ in (9-238)). The convergence rate is also an $O_p(F^{-1/2})$.

Appendix 9.R Subspace Algorithm for Discrete-Time Systems (Algorithm 9.24)

We first discuss the three basic steps of the subspace algorithm in more details. Next, we present a numerically efficient implementation.

9.R.1 Basic Subspace Algorithm.
The three basic steps of the subspace algorithm are (i) estimation of the range space of O_r, (ii) estimation of A and C given $\hat{O}_r$, and (iii) estimation of B and D given $\hat{A}$ and $\hat{C}$.

FIRST STEP. As O_r is known only within a right invertible transformation matrix T, see (9-118), it is sufficient to estimate the range space of O_r. If the $F > n_a$ frequencies are distinct, then the matrix $\mathbf{X}$ has rank n_a (see Appendix 9.P), and range($O_r\mathbf{X}$) = range(O_r). Hence, we can estimate the range of O_r using (9-120) if we can eliminate $S_r\mathbf{U}$ and suppress the influence of the noise $\mathbf{N_Y} - S_r\mathbf{N_U}$.

Because O_r is a real matrix, we are interested in a real range space. Therefore, we convert (9-120) into a set of real equations as

$$\mathbf{Y}^{\mathrm{re}} = O_r\mathbf{X}^{\mathrm{re}} + S_r\mathbf{U}^{\mathrm{re}} + \mathbf{N}_{\mathbf{Y}}^{\mathrm{re}} - S_r\mathbf{N}_{\mathbf{U}}^{\mathrm{re}} \qquad (9\text{-}235)$$

where $(\ )^{\mathrm{re}}$ locates the real and imaginary parts beside each other, for example,

$$\mathbf{Y}^{\mathrm{re}} = [\operatorname{Re}(\mathbf{Y})\ \operatorname{Im}(\mathbf{Y})] \qquad (9\text{-}236)$$

The operator $(\ )^{\mathrm{re}}$ should not be confused with $(\ )_{\mathrm{re}}$, which stacks the real and imaginary parts on top of each other. Both operators are related by $X^{\mathrm{re}} = ((X^T)_{\mathrm{re}})^T$.

The term $S_r\mathbf{U}^{\mathrm{re}}$ in (9-235) is eliminated by right multiplication of (9-235) with an orthogonal projection Π

$$\Pi = I_{2F} - \mathbf{U}^{\mathrm{re}\,T}(\mathbf{U}^{\mathrm{re}}\mathbf{U}^{\mathrm{re}\,T})^{-1}\mathbf{U}^{\mathrm{re}} \qquad (9\text{-}237)$$

which has the property $\mathbf{U}^{\mathrm{re}}\Pi = 0$. We get

$$\mathbf{Y}^{\text{re}}\Pi = O_r \mathbf{X}^{\text{re}}\Pi + \mathbf{N}$$
$$\mathbf{N} = (\mathbf{N}_\mathbf{Y}^{\text{re}} - S_r \mathbf{N}_\mathbf{U}^{\text{re}})\Pi \qquad (9\text{-}238)$$

If $\mathbf{U}^{\text{re}}\mathbf{U}^{\text{re}T}/F > cI_r$ with $0 < c < \infty$ and where c is independent of F, for any F, ∞ included, the frequencies are distinct, and $F \geq n_a + r$, then range$(O_r \mathbf{X}^{\text{re}}\Pi)$ = range(O_r) for any F, ∞ included (see Appendix 9.P).

From (9-238) it follows that the range of O_r can be estimated as range$(\mathbf{Y}^{\text{re}}\Pi)$. Since the range of a matrix equals the span of the left singular vectors corresponding to the nonzero singular values (see Section 15.4.1), range$(\mathbf{Y}^{\text{re}}\Pi)$ is calculated via a singular value decomposition (SVD) of $\mathbf{Y}^{\text{re}}\Pi$. The left singular vectors of $\mathbf{Y}^{\text{re}}\Pi$ are consistently estimated if

$$\underset{F \to \infty}{\text{a.s.lim}} \ \mathbf{X}^{\text{re}}\Pi \mathbf{N}^T/F = 0 \text{ and } \underset{F \to \infty}{\text{a.s.lim}} \ \mathbf{N}\mathbf{N}^T/F = C_\mathbf{N} \text{ with } C_\mathbf{N} = \sigma^2 I_r \qquad (9\text{-}239)$$

(see Appendix 9.Q). The second condition in (9-239) is in general not satisfied and, therefore, the noise $\mathbf{N}$ in (9-238) is whitened by left multiplication of (9-238) with $C_\mathbf{N}^{-1/2}$,

$$C_\mathbf{N}^{-1/2}\mathbf{Y}^{\text{re}}\Pi = C_\mathbf{N}^{-1/2}O_r \mathbf{X}^{\text{re}}\Pi + C_\mathbf{N}^{-1/2}\mathbf{N} \qquad (9\text{-}240)$$

where $C_\mathbf{N}^{1/2}$ is a square root of $C_\mathbf{N}$ (see Section 15.4.4 for the calculation of the square root of a positive (semi-)definite matrix). Because $\mathbb{E}\{C_\mathbf{N}^{-1/2}(\mathbf{N}\mathbf{N}^T/F)C_\mathbf{N}^{-T/2}\} \to I_r$ for $F \to \infty$, the left singular vectors of $C_\mathbf{N}^{-1/2}\mathbf{Y}^{\text{re}}\Pi$ are consistently estimated. From the SVD

$$C_\mathbf{N}^{-1/2}\mathbf{Y}^{\text{re}}\Pi = U\Sigma V^T \qquad (9\text{-}241)$$

we estimate the extended observability matrix O_r as

$$C_\mathbf{N}^{-1/2}\hat{O}_r = U_{[:, 1:n_a]} \text{ or } \hat{O}_r = C_\mathbf{N}^{1/2}U_{[:, 1:n_a]} \qquad (9\text{-}242)$$

The problem with the proposed algorithm is that $\mathbf{N}$ is a function of the unknown state space parameters, via S_r (see (9-114) and (9-238)), and, hence, $C_\mathbf{N}$ cannot be calculated. If the input is exactly known, $\mathbf{N}_\mathbf{U} = 0$, then $\mathbf{N}$ is independent of S_r and $C_\mathbf{N}$ can be calculated. If the input observations are disturbed by noise, $\mathbf{N}_\mathbf{U} \neq 0$, then we replace $U(k)$, $Y(k)$, $\sigma_U^2(k)$, and $\sigma_Y^2(k)$ everywhere by, respectively, 1, $G(\Omega_k) = Y(k)/U(k)$, 0, and σ_G^2 (9-54). If the worst case input and output signal-to-noise ratio is larger than 10 dB, then the bias on $\mathbb{E}\{Y(k)/U(k)\}$ can be neglected and the variance of the truncated ratio $\underline{G}(\Omega_k)$ is given by (9-54) (see Section 9.9 for an elaborated discussion). We conclude that from a practical point of view $Y(k)/U(k)$ acts as a zero mean random variable with variance (9-54). From now on, we will assume that $\mathbf{N}_\mathbf{U} = 0$. The matrix $C_\mathbf{N}$ is then asymptotically ($F \to \infty$) given by

$$C_\mathbf{N} = \lim_{F \to \infty} C_\mathbf{Y}/F \text{ with } C_\mathbf{Y} = \mathbb{E}\{\mathbf{N}_\mathbf{Y}^{\text{re}}\mathbf{N}_\mathbf{Y}^{\text{re}T}\} = \text{Re}(\sum_{k=1}^{F} \sigma_Y^2(k) W_r(k) W_r^H(k)) \qquad (9\text{-}243)$$

(see Appendix 9.Q). We conclude that the estimate $\hat{O}_r$ converges w.p. 1 to the true solution O_{r0} under the assumptions of Section 9.6.6 and Assumption 9.14 and that it convergences w.p. 1 to the noiseless solution O_{r*} under the assumptions of Section 9.6.1 and Assumption 9.14. Moreover, the convergence rate is an $O_p(F^{-1/2})$ (see Appendix 9.Q).

SECOND STEP.   Using the estimate $\hat{O}_r$ we calculate from (9-117) and (9-114),

$$\hat{A} = \hat{O}_{r[1:r-1,:]}^+ \hat{O}_{r[2:r,:]} \text{ and } \hat{C} = \hat{O}_{r[1,:]} \tag{9-244}$$

Note that $\hat{A}$, $\hat{C}$, and their derivatives w.r.t. $\hat{O}_r$ are continuous functions of $\hat{O}_r$ in a closed and bounded neighborhood of the true value O_{r0} or the noiseless solution O_{r*}. Because $\hat{O}_r$ converges w.p. 1 to the true solution or to the noiseless solution, the estimates $\hat{A}$ and $\hat{C}$ also converge w.p. 1 to the true solution or to the noiseless solution (Lemma 17.31). Since the convergence rate of $\hat{O}_r$ is an $O_p(F^{-1/2})$, the convergence rate of $\hat{A}$ and $\hat{C}$ is also an $O_p(F^{-1/2})$ (Lemma 17.34).

THIRD STEP.   We choose $W(\xi_k) = \sigma_Y^{-1}(k)$ in the cost function $V_{SUB}(B, D, \hat{A}, \hat{C}, Z)$ (9-121). If $N_U(k) \neq 0$ then we replace $Y(k)$, $U(k)$ by $Y(k)/U(k)$, 1 (see first step) and put $W(\xi_k) = \sigma_G^{-1}(k)$ (see (9-54)). This choice would give the smallest uncertainty on the estimates $\hat{B}$ and $\hat{D}$ if $\hat{A}$ and $\hat{C}$ were nonrandom. Under Assumptions 9.14 and 9.16 the linear least squares problem (9-121) is identifiable if and only if the state space realization (9-111) is observable (McKelvey et al., 1996).

Under the assumptions of Sections 9.6.5 and 9.6.6, the estimates $\hat{B}$ and $\hat{D}$ are strongly consistent because $\hat{A}$ and $\hat{C}$ are strongly consistent. To prove this statement, it is sufficient to apply Theorem 18.7 with $w(\theta, \eta(Z), Z) = 0$, $\eta(Z) = [(\text{vec}(\hat{A}))^T \hat{C}^T]^T$, and $\eta_* = [(\text{vec}(A_0))^T C_0^T]^T$, and to verify that $\mathbb{E}\{V_{SUB}(B, D, A_0, C_0, Z)\}$ is minimal in the true parameters C_0, D_0. This last condition is satisfied because for $W(\xi_k) = \sigma_Y^{-1}(k)$,

$$\mathbb{E}\{V_{SUB}(B, D, A_0, C_0, Z)\} = V_{SUB}(B, D, A_0, C_0, Z_0) + F \tag{9-245}$$

with $V_{SUB}(B_0, D_0, A_0, C_0, Z_0) = 0$. Similarly, under the assumptions of Sections 9.6.1 and 9.6.5, the estimates $\hat{B}$ and $\hat{D}$ converge w.p. 1 the noiseless solution (see Theorem 18.5).

Under the assumptions of Section 9.6.2 and 9.6.5 the estimates $\hat{B}$ and $\hat{D}$ converge in probability at the rate $O_p(F^{-1/2})$ to their limit value. To prove this, it is sufficient to note that the convergence rate of $\hat{A}$ and $\hat{C}$ is an $O_p(F^{-1/2})$ and to verify that all the conditions of Theorem 18.16 are satisfied.

9.R.2 Numerically Efficient Implementation.   The matrix $\mathbf{Y}^{re}\Pi$ can be calculated without forming the huge $2F$ by $2F$ matrix Π. This is done as follows. Form the matrix $Z = [\mathbf{U}^{reT} \ \mathbf{Y}^{reT}]^T$ and calculate the QR factorization (see Section 15.4.3) $Z^T = QR$ with $Q^T Q = I_{4F}$ and R a $2r$ by $2r$ upper triangular matrix. This factorization can be written as

$$Z = R^T Q^T \text{ or } \begin{bmatrix} \mathbf{U}^{re} \\ \mathbf{Y}^{re} \end{bmatrix} = \begin{bmatrix} R_{11}^T & 0 \\ R_{12}^T & R_{22}^T \end{bmatrix} \begin{bmatrix} Q_1^T \\ Q_2^T \end{bmatrix} \tag{9-246}$$

with R_{11} a regular r by r matrix. Using the property $Q_2^T Q_1 = 0$, it is easy to verify that $\mathbf{Y}^{re}\Pi = R_{22}^T Q_2^T$. The left singular vectors of $\mathbf{Y}^{re}\Pi$ are the eigenvectors of $\mathbf{Y}^{re}\Pi(\mathbf{Y}^{re}\Pi)^T$ (see Exercise 15.16). Using $Q_2^T Q_2 = I_{2F}$ we find that $\mathbf{Y}^{re}\Pi(\mathbf{Y}^{re}\Pi)^T = R_{22}^T R_{22}$ is independent of Q_2. It follows that the left singular vectors of $\mathbf{Y}^{re}\Pi$ and the (asymptotic) covariance matrices C_N and C_Y (9-243) are not influenced by Q_2. Hence, we can calculate (9-241) as

$$C_Y^{-1/2} R_{22}^T = U\Sigma V^T \tag{9-247}$$

Appendix 9.S Subspace Algorithm for Continuous-Time Systems (Algorithm 9.25)

The algorithm is a three-step procedure. The main differences from Algorithm 9.24 for discrete-time systems are (i) the orthogonalization of the input and output data, (ii) the estimation of a generalized extended observability matrix $O_{r\perp}$ in the first step, and (iii) the estimation of $\hat{A}$ from a generalized shift property of $O_{r\perp}$ in the second step. The third step remains exactly the same. We first explain the orthogonalization, next we discuss the impact of the orthogonalization on the model equation (9-120), and finally we prove the generalized shift property of $O_{r\perp}$. The appendix is concluded with a discussion of the stochastic properties and the numerical implementation.

9.S.1 Orthogonalization Procedure.
The data matrices $\mathbf{Y}$ and $\mathbf{U}$ in (9-116) depend on the scalar polynomial basis s^n, $n = 0, 1, \ldots, r-1$. We will construct two scalar orthogonal polynomial bases $p_n(s)$, $n = 0, 1, \ldots, r-1$ and $q_n(s)$, $n = 0, 1, \ldots, r-1$ such that the data matrices $\mathbf{Y}_\perp$, constructed using $p_n(s)$, and $\mathbf{U}_\perp$, constructed using $q_n(s)$, satisfy, respectively, $\mathrm{Re}(\mathbf{Y}_\perp \mathbf{Y}_\perp^H) = I_r$ and $\mathrm{Re}(\mathbf{U}_\perp \mathbf{U}_\perp^H) = I_r$. The matrix $Z_\perp = [\mathbf{U}_\perp^{\mathrm{re}T} \; \mathbf{Y}_\perp^{\mathrm{re}T}]^T$, where $(\;)^{\mathrm{re}}$ locates the real and imaginary parts of the matrix beside each other (see (9-236)), thus has the property

$$Z_\perp Z_\perp^T = \begin{bmatrix} I_r & C_1 \\ C_1^T & I_r \end{bmatrix} \qquad (9\text{-}248)$$

In Rolain et al. (1995) it has been shown that no other two scalar polynomial bases resulting in a smaller condition number of $Z_\perp Z_\perp^T$ can be found. Hence, $Z_\perp$ is best conditioned for scalar bases.

The matrices $\mathrm{Re}(\mathbf{Y}\mathbf{Y}^H)$ and $\mathrm{Re}(\mathbf{U}\mathbf{U}^H)$ each define an inner product,

$$\begin{aligned}\mathrm{Re}(\mathbf{Y}\mathbf{Y}^H) &\Rightarrow \langle x(s), y(s)\rangle_Y = \mathrm{Re}(\sum_{k=1}^{F} x(s_k)\bar{y}(s_k)|Y(k)|^2) \\ \mathrm{Re}(\mathbf{U}\mathbf{U}^H) &\Rightarrow \langle x(s), y(s)\rangle_U = \mathrm{Re}(\sum_{k=1}^{F} x(s_k)\bar{y}(s_k)|U(k)|^2)\end{aligned} \qquad (9\text{-}249)$$

that is used to calculate, respectively, the bases $p_n(s)$, $n = 0, 1, \ldots, r-1$, and $q_n(s)$, $n = 0, 1, \ldots, r-1$, via a Gram-Schmidt orthogonalization procedure (see Section 15.11). Applying this procedure with inner product $\langle\;,\;\rangle_Y$ gives

1. Initialization:
$$\begin{aligned} p_0(s) &= 1/\alpha_1 \quad \text{with} \quad \alpha_1 = \|1\| \\ p_1(s) &= sp_0(s)/\alpha_2 \quad \text{with} \quad \alpha_2 = \|sp_0(s)\| \end{aligned} \qquad (9\text{-}250)$$

2. Recursion: for $n = 2$ to $r-1$
$$\begin{aligned} p_n(s) &= (sp_{n-1}(s) + \alpha_n p_{n-2}(s))/\alpha_{n+1} \quad \text{with} \\ \alpha_{n+1} &= \|sp_{n-1}(s) + \alpha_n p_{n-2}(s)\| \end{aligned} \qquad (9\text{-}251)$$

where $\|t(s)\|^2 = \langle t(s), t(s)\rangle_Y$. The resulting polynomial basis has the property

$$\langle p_n(s), p_m(s)\rangle_Y = \delta(n-m) \qquad (9\text{-}252)$$

(see Section 15.11). Now define the vector $W_Y(k)$

$$W_Y(k) = [p_0(s_k) \; p_1(s_k) \; \ldots \; p_{r-1}(s_k)]^T \tag{9-253}$$

and construct $\mathbf{Y}_\perp$ as

$$\mathbf{Y}_\perp = [W_Y(1)Y(1) \; W_Y(2)Y(2) \; \ldots \; W_Y(F)Y(F)] \tag{9-254}$$

Using (9-252), it is easy to verify that

$$\text{Re}(\mathbf{Y}_\perp \mathbf{Y}_\perp^T) = \text{Re}(\sum_{k=1}^{F} |Y(k)|^2 W_Y(k) W_Y^H(k)) = I_r \tag{9-255}$$

Making the same calculations with the inner product $\langle \; , \; \rangle_U$, gives the scalar orthogonal polynomial basis $q_n(s)$ and the numbers β_n, $n = 0, 1, \ldots, r-1$. Similar to (9-253) and (9-254), we define

$$W_U(k) = [q_0(s_k) \; q_1(s_k) \; \ldots \; q_{r-1}(s_k)]^T \tag{9-256}$$

$$\mathbf{U}_\perp = [W_U(1)U(1) \; W_U(2)U(2) \; \ldots \; W_U(F)U(F)] \tag{9-257}$$

where $\text{Re}(\mathbf{U}_\perp \mathbf{U}_\perp^T) = I_r$.

9.S.2 Impact of the Orthogonalization on the Model Equation. From the Gram-Schmidt procedure it follows that the orthogonal polynomial bases $p_n(s)$, $n = 0, 1, \ldots, r-1$, and $q_n(s)$, $n = 0, 1, \ldots, r-1$, are related to the basis s^n, $n = 0, 1, \ldots, r-1$, via a lower triangular r by r matrix

$$W_Y(k) = L_Y W_r(k), \quad W_U(k) = L_U W_r(k) \text{ with } L_Y, L_U \in \mathbb{R}^{r \times r} \tag{9-258}$$

Applying (9-258) to (9-254) and (9-257) gives, using (9-116),

$$\begin{aligned} \mathbf{Y}_\perp &= L_Y \mathbf{Y} & \mathbf{Y}_\perp^{\text{re}} &= L_Y \mathbf{Y}^{\text{re}} \\ \mathbf{U}_\perp &= L_U \mathbf{U} & \mathbf{U}_\perp^{\text{re}} &= L_U \mathbf{U}^{\text{re}} \end{aligned} \tag{9-259}$$

Left multiplication of (9-235) with $\mathbf{N}_U = 0$, by L_Y, gives, using (9-259),

$$\mathbf{Y}_\perp^{\text{re}} = O_{r\perp} \mathbf{X}^{\text{re}} + L_Y S_r L_U^{-1} \mathbf{U}_\perp^{\text{re}} + L_Y \mathbf{N}_Y^{\text{re}} \tag{9-260}$$

with $O_{r\perp} = L_Y O_r$ the generalized extended observability matrix. Constructing the orthogonal projection $\Pi_\perp$ as in (9-237), where $\mathbf{U}^{\text{re}}$ is replaced by $\mathbf{U}_\perp^{\text{re}}$, makes it possible to eliminate the input term in (9-260)

$$\mathbf{Y}_\perp^{\text{re}} \Pi_\perp = O_{r\perp} \mathbf{X}^{\text{re}} \Pi_\perp + \mathbf{N}_\perp \tag{9-261}$$

with $\mathbf{N}_\perp = L_Y \mathbf{N}_Y^{\text{re}} \Pi_\perp$. Using the results of Appendix 9.R and (9-254) it follows that $\mathbf{N}_\perp \mathbf{N}_\perp^T / F$ converges w.p. 1 to

Section 9.24 ■ Appendixes

$$\frac{1}{F} C_{\mathbf{Y}_\perp} = \frac{1}{F} \mathbb{E}\{\mathrm{Re}(\mathbf{N}_{\mathbf{Y}_\perp} \mathbf{N}_{\mathbf{Y}_\perp}^H)\} = \frac{1}{F} \mathrm{Re}(\sum_{k=1}^{F} \sigma_Y^2(k) W_Y(k) W_Y^H(k)) \qquad (9\text{-}262)$$

The range space of $O_{r\perp}$ is estimated as in Appendix 9.R: from $C_{\mathbf{Y}_\perp}^{-1/2} \mathbf{Y}_\perp^{\mathrm{re}} \Pi_\perp = U\Sigma V^T$ we get $\hat{O}_{r\perp} = C_{\mathbf{Y}_\perp}^{1/2} U_{[:, 1:n_a]}$.

9.S.3 Generalized Shift Property. Because $O_{r\perp} = L_Y O_r$ and $\mathbf{Y}_\perp = L_Y \mathbf{Y}$, it follows that the rows of $O_{r\perp}$ and $\mathbf{Y}_\perp$ can be derived from the respective rows of O_r and $\mathbf{Y}$, using the same linear combinations

$$\begin{aligned} \mathbf{Y}_{\perp[n,:]} &= \sum_{m=1}^{n} \gamma_m \mathbf{Y}_{[m,:]} \\ O_{r\perp[n,:]} &= \sum_{m=1}^{n} \gamma_m O_{r[m,:]} \end{aligned} \qquad (9\text{-}263)$$

First, we establish the relationship between the rows of $\mathbf{Y}_\perp$. Next, using (9-263), we show that a similar relationship exists between the rows of $O_{r\perp}$.

Multiplying (9-250) and (9-251), evaluated at s_k, by $Y(k)$ for $k = 1, 2, ..., F$, gives the following relationship between the rows of $\mathbf{Y}_\perp$

1. Initialization:

$$\mathbf{Y}_{\perp[1,:]} = \mathbf{Y}_{[1,:]}/\alpha_1, \; \mathbf{Y}_{\perp[2,:]} = \mathbf{Y}_{\perp[1,:]} D_s/\alpha_2 \qquad (9\text{-}264)$$

with $Y_{[1,:]} = [Y(1) \; Y(2) \; ... \; Y(F)]$ and $D_s = \mathrm{diag}(s_1, s_2, ..., s_F)$.

2. Recursion: for $n = 3$ to r

$$\mathbf{Y}_{\perp[n,:]} = (\mathbf{Y}_{\perp[n-1,:]} D_s + \alpha_{n-1} \mathbf{Y}_{\perp[n-2,:]})/\alpha_n \qquad (9\text{-}265)$$

Using the definition (9-249) of $\langle \, , \, \rangle_Y$, it follows that

$$\alpha_1 = \|\mathbf{Y}_{[1,:]}\|_2, \; \alpha_2 = \|\mathbf{Y}_{\perp[1,:]} D_s\|_2, \; \alpha_n = \|\mathbf{Y}_{\perp[n-1,:]} D_s + \alpha_{n-1} \mathbf{Y}_{\perp[n-2,:]}\|_2 \qquad (9\text{-}266)$$

Using (9-263) we can rewrite (9-265) as

$$\begin{aligned} \mathbf{Y}_{\perp[n,:]} &= (\sum_{m=1}^{n-1} \gamma_m \mathbf{Y}_{[m,:]} D_s + \alpha_{n-1} \sum_{m=1}^{n-2} \gamma_m \mathbf{Y}_{[m,:]})/\alpha_n \\ &= (\sum_{m=1}^{n-1} \gamma_m \mathbf{Y}_{[m+1,:]} + \alpha_{n-1} \sum_{m=1}^{n-2} \gamma_m \mathbf{Y}_{[m,:]})/\alpha_n \end{aligned} \qquad (9\text{-}267)$$

where the last equality is due to the property $\mathbf{Y}_{[m+1,:]} = \mathbf{Y}_{[m,:]} D_s$ (see (9-116)). The second equation of (9-267) is just another way of writing the first equation of (9-263). Hence, the rows of $O_{r\perp}$ should satisfy the same expression

$$\begin{aligned} O_{r\perp[n,:]} &= (\sum_{m=1}^{n-1} \gamma_m O_{r[m+1,:]} + \alpha_{n-1} \sum_{m=1}^{n-2} \gamma_m O_{r[m,:]})/\alpha_n \\ &= (\sum_{m=1}^{n-1} \gamma_m O_{r[m,:]} A + \alpha_{n-1} \sum_{m=1}^{n-2} \gamma_m O_{r[m,:]})/\alpha_n \end{aligned} \qquad (9\text{-}268)$$

where the last equality is due to the shift property $O_{r[m+1,:]} = O_{r[m,:]}A$ (see (9-114)). Using (9-263), the last equation of (9-268) becomes

$$O_{r\perp[n,:]} = (O_{r\perp[n-1,:]}A + \alpha_{n-1}O_{r\perp[n-2]})/\alpha_n \tag{9-269}$$

for $n = 3, 4, \ldots, r$. Following the same lines we get from (9-264)

$$O_{r\perp[1,:]} = O_{r[1,:]}/\alpha_1 \tag{9-270}$$

$$O_{r\perp[2,:]} = O_{r\perp[1,:]}A/\alpha_2 \tag{9-271}$$

From (9-270) it follows that $C = \alpha_1 O_{r\perp[1,:]}$. Writing (9-269) and (9-271) under matrix notation gives the generalized shift property of $O_{r\perp}$

$$[D_1 O_{r\perp[1:r-1,:]}]A = [O_{r\perp[2:r,:]} - b] \tag{9-272}$$

with

$$b = \begin{bmatrix} 0 \\ D_2 O_{r\perp[1:r-2,:]} \end{bmatrix}, \quad \begin{aligned} D_1 &= \mathrm{diag}(1/\alpha_2, 1/\alpha_3, \ldots, 1/\alpha_r) \\ D_2 &= \mathrm{diag}(\alpha_2/\alpha_3, \alpha_3/\alpha_4, \ldots, \alpha_{r-1}/\alpha_r) \end{aligned}$$

9.S.4 Discussion. The range space estimation of $O_{r\perp}$ follows exactly the same lines as the range space estimation of O_r. Therefore, $\hat{O}_{r\perp}$ has the same asymptotic $(F \to \infty)$ properties as $\hat{O}_r$ in Appendix 9.R. Because $\hat{A}$ and $\hat{C}$ are continuous, differentiable functions of $\hat{O}_{r\perp}$, they have the same asymptotic $(F \to \infty)$ properties as $\hat{A}$ and $\hat{C}$ in Appendix 9.R. The third step of both algorithms is identical and, therefore, the properties of $\hat{B}$ and $\hat{D}$ also remain the same.

To orthogonalize the data matrices, we use formulas (9-264) to (9-266) instead of (9-250) and (9-251). Note that the orthogonalization is done without calculating, explicitly, the matrices L_Y and L_U in (9-258). Because $\mathbf{Y}^{\mathrm{re}}\mathbf{Y}^{\mathrm{re}T} = (L_Y^T L_Y)^{-1}$ and $\mathbf{U}^{\mathrm{re}}\mathbf{U}^{\mathrm{re}T} = (L_U^T L_U)^{-1}$, the matrices L_Y and L_U have the same condition numbers as $\mathbf{Y}^{\mathrm{re}}$ and $\mathbf{U}^{\mathrm{re}}$, respectively. Therefore, L_Y and L_U should never be computed. They are used for theoretical derivations only. As $W_Y(k)$ is not explicitly calculated in (9-264) to (9-265), the covariance matrix $C_{\mathbf{Y}_\perp}$ cannot be calculated using (9-262). This is done as follows. Using $\mathbf{N}_{\mathbf{Y}_\perp} = L_Y \mathbf{N}_\mathbf{Y}$ we find

$$C_{\mathbf{Y}_\perp} = L_Y C_\mathbf{Y} L_Y^T = L_Y \, \mathrm{Re}(\mathbf{CC}^H) \, L_Y^T = \mathrm{Re}(\mathbf{C}_\perp \mathbf{C}_\perp^H) \tag{9-273}$$

where $\mathbf{C} = [W_r(1)\sigma_Y(1) W_r(2)\sigma_Y(2) \ldots W_r(F)\sigma_Y(F)]$ and $\mathbf{C}_\perp = L_Y \mathbf{C}$. Because $\mathbf{C}$ has the shift property $\mathbf{C}_{[n+1,:]} = \mathbf{C}_{[n,:]}D_s$ and $\mathbf{C}_\perp = L_Y \mathbf{C}$, the relationship between the rows of $\mathbf{C}_\perp$ is given by (9-264), (9-265).

1. Initialization:

$$\mathbf{C}_{\perp[1,:]} = \mathbf{C}_{[1,:]}/\alpha_1, \quad \mathbf{C}_{\perp[2,:]} = \mathbf{C}_{\perp[1,:]}D_s/\alpha_2 \tag{9-274}$$

2. Recursion: for $n = 3$ to r

$$\mathbf{C}_{\perp[n,\,:]} = (\mathbf{C}_{\perp[n-1,\,:]}D_s + \alpha_{n-1}\mathbf{C}_{\perp[n-2,\,:]})/\alpha_n \tag{9-275}$$

The proof follows exactly the same lines as for $O_{r\perp}$ in Section 9.S.3 of this appendix.

The matrix $\mathbf{Y}_\perp^{\text{re}}\Pi_\perp$ is calculated using a QR decomposition of $Z_\perp^T = [\mathbf{U}_\perp^{\text{re}T}\ \mathbf{Y}_\perp^{\text{re}T}]$ as explained in Section 9.R.2 of Appendix 9.R.

Appendix 9.T Sensitivity Estimates to Noise Model Errors

We study the influence of noise model errors on the bias of the GTLS, ML, and BTLS estimates assuming that a true plant model exists and that it belongs to the considered model set (Assumptions 9.16 and 9.17).

9.T.1 Bias of the ML, GTLS, and BTLS Estimators. If the wrong noise (co)variances are used, then Theorem 9.21, except properties 6 to 8 (consistency, bias, and efficiency), is still valid for the GTLS, ML, and BTLS estimates. The estimates are no longer consistent, $\tilde{\theta}(Z_0) \neq \theta_0$ ($\theta_* \neq \theta_0$), because $\mathbb{E}\{v_F(\theta, N_Z)\}$ is no longer θ independent (use quick analysis tool number 2 of Section 9.5).

An explicit expression of the bias $\tilde{\theta}(Z_0) - \theta_0$ can be found for models (9-7) ($\Omega = z^{-1}$, s, $\sqrt{s}$, and $\tanh(\tau_R s)$) and (9-8) ($\Omega = z^{-1}$) through a Taylor series expansion of the expected value of the cost function at θ_0

$$V_F'^T(\theta) = V_F'^T(\theta_0) + V_F''(\widehat{\theta})(\theta - \theta_0) \tag{9-276}$$

with $\widehat{\theta} = t\theta + (1-t)\theta_0$ and $t \in [0, 1]$. Because $V_F'^T(\tilde{\theta}(Z_0)) = 0$, it follows from (9-276) that

$$\tilde{\theta}(Z_0) - \theta_0 = -V_F''^{-1}(\widehat{\theta})V_F'^T(\theta_0) \tag{9-277}$$

Under Assumptions 9.16 and 9.17 we have $\mathbb{E}\{V_F'(\theta_0, Z_0)\} = 0$ even if the wrong noise (co)variances are used (see (9-72), (9-80), and (9-98)). Together with (9-20), it facilitates rewriting (9-277) as

$$\tilde{\theta}(Z_0) - \theta_0 = -V_F''^{-1}(\widehat{\theta})\,\mathbb{E}\{v_F'(\theta_0, N_Z)\} \tag{9-278}$$

The GTLS, ML, and BTLS estimates are inconsistent because $\mathbb{E}\{v_F'(\theta_0, N_Z)\} \neq 0$ if the wrong noise (co)variances are used. The bias term (9-278) will be calculated explicitly for the ML estimator. From the ML cost (9-80) it follows that

$$\mathbb{E}\{v_F(\theta, N_Z)\} = \frac{1}{F}\sum_{k=1}^{F}\frac{\sigma_e^2(\Omega_k, \theta)}{\hat{\sigma}_e^2(\Omega_k, \theta)} = \frac{1}{F}\sum_{k=1}^{F}\frac{\sigma_Y^2(\Omega_k, \theta)}{\hat{\sigma}_Y^2(\Omega_k, \theta)} \tag{9-279}$$

where $\sigma_e^2(\Omega_k, \theta)$, $\sigma_Y^2(\Omega_k, \theta)$, and $\hat{\sigma}_e^2(\Omega_k, \theta)$, $\hat{\sigma}_Y^2(\Omega_k, \theta)$ are calculated as in (9-35), (9-45) using, respectively, the true and wrong noise (co)variances. The second equality in (9-279) is obtained via $\sigma_e^2(\Omega_k, \theta) = |A(\Omega_k, \theta)|^2 \sigma_Y^2(\Omega_k, \theta)$. Using

$$G(\Omega_k, \theta_0) = G_0(\Omega_k) = Y_0(k)/U_0(k)$$

$$\frac{\partial |G(\Omega_k, \theta)|^2}{\partial \theta_0} = 2|G_0(\Omega_k)|^2 \text{Re}(\frac{\partial \ln(G(\Omega_k, \theta))}{\partial \theta_0})$$

$$\text{Re}(\sigma_{YU}^2(k)\frac{\partial \overline{G}(\Omega_k, \theta)}{\partial \theta_0}) = \text{Re}(\overline{\sigma}_{YU}^2(k) G_0(\Omega_k)\frac{\partial \ln(G(\Omega_k, \theta))}{\partial \theta_0})$$

we get

$$\frac{\partial \sigma_Y^2(\Omega_k, \theta)}{\partial \theta_0} = 2|Y_0(k)|^2 \text{Re}(V_U(k)\frac{\partial \ln(G(\Omega_k, \theta))}{\partial \theta_0}) \quad (9\text{-}280)$$

$$V_U(k) = \sigma_U^2(k)/|U_0(k)|^2 - \overline{\sigma}_{YU}^2/(\overline{Y}_0(k)U_0(k))$$

Using (9-280) and

$$\sigma_Y^2(\Omega_k, \theta_0) = |Y_0(k)|^2(V_U(k) + V_Y(k))$$
$$V_Y(k) = \sigma_Y^2(k)/|U_0(k)|^2 - \sigma_{YU}^2/(Y_0(k)\overline{U}_0(k)) \quad (9\text{-}281)$$

the derivative of (9-279) w.r.t. θ at θ_0 equals

$$\mathbb{E}\{v_F'(\theta_0, N_Z)\} = \frac{2}{F}\sum_{k=1}^{F} \text{Re}\left(\frac{\partial \ln(G(\Omega_k, \theta))}{\partial \theta_0}\frac{V_U(k)\hat{V}_Y(k) - \hat{V}_U(k)V_Y(k)}{(\hat{V}_U(k) + \hat{V}_Y(k))^2}\right) \quad (9\text{-}282)$$

where $\hat{V}_U(k)$, $\hat{V}_Y(k)$ equal $V_U(k)$, $V_Y(k)$ evaluated with the wrong noise (co)variances.

From (9-278) and (9-282), it follows that the bias is a function of the difference between the actual and the true noise (co)variances. The same is true for the GTLS and BTLS estimators (see Exercise 9.9). If the noise covariance matrix used $\hat{C}_{N_Z(k)}$ satisfies (9-129), then $\hat{V}_Y(k) = f(k)V_Y(k)$, $\hat{V}_U(k) = f(k)V_U(k)$ and the bias (9-278) is zero.

For model (9-8) with $\Omega = s$, the bias $\theta_* - \theta_0$ is calculated via a Taylor series expansion of $V_*'(\theta)$ at θ_0. Following the same lines as in the previous paragraph, we find

$$\theta_* - \theta_0 = -V_*''^{-1}(\widehat{\theta})\lim_{F \to \infty}\mathbb{E}\{v_F'(\theta_0, N_Z)\} \quad (9\text{-}283)$$

with $\widehat{\theta} = t\theta_* + (1-t)\theta_0$ and $t \in [0, 1]$. Comparing this expression with (9-278), it follows that the conclusions of the previous paragraph also apply to the bias $\theta_* - \theta_0$ of model (9-8) with $\Omega = s$.

9.T.2 Bias of the NLS-FRF and LOG Estimators. The NLS-FRF (9-47) and LOG estimators (9-55) apply to model (9-7). The bias $\tilde{\theta}(Z_0) - \theta_0$ calculation follows the same lines as in the previous section. Therefore, expression (9-277) for the bias is valid with

$$V_F'(\theta_0) = -\frac{2}{F}\sum_{k=1}^{F}\text{Re}(\frac{\partial \ln(G(\Omega_k, \theta))}{\partial \theta_0}\overline{b}(k)g(k)) \quad (9\text{-}284)$$

where $g(k) = 1$, $b(k) = \mathbb{E}\{N(k)\}$ and $N(k)$ is defined in (9-56) for the LOG estimator, and $g(k) = |G_0(\Omega_k)|^2$, $b(k) = \mathbb{E}\{N_G(k)\}/G_0(\Omega_k)$ and $N_G(k)$ is defined in (9-48) for the NLS-FRF estimator. For circular complex normally distributed input-output errors, the bias $b(k)$ is given by (9-50) and (9-57). For circular complex noise with even pdf, the bias is a function of the fourth-order moments of the noise (see Appendix 9.G). If the noise is not circular complex then $b(k)$ is a function of the second-order moments of the noise (put $\mathbb{E}\{z^2\} \neq 0$ in (9-178)). □

Appendix 9.U IWLS Solution in Case of Vector Orthogonal Polynomials

Using the inner product (9-138), the orthogonality condition (9-139), and

$$\begin{bmatrix} A(\Omega, \theta^{(i)}) \\ B(\Omega, \theta^{(i)}) \end{bmatrix} = \sum_{r=0}^{n_a+n_b+1} a_r \begin{bmatrix} p_r(\Omega) \\ q_r(\Omega) \end{bmatrix} \tag{9-285}$$

with $a_{n_a+n_b+1} = 1$, the cost function (9-136) can be written as

$$\left\langle \begin{bmatrix} A(\Omega, \theta^{(i)}) \\ B(\Omega, \theta^{(i)}) \end{bmatrix}, \begin{bmatrix} A(\Omega, \theta^{(i)}) \\ B(\Omega, \theta^{(i)}) \end{bmatrix} \right\rangle = 1 + \sum_{r=0}^{n_a+n_b} a_r^2 \tag{9-286}$$

It follows directly that (9-286) is minimal for $a_0 = a_1 = \ldots = a_{n_a+n_b} = 0$. □

Appendix 9.V Asymptotic Properties in the Presence of Nonlinear Distortions

Consider the errors-in-variables model (9-146) where $M_U(k)$, $M_Y(k)$, and $N_P(k)$ satisfy the noise assumptions of a time-domain experiment in Section 9.6. Multiplying (9-146) by $e^{-j\angle U_0(k)}$ gives

$$\begin{aligned} Y(k)e^{-j\angle U_0(k)} &= G_{\text{BLA}}(s_k)|U_0(k)| + N_Y(k)e^{-j\angle U_0(k)} \\ U(k)e^{-j\angle U_0(k)} &= |U_0(k)| + N_U(k)e^{-j\angle U_0(k)} \end{aligned} \tag{9-287}$$

for $k = 1, 2, \ldots, F$. Note that this phase shift, applied to model (9-7) or (9-9), does not change any of the cost functions of Sections 9.8 to 9.14. To prove that Theorem 9.21, with $G_0(s)$ replaced by $G_{\text{BLA}}(s)$, is valid, it is sufficient to show that the noisy part $N_Z(k)e^{-j\angle U_0(k)}$ of (9-287) satisfies all the assumptions of Section 9.6. First, note that $N_Z(k)e^{-j\angle U_0(k)}$ is independent of the true unknown excitation $|U_0(k)|$. Because $M_U(k)$, $M_Y(k)$ and $N_P(k)$ are independent of $Y_S(k)$ and $U_0(k)$, it follows that $M_U(k)e^{-j\angle U_0(k)}$, $M_Y(k)e^{-j\angle U_0(k)}$, and $N_P(k)e^{-j\angle U_0(k)}$ satisfy the noise assumptions of Section 9.6. $Y_S(k)e^{-j\angle U_0(k)}$ has the same phase as $G_S(k)$ and, therefore, has the same stochastic properties as $G_S(k)$ in Theorems 3.10 (mixing of order infinity) and 3.11 (asymptotic normality). As $Y_S(k)e^{-j\angle U_0(k)}$ is mixing of order four (infinity), all properties of Theorem 9.21 remain valid (proof: see introduction of Appendix 9.E). □

Appendix 9.W Consistency of the Missing Data Problem

The consistency proof follows the lines of the proof of Theorem 9.21 (Appendix 9.E). We first show the result for discrete-time systems ((6-46) with $\Omega = z^{-1}$) and afterward for continuous-time systems ((6-46) with $\Omega = s, \sqrt{s}$). Because the output error $Y^m(k) - Y^m(\Omega_k, \Theta)$ is linear in the missing samples ψ, we can rewrite the cost function (9-148) as

$$V(\theta, \psi, Z^m) = \frac{1}{2}(\varepsilon_1(\theta, Z^m) + \varepsilon_2(\theta)\psi)^T(\varepsilon_1(\theta, Z^m) + \varepsilon_2(\theta)\psi) \qquad (9\text{-}288)$$

where $\varepsilon_1(\theta, Z^m) \in \mathbb{R}^N$ is a linear function of the missing data set Z^m, and $\varepsilon_2(\theta) \in \mathbb{R}^{N \times (M_U + M_y)}$ is independent of Z^m. Elimination of ψ in (9-288) gives

$$\psi(\theta, Z^m) = -(\varepsilon_2^T(\theta)\varepsilon_2(\theta))^{-1}\varepsilon_2^T(\theta)\varepsilon_1(\theta, Z^m) \qquad (9\text{-}289)$$

$$V(\theta, Z^m) = \frac{1}{2}\varepsilon_1^T(\theta, Z^m)P(\theta)\varepsilon_1(\theta, Z^m) \qquad (9\text{-}290)$$

with $P(\theta) = I_N - \varepsilon_2(\theta)(\varepsilon_2^T(\theta)\varepsilon_2(\theta))^{-1}\varepsilon_2^T(\theta)$ a symmetric idempotent matrix of rank $N - M_u - M_y$. As $\varepsilon_2(\theta)$ lies in the null space of $P(\theta)$, (9-290) can be written as

$$V(\theta, Z^m) = \frac{1}{2}(\varepsilon_1(\theta, Z^m) + \varepsilon_2(\theta)\varphi_Z)^T P(\theta)(\varepsilon_1(\theta, Z^m) + \varepsilon_2(\theta)\varphi_Z) \qquad (9\text{-}291)$$

with φ_Z the vector ψ (6-45) where the missing input and output samples are replaced by the disturbing noise on these samples. To simplify the notations, we will assume without any loss of generality that the excitation is deterministic. Because the noisy part of $\varepsilon_1(\theta, Z^m) + \varepsilon_2(\theta)\varphi_Z$ contains the DFT spectra of the complete disturbing noise sequences (no missing samples) and the output error $Y^m(k) - Y^m(\Omega_k, \Theta)$ in (9-148) is divided by $\sigma_Y(\Omega_k, \theta)$ (9-45), which contains the (co)variances of the complete noise sequences, we have $\text{Cov}(\varepsilon_1(\theta, Z^m) + \varepsilon_2(\theta)\varphi_Z) = I_N$. Using this result together with $\text{trace}(P(\theta)) = N - M_u - M_y$, the expected value of (9-291) equals (see Exercise 17.2)

$$\mathbb{E}\{V(\theta, Z^m)\} = \frac{1}{2}\varepsilon_1^T(\theta, Z_0^m)P(\theta)\varepsilon_1(\theta, Z_0^m) + \frac{1}{2}(N - M_u - M_y) \qquad (9\text{-}292)$$

with Z_0^m the true (noiseless) missing data set. Under Assumption 9.16 we have

$$P(\theta_0)\varepsilon_1(\theta_0, Z_0^m) = \varepsilon_1(\theta_0, Z_0^m) + \varepsilon_2(\theta_0)\psi(\theta_0, Z_0^m) = \varepsilon_1(\theta_0, Z_0^m) + \varepsilon_2(\theta_0)\psi_0 = 0 \qquad (9\text{-}293)$$

so that θ_0 minimizes the expected value of the cost function (9-292). Since $P(\theta) = V\Sigma V^T$, with V an orthogonal matrix and

$$\Sigma = \begin{bmatrix} I_{N-M_u-M_y} & 0 \\ 0 & 0 \end{bmatrix}$$

(see Exercise 15.19), the cost function (9-291) can be written as

$$V(\theta, Z^m) = \frac{1}{2}\varepsilon_3^T(\theta, Z^m)\varepsilon_3(\theta, Z^m)$$
$$\varepsilon_3(\theta, Z^m) = [I_{N-M_U-M_Y}\ 0]V^T(\varepsilon_1(\theta, Z^m) + \varepsilon_2(\theta)\varphi_Z) \quad (9\text{-}294)$$

Under Assumptions 9.18 and 9.19, the entries of the vector $\varepsilon_3(\theta, Z^m)$ are independent Gaussian random variables with variance 1, so that $V(\theta, Z^m)/N$ converges ($N \to \infty$) w.p. 1 to its expected value (see Section 16.9, version 2 of the strong law of large numbers). Under Assumption 9.6, applied to $V(\theta, Z^m)$ (9-294), this convergence is uniform in a closed and bounded neighborhood of θ_0. Under Assumption 9.7, applied to $V(\theta, Z^m)$ (9-294), this implies the strong convergence of the minimizer $\hat{\theta}_{WNLS}(Z^m)$ of (9-294) to the minimizer, θ_0, of (9-292) (see Appendix 17.B). The true coefficients of the $T_G(\Omega, \theta)$ polynomial in (9-147) are asymptotically zero (Lemma 6.5), thus only the plant model parameters $a_0 a_1 \ldots a_{n_a} b_0 b_1 \ldots b_{n_b}$ in θ are consistently estimated. Note also that the estimate of the missing data $\hat{\psi}_{WNLS} = \psi(\hat{\theta}_{WNLS}(Z^m), Z^m)$ (see (9-289)) is inconsistent.

The proof for continuous-time systems (6-46) follows exactly the same lines. The only differences are that θ_0 minimizes the limit cost function $\lim_{N \to \infty} \mathbb{E}\{V(\theta, Z^m)\}/N$ instead of (9-292) and that $V(\theta, Z^m)/N$ convergences weakly, instead of strongly, to its expected value. This is due to the presence of the alias term $\delta(s_k)$ in the true output observations (see model (6-46)) which is only in probability ($N \to \infty$) zero (Lemma 6.6). □

Appendix 9.X   Normal Equation for Complex Parameters and Analytic Residuals

Because $\varepsilon(\theta, Z)$ is an analytic function of θ we have

$$\frac{\partial \varepsilon(\theta, Z)}{\partial \text{Re}(\theta)} = \frac{\partial \varepsilon(\theta, Z)}{\partial \theta} \text{ and } \frac{\partial \varepsilon(\theta, Z)}{\partial \text{Im}(\theta)} = j\frac{\partial \varepsilon(\theta, Z)}{\partial \theta} \quad (9\text{-}295)$$

so that

$$\frac{\partial \text{Re}(\varepsilon(\theta, Z))}{\partial \text{Re}(\theta)} = \text{Re}\left(\frac{\partial \varepsilon(\theta, Z)}{\partial \theta}\right)$$
$$\frac{\partial \text{Im}(\varepsilon(\theta, Z))}{\partial \text{Re}(\theta)} = \text{Im}\left(\frac{\partial \varepsilon(\theta, Z)}{\partial \theta}\right)$$
$$\frac{\partial \text{Re}(\varepsilon(\theta, Z))}{\partial \text{Im}(\theta)} = \text{Re}\left(j\frac{\partial \varepsilon(\theta, Z)}{\partial \theta}\right) = -\text{Im}\left(\frac{\partial \varepsilon(\theta, Z)}{\partial \theta}\right) \quad (9\text{-}296)$$
$$\frac{\partial \text{Im}(\varepsilon(\theta, Z))}{\partial \text{Im}(\theta)} = \text{Im}\left(j\frac{\partial \varepsilon(\theta, Z)}{\partial \theta}\right) = \text{Re}\left(\frac{\partial \varepsilon(\theta, Z)}{\partial \theta}\right)$$

Equations (9-296) are known as the Cauchy-Riemann conditions of an analytic function (Henrici, 1974). Using (9-296) and definition (15-40), we find

$$\frac{\partial \varepsilon_{\text{re}}(\theta, Z)}{\partial \theta_{\text{re}}} = \begin{bmatrix} \dfrac{\partial \text{Re}(\varepsilon(\theta, Z))}{\partial \text{Re}(\theta)} & \dfrac{\partial \text{Re}(\varepsilon(\theta, Z))}{\partial \text{Im}(\theta)} \\ \dfrac{\partial \text{Im}(\varepsilon(\theta, Z))}{\partial \text{Re}(\theta)} & \dfrac{\partial \text{Im}(\varepsilon(\theta, Z))}{\partial \text{Im}(\theta)} \end{bmatrix} = \left(\frac{\partial \varepsilon(\theta, Z)}{\partial \theta} \right)_{\text{Re}} \qquad (9\text{-}297)$$

Applying (9-297) and Lemma 15.4 to the right-hand side of (9-153) gives

$$J_{\text{re}}(\theta_{\text{re}}^{(i-1)}, Z) \Delta \theta_{\text{re}}^{(i)} = (J(\theta^{(i-1)}, Z))_{\text{Re}} \Delta \theta_{\text{re}}^{(i)} = (J(\theta^{(i-1)}, Z) \Delta \theta^{(i)})_{\text{re}} \qquad (9\text{-}298)$$

(9-298) together with $\varepsilon_{\text{re}}(\theta_{\text{re}}^{(i-1)}, Z) = (\varepsilon(\theta^{(i-1)}, Z))_{\text{re}}$ shows that (9-153) is equivalent with (9-154). $\square$

Appendix 9.Y Total Least Squares for Complex Parameters

From Appendix 9.X, it follows that

$$J_{\text{re}}(Z) = \frac{\partial \varepsilon_{\text{re}}(\theta_{\text{re}}, Z)}{\partial \theta_{\text{re}}} = \left(\frac{\partial \varepsilon(\theta, Z)}{\partial \theta} \right)_{\text{Re}} = (J(Z))_{\text{Re}} \qquad (9\text{-}299)$$

Using (9-299) and Lemma 15.4 we find

$$\begin{aligned} W_{\text{Re}} J_{\text{re}}(Z) &= (WJ(Z))_{\text{Re}} \\ C^T C &= \mathbb{E}\{ W_{\text{Re}}^T j_{\text{re}}^T(N_Z) j_{\text{re}}(N_Z) W_{\text{Re}} \} = (\mathbb{E}\{ W^H j^H(N_Z) j(N_Z) W \})_{\text{Re}} = (C_c^H C_c)_{\text{Re}} \end{aligned} \qquad (9\text{-}300)$$

with $j(N_Z) = J(Z) - J(Z_0)$, and a possible choice for C is $C = (C_c)_{\text{Re}}$. The total least squares solution is calculated via the GSVD of the real matrix pair $(W_{\text{Re}} J_{\text{re}}(Z), C)$ (see Section 9.10). Because $W_{\text{Re}} J_{\text{re}}(Z) = (WJ(Z))_{\text{Re}}$ and $C = (C_c)_{\text{Re}}$, it can also be calculated via the GSVD of the complex matrix pair $(WJ(Z), C_c)$ (see Section 15.8). $\square$

10

Estimation with Unknown Noise Model – Standard Solutions

Abstract: In the identification schemes that were presented in the previous chapters, it was assumed that the covariance matrix of the noise is known a priori. In practice this information should also be extracted from the experimental data. In this chapter, it is shown that a utilizable nonparametric frequency domain noise model can be obtained from a very small number M of independent repeated experiments. Under these conditions the consistency of the estimates is maintained, while the loss in efficiency is small. In practice, $M \geq 4$ consecutive periods of the steady-state response to a periodic excitation are taken as independent repeated experiments. As such it is assumed that the correlation length of the disturbing noise is small compared with the period length of the excitation. Also the classical solution for identifying a parametric noise model together with the plant model is discussed. Both the discrete-time and continuous-time cases are handled.

10.1 INTRODUCTION

In Chapter 9 a large variety of estimators were discussed, ranging from unweighted linear least squares methods to maximum likelihood estimators. The more advanced estimators such as Markov, GTLS, BTLS, and ML estimators require knowledge of the covariance matrix with the disturbing noise as a function of the frequency. For example, the ML estimator was given as the minimizer of (9-80), which reduces to (10-1) if no transients are added to the model:

$$V_{\mathrm{ML}}(\theta, Z) = \sum_{k=1}^{F} \frac{|e(\Omega_k, \theta, Z(k))|^2}{\sigma_e^2(\Omega_k, \theta)} = \sum_{k=1}^{F} |\varepsilon(\Omega_k, \theta, Z(k))|^2 \quad (10\text{-}1)$$

with $e(\Omega_k, \theta, Z(k)) = A(\Omega_k, \theta)Y(k) - B(\Omega_k, \theta)U(k)$, $\sigma_e^2(\Omega_k, \theta) = \mathrm{var}(e(\Omega_k, \theta, N_Z(k)))$,

$$\sigma_e^2(\Omega_k, \theta) = \sigma_Y^2(k)|A(\Omega_k, \theta)|^2 + \sigma_U^2(k)|B(\Omega_k, \theta)|^2 - 2\mathrm{Re}(\sigma_{YU}^2(k)A(\Omega_k, \theta)\bar{B}(\Omega_k, \theta)) \quad (10\text{-}2)$$

and $\varepsilon(\Omega_k, \theta, Z(k)) = e(\Omega_k, \theta, Z(k))/\sigma_e(\Omega_k, \theta)$. The noise (co)variances $\sigma_U^2(k)$, $\sigma_Y^2(k)$, and $\sigma_{YU}^2(k)$ were assumed to be known exactly, and under these conditions the properties of the estimators were studied. In practice, this information is not available but should be extracted from the experimental data. In this chapter we will replace the exact noise (co)variances by their sample values. This is possible only if independent, repeated experiments are available. A practical solution consists of applying periodic excitations to the plant and observing M consecutive periods of the steady-state response. Therefore, the simple plant model (9-7) will be used throughout this chapter

$$Y(\Omega_k, \theta) = G(\Omega_k, \theta) U(k) \tag{10-3}$$

with $\Omega = z^{-1}$, s, $\sqrt{s}$, or $\tanh(\tau_R s)$. The M experiments are processed and their DFT spectra

$$U^{[l]}(k), Y^{[l]}(k), \ l = 1, \ldots, M \text{ and } k = 1, \ldots, F \tag{10-4}$$

are calculated as explained in Chapter 2. The sample (co)variances are obtained directly from these measurements, e.g.,

$$\hat{\sigma}_U^2(k) = \frac{1}{M-1} \sum_{l=1}^{M} |U^{[l]}(k) - \hat{U}(k)|^2, \text{ with } \hat{U}(k) = \frac{1}{M} \sum_{l=1}^{M} U^{[l]}(k) \tag{10-5}$$

(see (2-36) and (2-37)). The underlying assumption made here is that the correlation length of the input-output noise is small compared with the period length of the excitation; otherwise the DFT spectra of the consecutive periods cannot be considered to be independently distributed. The values (10-5) are used in (10-1) and (10-2) instead of the exact values.

To compare this approach with the classical framework that deals with arbitrary excitations (Ljung, 1999), we have to simplify the errors-in-variables framework to a weighted output error problem. This means that only process (open and closed loop setup) and output measurement (open loop only) noise is considered; the input measurement noise is assumed to be zero ($\sigma_U^2(k) = 0$ and also $\sigma_{YU}^2(k) = 0$) so that the cost function (10-1) reduces to

$$V_{\text{ML}}(\theta, Z) = \sum_{k=1}^{F} \frac{\left| Y(k) - \frac{B(\Omega_k, \theta)}{A(\Omega_k, \theta)} U(k) \right|^2}{\sigma_Y^2(k)} = \sum_{k=1}^{F} \frac{|Y(k) - G(\Omega_k, \theta) U(k)|^2}{\sigma_Y^2(k)} \tag{10-6}$$

Because in this classical framework no repeated measurements are imposed, the sample variance $\hat{\sigma}_Y^2(k)$ cannot be calculated. Instead a parametric noise model $\sigma_Y^2(k) = \sigma^2 |H(z_k^{-1}, \theta)|^2$ is used and the additional noise model parameters are estimated together with the plant model parameters (Ljung, 1999). This poses the question of what approach should be preferred: the parametric or the nonparametric (sample (co)variances) noise modeling approach?

The major advantages of the parametric modeling approach are its applicability to arbitrary excitations, the ability to identify the plant dynamics from output observations only (ARMA models), and the fact that the noise model helps to identify the plant poles for some model structures (ARX, ARMAX). Its major disadvantages are the need for a double model selection problem (plant and noise models), the more complex optimization problem, and the fact that the identified noise model is inconsistent in the presence of plant model errors. The reader is referred to Ljung (1999) for a comprehensive discussion of these techniques. A frequency domain solution to this problem is also given in Section 10.9 of this chapter. Both the discrete-time and continuous-time cases are handled.

The major disadvantages of the nonparametric approach *presented in this chapter* are the restriction to periodic excitations under steady-state conditions, the loss in frequency resolution of a factor $M \geq 4$ w.r.t. the parametric approach, and the (weak) correlation over the consecutive periods of the DFT spectra due to the coloring of the disturbing noise. However, whenever periodic excitations can be applied, significant advantages appear: the nonparametric model is generated automatically, without any user interaction; the errors-in-variables problem can be solved straightforwardly (no equivalent solution is available in the classical approach); the cost function is absolutely interpretable, which simplifies the validation process significantly (see Chapter 11); and the quality of the nonparametric noise model is independent of the identified parametric plant model. For these reasons, we prefer to use the nonparametric noise models whenever it is possible to apply periodic excitations, independent of the fact that a time or frequency domain method will be used later on.

Note that the major disadvantages of the nonparametric approach presented here are tackled in Chapter 12 via a local polynomial approximation of the frequency response function and the (noise) transient (leakage) errors: the correlation between consecutive signal periods is suppressed by removing nonparametrically the noise transients in the input-output data; the frequency resolution is increased by handling the first two periods of the transient response of the plant to a periodic excitation; and the nonparametric noise analysis is generalized to arbitrary excitations.

10.2 DISCUSSION OF THE DISTURBING NOISE ASSUMPTIONS

10.2.1 Assuming Independent Normally Distributed Noise for Time Domain Experiments

Actually, we will prove the theorems under the frequency domain experiment Assumption 9.18 and Assumption 9.19 (assuming independently normally distributed noise in the frequency domain) while in practice the data are obtained from a time domain experiment. In the latter it is assumed that the disturbing noise is described by a filtered white noise source, which introduces a (weak) correlation over the oberserved signal periods. Switching to the idealized frequency domain assumptions makes it possible to set up a formal theory to analyze the replacement of the exact variances by their sample values. Moreover, this mixed use of both assumptions is supported by Theorem 16.25 and will be further discussed later.

The following assumptions are necessary to study the asymptotic behavior ($F \to \infty$) of the estimators. First, we require that M independent repeated experiments are available. Next, we make an assumption about the disturbing errors of the lth experiment.

Assumption 10.1 (M Independent Repeated Experiments): The measured input-output DFT spectra $U^{[l]}(k)$, $Y^{[l]}(k)$, $k = 1, 2, \ldots, F$ and $l = 1, 2, \ldots, M$, satisfy

$$Y^{[l]}(k) = Y_0(k) + N_Y^{[l]}(k)$$
$$U^{[l]}(k) = U_0(k) + N_U^{[l]}(k)$$
(10-7)

where the true unknown deterministic values $U_0(k)$, $Y_0(k)$ are independent of l and where the disturbing input-output errors $N_U^{[l]}(k)$, $N_Y^{[l]}(k)$ are independent over l.

Following the lines of Chapter 9 we introduce the data vector $Z^{[l]}$ (see (9-3))

$$Z^{[l]T} = [Z^{[l]T}(1)Z^{[l]T}(2)...Z^{[l]T}(F)] \text{ with } Z^{[l]T}(k) = [Y^{[l]}(k) \ U^{[l]}(k)] \quad (10\text{-}8)$$

and similarly for $N_Z^{[l]}$. It is related to the true values by $Z^{[l]} = Z_0 + N_Z^{[l]}$.

Assumption 10.2 (Zero Mean Normally Distributed Errors): The noise $N_Z^{[l]}(k)$ is independent over the frequency k, has zero mean, and is circular complex normally distributed with covariance matrix

$$C_{N_Z}(k) = \mathbb{E}\{N_Z^{[l]}(k)(N_Z^{[l]}(k))^H\} = \begin{bmatrix} \sigma_Y^2(k) & \sigma_{YU}^2(k) \\ \bar{\sigma}_{YU}^2(k) & \sigma_U^2(k) \end{bmatrix} \quad (10\text{-}9)$$

In the theory that is set up below, the normal distribution of the noise will be a kernel property. It is asymptotically guaranteed by Theorem 16.25. It is assumed that the noise is independent from one frequency to the other. Again, this property is only asymptotically met in practice, so that in principle a full covariance matrix should be used, including the covariance over different frequencies. However, this would make the nonparametric approach very intractable because the large full matrix has to be inverted to calculate the cost function. It is shown that under the time domain experiment Assumption 9.3, the nondiagonal terms can be omitted without affecting the asymptotic properties of the estimates, so that the cost function (10-1) still can be used (Schoukens et al., 1999a).

10.2.2 Considering Successive Periods as Independent Realizations

The noise behavior is characterized using the sample mean and sample variance, obtained from a set of repeated measurements. In practice we often obtain these repeated measurements by measuring M successive periods in one record. For each period, we calculate the Fourier coefficients and consider them as independent experiments from one period to the other as formalized in Assumption 10.1. Again, this is only approximately met in practice because some correlation exists between neighboring periods. Because the correlation of filtered white noise (time domain experiment assumption) decays exponentially, the correlation between two neighboring periods disappears in inverse proportion to the length of the period. In practice it can be neglected if the period length is large compared with the correlation length of the noise.

10.3 PROPERTIES OF THE ML ESTIMATOR USING A SAMPLE COVARIANCE MATRIX

10.3.1 The Sample Maximum Likelihood Estimator: Definition of the Cost Function

A new cost function is defined putting $\hat{U}(k)$, $\hat{Y}(k)$ and $\hat{\sigma}_U^2(k)$, $\hat{\sigma}_Y^2(k)$, $\hat{\sigma}_{YU}^2(k)$ as the measurements and the variances, respectively, into the cost function (10-1):

Section 10.3 ■ Properties of the ML Estimator Using a Sample Covariance Matrix

$$V_{\text{SML}}(\theta, Z) = \sum_{k=1}^{F} \frac{|\hat{e}(\Omega_k, \theta, \hat{Z}(k))|^2}{\hat{\sigma}_{\hat{e}}^2(\Omega_k, \theta)} = \sum_{k=1}^{F} |\hat{\varepsilon}(\Omega_k, \theta, \hat{Z}(k))|^2 \quad (10\text{-}10)$$

with $\hat{\varepsilon}(\Omega_k, \theta, \hat{Z}(k)) = \hat{e}(\Omega_k, \theta, \hat{Z}(k)) / \hat{\sigma}_{\hat{e}}(\Omega_k, \theta)$ and

$$\hat{e}(\Omega_k, \theta, \hat{Z}(k)) = A(\Omega_k, \theta)\hat{Y}(k) - B(\Omega_k, \theta)\hat{U}(k)$$
$$\hat{\sigma}_{\hat{e}}^2(\Omega_k, \theta) = \hat{\sigma}_e^2(\Omega_k, \theta)/M \quad (10\text{-}11)$$
$$\hat{\sigma}_e^2(\Omega_k, \theta) = \hat{\sigma}_Y^2(k)|A(\Omega_k, \theta)|^2 + \hat{\sigma}_U^2(k)|B(\Omega_k, \theta)|^2 - 2\text{Re}(\hat{\sigma}_{YU}^2(k)A(\Omega_k, \theta)\bar{B}(\Omega_k, \theta))$$

$\hat{\sigma}_e^2(\Omega_k, \theta) = \text{var}(\hat{e}(\Omega_k, \theta, N_Z^{[l]}(k)))$ stands for the variance of the equation error of one experiment, while $\hat{\sigma}_{\hat{e}}^2(\Omega_k, \theta) = \text{var}(\hat{e}(\Omega_k, \theta, \hat{N}_Z(k)))$ is the variance of the sample mean of the equation error.

10.3.2 Properties of the Sample Maximum Likelihood Estimator

The most important concern, when replacing the exact noise (co)variances by their sample values, is the loss in quality of the new estimator $\hat{\theta}_{\text{SML}}(Z)$ with respect to the original estimate $\hat{\theta}_{\text{ML}}(Z)$ due to this change. It turns out that this loss is small, even for a very small number of periods, typically 4 or 7. It will be shown that the sample estimate $\hat{\theta}_{\text{SML}}(Z)$ converges asymptotically $(F \to \infty)$ to $\hat{\theta}_{\text{ML}}(Z)$ (the estimate obtained with the exact noise (co)-variances). Also the loss in efficiency is small. The covariance matrix of the estimates grows with a factor $(M-2)/(M-3)$. These results are formulated precisely in the next two theorems. The first theorem gives a precise formulation of the properties of the sample estimate. The second describes the relationship between the "sample" estimate and the "exact" estimate.

Theorem 10.3 (Asymptotic Properties $\hat{\theta}_{\text{SML}}(Z)$): Consider model (10-3) with any identifiable parameterization of Section 6.2. Under the assumptions of Section 9.6, Assumptions 10.1 ($M \geq 4$), and Assumption 10.2 the minimizer $\hat{\theta}_{\text{SML}}(Z)$ of (10-10) has the asymptotic properties of Theorem 9.21 with $V_F(\theta, Z) = V_{\text{SML}}(\theta, Z)/F$. For

1. $M \geq 4$ the stochastic and the deterministic convergence (properties 1, 5, and 6 of Theorem 9.21) are valid.
2. $M \geq 6$ the stochastic convergence rate (properties 2 and 6 of Theorem 9.21) are valid.
3. $M \geq 7$ the systematic and stochastic errors, the asymptotic normality, and the asymptotic bias (properties 3, 4, 6 and 7 of Theorem 9.21) are valid.

Proof. Apply Theorem 9.21 to (10-10), using the results of Appendix 10.D (which guarantees that all moments that appear in the proof exist). □

Theorem 10.4 (Relationship between $\hat{\theta}_{\text{SML}}(Z)$ and $\hat{\theta}_{\text{ML}}(Z)$): Under the conditions of Theorem 10.3, the estimates based on the true (10-1) and the sample (10-10) noise (co)variances are related to each other by:

1. For $M \geq 3$, the expected value of the cost functions,

$$V_{\text{SML}}(\theta) = \frac{M-1}{M-2} V_{\text{ML}}(\theta), \qquad (10\text{-}12)$$

2. For $M \geq 4$, the asymptotic value of the cost functions,

$$V_{*\text{SML}}(\theta) = \frac{M-1}{M-2} V_{*\text{ML}}(\theta) \qquad (10\text{-}13)$$

the minimizer of the expected value of the cost functions,

$$\tilde{\theta}_{\text{SML}}(Z_0) = \tilde{\theta}_{\text{ML}}(Z_0) \qquad (10\text{-}14)$$

and the minimizer of the asymptotic value of the cost functions,

$$\theta_{*\text{SML}} = \theta_{*\text{ML}} \qquad (10\text{-}15)$$

3. For $M \geq 7$, the parameter uncertainty in the absence of modeling errors, $\tilde{\theta}_{\text{SML}}(Z_0) = \theta_0$,

$$\text{Cov}(\delta_{\theta\text{SML}}(Z)) \approx \frac{M-2}{M-3} \text{Cov}(\delta_{\theta\text{ML}}(Z)), \qquad (10\text{-}16)$$

where $\hat{\theta}(Z) = \theta_0 + \delta_\theta(Z) + O_p(F^{-1})$ with $\mathbb{E}\{\delta_\theta(Z)\} = 0$ and $\delta_\theta(Z) = O_p(F^{-1/2})$, and where $\sqrt{F}\delta_\theta(Z)$ is asymptotically normally distributed.

Proof. See the proof of Theorem 10.3 and Appendices 10.B, 10.F. □

The full proofs of both theorems are in the appendices, but the basic idea is easy to grasp. Consider

$$V_{\text{SML}}(\theta, Z) = \sum_{k=1}^{F} \frac{|e(\Omega_k, \theta, \hat{Z}(k))|^2}{\hat{\sigma}_e^2(\Omega_k, \theta)} = \sum_{k=1}^{F} c_k(\theta) d_k(\theta)$$

$$c_k(\theta) = \sigma_e^2(\Omega_k, \theta)/\hat{\sigma}_e^2(\Omega_k, \theta) \qquad (10\text{-}17)$$

$$d_k(\theta) = |e(\Omega_k, \theta, \hat{Z}(k))|^2 / \sigma_e^2(\Omega_k, \theta)$$

Observe that $c_k(\theta)$ and $d_k(\theta)$ are independently distributed: the first depends on the sample variance, while the second depends on the sample mean. It is well known that these are independent random variables for normally distributed noise (Stuart and Ord, 1987). In Appendix 10.B it is shown that $\mathbb{E}\{c_k(\theta)\} = (M-1)/(M-2)$, for any $\theta \in \mathbb{P}_r$, so that

$$\mathbb{E}\{V_{\text{SML}}(\theta, Z)\} = \mathbb{E}\{c_k(\theta)\} \mathbb{E}\{V_{\text{ML}}(\theta, Z)\} = \frac{M-1}{M-2} \mathbb{E}\{V_{\text{ML}}(\theta, Z)\} \qquad (10\text{-}18)$$

This shows that the minimizer of the expected value of the cost is not changed by introducing the sample variance, and this result is the kernel of the classical consistency proof.

10.3.3 Discussion

From Theorem 10.3, it follows that the estimate $\hat{\theta}_{SML}(Z)$ is consistent and that $\tilde{\theta}_{SML}(Z_0)$, θ_{*SML} are the noiseless solutions when model errors are present (apply quick analysis tools number 2 and 3 of Section 9.5).

Because $V_{SML}(\lambda\theta, Z) = V_{SML}(\theta, Z)$, the estimate $\hat{\theta}_{SML}(Z)$ is independent of the particular constraint $a_i = 1$, $b_i = 1$, or $\|\theta\|_2^2 = 1$ chosen (quick tool number 4).

If the input and output measurements are noisy, the sample covariance $\hat{\sigma}_{YU}^2(k)$ must be calculated, even if it is known that the input and output errors are uncorrelated $\sigma_{YU}^2(k) = 0$. Otherwise the properties of Theorems 10.3 and 10.4 are no longer valid (see the proof of Theorem 10.3). Both theorems show that in practice the unknown exact noise (co)variances can be replaced by the sample noise (co)variances without any problem.

It is not necessary to make a precise measurement, $M = 4$ independent repeated measurements suffice to get consistency, and $M = 7$ independent repeated measurements are enough to guarantee the existence of the covariance matrix of the limiting parameter distribution.

The loss in efficiency is not large (below 12% for $M = 7$) so that this is a very acceptable solution.

In practice it can sometimes be a problem to measure seven periods, especially when the period length becomes very long as is necessary to access very low frequencies. In that case the number of periods can be restricted to $M = 2$, but at that moment the variances should be averaged over seven neighboring frequency lines.

10.3.4 Estimation of Covariance Matrix of the Model Parameters

Equation (10-16) quantifies the loss in efficiency due to the use of the sample variances. However, it does not give an answer about how to calculate $\hat{C}_{\hat{\theta}}$ from the available information. C_θ is approximated by

$$\text{Cov}(\hat{\theta}_{ML}(Z)) \approx [2\text{Re}((\varepsilon'(\hat{\theta}_{ML}(Z), Z))^H (\varepsilon'(\hat{\theta}_{ML}(Z), Z)))]^{-1}$$

(see Section 9.11.4) and in practice, during the calculations of the covariance matrix, the exact variances in ε' are again replaced by the sample variances, and only $\hat{\varepsilon}'$ is available. Using similar calculations as in (10-18), it turns out that

$$[2\text{Re}((\varepsilon'(\hat{\theta}_{ML}(Z), Z))^H (\varepsilon'(\hat{\theta}_{ML}(Z), Z)))]^{-1} \approx$$
$$\frac{M-1}{M-2}[2\text{Re}((\hat{\varepsilon}'(\hat{\theta}_{SML}(Z), \hat{Z}))^H (\hat{\varepsilon}'(\hat{\theta}_{SML}(Z), \hat{Z})))]^{-1} \quad (10\text{-}19)$$

so that (10-16) is replaced by

$$\hat{C}_{\hat{\theta}} \approx \frac{M-1}{M-3}[2\text{Re}((\hat{\varepsilon}'(\hat{\theta}_{SML}(Z), \hat{Z}))^H (\hat{\varepsilon}'(\hat{\theta}_{SML}(Z), \hat{Z})))]^{-1} \quad (10\text{-}20)$$

10.3.5 Properties of the Cost Function in Its Global Minimum

Due to the availability of the nonparametric noise model it is possible to give a prediction of the value of the cost function that is expected to be observed at the end of the identification process if no model errors are present. To judge the difference between the actual, observed value and the expected cost, the variance of the cost should also be known. Consequently, this information can be used as an undermodeling detection tool during the model selection and validation process by comparing the actual value of the cost function with its expected value (see also Chapter 11). Undermodeling occurs if the orders of the numerator and/or denominator polynomials of the transfer function model are too small (unmodeled dynamics), or if a true linear time-invariant model simply does not exist (for example, nonlinear distortions).

Theorem 10.5 (Mean and Variance of the Global Minimum of the Cost Function): Under the conditions of Theorem 10.3, and for $M \geq 6$, the mean and variance of the global minimum of the cost function $\text{var}(V_{\text{SML}}(\hat{\theta}_{\text{SML}}(Z), Z))$ are given by

$$\mathbb{E}\{V_{\text{SML}}(\hat{\theta}_{\text{SML}}(Z), Z)\} \approx \frac{M-1}{M-2} \mathbb{E}\{V_{\text{ML}}(\hat{\theta}_{\text{ML}}(Z), Z)\} - \frac{M-1}{(M-3)(M-2)} n_\theta / 2$$

$$\text{var}(V_{\text{SML}}(\hat{\theta}_{\text{SML}}(Z), Z)) \approx \frac{(M-1)^2}{(M-2)(M-3)} \text{var}(V_{\text{ML}}(\hat{\theta}_{\text{ML}}(Z_0), Z)) \quad (10\text{-}21)$$

$$+ \frac{(M-1)^2}{(M-2)^2(M-3)} \sum_{k=1}^{F} (\mathbb{E}\{|\hat{\varepsilon}(\Omega_k, \tilde{\theta}_{\text{ML}}(Z_0), \hat{Z}(k))|^2\})^2$$

in the presence of model errors ($\tilde{\theta}_{\text{ML}}(Z_0) \neq \theta_0$), and

$$\mathbb{E}\{V_{\text{SML}}(\hat{\theta}_{\text{SML}}(Z), Z)\} \approx \frac{M-1}{M-2}(F - n_\theta/2)$$

$$\text{var}(V_{\text{SML}}(\hat{\theta}_{\text{SML}}(Z), Z)) \approx \frac{(M-1)^3}{(M-2)^2(M-3)} F \quad (10\text{-}22)$$

if no model errors are present ($\tilde{\theta}_{\text{ML}}(Z_0) = \theta_0$).

Proof. See Appendix 10.G. □

10.4 PROPERTIES OF THE GTLS ESTIMATOR USING A SAMPLE COVARIANCE MATRIX

The general form of the cost function of the GTLS estimator is given by (9-72):

$$V_{\text{GTLS}}(\theta, Z) = \frac{\sum_{k=1}^{F} |e(\Omega_k, \theta, Z(k))|^2}{\sum_{k=1}^{F} \sigma_e^2(\Omega_k, \theta)} \quad (10\text{-}23)$$

Replacing $Z(k)$ in this expression by the sample mean $\hat{Z}(k)$ and the exact noise (co)variances by the sample noise (co)variances gives the sample GTLS (SGTLS) cost function

$$V_{\text{SGTLS}}(\theta, Z) = \frac{\sum_{k=1}^{F} |\hat{e}(\Omega_k, \theta, \hat{Z}(k))|^2}{\sum_{k=1}^{F} \hat{\sigma}_e^2(\Omega_k, \theta)} \quad (10\text{-}24)$$

Section 10.4 ■ Properties of the GTLS Estimator Using a Sample Covariance Matrix

where $\hat{e}(\Omega_k, \theta, \hat{Z}(k))$, and $\hat{\sigma}_{\hat{e}}^2(\Omega_k, \theta)$ are defined in (10-11). The minimizer $\hat{\theta}_{\text{SGTLS}}(Z)$ of (10-24) is not calculated using the iterative Newton-Gauss scheme (9-16) or (9-18) but via the generalized singular value decomposition of the matrix pair $(\hat{J}_{\text{re}}(Z), \hat{C})$ with $\hat{J}(Z) = \partial \hat{e}(\theta, \hat{Z})/\partial \theta$ and $\hat{C}$ a square root of the column covariance matrix of $\hat{J}_{\text{re}}(N_Z)$, with $\hat{J}(N_Z) = \hat{J}(Z) - \hat{J}(Z_0)$, calculated using the sample noise (co)variances (see Section 9.10). Just like the GTLS estimate, $\hat{\theta}_{\text{SGTLS}}(Z)$ suffers from the amplification of the high-frequency errors (see Section 9.10.3). To cope with this problem weighted SGTLS versions can be constructed as in Sections 9.10.3 and 9.12.4.

Theorem 10.6 (Asymptotic Properties $\hat{\theta}_{\text{SGTLS}}(Z)$): Consider model (10-3) with any identifiable parameterization of Section 6.2. Under the assumptions of Section 9.6 and Assumption 10.1 ($M \geq 2$) the minimizer $\hat{\theta}_{\text{SGTLS}}(Z)$ of (10-24) has the asymptotic properties of Theorem 9.21 with $V_F(\theta, Z) = V_{\text{SGTLS}}(\theta, Z)$, where

1. $V_F(\theta)$ and $V_*(\theta)$ are given by

$$V_F(\theta) = \frac{\sum_{k=1}^{F} |e(\Omega_k, \theta, Z_0(k))|^2}{\frac{1}{M}\sum_{k=1}^{F} \sigma_e^2(\Omega_k, \theta)} + 1, \quad V_*(\theta) = \frac{\int_{f_{\min}}^{f_{\max}} |e(\Omega(f), \theta, Z_0(f))|^2 n(f) df}{\frac{1}{M}\int_{f_{\min}}^{f_{\max}} \sigma_e^2(\Omega(f), \theta) n(f) df} + 1 \quad (10\text{-}25)$$

2. $\mathbb{E}\{\delta_\theta(Z)\}$ in (9-26) is not necessarily zero or even may not exist.

3. $\delta_\theta(Z)$ in the expression of the covariance matrix (9-27) is replaced by $d_\theta(Z)$ as in Theorem 18.25.

Proof. See Appendix 10.H. □

From Theorem 10.6 it follows that the estimate $\hat{\theta}_{\text{SGTLS}}(Z)$ is consistent and $\tilde{\theta}_{\text{SGTLS}}(Z_0)$, $\theta_{*\text{SGTLS}}$ are the noiseless solutions in case model errors are present (apply quick analysis tools number 2 and 3 of Section 9.5 to $V(\theta)$ and $V_*(\theta)$ in (10-25)). Because $V_{\text{SGTLS}}(\lambda\theta, Z) = V_{\text{SGTLS}}(\theta, Z)$ the estimate $\hat{\theta}_{\text{SGTLS}}(Z)$ is independent of the particular constraint chosen, for example, $a_i = 1$, $b_i = 1$, or $\|\theta\|_2^2 = 1$ (quick tool number 4). The relationship between the asymptotic behavior of the estimates, based on the true (10-23) and the sample (10-24) noise (co)variances, is established in the following theorem.

Theorem 10.7 (Relationship between $\hat{\theta}_{\text{SGTLS}}(Z)$ and $\hat{\theta}_{\text{GTLS}}(Z)$): Under the conditions of Theorem 10.6, the estimates based on the true (10-23) and the sample (10-24) noise (co)variances are related to each other by

$$V_{\text{SGTLS}}(\theta) = V_{\text{GTLS}}(\theta) \text{ and } V_{*\text{SGTLS}}(\theta) = V_{*\text{GTLS}}(\theta) \quad (10\text{-}26)$$

$$\tilde{\theta}_{\text{SGTLS}}(Z_0) = \tilde{\theta}_{\text{GTLS}}(Z_0) \text{ and } \theta_{*\text{SGTLS}} = \theta_{*\text{GTLS}} \quad (10\text{-}27)$$

In the absence of model errors, $\tilde{\theta}_{\text{SGTLS}}(Z_0) = \theta_0$, we have

$$\text{Cov}(\delta_{\hat{\theta}_{\text{SGTLS}}}(Z)) = \text{Cov}(\delta_{\hat{\theta}_{\text{GTLS}}}(Z)) \tag{10-28}$$

where $\hat{\theta}(Z) = \theta_0 + \delta_\theta(Z) + O_p(F^{-1})$ with $\mathbb{E}\{\delta_\theta(Z)\} = 0$ and $\delta_\theta(Z) = O_p(F^{-1/2})$, and where $\sqrt{F}\delta_\theta(Z)$ is asymptotically normally distributed.

Proof. See Appendix 10.I. □

Contrary to the SML solution, it is not necessary to calculate the sample covariance $\hat{\sigma}_{YU}^2(k)$ if it is known that the input and output errors are uncorrelated $\sigma_{YU}^2(k) = 0$. From both theorems it follows that $\hat{\theta}_{\text{SGTLS}}(Z)$ has asymptotic ($F \to \infty$) properties similar to those of $\hat{\theta}_{\text{GTLS}}(Z)$, even if the sample (co)variances are calculated using only $M = 2$ independent repeated experiments. For example, in the absence of model errors, the asymptotic uncertainty of the SGTLS equals that of the GTLS. This is no longer true if model errors are present. The basic reason for the similar asymptotic behavior of $\hat{\theta}_{\text{SGTLS}}(Z)$ and $\hat{\theta}_{\text{GTLS}}(Z)$ is that the "poor quality" sample (co)variances are averaged over the frequency in the cost function (10-24), resulting in a "high quality" estimate of the denominator of the cost function.

10.5 PROPERTIES OF THE BTLS ESTIMATOR USING A SAMPLE COVARIANCE MATRIX

The general form of the cost function of the BTLS estimator is given by (9-98)

$$V_{\text{BTLS}}(\theta^{(i)}, Z) = \frac{\sum_{k=1}^{F} \frac{|e(\Omega_k, \theta^{(i)}, Z(k))|^2}{\sigma_e^{2r}(\Omega_k, \theta^{(i-1)})}}{\sum_{k=1}^{F} \frac{\sigma_e^2(\Omega_k, \theta^{(i)})}{\sigma_e^{2r}(\Omega_k, \theta^{(i-1)})}} \tag{10-29}$$

We recall that $\sigma_e^{2r}(\Omega_k, \theta^{(i-1)})$ and $\sigma_e^2(\Omega_k, \theta^{(i)})$ stem, respectively, from the left W and right C weighting matrices in the total least squares problem (see Sections 9.10 and 9.12.3). Following along the lines of Section 10.4, one could think of replacing the true noise (co)variances everywhere in (10-29) by the sample noise (co)variances

$$\frac{\sum_{k=1}^{F} \frac{|\hat{e}(\Omega_k, \theta^{(i)}, \hat{Z}(k))|^2}{\hat{\sigma}_{\hat{e}}^{2r}(\Omega_k, \theta^{(i-1)})}}{\sum_{k=1}^{F} \frac{\hat{\sigma}_{\hat{e}}^2(\Omega_k, \theta^{(i)})}{\hat{\sigma}_{\hat{e}}^{2r}(\Omega_k, \theta^{(i-1)})}} \tag{10-30}$$

Proceeding in that way, we violate the assumptions of the framework developed in Chapter 18. Indeed, the theorems of Chapter 18 (strong consistency, convergence rate, asymptotic bias, and asymptotic normality) are valid only if the number of stochastic parameters in the weighting remains finite for finite M and $F \to \infty$, and if these parameters converge strongly ($F \to \infty$) to a nonrandom limit value. This is certainly not the case for the left weighting $\hat{\sigma}_{\hat{e}}^{2r}(\Omega_k, \theta^{(i-1)})$ in (10-30). Therefore, to preserve the strong consistency, the noise (co)variances in the left weighting matrix W ($\hat{\sigma}_{\hat{e}}^{2r}(\Omega_k, \theta^{(i-1)})$) in (10-30) should be modeled over the frequency using a finite (F-independent) number of parameters α. The estimates $\hat{\alpha}(Z)$ should strongly converge to some nonrandom value α_*. As it is the case for the SGTLS estimator, the right weight-

Section 10.5 ■ Properties of the BTLS Estimator Using a Sample Covariance Matrix

ing C ($\hat{\sigma}_{\hat{e}}^2(\Omega_k, \theta^{(i)})$) in (10-30) must still be calculated using the original sample noise (co)variances.

For computational reasons, only noise models that are linear in the parameters are considered. For example,

$$\sigma_n^2(k, \alpha) = \sum_{r=1}^{p_n} \alpha_{nr} h_{nr}(\Omega_k) \qquad n = 1, 2, 3 \qquad (10\text{-}31)$$

with $\sigma_1^2 = \sigma_U^2$, $\sigma_2^2 = \sigma_Y^2$, and $\sigma_3^2 = \sigma_{YU}^2$ and where $h_{nr}(\Omega)$, $r = 1, 2, ..., p_n$, are linear independent basis functions with p_n independent of F. The choice of the parametric noise model is not critical because it does not influence the consistency property. It, however, influences the uncertainty of the estimated plant model parameters. Under Assumption 9.3 or 9.4 and Assumption 10.1 the linear least squares estimate $\hat{\alpha}^T(Z) = [\hat{\alpha}_1^T(Z) \; \hat{\alpha}_2^T(Z) \; \hat{\alpha}_3^T(Z)]$

$$\hat{\alpha}_n(Z) = (H_n^T H_n)^{-1} H_n^T [\hat{\sigma}_n^2(1), ..., \hat{\sigma}_n^2(F)]^T \qquad n = 1, 2, 3 \qquad (10\text{-}32)$$

with $H_{n[k, r]} = h_{nr}(\Omega_k)$ and $\hat{\sigma}_n^2(k)$ the sample noise (co)variances ($\hat{\sigma}_1^2 = \hat{\sigma}_U^2$, $\hat{\sigma}_2^2 = \hat{\sigma}_Y^2$, and $\hat{\sigma}_3^2 = \hat{\sigma}_{YU}^2$) converges strongly ($F \to \infty$) to a nonrandom value α_*. Note that the estimated parametric noise models $\sigma_n^2(k, \hat{\alpha}(Z))$, $n = 1, 2, 3$, represent a linear projection of an F-dimensional space onto a p_n-dimensional space.

Replacing the true noise (co)variances $\sigma_n^2(k)$ in (10-29) by the estimated parametric noise model $\sigma_n^2(k, \hat{\alpha}(Z))$ in the left weighting W, $\sigma_n^2(k)$ by the sample noise (co)variances $\hat{\sigma}_n^2(k)$ in the right weighting C, and the measurements Z by the sample mean $\hat{Z}$ gives the sample BTLS (SBTLS) cost function

$$V_{\text{SBTLS}}(\theta^{(i)}, Z) = \frac{\sum_{k=1}^{F} \frac{|\hat{e}(\Omega_k, \theta^{(i)}, \hat{Z}(k))|^2}{\sigma_{\hat{e}}^{2r}(\Omega_k, \theta^{(i-1)}, \hat{\alpha}(Z))}}{\sum_{k=1}^{F} \frac{\hat{\sigma}_{\hat{e}}^2(\Omega_k, \theta^{(i)})}{\sigma_{\hat{e}}^{2r}(\Omega_k, \theta^{(i-1)}, \hat{\alpha}(Z))}} \qquad (10\text{-}33)$$

where $\hat{e}(\Omega_k, \theta, \hat{Z}(k))$ and $\hat{\sigma}_{\hat{e}}^2(\Omega_k, \theta)$ are defined in (10-11) and

$$\sigma_{\hat{e}}^2(\Omega_k, \theta, \hat{\alpha}(Z)) = \sigma_e^2(\Omega_k, \theta, \hat{\alpha}(Z))/M$$
$$\sigma_e^2(\Omega_k, \theta, \hat{\alpha}(Z)) = \sigma_Y^2(k, \hat{\alpha}(Z))|A(\Omega_k, \theta)|^2 + \sigma_U^2(k, \hat{\alpha}(Z))|B(\Omega_k, \theta)|^2 \qquad (10\text{-}34)$$
$$-2\text{Re}(\sigma_{YU}^2(k, \hat{\alpha}(Z))A(\Omega_k, \theta)\bar{B}(\Omega_k, \theta))$$

Likewise, the SGTLS estimator the minimizer $\hat{\theta}_{\text{SBTLS}}(Z)$ of (10-33) is calculated via the generalized singular value decomposition of the matrix pair $(W_{\text{Re}} \hat{J}_{\text{re}}(Z), \hat{C})$ with $\hat{J}(Z) = \partial \hat{e}(\theta, \hat{Z})/\partial \theta$, W a diagonal matrix with $W_{[k, k]} = \sigma_{\hat{e}}^{-r}(\Omega_k, \theta^{(i-1)}, \hat{\alpha}(Z))$, and $\hat{C}$ a square root of the column covariance matrix of $W_{\text{Re}} \hat{J}_{\text{re}}(N_Z)$, with $\hat{J}(N_Z) = \hat{J}(Z) - \hat{J}(Z_0)$, calculated using the sample noise (co)variances (see Section 9.10). The asymptotic properties of the first step of the iterative procedure (10-33) are analyzed in the following theorem.

Theorem 10.8 (Asymptotic Properties $\hat{\theta}_{\text{SBTLS}}(Z)$): Consider model (10-3) with any identifiable parameterization of Section 6.2. Under the assumptions of Section 9.6 and Assumption 10.1 ($M \geq 2$) the minimizer $\hat{\theta}_{\text{SBTLS}}(Z) = \theta^{(1)}$ of (10-33), with parametric noise model (10-31) and initial guess $\theta^{(0)} = \hat{\theta}(Z)$ satisfying Theorem 9.21, has the asymptotic properties of Theorem 9.21 with $V_F(\theta, Z) = V_{\text{SBTLS}}(\theta, Z)$ and where

1. $V_F(\theta)$ and $V_*(\theta)$ are given by

$$V(\theta) = \frac{\sum_{k=1}^{F} \frac{|e(\Omega_k, \theta, Z_0(k))|^2}{\sigma_e^{2r}(\Omega_k, \theta_*, \alpha_*)}}{\frac{1}{M}\sum_{k=1}^{F} \frac{\sigma_e^2(\Omega_k, \theta)}{\sigma_e^{2r}(\Omega_k, \theta_*, \alpha_*)}} + 1, \quad V_*(\theta) = \frac{\int_{f_{\min}}^{f_{\max}} \frac{|e(\Omega(f), \theta, Z_0(f))|^2}{\sigma_e^{2r}(\Omega(f), \theta_*, \alpha_*)} n(f)df}{\frac{1}{M}\int_{f_{\min}}^{f_{\max}} \frac{\sigma_e^2(\Omega(f), \theta)}{\sigma_e^{2r}(\Omega(f), \theta_*, \alpha_*)} n(f)df} + 1 \quad (10\text{-}35)$$

2. $\mathbb{E}\{\delta_\theta(Z)\}$ in (9-26) is not necessarily zero or even may not exist,
3. $\delta_\theta(Z)$ in the expression of the covariance matrix (9-27) is replaced by $d_\theta(Z)$ as in Theorem 18.25.

Proof. See Appendix 10.J. □

From Theorem 10.8 it follows that the estimate $\hat{\theta}_{\text{SBTLS}}(Z)$ is consistent and $\tilde{\theta}_{\text{SBTLS}}(Z_0)$, $\theta_{*\text{SBTLS}}$ are the noiseless solutions in case model errors are present (apply quick analysis tools number 2 and 3 of Section 9.5 to $V(\theta)$ and $V_*(\theta)$ in (10-35)). Because $V_{\text{SBTLS}}(\lambda\theta, Z) = V_{\text{SBTLS}}(\theta, Z)$, the estimate $\hat{\theta}_{\text{SBTLS}}(Z)$ is independent of the particular constraint chosen, for example, $a_i = 1$, $b_i = 1$ or $\|\theta\|_2^2 = 1$ (quick tool number 4). The relationship between the asymptotic behavior of $\hat{\theta}_{\text{SBTLS}}(Z)$ and $\hat{\theta}_{\text{BTLS}}(Z)$ is established in the following theorem.

Theorem 10.9 (Relationship between $\hat{\theta}_{\text{SBTLS}}(Z)$ and $\hat{\theta}_{\text{BTLS}}(Z)$): Under the conditions of Theorem 10.8, and assuming that the parametric noise models (10-31), $\sigma_n^2(k, \hat{\alpha}(Z))$ with $n = 1, 2, 3$, are consistent estimates of the true noise (co)variances, $\sigma_n^2(k)$ with $n = 1, 2, 3$, the estimates based on the true (10-29) and the sample (10-33) noise (co)variances are related to each other by

$$V_{\text{SBTLS}}(\theta) = V_{\text{BTLS}}(\theta) \quad \text{and} \quad V_{*\text{SBTLS}}(\theta) = V_{*\text{BTLS}}(\theta) \quad (10\text{-}36)$$

$$\tilde{\theta}_{\text{SBTLS}}(Z_0) = \tilde{\theta}_{\text{BTLS}}(Z_0) \quad \text{and} \quad \theta_{*\text{SBTLS}} = \theta_{*\text{BTLS}} \quad (10\text{-}37)$$

In the absence of model errors, $\tilde{\theta}_{\text{SBTLS}}(Z_0) = \theta_0$, we have

$$\text{Cov}(\delta_{\theta\text{SBTLS}}(Z)) = \text{Cov}(\delta_{\theta\text{BTLS}}(Z)) \quad (10\text{-}38)$$

where $\hat{\theta}(Z) = \theta_0 + \delta_\theta(Z) + O_p(F^{-1})$ with $\mathbb{E}\{\delta_\theta(Z)\} = 0$ and $\delta_\theta(Z) = O_p(F^{-1/2})$, and where $\sqrt{F}\delta_\theta(Z)$ is asymptotically normally distributed.

Proof. See Appendix 10.K. □

As with the SGTLS solution, it is not necessary to calculate the sample covariance $\hat{\sigma}_{YU}^2(k)$ if it is known that $\sigma_{YU}^2(k) = 0$. From both theorems it follows that $\hat{\theta}_{\text{SBTLS}}(Z)$ has asymptotic $(F \to \infty)$ properties similar to those of $\hat{\theta}_{\text{BTLS}}(Z)$, even for $M = 2$. The price to be paid is the construction of a consistent parametric model for the noise (co)variances in the left weighting. If the parametric noise model is a poor approximation of the true noise model, then the estimated plant model parameters are still strongly consistent (Theorem 10.8 is still valid), but their uncertainty may increase (Theorem 10.9 is no longer valid).

10.6 PROPERTIES OF THE SUB ESTIMATOR USING A SAMPLE COVARIANCE MATRIX

Subspace Algorithms 9.24 and 9.25 identify model (10-3) where $G(\Omega, \theta)$ is parameterized in the state space parameters (A, B, C, D)

$$G(\xi, \theta) = C(\xi I_{n_a} - A)^{-1} B + D \tag{10-39}$$

with $\xi = z$ for discrete-time systems and $\xi = s$ for continuous-time systems. This is done in three steps: (i) estimation of the (generalized) extended observability matrix O_r or $O_{r\perp}$ using the input-output spectra, (ii) estimation of A and C using $\hat{O}_r$ or $\hat{O}_{r\perp}$, and (iii) estimation of B and D using $\hat{A}$, $\hat{C}$, and the input-output spectra. The procedure has been developed for problems where the input is exactly known, $N_U(k) = 0$. If the input observations are noisy, $N_U(k) \neq 0$, then the errors-in-variables problem (10-7) is transformed into an equivalent frequency response function measurement problem

$$Y(k)/U(k) = G_0(\Omega_k) + N_G^{[l]}(k) \text{ with } \mathbb{E}\{N_G^{[l]}(k)\} = 0 \tag{10-40}$$

where $\text{var}(N_G^{[l]}(k)) = \sigma_G^2(k)$ is related to the input-output noise (co)variances by (9-54). For worst case input and output signal-to-noise ratios $|U_0(k)|/\sigma_U(k)$ and $|Y_0(k)|/\sigma_Y(k)$ larger than 10 dB, (10-40) is a very good approximation (see Section 9.9 for an elaborated discussion).

Knowledge of the noise variance $\sigma_Y^2(k)$ or $\sigma_G^2(k)$ is required in the first and third steps of the subspace algorithms. In the first step we need the covariance matrix C_Y or $C_{Y_\perp}$ for, respectively, discrete-time or continuous-time modeling,

$$\begin{aligned} C_Y &= \text{Re}(\sum_{k=1}^{F} \sigma_Y^2(k) W_r(k) W_r^H(k)) \text{ with } W_r(k) = [1 \ z_k \ \ldots \ z_k^{r-1}]^T \\ C_{Y_\perp} &= \text{Re}(\sum_{k=1}^{F} \sigma_Y^2(k) W_Y(k) W_Y^H(k)) \text{ with } W_Y(k) = [p_0(s_k) \ p_1(s_k) \ \ldots \ p_{r-1}(s_k)]^T \end{aligned} \tag{10-41}$$

where $p_n(s)$ are scalar orthogonal polynomials of order $n = 0, 1, \ldots, r-1$ (see Appendix 9.S). In the third step the estimates $\hat{B}$ and $\hat{D}$ are obtained by minimizing

$$V_{\text{SUB}}(B, D, \hat{A}, \hat{C}, Z) = \sum_{k=1}^{F} \frac{|Y(k) - (\hat{C}(\xi I_{n_a} - \hat{A})^{-1} B + D) U(k)|^2}{\sigma_Y^2(k)} \tag{10-42}$$

w.r.t. B and D. Replacing $Z(k)$ by the sample mean $\hat{Z}(k)$, and the exact (co)variances by the sample (co)variances, in the first and the third step of Algorithms 9.24 and 9.25, defines the sample subspace (SSUB) algorithms. Expressions (10-41) and (10-42) then become

$$\begin{aligned} \hat{C}_Y &= \text{Re}(\sum_{k=1}^{F} M^{-1} \hat{\sigma}_Y^2(k) W_r(k) W_r^H(k)) \\ \hat{C}_{Y_\perp} &= \text{Re}(\sum_{k=1}^{F} M^{-1} \hat{\sigma}_Y^2(k) W_Y(k) W_Y^H(k)) \\ V_{\text{SSUB}}(B, D, \hat{A}, \hat{C}, Z) &= \sum_{k=1}^{F} \frac{|\hat{Y}(k) - (\hat{C}(\xi I_{n_a} - \hat{A})^{-1} B + D) \hat{U}(k)|^2}{\hat{\sigma}_Y^2(k)/M} \end{aligned} \tag{10-43}$$

For noisy inputs, $\hat{Y}(k)$, $\hat{U}(k)$, and $\hat{\sigma}_Y^2(k)$ are replaced by $\hat{Y}(k)/\hat{U}(k)$, 1, and $\hat{\sigma}_G^2(k)$.

Theorem 10.10 (Asymptotic Properties $\hat{\theta}_{\text{SSUB}}(Z)$): Consider transfer function model (10-39), parameterized in its state space representation, and Algorithms 9.24 and 9.25, where the input-output spectra are replaced by their sample mean, and the exact noise (co)variances by the sample noise (co)variances. The resulting estimate $\hat{\theta}_{\text{SSUB}}(Z)$ has the asymptotic ($F \to \infty$) properties of Theorem 9.28, where

1. $M \geq 2$ for the stochastic convergence and stochastic convergence rate (properties 1, 2, and 3 of Theorem 9.28) of $\hat{A}$ and $\hat{C}$ (assumptions of Sections 9.6.1 and 9.6.5 and Assumptions 9.26, 9.27 and 10.1).
2. $M \geq 4$ for the stochastic convergence and the stochastic convergence rate (properties 1, 2, and 3 of Theorem 9.28) of $\hat{B}$ and $\hat{D}$ (assumptions of Sections 9.6.1 and 9.6.5, and Assumptions 9.26, 9.27, 10.1, and 10.2).

Proof. See Appendix 10.L. □

From Theorem 10.10, it follows that $\hat{\theta}_{\text{SSUB}}(Z)$ is consistent and that $\theta_{*\text{SSUB}}$ is the noiseless solution in case model errors are present. Note that estimation of the poles, which depend on $\hat{A}$ only, has a fundamentally different stochastic behavior than the estimation of the zeros, which depend on $\hat{A}$, $\hat{B}$, $\hat{C}$, and $\hat{D}$. Indeed, the poles can be estimated consistently under the same noise assumptions as those when the noise (co)variances are known, while consistent estimation of the zeros requires that the disturbing noise is independent (over the frequency) and normally distributed (Assumption 10.2). The relationship between the asymptotic behavior of $\hat{\theta}_{\text{SSUB}}(Z)$ and $\hat{\theta}_{\text{SUB}}(Z)$ is established in the following theorem.

Theorem 10.11 (Relationship between $\hat{\theta}_{\text{SSUB}}(Z)$ and $\hat{\theta}_{\text{SUB}}(Z)$): Under the conditions of Theorem 10.10, the estimates based on the true (10-41), (10-42), and the sample (10-43) noise (co)variances are related to each other by

1. For $M \geq 2$, $A_{*\text{SSUB}} = A_{*\text{SUB}}$ and $C_{*\text{SSUB}} = C_{*\text{SUB}}$
2. For $M \geq 4$, $B_{*\text{SSUB}} = B_{*\text{SUB}}$ and $D_{*\text{SSUB}} = D_{*\text{SUB}}$

Proof. See Appendix 10.L. □

10.7 IDENTIFICATION IN THE PRESENCE OF NONLINEAR DISTORTIONS

We recall that the steady-state response of the nonlinear system $y(t) = G[u_0(t)]$ to a random phase multisine (see Definition 3.2) $u_0(t)$ is given by

$$Y(k) = G_{\text{BLA}}(s_k)U_0(k) + Y_S(k) \quad (10\text{-}44)$$

with $Y_S(k)$ the stochastic nonlinear contributions (see Section 6.8) and where $y_S(t) = \text{IDFT}(Y_S(k))$ has the same periodicity as the input signal $u_0(t)$. Calculating the sample (co)variances over several periods of the output signal $y(t)$ will, hence, give no information about the stochastic nonlinear contributions $Y_S(k)$. Experiments with different realizations of the random phases $\varphi_k = \angle U_0(k)$ of the input signal are necessary to get the contribution of $Y_S(k)$ to the sample (co)variances. To distinguish between the signal part $G_{\text{BLA}}(s_k)U_0(k)$ and the noise part $Y_S(k)$ in (10-44), the input and output DFT spectra of each experiment must be turned back with the corresponding phases $\varphi_k = \angle U_0(k)$ of the input. Because the observations of the input are in general corrupted by measurement noise, a reference signal $r(t)$ is required to perform this operation (see Figure 10-1). This leads to the following robust measurement strategy (see also Section 4.3.1):

1. Choose the amplitude spectrum of the random phase multisine (see Definition 3.2).
2. Make a random choice of the phases φ_k of the random phase multisine (see Definition 3.2) and calculate the corresponding time signal $r(t)$.
3. Apply the excitation to the plant and measure $P \geq 1$ periods of the steady-state response $u(t)$, $y(t)$.
4. Repeat steps 2 and 3 $M \geq 4$ times.
5. Calculate the DFT spectra of the input $u(t)$, output $y(t)$, and reference $r(t)$ signals for each experiment at the excited DFT frequencies. This gives M sets of the reference $R^{[m]}(k)$, the noisy input $U^{[m]}(k)$, and the noisy output $Y^{[m]}(k)$ spectra, $k = 1, 2, ..., F$ and $m = 1, 2, ..., M$.
6. Divide the input and output spectra by $e^{j\angle R^{[m]}(k)}$

$$X_R^{[m]}(k) = X^{[m]}(k)/e^{j\angle R^{[m]}(k)} \tag{10-45}$$

and finally calculate the sample mean and sample (co)variances

$$\hat{X}(k) = \frac{1}{M}\sum_{m=1}^{M} X_R^{[m]}(k) \tag{10-46}$$

$$\hat{\sigma}_{XL}^2(k) = \frac{1}{M-1}\sum_{m=1}^{M}(\hat{X}(k) - X_R^{[m]}(k))\overline{(\hat{L}(k) - L_R^{[m]}(k))} \tag{10-47}$$

where $X, L = Y, U$.

Using the sample means (10-46) and sample (co)variances (10-47), we can calculate the SML (10-11), SGTLS (10-24), and SBTLS (10-33) estimates of the best linear approximation $G_{BLA}(s)$. Because Assumption 10.2 is asymptotically valid ($F \to \infty$) for $Y_S(k)$ (see Section 6.8 and Theorems 3.10 and 3.11) it is tempting to study the properties of $\hat{\theta}_{SML}(Z)$ under the idealized assumptions of Theorems 10.3 and 10.4, where $G_0(s)$ is replaced by $G_{BLA}(s)$. However, one should not forget that $Y_S(k)$ is uncorrelated with – but not independent of – the input DFT spectrum $U(k)$ (see Chapter 3). It follows that Theorems 10.3 and 10.4 remain

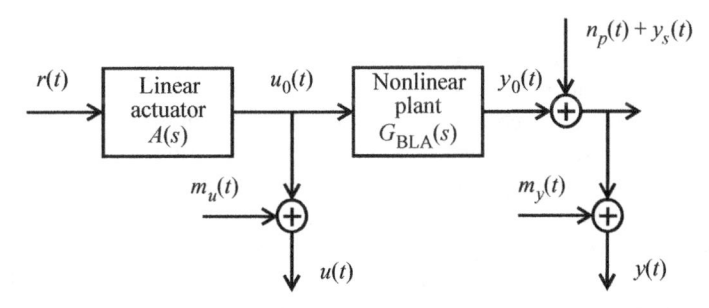

Figure 10-1. Measurement of the best linear approximation $G_{BLA}(s)$ of a nonlinear device: $u_0(t)$, $y_0(t)$ are the true input/output signals, $m_u(t)$, $m_y(t)$ are the input-output measurement errors, $y_s(t)$ is the zero mean stochastic nonlinear contribution, and $r(t)$ is the reference signal (typically the waveform stored in the arbitrary waveform generator).

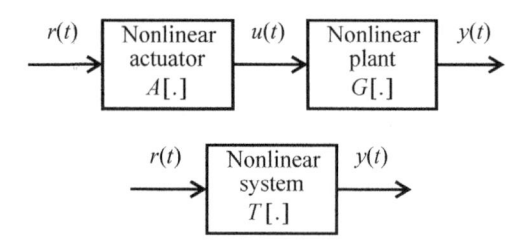

```
                r(t)  ┌───────────┐ u(t) ┌───────────┐ y(t)
                ────→ │ Nonlinear │ ───→ │ Nonlinear │ ────→
                      │ actuator  │      │  plant    │
                      │   A[.]    │      │   G[.]    │
                      └───────────┘      └───────────┘

                r(t)  ┌───────────┐ y(t)
                ────→ │ Nonlinear │ ────→
                      │  system   │
                      │   T[.]    │
                      └───────────┘
```

Figure 10-2. Schematic representation of a nonlinear device loading nonlinearly the generator: the input $u(t)$ of the device is nonlinearly related to the reference signal $r(t)$. The overall system from the reference $r(t)$ to the output of the plant is denoted by $T[.]$.

valid except that the covariance expression (10-16) underestimates the true covariance (see Section 12.4 or Schoukens and Pintelon, 2010b, for a detailed explanation). Because $Y_S(k)$ has similar second order properties as the measurement and process noise in a time domain experiment (see Sections 3.4.4 and 9.6), Theorems 10.6 and 10.7 and Theorems 10.8 and 10.9, where $G_0(s)$ is replaced by $G_{BLA}(s)$, remain valid for, respectively, $\hat{\theta}_{SGTLS}(Z)$ and $\hat{\theta}_{SBTLS}(Z)$ (proof: similar to Appendix 9.V; see also Section 12.4). Similar conclusions hold for the SSUB algorithms of Section 10.6.

It may happen that the nonlinear plant loads the generator nonlinearly or that the actuator itself is nonlinear, creating nonlinear distortions at the input of the plant. Figure 10-2 shows the corresponding block diagram. Applying the measurement strategy to this situation gives an estimate $\hat{G}_{BLA}(s_k) = \hat{Y}(k)/\hat{U}(k)$ that converges strongly ($M \to \infty$) to $\mathbb{E}\{Y_R(k)\}/\mathbb{E}\{U_R(k)\} = T_R(s_k)/A_R(s_k)$ where $T_R(s_k)$ and $A_R(s_k)$ are the best linear approximations of the nonlinear systems $T[.]$ and $A[.]$, respectively (see Appendix 10.M). Note that in general $T_R(s_k)/A_R(s_k) \neq G_{BLA}(s_k)$; however, if the nonlinear distortions at the input are small, $\text{var}(U_S(k)) \ll |A_R(s_k)R(k)|^2$, or if the best linear approximation $G_{BLA}(s_k)$ is not very sensitive to (small) variations of the input power spectrum, then $T_R(s_k)/A_R(s_k) \approx G_{BLA}(s_k)$. The same is true for the parametric estimate $G(s, \hat{\theta})$ because it converges strongly ($F \to \infty$) to the best linear approximation.

10.8 ILLUSTRATION AND OVERVIEW OF THE PROPERTIES

10.8.1 Real Measurement Example

We illustrate the SML (10-10), SGTLS (10-24), "full" ($r = 1$) SBTLS (10-33), and SSUB (Algorithm 9.25 with $\hat{\sigma}_Y^2(k)$ and $r = 70$) estimators on the second measurement example (flight flutter data) of Section 9.15.3. The sample noise (co)variances are calculated using three independent burst swept-sine experiments (see Figure 10-3). Because the three experiments were not synchronized, a postsynchronization was first executed before calculating the sample noise (co)variances. The postsynchronization consists of estimating the delay of the second and third experiments with respect to the first experiment and adding a corresponding phase shift $e^{j\omega \hat{\tau}^{[l]}}$, $l = 1, 2$, to DFT spectra of the second and the third experiment. Although $M = 3$ independent repeated experiments are not sufficient to use the sample noise (co)variances within the SML and SSUB framework (see Theorem 10.3 and Theorem 10.10), we still calculated the SML and SSUB estimates to show that they are rather robust w.r.t. to this condition. For SBTLS the noise (co)variances in the left weighting are modeled by a constant: put $h_{nr}(\Omega) = 1$ and $p_n = 1$, $n = 1, 2, 3$, in (10-31). Figure 10-4 shows the estimation results for a rational form in s of order $n_b = 11$ over $n_a = 10$. It can

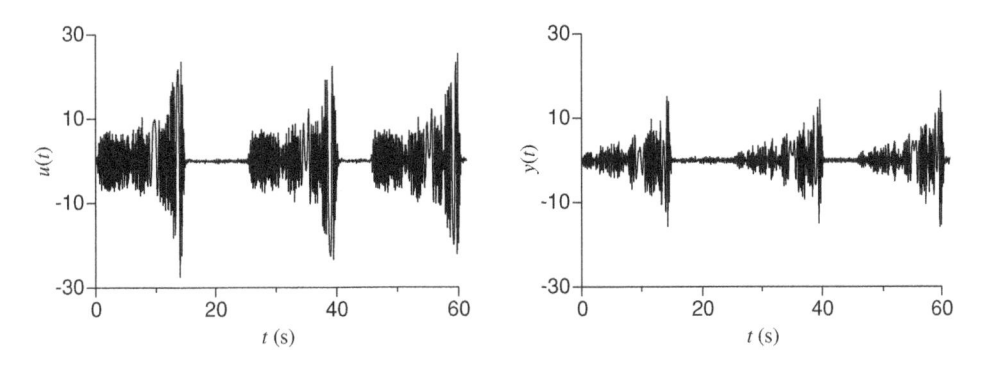

Figure 10-3. Input and output signals of the flight flutter data showing the three independent burst swept-sine experiments.

be seen that the SBTLS and SSUB estimates have SML quality. The SGTLS estimate misses the second resonance peak, which can be explained by its inappropriate frequency weighting (see also Section 9.10.3). Comparing Figure 9-13 on page 338 and Figure 10-4, it follows that SGTLS, SBTLS, and SSUB perform (much) better than GTLS, BTLS, and SUB. The basic reason for this is that in Figure 9-13 the three burst swept-sine excitations were treated as independent nonsynchronized experiments, whereas in Figure 10-4 the set of three postsynchronized input-output DFT spectra have been averaged (see also Section 14.3.5.5). The averaging improves the signal-to-noise ratio (compare the measured frequency response functions in both figures), which explains the better behavior of SGTLS, SBTLS, and SSUB.

10.8.2 Overview of the Properties

The SML, SGTLS, SBTLS, and SSUB estimators have the same basic properties as the ML, GTLS, BTLS, and SUB estimators (see Table 9-5 on page 340), except that no prior noise information is required. Table 10-1 gives an overview of the differences and the similarities between the estimates using the sample and the true noise (co)variances

TABLE 10-1 Comparison of the Estimates Using the Sample (Subscript S) and the True Noise (Co)Variances in the General Case of Input-Output Errors

Estimator	$\theta_{*S} = \theta_*$?	$R = \dfrac{\operatorname{var}(f(\hat{\theta}_S))}{\operatorname{var}(f(\hat{\theta}))}$ (i)	Same Noise Assumption	Estimation $\hat{\sigma}_{YU}^2$ if $\sigma_{YU}^2 = 0$	Sensitivity to Noise Model Errors
SML	Yes[ii]	$\dfrac{M-2}{M-3}$ (iii)	No[iv]	Yes[v]	Excellent[vi]
SGTLS	Yes[ii]	1[iii]	Yes[iv]	No	Excellent[vi]
SBTLS	Yes[ii]	1[iii]	Yes[iv]	No	Very good[vi]
SSUB	Yes[ii]	—	Yes/no[iv]	No/yes[v]	Excellent[vi]

See Table 9-5 on page 340 for an overview of the basic properties of ML, GTLS, BTLS, and SUB. The roman numbers in the table refer to the property numbers in the text.

(i) $f(.)$ represents any invariant of the model $G(\Omega, \theta)$, for example, the frequency response function, the poles, or the zeros. In order to ensure the existence of the variance, the function $f(.)$ is truncated as $\hat{\theta}(Z)$ in (9-25).

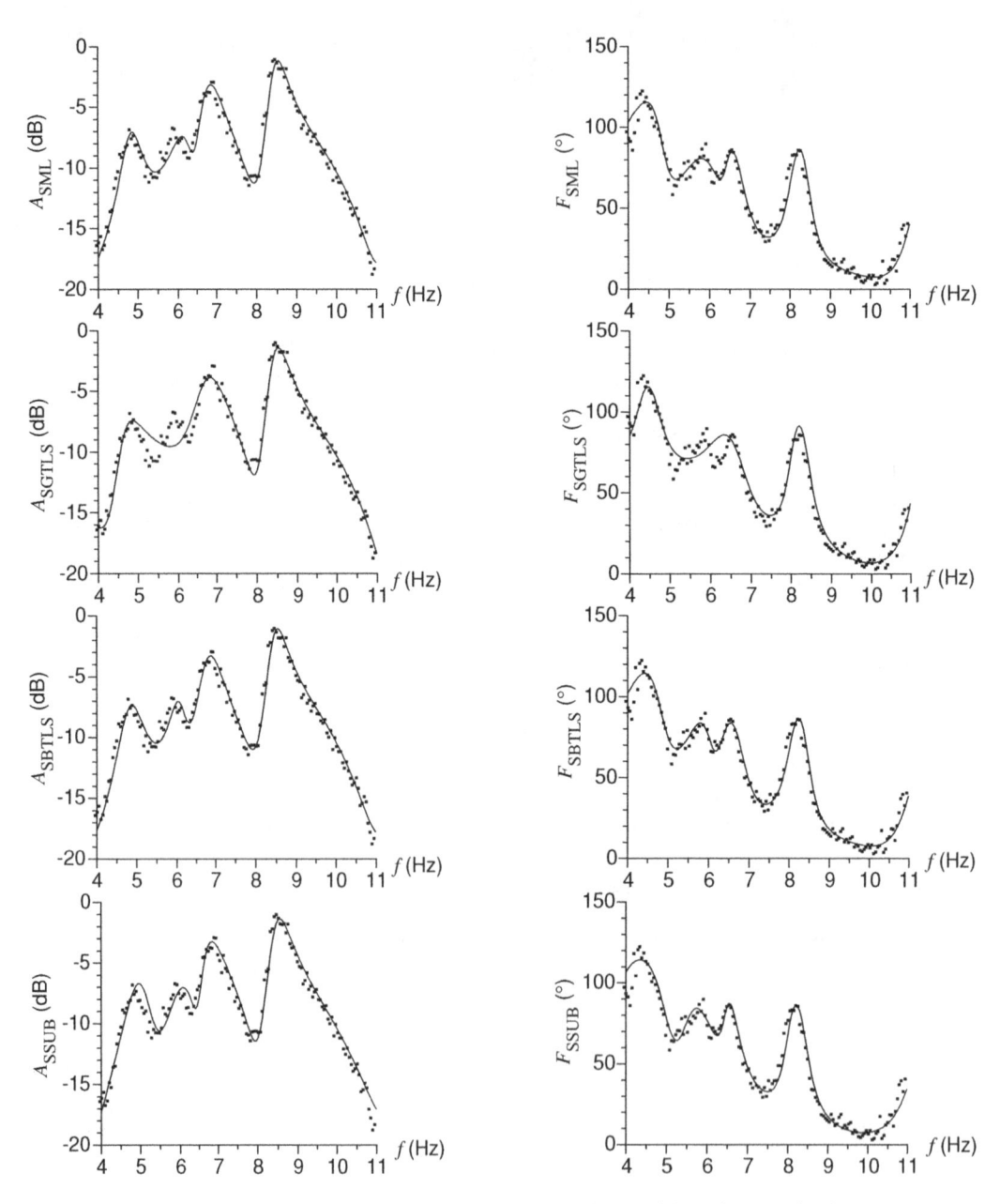

Figure 10-4. Comparison between the measurements (dots) and the estimates using the sample noise (co)variances (solid line) of the flight flutter data (model $n_a = 10$, $n_b = 11$). From left to right amplitude and phase.

(ii) The required number of independent repeated experiments is $M \geq 4$ for SML and $M \geq 2$ for SGTLS and SBTLS. The SSUB estimator requires $M \geq 2$ for the poles and $M \geq 4$ for the zeros. If the input is observed with errors, then SSUB uses $G(\Omega_k) = Y(k)/U(k)$ as primary data, and $\theta_{*S} = \theta_*$ is "practically valid" if the worst case input-output signal-to-noise ratio is larger than 10 dB (see Section 9.9).

(iii) The ratio R is equal to 1 if no model errors are present, otherwise the it is larger. For SBTLS the parametric noise model in the left weighting should be a consistent estimate of the true noise model. The required number of independent repeated experiments is $M \geq 6$ for SML and $M \geq 2$ for SGTLS and SBTLS.

(iv) SML requires $M \geq 6$ and independent (over the frequency k), normally distributed errors $N_Z(k)$, whereas SGTLS and SBTLS require $M \geq 2$ and the noise assumptions are exactly the same as when the noise model is known. The picture is somewhat more complicated for SSUB: the estimation of the poles requires $M \geq 2$ and the noise assumptions are the same as when the noise model is known, while the estimation of the zeros requires $M \geq 4$ and independent (over the frequency k), normally distributed errors $N_Z(k)$.

(v) Even if it is known that the input-output errors are uncorrelated, the sample noise (co)variance must be estimated, otherwise SML is no longer consistent. This is also the case for the SSUB estimates of the zeros, but not of the poles.

(vi) All the estimates are sensitive to the assumption that the repeated experiments are independent. On top of that, the uncertainty of the SBTLS estimate is sensitive to the quality of the parametric noise model in the left weighting.

If the sample noise (co)variances are regularized for $M = 2, 3$, then the estimate $\hat{\theta}_{\text{SML}}(Z)$ satisfies Theorem 9.21 and is "practically consistent." The same can be done for the SSUB estimates of B and D.

10.9 IDENTIFICATION OF PARAMETRIC NOISE MODELS

Without an additional piece of information, it is impossible to identify (non)parametric noise models within an errors-in-variables stochastic framework. There are four possibilities:

1. The unknown excitation signal is periodic (see Sections 10.1 to 10.8).
2. A noiseless reference signal (typically the signal stored in the arbitrary waveform generator) is available (see Chapter 12).
3. The unknown excitation can be written as filtered white noise, and parametric transfer function models for the plant, the input noise, the output noise, and the excitation signal are identified simultaneously (see, for example, Söderström, 2007, and Pintelon and Schoukens, 2007).
4. The arbitrary input is exactly known (see, for example, Ljung, 1999, and Söderström and Stoica, 1981).

In this section we discuss the fourth possibility, which is often applicable in control applications. It also includes problems where the plant dynamics are identified from output observations only (the unobserved input is assumed to be white). We first present the classical time domain approach (prediction error framework), and next develop a frequency domain maximum likelihood solution to the problem. The properties of the frequency domain estimator are studied in detail and an in-depth comparison with the prediction error approach is made.

10.9.1 Generalized Output Error Stochastic Framework

Within a generalized output error stochastic framework the input is exactly known $u(t) = u_0(t)$, and the output is disturbed by filtered (band-limited) white noise $n_y(t) = n_p(t) + m_y(t)$ (see Figure 10-5). A parametric plant model and output noise model are identified simultaneously from known input, noisy output samples $u(t)$ and $y(t)$, $t = 0, 1, \ldots, N-1$. The classical time domain solution to this problem assumes that the input is piecewise constant, and a discrete-time plant and noise model are estimated by minimizing the prediction error (PE) cost function

$$V_{\text{PE}}(\theta, Z) = \sum_{t=0}^{N-1} \varepsilon^2(t, \theta) \tag{10-48}$$

w.r.t. the plant and noise model parameters θ, where $\varepsilon(t, \theta)$ is the one-step-ahead prediction error

$$\varepsilon(t, \theta) = H^{-1}(q, \theta)(y(t) - G(q, \theta)u(t)) \tag{10-49}$$

with q the backward shift operator $qx(t) = x(t-1)$. A comprehensive study of the properties of the minimizer $\hat{\theta}_{\text{PE}}(Z)$ of (10-48) can be found in Ljung (1999) and Söderström and Stoica (1981).

Applying Parceval's equality to (10-48) gives a frequency domain interpretation of the prediction error method

$$V_{\text{PE}}(\theta, Z) = \sum_{t=0}^{N-1} \varepsilon^2(t, \theta) = \sum_{k=0}^{N-1} |\varepsilon(z_k^{-1}, \theta)|^2 \tag{10-50}$$

where $\varepsilon(z_k^{-1}, \theta) = \text{DFT}(\varepsilon(t, \theta))$ is related to the input-output DFT spectra as

$$\varepsilon(z_k^{-1}, \theta) = H^{-1}(z_k^{-1}, \theta)(Y(k) - G(z_k^{-1}, \theta)U(k) - T_G(z_k^{-1}, \theta) - T_H(z_k^{-1}, \theta)) \tag{10-51}$$

with $T_G(z_k^{-1}, \theta)$ and $T_H(z_k^{-1}, \theta)$, respectively, the plant (6-35) and noise (6-66) transient terms (proof: solve the combined plant-noise transfer function model (6-69) for $E(k)$). It follows that $|H(z_k^{-1}, \theta)|^2$ plays the role of the nonparametric noise weighting $\sigma_Y^2(k)$ in the ML solution (10-6).

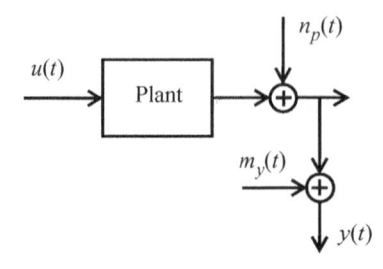

Figure 10-5. General output error framework: the arbitrary input is exactly known, $u(t) = u_0(t)$, and the output is disturbed by process noise $n_p(t)$ and measurement noise $m_y(t)$.

Section 10.9 ■ Identification of Parametric Noise Models

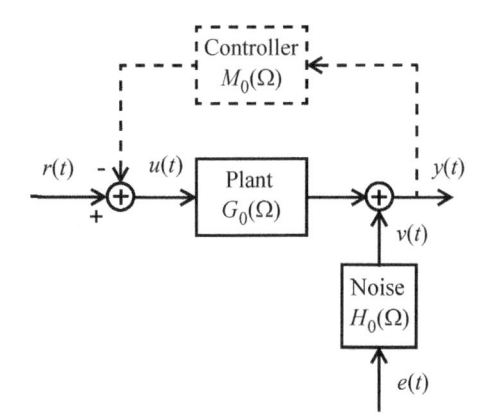

Figure 10-6. Plant operating in open (solid lines only) or closed (solid and dashed lines) loop. $G_0(\Omega)$, $H_0(\Omega)$, and $M_0(\Omega)$ are, respectively, the plant, the noise, and the controller transfer functions. The noise $v(t)$ is the sum of the process noise (open and closed loop) and the measurement noise (open loop only).

10.9.2 A Frequency Domain Solution

In this section we present a frequency domain solution to the parametric noise modeling problem within a generalized output error stochastic framework. The plant may operate in open or closed loop (see Figure 10-6), and both the discrete-time (DT) and continuous-time (CT) cases are handled. First, the input-output DFT spectra $U(k)$ and $Y(k)$ of the exactly known input and noisy output samples $u(t)$ and $y(t)$, $t = 0, 1, ..., N-1$, are calculated. Next, the DFT frequencies in the frequency band(s) of interest are selected: $f_k = kf_s/N$ with $k \in \mathbb{K}$ and where

$$\mathbb{K} \subseteq \{0, 1, ..., N/2\} \tag{10-52}$$

Finally, starting from $U(k)$ and $Y(k)$, $k \in \mathbb{K}$, a Gaussian maximum likelihood estimator for the plant $G(\Omega, \theta)$ and noise $H(\Omega, \theta)$ models is constructed. It requires the following assumptions.

Since the plant, noise, and controller transient terms decrease as an $O(N^{-1/2})$ in the relationship between the input-output DFT spectra (see Section 6.3.2), the Gaussian maximum likelihood (ML) estimator is constructed assuming that the transient terms are zero. Afterwards the finite sample behavior of the ML estimator is improved by adding the plant and noise transient terms in the ML cost function.

Assumption 10.12 (Generalized Output Error Framework): The input $u(t)$ is observed without measurement errors $m_u(t) = 0$. The output is disturbed by process noise $n_p(t)$ (open and closed loop) and measurement noise $m_y(t)$ (open loop only).

Assumption 10.13 (Relationship Input-Output DFT Spectra): The input-output DFT spectra $U(k)$ and $Y(k)$ are related to the DFT spectrum of the reference signal $R(k)$ and the driving (band-limited) white noise source $E(k)$ as

$$\begin{aligned} Y(k) &= G_0(\Omega_k)U(k) + V(k) \\ V(k) &= H_0(\Omega_k)E(k) \end{aligned} \tag{10-53}$$

where $U(k) = R(k)$ for the open loop setup and

$$U(k) = R(k) - M_0(\Omega_k)Y(k) \tag{10-54}$$

for the closed loop setup. $G_0(\Omega)$, $H_0(\Omega)$, and $M_0(\Omega)$ are rational functions of Ω.

Assumption 10.14 (Excitation Signal): The reference signal $r(t)$ satisfies Assumptions 9.1 and 9.11 ($r(t) = u(t)$ in open loop).

Assumption 10.15 (Noise Moments of Order m): $E(k)$ in (10-53) is independent (over k), circular complex distributed noise ($\mathbb{E}\{E^2(k)\} = 0$), with zero mean, variance λ, and finite moments of order m. $E(k)$ is independent of the input $U(k)$ (open loop, $M_0(\Omega) = 0$) or the reference signal $R(k)$ (closed loop, $M_0(\Omega) \neq 0$).

Assumption 10.16 (Noise Probability Density Function): $E(k)$ in (10-53) is normally distributed.

Assumption 10.17 (Known Controller): The controller transfer function $M_0(\Omega)$ is known.

Note that Assumptions 10.13-10.16 are asymptotically ($N \to \infty$) valid (see Sections 6.3.2 and 16.16, respectively). Note also that under Assumption 10.12 the controller is known (Assumption 10.17) if and only if the reference signal is known.

Theorem 10.18 (Gaussian Log-Likelihood Function): Under Assumptions 10.12-10.14, 10.15 ($m = 2$), 10.16, and 10.17 the negative Gaussian log-likelihood function is, within a constant, given by

$$\sum_{k \in \mathbb{K}} \ln(\lambda |S(\Omega_k, \theta)|^2) + \sum_{k \in \mathbb{K}} \lambda^{-1} |\varepsilon(\Omega_k, \theta)|^2 \qquad (10\text{-}55)$$

with $\mathbb{K}$ defined in (10-52), $\lambda = \text{var}(E(k))$, $S(\Omega, \theta)$ the following rational function of Ω

$$S(\Omega, \theta) = \frac{H(\Omega, \theta)}{1 + G(\Omega, \theta) M_0(\Omega)} \qquad (10\text{-}56)$$

and $\varepsilon(\Omega_k, \theta)$ the prediction error

$$\varepsilon(\Omega_k, \theta) = H^{-1}(\Omega_k, \theta)(Y(k) - G(\Omega_k, \theta) U(k)) \qquad (10\text{-}57)$$

At DC ($k = 0$) and Nyquist ($k = N/2$) the sums in (10-55) are multiplied by $1/2$.

Proof. See Appendix 10.O. □

For $M_0(\Omega) = 0$, the log-likelihood function (10-55) reduces to the open loop result in Ljung (1999, p. 230). By eliminating λ, log-likelihood function (10-55) can be simplified to a quadratic form.

Corollary 10.19 (Gaussian Maximum Likelihood Cost Function): Under the assumptions of Theorem 10.18 the Gaussian maximum likelihood (ML) cost function $V_F(\theta, Z)$, where Z represents the data, is given by

$$V_F(\theta, Z) = F^{-1} \sum_{k \in \mathbb{K}} |\varepsilon(\Omega_k, \theta) g_F(\theta)|^2, \qquad (10\text{-}58)$$

Section 10.9 ■ Identification of Parametric Noise Models

with $\mathbb{K}$ defined in (10-52), F the number of frequencies in the set $\mathbb{K}$ (DC, $k = 0$, and Nyquist, $k = N/2$, count for 1/2), $\varepsilon(\Omega_k, \theta)$ defined in (10-57), and

$$g_F(\theta) = \left(\prod_{k \in \mathbb{K}} S(\Omega_k, \theta)\right)^{1/F} = \exp\left(F^{-1} \sum_{k \in \mathbb{K}} \ln S(\Omega_k, \theta)\right), \quad (10\text{-}59)$$

where $S(\Omega, \theta)$ is defined in (10-56). The variance of the driving white noise source equals

$$\lambda(\theta) = F^{-1} \sum_{k \in \mathbb{K}} |\varepsilon(\Omega_k, \theta)|^2 \quad (10\text{-}60)$$

At DC ($k = 0$) and Nyquist ($k = N/2$) the sums in (10-58) to (10-60) are multiplied by 1/2.

Proof. See Appendix 10.P. □

As a result the minimizer of (10-58) can be calculated in a numerical stable way via the iterative Newton-Gauss and Levenberg-Marquardt (see Pintelon et al., 2006a, for the details). To improve the finite sample behavior of the ML estimator (10-58), the plant $T_G(\Omega_k, \theta)$ (6-35) and noise $T_H(\Omega_k, \theta)$ (6-66) transient terms are added in the prediction error (10-58), giving

$$\varepsilon(\Omega_k, \theta) = H^{-1}(\Omega_k, \theta)(Y(k) - G(\Omega_k, \theta)U(k) - T_G(\Omega_k, \theta) - T_H(\Omega_k, \theta)) \quad (10\text{-}61)$$

Note, however, that the numerator coefficients of $T_G(\Omega_k, \theta)$ and $T_H(\Omega_k, \theta)$ are not consistently estimated. For discrete-time models and frequency sets covering uniformly the unit circle, (10-58) can be simplified further.

Corollary 10.20 (Gaussian Maximum Likelihood Cost Function for Discrete-Time Models over the Full Unit Circle): If the frequencies cover uniformly the whole unit circle ($z_k = \exp(j2\pi k/N)$ with $k \in \mathbb{K} = \{0, 1, ..., N/2\}$), and if $S(z^{-1}, \theta)$ (10-56) and $S^{-1}(z^{-1}, \theta)$ are stable and satisfy

$$\lim_{z \to \infty} S(z^{-1}, \theta) = 1, \quad (10\text{-}62)$$

then the ML cost function (10-58) simplifies to

$$N^{-1} \sum_{k=0}^{N-1} |\varepsilon(z_k^{-1}, \theta)|^2 + O(|\lambda_{\max}|^N / N) \quad (10\text{-}63)$$

where $\lambda_{\max}$ is the dominant pole of $\ln S(z^{-1}, \theta)$.

Proof. See Appendix 10.Q. □

It can be concluded that under the assumptions of Corollary 10.20, the ML cost function (10-58) converges ($N \to \infty$) at the rate $O(|\lambda_{\max}|^N / N)$ to the classical prediction error cost function (10-50). However, they are different in all other cases: discrete-time models with non-monic noise model ($c_0 \neq 1$ or $d_0 \neq 1$ in (6-59)), discrete-time models with non-uniform frequency grid, or continuous-time models.

Condition (10-62) is fulfilled if $c_0 = d_0 = 1$ in (6-59) (monic noise model) and if the plant and/or controller transfer functions have a delay of at least one sample, for example,

$b_0 = 0$, $a_0 = 1$ in (6-20) ($G(0, \theta) = 0$) and $M_0(0) \neq \infty$. Note that the step-invariant transfer function model $G(z^{-1}, \theta) = (1 - z^{-1})Z\{G(s)/s\}$ (see Example 6.3 on page 179) of most physical continuous-time plants $G(s)$ satisfies $G(0, \theta) = 0$.

10.9.3 Asymptotic Properties of the Gaussian Maximum Likelihood Estimator

To study the asymptotic ($F = O(N) \to \infty$) behavior of the minimizer of the maximum likelihood cost function (10-58), we need an identifiable parameterization of the model structure

$$Y(k, \theta) = G(\Omega_k, \theta)U(k) + H(\Omega_k, \theta)E(k) + T_G(\Omega_k, \theta) + T_H(\Omega_k, \theta) \qquad (10\text{-}64)$$

with $G(\Omega, \theta)$ and $T_G(\Omega_k, \theta)$ defined in, respectively, (6-20) and (6-35), and $H(\Omega, \theta)$ and $T_H(\Omega_k, \theta)$ defined in, respectively, and (6-59), (6-62) and (6-66). Model structure (10-64) is overparameterized: multiplying the numerator b and denominator a coefficients of the plant model by the same non-zero real number leaves $G(\Omega, \theta)$ unchanged, and similarly for the noise model $H(\Omega, \theta)$. Further, the fact that $E(k)$ in (10-64) is not observed, and that the term $H(\Omega_k, \theta)E(k)$ remains the same when multiplying $H(\Omega_k, \theta)$ and dividing $E(k)$ by the same non-zero real number, imposes an additional constraint on the noise model parameters. Hence, according to the particular model structure, one (OE), two (ARMA, ARMAX), or three (BJ, hybrid BJ) parameter constraints are needed (see Table 10-2). For example, in discrete-time BJ modeling usually the choice $a_0 = c_0 = d_0 = 1$ is made in (6-20) and (6-59). Other choices are, however, possible (see Pintelon et al., 2006a). The cost function $V_F(\theta, Z)$ (10-58) contains exactly the same parameter ambiguities as model structure (10-64) and, therefore, the estimated models $G(\Omega, \hat{\theta}_{ML}(Z))$ and $\hat{\lambda}^{1/2}H(\Omega, \hat{\theta}_{ML}(Z))$, with $\hat{\theta}_{ML}(Z)$ the minimizer of (10-58) and $\hat{\lambda} = \lambda(\hat{\theta}_{ML}(Z))$ (10-60), are independent of the particular parameter constraint(s) chosen (see Chapter 20).

Since cost function (10-58) only depends on the magnitude of the noise model there is a global identifiability problem. For BJ model structures it is avoided by restricting the allowable poles/zeros positions of the noise model to the stable region of the Ω-domain. This is not necessary for the poles of ARMAX model structures (the plant and noise models have the same poles) that are determined by the plant dynamics. Both observations lead to the following standard assumption.

Assumption 10.21 (Constraint on the Noise Model): $H^{-1}(\Omega, \theta)$ is a stable transfer function. The poles of $H(\Omega, \theta)$ that are not in common with $G(\Omega, \theta)$ are stable.

TABLE 10-2 Possible Model Parameter Constraints (for Continuous-Time Models the Constraints Are Imposed on the Normalized Parameters)

Model Structure	Constraints on θ for Discrete-Time Models	Constraints on θ for Continuous-Time Models
ARMA ((10-64) with $G = 0$ and $T_G = 0$)	$c_0 = d_0 = 1$	$c_{n_c} = d_{n_d} = 1$
OE ((10-64) with $H = 1$ and $T_H = 0$)	$a_0 = 1$	$a_{n_a} = 1$
ARMAX ((10-64) with $D = A$) [a]	$a_0 = c_0 = 1$	$a_{n_a} = c_{n_c} = 1$
BJ	$a_0 = c_0 = d_0 = 1$	$a_{n_a} = c_{n_c} = d_{n_d} = 1$

a. For ARMAX models $T = T_G + T_H$ represents the sum of the plant and the noise transient terms.

These results are summarized in the following theorem.

Theorem 10.22 (Gaussian Maximum Likelihood Estimator $\hat{\theta}_{ML}(Z)$): Under the assumptions of Theorem 10.18 and Assumption 10.21, the Gaussian maximum likelihood estimator $\hat{\theta}_{ML}(Z)$ of the plant and noise model parameters minimizes (10-58) subject to the constraints in Table 10-2.

To study the asymptotic ($F = O(N) \to \infty$) behavior of the Gaussian maximum likelihood estimator $\hat{\theta}_{ML}(Z)$ defined in Theorem 10.22, we need an assumption concerning the true plant and noise models.

Assumption 10.23 (Existence of a True Plant/Noise Model): The true plant $G_0(\Omega)$ and noise $H_0(\Omega)$ transfer functions belong to the considered model set. The common poles of G_0 and H_0 are not common zeros of G_0 and H_0, the private poles of G_0 are not zeros of G_0, and the private poles of H_0 are not zeros of H_0.

Under the assumptions of Theorem 10.22, Assumption 10.23, and Assumptions 9.6–9.10, the Gaussian maximum likelihood (ML) estimator $\hat{\theta}_{ML}(Z)$ satisfies the standard conditions, among others, (i) the likelihood function is based on independent and identically distributed random variables $E(k)$, $k = (1, 2, ..., F)$, and (ii) the number of model parameters $\dim(\theta)$ does not increase with the amount of data F. Therefore, $\hat{\theta}_{ML}(Z)$ is strongly consistent ($\hat{\theta}_{ML}(Z) \to \theta_0$ with probability one as $F \to \infty$), asymptotically efficient (the asymptotic covariance matrix equals the Cramér-Rao lower bound), and asymptotically normally distributed (Caines, 1988). One may wonder now how sensitive these asymptotic properties are w.r.t. the basic assumptions made to construct the Gaussian ML estimator. For example, what if the errors are not normally distributed (Assumption 10.16), what if the true model does not belong to the considered model set (Assumption 10.23), or what if the independence assumption is violated (Assumption 10.15)? To analyze the robustness of the asymptotic properties of $\hat{\theta}_{ML}(Z)$ the standard assumptions of Section 9.6 are made. In addition Assumption 10.15 is replaced by the following mixing condition.

Assumption 10.24 (Noise Mixing Condition of Order P): The process noise $V(k)$ satisfies (10-53) where $E(k)$ is circular complex distributed ($\mathbb{E}\{E^2(k)\} = 0$), with zero mean and variance λ. $E(k)$ is independent of the input $U(k)$ (open loop, $M_0(\Omega) = 0$) or the reference signal $R(k)$ (closed loop, $M_0(\Omega) \neq 0$). $V(k)$ is mixing over k of order P.

Note that the stochastic nonlinear contributions generated by a nonlinear system excited by Gaussian noise is mixing over the frequency of order infinity (Theorem 3.16 on page 86). This is a quite important example since the process noise in a linear identification framework is mostly dominated by the stochastic nonlinear distortions (Schoukens et al., 2005). The following theorem proves the asymptotic properties of the Gaussian ML estimate $\hat{\theta}_{ML}(Z)$ under non-standard conditions.

Theorem 10.25 (Asymptotic Properties $\hat{\theta}_{ML}(Z)$): Consider model structure (10-64) where $G(\Omega, \theta)$, $T_G(\Omega, \theta)$, $H(\Omega, \theta)$, and $T_H(\Omega, \theta)$ are subject to the constraints of Table 10-2. Consider the assumptions of Section 9.6, where Assumptions 9.1–9.5 are replaced by Assumptions 10.12–10.14, 10.17, 10.21, and 10.15 ($m = 4$) or 10.24 (stochastic convergence); Assumptions 9.11–9.13 by Assumptions 10.14, and 10.15 ($m = 4 + \delta$) or 10.24 (asymptotic normality); Assumption 9.16 by Assumption 10.23 (consistency); and Assumptions 9.18–9.20 by Assumptions 10.15 ($m = 2$) and 10.16 (asymptotic efficiency).

Under these conditions the minimizer $\hat{\theta}_{\mathrm{ML}}(Z)$ of (10-58) has the asymptotic ($F = O(N) \to \infty$) properties of Theorem 9.21 on page 298 with the following modifications:

1. The asymptotic covariance matrix in property 4 of Theorem 9.21 is given by

$$\mathrm{Cov}(\delta_\theta(Z)) = (F_1 + F_2)^{-1}(F_1 + (k_u^c - 1)F_2)(F_1 + F_2)^{-1} \qquad (10\text{-}65)$$

with $k_u^c = (k_u + 1)/2$ and k_u the kurtosis factor of the real and imaginary parts of the noise (e.g., $k_u^c = 2$ for Gaussian noise),

$$
\begin{aligned}
F_1 &= 2\sum\nolimits_{k \in \mathbb{K}} \frac{|S_0(\Omega_k)|^2}{|H_0(\Omega_k)|^4} \frac{\mathbb{E}\{|R(k)|^2\}}{\lambda_0} \mathrm{Re}\!\left(\!\left(\frac{\partial G(\Omega_k, \theta)}{\partial \theta_0}\right)^{\!H}\!\left(\frac{\partial G(\Omega_k, \theta)}{\partial \theta_0}\right)\!\right) \\
F_2 &= \sum\nolimits_{k \in \mathbb{K}} (\chi_k - F^{-1}\sum\nolimits_{k=1}^F \chi_k)^T (\chi_k - F^{-1}\sum\nolimits_{k=1}^F \chi_k) \\
\chi_k &= \frac{\partial}{\partial \theta_0} \ln |S(\Omega_k, \theta)|^2
\end{aligned}
\qquad (10\text{-}66)
$$

where $\mathbb{K}$ and $S(\Omega, \theta)$ are defined in, respectively, (10-52) and (10-56), and with F the number of frequencies in the set $\mathbb{K}$ (DC, $k = 0$, and Nyquist, $k = N/2$, count for 1/2). At DC and Nyquist the sums in (10-66) are multiplied by 1/2. For BJ model structures (independently parameterized G and H) identified in open loop ($M_0 = 0$), (10-66) simplifies to

$$\mathrm{Cov}(\delta_{\theta_G}(Z)) = F_1^{-1} \text{ and } \mathrm{Cov}(\delta_{\theta_H}(Z)) = (k_u^c - 1)F_2^{-1} \qquad (10\text{-}67)$$

with θ_G the plant model parameters, for example, $\theta_G^T = [a^T, b^T]$, and θ_H the noise model parameters, for example, $\theta_H^T = [c^T, d^T]$.

2. $V_*(\theta)$ in property 5 of Theorem 9.21 is replaced by ($S(\Omega, \theta)$ is defined in (10-56))

$$
\begin{aligned}
V_*(\theta) &= |g_*(\theta)|^2 \int_{f_{\min}}^{f_{\max}} \mathbb{E}\{|\varepsilon(\Omega(f), \theta)|^2\} n(f) df \\
g_*(\theta) &= \exp\!\left(\int_{f_{\min}}^{f_{\max}} \ln(S(\Omega(f), \theta)) n(f) df\right)
\end{aligned}
\qquad (10\text{-}68)
$$

3. $\hat{\theta}_{\mathrm{ML}}(Z)$ in property 6 of Theorem 9.21 is replaced by $\hat{\theta}_{\mathrm{ML}}(Z)$ and $\hat{\lambda}_{\mathrm{ML}} = \lambda(\hat{\theta}_{\mathrm{ML}}(Z))$, where $\lambda(\theta)$ is defined in (10-60).

4. Under the constraints of Table 10-2, the Cramér-Rao lower bound in property 8 of Theorem 9.21 equals

$$\mathrm{Cov}(\delta_\theta(Z)) = (F_1 + F_2)^{-1} \qquad (10\text{-}69)$$

with $F_1 + F_2$ the Fisher information matrix.

Proof. See Appendix 10.R. □

10.9.4 Discussion

- A surprising consequence of Theorem 10.18 is that the knowledge of the controller contributes to the knowledge of the plant and noise models ($M(\Omega) \neq M_0(\Omega)$ in (10-58) implies $S(\Omega, \theta_0) \neq S_0(\Omega)$ giving biased estimates: see Appendix 10.R), which is not the case for the time domain prediction error method (see Ljung, 1999, and Corollary 10.20). This has been mentioned for the first time in McKelvey (2000). The apparent contradiction is explained by the fact that cutting out a part of the unit circle corresponds to non-causal filtering in the time domain (e.g., convolution with a sinc-function). The latter invalidates the classical construction of the likelihood function based on time domain data captured in feedback (Caines, 1988).

- Cost function (10-58), with zero transient terms (see (10-57)), is called the *conditional* maximum likelihood function in the literature (Söderström and Stoica, 1989). By adding a posteriori the plant and noise transient terms to the maximum likelihood cost function as in (10-61), we assume that the numerator coefficients of $T_G(\Omega, \theta)$ and $T_H(\Omega, \theta)$ are deterministic parameters. This is called the *approximate* maximum likelihood solution in the statistics literature. If we construct the likelihood function assuming that the numerator coefficients of $T_G(\Omega, \theta)$ and $T_H(\Omega, \theta)$ are random parameters that are correlated with the excitation $u(t)$ and the driving (band-limited) white noise source $e(t)$, then we get the *exact* maximum likelihood solution (Söderström and Stoica, 1989; Agüero et al., 2010). Note that similar (conditional, approximate, and exact) time domain maximum likelihood (ML) solutions can be constructed. While the equivalence of the time and frequency domain solutions is asymptotic in the data length N for the *conditional* and *approximate* ML estimators (see Corollary 10.20), the *exact* time and frequency domain ML estimates are equal for finite values of N (Agüero et al., 2010).

- In contrast to the time domain prediction error method, consistent estimation of the plant model parameters in open loop ALWAYS requires the correct noise model structure (if $H(\Omega, \theta_*) \neq H_0(\Omega)$, with θ_* the minimizer of $V_*(\theta)$ (10-68), then $S(\Omega, \theta_*) \neq S_0(\Omega)$, and, hence, the estimates are biased: see Appendix 10.R).

- The asymptotic uncertainty (10-69) of the time domain prediction error method is also valid for non-Gaussian errors $e(t)$ (see Ljung, 1999). This is not in contradiction with (10-65), showing that $(F_1 + F_2)^{-1}$ is valid for Gaussian frequency domain noise only ($k_u^c = 2$). Indeed, the DFT of filtered iid noise with existing moments of any order is asymptotically (number of time domain samples $N \to \infty$) independent, circular complex normally distributed (see Theorem 16.25).

- Theorem 10.25 is valid for any other identifiable parameterization of the plant and noise transfer function models: ratio (orthogonal) polynomials where c_0 (DT) or c_{n_c} (CT) is not constrained (λ is not estimated), partial fraction expansion, and state space representation. However, the particular expression of the asymptotic covariance matrix (10-65) and the Cramér-Rao lower bound (10-69) can change. For example, the constraint $c_0 = 1$ (DT) or $c_{n_c} = 1$ (CT) can be replaced by $\lambda = 1$, which eases the uncertainty calculation of the estimated noise model. The expressions for F_1 and F_2 in (10-66) then become

$$F_1 = 2\sum_{k \in \mathbb{K}} \frac{|S_0(\Omega_k)|^2}{|H_0(\Omega_k)|^4} \mathbb{E}\{|R(k)|^2\} \operatorname{Re}\left(\left(\frac{\partial G(\Omega_k, \theta)}{\partial \theta_0}\right)^H \left(\frac{\partial G(\Omega_k, \theta)}{\partial \theta_0}\right)\right)$$

$$F_2 = \sum_{k \in \mathbb{K}} \chi_k^T \chi_k$$

(10-70)

where χ_k is defined in (10-66) (proof: see Pintelon et al., 2007a).

- Theorem 10.25 is also valid for hybrid Box-Jenkins (BJ) models (6-75) identified in open loop or closed loop (proof: replace everywhere $H(\Omega_k, \theta)$ by $H(z_k^{-1}, \theta)$ in Appendix 10.R). Note that the limit of $g_F(\theta)$ (10-59) for $F \to \infty$ is always different from 1 for hybrid BJ models identified in closed loop, even if the frequencies cover uniformly the unit circle.

- Since the identified transfer function models $G(\Omega, \hat{\theta}_{ML}(Z))$ and $H(\Omega, \hat{\theta}_{ML}(Z))$ are independent of the particular constraint chosen, the minimizer of the cost function (10-58) is in practice not calculated using the constraints of Table 10-2. Instead, all parameters are left free and the pseudo-inverse of the Jacobian matrix is calculated taking into account its rank (= $\dim(\theta)$ - the number of constraints in Table 10-2). The reader is referred to Pintelon et al. (2006a) and Appendix 9.L.4 for the algorithmic details.

- If the reference signal or the controller is unknown (Assumption 10.17 is violated), then the estimated plant and noise models are biased (Theorem 10.25). This bias can be avoided at the cost of modeling simultaneously the plant, the noise, the controller, and the reference signal transfer functions (see Pintelon and Schoukens, 2006). In the time domain the latter is known as the joint input-output method (Ljung, 1999; Söderström and Stoica, 1989).

- The results of Sections 10.9.2 and 10.9.3 can be extended to multivariable systems (see Pintelon et al., 2007a).

- A disadvantage of the parametric noise model is that its quality (strongly) depends on quality of the estimated plant model (the noise model tries to follow every systematic deviation from the true plant model). This is not the case for the sample noise (co)variances because they are independent of the estimated plant model.

10.9.5 Experimental Illustration

The nonlinear electrical circuit of Figure 10-7 simulates a second order nonlinear mechanical system with a hardening spring (see (3-25)). Using normally distributed excitations, the main error source within a linear system identification framework is the stochastic nonlinear contribution that satisfies the mixing condition of Assumption 10.24. Two experiments have been performed, the first with a white band-limited normally distributed signal and the second with a white band-limited periodic signal with random phases. Both signals have the same power (16 mV rms) and bandwidth (200 Hz). For both experiments the sampling frequency equals $f_s = 20 \text{ MHz}/2^{15} \approx 600 \text{ Hz}$, and the input/output signals are lowpass filtered before sampling. The random excitation data is used for identifying the Box-Jenkins models, while the periodic excitation data is used for cross-validation.

Starting from $N = 10400$ input-output samples of the random excitation experiment the following three model structures are identified: (i) hybrid Box-Jenkins (CT plant and DT noise model (6-75)) over the full unit circle without DC and Nyquist ($F = 5199$), (ii) hybrid Box-Jenkins in the band (0 Hz, 200 Hz] ($F = 3406$), and (iii) continuous-time Box-Jenkins

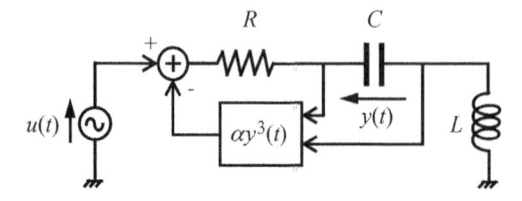

Figure 10-7. Block diagram of the nonlinear electrical circuit with $R = 16 \, \Omega$, $C = 9.4 \, \mu\text{F}$, $L = 1 \, \text{H}$, and $\alpha \sim 1000 \, \text{V}^{-2}$.

(CT plant and CT noise model) in the band (0 Hz, 200 Hz] ($F = 3406$). From the validation tests in Figures 10-8 and 10-9 it can be concluded that the random excitation measurements are explained very well by a second-order CT plant model ($n_a = 2$, $n_b = 0$) combined with, respectively, the following noise models: (i) eight-order DT noise model ($n_c = n_d = 8$, $n_j = 7$, and $n_i = 4$), (ii) sixth-order DT noise model ($n_c = 6$, $n_d = 5$, $n_j = 5$, and $n_i = 2$), and (iii) fifth-order CT noise ($n_c = 4$, $n_d = 5$, $n_j = 0$, and $n_i = 2$). Since the quality of the identified plant model is identical for the three model structures, only the result for the CT-BJ model structure is shown in Figure 10-8. In the cross-validation tests of Figures 10-8 and 10-9 the estimated BJ models are compared with the measured frequency response function (FRF) $\hat{G}(j\omega_k)$ and the measured (variance of the) output residual $\hat{V}(k)$ obtained from the periodic data. It follows that the BJ models also explain the periodic excitation data very well. Note that the reduced order noise models perform somewhat less in the neighborhood of 200 Hz (see Figure 10-9, right column, middle and bottom rows). This is due to the resonance peak around 210 Hz in the noise power spectrum (see Figure 10-9, top left figure), which cannot be identified from the data in the band (0 Hz, 200 Hz].

The measured FRF $\hat{G}(j\omega_k)$ and measured (variance of the) output residual $\hat{V}(k)$ in Figures 10-8 and 10-9 are calculated as

$$\hat{G}(j\omega_k) = (Y(k) - T_G(\Omega_k, \hat{\theta}_{ML}(Z)) - T_H(\Omega_k, \hat{\theta}_{ML}(Z)))/U(k)$$
$$\hat{V}(k) = Y(k) - G(\Omega_k, \hat{\theta}_{ML}(Z))U(k) - T_G(\Omega_k, \hat{\theta}_{ML}(Z)) - T_H(\Omega_k, \hat{\theta}_{ML}(Z))$$
(10-71)

for the random excitation, and as

$$\hat{G}(j\omega_k) = \hat{Y}(k)/\hat{U}(k) \text{ and } \hat{\sigma}_V^2(k) = \hat{\sigma}_G^2(k)|\hat{U}(k)|^2$$
(10-72)

for the periodic excitation. In (10-72) $\hat{G}(j\omega_k)$ and $\hat{\sigma}_G^2(k)$ are, respectively, the sample mean and sample variance obtained by analyzing consecutive periods of the input-output signals. The uncertainty of $\lambda^{1/2}(\hat{\theta}_{ML}(Z))|H(\Omega_k, \hat{\theta}_{ML}(Z))|$ and $G(j\omega_k, \hat{\theta}_{ML}(Z))$ is calculated using, respectively, (16-23) and (16-24), where $Cov(\hat{\theta}_{ML}(Z))$ is given by (10-67) with $k_u^c = 2$.

Figure 10-9 nicely illustrates the benefit of modeling the process noise in the frequency band of interest: in the band (0 Hz, 200 Hz] a DT noise model of order 6/5 is sufficient to describe the noise power spectrum, while over the full unit circle an order 8/8 is needed. It also shows that the CT noise model (order 4/5) is a valid alternative for the classical DT description (order 6/5).

10.10 IDENTIFICATION IN FEEDBACK

Consider the linear feedback experiment of Figure 9-14 on page 345. For periodic reference signals $r(t)$, the SML (10-10), SGTLS (10-23), SBTLS (10-33), and SSUB (Section 10.6) estimates remain strongly consistent (see Theorems 10.3, 10.6, 10.8, and 10.10). The basic reason for this is that the sample mean and sample (co)variance calculation makes a natural separation between the random and the periodic parts of the input $u(t)$ and output $y(t)$ signals without requiring knowledge of $r(t)$.

For arbitrary reference signals $r(t)$, the technical difficulty is that the input is correlated with the process noise (Ljung, 1999; Söderström and Stoica, 1989). The prediction error estimate (10-48) of the plant model parameters is consistent if the input and output signals are observed without errors ($m_u(t) = 0$, $m_y(t) = 0$), and if the true noise model belongs to the

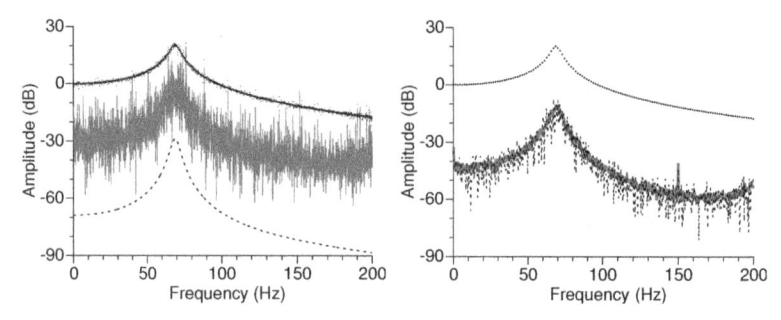

Figure 10-8. (Cross-)validation of the identified CT plant models – nonlinear electrical circuit. Left: validation with the random excitation measurement. Right: cross-validation with the periodic excitation measurement. Dots: measured FRF $\hat{G}(j\omega_k)$. Gray line: complex difference between the measured and modeled FRF. Dashed line: standard deviation of the model (left) or the measurement (right).

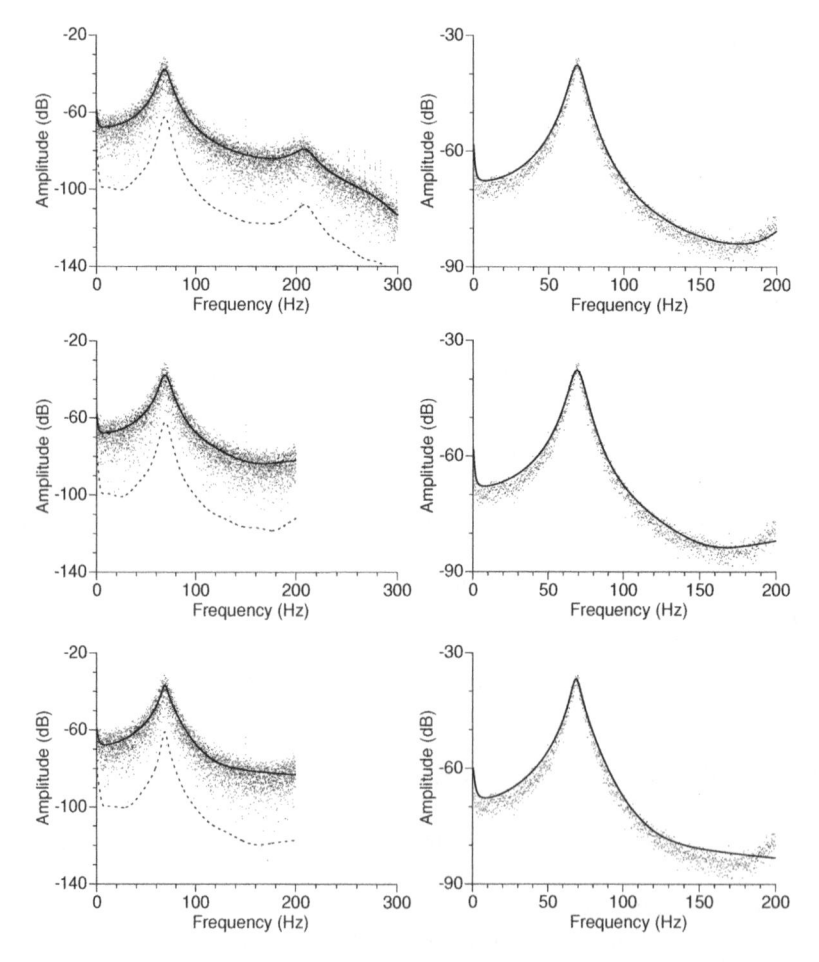

Figure 10-9. (Cross-)validation of the identified noise models – nonlinear electrical circuit. Left column: validation with the output residuals $\hat{V}(k)$ of the random excitation measurement. Right column: cross-validation with variance $\hat{\sigma}_V^2(k)$ of the output residuals of the periodic excitation measurement. Top row: DT model of order 8/8 over the full unit circle. Middle row: DT model of order 6/5 in the band $(0\,\text{Hz}, 200\,\text{Hz}]$. Bottom row: CT model of order 4/5 in the band $(0\,\text{Hz}, 200\,\text{Hz}]$. Dots: measured output residuals. Solid line: noise model times standard deviation driving white noise source. Dashed line: standard deviation model.

considered model set. For the frequency domain ML solution (10-58) in addition the controller or reference signal should be known (see Theorem 10.25). This additional requirement is the price to be paid for modeling the plant and the process noise in the relevant frequency band(s).

Other solutions have been developed that do not require the construction of a consistent parametric noise model: see Forssell and Ljung (1999) and Van den Hof and Schrama (1995) for an overview of the classical time domain methods, which all require that the input and output signals are observed without errors ($m_u(t) = 0$, $m_y(t) = 0$), and see Chapter 12 for the full errors-in-variables problem ($m_u(t) \neq 0$, $m_y(t) \neq 0$). Several of the solutions assume that the reference signal is exactly known and consider the controller and process noise as disturbing errors. The influence of these errors is eliminated by projection of the input and output signals on the reference signal. Note that in the periodic case the distinction between the signal originating from $r(t)$ and the disturbing noise can be made without explicit knowledge of the reference signal.

10.11 APPENDIXES

Appendix 10.A Expected Value and Variance of the Inverse of Chi-Square Random Variable

Let x be a $\chi^2(n)$ distributed random variable. Starting from the results of an F-distribution, it is found that

$$\mathbb{E}\{x^{-1}\} = \frac{1}{n-2} \quad \text{and} \quad \text{var}(x^{-1}) = \frac{2}{(n-2)^2(n-4)} \quad (10\text{-}73)$$

Appendix 10.B First and Second Moments of the Ratio of the True and the Sample Variance of the Equation Error

Consider $c_k(\theta)$ defined in (10-17). Under Assumptions 10.1 and 10.2, the expected value $\mathbb{E}\{c_k(\theta)\}$ is independent of θ and equals:

$$\mathbb{E}\{c_k(\theta)\} = \frac{M-1}{M-2}, \text{ for } M > 2 \quad \text{and} \quad \text{var}(c_k(\theta)) = \frac{(M-1)^2}{(M-2)^2(M-3)}, \text{ for } M > 3 \quad (10\text{-}74)$$

Proof. Consider

$$\begin{aligned}
2(M-1)/c_k(\theta) &= 2(M-1)\hat{\sigma}_{\hat{e}}^2(\Omega_k, \theta)/\sigma_{\hat{e}}^2(\Omega_k, \theta) \\
&= 2(M-1)\frac{\frac{1}{M}\frac{1}{M-1}\sum_{l=1}^{M}\left|e(\Omega_k, \theta, Z^{[l]}(k)) - \hat{e}(\Omega_k, \theta, \hat{Z}(k))\right|^2}{\sigma_{\hat{e}}^2(\Omega_k, \theta)/M} \\
&= \sum_{l=1}^{M}(\text{Re}(z_k^{[l]}(\theta)))^2 + (\text{Im}(z_k^{[l]}(\theta)))^2
\end{aligned} \quad (10\text{-}75)$$

with $z_k^{[l]}(\theta) = (e(\Omega_k, \theta, Z^{[l]}(k)) - \hat{e}(\Omega_k, \theta, \hat{Z}(k)))/(\sqrt{2}\sigma_e(\Omega_k, \theta))$. It follows that (10-75) consist of the sum of two independent central χ^2-distributed variables with $M-1$ degrees of freedom resulting in a χ^2-distribution with $2M-2$ degrees of freedom. Using the results of Appendix 10.A, we find

$$\mathbb{E}\{c_k(\theta)\} = \frac{M-1}{M-2} \quad \text{and} \quad \text{var}(c_k(\theta)) = \frac{(M-1)^2}{(M-2)^2(M-3)} \tag{10-76}$$

Note that the moments of $c_k(\theta)$ are θ-independent due to the fact that $\text{Re}(z_k^{[l]}(\theta))$ and $\text{Im}(z_k^{[l]}(\theta))$ have a θ-independent distribution.

Corollary 10.26: $\hat{\sigma}_{\hat{e}}^2/\sigma_{\hat{e}}^2$ has a $\chi^2(2M-2)$ distribution. The moments $\mathbb{E}\{1/\hat{\sigma}_{\hat{e}}^k\}$ are finite if $M \geq k/2 + 2$.

Proof. The $\chi^2(2M-2)$ distributions follows directly from (10-75). The second claim is a direct result of the $\chi^2(\nu)$ distribution:

$$\chi^2(\nu) \sim \left(\frac{1}{2}x^2\right)^{\nu/2-1} e^{-x^2/2} \tag{10-77}$$

Appendix 10.C Calculation of Some First- and Second-Order Moments

The results in this appendix depend strongly on the number of degrees of freedom that appear in the χ^2 of the sample variance. M measured periods result in M observations of the Fourier coefficient at a given frequency. After subtracting the mean value, this results in $M-1$ degrees of freedom for the sample variance on real or imaginary parts, and those are quadratically combined so that finally the sample variance has $2(M-1)$ degrees of freedom.

For notational simplicity, the dependence of $c_k(\theta)$ and $\hat{\sigma}_{\hat{e}}^2(\Omega_k, \theta)$ on θ, k, and Ω_k is omitted.

(i) c has finite first- and second-order moments for $M > 3$. From (10-74):

$$\mathbb{E}\{c\} = \frac{M-1}{M-2}$$

$$\mathbb{E}\{c^2\} = (\mathbb{E}\{c\})^2 + \text{var}(c) = \left(\frac{M-1}{M-2}\right)^2 + \frac{(M-1)^2}{(M-2)^2(M-3)} = \frac{(M-1)^2}{(M-2)(M-3)}$$

(ii) c' has finite first- and second-order moments for $M \geq 5$.

Because $\mathbb{E}\{c\}$ is independent of θ, it follows directly that $\mathbb{E}\{c'\} = 0$.

$\mathbb{E}\{(\partial c/\partial\theta_{[l]})^2\} = \mathbb{E}\{[\hat{\sigma}_{\hat{e}}^{-2}\partial\sigma_{\hat{e}}^2/\partial\theta_{[l]} - (\sigma_{\hat{e}}^2/\hat{\sigma}_{\hat{e}}^4)\partial\hat{\sigma}_{\hat{e}}^2/\partial\theta_{[l]}]^2\}$ with $\theta_{[l]}$ the lth element of θ. This is an expression of the form $\mathbb{E}\{(a-b)^2\}$. Applying the inequality $-\mathbb{E}\{a^2\} - \mathbb{E}\{b^2\} \leq 2\mathbb{E}\{ab\} \leq \mathbb{E}\{a^2\} + \mathbb{E}\{b^2\}$ shows that it is sufficient to prove that the second-order moments of a and b are finite. This is obvious for a because c_k has finite moments. Bounding $\mathbb{E}\{b^2\}$ is more involved. Calculating the derivative of $\hat{\sigma}_{\hat{e}}^2$ w.r.t. $\theta_{[l]}$, a parameter of the numerator $B(\Omega, \theta)$, gives

$$\partial\hat{\sigma}_{\hat{e}}^2/\partial\theta_{[l]} = \frac{2}{M}\text{Re}((B\hat{\sigma}_U^2 - A\hat{\sigma}_{YU}^2)\overline{\Omega}^l)$$

and

$$(\partial \hat{\sigma}_{\hat{e}}^2 / \partial \theta_{[l]})^2 = 4[\text{Re}(B\hat{\sigma}_U^2 - A\hat{\sigma}_{YU}^2)\overline{\Omega}^l]^2 / M^2$$

$$\leq 4|\Omega|^{2l}|B\hat{\sigma}_U^2 - A\hat{\sigma}_{YU}^2|^2 / M^2$$

$$\leq 4|\Omega|^{2l}\hat{\sigma}_U^2(|B|^2\hat{\sigma}_U^2 + |A|^2|\hat{\sigma}_{YU}^2|^2 / \hat{\sigma}_U^2 - 2\text{Re}(A\overline{B}\hat{\sigma}_{YU}^2)) / M^2$$

$$\leq 4|\Omega|^{2l}\hat{\sigma}_U^2\hat{\sigma}_{\hat{e}}^2 / M$$

where the last inequality is due to $|\hat{\sigma}_{YU}^2|^2 \leq \hat{\sigma}_U^2 \hat{\sigma}_Y^2$ and (10-11). Following the same lines, we find $(\partial \hat{\sigma}_{\hat{e}}^2 / \partial \theta_{[l]})^2 \leq 4|\Omega|^{2l}\hat{\sigma}_Y^2\hat{\sigma}_{\hat{e}}^2 / M$ if θ_l is a parameter of the denominator $A(\Omega, \theta)$. Without loss of generality, the first situation will be considered here and it will be shown that the second moment of

$$b^2 = \sigma_{\hat{e}}^4 (1/\hat{\sigma}_{\hat{e}}^2)^4 (\partial \hat{\sigma}_{\hat{e}}^2 / \partial \theta_{[l]})^2 \leq \frac{4}{M} \sigma_{\hat{e}}^4 |\Omega|^{2l} \hat{\sigma}_U^2 / (\hat{\sigma}_{\hat{e}}^2)^3$$

is bounded. For notational simplicity, the dependence on k will be omitted.

The sample variances $(\hat{\sigma}_U^2, \hat{\sigma}_{\hat{e}}^2)$ have a Wishart distribution that also depends on the cross-correlation $\hat{\sigma}_{U\hat{e}}^2$ and is noticed as $dF(\hat{\sigma}_U^2, \hat{\sigma}_{\hat{e}}^2, \rho)$ with $\rho = \sigma_{U\hat{e}}^2 / \sqrt{\sigma_U^2 \sigma_{\hat{e}}^2}$ (Kendall and Stuart, 1979). The domain of the integral to calculate the expected value of b^2 can be split into two parts: with respect to the variable $\hat{\sigma}_{\hat{e}}^2$: $D_\varepsilon = \{\hat{\sigma}_{\hat{e}}^2 | \hat{\sigma}_{\hat{e}}^2 < \varepsilon\}$ and $D_r = \{\hat{\sigma}_{\hat{e}}^2 | \hat{\sigma}_{\hat{e}}^2 \geq \varepsilon\}$. The integral over D_r will be finite because σ_U^2 has finite moments and the denominator is bounded. It can be shown after some calculations that on D_ε the marginal density function $dH(\hat{\sigma}_U^2, \hat{\sigma}_{\hat{e}}^2) = \int_{D_\rho} dF(\hat{\sigma}_U^2, \hat{\sigma}_{\hat{e}}^2, \rho)$ is bounded by

$$dH(\hat{\sigma}_U^2, \hat{\sigma}_{\hat{e}}^2) \leq C(\nu) e^{(|\rho|\sigma_U\sqrt{\varepsilon})(\nu+1)/2} dG(\nu, \hat{\sigma}_U^2) dG(\nu, \hat{\sigma}_{\hat{e}}^2) \qquad (10\text{-}78)$$

Here $dG(\nu, \hat{\sigma}_U^2)$ is a χ^2 distribution with respect to $\hat{\sigma}_U^2$ with ν degrees of freedom and similarly for $G(\nu, \hat{\sigma}_{\hat{e}}^2)$, and $C(\nu)$ a constant with respect to $(\hat{\sigma}_U^2, \hat{\sigma}_{\hat{e}}^2)$, depending on ν. Because the density function $dG(\nu, \hat{\sigma}_{\hat{e}}^2)$ of a χ^2 distribution is proportional to $(\hat{\sigma}_{\hat{e}}^2 / 2)^{\nu/2-1} e^{-\hat{\sigma}_{\hat{e}}^2/2}$ it is clear that the expected value of b^2 over D_ε will be bounded if $M \geq 5$ (Corollary 10.26).

Remark. To obtain (10-78), it had to be assumed that $|\rho| < 1 - \delta$, with $\delta > 0$. For the singular case that $|\rho| = 1$, it is straightforwardly shown that c_k becomes θ-independent and, hence, the derivatives are zero.

(iii) $c'' = \sigma_{\hat{e}}^{2\prime\prime} / \hat{\sigma}_{\hat{e}}^2 - (\hat{\sigma}_{\hat{e}}^{2\prime T}\hat{\sigma}_{\hat{e}}^{2\prime} + \hat{\sigma}_{\hat{e}}^{2\prime T}\hat{\sigma}_{\hat{e}}^{2\prime}) / \hat{\sigma}_{\hat{e}}^4 - \sigma_{\hat{e}}^2 \hat{\sigma}_{\hat{e}}^{2\prime\prime} / \hat{\sigma}_{\hat{e}}^4 + 2(\hat{\sigma}_{\hat{e}}^{2\prime T}\hat{\sigma}_{\hat{e}}^{2\prime})\sigma_{\hat{e}}^2 / \hat{\sigma}_{\hat{e}}^6$ has finite first- and second-order moments for $M \geq 6$.

The proof is completely similar to the previous one, noticing that $(\partial \hat{\sigma}_{\hat{e}}^2 / (\partial \theta_{[r]} \partial \theta_{[s]})) = 2\hat{\sigma}_U^2 \text{Re}(\Omega^r \overline{\Omega}^s)$, $2\hat{\sigma}_{YU}^2 \text{Re}(\Omega^r \overline{\Omega}^s)$ or $2\hat{\sigma}_Y^2 \text{Re}(\Omega^r \overline{\Omega}^s)$, and $|\hat{\sigma}_{YU}^2|^2 \leq (\hat{\sigma}_U^4 + \hat{\sigma}_Y^4)/2$. This time contributions of $(1/\hat{\sigma}_{\hat{e}}^2)^4$ should be bounded, requiring that $M \geq 6$ (Corollary 10.26).

Appendix 10.D Proof of Theorem 10.3

Because the noise is, by assumption, independent over the frequency (Assumption 10.2), the proof of Theorem 9.21 for a frequency domain experiment can be applied to the cost function (10-10) provided that the necessary moments of the cost function and its derivatives w.r.t. θ are finite in $\mathbb{P}_r$. We need the first- and second-order moments of $V_{\text{SML}}(\theta, Z)$ for properties 1, 5, and 6 of Theorem 9.21; in addition, the first- and second-order moments of $V_{\text{SML}}'(\theta, Z)$ and $V_{\text{SML}}''(\theta, Z)$ for properties 2 and 6 of Theorem 9.21; in addition, the first- and second-order moments of $V_{\text{SML}}'''(\theta, Z)$ for properties 3, 6, and 7 of Theorem 9.21; and in addition, the third-order moments of $V_{\text{SML}}(\theta, Z)$ and $V_{\text{SML}}'(\theta, Z)$ for property 4 of Theorem 9.21 (the moment $2 + \delta$ in Assumption 9.13 is bounded above by a third-order moment). The moments of the cost function $V_{\text{SML}}(\theta, Z)$ and its derivatives w.r.t. θ exist, if the moments of $|\hat{\varepsilon}(\Omega_k, \theta, \hat{Z}(k))|^2$ and its derivatives w.r.t. θ exist. These moments are calculated in the following. For notational simplicity, we dropped the arguments of $|\hat{\varepsilon}|^2$, $|\hat{e}|^2$ and $\hat{\sigma}^2$.

1. The moments $\mathbb{E}\{|\hat{\varepsilon}|^{2l}\}$ are finite with $l = 1, M \geq 3$; $l = 2, M \geq 4$; $l = 3, M \geq 5$.
   Proof: It follows directly from Corollary 10.26 in Appendix 10.B.

2. The moments $\mathbb{E}\{(\partial|\hat{\varepsilon}|^2/\partial\theta_{[r]})^l\}$ are finite with $l = 1, M \geq 4$; $l = 2, M \geq 5$; $l = 3, M \geq 6$.
   Proof:
   $$\frac{\partial|\hat{\varepsilon}|^2}{\partial\theta_{[r]}} = \frac{1}{\hat{\sigma}_{\hat{e}}^2}\frac{\partial|\hat{e}|^2}{\partial\theta_{[r]}} - |\hat{e}|^2\frac{1}{\hat{\sigma}_{\hat{e}}^4}\frac{\partial\hat{\sigma}_{\hat{e}}^2}{\partial\theta_{[r]}} \qquad (10\text{-}79)$$

   In this expression, it is the last term that is critical in bounding its expected value. Using the same technique as in Appendix 10.C, and noticing that $\hat{e}$ is independent of $\hat{\sigma}_{\hat{e}}$, it turns out that it is enough to bound $\mathbb{E}\{1/\hat{\sigma}_{\hat{e}}^{3l}\}$. The claim then follows directly from Corollary 10.26.

3. The moments $\mathbb{E}\{[\partial^2|\hat{\varepsilon}|^2/(\partial\theta_{[r]}\partial\theta_{[s]})]^l\}$ are finite with $l = 1, M \geq 4$; $l = 2, M \geq 6$.
   Proof: This time the term that is critical in bounding the expected value becomes:
   $$|\hat{e}|^2\frac{1}{\hat{\sigma}_{\hat{e}}^6}\frac{\partial\hat{\sigma}_{\hat{e}}^2}{\partial\theta_{[r]}}\frac{\partial\hat{\sigma}_{\hat{e}}^2}{\partial\theta_{[s]}} \qquad (10\text{-}80)$$

   Its lth order moment is again bounded if $\mathbb{E}\{1/\hat{\sigma}_{\hat{e}}^{4l}\}$ is finite. The claim then follows again from Corollary 10.26.

4. The moments $\mathbb{E}\{(\partial^3|\hat{\varepsilon}|^2/(\partial\theta_{[r]}\partial\theta_{[s]}\partial\theta_{[t]}))^l\}$ are finite with $l = 1, M \geq 5$; $l = 2, M \geq 7$.

   Proof: This time the critical term in bounding the expected value becomes:
   $$|\hat{e}|^2\frac{1}{\hat{\sigma}_{\hat{e}}^8}\frac{\partial\hat{\sigma}_{\hat{e}}^2}{\partial\theta_{[r]}}\frac{\partial\hat{\sigma}_{\hat{e}}^2}{\partial\theta_{[s]}}\frac{\partial\hat{\sigma}_{\hat{e}}^2}{\partial\theta_{[t]}} \qquad (10\text{-}81)$$

   Its lth order moment is again bounded if $\mathbb{E}\{1/\hat{\sigma}_{\hat{e}}^{5l}\}$ is finite. The claim then follows again from Corollary 10.26.

Appendix 10.E Approximation of the Derivative of the Cost Function

Replacing N_Z by υN_Z (C_{N_Z} by $\upsilon^2 C_{N_Z}$) and $\hat{e}(\Omega_k, \tilde{\theta}(Z_0), \hat{Z}_0(k))$ by $\mu\hat{e}(\Omega_k, \tilde{\theta}(Z_0), \hat{Z}_0(k))$ makes it possible to analyze $V_{\text{SML}}'(\theta, Z)$ for small noise levels $\upsilon \to 0$

Section 10.11 ■ Appendixes

(large signal-to-noise ratios) and small model errors $\mu \to 0$ (see Section 9.5, quick analysis tool 6)

$$V_{\text{SML}}'(\tilde{\theta}(Z_0), Z) \approx \sum_{k=1}^{F} c_k(\tilde{\theta}(Z_0)) d_k'(\tilde{\theta}(Z_0)) \tag{10-82}$$

Proof. Using (10-17), $V_{\text{SML}}'(\theta, Z)$ can be written as

$$\frac{1}{F}V_{\text{SML}}'(\tilde{\theta}(Z_0), Z) = \frac{1}{F}\sum_{k=1}^{F} c_k'(\tilde{\theta}(Z_0)) d_k(\tilde{\theta}(Z_0)) + \frac{1}{F}\sum_{k=1}^{F} c_k(\tilde{\theta}(Z_0)) d_k'(\tilde{\theta}(Z_0)) \tag{10-83}$$

Because $c_k(\tilde{\theta}(Z_0)) = O(\upsilon^0 \mu^0)$, $c_k'(\tilde{\theta}(Z_0)) = O(\upsilon^0 \mu^0)$, $d_k(\tilde{\theta}(Z_0)) = \upsilon^{-2} O(\mu^2 + \mu\upsilon + \upsilon^2)$, and $d_k'(\tilde{\theta}(Z_0)) = \upsilon^{-2} O(\mu + \upsilon)$, it follows that for small model errors $\mu \to 0$ and large signal-to-noise ratios $\upsilon \to 0$, the term $c_k(\tilde{\theta}(Z_0)) d_k'(\tilde{\theta}(Z_0))$ in (10-83) dominates over $c_k'(\tilde{\theta}(Z_0)) d_k(\tilde{\theta}(Z_0))$. Summing over the frequencies will not change this behavior because the two sums in the right-hand side of (10-83) are both an $O_p(F^{-1/2})$. The last statement follows from the fact that the terms in both sums are independent zero mean random variables with bounded first- and second-order moments (see Section 16.9, version 2 of the law of large numbers).

Appendix 10.F Loss in Efficiency of the Sample Estimator

In both sections we use the following properties: (i) $V_{\text{ML}}(\theta, Z) = \sum_{k=1}^{F} d_k(\theta)$ and (ii) $c_k(\theta)$ and $d_k(\theta)$ are mutually independent random variables (see (10-17)). Property (i) follows directly from the fact that $d_k(\theta)$ contains the true noise (co)variances (see (10-17)). Property (ii) is shown as follows: $c_k(\theta)$ depends on the sample variance, while $d_k(\theta)$ depends on the sample mean. It is well known that the sample variance and the sample mean are independent random variables for normally distributed noise (Stuart and Ord, 1987). Hence, this is also the case for $c_k(\theta)$ and $d_k(\theta)$.

10.F.1 Approximate Expression for the Parameter Deviations. Under the assumptions of Theorem 10.3, formula (9-26) of Theorem 9.21 is valid for $M \geq 7$ (Theorem 10.3)

$$\begin{aligned}\hat{\theta}_{\text{SML}}(Z) &= \tilde{\theta}_{\text{SML}}(Z_0) + \delta_{\theta\text{SML}}(Z) + O_p(F^{-1}) \\ \delta_{\theta\text{SML}}(Z) &= -V_{\text{SML}}''^{-1}(\tilde{\theta}_{\text{SML}}(Z_0)) V_{\text{SML}}'^T(\tilde{\theta}_{\text{SML}}(Z_0), Z) \end{aligned} \tag{10-84}$$

Using notation (10-17), the Hessian $V_{\text{SML}}''(\theta)$ can be written as

$$V_{\text{SML}}''(\theta) = \sum_{k=1}^{F} \mathbb{E}\{c_k''(\theta) d_k(\theta) + 2\text{herm}(c_k'^T(\theta) d_k'(\theta)) + c_k(\theta) d_k''(\theta)\} \tag{10-85}$$

with $\text{herm}(x) = (x + x^H)/2$. Because $d_k(\theta)$ is independent of $c_k(\theta)$, $\mathbb{E}\{c_k(\theta)\} = (M-1)/(M-2)$, $\mathbb{E}\{c_k'(\theta)\} = 0$ and $\mathbb{E}\{c_k''(\theta)\} = 0$ (see Appendix 10.C), and $V_{\text{ML}}(\theta, Z) = \sum_{k=1}^{F} d_k(\theta)$, (10-85) becomes

$$V_{\text{SML}}''(\theta) = \frac{M-1}{M-2}\sum_{k=1}^{F} \mathbb{E}\{d_k''(\theta)\} = \frac{M-1}{M-2} V_{\text{ML}}''(\theta) \tag{10-86}$$

Using $\tilde{\theta}_{SML}(Z_0) = \tilde{\theta}_{ML}(Z_0)$ (Theorem 10.4), approximation (10-82) (see Appendix 10.E), we finally get

$$\delta_{\theta SML}(Z) \approx -\frac{M-2}{M-1} V_{ML}''^{-1}(\tilde{\theta}_{ML}(Z_0)) \sum_{k=1}^{F} c_k(\tilde{\theta}_{ML}(Z_0)) d_k'^T(\tilde{\theta}_{ML}(Z_0))$$

$$\mathbb{E}\{\delta_{\theta SML}(Z)\} = -\frac{M-2}{M-1} V_{ML}''^{-1}(\tilde{\theta}_{ML}(Z_0)) V_{SML}'^T(\tilde{\theta}_{ML}(Z_0)) = 0$$
(10-87)

Note that (10-87) is also valid in the presence of model errors.

10.F.2 An Approximate Expression for the Covariance Matrix. Using (10-87), $\mathbb{E}\{c_k^2(\theta)\} = (M-1)^2/((M-2)(M-3))$ for $M \geq 6$ (see Appendix 10.C), and the fact that $c_k(\theta)$ and $d_k(\theta)$ are mutually independent and independent over the frequency k, the following asymptotic ($F \to \infty$) expression is found:

$$\hat{C}_\theta = \mathbb{E}\{\delta_{\theta SML}(Z) \delta_{\theta SML}^T(Z)\}$$

$$= \frac{M-2}{M-3} V_{ML}''^{-1}(\tilde{\theta}_{ML}(Z_0)) \sum_{k=1}^{F} \mathbb{E}\{d_k'^T(\tilde{\theta}_{ML}(Z_0)) d_k'(\tilde{\theta}_{ML}(Z_0))\} V_{ML}''^{-1}(\tilde{\theta}_{ML}(Z_0))$$
(10-88)

For independent (over the frequency) distributed errors, (9-27) reduces to

$$C_\theta = \mathbb{E}\{\delta_{\theta ML}(Z) \delta_{\theta ML}^T(Z)\}$$

$$= V_{ML}''^{-1}(\tilde{\theta}_{ML}(Z_0)) \sum_{k=1}^{F} \mathbb{E}\{d_k'^T(\tilde{\theta}_{ML}(Z_0)) d_k'(\tilde{\theta}_{ML}(Z_0))\} V_{ML}''^{-1}(\tilde{\theta}_{ML}(Z_0))$$
(10-89)

where $V_{ML}(\theta, Z) = \sum_{k=1}^{F} d_k(\theta)$. Hence, from (10-88) and (10-89) it follows that

$$\hat{C}_\theta \approx \frac{M-2}{M-3} C_\theta$$
(10-90)

for $M \geq 7$ ((10-87) is valid for $M \geq 7$), with C_θ the covariance matrix of the parameters when the noise (co)variances are exactly known.

Appendix 10.G Mean and Variance of the Sample Cost in Its Global Minimum

Some precautions should be taken when calculating the expected value and the variance of $V_{SML}(\hat{\theta}_{SML}(Z), Z)$ since $c_k(\hat{\theta}_{SML}(Z))$ and $d_k(\hat{\theta}_{SML}(Z))$ (see (10-17)) are no longer independent because they both depend on $\hat{\theta}_{SML}(Z)$. To get around this problem, a Taylor series expansion is made around the asymptotic value $\tilde{\theta}_{SML}(Z_0) = \tilde{\theta}_{ML}(Z_0)$ (Theorem 10.4), which is denoted as $\tilde{\theta}$ for simplicity of notation

$$V_{SML}(\hat{\theta}_{SML}(Z), Z) \approx V_{SML}(\tilde{\theta}, Z) + V_{SML}'(\tilde{\theta}, Z)\delta + \frac{1}{2}\delta^T V_{SML}''(\tilde{\theta}, Z)\delta$$
(10-91)

with $\delta = \hat{\theta}_{SML}(Z) - \tilde{\theta}_{ML}(Z_0)$. Under the assumptions of Theorem 10.3, the Hessian $V_{SML}''(\tilde{\theta}, Z)/F$ converges w.p. 1 at the rate $O_p(F^{-1/2})$ to its expected value $V_{SML}''(\tilde{\theta})/F$ (proof: apply version 2 of the strong law of large numbers (16-69)). Hence, using (10-86), (10-91) can be written as

$$V_{SML}(\hat{\theta}_{SML}(Z), Z) \approx V_{SML}(\tilde{\theta}, Z) + V_{SML}'(\tilde{\theta}, Z)\delta + \frac{1}{2}\frac{M-1}{M-2}\delta^T V_{ML}''(\tilde{\theta})\delta \qquad (10\text{-}92)$$

Substituting approximation (10-87) for δ in (10-92) gives, using (10-86),

$$\begin{aligned} V_{SML}(\hat{\theta}_{SML}(Z), Z) &\approx V_{SML}(\tilde{\theta}, Z) - \Delta V(\tilde{\theta}, Z) \\ \Delta V(\tilde{\theta}, Z) &= \frac{1}{2}\sum_{k=1}^{F} c_k(\tilde{\theta}) d_k'(\tilde{\theta}) \frac{M-2}{M-1} V_{ML}''^{-1}(\tilde{\theta}) \sum_{k=1}^{F} c_k(\tilde{\theta}) d_k'^T(\tilde{\theta}) \end{aligned} \qquad (10\text{-}93)$$

10.G.1 Calculation of the Mean Value. Using $\mathbb{E}\{c_k^2(\tilde{\theta})\} = (M-1)^2/((M-2)(M-3))$ for $M \geq 6$ (see Appendix 10.C), and the fact that $c_k(\tilde{\theta})$ and $d_k(\tilde{\theta})$ are mutually independent over the frequency k, the expected value of the second term in the right-hand side of (10-93) equals

$$\begin{aligned} \Delta V(\tilde{\theta}, Z) &= \frac{1}{2}\frac{M-1}{M-3}\sum_{k=1}^{F} \mathbb{E}\{d_k'(\tilde{\theta}) V_{ML}''^{-1}(\tilde{\theta}) d_k'^T(\tilde{\theta})\} \\ &= \frac{1}{2}\frac{M-1}{M-3}\text{trace}(V_{ML}''^{-1}(\tilde{\theta})\sum_{k=1}^{F} \mathbb{E}\{d_k'^T(\tilde{\theta}) d_k'(\tilde{\theta})\}) \end{aligned} \qquad (10\text{-}94)$$

For small model errors ($\mu \to 0$) and large signal-to-noise ratios ($\upsilon \to 0$) we have

$$V_{ML}''(\tilde{\theta}) \approx \mathbb{E}\{\sum_{k=1}^{F} d_k'^T(\tilde{\theta}) d_k'(\tilde{\theta})\} \Rightarrow \Delta V(\tilde{\theta}, Z) \approx \frac{1}{2}\frac{M-1}{M-3} n_\theta \qquad (10\text{-}95)$$

with n_θ the number of free model parameters (apply quick analysis tool 6 of Section 9.5). Collecting (10-93) and (10-95), using $\mathbb{E}\{c_k(\tilde{\theta})\} = (M-1)/(M-2)$ (see Appendix 10.C), gives

$$\mathbb{E}\{V_{SML}(\hat{\theta}_{SML}(Z), Z)\} \approx \frac{M-1}{M-2}\mathbb{E}\{V_{ML}(\tilde{\theta}, Z)\} - \frac{M-1}{M-3} n_\theta/2 \qquad (10\text{-}96)$$

Because $\mathbb{E}\{V_{ML}(\tilde{\theta}, Z)\} = \mathbb{E}\{V_{ML}(\tilde{\theta}, Z_0)\} + F$, $\tilde{\theta} = \tilde{\theta}_{ML}(Z_0)$ and

$$\mathbb{E}\{V_{ML}(\hat{\theta}_{ML}(Z), Z)\} \approx \mathbb{E}\{V_{ML}(\tilde{\theta}_{ML}(Z_0), Z_0)\} + F - n_\theta/2 \qquad (10\text{-}97)$$

(see Theorem 19.12), (10-96) can be written as

$$\mathbb{E}\{V_{SML}(\hat{\theta}_{SML}(Z), Z)\} \approx \frac{M-1}{M-2}\mathbb{E}\{V_{ML}(\hat{\theta}_{ML}(Z), Z)\} - \frac{M-1}{(M-3)(M-2)} n_\theta/2 \qquad (10\text{-}98)$$

If no model errors are present ($\tilde{\theta}_{ML}(Z_0) = \theta_0$), then $\mathbb{E}\{V_{ML}(\hat{\theta}_{ML}(Z), Z)\} = F - n_\theta/2$ (see Theorem 19.12).

10.G.2 Calculation of the Variance. For the variance, it is mostly sufficient to have a rough estimate, which can be obtained by neglecting the influence of δ and calculating

$$\text{var}(V_{SML}(\hat{\theta}_{SML}(Z), Z)) \approx \text{var}(V_{SML}(\tilde{\theta}_{SML}(Z_0), Z)) \tag{10-99}$$

Using $\text{var}(xy) = \sigma_x^2 \sigma_y^2 + \sigma_x^2 (\mathbb{E}\{y\})^2 + (\mathbb{E}\{x\})^2 \sigma_y^2$, we find after some calculations

$$\text{var}(V_{SML}(\hat{\theta}_{SML}(Z), Z)) \approx \frac{(M-1)^2}{(M-2)(M-3)} \text{var}(V_{ML}(\tilde{\theta}_{ML}(Z_0), Z)) \\ + \frac{(M-1)^2}{(M-2)^2(M-3)} \sum_{k=1}^{F} (\mathbb{E}\{|\hat{\varepsilon}(\Omega_k, \tilde{\theta}_{ML}(Z_0), \hat{Z}(k))|^2\})^2 \tag{10-100}$$

In the absence of model errors, we have $\tilde{\theta}_{ML}(Z_0) = \theta_0$, $\mathbb{E}\{|\hat{\varepsilon}(\Omega_k, \theta_0, \hat{Z}(k))|^2\} = 1$ and $\text{var}(V_{ML}(\theta_0, Z)) = F$, so that (10-100) reduces to (10-22).

Appendix 10.H Asymptotic Properties of the SGTLS Estimator (Theorem 10.6)

Following the same lines as in the proof of Theorem 9.21 (see Appendix 9.E), it is sufficient to verify that all the conditions of the theorems in Chapter 18 are fulfilled. The cost function (10-24) is of the form

$$V_{SGTLS}(\theta, Z) = f_F(\theta, w(\theta, Z), Z) \tag{10-101}$$

where $w(\theta, Z) = F^{-1} \sum_{k=1}^{F} \hat{\sigma}_e^2(\Omega_k, \theta)$. Because $w(\theta, Z)$ is quadratic in the measurements Z and quadratic in the model parameters θ, the theorems of Chapter 17 are valid under the assumptions of Section 9.6: $w^{(k)}(\theta, Z)$, $k = 0, 1, 2$, converge uniformly w.p. 1 (in prob.) to their expected value at the rate $O_p(F^{-1/2})$ in $\mathbb{P}_r$; and $w^{(k)}(\theta, Z)$, $k = 0, 1$, converge in law to a Gaussian random variable at the rate $F^{-1/2}$. Hence, $w(\theta, Z)$ satisfies all the assumptions of Chapter 18. The cost function $f_F(\theta, w(\theta), Z)$, with $w(\theta) = \mathbb{E}\{w(\theta, Z)\}$, is quadratic in Z and also satisfies all the assumptions of Chapter 18. We conclude that all the theorems of Chapter 18 apply to the SGTLS cost function (10-24). From Theorems 18.5 and 18.6 follows that $\tilde{\theta}_{SGTLS}(Z_0)$ and θ_{*SGTLS} are the minimizers of $V_F(\theta) = \mathbb{E}\{f_F(\theta, w(\theta), Z)\}$ and $V_*(\theta) = \lim_{F \to \infty} \mathbb{E}\{f_F(\theta, w(\theta), Z)\}$, respectively. Therefore, Theorem 9.21 is valid with three modifications: (i) to calculate $V_F(\theta)$ and $V_*(\theta)$ we first replace $w(\theta, Z)$ by its expected value $w(\theta) = F^{-1} \sum_{k=1}^{F} \sigma_e^2(\Omega_k, \theta)/M$ before taking the expected value of the cost function; (ii) the expected value $\mathbb{E}\{\delta_\theta(Z)\}$ is, in general, not zero as $\tilde{\theta}_{SGTLS}(Z_0)$ is not the minimizer of $\mathbb{E}\{f_F(\theta, w(\theta, Z), Z)\}$ (see Theorem 18.18); and (iii) $\delta_\theta(Z)$ is replaced by $d_\theta(Z)$ in the expression (9-27) of the covariance matrix

$$d_\theta(Z) = -V_F''^{-1}(\tilde{\theta})d_F(Z)$$

$$d_F(Z) = g_F(\tilde{\theta}, w(\tilde{\theta}), w'(\tilde{\theta}), Z) + \left.\frac{\partial g_F(\tilde{\theta}, w(\tilde{\theta}), w'(\tilde{\theta}))}{\partial x}\right|_{x=x_*}(\tilde{x}(Z) - x_*) \quad (10\text{-}102)$$

with $x^T = [w(\tilde{\theta})\ w'(\tilde{\theta})]$, $x_*^T = [w(\tilde{\theta})\ w'(\tilde{\theta})]$, $\tilde{x}^T(Z) = [\tilde{w}(\tilde{\theta}, Z)\ \tilde{w}'(\tilde{\theta}, Z)]$,

$$g_F(\theta, w(\theta, Z), w'(\theta, Z), Z) = f_F'(\theta, w(\theta, Z), Z)$$
$$g_F(\theta, w, w_1) = \mathbb{E}\{g_F(\theta, w, w_1, Z)\} \quad (10\text{-}103)$$

and $\tilde{\theta} = \tilde{\theta}_{\text{SGTLS}}(Z_0)$, and where w and w_1 are deterministic variables that replace the random variables $w(\theta, Z)$ and $w'(\theta, Z)$, respectively (see Theorem 18.25). □

Appendix 10.I Relationship between the GTLS and the SGTLS Estimates (Theorem 10.7)

Formulas (10-26) and (10-27) follow directly from (10-25), (9-72), Exercise 9.6 and the fact that the (co)variances of the mean value equal the (co)variances of one realization divided by M.

In the absence of model errors, $\tilde{\theta}(Z_0) = \theta_0$, we have $e(\Omega_k, \theta_0, Z_0(k)) = 0$. Calculating (10-103), where $f_F(\theta, w(\theta, Z), Z)$ and $w(\theta, Z)$ are defined in (10-101), gives

$$g_F(\theta, w, w_1) = \frac{2M}{Fw}\sum_{k=1}^{F}\text{Re}(e'(\Omega_k, \theta, Z_0(k))\bar{e}(\Omega_k, \theta, Z_0(k))) \\ -\frac{Mw_1}{Fw^2}\sum_{k=1}^{F}|e(\Omega_k, \theta, Z_0(k))|^2 \quad (10\text{-}104)$$

It follows that $g_F(\theta_0, w, w_1) = 0$ for any w and w_1. Hence, the derivatives of $g_F(\theta_0, w, w_1)$ w.r.t. w and w_1 are zero, so that the second term in the right-hand side of (10-102) is zero. The remaining term in $d_F(Z)$ is the term one would have if the true noise (co)variances were used instead of the sample noise (co)variances. This concludes the proof of (10-28). □

Appendix 10.J Asymptotic Properties of SBTLS Estimator (Theorem 10.8)

The proof is similar to that of Theorem 10.6 (see Appendix 10.H). The cost function (10-33), with $\theta^{(0)} = \hat{\theta}(Z)$ and $\theta = \theta^{(1)}$, is of the form

$$V_{\text{SBTLS}}(\theta, Z) = f_F(\theta, \eta(Z), w(\theta, \eta(Z), Z), Z) \quad (10\text{-}105)$$

where

$$\eta^T(Z) = [\hat{\theta}^T(Z) \; \hat{\alpha}^T(Z)]$$

$$w(\theta, \eta(Z), Z) = \frac{1}{F}\sum_{k=1}^{F} \frac{\hat{\sigma}_{\tilde{e}}^2(\Omega_k, \theta)}{\sigma_{\tilde{e}}^{2r}(\Omega_k, \hat{\theta}(Z), \hat{\alpha}(Z))} \tag{10-106}$$

satisfy all the conditions of the theorems in Chapter 18. □

Appendix 10.K Relationship between the BTLS and the SBTLS Estimates (Theorem 10.9)

The proof is similar to that of Theorem 10.7 (see Appendix 10.I). Because $\sigma_n^2(k, \hat{\alpha}(Z))$, $n = 1, 2, 3$, is a consistent estimate of the true noise model $\sigma_n^2(k)$, $n = 1, 2, 3$, we have that $\sigma_{\tilde{e}}^2(\Omega_k, \theta_*, \alpha_*) = \sigma_e^2(\Omega_k, \theta_*)$, which proves (10-36) and (10-37).

In the absence of model errors, $\tilde{\theta}(Z_0) = \theta_0$, we have $e(\Omega_k, \theta_0, Z_0(k)) = 0$. Calculating

$$\begin{aligned} g_F(\theta, \eta(Z), w(\theta, \eta(Z), Z), w'(\theta, \eta(Z), Z), Z) &= f_F'(\theta, \eta(Z), w(\theta, \eta(Z), Z), Z) \\ g_F(\theta, \eta, w, w_1) &= \mathbb{E}\{g_F(\theta, \eta, w, w_1, Z)\} \end{aligned} \tag{10-107}$$

where $f_F(.)$, $\eta(Z)$, and $w(\theta, \eta(Z), Z)$ are defined in (10-105) and where η, w, and w_1 are deterministic variables that replace the random variables $\eta(Z)$, $w(\theta, \eta(Z), Z)$, and $w'(\theta, \eta(Z), Z)$, respectively (see Lemma 18.14), gives

$$\begin{aligned} g_F(\theta, \eta, w, w_1) = &\frac{2M}{Fw}\sum_{k=1}^{F} \frac{\mathrm{Re}(e'(\Omega_k, \theta, Z_0(k))\bar{e}(\Omega_k, \theta, Z_0(k)))}{\sigma_e^{2r}(\Omega_k, \eta)} \\ &-\frac{Mw_1}{Fw^2}\sum_{k=1}^{F} \frac{|e(\Omega_k, \theta, Z_0(k))|^2}{\sigma_e^{2r}(\Omega_k, \eta)} \end{aligned} \tag{10-108}$$

It follows that $g_F(\theta_0, \eta, w, w_1) = 0$ for any η, w, and w_1. Hence, the derivatives of $g_F(\theta_0, \eta, w, w_1)$ w.r.t. η, w, and w_1 are zero, so that the second term in $d_F(Z)$ (18-17) is zero. Because $\sigma_{\tilde{e}}^2(\Omega_k, \theta_*, \alpha_*) = \sigma_e^2(\Omega_k, \theta_*)$, the remaining term in $d_F(Z)$ is the term one would have if the true noise (co)variances were used instead of the sample noise (co)variances. This concludes the proof of (10-38). □

Appendix 10.L Asymptotic Properties of SSUB Algorithms (Theorem 10.10)

Because the second step of the subspace Algorithms 9.24 and 9.25 does not require any noise information, it is sufficient to analyze the first and the third steps only.

10.L.1 Step 1: Estimation of the Extended Observability Matrix. The results of Appendix 9.Q, estimation of the extended observability matrix O_r, remain valid if we can show that $\hat{C}_Y/F$ converges w.p. 1 to its expected value C_Y/F and if the convergence rate is an $O_p(F^{-1/2})$. From this result it follows directly that the limit value O_{r*}, calculated with the sample variance $\hat{\sigma}_Y^2(k)$, equals that calculated with the true variance $\sigma_Y^2(k)$.

Section 10.11 ■ Appendixes

Under the assumptions of Section 9.6.1, $|N_Y(k)|^2$ is independently distributed for a frequency domain experiment and converges for $F \to \infty$ to a random variable that is mixing of order 2 for time domain experiment. This is also valid for

$$\hat{\sigma}_Y^2(k) = \frac{1}{M-1}\sum_{l=1}^{M}|Y^{[l]}(k) - \hat{Y}(k)|^2 = \frac{1}{M-1}\sum_{l=1}^{M}\left|N_Y^{[l]}(k) - \frac{1}{M}\sum_{m=1}^{M}N_Y^{[m]}(k)\right|^2 \quad (10\text{-}109)$$

and, therefore, $\hat{C}_\mathbf{Y}/F$

$$\hat{C}_\mathbf{Y}/F = \mathrm{Re}\left(\frac{1}{F}\sum_{k=1}^{F}\frac{1}{M}\hat{\sigma}_Y^2(k)W_r(k)W_r^H(k)\right) \quad (10\text{-}110)$$

converges w.p. 1 to its expected value at the rate $O_p(F^{-1/2})$ (see Section 16.9, versions 2 and 3 of the law of large numbers). Because $\mathbb{E}\{\hat{\sigma}_Y^2(k)\} = \sigma_Y^2(k)$, the expected value of $\hat{C}_\mathbf{Y}$ equals $C_\mathbf{Y}$.

The proof for $\hat{C}_{\mathbf{Y}_\perp}$ is somewhat more complicated because the scalar orthogonal basis $p_n(s)$, $n = 0, 1, ..., r-1$ depends on the measurements $\hat{Y}(k)$, $k = 1, 2, ..., F$. Indeed, the orthogonal basis is calculated using the inner product (see Appendix 9.S)

$$\langle x(s), y(s)\rangle_Y = \mathrm{Re}(\sum_{k=1}^{F}|\hat{Y}(k)|^2 x(s_k)\bar{y}(s_k)) \quad (10\text{-}111)$$

which clearly depends on $\hat{Y}(k)$. Under the assumptions of Section 9.6.1, the inner product $\langle x(s), y(s)\rangle_Y/F$ converges w.p. 1 at the rate $O_p(F^{-1/2})$ to its expected value

$$\mathbb{E}\{\langle x(s), y(s)\rangle_Y/F\} = \mathrm{Re}\left(\frac{1}{F}\sum_{k=1}^{F}\left(|Y_0(k)|^2 + \frac{1}{M}\sigma_Y^2(k)\right)x(s_k)\bar{y}(s_k)\right) \quad (10\text{-}112)$$

Hence, the polynomials $p_n(s)$ also converge w.p. 1 at the rate $O_p(F^{-1/2})$ to its limit value $\tilde{p}_n(s)$ (see recursion formulas (9-250) and (9-251)). We conclude that $W_Y(k)$ converges w.p. 1 at the rate $O_p(F^{-1/2})$ to $\tilde{W}_Y(k)$. Applying Corollary 17.35 to

$$\hat{C}_{\mathbf{Y}_\perp}/F = \mathrm{Re}\left(\frac{1}{F}\sum_{k=1}^{F}\frac{1}{M}\hat{\sigma}_Y^2(k)W_Y(k)W_Y^H(k)\right) \quad (10\text{-}113)$$

shows that $\hat{C}_{\mathbf{Y}_\perp}/F$ converges w.p. 1 at the rate $O_p(F^{-1/2})$ to

$$\mathrm{Re}\left(\frac{1}{F}\sum_{k=1}^{F}\frac{1}{M}\sigma_Y^2(k)\tilde{W}_Y(k)\tilde{W}_Y^H(k)\right) \quad (10\text{-}114)$$

This concludes the proof because (10-114) is the asymptotic expression of the covariance matrix when the noise variance is known.

10.L.2 Step 3: Estimation of B and D. From step 1 it follows that $\hat{A}$ and $\hat{C}$ converge w.p. 1 at the rate $O_p(F^{-1/2})$ to their limit values A_0, C_0 or, in case of model errors, to A_*, C_* (see Appendix 9.R). The cost function

$$V_{\text{SSUB}}(B, D, A_1, C_1, Z) = \sum_{k=1}^{F} \frac{|\hat{Y}(k) - (C_1(\xi_k I_{n_a} - A_1)^{-1} B + D)\hat{U}(k)|^2}{\hat{\sigma}_Y^2(k)/M} \quad (10\text{-}115)$$

where A_1 and C_1 are deterministic and where, by assumption, the input is known ($\hat{U}(k) = U_0(k)$), has exactly the same stochastic structure as the SML cost function (10-10). Therefore, under the same assumptions $V_{\text{SSUB}}(B, D, A_1, C_1, Z)$ has the same asymptotic properties as $V_{\text{SML}}(\theta, Z)$. The only difference is that the denominator in (10-115) is independent of θ, which decreases the value of M from 6 to 4 for the convergence rate (see Appendix 10.D). Hence, for $M \geq 4$, (10-115) converges w.p. 1 at the rate $O_p(F^{-1/2})$ to its expected value (see proof of Theorem 10.3). We conclude that $V_{\text{SSUB}}(B, D, \hat{A}, \hat{C}, Z)$ satisfies all the assumptions of Theorems 18.5, 18.7, and 18.16. Hence, for $M \geq 4$, $\hat{B}$ and $\hat{D}$ converge w.p. 1 to their limit values B_0, D_0 (Theorem 18.7) or, in case of model errors, to B_*, D_* (Theorem 18.5), and the convergence rate is an $O_p(F^{-1/2})$ (Theorem 18.16). Note that the limit values A_*, C_*, B_*, and D_*, calculated with the sample variance $\hat{\sigma}_Y^2(k)$, equal those calculated with the true variance $\sigma_Y^2(k)$ since this is also the case for O_{r^*}.

Appendix 10.M Best Linear Approximation of a Cascade of Nonlinear Systems

Applying (10-44) to the nonlinear operators $T[.]$ and $A[.]$ gives, taking into account the measurement errors $M_U(k)$, $M_Y(k)$ and the process noise $N_P(k)$,

$$\begin{aligned} Y(k) &= T_R(s_k)R(k) + N_Y(k) \\ U(k) &= A_R(s_k)R(k) + N_U(k) \end{aligned} \quad (10\text{-}116)$$

with $N_Y(k) = N_P(k) + Y_S(k) + M_Y(k)$, $N_U(k) = U_S(k) + M_U(k)$, and $Y_S(k)$, $U_S(k)$ the zero mean nonlinear distortions, which are independent of $R(k)$. Dividing both sides of (10-116) by $R(k)$ and taking the expected value w.r.t. to the measurement noise and the random phase φ_k of $R(k)$ shows that $\mathbb{E}\{Y_R(k)\}/\mathbb{E}\{U_R(k)\} = T_R(s_k)/A_R(s_k)$, where $Y_R(k)$, $U_R(k)$ are defined as in (10-45). □

Appendix 10.N Sum of Analytic Function Values over a Uniform Grid of the Unit Circle

Lemma 10.27 (Sum of Analytic Function Values over the Unit Circle): Let $F(z^{-1})$ be a rational function of z^{-1} that is analytic outside the unit circle ($|z| \geq 1$) and zero at $z = \infty$ ($F(0) = 0$). Consider a uniform grid $z_k = \exp(j2\pi k/N)$, $k = 0, 1, \ldots, N-1$, on the unit circle. The sum $\sum_{k=0}^{N-1} F(z_k^{-1})$ converges ($N \to \infty$) then to zero at exponential rate

$$\sum_{k=0}^{N-1} F(z_k^{-1}) = O(N|\lambda_{\max}|^N) \text{ with } |\lambda_{\max}| < 1 \quad (10\text{-}117)$$

and where $\lambda_{\max}$ is the dominant pole of $F(z^{-1})$ (= the pole of $F(z^{-1})$ closest to the unit circle).

Proof. The Taylor series of $F(z^{-1})$ w.r.t. z^{-1} at $z^{-1} = 0$ equals

Section 10.11 ■ Appendixes

$$F(z^{-1}) = \sum_{r=1}^{\infty} f_r z^{-r} \text{ for any } |z| \geq 1 \tag{10-118}$$

Using

$$\sum_{k=0}^{N-1} z_k^{-r} = \begin{cases} 0 & \text{for } r \neq nN \\ N & \text{for } r = nN \end{cases} \text{ with } n = 0, 1, \ldots \tag{10-119}$$

the sum of $F(z^{-1})$ over can be written as

$$\sum_{k=0}^{N-1} F(z_k^{-1}) = \sum_{r=1}^{\infty} f_r \sum_{k=0}^{N-1} z_k^{-r} = N \sum_{n=1}^{\infty} f_{nN} \tag{10-120}$$

Since $|f_r| \leq K |\lambda_{\max}|^r$, with $z = \lambda_{\max}$ the pole of $F(z^{-1})$ closest to the unit circle ($|\lambda_{\max}| < 1$) and K a constant, the absolute value of (10-120) can be bounded above as

$$\left| \sum_{k=0}^{N-1} F(z_k^{-1}) \right| \leq N \sum_{n=1}^{\infty} K |\lambda_{\max}|^{nN} \leq O(N |\lambda_{\max}|^N) \tag{10-121}$$

which proves (10-117). □

Lemma 10.28 (Sum of the Logarithm of Analytic Function Values over the Unit Circle): Let $H(z^{-1})$ be a stable and inversely stable monic rational form in z^{-1}. Consider a uniform grid $z_k = \exp(j2\pi k/N)$, $k = 0, 1, \ldots, N-1$, on the unit circle. The sum $\sum_{k=0}^{N-1} \ln(H(z_k^{-1}))$ converges ($N \to \infty$) then to zero at exponential rate

$$\sum_{k=0}^{N-1} \ln(H(z_k^{-1})) = O(|\lambda_{\max}|^N) \text{ with } |\lambda_{\max}| < 1 \tag{10-122}$$

and where $\lambda_{\max}$ is the dominant pole of $d\ln(H(z^{-1}))/dz$ (= pole or zero of $H(z^{-1})$ closest to the unit circle).

Proof. The natural logarithm of $H(z^{-1})$ equals

$$\ln(H(z^{-1})) = \sum_r \ln(1 - \beta_r z^{-1}) - \sum_l \ln(1 - \alpha_l z^{-1}) \tag{10-123}$$

with α_l, β_r the poles and zeros of $H(z^{-1})$ satisfying $|\alpha_l| < 1$ and $|\beta_r| < 1$. Using (10-119) and the Taylor series expansion

$$\ln(1 - \lambda z^{-1}) = -\sum_{r=1}^{\infty} (\lambda z^{-1})^r / r, \tag{10-124}$$

the sum of $\ln(1 - \lambda z^{-1})$ over a uniform grid on the unit circle $z_k = \exp(j2\pi k/N)$, $k = 0, 1, \ldots, N-1$, can be written as

$$\sum_{k=0}^{N-1} \ln(1 - \lambda z_k^{-1}) = -\sum_{r=1}^{\infty} \frac{\lambda^r}{r} \sum_{k=0}^{N-1} z_k^{-r} = -N \sum_{n=1}^{\infty} \frac{\lambda^{nN}}{nN} \tag{10-125}$$

The absolute value of (10-125) can be bounded above as

$$\left|\sum_{k=0}^{N-1} \ln(1 - \lambda z_k^{-1})\right| \leq \sum_{n=1}^{\infty} \frac{|\lambda|^{nN}}{n} \leq O(|\lambda|^N) \qquad (10\text{-}126)$$

Collecting (10-123) and (10-126) gives

$$\left|\sum_{k=0}^{N-1} \ln(H(z_k^{-1}))\right| \leq O(|\lambda_{max}|^N) \qquad (10\text{-}127)$$

where $z = \lambda_{max}$ is the pole or zero of $H(z^{-1})$ closest to the unit circle ($|\lambda_{max}| < 1$). □

Appendix 10.O Gaussian Log-likelihood Function (Theorem 10.18)

Under Assumptions 10.12–10.16 $Y(k)$ (10-53) is independent (over k), circular complex normally distributed. To construct the likelihood function ($U(k)$ is exactly known) it is sufficient to calculate the mean and variance of $Y(k)$ given the plant and noise model parameters θ and the variance λ of the driving (band-limited) white noise source. In closed loop the process noise $V(k)$ is correlated with the input of the plant $U(k)$ (Assumption 10.13), and is independent of the reference signal $R(k)$ (Assumption 10.15). Therefore, the expected values in the mean and variance calculation of $Y(k)$ should be conditioned on $R(k)$. The latter is known since the controller $M_0(\Omega)$ is known (Assumption 10.17) and since $Y(k)$ and $U(k)$ are observed without errors (Assumption 10.12). Using (10-53), where $G_0(\Omega)$ and $H_0(\Omega)$ are replaced by, respectively, $G(\Omega, \theta)$ and $H(\Omega, \theta)$, we find

$$\mathbb{E}\{Y(k)|R(k), \theta, \lambda\} = \frac{G(\Omega_k, \theta)}{1 + G(\Omega_k, \theta)M_0(\Omega_k)} R(k) \equiv Y_0(k, \theta) \qquad (10\text{-}128)$$

$$\text{var}(Y(k)|R(k), \theta, \lambda) = \lambda |S(\Omega_k, \theta)|^2$$

where $S(\Omega, \theta)$ is defined in (10-56). Hence, the probability density function of $Y(k)$ equals

$$f_{Y(k)}(Y(k)|R(k), \theta, \lambda) = \begin{cases} \dfrac{1}{\pi \lambda |S(\Omega_k, \theta)|^2} \exp\left(-\dfrac{|Y(k) - Y_0(k, \theta)|^2}{\lambda |S(\Omega_k, \theta)|^2}\right) & k \neq 0, \dfrac{N}{2} \\[2ex] \dfrac{1}{\sqrt{2\pi \lambda |S(\Omega_k, \theta)|^2}} \exp\left(-\dfrac{|Y(k) - Y_0(k, \theta)|^2}{2\lambda |S(\Omega_k, \theta)|^2}\right) & k = 0, \dfrac{N}{2} \end{cases} \qquad (10\text{-}129)$$

because $Y(k)$ is real at DC ($k = 0$) and Nyquist ($k = N/2$) and circular complex elsewhere (see (16-14) on page 569). Using the independence of $Y(k)$ over k we get

$$f_Y(Y|R, \theta, \lambda) = \prod_{k \in \mathbb{K}} f_{Y(k)}(Y(k)|R(k), \theta, \lambda) \qquad (10\text{-}130)$$

with f_Y the likelihood of the output data $Y(k)$, $k \in \mathbb{K}$. Elaborating the exponent in (10-129),

$$S^{-1}(\Omega_k, \theta)(Y(k) - Y_0(k, \theta)) = H^{-1}(\Omega_k, \theta)(Y(k) - G(\Omega_k, \theta)U(k)) \qquad (10\text{-}131)$$

finally proves (10-55). The factor 1/2 at DC and Nyquist in the sums of (10-55) stems from (10-129). □

Appendix 10.P Proof of Corollary 10.19

Minimizing (10-55) w.r.t. λ gives (10-60). Using (10-60) λ is eliminated in (10-55)

$$\sum_{k \in \mathbb{K}} \ln|S(\Omega_k, \theta)|^2 + F\ln(F^{-1}\sum_{k \in \mathbb{K}} |\varepsilon(\Omega_k, \theta)|^2) + F \qquad (10\text{-}132)$$

Dividing (10-132) by F, subtracting one, and taking the exponential function finally gives (10-58). □

Appendix 10.Q Proof of Corollary 10.20

For frequency sets covering uniformly the unit circle, $z_k = \exp(j2\pi k/N)$ with $\mathbb{K} = \{0, 1, \ldots, N/2\}$, the sums in (10-58) and (10-59) are replaced by

$$\sum_{k \in \mathbb{K}} \ldots = 0.5\sum_{k=0}^{N-1} \ldots \qquad (10\text{-}133)$$

where the additional factor 1/2 accounts for DC ($k = 0$), Nyquist ($k = N/2$), and the fact that each frequency $k = 1, 2, \ldots, N/2 - 1$ appears twice in the sum ($z_k = \overline{z_{N-k}}$). In the sequel of the appendix we study the sum in the exponent of $g_N(\theta)$ for $N \to \infty$.

Since by assumption $S(0, \theta) = 1$, and $S(z^{-1}, \theta)$ and $S^{-1}(z^{-1}, \theta)$ are stable, it follows from Lemma 10.28 of Appendix 10.N that

$$\left|\sum_{k=0}^{N-1} \ln S(z_k^{-1}, \theta)\right| \leq O(|\lambda_{\max}|^N) \qquad (10\text{-}134)$$

where $z = \lambda_{\max}$ is the pole or zero of $S(z^{-1}, \theta)$ closest to the unit circle (= dominant pole of $\ln S(z^{-1}, \theta)$). Since $|\lambda_{\max}| < 1$ we conclude from (10-59), (10-133), (10-134), and $N = 2F$ that $g_N(\theta) = (1 + O(|\lambda_{\max}|^N/N))$, which proves the corollary. □

Appendix 10.R Proof of Theorem 10.25

The stochastic convergence, stochastic convergence rate, systematic and stochastic errors, asymptotic normality, deterministic convergence, and asymptotic bias are proven in, respectively, properties 1–5 and 7 of Theorem 9.21, for the independent frequency domain noise (Assumption 10.15), and in Chapter 17 for the mixing frequency domain noise (Assumption 10.24), while the asymptotic efficiency immediately follows from the ML properties under standard conditions (Caines, 1988). Only the consistency (Theorem 9.21, property 6), and the particular form of the asymptotic covariance matrix (Theorem 9.21, property 4), the limit cost function $V_*(\theta)$ (Theorem 9.21, property 5), and the Cramér-Rao bound (Theorem 9.21, property 8) remain to be proven.

10.R.1 Limit cost function. Eq. (10-68) is the result of the convergence of the Riemann sum (10-58) to the corresponding Riemann integral (see Theorem 7.21, property 5), and only the particular expression for $g_*(\theta)$ should be clarified. The latter follows immediately from the convergence of the Riemann sum in the exponent of $g_F(\theta)$ (10-59).

10.R.2 Consistency. To prove the strong consistency of $\hat{\theta}_{\mathrm{ML}}(Z)$ it is sufficient to show that θ_0 minimizes the expected value of $V_F(\theta, Z)$ (10-58), with $\varepsilon(\Omega_k, \theta)$ defined in (10-57). The strong consistency of $\hat{\lambda} = \lambda(\hat{\theta}_{\mathrm{ML}}(Z))$ follows then from $\lambda_0 = \lambda(\theta_0)$. Since the minimizers of (10-55) and (10-58) are by construction exactly the same (see Appendix 10.P), the expected value of (10-58) is minimal in $\theta = \theta_0$ if and only if the expected value of (10-55), $V(\theta, \lambda)$, is minimal in $\theta = \theta_0$ and $\lambda = \lambda_0$. Under Assumptions 10.15 ($m = 2$) and 10.23, $V(\theta, \lambda)$ equals

$$V(\theta, \lambda) = \sum_{k \in \mathbb{K}} \ln |S(\Omega_k, \theta)|^2 + \frac{1}{\lambda} \sum_{k \in \mathbb{K}} \left|\frac{S_0(\Omega_k)}{H(\Omega_k, \theta)}\right|^2 \frac{|\Delta G(\Omega_k, \theta)|^2 \mathbb{E}\{|R(k)|^2\}}{|H_0(\Omega_k)|^2} + F \ln \lambda + \frac{\lambda_0}{\lambda} \sum_{k \in \mathbb{K}} \left|\frac{S_0(\Omega_k)}{S(\Omega_k, \theta)}\right|^2 \qquad (10\text{-}135)$$

where $S(\Omega, \theta)$ is defined in (10-56), $S_0 = H_0/(1 + G_0 M_0)$, and $\Delta G(\Omega_k, \theta) = G_0(\Omega_k) - G(\Omega_k, \theta)$. Calculating the derivative of (10-135) w.r.t. θ and λ gives

$$\frac{\partial V(\theta, \lambda)}{\partial \theta} = \sum_{k \in \mathbb{K}} \frac{1}{|S(\Omega_k, \theta)|^2} \frac{\partial |S(\Omega_k, \theta)|^2}{\partial \theta} + \frac{1}{\lambda} \sum_{k \in \mathbb{K}} \left|\frac{S_0(\Omega_k)}{H(\Omega_k, \theta)}\right|^2 \frac{\mathbb{E}\{|R(k)|^2\}}{|H_0(\Omega_k)|^2} \frac{\partial |\Delta G(\Omega_k, \theta)|^2}{\partial \theta}$$
$$- \frac{1}{\lambda} \sum_{k \in \mathbb{K}} \frac{|S_0(\Omega_k)|^2 |\Delta G(\Omega_k, \theta)|^2 \mathbb{E}\{|R(k)|^2\}}{|H(\Omega_k, \theta)|^4 |H_0(\Omega_k)|^2} \frac{\partial |H(\Omega_k, \theta)|^2}{\partial \theta} \qquad (10\text{-}136)$$
$$- \frac{\lambda_0}{\lambda} \sum_{k \in \mathbb{K}} \frac{|S_0(\Omega_k)|^2}{|S(\Omega_k, \theta)|^4} \frac{\partial |S(\Omega_k, \theta)|^2}{\partial \theta}$$

$$\frac{\partial V(\theta, \lambda)}{\partial \lambda} = \frac{F}{\lambda} - \frac{1}{\lambda^2} \left(\sum_{k \in \mathbb{K}} \left|\frac{S_0(\Omega_k)}{H(\Omega_k, \theta)}\right|^2 \frac{|\Delta G(\Omega_k, \theta)|^2 \mathbb{E}\{|R(k)|^2\}}{|H_0(\Omega_k)|^2} + \lambda_0 \sum_{k \in \mathbb{K}} \left|\frac{S_0(\Omega_k)}{S(\Omega_k, \theta)}\right|^2 \right)$$

Evaluating (10-136) in $\theta = \theta_0$ and $\lambda = \lambda_0$, using $H(\Omega_k, \theta_0) = H_0(\Omega_k)$, $\Delta G(\Omega_k, \theta_0) = 0$, and

$$\frac{\partial |\Delta G(\Omega_k, \theta)|^2}{\partial \theta} = 2\mathrm{Re}\left(\frac{\partial \Delta G(\Omega_k, \theta)}{\partial \theta} \overline{\Delta G(\Omega_k, \theta)}\right)$$

where $\bar{x}$ denotes the complex conjugate of x, gives $\partial V(\theta, \lambda_0)/\partial \theta_0 = 0$ and $\partial V(\theta_0, \lambda)/\partial \lambda_0 = 0$ which concludes the proof.

10.R.3 Cramér-Rao lower bound. Evaluating the derivatives of (10-136) w.r.t. θ, λ in θ_0, λ_0 gives

$$M_{22} = \left.\frac{\partial^2 V(\theta, \lambda)}{\partial \lambda^2}\right|_{\theta = \theta_0, \lambda = \lambda_0} = \frac{F}{\lambda_0^2}$$

Section 10.11 ■ Appendixes 429

$$M_{11} = \left.\frac{\partial^2 V(\theta, \lambda)}{\partial \theta^2}\right|_{\theta = \theta_0, \lambda = \lambda_0} = \sum_{k \in \mathbb{K}} \frac{1}{|S_0(\Omega_k)|^4}\left(\frac{\partial |S(\Omega_k, \theta)|^2}{\partial \theta_0}\right)^T\left(\frac{\partial |S(\Omega_k, \theta)|^2}{\partial \theta_0}\right)$$

$$+ \sum_{k \in \mathbb{K}} \frac{|S_0(\Omega_k)|^2}{|H_0(\Omega_k)|^4}\frac{\mathbb{E}\{|R(k)|^2\}}{\lambda_0}\frac{\partial^2 |\Delta G(\Omega_k, \theta)|^2}{\partial \theta_0^2}$$

$$M_{12} = \left.\frac{\partial^2 V(\theta, \lambda)}{\partial \theta \partial \lambda}\right|_{\theta = \theta_0, \lambda = \lambda_0} = \frac{1}{\lambda_0}\sum_{k \in \mathbb{K}}\frac{1}{|S_0(\Omega_k)|^2}\left(\frac{\partial |S(\Omega_k, \theta)|^2}{\partial \theta_0}\right)^T$$

Using the inverse of a 2-by-2 block matrix (see Exercise 15.10), the inverse of the Cramér-Rao lower bound $CR(\theta_0)$ of the model parameters θ is given by $CR^{-1}(\theta_0) = (M_{11} - M_{12}M_{22}^{-1}M_{12}^T)$. After some straightforward calculations one finds $CR^{-1}(\theta_0) = F_1 + F_2$ with F_1, F_2 given in (10-66).

10.R.4 Asymptotic covariance matrix. Using (10-53) and (10-54), cost function (10-58) can be written as

$$FV_F(\theta, Z) = |g_F(\theta)|^2 \sum_{k \in \mathbb{K}}\left(\left|\frac{\Delta G(\Omega_k, \theta)}{H(\Omega_k, \theta)}\right|^2\left|\frac{S_0(\Omega_k)}{H_0(\Omega_k)}\right|^2|R(k)|^2 + \left|\frac{S_0(\Omega_k)}{S(\Omega_k, \theta)}\right|^2|E(k)|^2 \right.$$

$$\left. + 2\text{Re}\left(\frac{\Delta G(\Omega_k, \theta)}{H(\Omega_k, \theta)}\frac{|S_0(\Omega_k)|^2}{\bar{S}(\Omega_k, \theta)}\frac{R(k)\bar{E}(k)}{H_0(\Omega_k)}\right)\right) \quad (10\text{-}137)$$

After some calculations we find

$$FV_F''(\theta_0) = \lambda_0|g_F(\theta_0)|^2(F_1 + F_2)$$

$$FV_F'(\theta_0, Z) = -2|g_F(\theta_0)|^2\sum_{k \in \mathbb{K}}\text{Re}\left(\frac{\partial G(\Omega_k, \theta)}{\partial \theta_0}\frac{S_0(\Omega_k)}{H_0^2(\Omega_k)}R(k)\bar{E}(k)\right) \quad (10\text{-}138)$$

$$- |g_F(\theta_0)|^2\sum_{k \in \mathbb{K}}(\chi_k - F^{-1}\sum_{k=1}^{F}\chi_k)|E(k)|^2$$

with $V_F(\theta) = \mathbb{E}\{V_F(\theta, Z)\}$, and where F_1, F_2 and χ_k are defined in (10-66). Using

$$\mathbb{E}\{|E(k)|^2|E(l)|^2\} = \begin{cases} \lambda_0^2 & k \neq l \\ k_u^c\lambda_0^2 & k = l \end{cases} \text{ and } \mathbb{E}\{\text{Re}(Z_1^T)\text{Re}(Z_1)\} = \frac{1}{2}\text{Re}(\mathbb{E}\{Z_1^H Z_1\}) \quad (10\text{-}139)$$

with $E(k)$ independent over k (Assumption 10.15), k_u^c the kurtosis factor $\mathbb{E}\{|E(k)|^4\}/(\mathbb{E}\{|E(k)|^2\})^2$, and Z_1 zero mean circular complex noise $\mathbb{E}\{Z_1^T Z_1\} = 0$, together with $\mathbb{E}\{R(k)\} = 0$ (under Assumption 10.14, $R(k)$ has zero mean: use $R(k) = H_u(z_k^{-1})E_u(k)$ and apply Lemma 16.24), it can be verified that

$$\mathbb{E}\{V_F'^T(\theta_0, Z)V_F'(\theta_0, Z)\} = \lambda_0^2|g_F(\theta_0)|^4(F_1 + (k_u^c - 1)F_2)/F^2 \quad (10\text{-}140)$$

Combining (9-27), (10-138), and (10-140) proves (10-65).

11

Model Selection and Validation

Abstract: A critical step in the identification process is the quality assessment of the identified model. A model without error bounds has no value. For this reason, we need tools to check whether all linear dynamics in the raw data are captured and tools to quantify the remaining model errors. Also, the presence of nonlinear distortions should be detected, qualified, and quantified. Finally, the validity of the disturbing noise models should be tested.

This chapter provides dedicated tools to test for over- and undermodeling. This information is used not only to validate the final model but also to guide the model selection process during the identification. The methods vary from a simple visual inspection (does the transfer function fit the FRF measurements well enough for the intended application?) to an advanced statistical analysis of the residuals. In the case of undermodeling, the remaining model error is quantified so that the user can decide whether the final model is acceptable for his/her application.

11.1 INTRODUCTION

At the end of an identification run, two important questions remain to be answered. What is the quality of the model? Can this model be used to solve my problem? Whereas the first question is an absolute one, the second question shows that, in practice, the applicability of an identified model strongly depends on the intended application. Each model is only an approximation of reality, and often the existence of a "true" model is only a fiction, in the mind of the experimenter. The deviations between the model and the system that generated the measurements are partitioned in two parts following their nature: systematic errors and stochastic errors. If the experiment is repeated under the same conditions, the systematic errors will be the same, but the stochastic errors vary from one realization to the other. Model validation is directed toward the quantification of the remaining model errors. Once the level of the systematic errors is known, the user should decide whether or not they are acceptable. It is not evident at all that one is looking for the lowest error level; often it is sufficient to push them below a given upper bound. In order to decide whether the errors are systematic, it is necessary to know the uncertainty on the estimated model. In this book we use probabilistic uncertainty bounds (e.g., 95% bounds) that describe how the individual realizations are scattered around their mean values. Errors that are outside this bound are considered to be unlikely, so that they are most probably due to systematic deviations.

This short discussion shows, clearly, that model validation starts with the generation of good uncertainty bounds. These bounds can be used in a second step to check for the presence of significant (from a statistical point of view) systematic errors. This two-step approach is developed in the course of this chapter. First, it is shown how error bounds on the transfer function and the poles can be generated starting from the covariance matrix of the parameters C_θ (generated by the estimation algorithm). Next, it will be explained how the presence of systematic errors can be detected. This information will be used to develop an automatized model selection procedure.

Note that using the measured frequency response function (FRF) in the validation step is theoretically not the best choice in an errors-in-variables concept (noise on the input and output measurements). However, as shown in Chapter 2, the FRF measurements are very usable for sufficient high signal-to-noise ratios (SNRs). Although we still prefer to do the identification from the measured input-output data (no user decision needed to check whether the measurements are good enough to jump to the FRF; no increase in complexity for the user; nonincreased computation time), using the FRF simplifies the validation process significantly because it is much easier to interpret, allows a simple visual inspection, and is often one of the major results that is needed.

11.2 ASSESSING THE MODEL QUALITY: QUANTIFYING THE STOCHASTIC ERRORS

As mentioned in the introduction, the first step in the validation process is the partitioning into stochastic and systematic errors. The stochastic error bounds are not only a tool to detect systematic errors, they are also intensively used to describe the overall quality of the model once it is known that systematic errors are no longer dominating. The basic "uncertainty" information is delivered under the form of the covariance matrix on the estimated parameters. The actual covariance matrix is mostly too difficult to calculate. But in most cases the Cramér-Rao lower bound (see Sections 1.3.2 and 16.12) can be used for asymptotically efficient estimators. Also, for weighted least squares estimators, approximative expressions to calculate the covariance matrix are available (see, for example, Theorem 9.21). An approximation of both expressions can be calculated easily at the end of the identification process. However, in many applications the user is not interested in the estimated parameters and their uncertainty but wants to calculate from these parameters other system characteristics such as the transfer function or the pole positions. In Section 16.12 it is shown that the Cramér-Rao lower bound of these derived quantities is generated by simple transformation laws, obtained from the first-order derivatives of the actual transformation. Remark that the same laws also apply to the approximated covariance matrices such as those obtained in Theorem 9.21:

$$\text{Cov}(f(x)) \approx \left.\frac{\partial f(x)}{\partial x}\right|_{x=\mu_x} \text{Cov}(x) \left(\left.\frac{\partial f(x)}{\partial x}\right|_{x=\mu_x}\right)^H \tag{11-1}$$

In practice, this works very well as long as the transformations are not heavily nonlinear (e.g., transfer function calculation), but sometimes it fails. A typical example of such a failure is the generation of the uncertainty regions on the estimated poles/zeros. Although the Cramér-Rao bounds (or the approximate covariance matrix) are correct, the actual uncertainties can significantly differ due to the fact that the asymptotic properties on these estimates are not yet reached for practical signal-to-noise ratios. For this case, we present a more precise numerical method to generate these bounds. In the following, the uncertainty on the transfer function, the transfer function residuals (difference between the measured and the estimated transfer function), and the poles are studied in detail. It is shown how to calculate these quantities, starting from the covariance matrix on the parameters C_θ. To evaluate the expression in practice, C_θ is replaced by the covariance matrix estimate, delivered by the identification algorithm, and the estimated parameter values are used instead of the limiting values parameters.

11.2.1 Uncertainty Bounds on the Calculated Transfer Functions

The transfer function interpretation of an identified model $G(\Omega, \hat{\theta})$ is used very intensively. It is also important to know the reliability of the estimated transfer function as a function of the frequency. Applying (11-1) gives the variance of the transfer function due to the noise sensitivity of the parameter estimates ($C_\theta = \text{Cov}(\hat{\theta})$):

$$\text{var}(G(\Omega, \hat{\theta})) \approx \left(\frac{\partial G(\Omega, \theta)}{\partial \theta}\bigg|_{\theta = \hat{\theta}}\right) C_\theta \left(\frac{\partial G(\Omega, \theta)}{\partial \theta}\bigg|_{\theta = \hat{\theta}}\right)^H \tag{11-2}$$

11.2.2 Uncertainty Bounds on the Residuals

A very simple, but popular, validation test is to compare the differences between the measured FRF, $G(\Omega_k)$, and the modeled transfer function, $G(\Omega_k, \hat{\theta})$. In order to decide whether these residuals $G(\Omega_k) - G(\Omega_k, \hat{\theta})$ are significantly different from zero, their variance should be calculated. Equation (11-2) of the previous section cannot be applied here directly because $G(\Omega_k) - G(\Omega_k, \hat{\theta})$ depends now not only on $\hat{\theta}$ but also on the raw data $G(\Omega_k)$. Note that $\hat{\theta}$ and $G(\Omega_k)$ are correlated stochastic variables because they both depend on the same noise distortions N_Z. The extended expression (19-37) is repeated here for the readers' convenience for complex-valued $f(Z, \theta)$ and Z, and real-valued $\hat{\theta}$:

$$\text{Cov}(f(Z, \hat{\theta})) \approx \left(\frac{\partial f(Z, \hat{\theta})}{\partial Z}\right) C_{N_Z} \left(\frac{\partial f(Z, \hat{\theta})}{\partial Z}\right)^H + \left(\frac{\partial f(Z, \theta)}{\partial \hat{\theta}}\right) \text{Cov}(\hat{\theta}(Z)) \left(\frac{\partial f(Z, \theta)}{\partial \hat{\theta}}\right)^H$$

$$+ 2\text{herm}\left(\left(\frac{\partial f(Z, \hat{\theta})}{\partial Z}\right) \text{Cov}(N_Z, \hat{\theta} - \tilde{\theta}) \left(\frac{\partial f(Z, \theta)}{\partial \hat{\theta}}\right)^H\right) \tag{11-3}$$

$$\text{Cov}(N_Z, \hat{\theta} - \tilde{\theta}) \approx -C_{N_Z} \left(\frac{\partial \varepsilon(\hat{\theta}, Z)}{\partial Z}\right)^H \left(\frac{\partial \varepsilon(\theta, Z)}{\partial \hat{\theta}}\right) \text{Cov}(\hat{\theta})$$

(proof: see Appendix 11.A). We apply this to the residual $G(\Omega_k) - G(\Omega_k, \hat{\theta})$, for deterministic excitations (the expected value with respect to a random input disappears), assuming that there is no input noise, and $\hat{\theta} = \hat{\theta}_{\text{ML}}(Z)$ considering $G(\Omega_k)$ as the raw data (see 8-1). The following expression is obtained:

$$\text{var}(G(\Omega_k) - G(\Omega_k, \hat{\theta})) = \text{var}(G(\Omega_k)) + \left(\frac{\partial G(\Omega_k, \theta)}{\partial \hat{\theta}}\right) \text{Cov}(\hat{\theta}) \left(\frac{\partial G(\Omega_k, \theta)}{\partial \hat{\theta}}\right)^H$$

$$- 2\text{herm}\left(\left(\frac{\partial G(\Omega_k, \theta)}{\partial \hat{\theta}}\right) \text{Cov}(\hat{\theta}) \left(\frac{\partial G(\Omega_k, \theta)}{\partial \hat{\theta}}\right)^H\right) \tag{11-4}$$

(proof: see Appendix 11.B). Using $\text{herm}(xCx^H) = xCx^H$ for Hermitian matrices C ($C^H = C$), (11-4) can be simplified as

$$\text{var}(G(\Omega_k) - G(\Omega_k, \hat{\theta})) = \sigma_G^2(k) - \sigma_G^2(\Omega_k, \hat{\theta}) \qquad (11\text{-}5)$$

where

$$\sigma_G^2(k) = \text{var}(G(\Omega_k)) \text{ and } \sigma_G^2(\Omega_k, \hat{\theta}) = \left(\frac{\partial G(\Omega_k, \theta)}{\partial \hat{\theta}}\right) \text{Cov}(\hat{\theta}) \left(\frac{\partial G(\Omega_k, \theta)}{\partial \hat{\theta}}\right)^H \qquad (11\text{-}6)$$

Practical Application. In general, $\sigma_G^2(\Omega_k, \hat{\theta}) \ll \sigma_G^2(k)$ so that the compensation in (11-5) can be neglected. $\sigma_G^2(\Omega_k, \hat{\theta})$ can become of the same order as $\sigma_G^2(k)$ only at those frequencies where the model is very flexible and depends only on a few data points (e.g., very sharp resonances). Because in this situation both terms in (11-5) cancel each other, the result becomes extremely sensitive to model errors. Expression (11-5) can even become negative! Hence, if $\sigma_G^2(\Omega_k, \hat{\theta}) \approx \sigma_G^2(k)$ the user should accept that in that region the presence of model errors cannot be detected because there is no reliable estimate of the residual uncertainty to decide whether or not they are significantly different from zero.

Example 11.1 (Calculation of the Uncertainty of Transfer Function Residuals): A very popular validation test is to compare the estimated transfer function $G(\Omega, \hat{\theta}_{\text{ML}}(Z))$ with the measured transfer function, obtained directly from the measured input-output spectra: $\hat{G}(\Omega_k) = Y(k)/U(k)$. In order to decide whether the errors $\hat{G}(\Omega_k) - G(\Omega_k, \hat{\theta}_{\text{ML}}(Z))$ are significantly different from zero, the variance of these residuals is calculated. Although the raw data were U and Y, we still use expression (11-5) with σ_G^2 calculated from the (co)variances σ_U^2, σ_Y^2, and σ_{YU}^2 with (2-25). A simulation is made on the system:

$$G(z^{-1}) = \frac{5.619 \times 10^{-3} + 2.248 \times 10^{-2} z^{-1} + 3.371 \times 10^{-2} z^{-2} + 2.248 \times 10^{-2} z^{-3} + 5.619 \times 10^{-3} z^{-4}}{1 - 1.585 z^{-1} + 2.124 z^{-2} - 1.544 z^{-3} + 0.9034 z^{-4}}$$

It is excited at the frequencies $k f_s / 128$, $k = 1, 2, \ldots, 44$, with $|U(k)| = 1$, and $\sigma_U(k) = 0.01\sqrt{2}$, $\sigma_Y(k) = 0.005\sqrt{2}$. The number of frequencies was kept very small in order to illustrate the effect of model errors on the uncertainty bounds. A thousand simulations are made and processed for a model order of $G(\Omega, \theta)$ equal to 4/4, 3/4, and 2/4. The results are given in Figure 11-1. The figures in the upper row compare the predicted uncertainty on the transfer function with the actual observed uncertainty. As can be seen, very good agreement is obtained as long as the model errors are small (models 3/4 and 4/4), while deviations become visible for model 2/4 at the second resonance peak. In this case, significant model errors are present. Observe that the uncertainty $\text{std}(G(\Omega, \hat{\theta}))$ on the estimated parametric model can become even larger than the measurement uncertainty $\sigma_G(k)$. The lower row shows the uncertainty on the residuals. For models 3/4 and 4/4 the theoretical values (11-5) and the observations are again in very good agreement. Note also that at most frequencies the standard deviation of the residual is almost equal to the measurement uncertainty (compare $\sigma_G(k)$ of the upper plot with $\text{std}(G(\Omega_k) - G(\Omega_k, \hat{\theta}))$ of the lower plot). Only at the second resonance peak (most important frequency band!) is there a significant drop. This is due to the fact that only two data points are put at the resonance peak so that the model can follow the raw data almost completely, leading to small residuals. Because of the errors for model 2/4, the predicted residual variance even becomes negative (the compensation in (11-5) fails), so that it loses all value. □

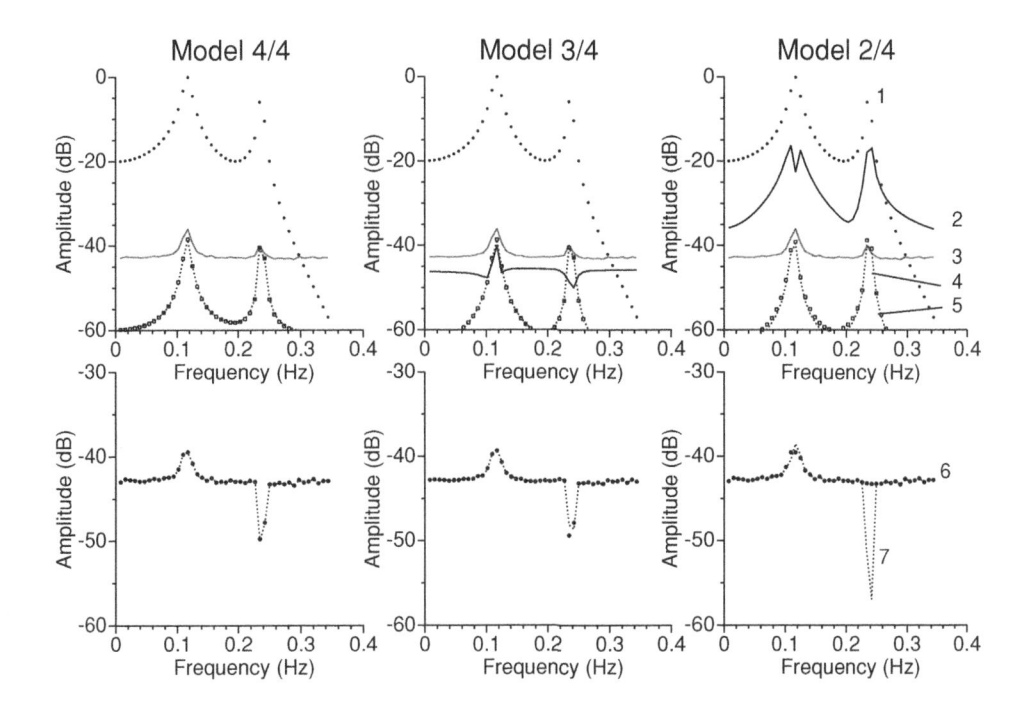

Figure 11-1. Study of the uncertainty bounds of the residuals for different levels of the model error. 1: $G_0(z_k^{-1})$, 2: model error, 3: measurement uncertainty $\sigma_G(k)$, 4: theoretic value of $\text{std}(G(z_k^{-1}, \hat{\theta}))$, 5: sample value of $\text{std}(G(z_k^{-1}, \hat{\theta}))$, 6: sample value of $\text{std}(G(\Omega_k) - G(\Omega_k, \hat{\theta}))$, 7: theoretic value of $\text{std}(G(\Omega_k) - G(\Omega_k, \hat{\theta}))$.

11.2.3 Uncertainty Bounds on the Poles/Zeros

The dispersion of the estimated parameters $\hat{\theta}$ around their mean value is given by the covariance matrix C_θ. Assuming that the estimates are normally distributed, the most compact uncertainty regions are ellipses. Practice has shown that this is a very usable description for realistic signal-to-noise ratios if θ are the coefficients of the numerator and denominator polynomials of the transfer function model. In the previous section it was shown how to calculate the covariance matrix of related system characteristics using linear approximations. However, if the user is interested in the uncertainty of the poles/zeros of the estimated system, it turns out that this linearization may fail. Even for high signal-to-noise ratios, the uncertainty ellipses calculated for the poles and zeros may not cover the true uncertainty regions. This is illustrated in the following simulation example. Consider the system $G(s)$ with zeros $-1.4355 \pm j4.0401$ and poles $-1.3010 \pm j4.8553$, $-3.5543 \pm j3.5543$, $-4.8553 \pm j1.3010$. The system has one dominating pole-zero pair and two pole pairs that have a smaller impact on the system. The transfer function is measured in 101 equidistant points between 0 and 1.25 rad/s with a signal-to-noise ratio of 40 dB ($\sigma_G(k) = |G(j\omega_k)|/100$). Although we specified all characteristics in the frequency domain, the results are completely independent of the method that is used to identify the system (time or frequency domain identification). The only important information that is used are the model parameters and their co-

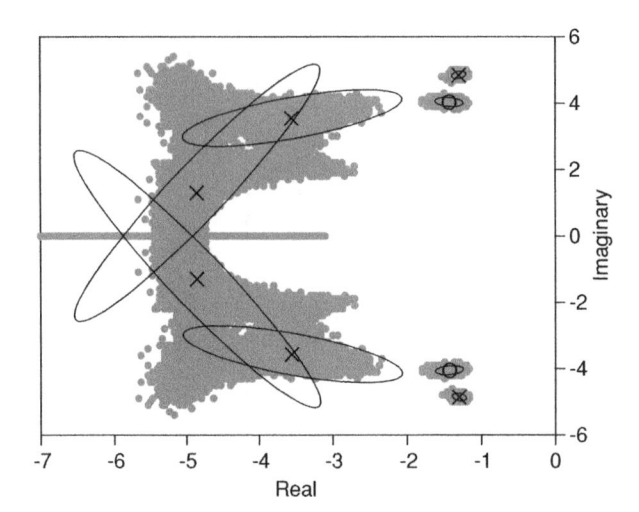

Figure 11-2. 95% confidence ellipsoids compared with the estimated poles and zeros of 10,000 simulations.

variance matrix. Ten thousand realizations were generated and for each of them the poles/zeros were calculated and are shown in Figure 11-2. Also, the "classical" 95% confidence ellipsoids calculated using (11-1) are shown (see Guillaume et al., 1989). In this figure it is clearly seen that the shape of the uncertainty regions differs significantly from the ellipsoids (for the nondominating poles); and that even for the dominating pole/zeros the uncertainties are significantly underestimated. This is an unacceptable result because it is used as an input to many design procedures. Consequently, there is a need for more precise techniques to produce reliable uncertainty regions. The basic idea behind the improved technique is explained in the next section, and the precise mathematical description is given in Appendix 11.D.

11.2.3.1 Improved Method—Practical Calculation. The basic idea is to consider one pole (or zero) as a parameter and to move it away from its estimated position. The position of the remaining poles/zeros is shifted such that the total impact of the movement on the cost function is minimized. This step is the major difference from the method presented by Walter and Pronzato (1997). In Appendix 11.D it is shown that it is sufficient to observe the quadratic form

$$\Delta \theta^T C_\theta^{-1} \Delta \theta \in As\chi^2(n_\theta) \tag{11-7}$$

Once this form reaches its maximum acceptable level given by the $p\%$ percentile $\chi_p^2(n_\theta)$, the border of the confidence region is found. The subsequent steps will follow that border so that the boundary is constructed (Vuerinckx et al., 1998).

11.2.3.2 Example. The improved method is applied to the previous example and the results are shown in Figure 11-3. It can be observed that there is now a very good match between the observed and the calculated uncertainty regions. In order to show how the confidence regions start to deviate from an ellipsoidal form, the 95% bounds are drawn in Figure 11-4 for increasing SNR. Starting from the same conditions as in the previous simulation (SNR = 40 dB), the signal-to-noise ratio is increased in steps of 6 dB to 64 dB. Note that even for a high SNR the ellipsoidal form is not followed

Section 11.3 ■ Avoiding Overmodeling

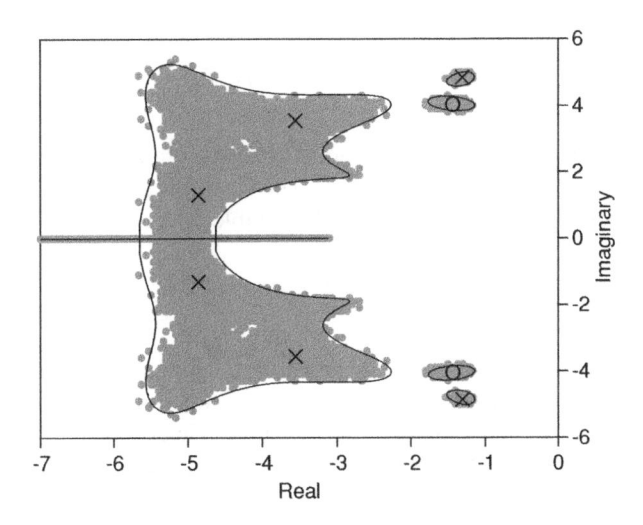

Figure 11-3. 95% confidence regions of the test system, calculated by perturbing the zeros and poles, using the coefficient covariance matrix.

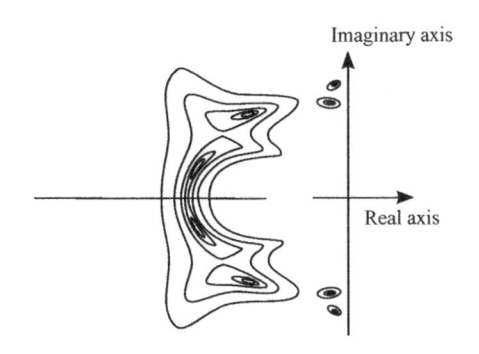

Figure 11-4. Evolution of the confidence regions as a function of the SNR (40 dB, 46 dB, 52 dB, 58 dB, 64 dB).

11.3 AVOIDING OVERMODELING

11.3.1 Introduction: Impact of an Increasing Number of Parameters on the Uncertainty

In this section we look into the dependence of the model variability on the model complexity. During the modeling process it is often quite difficult to decide whether or not the introduction of a new parameter is meaningful. A simple strategy would be to fit all the parameters that could be of possible interest, but this is not a good idea because the uncertainty on the estimates will then be increased. Consider a model with a partitioned set of parameters $\theta = (\theta_1, \theta_2)$. What is the impact on the model uncertainty if the simple model $G(\theta_1)$ is extended to the more complex one $G(\theta_1, \theta_2)$? In Example 1.5, it was illustrated that the uncertainty will increase. Here it is shown that this is a general result. Consider the information matrix of the full model:

$$Fi(\theta_1, \theta_2) = \begin{bmatrix} Fi_{11} & Fi_{12} \\ Fi_{21} & Fi_{22} \end{bmatrix} \quad (11\text{-}8)$$

The covariance matrix of the simple model is $C(\theta_1) = Fi_{11}^{-1}$, while the covariance matrix of the complete model is $C(\theta_1, \theta_2) = Fi^{-1}$. The covariance matrix C_{θ_1} of the subset θ_1 is related to the covariance matrix $C(\theta_1)$ of the complete set by

$$C_{\theta_1} = Fi_{11}^{-1} + Fi_{11}^{-1} Fi_{12}(Fi_{22} - Fi_{21} Fi_{11}^{-1} Fi_{12})^{-1} Fi_{21} Fi_{11}^{-1} = C(\theta_1) + \Delta \quad (11\text{-}9)$$

(see (15-8)). Because $\Delta \geq 0$ it is clear that adding additional parameters to a model increases its uncertainty. A similar result is available for transfer function estimation. Ljung (1985) has shown that in case of output noise only, the asymptotic expression (for the order increasing to ∞) for the variance $\sigma_G^2(\Omega, \hat{\theta}(Z))$ on the estimated transfer function becomes

$$\sigma_G^2(\Omega, \hat{\theta}(Z)) \sim \frac{n_\theta}{N} \text{SNR}^{-1}(\omega) \quad (11\text{-}10)$$

This expression gives a great deal of insight: the uncertainty on the parametric model is proportional to that of the nonparametric estimate, but due to the averaging effect (over the frequency) of the parametric model an additional noise reduction of n_θ/N appears. The dependence on n_θ is illustrated in Figure 11-5. A 5th-order FIR system is identified, the first time using a 5th-order model ($n_\theta = 5$) and the second time with a 50th-order model ($n_\theta = 50$). From (11-10) it is expected that the standard deviation should increase about 9 dB, which is in agreement with the simulation results.

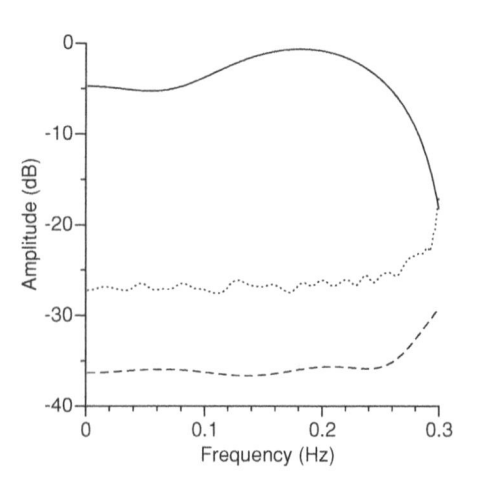

Figure 11-5. Dependence of $\sigma_G(\Omega, \hat{\theta}(Z))$ on the model order. ___ $G_0(z_k^{-1})$, --- $\sigma_G(\Omega, \hat{\theta}(Z))$ of 5th-order system, ... $\sigma_G(\Omega, \hat{\theta}(Z))$ of 50th-order system.

11.3.2 Balancing the Model Complexity versus the Model Variability

In the previous section it was illustrated that the systematic errors decrease with increasing model complexity. However, at the same time the model variability increases as shown in (11-9) and (11-10). In practice, the optimal complexity should be selected from the available information. Usually this choice is based on the evolution of the cost function. As

Section 11.3 ■ Avoiding Overmodeling

explained in Sections 19.6 and 19.7, it is not a good idea to select the model with the smallest cost function because it will continue to decrease if additional parameters are added. From a given complexity, the additional parameters no longer reduce the systematic errors but are used only to follow the actual noise realization on the data. As these vary from measurement to measurement, they increase only the model variability. Many techniques were proposed to avoid this unwanted behavior. These are based on extending the cost function with a model complexity term that estimates and compensates for the unwanted increasing model variability. Two popular methods are actually in use, the AIC (Akaike information criterion) and the MDL (minimum description length):

$$\text{AIC:} \quad V_{\text{ML}}(\hat{\theta}_{\text{ML}}(Z), Z)\left(1 + \frac{n_\theta}{F}\right)$$

$$\text{MDL:} \quad V_{\text{ML}}(\hat{\theta}_{\text{ML}}(Z), Z)\left(1 + \frac{n_\theta}{2F}\ln(2\alpha F)\right)$$

(11-11)

with n_θ the number of identifiable (free) model parameters (= total number of parameters minus the number of constraints), and F the number of frequencies (see Section 19.7 and Schoukens et al., 2002). $\alpha = 1$ for output error problems, and $\alpha = 2$ for the errors-in-variables problem. AIC minimizes the prediction error (Shibata, 1980), while MDL is best suited for physical modeling. The latter is illustrated in the following simulations.

11.3.2.1 Simulation of a Second-Order System. In the first simulation, a discrete-time second-order system ($n_a = n_b = 2$) with independently uniformly distributed coefficients, $a_r, b_r \in [0, 1]$ for $r = 0, 1, 2$, was considered. Because the estimation is performed in the frequency domain, the stability of the system was not an issue, and we also kept the unstable systems in the simulation. The system is excited over the full frequency band $(0, f_s/2)$ with 200 equidistantly distributed frequencies, $U_0(k) = 1$ and $Y_0(k) = G_0(z_k^{-1})$. The output is disturbed with zero mean white Gaussian noise with variance $\sigma_Y^2(k) = 2 \times 10^{-4}$. Then 1,000 random realizations of the random system were generated and identified. Each time all the models between 1/1 to 3/3 were tested and the best one was selected following the AIC and MDL rule (11-11). The results are shown in Table 11-1, giving how many times each model is selected. The MDL rule selects the correct model almost every time, while the AIC rule has a strong tendency to select models that are too complex.

11.3.2.2 Simulation of a Sixth-Order System. In the second simulation, a system with well-separated resonances is identified (see Figure 11-6, thin black line), namely a sixth-order discrete-time Chebyshev filter ($n_a = n_b = 6$) with a passband ripple of 6 dB, a cutoff frequency of 0.225 Hz, and $f_s = 1$ Hz. The user is actually interested only in a model for the middle resonance. Therefore, most input power is focused around this resonance (see Fig-

TABLE 11-1 Selection of the Model Order Using the AIC (left columns) and the MDL (right columns) Model Selection Rules – Thousand Monte Carlo Runs

n_b \ n_a	1	2	3	n_b \ n_a	1	2	3
1	1	1	11	$n_b = 1$	2	5	15
2	1	633	74	$n_b = 2$	3	942	4
3	10	66	203	$n_b = 3$	11	8	10
	AIC				MDL		

ure 11-6, black dashed line): $F = 200$ frequencies are equidistantly distributed in the band (0 Hz, 0.3 Hz) with $|U(k)| = 1$ in the band [0.145 Hz, 0.225 Hz] and $|U(k)| = 0.02$ elsewhere. Noise was only added to the output ($\sigma_Y^2(k) = 1 \times 10^{-4}$). All models with $4 \leq n_a, n_b \leq 8$ are scanned, and the best one is selected following the AIC-rule (11-11), the MDL-rule (11-11), and the cost function–based rule (select the model corresponding to the smallest value of $V_{ML}(\hat{\theta}_{ML}(Z), Z)$). This is repeated for 1,000 independent output noise realizations. The results are given in Tables 11-2 and 11-3. It can be seen that none of the methods is able to select the correct 6/6 model. For AIC and MDL this is mainly due to the fact that only a part of the band is properly excited so that the concept "exact model" loses its value as is commonly experienced in practice. In order to get a better appreciation of the results, we plotted the rms errors of the 1,000 AIC, MDL, and cost function selected models (Figure 11-6). It can be seen that the rms errors on the MDL models are the smallest, and that those of the AIC and cost function–based models almost coincide. Note also that the rms errors are small inside the frequency band of interest (band with the highest input power). This suggests that the additional model flexibility, used to model out-of-band effects, does not have a great impact on the in-band uncertainty. This is further analyzed in Section 11.4.1.

TABLE 11-2 Selection of the Model Order Using the AIC (left columns) and the MDL (right columns) Model Selection Rules – Thousand Monte Carlo Runs

n_b \ n_a	4	5	6	7	8	n_b \ n_a	4	5	6	7	8
4	0	0	0	127	46	4	0	0	15	404	94
5	0	0	97	20	28	5	0	0	352	24	11
6	0	0	13	20	70	6	0	0	4	2	5
7	0	0	36	31	291	7	0	0	64	4	17
8	0	0	3	15	203	8	0	0	0	0	4
			AIC						MDL		

Remarks

(i) Although the cost function–based model selection rule should always select the most complex model 8/8, it can be seen from Table 11-3 that this is not the case here. This is due to the overmodeling, which increases the probability that the search algorithm gets stuck in a local minimum. Nevertheless, it can still be observed that without an additional model complexity term, there is a strong tendency to select too complex models. Although this is not really a disaster from a model variability point of view (the uncertainty on the transfer function does not really explode), it still represents a lot of wasted work.

(ii) Although the selection of the model complexity is not that critical, it is important not to exaggerate the order (e.g., just doubling it), in order to avoid the appearance of coinciding pole-zero pairs that can create sharp resonances between two frequency points, resulting in a locally increased variance of the model.

Conclusions. Using the MDL rule, it is possible to balance the model complexity versus the model variability without requiring prior knowledge that would not be available in practice. The major contribution of the MDL rule is to give a data-based restriction on the maximum complexity of the models that need to be checked. As it is precisely these too complex models that require a lot of computation time, it can be concluded that the MDL rule essentially helps to save time.

Section 11.4 ■ Detection of Undermodeling

TABLE 11-3 Cost Function–Based Model Selection – Thousand Monte Carlo Runs

n_b \ n_a	4	5	6	7	8
4	0	0	0	0	0
5	0	0	0	0	2
6	0	0	0	0	39
7	0	0	1	17	315
8	0	0	2	39	585

Figure 11-6. Comparison of the rms error of AIC (light gray), MDL (dark gray), and cost function–based (black) selected models. Thin black line: true FRF $G_0(z_k^{-1})$, and black dashed line: input DFT spectrum $|U(k)|$.

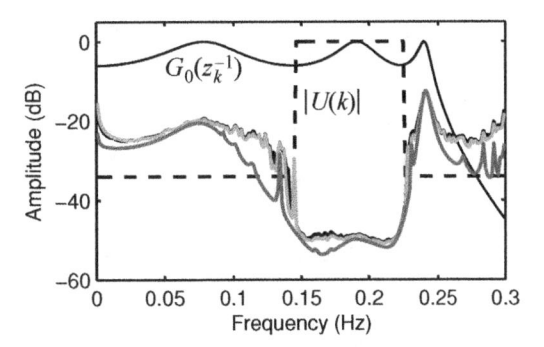

Using the MDL rule, it is possible to go for the best model making an exhaustive scan over all possible models in a predefined set (e.g., $n_{\min} \leq n_a, n_b \leq n_{\max}$) and picking out the model with the smallest MDL cost function (11-11). The major disadvantage of this approach is that many models should be evaluated, and fine tuning is very time consuming. Consequently, a top-down approach will be presented in Section 11.5, where initially a model that is too complex is estimated and, next, the complexity is reduced by stripping off the superfluous parts, using the MDL rule again as a decision criterion. Special actions will be needed to avoid the numerical conditioning problems and to guarantee good convergence.

To solve the general model selection problem, it is also necessary to detect the presence of model errors so that unmodeled dynamics or nonlinear distortions can be detected. This will be discussed in the next section.

11.4 DETECTION OF UNDERMODELING

In the previous section we were mainly concerned to restrict the model complexity in order to avoid a noise sensitivity of the model that is too large. This might suggest that it is a good idea to select too simple models deliberately, hoping for a significant reduction of the noise sensitivity. This idea is checked by means of a simple simulation. Next we will analyze how can we detect, qualify, and quantify systematic errors. The detection step should indicate the presence of model errors. In the qualification step, it is checked whether the model error is either due to too low a model order, so that there remain unmodeled dynamics, or due to nonlinear distortions. Finally, an idea is given about the average level of the model errors.

11.4.1 Undermodeling: A Good Idea?

In some applications, the users are not interested in a complete model covering the full frequency band. They want only a good description in the frequency band of interest, which might be covered by a low-order model. Equation (11-10) suggests that a high-order model would suffer from a larger variance than the low-order model. A simulation is set up to analyze this problem. The same setup as in Section 11.3.2, simulation 11.3.2.2, is used. A hundred runs were made, and the results are shown in Figure 11-7. From the left side it is seen that in the frequency band of interest (where the input amplitude $|U(k)|$ is high) the standard deviations are about the same. However, on the right side it is seen that for the simple model, there remain significant model errors, even in the frequency band of interest. This shows that it is seemingly better to go for a sufficiently complex model that pushes the systematic model errors down to the noise level. This result conflicts with the previous asymptotic result where the model variability increased proportionally with n_θ. Note here also that the model variability is increased, but only outside the frequency band of interest, where the simple model is actually not applicable.

This brings the model complexity question to a mature level: how should the model complexity be chosen to balance the model variability versus its systematic errors? We advise the reader to increase the model complexity in the identification step until model errors can no longer be detected. If this model is too complex for a user's final goal, a model reduction step can be applied next. This offers the advantage that users know exactly what model errors they introduced themselves.

11.4.2 Detecting Model Errors

In "classical" identification (Ljung, 1999), two tests are very popular to detect model errors. Both are based on the residuals that are given as the difference between the model-based predicted output and the actual measured output. In the most general case, a plant model and a noise model are estimated. If the "true models" are reached, it is shown that the residuals should be white. If one of both is wrong, correlation will be detected. In practice, the plant model is more important than the noise model. Many times the user is not really interested in modeling the noise characteristics. Some methods, such as the output-error method (Ljung, 1999), do not even estimate the noise model at the price of a poorer efficiency, but they are still consistent in open loop identification without input noise. In that case, the residuals will basically mimic the colored process noise, and, hence, they should not be white. Consequently, a whiteness test loses its applicability for these frequently occurring situations.

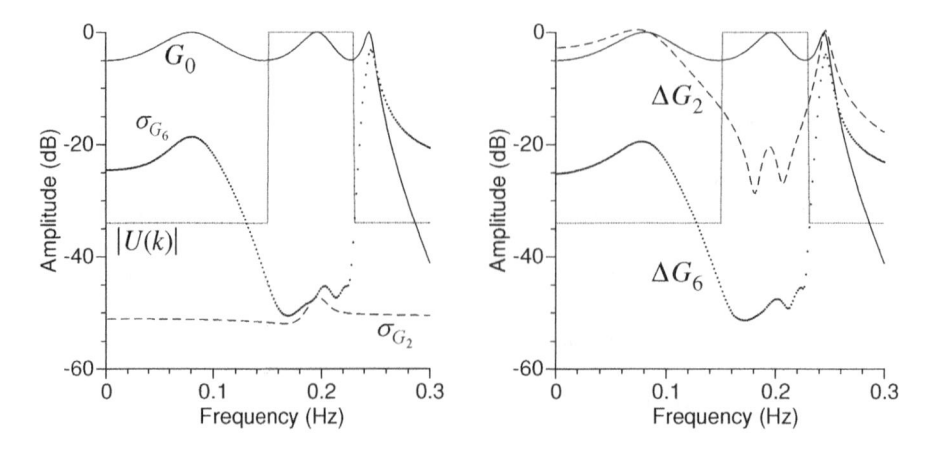

Figure 11-7. Impact of undermodeling on $\sigma_G(z_k^{-1}, \hat{\theta}(Z))$. Left: $\sigma_G(z_k^{-1}, \hat{\theta}(Z))$ for a second-order ($\sigma_{G_2}(z_k^{-1}, \hat{\theta}(Z))$) and a sixth-order ($\sigma_{G_6}(z_k^{-1}, \hat{\theta}(Z))$) system, and right: th mean model error of the transfer function estimate $\mathbb{E}\{|G(z_k^{-1}, \hat{\theta}(Z)) - G_0(z_k^{-1})|\}$.

Section 11.4 ■ Detection of Undermodeling

The other test checks for the presence of unmodeled dynamics by looking for cross-correlations between the residuals and the input. If all linear relations are modeled, the cross-correlation should not be significantly different from zero and this leads to a statistical validation test.

It is essential to note that in both tests no check of the absolute level of the residuals is made. This is intrinsically due to the fact that the variance of the disturbing noise is not estimated a priori from the raw data; it is estimated together with the plant and noise model. This is the major difference from the framework that is set up in this book. As explained before, periodic excitations make it possible to separate the signal and the noise before starting the identification process. So, an estimate of the noise model is obtained using the sample (co)variances $\hat{\sigma}_U^2(k)$, $\hat{\sigma}_Y^2(k)$, and $\hat{\sigma}_{YU}^2(k)$ as explained in Section 2.5.1. The knowledge of this noise model, which is obtained directly from the raw data independent of the identification process and the selected model, opens completely new possibilities. It will become possible to check for the amplitude of the residuals, which is a more direct measure of remaining model errors. Because the cost function is nothing other than the sum of the squared amplitudes of the normalized residuals, we will use its value as the primary check for the detection of model errors. The properties of the cost function are studied in detail in Section 19.6. It is shown that the expected value of the cost function (based on a nonparametric noise model) can be split in two parts, the first one accounting for the noise contributions and the second one being due to modeling errors (Theorem 19.12). Consider the cost function evaluated in the estimated parameters: $V_{ML}(\hat{\theta}_{ML}(Z), Z)$ (see Section 9.11); then the following result holds:

Theorem 11.2 (Properties Global Minimum ML Cost Function): The global minimum $V_{ML}(\hat{\theta}_{ML}(Z), Z)$ of the maximum likelihood cost function (9-83) has the following properties:

1. In the presence of model errors, deterministic inputs (Z_0 is deterministic), and circular complex normally distributed noise N_Z, $V_{ML}(\hat{\theta}_{ML}(Z), Z)$ is asymptotically ($F \to \infty$) normally distributed with mean and variance

$$\mathbb{E}\{V_{ML}(\hat{\theta}_{ML}(Z), Z)\} \approx V_{noise} + V_{model}$$
$$\text{var}(V_{ML}(\hat{\theta}_{ML}(Z), Z)) \approx V_{noise} + 2V_{model} \tag{11-12}$$

(assumptions of Section 9.6.4),

2. In the absence of model errors, deterministic or random inputs (Z_0 is deterministic or random), and circular complex normally distributed noise N_Z, $V_{ML}(\hat{\theta}_{ML}(Z), Z)$ is asymptotically ($F \to \infty$) normally distributed with mean and variance

$$\mathbb{E}\{V_{ML}(\hat{\theta}_{ML}(Z), Z)\} \approx V_{noise}$$
$$\text{var}(V_{ML}(\hat{\theta}_{ML}(Z), Z)) \approx V_{noise} \tag{11-13}$$

(assumptions of Sections 9.6.4 and 9.6.6),
with $V_{noise} = F - n_\theta/2$ and $V_{model} = \varepsilon^T(\tilde{\theta}_{ML}(Z_0), Z_0)\varepsilon(\tilde{\theta}_{ML}(Z_0), Z_0)$.

Proof. See Appendix 11.C. □

This result gives an extremely simple test to check for model errors. If the actual cost function is significantly larger than the expected value, model errors are present. Otherwise, it can be decided that no significant model errors are detected. This conclusion cannot be made within the "classical" identification schemes because those algorithms estimate the noise model and its variance. They cannot recognize the presence of white residuals that are too large. Here, these errors are detected because the noise (co)variances are known a priori.

The impact of replacing the exact variance by its sampling value is studied in Theorem 10.5, where it is shown that similar rules still apply. If no model errors nor nonlinear distortions are present, then (11-13) becomes

$$\mathbb{E}\{V_{\text{SML}}(\hat{\theta}_{\text{SML}}(Z), Z)\} \approx \frac{M-1}{M-2} V_{\text{noise}}$$

$$\text{var}(V_{\text{SML}}(\hat{\theta}_{\text{SML}}(Z), Z)) \approx \frac{(M-1)^3}{(M-2)^2(M-3)} V_{\text{noise}}$$

(11-14)

(see Theorem 10.5). If the sample noise (co)variances in the cost function (10-10) are replaced by the sum of the sample noise (co)variances and the sample (co)variances of the stochastic nonlinear distortions, then (11-14) remains valid in the presence of nonlinear distortions. The reader is referred to Section 12.4 for a detailed discussion.

11.4.3 Qualifying and Quantifying the Model Errors

In this section we will develop the theory explicitly using the output error framework instead of the errors-in-variables viewpoint. The major reason for the choice is that we perform the analysis on the FRF residuals. In practice, we first do the identification in the errors-in-variables framework and next make the validation on the measured transfer function $G(\Omega_k)$ obtained from the raw data by (2-17) and the estimator $G(\Omega_k, \hat{\theta}_{\text{ML}}(Z))$, assuming that the following assumptions are valid:

Assumption 11.3 (pdf FRF Measurement Errors): The noise $N_G(k)$ on the FRF measurement is independent (over k), circular complex normally distributed.

Assumption 11.4 (FRF Estimate): The estimate $\hat{\theta}(Z) \approx \hat{\theta}_{\text{ML}}(Z)$, with

$$\hat{\theta}(Z) = \arg\min_{\theta} \sum_{k=1}^{F} \frac{|G(\Omega_k) - G(\Omega_k, \theta)|^2}{\sigma_G^2(k)}$$

(11-15)

The residuals are the difference between the measured and the modeled transfer function, weighted by the standard deviation on the FRF measurement:

$$\varepsilon(\Omega_k, \hat{\theta}(Z)) = (G(\Omega_k) - G(\Omega_k, \hat{\theta}(Z)))/\sigma_G(k)$$

(11-16)

These residuals will be used to qualify the nature of the error, once model errors are detected (the cost function is too large). Because we also want to include nonlinear distortions in this analysis, we have to obey the assumptions and restrictions put forward in Chapter 3: normalized excitations $x_F \in \mathbb{E}_F$ (Definition 3.2); the class of systems from Definition 3.5; and the noise properties of a frequency domain experiment with $\sigma_U^2(k) = 0$ (see Section 9.6). The measured FRF can then be written as the sum of three parts:

$$G(\Omega_k) = G_{\text{BLA}}(\Omega_k) + G_S(\Omega_k) + N_G(k) \qquad (11\text{-}17)$$

with $G_{\text{BLA}}(\Omega_k)$ the best linear approximation to the overall system, $G_S(\Omega_k)$ the stochastic nonlinear contributions, and $N_G(k)$ the errors due to the output noise. The related dynamic system $G_{\text{BLA}}(\Omega_k)$ consists of two parts:

$$G_{\text{BLA}}(\Omega_k) = G_0(\Omega_k) + G_B(\Omega_k) \qquad (11\text{-}18)$$

with $G_0(\Omega_k)$ the underlying linear system and $G_B(\Omega_k)$ the bias or systematic errors due to the nonlinear distortions. $G_S(\Omega_k)$ is called a stochastic contribution; it behaves as uncorrelated (over the frequencies) noise, although the reader should be aware that it is not really a noise component. Its properties were explicitly stated in Theorem 3.9. Due to this noisy behavior, the presence of nonlinear distortions is often not recognized, although it is exactly this noisy behavior that will make it possible to detect their presence. The linear model will converge to the best linear approximation $G_{\text{BLA}}(\Omega_k)$ if the model complexity is high enough. So, depending on the nonlinear distortion and the nature of the excitation, but opposed to the classical validation techniques, the user gets a warning about their presence.

The model errors can be written as:

$$\begin{aligned} G(\Omega_k) - G(\Omega_k, \hat{\theta}(Z)) &= G_E(\Omega_k) - G_v(\Omega_k, \hat{\theta}(Z)) + q_k \\ G_E(\Omega_k) &= G_{\text{BLA}}(\Omega_k) - G(\Omega_k, \theta_*) \\ G_v(\Omega_k, \hat{\theta}(Z)) &= G(\Omega_k, \hat{\theta}(Z)) - G(\Omega_k, \theta_*) \end{aligned} \qquad (11\text{-}19)$$

$G_E(\Omega_k)$ is the bias error due to undermodeling (unmodeled dynamics and approximation of the nonlinear system), $G_v(\Omega_k, \hat{\theta}(Z))$ is the model uncertainty contribution (the estimated parameters $\hat{\theta}(Z)$ are different from θ_* due to the noise), and $q_k = N_G(k) + G_S(\Omega_k)$ are the stochastic errors (see Chapter 3).

The basic idea to qualify the model errors is based on the sample correlation analysis of the transfer function residuals. Consider $\hat{R}_{\varepsilon\varepsilon}(m)$:

$$\hat{R}_{\varepsilon\varepsilon}(m) = \frac{1}{F - |m|} \sum_{k=1}^{F - |m|} \frac{(G(\Omega_k) - G(\Omega_k, \hat{\theta}(Z)))\overline{(G(\Omega_{k+m}) - G(\Omega_{k+m}, \hat{\theta}(Z)))}}{\sigma_G(k)\sigma_G(k+m)} \qquad (11\text{-}20)$$

In the following theorem it is shown that the $\hat{R}_{\varepsilon\varepsilon}(m)$ converges to zero, except at the origin ($m = 0$), if the selected model includes the BLA model structure ($G_E(\Omega_k) = 0$).

Theorem 11.5 (Properties Sample Correlation in the Absence of Unmodeled Dynamics): Consider a system belonging to the set $\mathbb{S}$ (see Definition 3.5), excited with a random multisine $u_F \in \mathbb{E}_F$. If no unmodeled dynamics are present ($G_E(j\omega_k) = 0$), then under Assumption 11.3,

$$\hat{R}_{\varepsilon\varepsilon}(m)\big|_{m \neq 0} = O_p(F^{-1/2}) \qquad \hat{R}_{\varepsilon\varepsilon}(0) = \frac{1}{F}\sum_{k=1}^{F}\frac{\sigma_q^2(k)}{\sigma_G^2(k)} + O_p(F^{-1/2})$$

$$\text{var}(\hat{R}_{\varepsilon\varepsilon}(m)) \approx \frac{1}{F-|m|} + \frac{1}{(F-|m|)^2}\left[\sum_{k=1}^{F-|m|}\frac{|G_S(j\omega_k)|^2|G_S(j\omega_{k+m})|^2}{\sigma_G^2(k)\sigma_G^2(k+m)} + \right. \tag{11-21}$$

$$\left.\sum_{k=1}^{F-|m|}\frac{|G_S(j\omega_k)|^2}{\sigma_G^2(k)} + \sum_{k=1}^{F-|m|}\frac{|G_S(j\omega_{k+m})|^2}{\sigma_G^2(k+m)}\right]$$

where $\sigma_q^2(k) = \text{var}(q_k) = \text{var}(N_G(k) + G_S(\Omega_k)) = \sigma_G^2(k) + \sigma_{G_S}^2(k)$.

Proof. See, respectively, Appendices 11.E, 11.F, and 11.G. □

Note that this theorem gives an alternative interpretation of the cost function: $\mathbb{E}\{V_{ML}(\hat{\theta}_{ML}(Z), Z)\} \approx F\hat{R}_{\varepsilon\varepsilon}(0)$. It also allows us to check whether modeling errors are present or not: if $100 \times p\%$ of the $\hat{R}_{\varepsilon\varepsilon}(m)$ samples are below the $100 \times p\%$ confidence bound $\sqrt{-\ln(1-p)}\text{std}(\hat{R}_{\varepsilon\varepsilon}(m))$, then the hypothesis $G_E(j\omega_k) = 0$ (no unmodeled dynamics) is not rejected; otherwise it is.

In order to separate the stochastic nonlinear distortions from the unmodeled linear dynamics, it is assumed that the latter have a smooth behavior. For a fixed bandwidth of the experiments, the density of the frequency grid increases in proportion to F. The neighboring model errors will be almost equal, for F large enough. Moreover, we assume that the model errors are bounded and that the derivative with respect to the parameters behaves well. This leads to the following formal assumptions:

Assumption 11.6 (Smooth Errors): For F sufficiently large

  (i) Smooth unmodeled dynamics: $G_E(\Omega_k)\overline{G_E(\Omega_{k+1})} = |G_E(\Omega_k)|^2 + O(F^{-1})$.
  (ii) Smooth model variability: $\mathbb{E}\{G_v(\Omega_k)\overline{G_v(\Omega_{k+1})}\} = \mathbb{E}\{|G_v(\Omega_k)|^2\}$.
  (iii) Smooth disturbing noise spectrum: $\sigma_G(k) \approx \sigma_G(k+1)$.
  (iv) The normalized model errors $G_E(\Omega_k)/\sigma_G(k)$ are bounded.

Under these assumptions, it can be shown that $\hat{R}_{\varepsilon\varepsilon}(m)$ is significantly different from zero for $m \neq 0$ in the presence of unmodeled linear dynamics, and a hypothesis test is set up to check this. Under Assumption 11.6 it is also possible to bound the unmodeled dynamics. A similar attempt has already been made, starting from the value of the cost function (Schoukens and Pintelon, 1991), but this idea cannot be applied directly in this nonlinear context because the cost function is too large not only due to the model errors of the related dynamic system but also due to the stochastic nonlinear contributions. In order to separate both effects, $\hat{R}_{\varepsilon\varepsilon}(1)$ is considered here.

Theorem 11.7 (Properties Sample Correlation at Lag One): Consider a system belonging to the system set $\mathbb{S}$, excited with a random multisine $x_F \in \mathbb{E}_F$. Under Assumptions 11.3 and 11.6, $\hat{R}_{\varepsilon\varepsilon}(1)$ depends only on the unmodeled dynamics:

$$\hat{R}_{\varepsilon\varepsilon}(1) = \frac{1}{F-1}\sum_{k=1}^{F-1}|G_E(\Omega_k)|^2/\sigma_G^2(k) + O_p(F^{-1/2}) \tag{11-22}$$

Proof. See Appendix 11.H. □

Section 11.4 ■ Detection of Undermodeling

A stochastic bound on the bias error can be given starting from the sample value:

$$\text{Prob}\left(\frac{1}{F-1}\sum_{k=1}^{F-1}\frac{|G_E(\Omega_k)|^2}{\sigma_G^2(k)} \leq |\hat{R}_{\varepsilon\varepsilon}(1)| + \alpha\,\text{std}(\hat{R}_{\varepsilon\varepsilon}(1))\right) = P_\alpha \qquad (11\text{-}23)$$

(proof: see Appendix 11.G). In this expression the norm of $\hat{R}_{\varepsilon\varepsilon}(1)$ is considered because in general it is a complex value. Hence, chi-square tables should be used to determine the value P_α, because $\hat{R}_{\varepsilon\varepsilon}(1)$ is asymptotically circular complex normally distributed (see Section 19.5.2). For example, the 95% level is given by the bound $\sqrt{3}\,\text{std}(\hat{R}_{\varepsilon\varepsilon}(1))$ (see Appendix 11.G for an explicit expression of $\text{std}(\hat{R}_{\varepsilon\varepsilon}(1))$). The reader should be aware that a hypothesis test does not guarantee that no model errors are present. It only makes a statement on the probability that there are no significant (with respect to the noise level $\sigma_G(k)$) model errors left.

Conclusion. The following qualification rules can be used:

1. The cost function equals the noise value within the uncertainty bounds, for example, with 95% confidence

$$V_{\text{ML}}(\hat{\theta}_{\text{ML}}(Z), Z) \in [V_{\text{noise}} - 2V_{\text{noise}}^{1/2},\ V_{\text{noise}} + 2V_{\text{noise}}^{1/2}] \qquad (11\text{-}24)$$

with $V_{\text{noise}} = F - n_\theta/2$ (see (11-13)). If the sample (co)variances are used, then (11-24) should be adapted according to (11-14). No model errors are detected. This test should be confirmed by the fact that $\hat{R}_{\varepsilon\varepsilon}(m) \approx 0$. If this is not the case, another error source, not discussed in this section, should be present.

2. The cost function is significantly larger than the noise value, for example, with 95% confidence

$$V_{\text{ML}}(\hat{\theta}_{\text{ML}}(Z), Z) > V_{\text{noise}} + 2V_{\text{noise}}^{1/2} \qquad (11\text{-}25)$$

(see (11-13)). If the sample (co)variances are used, then (11-25) should be adapted according to (11-14). Model errors are present. These can then be qualified by checking the correlation between the transfer function residuals:

2a. $\hat{R}_{\varepsilon\varepsilon}(m) \neq 0$ for $m \neq 0$: there are still unmodeled dynamics.

2b. $\hat{R}_{\varepsilon\varepsilon}(m) \approx 0$ for $m \neq 0$: no unmodeled dynamics can be detected. This behavior can be explained, assuming the presence of nonlinear distortions.

In order to test whether $\hat{R}_{\varepsilon\varepsilon}(m) \approx 0$, a percentile test can be used. In such a test it is checked if, e.g., $100 \times p\%$ of the $\hat{R}_{\varepsilon\varepsilon}(m)$ samples have an amplitude below the predicted $100 \times p\%$ confidence bound $\sqrt{-\ln(1-p)}\,\text{std}(\hat{R}_{\varepsilon\varepsilon}(m))$, where $\text{std}(\hat{R}_{\varepsilon\varepsilon}(m))$ can be calculated from the variance expression given in (11-21). It turned out from our experience that this is a very sensitive test to indicate the presence of unmodeled dynamics.

3. If nonlinear distortions are detected, a new experiment can be set up to measure the variances of the nonlinear distortions and use them in a slightly modified estimation procedure (see Section 10.7 for a detailed description of the procedure). This will result in a model with a smaller variance.

In practice, the sample correlation is calculated using not the exact variances but the sample variances (see (10-10), Section 10.3.1), giving

$$\hat{R}_{\hat{\varepsilon}\hat{\varepsilon}}(m) = \frac{\alpha_1(m)}{F-|m|} \sum_{k=1}^{F-|m|} \frac{\overline{(G(\Omega_k) - G(\Omega_k, \hat{\theta}(Z)))(G(\Omega_{k+m}) - G(\Omega_{k+m}, \hat{\theta}(Z)))}}{\hat{\sigma}_G(k)\hat{\sigma}_G(k+m)} \quad (11\text{-}26)$$

where $\alpha_1(m)$

$$\alpha_1(0) = \frac{M-2}{M-1} \text{ and } \alpha_1(m)\big|_{m \neq 0} = \frac{M-5/3}{M-11/12} \quad (11\text{-}27)$$

corrects for the bias introduced by the sample variances (proof: see Appendix 11.I). In the absence of model errors $(G_E(\Omega_k) = 0)$ and nonlinear distortions $(G_S(j\omega_k) = 0)$, the standard deviation of (11-26) is related to that of (11-20) as

$$\text{std}(\hat{R}_{\hat{\varepsilon}\hat{\varepsilon}}(m)) = \alpha_2(m)\text{std}(\hat{R}_{\varepsilon\varepsilon}(m)) = \alpha_2(m)/\sqrt{F-|m|} \quad (11\text{-}28)$$

where $\alpha_2(m)$

$$\alpha_2(0) = \alpha_1(0) \frac{(M-1)^{3/2}}{(M-2)(M-3)^{1/2}} \text{ and } \alpha_2(m)\big|_{m \neq 0} = \alpha_1(m) \frac{M-1}{M-2} \quad (11\text{-}29)$$

accounts for the increase in uncertainty introduced by the sample variances (proof: see Appendix 11.J). If the sample noise variance $\hat{\sigma}_G^2(k)$ in (11-26) is replaced by the sum of the sample noise variance and the sample variance of the stochastic nonlinear distortions $\hat{\sigma}_G^2(k) + \hat{\sigma}_{G_S}^2(k)$, then (11-26)–(11-29) remain valid in the presence of nonlinear distortions $(G_E(\Omega_k) \neq 0)$. The reader is referred to Section 12.4 for a detailed discussion.

11.4.4 Illustration on a Mechanical System

The previous ideas are illustrated on a vibrating robot arm. Jan Swevers and Dirk Torfs (Department PMA of the Katholieke Universiteit Leuven, Belgium) have provided us with the experimental data (Torfs et al., 1998). As the input, the driving couple is measured, while the output is the acceleration at the tip of the robot arm. Ten periods are measured, each period consisting of 4096 points sampled at a frequency of 500 Hz. Only the odd harmonics (1, 3,..., 199) are excited. The identification results are shown in Figure 11-8 for models of order 4/4 and 6/6. Although the model 4/4 already gives quite a good fit, the correlation analysis clearly indicates that there are still significant (with respect to the noise level) unmodeled dynamics. Hence, it makes sense to increase the model order. The cost function is 4964.8 while a value of 95.5 is expected. A closer inspection of Figure 9-8(a) shows that the errors are, indeed, larger than the measurement uncertainty. Increasing the model order to 6/6 reduces the cost function to 220.5 (compared with an expected value of 93.5), still pointing to significant model errors. However, the correlation analysis cannot detect any more unmodeled dynamics. 58% of the correlation results are above the 50% percentile (98% for the 4/4 model) and 4% above the 95% percentile (92% for the 4/4 model). From this, we conclude that it makes no sense to increase the model order further, and most probably there are nonlinear distortions present. This was confirmed by more detailed tests.

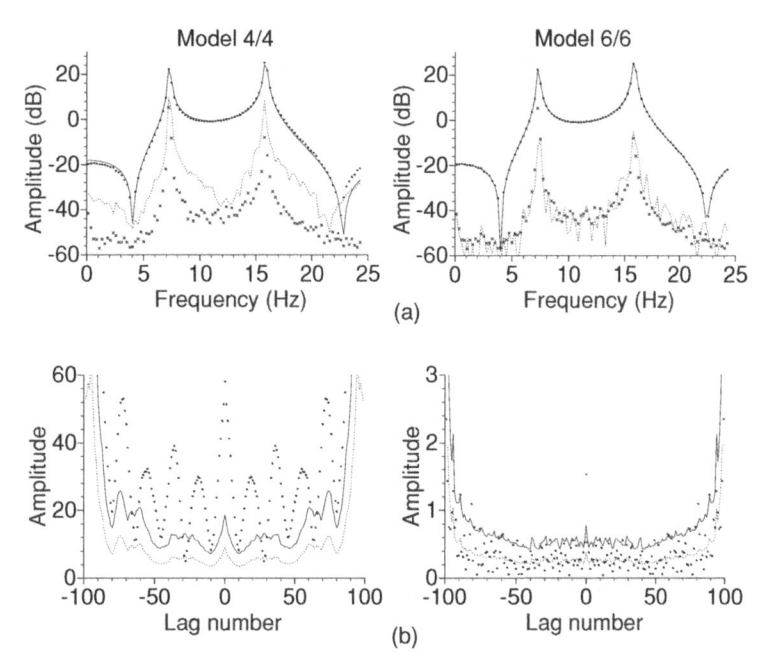

Figure 11-8. Illustration of model error detection and qualification on a vibrating robot arm. (a) The identified transfer function, dots: measurements, — model, ⋯ model errors, × measurement uncertainty σ_G; (b) $\hat{R}_{\varepsilon\varepsilon}(m)$, dots: measure-ment, ⋯ 50% bound, — 95% bound.

Remarks

(i) In practice, it is advisable not to use the correlation results at all the lags. The uncertainty on it increases very fast for the extreme lag numbers due to the small number of points that add to the sum. It is better to restrict the analysis to lag numbers that are smaller than half the number of frequencies.

(ii) The model validation is extremely simplified due to the presence of high-quality FRF measurements, so that even small model errors on the transfer function become visible.

11.5 MODEL SELECTION

In this section a "new" model selection procedure is proposed. In the classical approach (Stoica et al., 1986), a first guess of the model order is made directly from the raw data. A typical example is to plot the nonparametric frequency response function to get an initial idea about the required model complexity. A first trial is made and then the complexity is adapted to the results of the validation test. Usually, the model order is increased step-by-step until the point where acceptable validation results are obtained. The major reason for this cautious approach is the sensitivity of most algorithms to overmodeling. It usually results in very poor conditioning of the normal equations and convergence problems of the iterative schemes, so that the complexity can only be increased gradually. However, orthogonal parameterizations (see Section 9.16) guarantee good numerical conditioning, even in the case of extreme overmodeling, so that it becomes possible to reverse the previous sketched procedure (Rolain et al., 1997). This is the approach that we will explain here. It consists basically of three steps:

1. Make an initial guess of the maximum order based on a rank decision of a raw data matrix. This choice should be conservative (biased toward a too high model order) so that the best order is below this selection.
2. The parameters of this high-order model are estimated.
3. A model reduction is performed by eliminating poles, zeros, or pole-zero pairs that do not significantly contribute to the model. The validity of each reduction is checked. If no further reduction is possible, a new estimate with the reduced order is made and the model reduction procedure is restarted.

A more detailed description of each step is given in the following. The whole procedure can be automated, and from our experience it turns out that it results in reasonable and sometimes even better models than those selected by human operators.

11.5.1 Model Structure Selection Based on Preliminary Data Processing: Initial Guess

In order to start the model selection procedure automatically, an initial guess for the order is needed that is generated directly from the raw data. To do so, the user should specify an initial order $((n_a)_{init}, (n_b)_{init})$ that is definitely too high, so that the best model order is guaranteed to be included. Next, the number of possible pole/zero cancellations is estimated. An identification method that is linear in the parameters is used. So, an improved order selection boils down to a rank detection problem on the raw data matrices. Consider the following equation error formulation (Section 9.8.2):

$$e(\Omega_k, \theta, Z(k)) = A(\Omega_k, \theta)Y(k) - B(\Omega_k, \theta)U(k) \qquad (11\text{-}30)$$

where A and B are polynomials of order $(n_a)_{init}$ and $(n_b)_{init}$, respectively. The Jacobian matrix $J(Z) = \partial e(\theta, Z)/\partial \theta$ is parameter independent and its rank is, at most, $(n_a)_{init} + (n_b)_{init} + 1$ in the noiseless case and no model errors present. In case of model errors, the rank is at most $(n_a)_{init} + (n_b)_{init} + 1$. If there are common pole-zero pairs in the system, for the given model orders, degenerations will appear. Their number equals the dimension of the null space of $J(Z)$ minus 1 (to account for the structural degeneration of a transfer function model). The initial estimate of the model order is then given by

$$\hat{n}_a = (n_a)_{init} + 1 - \dim(\text{null}(J(Z))), \quad \hat{n}_b = (n_b)_{init} + 1 - \dim(\text{null}(J(Z))) \qquad (11\text{-}31)$$

In practice, only noisy data are available and this simple principle fails to work because the noise and model errors increase the rank of $J(Z)$. To reduce the noise sensitivity, the following extensions are made

- Add a frequency weighting to (11-30) to get as close as possible to the maximum likelihood weighting. This is exactly the problem that is solved in the starting values generating methods (see Section 9.12.4 on starting values) where $J(Z) \to WJ(Z)$ with the diagonal matrix (9-69).
- The column space of $W_{Re}J_{re}(Z)$ (see Section 9.10.3) is weighted with a square root of the column covariance matrix C_{WJ} (9-75) of $W_{Re}j_{re}(N_Z)$, with $j(N_Z) = J(Z) - J(Z_0)$ (see (9-68)). The whitened Jacobian is given by

$$W_{\text{Re}}J_{\text{re}}(Z)C_{WJ}^{-1/2} \qquad (11\text{-}32)$$

For deterministic weighting matrices W it is shown in Rolain et al. (1997) that the "noise" singular values σ_k^2 converge strongly to those of the noiseless matrix + 1:

$$\underset{F \to \infty}{\text{a.s.lim}}\ \sigma_k^2 = \sigma_{k0}^2 + 1 \qquad (11\text{-}33)$$

The variance of the noise singular values can be calculated (Rolain et al., 1997), but this requires unacceptably long computation time. Consequently, the dimension of the null space of $W_{\text{Re}}J_{\text{re}}(Z)C_{WJ}^{-1/2}$ is estimated by the number of singular values between zero and one, $\#\{\sigma_k | 0 < \sigma_k \leq 1\}$. This results in an overestimate of the rank of $W_{\text{Re}}J_{\text{re}}(Z)C_{WJ}^{-1/2}$. This is a desirable property because the initial estimate of the model order should be high enough so that the peeling process can be started.

Remarks

(i) In practice, the whitening (11-32) is not explicitly calculated as it is not guaranteed that C_{WJ} is of full rank (see Assumption 9.20(i) or (ii) and Appendix 9.K). Instead, the generalized singular value decomposition (GSVD) is used (Section 15.4.2; Paige, 1986; Bai and Demmel, 1993) to calculate the singular values of the whitened matrix directly from the matrix pair $(W_{\text{Re}}J_{\text{re}}(Z), C_{WJ}^{1/2})$ without calculating the inverse $C_{WJ}^{-1/2}$.

(ii) As we are dealing here with extremely high orders, the numerical conditioning can be cumbersome. In order to avoid these problems, an orthogonal parameterization is used.

11.5.2 "Postidentification" Model Structure Updating

The input to this second step consists of an initial guess of both the model parameters and the model order as obtained, for example, in the coarse step. The methods discussed below are in principle applicable to any estimator, as long as the cost function is absolutely interpretable (this means that it should be possible to predict and calculate its expected value in case there are no model errors). This facilitates validating the intermediate models that are obtained when we reduce the model complexity.

The model reduction procedure consists of the following steps:

- A full identification is performed starting from the initial parameters and it is checked whether the resulting model passes the validation test. After a positive validation, the reduction step can be started, otherwise the order should be increased until the validation test is successful.

- The poles, zeros, and pole-zero pairs are ranked with respect to their impact on the transfer function in the frequency band of interest. Possible candidates for elimination are poles or zeros that are far away from the modeled frequency band or almost coinciding pole-zero pairs. Next, these roots are eliminated one after another without changing the remaining poles and zeros. Each time it is checked whether the remaining model is still acceptable, using a simplified validation test. For example, the MDL test is a good method to compensate for the reduced order.

- Once it is no longer possible to continue the peeling process without violating the validation test, a new estimate is calculated, starting from the last accepted model. This optimizes the positions of the remaining poles and zeros, again reducing the cost function. Before starting a new peeling step, a full validation is performed (for example, a cost function test and a whiteness test of the residuals).

- Repeating these steps a few times results eventually in a "simple" model that still passes the validation tests. Often, this model is too complex for practical use. If the user can specify an acceptable level of model errors, a further reduction can be made until the user-imposed restrictions are violated.

Remarks

(i) This procedure does not guarantee that the optimal model is found. However, from our experience, it turned out that the in most cases the resulting model is very reasonable.

(ii) Because the procedure is controlled by a series of mechanical rules, it is very suited for a fully automated model selection. The only required user interaction is the definition of the maximum, acceptable level of model errors.

(iii) In practice, it is never guaranteed that the global minimum of the nonlinear cost function is reached, especially when dealing with more complex systems. Mostly a "good" local minimum is reached. We observed that with the top-down approach we sometimes ended in a better local minimum than the one obtained by a bottom-up approach on the same model order.

11.6 GUIDELINES FOR THE USER

In Table 11-4, we summarize the actions to be followed during the model selection process. This will help the less experienced reader to select a good model. We strongly advise comparing the estimated model with the nonparametric FRF at the end of the process to look for undetected anomalies (e.g., large errors in some frequency bands), strange behavior of the residuals (e.g., strong correlation in a subband), or undesired behavior of the model in frequency bands that were not well excited.

TABLE 11-4 Recommendations for the Model Selection Process

	White Residuals	Colored Residuals
The cost function weighted with the noise (co)variances is too large	- Best linear approximation - Nonlinear distortions present - It makes no sense to increase the model order	- There are still unmodeled dynamics (model errors). Increase the model order to reduce them
The cost function weighted with the noise (co)variances is not significantly different from the expected value	- This is the ideal situation - Best linear model - No unmodeled dynamics detectable	- Good linear approximation - Check the noise analysis
The cost function weighted with the noise (co)variances is too small	- Good linear approximation - Check the noise analysis or reduce the model order	- Good linear approximation - Check the noise analysis

11.7 EXERCISES

11.1. Consider a polynomial model:

$$y_0(k) = \sum_{p=1}^{5} a_p u^p(k) \qquad (11\text{-}34)$$

that is identified from a set of measurements $y(k) = y_0(k) + n_y(k)$, with $u(k) = [-N:N]/N$ and $n_y(k)$ zero mean iid distributed noise with variance σ_y^2. Set up the least squares estimator for this problem, and calculate the covariance matrix C_a, and the uncertainty of the model output $\sigma_y(u)$.

11.2. Check the previous results on simulations.

11.3. Set up a weighted least squares estimator to identify a parametric transfer function mode $G(\Omega, \theta)$ from measured values $G(\Omega_k) = G_0(\Omega_k) + n_G(k)$ with $n_G(k)$ zero mean iid distributed noise with variance σ_G^2.

11.4. Apply the estimator of Exercise 11.3 to simulation data. Select a second-order system to generate the data. Repeat the simulation 100 times, and compare the mean value with the exact system $G_0(\Omega_k)$.

11.5. Use the results of Exercise 11.4 to calculate the covariance matrix of the transfer function parameter C_θ, and predict from these results the uncertainty on the transfer function $\text{var}(G(\Omega, \theta))$. Compare the predicted variance with the simulation results.

11.6. Calculate $\text{var}(G(\Omega, \theta))$ after putting the nondiagonal terms in the covariance matrix to zero, and discuss your results. Start from the results of Exercise 11.5.

11.7. Make a residual analysis on the results of Exercise 11.4 (use only one simulation), and discuss the results as a function of the selected model order.

11.8. Select a fourth-order system with two well-separated resonances, and identify this system (use, for example, the estimator of Exercise 11.3). Apply the AIC and MDL model selection rules. Do this first for a simulation using a broadband excitation that covers the complete passband of the system, and then repeat the exercise with an excitation that concentrates most of its power on one of both resonances. Analyze the results.

11.9. Set up a simulation, using a Wiener-Hammerstein system as plant. Use an FIR structure for the linear dynamic parts and a third-order polynomial for the static nonlinearities. Scale the excitation signals so that the energy of the second-degree nonlinearity is 10% of that of the linear part, and the third-degree nonlinearity contributes about 1% to the output energy. For this structure determine the underlying linear system and the best linear approximation (see also Chapter 3).

11.10. Identify the best linear approximation of Exercise 11.9, and make a full model validation using the tools developed in this chapter.

11.11. Study the impact of the excitation signal on the quality of the identified model of Exercise 11.10 (cost function, residue analysis, model uncertainty). Analyze 10 realizations in each run.

11.8 APPENDIXES

Appendix 11.A Proof of Equation (11-3)

Eq. (19-31) remains valid for complex-valued measurements Z and real-valued estimates $\hat{\theta} = \hat{\theta}_{\text{ML}}(Z)$ if ε, Z, and N_Z are replaced by, respectively, ε_{re}, Z_{re}, and $(N_Z)_{\text{re}}$; where x_{re} puts the real and imaginary parts of x on top of each other (see (15-48)). Using Lemma 15.4, (19-31) can be rewritten as

$$\Delta_\theta(Z) = -\left[\text{Re}\left(\left(\frac{\partial \varepsilon(\theta, Z_0)}{\partial \tilde{\theta}}\right)^H \left(\frac{\partial \varepsilon(\theta, Z_0)}{\partial \tilde{\theta}}\right)\right)\right]^{-1} \text{Re}\left(\left(\frac{\partial \varepsilon(\theta, Z_0)}{\partial \tilde{\theta}}\right)^H \left(\frac{\partial \varepsilon(\tilde{\theta}, Z)}{\partial Z}\right) N_Z\right) \qquad (11\text{-}35)$$

with $\tilde{\theta} = \tilde{\theta}(Z_0)$ and where

$$\text{Cov}(\hat{\underline{\theta}}) \approx \left[2\,\text{Re}\left(\left(\frac{\partial \varepsilon(\theta, Z_0)}{\partial \tilde{\theta}}\right)^H \left(\frac{\partial \varepsilon(\theta, Z_0)}{\partial \tilde{\theta}}\right)\right)\right]^{-1} \qquad (11\text{-}36)$$

(see (19-22)). Hence, the covariance between the complex-valued noise N_Z on the measurements and the real-valued noise $\Delta_\theta(Z)$ on the model parameters is given by

$$\begin{aligned}
\mathbb{E}\{N_Z \Delta_\theta^T(Z)\} &= -\mathbb{E}\left\{N_Z\, 2\text{Re}\left(N_Z^H \left(\frac{\partial \varepsilon(\tilde{\theta}, Z)}{\partial Z}\right)^H \left(\frac{\partial \varepsilon(\theta, Z_0)}{\partial \tilde{\theta}}\right)\right)\right\} \text{Cov}(\hat{\underline{\theta}}) \\
&= -C_{N_Z}\left(\frac{\partial \varepsilon(\tilde{\theta}, Z)}{\partial Z}\right)^H \left(\frac{\partial \varepsilon(\theta, Z_0)}{\partial \tilde{\theta}}\right) \text{Cov}(\hat{\underline{\theta}})
\end{aligned} \qquad (11\text{-}37)$$

where the last equality uses $2\text{Re}(x) = x + \bar{x}$ and $\mathbb{E}\{N_Z N_Z^T\} = 0$ (N_Z is circular complex distributed, see (16-12)). Finally, $\Delta_\theta(Z) \approx \hat{\theta} - \tilde{\theta}$ (see Theorem 19.2), $Z \approx Z_0$, and $\hat{\theta} \approx \tilde{\theta}$, which concludes the proof. □

Appendix 11.B Proof of Equation (11-4)

Since $f(Z, \hat{\theta}) = G(\Omega_k) - G(\Omega_k, \hat{\theta})$, $\varepsilon_{[k]}(\hat{\theta}, Z) = (G(\Omega_k) - G(\Omega_k, \hat{\theta}))/\sigma_G(k)$, and $Z_{[k]} = G(\Omega_k)$, the partial derivatives in (11-3) equal

$$\begin{aligned}
\left.\frac{\partial f(Z, \hat{\theta})}{\partial Z_{[r]}}\right|_{r \neq k} &= 0 \quad \text{and} \quad \frac{\partial f(Z, \hat{\theta})}{\partial Z_{[k]}} = 1 \\
\frac{\partial f(Z, \hat{\theta})}{\partial \hat{\theta}} &= -\frac{\partial G(\Omega_k, \theta)}{\partial \hat{\theta}} \\
\frac{\partial \varepsilon(\hat{\theta}, Z)}{\partial Z} &= \text{diag}(\sigma_G^{-1}(1) \ldots \sigma_G^{-1}(k) \ldots \sigma_G^{-1}(F)) \\
\frac{\partial \varepsilon_{[k]}(\hat{\theta}, Z)}{\partial \hat{\theta}} &= -\frac{1}{\sigma_G(k)} \frac{\partial G(\Omega_k, \theta)}{\partial \hat{\theta}}
\end{aligned} \qquad (11\text{-}38)$$

where $\hat{\theta} = \hat{\theta}(Z)$. Combining (11-38) with $C_{N_Z} = \text{diag}(\sigma_G^2(1) \ldots \sigma_G^2(k) \ldots \sigma_G^2(F))$, it can easily be verified that

$$\left(\frac{\partial f(Z, \hat{\theta})}{\partial Z}\right) \text{Cov}(N_Z, \hat{\theta} - \tilde{\theta}) \left(\frac{\partial f(Z, \theta)}{\partial \hat{\theta}}\right)^H = -\left(\frac{\partial G(\Omega_k, \theta)}{\partial \hat{\theta}}\right) \text{Cov}(\hat{\underline{\theta}}(Z)) \left(\frac{\partial G(\Omega_k, \theta)}{\partial \hat{\theta}}\right)^H$$

which concludes the proof. □

Appendix 11.C Properties of the Global Minimum of the Maximum Likelihood Cost Function (Theorem 11.2)

Theorem 11.2 would follow directly from Theorem 19.12 if the frequency domain errors of the time and frequency domain experiment (see Section 9.6) were mixing of order four (infinity). For the frequency domain experiment the frequency domain errors are mixing of order four (Assumption 9.4) but not of order infinity (moments of order higher than $4 + \varepsilon$ do not necessarily exist, see Assumption 9.13). Hence, Lemma 19.11 is valid and only the asymptotic normality of $V_{\mathrm{ML}}(\hat{\theta}_{\mathrm{ML}}(Z), Z)$ in Theorem 19.12 remains to be proved. Because after a DFT the noise is not mixing of order four (infinity) (see Section 16.16), all the properties of $V_{\mathrm{ML}}(\hat{\theta}_{\mathrm{ML}}(Z), Z)$ in Theorem 19.12 remain to be proved for the time domain experiment. Fortunately, following the same lines as in the proof of Theorem 9.21 (see Appendix 9.E and Appendix 9.D), the resulting technical difficulties in the proofs can easily be solved using the results of Section 16.16. □

Appendix 11.D Calculation of Improved Uncertainty Bounds for the Estimated Poles and Zeros

This appendix gives the theoretical foundation of the method explained in Section 11.2.3. In order to keep the application field as general as possible, we emphasize that this result is independent of the specific identification scheme that is used as long as it meets some minimum requirements.

Consider an identification scheme that extracts the model parameters $\theta \in \mathbb{R}^{n_\theta}$ from the measurements $z \in \mathbb{R}^N$ (note that we no longer specify that it is time or frequency domain measurements),

$$\hat{\theta}(z) = \arg\min_{\theta \in \mathbb{P}_r} V_N(\theta, z) \tag{11-39}$$

$V_N(\theta, z) = V(\theta, z)/N$ is a well-designed cost function, such that

1. $V''(\tilde{\theta}(z_0), z_0) = C_\theta^{-1}$, with $\tilde{\theta}(z_0) = \arg\min_{\theta \in \mathbb{P}_r} \mathbb{E}\{V_N(\theta, z)\}$ and z_0 the noiseless data.
2. $\mathrm{a.s.}\lim_{N \to \infty} \hat{\theta}(z) = \tilde{\theta}(z_0)$, and $\Delta\theta = \hat{\theta}(z) - \tilde{\theta}(z_0) \in AsN(0, C_\theta)$.

Note that these assumptions are met for maximum likelihood and Markov estimators. Even linear least squares estimators can be used if, in a second step, a correct estimation of C_θ is made.

Note that

$$\Delta\theta^T C_\theta^{-1} \Delta\theta \in As\chi^2(n_\theta) \tag{11-40}$$

This result is also valid in the presence of model errors. At that moment (11-40) describes, under well-known conditions, the behavior of $\hat{\theta}(z)$ around the parameters that would be obtained in the noiseless case. The p-percentages uncertainty ellipsoids on $\hat{\theta}(z)$ are then given by

$$\mathbb{S}_\theta = \{\theta | \Delta\theta^T C_\theta^{-1} \Delta\theta \leq \chi_p^2(n_\theta)\} \qquad (11\text{-}41)$$

with $\chi_p^2(n_\theta)$ the p-percentile of a χ^2 distribution with n_θ degrees of freedom. Starting from (11-41), an uncertainty set for the poles/zeros $\rho = \rho(\tilde{\theta}) + \Delta\rho$ around $\rho(\tilde{\theta})$ is defined, where $\rho(\tilde{\theta})$ are the poles and zeros corresponding to the parameters $\tilde{\theta}$.

$$\mathbb{S}_\rho = \{\rho | ([\Delta\theta(\Delta\rho)]^T C_\theta^{-1} [\Delta\theta(\Delta\rho)] \leq \chi_p^2(n_\theta))\} \qquad (11\text{-}42)$$

where $\Delta\theta(\Delta\rho)$ is the parameter variation due to the variation of the poles/zeros. Because the transformation $\theta \to \rho$ is highly nonlinear, linear approximations fail and an alternative method is formulated. In the following, we discuss the individual steps of the method that was explained in Section 11.2.3 in more detail. The basic idea was to move a pole (or zero) and to minimize the impact of this movement on the quality of the fit. First, a reparameterization is described to introduce a pole or zero as a parameter. For generality, a complex pair is considered but the results can be reduced without any problem to the situation of a real pole or zero. Next it is shown how the impact of a movement is minimized. Finally, the criterion $[\Delta\theta(\Delta\rho)]^T C_\theta^{-1} [\Delta\theta(\Delta\rho)] \leq \chi_p^2(n_\theta)$ in (11-42) is used to accept or reject the movement. Remark that the poles and zeros do not fully determine θ because variation of the gain of the transfer function does not change the pole/zero positions. This additional free parameter will be set such that the impact of a pole or zero movement on the criterion is minimized.

11.D.1 Reparameterization. To focus the ideas, we consider a complex pole pair $\pi_1, \pi_2 = \bar{\pi}_1$ as parameter. The original transfer function $G(\Omega, \theta)$ is partitioned into two subsystems:

$$\underset{\sim}{G}(\Omega, \underset{\sim}{\theta}) = \frac{\pi_1 \pi_2}{(\Omega - \pi_1)(\Omega - \pi_2)} = \frac{1}{\underset{\sim}{a_2}\Omega^2 + \underset{\sim}{a_1}\Omega + 1} \text{ and } \tilde{G}(\Omega, \tilde{\theta}) = \frac{\sum_{i=0}^{n_b} b_i \Omega^i}{\sum_{i=0}^{n_a - 2} \tilde{a}_i \Omega^i} \qquad (11\text{-}43)$$

such that $G(\Omega, \theta) = \underset{\sim}{G}(\Omega, \underset{\sim}{\theta}) \tilde{G}(\Omega, \tilde{\theta})$.

Remarks

$\tilde{G}(\Omega, \tilde{\theta})$ includes the gain variations that were mentioned before.

Note that there exists a bilinear relationship between the old and the new parameterization:

$$\theta = T(\underset{\sim}{\theta})\tilde{\theta} \qquad (11\text{-}44)$$

with

A similar transformation can be set up when a complex zero pair or a real pole or zero is selected as parameter.

11.D.2 Minimizing the Impact of a Movement, Accepting or Rejecting a Movement. The impact of pole (zero) movement on the cost function is minimized by changing the remaining parameters $\tilde{\theta}$. This is done by minimizing

$$T(\underset{\sim}{\theta}) = \begin{bmatrix} I_{n_b+1} & 0 & 0 & \cdots & 0 & 0 & 0 \\ \hdashline 0 & 1 & 0 & \cdots & 0 & 0 & 0 \\ 0 & \underset{\sim}{a}_1 & 1 & \cdots & 0 & 0 & 0 \\ 0 & \underset{\sim}{a}_2 & \underset{\sim}{a}_1 & \cdots & 0 & 0 & 0 \\ 0 & 0 & \underset{\sim}{a}_2 & \cdots & 0 & 0 & 0 \\ \vdots & \vdots & \vdots & & \vdots & \vdots & \vdots \\ 0 & 0 & 0 & \cdots & \underset{\sim}{a}_2 & \underset{\sim}{a}_1 & 1 \\ 0 & 0 & 0 & \cdots & 0 & \underset{\sim}{a}_2 & \underset{\sim}{a}_1 \\ 0 & 0 & 0 & \cdots & 0 & 0 & \underset{\sim}{a}_2 \end{bmatrix} \begin{matrix} \updownarrow n_b+1 \\ \\ \\ \\ \updownarrow n_a+1 \\ \\ \\ \\ \end{matrix}$$

$$\underset{n_b+1}{\longleftrightarrow} \underset{n_a-1}{\longleftrightarrow}$$

$$[\Delta\theta(\Delta\rho)]^T C_\theta^{-1} [\Delta\theta(\Delta\rho)] = [T(\underset{\sim}{\theta})\tilde{\theta} - \tilde{\theta}(z_0)]^T C_\theta^{-1} [T(\underset{\sim}{\theta})\tilde{\theta} - \tilde{\theta}(z_0)] \qquad (11\text{-}45)$$

with respect to $\tilde{\theta}$. The solution is found by solving

$$T^T(\underset{\sim}{\theta}) C_\theta^{-1} T(\underset{\sim}{\theta}) \tilde{\theta} = T^T(\underset{\sim}{\theta}) C_\theta^{-1} \tilde{\theta}(z_0) \qquad (11\text{-}46)$$

Plugging this solution back into (11-42) allows us to verify whether $\rho \in \mathbb{S}_\rho$, using relation (11-42). The corresponding pole/zero positions can be calculated from this "improved" parameter set and be used to construct all uncertainty regions at once, instead of looking for the extreme positions for all pole-zero pairs. In practice, it might be necessary to repeat the whole process for a few poles/zeros in order to get a precise description of the uncertainty regions as there is no guarantee that all pole/zeros reach their extreme positions at the same time.

Remark. In practice, the following approximations are made to evaluate the solutions: $\tilde{\theta}(z_0) \to \hat{\theta}(z)$ and $C_\theta^{-1} \to V''(\hat{\theta}(z), z)$, as the exact values are unknown.

Appendix 11.E Sample Correlation at Lags Different from Zero (Proof of Theorem 11.5)

Consider a system belonging to the set $\mathbb{S}$ (see Definition 3.5), excited with a random multisine $x_F \in \mathbb{E}_F$. If no unmodeled dynamics are present ($G_E(\Omega_k) = 0$), then under the assumptions of Section 9.6.5 (frequency domain experiment with $\sigma_U^2(k) = 0$)

$$\hat{R}_{\varepsilon\varepsilon}(m) = O_p(F^{-1/2}) \qquad m \neq 0 \qquad (11\text{-}47)$$

Proof. In this proof we use the more compact notation $G_{vk} = G_v(\Omega_k, \hat{\theta}(Z))$. From (11-19) it follows that in the absence of model errors ($G_E(\Omega_k) = 0$)

$$\hat{R}_{\varepsilon\varepsilon}(m) = \frac{1}{F - |m|} \sum_{k=1}^{F-|m|} \frac{(q_k - G_{vk})(\bar{q}_{k+m} - \overline{G}_{v(k+m)})}{\sigma_G(k)\sigma_G(k+m)}$$

$$= \frac{1}{F-|m|}\sum_{k=1}^{F-|m|}\frac{q_k \bar{q}_{k+m}}{\sigma_G(k)\sigma_G(k+m)} + \frac{1}{F-|m|}\sum_{k=1}^{F-|m|}\frac{G_{vk}\bar{G}_{v(k+m)}}{\sigma_G(k)\sigma_G(k+m)}$$
$$-\frac{1}{F-|m|}\sum_{k=1}^{F-|m|}\frac{q_k \bar{G}_{v(k+m)}}{\sigma_G(k)\sigma_G(k+m)} - \frac{1}{F-m}\sum_{k=1}^{F-|m|}\frac{q_{k+m}\bar{G}_{vk}}{\sigma_G(k)\sigma_G(k+m)} \qquad (11\text{-}48)$$

Each of these terms converges for $m \neq 0$ to zero, at least, as an $O_p(F^{-1/2})$. Essential in the proof is that m does not tend to F: we require that $F - |m| = O(F)$ so that the results are also valid for a constant fraction $m = \alpha F$ with $\alpha < 1$.

(i) $s_1 = \frac{1}{F-|m|}\sum_{k=1}^{F-|m|}\frac{q_k \bar{q}_{k+m}}{\sigma_G(k)\sigma_G(k+m)}$ is an $O_{\text{m.s.}}((F-|m|)^{-1/2})$

Using $\text{var}(s_1) \leq \mathbb{E}\{s_1^2\}$, we show the mean square convergence of s_1

$$\mathbb{E}\{|s_1|^2\} = \frac{1}{(F-|m|)^2}\sum_{k=1}^{F-|m|}\sum_{l=1}^{F-|m|}\mathbb{E}\left\{\frac{q_k \bar{q}_{k+m}}{\sigma_G(k)\sigma_G(k+m)}\overline{\frac{q_l \bar{q}_{l+m}}{\sigma_G(l)\sigma_G(l+m)}}\right\} \qquad (11\text{-}49)$$

By careful examination of the right side and using the results of Theorem 3.9 and the noise assumptions of a frequency domain experiment (see Section 9.6.5), it can be shown that the double sum contains the following contributions:

- $k \neq l$: $O((F-|m|)^2)$ contributions of $O(F^{-2})$
- $k = l$: $O((F-|m|))$ contributions of $O(F^0)$

Hence, $\mathbb{E}\{|s_1|^2\} = O((F-|m|)^{-1})$ and $s_1 = O_{\text{m.s.}}((F-|m|)^{-1/2})$.

(ii) $s_2 = \frac{1}{F-|m|}\sum_{k=1}^{F-|m|}\frac{G_{vk}\bar{G}_{v(k+m)}}{\sigma_G(k)\sigma_G(k+m)}$ is an $O_p(F^{-1})$.

$G(\Omega_k, \hat{\theta}(Z))$ is a consistent estimate obtained under the standard conditions for output error estimates, which is a special case of the errors-in-variables formulation. For this class of estimators it is known that $\hat{\theta}(Z) - \theta_*$ is an $O_p(F^{-1/2})$ (Theorem 9.21, properties 2 and 5). Applying the mean value theorem to $G(\Omega_k, \hat{\theta}(Z))$ gives

$$G_{vk} = \frac{\partial G(\Omega_k, \widehat{\theta})}{\partial \widehat{\theta}}(\hat{\theta}(Z) - \theta_*) \qquad (11\text{-}50)$$

with $\widehat{\theta} = (1-t)\theta_* + t\hat{\theta}(Z)$, and $t \in [0, 1]$. Under Assumption 9.8, the derivatives of G are uniformly bounded, so that G_{vk} is $O_p(F^{-1/2})$. Hence, s_2 is an $O_p(F^{-1})$.

(iii) $s_3 = \frac{1}{F-|m|}\sum_{k=1}^{F-|m|}\frac{q_k \bar{G}_{v(k+m)}}{\sigma_G(k)\sigma_G(k+m)}$ or $\frac{1}{F-|m|}\sum_{k=1}^{F-|m|}\frac{q_{k+m}\bar{G}_{vk}}{\sigma_G(k)\sigma_G(k+m)}$ are an $O_p(F^{-1/2})$.

We prove the result for the first sum; that of the second sum follows exactly the same lines. Taking the absolute value of s_3 gives

Section 11.8 ■ Appendixes

$$|s_3| \leq \frac{1}{F-|m|} \sum_{k=1}^{F-|m|} \frac{|q_k||G_{v(k+m)}|}{\sigma_G(k)\sigma_G(k+m)} \leq \frac{O_p(F^{-1/2})}{F-|m|} \sum_{k=1}^{F-m} |q_k| \qquad (11\text{-}51)$$

Because q_k is an $O_{\text{m.s.}}(F^0)$, the conclusion follows directly. □

Appendix 11.F Sample Correlation at Lag Zero (Proof of Theorem 11.5)

$$\hat{R}_{\varepsilon\varepsilon}(0) = \frac{1}{F} \sum_{k=1}^{F} \frac{|q_k|^2}{\sigma_G^2(k)} + O_p(F^{-1/2})$$

Proof. Putting $m = 0$ in (11-48) gives

$$\hat{R}_{\varepsilon\varepsilon}(0) = \frac{1}{F}\sum_{k=1}^{F}\frac{|q_k|^2}{\sigma_G^2(k)} + \frac{1}{F}\sum_{k=1}^{F}\frac{|G_{vk}|^2}{\sigma_G^2(k)} - \frac{2}{F}\text{Re}\left(\sum_{k=1}^{F}\frac{q_k\overline{G}_{vk}}{\sigma_G^2(k)}\right) \qquad (11\text{-}52)$$

The proof of convergence of the second and third terms at the right-hand side in (11-52) is similar to that of the previous appendix. The first sum can be written as:

$$s_1 = \frac{1}{F}\sum_{k=1}^{F}\frac{|q_k|^2}{\sigma_G^2(k)} = \frac{1}{F}\sum_{k=1}^{F}\frac{\sigma_q^2(k)}{\sigma_G^2(k)} + \frac{1}{F}\sum_{k=1}^{F}\frac{|q_k|^2 - \sigma_q^2(k)}{\sigma_G^2(k)} \qquad (11\text{-}53)$$

From Theorem 3.9(iv) and the noise assumptions of a frequency domain experiment (see Section 9.6), it follows directly that the last sum in (11-53) is of order $O_{\text{m.s.}}(F^{-1/2})$, which concludes the proof. □

Appendix 11.G Variance of the Sample Correlation (Proof of Theorem 11.5)

In this section the variance of $\hat{R}_{\varepsilon\varepsilon}(m)$ is calculated under Assumptions 1 and 2 and putting $G_v = 0$. The variance of $\hat{R}_{\varepsilon\varepsilon}(m)$ will be calculated assuming that there are no unmodeled dynamics left ($G_E = 0$). Hence, this result can be used to check whether or not this hypothesis is valid. Because $\mathbb{E}\{\hat{R}_{\varepsilon\varepsilon}(m)\} = O((F-m)^{-1})$ for $m \neq 0$, we have $\text{var}(\hat{R}_{\varepsilon\varepsilon}(m)) \approx \mathbb{E}\{|\hat{R}_{\varepsilon\varepsilon}(m)|^2\}$ so that

$$\text{var}(\hat{R}_{\varepsilon\varepsilon}(m)) \approx \frac{1}{(F-|m|)^2}\mathbb{E}\left\{\sum_{k=1}^{F-|m|}\sum_{l=1}^{F-|m|}\frac{q_k\bar{q}_{k+m}\bar{q}_l q_{l+m}}{\sigma_G(k)\sigma_G(k+m)\sigma_G(l)\sigma_G(l+m)}\right\} \qquad (11\text{-}54)$$

for $m \neq 0$ and $F \to \infty$. Replacing $q_k = N_G(k) + G_S(j\omega_k)$ and using the properties of $G_S(j\omega_k)$, we find asymptotically ($F - |m| \to \infty$) the variance expression in (11-21). A detailed analysis shows that this expression is also valid for $m = 0$.

Remarks

(i) Expression (11-21) cannot be calculated directly because only q is available and not G_S. Replacing $G_S(j\omega_k)$ by q results in an overestimate of the uncertainty bounds. This can be compensated by substituting $|G_S(j\omega_k)|^2/\sigma_G^2(k)$ in the variance expressions by $|q_k|^2/\sigma_G^2(k) - 1$.

(ii) During the validation tests, graphical representations of the amplitude of $\hat{R}_{\varepsilon\varepsilon}(m)$ are used. Hence, the complex variance should be transferred into a bound on the amplitude. Because the variance dominates the bias error, it follows that $\hat{R}_{\varepsilon\varepsilon}(m)$ is asymptotically zero mean complex normally distributed (The real and imaginary parts of the individual contributions to the $\hat{R}_{\varepsilon\varepsilon}(m)$ are uncorrelated and have equal variance). So the amplitude is chi-squared distributed with two degrees of freedom. The $\alpha \times 100\%$ confidence bound is given by $\sqrt{-\ln(1-\alpha)}\text{std}(\hat{R}_{\varepsilon\varepsilon}(m))$ (proof: see Appendix 2.A). For example, for $\alpha = 0.95$ we get $\sqrt{3}\text{std}(\hat{R}_{\varepsilon\varepsilon}(m))$.

(iii) If G_v is not zero in the previous calculations, the variance expression is still valid but only a weaker statement about convergence in distribution can be made because the expected value $\mathbb{E}\{G_{vk}G_{vl}\}$ is not guaranteed to exist. The sample correlation $\hat{R}_{\varepsilon\varepsilon}(m)$ converges in distribution to a random variable with zero mean and variance $\text{var}(\hat{R}_{\varepsilon\varepsilon}(m))$.

(iv) In Section 10.7 it is shown how to identify the best linear approximation for systems that are disturbed by nonlinear distortions. In that case not only the measurement noise but also the stochastic nonlinearities are considered as disturbing noise. The proposed procedure accounts for both effects during the extraction of the (co)variances from the raw data. As a consequence, the presence of nonlinear distortions will not be detected during the validation tests because under these conditions, it just acts as an additional noise source. In this case $\text{var}(\hat{R}_{\varepsilon\varepsilon}(m))$ (11-21) reduces to $1/(F-|m|)$.

Appendix 11.H Study of the Sample Correlation at Lag One (Proof of Theorem 11.7)

$$\hat{R}_{\varepsilon\varepsilon}(1) = \frac{1}{F-1}\sum_{k=1}^{F-1}\frac{G_E(\Omega_k)\overline{G}_E(\Omega_{k+1})}{\sigma_G(k)\sigma_G(k+1)} + O_p(F^{-1/2})$$

Proof. In this proof we use the more compact notation $G_{Ek} = G_E(\Omega_k)$.

$$\hat{R}_{\varepsilon\varepsilon}(1) = \frac{1}{F-1}\sum_{k=1}^{F-1}\frac{[G_{Ek}+q_k-G_{vk}][\overline{G}_{E(k+1)}+\bar{q}_{k+1}-\overline{G}_{v(k+1)}]}{\sigma_G(k)\sigma_G(k+1)}$$

$$= \frac{1}{F-1}\sum_{k=1}^{F-1}\frac{[G_{Ek}-d_k][G_{E(k+1)}-\bar{d}_{k+1}]}{\sigma_G(k)\sigma_G(k+1)}$$

(11-55)

with $d_k = G_{vk} - q_k$. Compared with Appendix 11.E, a new term G_{Ek} appeared, raising new contributions of the type

$$\frac{1}{F-1}\sum_{k=1}^{F-1}\frac{G_{Ek}\bar{d}_{k+1}}{\sigma_G(k)\sigma_G(k+1)} = \frac{1}{F-1}\sum_{k=1}^{F-1}\frac{G_E\bar{q}_{k+1}}{\sigma_G(k)\sigma_G(k+1)} - \frac{1}{F-1}\sum_{k=1}^{F-1}\frac{G_{Ek}\bar{G}_{v(k+1)}}{\sigma_G(k)\sigma_G(k+1)}$$

The last sum at the right-hand side is an $O_p(F^{-1/2})$ because $\bar{G}_{v(k+1)}$ is an $O_p(F^{-1/2})$. Using $\mathrm{var}(x) \leq \mathbb{E}\{x^2\}$, the mean square convergence of the first sum is shown

$$\frac{1}{(F-1)^2}\sum_{k=1}^{F-1}\sum_{l=1}^{F-1}\frac{G_{Ek}\bar{G}_{El}\mathbb{E}\{\bar{q}_{k+1}q_{l+1}\}}{\sigma_G(k)\sigma_G(k+1)\sigma_G(l)\sigma_G(l+1)} \leq O(F^{-1}) \qquad (11\text{-}56)$$

The last inequality follows from the fact that $G_{Ek}/\sigma_G(k)$ is uniformly bounded by assumption and because $\mathbb{E}\{\bar{q}_{k+1}q_{l+1}\} = O(F^{-1})$ for $k \neq l$, and $\mathbb{E}\{|q_{k+1}|^2\} = \sigma_q^2(k)$ for $k = l$. □

Appendix 11.I Expected Value Sample Correlation

The scaling factor $\alpha_1(m)$ in the sample correlation $\hat{R}_{\hat{\varepsilon}\hat{\varepsilon}}(m)$ (11-26) accounts for the bias introduced by the sample variance in de denominator. The expected value of the sample correlation requires the knowledge of $\mathbb{E}\{\hat{\sigma}^{-2}\}$ ($m = 0$) and $(\mathbb{E}\{\hat{\sigma}^{-1}\})^2$ ($m \neq 0$). In Appendix 10.B it has been shown that

$$\mathbb{E}\{\sigma^2/\hat{\sigma}^2\} = \frac{M-1}{M-2} \qquad (11\text{-}57)$$

Extensive MATLAB® simulations for different values of M indicate that

$$(\mathbb{E}\{\sigma/\hat{\sigma}\})^2 = \frac{M-11/12}{M-5/3} \qquad (11\text{-}58)$$

Using (11-57) and (11-58) it can easily be shown that $\mathbb{E}\{\hat{R}_{\hat{\varepsilon}\hat{\varepsilon}}(m)\} = \mathbb{E}\{\hat{R}_{\varepsilon\varepsilon}(m)\}$.

Appendix 11.J Standard Deviation Sample Correlation

For $m \neq 0$ and in the absence of model errors $\mathbb{E}\{\hat{R}_{\hat{\varepsilon}\hat{\varepsilon}}(m)\} = 0$ and, hence, the variance of the sample correlation $\hat{R}_{\hat{\varepsilon}\hat{\varepsilon}}(m)$ (11-26) requires the knowledge of $(\mathbb{E}\{\sigma^2/\hat{\sigma}^2\})^2$, which is given by the square of (11-57). For $m = 0$, the variance of one term of the sum in (11-26) equals the variance of the cost function $V_{\mathrm{SML}}(\hat{\theta}_{\mathrm{SML}}(Z), Z)$ divided by F. In the absence of model errors the latter is given by (11-14). Taking the square root of both results concludes the proof.

12

Estimation with Unknown Noise Model – The Local Polynomial Approach

Abstract: The identification methods of Chapter 10 using standard nonparametric estimates of the input-output noise models suffer from the following shortcomings: due to the noise coloring consecutive signal periods are not independently distributed, the non-steady state part of the input-output signals should be removed, arbitrary excitations cannot be handled, and the frequency resolution is limited by the required minimal number of consecutive periods (at least four periods for the single-input, single output case). This chapter describes frequency domain estimators for parametric plant transfer function models of multivariable n_u input, n_y output systems excited by periodic or random signals. The aforementioned problems are solved by using nonparametric noise models estimated via the local polynomial approach of Chapter 7. The key idea is to calculate a generalized sample mean and sample covariance of the input-output DFT spectra via the local polynomial approximation of the frequency response matrix.

12.1 INTRODUCTION

The nonparametric noise models used in Chapter 10 are obtained by calculating the sample mean and the sample covariances over the DFT spectra of consecutive periods of the steady state response to a periodic input. The basic property needed to prove the consistency of the SGTLS, SBTLS, and SML estimators is that the sample mean and sample covariances are independently distributed (see Section 10.3.2 on page 387). Due to the noise coloring the consecutive signal periods are (weakly) correlated and, hence, the independence property of the sample mean and sample covariances is only approximately true. Moreover, transient data and random excitations cannot be handled. These problems are solved via the local polynomial approach of Chapter 7: it suppresses the plant and noise transient (leakage) errors in the nonparametric estimates of the frequency response matrix (FRM) and the noise covariances and, as such, decorrelates consecutive signal periods (suppression of the noise transients), eliminates the non-steady state part in the transient response to periodic inputs (suppression of the plant transients), and allows for non-periodic excitations (suppression of the plant transients). The key idea consists of calculating generalized sample means and sample covariances of the input-output DFT spectra via the nonparametric local polynomial estimates of the FRM and the noise covariances. Since at most two signal periods are needed, the frequency resolution of the presented approach is at least two times larger than the standard solution.

Using the generalized sample means and sample covariances we develop in this chapter a multivariable version of the sample maximum likelihood (SML) estimator (10-10). There are two major technical differences with the standard solution (10-10). The first difference is that the generalized sample means and sample covariances – by construction – are correlated over the frequency with a finite correlation length, while the standard sample means and sample covariances – if the noise transients are neglected – are uncorrelated over the frequency. This has implications on the calculation of the covariance of the estimated model parameters, and on the model validation and model selection tools. The second difference is that the generalized sample means and sample covariances are asymptotically ($N \to \infty$) independently distributed for normally distributed input-output errors, while the standard sample means and sample covariances – if the noise transients are neglected – are independently distributed.

Since the nonparametric noise covariance estimates are obtained by averaging squared residuals over the frequency rather than over independent experiments, the variability of the covariance estimates will be characterized by the degrees of freedom dof of the local polynomial approximations. For example, q in (7-13) for arbitrary excitations within an output error framework; $dof_{\text{robust}}^{\text{noise}}$ and dof_{robust} in (7-91) for the robust method using full random orthogonal multisines; and $dof_{\text{fast}}^{\text{noise}}$ and dof_{fast} in (7-103) for the fast method using uncorrelated random phase multisines. Finally, the number of independent repeated experiments M (in practice: the number of consecutive periods) in the standard solution (10-10) is related to the degrees of freedom as $dof = M - 1$.

Replacing everywhere the noise covariances by the total covariances (= sum noise covariance and covariance of the stochastic nonlinear distortions), it is shown in this chapter that all results for the identification of linear systems – except one – are also valid for the identification of the best linear approximation (BLA) of a nonlinear system. The difference with the linear case is that only the order of magnitude of the covariance matrix of the estimated model parameters of the BLA can be given. The reason for this is that the stochastic nonlinear distortions are uncorrelated with – but not independent of – the input (open loop case) or the reference signal (closed loop case).

Finally, notice that the quality (bias, variance) of the nonparametric noise models in Chapter 7 is solely determined by the frequency width $2n + 1$, the degrees of freedom q, and the degree R of the local polynomial approximation. Since these noise models are obtained in a preprocessing step prior to the parametric plant modeling, there is no link between the plant modeling errors and the estimated noise models. This is a major advantage w.r.t. the prediction error framework.

12.2 GENERALIZED SAMPLE MEAN AND SAMPLE COVARIANCE

In Chapter 10 the input-output signals are separated from the noise by calculating the sample means and sample (co)variances of the DFT spectra over consecutive signal periods (linear plants) and/or independent experiments (nonlinear plants). The sample (co)variances are then used as weighting in the sample maximum likelihood cost function. This idea is generalized to random excitations via nonparametric local polynomial estimates of the input-output spectra and the (noise) covariances.

In this section we consider the setup of Figure 7-4 on page 240 for measuring/identifying the frequency response matrix (FRM) of a linear multivariable system operating in open or closed loop. We assume that both the actuator and controller are linear. Since the plant behaves linearly, the total sample covariances of the robust and fast methods and the sample covariance of the local polynomial method for random excitations depend on the disturbing noise only. For nonlinear plants, or linear plants and nonlinear actuators or controllers, these sample covariances depend on the disturbing noise as well as the stochastic nonlinear distortions. These cases are handled in Section 12.4.

12.2.1 Arbitrary Excitations

12.2.1.1 Generalized Output Error Framework. Starting from one experiment with n_u uncorrelated random excitations the local polynomial method of Section 7.2.2 on page 228 estimates the frequency response function, and the noise sample covariance $\hat{C}_V(k)$. At first glance a natural solution would be to combine the known input $U(k)$ and noisy output $Y(k)$ DFT spectra with $\hat{C}_V(k)$ (7-13). However, the resulting parametric transfer function estimate is inconsistent because for normally distributed output noise $v(t)$ (see Figure 7-1 on page 227), $Y(k)$ and $\hat{C}_V(k)$ are uncorrelated – but not independently distributed – random variables. A better solution consists in combining the known input $U(k)$, with the local polynomial estimate of the output DFT spectrum $Y(k)$ (= generalized sample mean) and its corresponding covariance (= generalized sample covariance), because these will be shown to be asymptotically ($N \to \infty$) independent. The generalized sample mean $\hat{Y}(k)$ of $Y(k)$ is calculated as

$$\hat{Y}(k) = \hat{G}(\Omega_k)U(k) + \hat{T}(\Omega_k) \quad (12\text{-}1)$$

with $\hat{G}(\Omega_k) = \hat{\Theta}_{[:,1:n_u]}$ (7-16) and $\hat{T}(\Omega_k) = \hat{\Theta}_{[:,n_u(R+1)+1]}$ the local polynomial estimates of, respectively, the frequency response matrix and the transient term, and where $\hat{\Theta}$ is given in (7-10). We define the generalized sample covariance of $\hat{Y}(k)$ as

$$\hat{C}_{\hat{Y}}(k) = Q_{n[n+1,n+1]}\hat{C}_V(k) \text{ with } Q_n = K_n^H(K_n K_n^H)^{-1} K_n, \quad (12\text{-}2)$$

and where K_n is defined in (7-7) to (7-9). The asymptotic ($N \to \infty$) properties of the generalized sample mean (12-1) and the generalized sample covariance (12-2) are established in the following lemmas.

Lemma 12.1 (Sample Covariance Sample Mean): The generalized sample covariance $\hat{C}_{\hat{Y}}(k)$ (12-2) is the asymptotic ($N \to \infty$) sample covariance of the generalized sample mean $\hat{Y}(k)$ (12-1). The expected values of (12-1) and (12-2) equal

$$\begin{aligned} \mathbb{E}\{\hat{Y}(k)\} &= Y_0(k) + O((n/N)^{(R+1)}) \\ \mathbb{E}\{\hat{C}_{\hat{Y}}(k)\} &= C_{\hat{Y}}(k) + O(n/N) \end{aligned} \quad (12\text{-}3)$$

with $Y_0(k)$ the noiseless output DFT spectrum, and n, R, respectively, the frequency width and the order of the local polynomial approximation.

Proof. See Appendix 12.A. □

Lemma 12.2 (Independence Sample Mean and Sample Covariance): For normally distributed output noise $v(t)$ (see Figure 7-1 on page 227), the generalized sample mean $\hat{Y}(k)$

(12-1) and sample covariance $\hat{C}_{\hat{Y}}(k)$ (12-2) are asymptotically ($N \to \infty$) independently distributed, and

$$\mathbb{E}\{F_1(\hat{Y}(k))F_2(\hat{C}_{\hat{Y}}(k))\} = \mathbb{E}\{F_1(\hat{Y}(k))\}\mathbb{E}\{F_2(\hat{C}_{\hat{Y}}(k))\} + O(n/N) \quad (12\text{-}4)$$

where F_1 and F_2 are (non)linear functions. For non-Gaussian noise $v(t)$ satisfying Theorem 16.25, (12-4), where $O(n/N)$ is replaced by $O(N^{-1/2})$, remains valid.

Proof. See Appendix 12.B. □

These two properties are crucial for proving the consistency of the sample maximum likelihood estimator (see Section 12.3.2).

Removing the transient term in the sample mean (12-1) is mostly beneficial for the parametric transfer function estimate (see Section 12.3.1 for a detailed discussion). The generalized sample mean and sample covariance are then calculated as

$$\hat{Y}(k) = \hat{G}(\Omega_k)U(k) \quad (12\text{-}5)$$

$$\hat{C}_{\hat{Y}}(k) = \|q_n\|_2^2 \hat{C}_V(k) \text{ with } q_n = K_n^H(K_n K_n^H)^{-1}\begin{bmatrix} U(k) \\ 0 \end{bmatrix} \quad (12\text{-}6)$$

and where K_n is defined in (7-7) to (7-9). As shown in the following lemmas, they have the same asymptotic ($N \to \infty$) properties as the generalized sample mean and sample covariance without transient removal (Lemmas 12.1 and 12.2).

Lemma 12.3 (Sample Covariance Sample Mean – Transient Removed): The generalized sample covariance $\hat{C}_{\hat{Y}}(k)$ (12-6) is the asymptotic ($N \to \infty$) sample covariance of the generalized sample mean $\hat{Y}(k)$ (12-5). The expected values of (12-5) and (12-6) satisfy (12-3).

Proof. See Appendix 12.C. □

Lemma 12.4 (Independence Sample Mean Sample Covariance – Transient Removed): For normally distributed output noise $v(t)$ (see Figure 7-1 on page 227), the generalized sample mean $\hat{Y}(k)$ (12-5) and sample covariance $\hat{C}_{\hat{Y}}(k)$ (12-6) are asymptotically ($N \to \infty$) independently distributed, and satisfy (12-4). For non-Gaussian noise $v(t)$ satisfying Theorem 16.25, (12-4), where $O(n/N)$ is replaced by $O(N^{-1/2})$, remains valid.

Proof. See Appendix 12.D. □

Remarks

(i) The results for the generalized sample mean and sample covariance with transient removal (Lemmas 12.3 and 12.4) can easily be extended to concatenated data records. Instead of removing one transient term, as many transient terms as concatenated records are removed in the output DFT spectrum (see Section 7.2.9 on

Section 12.2 ■ Generalized Sample Mean and Sample Covariance

page 243 for the details). The generalized sample mean $\hat{Y}^c(k)$ is calculated as in (12-5), while the corresponding generalized sample covariance is obtained as

$$\hat{C}_{\hat{Y}^c}(k) = \|q_n^c\|_2^2 \hat{C}_V(k) \text{ with } q_n^c = K_n^H(K_n K_n^H)^{-1} \begin{bmatrix} U^c(k) \\ 0 \end{bmatrix} \quad (12\text{-}7)$$

with $U^c(k)$ the input DFT spectrum of the concatenated data records, and with K_n the matrix defined in (7-7) to (7-9) where $U(k)$ is replaced by $\tilde{U}^c(k)$ (7-58) (proof: see Appendix 12.E).

(ii) The numerically stable implementation of Q_n in (12-2), q_n in (12-6), and q_n^c in (12-7) uses the singular value decomposition $K_n^H = U_K \Sigma_K V_K^H$, giving, respectively,

$$Q_{n[n+1, n+1]} = U_{K[n+1, :]} U_{K[n+1, :]}^H$$
$$q_n = U_K \Sigma_K^{-1} V_{K[1:n_u, :]}^H U(k)$$
$$q_n^c = U_K \Sigma_K^{-1} V_{K[1:n_u, :]}^H U^c(k)$$

(iii) Since the local polynomial estimate of the frequency response matrix is correlated over the frequency (finite correlation length of $\pm 2n$; see Section 7.2.2 on page 228), this is also the case for the sample means $\hat{Y}(k)$ (12-1) and (12-5), and the sample covariances $\hat{C}_{\hat{Y}}(k)$ (12-2) and (12-6). Hence, some information is lost as the sample covariances (12-2) and (12-6) do not contain the correlation information over the frequency of the sample means (12-1) and (12-5).

(iv) Formulas (12-1) to (12-6) assume that n frequencies are available at the left- and right-hand sides of the DFT frequency k, which is not the case at the left and right borders of the frequency band. At those borders, $k + r$ with $r = -n, ..., n$ in (12-1) to (12-6) is replaced by the r-values defined in (7-29), and n/N in the bias error (12-3) and (12-4) is replaced by $(n + |p|)/N$, $p = \pm 1, \pm 2, ..., \pm n$.

(v) If the transient term $T(\Omega_k) = T_G(\Omega_k) + T_H(\Omega_k)$ in (7-1) is smaller than the noise term $V(k) = H(\Omega_k)E(k)$ ($\|T(\Omega_k)\|_2 < \|H(\Omega_k)E(k)\|_2$), then the transient (leakage) removal in the sample mean (12-5) will slightly increase the covariance of the estimate (12-6 is typically 1 dB larger than 12-2). However, if $\|T(\Omega_k)\|_2 > \|H(\Omega_k)E(k)\|_2$, then the transient removal results in a (significant) reduction of the output covariance.

12.2.1.2 Errors-in-Variables Framework. If both the input $u(t)$ and output $y(t)$ signals are noisy (see Figure 7-4 on page 240), then the reference signal $r(t)$ should be known. The known reference $r(t)$ is taken as input and the noisy outputs $y(t)$ and inputs $u(t)$ stacked on top of each other $z(t) = [y^T(t) \ u^T(t)]^T$ as noisy output. Proceeding in this way the original errors-in-variables problem with n_u noisy inputs and n_y noisy outputs is transformed into a generalized output error problem with n_u known inputs and $n_y + n_u$ noisy outputs (see Section 7.2.7 on page 240 for the details). Following the same lines of Section 12.2.1.1 we obtain then the generalized sample mean $\hat{Z}(k)$ of the input-output DFT spectra and the corresponding generalized sample covariance matrix $\hat{C}_{\hat{Z}}(k)$; both with or without transient removal.

12.2.2 Periodic Excitations

12.2.2.1 Robust Method. The robust procedure of Section 7.3.6 on page 254 delivers the $(n_y + n_u) \times n_u$ sample means $\hat{\mathbf{Z}}_R(kP)$ (7-87) of the input-output DFT spectra of $P \geq 2$ periods of n_u experiments with (full) random orthogonal multisines, repeated for $M \geq 2$ random phase realizations; and the corresponding n_u noise $\hat{C}_{\hat{Z}_R}^{\text{noise}[e]}(kP)$ (7-89) and n_u total $\hat{C}_{\hat{Z}_R}^{[e]}(kP)$ (7-88) sample covariances

$$\hat{Z}^{[e]}(k) = (\hat{\mathbf{Z}}_R(kP))_{[:,e]} \tag{12-8}$$

$$\hat{C}_{\hat{Z}}^{\text{noise}[e]}(k) = \hat{C}_{\hat{Z}_R}^{\text{noise}[e]}(kP) \text{ and } \hat{C}_{\hat{Z}}^{[e]}(k) = \hat{C}_{\hat{Z}_R}^{[e]}(kP) \tag{12-9}$$

($e = 1, 2, ..., n_u$). Note that the robust procedure of Section 7.3.3 on page 252 only provides the noise sample covariance. The sample means and sample covariances have the following properties.

Lemma 12.5 (Properties Sample Mean and Sample Covariance – Robust Method): The sample covariances $\hat{C}_{\hat{Z}}^{\text{noise}[e]}(k)$ and $\hat{C}_{\hat{Z}}^{[e]}(k)$ in (12-9) are the asymptotic ($N \to \infty$) noise and total sample covariances of the sample mean $\hat{Z}^{[e]}(k)$ (12-8), $e = 1, 2, ..., n_u$. The expected values of (12-8) and (12-9) equal

$$\mathbb{E}\{\hat{Z}^{[e]}(k)\} = Z_0^{[e]}(k)$$
$$\mathbb{E}\{\hat{C}_{\hat{Z}}^{\text{noise}[e]}(k)\} = C_{\hat{Z}}^{\text{noise}[e]}(k) + O((r_n/(PN))^2) \tag{12-10}$$
$$\mathbb{E}\{\hat{C}_{\hat{Z}}^{[e]}(k)\} = C_{\hat{Z}}(k) + O((r_{n_E}/N)^\alpha)$$

with $\alpha = 1, 2$ for, respectively, non-uniformly and uniformly distributed excited harmonics. For normally distributed input-output errors (see Figure 7-4 on page 240), the sample means (12-8) and the sample noise and total covariances (12-9) are asymptotically ($N \to \infty$) independently distributed, and satisfy (12-4). For non-Gaussian input-output errors satisfying Theorem 16.25, (12-4), where $O(n/N)$ is replaced by $O(N^{-1/2})$, remains valid.

Proof. See Appendix 12.F. □

12.2.2.2 Fast Method. Starting from $P \geq 2$ periods of one experiment with n_u uncorrelated random phase multisines, the fast method of Section 7.3.7 estimates the frequency response function, and the noise and total sample covariances. The noise information is obtained via an analysis over the periods, while the total covariance is obtained via an analysis over the excited frequencies. Therefore, two different sample means are defined: the first being the average over the periods, and the second the average over both the periods and neighboring excited frequencies.

The sample mean over the periods and the corresponding noise sample covariance are given by, respectively, $\hat{Z}(kP)$ (7-73) and $\hat{C}_{\hat{Z}}^{\text{noise}}(kP)$ (7-74)

$$\hat{Z}(k) = \hat{Z}(kP) \tag{12-11}$$

$$\hat{C}_{\hat{Z}}^{\text{noise}}(k) = \hat{C}_{\hat{Z}}^{\text{noise}}(kP) \tag{12-12}$$

They satisfy the properties of Lemma 12.5 since the first (noise analysis) step of the robust method is exactly the same as the first (noise analysis) step of the fast method.

The sample mean over both the periods and the neighboring excited frequencies is based on the local polynomial estimate of the frequency response function $\hat{G}_{rz}(\Omega_k)$ from reference to both input and output

$$\hat{Z}(k) = \hat{G}_{rz}(\Omega_k) R(k) \tag{12-13}$$

with $\hat{G}_{rz}(\Omega_k) = \hat{\Psi}_{[:,\,1:n_u]}$ (7-99) and $R(k)$ the DFT spectrum of the known reference signal. The corresponding total sample covariance is calculated as

$$\hat{C}_{\hat{Z}}(k) = T_{n[n+1,\,n+1]} \hat{C}_{\hat{Z}}(kP) \text{ with } T_n = L_{n_E}^H (L_{n_E} L_{n_E}^H)^{-1} L_{n_E}, \tag{12-14}$$

with L_{n_E} defined in (7-100) to (7-101). Note that T_n can be calculated in a numerically stable way via the singular value decomposition of L_{n_E} (see Remark ii of Section 12.2.1.1). The sample mean (12-13) and total sample covariance (12-14) satisfy the properties of Lemmas 12.1, 12.2, and 12.5 because the same local polynomial algorithms are used, except that the transient term is not estimated in $\hat{\Psi}$ (7-99) (it has been removed in the first step of the fast method, which is equal to the first step of the robust method). Summarized we have proven the following properties:

Lemma 12.6 (Properties Sample Mean and Sample Covariance – Fast Method): The covariances $\hat{C}_{\hat{Z}}^{\text{noise}}(k)$ (12-12) and $\hat{C}_{\hat{Z}}(k)$ (12-14) are the asymptotic ($N \to \infty$) noise and total sample covariances of the sample means $\hat{Z}(k)$ (12-11) and (12-13), respectively. The expected values of (12-11) and (12-12) satisfy (12-10), while the expected values of (12-13) and (12-14) are given by (12-3) where n is replaced by r_{n_E}. For normally distributed input-output errors (see Figure 7-4 on page 240), the sample means (12-11) and (12-13) are asymptotically ($N \to \infty$) independently distributed of the noise (12-12) and total (12-14) sample covariances, respectively, and they satisfy (12-4), where n is replaced by n_E for the total covariance. For non-Gaussian input-output errors satisfying Theorem 16.25, (12-4), where $O(n/N)$ is replaced by $O(N^{-1/2})$, remains valid.

12.2.3 Choice Frequency Width of the Local Polynomial Approach

At each frequency k, the frequency width parameter n and/or n_E of the local polynomial method can be chosen such that an optimal bias-variance trade-off is made for the frequency response matrix estimate $\hat{G}(\Omega_k)$ (see Section 7.2.6). As such, the degrees of freedom of the noise and the total sample covariance estimates of $\hat{G}(\Omega_k)$ will depend on the frequency k.

Another choice is made for the generalized sample means and the generalized noise and total sample covariances of the input-output DFT spectra defined in Sections 12.2.1 and 12.2.2. Since the generalized sample covariances are used as weighting in the sample maximum likelihood cost function (see Section 12.3.1), the frequency width parameter n and/or n_E in (7-13), (7-75), (7-91), and (7-103) is chosen to be frequency independent and such that a given minimum number of degrees of freedom is obtained for the sample covariance estimates. The required minimum number of degrees of freedom is dictated by the asymptotic ($N \to \infty$) properties of the sample maximum likelihood estimator (see Section 12.3.2). Keeping the frequency width parameter n and/or n_E as small as possible has the following consequences on the generalized sample means and sample covariances: it (i) minimizes the bias error (see, for example, (12-3)), (ii) maximizes the variance, and (iii) minimizes the correlation length over the frequency. Since the variance increase goes along with a correlation length decrease, the information content for the parametric estimate remains the same (see Appendix 12.H, second step). Therefore, the minimum width solution is optimal for the parametric transfer function estimate.

12.2.4 Overview of the Properties

The generalized sample means and sample covariances have similar properties for arbitrary (arb method) and periodic (robust and fast methods) excitations: they are asymptotically ($N \to \infty$) independently distributed and are asymptotically ($N \to \infty$) unbiased. Table 12-1 gives an overview of the rate at which the bias converges to zero. The reader is referred to Table 7-1 on page 249 for an overview of the experimental conditions and the approximations made.

TABLE 12-1 Asymptotic Bias of the Generalized Sample Means and Sample Covariances for Arbitrary (Arb) and Periodic (Robust, Fast) Excitations (R is the Degree of the Local Polynomial Approximation, $2n + 1$ the Frequency Width, and where r_n and r_{n_E} are Defined in, Respectively, (7-72) and (7-93))

Algorithm	Bias Total Sample Covariance	Bias Noise Sample Covariance	Bias Sample Mean
Arb	$O(n/N)$	—	$O((n/N)^{R+1})$
Robust	$O((r_{n_E}/N)^\alpha)$ [a]	$O(r_n/(PN)^2)$	0
Fast	$O(r_{n_E}/N)$	$O(r_n/(PN)^2)$	$O((r_{n_E}/N)^{R+1})$

a. $\alpha = 1, 2$ for, respectively, non-uniformly and uniformly distributed excited harmonics.

12.3 SAMPLE MAXIMUM LIKELIHOOD ESTIMATOR

12.3.1 Sample Maximum Likelihood Cost Function

We start from the generalized sample means $\hat{U}^{[l]}(k)$, $\hat{Y}^{[l]}(k)$ and noise or total sample covariances $\hat{C}_U^{[l]}(k)$, $\hat{C}_Y^{[l]}(k)$, and $\hat{C}_{YU}^{[l]}(k)$ ($l = 1, 2, ..., n_{\exp}$), where the transient (leakage) errors have been suppressed. They originate from a single experiment ($n_{\exp} = 1$) with n_u uncorrelated random excitations ("arb" method: see Section 12.2.1) or n_u uncorrelated random phase multisines ("fast" method: see Section 12.2.2.2); or from multiple experiments ($n_{\exp} = n_u$) with orthogonal or (full) random orthogonal multisines ("robust" method: see Section 12.2.2.1). Since the sample means $\hat{Z}^{[l]}(k)$ (12-8) are either uncorrelated (robust procedure of Section 7.3.6: see Appendix 7.K) or independently distributed (robust procedure of Section 7.3.3) over the experiments l, a natural multivariable extension of the sample maximum likelihood (SML) cost function (10-10) is given by

Section 12.3 ■ Sample Maximum Likelihood Estimator

$$V_{\text{SML}}(\theta, Z) = \sum_{l=1}^{n_{\text{exp}}} \sum_{k \in \mathbb{K}} e^H(\Omega_k, \theta, \hat{Z}^{[l]}(k))(\hat{C}_{\hat{e}}^{[l]}(\Omega_k, \theta))^{-1} e(\Omega_k, \theta, \hat{Z}^{[l]}(k)) \qquad (12\text{-}15)$$

with $\mathbb{K}$ the set of $F = O(N)$ excited DFT frequencies lying in the frequency band(s) of interest, Z the vector of the input-output DFT spectra at all frequencies, n_{exp} the number of experiments ($n_{\text{exp}} = 1$ for the "arb" and "fast" methods, and $n_{\text{exp}} = n_u$ for the "robust" method), $e(\Omega_k, \theta, \hat{Z}^{[l]}(k))$ the $n_y \times 1$ equation error of experiment l

$$e(\Omega_k, \theta, \hat{Z}^{[l]}(k)) = \hat{Y}^{[l]}(k) - G(\Omega_k, \theta)\hat{U}^{[l]}(k) = \begin{bmatrix} I_{n_y} & -G(\Omega_k, \theta) \end{bmatrix} \hat{Z}^{[l]}(k) \qquad (12\text{-}16)$$

and $\hat{C}_{\hat{e}}^{[l]}(\Omega_k, \theta)$ the corresponding $n_y \times n_y$ noise or total sample covariance matrix

$$\hat{C}_{\hat{e}}^{[l]}(\Omega_k, \theta) = \begin{bmatrix} I_{n_y} & -G(\Omega_k, \theta) \end{bmatrix} \hat{C}_{\hat{Z}}^{[l]}(k) \begin{bmatrix} I_{n_y} & -G(\Omega_k, \theta) \end{bmatrix}^H \qquad (12\text{-}17)$$

The rational transfer function model $G(\Omega, \theta)$ can be parameterized as a left or right matrix fraction, as a common denominator model, or as a function of the state space matrices (see Section 6.6 on page 193). Minimizing (12-15) w.r.t. θ gives the SML estimate $\hat{\theta}_{\text{SML}}(Z)$ (see Section 12.3.3 for the computational details).

Remarks

(i) Since the local polynomial estimates of the frequency response matrix (FRM), the transient (leakage) error, and the noise covariance matrix are correlated over the frequency with a finite correlation length (see Chapter 7), this is also the case for the generalized sample means and sample covariances of Section 12.2. Hence, replacing in (12-15) the sample covariances by the true noise covariances does not give the true maximum likelihood (ML) cost function because the correlation over the frequency has not been accounted for in the weighting of the residuals. This is a major technical difference with the standard SML solution (10-10).

(ii) The sample maximum likelihood (SML) solution (12-15) assumes that the transient (leakage) errors have been suppressed nonparametrically in the input-output DFT spectra and their covariances. This is the recommended default choice. If at all frequencies the transient (leakage) error term is smaller than the noise contribution (e.g., $\|T(\Omega_k)\|_2 < \|H(\Omega_k)E(k)\|_2$ in (7-1)), then the nonparametric transient removal will slightly increase the noise standard deviation (typically 1 dB) of the generalized sample means and, hence, also that of the estimated transfer function model $G(\Omega, \hat{\theta}_{\text{SML}}(Z))$. To avoid this potential increase in uncertainty one can omit the nonparametric transient suppression (e.g., use (12-1) and (12-2) instead of (12-5) and (12-6)) add a parametric transient term in (12-16)

$$e(\Omega_k, \theta, \hat{Z}^{[l]}(k)) = \hat{Y}^{[l]}(k) - G(\Omega_k, \theta)\hat{U}^{[l]}(k) - T_G(\Omega_k, \theta) \qquad (12\text{-}18)$$

where $T_G(\Omega_k, \theta)$ has the same poles (denominator) as $G(\Omega_k, \theta)$ (see Section 6.3 on page 184). This works well so long as the system transient T_G is dominant over the noise transient T_H, or if the system and noise transients have the same poles (e.g., ARX and ARMAX models, see Section 6.7.3.5 on page 200). In all the other cases one would need to add explicitly a noise transient term $T_H(\Omega_k, \theta)$ to (12-18), which complicates the minimization and the model selection.

(iii) A similar discussion as in (ii) can be made for concatenated data sets. The recommended default choice is to combine (12-15) with the generalized sample mean (12-5) and sample covariance (12-7) where the transient (leakage) errors due to the concatenation have been suppressed nonparametrically. An alternative approach consists in omitting the nonparametric transient suppression and to add delayed parametric transient terms to the equation error (12-18) (see (6-48)). This works well if $\|T_G\|_2 > \|T_H\|_2$, or if T_G and T_H have the same poles.

(iv) If the identification starts from frequency response matrix measurements $\hat{G}(\Omega_k)$, then we can still use (12-15) with $n_E = 1$, $\hat{Y}^{[l]}(k) = \text{vec}(\hat{G}(\Omega_k))$, $\hat{U}^{[l]}(k) = 1$, $\hat{C}_e^{[l]}(\Omega_k, \theta) = \hat{C}_{\text{vec}\hat{G}}(k)$, and where $G(\Omega_k, \theta)$ is replaced by $\text{vec}(G(\Omega_k, \theta))$.

(v) If the arbitrary input is exactly known, then the SML estimator (12-15) using the "arb" method is a valid alternative for MIMO Box-Jenkins modeling. The advantages of (12-15) are: (i) no parametric noise model must be estimated (simplified minimization cost function and simplified model selection), and (ii) no symbolic calculation is needed for calculating the matrix inverse of the noise model.

12.3.2 Asymptotic Properties

Compared with the standard sample maximum likelihood (SML) solution (10-10), there are three additional technical difficulties in the analysis of the asymptotic ($N \to \infty$) properties of the SML estimator (12-15): the generalized sample means and sample covariances are (i) correlated over the frequency with a finite correlation length, (ii) asymptotically ($N \to \infty$) independently distributed, and (iii) asymptotically ($N \to \infty$) unbiased. Nevertheless, it is shown that the asymptotic properties of $\hat{\theta}_{\text{SML}}(Z)$ remain the same for the class of excitations considered in Section 12.2. Since we handle the multivariable case, the required minimal values of the degrees of freedom *dof* of the sample covariances depend here on the number of outputs n_y. Finally, we compare the SML estimate (12-15) with the true maximum likelihood (ML) solution ((12-15) with the exact noise covariances and the original input-output DFT spectra $U(k)$ and $Y(k)$ of $u(t)$ and $y(t)$).

Theorem 12.7 (Asymptotic Properties $\hat{\theta}_{\text{SML}}(Z)$): Consider the transfer function model $G(\Omega, \theta)$ with any identifiable parameterization of Section 6.6. Under the assumptions of Sections 9.6 and 12.2 the minimizer $\hat{\theta}_{\text{SML}}(Z)$ of (12-15) has the asymptotic ($F = O(N) \to \infty$) properties of Theorem 9.21 on page 298 with $V_F(\theta, Z) = V_{\text{SML}}(\theta, Z)/F$:

1. For $dof \geq n_y + 2$, the stochastic and the deterministic convergence (properties 1, 5, and 6 of Theorem 9.21) are valid.

2. For $dof \geq n_y + 7$, the stochastic convergence rate (properties 2 and 6 of Theorem 9.21) are valid.

3. For $dof \geq n_y + 8$, the systematic and stochastic errors, the asymptotic normality, and the asymptotic bias (properties 3, 4, 6 and 7 of Theorem 9.21) are valid. In the absence of modeling errors the bias $b_\theta(Z)$ is an $O(F^{-1/2})$ for non-Gaussian noise satisfying Theorem 16.25.

Proof. See Appendix 12.G. □

Section 12.3 ■ Sample Maximum Likelihood Estimator

Theorem 12.8 (Relationship between $\hat{\theta}_{\text{SML}}(Z)$ and $\hat{\theta}_{\text{ML}}(Z)$): Under the conditions of Theorem 12.8, the sample maximum likelihood estimate $\hat{\theta}_{\text{SML}}(Z)$ (minimizer of (12-15)) and the maximum likelihood estimate $\hat{\theta}_{\text{ML}}(Z)$ (minimizer of (12-15) with the exact noise covariances and the original input-output DFT spectra $U(k)$ and $Y(k)$) are related to each other by:

1. For $dof \geq n_y + 1$, the expected value of the cost functions,

$$\frac{V_{\text{SML}}(\theta)}{F} = \frac{dof}{dof - n_y} \frac{V_{\text{ML}}(\theta)}{F} + O(F^{-\alpha}) \qquad (12\text{-}19)$$

with $\alpha = 1$ for normally distributed input-output errors, and $\alpha = 0.5$ for non-Gaussian noise satisfying Theorem 16.25.

2. For $dof \geq n_y + 2$, the asymptotic value of the cost functions,

$$V_{*\text{SML}}(\theta) = \frac{dof}{dof - n_y} V_{*\text{ML}}(\theta) \qquad (12\text{-}20)$$

the minimizer of the expected value of the cost functions,

$$\tilde{\theta}_{\text{SML}}(Z_0) = \tilde{\theta}_{\text{ML}}(Z_0) + O(F^{-\alpha}) \qquad (12\text{-}21)$$

with α defined in (12-19), and the minimizer of the asymptotic value of the cost functions,

$$\theta_{*\text{SML}} = \theta_{*\text{ML}} \qquad (12\text{-}22)$$

3. For $dof \geq n_y + 8$, the parameter uncertainty in the absence of modeling errors, $\tilde{\theta}_{\text{SML}}(Z_0) = \theta_0$,

$$\text{Cov}(\delta_{\theta\text{SML}}(Z)) \geq \lambda_1(dof) \, \text{Cov}(\delta_{\theta\text{ML}}(Z))$$
$$\lambda_1(dof) = \frac{dof \, (dof - n_y)}{(dof - n_y + 1)(dof - n_y - 1)} \qquad (12\text{-}23)$$

where $\hat{\theta}(Z) = \theta_0 + \delta_\theta(Z) + O_p(F^{-1})$ with $\mathbb{E}\{\delta_\theta(Z)\} = 0$ and $\delta_\theta(Z) = O_p(F^{-1/2})$, and where $\sqrt{F}\delta_\theta(Z)$ is asymptotically normally distributed. For non-Gaussian input-output noise satisfying Theorem 16.25 the bias $b_\theta(Z)$ is an $O_p(F^{-1/2})$. The robust method reaches the equality in (12-23).

with dof the degrees of freedom of the sample covariances (see Chapter 7); $V_X(\theta) = \mathbb{E}\{V_X(\theta)\}$, with X = SML or ML; and $V_{*X}(\theta) = \lim_{F \to \infty} V_X(\theta)/F$.

Proof. See Appendix 12.H. □

Remarks

(i) The randomness of the excitation (filtered white noise or random phase multisines) is explicitly needed for the "arb" (see Section 12.2.1) and "fast" (see Section 12.2.2.2) local polynomial estimates of the generalized sample means and sample covariances (see Section 7.2). It is also explicitly used to establish

(12-23). This is a difference with the standard SML method where the randomness of the input over the frequency is not required at all (see Assumption 10.1).

(ii) For single output systems ($n_y = 1$) a tighter lower bound on the degrees of freedom *dof* can be derived for properties 2 and 3 in Theorem 12.7, and property 3 in Theorem 12.8 (see Theorems 10.3 and 10.4). For example, in the single output case, $dof \geq 6$ instead of $dof \geq 9$ is enough for properties 3.

(iii) Via regularization of the matrix inverse $(\hat{C}_{\hat{e}}^{[l]}(\Omega_k, \theta))^{-1}$ in (12-15) (a positive number is added to the smallest eigenvalues of $\hat{C}_{\hat{e}}^{[l]}(\Omega_k, \theta)$ such that their inverse cannot exceed a certain user-defined threshold), the conditions on the degrees of freedom in Theorems 12.7 and 12.8 can be relaxed to $dof \geq n_y + 2$.

(iv) The asymptotic properties of the SML estimator remain valid for noisy input-output observations of systems operating in closed loop (see Figure 7-4 on page 240), provided the reference signal $r(t)$ is available (only the robust method of Section 7.3.3 does not require the knowledge of $r(t)$). An asymptotically ($F = O(N) \to \infty$) unbiased estimate of the generalized sample means and sample covariances is then obtained via the indirect method of Section 7.2.7.

(v) Even if it is known that the input-output errors are uncorrelated, it is necessary to estimate the input-output noise covariance; otherwise $\hat{C}_{\hat{e}}^{[l]}(\Omega_k, \theta)$ is no longer the (asymptotic) sample covariance of $e(\Omega_k, \theta, \hat{Z}^{[l]}(k))$ and the SML estimate (12-15) is inconsistent.

12.3.3 Computational Issues

Although analytic calculation of the derivative of a square root of $(\hat{C}_{\hat{e}}^{[l]}(\Omega_k, \theta))^{-1}$ w.r.t. θ is practically impossible, an iterative Gauss-Newton minimization scheme to compute the minimizer of (12-15) can still be constructed via a pseudo-Jacobian matrix $J_+(\theta, Z)$

$$J_+(\theta, Z) = \begin{bmatrix} J_+^{[1]}(\theta, Z) \\ J_+^{[2]}(\theta, Z) \\ \ldots \\ J_+^{[n_{\exp}]}(\theta, Z) \end{bmatrix} \text{ with } J_+^{[l]}(\theta, Z) = \begin{bmatrix} J_+^{[l,1]}(\theta, Z) \\ J_+^{[l,2]}(\theta, Z) \\ \ldots \\ J_+^{[l,F]}(\theta, Z) \end{bmatrix} \quad (12\text{-}24)$$

and

$$J_{+[:,r]}^{[l,k]}(\theta, Z) = (\hat{C}_{\hat{e}}^{[l]}(\Omega_k, \theta))^{-1/2} \Big(\frac{\partial e(\Omega_k, \theta, \hat{Z}^{[l]}(k))}{\partial \theta_{[r]}} - \ldots \\ \frac{1}{2} \frac{\partial \hat{C}_{\hat{e}}^{[l]}(\Omega_k, \theta)}{\partial \theta_{[r]}} (\hat{C}_{\hat{e}}^{[l]}(\Omega_k, \theta))^{-1} e(\Omega_k, \theta, \hat{Z}^{[l]}(k)) \Big) \quad (12\text{-}25)$$

with $X_{[:,r]}$ the rth column of X, and $C^{1/2}$ a square root of the positive definite matrix C (see Section 15.4.4 on page 550)

$$C = C^{1/2} C^{H/2} \Rightarrow C^{-1} = C^{-H/2} C^{-1/2} \quad (12\text{-}26)$$

(proof: see Appendix 12.I and Guillaume and Pintelon, 1996). The parameter update in the iterative algorithm is found by solving the overdetermined set of equations

$$J_{+\mathrm{re}}(\theta, Z)\Delta\theta = -\varepsilon_{\mathrm{re}}(\theta, Z) \qquad (12\text{-}27)$$

using the singular value decomposition (SVD) of $J_{+\mathrm{re}}(\theta, Z)$, where X_{re} puts the real and imaginary parts of X on top of each other, and where $\varepsilon(\theta, Z)$ has the same structure as $J_+(\theta, Z)$ in (12-24) with

$$\varepsilon^{[l,k]}(\theta, Z) = (\hat{C}_{\hat{e}}^{[l]}(\Omega_k, \theta))^{-1/2} e(\Omega_k, \theta, \hat{Z}^{[l]}(k)) \qquad (12\text{-}28)$$

for $l = 1, 2, \ldots, n_{\exp}$ and $k = 1, 2, \ldots, F$. To increase the numerical stability of the calculations, the angular frequencies are normalized by their median for continuous-time models ($\Omega = s, \sqrt{s}$), and each column of the pseudo-Jacobian matrix (12-24) is divided by its 2-norm (see Section 9.4 on page 289 for the details).

To avoid switching from one identifiable parameterization to another during the minimization procedure (Gevers and Wertz, 1984), all parameters in the transfer function model (e.g., common denominator, left matrix fraction, state space description, ...) are left free. Since the pseudo-Jacobian matrix of the overparameterized model is rank deficient, (12-27) is solved by calculating the pseudo-inverse of $J_{+\mathrm{re}}(\theta, Z)$ (the number of zero singular values depends on the overparameterization and is known beforehand). Next, the parameter constraint is imposed on the updated parameter vector $\theta + \Delta\theta$ (see, for example, Section 9.L.4 on page 365 for the single-input, single-output case using the 2-norm constraint). The justification of this procedure can be found in Pintelon et al. (1999), McKelvey et al. (2004), and Wills and Ninness (2008).

12.3.4 Calculation of the Asymptotic Covariance Matrix

Inequality (12-25) shows that an estimate of the asymptotic ($F \to O(N) \to \infty$) covariance matrix of $\hat{\theta}_{\mathrm{SML}}(Z)$ is obtained via $\mathrm{Cov}(\hat{\theta}_{\mathrm{ML}}(Z))$ which can be approximated as

$$\mathrm{Cov}(\hat{\theta}_{\mathrm{ML}}(Z)) \approx (2\mathrm{Re}(J_{\mathrm{ML}+}^H J_{\mathrm{ML}+}))^{-1} \qquad (12\text{-}29)$$

with $J_{\mathrm{ML}+}$ the pseudo-Jacobian (12-24) evaluated at $\theta = \hat{\theta}_{\mathrm{ML}}(Z)$, and where the generalized sample mean $\hat{Z}(k)$ and sample covariance $\hat{C}_{\hat{Z}}(k)$ are replaced by the original input-output DFT spectra $Z(k)$ and the corresponding true noise covariance $C_Z^{\mathrm{noise}}(k)$ (proof of (12-29): follow the same lines of Section 9.11.4 on page 318). Since the ML estimate $\hat{\theta}_{\mathrm{ML}}(Z)$ and the true noise covariance $C_Z^{\mathrm{noise}}(k)$ are unknown, we replace them by, respectively, the SML estimate $\hat{\theta}_{\mathrm{SML}}(Z)$ and the sample noise covariance $\hat{C}_Z^{\mathrm{noise}}(k)$ of the original input-output DFT spectra $Z(k)$ ((7-13) and (7-71) for, respectively, the "arb" and "fast" methods; and (7-77) for the "robust" method). Using the singular value decomposition $U_J \Sigma_J V_J^T$ of the pseudo-Jacobian $J_{+\mathrm{re}}(\hat{\theta}_{\mathrm{SML}}(Z), Z)$ (12-24), where $\hat{Z}(k)$ and $\hat{C}_{\hat{Z}}(k)$ are replaced by $Z(k)$ and $\hat{C}_Z^{\mathrm{noise}}(k)$, we finally get

$$\mathrm{Cov}(\hat{\theta}_{\mathrm{SML}}(Z)) \approx 0.5 \lambda_2(dof)(V_J \Sigma_J^+)(V_J \Sigma_J^+)^T$$
$$\lambda_2(dof) = \frac{dof^2}{(dof - n_y + 1)(dof - n_y - 1)} \qquad (12\text{-}30)$$

with Σ_J^+ the pseudo-inverse of Σ_J, and where $\lambda_2(dof)$ accounts for the fact that the pseudo-Jacobian is calculated using the sample noise covariance instead of the true noise covariance (proof: see Appendix 12.J).

12.3.5 Generation of Starting Values

For transfer function matrices $G(\Omega, \theta)$ parameterized as a *common denominator model* or a *left matrix fraction description* (see (6-53) and (6-54)), an equation error that is linear in the transfer function coefficients can be obtained by multiplying (12-16) with the denominator (matrix) polynomial $A(\Omega_k, \theta)$

$$A(\Omega_k, \theta)\hat{Y}^{[l]}(k) - B(\Omega_k, \theta)\hat{U}^{[l]}(k) \approx 0 \quad l = 1, 2, \ldots, n_{\exp} \text{ and } k = 1, 2, \ldots, F \quad (12\text{-}31)$$

Eq. (12-31) can be written as

$$J_{\text{LS}}(Z)\theta \approx 0 \qquad (12\text{-}32)$$

where $J_{\text{LS}}(Z)$ is an $n_y n_{\exp} F \times n_\theta$ matrix, and

$$\theta = \left[\text{vec}^T(A_0) \; \text{vec}^T(A_1) \; \ldots \; \text{vec}^T(A_{n_a}) \; \text{vec}^T(B_0) \; \ldots \; \text{vec}^T(B_{n_b})\right]^T \qquad (12\text{-}33)$$

with B_r the $n_y \times n_u$ numerator matrix coefficients, and A_r the $n_p \times n_p$ denominator (matrix) coefficients, with $n_p = 1$ or n_u for, respectively, the common denominator model and the left matrix fraction description.

Since (12-32) is similar to the scalar case (9-60), it can be used to construct (iterative) weighted linear least squares (Bayard, 1994a; Verboven et al., 2005; de Callafon et al., 1996; Gaikwad and Rivera, 1997), sample weighted generalized total least squares (Verboven et al., 2004; Pintelon et al., 1998), and sample bootstrapped total least squares (Pintelon et al., 1998) estimators (follow the same lines of Sections 9.8, 9.10, and 9.12.3). However, there is a subtle technical difference for the *left matrix fraction description*. Indeed, since the numerator B_r and denominator A_r matrix coefficients can be multiplied with a regular matrix $\Lambda \in \mathbb{R}^{n_y \times n_y}$ without changing the transfer function $G(\Omega, \theta)$ (6-54), the one-dimensional constraint $\|\theta\|_2 = 1$ is not sufficient to remove the parameter redundancy (n_y^2 constraints are needed). Therefore, one denominator matrix coefficient should be fixed (for example, $A_0 = I_{n_y}$) leading to a parameter vector θ of reduced dimension. Consequently, contrary to the scalar case, the one-dimensional weighted generalized and bootstrapped total least squares solutions will depend on this particular choice. Notice that a multi-dimensional (generalized) total least squares solution can be constructed that does not suffer from this problem (see Pintelon et al., 1998 for the details).

For a *state space parameterization* of $G(\Omega, \theta)$ (see (6-26) and (6-27)) starting values can be generated via multivariate subspace algorithms (see McKelvey et al., 1996 and Van Overschee and De Moor, 1996b for the details).

If the identification starts from measured frequency response data then a *right matrix fraction description* of $G(\Omega, \theta)$ (see (6-55)) can also be used for generating initial estimates (see, for example, de Callafon et al., 1996). Similar to the left matrix fraction description, one- and multi-dimensional (generalized) total least squares solutions can be constructed.

12.3.6 Model Selection and Validation

This section provides tools for answering the following two questions: "Is the identified model complex enough?" and "Is the identified model not too complex?". To handle the first question (detection of undermodeling) we perform three tests: (i) comparison of the identified transfer function model $G(\Omega_k, \hat{\theta}_{SML}(Z))$ with the nonparametric frequency response matrix (FRM) estimate $\hat{G}(\Omega_k)$, (ii) comparison of the minimum of the SML cost function with its expected value assuming that no modeling errors are present, and (iii) a whiteness test on the FRM residuals $\hat{G}(\Omega_k) - G(\Omega_k, \hat{\theta}_{SML}(Z))$. The second question (detection of overmodeling) is tackled by adding a penalty term for the model complexity to the cost function. The tools discussed here are the same as those in Chapter 11 but modified where necessary to handle the multivariable SML estimates.

12.3.6.1 Comparison with the Nonparametric FRM Estimate. As a straightforward extension of the single-input, single-output case, each entry of the identified transfer function model $G_{[r,s]}(\Omega_k, \hat{\theta}_{SML}(Z))$ is compared with the nonparametric FRM estimate $\hat{G}_{[r,s]}(\Omega_k)$ taking into account its uncertainty. If no modeling errors are present, then the inequality

$$\left| G_{[r,s]}(\Omega_k, \hat{\theta}_{SML}(Z)) - \hat{G}_{[r,s]}(\Omega_k) \right| \leq \sqrt{F_p(2, 2dof)} \, \hat{\sigma}_{\hat{G}_{[r,s]}}(k) \qquad (12\text{-}34)$$

should be satisfied for about $100 \times p\%$ of the F frequencies, where dof are the degrees of freedom of the estimated variance $\hat{\sigma}^2_{\hat{G}_{[r,s]}}(k)$ of the FRM $\hat{G}_{[r,s]}(\Omega_k)$, and with $F_p(2, 2dof)$ the $100 \times p\%$ percentile of an $F(2, 2dof)$-distributed random variable (proof: use (2-40) on page 51 with $M = dof + 1$).

A multivariate extension of (12-34) is given by

$$\text{vec}^H(G_{\hat{\theta}}(\Omega_k) - \hat{G}(\Omega_k)) \hat{C}^{-1}_{\text{vec}\hat{G}}(k) \text{vec}(G_{\hat{\theta}}(\Omega_k) - \hat{G}(\Omega_k)) \leq \frac{n_1 dof}{n_2} F_p(n_1, n_2) \qquad (12\text{-}35)$$

with $G_{\hat{\theta}}(\Omega_k) = G(\Omega_k, \hat{\theta}_{SML}(Z))$, dof the degrees of freedom of the covariance estimate estimate $\hat{C}_{\text{vec}\hat{G}}(k)$ of $\hat{G}(\Omega_k)$, $n_1 = 2n_u n_y$, $n_2 = 2dof - 2n_u n_y + 2$, and $F_p(n_1, n_2)$ the $100 \times p\%$ percentile of a $F(n_1, n_2)$-distributed random variable (proof: use (7-44) on page 237 with $X = G_{\hat{\theta}}(\Omega_k)$).

To test whether modeling errors are present or not for the "arb" (Section 12.2.1) and "fast" (Section 12.2.2) methods, one cannot just compare p with the fraction of the F frequencies satisfying (12-34) or (12-35). Indeed, the "arb" and "fast" FRM estimates $\hat{G}(\Omega_k)$ are correlated over the frequency with a finite correlation length n_{CL} of, respectively, $\pm 2n$ and $\pm 2n_E$. Therefore, the fraction should be counted on a subset of $\mathbb{K}$, for example,

$$k_r = (r-1)(n_{CL}+1) + 1 \qquad (12\text{-}36)$$

with $r = 1, 2, \ldots, F_1$, $F_1 = \lfloor (F-1)/(n_{CL}+1) \rfloor + 1$, and $n_{CL} = 2n$ or $2n_E$ ($\lfloor x \rfloor$ is the largest integer number smaller than or equal to x).

12.3.6.2 Analysis SML Cost Function. A second test for detecting modeling errors consists in comparing the global minimum of the SML cost function $V_{SML}(\hat{\theta}_{SML}(Z), Z)$ to the expected value assuming that no modeling errors are present. To decide whether the actual value of the cost function coincides with this expected value, we also need the variance of the cost function in the absence of modeling errors. While the correlation over the frequency of the generalized sample means does not affect the expected value of the cost function, it does influence its variance. This is the major technical difference with the standard SML result in (11-14).

Theorem 12.9 (Properties Global Minimum SML Cost Function): Under the conditions of Theorem 12.7 and assuming that no modeling errors are present, the global minimum $V_{\text{SML}}(\hat{\theta}_{\text{SML}}(Z), Z)$ of the sample maximum likelihood (SML) cost function (12-15) is asymptotically ($F = O(N) \to \infty$) normally distributed with mean and variance given by:

1. For $dof \geq n_y + 2$ the expected value of the global minimum of (12-15) equals

$$\mathbb{E}\{V_{\text{SML}}(\hat{\theta}_{\text{SML}}(Z), Z)\} \approx \frac{dof}{dof - n_y}(n_{\exp}n_y F - n_\theta/2) \qquad (12\text{-}37)$$

with n_θ the number of free model parameters, and where $n_{\exp} = 1$ for the "arb" (Section 12.2.1) and "fast" (Section 12.2.2.2) methods and $n_{\exp} = n_u$ for the "robust" (Section 12.2.2.1) method.

2. For $dof \geq n_y + 4$ the variance of the global minimum of (12-15) lies in the interval

$$\sigma_1^2 \leq \text{var}(V_{\text{SML}}(\hat{\theta}_{\text{SML}}(Z), Z)) \leq 3\sigma_1^2$$
$$\sigma_1^2 = \frac{dof^3}{(dof - n_y)^2(dof - n_y - 1)}(n_{\exp}n_y F - n_\theta/2) \qquad (12\text{-}38)$$

where the upper bound is reached for the "arb" and "fast" methods (strongly correlated over the excited frequencies), and the lower bound for the "robust" method (weakly correlated over the excited frequencies).

Proof. See Appendix 12.K. □

Remark. The upper bound in (12-38) is supported by MATLAB® simulations. However, since no formal proof is given, it should be used as an order of magnitude.

12.3.6.3 Whiteness Test of the Residuals. If no modeling errors are present, then the weighted residuals

$$\hat{\varepsilon}_{G_{[r,s]}}(k) = \frac{\hat{G}_{[r,s]}(\Omega_k) - \hat{G}_{[r,s]}(\Omega_k, \hat{\theta}_{\text{SML}}(Z))}{\hat{\sigma}_{\hat{G}_{[r,s]}}(k)} \qquad (12\text{-}39)$$

with $\hat{\sigma}^2_{\hat{G}_{[r,s]}}(k)$ the sample variance of $\hat{G}_{[r,s]}(\Omega_k)$, should be uncorrelated over the frequencies k for the "robust" method and the subset of frequencies (12-36) for the "arb' and "fast" methods. This is tested via the sample correlation (11-26) and its standard deviation (11-28) at all frequencies $k \in \mathbb{K}$ for the "robust" method, and at the subset (12-36) for the "arb" and "fast" methods

Section 12.4 ■ Identification in the Presence of Nonlinear Distortions 479

$$\hat{R}_{[r,s]}(m) = \frac{\alpha_1(m)}{F_1 - |m|} \sum_{r=1}^{F_1 - |m|} \hat{\varepsilon}_{G_{[r,s]}}(k_r) \overline{\hat{\varepsilon}_{G_{[r,s]}}(k_{r+m})} \text{ and } \text{std}(\hat{R}_{[r,s]}(m)) \approx \frac{\alpha_2(m)}{\sqrt{F_1 - |m|}} \quad (12\text{-}40)$$

where $\alpha_1(m)$ and $\alpha_2(m)$ are defined in, respectively, (11-27) and (11-29) with $M = dof + 1$. Since $\hat{R}_{[r,s]}(m)$ is asymptotically circular complex normally distributed (see Section 19.5.2), a $100 \times p\%$ confidence bound on $\hat{R}_{[r,s]}(m)$ is given by $\sqrt{-\ln(1-p)} \text{ std}(\hat{R}_{[r,s]}(m))$ (proof: see Appendix 2.A).

12.3.6.4 Detection of Overmodeling. If the identified model passes the three validation tests (comparison with the nonparametric FRM, analysis of the SML cost function, and whiteness test of the FRM residuals), then one should verify whether the model is not too complex. This can be done via the AIC (minimizes the prediction error) or MDL (selects the true model order) model selection criteria of Section 19.7.2

$$\begin{aligned} \text{AIC:} \quad & V_{\text{SML}}(\hat{\theta}_{\text{SML}}(Z), Z)\left(1 + \frac{n_\theta}{n_{\exp}Fn_y}\right) \\ \text{MDL:} \quad & V_{\text{SML}}(\hat{\theta}_{\text{SML}}(Z), Z)\left(1 + \frac{n_\theta}{2n_{\exp}Fn_y}\ln(2\alpha n_{\exp}F)\right) \end{aligned} \quad (12\text{-}41)$$

with $\alpha = n_y$ for output error problems, and $\alpha = n_y + n_u$ for an errors-in-variable problem.

12.4 IDENTIFICATION IN THE PRESENCE OF NONLINEAR DISTORTIONS

In this section we consider the setup of Figure 7-5 on page 242 for measuring the best linear approximation (BLA) of a nonlinear system operating in open or closed loop. Replacing the nonlinear plant in Figure 7-5 with its BLA plus an additive output source representing the stochastic nonlinear distortions $y_s(t)$ (see Section 6.8), gives the input-output relationship (7-53) where $y_s(t)$ is independent of the input-output measurement noise $m_u(t)$, $m_y(t)$ (open and closed loop) and the process noise $n_p(t)$ (open loop only). If generator noise (open and closed loop) or process noise (closed loop only) is present in the setup, then $y_s(t)$ depends on the input-output disturbances. We explicitly exclude this situation.

Due to the nonlinear behavior of the plant, the generalized sample covariance of the "arb" method (Section 12.2.1) and the generalized total sample covariances of the "robust" (Section 12.2.2.1) and "fast" (Section 12.2.2.2) methods depend on the disturbing input-output noise and the stochastic nonlinear distortions. To estimate a parametric model for the BLA of the nonlinear plant, we replace the sample noise covariances in the SML cost function (12-15) by the sample total covariances. This is a natural choice because the noise covariances describe only a part of the total variability of the nonparametric BLA measurement.

Since the properties of the stochastic nonlinear distortions $Y_S(k)$, are similar to those of the disturbing noise (compare Theorems 3.16, 3.17, 3.20 and 3.22 to Theorem 16.25) it can easily be shown that Lemma's 12.1 to 12.6 are still correct for the setup of Figure 7-5, except that $C_{\hat{X}}$ ($X = Y, Z$) in the expected values of the sample covariances (12-3) and (12-10) should be replaced by

$$C_{\hat{X}} = C_{\hat{X}}^{\text{noise}}(k) + C_{\hat{X}_S}(k) \quad (12\text{-}42)$$

with $C_{\hat{X}}^{\text{noise}}(k)$ and $C_{\hat{X}_S}(k)$ the contribution of the disturbing noise and the stochastic nonlinear distortions (proof: see Appendix 12.L). Therefore, Theorems 12.7 to 12.9 and the whiteness test on the residuals (12-40) remain valid except that the covariance expressions (12-23) and (12-30) are approximations (see Appendix 12.M) that underestimate the true covariance (see Schoukens and Pintelon, 2010b). The technical reason for this difference with the linear case is that – contrary to the disturbing noise – $Y_S(k)$ is not independently distributed of $U_0(k)$ (open loop) or $R(k)$ (closed loop). If the degree of the nonlinearity is known as well as the contribution of the stochastic nonlinear distortions to the sample total covariances, then a scaling factor compensating for the underestimation can be calculated (see Schoukens and Pintelon, 2010b, for the details). This scaling factor on the covariance (12-30) can be as large as 7 (8.5 dB) or more depending on the type of excitation (Gaussian noise or random phase multisine) and the degree of the nonlinearity.

Remark. As a special case we can also consider the setup of Figure 7-4 on page 240 where either the actuator or the controller is nonlinear. Although the plant behaves linearly, the input-output relationship (7-51) depends on the nonlinear distortions of either the actuator or the controller (see Section 7.2.7.3). Therefore, the covariances of the measured input-output DFT spectra depend on both the disturbing noise and the nonlinear distortions. All results of Section 12.3 remain valid, the covariance expressions (12-23) and (12-30) included (proof: see Appendix 12.N).

12.5 EXPERIMENTAL ILLUSTRATION

In this section we illustrate the parametric transfer function modeling on the measurement example of Section 7.4.2 (aluminum tooling plate excited by two mini-shakers). Starting from two consecutive periods of the transient response to one set of $n_u = 2$ uncorrelated random phase multisines, the "fast" local polynomial estimates of the generalized sample means (12-13) and sample total covariances (12-14) of the input-output DFT spectra are calculated. The "fast" local polynomial method (see Section 7.3.7.1) uses here an $R = 4$th order local polynomial approximation of the frequency response matrix (FRM) over $2n_E + 1$ neighboring excited frequencies, where n_E is chosen such that $dof_{\text{fast}} = 9$ (see (7-103)). Hence, the theoretical correlation length n_{CL} of the generalized sample means (12-13) and sample total covariances (12-14) is ± 18.

Figure 7-15 on page 265 shows the "fast" nonparametric FRM estimate at all excited frequencies. In this section we model the 2×2 FRM in the band [247 Hz, 254 Hz] ($F = 188$ excited frequencies) using a common denominator transfer function (6-53). Since the aluminum plate behaves nonlinearly (see Figure 7-17 on page 267), the sample total covariances are used in the SML cost function (12-15) and in the model validation/selection tools of Section 12.3.6. The orders n_a and n_b of, respectively, the denominator polynomial and the numerator 2×2 matrix polynomial are obtained via the model validation/selection tools of Section 12.3.6. Since two resonances and one anti-resonance are visible in the band [247 Hz, 254 Hz] (see Figure 7-17), and two anti-resonances are adjacent to the band [247 Hz, 254 Hz] (see Figure 7-15), it is reasonable to start with a minimal model order n_b/n_a of 6/4. Table 12-2 shows the SML cost function and the value of the MDL and AIC criteria for increasing model complexity. It can be seen that for model orders n_b/n_a equal to 6/4 and 6/6 the actual value of the SML cost function $V_{\text{SML}} = V_{\text{SML}}(\hat{\theta}_{\text{SML}}(Z), Z)$ is much larger than its expected value $\mathbb{E}\{V_{\text{SML}}\}$ in the absence of modeling errors, while for model orders 6/6 to 8/7 the actual value V_{SML} equals the expected value $\mathbb{E}\{V_{\text{SML}}\}$ within its 95% uncertainty bound ($\pm 2\text{std}(V_{\text{SML}})$). To decide whether the differences between the cost function values of models 6/6 to 8/7 are significant or not, we look at the corresponding MDL and AIC criteria

Section 12.5 ■ Experimental Illustration

TABLE 12-2 SML Cost Function and the AIC and MDL Rules for Increasing Order n_a, n_b of the Common Denominator Transfer Function Model

n_b/n_a	V_{SML}	$\mathbb{E}\{V_{\text{SML}}\}$ (12-37)	std(V_{SML}) (12-38)	MDL (12-41)	AIC (12-41)
6/4	9855	462.5	[29.9, 51.7]	12920	10690
6/5	9828	462.2	[29.9, 51.7]	12980	10690
6/6	527.4	461.6	[29.8, 51.7]	701.9	575.1
7/6	394.4	459.0	[29.7, 51.5]	540.3	434.4
8/6	365.2	456.4	[29.7, 51.4]	514.5	406.0
8/7	363.0	455.8	[29.6, 51.4]	515.0	404.6

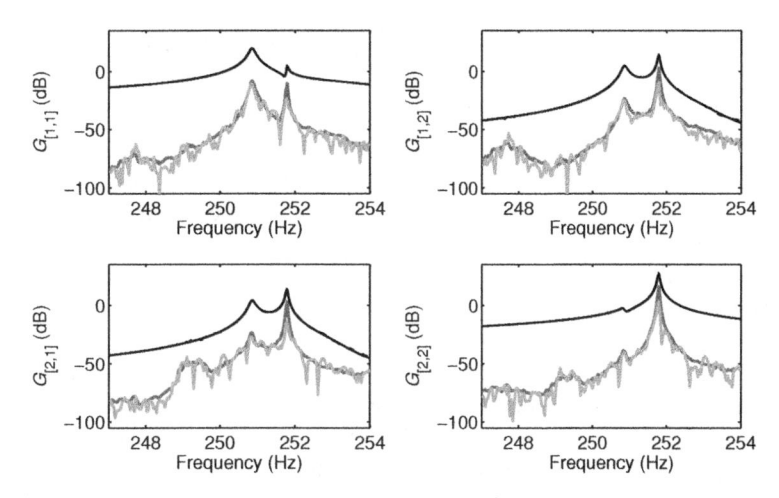

Figure 12-1. Comparison between the nonparametric FRM $\hat{G}(\Omega_k)$ (black line) and the parametric transfer function model $G(\Omega_k, \hat{\theta}_{\text{SML}}(Z))$ of order 8/6 (black dashed line; coincides with the black line) – aluminum plate. Dark gray line: $\text{var}(\hat{G}(\Omega_k))$, and light gray line $|\hat{G}(\Omega_k) - G(\Omega_k, \hat{\theta}_{\text{SML}}(Z))|$.

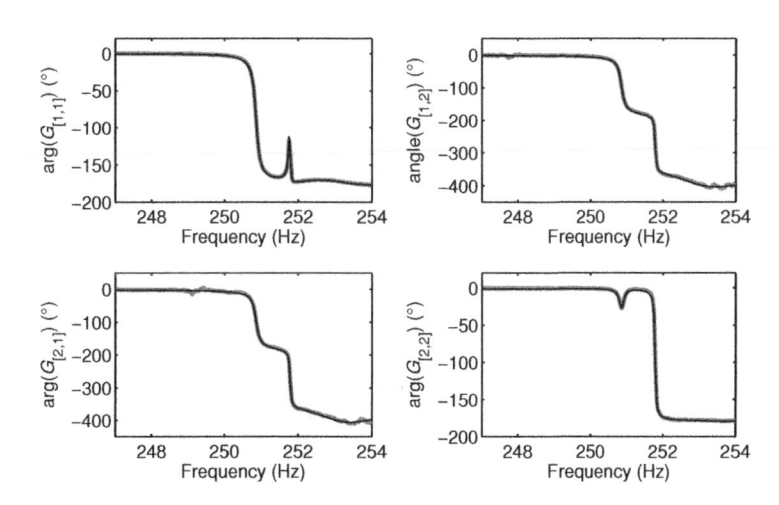

Figure 12-2. Phase of the nonparametric FRM $\hat{G}(\Omega_k)$ (gray line; coincides with the black line) and the parametric transfer function model $G(\Omega_k, \hat{\theta}_{\text{SML}}(Z))$ of order 8/6 (black line).

(see Table 12-2). While AIC decreases monotonically for increasing model complexity, MDL is minimal for model order 8/6. Hence, according to MDL, model 8/6 is the best choice.

Figures 12-1 and 12-2 compare the "fast" nonparametric FRM estimate $\hat{G}(\Omega_k)$ to the estimated transfer function model $G(\Omega_k, \hat{\theta}_{SML}(Z))$ of order 8/6. It can be seen that the difference $\hat{G}(\Omega_k) - G(\Omega_k, \hat{\theta}_{SML}(Z))$ between both estimates is at the level of the standard deviation of the nonparametric estimate. Finally, a whiteness test on the FRM residuals is performed. Although the theoretical correlation length over the frequency of the FRM estimate $\hat{G}(\Omega_k)$ is ±18, the observed correlation length is ±7. Therefore, the sample correlation (12-40) is calculated over the frequency set (12-36) with $n_{CL} = 7$. Figure 12-3 shows the result: about 42% and 2.2% of the sample correlation values lie outside, respectively, the 50% and 95% uncertainty bounds. We conclude that model 8/6 passes the model validation/selection tests of Section 12.3.6. Table 12-3 shows the identified resonance frequencies and damping ratios of the two visible resonances in Figure 12-1. The standard deviations are calculated via a linearization (16-22) and (16-23), where $x = \hat{\theta}_{SML}(Z)$ and $Cov(x) = Cov(\hat{\theta}_{SML}(Z))$ (12-30) (see Pintelon et al., 2007b, for the details).

Since the MDL value of model 7/6 is quite close to that of model 8/6 (see Table 12-2), we could – using the parsimonious principle – also select model 7/6. Indeed, model 7/6 passes all the other model selection/validation tests: (i) the actual value of the cost function equals the expected value in the absence of modeling errors within its uncertainty (see Table 12-2), (ii) the identified parametric model equals the nonparametric FRM estimate within its uncertainty (not shown here), and (iii) about 45% and 2.2% of the sample correlation values (12-40) lie outside, respectively, the 50% and 95% uncertainty bounds (not shown here). Table 12-3 compares the identified resonance frequencies and damping ratios of model 7/6 to that of model 8/6. It can be seen that the estimates of both models agree fairly well: the damping ratios coincide within their standard deviation, while the differences between the resonance frequencies (a few hundred μHz) are a factor 10 larger than their standard deviation.

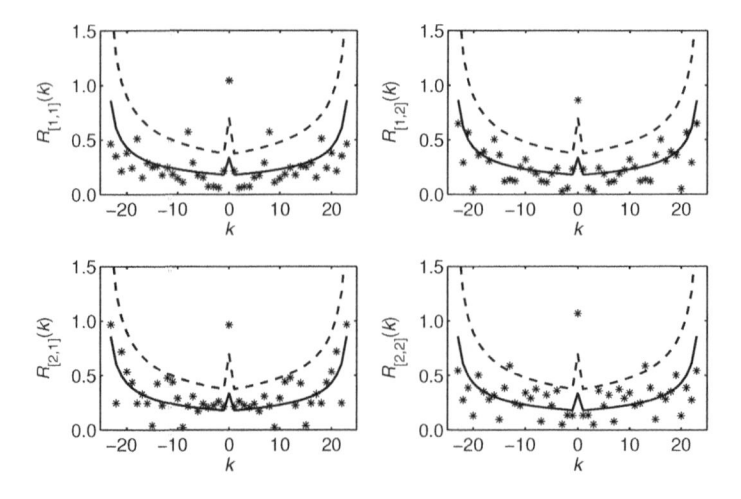

Figure 12-3. Sample correlation of the FRM residuals ('*') for the estimated transfer function model of order 8/6. Solid line: 50% confidence bound, and dashed line: 95% confidence bound.

TABLE 12-3 Identified Resonance Frequencies f_0 and Damping Ratios ζ.

	$n_b/n_a = 8/6$	$n_b/n_a = 7/6$	Difference Estimates
$f_0 \pm \text{std}(f_0)$ (Hz)	$250.8552 \pm 2.0 \times 10^{-5}$	$250.8550 \pm 2.0 \times 10^{-5}$	2.0×10^{-4}
$\zeta \pm \text{std}(\zeta)$	$2.96 \times 10^{-4} \pm 3.2 \times 10^{-6}$	$2.97 \times 10^{-4} \pm 3.1 \times 10^{-6}$	-1.4×10^{-6}
$f_0 \pm \text{std}(f_0)$ (Hz)	$251.7844 \pm 3.8 \times 10^{-5}$	$251.7840 \pm 3.7 \times 10^{-5}$	4.0×10^{-4}
$\zeta \pm \text{std}(\zeta)$	$9.0 \times 10^{-5} \pm 6.0 \times 10^{-6}$	$9.2 \times 10^{-5} \pm 5.9 \times 10^{-6}$	-1.5×10^{-6}

12.6 GUIDELINES FOR PARAMETRIC TRANSFER FUNCTION MODELING

- *Guideline 1: Select the Smallest Bandwidth for the Local Polynomial Approximation.* The frequency width n (or n_E) of the local polynomial approximation can be tuned as a function of the frequency to minimize the mean square error (optimal bias-variance trade-off) of the nonparametric frequency response matrix estimate (see Section 7.2.6). A totally different choice is made when calculating the generalized sample means and sample covariances used in the SML cost function. Indeed, for parametric transfer function modeling the bias error in the generalized sample means and sample covariances should be minimized, because it cannot be reduced by the parametric step. Therefore, the frequency width n (or n_E) is chosen as small as possible such that the minimum requirement on the degrees of freedom *dof* of the sample covariance estimate (see Theorems 12.7 and 12.8) is satisfied. Although this choice maximizes the variance of the generalized sample means, it also minimizes the correlation length over the frequency of the nonparametric estimates and, hence, it does not affect the variability of the estimated transfer function model.

- *Guideline 2: Suppress Nonparametrically the Transient (Leakage) Errors in the Generalized Sample Means.* It is strongly recommended to suppress nonparametrically the leakage (transient) errors in the generalized sample means, because it decreases significantly the sensitivity of the parametric transfer function estimate w.r.t. (unexpected) plant and/or noise transients in the data. However, if the leakage errors can be neglected over the whole frequency band, then the nonparametric transient suppression increases the variability of the estimated parametric transfer function model by about 1 dB. Only if it is known beforehand that either the noise and plant models have the same poles (ARMAX model structure), or the noise transient can be neglected w.r.t. the plant transient, one can combine the generalized sample means without transient removal (e.g., (12-1)) with the SML cost function where a plant transient term is added to the parametric model (see (12-18)).

- *Guideline 3: Use the Sample Total Covariances in the SML Cost Function.* Since most real-life systems behave to some extent nonlinearly, the best linear approximation is estimated rather than the true underlying linear system. Therefore, it is more natural/appropriate to use the sample total covariances instead of the sample noise covariances in the SML cost function (12-15), and in the model validation/selection tools of Section 12.3.6.

Concluding remarks. The required frequency resolution (measurement time) of the identification experiment increases linearly with the number of inputs n_u and outputs n_y. Indeed, to keep the bias errors of the generalized sample means and sample covariances small, the ratio n/N, with $2n+1$ the bandwidth of the local polynomial approximation and N the number of time domain samples, should remain small (see, for example, (12-3)). Further, from the lower bound on the degrees of freedom of the sample covariance estimate, for example, $dof \geq n_y + 8$ (see Theorems 12.7 and 12.8), and $dof = 2n + 1 - (n_u + 1)(R + 1)$, with R the order of the local polynomial approximation (see (7-13)), it follows that n increases linearly with n_u and n_y

$$2n + 1 \geq n_y + 8 + (n_u + 1)(R + 1)$$

Hence, to keep n/N small, N should also increase linearly with n_u and n_y.

12.7 APPENDICES

Following the lines of Chapter 7, $T(\Omega)$ represents for continuous-time systems the sum of the transient (leakage) term and the residual alias term. This is permitted since the transient and the residual alias terms have exactly the same properties (see Lemmas 6.5 and 6.6 and Theorems 6.15 and 6.16).

Appendix 12.A Proof of Lemma 12.1

Defining $Q_n = K_n^H(K_n K_n^H)^{-1} K_n$, we deduce from (7-10) that (12-1) can be written as

$$\hat{Y}(k) = (\hat{\Theta} K_n)_{[:, n+1]} = (Y_n Q_n)_{[:, n+1]} \qquad (12\text{-}43)$$

Using $Y_n = \Theta K_n + V_n$ (7-8) and $K_n Q_n = K_n$, we find $Y_n Q_n = \Theta K_n + V_n Q_n$. Combining this result with (12-43) gives

$$\hat{Y}(k) = \Theta K_n 1_{n+1} + V_n Q_n 1_{n+1} \qquad (12\text{-}44)$$

with 1_m a vector containing everywhere zeroes except at entry m where it is one. From (12-44) and $Q_n^H = Q_n$ it follows that entry $[i, j]$ of the covariance matrix of $\hat{Y}(k)$, given the input and the initial and final conditions, equals

$$\begin{aligned}
(\text{Cov}(\hat{Y}(k)))_{[i,j]} &= \mathbb{E}\{V_{n[i,:]} Q_n 1_{n+1} 1_{n+1}^T Q_n^H V_{n[j,:]}^H\} \\
&= \text{trace}(Q_n 1_{n+1} 1_{n+1}^T Q_n \mathbb{E}\{V_{n[j,:]}^H V_{n[i,:]}\}) \qquad (12\text{-}45) \\
&= \text{trace}(Q_n 1_{n+1} 1_{n+1}^T Q_n C_n)
\end{aligned}$$

where the diagonal matrix C_n is defined in (7-114). Combining (7-114), (7-115), and (12-45) gives, using $Q_n^2 = Q_n$,

$$(\text{Cov}(\hat{Y}(k)))_{[i,j]} = (C_V(k))_{[i,j]} \text{trace}(1_{n+1}^T Q_n 1_{n+1}) + (C_V^{(1)}(k))_{[i,j]} O_{\text{int}H}(n/N) \qquad (12\text{-}46)$$

where $\text{trace}(1_{n+1}^T Q_n 1_{n+1}) = Q_{n[n+1, n+1]}$. From (7-17), (12-2), and (12-46) we deduce that

$$\lim_{N \to \infty} \text{Cov}(\hat{Y}(k)) = \lim_{N \to \infty} \mathbb{E}\{\hat{C}_{\hat{Y}}(k)\} \tag{12-47}$$

It proves that $\hat{C}_{\hat{Y}}(k)$ (12-2) is the asymptotic ($N \to \infty$) sample covariance of $\hat{Y}(k)$ (12-1).

Combining (7-17) with (12-1) and (12-46) proves immediately the expected value of the generalized sample covariance (12-3). First note that the expected value of the estimated transient term $\hat{T}(\Omega_k)$ satisfies

$$\mathbb{E}\{\hat{T}(\Omega_k)\} = T(\Omega_k) + O_{\text{intG}}((n/N)^{(R+1)}) \tag{12-48}$$

(proof: combine (7-6) and (7-124)). Collecting (7-18), (12-1), and (12-48) proves the expected value of the generalized sample mean. □

Appendix 12.B Proof of Lemma 12.2

If $v(t)$ in Figure 7-1 on page 227 is normally distributed, then, $V(k)$ (7-2) is circular complex normally distributed. In this appendix we will show that the generalized sample mean $\hat{Y}(k)$ (12-1) and the generalized sample covariance $\hat{C}_{\hat{Y}}(k)$ (12-2) are asymptotically ($N \to \infty$) independently distributed for circular complex normally distributed noise $V(k)$ (7-2). Therefore, it is sufficient to show that $\hat{Y}_n = Y_n Q_n$ in (12-43) is asymptotically ($N \to \infty$) independently distributed of the noise residual $\hat{V}_n = V_n P_n$ (7-12). Indeed, if $\hat{Y}_n$ and $\hat{V}_n$ are asymptotically ($N \to \infty$) independently distributed, then $F_1(\hat{Y}_n)$ and $F_2(\hat{V}_n)$, with F_1 and F_2 (non)linear functions, are also asymptotically ($N \to \infty$) independently distributed (Stuart and Ord, 1987). Taking as a special case of F_1 and F_2 the sample mean $\hat{Y}(k)$ (12-1) and the sample covariance $\hat{C}_{\hat{Y}}(k)$ (12-2) proves the result.

Note that $\hat{Y}(k)$ and $\hat{V}_n$ are jointly circular complex normally distributed because they are both analytic functions of V_n. Hence, it is sufficient to show that $\text{vec}(\hat{Y}_n)$ and $\text{vec}(\hat{V}_n)$ are asymptotically ($N \to \infty$) uncorrelated. Using $Y_n = \Theta K_n + V_n$ (7-8), $\text{vec}(ABC) = (C^T \otimes A)\text{vec}(B)$, $(A \otimes B)^H = A^H \otimes B^H$, and $P_n^H = P_n$, we find

$$\begin{aligned}\mathbb{E}\{\text{vec}(\hat{Y}_n)\text{vec}^H(\hat{V}_n)\} &= (Q_n^T \otimes I_{n_y})\mathbb{E}\{\text{vec}(V_n)\text{vec}^H(V_n)\}(P_n^T \otimes I_{n_y}) \\ &= (Q_n^T \otimes I_{n_y})D_n(P_n^T \otimes I_{n_y})\end{aligned} \tag{12-49}$$

with D_n the following $(2n+1)n_y \times (2n+1)n_y$ block diagonal matrix

$$D_n = \text{blockdiag}(C_V(k-n), \ldots, C_V(k+n)) \tag{12-50}$$

Combining (7-114), (12-49), and (12-50), using $(A \otimes B)(C \otimes D) = (AC \otimes BD)$, gives

$$\mathbb{E}\{\text{vec}(\hat{Y}_n)\text{vec}^H(\hat{V}_n)\} = (P_n Q_n)^T \otimes C_V(k) + O_{\text{intH}}(n/N) \tag{12-51}$$

Using $Q_n^2 = Q_n$ ($Q_n = K_n^H(K_n K_n^H)^{-1}K_n$) and $P_n = I_{2n+1} - Q_n$, it can easily be verified that $P_n Q_n = 0$. Hence,

$$\mathbb{E}\{\text{vec}(\hat{Y}_n)\text{vec}^H(\hat{V}_n)\} = O_{\text{intH}}(n/N) \tag{12-52}$$

which proves the asymptotic ($N \to \infty$) independence of the sample mean $\hat{Y}(k)$ and sample covariance $\hat{C}_{\hat{Y}}(k)$. From (12-52) it follows that

$$\mathbb{E}\{F_1(\hat{Y}_n)F_2(\hat{V}_n)\} = \mathbb{E}\{F_1(\hat{Y}_n)\}\mathbb{E}\{F_2(\hat{V}_n)\} + O(n/N) \qquad (12\text{-}53)$$

where F_1 and F_2 are (non)linear functions, which proves (12-4) because it is a special case of (12-53).

If $v(t)$ in Figure 7-1 satisfies the conditions of Theorem 16.25, then $V(k)$ (7-2) is asymptotically ($N \to \infty$) circular complex normally distributed (convergence in law at the rate $O(N^{-1/2})$). Since (12-52) has been derived without using the actual distribution of $V(k)$, (12-53), where $O(n/N)$ is replaced by $O(N^{-1/2})$, remains valid for filtered white noise satisfying Theorem 16.25.

Appendix 12.C Proof of Lemma 12.3

Using $Y_n = \Theta K_n + V_n$ (7-8) and $\hat{\Theta} = Y_n S_n$ (7-10), with $S_n = K_n^H(K_n K_n^H)^{-1}$, we find $\hat{\Theta} = \Theta + V_n S_n$. Hence, the local polynomial estimate $\hat{T}(\Omega_k)$ of the transient (leakage error) term can be written as

$$\hat{T}(\Omega_k) = \Theta 1_p + V_n S_n 1_p \qquad (12\text{-}54)$$

with $p = n_u(R+1) + 1$, and where 1_p is a vector containing everywhere zeroes except at entry p where it is one. Subtracting the estimated transient (12-54) from the estimated output DFT spectrum (12-44) gives an expression for the sample mean (12-5) without transient

$$\hat{Y}(k) = \Theta(K_n 1_{n+1} - 1_p) + V_n q_n \qquad (12\text{-}55)$$

where $q_n = Q_n 1_{n+1} - K_n^H(K_n K_n^H)^{-1} 1_p = K_n^H(K_n K_n^H)^{-1}(K_n 1_{n+1} - 1_p)$. Using (7-7) to (7-9) and the definition of 1_p it can easily be verified that

$$K_n 1_{n+1} - 1_p = \begin{bmatrix} U(k) \\ 0 \end{bmatrix} \qquad (12\text{-}56)$$

which proves the expression for q_n in (12-6). The rest of the proof follows the same lines of Appendix 12.A, where $Q_n 1_{n+1}$ is replaced by q_n, and $\text{trace}(Q_n 1_{n+1} 1_{n+1}^T Q_n^H)$ by $\|q_n\|_2^2$. $\square$

Appendix 12.D Proof of Lemma 12.4

The proof follows the same lines of Appendix 12.B, where $Q_n 1_{n+1}$ is replaced by $q_n = Q_n 1_{n+1} - S_n 1_p$, with $S_n = K_n^H(K_n K_n^H)^{-1}$ and $p = n_u(R+1) + 1$ (compare (12-44) and (12-55)). The asymptotic ($N \to \infty$) independence of the sample mean (12-1) and sample covariance (12-2) without transient removal is a consequence of the property that $P_n Q_n = 0$, where $P_n = I_{2n+1} - Q_n$ and $Q_n = K_n^H(K_n K_n^H)^{-1} K_n$ (see Appendix 12.B). Hence, to prove the asymptotic ($N \to \infty$) independence of the sample mean (12-5) and sample covariance (12-6) with transient removal it is sufficient to note that also $P_n S_n = 0$. $\square$

Appendix 12.E Proof of Equation (12-7)

Using $\hat{Y}(k) = \text{vec}(\hat{Y}(k)) = \text{vec}(\hat{G}(\Omega_k)U^c(k)) = (U^{cT}(k) \otimes I_{n_y})\text{vec}(\hat{G}(\Omega_k))$ we get

$$\text{Cov}(\hat{Y}(k)) = (U^{cT}(k) \otimes I_{n_y})\text{Cov}(\text{vec}(\hat{G}(\Omega_k)))(\overline{U^c}(k) \otimes I_{n_y}) \tag{12-57}$$

Replacing $\text{Cov}(\text{vec}(\hat{G}(\Omega_k)))$ in (12-57) by its sample estimate gives

$$\hat{C}_{\hat{Y}}(k) = (U^{cT}(k) \otimes I_{n_y})\hat{C}_{\text{vec}G}(k)(\overline{U^c}(k) \otimes I_{n_y}) \tag{12-58}$$

where $\hat{C}_{\text{vec}G}(k)$ is the upper $n_y n_u \times n_y n_u$ block of the estimated covariance (7-21) of $\tilde{G}(\Omega_k)$ defined in (7-58). $\hat{C}_{\text{vec}G}(k)$ can be elaborated as

$$\hat{C}_{\text{vec}G}(k) = (\overline{S^H S} \otimes \hat{C}_V(k))_{[1:n_y n_u, 1:n_y n_u]} = (\overline{S^H S})_{[1:n_u, 1:n_u]} \otimes \hat{C}_V(k) \tag{12-59}$$

with S the matrix defined in (7-20) where $U(k)$ and I_{n_u} are replaced by, respectively, $\tilde{U}^c(k)$ (7-58) and $I_{n_u + M_c - 1}$. Combining (12-58) and (12-59) using $(A \otimes B)(C \otimes D) = AC \otimes BD$,

$$(S^H S)_{[1:n_u, 1:n_u]} = S_1^H S_1 \text{ with } S_1 = K_n^H (K_n K_n^H)^{-1} \begin{bmatrix} I_{n_u} \\ 0 \end{bmatrix},$$

and $S_1 U^c(k) = q_n^c$, gives $\hat{C}_{\hat{Y}}(k) = \overline{U^{cH}(k) S_1^H S_1 U^c(k)} \otimes \hat{C}_V(k) = \|q_n^c\|_2^2 \hat{C}_V(k)$, which proves (12-7). □

Appendix 12.F Proof of Lemma 12.5

In Appendix 7.H it is shown that the sample noise covariance $\hat{C}_{\hat{Z}}^{\text{noise}[e]}(k)$ is the asymptotic ($N \to \infty$) noise covariance of the sample mean $\hat{Z}^{[e]}(k)$, while the sample covariance $\hat{C}_{\hat{Z}}^{[e]}(k)$ is by construction the total sample covariance of $\hat{Z}^{[e]}(k)$ (see (7-87) and (7-88)).

To prove the asymptotic ($N \to \infty$) independence of the sample mean $\hat{Z}^{[e]}(k)$ (12-8) and the noise sample covariance $\hat{C}_{\hat{Z}}^{\text{noise}[e]}(k)$ (12-9), it is sufficient to show that the estimated noise transient $\hat{T}_{H_Z}(\Omega_{kP})$ (7-71) is asymptotically ($N \to \infty$) independently distributed of the noise residual

$$\hat{V}_n = Z_n - \hat{\Theta} K_n = Z_n P_n \text{ with } P_n = I_{2n} - K_n^H (K_n K_n^H)^{-1} K_n \tag{12-60}$$

Since the estimated noise transient $\hat{T}_{H_Z}(\Omega_{kP})$ is related to the true value $T_{H_Z}(\Omega_{kP})$ and the noise V_n as

$$\hat{T}_{H_Z}(\Omega_{kP}) = T_{H_Z}(\Omega_{kP}) + V_n S_n 1_1 \text{ with } S_n = K_n^H (K_n K_n^H)^{-1} \tag{12-61}$$

and where 1_1 is a vector containing everywhere zeroes except at the first entry where it is one (see (7-137) and (7-138)). Since $P_n S_n = 0$, it can be concluded that $\hat{T}_{H_Z}(\Omega_{kP})$ and $\hat{V}_n$ are asymptotically ($N \to \infty$) independently distributed for normally distributed input-output errors (follow the same lines of Appendix 12.B). Hence, $\hat{C}_{\hat{Z}}^{\text{noise}[e]}(k)$ and $\hat{Z}^{[e]}(k)$ are asymptotically ($N \to \infty$) independently distributed.

The only difference between the total sample covariance $\hat{C}_{\hat{Z}}^{[e]}(k)$ (7-88) and the standard sample covariance is the averaging of the squared residuals over neighboring frequencies. This introduces a bias error in the covariance estimate that is asymptotically ($N \to \infty$) zero (see (7-93)). Hence, for normally distributed input-output errors, $\hat{Z}^{[e]}(k)$ and $\hat{C}_{\hat{Z}}^{[e]}(k)$ are asymptotically ($N \to \infty$) independently distributed.

The expected values (12-10) follow immediately from (7-76) (sample mean) and (7-93) (sample covariances).

Appendix 12.G Proof of Theorem 12.7

The proof follows the same lines of the proof of Theorem 10.3, but extended to handle the three technical differences of the generalized sample means and sample covariances w.r.t. the standard solution: (i) the finite correlation length over the frequency, (ii) the asymptotic ($N \to \infty$) normality and unbiasedness, and (iii) the multiple-output case.

(i) *Correlation over the frequency*: the (strong) laws of large numbers, and the central limit theorems all remain valid for random variables with a finite correlation length over the frequency.

(ii) *Asymptotic normality and unbiasedness*: the expected value of the product of a (non)linear function of the generalized sample mean and a (non)linear function of the generalized sample covariance equals

$$\mathbb{E}\{F_1(\hat{Z}(k))F_2(\hat{C}_{\hat{Z}}(k))\} = \mathbb{E}\{F_1(\hat{Z}(k))\}\mathbb{E}\{F_2(\hat{C}_{\hat{Z}}(k))\} + O(N^{-\alpha}) \quad (12\text{-}62)$$

with $\alpha = 1$ for normally distributed input-output errors, and $\alpha = 0.5$ for non-Gaussian input-output noise satisfying Theorem 16.25. Since the bias errors of $\hat{Z}(k)$ and $\hat{C}_{\hat{Z}}(k)$ are an $O(N^{-1})$ or smaller, they will not increase the $O(N^{-\alpha})$ term in the expected value (12-62).

(iii) *Multiple-outputs*: Because the input-output noise is (asymptotically) circular complex normally distributed, the sample covariance $\hat{C}_{\hat{Z}}(k)$ (and hence also $\hat{C}_{\hat{e}}^{[l]}(\Omega_k, \theta)$ (12-17)) is (asymptotically) complex Wishart distributed of dimension n_y and degrees of freedom *dof* (Goodman, 1963, Brillinger, 1981). To guarantee the existence of the m th order moment of the SML cost function (12-15), the m th order moment of $(\hat{C}_{\hat{e}}^{[l]}(\Omega_k, \theta))^{-1}$ should be finite. This puts a lower bound on the degrees of freedom of the sample covariance (Maiwald and Kraus, 2000). For example, if $\hat{S}$, with $\mathbb{E}\{\hat{S}\} = S$, is complex Wishart distributed of dimension n_y and degrees of freedom *dof*, then the first two moments of $\hat{T} = \hat{S}^{-1}$ ($T = S^{-1}$) are given by

$$\mathbb{E}\{\hat{S}^{-1}\} = \frac{dof}{dof - n_y}S^{-1} \quad (12\text{-}63)$$

for $dof \geq n_y + 1$, and

$$\mathbb{E}\{\hat{T}_{[i,j]}\hat{T}_{[k,l]}\} = \frac{dof^2}{(dof - n_y)^2 - 1}\left(T_{[i,j]}T_{[k,l]} + \frac{1}{dof - n_y}T_{[i,k]}T_{[l,j]}\right) \quad (12\text{-}64)$$

for $dof \geq n_y + 2$. In general, the mth order moment of $\hat{T}$ exists if $dof \geq n_y + m$ (Maiwald and Kraus, 2000). Following the same lines of Appendix 10.D and Mahata et al. (2006) we find then the lower bounds on dof in Theorem 12.7.

Appendix 12.H Proof of Theorem 12.8

The proof follows exactly the same lines of Appendix 12.G, and only the particular expression of the covariance matrix of the model parameters (12-23) must be shown. This is done in two steps. First, for filtered white noise and random phase multisine excitations it is shown that the correlation over the frequency does not affect the covariance expression of the estimated model parameters. Next, the link with the maximum likelihood (ML) estimate is established.

FIRST STEP.   The asymptotic ($F = O(N) \to \infty$) covariance matrix of $\hat{\theta}_{\text{SML}}(Z)$ is exactly given by

$$\text{Cov}(\sqrt{F}\delta_{\theta\text{SML}}(Z)) = V_F''^{-1}(\theta_0) Q_F(\theta_0) V_F''^{-1}(\theta_0) \tag{12-65}$$

with $V_F = V_{\text{SML}}/F$, $V_F''(\theta)$ the second order derivative of $V_F(\theta) = \mathbb{E}\{V_F(\theta, Z)\}$ w.r.t. θ,

$$Q_F(\theta_0) = F \mathbb{E}\{V_F'^T(\theta_0, Z) V_F'(\theta_0, Z)\} \tag{12-66}$$

and $V_F'(\theta, Z)$ the derivative of $V_F(\theta, Z)$ w.r.t. θ (see Theorem 9.21 on page 298). Assume for simplicity of notation that $n_{\text{exp}} = 1$ in the SML cost function (12-15). The equation error can then be written as

$$e(\Omega_k, \theta, \hat{Z}(k)) = e_0(\Omega_k, \theta) + N_e(\Omega_k, \theta) + O(N^{-(R+1)}) \tag{12-67}$$

with $e_0(\Omega_k, \theta) = Y_0(k) - G(\Omega_k, \theta)U_0(k)$, $N_e(\Omega_k, \theta) = V_Y(k) - G(\Omega_k, \theta)V_U(k)$, $O(N^{-(R+1)})$ the bias of the generalized sample mean (zero for the robust method), R the order of the local polynomial approximation, and where the input-output noise contributions $V_U(k)$, $V_Y(k)$ are defined in (7-48). If the transient (leakage) error has not been suppressed nonparametrically, then the equation error is extended with a parametric transient term.

The derivative of the cost function $V_F'(\theta_0, Z)$ depends on the equation error (12-67) and its derivative evaluated at the true parameter value θ_0

$$\begin{aligned} e(\Omega_k, \theta_0, \hat{Z}(k)) &= N_e(\Omega_k, \theta_0) + O(N^{-(R+1)}) \\ e'^{,r}(\Omega_k, \theta_0, \hat{Z}(k)) &= -G'^{,r}(\Omega_k, \theta_0)U_0(k) + N_e'^{,r}(\Omega_k, \theta_0) + O(N^{-(R+1)}) \end{aligned} \tag{12-68}$$

with $e_0(\Omega_k, \theta_0) = 0$, and where $x'^{,r}(\theta)$ is the derivative of $x(\theta)$ w.r.t. $\theta_{[r]}$. Neglecting the terms in $V_F'(\theta_0, Z)$ depending on the product of two noise contributions we find

$$V_F'^{,r}(\theta_0, Z) \approx \frac{1}{F} \sum_{k \in \mathbb{K}} -2\text{Re}(N_e^H(\Omega_k, \theta_0)\hat{C}_e^{-1}(\Omega_k, \theta_0)G'^{,r}(\Omega_k, \theta_0)U_0(k)) + O(N^{-(R+1)}) \tag{12-69}$$

For filtered white noise and random phase multisine excitations the noiseless input DFT spectrum is asymptotically ($F = O(N) \to \infty$) uncorrelated over the frequency

$$\mathbb{E}\{U_0(k)\overline{U_0(l)}\} = \begin{cases} O(N^{-(R+3/2)}) & \text{transient suppressed} \\ O(N^{-1}) & \text{no transient suppression} \end{cases} \text{ for } k \neq l \quad (12\text{-}70)$$

and similarly for $\mathbb{E}\{U_0(k)U_0(l)\}$ (proof: use (12-48) and Theorem 6.17). Using (12-69), (12-70), and the finite correlation length over the frequency of $N_e(\Omega_k, \theta_0)$, it follows that the double sum over the frequency in $Q_F(\theta_0)$ (12-66) reduces to a single sum

$$Q_{F[r,s]}(\theta_0) = \frac{4}{F}\sum_{k \in \mathbb{K}} \mathbb{E}\{\text{Re}(N_e^H(\Omega_k, \theta_0)\hat{C}_{\hat{e}}^{-1}(\Omega_k, \theta_0)G^{\prime, r}(\Omega_k, \theta_0)U_0(k)) \dots \\ \text{Re}(N_e^H(\Omega_k, \theta_0)\hat{C}_{\hat{e}}^{-1}(\Omega_k, \theta_0)G^{\prime, s}(\Omega_k, \theta_0)U_0(k))\} + O(N^{-\beta}) \quad (12\text{-}71)$$

with $\beta = R + 3/2$ if the transient has been suppressed nonparametrically, and $\beta = 1$ otherwise. Eq. (12-71) shows that the correlation over the frequency of the generalized sample mean $\hat{Z}(k)$ and sample covariance $\hat{C}_{\hat{Z}}(k)$ does not influence the asymptotic ($F = O(N) \to \infty$) expression of the covariance matrix of the model parameters (12-65).

SECOND STEP. We have to distinguish here between on the one hand the "robust" method (Section 12.2.2.1) and the other hand the "fast" method (Section 12.2.2.2) and the local polynomial method for arbitrary excitations (= "arb" method, see Section 12.2.1). Indeed, in both the "fast" and "arb" methods a local polynomial approximation of the frequency response matrix is made which is not the case for the "robust" method. As a consequence, the correlation over the frequency of the generalized sample mean in the "robust" method is only due to the transient suppression. Since we study the asymptotic ($F = O(N) \to \infty$) covariance of the model parameters, the correlation over the frequency of the generalized sample mean is asymptotically zero for the "robust" method, which is not the case for the "fast" and "arb" methods. Although this result is not valid for the generalized sample covariance of the "robust" method (it remains correlated due to the averaging over neighboring frequencies, see (7-88)), we conclude that (12-23) is the multi-output extension of the single-output case (10-16) where $M = dof + 1$

$$\text{Cov}(\hat{\theta}_{\text{SML}}(Z)) = \lambda_1(dof)\text{Cov}(\hat{\theta}_{\text{ML}}(Z)) \quad (12\text{-}72)$$

The proof of the particular expression (12-72) follows the same lines of Appendix 10.F, using the first (12-63) and second (12-64) order moments of a complex inverse Wishart distribution (see Mahata et al., 2006 for a detailed derivation within a generalized output error framework).

The reasoning is somewhat more complicated for the "arb" and "fast" methods. Since we study the asymptotic ($F = O(N) \to \infty$) covariance of the model parameters, we neglect the transient (leakage) errors in the analysis. The "arb" and "fast" methods both define a linear transformation between the $F(n_y + n_u) \times 1$ vector Z of the measured input-output DFT spectra, which are uncorrelated over the frequency, and the $F(n_y + n_u) \times 1$ vector $\hat{Z}$ of the generalized sample means of the input-output DFT spectra (see (12-5) and (12-13)), which are correlated over the frequency with a correlation width of $\pm 2n$ ("arb" method) or $\pm 2n_E$ ("fast" method). Hence, within a Gaussian stochastic framework, the information content of $(\hat{Z}, \hat{C}_{\hat{Z}})$, with $\hat{C}_{\hat{Z}}$ the $F(n_y + n_u) \times F(n_y + n_u)$ sample covariance of $\hat{Z}$, is exactly the same

Section 12.7 ■ Appendices

as $(Z, \hat{C}_Z)$, with $\hat{C}_Z$ the $F(n_y + n_u) \times F(n_y + n_u)$ sample covariance of Z (the sample mean and sample covariances are sufficient statistics for the parameters of a normal distribution Kendall and Stuart, 1979). Hence, the covariance matrices of the corresponding sample maximum likelihood estimators $\hat{\theta}_{\hat{Z}}$ and $\hat{\theta}_Z$ are the same

$$\text{Cov}(\hat{\theta}_{\hat{Z}}) = \text{Cov}(\hat{\theta}_Z) \tag{12-73}$$

Since the correlation over the frequency has been neglected in the SML cost function (12-15), the covariance matrix of the minimizer $\hat{\theta}_{\text{SML}}(Z)$ of (12-15) is larger than that of the SML estimate $\hat{\theta}_{\hat{Z}}$ using the correlation over the frequency

$$\text{Cov}(\hat{\theta}_{\text{SML}}(Z)) \geq \text{Cov}(\hat{\theta}_{\hat{Z}}) \tag{12-74}$$

Combining (12-73) and (12-74) gives

$$\text{Cov}(\hat{\theta}_{\text{SML}}(Z)) \geq \text{Cov}(\hat{\theta}_Z) = \lambda_1(dof)\text{Cov}(\hat{\theta}_{\text{ML}}(Z)) \tag{12-75}$$

with $\lambda_1(dof)$ defined in (12-23), and $\hat{\theta}_{\text{ML}}(Z)$ the maximum likelihood estimator using the original data Z and its true covariance $\text{Cov}(Z)$. The equality in (12-75) follows from (12-72).

Appendix 12.I Proof of the Pseudo-Jacobian (12-25)

To simplify the notations we take a cost function of the form

$$V(\theta) = e^H(\theta)C_e^{-1}(\theta)e(\theta) \tag{12-76}$$

with corresponding pseudo-Jacobian matrix $J_+(\theta)$

$$J_{+[:,r]}(\theta) = C_e^{-1/2}(\theta)\left(\frac{\partial e(\theta)}{\partial \theta_{[r]}} - \frac{1}{2}\frac{\partial C_e(\theta)}{\partial \theta_{[r]}}C_e^{-1}(\theta)e(\theta)\right) \text{ for } r = 1, 2, ..., n_\theta \tag{12-77}$$

and where $C_e^{1/2}$ is a square root of C_e ($C_e = C_e^{1/2}C_e^{H/2}$). In the sequel of this appendix we will show that the derivative of the cost function (12-76) w.r.t. θ is given by

$$V'^T(\theta) = 2\text{Re}(J_+^H(\theta)\varepsilon(\theta)) = 2J_{+\text{re}}^T(\theta)\varepsilon_{\text{re}}(\theta) \text{ with } \varepsilon(\theta) = C_e^{-1/2}(\theta)e(\theta) \tag{12-78}$$

It shows that the pseudo-Jacobian matrix (12-77) can be used to construct a Gauss-Newton like iterative scheme

$$2J_{+\text{re}}^T(\theta)J_{+\text{re}}(\theta)\Delta\theta = -2J_{+\text{re}}^T(\theta)\varepsilon_{\text{re}}(\theta) \tag{12-79}$$

where $2J_{+\text{re}}^T(\theta)J_{+\text{re}}(\theta)$ is an approximation of the second order derivative $V''(\theta)$.

Using $\partial C_e^{-1}/\partial \theta_{[r]} = -C_e^{-1}\partial C_e/\partial \theta_{[r]}C_e^{-1}$, the derivative of the cost function $V(\theta)$ w.r.t. $\theta_{[r]}$ can be elaborated as

$$\frac{\partial V(\theta)}{\partial \theta_{[r]}} = 2\text{Re}\left(\left(\frac{\partial e(\theta)}{\partial \theta_{[r]}}\right)^H C_e^{-1}(\theta)e(\theta) - \frac{1}{2}e^H(\theta)C_e^{-1}(\theta)\frac{\partial C_e(\theta)}{\partial \theta_{[r]}}C_e^{-1}(\theta)e(\theta)\right)$$

$$= 2\text{Re}\left(\left(\left(\frac{\partial e(\theta)}{\partial \theta_{[r]}}\right)^H C_e^{-H/2}(\theta) - \frac{1}{2}e^H(\theta)C_e^{-1}(\theta)\frac{\partial C_e(\theta)}{\partial \theta_{[r]}}C_e^{-H/2}(\theta)\right)C_e^{-1/2}(\theta)e(\theta)\right) \quad (12\text{-}80)$$

$$= 2\text{Re}(J_{+[:,\,r]}^H(\theta)\varepsilon(\theta))$$

Putting the contributions (12-80), $r = 1, 2, \ldots, n_\theta$, on top of each other gives (12-78). □

Appendix 12.J Proof of (12-30)

Retaining the dominant term in the pseudo-Jacobian $J_{+[:,\,r]}^{[l,\,k]}(\theta, Z)$ (12-25)

$$(\hat{C}_{\hat{e}}^{[l]}(\Omega_k, \theta))^{-1/2}\frac{\partial e(\Omega_k, \theta, \hat{Z}^{[l]}(k))}{\partial \theta_{[r]}}, \quad (12\text{-}81)$$

entry $[r, s]$ of $J_+^H J_+$ can be approximated as

$$(J_+^H J_+)_{[r,\,s]} \approx \sum_{k \in \mathbb{K},\,l} \left(\frac{\partial e(\Omega_k, \theta, \hat{Z}^{[l]}(k))}{\partial \theta_{[r]}}\right)^H (\hat{C}_{\hat{e}}^{[l]}(\Omega_k, \theta))^{-1}\left(\frac{\partial e(\Omega_k, \theta, \hat{Z}^{[l]}(k))}{\partial \theta_{[s]}}\right) \quad (12\text{-}82)$$

with $\text{ord}(\mathbb{K}) = F$, and where the sum over l runs from 1 to n_{exp}. Since $(J_+^H J_+)_{[r,\,s]}/F$ converges to its expected value (law of large numbers), (12-82) can be elaborated as

$$(J_+^H J_+)_{[r,\,s]} \approx \sum_{k \in \mathbb{K},\,l} \mathbb{E}\left\{\left(\frac{\partial e(\Omega_k, \theta, \hat{Z}^{[l]}(k))}{\partial \theta_{[r]}}\right)^H (\hat{C}_{\hat{e}}^{[l]}(\Omega_k, \theta))^{-1}\left(\frac{\partial e(\Omega_k, \theta, \hat{Z}^{[l]}(k))}{\partial \theta_{[s]}}\right)\right\}$$

$$\approx \sum_{k \in \mathbb{K},\,l} \mathbb{E}\left\{\left(\frac{\partial e(\Omega_k, \theta, \hat{Z}^{[l]}(k))}{\partial \theta_{[r]}}\right)^H \mathbb{E}\{(\hat{C}_{\hat{e}}^{[l]}(\Omega_k, \theta))^{-1}\}\left(\frac{\partial e(\Omega_k, \theta, \hat{Z}^{[l]}(k))}{\partial \theta_{[s]}}\right)\right\} \quad (12\text{-}83)$$

$$\approx \frac{dof}{dof - n_y}\sum_{k \in \mathbb{K},\,l} \mathbb{E}\left\{\left(\frac{\partial e(\Omega_k, \theta, \hat{Z}^{[l]}(k))}{\partial \theta_{[r]}}\right)^H (C_e^{[l]}(\Omega_k, \theta))^{-1}\left(\frac{\partial e(\Omega_k, \theta, \hat{Z}^{[l]}(k))}{\partial \theta_{[s]}}\right)\right\}$$

$$\approx \frac{dof}{dof - n_y}(J_{\text{ML}+}^H J_{\text{ML}+})_{[r,\,s]}$$

where the second equality uses the asymptotic ($F = O(N) \to \infty$) independence of the generalized sample mean and sample covariance (Lemmas 12.2, and 12.4–12.6), and the third equality (12-63).

Eq. (12-83) shows that $\lambda_1(dof)$ in (12-23) must be multiplied by $dof/(dof - n_y)$ when evaluating the right-hand side in the generalized sample covariances. This explains the factor $\lambda_2(dof)$ in (12-30). Using $\text{Re}(J_+^H J_+) = J_{+\text{re}}^T J_{+\text{re}}$ (see Lemma 15.4) and $J_{+\text{re}} = U_J \Sigma_J V_J^T$ explains the other factor $(V_J \Sigma_J^+)(V_J \Sigma_J^+)^T$. □

Appendix 12.K Proof of Theorem 12.9

We first assume that the generalized sample mean $\hat{Z}(k)$ and sample covariance $\hat{C}_{\hat{Z}}(k)$ are not correlated over the frequency. Although this assumption does not influence the expected value of the global minimum of the cost function it has an impact on the variance. Afterwards, a *rationale* for the influence of the correlation over the frequency on the variance of the cost function is given. The analysis is made for "F sufficiently large" such that (i) the bias of the generalized sample means and sample covariances can be neglected, (ii) the generalized sample means and sample covariances are independently distributed (mutually and over the frequency), and (iii) the noise influence on the estimated model parameters $\hat{\theta}_{\text{SML}}(Z)$ can be quantified via a linear approximation.

Let X_r denote one term in the double sum of the global minimum of (12-15), and assume first that the noise on $\hat{\theta}_{\text{SML}}(Z)$ can be neglected. For "F sufficiently large" it is approximately complex Hotelling's T^2-distributed

$$X_r \in \frac{n_y \, dof}{dof - n_y + 1} F(2n_y, 2(dof - n_y + 1)) \tag{12-84}$$

((16-32b) with $R = dof + 1$, and $n = n_y$). Using the mean value and variance of an $F(n_1, n_2)$-distributed random variable Y

$$\mathbb{E}\{Y\} = \frac{n_2}{n_2 - 2} \quad \text{and} \quad \text{var}(Y) = \frac{2n_2^2(n_1 + n_2 - 2)}{n_1(n_2 - 2)^2(n_2 - 4)} \tag{12-85}$$

we obtain the mean and variance of X_r (12-84)

$$\mathbb{E}\{X_r\} = \frac{n_y \, dof}{dof - n_y} \quad \text{and} \quad \text{var}(X_r) = \frac{n_y \, dof^3}{(dof - n_y)^2 (dof - n_y - 1)} \tag{12-86}$$

Neglecting the correlation over the frequency, the mean value and the variance of the sum of the $n_{\text{exp}} F$ terms X_r equal

$$\mathbb{E}\left\{\sum_{r=1}^{n_{\text{exp}} F} X_r\right\} = n_{\text{exp}} F \, \mathbb{E}\{X_r\} \quad \text{and} \quad \text{var}\left(\sum_{r=1}^{n_{\text{exp}} F} X_r\right) \approx n_{\text{exp}} F \, \text{var}(X_r) \tag{12-87}$$

Combining (12-86) and (12-87) gives

$$\mathbb{E}\{V_{\text{SML}}(\hat{\theta}_{\text{SML}}(Z), Z)\} \approx \frac{n_{\text{exp}} F n_y \, dof}{dof - n_y}$$

$$\text{var}(V_{\text{SML}}(\hat{\theta}_{\text{SML}}(Z), Z)) \approx \frac{n_{\text{exp}} F n_y \, dof^3}{(dof - n_y)^2 (dof - n_y - 1)} \tag{12-88}$$

The noise on the estimated model parameters $\hat{\theta}_{\text{SML}}(Z)$ introduces a correlation among the $2n_{\text{exp}} F n_y$ real residuals that decreases the real degrees of freedom to $2n_{\text{exp}} n_y F - n_\theta$ (proof: follow the same lines of Appendix 19.I). Hence, replacing $n_{\text{exp}} F n_y$ by $n_{\text{exp}} F n_y - n_\theta / 2$ in

(12-88) finally proves (12-37) for the "arb", "fast" and "robust" methods and the lower bound of (12-38) for the "robust" method.

For the "arb" and "fast" methods the correlation of the generalized sample means over $\pm n_{CL}$ neighboring frequencies increases the variance. If we concentrate the correlation over the two nearest frequencies with a correlation coefficient equal to 1, then we get the upper bound σ_1^2 in (12-38). Numerous MATLAB® simulations confirm this bound.

To prove the asymptotic normality of $V_{SML}(\hat{\theta}_{SML}(Z), Z)$ using version 2 of the central limit theorem (see Section 16.10), it is sufficient to note that (i) each term X_r of the cost function converges for $F = O(N) \to \infty$ to a complex Hotelling's T^2-distribution (12-84) whose third order moments are finite if $dof > n_y + 2$, and (ii) the finite correlation length over the frequency does not affect the central limit theorem. □

Appendix 12.L Properties Generalized Sample Means and Sample Covariances in the Presence of Nonlinear Distortions

The major differences between the stochastic nonlinear distortions $Y_S(k)$ and the disturbing noise is that (i) $Y_S(k)$ is uncorrelated with – but not independent of – the input $U_0(k)$ (open loop) or the reference signal $R(k)$ (closed loop), and (ii) $Y_S(k)$ is mixing of order infinity over the frequency with $\mathbb{E}\{Y_S(k)\overline{Y_S}(l)\} = O(N^{-1})$ for $k \neq l$ (compare Theorems 3.16, 3.17, 3.20 and 3.22 to Theorem 16.25). Hence, the proofs in Appendices 12.A to 12.D and 12.F remain valid in the presence of stochastic nonlinear distortions, except that the matrices C_n and D_n in (12-45) and (12-49) are no longer (block) diagonal

$$\begin{aligned} C_n &= \text{diag}(C_V(k-n)_{[i,j]}, ..., C_V(k+n)_{[i,j]}) + O_1(N^{-1}) \\ D_n &= \text{blockdiag}(C_V(k-n), ..., C_V(k+n)) + O_2(N^{-1}) \end{aligned} \quad (12\text{-}89)$$

with $C_V(k) = \text{Cov}(H(\Omega_k)E(k)) + \text{Cov}(Y_S(k))$, and where $O_1(N^{-1})$ and $O_2(N^{-1})$ are full matrices of size $(2n+1) \times (2n+1)$ and $(2n+1)n_y \times (2n+1)n_y$, respectively, whose elements decrease as an $O(N^{-1})$. Moreover, due to the mixing property of $Y_S(k)$ over k, the correlation of $Y_S(k)$ over k satisfies (16-36) with $K = F$, so that

$$\sum_{r,s} |\alpha_{[s,r]}||O_{i[r,s]}(N^{-1})| = O(n/N) \text{ for } i = 1, 2 \quad (12\text{-}90)$$

where $\alpha_{[s,r]}$ is independent of n and N.

Combining (7-115), (12-45), (12-89), and (12-90) shows that an additional bias error term $O(n/N)$ appears in (12-46) originating from $O_1(N^{-1})$ in (12-89). It proves already (12-3) with $C_{\hat{y}}(k)$ defined in (12-42) for the "arb" and "fast" methods. The proof of (12-10) with $C_{\hat{z}}(k)$ defined in (12-42) for the "robust" method follows immediately from Appendix 7.L.

Combining (7-115), (12-49), (12-89), and (12-90) shows that an additional bias error term $O(n/N)$ appears in (12-51) originating from $O_2(N^{-1})$ in (12-89). Hence, (12-51) and (12-52) remain valid in the presence of nonlinear distortions.

Appendix 12.M Covariance Model Parameters and Variance SML Cost Function in the Presence of Nonlinear Distortions

In this appendix we use the properties of the stochastic nonlinear distortions $Y_S(k)$ described in Theorems 3.16, 3.17, 3.20, and 3.22. In the sequel of this appendix we denote by $U_0(k)$ that part of the input DFT spectrum that depends on the reference signal $R(k)$ only.

12.M.1 Asymptotic Covariance of the Model Parameters. For simplicity of notation we assume here that only the output is affected by the nonlinear distortions. Eqs. (12-65) to (12-70), where the true linear system $G_0(\Omega)$ is replaced by the BLA $G_{\text{BLA}}(\Omega)$, and $N_e(\Omega_k, \theta)$ by $N_e(\Omega_k, \theta) + Y_S(k)$, remain valid in the presence of nonlinear distortions. What changes is the expression for $Q_{F[r,s]}(\theta_0)$ (12-71). Indeed, the double sum over the frequencies in (12-66) contains contributions of the form $\mathbb{E}\{\overline{U_{0[r]}(k)}Y_{S[p]}(k)\overline{Y_{S[q]}(l)}U_{0[s]}(l)\}$ that can be elaborated as

$$\mathbb{E}\{\overline{U_{0[r]}(k)}Y_{S[p]}(k)\overline{Y_{S[q]}(l)}U_{0[s]}(l)\} = \mathbb{E}\{Y_{S[p]}(k)\overline{Y_{S[q]}(k)}\}\mathbb{E}\{\overline{U_{0[r]}(k)}U_{0[s]}(l)\}\delta(k-l) + O_{[k,l]}(N^{-1}) \qquad (12\text{-}91)$$

with $\delta(k-l)$ the Kronecker delta, and where the $O_{[k,l]}(N^{-1})$ term accounts for the weak dependency between $Y_S(k)$ and $U_0(k)$ (see Appendix 3.L and Schoukens and Pintelon, 2010b). The double sum over the first term in (12-91) reduces to a single sum of the form (12-71), while the double sum over the second term in (12-91) gives an $O(F^0)$ contribution to $Q_{F[r,s]}(\theta_0)$ that cannot be neglected w.r.t. the first contribution. Therefore, (12-23) and (12-30) only indicate the order of magnitude of the asymptotic covariance of the model parameters.

12.M.2 Mean Value of the SML Cost Function. To prove (12-37) in the presence of nonlinear distortions it is sufficient to note that $Y_S(k)$ and $U_0(k)$ are uncorrelated.

12.M.3 Variance of the SML Cost Function. The noise on the estimate $\hat{\theta}_{\text{SML}}(Z)$ is for $F = O(N) \to \infty$ a second order effect in the variance of the SML cost function (12-15) and, therefore, it will be neglected here. To prove (12-38) it is sufficient to note that (i) $V_{\text{SML}}(\theta_0, Z)$ is independent of $U_0(k)$, and (ii) due to the mixing of order ∞ property of $Y_S(k)$ and property (iv) of Theorem 3.9, where G_S is replaced by Y_S, the correlation over the frequency of $Y_S(k)$ has an $O(F^0)$ contribution to $\text{var}(V_{\text{SML}}(\theta_0, Z))$.

12.M.4 Whiteness Test of the Residuals. To simplify the notations we consider the single-input, single-output case. Using Theorems 3.9 and 3.25 and property (12-4) we will show that the expected value and the variance of the sample correlation $\hat{R}(m)$ (12-40) is asymptotically ($F = O(N) \to \infty$) not affected by the correlation of $G_S(k)$ over the frequency. For N sufficiently large the numerator of the residual (12-39) can be approximated by $G_S(k) + N_G(k)$, where $N_G(k)$ quantifies the noise on the BLA estimate $\hat{G}(\Omega_k)$. Using property (ii) of Theorem 3.9 and (12-4), and taking into account that $N_G(k)$ satisfies Assumption 3.6, the expected value of $\hat{R}(m)$, $m \neq 0$, is given by

$$\mathbb{E}\{\hat{R}(m)\} = \frac{\alpha_1(m)}{F-|m|}\sum_{k=1}^{F-|m|}\mathbb{E}\{\hat{\varepsilon}_G(k)\overline{\hat{\varepsilon}_G(k+m)}\}$$

$$= \frac{\alpha_1(m)}{F-|m|} \sum_{k=1}^{F-|m|} (\mathbb{E}\{\hat{\varepsilon}_G(k)\}\mathbb{E}\{\overline{\hat{\varepsilon}_G(k+m)}\} + O(N^{-1})) \qquad (12\text{-}92)$$

$$= O(N^{-1})$$

(to simplify the notations we use all frequencies and neglect the finite correlation length over the frequency). The variance of $\hat{R}(m)$, $m \neq 0$, can be elaborated as

$$\text{var}(\hat{R}(m)) = \frac{\alpha_1^2(m)}{(F-|m|)^2} \Bigg(\sum_{k=1}^{F-|m|} \mathbb{E}\{|\hat{\varepsilon}_G(k)|^2 |\hat{\varepsilon}_G(k+m)|^2\} + \ldots$$

$$\sum_{k,l=1, k\neq l}^{F-|m|} \mathbb{E}\{\hat{\varepsilon}_G(k)\overline{\hat{\varepsilon}_G(k+m)}\overline{\hat{\varepsilon}_G(l)}\hat{\varepsilon}_G(l+m)\} \Bigg) \qquad (12\text{-}93)$$

where

$$\mathbb{E}\{|\hat{\varepsilon}_G(k)|^2 |\hat{\varepsilon}_G(k+m)|^2\} = \mathbb{E}\{|\hat{\varepsilon}_G(k)|^2\}\mathbb{E}\{|\hat{\varepsilon}_G(k+m)|^2\} + O(N^{-1}) \qquad (12\text{-}94)$$

(use property (iv) of Theorem 3.9 and (12-4)), and

$$\mathbb{E}\{\hat{\varepsilon}_G(k)\overline{\hat{\varepsilon}_G(k+m)}\overline{\hat{\varepsilon}_G(l)}\hat{\varepsilon}_G(l+m)\} = O(N^{-2}) \qquad (12\text{-}95)$$

(use property 5 of Theorem 3.25 and (12-4)). We conclude from (12-94) and (12-95) that the contributions of the correlation of $G_S(k)$ over the frequency to (12-93) are, respectively, an $O(1/(N(F-|m|)))$ and $O(N^{-2})$, while the main contribution (the first term in (12-94)) is an $O(1/(F-|m|))$.

Appendix 12.N Linear Plant and Nonlinear Actuator or Controller

The stochastic nonlinear distortions produced by either the actuator or the controller satisfy Theorems 3.16, 3.17, 3.20 and 3.22 and, therefore, the proofs of Appendices 12.L and 12.M remain valid, except that the covariance expression (12-23) and (12-30) are still correct as in the linear case. The latter is proven as follows. For a linear plant and a nonlinear actuator or controller, $N_e(\Omega_k, \theta)$ in (12-67) equals

$$N_e(\Omega_k, \theta) = V_Y(k) - G(\Omega_k, \theta)V_U(k) + Y_S(k) - G(\Omega_k, \theta)U_S(k) \qquad (12\text{-}96)$$

where $Y_S(k)$ and $U_S(k)$ satisfy (7-52). Hence, evaluating (12-96) at $\theta = \theta_0$ using (7-52) shows that $N_e(\Omega_k, \theta_0) = V_Y(k) - G(\Omega_k, \theta_0)V_U(k)$ is independent of the nonlinear distortions. Therefore, $N_e(\Omega_k, \theta_0)$ in (12-69) is independent of $U_0(k)$, which is the basic property used to obtain (12-71).

13

Basic Choices in System Identification

Abstract: In this chapter we discuss five fundamental questions regarding very basic aspects of the identification process. The first one deals with the signal assumption that is made to reconstruct the intersample behavior. Two possibilities are considered: zero-order-hold reconstruction and band-limited reconstruction. The second question handles the choice between nonparametric and parametric noise models. The third question looks into the selection of the excitation signal, dealing mainly with the choice between periodic or nonperiodic excitations. The choice between time domain or frequency domain identification is the topic of the fourth part of this chapter. Finally, the fifth and final part discusses the issue of imposing physical constraints on the identified model such as stability, reciprocity, and passivity.

13.1 INTRODUCTION

At the beginning of the 1970s, the identification field gave a quite disordered impression. Many methods were proposed to identify linear dynamic systems. However, due to the lack of integration, the whole field looked more like a "bag of tricks" than a consistent scientific discipline. In the 1980s the field became well ordered, pointing out the relations and the differences between the widely scattered methods (Eykhoff, 1974; Ljung, 1999; Norton, 1986; Söderström and Stoica, 1989). The major part of this work was done in the time domain, leading to a complete dominance of these methods over frequency domain techniques. Since then, new methods have popped up, some of them being applicable to time domain or frequency domain identification (for example, subspace methods: McKelvey et al., 1996; Van Overschee and De Moor, 1994 and 1996b; Verhaegen, 1994; Viberg et al., 1997), others being completely focused on frequency domain identification. This does not lead us back to chaos, because the clear insight of the 1980s still applies to the new situations. Rather, these reviving approaches just complete the puzzle, making the picture better balanced. As explained in Chapter 1, the identification process is mainly determined by the answers to three basic questions: (i) what data will be used (experiment design), (ii) what model will be used (model selection), and (iii) how will the model be matched to the data (choice of a cost function)? These questions have to be answered, independent of the user's intention to work in the time domain or in the frequency domain. Perpendicular to these questions, two other important choices have to be made. (i) Choice of the intersample behavior. As identification starts mostly from discrete data, an assumption is needed to make precise what is going on between the samples. This choice has a major impact on the experimental setup, the model choice, and the selection of the cost function. (ii) Periodic versus arbitrary excitation. This question is not linked to the choice between time or frequency domain identification. Periodic excitations offer significant advantages whenever they can be applied, and this is almost independent of the domain (time or frequency) that is selected to process the measurements.

In this chapter we analyze the consequences of these two basic choices that have to be made. Too often, no conscious selection is made, although the consequences maintain their full impact on the users' result. Therefore, it is important to make a well-considered selection, at the beginning of the process, to avoid undesired surprises at the end. The discussions are made without any prejudices to a specific application, so that for some fields the risk exists that some parts of this chapter are not relevant. For that reason we separated the objective facts, which are true without any discussion, from the interpretation of these facts, where we look for their (un)importance for specific fields. The latter is much more subjective as it is strongly influenced by our personal experiences. Hence, we strongly advise readers to test these sections according to their own experiences and to draw their own conclusions.

Finally we also deal with the choice between time and frequency domain methods. Too often this selection is presented as conflicting options. To address this, we first point to the (sometimes even unexpected) equivalences between time and frequency domain methods. Eventually, we zoom in on the differences so that at the end of the chapter the user should be able to select the most dedicated method for his/her problem.

13.2 INTERSAMPLE ASSUMPTIONS: FACTS

Nowadays, almost every identification scheme is applied to sampled data. In a first step, the continuous-time signals are sampled in time and stored in the computer. The discrete samples do not carry all information contained in the original signals, unless additional assumptions are made on the intersample behavior. What is going on between the samples? As we did not measure this information, we do not know. We can only make a guess, formalized as an assumption, and hope that in practice the real behavior is close to the assumed one. Two assumptions are very popular. The zero-order-hold (ZOH) assumption considers the signal to be constant between consecutive samples, while under the band-limited (BL) assumption we suppose that the power spectrum of the signal is zero above half the sampling frequency $f_{max} < f_s/2$. We discuss, in detail, the impact of this choice on the experimental setup, the model, and the identification process. We also analyze what happens if the wrong assumption is applied to a given set of data, for example, band-limited data are processed under the ZOH assumption.

13.2.1 Formal Description of the Zero-Order-Hold and Band-Limited Assumptions

Consider a discrete-time signal $u_d(kT_s)$. Notice that in this section we will sometimes, explicitly, mention the sampling period T_s. This is to indicate that these samples are generated at the time instances kT_s and that the spectrum of this signal is periodic with period $f_s = 1/T_s$ as shown in Figure 13-1.

Assumption 13.1 (Zero-Order-Hold Assumption): The zero-order-hold (ZOH) reconstruction of a discrete-time signal $u_d(kT_s)$ is

$$u_{ZOH}(t) = \sum_{k=-\infty}^{\infty} u_d(kT_s)\text{zoh}(t - kT_s) \qquad (13\text{-}1)$$

Section 13.2 ■ Intersample Assumptions: Facts

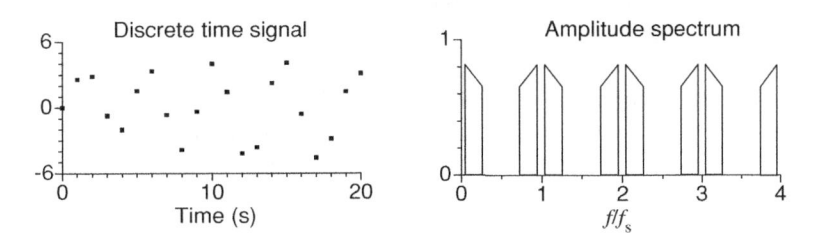

Figure 13-1. A discrete-time signal and part of its periodic spectrum.

with $zoh(t) = 1$ for $0 \leq t < T_s$ and 0 elsewhere. The spectrum, after a ZOH reconstruction, is (see Exercise 13.1)

$$X_{ZOH}(j\omega) = U_d(j\omega)ZOH(\omega/\omega_s) \qquad (13\text{-}2)$$

with $ZOH(x) = T_s e^{-j\pi x}\sin(\pi x)/(\pi x)$.

Assumption 13.2 (Band-Limited Assumption): A signal $u(t)$ with power spectrum $\Phi(\omega)$ is called band-limited (BL) if there exists a value ω_{max} such that $\Phi(\omega) = 0$ for $\forall \, |\omega| > \omega_{max}$.

In Figure 13-2(a), both reconstructions are illustrated. The reconstructed signals and their spectra differ considerably. The steps in a ZOH reconstruction create high-frequency components far above the sampling frequency, but this is not the case for the BL reconstruction.

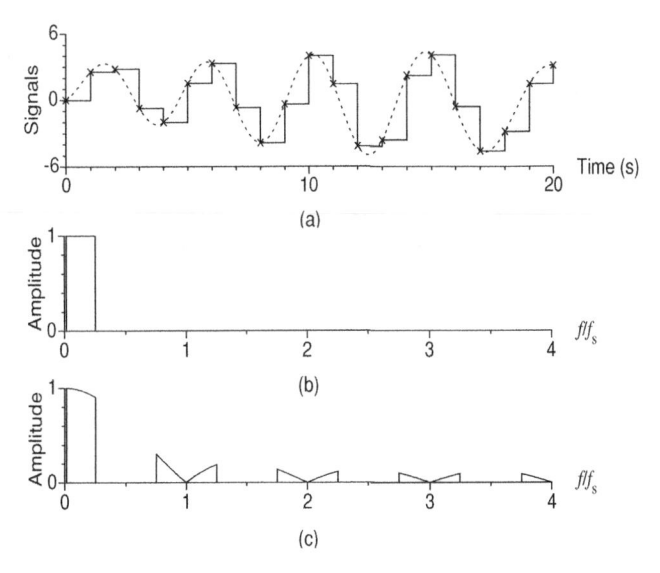

Figure 13-2. Reconstruction of a discrete-time signal under the BL and ZOH assumptions. (a) The time signals: x samples, --- BL, —— ZOH; (b) spectrum of the BL reconstruction; and (c) part of the spectrum of the ZOH reconstruction.

13.2.2 Relation between the Intersample Behavior and the Model

13.2.2.1 Zero-Order-Hold Assumption. In this setup, the transfer function between the discrete-time signal $u_d(k)$ and the output $y(k)$ is measured (Figure 13-3). This means that besides the linear system itself, the actuator and measurement channel are also modeled as indicated by the gray area in Figure 13-3. Assume, for simplicity, that the disturbing noise sources are zero. The overall continuous-time system transfer function $G_c(j\omega)$ comprises the actuator $G_{act}(j\omega)$, the process $G(j\omega)$, and the data acquisition $G_y(j\omega)$ transfer function. Under the ZOH setup, it is modeled as a discrete-time system with impulse response $g_{ZOH}(k)$ that links the discrete-time input $u_d(k)$ to the discrete-time output $y(k)$ (Ljung, 1999):

$$y(m) = \sum_{k=1}^{\infty} g_{ZOH}(k) u_d(m-k) \text{ with } g_{ZOH}(k) = \int_{(k-1)T_s}^{kT_s} g_c(\tau) d\tau \quad (13\text{-}3)$$

where $g_c(\tau)$ is the impulse response of the continuous-time system, between $u_{ZOH}(t)$ and the continuous-time output $y_{AA}(t)$.

$G_{ZOH}(z^{-1})$ and $G_c(s)$ are linked by the step-invariant transformation for ZOH excitations (Middleton and Goodwin, 1990; see also Example 6.3):

$$G_{ZOH}(z^{-1}) = (1 - z^{-1}) Z\{G_c(s)/s\} \quad (13\text{-}4)$$

If $f_s/2 > \text{Im(poles)}/(2\pi)$, then the original continuous-time parameters can be retrieved from the discrete-time model, using an inverse transformation (Ljung, 1999). The poles are found using the impulse invariant transformation, but the transformation of the zeros is much more complex (Åström et al., 1984). The influence of $G_y(s)$ in (13-4) can be eliminated via an absolute amplitude and phase calibration of the data acquisition channel. While the amplitude characteristic $|G_y(j\omega)|$ can easily be measured using a calibrated power meter, the measurement of the phase characteristic $\angle G_y(j\omega)$ requires a phase calibrated excitation signal. No standard (commercial) solutions are available for the latter.

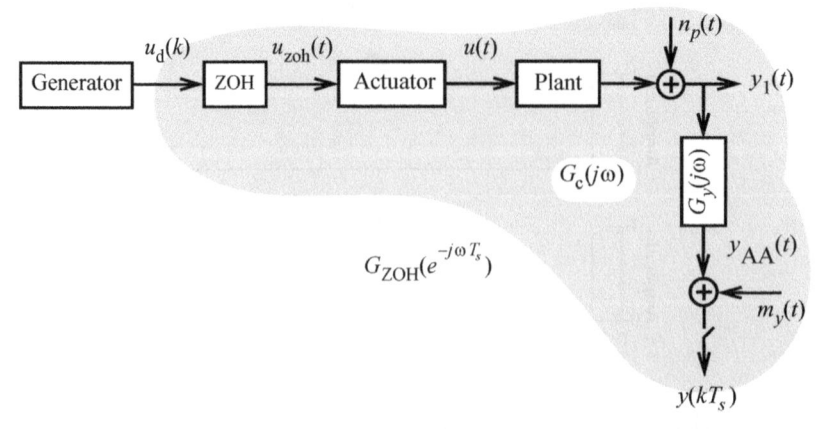

Figure 13-3. Basic setup for the ZOH assumption, interpretation of the continuous-time ($G_c(s)$) and the equivalent discrete-time ($G_{ZOH}(z^{-1})$) system.

For $|\omega| < \omega_s/2$, the final relation between the discrete input spectrum and the spectrum of the sampled output signal is given by (see Exercise 13.2)

$$Y(e^{j\omega T_s}) = U_d(e^{j\omega T_s}) \sum_{k=-\infty}^{\infty} G(j\Omega_k)A(j\Omega_k)G_y(j\Omega_k)\text{ZOH}(\Omega_k/\omega_s)\Big|_{\Omega_k = \omega - k\omega_s} \quad (13\text{-}5)$$

where $A(j\omega)$ represents the actuator dynamics. The sum in (13-5) is due to the repeated spectra, as they appear in the spectrum of $u_{\text{ZOH}}(k)$ (see Figure 13-2).

At the sampling instances the output noise $n_y(t) = n_p(t) + m_y(t)$ can be modeled exactly as discrete-time filtered white noise if the unobserved continuous-time driving noise source is either a Wiener stochastic process or zero-order-hold noise (see Theorem 6.11). Since the prediction error (10-50) and maximum likelihood (10-58) cost functions do not depend on the phase of the noise filter, an absolute amplitude calibration is sufficient to eliminate the influence of the data acquisition transfer function $G_y(s)$ on the identified parametric noise model.

13.2.2.2 Band-Limited Assumption. The BL setup is given in Figure 13-4. Only the gray box is directly involved in the identification process. Starting from the spectra of the continuous-time signals $U_1(j\omega) = F\{u_1(t)\}$ and $Y_1(j\omega) = F\{y_1(t)\}$, it can easily be shown that the following relations exist between the spectra of the sampled signals $U(e^{j\omega T_s}) = F\{u(k)\}$, $Y(e^{j\omega T_s}) = F\{u(k)\}$:

$$G_{\text{BL}}(j\omega) = \frac{Y(e^{j\omega T_s})}{U(e^{j\omega T_s})} = \frac{G_y(j\omega)Y_1(j\omega) + \sum_{k=-\infty, k\neq 0}^{\infty} G_y(j\Omega_k)Y_1(j\Omega_k)\Big|_{\Omega_k = \omega - k\omega_s}}{G_u(j\omega)U_1(j\omega) + \sum_{k=-\infty, k\neq 0}^{\infty} G_u(j\Omega_k)U_1(j\Omega_k)\Big|_{\Omega_k = \omega - k\omega_s}} \quad (13\text{-}6)$$

(see Exercise 13.3).

The sum terms in this expression are due to the alias effect of the sampling process (see Section 2.2.1). If the measurement channels are provided with good anti-alias filters with a cutoff frequency below $\omega_s/2$, the band-limited assumption holds and (13-6) becomes

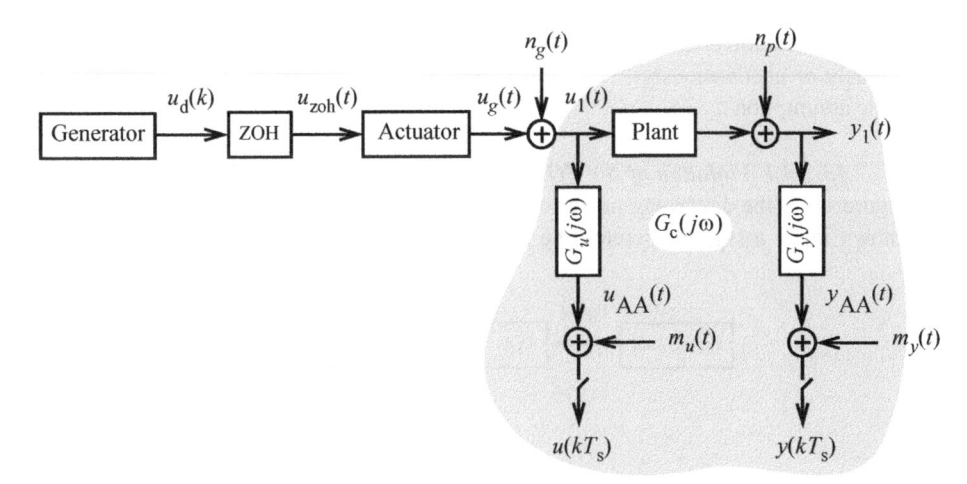

Figure 13-4. BL measurement setup.

$$G_{BL}(j\omega) = \frac{G_y(j\omega)Y_1(j\omega)}{G_u(j\omega)U_1(j\omega)} = G(j\omega)\frac{G_y(j\omega)}{G_u(j\omega)} \text{ for } |\omega| < \omega_s/2 \qquad (13\text{-}7)$$

which shows that $G_{BL}(j\omega) = G(j\omega)$ for $|\omega| < \omega_s/2$ if $G_y(j\omega) = G_u(j\omega)$ in this frequency band. The influence of the data acquisition channels in (13-7) is eliminated via a relative calibration: a periodic signal exciting the frequency band $(0, f_s/2]$ is applied simultaneously to both acquisition channels and the frequency response function from the input channel to the output channel is measured, giving $G_y(j\omega)/G_u(j\omega)$.

Note that the model $G_{BL}(j\omega) = G(j\omega)$ is the continuous-time representation of the plant. So the model equations are given by differential equations in the time domain and algebraic equations in the frequency domain. The latter are given by the transfer function model formulated in the Laplace domain ($s = j\omega$ is used as frequency variable in the transfer function).

If the input is known then a continuous-time parametric noise can be identified for the output $n_y(t)$ noise if the unobserved driving noise source is continuous-time band-limited white noise with a bandwidth of at least $f_s/2$ (see Theorem 6.14). Since the maximum likelihood (10-58) cost function does not depend on the phase of the noise filter, an absolute amplitude calibration is sufficient to eliminate the influence of the data acquisition transfer function $G_y(s)$ on the identified parametric noise model.

13.2.2.3 Conclusion. The ZOH assumption imposes an experimental condition on the excitation signal; it is generated from a discrete-time sequence using a piecewise constant interpolation. Under these conditions, a discrete-time model is obtained between the discrete-time input and the sampled output.

The BL assumption is a condition on the observation of the signals and does not impose constraints on the applied excitation (e.g., BL observations of ZOH signals can be made). It results in a continuous-time model of the plant in the observed frequency band.

13.2.3 Mixing the Intersample Behavior and the Model

In the previous section it was found that the ZOH assumption leads, naturally, to a discrete-time model, and the BL assumption results in a continuous-time model. If a discrete-time model is combined with non-ZOH inputs, or continuous-time models are identified under the ZOH assumption (without applying anti-alias filters so that the BL condition is violated), systematic errors appear. In practice, these wrong combinations are often made (consciously or unconsciously). Hence, it is important to understand the impact of violating the basic assumption.

13.2.3.1 Violation of the ZOH Assumption. Consider the generalized setup of Figure 13-5 (the disturbing noise sources are not shown for simplicity). Instead of using the known input $u_d(k)$, a discrete-time model is built between the measured input $u(kT_s)$ and

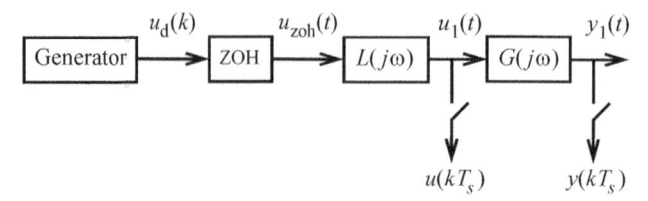

[Generator] → $u_d(k)$ → [ZOH] → $u_{zoh}(t)$ → [$L(j\omega)$] → $u_1(t)$ → [$G(j\omega)$] → $y_1(t)$
                                                                    ↓   ↓
                                                                 $u(kT_s)$ $y(kT_s)$

Figure 13-5. Violation of the ZOH assumption.

output $y(kT_s)$. Because the excitation signal passed through a first subsystem $L(j\omega)$, the signal $u_1(t)$ is no longer ZOH. The discrete-time transfer function $G_L(z^{-1})$ that relates the input samples $u(kT_s)$ to the output samples $y(kT_s)$ is found directly, applying (13-4) twice

$$G_L(z^{-1}) = \frac{(1-z^{-1})Z\{L(s)G(s)/s\}}{(1-z^{-1})Z\{L(s)/s\}} = \frac{Z\{L(s)G(s)/s\}}{Z\{L(s)/s\}} \quad (13\text{-}8)$$

(see Example 6.3 on pages 179-180). Assuming that the sums converge, $G_L(e^{-j\omega T_s})$ can also be written as

$$G_L(e^{-j\omega T_s}) = \frac{\sum_{k=-\infty}^{\infty} L(j\Omega_k)G(j\Omega_k)\text{ZOH}(\Omega_k/\omega_s)\big|_{\Omega_k = \omega - k\omega_s}}{\sum_{k=-\infty}^{\infty} L(j\Omega_k)\text{ZOH}(\Omega_k/\omega_s)\big|_{\Omega_k = \omega - k\omega_s}} \quad (13\text{-}9)$$

for $|\omega| < \omega_s/2$. Note that the result is still independent of the input $U_d(e^{j\omega T_s})$, but it depends on the preceding system $L(s)$. If the same subsystem $G(j\omega)$ is measured in another environment ($L(j\omega) \to \hat{L}(j\omega)$), the resulting model will change. Under these conditions, the model is no longer independent of the measurement environment, and the results cannot be transferred from one setup to the other. However, as long as the setup is not changed, a good description of the measurements is given. If $L(j\omega) = 1$, $\forall \omega$, the original ZOH setup is retrieved (see Exercise 13.4). If L is chosen as a perfect reconstruction filter, $L(j\omega) = 1$ for $|\omega| < \omega_s/2$ and $L(j\omega) = 0$ elsewhere, the BL setup is retrieved and $G_L(e^{-j\omega T_s})$ (13-9) equals $G(j\omega)$ instead of $G_{\text{ZOH}}(e^{-j\omega T_s})$. Assume now that a discrete-time model $G(z^{-1}, \theta)$ is fitted to these BL measurements such that $G(e^{-j\omega T_s}, \theta) = G(j\omega)$ for $|\omega| < \omega_s/2$. If the model $G(z^{-1}, \theta)$, based on BL measurements, is used later under ZOH conditions, the ratio $\gamma(j\omega)$ between the predicted and the actual output is

$$\begin{aligned}\gamma(j\omega) &= \frac{G(e^{-j\omega T_s}, \theta)U_1(e^{j\omega T_s})}{G_{\text{ZOH}}(e^{-j\omega T_s})U_1(e^{j\omega T_s})} \\ &= \frac{G(j\omega)}{G_{\text{ZOH}}(e^{-j\omega T_s})} \qquad (13\text{-}10) \\ &= \left(\text{ZOH}(\omega/\omega_s) + \sum_{k=-\infty, k \neq 0}^{\infty} \frac{G(j\omega - kj\omega_s)}{G(j\omega)}\text{ZOH}(\omega/\omega_s - k)\right)^{-1}\end{aligned}$$

for $|\omega| < \omega_s/2$. $\gamma(j\omega)$ will be close to 1 only if in the frequency band of interest $|G(j\omega - kj\omega_s)| \ll |G(j\omega)|$, $k \neq 0$ and $|\omega/\omega_s| \ll 1$.

13.2.3.2 Applying Continuous Models to ZOH Measurements. The second possibility is to fit a continuous-time model $G(s, \theta)$ to the ZOH measurements such that $G(j\omega, \theta) = G_{\text{ZOH}}(e^{-j\omega T_s})$ for $|\omega| < \omega_s/2$. If the model $G(s, \theta)$, based on ZOH measurements, is used under BL conditions, then the ratio between the predicted and the actual output is

$$\frac{G(j\omega, \theta)U_1(j\omega)}{G(j\omega)U_1(j\omega)} = \frac{G_{\text{ZOH}}(e^{-j\omega T_s})}{G(j\omega)} = \frac{1}{\gamma(j\omega)} \quad (13\text{-}11)$$

where $\gamma(j\omega)$ is defined in (13-10). Again, the same conclusions can be drawn.

13.2.3.3 Conclusion.
In the previous sections it was shown that a continuous-time system can be modeled, without systematic errors, by selecting the proper experimental conditions so that the assumptions that describe the intersample behavior are met. If they are violated, it is still possible to get a good model for the observations, but this model is no longer independent of the measurement environment. The intersample behavior becomes an intrinsic part of the model (see Exercise 13.6 for a feedback example).

13.2.4 Experimental Illustration

The goal of the measurement example is to illustrate the consequences of mixing the intersample behavior and the model. To make the errors (13-10) and (13-11) apparent, a first order continuous-time system is chosen such that the magnitude of the first term in the infinite sum, $|G(j\omega - j\omega_s)/G(j\omega)|$, is not much smaller than one. For higher order systems where the sampling frequency is chosen to be at least ten times the bandwidth of the plant, the factor $\gamma(j\omega)$ is close to one for all frequencies within the system bandwidth.

13.2.4.1 Measurement Setup.
Figure 13-6 shows the block schematic of an RC- circuit. This first order continuous-time plant has been measured using the zero-order-hold (ZOH) and band-limited (BL) setups of Figures 13-3 (ZOH) and 13-4 (BL), where the actuator dynamics satisfies $G_{\text{act}}(j\omega) = 1$, and where the input-output acquisition channels have been calibrated ($G_u(j\omega) = G_y(j\omega)$). The continuous-time transfer function and the corresponding step-invariant transformation (13-4) are given by

$$\text{BL setup:} \quad G(s) = \frac{1}{1 + RCs}$$

$$\text{ZOH setup:} \quad G_{\text{ZOH}}(z^{-1}) = \frac{(1 - e^{-T_s/(RC)})z^{-1}}{1 - e^{-T_s/(RC)}z^{-1}} \tag{13-12}$$

where $RC = 142.7$ μs ± 0.4 μs. For both measurements the same Schroeder multisine (5-10) has been downloaded under digital form in a 12 bit arbitrary waveform generator

$$u_d(t) = \sum_{k=1}^{20} A \sin(2\pi kt/N + \phi_k) \tag{13-13}$$

with $t = 0, 1, ..., N-1$, $N = 1048$, and ϕ_k the Schroeder phases (5-10). The signal (13-13) is generated at the clock frequency $f_c = 10$ kHz, and the acquisition channels are synchronized with the generator ($f_s = f_c$).

The steady state response to the periodic excitation (13-13) is measured. For the BL setup the modeling starts from the noisy input-output DFT spectra $U(k)$, $Y(k)$, $k = 1, 2, ..., 20$, while for the ZOH setup it starts from the DFT $U_d(k)$ of the signal $u_d(t)$ stored in the arbitrary waveform generator and the noisy output DFT spectrum $Y(k)$. To compensate for the gain from the digital samples to the analog output of the waveform generator and the delay between the start of the acquisition and the zero time reference of the generator,

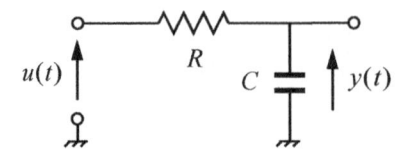

Figure 13-6. Electrical circuit consisting of a series resistor and a parallel capacitor.

the input spectrum of the ZOH measurements is divided by 5.56 and delayed over 31.40 samples (multiplication of $U_d(k)$ by $e^{-31.40 \times j2\pi k/N}/5.56$).

13.2.4.2 Band-limited Measurements – Continuous-Time Model.
Using the prior knowledge (13-12) about the continuous-time plant, the band-limited (BL) measurements are modeled with a first order continuous-time (CT) transfer function of the form

$$G(s, \theta) = \frac{b_0}{a_0 + a_1 s} \qquad (13\text{-}14)$$

The results are shown in Figure 13-7. It follows that the first order CT-model (13-14) explains very well the BL-measurements (amplitude and phase errors of about 0.1%). The RC- time constant calculated from the estimated a_0 and a_1 coefficients equals 142.6 µs.

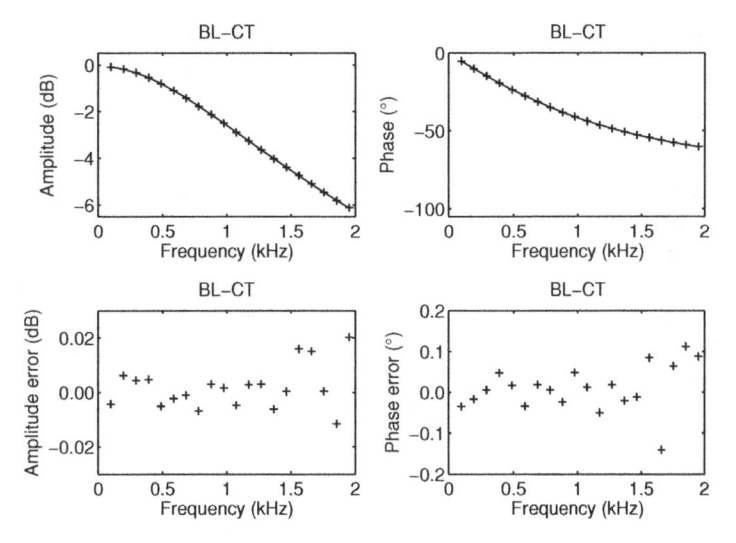

Figure 13-7. Band-limited (BL) measurements modeled with a first order continuous-time (CT) transfer function (13-14). Top row: measured ('+') and modeled (black line) FRF. Bottom row: difference between measurement and estimate.

13.2.4.3 Zero-Order-Hold Measurements – Discrete-Time Model.
Using the prior knowledge (13-12) about the zero-order-hold (ZOH) setup, the ZOH-measurements are modeled with the following first order discrete-time (DT) transfer function

$$G(z, \theta) = \frac{b_1 z^{-1}}{a_0 + a_1 z^{-1}} \qquad (13\text{-}15)$$

The results are shown in Figure 13-8. It can be seen that the first order DT-model (13-15) explains very well the ZOH-measurements (amplitude and phase errors of about 0.1%). Note that the phase error at the first two frequencies is much higher than in Figure 13-7. This indicates that it more difficult to realize a perfect ZOH-setup than a perfect BL-setup. The RC- time constant calculated from the estimated a_0 and a_1 coefficients equals 142.9 µs.

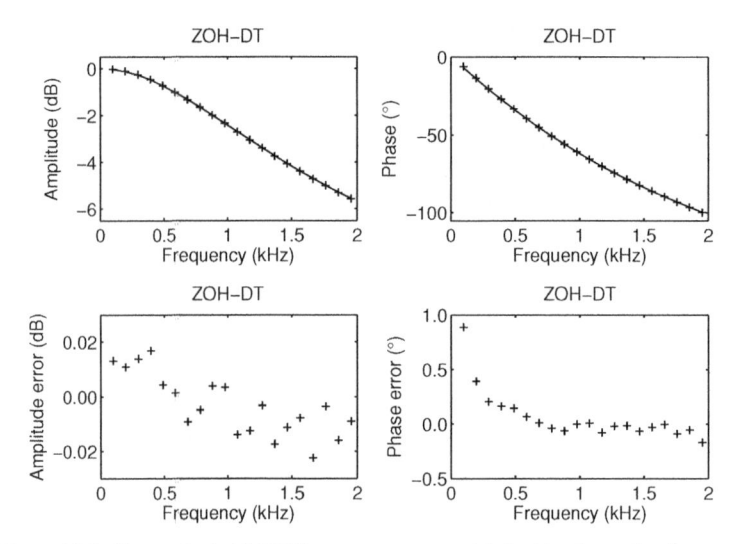

Figure 13-8. Zero-order-hold (ZOH) measurements modeled with a first order discrete-time (DT) transfer function (13-15). Top row: measured ('+') and modeled (black line) FRF. Bottom row: difference between measurement and estimate.

13.2.4.4 Mixing the Intersample Behavior and the Model. What happens if we mix the intersample behavior and the model is shown in Figure 13-10: the band-limited (BL) measurements are modeled with the first order discrete-time (DT) model (13-15), and the zero-order-hold (ZOH) measurements are modeled with the first order continuous-time (CT) model (13-14). It can be seen that in both cases the model is unable to explain the measurements. The corresponding RC-time constants are 60 µs (DT-model (13-15)) and 228 µs (CT-model (13-14)). The mismatch between measurements and model and the large bias in the estimated RC-time constant are due to the discrepancy between the true intersample be-

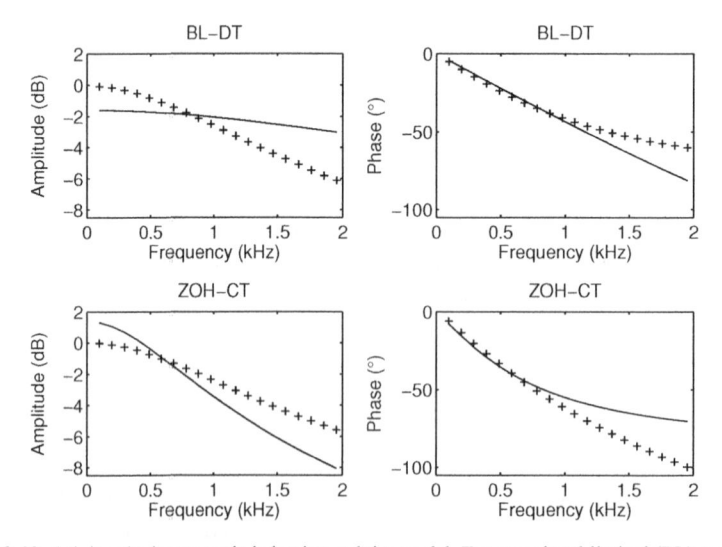

Figure 13-10. Mixing the intersample behavior and the model. Top row: band-limited (BL) measurements ('+') modeled with the first order discrete-time (DT) transfer function (black line) of (13-15). Bottom row: zero-order-hold (ZOH) measurements ('+') modeled with the first order continuous-time (CT) transfer function (black lines) of (13-14).

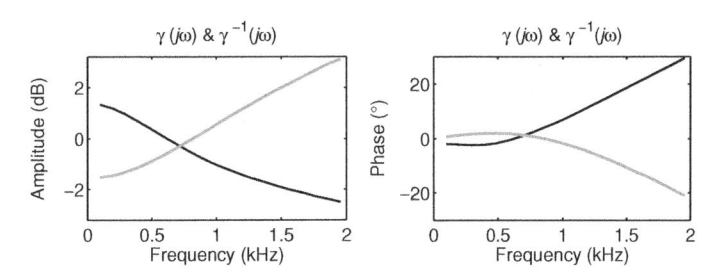

Figure 13-11. Mixing the intersample behavior and the model: ratio of the predicted to the actual output of the system. Black lines: $\gamma(j\omega)$ (13-10) of the first order DT model identified using the BL measurements (see Figure 13-10, top row). Light gray lines: $\gamma^{-1}(j\omega)$ (13-11) of the first order CT model identified using the ZOH measurements (see Figure 13-10, bottom row).

havior of the measurements (BL and ZOH) and the one assumed by the model (ZOH and BL for, respectively, the DT and CT models). Figure 13-11 shows the $\gamma(j\omega)$ (13-10) and $\gamma^{-1}(j\omega)$ (13-11) factors of the identified first order models. As predicted by the theory, the mismatch between the DT-model and the BL-measurements is - in good approximation - one over the mismatch between the CT-model and the ZOH-measurements.

If we increase the model complexity of the DT-model to a second order rational form with fractional sample delay τ,

$$G(z, \theta) = z^{-\tau} \frac{b_0 + b_1 z^{-1} + b_2 z^{-2}}{a_0 + a_1 z^{-1} + a_2 z^{-2}} \quad (13\text{-}16)$$

then the difference between the BL-measurements and the DT-model is at the noise level (compare the top row of Figure 13-12 with the bottom row of Figure 13-7). Similarly, a second order CT-model with time delay

$$G(s, \theta) = e^{-\tau s} \frac{b_0 + b_1 s + b_2 s^2}{a_0 + a_1 s + a_2 s^2} \quad (13\text{-}17)$$

explains the ZOH-measurements as well as a first order discrete-time model (compare the bottom row of Figure 13-12 with the bottom row of Figure 13-8).

Note that the additional poles and zeroes and time delay in models (13-16) and (13-17) model the difference between the actual intersample behavior, and the one assumed by the model. One of the poles of each transfer function model can be linked to the physical pole of the RC-circuit shown in Figure 13-6. The corresponding estimated RC-time constants are 142.5 μs (DT-model (13-16)) and 137.1 μs (CT-model (13-17)). Although these values are much closer to the true value (142.7 μs ± 0.4 μs) than those of the first order models in Figure 13-10 (60 μs and 228 μs), the physical interpretation of the identified models requires prior knowledge to separate the physical plant poles and zeroes from the mathematical poles and zeroes introduced by the missmatch between the actual and the assumed intersample behavior. Hence, it should be done with (great) care. However, the identified models (13-16) and (13-17) are well suited to predict the output of the system at the sampling instances for a new input signal, so long as the intersample behavior of the new input matches that of the input in the identification experiment.

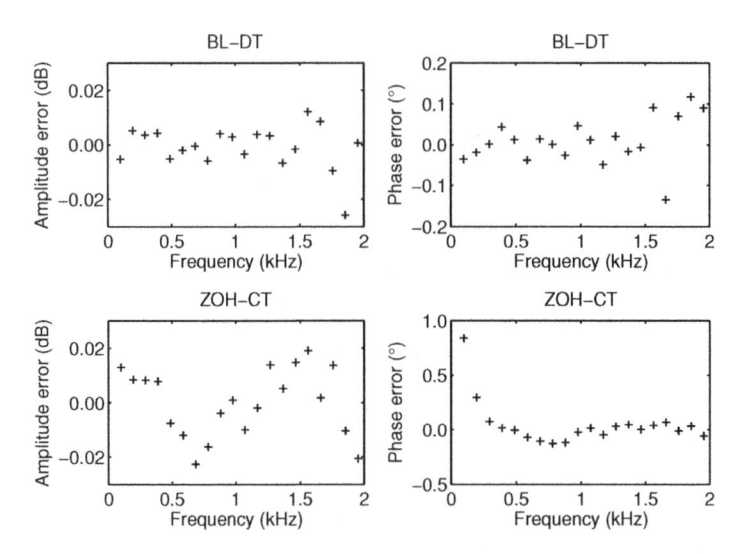

Figure 13-12. Mixing the intersample behavior and the model. Top row: difference between the band-limited (BL) measurements and the estimated second order discrete-time (DT) model (13-16). Bottom row: difference between the zero-order-hold (ZOH) measurements estimated the second order continuous-time (CT) model (13-17).

Finally, the DT-model (13-15) with f_s = 100 kHz instead of f_s = 10 kHz has been identified using the BL-measurements (no new measurements were performed; only the sampling frequency of the discrete-time model has been changed). The estimated RC-time constant equals now 142.6 µs. This can be explained by the fact that at the higher sampling frequency the corresponding mismatch factor $\gamma(j\omega)$ (13-10) is much closer to one.

13.2.4.5 Summary. The estimated RC-time constant of the different CT- and DT-models are given in Table 13-1. The estimates obtained with the exact approach (Sections 13.2.4.2 and 13.2.4.3) equal the true value within its measurement uncertainty. When mixing the intersample behavior and the plant model (Section 13.2.4.4) the errors on the estimates are very large. These errors are significantly reduced by increasing the model complexity. As could be expected the estimate at f_s = 100 kHz equals the true value.

TABLE 13-1 Estimation Results RC-time constant (true value = 142.8 µs ± 0.4 µs)

Measurement setup	Plant model	RC (µs)	Remarks
BL	CT (13-14)	142.6	exact approach
ZOH	DT (13-15)	142.9	exact approach
BL	DT (13-15)	60	–
ZOH	CT (13-14)	228	–
BL	DT (13-16)	142.5	unstable model
ZOH	CT (13-17)	137.1	non-minimum phase model
BL	DT (13-15)	142.6	f_s = 100 kHz

13.3 THE INTERSAMPLE ASSUMPTION: APPRECIATION OF THE FACTS

In Sections 13.2.1 to 13.2.3 we have given objective facts. These are true without any discussion. However, the importance of these facts can be very different from one application field to another. As mentioned before, it is the responsibility of the reader to judge their impact on his/her application. Here, we give some thoughts on possible implications, but it should be clear that these are influenced by personal experiences. As such, we advise the reader to consider them critically.

13.3.1 Intended Use of the Model

Although from an information point of view there is no fundamental difference between the BL and the ZOH assumption (besides the fact that the ZOH setup provides more high-frequency information), it is still advisable to match the choice for the basic assumption with the intended use of the model. In theory, it is possible to relate a discrete-time and a continuous-time model, using (13-4), if the basic assumption is met, but in practice, additional errors are added due to the nonideal experimental conditions. For some applications, the signal choice is not really critical, but for others, the application leads to a natural choice. The following applications are discussed in more detail next: controller design, physical interpretation, simulation, and modeling of subsystems.

13.3.1.1 Controller Design. In model-based control design, a mathematical model of the device under test is required. For discrete controllers, it is clear that a discrete-time model is the best choice. The digital controller generates a ZOH excitation that is exactly known (it is available in the memory of the controller). Everything between the controller output and the observed system output (including the actuator and the noise reduction filters) should be modeled because it is part of the control problem. This is the standard setup for ZOH modeling; in fact, the whole ZOH theory originated from this problem. Even if the ZOH reconstruction is poor, the behavior can still be included in the characteristic by increasing the model order. The drawback of this approach is that the models are not portable from one setup to the other because plant model and signal reconstruction are mixed up in one single model. Another nonideal ZOH characteristic results in another model.

Conclusion: for digital control design the ZOH assumption is well suited. Even if the ideal ZOH reconstruction is not closely matched, the nonideal ZOH characteristic can be included in the model.

13.3.1.2 Physical Interpretation. In some applications, users are not really interested in the identified model but use it only as an intermediate step to get deeper physical insight into a problem. They want to measure model parameters that are not directly accessible with classical instruments, for example, time constants, diffusion constants, or the values of some components in an electrical circuit. Mostly, it is not a good idea to try to identify these parameters directly as they are linked to the measurements through highly nonlinear relations, complicating the identification significantly, and they may not even be uniquely identifiable. It is easier to identify, first, an intermediate model, and to extract the physical parameters from this result. Usually, the coefficients of the differential equations are closer linked to these parameters than those of the approximating difference equations, so that continuous-time models are preferably used.

Although it is possible to identify continuous-time models under the ZOH conditions using a direct continuous-time parameterization (Ljung, 1999) or using dedicated prepro-

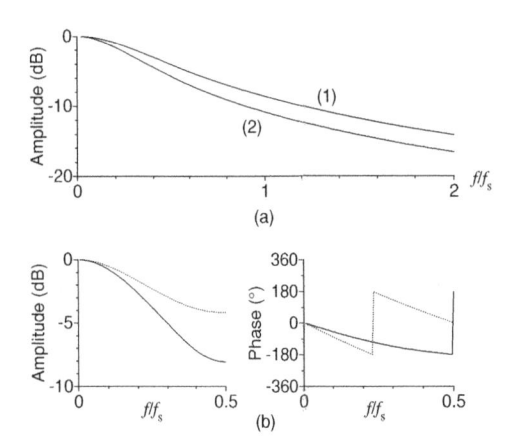

Figure 13-13. Illustration of the cascading error of ZOH models. (a) Transfer function of the original systems $L(j\omega)$ (1) and $G(j\omega)$ (2) and (b) comparison of the ZOH model of the cascaded system (solid line) with the cascade of the ZOH models (dots).

cessing methods based on block-pulse functions (Sinha and Rao, 1991) or delta operators (Ninness and Goodwin, 1991), we advise starting from BL measurements. These techniques can be applied in the time domain (Van hamme et al., 1991) or in the frequency domain, even for arbitrary excitations, using the extended models as explained in Section 6.3.2. The preference for the BL setup might be surprising because there is a formal relation (13-4) between both approaches. However, this relation is valid only if the experimental conditions were in perfect agreement with the underlying assumptions (perfect ZOH, perfect BL), and this can be very hard to realize, especially for broadband ZOH excitations. In many cases the ZOH reconstruction is disturbed due to the load of the output impedance $Z_{out}(j\omega)$ of the ZOH reconstructor by the input impedance of the actuator $Z_{in}(j\omega)$, so that the actual generated spectrum differs from the theoretical one by $Z_{in}(j\omega)/(Z_{in}(j\omega) + Z_{out}(j\omega))$. On the other hand, it is quite easy to get a set of two identical, good anti-aliasing filters, so that the BL assumption is matched well.

13.3.1.3 Simulation and Modeling of Subsystems. Building a model for a very complex system is a tedious task. Instead of catching the system in one extreme complex model, it is much more feasible to split the problem into a series of subtasks, each modeling a subsystem. In principle, for each of these subsystems we can build a discrete-time model under the ZOH assumption. However, even if these submodels are perfect, they will not describe the actual signals in the cascaded system because the subsystems are not excited by ZOH excitations. This is very similar to the setup given in Figure 13-5. Assume that perfect ZOH models $L_{ZOH}(z^{-1}) = (1-z^{-1})Z\{L(s)/s\}$ and $G_{ZOH}(z^{-1}) = (1-z^{-1})Z\{G(s)/s\}$ are available for $L(j\omega)$ and $G(j\omega)$; then the cascaded system $L(j\omega)G(j\omega)$ is described not by $L_{ZOH}(z^{-1})G_{ZOH}(z^{-1})$ but by $(1-z^{-1})Z\{L(s)G(s)/s\}$, so that an error appears:

$$\frac{(1-z^{-1})Z\{L(s)/s\}(1-z^{-1})Z\{G(s)/s\}}{(1-z^{-1})Z\{L(s)G(s)/s\}} \tag{13-18}$$

Note that this error is independent of the order of cascading. In Figure 13-13, the error due to this wrong combination is shown for the cascade of two first-order systems:

$$L(s) = 1/(1 + s/(0.6\pi)) \quad \text{and} \quad G(s) = 1/(1 + s/(0.8\pi)) \tag{13-19}$$

In this case severe errors appear because we considered systems with a bandwidth that is large compared with the sampling frequency, so that the repeated spectra of the ZOH excitation (see Figure 13-2) are not filtered out by the plant. If this were the case, the errors would be much lower. However, for a general approach this is an undesired restriction.

A first possibility would be to transform the ZOH models to continuous-time models, using the inverse relation, and next apply (13-4) again to the cascaded continuous-time models (assuming that $f_s/2 > \text{Im}(\text{poles})/(2\pi)$). As mentioned before, this approach relies heavily on the ideal ZOH behavior, which may be difficult to obtain. The sound approach is to select the BL assumption and combine it with discrete-time models. Although the resulting models lose their physical interpretation, they are perfectly suited for simulation. By increasing the complexity, an arbitrary precision can be obtained. Moreover, cascading of these models is allowed as long as the signals in the simulator obey the BL assumption.

Note that by using the same arguments, it is also possible to identify continuous or discrete-time models of an arbitrary subsystem of a complex system. Hooking the probes of the measurement device at the input and output of the subsystem makes it possible to zoom in on each accessible part of the overall system, as shown in Figure 13-4. The BL assumption is realized using good anti-alias filters. Because it is almost impossible to impose a ZOH excitation in the middle of a complex process, the ZOH assumption is not well suited to solve this kind of problem.

Conclusion: under the BL assumption it is possible to build continuous or discrete-time models of subsystems of a complex plant, even if they are preceded by nonlinear systems. These can be used, for example, as portable building blocks for simulators.

13.3.2 Impact of the Intersample Assumption on the Setup

The intersample assumption has a significant impact on the experimental aspects and the actual quality of the measurements. In each measurement setup it is important to reduce the errors. The identification methods take care of the stochastic errors but cannot cope with systematic errors. These should be removed in an appropriate calibration procedure. This is relevant only if accuracy is important, but why bother about consistency and efficiency if the systematic instrumentation errors dominate? Therefore, it is always necessary to check the quality of the measurement setup and to verify its impact on the quality of the final models. Typical errors that appear in many data acquisition channels are DC offsets and dynamic distortions due to the measurement channel characteristics $G_u(j\omega)$ and $G_y(j\omega)$. The offset errors can often be eliminated by excluding the DC information, while the compensation of the channel characteristics requires a calibration.

13.3.2.1 Perfect BL Setup. Under the BL assumption, two channels measuring the input and the output are needed. Because in (13-7) only the ratio $G_y(j\omega)/G_u(j\omega)$ appears, a relative calibration that measures this ratio will do, and this for $|\omega| < \omega_s/2$. In order to guarantee that the BL assumption is met, the acquisition channels should be equipped with Antialias filters. An alternative is to filter the excitation signal so that no power is injected above $\omega_s/2$. If the plant is guaranteed to be linear, the measured input and output also obey the BL assumption. The advantage of this approach is that only one filter is required, so it is easier to get two identical measurement channels.

13.3.2.2 Perfect ZOH Setup. Under the ZOH assumption, the situation changes drastically. In this case, only the output is measured. From (13-5) it is seen that the acquisition channel should have a transfer function $G_y(j\omega) = 1$ in a frequency band that covers ω_s many times in order to pass the high-frequency components created in the ZOH reconstruction. This is a difficult constraint. An absolute calibration is required in this case to measure and compensate the channel characteristics. This will be a tedious task, especially if arbitrary excitations are used, since in that case the compensation should be done in the time domain using inverse filtering techniques (Pintelon et al., 1990; Kollár et al., 1991). The alternative is to select an instrument with a very large bandwidth compared with the sampling frequency and hope that the roll off and phase distortion will be small in the frequency band of interest. Notice that in this setup it is NOT allowed to use an anti-alias filter as this would eliminate all the repeated spectral contributions of the ZOH. These results are grouped in Table 13-2. From this table we conclude that it is easier to approach the ideal measurement setup for the BL assumption compared with the ZOH requirements. This suggests that, due to the experimental constraints, the BL setup is best suited for accurate measurements of the system.

TABLE 13-2 Implications of the BL Assumption and the ZOH Assumption for the Ideal Measurement Setup

BL	ZOH
Two-channel measurement	Single-channel measurement
Relative amplitude and phase calibration[a]	Absolute amplitude and phase calibration[a]
No flat amplitude/linear phase required	Flat amplitude/linear phase required
Instrument bandwidth $\leq \omega_s/2$	Instrument bandwidth $\geq$ many times ω_s
Anti-alias filters required	Anti-alias filtering not allowed

a. Parametric noise models only require an absolute amplitude calibration.

13.3.2.3 ZOH Setup for Control. For many control applications the situation is, luckily, not that bad. Often, the bandwidth of the plant is not very large (for example a few kHz or lower) and the sample frequency is typically chosen 10 times larger. This makes it possible to filter the output before feeding it back to the controller in order to reduce the out-of-band process noise without adding too much delay to the system. Under these conditions, the previously mentioned problems with the ZOH setup become less pronounced. Moreover, in this field, the desired accuracy is also much lower than the accuracy that is typically required in many measurement applications. This leads to the conclusion that the ZOH setup is the natural choice for digital prediction/control design, where the actuator is an intrinsic part of the modeling problem and high accuracy is not the first requirement.

13.3.3 Impact of the Intersample Behavior Assumption on the Identification Methods

In the stochastic approach to system identification, the cost function is completely set by the noise model. From Figures 13-3 and 13-4, it is seen that under the ZOH setup, the input is assumed to be known, whereas under the BL assumption the input is measured. As each measurement is disturbed by noise, two (correlated) noise sources are needed for the stochastic model under the BL assumption, whereas only one noise source on the output measurements is needed under the ZOH assumption. This has a significant impact on the general structure of the identification scheme: the BL assumption leads to the errors-in-variables approach (see Chapter 9), while the ZOH assumption is the basis for the prediction error methods (see Ljung, 1999 and Sections 10.9 and 10.10).

13.4 NONPARAMETRIC NOISE MODELS: FACTS

In the next section we study the impact of selecting a nonparametric or a parametric noise model. Here, we give an enumeration of the facts connected to nonparametric noise models estimated from experiments with periodic or random excitation signals. In the next section, we give a more detailed discussion, including the appreciation, of these facts. The most important facts about nonparametric noise models are:

1. The quality of the noise model is independent of the parametric plant model
2. Improved/simplified model validation
3. Simplified model selection/minimization cost function
4. Errors-in-variables identification and identification in feedback are as easy as the generalized output error problem
5. Increased uncertainty of the parametric plant model
6. Not suitable for output data only

Facts 5 and 6 are a drawbacks of nonparametric noise models compared with parametric noise models.

13.5 NONPARAMETRIC NOISE MODELS: DETAILED DISCUSSION AND APPRECIATION OF THE FACTS

The choice between nonparametric and parametric noise models is one of the most important selections to be made when designing a system identification experiment. Nonparametric noise models simplify considerably the identification of parametric plant models, for example, identification in feedback and errors-in-variables problems become as easy as generalized output error problems.

13.5.1 The Quality of the Noise Model

With periodic excitations, all nonperiodic variations are assigned to the disturbing noise (Schoukens et al., 1997b; Pintelon et al., 2011a) because the signal repeats itself from period to period. This is a very general technique that can separate the noise contribution from the nonlinear distortions. It fails in two situations:

- In the presence of periodic noise that is synchronous with the periodic excitation.
- In the presence of chaos, where a periodic input does not necessarily result in a periodic output.

Note that measuring two periods of the transient response to a periodic excitation is sufficient to fully characterize the nonparametric frequency response function estimate (noise level and level nonlinear distortions), even for multivariable systems (see Section 7.3, and Pintelon et al., 2011a, b).

With random excitations it is also possible to extract nonparametric noise models from the data under the conditions that the input is known and that the plant transfer function can locally be approximated by a polynomial (see Section 7.2, and Pintelon et al., 2010a). If the input is also noisy then a known reference signal (typically the signal stored in the arbitrary waveform generator) is needed (see Section 7.2, and Pintelon et al., 2010b). While the noise cannot be distinguished from the nonlinear distortions for stationary random excitations, the nonlinear behavior can be detected and an upper bound on the noise level can be estimated using nonstationary random inputs (Zhang et al., 2010).

In practice, the signal is separated from the noise by calculating (generalized) sample means and (generalized) sample (co)variances (see Section 4.3 for the classical methods and Section 12.2 for the local polynomial methods). The sample mean carries the signal information, while the sample (co)variances can be used as a nonparametric noise model. This analysis is done before starting the identification process; for example, no parametric plant model is selected yet. Consequently, the nonparametric noise model is independent of the plant model errors. This noise model can be used as a weighting in the estimation step, even during the generation of starting values.

Example 13.3. Figure 13-14 shows the results of a nonparametric noise analysis of electrical machine measurements. It not only gives the noise levels but also makes it possible to make a quality check of the measurements (e.g., What is the SNR? Is the system well excited in the frequency band of interest?) before starting the identification procedure. □

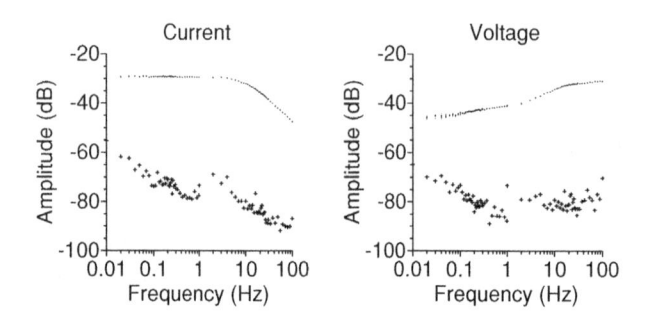

Figure 13-14. Separation of the signal and the noise using 10 repeated experiments on an electrical machine. − the raw measurements, + the nonparametric noise model.

13.5.2 Improved/Simplified Model Validation

The independent nonparametric noise characterization (see Section 13.5.1) allows an absolute interpretation of the cost function, leading to significant advantages during the model selection and validation process. Not only is it possible to make an absolute detection of model errors, starting from the value of the cost function, but also the presence of unmodeled dynamics (and nonlinear distortions) can be checked (see Chapter 11 for the classical methods and Chapter 12 for the local polynomials methods).

Besides these global qualifications, the direct comparison of the measured FRF (see Section 13.7.3) with the modeled transfer function shows in which frequency bands the model fails, as illustrated next on a mechanical system (Figure 13-15). Using $M = 34$ measured periods, the mean value and the standard deviation (complex error) of the FRF were measured and compared with a parametric model. From this simple test we can conclude that the errors mainly appear at the resonance frequencies.

13.5.3 Simplified Model Selection/Minimization Cost Function

Within the prediction error framework the model orders and the coefficients of the plant and the noise transfer functions must be estimated (see Section 10.9, and Ljung, 1999). For nonparametric noise models this reduces to the identification of the plant transfer function (model order + coefficients) only. This simplification also decreases the risk of getting trapped in a local minimum of the cost function (Schoukens et al., 2011).

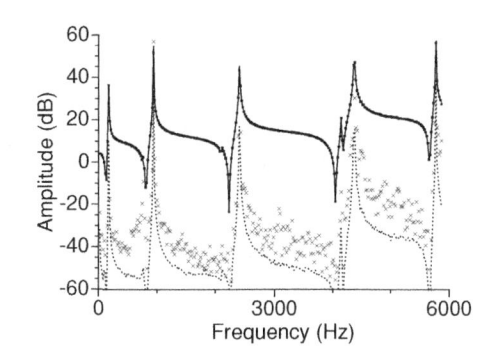

Figure 13-15. Measurement of a mechanical system (acceleration as a function of force): • measurement, ... noise standard deviation σ_G, ___ model, × difference between model and measurement.

13.5.4 Errors-in-Variables Identification and Identification in Feedback

The errors-in-variables problem (all observations are noisy) can be solved using parametric input-output noise models provided (i) the system operates in open loop, and (ii) the unknown excitation can be written as filtered white noise. The solution requires the simultaneous identification of the plant, the input noise, the output noise, and the excitation signal transfer functions (see, for example, Söderström, 2007, and Pintelon and Schoukens, 2007). Beside some identifiability conditions (Aguero and Goodwin, 2008), the major difficulties are (i) the generation of (high quality) starting values for all transfer functions, (ii) the model order selection, and (iii) the model validation.

Using nonparametric noise models, the identification of a plant operating in feedback from noisy input-output data (= most general errors-in-variables problem, see Figure 7-4 on page 240) has exactly the same complexity as the identification of a system operating in open loop from known input, noisy output data (see Chapters 10 and 12). If the best linear approximation of a nonlinear system is identified and/or the excitation is arbitrary, then, besides the noisy input-output data, also a known reference signal should be available.

13.5.5 Increased Uncertainty of the Plant Model

The uncertainty of the estimated parametric plant model using a nonparametric noise model is larger than that using a parametric noise model because,

1. Replacing in the maximum likelihood cost function the true input-output noise (co)variances factor by their (generalized) sample estimates increases the uncertainty of the estimates by the factors given in (10-16) and (12-23).

2. For ARX and ARMAX model structures (see (6-70) and (6-71)), the disturbing noise contributes to the identification of the plant poles via the common denominator in the plant and noise models. This information is lost when using a nonparametric noise model.

3. The (generalized) sample mean (see Sections 4.3 and 12.2) captures that part of the plant input and output that is correlated with the reference signal. Hence, it suppresses that part of the excitation that is not correlated with the reference signal, for example, the generator noise (open and closed loop) and/or the process noise (closed loop only). This results in some loss of information.

While it is easy to quantify the increase in uncertainty due to the variability of the sample noise (co)variances (cause 1), the increase due to causes 2 and 3 is strongly case dependent.

13.5.6 Not Suitable for Output Data Only

In some applications neither the plant can be excited nor is it possible to measure the operational perturbations. Think, for example, of the operational modal analysis of civil engineering structures like bridges, buildings, and windmills. In those cases the plant dynamics should be identified from output data only. Assuming that the unobserved operational perturbation is white in the frequency band of interest, the problem can be solved by estimating a parametric noise model (see Section 10.9, and Ljung, 1999). Nonparametric noise models do not offer a solution here.

13.6 PERIODIC EXCITATIONS: FACTS

In the next sections, we study the impact of selecting periodic or arbitrary excitations. The impact of this choice is less dependent on the application field than the selection of the intersample assumption. Here, we give an enumeration of the facts connected to periodic excitations. In the next section, we give a more detailed discussion, including the appreciation, of these facts. Finally, we look into some user aspects of periodic excitations. The most important facts about periodic excitations are:

1. Data reduction linked to an improved signal-to-noise ratio of the raw data
2. Elimination of nonexcited frequencies
3. Improved frequency response function measurements
4. Detection, qualification, and quantification of nonlinear distortions
5. Detection and removal of trends
6. Reduced frequency resolution
7. Increased uncertainty if the nonlinear distortions dominate over the noise

Facts 6 and 7 are drawbacks of periodic signals compared with random noise excitations.

13.7 PERIODIC EXCITATIONS: DETAILED DISCUSSION AND APPRECIATION OF THE FACTS

The choice between periodic and nonperiodic excitations is an important selection to be made during the experiment design. Many times, it is incorrectly linked to the selection between time and frequency domain identification. Periodic excitations open up a number of possibilities that are not accessible with arbitrary excitations and this for time and frequency domain identification.

13.7.1 Data Reduction Linked to an Improved Signal-to-Noise Ratio of the Raw Data

When periodic excitations are applied, it is possible to collect P successive periods (with length N_p) and to average the measurements in the time domain over these repeated periods, for example, for the output measurement (Figure 13-16):

$$\hat{y}(k) = \frac{1}{P}\sum_{l=1}^{P} y(k+(l-1)N_p) = \frac{1}{P}\sum_{l=1}^{P} y^{[l]}(k) \qquad (13\text{-}20)$$

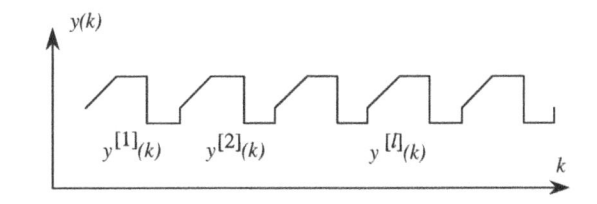

Figure 13-16. Making use of the periodic nature to improve the SNR.

with $y^{[l]}(k) = y(k + (l-1)N_p)$. It is clear that due to the averaging process, the noise is reduced in $P^{-1/2}$ under very weak conditions (the total measurement time should be much larger than the correlation length of the noise), and $\lim_{P \to \infty} \hat{y}(k) = y_0(k)$ w.p. 1. Many dynamic signal analyzers offer this measurement option; for example, $P = 128$ averages are made over $N_p = 2048$ data points. As this reduces the record length at a very low computational cost, it is strongly advised to make full use of this option. Why should we restrict ourselves to 2048 data samples if we can get 128×2048 data samples almost for free? In practice, P is determined by the maximum measurement time T and the minimum required frequency resolution f_0: $P = f_0 T$. Note that for a fixed experiment time, the frequency resolution dropped by a factor P. Another interpretation of using consecutive periods is given in Section 13.7.2.

13.7.2 Elimination of Nonexcited Frequencies

When a periodic excitation is applied, the user knows mostly what spectral lines are present. Often, not all lines are excited and it is possible to eliminate the "zero lines." This operation offers many new possibilities: generation of improved starting values and data reduction.

13.7.2.1 Improved Starting Values. Eliminating nonexcited lines does not change the asymptotic properties of well-designed estimators because the information matrix is not affected by removing zero lines. However, during the generation of starting values, simplified schemes such as ARX (linear least squares) methods are used. These are sensitive to the noise on the zero lines, so their elimination results in improved starting values (Schoukens et al., 1994). The risk of getting stuck in a local minimum is much larger if all spectral lines are retained, including the nonexcited lines.

13.7.2.2 Data Reduction. If a very wide frequency band has to be covered, fine resolution is needed at the low frequencies, whereas in the higher frequency bands the resolution can be reduced. For this reason, many systems are shown on a logarithmic frequency axis in a Bode plot. In these situations, it is advisable to excite the system with a semilogarithmic multisine, exciting the system at logarithmically spaced frequencies (on an equidistant grid of a DFT), so that a constant relative frequency resolution is obtained. This results in a sparsely filled spectrum, where only a small fraction of all lines is excited, for example, 200 out of 8192 lines. Only these frequencies should be retained so that the amount of raw data to be stored can be reduced further.

13.7.2.3 Special Case: Measuring P Consecutive Periods. In Section 2.2.3, it was shown that the spectrum of a signal consisting of P measured periods is sparse with nonzero lines at the multiples of P. The other lines are different from zero only due to the presence of noise. By putting them to zero, the spectrum of the averaged signal (e.g., $\hat{y}(k)$ in (13-20)) repeated over P periods is obtained.

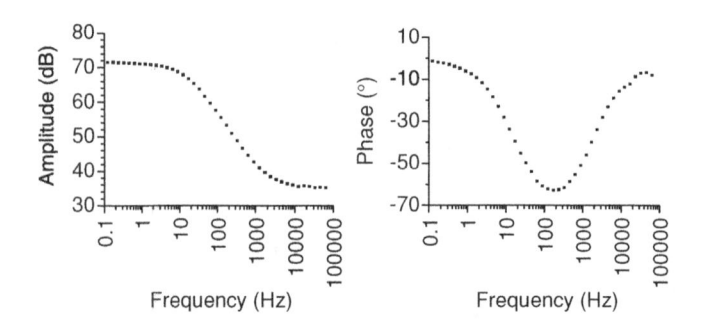

Figure 13-17. Measurement of the impedance of an electrochemical reaction in a wide frequency range on a semilogarithmic frequency grid.

13.7.2.4 Examples

Example 13.4. In Figure 13-17 the measured impedance of an electrochemical reaction $Fe^{3+} + e^- \to Fe^{2+}$ is shown in a frequency band from 0.123 Hz to 64 kHz. Using a semilogarithmic multisine, this very wide frequency range is covered with a small number of frequency points. □

Example 13.5. The effect of removing the nonexcited lines on the noise level is illustrated in Figure 13-18 , where the impact of averaging and filtering (removing the zero lines) is shown on a signal with a semilogarithmic spectrum. □

Remark. In some special cases such as ARX modeling, the disturbing noise also contributes to the plant knowledge through the common denominator of the noise model and the plant model. Some modifications are needed to restore the information lost during the averaging (Gustafsson and Schoukens, 1998).

13.7.3 Improved Frequency Response Function Measurements

When an integer number of periods is measured in steady-state conditions, the spectra of the signals calculated using the DFT are free of leakage errors due to the plant dynamics. High-quality FRF measurements are obtained by first averaging the output and input spectra and next making the division (see Chapter 2). From the nonparametric noise analysis, the uncertainty on this estimate is obtained directly. The availability of the FRF measurements not only simplifies the model validation significantly (compare the FRF of the estimated transfer function with the measured one), it also gives a prior view of the required model complexity so that the model selection process is speeded up.

In some applications (e.g., vibrating mechanical structures) the leakage error due to the noise dynamics cannot be neglected in the steady state response of the plant to a periodic excitation. Using the smooth frequency behavior of the leakage error, it can be suppressed nonparameterically in the FRF estimate via a local polynomial approximation (see Section 7.3).

Under the condition that the FRF and the leakage error due to the plant and noise dynamics can locally be approximated by a polynomial, high quality FRF estimates can also be obtained using random excitations (see Section 7.2).

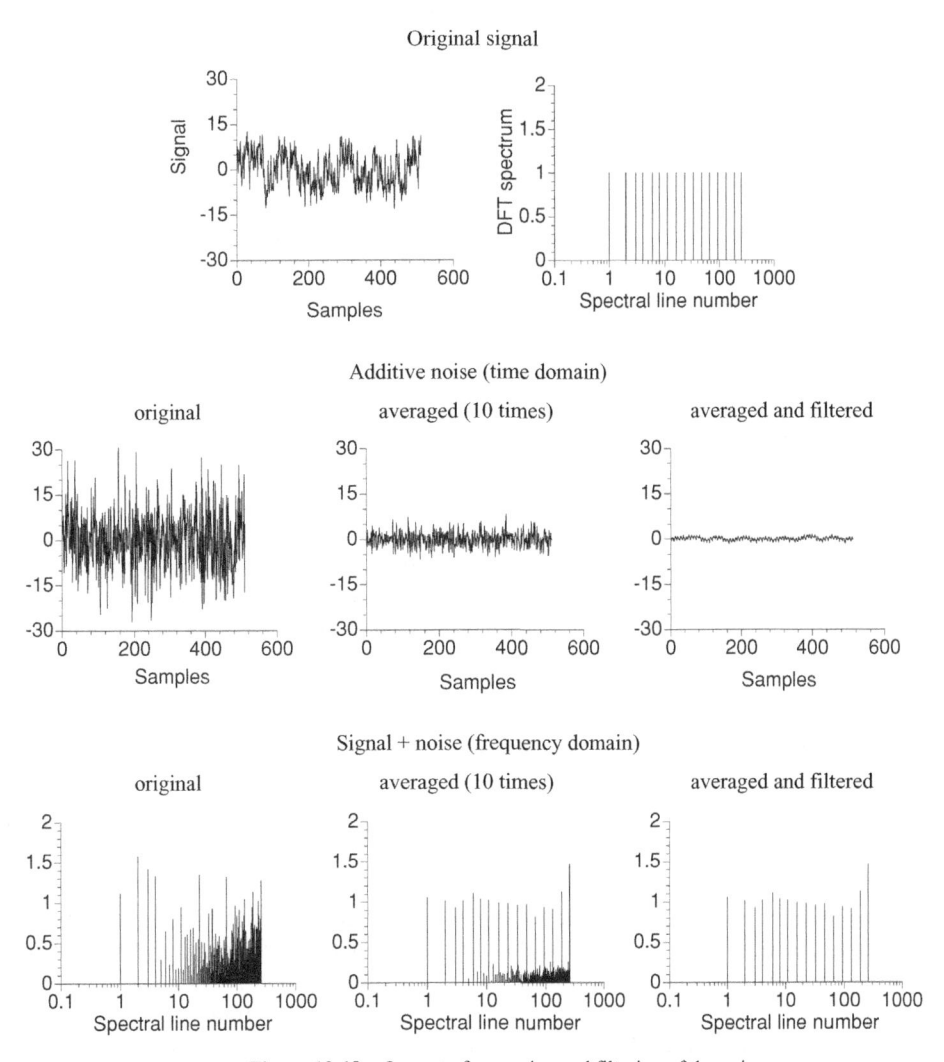

Figure 13-18. Impact of averaging and filtering of the noise.

13.7.4 Detection, Qualification, and Quantification of Nonlinear Distortions

Using, for example, an odd random phase multisine with random harmonic grid gives a great deal of insight into the nonlinear behavior of the plant (see Section 4.3.2). The even nonlinearities become visible at the even frequencies, while the odd nonexcited frequencies can be used to detect and quantify the level of the odd nonlinear distortions. Again, this is a nonparametric test that can be applied directly to the raw data before starting the identification process. These methods are extensively discussed in Chapter 4.

Using stationary excitations the noise cannot be distinguished from the nonlinear distortions in FRF measurements. If nonstationary random inputs are used, then it is possible to detect the presence of nonlinear distortions and to estimate un upper bound on the noise level (see Zhang et al., 2010). Classification in even and odd nonlinear distortions is, however, impossible.

13.7.5 Detection and Removal of Trends

In this book, we consider time-invariant systems. Periodic excitations facilitate testing for this assumption by calculating the sample mean of each period

$$\mu_y(l) = \sum_{k=1}^{N} y^{[l]}(k)/N,$$

and checking that it does not vary systematically from one period to the other. This test makes it possible to detect very small variations of the mean value revealing the presence of a (weak) trend. The trend can be removed by fitting the sum of sinewaves and a polynomial to the data. A time efficient implementation of this linear least squares problem can be found in Peirlinckx et al. (1996).

Similarly, for random excitations polynomials or curve segments can be fitted to detrend the data.

13.7.6 Reduced Frequency Resolution

To distinguish the noise from the nonlinear distortions, at least two periods of the input-output signals should be measured (see Sections 4.3 and 7.3). Therefore, the frequency resolution of the frequency response function measurement is at least two times smaller than that using random excitations (see Section 7.2).

13.7.7 Increased Uncertainty if the Nonlinear Distortions Dominate over the Noise

If the nonlinear distortions are the dominant error source, then the averaging of the stochastic nonlinear distortions should be maximized for a given measurement time T. Since the stochastic nonlinear distortions have the same periodicity as the excitation, their contribution is not decreased by averaging the input-output signals over the P periods. Therefore, the optimal choice is one experiment with a random excitation, or one experiment with a random phase multisine with period length T, or M experiments with random phase multisines with period length T/M (the frequency resolution and the variance of the stochastic nonlinear distortions of the latter are M times smaller than those of the single experiments and, therefore, the Fisher information matrices of these three measurements are the same). The variance of the estimated parametric plant model is then P times smaller than that of the periodic experiment with P measured periods.

13.8 PERIODIC VERSUS RANDOM EXCITATIONS: USER ASPECTS

Three aspects are discussed in detail: how difficult it is to design a signal; what about the frequency resolution; and finally, how flexible the signal characteristics can be set.

13.8.1 Design Aspects: Required User Interaction

Although periodic excitations offer a number of extended possibilities, they are still not very popular. One of the basic reasons for the unpopularity of periodic signals might be the user interaction required to design them. The user should specify the power spectrum (what frequencies are excited) and the period T_0 (determining the frequency resolution $f_0 = 1/T_0$ that is obtained in the frequency domain) before the periodic signal can be generated. For arbitrary signals, this information is seemingly not required. For example, the user can decide to generate a random sequence of length $T_0 = NT_s$ without even bothering about all these boring questions. However, this is a misleading impression. Also, in the latter case users should realize that even if they do not select a power spectrum and a frequency resolution consciously, their choices fix (unconsciously) these parameters, as shown below.

13.8.1.1 Frequency Resolution.
A periodic signal, with period T_0, probes the plant only at frequencies kf_0 with $f_0 = 1/T_0$, and it is completely insensitive to what happens between these frequencies, so that very narrow (compared with f_0) resonance peaks can be missed. However, although arbitrary excitations have a continuous spectrum, they do not offer an unrestricted resolution. As the signals are measured only in a finite time interval T_0, the windowing (or leakage) effect smears the spectra so that details below a resolution f_0 are also lost.

13.8.1.2 Imposing the Power Spectrum.
A good excitation signal should excite the process in the frequency band of interest. For periodic excitations, the user has full control over the power spectrum, so it is easy to inject all the available power in this band. Moreover, because these are deterministic signals, it is guaranteed that after one period this power spectrum is realized exactly. The situation is completely different for arbitrary excitations. Only indirect control, using digital filters, is possible. Starting from a white noise generator, a colored noise process is generated. Because this is a stochastic signal, its power spectrum is reached only asymptotically, and for short data records significant differences between the actual power spectrum and the desired one can appear (see Section 2.6).

Example 13.6. In Figure 13-19 a typical example of an arbitrary (white uniform noise) and a periodic (a flat multisine with 31 components, period 64 samples) ZOH excitation is shown. Both signals have a flat power spectrum filling the complete frequency band. They were generated in 64 samples. It can be seen that a perfect realization of the flat power spectrum is obtained for the multisine, but the spectrum of the considered noise realization is not flat at all. The spectrum drops at some frequencies more than 10 dB, resulting in a poor SNR at those frequencies. □

13.8.1.3 Small Crest Factors.
A second important aspect is the crest factor, measuring the ratio between the peak value and the rms value. Signals with a low crest factor make

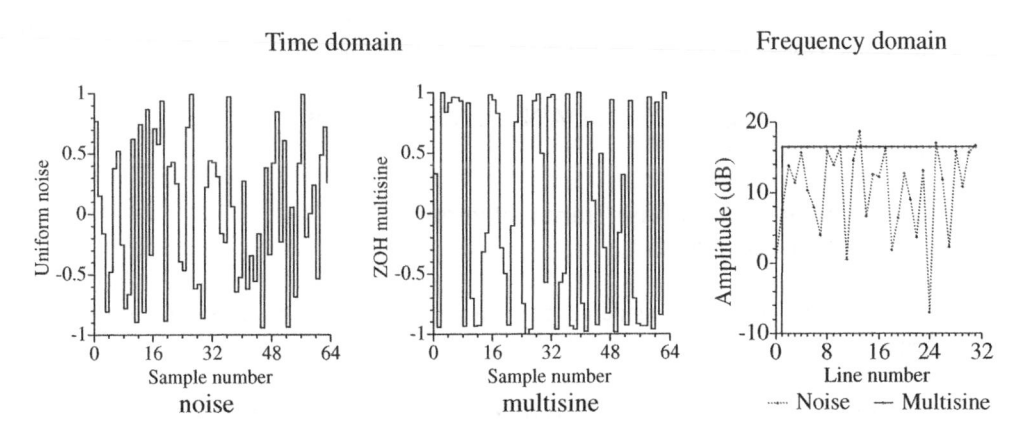

Figure 13-19. Comparison of an arbitrary (white noise) excitation with a periodic (flat multisine) excitation in the time and frequency domain.

it possible to inject more power into the system for a restricted peak value of the excitation, resulting in a better SNR. Algorithms are available to minimize the crest factor of periodic signals for a specified power spectrum. Also for arbitrary excitations, the crest factor can be reduced, but at the cost of a distorted power spectrum: the user does not have full control over the power spectrum and the crest factor at the same time (see Chapter 5).

13.9 TIME AND FREQUENCY DOMAIN IDENTIFICATION

Time and frequency domain identification were considered as competing methods for a long time. However, in most cases, the frequency domain data are obtained by a DFT from the raw time domain data. Note that there is a one-to-one relation between the time and the frequency domain. The only difference is that some information is more easily accessible in one domain than in the other. This is also reflected in the methods themselves. Assuming that all transients have died out, it can be shown by Parseval's relation that the least squares cost functions are identical (Ljung, 1993). Moreover, it is shown in Section 6.3.2 that the leakage effect on transfer function modeling in the frequency domain data is described by exactly the same transient model structure as in the time domain, where it is used to include the initial conditions. So there is full equivalence between time and frequency domain identification. The only difference is how the available information is formulated, but even here mixed algorithms popped up that combine the time and frequency domain representation in one algorithm, for example, the use of a nonparametric noise model in the time domain (Gustafsson and Schoukens, 1998) or mixed implementations of ARX methods (Schoukens et al., 1998b). This results in the (surprising) conclusion that it is not possible to give a clear formal definition of what time and frequency domain identification schemes are. Nevertheless, we still like to use the term time domain identification for algorithms that mainly operate on time domain data and frequency domain identification for algorithms that work on frequency domain data.

Many advantages that are often claimed for frequency domain identification are intrinsically due to the periodic nature of the excitation. So, the prime question is not to choose time or frequency domain methods but to select periodic or nonperiodic excitations! As explained before, many of the advantages of periodic excitations can be used in both domains. Some of the equivalences between both domains are not directly visible and are discussed in more detail later. Next, we also deal with some important differences between time and frequency domain identification.

Again, the text is organized along the same lines. First, a series of facts is stated, in this case equivalences and differences. Next, a more personal interpretation of the methods is given in the section on the natural choices in identification.

13.10 TIME AND FREQUENCY DOMAIN IDENTIFICATION: EQUIVALENCES

13.10.1 Initial Conditions: Transient versus Leakage Errors

Two situations are considered: the nonparametric measurements (impulse response or frequency response) and the parametric transfer function model.

13.10.1.1 Nonparametric Measurement. Usually, the frequency domain is cursed because the time-frequency transform is prone to leakage errors, so that the frequency response function obtained by dividing these spectra will also be wrong. In the time domain, the standard nonparametric impulse response measurements are based on a correlation analysis; for example, for discrete-time systems we have

$$R_{yu}(t) = g(t) * R_{uu}(t) \tag{13-21}$$

with $R_{yu}(t)$ and $R_{uu}(t)$ the cross- and the autocorrelation (Bendat and Piersol, 1980). These have to be estimated from the finite set of available data, e.g.,

$$\hat{R}_{yu}(k) = \sum_{l=0}^{N-1} y(l)u(l-k)/N,$$

where the data outside the window are put equal to zero. This shows that in the time domain windowing problems occur also, hence we can conclude that nonparametric measurements are prone to window errors in both domains. In the frequency domain these appear as leakage errors.

Remark. The Wiener-Hopf equations (13-21) are usually solved in the frequency domain (see (2-51)), emphasizing even more the time-frequency equivalence (Bendat and Piersol, 1980).

13.10.1.2 Parametric Models. For parametric system identification the time and frequency domain problem (initial conditions versus leakage) is cured in exactly the same way for both domains: the model is extended with a transient term that is linear-in-the-parameters (see Chapter 6) so that the additional computational cost is low. This solves the problem completely.

13.10.1.3 Impact on Whiteness Tests. At the end of the identification process the model is validated. A very popular test is to check the whiteness of the residuals using a correlation analysis. It turns out that this test is very sensitive to unmodeled initial conditions because these appear as model errors. If they are not recognized as such, the test leads to a too complex model structure. For that reason it is strongly advised to add an initial conditions estimate (keeping the model parameters fixed) for the validation data at the beginning of the validation process or to wait until the transients have died out.

13.10.2 Windowing in the Frequency Domain, (Noncausal) Filtering in the Time Domain

Sometimes we want to emphasize or deemphasize some spectral bands, expressing our belief in the quality of these measurements. Eliminating frequency lines, as explained in Section 13.7.2, is an extreme example of this method. Weighting in the frequency domain (multiplication with a frequency weighting $W(\Omega)$) corresponds to a filter operation in the time domain (convolution of the measured input and output with the impulse response $w(t) = F^{-1}\{W(\Omega)\}$). Removing some frequencies from the data set corresponds to a rectangular window. This is a noncausal filter with an impulse response of the form $\sin(\alpha t)/(\alpha t)$, $t \in \,]-\infty, \infty[$. Its absolute value is not summable, as should be for a stable system, so there are no simple alternative formulations in the time domain. Moreover, the maximum likelihood interpretation of the classical prediction error scheme is also lost because the transformation matrix, as it is introduced in Söderström and Stoica (1989, pp. 251), is no longer triangular due to the noncausal filter operation. The filtered output depends not only on the past but also on the future data.

Remarks

(i) Prefiltering the raw data is somehow cheating because it also changes the noise model. For Box-Jenkins methods, prefiltering does not change anything because it is completely compensated by a similar change in the noise model. However, in practice, the identification process is continued with the simple noise model, so that the efficiency can be affected, or a bias can even appear in closed loop identification.

(ii) Removing undesired frequencies (non-causal filtering) is possible within the prediction error framework at the cost of an additional term in the cost function (see Sections 10.9.2 and 13.10.3.1).

13.10.3 Cost Function Interpretation

The cost function that measures the goodness of the fit can be expressed in the time domain or in the frequency domain. It is clear that there should again be a full equivalence. However, a detailed study reveals some small differences depending on the practical implementation.

13.10.3.1 Extra Term in the Frequency Domain. If not only the plant model $G(z^{-1}, \theta)$ but also the noise model $H(z^{-1}, \theta)$ is identified, an additional term appears in the frequency domain interpretation of the cost function that was seemingly missing in the time domain expressions. Consider the maximum likelihood formulation for the generalized output error situation of a system operating in open loop, assuming normally distributed noise and neglecting the plant and noise transient terms (see Ljung, 1993 and Theorem 10.18 with $\Omega = z^{-1}$ and $M_0 = 0$),

$$\frac{1}{2}\sum_{k=0}^{N-1} \ln(|H(z_k^{-1}, \theta)|^2) + \left(\frac{N}{2}\ln(\lambda) + \frac{1}{2\lambda}\sum_{k=0}^{N-1}|\varepsilon(z_k^{-1}, \theta)|^2\right) \qquad (13\text{-}22)$$

with $\varepsilon(z_k^{-1}, \theta) = H^{-1}(z_k^{-1}, \theta)(Y(k) - G(z_k^{-1}, \theta)U(k))$ and $\lambda = \text{var}(e(t)) = \text{var}(E(k))$. The first term in (13-22) does not appear explicitly in the frequency domain interpretation (10-50) of the classical time domain expressions. However, if the frequencies z_k are equidistantly distributed on the unit circle and if $H(z^{-1}, \theta)$ is monic ($c_0 = d_0 = 1$), then this extra term converges for $N \to \infty$ to zero at exponential rate (see Corollary 10.20), and the difference between the time and frequency domain cost function disappears. This also reveals an additional condition on time domain identification: to get consistent noise models, it is not allowed to eliminate some frequency lines, nor is it allowed to restrict the identification to a subband on the unit circle.

The previous discussion is irrelevant if a prior known noise model is used; for example, the nonparametric model obtained from independent repeated measurements ($\lambda |H(z_k^{-1}, \theta)|^2$ in the third term of (13-22) is then replaced by the sample variance $\hat{\sigma}_Y^2(k)$ (2-37)). At that moment the noise model is fixed, and the additional term only adds a parameter-independent constant to the cost.

Remark. If the *exact* time and frequency domain maximum likelihood estimators are constructed (the exact likelihoods consider the initial and final conditions as random parameters that are correlated with the input and the disturbing noise), then both estimates are equal for finite values of the data length N (Agüero et al., 2010).

13.10.3.2 Optimization Aspects. When time and frequency domain identification lead to the same cost function, the only remaining difference is the optimization technique that is used to minimize the cost function. This can sometimes lead to tricky situations. Consider, for example, the generalized output error problem in the frequency domain formulation as discussed in Chapter 9 (see (9-83) with $\Omega = z^{-1}$, $\sigma_U^2(k) = 0$ and $\sigma_{YU}^2(k) = 0$):

$$\sum_{k=0}^{N-1} |\varepsilon(z_k^{-1}, \theta, Z(k))|^2 \qquad (13\text{-}23)$$

where $\varepsilon(z_k^{-1}, \theta, Z(k))$ can be written as

$$\varepsilon(z_k^{-1}, \theta, Z(k)) = \frac{e(z_k^{-1}, \theta, Z(k))}{\sigma_Y(k)|A(z_k^{-1}, \theta)|} \quad \text{or} \quad \varepsilon(z_k^{-1}, \theta, Z(k)) = \frac{e(z_k^{-1}, \theta, Z(k))}{\sigma_Y(k)A(z_k^{-1}, \theta)} \qquad (13\text{-}24)$$

with $e(z_k^{-1}, \theta, Z(k)) = A(z_k^{-1}, \theta)Y(k) - B(z_k^{-1}, \theta)U(k)$ (see (9-84)). Although both expressions in (13-24) lead to the same cost function (13-23), it turns out that the first form creates less problems with local minima. It also has a wider convergence region if a Gauss-Newton optimization method is used. In this method, the second-order derivatives are approximated from the first-order derivatives, and, seemingly, this approximation is better for the first expression (where some phase dependence is eliminated) than for the second. The disadvantage is that more calculations are needed to deal with the derivative of the absolute value, and slower convergence is obtained in the close neighborhood of the solution.

13.11 TIME AND FREQUENCY DOMAIN IDENTIFICATION: DIFFERENCES

13.11.1 Choice of the Model

Discrete-time models are the natural model class to be used in combination with time domain methods. Generalizing to other classes such as continuous-time models is not completely excluded, but it turns out from the literature that it is quite a complicated task (Sinha and Rao, 1991), and unexpected problems can appear (Söderström et al., 1997a, 1997b; Söderström and Mossberg, 2000).

In the frequency domain, the choice is more general. This is basically due to the fact that the differential (or difference) equations are replaced by algebraic equations in the related frequency variable. For continuous-time systems, the Laplace representation (transfer function) is used and evaluated on the imaginary axis ($s = j\omega$). For discrete-time systems, z-domain models are used and evaluated along the unit circle ($z = e^{j\omega T_s}$). Also, other frequency variables can be chosen; for example, $\sqrt{j\omega}$ is the natural representation for diffusion phenomena (e.g., used to model electrochemical processes) and $\tanh(\tau_R j\omega)$ is the logical choice to model commensurate microwave structures.

13.11.2 Unstable Plants

Prediction error techniques are mostly used to identify discrete-time models. These, typically, consider the following model structure:

$$y(t) = G(q, \theta)u(t) + H(q, \theta)e(t) \qquad (13\text{-}25)$$

with q, the unit delay operator ($qu(t) = u(t-1)$). The plant $G(z^{-1}, \theta)$ and the noise $H(z^{-1}, \theta)$ models are rational functions of z^{-1}, parameterized in θ. $e(t)$ is white noise, and $|H(z^{-1}, \theta)|^2$ models the power spectrum of the disturbing noise. The parameters θ are estimated by minimizing the prediction errors

$$\varepsilon(t, \theta) = H^{-1}(q, \theta)(y(t) - G(q, \theta)u(t)) \tag{13-26}$$

in least squares sense. It is clear that $H^{-1}(z^{-1}, \theta)$ and $H^{-1}(z^{-1}, \theta)G(z^{-1}, \theta)$ and their derivatives with respect to θ should be stable in order to be able to calculate (13-26). A stable plant and noise model is a sufficient condition to guarantee stability. Recently, a less restrictive solution was proposed for this problem (Forssell and Ljung, 2000a) by adding an all-pass section to the noise filter that cancels the unstable plant poles.

In the frequency domain, there is no problem to model unstable plants because these methods calculate the transfer function only at a discrete grid on the unit circle (or imaginary axis). As long as a pole does not coincide with one of these grid points, the cost function remains well defined; otherwise, regularization procedures can be used.

13.11.3 Noise Models: Parametric or Nonparametric

The efficiency of the estimates is improved using a well-chosen weighting function. The best option is to choose it inversely proportional to the power spectrum of the disturbing noise. Time domain methods apply filtering techniques to realize this weighting. Without these, the full covariance matrix of the noise should be inverted and next used in each iteration step, leading to more calculation work. For prediction error methods (time domain), these noise filters are an intrinsic part of the method (see (13-26)), and the noise model $H(z^{-1}, \theta)$ is estimated together with the plant model $G(z^{-1}, \theta)$. The obvious advantage is that no constraints are imposed on the excitation at a cost of a second model selection problem. Moreover, the convergence is significantly slowed down.

A nonparametric noise model can be generated automatically without user interaction for periodic and random excitations, leaving the complexity of the methods unaffected (see Section 13.5.1).

13.11.4 Extended Frequency Range: Combination of Different Experiments

The number of measured data points N in an experiment is directly linked to the record length T and the sample frequency f_s, as $N = Tf_s$. The minimum record length is mainly imposed by the lowest frequency of interest (or the spectral resolution) $T = 1/f_0$. The sample frequency is imposed by the highest frequencies, which have to be chosen so that the frequency band of interest of the plant is covered $f_s \geq 2f_{max}$. Hence, the minimum number of samples that should be measured and processed is $N > 2f_{max}/f_0$. It is obvious that the number of measurements increases drastically if a large frequency range should be covered. This leads to impractical situations if N becomes very large, for example, 1 million points.

If periodic excitations are applied and combined with frequency domain identification, two significant simplifications can be made, the first being the data compression, as explained in Section 13.7.1, and the second consisting of a simplification of the experimental conditions. The latter is obtained by splitting the experiment into a number of subexperiments, each covering another frequency range. For each of these subexperiments a much shorter record length can be used while it is still possible to measure all the required Fourier coefficients. A similar approach cannot be applied to the ZOH models because they strongly depend on the sample frequency and, hence, combination of the different records is much more complicated. An alternative might be to use multirate systems (Crochiere and Rabiner, 1983).

Example 13.7. In the measurement of the electrochemical process (Figure 13-17) a wide frequency band [0.123 Hz, 64 kHz] had to be covered. To do this in one experiment, at least 1 million points are needed. The actual measurements were obtained in two experiments covering [0.123 Hz, 100 Hz] and [100 Hz, 64 kHz], using 4096 points each time. □

13.11.5 The Errors-in-Variables Problem

The errors-in-variables concept is a more general approach than the classical feedback situation as shown in Figure 13-20. The basic structure is captured in the gray area and it can be part of a larger structure, for example, a feedback system. However, this additional information is not used in the algorithms developed in this book. Starting from the measured input $u(t)$ and output $y(t)$, the plant model $G(\Omega, \theta)$ is identified. The noise sources can be correlated with each other but are assumed to be independent of the driving signal $r(t)$. In general, this is an unidentifiable problem (Anderson and Deistler, 1984; Bohlin, 1971) unless additional constraints are imposed on the excitation, and/or the noise models, and/or the plant model (Söderström, 2007; Aguero and Goodwin, 2008). The following four situations can be solved: (i) the input is known exactly and the plant and output noise transfer function models are identified simultaneously (Ljung, 1999); or (ii) the excitation signal can be written as filtered white noise and the plant, signal, input noise, and output noise transfer function models are identified simultaneously (Söderström, 2007; Pintelon and Schoukens, 2007); or (iii) the signals are known to be periodic (Schoukens et al., 1997b); or (iv) an exactly known external reference signal is available (Forssell and Ljung, 2000c; Pintelon et al., 2010b). The first situation is the classical setup for time domain identification; while the second is the most difficult linear system identification problem concerning the generation of "high quality" starting values, the minimization of the cost function, and the model selection/validation. Solutions (iii) and (iv) are very attractive in combination with frequency domain identification and non-parametric noise models. Actually, this is the standard setup we consider in this book (see Chapters 10 and 12).

Note that the major difference between this setup and the ZOH setup is the information that is used to get the input signal $u_1(t)$. In the ZOH setup, the user relies on the validity of the ZOH assumption and the exact knowledge of the generator signal $u_d(k)$, whereas in this framework the assumption is replaced by a BL measurement.

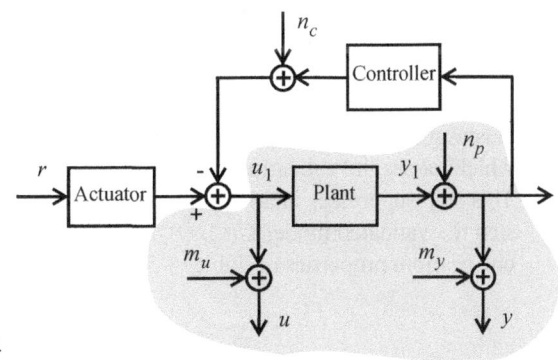

Figure 13-20. The errors-in-variables concept.

13.12 IMPOSING CONSTRAINTS ON THE IDENTIFIED MODEL

In applications like prediction and simulation (finite element programs, network simulation, virtual prototyping) the identified transfer function model should satisfy some constraints like, for example, reciprocity, (strict) stability, (strict) passivity (positive real), or bounded real (see Table 13-3). While the reciprocity constraint is linear in the transfer function model parameters θ, the others are strongly nonlinear functions of θ. For example, the matrix inequality constraint $T(j\omega) > 0$ in Table 13-3 imposes the quadrant symmetry of the zeroes of $\det(T(s, \theta))$: if $s_0 \in \mathbb{C}$, $\sigma_0 \in \mathbb{R}$, and $j\omega_0 \in j\mathbb{R}$ are, respectively, complex, real, and imaginary zeroes of $\det(T(s, \theta))$, then $\bar{s}_0, -s_0, -\bar{s}_0, -\sigma_0$, and $-j\omega_0$ are also zeroes of $\det(T(s, \theta))$, and $j\omega_0$ has an even multiplicity (for the discrete-time case $s_0, -s_0, j\omega_0$ are replaced by $z_0, z_0^{-1}, e^{j\omega_0 T_s}$).

TABLE 13-3 Possible Properties of an $n_u \times n_u$ Transfer Function Model $G(\Omega, \theta)$.

Property	s-domain	z-domain
Reciprocity	$G^T(s, \theta) = G(s, \theta)$	$G^T(z^{-1}, \theta) = G(z^{-1}, \theta)$
Strictly stable	poles $G(s, \theta)$ in $\text{Re}(s) < 0$	poles $G(z^{-1}, \theta)$ in $|z| < 1$
Strictly passive (positive real)[a]	$G(s, \theta)$ strictly stable and $\forall \omega \in \mathbb{R}$: $T(j\omega, \theta) = G(j\omega, \theta) + G^T(-j\omega, \theta) > 0$	$G(z^{-1}, \theta)$ strictly stable and $\forall \omega \in \mathbb{R}$: $T(e^{-j\omega T_s}, \theta) = G(e^{-j\omega T_s}, \theta) + G^T(e^{j\omega T_s}, \theta) > 0$
Bounded real[a]	$G(s, \theta)$ strictly stable and $\forall \omega \in \mathbb{R}$: $T(j\omega, \theta) = I_{n_u} - G^T(-j\omega, \theta) G(j\omega, \theta) > 0$	$G(z^{-1}, \theta)$ strictly stable and $\forall \omega \in \mathbb{R}$: $T(e^{-j\omega T_s}, \theta) = I_{n_u} - G^T(e^{j\omega T_s}, \theta) G(e^{-j\omega T_s}, \theta) > 0$

a. The matrix inequality constraint is equivalent with: the zeroes of $\det(T(\Omega, \theta))$ are quadrant symmetric and there exists at least one ω such that $T(j\omega, \theta) > 0$ (s- domain) or $T(e^{-j\omega T_s}, \theta) > 0$ (z- domain).

The constraints mentioned in Table 13-3 are often imposed during the estimation of the transfer function model from noisy data (Baratchard et al., 1997; Van Gestel et al., 2001; Goethals et al., 2003; Grivet-Talocia and Bandinu, 2006). The disadvantage of this approach is that conflicting demands are imposed, which results in noncontrollable bias errors. Indeed, to suppress the noise on the data, the cost function should be weighted with the inverse of the noise variance. Imposing a constraint of Table 13-3 might introduce model errors because, for example, there is no a priori reason why the best linear approximation of a nonlinear system should satisfy the properties of Table 13-3. These model errors should be distributed over the frequency band of interest according to some user-defined criteria. Since the corresponding frequency weighting will be different from the inverse of the noise variance, optimal noise removal and stability (reciprocity, passivity, …) are contradictory demands.

Another approach consists in imposing the constraints in a post-processing step (D'haene et al., 2006; D'haene and Pintelon, 2008; Grivet-Talocia and Ubolli, 2007; Gustavsen, 2008). In a first step an unconstrained (high-order) model is estimated $G(\Omega, \psi)$, which passes the validation tests (e.g., analysis of the cost function, whiteness test residuals). This step suppresses in an optimal way the noise without introducing bias errors. In a second step the validated model $G(\Omega, \hat{\psi})$ is approximated by a constrained one $G_c(\Omega, \theta)$ satisfying one or more properties of Table 13-3. This is done by minimizing the following cost function

$$\sum_{k \in \mathbb{K}} w_k^2 \| G(\Omega, \hat{\psi}) - G_c(\Omega, \theta) \|_F^2 \qquad (13\text{-}27)$$

w.r.t. θ in a user-specified frequency band $\mathbb{K}$, and with user-defined weights w_k. To keep the approximation error of the constrained model below a user-imposed level, the order of the constrained model might be larger than that of the validated model (D'haene et al., 2006; D'haene and Pintelon, 2008). The big advantage of this two-step procedure is that it provides models with uncertainty bounds (in the first step) and bias error bounds (in the second step), which is not the case when the constraints are imposed during the noise removal.

13.13 CONCLUSIONS

In this chapter we have refined the order in the identification field by putting forward four basic questions that should be answered before starting the identification process: (i) What signal assumption should be used, zero-order-hold (ZOH) or band-limited (BL)?; (ii) What excitation should be preferred, arbitrary or periodic excitations?; (iii) What kind of noise model should be used, parametric or nonparametric? and (iv) Finally, a last choice that should be made is the criterion to match the model and the data, a generalized output error or an errors-in-variables cost. This leads to the following major (interrelated) steps in the design of the identification process:

1. Experiment design

   1a. Select the ZOH or BL signal assumption

   1b. Choose between arbitrary or periodic excitations

2. Model: discrete- or continuous-time model?

   2a. ZOH $\rightarrow$ discrete-time models

   2b. BL $\rightarrow$ continuous-time models

   If the signal assumption is violated, the choice is free but a more complex model is needed to describe the measurements. This model can be sued for prediction purposes so long as the setup is not changed.

3. Cost function

   3a. Noise on input and output: errors-in-variables method using nonparametric noise models

   3b. Noise on the input negligible: generalized output error method using a nonparametric noise model (default choice) or a parametric noise model (ARX and ARMAX model structures) if the output is disturbed by unobserved random inputs

   3c. Output observations only: prediction error or maximum likelihood estimation of a parametric noise model

And what about time or frequency domain identification? For some selections among the preceding choices, frequency domain identification methods seem to be preferred, for example, nonparametric noise models, very wide frequency ranges, continuous-time models, modeling subsystems for simulation, and errors-in-variables with periodic or random excitations. For online identification (Ljung and Söderström, 1983), or identification in the presence of nonstationary noise, time domain identification is the natural choice. For the other situations, both domains are equivalent, and the user can make the choice by using other criteria such as familiarity with one domain.

13.14 EXERCISES

13.1. The ZOH reconstruction of a discrete-time signal $u_d(kT_s)$ is

$$u_{ZOH}(t) = \sum_{k=-\infty}^{\infty} u_d(kT_s)\text{zoh}(t - kT_s)$$

with $\text{zoh}(t) = 1$ for $0 \le t < T_s$ and $\text{zoh}(t) = 0$ elsewhere. Show that the spectrum after a ZOH reconstruction is $X_{ZOH}(\omega) = U_d(e^{j\omega T_s})\text{ZOH}(\omega/\omega_s)$ with $\text{ZOH}(x) = T_s \sin(\pi x)/(\pi x)e^{-j\pi x}$.

13.2. Consider the setup in Figure 13-3 and prove relation (13-5) (hint: first calculate the spectrum of $u_{zoh}(t)$, and $y(t)$. Next, apply the sampling theorem).

13.3. Consider the setup in Figure 13-4 and prove relation (13-6) (hint: first calculate the spectrum of $u_{zoh}(t)$, and $u(t), y(t)$. Next, apply the sampling theorem).

13.4. Consider the setup in Figure 13-5 and show that (13-8) reduces to (13-5) for $L(j\omega) = 1$. Note that this result is different from what would be found starting directly from (13-9) using a Taylor series expansion of $\text{cosec}(x)$ (hint: check the convergence of the series expansions carefully).

13.5. Reproduce the transfer characteristics of Figure 13-13 for the systems given in (13-19).

13.6. Consider the continuous-time system of Figure 13-21, with $G_{act}(s)$, $G(s)$, and $M(s)$ the continuous-time actuator, plant and controller characteristics, respectively, and with $r_{zoh}(t)$ a zero-order-hold reference signal. Prove that the response of all signals $p(t)$, $u(t)$, $y(t)$, and $w(t)$ in the continuous-time system to the ZOH reference $r_{zoh}(t)$, at the sampling instances, can exactly be calculated via the discrete-time equivalent in Figure 13-21, where

$$G_{act}^d(z^{-1}) = (1 - z^{-1})Z\{G_{act}(s)/s\} \qquad G^d(z^{-1}) = \frac{Z\{G_{act}(s)G(s)/(s(1 + G_{act}(s)M(s)))\}}{Z\{G_{act}(s)/(s(1 + G_{act}(s)M(s)))\}}$$

$$M^d(z^{-1}) = \frac{Z\{G_{act}(s)G(s)M(s)/(s(1 + G_{act}(s)M(s)))\}}{Z\{(G_{act}(s)G(s))/(s(1 + G_{act}(s)M(s)))\}}$$

What do you conclude? Is physical interpretation of the identified discrete-time plant model $G^d(z^{-1})$ possible? (hint: apply (13-4) from the reference signal to each internal signal, and calculate the appropriate ratios)

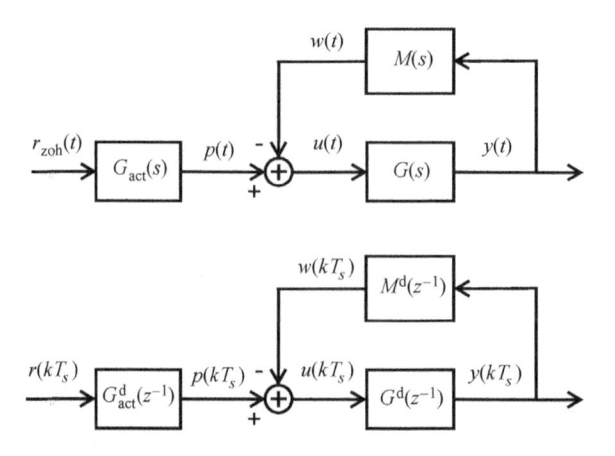

Figure 13-21. Continuous-time system (top) excited by a zero-order-hold signal $r_{zoh}(t)$, and its exact discrete-time equivalent (bottom).

14

Guidelines for the User

Abstract: A guideline for the user is provided in this chapter. It not only gives an overview of the complete identification process, but also discusses the decisions that should be made at each stage. So, inexperienced users have a road map that reduces the risk of getting trapped and increases their chances of arriving at a good model for their problem.

14.1 INTRODUCTION

From the previous chapters, it became clear that identification is a complex task, bringing together many different skills. It is not enough to know the specific application field well (e.g., automotive, acoustics, electrochemistry), but the user is also expected to be familiar with measurement techniques, statistical theories, and numerical methods. As it is quite unlikely that all these skills are found in one person, the risk of making a serious mistake during one of the identification steps is always present. The aim of this chapter is not to turn all readers into absolute specialists but to offer some guidance to inexperienced users in order to increase their chances of a successful identification. To do so, we present two tools for readers to select the best solution for their problem. First, we provide a table that will guide readers to a good identification scheme (experiment setup, noise model, estimator) for their problem, starting from a few simple questions. Second, we provide some rules of thumb that may help the reader to avoid frequently appearing problems and pitfalls.

14.2 SELECTION OF AN IDENTIFICATION SCHEME

The aim of this section is to make a proper selection among possible identification schemes. By answering two questions, we will guide users to a good choice of the measurement setup, the noise and the plant model, and the identification method that can solve their problem. Of course, these guidelines are strongly influenced by our personal background and experiences. For these reasons, we strongly urge readers to judge them critically and combine our advice with their own experience.

14.2.1 Questions – Proposed Solutions

There are basically two questions that set the complete identification scheme: the stochastic framework and the particular application. In the sequel we discuss in detail these questions and the proposed solutions.

TABLE 14-1 Selection of the Noise Model and the Identification Method

Stochastic Framework?	Output Observations only/ Known Input – Output Disturbed by Unobserved Input(s)	Other[a]
Noise Model	Parametric	Nonparametric
Identification Method	Prediction error/Frequency domain maximum likelihood	Sample maximum likelihood

a. Generalized output error with independent plant and noise dynamics, and errors-in-variables.

14.2.1.1 Stochastic Framework? How does the disturbing noise sneak into the process: As process noise? On the output measurements? Or on the input and output measurements? Is the output perturbed by unobserved random inputs? Can the input be observed? For some methods, this is not important at all, but for other methods it is a critical issue. Identification in feedback is a more tedious problem in control design than solving the same problem in open loop. The most general situation is that we consider (correlated) noise on the input and output measurements, which also includes the feedback problem. In these cases nonparametric noise models combined with the sample maximum likelihood estimator (see Sections 10.3 and 12.3) is the prime choice. However, the case where the identification starts from output observations only can only be solved using parametric noise models. If the input is known and the output is disturbed by unobserved random plant inputs, then parametric noise models where (some of) the poles are common to the plant model (e.g., ARMAX model structure) give the best results. Either a time domain prediction error method (Ljung, 1999) or a frequency domain maximum likelihood estimator can be used (see Section 10.9).

14.2.1.2 Application? What is the problem to be solved? For what reason do I need a model? We consider three possible answers. A first possibility is that a model-based digital controller will be designed on the basis of the identified model. In this case a discrete-time model combined with a zero-order-hold setup is optimal. Another possibility is modeling for physical interpretation (for example, estimation of the resonance frequencies, damping ratios, and mode shapes of a vibrating structure) or for analog simulation (for example, identification of an equivalent electrical scheme for use in an electrical network simulator such as SPICE). For these applications a continuous-time model combined with a band-limited setup is the prime choice. A final possibility is the modeling of analog (sub)structures for implementation in a digital simulator. In that case it is most suited to describe the analog characteristics by discrete-time models identified from band-limited data.

14.2.1.3 Proposed Solutions. The optimal choices of the experimental setup, the noise model, the domain of the parametric model, and the identification method are summarized in Tables 14-1 and 14-2. While most calculations can be done either in the time domain or in the frequency domain, some are easier performed in the time domain, others in the frequency domain. Therefore, the continuous-time models and the nonparametric noise models are solely handled in the frequency domain. Focusing the identification in a particular frequency band is also easier in the frequency domain than in the time domain.

TABLE 14-2 Selection of the Setup and the Domain of the Parametric Model

Application?	Digital Control	Physical Interpretation/ Analog Simulation	Digital Simulation
Experimental Setup	Zero-order-hold	Band-limited	Band-limited
Domain	Discrete-time	Continuous-time	Discrete-time

14.3 IDENTIFICATION STEP-BY-STEP

In this section, a series of general advice is formulated covering the different phases of the identification process as listed:

Check and selection of the experimental setup

Design of an experiment

Choice noise model

Preprocessing of the data

The identification step

Validation of the results

Each of these topics will be visited shortly, resulting sometimes in an overlap with earlier or later material in this book. However, we chose to bring it all together here in order to optimize the global overview and insight of the reader, to minimize the risk of making bad decisions.

14.3.1 Check and Selection of the Experimental Setup

In many cases, an identification run starts from data that were made available at some place. From our experience, it is definitely no loss of time to inspect the experimental setup and to check how the data were collected. Quite often, significant improvements can be obtained by very simple changes in the setup. Are the amplifiers properly set? What preprocessing is done on the raw data? What are the properties of the sensors that are used to get the raw data? Is the process operating under stationary conditions? ... Each of these aspects can have a considerable impact on the overall quality of the data. A short visit to the experimental site is very informative in revealing unexpected complications that would be detected only after wasting a lot of time and effort. For example, the data can be collected with a specific goal in mind (e.g., quality control), paying no attention to disturbing effects or bad settings that eventually make the data useless for the intended modeling purposes.

When looking at a measurement setup, two levels can be distinguished. A typical instrumentation configuration consists of a signal generator, a data acquisition arrangement, and a data-processing part, which extracts the parameters of interest from the raw data. Understandably, the sensor and actuator technology of the setup are closely connected to the application, whereas the actual data acquisition (amplification, attenuation, filtering, sampling, and quantization) is only loosely coupled to a specific application.

It is not easy to give general rules on the sensor or actuator part, although it is always worthwhile to check for the linearity, offsets, and drifts of these devices. These questions are closely linked to the calibration of the setup. A good identification scheme makes it possible to reduce the impact of stochastic errors, but systematic errors should be eliminated, either by a proper calibration procedure that minimizes these errors or by extending the model to include them as unknown parameters. What choice is optimal depends strongly on the effort that is needed to go for one of these solutions. In general, the quality of the model improves with the quality of the measurements. Identification should be no excuse to do sloppy measurements, although it can open new possibilities to extract the desired information under worse operational conditions.

Because the acquisition part is quite similar for many instruments, more general advice can be formulated. A first general choice is to select between the ZOH and the BL setup. Although it is not critical in every situation, it is better to match this choice with the application in mind (discrete-time versus continuous-time model, digital simulation versus physical model, control application, etc.). The BL setup is assured to be valid by putting proper anti-alias filters in place before sampling the signals (check for the cutoff frequency, the stopband rejection, the linearity).

A second, very important, aspect is the synchronization between the generator and the acquisition. If periodic signals will be applied and explicit advantage will be taken of the periodic nature (averaging, plant leakage suppression, etc.), it is extremely important that the generator is synchronized with the acquisition. Otherwise, it is more complicated to use the redundancy induced by the periodic behavior.

For critical applications, it is also necessary to check the stability of the master clock and the triggering in order to assure the best quality. Jitter (see Section 2.5.2) decreases the signal-to-noise ratio of the measurements, and clock instabilities (Schoukens et al., 1996a) can induce systematic errors.

Finally, the signal stored in the arbitrary waveform generator (= known reference signal) should always be saved together with the input-output data, because it can significantly simplify the system identification task. Indeed, if a known reference signal is available then, for example, (i) the noisy input, noisy output problem can be solved for arbitrary excitations using nonparametric noise models (see Section 7.2.7); (ii) exact filtering of the input-output signals becomes possible within a frequency domain prediction error framework (see Section 10.9); and (iii) the noise can be separated from the nonlinear distortions in the transient response of a plant to a periodic excitation (see Section 7.3.8).

Advice

Visit the site of the experimental setup and talk to the operators to learn from their experience.

Check the systematic errors of the complete data acquisition – calibrate the setup.

Check the validity of the signal assumptions (ZOH or BL, anti-alias filters).

Pay attention to the synchronization of the setup.

Always save the signal stored in the arbitrary waveform generator together with the noisy input-output data.

14.3.2 Design of an Experiment

The second phase of the identification process is the design of the excitation signals. Sometimes the user cannot influence the process at all. But even then, it should be checked whether the natural fluctuations (operational perturbations) carry enough information to give, at least, a chance of a successful identification. In all the other cases an excitation signal should be selected. This raises a series of questions immediately: what excitation level should be applied? What frequency band should be excited? In the initial phase of the identification process, we can only use prior information to set these values. For operator-controlled processes, the operators should have good knowledge of acceptable values. For other devices the nominal values given in the user manual might give some indications. And if none of this information is available, initial tests should give the required information. In this case we can only hope that our experience will help us to protect the device under test against dangerous overloads.

A second question is the linearity of the device. In this book we deal with methods to model linear systems or the best linear approximation of nonlinear systems. So it is important to know whether or not the linearity assumption is met. If the user is very confident, it is not necessary to check for nonlinear distortions; otherwise it is better to use excitations that make it possible to detect their presence. According to the available prior knowledge, some precise guidelines can be given:

- If it is known that the odd (even) nonlinear distortions are dominant over the even (odd) distortions, then the optimal choice is measuring (at least) two periods of the response to a full (odd) random phase multisine.

- If it is known beforehand that the nonlinear distortions are dominant over the disturbing noise, then one period of a full random phase multisine or a random noise excitation should be used.

- If nothing is known at all, then the recommended default choice is measuring (at least) two periods of the response to an odd random phase multisine.

If significant nonlinearities are detected, it is also important to reflect carefully on the goal of the modeling process. Do you intend to extract the underlying linear system or are you interested in a best linear approximation? An appropriate excitation design in agreement with the previous selection should be made. In the first case, the amplitude should be made as small as possible (although some nonlinearities such as stick slip are pronounced in that case). In the latter case, the excitation should be representative for the class of excitations that will be applied later on to the device (e.g., same amplitude distribution and same power spectrum; see Chapters 3 and 4).

Next, it is necessary to check whether the experiments have to be done under feedback conditions. These can be explicitly visible (a controller is in place) or can be implicitly present (e.g., loading of the non-ideal actuator by the plant). An example of the latter is a mechanical device that interacts with the output of the shaker. In these cases, additional care is needed because many identification methods fail when proper action is not taken (see Section 10.10).

Finally, the question of the required frequency resolution should be answered. For lightly damped systems (e.g., mechanical vibrating systems) one should minimize the risk of missing a resonance and, hence, maximize the frequency resolution for a given measurement time T: use random noise excitations (or $P = 1$ period of a full random phase multisine) if the nonlinear distortions are dominant (see Section 7.2.8); otherwise measure $P = 2$ periods of the transient response to a full random phase multisine (see Section 7.3.7). If the required frequency resolution f_0 is smaller than that of the total experiment time T ($f_0 > 1/T$), then one can maximize the signal-to-noise and signal-to-distortion ratios of the FRF measurement using random phase multisines: perform $M = \lfloor f_0 T \rfloor$ ($\lfloor x \rfloor$ is the largest integer smaller than or equal to x) experiments and measure $P = 1$ period if the nonlinear distortions are dominant (see Section 7.3.6); otherwise perform $M = \lfloor f_0 T/2 \rfloor$ experiments and measure $P = 2$ periods (see Section 7.3.6).

Advice

Choose the excitation (random or periodic) that best fits your specific needs.

Select the amplitude range and frequency band of the excitation signal to cover the frequency band of interest.

Check for the presence of nonlinear distortions.

Check whether the device is captured in a feedback loop.

Keep your application in mind.

14.3.3 Choice Noise Model

A third important decision to be made concerns the choice between a parametric or a nonparametric noise model. As will become clear from the discussion, the choice of the noise model has an important impact on the complexity of the parametric plant modeling step.

If the identification starts from output observations only, then parametric noise models are the only option. This is known as time series analysis in econometrics, spectral analysis in signal processing, and operational modal analysis in mechanical and civil engineering. Parametric noise models are also very useful in all cases where the input is known, and where the output noise is solely due to unobserved random inputs filtered by the plant dynamics, because they help identifying the plant dynamics (e.g., ARMAX model structure). The major disadvantages of parametric noise models are: (i) the quality of the estimated noise model depends on the quality of the estimated plant model, (ii) the increased complexity of the model selection/validation and minimization of the cost function, and (iii) they offer no practical solution for errors-in-variables problems.

Nonparametric noise models are obtained in a preprocessing step and, hence, their quality is independent of the quality of the estimated parametric plant model, and the minimization of the cost function is less complex. As such, they simplify the model selection/validation procedure. Moreover, errors-in-variables problems and identification in feedback become as easy as a generalized output error problem of a system operating in open loop. A last advantage is that using periodic excitations the noise can be distinguished from the nonlinear distortions. The major drawbacks of nonparametric noise models are: (i) the increased variability of the identified plant model due to the variability of the noise model and the suppression of that part of the input that is uncorrelated with the reference signal, (ii) unobserved inputs do not contribute to the estimation of the plant model, and (iii) they offer no solution for output data only problems.

Advice

Default choice: nonparametric models.

Use parametric noise models for output data only problems, and problems where the input is known and the output is disturbed by an unobserved random input.

14.3.4 Preprocessing

The raw data, collected during the experiment, need to be preprocessed before starting the more demanding identification step. This not only facilitates checking for anomalies in the data (e.g., outliers or missing data) and bad experiments (poor signal-to-noise ratio) in an early phase of the identification process but also provides more insight into the complexity of the problem (look to the FRF), and makes it possible to separate different side aspects (such as trends or sensor drift) from the main task, which is to extract a linear parametric model from the data.

14.3.4.1 Removal of Trends, Drifts, and Offsets. In many problems, a linear model is used as a local linearization of a nonlinear system, around a given operating point, that might be slowly varying as a function of an uncontrolled input (e.g., temperature). If the user is not interested in building a full-blown nonlinear model that accounts for all these effects, it is important to eliminate their impact on the data as much as possible. A whole bunch of methods, ranging from very simple to complex procedures, can be used to eliminate these undesired effects. The simplest technique is to eliminate the DC offset from the measurements. This can be done effectively, under periodic operating conditions, by putting the DC line to zero after the DFT (or just do not use the DC line during the identification). In that case it is also very simple to observe slow drifts of the offset signals: calculate the mean value for a series of successive periods and check for systematic variations as a function of the period number. Next, simple correction methods such as linear interpolation between the successive DC values can be used to remove the first-order effects of these variations. If the variations are large, then more sophisticated trend-removing algorithms are recommended (McCormack et al., 1994a; Peirlinckx et al., 1996). An alternative is to disregard the spectral contributions at the low frequencies that are well below the reverse of the dominating time constants of the system. These techniques are also applicable to reduce the impact of sensor drift.

Advice

Check for trends by calculating the mean value over the successive periods.

Do not use the DC information during the identification.

14.3.4.2 Dealing with Outliers and Missing Data. Sometimes the measurements are very disturbed during a short interval (e.g., the presence of spikes or loss of data in a transmission link). This results in a few data that are very unreliable or even completely missing. In such a case the first advice is to repeat the experiment at a reasonable cost, if possible. Only if this is excluded do we advise restoring the data by trying to remove the artifacts. In case of missing data in highly oversampled signals, simple interpolation methods can help a lot (Rolain et al., 1998). In more complex cases, where the oversampling is low, the missing or heavily disturbed data can be replaced, considering them as missing data that also need to be identified (Pintelon and Schoukens, 2000; see also Section 9.20). This increases the complexity of the algorithms considerably and should be regarded as a last resort. Another possibility consists in concatenating the data records (see Section 14.3.5.5).

Advice

Perform new experiments.

If this is not possible, use simple interpolation methods if $f_{max} < 0.1 f_s$.

Last resort: estimate the missing data.

14.3.4.3 Estimate the Nonparametric FRF. We strongly advise calculating, always, the nonparametric FRF estimate before starting the parametric modeling step. This additional effort is negligible (see Chapters 2–4, 7). Simple visual inspection of the FRF not only gives a first impression of the model complexity but also allows a first evaluation of the quality of the data (noise level + level nonlinear distortions) and reveals, in a very early phase of the process, many problems. Sensor failure, saturated amplifiers, and acquisition overloads all result in an unexpected but mostly conspicuous distortions of the FRF. Finally, the user can check whether the appropriate frequency band is excited.

Advice

Calculate the FRF, the noise level, and the level of the nonlinear distortions, and make a visual inspection.

Select the frequency band of interest.

14.3.4.4 Check Whether the System Is Time Invariant. Slow-varying trends and offsets not only disturb the measurements but also can change the linearized behavior of the system intrinsically. Under these conditions, users should carefully reflect on the value of their models and the aim of the experiments. A useful idea for the variability is a necessary condition to make a ripe decision. In case of periodic excitations, the FRF can be calculated for each individual period and, again, a simple visual inspection will give good insight into the significance of the problem.

Another way to detect slow time-variations of the system consists in taking the DFT of all output periods at once and checking whether or not skirts are present around the excited frequencies (Lataire and Pintelon, 2009). Before performing this test one should first verify that the acquisition and generator units are synchronized, and that the actuator(s) and sensor(s) are time-invariant. This can be done by measuring with the same setup a dynamic system that is known to be time-invariant.

Advice

Check the time invariance of the system by calculating the FRF over successive periods after trend removal (see 14.3.4.1); or by looking for the presence of skirts in the output DFT spectrum of all output samples at once.

14.3.4.5 Extract the Nonparametric Noise Model. It is very easy to extract the nonparametric noise model from periodic (Chapter 2–4, 7) or random (Chapter 7) records. Again, this information is very revealing.

Observing the SNR of the input and output measurements not only gives a good impression of the overall quality of the data but also shows where the noise sneaks into the measurements. A low SNR at the input (or the output) points to high noise levels at the input (or output). Low SNR values at the input and the output in combination with a high input-output correlation indicate dominating generator noise or process noise that is turning around in a feedback loop.

The noise levels are known as a function of frequency. So, the user can check whether or not the frequency band of interest is affected too much by the noise. It also gives feedback in an early stage of the identification process for the design of improved experiments, such as putting more power in the frequency bands with a too low SNR. Of course, the noise information can also reveal problems in the measurement setup and alert the user to their presence. For example, bad grounding can be denoted by the presence of high disturbing components at the harmonics of the mains frequency.

Advice

Make a nonparametric noise analysis.

Check for anomalies.

Judge the quality of the experiment.

Improve the experiment if necessary and possible.

14.3.4.6 Check for the Presence of Nonlinear Distortions. A final, but important, check is to look for the presence of nonlinear distortions. From Chapter 3 it is known that such distortions can be masked completely as filtered white noise in the case of random excitations. All classical validation tests at the end of the identification process will fail to indicate their presence. This may lead to dangerous situations in which users erroneously believe they captured a good model. A significant change in the excitation signal, in a later phase of the design process, would completely fool the quality of the predicted output. Moreover, the noisy behavior of the measurements is actually not due to the noise but should be attributed to the stochastic nonlinear distortions. For this reason, early detection of the presence and the level of nonlinear distortions is very valuable. It gives users, from the very beginning, an idea of the best quality that can be obtained through linear modeling. This makes it possible to make a conscious decision to go on or to stop with the modeling effort before wasting a lot of time in the identification step.

If the goal is to maximize the nonlinear detection sensitivity of the measurement for a given excitation rms value, frequency resolution f_0, and experiment time T, then one should measure $P = \lfloor pf_0T \rfloor$ or $P = \lfloor pf_0T/2 \rfloor$ ($\lfloor x \rfloor$ is the largest integer smaller than or equal to x) periods of the (transient) response to a full or odd random phase multisine with random harmonic grid where $100 \times p\%$ of the (odd) harmonics are excited. If the goal is to minimize the total uncertainty (noise + nonlinear distortions) on the modeled best linear approximation, while maintaining the ability to distinguish the noise from the nonlinear distortions, then one should measure $P = 2$ periods of the response to a full (the odd distortions are dominant over the even) or odd (the even distortions are larger than the odd) random phase multisine.

Advice

Use specially designed periodic excitation to check for the presence of nonlinear distortions (see Sections 4.4 and 14.3.2).

If the nonlinear distortions cannot be neglected, use the total (co)variances (sum noise (co)variances and (co)variances of the nonlinear distortions) for generating uncertainty bounds on the FRF, and for identifying the parametric plant model.

14.3.5 Identification

Only at the fifth step do we arrive, finally, at the kernel of the identification procedure where a parametric model is extracted from the (preprocessed) data. Just as for the previous phases, a number of user decisions have to be made. Among them, we will discuss the choice of a model class; the selection of the model complexity; the impact of initial conditions or transients; dealing with time delays; and finally, we spend a few moments on the problem of local minima.

14.3.5.1 Choice of a Model Class. In a first step the desired model class should be selected: do you want to get a continuous-time model (e.g., physical interpretation of the results, or synthesis of an electrical network equivalent of the model for implementation in an electrical network simulator), a discrete-time model (e.g., for control design or to set up a simulator), or one of the special models such as $\sqrt{s}$ to model diffusion processes or distributed systems. Remember that this choice should be matched with the selected experimental setup. Otherwise, more complex models will be needed to capture the difference between the actual intersample behavior and the one assumed by the model.

Advice

Select the model class that best fits your application.

14.3.5.2 Selection of the Model Complexity. During the identification process, not only do the parameters need to be estimated but also the model order should be selected. There exist a series of simplified estimators, with increased noise sensitivity (e.g., linear least squares), that make it possible to estimate a whole bench of models in one step. This result can be used to get an initial guess of the required complexity. An alternative is to calculate the FRF (for input-output data problems) or the power spectrum of the data (for output data only problems) to get an initial idea (e.g., counting the number of resonances for vibrating mechanical structures). Next, this guess should be refined using more advanced estimators. Two strategies are possible. The first one is conservative, starts from a simple model, and searches gradually for more complex models. The alternative is to go for a very complex model and check next what poles and zeros can be eliminated. This method can be used only if the estimator is robust for numerical singularities (common pole-zero pairs). In both cases, the cost function and a residue analysis are very valuable tools to guide one in the selection process (see Chapters 11 and 12).

Advice

Calculate the FRF (input-output data) or power spectrum (output data only) to get an initial idea of the complexity of the problem.

Check the phase of the FRF and the cross-correlation between the input and output signals to detect the presence of a delay.

Analyze the value of the cost function.

Perform a whiteness test of the residuals.

Detect overmodeling via AIC (prediction) or MDL (physical interpretation).

14.3.5.3 Impact of Initial Conditions or Transients. During the identification and validation it is important to safeguard against the impact of initial conditions (time domain) or leakage effects (frequency domain) induced by the plant and noise dynamics on the data. While the leakage errors due to the plant dynamics can easily be included in the parametric plant model by adding an additional transient term (see Chapter 6), this is somewhat more tricky for the noise leakage errors because it is more difficult to generate starting values. The leakage/transient errors due to the plant and the noise dynamics can easily be suppressed nonparametrically in the input-output data for both random and periodic excitations (see Chapter 7). Using periodic excitations, the plant leakage errors can be avoided if an integer number of periods of the steady state response is observed. Note, however, that the noise leakage errors always remain present in these measurements.

Even if these plant and noise leakage effects mostly have a second-order impact on the quality of the identified model, they become dominant during the analysis of the residuals. This can lead the user to very complex models because the correlation test of the residuals is very sensitive to these effects. For this reason, the user is advised to add these additional terms to his/her model or to suppress it nonparametrically in the data.

For lightly damped systems (e.g., mechanical vibrating structures) the leakage errors cannot be neglected in the parametric plant modeling; even in the steady state response to a periodic excitation (noise leakage). Hence, nonparametric suppression of the leakage errors in the data is recommended here. It also allows us to handle the transient response to periodic excitations, which either significantly reduces the experiment time for a given frequency resolution, or increases the frequency resolution for a given experiment time.

Advice

Suppress nonparametrically the leakage errors in the input-output data (random and periodic excitations).

If the transient time can be neglected, measure the steady-state response to a periodic input.

Always add a plant and noise transient term to the model structure when identifying parametric noise models.

Suppress (non)parametrically the transient/leakage errors in the validation data set.

14.3.5.4 Dealing with Time Delays. Some systems, such as transmission lines or transport phenomena, cannot be modeled as a lumped system. An additional delay term becomes explicitly visible and should be added to the model. The presence of such terms can be recognized from the impulse response (where a delay is explicitly visible) or from the FRF (by looking for a rapidly varying (linear) phase). When a delay is present, we advise that all information is used to get a good initial estimate. It reduces the risk of stumbling on a local minimum during the optimization. Delay systems have many local minima, and it is very hard to find the global minimum.

Advice

Add an explicit delay term to the model and use all prior information available to get an initial value.

Restart the search, using different starting values, to make sure that you are not trapped in a poor local minimum.

14.3.5.5 Combining Experiments. It may happen that data sets of different experiments on the same plant are available. These data sets may originate from time domain experiments, frequency domain experiments, or time and frequency domain experiments. The basic question that arises then is how to combine these data sets in an optimal way. The solution to this problem depends on the prior knowledge and the type of experiments performed. We distinguish between the following three cases.

1. It is known only that the time and/or frequency domain experiments are independent (the noise in one experiment has nothing to do with the noise in the other experiments).
2. The independent time or frequency domain experiments are synchronized.
3. The time domain experiments stem from one experiment where at several time instances a (large) number of consecutive input and output samples are missing; or separate input-output data sets measured under the same operational conditions are available.

The following solutions are recommended for each of these situations:

1. To solve case 1 we apply the Gaussian maximum likelihood (ML) principle to independent experiments. The only thing to do is to extend the frequency domain data vector Z as $Z^T = [Z^{[1]T} \ Z^{[2]T} ... Z^{[M]T}]$ with $Z^{[r]}$ the input-output DFT spectra of the rth experiment (see (9-3)), and similarly for the noise (co)variances. If no periodic excitations are used, the equivalent initial conditions are different for each experiment and they should be added to the model parameters.
2. For synchronized experiments, the sample mean and sample (co)variance should be calculated, and these data are then considered as the raw input data for the identification process. Note that the improved signal-to-noise ratio also relaxes the starting value generating problem, resulting in a wider convergence rate of the search algorithms. The single experiment software can be used without any modi-

fication to handle the synchronized experiments, even in case of arbitrary excitations (see Appendix 14.B).

3. A first possibility to tackle case 3 is to handle the complete input-output data sets as independent experiments. However, it is better to express that the data sets stem from the same experiment. The identification starts, then, from the DFT spectra of the concatenated data sets (the missing data points are just taken out). The recommended default choice is to suppress nonparametrically the transient (leakage) errors in the concatenated data sets and to use the steady state input-output model (see Section 12.3.1). An alternative approach consists of combining the concatenated data sets without nonparametric transient suppression with an extended parametric model (see Section 6.3.4, (6-48) for the concatenation of two data sets; and see Exercise 6.8, (6-81) for the general case). This works well so long as the noise leakage error can be neglected or if the plant and noise transient terms have the same poles.

14.4 VALIDATION

At the end of the identification process, it should be checked whether the identified model is a valid one. Ideally, the estimated model should be close to the exact one, but the reader should realize that the "exact" model is only an idealized concept. Most real-life systems cannot be described exactly by a rational transfer function. Moreover, because the exact system will always be unknown, we will never be able to answer that question. For this reason, we should focus on more realistic questions such as: Does the model describe the data well? Does the model fit my needs? These questions can be properly answered. A set of tools is available to check whether all information is extracted from the data. We briefly repeat them here; for more details, the reader is referred to Chapters 11 and 12.

A *global test* is based on the value of the cost function and a correlation test of the frequency response function (FRF) residuals (difference between the nonparametric FRF estimate and the identified transfer function model):

(i) The cost is too small: check the noise and/or nonlinear distortion analysis, are the (co)variances correct?

(ii) The value equals the expected value within the uncertainty bands: no information is left in the data (this should be confirmed by the residual analysis). If the noise (co)variances are used as weighting in the cost function, then the system behaves linearly and no unmodeled dynamics can be detected. If the total (co)variances (noise + nonlinear distortions) are used, then no unmodeled dynamics can be detected in the best linear approximation.

(iii) The value is too large: there are still model errors present. Their nature can be determined from an analysis of the FRF residuals. Correlated residuals point to unmodeled dynamics and demand increasing the model complexity. White residuals point to nonlinear distortions if noise (co)variances are used as weighting, and increasing the model complexity will not help.

Analyzing the cost function reveals the presence or absence of model errors, but gives no indication about the frequency location of the unmodeled dynamics. A *local test* obtains this information by comparing the frequency response function (FRF) residuals to the uncertainty of the nonparametric FRF estimate:

(i) About $100 \times p\%$ of the FRF residuals lie within the $100 \times p\%$ noise uncertainty bound: the system behaves linearly and no unmodeled dynamics can be detected (to be confirmed by a correlation test on the FRF residuals).

(ii) Too much ($> 100(1-p)\%$) of the FRF residuals lie outside the $100 \times p\%$ noise uncertainty bound: unmodeled dynamics (correlated FRF residuals) and/or nonlinear distortions (uncorrelated FRF residuals).

(iii) About $100 \times p\%$ of the FRF residuals lie within the $100 \times p\%$ total uncertainty bound (noise + nonlinear distortions): the best linear approximation is identified and no unmodeled dynamics can be detected (to be confirmed by a correlation test on the FRF residuals).

(iv) Too much ($> 100(1-p)\%$) of the FRF residuals lie outside the $100 \times p\%$ total uncertainty bound (noise + nonlinear distortions): unmodeled dynamics in the best linear approximation (correlated FRF residuals).

Sometimes it is not necessary to extract all information from the data, especially when this would lead to very complex models. At that stage the reader should specify an acceptable error level (e.g., no model errors larger than 10%), and once this level is reached the model complexity is no longer increased. The choice between these options depends on the intended use of the model completely.

In the classical identification approach (see, for example, Ljung, 1999), it is strongly advised to split the available data in two sets: an identification set, used to identify the model, and a validation set to check for the model. Although this is a very robust check of the quality of the model, we prefer to use all the available experimental time and data to identify the model. The availability of the nonparametric (high-quality) FRF and a nonparametric noise model turns out to be a good alternative for the validation set. Of course, it always makes sense to perform a second experiment with another excitation signal, but then we advise using this information also during the identification step. Before starting the parametric identification, the two nonparametric FRFs can be compared with each other to check whether one model can be used to describe both experiments.

Advice

Compare the parametric model with the nonparametric FRF and its noise (and total) uncertainty bounds.

Check the value of the cost function.

Perform a correlation test on FRF residuals.

14.5 CONCLUSION

The system identification task can be simplified significantly if the following two recommendations are followed:

- Always save the signal stored in the arbitrary waveform generator together with the input-output signals.
- Use nonparametric noise models.

It is also a good practice to make a nonlinearity test and to quantify the level of the nonlinear distortions, because it sets the limits of the linear system identification framework.

14.6 APPENDIXES

Appendix 14.A Independent Experiments

Because the Gaussian negative log-likelihood function (9-78) of the union of independent experiments is the sum of the contributions of the individual experiments separately, it follows that the ML solution (9-83) equals the sum of the ML cost functions the data sets separately. The only thing to do is to extend the frequency domain data vector Z as $Z^T = [Z^{[1]T} Z^{[2]T} ... Z^{[M]T}]$ with $Z^{[r]}$ the input-output DFT spectra of the rth experiment (see (9-3)), and similarly for the noise (co)variances. For arbitrary excitations the equivalent initial conditions are different for each experiment and they should be added to the model parameters. To see this it is sufficient to note that the weighted residual $\varepsilon(\theta, Z)$ of the ML solution (9-83) of the combined experiments can be written as

$$\varepsilon^T(\theta, Z) = [\varepsilon^{[1]T}(\theta, Z^{[1]}) \; \varepsilon^{[2]T}(\theta, Z^{[2]}) \; ... \varepsilon^{[M]T}(\theta, Z^{[M]})] \qquad (14\text{-}1)$$

with $\varepsilon^{[r]}(\theta, Z^{[k]})$ the weighted residual vector of the rth experiment.

Appendix 14.B Relationship between Averaged DFT Spectra and Transfer Function for Arbitrary Excitations

The experiments are synchronized (case number 2) if the phases of the true input DFT spectra are the same

$$\angle U_0^{[1]}(k) = \angle U_0^{[2]}(k) = \cdots = \angle U_0^{[M]}(k), \; k = 1, 2, ..., F \qquad (14\text{-}2)$$

The solution consists of averaging the data $Z = M^{-1} \sum_{r=1}^{M} Z^{[r]}$ and changing the noise (co)variances accordingly $\sigma^2 = M^{-2} \sum_{r=1}^{M} \sigma^{[r]2}$ with $\sigma = \sigma_U$, σ_Y and σ_{YU}. The input-output DFT spectra of each experiment satisfy

$$\begin{aligned} A(s_k, \theta^{[r]}) Y^{[r]}(k) &= B(s_k, \theta^{[r]}) U^{[r]}(k) + I(s_k, \theta^{[r]}) + \Delta^{[r]}(s_k) \\ A(z_k^{-1}, \theta^{[r]}) Y^{[r]}(k) &= B(z_k^{-1}, \theta^{[r]}) U^{[r]}(k) + I(z_k^{-1}, \theta^{[r]}) \end{aligned} \qquad (14\text{-}3)$$

with $\theta^{[r]} = [a_0 \, a_1 ... a_{n_a} \, b_0 \, b_1 ... b_{n_b} \, i_0^{[r]} \, i_1^{[r]} \, ... i_{n_i}^{[r]}]^T$, $r = 1, 2, ..., M$ (see (6-33)). Averaging (14-3) over all experiments gives

$$\begin{aligned} A(s_k, \theta) Y(k) &= B(s_k, \theta) U(k) + I(s_k, \theta) + \Delta(s_k) \\ A(z_k^{-1}, \theta) Y(k) &= B(z_k^{-1}, \theta) U(k) + I(z_k^{-1}, \theta) \end{aligned} \qquad (14\text{-}4)$$

where $\theta = [a_0 \, a_1 ... a_{n_a} \, b_0 \, b_1 ... b_{n_b} \, i_0 \, i_1 ... i_{n_i}]^T$ with $i_s = M^{-1} \sum_{r=1}^{M} i_s^{[r]}$, $s = 0, 1, ... \; n_i$; $X(k) = M^{-1} \sum_{r=1}^{M} X^{[r]}(k)$ with $X = Y$ and U; and $\Delta(s_k) = M^{-1} \sum_{r=1}^{M} \Delta^{[r]}(s_k)$. Because the experiments are synchronized, the averaged DFT spectra will not tend to zero as $M \to \infty$.

15

Some Linear Algebra Fundamentals

Abstract: This chapter states and reviews linear algebra notations and basic concepts that are used throughout this book. In order to promote familiarity with these concepts, many exercises are provided at the end of the chapter. Elaborated discussions and proofs on the topic can be found in Gantmacher (1990), Golub and Van Loan (1996), Lancaster and Tismenetsky (1985), and Wilkinson (1988). Elementary matrix operations such as the sum, the inverse, the transpose, and the determinant are assumed to be known.

15.1 NOTATIONS AND DEFINITIONS

The entries of a matrix $A \in \mathbb{C}^{n \times m}$ are denoted by $A_{[i,j]}$

$$A = \begin{bmatrix} A_{[1,1]} & \cdots & A_{[1,m]} \\ \cdots & \cdots & \cdots \\ A_{[n,1]} & \cdots & A_{[n,m]} \end{bmatrix} \tag{15-1}$$

$A_{[:,k]}$ ($A_{[k,:]}$) stands for the kth column (row) of A. $A_{[i:j,k:l]}$, with $j \geq i$ and $l \geq k$, selects a $(j-i+1) \times (k-l+1)$ block of A containing rows i to j and columns k to l. Superscript T (H) is for the matrix transpose (complex conjugate transpose) and superscript $-T$ ($-H$) denotes the transpose (complex conjugate transpose) of the inverse matrix. A matrix A is *Hermitian (skew Hermitian)* if $A^H = A$ ($A^H = -A$) and it is *symmetric (skew-symmetric)* if $A^T = A$ ($A^T = -A$). I_n (O_n) denotes the $n \times n$ identity (zero) matrix.

The *row (column) rank* of a matrix is the maximum number of linearly independent rows (columns). A matrix $A \in \mathbb{C}^{n \times m}$ has a *full row (column) rank* if its row (column) rank is n (m). For any matrix the column rank equals the row rank (Lancaster and Tismenetsky, 1985). This motivates the following definition of the *rank* of a matrix A:

$$\text{rank}(A) = \text{column rank of } A = \text{row rank of } A \tag{15-2}$$

A square matrix is called *regular* if it is of full rank.

For $A \in \mathbb{C}^{n \times m}$, null($A$) is the linear subspace of $\mathbb{C}^m$ defined by $Ax = 0$

$$\text{null}(A) = \{x \in \mathbb{C}^m | Ax = 0\} \tag{15-3}$$

The *range (column space)* of a matrix $A \in \mathbb{C}^{n \times m}$ is the linear subspace of $\mathbb{C}^n$ that is obtained by making all possible linear combinations of the columns of A

$$\text{range}(A) = \{y \in \mathbb{C}^n | y = Ax, x \in \mathbb{C}^m\} \tag{15-4}$$

Note that $\text{range}(A) = \text{range}(AA^H) = (\text{null}(A^H))^\perp$ where superscript $\perp$ stands for the orthogonal complement of a subspace (proof: see Exercise 15.1).

The *span* of m vectors $a_1, a_2, \ldots, a_m \in \mathbb{C}^n$ is the linear subspace of $\mathbb{C}^n$ obtained by making all possible linear combinations of $a_1, a_2, \ldots, a_m$

$$\text{span}\{a_1, a_2, \ldots, a_m\} = \{x \in \mathbb{C}^n | x = \sum_{i=1}^{m} \alpha_i a_i, \alpha_i \in \mathbb{C}\} \tag{15-5}$$

The *eigenvalues* $\lambda(A)$ of a matrix $A \in \mathbb{C}^{n \times n}$ are the roots of the *characteristic polynomial* $\det(A - \lambda I_n) = 0$, where $\det(\ )$ denotes the determinant. The nonzero vectors $x \neq 0$ that satisfy $Ax = \lambda x$ are the corresponding *eigenvectors*. The eigenvalues are invariant with respect to a regular transformation $T \in \mathbb{C}^{n \times n}$ (Golub and Van Loan, 1996):

$$B = TAT^{-1} \text{ with } \det(T) \neq 0 \tag{15-6}$$

whence, after ordering of the eigenvalues, $\lambda_k(B) = \lambda_k(A)$, $(k = 1, 2, \ldots, n)$. We note that B and A are *similar,* and T is called a *similarity transformation.* Hermitian matrices have real eigenvalues (Wilkinson, 1988).

By definition, a real matrix $A \in \mathbb{R}^{n \times n}$ is *positive (semi-)definite* if for any $x \in \mathbb{R}_0^n$, the *quadratic form* $x^T A x$ is strictly positive (positive): $x^T A x > 0$ ($x^T A x \geq 0$). Similarly, a matrix $A \in \mathbb{C}^{n \times n}$ is positive (semi-)definite if for any $x \in \mathbb{C}_0^n$, $x^H A x > 0$ ($x^H A x \geq 0$). These conditions are satisfied if and only if all the eigenvalues of A are real and $\lambda_k(A) > 0$ ($\lambda_k(A) \geq 0$), $k = 1, 2, \ldots, n$. Note that no symmetry is required in the real case while in the complex case the positive (semi-)definite condition implies that A is Hermitian. We shall write $A > 0$ for positive definite and $A \geq 0$ for positive semidefinite matrices.

The *right singular vectors* v (*left singular vectors* u) of a matrix $A \in \mathbb{C}^{n \times m}$ are the eigenvectors of the matrix $A^H A$ (AA^H). The *singular values* $\sigma_k(A)$, $k = 1, 2, \ldots$ $\min(n, m)$, are the positive square roots of the eigenvalues of $A^H A$ (AA^H) and are usually ordered from large to small values.

A matrix $U \in \mathbb{R}^{n \times n}$ ($U \in \mathbb{C}^{n \times n}$) is said to be *orthogonal (unitary)* if $U^T U = I_n$ ($U^H U = I_n$). Orthogonal (unitary) matrices have the property $\det(U) = \pm 1$ ($|\det(U)| = 1$) and $U^{-1} = U^T$ ($U^{-1} = U^H$).

15.2 OPERATORS AND FUNCTIONS

Let $A_i \in \mathbb{C}^{n \times m}$, $i = 1, 2, \ldots, K$, then $\text{diag}(A_1, A_2, \ldots, A_K) \in \mathbb{C}^{nK \times mK}$ is a block diagonal matrix

$$\text{diag}(A_1, A_2, \ldots, A_K) = \begin{bmatrix} A_1 & 0 & \ldots & 0 \\ 0 & A_2 & \ldots & 0 \\ \ldots & \ldots & \ldots & \ldots \\ 0 & 0 & \ldots & A_K \end{bmatrix} \qquad (15\text{-}7)$$

The *Hermitian part* (*skew Hermitian part*) of $A \in \mathbb{C}^{n \times n}$ is $\text{herm}(A) = (A + A^H)/2$ ($(A - A^H)/2$). Any matrix can be written as the sum of a Hermitian and a skew Hermitian matrix: $A = (A + A^H)/2 + (A - A^H)/2$.

Inverse of block matrices: if $\begin{bmatrix} A & D \\ C & B \end{bmatrix}^{-1}$ and A^{-1} exist, then (Kailath, 1980)

$$\begin{bmatrix} A & D \\ C & B \end{bmatrix}^{-1} = \begin{bmatrix} A^{-1} + E\Delta^{-1}F & -E\Delta^{-1} \\ -\Delta^{-1}F & \Delta^{-1} \end{bmatrix} \qquad (15\text{-}8)$$

where $\Delta = B - CA^{-1}D$, $E = A^{-1}D$, and $F = CA^{-1}$.

The *trace* of $A \in \mathbb{C}^{n \times n}$ is defined as $\text{tr}(A) = \sum_{k=1}^{n} A_{[k,k]}$. It is circular shift invariant: for any $A \in \mathbb{C}^{n \times m}$, $B \in \mathbb{C}^{m \times p}$, and $C \in \mathbb{C}^{p \times n}$, $\text{tr}(ABC) = \text{tr}(BCA)$.

For $A \in \mathbb{C}^{n \times m}$, $\text{vec}(A) \in \mathbb{C}^{nm}$ is a column vector obtained by stacking the columns of A on top of each other

$$\text{vec}(A) = \begin{bmatrix} A_{[:,1]} \\ A_{[:,2]} \\ \ldots \\ A_{[:,m]} \end{bmatrix} \qquad (15\text{-}9)$$

15.3 NORMS

$\| \ \|$ is a matrix norm if the following properties are satisfied for all $A, B \in \mathbb{C}^{n \times m}$ and $\alpha \in \mathbb{C}$:

1. $\|A\| \geq 0$ and $\|A\| = 0 \Leftrightarrow A = 0$
2. $\|A + B\| \leq \|A\| + \|B\|$
3. $\|\alpha A\| = |\alpha| \|A\|$

The following matrix norms ($A \in \mathbb{C}^{n \times m}$) are used frequently: the Frobenius norm,

$$\|A\|_F = \sqrt{\text{tr}(A^H A)} = \sqrt{\sum_{k=1}^{n} \sum_{l=1}^{m} |A_{[k,l]}|^2} \qquad (15\text{-}10)$$

the 1-norm,

$$\|A\|_1 = \max_{1 \leq l \leq m} \sum_{k=1}^{n} |A_{[k,l]}| \qquad (15\text{-}11)$$

the 2-norm,

$$\|A\|_2 = \max_{1 \leq k \leq m} \sigma_k(A) = \sigma_1(A) \tag{15-12}$$

and the ∞-norm,

$$\|A\|_\infty = \max_{1 \leq k \leq n} \sum_{l=1}^{m} |A_{[k,l]}| \tag{15-13}$$

The Frobenius, 1-, 2- and ∞-norms satisfy the submultiplicative property

$$\|AB\| \leq \|A\|\|B\| \qquad \forall A \in \mathbb{C}^{n \times m}, \forall B \in \mathbb{C}^{m \times p} \tag{15-14}$$

Note that not all matrix norms satisfy (15-14). We also have

$$\|A\|_2 \leq \|A\|_F \tag{15-15}$$

Perturbations and the inverse (Theorem 2.3.4 of Golub and Van Loan, 1996): take $A, E \in \mathbb{C}^{n \times n}$, if $\|A^{-1}E\| = r < 1$, then $\det(A + E) \neq 0$ and

$$\|(A+E)^{-1} - A^{-1}\| \leq \frac{\|E\|\|A^{-1}\|^2}{1-r} \tag{15-16}$$

with $\|\ \|$ any matrix norm that satisfies the submultiplicative property (15-14).

15.4 DECOMPOSITIONS

15.4.1 Singular Value Decomposition

For any $A \in \mathbb{C}^{n \times m}$ with $n \geq m$ there exist $U \in \mathbb{C}^{n \times m}$ and $\Sigma, V \in \mathbb{C}^{m \times m}$ such that (Golub and Van Loan, 1996)

$$A = U\Sigma V^H \tag{15-17}$$

where $V^H V = VV^H = U^H U = I_m$ and $\Sigma = \text{diag}(\sigma_1, \sigma_2, \cdots, \sigma_m)$ with $\sigma_1 \geq \sigma_2 \geq \cdots \geq \sigma_m \geq 0$. The nonnegative real numbers σ_k are the *singular values* of A, and the columns $V_{[:,k]}$ and $U_{[:,k]}$ are the corresponding right and left singular vectors. (15-17) is called the *singular value decomposition* (SVD) of the matrix A. It can be expanded as

$$A = \sum_{k=1}^{m} \sigma_k U_{[:,k]} V_{[:,k]}^H \tag{15-18}$$

Taking the Hermitian transpose of (15-17) covers the case $n \leq m$. A numerically stable calculation of the singular value decomposition is available in standard mathematical software packages.

Section 15.4 ■ Decompositions

The singular value decomposition (15-17) contains a lot of information about the structure of the matrix. Indeed, if $\sigma_1 \geq \sigma_2 \geq \cdots \geq \sigma_r > \sigma_{r+1} = \cdots = \sigma_m = 0$, then

$$\begin{aligned}
\text{rank}(A) &= r \\
\text{null}(A) &= \text{span}\{V_{[:,r+1]}, V_{[:,r+2]}, \ldots, V_{[:,m]}\} \\
\text{range}(A) &= \text{span}\{U_{[:,1]}, U_{[:,2]}, \ldots, U_{[:,r]}\}
\end{aligned} \qquad (15\text{-}19)$$

If $A \in \mathbb{C}^{m \times n}$ with $n \geq m$, then it can be decomposed into singular values as

$$A = V \Sigma U^H \qquad (15\text{-}20)$$

with $V \in \mathbb{C}^{m \times m}$, $\Sigma = \text{diag}(\sigma_1, \ldots, \sigma_m)$, and $U \in \mathbb{C}^{n \times m}$ (proof: apply (15-17) to A^H). If rank$(A) = r$ then

$$\begin{aligned}
\text{null}(A) &= (\text{span}\{U_{[:,1]}, U_{[:,2]}, \ldots, U_{[:,r]}\})^\perp \\
\text{range}(A) &= \text{span}\{V_{[:,1]}, V_{[:,2]}, \ldots, V_{[:,r]}\}
\end{aligned} \qquad (15\text{-}21)$$

If rank$(A) = n$ then null$(A) = \{x \in \mathbb{C}^n | x = U_\perp y, y \in \mathbb{C}^{n-m}\}$ with $U_\perp \in \mathbb{C}^{n \times (n-m)}$ the orthogonal complement of U: $U^H U_\perp = 0$ and $U_\perp^H U_\perp = I_{n-m}$.

The *condition number* of a matrix $A \in \mathbb{C}^{n \times m}$ is defined as the ratio of the largest singular value to the smallest singular value $\kappa(A) = \sigma_1/\sigma_m$. For regular square matrices $m = n$ it is a measure of the sensitivity of the solution of the linear system $Ax = b$, with $b \in \mathbb{C}^n$, to perturbations in A and b. It can be shown that (Golub and Van Loan, 1996)

$$\frac{\|\Delta x\|_2}{\|x\|_2} \leq \kappa(A) \left(\frac{\|\Delta A\|_2}{\|A\|_2} + \frac{\|\Delta b\|_2}{\|b\|_2} \right) \qquad (15\text{-}22)$$

where Δ denotes the perturbation. For rectangular matrices $m > n$ of full rank, it is a measure of the sensitivity of the least squares solution $x_{LS} = (A^H A)^{-1} A^H b$ of the overdetermined set of equations $Ax \approx b$, with $b \in \mathbb{C}^m$, to perturbations in A and b (see Section 15.13). For singular matrices $\kappa(A) = \infty$. If $\kappa(A)$ is large ($\log_{10}(\kappa(A))$ is of the order of the number of significant digits used in the calculations), then A is said to be *ill-conditioned*. Unitary (orthogonal) matrices are perfectly conditioned ($\kappa = 1$), while matrices with small condition number ($\kappa \sim 1$) are said to be *well-conditioned*.

15.4.2 Generalized Singular Value Decomposition

Let $A \in \mathbb{C}^{n \times m}$ with $n \geq m$, $B \in \mathbb{C}^{p \times m}$ with $p \geq m$ and rank$([A^T B^T]^T) = m$ then there exist $U_A \in \mathbb{C}^{n \times m}$, $U_B \in \mathbb{C}^{p \times m}$, and a regular $X \in \mathbb{C}^{m \times m}$ such that (Golub and Van Loan, 1996; Paige, 1986)

$$A = U_A \Sigma_A X^{-1} \qquad B = U_B \Sigma_B X^{-1} \qquad (15\text{-}23)$$

where $U_A^H U_A = U_B^H U_B = I_m$, $\Sigma_A = \text{diag}(\alpha_1, \alpha_2, \ldots, \alpha_m)$, and $\Sigma_B = \text{diag}(\beta_1, \beta_2, \ldots, \beta_m)$, with $\alpha_k \geq 0$, $\beta_k \geq 0$, and $\alpha_k^2 + \beta_k^2 = 1$. The ratios $\sigma_k(A, B) = \alpha_k/\beta_k$ are the *generalized singular values* of the matrix pair (A, B), and the columns $X_{[:,k]}$ are the corresponding *gen-*

eralized right singular vectors. If $B \in \mathbb{C}^{m \times m}$ is regular then the generalized singular values of (A, B) are equal to the singular values of AB^{-1}: $\sigma_k(A, B) = \sigma_k(AB^{-1})$.

The generalized singular value decomposition (15-23) can be used to solve the generalized eigenvalue problem

$$A^H A x = \lambda B^H B x \qquad (15\text{-}24)$$

without forming $A^H A$ and $B^H B$. It is easy to verify that $x = X_{[:, k]}$ and $\lambda = \alpha_k^2 / \beta_k^2$, $k = 1, 2, \ldots, m$, are the solutions of (15-24) (see Exercise 13.18). Fortran and C versions of the generalized singular value decomposition are available in public domain software (Anderson et al., 1992; Bai and Demmel, 1993). For $B^H B = I_m$ the generalized eigenvalue problem reduces to an ordinary eigenvalue problem that can be solved using the singular value decomposition (15-20) of A. $x = V_{[:, k]}$ and $\lambda = \sigma_k^2$, $k = 1, 2, \ldots, m$, are then the solutions of (15-24) (see Exercise 15.16).

15.4.3 The QR Factorization

The QR factorization of $A \in \mathbb{C}^{n \times m}$ with $n \geq m$ is given by

$$A = QR \qquad (15\text{-}25)$$

where $Q \in \mathbb{C}^{n \times n}$ satisfies $Q^H Q = I_n$, and R is an upper triangular matrix (Golub and Van Loan, 1996). If A is of full rank then the QR factorization has the following properties: Q and R are unique, the diagonal elements of R are positive, and range(A) = range(Q).

15.4.4 Square Root of a Positive (Semi-)Definite Matrix

Any positive (semi-)definite matrix $A \in \mathbb{C}^{n \times n}$ can be decomposed as

$$A = \Lambda^H \Lambda \quad \text{or} \quad A = SS^H \qquad (15\text{-}26)$$

where $\Lambda^H, S \in \mathbb{C}^{n \times m}$ and $m \geq \text{rank}(A)$. Λ, S are *square roots* of A that are not unique and often have no analytic solution. Numerical, $n \times n$, solutions can be calculated using the singular value decomposition. For example, if $A = V\Sigma V^H$ then any $\Lambda = T\Sigma^{1/2} V^H$ with $T \in \mathbb{C}^{n \times n}$ a unitary matrix satisfies (15-26). Choosing $T = V$ gives a Hermitian solution that motivates the following notation:

$$A = A^{1/2} A^{1/2} \qquad (15\text{-}27)$$

with $A^{1/2} = V\Sigma^{1/2} V^H$. In the real case, similar results apply to symmetric positive (semi-)definite matrices.

15.5 MOORE-PENROSE PSEUDOINVERSE

For any matrix $A \in \mathbb{C}^{n \times m}$ there exists a unique generalized inverse $A^+ \in \mathbb{C}^{m \times n}$, also called a Moore-Penrose pseudoinverse, that satisfies the four Moore-Penrose conditions (Ben-Israel and Greville, 1974)

1. $AA^+A = A$
2. $A^+AA^+ = A^+$
3. $(AA^+)^H = AA^+$
4. $(A^+A)^H = A^+A$

For regular square matrices it is clear that $A^+ = A^{-1}$. The pseudoinverse can be constructed using, for example, the singular value decomposition (Golub and Van Loan, 1996). If rank$(A) = r$ then

$$A^+ = V\Sigma^+U^H \text{ with } \Sigma^+ = \text{diag}(\sigma_1^{-1}, \sigma_2^{-1}, ..., \sigma_r^{-1}, 0, ..., 0) \qquad (15\text{-}28)$$

Using (15-28) it can easily be shown that for every matrix A, $(A^+)^+ = A$, $(A^+)^H = (A^H)^+$, and $A^+ = (A^HA)^+A^H = A^H(AA^H)^+$.

Although the properties of the pseudoinverse very much resemble those of the inverse, in general $(AB)^+ \neq B^+A^+$. If the matrices $A \in \mathbb{C}^{n \times r}$ and $B \in \mathbb{C}^{r \times m}$ with $r \leq \min(n, m)$ are of full rank then $(AB)^+ = B^+A^+$ (Ben-Israel and Greville, 1974).

Theorem 15.1: For any $C \in \mathbb{C}^{n \times m}$ with $n \geq m$ and rank$(C) = m$ and any $B \in \mathbb{C}^{m \times m}$ with rank$(B) = m$ we have $C^H(CBC^H)^+C = B^{-1}$.

Proof. Apply condition 1 with $A = CBC^H$

$$CBC^H(CBC^H)^+CBC^H = CBC^H \qquad (15\text{-}29)$$

Left multiplication with C^H and right multiplication with C of (15-29) results in

$$C^HCBC^H(CBC^H)^+CBC^HC = C^HCBC^HC \qquad (15\text{-}30)$$

where C^HC and B are, by assumption, regular matrices. Left division by C^HCB and right division by BC^HC of (15-30) proves the theorem. □

15.6 IDEMPOTENT MATRICES

By definition, an *idempotent matrix* $P \in \mathbb{C}^{n \times n}$ satisfies $P^2 = P$. If P is an idempotent matrix then (Lancaster and Tismenetsky, 1985)

1. $\lambda_k(P) = 1$, $k = 1, 2, ..., \text{rank}(P)$ and $\lambda_k(P) = 0$, $k = \text{rank}(P) + 1, ..., n$.
2. $I_n - P$ is also a idempotent matrix
3. range$(I_n - P) = $ null(P) and null$(I_n - P) = $ range(P)
4. null(P) + range$(P) = \mathbb{C}^n$ and null$(P) \cap $ range$(P) = \{0\}$

where $\mathbb{A} + \mathbb{B} = \mathbb{D}$ means that for each element $d \in \mathbb{D}$ there exist an element $a \in \mathbb{A}$ and an element $b \in \mathbb{B}$ such that $d = a + b$.

An idempotent matrix P can be interpreted geometrically as a projection on range(P) along null(P). This projection is orthogonal for Hermitian idempotent matrices P: range$(P) = ($null$(P))^\perp$. Note that a Hermitian idempotent matrix is positive (semi-)definite.

Theorem 15.2: Let $P, Q \in \mathbb{C}^{n \times n}$ be Hermitian idempotent matrices with rank$(P) = r$ and rank$(Q) = n - r$, respectively. If $QP = 0$ then $Q + P = I_n$.

Proof. Because P and Q are Hermitian, it follows from $QP = 0$ that $PQ = 0$. Take any eigenvector v_k of P with $\lambda_k(P) = 1$, $k = 1, 2, \ldots, r$. Right multiplication of $QP = 0$ by v_k gives $Qv_k = 0$ and, hence, $\lambda_k(Q) = 0$, $k = 1, 2, \ldots, r$. Using $PQ = 0$ it follows, similarly, that any eigenvector v_k of Q with $\lambda_k(Q) = 1$ is an eigenvector of P with $\lambda_k(P) = 0$, $k = r+1, \ldots, n$. Because the eigenvectors of a Hermitian positive (semi-)definite matrix form an orthonormal basis (Exercise 15.19), we have

$$P = V \begin{bmatrix} I_r & 0 \\ 0 & 0 \end{bmatrix} V^H \text{ and } Q = V \begin{bmatrix} 0 & 0 \\ 0 & I_{n-r} \end{bmatrix} V^H \qquad (15\text{-}31)$$

where $V = [v_1 v_2 \ldots v_n]$, so that $Q + P = I_n$. □

15.7 KRONECKER ALGEBRA

This section gives some basic properties of the Kronecker product of matrices. A complete overview and elaborated proofs can be found in Brewer (1978) and Lancaster and Tismenetsky (1985).

Consider the following matrices: $A, H \in \mathbb{C}^{p \times q}$, $B \in \mathbb{C}^{s \times t}$, $C \in \mathbb{C}^{r \times l}$, $D \in \mathbb{C}^{q \times s}$, $G \in \mathbb{C}^{t \times u}$, $N \in \mathbb{C}^{n \times n}$, and $M \in \mathbb{C}^{m \times m}$. The Kronecker product of two matrices is defined as

$$A \otimes B = \begin{bmatrix} A_{[1,1]}B & A_{[1,2]}B & \ldots & A_{[1,q]}B \\ A_{[2,1]}B & A_{[2,2]}B & \ldots & A_{[2,q]}B \\ \ldots & \ldots & \ldots & \ldots \\ A_{[p,1]}B & \ldots & \ldots & A_{[p,q]}B \end{bmatrix} \in \mathbb{C}^{ps \times qt} \qquad (15\text{-}32)$$

It has the following properties:

$$(A \otimes B) \otimes C = A \otimes (B \otimes C) \qquad (15\text{-}33)$$

$$(A + H) \otimes B = A \otimes B + H \otimes B \qquad (15\text{-}34)$$

$$(A \otimes B)^T = A^T \otimes B^T \qquad (15\text{-}35)$$

$$(A \otimes B)(D \otimes G) = (AD) \otimes (BG) \qquad (15\text{-}36)$$

$$(N \otimes M)^{-1} = N^{-1} \otimes M^{-1} \qquad (15\text{-}37)$$

$$\text{vec}(ADB) = (B^T \otimes A)\text{vec}(D) \qquad (15\text{-}38)$$

$$\|A \otimes B\| = \|A\| \|B\| \qquad (15\text{-}39)$$

where $\| \ \|$ denotes the 1-, 2-, ×-, and Frobenius norm.

15.8 ISOMORPHISM BETWEEN COMPLEX AND REAL MATRICES

Any complex matrix $A \in \mathbb{C}^{n \times m}$ can be transformed into a real matrix $A_{\text{Re}} \in \mathbb{R}^{(2n) \times (2m)}$ through

$$A_{\text{Re}} = \begin{bmatrix} \text{Re}(A) & -\text{Im}(A) \\ \text{Im}(A) & \text{Re}(A) \end{bmatrix} \tag{15-40}$$

Lemma 15.3

$$A = B + C \Leftrightarrow A_{\text{Re}} = B_{\text{Re}} + C_{\text{Re}} \tag{15-41}$$

$$A = BC \Leftrightarrow A_{\text{Re}} = B_{\text{Re}} C_{\text{Re}} \tag{15-42}$$

$$A = B^{-1} \Leftrightarrow A_{\text{Re}} = B_{\text{Re}}^{-1} \tag{15-43}$$

$$A = B^{+} \Leftrightarrow A_{\text{Re}} = B_{\text{Re}}^{+} \tag{15-44}$$

$$A = B^{H} \Leftrightarrow A_{\text{Re}} = B_{\text{Re}}^{T} \tag{15-45}$$

$$\det(A_{\text{Re}}) = |\det(A)|^2 \tag{15-46}$$

$$\text{rank}(A_{\text{Re}}) = 2\,\text{rank}(A) \tag{15-47}$$

provided that the matrix dimensions are appropriate. Moreover, if $A \in \mathbb{C}^{n \times n}$ is unitary (positive definite) then $A_{\text{Re}} \in \mathbb{R}^{(2n) \times (2n)}$ is orthogonal (symmetric and positive definite) and vice versa.

Proof. Exercises 15.34 to 15.38. □

Lemma 15.3 defines an isomorphism between the complex $n \times m$ matrices and the real $(2n) \times (2m)$ matrices. Using (15-42), (15-43), and (15-45), the relationships between the eigenvectors, eigenvalues, singular values, and singular vectors of A and A_{Re} are readily obtained. For example, if the singular value decomposition of A is given by $U\Sigma V^H$, then that of A_{Re} equals $U_{\text{Re}} \Sigma_{\text{Re}} V_{\text{Re}}^T$. Similarly, if the generalized singular value decomposition of the matrix pair (A, B) is given by $A = U_A \Sigma_A X^{-1}$, $B = U_B \Sigma_B X^{-1}$, then that of $(A_{\text{Re}}, B_{\text{Re}})$ is given by $A_{\text{Re}} = (U_A)_{\text{Re}} (\Sigma_A)_{\text{Re}} X_{\text{Re}}^{-1}$, $B_{\text{Re}} = (U_B)_{\text{Re}} (\Sigma_B)_{\text{Re}} X_{\text{Re}}^{-1}$, and vice versa.

Another transformation between complex ($A \in \mathbb{C}^{n \times m}$) and real ($A_{\text{re}} \in \mathbb{R}^{(2n) \times m}$) matrices is given by

$$A_{\text{re}} = \begin{bmatrix} \text{Re}(A) \\ \text{Im}(A) \end{bmatrix} \tag{15-48}$$

It has the following properties.

Lemma 15.4: Take any $A \in \mathbb{C}^{n \times m}$, $B \in \mathbb{C}^{n \times p}$, $X \in \mathbb{C}^{p \times m}$, and $Y \in \mathbb{R}^{p \times m}$

$$A = BX \Leftrightarrow A_{\text{re}} = B_{\text{Re}} X_{\text{re}} \tag{15-49}$$

$$A = BY \Leftrightarrow A_{\text{re}} = B_{\text{re}} Y \tag{15-50}$$

$$\text{Re}(A^H B) = A_{\text{re}}^T B_{\text{re}} \tag{15-51}$$

Proof. Exercise 15.39. □

Lemmas 15.3 and 15.4 are very useful to generalize results obtained for the real-valued case to the complex-valued case. This is illustrated in the following example.

Example 15.5: Consider the following expression:

$$\frac{1}{2} x^T C_x^{-1} x \tag{15-52}$$

where $x \in \mathbb{R}^n$ is a real-valued random vector and $C_x = \text{Cov}(x) \in \mathbb{R}^{n \times n}$ is the corresponding covariance matrix. To obtain the result for the complex-valued case ($x \in \mathbb{C}^n$), x and C_x are replaced in (15-52) by x_{re} and $C_{x_{\text{re}}}$, respectively. Assuming that $x \in \mathbb{C}^n$ is a circular complex random vector (see Section 16.1), $\text{Cov}(\text{Re}(x), \text{Im}(x)) = -\text{Cov}(\text{Im}(x), \text{Re}(x))$ and $\text{Cov}(\text{Re}(x)) = \text{Cov}(\text{Im}(x))$, and we have $C_{x_{\text{re}}} = 0.5(C_x)_{\text{Re}}$. Using Lemmas 15.3 and 15.4, we find

$$\begin{aligned}
\frac{1}{2} x_{\text{re}}^T C_{x_{\text{re}}}^{-1} x_{\text{re}} &= x_{\text{re}}^T (C_x^{-1})_{\text{Re}} x_{\text{re}} && \text{(property (15-43))} \\
&= x_{\text{re}}^T (C_x^{-1} x)_{\text{re}} && \text{(property (15-49))} \\
&= \text{Re}(x^H C_x^{-1} x) && \text{(property (15-51))} \\
&= x^H C_x^{-1} x && (C_x \text{ is positive definite})
\end{aligned}$$

We conclude that $\frac{1}{2} x_{\text{re}}^T C_{x_{\text{re}}}^{-1} x_{\text{re}} = x^H C_x^{-1} x$. □

15.9 DERIVATIVES

15.9.1 Derivatives of Functions and Vectors w.r.t. a Vector

The first- and second-order derivatives of an analytic function $f(x) \in \mathbb{C}$ (vector function $F(x) \in \mathbb{C}^n$) with respect to a vector $x \in \mathbb{C}^m$ are defined as

$$\frac{\partial f(x)}{\partial x} \in \mathbb{C}^{1 \times m} \text{ with } \left(\frac{\partial f(x)}{\partial x}\right)_{[1, k]} = \frac{\partial f(x)}{\partial x_{[k]}} \tag{15-53}$$

Section 15.9 ■ Derivatives

$$\frac{\partial^2 f(x)}{\partial x^2} \in \mathbb{C}^{m \times m} \text{ with } \left(\frac{\partial^2 f(x)}{\partial x^2}\right)_{[k,l]} = \frac{\partial^2 f(x)}{\partial x_{[k]} \partial x_{[l]}} \tag{15-54}$$

$$\frac{\partial F(x)}{\partial x} \in \mathbb{C}^{n \times m} \text{ with } \left(\frac{\partial F(x)}{\partial x}\right)_{[k,l]} = \frac{\partial F_{[k]}(x)}{\partial x_{[l]}} \tag{15-55}$$

Let $g(x) \in \mathbb{C}^q$ be an analytic vector function of $x \in \mathbb{C}^m$ and $A \in \mathbb{C}^{n \times m}$, $B \in \mathbb{C}^{m \times m}$. Using definitions (15-53), (15-54), and (15-55) it can be verified that

$$\frac{\partial Ax}{\partial x} = A \quad \frac{\partial}{\partial x}\left(\frac{x^T B x}{2}\right) = x^T\left(\frac{B + B^T}{2}\right) \quad \frac{\partial^2}{\partial x^2}\left(\frac{x^T B x}{2}\right) = \frac{B + B^T}{2} \tag{15-56}$$

$$\frac{\partial}{\partial x}\left(\frac{g^T(x) g(x)}{2}\right) = g^T(x) \frac{\partial g(x)}{\partial x} \tag{15-57}$$

$$\frac{\partial^2}{\partial x^2}\left(\frac{g^T(x) g(x)}{2}\right) = \left(\frac{\partial g(x)}{\partial x}\right)^T \left(\frac{\partial g(x)}{\partial x}\right) + \sum_{k=1}^{q} g_{[k]}(x) \frac{\partial^2 g_{[k]}}{\partial x^2} \tag{15-58}$$

The derivative of a real function $f(x, \bar{x}) \in \mathbb{R}$ with respect to the real and imaginary parts of the vector $x \in \mathbb{C}^m$ can be found using the chain rule and symbol derivation w.r.t. x and $\bar{x}$ ($f(x, \bar{x})$ is not an analytic function of x)

$$\begin{aligned}\frac{\partial f(x, \bar{x})}{\partial \mathrm{Re}(x)} &= \frac{\partial f(x, \bar{x})}{\partial x} \frac{\partial x}{\partial \mathrm{Re}(x)} + \frac{\partial f(x, \bar{x})}{\partial \bar{x}} \frac{\partial \bar{x}}{\partial \mathrm{Re}(x)} = \frac{\partial f(x, \bar{x})}{\partial x} + \frac{\partial f(x, \bar{x})}{\partial \bar{x}} \\ \frac{\partial f(x, \bar{x})}{\partial \mathrm{Im}(x)} &= \frac{\partial f(x, \bar{x})}{\partial x} \frac{\partial x}{\partial \mathrm{Im}(x)} + \frac{\partial f(x, \bar{x})}{\partial \bar{x}} \frac{\partial \bar{x}}{\partial \mathrm{Im}(x)} = j\left(\frac{\partial f(x, \bar{x})}{\partial x} - \frac{\partial f(x, \bar{x})}{\partial \bar{x}}\right)\end{aligned} \tag{15-59}$$

Because $f(x, \bar{x})$ is real, $\partial f(x, \bar{x})/\partial \bar{x} = \overline{\partial f(x, \bar{x})/\partial x}$, so that (15-59) can be written as

$$\left(\frac{\partial f(x, \bar{x})}{\partial x_{\mathrm{re}}}\right)^T = 2\left[\left(\frac{\partial f(x, \bar{x})}{\partial x}\right)^H\right]_{\mathrm{re}} \tag{15-60}$$

15.9.2 Derivative of a Function w.r.t. a Matrix

The derivative of an analytic function $f(A) \in \mathbb{C}$ with respect to a matrix $A \in \mathbb{C}^{n \times m}$ is defined as

$$\frac{\partial f(A)}{\partial A} \in \mathbb{C}^{n \times m} \text{ with } \left(\frac{\partial f(A)}{\partial A}\right)_{[k,l]} = \frac{\partial f(A)}{\partial A_{[k,l]}} \tag{15-61}$$

Using definition (15-61) it can be verified that

$$\frac{\partial \mathrm{tr}(BA)}{\partial A} = B^T \tag{15-62}$$

$$\frac{\partial \text{tr}(CABA^TC^T)}{\partial A} = C^TCA(B+B^T) \tag{15-63}$$

$$\frac{\partial \ln(\det(A))}{\partial A} = A^{-T} \tag{15-64}$$

$$\frac{\partial \text{tr}(BA^{-1})}{\partial A} = -(A^{-1}BA^{-1})^T \tag{15-65}$$

provided that the matrix dimensions are appropriate.

Following the same procedure as in Section 15.9.1, the derivative of a real function $f(A, \bar{A}) \in \mathbb{R}$ with respect to the real and imaginary parts of the matrix $A \in \mathbb{C}^{n \times m}$ can be found through symbolic derivation ($f(A, \bar{A})$ is *not* an analytic function of A)

$$\frac{\partial f(A, \bar{A})}{\partial A_{\text{re}}} = 2\left(\frac{\partial f(A, \bar{A})}{\partial \bar{A}}\right)_{\text{re}} \tag{15-66}$$

Take, for example, $f(A, \bar{A}) = \text{tr}(CABA^HC^H)$ with $B^H = B$, then

$$\frac{\partial \text{tr}(CABA^HC^H)}{\partial A_{\text{re}}} = 2(C^HCAB)_{\text{re}} \tag{15-67}$$

Note that the derivative of a function w.r.t. a vector in Section 15.9.1 corresponds to the derivative of a function w.r.t. a row in Section 15.9.2.

15.10 INNER PRODUCT

Consider a finite-dimensional linear space $\mathbb{L}$ over the field $\mathbb{F}$ of real or complex numbers ($\mathbb{F} = \mathbb{R}$ or $\mathbb{F} = \mathbb{C}$), and let $x, y \in \mathbb{L}$. The function $\langle x, y \rangle$ from $\mathbb{L} \times \mathbb{L}$ to $\mathbb{F}$ is an *inner product* on the linear space $\mathbb{L}$ if the following properties are satisfied for all $x, y \in \mathbb{L}$ and $\alpha, \beta \in \mathbb{F}$:

1. $\langle x, x \rangle \geq 0$ and $\langle x, x \rangle = 0 \Leftrightarrow x = 0$
2. $\langle \alpha x + \beta y, z \rangle = \alpha \langle x, z \rangle + \beta \langle y, z \rangle$
3. $\langle x, y \rangle = \overline{\langle y, x \rangle}$

These properties are known as the positivity, linearity in the first argument, and Hermitian symmetry ($\mathbb{F} = \mathbb{C}$) or symmetry ($\mathbb{F} = \mathbb{R}$), respectively. The inner product also defines a *norm* on the space $\mathbb{L}$: $\langle x, x \rangle^{1/2} = \|x\|$ (Lancaster and Tismenetsky, 1985). Two nonzero elements $x, y \in \mathbb{L}$ are *orthogonal* if $\langle x, y \rangle = 0$.

Lemma 15.6: Let $\mathbb{P}_m(\mathbb{R})$ be the linear space of real polynomials (= polynomials with real coefficients) of order smaller than or equal to m. Take any $p(x), q(x) \in \mathbb{P}_m(\mathbb{R})$ and define

$$\langle p(x), q(x) \rangle = \text{Re}(\sum_{k=1}^{n} [w_{1k}p(x_k) + w_{2k}\bar{p}(x_k)][\bar{w}_{1k}\bar{q}(x_k) + \bar{w}_{2k}q(x_k)]) \tag{15-68}$$

where $w_{1k}, w_{2k} \in \mathbb{C}$ are the weights and $x_k \in \mathbb{C}$ the grid points. (15-68) defines an inner product if and only if the matrix $J \in \mathbb{C}^{n \times (m+1)}$ with $J_{[k,r]} = w_{1k}x_k^{r-1} + w_{2k}\bar{x}_k^{r-1}$, $k = 1, 2, \ldots, n$ and $r = 1, 2, \ldots, m+1$ satisfies the rank condition

$$\text{rank}(J_{\text{re}}) = m + 1 \tag{15-69}$$

Proof. The linearity (only real linear combinations are considered for real polynomials) and symmetry of (15-68) follow directly. To show that the positivity condition is satisfied, we rewrite (15-68) using $p(x) = \sum_{r=0}^{m} p_r x^r$ and $p = [p_0 \, p_1 \cdots p_m]^T \in \mathbb{R}^{m+1}$ as

$$\langle p(x), p(x) \rangle = \text{Re}(p^T J^H J p) = p^T \text{Re}(J^H J) p = p^T J_{\text{re}}^T J_{\text{re}} p \tag{15-70}$$

The last equivalence is due to property (15-51). Under rank condition (15-69), the matrix $J_{\text{re}}^T J_{\text{re}}$ is positive definite and, hence, $\langle p(x), p(x) \rangle = 0$ if and only if $p = 0$. □

Lemma 15.7: Let $\mathbb{P}_m^2(\mathbb{R})$ be the linear space of real 2 by 1 vector polynomials of order smaller than or equal to m. If $p(x) \in \mathbb{P}_m^2(\mathbb{R})$, then $p_{[i]}(x) \in \mathbb{P}_m(\mathbb{R})$, $i = 1, 2$. Take any $p(x), q(x) \in \mathbb{P}_m^2(\mathbb{R})$ and define

$$\langle p(x), q(x) \rangle = \text{Re}(\sum_{k=1}^{n} q^H(x_k) W_k^H W_k p(x_k)) \tag{15-71}$$

with $W_k \in \mathbb{C}^2$ the weighting matrices and $x_k \in \mathbb{C}$ the grid points. Define the 2 by 1 vector polynomials $E_{2r}^T(x) = [x^r \; 0]$ and $E_{2r+1}^T(x) = [0 \; x^r]$, $r = 0, 1, \ldots, m$. (15-71) defines an inner product if and only if the matrix $J \in \mathbb{C}^{n \times (2m+2)}$ with $J_{[2k-1 \,:\, 2k, r]} = W_k E_{r-1}(x_k)$, $k = 1, 2, \ldots, n$ and $r = 1, 2, \ldots, 2m+2$ satisfies the rank condition

$$\text{rank}(J_{\text{re}}) = 2m + 2 \tag{15-72}$$

Proof. The linearity (only real linear combinations are considered for real polynomials) and symmetry of (15-71) follow directly. The proof of the positivity condition is along the lines of Lemma 15.6. Using $p(x) = \sum_{r=0}^{2m+1} p_r E_r(x)$ and $p = [p_0 \, p_1 \cdots p_{2m+1}]^T \in \mathbb{R}^{2m+2}$, (15-71) becomes $\langle p(x), p(x) \rangle = p^T J_{\text{re}}^T J_{\text{re}} p$. Under the rank condition (15-72) $J_{\text{re}}^T J_{\text{re}}$ is positive definite so that $\langle p(x), p(x) \rangle = 0$ if and only if $p = 0$. □

Lemma 15.8: Let $\mathbb{P}_m(\mathbb{C})$ be the linear space of complex polynomials (= polynomials with complex coefficients) of order smaller than or equal to m. Take any $p(x), q(x) \in \mathbb{P}_m(\mathbb{C})$ and define

$$\langle p(x), q(x) \rangle = \sum_{k=1}^{n} |w_k|^2 p(x_k) \bar{q}(x_k) \tag{15-73}$$

with $w_k \in \mathbb{C}$ the weights and $x_k \in \mathbb{C}$ the grid points. (15-73) is an inner product if and only if the matrix $J \in \mathbb{C}^{n \times (m+1)}$ with $J_{[k,r]} = w_k x_k^{r-1}$, $k = 1, 2, \ldots, n$ and $r = 1, 2, \ldots, m+1$ has rank $m + 1$.

Proof. Similar to Lemma 15.6. □

Lemma 15.9: Let $\mathbb{P}_m^2(\mathbb{C})$ be the linear space of complex 2 by 1 vector polynomials (= polynomials with complex coefficients) of order smaller than or equal to m. Take any $p(x), q(x) \in \mathbb{P}_m^2(\mathbb{C})$ and define

$$\langle p(x), q(x) \rangle = \sum_{k=1}^{n} q^H(x_k) W_k^H W_k p(x_k) \tag{15-74}$$

with $W_k \in \mathbb{C}^2$ the weighting matrices and $x_k \in \mathbb{C}$ the grid points. Define the 2 by 1 vector polynomials $E_{2r}^T(x) = [x^r \; 0]$ and $E_{2r+1}^T(x) = [0 \; x^r]$, $r = 0, 1, \ldots, m$. (15-73) is an inner product if and only if the matrix $J \in \mathbb{C}^{n \times (2m+2)}$ with $J_{[2k-1 \,:\, 2k, r]} = W_k E_{r-1}(x_k)$, $k = 1, 2, \ldots, n$ and $r = 1, 2, \ldots, 2m+2$ has rank $2m+2$.

Proof. Similar to Lemma 15.7. □

Note that Lemmas 15.7 and 15.9 can easily be generalized to vector polynomials with more than two entries.

15.11 GRAM-SCHMIDT ORTHOGONALIZATION

The *Gram-Schmidt orthogonalization* calculates an orthonormal set $\{y_1, y_2, \ldots, y_n\}$ from a given linear independent set $\{x_1, x_2, \ldots, x_n\}$ with the property

$$\text{span}\{y_1, y_2, \ldots, y_s\} = \text{span}\{x_1, x_2, \ldots, x_s\} \quad \text{for } s = 1, 2, \ldots, n \tag{15-75}$$

It works as follows. In the first step we assign $z_1 = x_1$ and calculate $y_1 = z_1/\|z_1\|$. In the second step we choose an element $z_2 \in \text{span}\{x_1, x_2\}$ that is orthogonal to y_1: $z_2 = x_2 + \alpha_{21} y_1$ and $\langle z_2, y_1 \rangle = 0$. We find $\alpha_{21} = -\langle x_2, y_1 \rangle$ and calculate $y_2 = z_2/\|z_2\|$. In the sth step we take an element $z_s \in \text{span}\{x_1, x_2, \ldots, x_s\}$ that is orthogonal to $y_1, y_2, \ldots, y_{s-1}$: $z_s = x_s + \sum_{r=1}^{s-1} \alpha_{sr} y_r$ and $\langle z_s, y_r \rangle = 0$, $r = 1, 2, \ldots, s-1$. We find $\alpha_{sr} = -\langle x_s, y_r \rangle$, $r = 1, 2, \ldots, s-1$, and calculate $y_s = z_s/\|z_s\|$.

It is well known that the Gram-Schmidt orthogonalization has poor numerical properties (Golub and Van Loan, 1996). There is, typically, a (severe) loss of orthogonality among the computed basis vectors. The method is, however, still very useful in applications where the orthogonality of the basis vectors is not explicitly taken into account during the calculations.

Example 15.10: Consider the space of real polynomials (Lemma 15.6) with inner product (15-68). Starting from the linear independent set $\{1, x, x^2, \ldots, x^m\}$ the $(s+1)$th step of the Gram-Schmidt method becomes

$$\begin{aligned} q_s(x) &= x p_{s-1}(x) - \sum_{l=0}^{s-1} \langle x p_{s-1}(x), p_l(x) \rangle p_l(x) \\ p_s(x) &= q_s(x)/\|q_s(x)\| \end{aligned} \tag{15-76}$$

If we take imaginary grid points, $\bar{x}_k = -x_k$, and $w_{2k} = 0$ for any k in the inner product (15-68), then (15-76) reduces to a three-term recursion formula

$$\begin{aligned} q_s(x) &= x p_{s-1}(x) + \|q_{s-1}(x)\| p_{s-2}(x) \\ p_s(x) &= q_s(x)/\|q_s(x)\| \end{aligned} \tag{15-77}$$

(proof: see Exercise 15.44 and Forsythe, 1957). Note that (15-77) generates only even $p_{2s}(x)$ and odd $p_{2s+1}(x)$ polynomials. □

Section 15.11 ■ Gram-Schmidt Orthogonalization

Example 15.11: Consider the space of real 2 by 1 vector polynomials (Lemma 15.7) with inner product (15-71). Applying the Gram-Schmidt method on the linear independent set $\{E_0(x), E_1(x), ..., E_{2m+1}(x)\}$, with $E_{2r}^T(x) = [x^r\ 0]$ and $E_{2r+1}^T(x) = [0\ x^r]$, $r = 0, 1, ..., m$, gives the following recursion formula:

$$q_s(x) = E_s(x) - \sum_{l=0}^{s-1} \langle E_s(x), p_l(x) \rangle p_l(x)$$
$$p_s(x) = q_s(x) / \|q_s(x)\| \tag{15-78}$$

Using $E_s(x) = xE_{s-2}(x)$ and $E_{s-2}(x) \in \text{span}\{p_0(x), p_1(x), ..., p_{s-2}(x)\}$, the recursion can be written as

$$q_s(x) = xp_{s-2}(x) - \sum_{l=0}^{s-1} \langle xp_{s-2}(x), p_l(x) \rangle p_l(x)$$
$$p_s(x) = q_s(x) / \|q_s(x)\| \tag{15-79}$$

If we take imaginary grid points, $\bar{x}_k = -x_k$ for any k, then (15-79) reduces to a five-term recursion formula

$$q_s(x) = xp_{s-2}(x) - \beta(p_{s-1}(x) - p_{s-3}(x)) + p_{s-4}(x)\|q_{s-2}(x)\|$$
$$p_s(x) = q_s(x) / \|q_s(x)\| \tag{15-80}$$

with $\beta = \langle xp_{s-2}(x), p_{s-1}(x) \rangle$ (proof: see Exercise 15.46). A numerically stable and time-efficient implementation of the orthogonalization can be found in Van Barel and Bultheel (1992) for real polynomials and real grid point $x_k \in \mathbb{R}$, and Van Barel and Bultheel (1994) for real polynomials and grid points on the unit circle, $|x_k| = 1$ for any k. □

Example 15.12: Consider the space of complex polynomials (Lemma 15.8) with inner product (15-73). Applying the Gram-Schmidt method on the linear independent set $\{1, x, x^2, ..., x^m\}$ gives the same full recursion formula (15-76). If we take imaginary grid points, $\bar{x}_k = -x_k$ for any k, then the orthogonalization reduces to

$$q_s(x) = (x - \alpha)p_{s-1}(x) - \beta p_{s-2}(x)$$
$$p_s(x) = q_s(x) / \|q_s(x)\| \tag{15-81}$$

with $\alpha = \langle xp_{s-1}(x), p_{s-1}(x) \rangle$ and $\beta = \langle xp_{s-1}(x), p_{s-2}(x) \rangle$ (proof: similar to Example 15.10). □

Example 15.13: Consider the space of complex 2 by 1 vector polynomials (Lemma 15.9) with inner product (15-74). Applying the Gram-Schmidt method on the linear independent set $\{E_0(x), E_1(x), ..., E_{2m+1}(x)\}$, with $E_{2r}^T(x) = [x^r\ 0]$ and $E_{2r+1}^T(x) = [0\ x^r]$, $r = 0, 1, ..., m$, gives the same full recursion formula (15-79). If we take imaginary grid points, $\bar{x}_k = -x_k$ for any k, then (15-79) reduces to a five-term recursion formula

$$q_s(x) = (x - \alpha)p_{s-2}(x) - \beta(p_{s-1}(x) + \bar{\beta}p_{s-3}(x)) - \gamma p_{s-4}(x)$$
$$p_s(x) = q_s(x) / \|q_s(x)\| \tag{15-82}$$

with $\alpha = \langle xp_{s-2}(x), p_{s-2}(x)\rangle$, $\beta = \langle xp_{s-2}(x), p_{s-1}(x)\rangle$, $\gamma = \langle xp_{s-2}(x), p_{s-4}(x)\rangle$ (proof: similar to Example 15.11). A numerically stable and time-efficient implementation of the orthogonalization can be found in Van Barel and Bultheel (1994) for complex polynomials and grid points on the unit circle, $|x_k| = 1$ for any k. □

Note that a particular value of the orthogonal polynomial $p_s(x_k)$ is calculated via the recursion formula used for the orthogonalization and not via the expansion of the orthogonal polynomials in powers of x. The last approach is numerically ill-conditioned for high-order polynomials. Similarly, the poles and zeros of orthogonal polynomials are calculated via a companion matrix based on the recursion formula and not via the expansion of the orthogonal polynomials in powers of x (see Section 15.12).

15.12 CALCULATING THE ROOTS OF POLYNOMIALS

15.12.1 Scalar Orthogonal Polynomials

In this section we study the problem of calculating the roots of a polynomial $A(x)$ that is written as a linear combination of scalar orthogonal polynomials $p_r(x)$

$$A(x) = \sum_{r=0}^{n_a} a_r p_r(x) \tag{15-83}$$

The coefficients a_r are known and the orthogonal basis $p_r(x)$, $r = 0, 1, \ldots, n_a$, is defined by the following recursion formula:

$$\begin{aligned} q_r(x) &= xp_{r-1} + \sum_{s=0}^{r-1} \alpha_{rs} p_s(x) \\ p_r(x) &= q_r(x)/\|q_r(x)\| \end{aligned} \tag{15-84}$$

with $q_0(x) = 1$ and $\alpha_{rs} = -\langle xp_{r-1}(x), p_s(x)\rangle$ (see Example 15.10).

To maintain good numerical conditioning, the calculation of the roots must use only the orthogonal decomposition of the polynomials and not their explicit form as a polynomial in powers of x. The eigenvalues of the modified companion matrix A,

$$A = A_1^{-1}(A_2 - A_3) \tag{15-85}$$

with A_1, A_2 and A_3, n_a by n_a matrices,

$$A_1 = \text{diag}(1/\|q_{n_a}(x)\|, 1/\|q_{n_a-1}(x)\|, \ldots, 1/\|q_1(x)\|)$$

$$A_2 = \begin{bmatrix} -\dfrac{a_{n_a-1}}{a_{n_a}} & -\dfrac{a_{n_a-2}}{a_{n_a}} & \cdots & -\dfrac{a_1}{a_{n_a}} & -\dfrac{a_0}{a_{n_a}} \\ 1 & 0 & \cdots & 0 & 0 \\ 0 & 1 & \cdots & 0 & 0 \\ \cdots & \cdots & \cdots & \cdots & \cdots \\ 0 & \cdots & \cdots & 1 & 0 \end{bmatrix}, A_3 = \begin{bmatrix} \dfrac{\alpha_{n_a(n_a-1)}}{\|q_{n_a}(x)\|} & \dfrac{\alpha_{n_a(n_a-2)}}{\|q_{n_a}(x)\|} & \cdots & \dfrac{\alpha_{n_a 0}}{\|q_{n_a}(x)\|} \\ 0 & \dfrac{\alpha_{(n_a-1)(n_a-2)}}{\|q_{n_a-1}(x)\|} & \cdots & \dfrac{\alpha_{(n_a-1)0}}{\|q_{n_a-1}(x)\|} \\ \cdots & \cdots & \cdots & \cdots \\ 0 & \cdots & 0 & \dfrac{\alpha_{10}}{\|q_1(x)\|} \end{bmatrix} \tag{15-86}$$

are the required roots of $A(x)$ (see Appendix 15.A). For real polynomials with grid points on the imaginary axis (see Example 15.10), (15-84) reduces to the three-term recursion (15-77), and A_3 in (15-86) contains only one nonzero diagonal.

15.12.2 Vector Orthogonal Polynomials

In this section we study the problem of finding the roots of the entries of a 2 by 1 polynomial vector $[A(x) \; B(x)]^T$ that is written as a linear combination of vector orthogonal polynomials

$$\begin{bmatrix} A(x) \\ B(x) \end{bmatrix} = \sum_{r=0}^{n_a+n_b+1} a_r \begin{bmatrix} p_r(x) \\ q_r(x) \end{bmatrix} \tag{15-87}$$

with n_a, n_b the orders of, respectively, $A(x)$, $B(x)$, and where the coefficients a_r are known. The vector orthogonal basis $[p_r(x) \; q_r(x)]^T$, $r = 0, 1, ..., n_a + n_b + 1$, stems from a rational approximation $B(x)/A(x)$ of a frequency response function $G(x)$ at the grid points x_k, $k = 1, 2, ..., F$ (see Section 9.16.2) and is calculated through a recursion formula (see, for example, the five-term recursion (15-80) in Example 15.11).

The calculation of the roots of $A(x) = 0$ and $B(x) = 0$ can be reduced to finding the roots of scalar orthogonal polynomials. This is done as follows.

1. Fit a scalar orthogonal polynomial $A_1(x) = \sum_{r=0}^{n_a} \alpha_r p_{1r}(x)$ of order n_a to the denominator $A(x)$ polynomial in (15-87) by minimizing

$$\sum_{k=1}^{F} \left| \frac{A_1(x_k) - A(x_k)}{B(x_k)} \right|^2 = \sum_{k=1}^{F} \left| \frac{\sum_{r=0}^{n_a} \alpha_r p_{1r}(x_k) - \sum_{r=0}^{n_a+n_b+1} a_r p_r(x_k)}{\sum_{r=0}^{n_a+n_b+1} a_r q_r(x_k)} \right|^2$$

w.r.t. $\alpha_0, \alpha_1, ..., \alpha_{n_a}$ (see Section 9.16.1).

2. Fit a scalar orthogonal polynomial $B_1(x) = \sum_{r=0}^{n_b} \beta_r q_{1r}(x)$ of order n_b to the numerator $B(x)$ polynomial in (15-87) by minimizing

$$\sum_{k=1}^{F} \left| \frac{B_1(x_k) - B(x_k)}{A(x_k)} \right|^2 = \sum_{k=1}^{F} \left| \frac{\sum_{r=0}^{n_b} \beta_r q_{1r}(x_k) - \sum_{r=0}^{n_a+n_b+1} a_r q_r(x_k)}{\sum_{r=0}^{n_a+n_b+1} a_r p_r(x_k)} \right|^2$$

w.r.t. $\beta_0, \beta_1, ..., \beta_{n_b}$ (see Section 9.16.1).

3. Calculate the roots of $A_1(x) = \sum_{r=0}^{n_a} \alpha_r p_{1r}(x)$ and $B_1(x) = \sum_{r=0}^{n_b} \beta_r q_{1r}(x)$ using the modified companion matrix approach of Section 15.12.1.

Note that the first and the second step of this procedure introduce no approximation errors because (i) there are no model errors (the true orders of $A(x)$ and $B(x)$ are, respectively, n_a and n_b), and (ii) there is no disturbing noise ($A(x)$, $B(x)$ are known exactly, within the numerical precision, at the grid points x_k, $k = 1, 2, ..., F$). Note also that the grid points x_k, $k = 1, 2, ..., F$, in the first and second steps are the same as those used to calculate the vector orthogonal basis $[p_r(x) \; q_r(x)]^T$, $r = 0, 1, ..., n_a + n_b + 1$.

15.13 SENSITIVITY OF THE LEAST SQUARES SOLUTION

Consider the overdetermined set of equations

$$Ax \approx b \tag{15-88}$$

with $A \in \mathbb{C}^{n \times m}$, $n > m$, regular and $b \in \mathbb{C}^m$. The least squares solution of (15-88) is

$$x_{LS} = (A^H A)^{-1} A^H b \tag{15-89}$$

The sensitivity of the least squares solution (15-89) to perturbations in A and b equals

$$\frac{\|\Delta x_{LS}\|_2}{\|x_{LS}\|_2} \leq \varepsilon \left(\frac{2\kappa(A)}{\cos(\alpha)} + \text{tg}(\alpha) \kappa^2(A) \right) \tag{15-90}$$

with $\sin(\alpha) = \|r_{LS}\|_2 / \|b\|_2$, $r_{LS} = A x_{LS} - b$, $\varepsilon = \max(\|\Delta A\|_2 / \|A\|_2, \|\Delta b\|_2 / \|b\|_2)$, and Δ the perturbation; that of the least square residual r_{LS} is given by

$$\frac{\|\Delta r_{LS}\|_2}{\|b\|_2} \leq 2\varepsilon \kappa(A) \tag{15-91}$$

(Golub and Van Loan, 1996). It shows that for nonzero residual problems ($r_{LS} \neq 0$) the sensitivity of x_{LS} depends on the square of $\kappa(A)$, while the sensitivity of r_{LS} just depends linearly on $\kappa(A)$.

The loss in numerical precision (high sensitivity) of the least squares solution (15-89) is basically due to the calculation of $A^H A$ ($\kappa(A^H A) = \kappa^2(A)$). There exist algorithms that calculate x_{LS} without forming the product $A^H A$ explicitly and, hence, have better sensitivity. For example, using the singular value decomposition $A = U \Sigma V^H$, (15-89) can be calculated as

$$x_{LS} = V \Sigma^{-1} U^H b \text{ or } x_{LS} = A^+ b, \tag{15-92}$$

while using the QR-factorization $A = QR$, (15-89) is calculated via back-substitution

$$R x_{LS} = Q^H b \tag{15-93}$$

The sensitivity of the SVD (15-92) and QR (15-93) solutions is approximately given by

$$\frac{\|\Delta x_{LS}\|_2}{\|x_{LS}\|_2} \leq \varepsilon \left(\frac{2\kappa(A)}{\cos(\alpha)} + \text{tg}(\alpha) \kappa^2(A) \right) 10^{-d} \tag{15-94}$$

where d is the number of significant digits used in the calculations (Golub and Van Loan, 1996).

15.14 EXERCISES

15.1. Prove that range(A) = (null(A^H))$^\perp$ (hint: take any $z_1 \in$ null(A^H), $z_2 \in$ range(A) and show that $z_1^H z_2 = 0$).

15.2. Show that the eigenvalues of a Hermitian matrix are real (hint: use the equality $x^H A x = \lambda x^H x$ valid for the eigenvalue, eigenvector pair λ, x and take the complex conjugate).

15.3. Show that the eigenvalues of A and A^T are the same (hint: use det(A) = det(A^T)).

15.4. Show that the eigenvalues of $A \in \mathbb{C}^{n \times n}$ are invariant w.r.t. a similarity transformation T (hint: use $I_n = TT^{-1}$).

15.5. Show that the eigenvalues of an upper or lower triangular matrix are the diagonal elements.

15.6. Consider the following matrix with a Vandermonde structure:

$$V(x_1, x_2, x_3) = \begin{bmatrix} 1 & x_1 & x_1^2 & \dots & x_1^n \\ 1 & x_2 & x_2^2 & \dots & x_2^n \\ 1 & x_3 & x_3^2 & \dots & x_3^n \end{bmatrix}$$

Show, via linear combinations, that it can be reduced to

$$f(x_1, x_2, x_3) \begin{bmatrix} 1 & x_1 & x_1^2 & \dots & x_1^n \\ 0 & 1 & x_1 + x_2 & \dots & \sum_{r_1 = 0}^{n-1} x_2^{n-1-r_1} x_1^{r_1} \\ 0 & 0 & 1 & \dots & \sum_{r_1 = 0}^{n-2} \sum_{r_2 = 0}^{n-2-r_1} x_3^{n-2-r_1-r_2} x_2^{r_2} x_1^{r_1} \end{bmatrix}$$

with $f(x_1, x_2, x_3) = (x_3 - x_1)(x_2 - x_1)(x_3 - x_2)$. Note that the matrix $V(x_1, x_2, x_3)$ has a full rank if and only if all x_i are different (hint: use $((y^n - x^n)/(y - x)) = \sum_{r=0}^{n-1} y^{n-1-r} x^r$).

15.7. Prove that only the symmetric part of a real matrix $A \in \mathbb{R}^{n \times n}$ contributes to $x^T A x$ (hint: use $A = (A + A^T)/2 + (A - A^T)/2$).

15.8. Show that for any $A \in \mathbb{C}^{n \times n}$ and $x \in \mathbb{C}^n$: Re($x^H A x$) = x^Hherm(A)x. Conclude that only the Hermitian part of a matrix contributes to the real value of $x^H A x$.

15.9. If $\begin{bmatrix} A & D \\ C & B \end{bmatrix}^{-1}$ and B^{-1} exist, show that

$$\det\left(\begin{bmatrix} A & D \\ C & B \end{bmatrix}\right) = \det(B)\det([A - DB^{-1}C]) \tag{15-95}$$

(hint: $\begin{bmatrix} A & D \\ C & B \end{bmatrix} = \begin{bmatrix} I & 0 \\ 0 & B \end{bmatrix} \begin{bmatrix} A & D \\ B^{-1}C & I \end{bmatrix}$ and det(AB) = det(A)det(B)).

15.10. If B^{-1} exists, show that

$$\begin{bmatrix} A & D \\ C & B \end{bmatrix}^{-1} = \begin{bmatrix} \Delta^{-1} & -\Delta^{-1}F \\ -E\Delta^{-1} & B^{-1} + E\Delta^{-1}F \end{bmatrix} \qquad (15\text{-}96)$$

where $\Delta = A - DB^{-1}C$, $E = B^{-1}C$, and $F = DB^{-1}$. Show, using Exercise 15.9, that Δ^{-1} exists.

15.11. Show that for any $A \in \mathbb{C}^{n \times m}$ and $B \in \mathbb{C}^{m \times n}$, $\text{tr}(AB) = \text{tr}(BA)$.

15.12. Show that for any $A \in \mathbb{C}^{n \times n}$, $\text{tr}(A) = \sum_{k=1}^{n} \lambda_k(A)$.

15.13. Show that for any $A, B \in \mathbb{C}^{n \times m}$, $\text{tr}(A^T B) = (\text{vec}(A))^T \text{vec}(B)$.

15.14. Prove the submultiplicative property (15-14) for the Frobenius norm.

15.15. Show that $\|A\|_2 \leq \|A\|_F$.

15.16. Using the definition of the singular value decomposition of a matrix $A \in \mathbb{C}^{n \times m}$ ($n \geq m$) show that the right singular vectors v_k and the squared singular values σ_k^2, $k = 1, 2, \ldots, m$, are the eigenvectors and eigenvalues of $A^H A$.

15.17. Using the definition of the singular value decomposition of a matrix $A \in \mathbb{C}^{n \times m}$ ($n \geq m$), show that the left singular vectors u_k and the squared singular values σ_k^2, $k = 1, 2, \ldots, n$, are the eigenvectors and eigenvalues of AA^H.

15.18. Let $A = U_A \Sigma_A X^{-1}$, $B = U_B \Sigma_B X^{-1}$ be the generalized singular value decomposition of the matrix pair (A, B). Show that $x = X_{[:, k]}$ and $\lambda = \alpha_k^2 / \beta_k^2$, $k = 1, 2, \ldots, m$, are the solutions of the generalized eigenvalue problem $A^H A x = \lambda B^H B x$.

15.19. Show that a Hermitian positive definite matrix $A \in \mathbb{C}^{n \times m}$ can be written as $A = V\Sigma V^H$ where $V^H V = VV^H = I_n$ and $\Sigma = \text{diag}(\lambda_1(A), \lambda_2(A), \ldots, \lambda_n(A))$ (hint: show that $\lambda_k(A) = \sigma_k(A)$ and that $U_{[:, k]} = V_{[:, k]}$).

15.20. Prove that a Hermitian matrix $A \in \mathbb{C}^{n \times m}$ is positive (semi-)definite if and only if $\lambda_k(A) > 0$ ($\lambda_k(A) \geq 0$) (hint: apply the results of Exercise 15.19, $A = V\Sigma V^H$ with $V^H = V^{-1}$, to the quadratic form $x^H A x$).

15.21. Let $Q \in \mathbb{C}^{n \times n}$ be a unitary matrix ($Q^H Q = I_n$). Show that $\kappa(Q) = 1$.

15.22. Prove inequality (15-16) (hint: first show that $(A+E)^{-1} - A^{-1} = -A^{-1}E(A+E)^{-1}$, next use $(I_n + A^{-1}E)^{-1} = \sum_{k=0}^{\infty} (-A^{-1}E)^k$).

15.23. Let $A \in \mathbb{C}^{n \times m}$ with $n \geq m$ and $B \in \mathbb{C}^{m \times m}$ nonsingular. Show that $\sigma(A, B) = \sigma(AB^{-1})$.

15.24. Let $x \in \mathbb{C}^n$. Show that $x^+ = x^H / (x^H x)$.

15.25. Show that expression (15-28) satisfies the four Moore-Penrose conditions.

15.26. Let $A \in \mathbb{C}^{n \times r}$ and $B \in \mathbb{C}^{r \times m}$ with $r \leq \min(n, m)$ and $\text{rank}(A) = \text{rank}(B) = r$. Show that $(AB)^+ = B^+ A^+$ (hint: verify that the four Moore-Penrose conditions are satisfied with $A^+ = (A^H A)^{-1} A^H$ and $B^+ = B^H (BB^H)^{-1}$).

15.27. Let $P \in \mathbb{C}^{n \times n}$ be a Hermitian idempotent matrix. Show that $P^+ = P$.

15.28. Show that the eigenvalues of an idempotent matrix P are one or zero (hint: left multiply $Px = \lambda x$ by P and work out).

15.29. Prove properties 2, 3, and 4 of the idempotent matrices (see Section 15.6).

15.30. Show that $(A \otimes B)(D \otimes G) = (AD) \otimes (BG)$.

15.31. Show that $\text{vec}(ADB) = (B^T \otimes A) \text{vec}(D)$ (hint: calculate the kth column of ADB and use $(XY)_{[:, k]} = XY_{[:, k]}$).

15.32. Show that $(N \otimes M)^{-1} = N^{-1} \otimes M^{-1}$ (hint: use (15-36)).

15.33. Take any $A \in \mathbb{C}^{n \times m}$ and $B \in \mathbb{C}^{p \times q}$. Show that $\|A \otimes B\| = \|A\|\|B\|$ where $\|\ \|$ denotes the 1-, 2-, ×-, and Frobenius norm (hint: for the 2- and Frobenius norm first show using (15-36) and (15-35) that $A \otimes B = (U_A \otimes U_B)(\Sigma_A \otimes \Sigma_B)(V_A \otimes V_B)^H$ with $A = U_A \Sigma_A V_A^H$ and $B = U_B \Sigma_B V_B^H$ the corresponding singular valued decompositions).

15.34. Prove properties (15-41), (15-42), (15-43), and (15-45) (hint: use $A = B^{-1} \Leftrightarrow AB = I_n$ for (15-43) and $A = B^+ \Leftrightarrow BAB = B$ for (15-44)).

15.35. Take a positive definite matrix $A \in \mathbb{C}^{n \times n}$. Show that the real matrix A_{Re} is symmetric and positive definite.

15.36. Take a unitary matrix A. Show that A_{Re} is orthogonal.

15.37. Prove property (15-46) (hint: using linear combinations of block rows and block columns show that

$$\det\left(\begin{bmatrix} \text{Re}(A) & -\text{Im}(A) \\ \text{Im}(A) & \text{Re}(A) \end{bmatrix}\right) = \det\left(\begin{bmatrix} A & 0 \\ \text{Im}(A) & \bar{A} \end{bmatrix}\right)$$

15.38. Prove property (15-47) (hint: first show that the singular value decompositions of A and A_{Re} are related to each other by $A = U\Sigma V^H$ and $A_{\text{Re}} = U_{\text{Re}} \Sigma_{\text{Re}} V_{\text{Re}}^T$).

15.39. Prove properties (15-49), (15-50), and (15-51).

15.40. Verify results (15-56), (15-57), (15-58), and (15-60).

15.41. Show that $\partial f(x, \bar{x})/\partial \bar{x} = \overline{\partial f(x, \bar{x})/\partial x}$ if $f(x, \bar{x}) \in \mathbb{R}$ (hint: use the limit defintion of the partial derivative operator and $\overline{f(x, \bar{x})} = f(x, \bar{x})$).

15.42. Verify results (15-62), (15-63), (15-64), (15-65), (15-66), and (15-67) (hint: use $A^{-1} = \text{adj}(A)/\det(A)$ for (15-64), where $\text{adj}(A)$ is the transposed matrix of cofactors of A; show first that $\partial A^{-1}/\partial A_{[k, l]} = -A^{-1}(\partial A/\partial A_{[k, l]})A^{-1}$ for (15-65)).

15.43. Prove Lemma 15.8.

15.44. Show that (15-76) reduces to a three-term recursion if for any k, $w_{2k} = 0$ and $\bar{x}_k = -x_k$ in (15-68) (hint: first use $\langle xp_{s-1}(x), p_l(x) \rangle = -\langle p_{s-1}(x), xp_l(x) \rangle$ and $xp_l(x) \in \text{span}\{p_0(x), p_1(x), \ldots, p_{l+1}(x)\}$ to prove the three-term recursion; next show that $\langle p_{s-1}(x), xp_{s-1}(x) \rangle = 0$ and $\langle p_{s-1}(x), xp_{s-2}(x) \rangle = \|q_{s-1}(x)\|$).

15.45. Consider Example 15.10 with real grid points ($x_k \in \mathbb{R}$) and $w_{2k} = 0$ for any k in (15-68). Show that (15-76) reduces to a three-term recursion formula (hint: use $\langle xp_{s-1}(x), p_l(x) \rangle = \langle p_{s-1}(x), xp_l(x) \rangle$, $xp_l(x) \in \text{span}\{p_0(x), p_1(x), \ldots, p_{l+1}(x)\}$).

15.46. Show that (15-79) reduces to a five-term recursion if $\bar{x}_k = -x_k$ for any k in (15-71) (hint: first use $\langle xp_{s-2}(x), p_l(x) \rangle = -\langle p_{s-2}(x), xp_l(x) \rangle$ and $xp_l(x) \in \text{span}\{p_0(x), p_1(x), \ldots, p_{l+2}(x)\}$ to prove the five-term recursion; next show that $\langle p_{s-2}(x), xp_{s-2}(x) \rangle = 0$, $\langle p_{s-2}(x), xp_{s-1}(x) \rangle = -\langle p_{s-2}(x), xp_{s-3}(x) \rangle$ and $\langle p_{s-2}(x), xp_{s-4}(x) \rangle = \|q_{s-2}(x)\|$).

15.15 APPENDIX

Appendix 15.A Calculation of the Roots of a Polynomial

The roots of the polynomial $A(x)$ are those values of x such that $A(x) = 0$ or

$$p_{n_a}(x) = -\frac{1}{a_{n_a}} \sum_{r=0}^{n_a - 1} a_r p_r(x) \tag{15-97}$$

(use (15-83)). Adding the equations $p_r(x) = p_r(x)$, $r = n_a - 1, n_a - 2, \ldots, 1$, to (15-97) gives the following set of n_a equations:

$$A_2 Z = Z_1 \tag{15-98}$$

with $Z_1^T = [p_{n_a}(x)\ p_{n_a-1}(x) \ldots p_1(x)]$, $Z^T = [p_{n_a-1}(x)\ p_{n_a-2}(x) \ldots p_0(x)]$ and where A_2 is defined in (15-86). Recursion (15-84) can be written in matrix form as

$$Z_1 = xA_1Z + A_3Z \qquad (15\text{-}99)$$

where A_1 and A_3 are n_a by n_a matrices defined in (15-86). Combining (15-98) and (15-99) shows that the roots x of $A(x) = 0$ are the solutions of the eigenvalue problem $A_1^{-1}(A_2 - A_3)Z = xZ$. □

16

Some Probability and Stochastic Convergence Fundamentals

Abstract: The goal of this chapter is to give insight into the way to analyze the stochastic properties of an estimator. Therefore, a great deal of attention is paid to the different concepts of stochastic convergence. The main ideas behind the stochastic convergence proofs used throughout this book are explained and some basic analysis tools are provided. More information on the topic can be found in the following textbooks: Billingsley (1995), Chow and Teicher (1988), Brillinger (1981), Lukacs (1975), Stout (1974), and Jazwinski (1970). The calculation of probabilities, expected values, and higher order (central) moments and the properties of standard distributions are assumed to be known. More information can be found in Anderson (1958), Stuart and Ord (1987), and Mathai and Provost (1992). This chapter also includes a study of the properties of the noise after a discrete Fourier transform, which is essential in frequency domain identification.

16.1 NOTATIONS AND DEFINITIONS

$\mathbb{E}\{\ \}$ and Prob( ) denote the expected value and the probability function, respectively. If $f_x(x)$ and $F(x)$ are the respective probability density function and distribution function of the random variable x, then the expected value of $g(x)$ is given by

$$\mathbb{E}\{g(x)\} = \int_X g(x)dF(x) = \int_X g(x)f_x(x)dx \tag{16-1}$$

where X is the domain of $F(x)$.

Let $x, y \in \mathbb{C}$ be complex random variables: then the *mean* μ_x and *variance* σ_x^2 of x and the *covariance* σ_{xy}^2 between x and y are defined as

$$\mu_x = \mathbb{E}\{x\} \qquad \sigma_x^2 = \text{var}(x) = \mathbb{E}\{|x - \mathbb{E}\{x\}|^2\} \tag{16-2}$$

$$\sigma_{xy}^2 = \text{covar}(x, y) = \mathbb{E}\{(x - \mathbb{E}\{x\})\overline{(y - \mathbb{E}\{y\})}\} \tag{16-3}$$

Let $x, y \in \mathbb{C}^n$ be complex random vectors: then the *covariance matrix* C_x of x and the *cross-covariance matrix* C_{xy} between x and y are given by

$$C_x = \text{Cov}(x) = \mathbb{E}\{(x - \mathbb{E}\{x\})(x - \mathbb{E}\{x\})^H\} \tag{16-4}$$

$$C_{xy} = \text{Cov}(x, y) = \mathbb{E}\{(x - \mathbb{E}\{x\})(y - \mathbb{E}\{y\})^H\} \tag{16-5}$$

Let $\hat{x} \in \mathbb{C}^n$ be an estimate of the true value x_0. The *bias* b_x and the *mean square error* $\text{MSE}(\hat{x})$ of the estimate are given, respectively, by

$$b_x = \mathbb{E}\{\hat{x}\} - x_0 \tag{16-6}$$

$$\text{MSE}(\hat{x}) = \mathbb{E}\{(\hat{x} - x_0)(\hat{x} - x_0)^H\} = \text{Cov}(\hat{x}) + b_x b_x^H \tag{16-7}$$

A stochastic process $x(t) \in \mathbb{C}^n$, $t \in \mathbb{Z}$, is *strictly stationary* if the joint distribution of $x(t_1 + t), x(t_2 + t), \ldots x(t_k + t)$ does not depend on t for every $t, t_1, \ldots, t_k \in \mathbb{Z}$ and $k = 1, 2, 3, \ldots$ For example, a series of independent, identically distributed random vectors is strictly stationary. A stochastic process $x(t) \in \mathbb{C}^n$, $t \in \mathbb{Z}$, is *second-order stationary* or *wide-sense stationary* if the first- and second-order moments are invariant under a common shift of the argument t

$$\mu_x(t_1 + t) = \mu_x(t_1) \tag{16-8}$$

$$\text{Cov}(x(t_1 + t), x(t_2 + t)) = \text{Cov}(x(t_1), x(t_2)) \tag{16-9}$$

for every $t, t_1, t_2 \in \mathbb{Z}$. A strictly stationary process with finite second-order moments is second-order stationary.

The stochastic process $x(t) \in \mathbb{C}^n$, $t \in \mathbb{Z}$, is *independent* if $x(t)$ and $x(s)$ are independent whenever $t - s \neq 0$. It is *m-dependent* if $x(t), x(t+1), \ldots, x(t+r)$ is independent of $x(t+r+n), x(t+r+n+1), \ldots, x(t+r+s)$ for any $n > m \geq 0$, with $r, s > 0$. It is uniformly bounded if $|x(t)| \leq C < \infty$ for any realization and for any t.

Let $x(t) \in \mathbb{C}^n$ and $y(t) \in \mathbb{C}^m$ be second-order stationary stochastic processes; then the *autocorrelation matrix* $R_{xx}(\tau)$ of $x(t)$ and the *cross-correlation matrix* $R_{xy}(\tau)$ between x and y are defined as

$$R_{xx}(\tau) = \mathbb{E}\{x(t)x^H(t - \tau)\} \tag{16-10}$$

$$R_{xy}(\tau) = \mathbb{E}\{x(t)y^H(t - \tau)\} \tag{16-11}$$

The *auto-* and *cross-power spectra* are the Fourier transforms of the auto- and cross-correlation matrices.

A *real normally distributed* random vector $x \in \mathbb{R}^n$ with mean μ_x and covariance matrix C_x, will be denoted as $x \in N_n(\mu_x, C_x)$. If subscript n is omitted then this is equivalent to $n = 1$. Similarly, $x \in E(\mu_x, \sigma_x^2)$, $x \in L(\mu_x, \sigma_x^2)$, and $x \in U(\mu_x, \sigma_x^2)$ denote *real exponential, Laplace,* and *uniform* random variables, respectively. The sum of the squares $y = \sum_{k=1}^{n} x_k^2$ of n independent and identically distributed normal random variables $x_k \in N(0, 1)$ is *chi-squared distributed* with n degrees of freedom $y \in \chi^2(n)$. The ratio $z = (n_2 y_1)/(n_1 y_2)$ of two independent chi-squared distributed random variables $y_1 \in \chi^2(n_1)$ and $y_2 \in \chi^2(n_2)$ is *F-distributed* with n_1 and n_2 degrees of freedom $z \in F(n_1, n_2)$. The matrix-valued random variable $y = \sum_{k=1}^{n} x_k x_k^T \in \mathbb{R}^{p \times p}$, formed by the sum of the product of n independent and identically distributed normal vectors $x_k \in N_p(0, C_x)$, is *Wishart distributed*, $y \in W_p(n, C_x)$, with n degrees of freedom and associated parameter matrix C_x (Anderson, 1958; Mathai and Provost, 1992). As a special case, we have $W_1(n, 1) = \chi^2(n)$.

A complex random vector $x \in \mathbb{C}^n$ is said to be *circular* if

$$\mathbb{E}\{(x - \mathbb{E}\{x\})(x - \mathbb{E}\{x\})^T\} = 0 \tag{16-12}$$

(Picinbono, 1993). If $x \in \mathbb{C}^n$ is a circular complex random vector and $A \in \mathbb{C}^{m \times n}$, then $y = Ax$ is also circular (Exercise 16.3). Condition (16-12) is equivalent to

$$\begin{aligned}\text{Cov}(\text{Re}(x)) &= \text{Cov}(\text{Im}(x)) \\ \text{Cov}(\text{Re}(x), \text{Im}(x)) &= -(\text{Cov}(\text{Im}(x), \text{Re}(x)))^T\end{aligned} \tag{16-13}$$

If (16-13) is valid, then it can be verified that $\text{Cov}(x_{re}) = 0.5(\text{Cov}(x))_{Re}$ (Exercise 16.4). The probability density function $f_x(x)$ of complex random variables $x \in \mathbb{C}^n$ is given by the joint probability density function $f_{x_{re}}(x_{re})$ of the real and imaginary parts of x. For *circular complex normally* distributed random vectors $x \in \mathbb{C}^n$, the probability density function is uniquely determined by the mean value μ_x and the covariance matrix C_x (Picinbono, 1993)

$$f_x(x) = f_{x_{re}}(x_{re}) = \frac{1}{\pi^n \det(C_x)} \exp(-(x - \mu_x)^H C_x^{-1} (x - \mu_x)) \tag{16-14}$$

If the complex normally distributed noise is not circular, then the other second-order moment $\mathbb{E}\{(x - \mathbb{E}\{x\})(x - \mathbb{E}\{x\})^T\}$ is also required to construct $f_{x_{re}}(x_{re})$. In the univariate case, (16-13) implies that the real and imaginary parts of $x \in \mathbb{C}$ have equal variance and have zero covariance. If $x \in \mathbb{C}$ is circular complex normally distributed, then its real and imaginary parts are independent and $\mathbb{E}\{(x - \mathbb{E}\{x\})^n\} = 0$ (see Exercise 16.8 and Schoukens and Pintelon, 1990).

A *circular complex normally distributed* random vector $x \in \mathbb{C}^n$ with mean μ_x and covariance matrix C_x, will be denoted as $x \in N_n^c(\mu_x, C_x)$. Similarly, $x \in E^c(\mu_x, \sigma_x^2)$, $x \in L^c(\mu_x, \sigma_x^2)$, and $x \in U^c(\mu_x, \sigma_x^2)$ denote *circular complex exponential, Laplace,* and *uniform* random variables with independent real and imaginary parts. The matrix-valued random variable $y = \sum_{k=1}^n x_k x_k^H \in \mathbb{C}^{p \times p}$ with $x_k \in N_p^c(0, C_x)$ is *complex Wishart distributed*, $y \in W_p^c(n, C_x)$, with n degrees of freedom and associated parameter matrix C_x (Goodman, 1963; Brillinger, 1981). Its probability density function is given by

$$f_y(y) = \frac{(\det(y))^{n-p} \exp(-\text{tr}(C_x^{-1} y))}{\pi^{p(p-1)/2} (\det(C_x))^n \prod_{k=1}^n (n-k)!} \tag{16-15}$$

for $n \geq p$ and $y \geq 0$ (Goodman, 1963; Brillinger, 1981). Note that the real part of a complex Wishart distributed random variable is, in general, not Wishart distributed. This is due to the fact that the real and imaginary parts of the $x_k \in N_p^c(0, C_x)$ are not necessarily independent.

Take n complex random variables $x_k \in \mathbb{C}$ with $\mathbb{E}\{|x_k|^n\} < \infty$, $k = 1, 2, \ldots, n$. The *joint cumulant* $\text{cum}(x_1, x_2, \ldots, x_n)$ of order n is given by

$$\text{cum}(x_1, x_2, \ldots, x_n) = \sum (-1)^{p-1} (p-1)! \prod_{m=1}^p \mathbb{E}\{\prod_{k_m \in \mathbb{V}_m} x_{k_m}\} \tag{16-16}$$

where the summation extends over all partitions $\{\mathbb{V}_1, \mathbb{V}_2, \ldots, \mathbb{V}_p\}$ of $\mathbb{I} = \{1, 2, \ldots, n\}$. If for any k, $x_k = x$, then the definition gives the *cumulant* of order n of x.

Example 16.1: Calculation of the third-order joint cumulant. All the partitions $\{\mathbb{V}_1, \mathbb{V}_2, \ldots, \mathbb{V}_p\}$ of the set $\{1, 2, 3\}$ are $\{\{1\}, \{2\}, \{3\}\}$, $\{\{1\}, \{2, 3\}\}$, $\{\{2\}, \{1, 3\}\}$, $\{\{3\}, \{1, 2\}\}$, and $\{\{1, 2, 3\}\}$. Hence, formula (16-16) becomes

$$\text{cum}(x_1, x_2, x_3) = (-1)^2 2! \,\mathbb{E}\{x_1\} \mathbb{E}\{x_2\} \mathbb{E}\{x_3\} + (-1)^1 1! \,\mathbb{E}\{x_1\} \mathbb{E}\{x_2 x_3\} + \\ (-1)^1 1! \,\mathbb{E}\{x_2\} \mathbb{E}\{x_1 x_3\} + (-1)^1 1! \,\mathbb{E}\{x_3\} \mathbb{E}\{x_1 x_2\} + (-1)^0 0! \,\mathbb{E}\{x_1 x_2 x_3\} \quad (16\text{-}17)$$

where $0! = 1$. □

The cumulants have the following properties for any random variables $x_k, y_l \in \mathbb{C}$ and constants $a_k, b_l \in \mathbb{C}$, $k, l = 1, 2, \ldots$ (Brillinger, 1981):

1. Symmetric in their arguments, for example, $\text{cum}(x_1, x_2, x_3) = \text{cum}(x_3, x_1, x_2)$.
2. Multilinear functions of their arguments, for example,
   $\text{cum}(\sum_{k=1}^n a_k x_k, \sum_{l=1}^m b_l y_l) = \sum_{k=1}^n \sum_{l=1}^m a_k b_l \text{cum}(x_k, y_l)$.
3. If any group of the x_k's, $k = 1, 2, \ldots, n$ are independent of the remaining x_k's, then $\text{cum}(x_1, x_2, \ldots, x_n) = 0$.
4. If the random variables $x_1, x_2, \ldots, x_n$ are independent of $y_1, y_2, \ldots, y_n$, then $\text{cum}(x_1 + y_1, x_2 + y_2, \ldots, x_n + y_n) = \text{cum}(x_1, x_2, \ldots, x_n) + \text{cum}(y_1, y_2, \ldots, y_n)$.
5. $\text{cum}(x_1, x_2, \ldots, x_{k-1}, a_1, x_{k+1}, \ldots, x_r) = 0$ for $r = 2, 3, \ldots, n$.
6. $\text{cum}(x_k) = \mathbb{E}\{x_k\}$, $\text{cum}(x_k, \bar{x}_k) = \text{var}(x_k)$ and $\text{cum}(x_k, \bar{y}_l) = \text{covar}(x_k, y_l)$.
7. For stationary random variables, $x(k) \in \mathbb{C}$, $k \in \mathbb{Z}$, we have
   $\text{cum}(x(k_1), x(k_2), \ldots, x(k_n)) = \text{cum}(x(k_1 - k_n), x(k_2 - k_n), \ldots, x(0))$
   for every $k_i \in \mathbb{Z}$, $i = 1, 2, \ldots, n$.

It can be shown that the joint cumulant (16-16) equals the coefficient of $j^r t_1 t_2 \ldots t_r$ in the Taylor series expansion of $\ln(\mathbb{E}\{\exp(j \sum_{k=1}^r x_k t_k)\})$ about the origin $t_1 = t_2 = \ldots = 0$ (Brillinger, 1981).

Example 16.2: Let $x \in \mathbb{R}^n$ be a multivariate normally distributed random variable with covariance matrix C_x and mean value μ_x. Its *characteristic function* $\phi(t)$ is given by (Stuart and Ord, 1987)

$$\phi(t) = \mathbb{E}\{\exp(j \sum_{k=1}^n x_{[k]} t_{[k]})\} = \exp(-\frac{1}{2} t^T C_x t + j t^T \mu_x) \quad (16\text{-}18)$$

with $t \in \mathbb{R}^n$. Note that this result remains valid if the covariance matrix C_x is not of full rank (Mathai and Provost, 1992). Taking the natural logarithm of (16-18) shows that all joint cumulants of order greater than two are zero

$$\text{cum}(x_{[k_1]}, x_{[k_2]}, \ldots, x_{[k_r]}) = 0 \qquad k_i \in \{1, 2, \ldots, r\}, r > 2 \quad (16\text{-}19)$$

If $x \in \mathbb{C}^n$ is a multivariate complex normal random variable, then $x_{\text{re}} \in \mathbb{R}^{2n}$ is a real-valued normal random variable for which result (16-19) applies. Because the cumulants of x can be written as a linear combination of the cumulants of x_{re} (multilinearity property 2), it follows that (16-19) is also valid for multivariate complex normal random variables. □

Chebyshev's inequality: Take two random variables x, y with finite second-order moments and ε an arbitrary positive real number, then (Billingsley, 1995)

$$\text{Prob}(|x-y| > \varepsilon) \le \frac{1}{\varepsilon^2} \mathbb{E}\{(x-y)^2\} \tag{16-20}$$

Markov's inequality: take two random variables x, y with finite absolute moments of order $p > 0$ and ε an arbitrary positive real number, then (Stuart and Ord, 1987)

$$\text{Prob}(|x-y| > \varepsilon) \le \frac{1}{\varepsilon^p} \mathbb{E}\{|x-y|^p\} \tag{16-21}$$

16.2 THE COVARIANCE MATRIX OF A FUNCTION OF A RANDOM VARIABLE

Let $x \in \mathbb{R}^n$ be a random vector with mean value μ_x. In general it is impossible to calculate the covariance matrix of the function $f(x) \in \mathbb{R}^m$. An approximation can be calculated through linearization of the function $f(x)$. Assuming that $f(x)$ has continuous derivative w.r.t. x, we find

$$f(x) \approx f(\mu_x) + \left.\frac{\partial f(x)}{\partial x}\right|_{x=\mu_x} (x - \mu_x) \tag{16-22}$$

and

$$\text{Cov}(f(x)) \approx \left.\frac{\partial f(x)}{\partial x}\right|_{x=\mu_x} \text{Cov}(x) \left(\left.\frac{\partial f(x)}{\partial x}\right|_{x=\mu_x}\right)^T \tag{16-23}$$

Under some conditions on x and $\partial f(x)/\partial x|_{x=\mu_x}$, the right-hand side of (16-22) is approximately Gaussian distributed (see Section 16.10). Then, (16-23) is very useful for calculating confidence levels and uncertainty bounds on $f(x)$, even if the second-order moments of $f(x)$ do not exist.

In general, if $x \in \mathbb{C}^n$ and/or $f(x) \in \mathbb{C}^m$, then x and/or $f(x)$ in (16-23) should be replaced, respectively, by x_{re} and/or $f_{\text{re}}(x)$. Some important special complex cases lead to simplified formulae. If $x \in \mathbb{C}^n$ is circular complex distributed and $f(x) \in \mathbb{C}^m$ is an analytic function of x, then the right-hand side of (16-22) is circular complex distributed and (16-23) becomes

$$\text{Cov}(f(x)) \approx \left.\frac{\partial f(x)}{\partial x}\right|_{x=\mu_x} \text{Cov}(x) \left(\left.\frac{\partial f(x)}{\partial x}\right|_{x=\mu_x}\right)^H \tag{16-24}$$

If $x \in \mathbb{C}^n$ is circular complex distributed and $f(x, \bar{x}) \in \mathbb{R}^m$, then

$$\text{Cov}(f(x, \bar{x})) \approx 2\text{Re}\left(\left.\frac{\partial f(x, \bar{x})}{\partial x}\right|_{x=\mu_x} \text{Cov}(x) \left(\left.\frac{\partial f(x, \bar{x})}{\partial x}\right|_{x=\mu_x}\right)^H\right) \tag{16-25}$$

(Proof: see Exercise 16.15).

16.3 SAMPLE VARIABLES

A particular realization of a stochastic process $x(t) \in \mathbb{C}^n$ is denoted by $x^{[k]}(t)$ where k can be a random variable itself. It indicates that the outcome of the process $x(t)$ for each value of t depends on the particular value of k. In most cases k is a positive integer number corresponding to the index of the realization.

The *sample mean* $\hat{x}(t)$ and *sample (cross-)covariance matrices* $\hat{C}_{xy}(t)$ and $\hat{C}_x(t)$ of R realizations of the stochastic processes $x(t), y(t) \in \mathbb{C}^n$ are defined as

$$\hat{x}(t) = \frac{1}{R} \sum_{k=1}^{R} x^{[k]}(t) \tag{16-26}$$

$$\hat{C}_x(t) = \frac{1}{R-1} \sum_{k=1}^{R} (x^{[k]}(t) - \hat{x}(t))(x^{[k]}(t) - \hat{x}(t))^H \tag{16-27}$$

$$\hat{C}_{xy}(t) = \frac{1}{R-1} \sum_{k=1}^{R} (x^{[k]}(t) - \hat{x}(t))(y^{[k]}(t) - \hat{y}(t))^H \tag{16-28}$$

The sample mean $\hat{x}(t)$ and sample (cross-)covariance matrices $\hat{C}_{xy}(t)$, $\hat{C}_x(t)$ of *independent realizations* are unbiased estimates of the mean $\mu_x(t)$ and (cross-)covariance matrices $C_{xy}(t)$, $C_x(t)$.

For real and circular complex normally distributed processes the sample mean $\hat{x}(t)$ and sample covariance matrix $\hat{C}_x(t)$ are independently distributed (see Anderson, 1958 for the real case; see Giri, 1965 for the circular complex case). Because $\hat{C}_x(t)$ and $\hat{x}(t)$ are only functions of, respectively, $\hat{C}_{x_{re}}(t)$ and $x_{re}(t)$, it can be seen that this result is also valid for noncircular complex normal processes. For real and circular complex normal processes we also have

$$x(t) \in N_n(\mu_x(t), C_x(t)) \Rightarrow \begin{cases} \hat{x}(t) \in N_n(\mu_x(t), C_x(t)/R) \\ \hat{C}_x(t) \in W_n((R-1), C_x(t)/(R-1)) \end{cases} \tag{16-29}$$

$$x(t) \in N_n^c(\mu_x(t), C_x(t)) \Rightarrow \begin{cases} \hat{x}(t) \in N_n^c(\mu_x(t), C_x(t)/R) \\ \hat{C}_x(t) \in W_n^c((R-1), C_x(t)/(R-1)) \end{cases} \tag{16-30}$$

(see Anderson, 1958 for the real case; see Giri, 1965 for the circular complex case).

The statistical performance of estimators is often compared through Monte Carlo simulations. In such simulations the true parameter value is known, while the true covariance of the estimates is unknown. Because the parameter estimates are mostly asymptotically normal, we may use Hotelling's T^2-statistic to test the bias of the estimates. For real and circular complex normally distributed, it becomes

$$b = (\hat{x} - \mu_x)^H (\hat{C}_x/R)^{-1} (\hat{x} - \mu_x) \tag{16-31}$$

with

Section 16.4 ■ Mixing Random Variables

$$x \in N_n(\mu_x, C_x) \Rightarrow b \in (R-1)\frac{n}{R-n}F(n, R-n) \quad \text{(a)}$$
$$x \in N_n^c(\mu_x, C_x) \Rightarrow b \in (R-1)\frac{2n}{2(R-n)}F(2n, 2(R-n)) \quad \text{(b)}$$
(16-32)

(see Hotelling, 1933 and Anderson, 1958 for the real case; see Giri, 1965 for the circular complex case). If for all realizations the same dependence exists between the estimated parameters, then $\hat{C}_x$ is rank deficient and the inverse in (16-31) should be replaced by the pseudoinverse. The statistic (16-32) is still valid if n is replaced by rank($\hat{C}_x$) (see Exercise 16.18). This rank deficiency problem often occurs when identifying models with a redundant number of parameters. For noncircular complex normal parameters $x \in \mathbb{C}^n$ the bias test (16-31), (16-32) is performed on x_{re} and $\hat{C}_{x_{\text{re}}}$ where n is replaced by $2n$ in (16-32-a).

16.4 MIXING RANDOM VARIABLES

16.4.1 Definition

An important requirement that will be imposed on the perturbing noise after sampling is that it has a limited span of dependence. This requirement is formalized by the mixing assumption for discrete-time noise. The definition for mixing random variables used throughout this book is slightly more general than the classical definition given in Brillinger (1981) in the sense that it is also valid for a certain class of nonstationary signals.

The real random vectors $x(t) \in \mathbb{R}^r$, $t = 0, 1, 2, \ldots$, are called *mixing of order P* if

$$\max_{t_k} \sum_{t_1, t_2, \ldots, t_{k-1} = 0}^{\infty} \left| \text{cum}(x_{[a_1]}(t_1), x_{[a_2]}(t_2), \ldots, x_{[a_{k-1}]}(t_{k-1}), x_{[a_k]}(t_k)) \right| < \infty \quad (16\text{-}33)$$

for every $a_i \in \{1, 2, \ldots, r\}$, $i = 1, 2, \ldots, k$, and $k = 1, 2, \ldots, P$. Because cumulants are symmetric in their arguments (see Section 16.1, property 1), this definition is independent of the particular choice of the index t_i, $i = 1, 2, \ldots k$, used to take the maximum. A mixing condition of order P assumes that all moments of order $k = 1, 2, \ldots, P$ exist and are uniformly bounded. The random vectors $x_1(t) \in \mathbb{R}^{r_1}$, $x_2(t) \in \mathbb{R}^{r_2}$, $\ldots x_q(t) \in \mathbb{R}^{r_q}$ are *jointly mixing of order P* if the random vector

$$y(t) = \left[x_1^T(t) \; x_2^T(t) \; \ldots \; x_q^T(t) \right]^T \in \mathbb{R}^{r_1 + r_2 + \ldots + r_q} \quad (16\text{-}34)$$

is mixing of order P. For complex random vectors $x(t) \in \mathbb{C}^r$ definition (16-33) is applied to $x_{\text{re}}(t) \in \mathbb{R}^{2r}$. For strictly stationary random variables $x(t) \in \mathbb{R}^r$, $t \in \mathbb{Z}$, definition (16-33) simplifies to the classical definition given in Brillinger (1981)

$$\max_{t_k} \sum_{t_1, t_2, \ldots, t_{k-1} = -\infty}^{\infty} \left| \text{cum}(x_{[a_1]}(t_1), x_{[a_2]}(t_2), \ldots, x_{[a_{k-1}]}(t_{k-1}), x_{[a_k]}(t_k)) \right|$$
$$= \sum_{u_1, u_2, \ldots, u_{k-1} = -\infty}^{\infty} \left| \text{cum}(x_{[a_1]}(u_1), x_{[a_2]}(u_2), \ldots, x_{[a_{k-1}]}(u_{k-1}), x_{[a_k]}(0)) \right|$$
(16-35)

where $u_i = t_i - t_k$, $i = 1, 2, \ldots, k-1$ (see Section 16.1, property 7). Note that the definition of mixing random variables (16-33) implies that

$$\sum_{t_1, t_2, \ldots, t_k = 1}^{K} \left| \text{cum}(x_{[a_1]}(t_1), x_{[a_2]}(t_2), \ldots, x_{[a_{k-1}]}(t_{k-1}), x_{[a_k]}(t_k)) \right| = O(K) \tag{16-36}$$

for $K \to \infty$. The converse is true only for strictly stationary random variables.

For strictly stationary random variables $x(t)$, the mixing condition (16-33) implies that the span of dependence over t is sufficiently small. That is, the random variables $x(t_1)$ and $x(t_2)$ become uncorrelated sufficiently fast (mixing of order 2) or independent (mixing of order $\times$) as $t_2 - t_1 \to \infty$. Indeed, if the zero mean noise $x(t)$ is mixing of order 2, then $\text{covar}(x(t_1), x(t_2)) = \text{cum}(x(t_1), \bar{x}(t_2)) \to 0$ as $t_2 - t_1 \to \infty$. As $\text{var}(x(t_1)) = \text{var}(x(t_2)) > 0$ are independent of t_1, t_2, we have

$$\frac{\text{covar}(x(t_1), x(t_2))}{\sqrt{\text{var}(x(t_1)) \text{var}(x(t_2))}} \to 0 \text{ as } t_2 - t_1 \to \infty \tag{16-37}$$

and thus $x(t_1)$ and $x(t_2)$ are asymptotically uncorrelated. Similarly, if the noise is mixing of order ∞, then all the higher order correlations tend to zero as $t_2 - t_1 \to \infty$ and, hence, $x(t_1)$ and $x(t_2)$ are asymptotically independent. Note that the converse is not necessarily true: if $x(t_1)$ and $x(t_2)$ are asymptotically ($t_2 - t_1 \to \infty$) independent, then this does not imply that $x(t)$ is mixing of order ∞. Indeed, the mixing condition (16-33) also imposes conditions on how fast the random variables should become independent. For general nonstationary noise, the mixing condition (16-33) is not sufficient to ensure that $x(t_1)$ and $x(t_2)$ are asymptotically uncorrelated or independent. To be asymptotically uncorrelated, we must assume, in addition, that the variance of the noise does not decrease to zero or decreases to zero more slowly than the covariance, so that (16-37) is fulfilled.

16.4.2 Properties

Besides its generality, the power of the mixing assumption lies in the property that a linear combination of powers of mixing variables is also mixing. These properties are formalized in the following lemmas.

Lemma 16.3: Let $x(t) \in \mathbb{C}$ be mixing of order P and $\alpha(t) \in \mathbb{C}$ nonrandom numbers. If $\max_t |\alpha(t)| \leq c < \infty$, then $\alpha(t) x(t)$ is mixing of order P.

Proof. Follows directly from definition (16-33). □

Lemma 16.4: Let $x_k(t) \in \mathbb{C}^r$, $k = 1, 2, \ldots, q$, be jointly mixing (over t) random variables of order P. The linear combination $y(t) = \sum_{k=1}^{q} \alpha_k z_k(t)$ with $z_k(t) = x_k(t)$ and/or $z_k(t) = \bar{x}_k(t)$ is mixing of order P.

Proof. Apply the multilinearity property 2 of the cumulants (see Section 16.1). □

Lemma 16.5: Let $x(t) \in \mathbb{C}$ be mixing of order P and $y(t) = \sum_{u=0}^{\infty} h(t,u)x(u)$. If $h(t,u) \in \mathbb{C}$ is absolutely summable with respect to t and u

$$\forall u: \sum_{t=0}^{\infty} |h(t,u)| \leq C < \infty \quad \text{(a)}$$
$$\forall t: \sum_{u=0}^{\infty} |h(t,u)| \leq C < \infty \quad \text{(b)}$$
(16-38)

with C a constant independent of t, u then, $y(t)$ is mixing of order P.

Proof. See Appendix 16.B. □

Lemma 16.5 says that mixing noise, filtered by a linear time-variant system with absolutely summable impulse response, remains mixing of the same order. Extension of Lemma 16.5 to the multivariable case is direct (Exercise 16.19). The real multivariable time-invariant version of Lemma 16.5 can be found in Brillinger (1981).

Example 16.6: Let $y(t)$ be generated by independent, identically distributed noise $e(t)$ passing through a stable discrete-time filter. If $e(t)$ has bounded Pth order moments, then $y(t)$ is mixing of order P. If $e(t)$ is uniformly bounded or belongs to the exponential family of distributions (for example, exponential, gamma, Laplace, Gaussian), then $y(t)$ is mixing of order ∞. Indeed, for independent noise, the mixing condition (16-33)

$$\max_{t_k} \sum_{t_1, t_2, \ldots, t_{k-1} = 0}^{\infty} |\text{cum}(e(t_1), \ldots, e(t_{k-1}), e(t_k))| = |\text{cum}(e(t_k), \ldots, e(t_k))| < \infty \quad (16\text{-}39)$$

boils down to the existence of the kth order moment. All higher order moments exist for uniformly bounded noise and noise belonging to the exponential family of distributions (Stuart and Ord, 1987). □

Corollary 16.7: Let $x \in \mathbb{C}^N$, $H \in \mathbb{C}^{N \times N}$, and $y = Hx$. If $x_{[t]}$ is mixing over t of order P for $N = 1, 2, \ldots, \infty$ and $\|H\|_1 \leq c < \infty$, $\|H\|_{\infty} \leq c < \infty$ for $N = 1, 2, \ldots, \infty$, with c a constant independent of N, then $y_{[t]}$ is mixing over t of order P for $N = 1, 2, \ldots, \infty$.

Proof. Follows directly from Lemma 16.5. □

Lemma 16.8: Let $x(t) \in \mathbb{C}$ be mixing of order qP and $y(t) \in \mathbb{C}$ defined as

$$y(t) = \sum_{m=0}^{q} \sum_{u_1, u_2, \ldots, u_m = 0}^{t} h_m(t-u_1, t-u_2, \ldots, t-u_m)x(u_1)x(u_2)\ldots x(u_m) \quad (16\text{-}40)$$

If the $h_m(t_1, t_2, \ldots, t_m) \in \mathbb{C}$, $m = 0, 1, \ldots, q$, are absolutely summable

$$\sum_{t_1, t_2, \ldots, t_m = 0}^{\infty} |h_m(t_1, t_2, \ldots, t_m)| \leq C_m < \infty \quad (16\text{-}41)$$

and $q < \infty$, then $y(t)$ is mixing of order P.

Proof. See Appendix 16.C. □

Lemma 16.8 says that mixing noise of order qP passing through a time-invariant nonlinear system that can be described by a Volterra functional of finite degree q with absolutely summable impulse responses is mixing of order P. Extension of Lemma 16.8 to the multivariable systems is direct. As a special case of the multivariable result, we have the following lemma.

Lemma 16.9: Let $x_k(t) \in \mathbb{C}$, $k = 1, 2, ..., q$, be jointly mixing (over t) of order qP. The product $y(t) = \prod_{k=1}^{q} x_k(t)$ is mixing of order P.

Special cases of Lemma 16.9 are $x^q(t)$, $x^r(t)\bar{x}^s(t)$ with $r + s = q$, and more generally, the product $\prod_{k=1}^{q} z_k(t)$ of q delayed and/or complex conjugate factors $z_k(t) = x(t - t_k)$ and/or $z_k(t) = \bar{x}(t - t_k)$.

16.5 PRELIMINARY EXAMPLE

Consider, again, the resistance measurement problem of Section 1.2. Assume that M independent experiments are made of N current and voltage measurements each,

$$i^{[r]}(k) = i_0 + n_i^{[r]}(k) \text{ and } u^{[r]}(k) = u_0 + n_u^{[r]}(k) \tag{16-42}$$

$k = 1, 2, ..., N$, $r = 1, 2, ..., M$. Assume, furthermore, that $i_0 = 1$ A, $u_0 = 1$ V, and that $n_i^{[r]}(k)$, $n_u^{[r]}(k)$ are independent (over the measurements k and the experiments r) uniformly distributed random variables in the intervals $[-0.5 \text{ A}, 0.5 \text{ A}]$ and $[-0.5 \text{ V}, 0.5 \text{ V}]$, respectively. Invoke the simple approach (1-1), least squares (1-2), and errors-in-variables (1-3) estimators, proposed in Section 1.2, for each experiment $r = 1, 2, ..., M$, giving M independent realizations of the estimates

$$\hat{R}_{SA}^{[r]}(N) = \frac{1}{N} \sum_{k=1}^{N} \frac{u^{[r]}(k)}{i^{[r]}(k)} \tag{16-43}$$

$$\hat{R}_{LS}^{[r]}(N) = \frac{\frac{1}{N} \sum_{k=1}^{N} u^{[r]}(k) i^{[r]}(k)}{\frac{1}{N} \sum_{k=1}^{N} (i^{[r]}(k))^2} \tag{16-44}$$

$$\hat{R}_{EV}^{[r]}(N) = \frac{\frac{1}{N} \sum_{k=1}^{N} u^{[r]}(k)}{\frac{1}{N} \sum_{k=1}^{N} i^{[r]}(k)} \tag{16-45}$$

Figure 16-1, top left plot, shows the evolution of errors-in-variables resistance estimate (16-45) as a function of the number of measurements N for the first experiment ($r = 1$). The basic question that arises now is: "Does the estimate converge along this particular realization?" That is,

$$\lim_{N \to \infty} \hat{R}_{EV}^{[1]}(N) = R_{EV}^{[1]} \tag{16-46}$$

Scrutiny of the errors-in-variables estimates of the first five experiments (see Figure 16-1, top right plot) may cause one to wonder to which realizations the estimates converge (also called *pointwise convergence*)

$$\lim_{N \to \infty} \hat{R}_{EV}^{[r]}(N) = R_{EV}^{[r]} \quad r = 1, 2, ..., \tag{16-47}$$

Section 16.5 ■ Preliminary Example

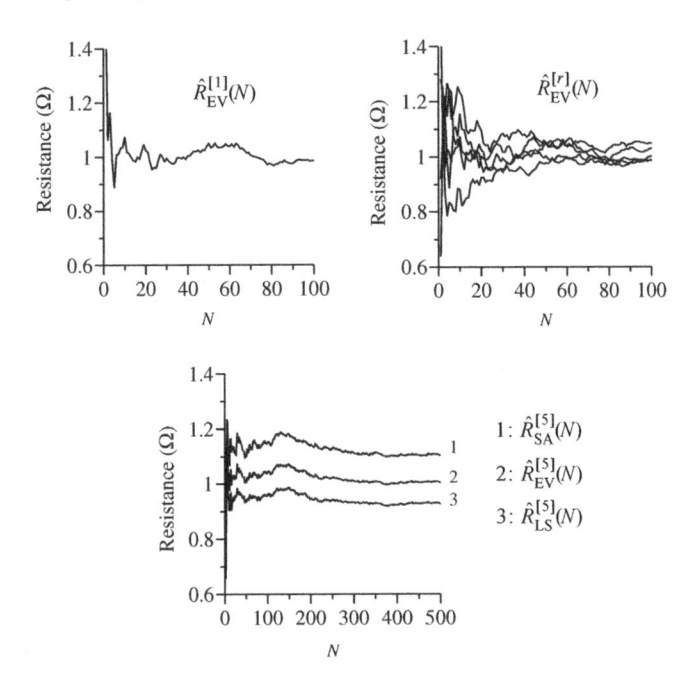

Figure 16-1. Measurement of a resistance. Top left: errors-in-variables estimates of the first experiment; top right: errors-in-variables estimates of the first five experiments ($r = 1, 2, ..., 5$); bottom: simple approach, $\hat{R}_{EV}^{[1]}(N)$ errors-in-variables and least squares estimate of the fifth experiment.

whether they converge to the same value,

$$R_{EV}^{[r]} = R_{EV} \qquad r = 1, 2, ... \tag{16-48}$$

and whether this value equals the true value R_0

$$R_{EV} = R_0 \tag{16-49}$$

If (16-47) is true for "almost all realizations," then we have *stochastic convergence* to a random number R_{EV} and it makes sense to write

$$\text{"lim"} \hat{R}_{EV}(N) = R_{EV} \tag{16-50}$$
$$N \to \infty$$

If (16-48) is true for "almost all realizations," then R_{EV} in (16-50) is a deterministic (nonrandom) number. Several definitions of "lim" can be given according to what is meant by "almost all realizations." The precise definitions of the stochastic limits, their properties, and their interrelations will be discussed later on in this chapter. If (16-49) is true, then the resistance estimates are called *consistent*, which means that they converge to the true value as the number of processed measurements N tends to infinity.

Figure 16-1, bottom plot, shows the simple approach (16-43), least squares (16-44), and errors-in-variables (16-45) estimates of the fifth experiment ($r = 5$). Clearly, the simple approach and least squares estimates seem to converge to a value that deviates significantly from the true resistance value of 1 Ω. Referring also to Figure 1-4 on page 5, we may ask the following questions: "Is the mean value of the estimates asymptotically different from the true value?", "Is the uncertainty of the estimates (asymptotically) minimal?", and "What is the asymptotic distribution of the estimates?" The first question is strongly related to the consistency problem but is not completely equivalent (see Section 16.14). The second question is handled in Section 16.12, where it is shown that the uncertainty is bounded below by the Cramér-Rao bound. The central limit theorems of Section 16.10 will be helpful to answer the third question. The knowledge of the asymptotic distribution makes it possible to calculate confidence intervals on the estimates.

16.6 DEFINITIONS OF STOCHASTIC LIMITS

Let $x(N)$, $N = 1, 2, \ldots$ be a scalar random sequence. There are several ways in which the sequence might converge to a (random) number x as $N \to \infty$. We will define four modes of stochastic convergence.

1. The sequence $x(N)$, $N = 1, 2, \ldots$ converges to x *in mean square* if, $\mathbb{E}\{|x|^2\} < \infty$, $\mathbb{E}\{|x(N)|^2\} < \infty$ for all N, and $\lim_{N \to \infty} \mathbb{E}\{|x(N) - x|^2\} = 0$. We write

$$\text{l.i.m.}_{N \to \infty} x(N) = x \Leftrightarrow \lim_{N \to \infty} \mathbb{E}\{|x(N) - x|^2\} = 0 \tag{16-51}$$

2. The sequence $x(N)$, $N = 1, 2, \ldots$ converges to x *with probability 1* (w.p. 1) or *almost surely* if, $\lim_{N \to \infty} x^{[\omega]}(N) = x^{[\omega]}$ for almost all realizations ω, except those $\omega \in \mathbb{A}$ such that $\text{Prob}(\mathbb{A}) = 0$. We write

$$\text{a.s.lim}_{N \to \infty} x(N) = x \Leftrightarrow \text{Prob}(\lim_{N \to \infty} x(N) = x) = 1 \tag{16-52}$$

This definition is equivalent to (Theorem 2.1.2 of Lukacs, 1975)

$$\text{a.s.lim}_{N \to \infty} x(N) = x \Leftrightarrow \forall \varepsilon > 0: \lim_{N \to \infty} \text{Prob}(\sup_{k \geq N}|x(k) - x| \leq \varepsilon) = 1 \tag{16-53}$$

3. The sequence $x(N)$, $N = 1, 2, \ldots$ converges to x *in probability* if, for every $\varepsilon, \delta > 0$ there exists an N_0 such that for every $N > N_0$: $\text{Prob}(|x(N) - x| \leq \varepsilon) > 1 - \delta$. We write

$$\text{plim}_{N \to \infty} x(N) = x \Leftrightarrow \forall \varepsilon > 0: \lim_{N \to \infty} \text{Prob}(|x(N) - x| \leq \varepsilon) = 1 \tag{16-54}$$

4. Let $F_N(x)$ and $F(x)$ be the distribution functions of, respectively, $x(N)$ and x. The sequence $x(N)$, $N = 1, 2, \ldots$ converges to x *in law* or *in distribution* if $F_N(x)$ converges weakly[1] to $F(x)$. We write

$$\text{Lim}_{N \to \infty} x(N) = x \Leftrightarrow \text{Lim}_{N \to \infty} F_N(x) = F(x) \tag{16-55}$$

---

1. This means at all continuity points of the limiting function and is denoted by "Lim."

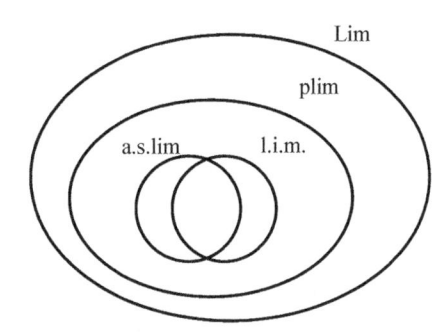

Figure 16-2. Convergence area of the stochastic limits.

16.7 INTERRELATIONS BETWEEN STOCHASTIC LIMITS

In the previous section we defined several modes of stochastic convergence. The connections between these concepts are

1. Almost sure convergence implies convergence in probability; the converse is not true (Theorem 2.2.1 of Lukacs, 1975; see also Appendix 16.D).
2. Convergence in mean square implies convergence in probability; the converse is not true (Theorem 2.2.2 of Lukacs, 1975; see also Appendix 16.E).
3. Convergence in probability implies convergence in law; the converse is not true (Theorem 2.2.3 of Lukacs, 1975).
4. There is no implication between almost sure and mean square convergence.
5. A sequence $x(N)$ converges in probability to x if and only if every subsequence $x(N_k)$ contains a sub-subsequence $x(N_{k_i})$ that converges ($i \to \infty$) almost surely to x (Theorem 2.4.4 of Lukacs, 1975).
6. A sequence converges in probability to a constant if and only if it converges in law to a degenerate distribution[1] (Corollary to Theorem 2.2.3 of Lukacs, 1975).

A graphical representation of the convergence area of the different stochastic limits is given in Figure 16-2. The interrelations between the concepts are summarized in Figure 16-3. As these allow a better understanding of the stochastic limits, some proofs are given in the appendixes. The importance of interrelation 5 is that any theorem proved for the almost sure limit is also valid for the limit in probability. Before illustrating some of the interrelations by (counter) examples, we cite the Borel-Cantelli and the Fréchet-Shohat lemmas, which are useful to establish, respectively, convergence w.p. 1 and convergence in distribution. The Borel-Cantelli lemma roughly says that if the convergence in probability or in mean square is sufficiently fast, this implies convergence with probability 1.

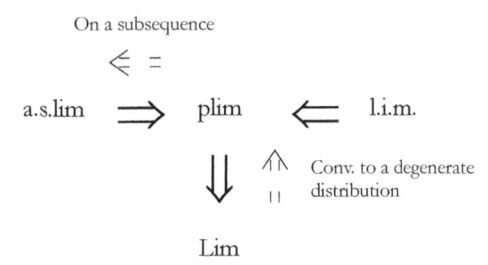

Figure 16-3. Interrelations between the stochastic limits.

---

1. $F(x)$ is degenerate if there exists an x_0 such that $F(x) = 0$ for $x < x_0$ and $F(x) = 1$ for $x \geq x_0$.

Lemma 16.10 (Borel-Cantelli Lemma): If

$$\sum_{N=1}^{\infty} \text{Prob}(|x(N) - x| > \varepsilon) < \infty \text{ or } \sum_{N=1}^{\infty} \mathbb{E}\{|x(N) - x|^2\} < \infty \qquad (16\text{-}56)$$

then $x(N)$ converges to x w.p. 1.

Proof. Theorems 2.1.1 and 2.1.3 of Stout (1974); see also Appendix 16.F. □

Lemma 16.11 (Fréchet-Shohat Lemma): Let x have a distribution function $F(x)$ that is uniquely determined by its moments (cumulants). If the moments (cumulants) of the sequence $x(N)$ converge for $N \to \infty$ to the moments (cumulants) of x, then $x(N)$ converges in distribution to x.

Proof. Theorem 1, Section 8.2 of Chow and Teicher (1988). □

Example 16.12: Convergence w.p. 1 and convergence in probability do not imply convergence in mean square (Example 2.1.1 of Stout). Take ω to be uniform in $[0, 1]$, and build the sequence $x(N)$ such that

$$x^{[\omega]}(N) = \begin{cases} N & \omega \in [0, 1/N) \\ 0 & \omega \in [1/N, 1] \end{cases} \qquad (16\text{-}57)$$

Two realizations of the sequence are, for example,

$$\{x^{[0.3]}(N)\} = \{1, 2, 3, 0, 0, 0, 0, 0, \ldots\}$$
$$\{x^{[0.15]}(N)\} = \{1, 2, 3, 4, 5, 6, 0, 0, \ldots\}$$

We see that $x^{[\omega]}(N)$ is zero for N sufficiently large, which suggests that it will converge to zero. Formally, $\underset{N \to \infty}{\text{plim}}\, x(N) = \underset{N \to \infty}{\text{a.s.lim}}\, x(N) = 0$ since

$$\text{Prob}(\sup_{k \geq N}|x(k)| \leq \varepsilon) = \text{Prob}(|x(N)| \leq \varepsilon) = \text{Prob}(x(N) = 0) = 1 - 1/N$$

is arbitrarily close to 1 for N sufficiently large. There is just one sequence, $x^{[0]}(N)$, that does not converge. This is not in contradiction with the previous results because the probability of getting this particular realization is zero: $\text{Prob}(\omega = 0) = 0$. The mean square limit $\underset{N \to \infty}{\text{l.i.m.}}\, x(N)$ does not exist because $\mathbb{E}\{x^2(N)\} = N$ is unbounded. Note that the Borel-Cantelli lemma cannot be used in this example to establish the almost sure convergence from the convergence in probability. Indeed, $\sum_{N=1}^{\infty} \text{Prob}(|x(N)| > \varepsilon) = \sum_{N=1}^{\infty} 1/N = \infty$. □

Example 16.13: Convergence in probability and convergence in mean square do not imply convergence w.p. 1 (Example 2.1.2 of Stout, 1974). Take ω to be uniform in $[0, 1)$, and build the sequence $T(n, k)$ such that

$$T^{[\omega]}(n, k) = \begin{cases} 1 & \omega \in [(k-1)/n, k/n) \\ 0 & \text{elsewhere} \end{cases}$$

Section 16.7 ■ Interrelations between Stochastic Limits

for $k = 1, 2, ..., n$ and $n \geq 1$. Let

$$\{x(N)\} = \{\{T(1, k)\}, \{T(2, k)\}, \{T(3, k)\}, ...\}$$

with $\{T(n, k)\} = \{T(n, 1), T(n, 2), ..., T(n, n)\}$ and $N = n(n-1)/2 + k$. Two realizations of the sequence are, for example,

$$\{x^{[0.27]}(N)\} = \{\{1\}, \{1, 0\}, \{1, 0, 0\}, \{0, 1, 0, 0\}, ...\}$$
$$\{x^{[0.85]}(N)\} = \{\{1\}, \{0, 1\}, \{0, 0, 1\}, \{0, 0, 0, 1\}, ...\}$$

We see that the length of each subsequence $\{T(n, k)\}$ of $\{x(N)\}$ increases with n and that it contains exactly one nonzero term. This suggests that $x(N)$ will converge in probability (the probability to get a 1 goes to zero), but not w.p. 1 (the supremum is 1 for any value of N). Formally, $\underset{N \to \infty}{\text{plim}}\, x(N) = 0$ since

$$\lim_{N \to \infty} \text{Prob}(|x(N)| \leq \varepsilon) = \lim_{N \to \infty} \text{Prob}(T(n, k) = 0) = \lim_{N \to \infty} (1 - 1/n) = 1$$

and $\underset{N \to \infty}{\text{l.i.m.}}\, x(N) = 0$ because

$$\lim_{N \to \infty} \mathbb{E}\{x^2(N)\} = \lim_{N \to \infty} \mathbb{E}\{T^2(n, k)\} = \lim_{N \to \infty} 1/n = 0$$

The almost sure limit $\underset{N \to \infty}{\text{a.s.lim}}\, x(N)$ does not exist since $\text{Prob}(\underset{r \geq N}{\text{sup}}|x(r)| > \varepsilon) = 1$. Note that the subsequence $T(n, k)$, with k fixed and $n \geq 1$, converges with probability one to zero. This is an illustration of interrelation 5. □

Example 16.14: Convergence in mean square and convergence w.p. 1 are compatible (Example 2.2.3 of Lukacs, 1975). Let $x(N)$ be a random variable that assumes only the values $1/N$ and $-1/N$ with equal probability. We find $\underset{N \to \infty}{\text{l.i.m.}}\, x(N) = 0$ since

$$\lim_{N \to \infty} \mathbb{E}\{x^2(N)\} = \lim_{N \to \infty} 1/N^2 = 0$$

Also $\underset{N \to \infty}{\text{a.s.lim}}\, x(N) = 0$ because $|x(k)| < |x(N)|$ for any $k > N$ so that

$$\text{Prob}(\sup_{k \geq N}|x(k)| \leq \varepsilon) = \text{Prob}(|x(N)| \leq \varepsilon)|_{N > 1/\varepsilon} = 1 \qquad \square$$

Example 16.15: Convergence in distribution does not imply convergence in probability (Example 2.2.4 of Lukacs, 1975). Let x be a random variable that can take only the values 0 and 1 with equal probability. Next, construct the sequence $x(N) = 1 - x$. We have $\underset{N \to \infty}{\text{Lim}}\, x(N) = x$ because $x(N)$ and x have the same distribution functions $F_N(x) = F(x)$. However, the limit in probability $\underset{N \to \infty}{\text{plim}}\, x(N)$ does not exist because $|x(N) - x| = 1$ so that $\text{Prob}(|x(N) - x| \leq \varepsilon) = 0$. □

16.8 PROPERTIES OF STOCHASTIC LIMITS

The properties of the stochastic limits are similar to those of the classical (deterministic) limit, but there are some subtle differences. The general properties are

1. A continuous function and the almost sure limit may be interchanged

$$\text{a.s.}\lim_{N \to \infty} f(x(N)) = f(x) \text{ with } x = \text{a.s.}\lim_{N \to \infty} x(N) \tag{16-58}$$

2. The almost sure limit and the expected value may be interchanged for uniformly bounded sequences (Theorem 5.4 of Billingsley, 1995)

$$\lim_{N \to \infty} \mathbb{E}\{x(N)\} = \mathbb{E}\{\text{a.s.}\lim_{N \to \infty} x(N)\} \tag{16-59}$$

A direct consequence of (16-59) is that

$$\mathbb{E}\{O_{\text{a.s.}}(N^{-k})\} = O(N^{-k}) \tag{16-60}$$

3. A continuous function and the limit in probability may be interchanged (Theorem 2.3.3 of Lukacs, 1975)

$$\text{plim}_{N \to \infty} f(x(N)) = f(x) \text{ with } x = \text{plim}_{N \to \infty} x(N) \tag{16-61}$$

4. The limit in probability and the expected value may be interchanged for uniformly bounded sequences (Theorem 5.4 of Billingsley, 1995)

$$\lim_{N \to \infty} \mathbb{E}\{x(N)\} = \mathbb{E}\{\text{plim}_{N \to \infty} x(N)\} \tag{16-62}$$

A direct consequence of (16-62) is that

$$\mathbb{E}\{O_p(N^{-k})\} = O(N^{-k}) \tag{16-63}$$

5. The mean square limit is linear (Theorem 3.1 of Jazwinski, 1970)

$$\text{l.i.m.}_{N \to \infty} (ax(N) + by(N)) = a \, \text{l.i.m.}_{N \to \infty} x(N) + b \, \text{l.i.m.}_{N \to \infty} y(N) \tag{16-64}$$

where a and b are deterministic (nonrandom) numbers.

6. The mean square limit and the expected value may be interchanged (Theorem 3.1 of Jazwinski, 1970),

$$\lim_{N \to \infty} \mathbb{E}\{x(N)\} = \mathbb{E}\{\text{l.i.m.}_{N \to \infty} x(N)\} \tag{16-65}$$

A direct consequence of (16-65) is that

$$\mathbb{E}\{O_{m.s.}(N^{-k})\} = O(N^{-k}) \qquad (16\text{-}66)$$

7. If l.i.m.$_{N \to \infty}$ $x(N) = x$ and $\mathbb{E}\{(x(N) - x)^2\} = O(N^{-k})$, with $k > 0$, then

$$x(N) = x + O_{m.s.}(N^{-k/2}) \text{ and } x(N) = x + O_p(N^{-k/2}) \qquad (16\text{-}67)$$

This is a direct consequence of (16-66) and interrelation 2, Section 16.7.

8. If the sequence $x(n)$ is deterministic (nonrandom), then the limit in mean square, the limit w.p. 1, and the limit in probability reduce to the deterministic limits.

Property 1 follows directly from the definition (16-52) of convergence w.p. 1, while property 3 follows from interrelation 5, Section 16.7, and property 1. Properties 1 and 3 require the continuity of the function at ALL values of the limit random variable x. If x is a constant (nonrandom), then continuity in a closed neighborhood of x is sufficient. Note that the limit in mean square and a continuous function may, in general, NOT be interchanged. Note also that the almost sure limit and the limit in probability, in general, do NOT commute with the expected value.

16.9 LAWS OF LARGE NUMBERS

The classical laws of large numbers are used to study the stochastic convergence of the partial sum $S(N) = \sum_{k=1}^{N} x(k)$ of a random sequence $x(k)$, with $x(k)$ independent of N. They state roughly that $S(N)$ converges to its expected value if the span of dependence of $x(k)$ is limited. In this book, we often need the more general case where the sequence $x(k)$ in the partial sum $S(N)$ depends on the number of samples N: $S(N) = \sum_{k=1}^{N} x_N(k)$. According to the stochastic limit used to establish the convergence, we speak about the *weak law of large numbers*,

$$\operatorname*{plim}_{N \to \infty} (S(N) - \mathbb{E}\{S(N)\})/N = 0 \qquad (16\text{-}68)$$

the *strong law of large numbers*,

$$\operatorname*{a.s.lim}_{N \to \infty} (S(N) - \mathbb{E}\{S(N)\})/N = 0 \qquad (16\text{-}69)$$

and the *law of large numbers*

$$\operatorname*{l.i.m.}_{N \to \infty} (S(N) - \mathbb{E}\{S(N)\})/N = 0 \qquad (16\text{-}70)$$

Note that the (strong) laws of large numbers (16-69) and (16-70) imply the weak law of large numbers (16-68) (see Section 16.7, interrelations 1 and 2). The analysis of the rate at which $S(N)/N$ converges to its expected value requires some additional assumptions. For the strong law of large numbers (16-69), this rate is given by the *law of the iterated logarithm*

$$S(N)/N = \mathbb{E}\{S(N)\}/N + O_{\text{a.s.}}(N^{-1/2}\sqrt{\ln(\ln(N))}) \tag{16-71}$$

For the law of large numbers (16-70) we have, typically,

$$S(N)/N = \mathbb{E}\{S(N)\}/N + O_{\text{m.s.}}(N^{-1/2}) \tag{16-72}$$

Some interesting versions of the laws of large numbers and their respective convergence rates are listed next. More versions can be found in Chow and Teicher (1988), Lukacs (1975), and Stout (1974).

1. If $x(k)$ is *independent and identically distributed (iid)*, then (16-69) applies if and only if $\mathbb{E}\{x(k)\} = x < \infty$ (Theorem 4.3.3 of Lukacs, 1975). If in addition $\text{var}(x(k)) = \sigma^2 < \infty$, then the convergence rate of (16-69) is given by (16-71) (Theorem 3.2.9 of Stout, 1974).
2. If $x(k)$ is *independent*, then

    2a. (16-68) and (16-69) are equivalent (Theorem 2.13.2 of Stout, 1974).

    2b. (16-69) applies if $\text{var}(x(k)) \le M < \infty$ for any k (Corollary 1 to Theorem 4.3.1 of Lukacs, 1975). If, in addition, $\text{var}(S(N)) = O(N)$ and for some $\delta > 0$,

    $$\sum_{N=1}^{\infty} \text{Prob}(|x(N) - \mathbb{E}\{x(N)\}| > \sqrt{\delta N \ln(\ln(N))}) < \infty \tag{16-73}$$

    then the convergence rate of (16-69) is given by (16-71) (Corollary 3, Section 10.2 of Chow and Teicher, 1988)

3. If $x_N(k)$, $N = 1, 2, \ldots, \infty$, is *mixing of order 2* and depends on N, then the law of large numbers (16-70) applies, and its convergence rate is given by (16-72) (proof: see Appendix 16.G). For the strong law of large numbers, the variations of the sequence $x_N(k)$ w.r.t. N should, in addition, be "small enough": if $\text{var}(\sum_{k=1}^{s} x_r(k) - x_s(k)) = O(r-s)$, $r \ge s$, then (16-69) applies (proof: see Appendix 16.G).

The uniformly boundedness condition on the variances in versions 2b and 3 of the law of large numbers is necessary to avoid any increase in the variance of the sequence $x(k)$ to infinity. Otherwise, the uncertainty on the partial sum would not decrease to zero, making it impossible, in general, for $S(N)/N$ to converge to its expected value.

Because the almost sure limit imposes some restrictions on the supremum (see (16-53)), the convergence rate of the strong law of large numbers depends upon the tails of the probability density functions (pdf's) of the random variables $x(k)$. Condition (16-73) dictates that the tails of the pdf's tend sufficiently fast to zero. It is satisfied for uniformly bounded random variables.

Example 16.16: Let $x \in \mathbb{C}^N$, $H \in \mathbb{C}^{N \times N}$, and $y = Hx$, where x, H satisfy the assumptions of Corollary 16.7 with $P = 2$, and where $H_{[i,j]}$ is independent of N for any i, j. The partial sum $S(N) = \sum_{k=1}^{N} y_N(k)$, with $y_N(k) = y_{[k]}$, satisfies the strong law of large numbers (16-69). Indeed,

$$\text{var}(\sum_{k=1}^{s} y_r(k) - y_s(k)) = \text{var}(\sum_{k=1}^{s}\sum_{l=s+1}^{r} H_{[k,l]} x_{[l]})$$

$$= \sum_{l_1, l_2 = s+1}^{r} \text{cum}(x_{[l_1]}, \bar{x}_{[l_2]}) \sum_{k_1=1}^{s} H_{[k_1, l_1]} \sum_{k_2=1}^{s} \bar{H}_{[k_2, l_2]}$$

$$\leq \left(\max_{l} \sum_{k=1}^{s} |H_{[k,l]}| \right)^2 \sum_{l_1, l_2 = s+1}^{r} |\text{cum}(x_{[l_1]}, \bar{x}_{[l_2]})|$$

$$\leq O(r-s)$$

The last inequality is due to the finite 1-norm of H and the second-order mixing property of $x_{[k]}$ (16-36). □

16.10 CENTRAL LIMIT THEOREMS

The classical central limit theorems pertain to the asymptotic distribution function of the partial sum $S(N) = \sum_{k=1}^{N} x(k)$ of a random sequence $x(k)$. They state, roughly, that $S(N)$ is asymptotically normally distributed

$$\lim_{N \to \infty} \frac{S(N) - \mathbb{E}\{S(N)\}}{\sqrt{\text{var}(S(N))}} \in N(0, 1) \tag{16-74}$$

if each $x(k)$ has high probability to be of the same order of magnitude and if the span of dependence of $x(k)$ is limited. Under some additional assumptions, the rate at which the distribution function of $S(N)$ converges to a normal distribution can be established. It is given by the Berry and Esseen theorem

$$\sup_{y} |F_N(y) - \Phi(y)| \leq O(N^{-1/2}) \tag{16-75}$$

with $F_N(y)$ the distribution function of $(S(N) - \mathbb{E}\{S(N)\})/\sqrt{\text{var}(S(N))}$ and $\Phi(y)$ the standard normal distribution function. In this book we often need the more general case where the sequence $x(k)$ in the partial sum $S(N)$ depends on the number of samples N: $S(N) = \sum_{k=1}^{N} x_N(k)$. Some interesting versions of the central limit theorem are listed next. More versions can be found in Billingsley (1995) and Feller (1968).

1. If $x(k)$ is *independent and identically distributed* with finite mean $\mu < \infty$ and finite nonzero variance $0 < \sigma^2 < \infty$, then (16-74) applies with $\mathbb{E}\{S(N)\} = N\mu$ and $\text{var}(S(N)) = N\sigma^2$ (Theorem 27.1 of Billingsley, 1995). If, in addition, $\mathbb{E}\{|x(k)|^3\} < \infty$, then the convergence rate of (16-74) is given by (16-75) (Theorem 9.1.3 of Chow and Teicher, 1988).

2. $x(k)$ is *independent*, with finite means $\mu_k \leq c_1 < \infty$ and finite variances $\sigma_k^2 \leq c_2 < \infty$. If for some $\varepsilon > 0$ $x(k)$ has uniformly bounded $2 + \varepsilon$ moments $\mathbb{E}\{|x(k)|^{2+\varepsilon}\} \leq C < \infty$ and if $N/(\sqrt{\text{var}(S(N))})^{2+\varepsilon} = o(N^0)$, then (16-74) applies with $\mathbb{E}\{S(N)\} = \sum_{k=1}^{N} \mu_k$ and $\text{var}(S(N)) = \sum_{k=1}^{N} \sigma_k^2$. (Theorem 27.3 of Billingsley, 1995). If, in addition, $\mathbb{E}\{|x(k)|^3\} \leq c_3 < \infty$, then the convergence rate of (16-74) is given by (16-75) (Theorem 9.1.3 of Chow and Teicher, 1988).

3. If $x(k)$ is m-*dependent*, then (16-74) is valid under the same conditions of version 2 of the central limit theorem (Orey, 1958; Rosén, 1967).
4. If $x_N(k)$, $N = 1, 2, ..., \infty$, is *mixing of order infinity* and if $\text{var}(S(N)) = O(N)$, then (16-74) and (16-75) apply (proof: see Appendix 16.H for the nonstationary case; see Theorem 4.4.1 of Brillinger (1981) for the stationary case).

The conditions $N/(\sqrt{\text{var}(S(N))})^{2+\varepsilon} = o(N^0)$ and $\text{var}(S(N)) = O(N)$ in, respectively, versions 2 and 3 and version 4 of the central limit theorem are necessary to avoid dominance of a few random variables over the partial sum $S(N)$. If not, the distribution function of $S(N)$ would be determined by the distribution functions of those few dominating random variables and would in general not be normal. Extension of these theorems to the complex and to the (complex) multivariate case is straightforward.

The central limit theorem should be interpreted with some care. The following example will illustrate this.

Example 16.17: Suppose that we take N independent samples of a uniformly distributed random variable $x(k) \in U(0, \sigma^2)$. According to version 1 of the central limit theorem (16-74), the mean value will be asymptotically normally distributed. Figure 16-4 compares the true probability density function of $S(N)/N$ (solid line) with the Gaussian pdf predicted by the central limit theorem (dashed line) for the case $\sigma = 5/\sqrt{3}$. It follows that even for small values of N the Gaussian approximation is remarkably within the interval $[-5, 5]$. Although the mean $S(N)/N$ cannot take values outside the interval $[-\sqrt{3}\sigma, \sqrt{3}\sigma]$, the central limit theorem predicts that this will happen with some (small) probability. Similarly, saying that the weight of newborn babies is normally distributed does not imply that there is a small risk of getting babies with a negative weight! We conclude that the central limit theorem describes, very well, the behavior of the distribution function around its mean value but not at its tails. □

16.11 PROPERTIES OF ESTIMATORS

What kind of properties do we expect from a "good" estimator? It would be nice that the estimate $\hat{\theta}(N)$ converges to the true value θ_0 as the number of noisy measurements N tends to infinity. We could also require that the expected value of $\hat{\theta}(N)$ equals the true value or that this is at least asymptotically ($N \to \infty$) valid. "Does the estimator have the smallest possible (asymptotic) mean square error?" and "Is its (asymptotic) distribution function known?" are also important issues. Besides, we would also like that most of (all) these properties remain valid if we do not satisfy some of (all) the basic assumptions made in constructing the estimator. The formal definitions are listed next.

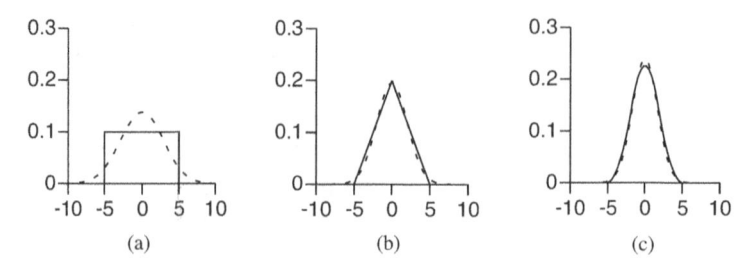

Figure 16-4. Comparison of the true pdf (solid line) and the Gaussian pdf predicted by the central limit theorem (dashed line) of $S(N)/N$ for zero mean, independent uniformly distributed random variables $x(k)$ with $\sigma = 5/\sqrt{3}$: (a) $N = 1$, (b) $N = 2$, and (c) $N = 3$.

Section 16.11 ■ Properties of Estimators

1. An estimator $\hat{\theta}(N)$ is *consistent* if it converges to the true value θ_0 as $N \to \infty$. According to the stochastic limit used, we say that $\hat{\theta}(N)$ is *weakly consistent* if

$$\operatorname*{plim}_{N \to \infty} \hat{\theta}(N) = \theta_0 \tag{16-76}$$

   strongly consistent if

$$\operatorname*{a.s.lim}_{N \to \infty} \hat{\theta}(N) = \theta_0 \tag{16-77}$$

   and *consistent* if

$$\operatorname*{l.i.m.}_{N \to \infty} \hat{\theta}(N) = \theta_0 \tag{16-78}$$

2. An estimator $\hat{\theta}(N)$ is *unbiased* if

$$\mathbb{E}\{\hat{\theta}(N)\} = \theta_0 \tag{16-79}$$

   It is *asymptotically unbiased* if (16-79) is valid for $N \to \infty$.

3. An estimator $\hat{\theta}(N)$ is *(statistically) efficient* if, for all θ_0-values, the mean square error matrix of any other estimator $\hat{\psi}(N)$ is not smaller than that of $\hat{\theta}(N)$

$$\mathrm{MSE}(\hat{\psi}(N)) \geq \mathrm{MSE}(\hat{\theta}(N)) \tag{16-80}$$

   It is *asymptotically efficient* if (16-80) is valid for $N \to \infty$. For unbiased estimators (16-80) becomes

$$\mathrm{Cov}(\hat{\psi}(N)) \geq \mathrm{Cov}(\hat{\theta}(N)) \tag{16-81}$$

4. The estimator $\hat{\theta}(N)$ is *(asymptotically) normally distributed*

$$\hat{\theta}(N) - \tilde{\theta}(N) \to \delta_\theta(N) \in N^{n_\theta}(0, \mathrm{Cov}(\delta_\theta(N))) \tag{16-82}$$

   with $\tilde{\theta}(N)$ a deterministic parameter vector depending on N.

5. An estimator $\hat{\theta}(N)$ is *robust* if one or more of the preceding properties remain unchanged when one or more of the basic assumptions made to construct the estimator are violated.

There is a fundamental difference between asymptotic unbiasedness and (weak or strong) consistency. Indeed, to be asymptotically unbiased, it is, for example, sufficient (but not necessary!) that the asymptotic probability density function $f_{\hat{\theta}}(\hat{\theta})$ of the estimate $\hat{\theta}$ satisfies $f_{\hat{\theta}}(\hat{\theta} - \theta_0) = f_{\hat{\theta}}(\theta_0 - \hat{\theta})$ (see Figure 16-5(a)), while (weak or strong) consistency requires that the asymptotic probability density function is a Dirac function (see Figure 16-5(b)).

The property that a continuous function may be interchanged with the almost sure limit and the limit in probability (see Section 16.8) explains why weak and strong consistency of

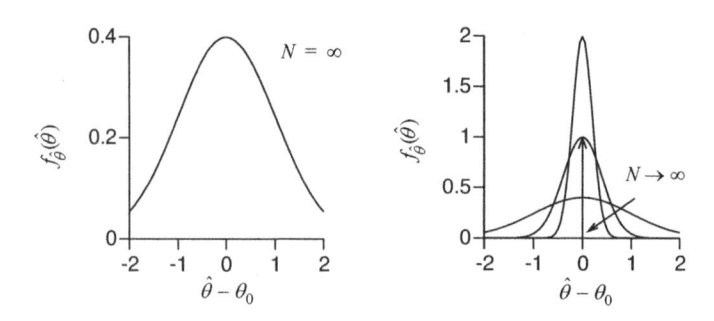

Figure 16-5. Asymptotic pdf of $\hat{\theta}$: (a) asymptotically unbiased estimator, (b) (weakly or strongly) consistent estimator for $N \to \infty$ (the limit pdf is a Dirac function).

the estimates are mostly proved. This is not the case for the limit in mean square, which is often used as an intermediate step in the consistency proofs (see Section 16.13). Note, however, that consistency (16-78) implies asymptotic unbiasedness (see Section 16.8, property 6), which is not the case for weak and strong consistency (see Section 16.14).

The practical importance of the efficiency property is that it makes no sense to look for estimators with a lower mean square error matrix. Although (asymptotic) efficiency is a highly desirable property, inefficient estimators with an acceptable accuracy may sometimes be the best practicable candidates (for example, if calculation time is important).

The existence of bias in the estimates is often the reason of the increased mean square error matrix compared with that of the unbiased estimates. However, simple examples of biased minimum mean square estimators exist that are statistically more efficient (have smaller MSE) than any other unbiased estimator (Kendall and Stuart, 1979; Stoica and Moses, 1990; see also Example 16.19 and Exercise 16.24). Minimum mean square error estimators have the following three drawbacks. First, they often require knowledge of the true (unknown) parameter values and, therefore, are not realizable (Kendall and Stuart, 1979; Norton, 1986). Next, if the estimation results are averaged in a second step, then the mean square error can only be reduced to the square of the bias as the number of averages tends to infinity (it can be reduced to zero for unbiased estimates). Finally, minimum mean square estimators are not robust w.r.t. the assumed underlying distribution function of the measurements. This explains why (asymptotically) unbiased estimators are usually preferred over minimum mean square estimators.

One should be very careful when interpreting the asymptotic normality property (16-82). It says that the difference $\hat{\theta}(N) - \tilde{\theta}(N)$, with $\tilde{\theta}(N)$ deterministic, converges in distribution to a zero mean random variable $\delta_\theta(N)$ that is (asymptotically) normally distributed. This does NOT imply the existence of the moments (expected value, variance, ...) of $\hat{\theta}(N)$ for any finite value of N. However, the asymptotic normality property makes it possible to calculate uncertainty bounds and confidence levels.

16.12 CRAMÉR-RAO LOWER BOUND

Consider the identification of the parameter vector $\theta \in \mathbb{R}^{n_\theta}$ using noisy measurements $z \in \mathbb{R}^N$. The quality of the estimator $\hat{\theta}(z)$ can be represented by its mean square error matrix

$$\text{MSE}(\hat{\theta}(z)) = \text{Cov}(\hat{\theta}(z)) + b_\theta b_\theta^T \qquad (16\text{-}83)$$

where θ_0 and b_θ denote, respectively, the true value and the bias on the estimates. We may wonder whether there exists a lower limit on the value of the mean square error (16-83) that can be obtained with various estimators. The answer is given by the *generalized Cramér-Rao lower bound*.

Theorem 16.18: Let $f_z(z, \theta_0)$ be the probability density function of the measurements $z \in \mathbb{R}^N$. Assume that $f_z(z, \theta_0)$ and its first- and second-order derivatives w.r.t. $\theta \in \mathbb{R}^{n_\theta}$ exist for all θ_0- values. Assume, furthermore, that the boundaries of the domain of $f_z(z, \theta_0)$ w.r.t. z are θ_0 independent. Then, the *generalized Cramér-Rao lower bound* on the mean square error of any estimator $\hat{G}(z)$ of the function $G(\theta) \in \mathbb{C}^r$ of θ is

$$\mathrm{MSE}(\hat{G}(\hat{\theta}(z))) \geq \left(\frac{\partial G(\theta_0)}{\partial \theta_0} + \frac{\partial b_G}{\partial \theta_0}\right) Fi^+(\theta_0) \left(\frac{\partial G(\theta_0)}{\partial \theta_0} + \frac{\partial b_G}{\partial \theta_0}\right)^H + b_G b_G^H \quad (16\text{-}84)$$

with $b_G = \mathbb{E}\{\hat{G}(z)\} - G(\theta_0)$ the bias that might be present in the estimate, and $Fi(\theta_0)$ the *Fisher information matrix* of the parameters θ_0

$$Fi(\theta_0) = \mathbb{E}\left\{\left(\frac{\partial \ln f_z(z, \theta_0)}{\partial \theta_0}\right)^T \left(\frac{\partial \ln f_z(z, \theta_0)}{\partial \theta_0}\right)\right\} = -\mathbb{E}\left\{\frac{\partial^2 \ln f_z(z, \theta_0)}{\partial \theta_0^2}\right\} \quad (16\text{-}85)$$

Equality holds in (16-84) if and only if there exists a nonrandom matrix Γ such that

$$\hat{G}(\hat{\theta}(z)) - \mathbb{E}\{\hat{G}(\hat{\theta}(z))\} = \Gamma \left(\frac{\partial \ln f_z(z, \theta_0)}{\partial \theta_0}\right)^T \quad (16\text{-}86)$$

The expectations in (16-84) and (16-85) are taken w.r.t. the measurements z.

Proof. See Appendix 16.J. □

Note that the calculation of the Cramér-Rao lower bound requires knowledge of the true parameters θ_0, which is often not available (except in simulations). An approximation can be calculated by replacing θ_0 by its estimated value $\hat{\theta}$ in (16-84). Two special cases of the Cramér-Rao inequality are worth mentioning.

If $G(\theta) = \theta$, $b_G = 0$, and $Fi(\theta_0)$ is regular, then we obtain the *Cramér-Rao lower bound for unbiased estimators* (abbreviated as UCRB)

$$\mathrm{Cov}(\hat{\theta}(z)) \geq Fi^{-1}(\theta_0) \quad (16\text{-}87)$$

If condition (16-86) is not satisfied, $\hat{\theta}(z) - \theta_0 \neq \Gamma(\partial \ln f_z(z, \theta_0)/\partial \theta_0)^T$, then the lower bound (16-87) is too conservative, and there may still be an unbiased estimator that has smaller variance than any other unbiased estimator. Better (larger) bounds exist when (16-87) is not attainable, but they are often (extremely) difficult to compute. An overview of tighter bounds can be found in Abel (1993).

If $G(\theta) = \theta$, $b_G \neq 0$, and $Fi(\theta_0)$ is regular, then we find the *Cramér-Rao lower bound on the mean square error of biased estimators* (abbreviated as CRB)

$$\text{MSE}(\hat{\theta}(z)) \geq \left(I_{n_\theta} + \frac{\partial b_\theta}{\partial \theta_0}\right) Fi^{-1}(\theta_0) \left(I_{n_\theta} + \frac{\partial b_\theta}{\partial \theta_0}\right)^T + b_\theta b_\theta^T \tag{16-88}$$

It follows that the Cramér-Rao lower bound for asymptotically unbiased estimators ($b_\theta \to 0$ as $N \to \infty$) is asymptotically given by (16-87) only if the derivative of the bias w.r.t. θ_0 is asymptotically zero. Likewise, in the unbiased case, the lower bound (16-88) may be too conservative and tighter bounds exist (Abel, 1993). Note that the first term in the right-hand side of (16-88) can be zero for biased estimators (see Example 16.20).

In general, it is impossible to show that the bias (and its derivative w.r.t. θ) of a weakly or strongly consistent estimator converges to zero as $N \to \infty$. However, the moments of the limiting random variable often exist. The (asymptotic) covariance matrix or mean square error of the limiting random variable is then compared with the UCRB. In this context, the concept of efficiency is also used for weakly or strongly consistent estimators.

Example 16.19: (Stoica and Moses, 1990) Let $z(k)$, $k = 1, 2, \ldots, N$ be zero mean iid Gaussian random variables, $z(k) \in N(0, \sigma^2)$. The sample variance $\hat{\sigma}_z^2 = \sum_{k=1}^{N} z^2(k)/N$ is an unbiased and efficient estimate of σ_z^2 (see Exercise 16.23). Now consider the estimator $\hat{s}^2 = a\hat{\sigma}_z^2$ where $a > 0$ is chosen to minimize the mean square error (16-7) of the estimate $\hat{s}^2$

$$\text{MSE}(\hat{s}^2) = a^2 \text{var}(\hat{\sigma}_z^2) + (a-1)^2 \sigma_z^4 \tag{16-89}$$

Minimizing (16-89) w.r.t. a gives $a = \sigma_z^4 / \mathbb{E}\{\hat{\sigma}_z^4\} = N/(N+2)$ with corresponding minimum mean square error

$$\min_a \text{MSE}(a) = 2\sigma_z^4/(N+2) \tag{16-90}$$

This should be compared with the UCRB

$$Fi^{-1}(\sigma_z^2) = \text{var}(\hat{\sigma}_z^2) = 2\sigma_z^4/N \tag{16-91}$$

which is clearly larger than the mean square error (16-90) of the biased estimate $\hat{\sigma}_z^2 N/(N+2)$. It can also be verified that $\hat{\sigma}_z^2 N/(N+2)$ is statistically efficient in the sense that its mean square error reaches the lower bound (16-88). We conclude that the lower bound on the mean square error matrix of biased estimators may be smaller than the lower bound on the covariance matrix of unbiased estimators. □

Example 16.20: Assume that we estimate the weight of a bread from N noisy measurements. The true weight of the bread is 800 g. Regardless of what we measure, we estimate the weight as 100 g. Clearly, the estimator is biased and has zero variance. This is not in contradiction with the lower bound (16-88). Indeed,

$$I_{n_\theta} + \partial b_\theta / \partial \theta = \partial \mathbb{E}\{\hat{\theta}(z)\} / \partial \theta_0 = 0$$

because the estimate $\hat{\theta}(z)$ of the weight is independent of the true value θ_0. □

16.13 HOW TO PROVE ASYMPTOTIC PROPERTIES OF ESTIMATORS?

Ideally, we would like to know everything about the finite sample behavior (N does not increase to infinity) of an estimator. In practice, however, we can prove only a few large sample ($N \to \infty$) properties and hope that an estimator with good asymptotic properties also behaves well for practical sample sizes. The goal of this section is to present the main ideas and techniques without going into the mathematical details. The exact technical conditions and assumptions can be found in Chapter 17. We distinguish two different situations: an explicit (analytic) expression for the estimates $\hat{\theta} \in \mathbb{R}^{n_\theta}$ as a function of the measurements $z \in \mathbb{R}^N$ is available,

$$\hat{\theta}(z) = f(z) \tag{16-92}$$

or $\hat{\theta}$ is implicitly known through the minimization of a cost function $V(\theta, z)$

$$\hat{\theta}(z) = \arg\min_\theta V(\theta, z) \tag{16-93}$$

The explicit case (16-92) is illustrated on the resistance measurement problem in Section 16.15, and the implicit case (16-93) is elaborated in Chapter 17 on cost functions that are quadratic-in-the-measurements. We assume that the measurements z are disturbed by additive noise n_z

$$z = z_0 + n_z \tag{16-94}$$

with z_0 the true unknown value.

The following tools are essential in the analysis of the asymptotic properties of an estimator: the law of large numbers (Section 16.9) for the convergence and the consistency, the convergence rate of the law of large numbers (Section 16.9) for the convergence rate of the estimates, the interchangeability of the stochastic limit and the expected value (Section 16.8) for the asymptotic bias, the central limit theorem (Section 16.10) or the Fréchet-Shohat lemma (Lemma 16.11) for the asymptotic normality, and in general the interchangeability of a continuous function and a stochastic limit (Section 16.8).

16.13.1 Convergence—Consistency

In both cases (16-92) and (16-93) the estimate $\hat{\theta}(z)$ converges to some nonrandom number $\tilde{\theta}(z_0)$ by averaging of the disturbing noise n_z. Therefore, we first locate in $f(z)$ and $V(\theta, z)$ the sums that average the measurements z, and next use one of the laws of large numbers of Section 16.9 to prove the convergence of the sums to their expected values. Further analysis is done with the limit in probability or the limit with probability one, as they have the nice property of being interchangeable with a continuous function (see Section 16.8). Putting the stochastic sums in the vector $w \in \mathbb{R}^p$, with p independent of N, we can write this down formally as

$$\begin{aligned} \hat{\theta}(z) &= f(z) = \tilde{f}(z_0, w(n_z, z_0)) & \text{(a)} \\ \hat{\theta}(z) &= \arg\min_\theta V(\theta, z) = \arg\min_\theta \tilde{V}(\theta, z_0, w(\theta, n_z, z_0)) & \text{(b)} \end{aligned} \quad (16\text{-}95)$$

The sums w converge for $N \to \infty$ in some sense (mean square, in probability, or w.p. 1) to their expected values

$$\begin{aligned} w(n_z, z_0) &\to \mathbb{E}\{w(n_z, z_0)\} = \mu_w(z_0) \\ w(\theta, n_z, z_0) &\to \mathbb{E}\{w(\theta, n_z, z_0)\} = \mu_w(\theta, z_0) \quad \text{uniformly in } \theta \end{aligned} \quad (16\text{-}96)$$

Note that the convergence of $w(\theta, n_z, z_0)$ must be uniform w.r.t. θ, otherwise $\mu_w(\theta, z_0)$ is not a continuous function of θ. The strong or weak convergence then follows directly from the interchangeability of a continuous function and the almost sure limit or the limit in probability

$$\begin{aligned} \hat{\theta}(z) &\to \tilde{\theta}(z_0) = \tilde{f}(z_0, \mu_w(z_0)) & \text{(a)} \\ \hat{\theta}(z) &\to \tilde{\theta}(z_0) = \arg\min_\theta \tilde{V}(\theta, z_0, \mu_w(\theta, z_0)) & \text{(b)} \end{aligned} \quad (16\text{-}97)$$

This is illustrated on the explicit case (16-97.a) using properties 1 and 8 of the almost sure limit (see Section 16.8),

$$\begin{aligned} \operatorname*{a.s.lim}_{N \to \infty} \hat{\theta}(z) &= \tilde{f}(\operatorname*{a.s.lim}_{N \to \infty} z_0, \operatorname*{a.s.lim}_{N \to \infty} w(n_z, z_0)) \\ &= \tilde{f}(\lim_{N \to \infty} z_0, \lim_{N \to \infty} \mu_w(z_0)) \\ &= \lim_{N \to \infty} \tilde{f}(z_0, \mu_w(z_0)) \\ &= \lim_{N \to \infty} \tilde{\theta}(z_0) \end{aligned} \quad (16\text{-}98)$$

The estimate $\hat{\theta}(z)$ is strongly or weakly consistent if the limit value

$$\theta_* = \lim_{N \to \infty} \tilde{\theta}(z_0) \quad (16\text{-}99)$$

equals the true value $\theta_* = \theta_0$. In most cases the stronger condition $\tilde{\theta}(z_0) = \theta_0$ is satisfied.

16.13.2 Convergence Rate

Suppose that we have estimated some parameters $\hat{\theta}$ using N data samples. We may wonder now how many additional samples we should measure in order to decrease the uncertainty on $\hat{\theta}$ by a factor of k. The answer is given by the convergence rate of the estimates. It, typically, obeys the so-called $\sqrt{N}$ law; for example, to decrease the uncertainty by a factor of 10 we need 100 times more data samples. The consistency property does not tell how quickly the estimates converge to the true value. An additional analysis is necessary to establish the convergence rate. It starts by analyzing how fast the variance of the stochastic sums w in (16-96) converges to zero. By properties 6 and 7 of Section 16.8, these convergence rates also apply for the limit in mean square and the limit in probability. For mixing sequences, this rate is at least $O_{m.s.}(N^{-1/2})$ (see Section 16.9, version 3 of the law of large numbers). The almost sure limit is not used in this context because it results in somewhat slower convergence rates (see Section 16.9). Hence, further analysis is done with the limit in probability as it is interchangeable with a continuous function. The main idea is to make a Taylor series expansion of $\hat{\theta}$ (16-95) as a function of the stochastic sums w. The implicit case (16-95.b) is somewhat involved because first an explicit expression of $\hat{\theta}$ as a function of w should be constructed. Therefore, the explicit case (16-95.a) is tackled first. To simplify the notations, in the sequel of the analysis we drop the dependence of w and μ_w on n_z and z_0.

16.13.2.1 Explicit Case.
The Taylor series expansion of the kth entry of $\tilde{f}(z_0, w)$ (16-95.a) w.r.t. w at the point $w = \mu_w$ gives

$$\tilde{f}_{[k]}(z_0, w) = \tilde{f}_{[k]}(z_0, \mu_w) + \left.\frac{\partial \tilde{f}_{[k]}(z_0, w)}{\partial w}\right|_{w = \mu_w} (w - \mu_w) + \\ \frac{1}{2}(w - \mu_w)^T \left.\frac{\partial^2 \tilde{f}_{[k]}(z_0, w)}{\partial w^2}\right|_{w = \widehat{w}} (w - \mu_w) \quad (16\text{-}100)$$

where $\widehat{w}$ is a point on the straight line connecting w to μ_w ($\widehat{w} = tw + (1-t)\mu_w$ with $t \in [0, 1]$). Suppose now that $\text{Cov}(w) = O(N^{-1})$ so that (property 7 of Section 16.8)

$$w = \mu_w + O_p(N^{-1/2}) \quad (16\text{-}101)$$

Using the definitions $\hat{\theta}(z) = \tilde{f}(z_0, w)$, $\tilde{\theta}(z_0) = \tilde{f}(z_0, \mu_w)$ and applying result (16-101) to (16-100), taking into account that the matrix dimensions of w, $\tilde{f}$, and the derivatives of $\tilde{f}$ w.r.t. w, are independent of N, gives

$$\hat{\theta}(z) = \tilde{\theta}(z_0) + \delta_\theta(z) + b_\theta(z) \quad (a)$$

$$\delta_\theta(z) = \left.\frac{\partial \tilde{f}_{[k]}(z_0, w)}{\partial w}\right|_{w = \mu_w} (w - \mu_w) = O_p(N^{-1/2}) \quad (b) \quad (16\text{-}102)$$

$$b_\theta(z) = O_p(N^{-1}) \quad (c)$$

From (16-102) it follows directly that the convergence rate of $\hat{\theta}(z)$ to the nonrandom value $\tilde{\theta}(z_0)$ is an $O_p(N^{-1/2})$.

16.13.2.2 Implicit Case.
Now we give an approximate analysis for the implicit case (16-93) (see Chapter 17 for the complete analysis). The implicit function that defines $\hat{\theta}(z)$ as a function of w is

$$\tilde{V}'(\hat{\theta}, z_0, w(\hat{\theta})) = 0 \quad (16\text{-}103)$$

where x' denotes the derivative of x w.r.t. θ. Taylor series expansion of (16-103) w.r.t. $\hat{\theta}$ at the point $\tilde{\theta}$ gives, neglecting the second and higher order terms,

$$\tilde{V}'^T(\hat{\theta}, z_0, w(\hat{\theta})) = \tilde{V}'^T(\tilde{\theta}, z_0, w(\tilde{\theta})) + \tilde{V}''(\tilde{\theta}, z_0, w(\tilde{\theta}))(\hat{\theta}(z) - \tilde{\theta}(z_0)) \qquad (16\text{-}104)$$

Because $\hat{\theta}(z)$ is the minimizing argument of $\tilde{V}(\hat{\theta}, z_0, w(\hat{\theta}))$ (16-104) reduces to

$$\hat{\theta}(z) - \tilde{\theta}(z_0) = -\tilde{V}''^{-1}(\tilde{\theta}, z_0, w(\tilde{\theta}))\tilde{V}'^T(\tilde{\theta}, z_0, w(\tilde{\theta})) \qquad (16\text{-}105)$$

Following the same lines as in the explicit case, the convergence rate of the first- and the second-order derivatives of the cost function in (16-105) is obtained. If (16-101) is valid, then

$$\begin{aligned}\tilde{V}'(\tilde{\theta}, z_0, w(\tilde{\theta})) &= \tilde{V}'(\tilde{\theta}, z_0, \mu_w(\tilde{\theta})) + O_p(N^{-1/2}) = O_p(N^{-1/2}) &\text{(a)}\\ \tilde{V}''(\tilde{\theta}, z_0, w(\tilde{\theta})) &= \tilde{V}''(\tilde{\theta}, z_0, \mu_w(\tilde{\theta})) + O_p(N^{-1/2}) &\text{(b)}\end{aligned} \qquad (16\text{-}106)$$

The last equality in (16-106a) is due to the fact that $\tilde{\theta}(z_0)$ is the minimizing argument of $\tilde{V}(\theta, z_0, \mu_w(\theta))$. Using the interchangeability of a continuous function and the limit in probability, (16-106b) becomes

$$\tilde{V}''^{-1}(\tilde{\theta}, z_0, w(\tilde{\theta})) = \tilde{V}''^{-1}(\tilde{\theta}, z_0, \mu_w(\tilde{\theta})) + O_p(N^{-1/2}) \qquad (16\text{-}107)$$

Collecting (16-105), (16-106), and (16-107) gives (16-102a) with

$$\begin{aligned}\delta_\theta(z) &= -\tilde{V}''^{-1}(\tilde{\theta}, z_0, \mu_w(\tilde{\theta}))\tilde{V}'^T(\tilde{\theta}, z_0, w(\tilde{\theta})) = O_p(N^{-1/2}) &\text{(a)}\\ b_\theta(z) &= O_p(N^{-1}) &\text{(b)}\end{aligned} \qquad (16\text{-}108)$$

Similarly to the explicit case, it follows from (16-108a) that the convergence rate of $\hat{\theta}(z)$ to $\tilde{\theta}(z_0)$ is an $O_p(N^{-1/2})$.

16.13.3 Asymptotic Bias

The asymptotic bias analysis is done for (weakly or strongly) consistent estimators. No explicit expression for the bias can be found except for some special examples. The best we can hope is to find how the bias behaves as $N \to \infty$. It is derived from the convergence rate analysis (16-102) and (16-108) by making the additional assumption that the disturbing noise is uniformly bounded and using the property that for such noise the expected value and the limit in probability may be interchanged (property 4 of Section 16.8). Taking the expected value of (16-102a) gives

$$\mathbb{E}\{\hat{\theta}(z)\} = \tilde{\theta}(z_0) + \mathbb{E}\{\delta_\theta(z)\} + O(N^{-1}) \qquad (16\text{-}109)$$

For the explicit case (16-102b) it is obvious that $\mathbb{E}\{\delta_\theta(z)\} = 0$, whereas this is true for the implicit case (16-108) only if

$$\mathbb{E}\{\tilde{V}'(\tilde{\theta}, z_0, w(\tilde{\theta}))\} = \frac{\partial \mathbb{E}\{\tilde{V}(\tilde{\theta}, z_0, w(\tilde{\theta}))\}}{\partial \tilde{\theta}} = 0 \qquad (16\text{-}110)$$

Condition (16-110) is often satisfied so that for both cases (16-109) can be written as

$$\mathbb{E}\{\hat{\theta}(z)\} = \tilde{\theta}(z_0) + O(N^{-1}) \tag{16-111}$$

For (weakly or strongly) consistent estimators we almost always have $\tilde{\theta}(z_0) = \theta_0$ so that the bias on $\hat{\theta}(z)$ behaves as an $O(N^{-1})$. If $\tilde{\theta}(z_0) \neq \theta_0$, then, the convergence rate of $\tilde{\theta}(z_0)$ to θ_0 should be added to the $O(N^{-1})$ bias term in (16-109).

16.13.4 Asymptotic Normality

From (16-102) and (16-108) it follows that $\sqrt{N}(\hat{\theta}(z) - \tilde{\theta}(z_0))$ converges in probability, and, hence, also in distribution (interrelation 3, Section 16.7), to $\sqrt{N}\delta_\theta(z)$. Hence, the study of the asymptotic distribution function of $\hat{\theta}(z)$ boils down to the study of the asymptotic distribution of, respectively, $w - \mu_w$, see (16-102.b), and $\tilde{V}'(\tilde{\theta}, z_0, w(\tilde{\theta}))$, see (16-108.a). Thereto we use the central limit theorems (Section 16.10) or the Fréchet-Shohat lemma (Lemma 16.11). If $w - \mu_w$ and $\tilde{V}'(\tilde{\theta}, z_0, w(\tilde{\theta}))$ are asymptotically normally distributed, then $\sqrt{N}\hat{\theta}(z)$ is also asymptotically normally distributed (a linear combination of Gaussian random variables is Gaussian) with mean $\sqrt{N}\tilde{\theta}(z_0)$ and covariance matrix $N\mathbb{E}\{\delta_\theta(z)\delta_\theta^T(z)\}$. Note that the analysis assumes only that the moments of $\delta_\theta(z)$ exist, not those of $\hat{\theta}(z)$.

16.13.5 Asymptotic Efficiency

The asymptotic efficiency analysis is done for (weakly or strongly) consistent estimators. It consists of comparing the covariance matrix of the limit random variable $\delta_\theta(z)$ to the Cramér-Rao lower bound for unbiased estimators (16-87). The consistent estimator is asymptotically efficient if

$$\lim_{N \to \infty} N(\mathbb{E}\{\delta_\theta(z)\delta_\theta^T(z)\} - Fi^{-1}(\theta_0)) = 0 \tag{16-112}$$

Note that (16-112) can be true while the moments of $\hat{\theta}(z)$ may not exist.

16.14 PITFALLS

Some erroneous statements such as "strong consistency implies asymptotic unbiasedness" or "the limit in mean square and a continuous function are interchangeable" are tempting to make. Therefore, a list of pitfalls is given, some of which are illustrated by means of counterexamples, namely:

1. Weak and strong consistency do not imply asymptotic unbiasedness.
2. Asymptotic unbiasedness does not imply any kind of consistency.
3. Weak and strong consistency do not imply that the limit of the variance is equal to the variance of the limit.
4. 1 and 3 are special cases of: the limit in probability and the almost sure limit are not interchangeable with the expected value. Similar $\mathbb{E}\{O_p(N^{-k})\} \neq O(N^{-k})$ and $\mathbb{E}\{O_{a.s.}(N^{-k})\} \neq O(N^{-k})$.
5. The limit in probability and the limit with probability one ($N \to \infty$) may not be interchanged with a continuous matrix function if its matrix dimensions vary with N. For example, let $J \in \mathbb{R}^{N \times p}$ then,

$$\text{plim}_{N\to\infty} J^T J \neq (\text{plim}_{N\to\infty} J)^T (\text{plim}_{N\to\infty} J)$$

Similarly, if $A, B \in \mathbb{R}^{N \times p}$ with $A = O_{\text{a.s.}}(N^{-s})$ and $B = O_{\text{a.s.}}(N^{-r})$ then $A^T B \neq O_{\text{a.s.}}(N^{-(r+s)})$.

6. The supremum (maximum) and the expected value may not be interchanged.
7. The limit in mean square and a continuous function are not interchangeable.

$$\text{l.i.m.}_{N\to\infty} f(x(N)) \neq f(\text{l.i.m.}_{N\to\infty} x(N))$$

Example 16.21: Weak consistency does not imply asymptotic unbiasedness. Let $\hat{\theta}(N)$ be an estimator of $\theta_0 = 1$ that takes the value 1 with probability $1 - 1/N$ and the value N with probability $1/N$. The estimator is weakly consistent,

$$\lim_{N\to\infty} \text{Prob}(|\hat{\theta}(N) - \theta_0| < \delta) = \lim_{N\to\infty} \text{Prob}(|\hat{\theta}(N) - \theta_0| = 0) = \lim_{N\to\infty} (1 - 1/N) = 1$$

and asymptotically biased $\lim_{N\to\infty} \mathbb{E}\{\hat{\theta}(N)\} = \lim_{N\to\infty} (1(1 - 1/N) + N(1/N)) = 2$. □

Example 16.22: Asymptotic unbiasedness does not imply consistency (16-78). Consider the squared magnitude of the DFT transform of a noise sequence $v(t)$

$$\hat{S}_V(k) = \left| \frac{1}{\sqrt{N}} \sum_{k=0}^{N-1} v(t) e^{-2\pi jkt/N} \right|^2 \tag{16-113}$$

In Kay (1988) it is shown that $\hat{S}_V(k)$ is an asymptotically unbiased estimate of the power spectral density $S_v(j\omega_k)$,

$$\lim_{N\to\infty} \mathbb{E}\{\hat{S}_V(k)\} = S_v(j\omega_k) \tag{16-114}$$

and that the variance of $\hat{S}_V(k)$ does not decrease to zero as $N \to \infty$

$$\lim_{N\to\infty} \text{var}(\hat{S}_V(k)) \approx S_v^2(j\omega_k) \qquad k \neq 0, k \neq N/2 \tag{16-115}$$

(see Appendix 4B of Kay, 1988). Hence, $\hat{S}_V(k)$ is an inconsistent estimate. □

16.15 PRELIMINARY EXAMPLE—CONTINUED

We retake the resistance measurement problem of Sections 1.2 and 16.5 and assume that one experiment consisting of N current and voltage measurements is made,

$$i(k) = i_0 + n_i(k) \text{ and } u(k) = u_0 + n_u(k) \tag{16-116}$$

$k = 1, 2, \ldots, N$. Unless mentioned otherwise, we assume that the current and voltage errors, $n_i(k)$ and $n_u(k)$, are mutually independent, zero mean iid random variables,

Section 16.15 ■ Preliminary Example—Continued

$n_u(k) \in U(0, \sigma_u^2)$ and $n_i(k) \in U(0, \sigma_i^2)$. The goal of this section is to predict, theoretically, the behavior of the three resistance estimators, the simple approach (1-1), the least squares method (1-2), and the errors-in-variables approach (1-3). The analysis follows the lines of Section 16.13. In a first step, we rewrite the estimates as a function of the stochastic sums w. We obtain

$$\hat{R}_{SA}(N) = \frac{1}{N} \sum_{k=1}^{N} \frac{u(k)}{i(k)} = w_{[1]}$$

$$\mathbb{E}\{w_{[1]}\} = \frac{1}{N} \sum_{k=1}^{N} \mathbb{E}\{u(k)\} \mathbb{E}\{1/i(k)\}$$

$$= \frac{u_0}{2\sqrt{3}\sigma_i} \int_{-\sqrt{3}\sigma_i}^{\sqrt{3}\sigma_i} (i_0 + z)^{-1} dz \qquad (16\text{-}117)$$

$$= R_0 \frac{i_0}{2\sqrt{3}\sigma_i} \ln\left(\frac{1 + \sqrt{3}\sigma_i/i_0}{1 - \sqrt{3}\sigma_i/i_0}\right) \qquad (\sqrt{3}\sigma_i < i_0)$$

for the simple approach,

$$\hat{R}_{LS}(N) = \frac{\frac{1}{N}\sum_{k=1}^{N} u(k)i(k)}{\frac{1}{N}\sum_{k=1}^{N} i^2(k)} = \frac{u_0 i_0 + w_{[1]}}{i_0^2 + w_{[2]}}$$

$$w_{[1]} = \frac{1}{N}\sum_{k=1}^{N}(u_0 n_i(k) + i_0 n_u(k) + n_i(k)n_u(k)) \quad \text{with} \quad \mathbb{E}\{w_{[1]}\} = 0 \qquad (16\text{-}118)$$

$$w_{[2]} = \frac{1}{N}\sum_{k=1}^{N}(2i_0 n_i(k) + n_i^2(k)) \quad \text{with} \quad \mathbb{E}\{w_{[2]}\} = \sigma_i^2$$

for the least squares method, and

$$\hat{R}_{EV}(N) = \frac{\frac{1}{N}\sum_{k=1}^{N} u(k)}{\frac{1}{N}\sum_{k=1}^{N} i(k)} = \frac{u_0 + w_{[1]}}{i_0 + w_{[2]}}$$

$$w_{[1]} = \frac{1}{N}\sum_{k=1}^{N} n_u(k) \quad \text{with} \quad \mathbb{E}\{w_{[1]}\} = 0 \qquad (16\text{-}119)$$

$$w_{[2]} = \frac{1}{N}\sum_{k=1}^{N} n_i(k) \quad \text{with} \quad \mathbb{E}\{w_{[2]}\} = 0$$

for the errors-in-variables approach. Note that under the condition $\sqrt{3}\sigma_i < i_0$, all the moments of the three resistance estimators exist for any N.

16.15.1 Consistency

Because by assumption $n_u(k)$ and $n_i(k)$ are mutually independent, iid random variables, each entry of w in (16-117) to (16-119) consists of the sum of iid random variables and converges to its expected value (see Section 16.9, version 1 of the law of large numbers). Hence, we find

$$\tilde{R}_{SA}(N) = \mathbb{E}\{w_{[1]}\} = R_0 \frac{i_0}{2\sqrt{3}\,\sigma_i} \ln\left(\frac{1+\sqrt{3}\,\sigma_i/i_0}{1-\sqrt{3}\,\sigma_i/i_0}\right) \quad \text{(a)}$$

$$\tilde{R}_{LS}(N) = \frac{u_0 i_0 + \mathbb{E}\{w_{[1]}\}}{i_0^2 + \mathbb{E}\{w_{[2]}\}} = \frac{R_0}{1+\sigma_i^2/i_0^2} \quad \text{(b)} \qquad (16\text{-}120)$$

$$\tilde{R}_{EV}(N) = \frac{u_0 + \mathbb{E}\{w_{[1]}\}}{i_0 + \mathbb{E}\{w_{[2]}\}} = R_0 \quad \text{(c)}$$

where $\sqrt{3}\,\sigma_i < i_0$ for $\tilde{R}_{SA}(N)$. The values $\tilde{R}(N)$ in (16-120) are independent of the number of samples N so that $R_* = \tilde{R}(N)$ for each estimator. We conclude that the simple approach and the least squares estimates are inconsistent, $R_{SA*} \neq R_0$, $R_{LS*} \neq R_0$, while the errors-in-variables estimate is strongly consistent, $R_{EV*} = R_0$. Note that $\tilde{R}_{SA}(N)$ and $\tilde{R}_{LS}(N)$ tend to R_0 as $i_0/\sigma_i \to 0$ (Exercise 16.25). Taking the same numerical example as in Section 16.5 ($i_0 = 1$ A, $\sigma_i = 1/\sqrt{12}$ A) gives $R_{SA*} = R_0 \ln 3 = 1.099 R_0$ and $R_{LS*} = R_0 12/13 = 0.923 R_0$.

16.15.2 Convergence Rate

According to Section 16.13.2, the convergence rate of $\hat{R}(N)$ to $\tilde{R}(N)$ equals the convergence rate of w to μ_w. Because the variance of each entry of w exists and is finite ($\sqrt{3}\,\sigma_i < i_0$ for (16-117)), the convergence rate of w to μ_w equals $O_{a.s.}(N^{-1/2}\ln(\ln(N)))$ or $O_p(N^{-1/2})$ (see Section 16.9, respectively versions 1 and 3 of the law of large numbers). Hence, for each of the three estimators, we have

$$\hat{R}(N) = \tilde{R}(N) + \delta(N) + b(N) \quad \text{(a)}$$

$$\delta(N) = \left.\partial \hat{R}(N)/\partial w\right|_{w=\mu_w} w - \mu_w = O_p(N^{-1/2}) \quad \text{(b)} \qquad (16\text{-}121)$$

$$b(N) = O_p(N^{-1}) \quad \text{(c)}$$

with $b_{SA}(N) = 0$ and

$$\left.\partial \hat{R}_{SA}(N)/\partial w\right|_{w=\mu_w} = 1$$

$$\left.\partial \hat{R}_{LS}(N)/\partial w\right|_{w=\mu_w} = \left[(i_0^2+\sigma_i^2)^{-1} \quad -u_0 i_0 (i_0^2+\sigma_i^2)^{-2}\right] \qquad (16\text{-}122)$$

$$\left.\partial \hat{R}_{EV}(N)/\partial w\right|_{w=\mu_w} = \left[1/i_0 \quad -u_0 i_0^{-2}\right]$$

16.15.3 Asymptotic Normality

The vector w consists of the sum of iid random variables with finite mean value and finite nonzero (co)variance matrix ($\sqrt{3}\,\sigma_i < i_0$ for (16-117)). According to the multivariable version of the central limit theorem (see Section 16.10, version 1), $w - \mu_w$ is asymptotically normally distributed at the rate $O(N^{-1/2})$. Hence, the estimates $\hat{R}(N)$ are asymptotically normally distributed (at the rate $O(N^{-1/2})$) with mean value $\tilde{R}(N)$ and variance $\mathbb{E}\{\delta^T \delta\}$ (see Section 16.13.4). We find

Section 16.15 ■ Preliminary Example—Continued

$$\text{var}(\delta_{SA}(N)) = \text{var}(w_{[1]}) \qquad (a)$$

$$\text{var}(\delta_{LS}(N)) = \frac{\text{var}(w_{[1]})}{(i_0^2 + \sigma_i^2)^2} + \frac{\text{var}(w_{[2]})}{(i_0^2 + \sigma_i^2)^4} u_0^2 i_0^2 - 2\frac{\text{covar}(w_{[1]}, w_{[2]})}{(i_0^2 + \sigma_i^2)^3} u_0 i_0 \qquad (b) \qquad (16\text{-}123)$$

$$\text{var}(\delta_{EV}(N)) = \frac{\text{var}(w_{[1]})}{i_0^2} + \frac{u_0^2}{i_0^4}\text{var}(w_{[2]}) \qquad (c)$$

where the stochastic vectors w in (a), (b), and (c) are defined in, respectively, (16-117), (16-118), and (16-119). For the numerical example of Section 16.5 ($u_0 = 1$ V, $i_0 = 1$ A, $\sigma_u = 1/\sqrt{12}$ V, and $\sigma_i = 1/\sqrt{12}$ A), we obtain $\text{var}(\delta_{SA}(N)) = 0.237N^{-1}$, $\text{var}(\delta_{LS}(N)) = 0.132N^{-1}$, and $\text{var}(\delta_{EV}(N)) = 0.167N^{-1}$. Note that $\text{var}(\delta_{LS}(N)) < \text{var}(\delta_{EV}(N))$, as observed in Section 1.2.2.3. Formulas (16-123b) and (16-123c) make it possible to predict the sample variances, obtained by Monte Carlo simulation, of the least squares and errors-in-variables estimates shown in Figure 1-7 on page 11 (Exercise 16.27).

16.15.4 Asymptotic Efficiency

The Cramér-Rao lower bound does not exist for uniformly distributed random variables because the uniform probability density function does not satisfy the regularity conditions of Theorem 16.18 (the derivatives of the pdf do not exist at the boundaries of the domain). Therefore, the asymptotic variance (16-123c) of the consistent estimator $\hat{R}_{EV}(N)$ is compared with the UCRB for Gaussian distributed errors (note that in opposition to the uniform case, the moments of $\hat{R}_{EV}(N)$ do not exist for Gaussian distributed errors).

Putting $\text{var}(w_{[1]}) = \sigma_u^2/N$ and $\text{var}(w_{[2]}) = \sigma_i^2/N$ in (16-123c) gives an explicit expression for the variance of the limiting random variable $\delta_{EV}(N)$

$$\text{var}(\delta_{EV}(N)) = \frac{R_0^2}{N}\left(\frac{\sigma_i^2}{i_0^2} + \frac{\sigma_u^2}{u_0^2}\right) \qquad (16\text{-}124)$$

To construct the UCRB we need the likelihood function of the measurements $z = [u(1)\ u(2)\ \ldots u(N)\ i(1)\ i(2)\ \ldots\ i(N)]^T$. As $u(k)$ and $i(k)$ are mutually independent, iid Gaussian random variables, it is given by ($u_0 = R_0 i_0$)

$$f_z(z, i_0, R_0) = \prod_{k=1}^{N} f_{u(k)}(u(k)) f_{i(k)}(i(k))$$

$$= \frac{1}{(2\pi\sigma_u\sigma_i)^N}\exp(-\frac{1}{2}\sum_{k=1}^{N}\frac{(u(k) - R_0 i_0)^2}{\sigma_u^2} + \frac{(i(k) - i_0)^2}{\sigma_i^2}) \qquad (16\text{-}125)$$

Two unknowns appear in (16-125), the true values R_0 and i_0 of, respectively, the resistance and the current. This means that, in maximum likelihood sense, the resistance as well as the current must be estimated. Applying (16-85) gives the UCRB on the current and resistance estimates

$$Fi^{-1}(i_0, R_0) = \frac{1}{N}\begin{bmatrix} \sigma_i^2 & -u_0\frac{\sigma_i^2}{i_0^2} \\ -u_0\frac{\sigma_i^2}{i_0^2} & R_0^2\left(\frac{\sigma_i^2}{i_0^2} + \frac{\sigma_u^2}{u_0^2}\right) \end{bmatrix} \qquad (16\text{-}126)$$

Finally, entry $[2, 2]$ of $Fi^{-1}(i_0, R_0)$ is the UCRB on the resistance estimate

$$Fi^{-1}(R_0) = \frac{R_0^2}{N}\left(\frac{\sigma_i^2}{i_0^2} + \frac{\sigma_u^2}{u_0^2}\right) \tag{16-127}$$

Comparing (16-127) and (16-124) shows that the errors-in-variable estimate $\hat{R}_{EV}(N)$ is asymptotically efficient.

16.15.5 Asymptotic Bias

Under the condition $\sqrt{3}\,\sigma_i < i_0$, the expected values of the three resistance estimators exist for any N. Applying property 4 of the limit in probability (see Section 16.8), it follows from (16-120) and (16-121) that the bias of the simple approach and the least squares estimates is an $O(N^0)$, while that of the errors-in-variables approach is an $O(N^{-1})$. For the numerical example of Section 16.5 we find $b_{SA} = 0.10\ \Omega$ and $b_{LS} = -0.08\ \Omega$.

16.15.6 Robustness

The simple approach (16-117) is not robust w.r.t. the underlying distribution function of the errors. Indeed, for Gaussian current measurement errors $n_i(k)$, neither the expected value nor the variance of $1/i(k)$ exists, so that estimate $\hat{R}_{SA}(N)$ does not converge w.p. 1, nor in probability, nor in mean square sense (see Section 16.9, versions 1 and 3 of the law of large numbers).

The properties of $\hat{R}_{EV}(N)$ have been analyzed, assuming that the errors $n_u(k)$ and $n_i(k)$ are mutually independent, iid uniform random variables. One may wonder now what happens with these properties when, for example, the measurement errors are no longer independent and/or are no longer identically distributed. Therefore, the analysis is redone, assuming that the errors $n_u(k)$ and $n_i(k)$ are mixing of order 2. For example, filtered independent noise with uniformly bounded variances satisfies this assumption. The errors may be correlated and their distribution function has not been specified. Stationarity is also no longer required. The following properties remain unchanged and are, hence, robust w.r.t. the independence and stationarity assumption.

1. Consistency: applying the strong law of large numbers for mixing sequences (see Section 16.9, version 3) shows that $w_{[1]}$ and $w_{[2]}$ in (16-119) still converge w.p. 1 to zero, which proves the consistency.
2. Convergence rate: applying the convergence rate of the law of large numbers for mixing sequences (see Section 16.9, version 3) to $w_{[1]}$ and $w_{[2]}$ in (16-119) shows that the convergence rate (16-121b) still applies.
3. Asymptotic normality: if the assumption is tightened to a mixing condition of order infinity, then the central limit theorem for mixing sequences (see Section 16.10, version 4) applied to $w_{[1]}$ and $w_{[2]}$ in (16-119) shows that $\hat{R}_{EV}(N)$ is still asymptotically normally distributed.
4. Asymptotic bias: if the class of allowable disturbances is restricted to uniformly bounded random variables, then the bias is still an $O(N^{-1})$.

The following property is not robust:

1. Asymptotic efficiency: the estimator $\hat{R}_{EV}(N)$ does not take into account the dependence between the measurement errors and the particular shape of their distribution function so that, in general, $\text{var}(\delta_{EV}(N))$ will not reach the UCRB.

16.16 PROPERTIES OF THE NOISE AFTER A DISCRETE FOURIER TRANSFORM

In this section we discuss the properties of discrete-time filtered white noise after a discrete Fourier transform (DFT). We first handle the scalar case and afterward generalize the results to the multivariable case. Taking the discrete Fourier transform of $v(t) = H(q)e(t)$ gives (see Section 6.7.3)

$$V(k) = H(z_k^{-1})E(k) + T_H(z_k^{-1}) \tag{16-128}$$

with $H(z^{-1}) = C(z^{-1})/D(z^{-1})$ the noise model, $E(k)$ and $V(k)$ the discrete Fourier transforms of, respectively, $e(t)$ and $v(t)$, and $T_H(z^{-1}) = J(z^{-1})/D(z^{-1})$ the initial and final conditions of the noise process. The transient term $T_H(z_k^{-1})$ is strongly correlated over the frequency and gives a nonmixing contribution to the noise $V(k)$. Fortunately, it can be shown that its influence decreases to zero with probability one.

Lemma 16.23: Consider filtered white noise $H(q)e(t)$, where $H(z^{-1})$ is stable and $e(t)$ has uniformly bounded absolute moments of order $2 + \delta$, with $\delta > 0$: $\mathbb{E}\{|e(t)|^{2+\delta}\} \leq c < \infty$, with c independent of t. The discrete Fourier transform of $H(q)e(t)$ converges w.p. 1 to $H(z_k^{-1})E(k)$. The convergence rate in probability is an $O_p(N^{-1/2})$.

Proof. See Appendix 16.K. □

Lemma 16.24: Let $e(t)$ be independent, identically distributed (iid) noise with existing moments of any order. The discrete Fourier transform $E(k)$ is asymptotically ($N \to \infty$) independent, circular complex normally distributed (convergence in law at the rate $O(N^{-1/2})$). $E(k)$ has zero mean except at $k = 0$ (DC).

Proof. See Appendix 16.L. □

Theorem 16.25 (Asymptotic Normality): The discrete Fourier transform $V(k)$ (16-128) of filtered iid noise $v(t) = H(q)e(t)$, where $H(z^{-1})$ is stable and $e(t)$ has existing moments of any order, is asymptotically ($N \to \infty$) independent, circular complex normally distributed (convergence in law at the rate $O(N^{-1/2})$). For any N, $V(k)$ has zero mean except at $k = 0$ (DC).

Proof. According to Lemma 16.24, $E(k)$ is asymptotically independent, circular complex normally distributed. This is also true for $H(z_k^{-1})E(k)$ because $|H(z_k^{-1})|$ is uniformly bounded (see Exercise 16.6). Applying Lemma 16.23 and interrelations 1 and 3 of the stochastic limits (see Section 16.7) proves the theorem. □

Due to the more restrictive noise assumption, this result is stronger than Theorem 4.4.1 of Brillinger (1981), which shows for mixing stationary time domain noise $v(t)$ that the DFT spectral lines $V(\zeta_1 N), V(\zeta_2 N), \ldots, V(\zeta_J N)$, at a set of *fixed* frequencies $f_r = \zeta_r f_s$,

$r = 1, 2, \ldots, J$, are asymptotically independent, circular complex normally distributed. As the number of time domain samples N increases, the number of DFT lines in between two consecutive spectral lines $V(\zeta_r N)$, $V(\zeta_{r+1} N)$, for which Theorem 4.4.1 of Brillinger (1981) applies, increases to infinity. This is not the case in Theorem 16.25.

Lemma 16.26 (Mixing of Order P): Let $e(t)$ be independent, identically distributed noise with finite moments of order $2P$ and discrete Fourier transform $E(k)$. The squared amplitude spectrum $|E(k)|^2$, DC not included, is mixing of order P.

Proof. See Appendix 16.M. □

Lemma 16.26 does not imply that $E(k)$ is mixing of order $2P$. On the contrary, it can only be proved that $E(k)$ is mixing of order 2 (Lemma 16.27).

Lemma 16.27: Let $e(t)$ be independently distributed noise with mean $\mu < \infty$, variance $\sigma^2 < \infty$, and uniformly bounded fourth-order moments. The discrete Fourier transform $E(k)$ of $e(t)$ and its squared amplitude spectrum $|E(k)|^2$, DC not included, are mixing of order 2.

Proof. See Appendix 16.N. □

Note that Lemma 16.27 requires only the stationarity of the first- and second-order moments of $e(t)$.

Theorem 16.28 (Strong Law of Large Numbers): Let $V(k)$ (16-128) be the DFT of filtered noise $v(t) = H(q)e(t)$, where $H(z^{-1})$ is stable, and where $e(t)$ is independently distributed noise with mean $\mu < \infty$, variance $\sigma^2 < \infty$, and uniformly bounded fourth-order moments: $\mu_4(t) \leq c < \infty$ with c independent of t. Consider the partial sums

$$S(F) = \sum_{k \in \mathbb{F}} W_k V(k) \text{ and } S(F) = \sum_{k \in \mathbb{F}} |W_k V(k)|^2 \tag{16-129}$$

with W_k a uniformly bounded deterministic weighting, $\mathbb{F}$ a subset of the DFT frequencies $k = 0, 1, \ldots, N/2$, and $F = O(N)$ the number of frequencies in the set $\mathbb{F}$. If DC ($k = 0$) belongs to the set $\mathbb{F}$, then $\mu = \mathbb{E}\{e(t)\}$ must be zero. The partial sums $S(F)$ in (16-129) satisfy the strong law of large numbers (16-69). The convergence rate of the partial sums $S(F)/F$ is an $O_p(F^{-1/2})$.

Proof. See Appendix 16.O. □

To study the asymptotic distribution of the estimates, we also need the following central limit theorem.

Theorem 16.29 (Central Limit Theorem): Let $V(k)$ (16-128) be the discrete Fourier transform of filtered iid noise $v(t) = H(q)e(t)$, where $H(z^{-1})$ is stable and $e(t)$ has existing moments of any order. Let $X(k)$ be the discrete Fourier transform of the deterministic signal $x(t)$. Define the sum

$$S(N) = \sum_{k=0}^{N-1} X(k) \bar{V}(k) \tag{16-130}$$

If $x(t)$ has constant power,

$$\frac{1}{N}\sum_{k=0}^{N-1}|X(k)|^2 = \frac{1}{N}\sum_{t=0}^{N-1}|x(t)|^2 = O(N^0) \qquad (16\text{-}131)$$

and has uniformly bounded peak value

$$\max_t |x(t)| \leq c < \infty \qquad (16\text{-}132)$$

for any N, ∞ included, with c a constant independent of N, then $N^{-1/2}S(N)$ is asymptotically normally distributed (convergence in law at the rate $O(N^{-1/2})$).

Proof. See Appendix 16.P. □

Theorem 16.29 is not valid if, for example, $X(k) = 1$, $k = 0, 1, ..., N-1$, because the corresponding time signal $x(t)$ is a pulse whose peak value increases as $O(\sqrt{N})$ (see also Exercise 16.28). This result can easily be understood by rewriting (16-130) as a circular DFT convolution

$$N^{-1/2}S(N) = N^{-1/2}\sum_{k=0}^{N-1} X(k)V(N-k) = \text{DFT}(x(t)v(t)) \qquad (16\text{-}133)$$

It shows that only a very few (independent of N) samples of $v(t)$ contribute to the statistics of $S(N)$ if $x(t)$ is a pulse-like signal, while the central limit theorem requires a large number (increasing with N) of samples.

Theorem 16.29 requires that the energy of the signal $x(t)$ is more or less equally distributed over all time samples and not concentrated in a few points. This is the case for the following classes of signals:

1. Periodic signals with a fixed number F of frequencies. The DFT spectrum $X(k)$ increases as $N^{1/2}$ at the F excited frequencies (power per frequency is an $O(N^0)$) and is zero (if an integer number of periods is observed) or decreases as $N^{-1/2}$ at the other frequencies. The peak value is independent of N.

2. Peak value optimized periodic signals with flat amplitude spectrum that excite all DFT lines $k = 0, 1, ..., N-1$ with $X(k) = \bar{X}(N-k)$ and $|X(k)| = 1$ (power per frequency is an $O(N^{-1})$). The phases of $X(k)$ can always be chosen such that the peak value of $x(t)$ is an $O(N^0)$ close to 1 (Kahane, 1980).

3. Peak value optimized periodic signals with flat amplitude spectrum that excite only DFT lines $r(k+1)$, $k = 0, 1, ..., N/r-2$ with $r \in \mathbb{N}_0$, $X(k) = \bar{X}(N-k)$ and $|X(k)| = 1$ (power per excited frequency is an $O(N^{-1})$). The signal $x(t)$ is an r-times periodic repetition of the signal based on N/r samples that excites all DFT lines (see signal class 2). The peak value is an $O(N^0)$ close to 1 because this is also the case for signal class 2.

4. Peak value optimized periodic signals with flat amplitude spectrum that excite K (independent of N) frequency bands. For example, $K = 2$ frequency bands $[N/16, N/8]f_0$ and $[N/5, N/3]f_0$ with $f_0 = f_s/N$ the DFT resolution. The signal $x(t)$ can be written as a linear combination of K modulated versions of class 2 and/or 3 signals, each with an $O(N^0)$ peak value (Schoukens et al., 1996b). Hence, the peak value of $x(t)$ is an $O(N^0)$ because the peak value of a linear combination of K signals with uniformly bounded peak value is uniformly bounded.

Only upper bounds on the peak value are available for periodic signals with nonflat amplitude spectra. The same is true for periodic signals that excite only logarithmically spaced DFT lines (lacunar multisines), for example, 2^k ($k = 0, 1, ..., \ln(N/2) - 1$, $X(k) = \bar{X}(N-k)$ and $|X(k)|^2 = O(N/\ln N)$). Fortunately, the upper bound increases slowly with N: for signals satisfying (16-131) the phases of $X(k)$ can always be chosen such that the peak value is bounded by $O(\sqrt{\ln N})$ (Theorem 4, Chapter 6 of Kahane, 1985), while for arbitrary phases the peak value is w.p. 1 an $O(\sqrt{\ln N})$

$$\text{Prob}(\max_t |x(t)| \geq O(\sqrt{\ln N})) \leq N^{-2} \qquad (16\text{-}134)$$

(Theorem 1 and Exercise 5, Chapter 6 of Kahane, 1985). This means that the risk of selecting phases with a peak value increasing faster than $O(\sqrt{\ln N})$ is very small. The following corollary can then be used.

Corollary 16.30: Let $V(k)$ (16-128) satisfy the assumptions of Theorem 16.29. If $x(t)$ satisfies (16-131) and has a peak value $\max_t |x(t)| = O(\sqrt{\ln N})$, then $N^{-1/2}S(N)$ (16-130) is asymptotically normally distributed (convergence in law at the rate $O(\sqrt{\ln N/N})$).

Proof. See Appendix 16.Q. □

The preceding theorems are useful for studying the asymptotic behavior of frequency domain estimators when using deterministic excitation signals. For signals with a stochastic behavior, such as filtered white noise, periodic noise, and random multisines, we need the following theorems.

Lemma 16.31 (Mixing of Order P): Let $H_1(z^{-1})$, $H_2(z^{-1})$ be stable filters and $E_1(k)$, $E_2(k)$ the DFT spectra of iid random variables $e_1(t)$, $e_2(t)$ with existing moments of order P. Let $X(k)$ be one of the following spectra, $H_2(z_k^{-1})E_2(k)$, or the DFT spectrum of an integer number of periods of a normalized random multisine (see Definition 3.2) or normalized periodic noise (see Definition 3.3), all with existing Pth order moments. If $X(k)$ is independent of $E_1(k)$, then $X(k)H_1(z_k^{-1})E_1(k)$ is mixing of order P.

Proof. See Appendix 16.R. □

Theorem 16.32 (Strong Law of Large Numbers): Let $V(k)$ (16-128) be the DFT of filtered noise $v(t) = H(q)e(t)$, where $H(z^{-1})$ is stable, and where $e(t)$ is independently distributed noise with uniformly bounded absolute moments of order $2 + \delta$, with $\delta > 0$. Let $X(k)$ be the DFT spectrum of one of the following signals, filtered white noise, or an integer number of periods of a normalized random multisine (see Definition 3.2) or normalized periodic noise (see Definition 3.3), all with uniformly bounded fourth-order moments. Assume, furthermore, that $X(k)$ is independent of $V(k)$. Consider the partial sums

$$S(F) = \sum_{k \in \mathbb{F}} W_k X(k) \bar{V}(k) \text{ and } S(F) = \sum_{k \in \mathbb{F}} |W_k X(k)|^2 \qquad (16\text{-}135)$$

with W_k a uniformly bounded deterministic weighting, $\mathbb{F}$ a subset of the DFT frequencies $k = 0, 1, ..., N/2$, and $F = O(N)$ the number of frequencies in the set $\mathbb{F}$. If DC ($k = 0$) belongs to the set $\mathbb{F}$, then $\mu = \mathbb{E}\{e(t)\}$ must be zero. The partial sums $S(F)$ (16-135) satisfy the strong law of large numbers (16-69). The convergence rate of $S(F)/F$ is an $O_p(F^{-1/2})$.

Proof. See Appendix 16.S. □

Theorem 16.33 (Central Limit Theorem): Let $V(k)$ (16-128) be the DFT of filtered iid noise $v(t) = H(q)e(t)$, where $H(z^{-1})$ is stable and $e(t)$ has existing moments of any order. Let $X(k)$ be the DFT spectrum of one of the following signals, filtered iid noise, an integer number of periods of a normalized random multisine (see Definition 3.2), or normalized periodic noise (see Definition 3.3), all with existing moments of any order. Assume, furthermore, that $X(k)$ is independent of $V(k)$. Consider the partial sums

$$S(F) = \sum_{k \in \mathbb{F}} W_k X(k) \bar{V}(k), \quad S(F) = \sum_{k \in \mathbb{F}} |W_k V(k)|^2 \text{ and } S(F) = \sum_{k \in \mathbb{F}} |W_k X(k)|^2 \quad (16\text{-}136)$$

with W_k a uniformly bounded deterministic weighting, $\mathbb{F}$ a subset of the DFT frequencies $k = 0, 1, \ldots, N/2$, and $F = O(N)$ the number of frequencies in the set $\mathbb{F}$. If DC ($k = 0$) belongs to the set $\mathbb{F}$, then $\mu = \mathbb{E}\{e(t)\}$ must be zero. The sums $F^{-1/2}S(F)$ (16-136) are asymptotically normally distributed (convergence in law at the rate $O(F^{-1/2})$).

Proof. See Appendix 16.T. □

The preceding theorems can easily be generalized to the multivariable case $v(t) = H(q)e(t)$ with $e(t) \in \mathbb{R}^p$, $v(t) \in \mathbb{R}^q$ and $H(q)$ a stable transfer function matrix. Formula (16-128) is still valid with $E(k) \in \mathbb{C}^p$, $V(k) \in \mathbb{C}^q$, $H(z^{-1}) = D^{-1}(z^{-1})C(z^{-1})$, $T_H(z^{-1}) = D^{-1}(z^{-1})J(z^{-1})$, $D(z^{-1})$ a q by q matrix, $C(z^{-1})$ a q by p matrix, and $J(z^{-1})$ a q by 1 vector (see Section 6.6).

16.17 EXERCISES

16.1. Let $x, y \in \mathbb{R}$ have finite second-order moments. Prove that $\text{var}(x+y) \leq 2\text{var}(x) + 2\text{var}(y)$ (hint: use $(a+b)^2 \leq (a+b)^2 + (a-b)^2 = 2a^2 + 2b^2$).

16.2. Show that conditions (16-13) are equivalent to condition (16-12).

16.3. Let $x \in \mathbb{C}^n$ be circular complex distributed and $A \in \mathbb{C}^{m \times n}$. Show that $y = Ax$ is circular complex distributed.

16.4. Show that $\text{Cov}(x_{\text{re}}) = 0.5(\text{Cov}(x))_{\text{Re}}$ for circular complex noise x.

16.5. Show that the probability density function of $x \in N_n^c(\mu_x, C_x)$ is given by (16-14) (hint: use $x_{\text{re}} \in N_{2n}((\mu_x)_{\text{re}}, 0.5(C_x)_{\text{Re}})$ and apply Lemma 15.3).

16.6. Let $x \in N^c(\mu, \sigma_x^2)$ and $a \in \mathbb{C}$ with $|a| < \infty$. Show that $ax \in N^c(a\mu, |a|^2\sigma_x^2)$.

16.7. Let $x \in N^c(0, \sigma_x^2)$. Show that $\mathbb{E}\{|x|^4\} = 2\sigma_x^4$.

16.8. Let $x \in N^c(0, \sigma_x^2)$. Show that $\mathbb{E}\{x^n\} = 0$ (hint: use $\mathbb{E}\{u^{2n}\} = \sigma_u^{2n}(2n)!/(2^n n!)$ for $u \in N(0, \sigma_u^2)$ (Stuart and Ord, 1987) and $\sum_{k=0}^{n \text{ div } 2} C_n^{2k} = \sum_{k=0}^{n \text{ div } 2} C_n^{2k+1} = 2^{n-1}$ (Gradshteyn and Ryzhik, 1980)).

16.9. Show that $x \in N_n^c(\mu_x, C_x)$ can be written as $x = \mu_x + Ay$ with $y \in N_p^c(0, I_p)$ and $p = \text{rank}(C_x)$ (hint: use $C_x = AA^H$ with $A \in \mathbb{C}^{n \times p}$ and $\text{rank}(A) = p$).

16.10. Let $x \in N_n(0, I_n)$ ($x \in N_n^c(0, I_n)$) and $P \in \mathbb{R}^{n \times n}$ ($P \in \mathbb{C}^{n \times n}$) a symmetric (Hermitian) idempotent matrix of rank p. Show that $x^T P x \in \chi^2(p)$ ($2x^H P x \in \chi^2(2p)$) (hint: use $P = U \Lambda U^H$ with $U^{-1} = U^H$ and $\Lambda = \text{diag}(I_p, O_{n-p})$ with $p = \text{rank}(P)$).

16.11. Let $x \in \chi^2(n)$. Show that $\mathbb{E}\{1/x\} = 1/(n-2)$ and $\text{var}(1/x) = 2/((n-2)^2(n-4))$ (hint: take $y \in F(n_1, n_2)$ and use the rules $\mathbb{E}\{y\} = \mathbb{E}\{x_1/n_1\}\mathbb{E}\{n_2/x_2\}$ and $\text{var}(x_1 x_2) = \sigma_{x_1}^2 \sigma_{x_2}^2 + \sigma_{x_1}^2 \mu_{x_2}^2 + \sigma_{x_2}^2 \mu_{x_1}^2$ with $\mathbb{E}\{x\} = n$, $\mathbb{E}\{y\} = n_2/(n_2 - 1)$ and $\text{var}(y) = 2n_2^2(n_1 + n_2 - 2)/(n_1(n_2-2)^2(n_2-4))$ (Stuart and Ord, 1987)).

16.12. Let $x \in \mathbb{C}^n$ be circular complex distributed and $f(x) \in \mathbb{C}^m$ an analytic function. Show that $\text{Cov}(f(x))$ is given by (16-24).

16.13. Let $y \in W_p^c(n, C_x)$ with $\text{Im}(C_x) = 0$. Show that $\text{Re}(y) \in W_p(2n, C_x/2)$.

16.14. Show using (16-16) that $\text{cum}(x_k) = \text{E}\{x_k\}$, $\text{cum}(x_k, \bar{x}_k) = \text{var}(x_k)$, and $\text{cum}(x_k, \bar{y}_l) = \text{covar}(x_k, y_l)$.

16.15. Let $x \in \mathbb{C}^n$ be circular complex distributed. Show that the covariance matrix of $f(x, \bar{x}) \in \mathbb{R}^m$ is given by (16-25) (hint: start from (16-23) with x replaced by x_{re}, and use $C_{x_{\text{re}}} = 0.5(C_x)_{\text{Re}}$, Lemma 15.4, and (15-60)).

16.16. Show that the sample mean (16-26) and sample (cross-)covariance matrices (16-27), (16-28) of independent realizations are unbiased estimates of the mean and (cross-)covariance matrices.

16.17. Let $x \in N(0, \sigma_x^2)$ ($x \in N^c(0, \sigma_x^2)$) and calculate the sample variance $\hat{\sigma}_x^2$ of R independent realizations. Show that $(R-1)\hat{\sigma}_x^2/\sigma_x^2 \in \chi^2(R-1)$ ($2(R-1)\hat{\sigma}_x^2/\sigma_x^2 \in \chi^2(2(R-1))$) (hint: first show that $(R-1)\hat{\sigma}_x^2 = X^H P X$ with $X_{[k]} = x[k]$, $P = I_R - U/R$ and $U_{[k,l]} = 1$, $k, l = 1, 2, ..., R$; next prove that P is a symmetric idempotent matrix of rank $R-1$ and apply the results of Exercise 16.10).

16.18. Let $x \in N_n(\mu_x, C_x)$ with $\text{rank}(C_x) = p < n$. Show that $b = (\hat{x} - \mu_x)^T (\hat{C}_x/R)^+(\hat{x} - \mu_x)$ is $p(R-1)/(R-p)F(p, R-p)$ distributed (hint: apply the results of Exercise 16.9 on $\hat{x}$ and $\hat{C}_x$, and use Theorem 15.1).

16.19. Let $x(t) \in \mathbb{C}^r$ be mixing of order P and $h(t, u) \in \mathbb{C}^{s \times r}$ the impulse response of a stable linear time-variant multivariable system. Show that $y(t) = \sum_{u=0}^{\infty} h(t, u)x(u)$ is mixing of order P.

16.20. Prove the linearity of the limit in mean square (16-64) (hint: use $(x+y)^2 \leq (x+y)^2 + (x-y)^2 = 2x^2 + 2y^2$).

16.21. Prove that the limit in mean square and the expected value commute (hint: first show that $\mathbb{E}\{x^2\} \geq (\mathbb{E}\{x\})^2$).

16.22. Prove that $\mathbb{E}\{O_{\text{m.s.}}(N^{-k})\} = O(N^{-k})$ (hint: show that $\lim_{N \to \infty} N^k \mathbb{E}\{O_{\text{m.s.}}(N^{-k})\} < \infty$).

16.23. Let $z(k) \in N(0, \sigma_z^2)$, $k = 1, 2, ..., N$. Show that the sample variance $\hat{\sigma}_z^2 = \sum_{k=1}^{N} z^2(k)/N$ is an unbiased and efficient estimate of σ_z^2.

16.24. Let $z(k) \in E(\mu_z, \mu_z^2)$, $k = 1, 2, ..., N$. Show that the sample mean $\hat{\mu}_z = \sum_{k=1}^{N} z(k)/N$ is an unbiased and efficient estimate of μ_z, with $\text{var}(\hat{\mu}_z) = \mu_z^2/N$. Now consider the estimator $\hat{m} = a\hat{\mu}_z$, where a is chosen to minimize the mean square error $\text{MSE}(\hat{m})$. Show that $a = N/(N+1)$ and that $\text{MSE}(\hat{m}) = \mu_z^2/(N+1)$. Conclude that the MSE of the biased estimate $\hat{m}$ is smaller than that of the unbiased estimate $\hat{\mu}_z$.

16.25. Show that the simple approach (16-117) and the least squares estimates (16-118) tend to the true value R_0 as the signal-to-noise ratio of the current measurements tends to infinity (hint: use (16-120) and let $i_0/\sigma_i \to \infty$).

16.26. Illustrate reasoning (16-98) on the errors-in-variables estimate (16-119).

16.27. Run a Monte Carlo simulation for the resistance measurement problem (16-116) with $u_0 = 1$ V, $i_0 = 1$ A, and iid errors $n_u(k) \in U(0, 12^{-1} \text{ V}^2)$, $n_i(k) \in U(0, 12^{-1} \text{ A}^2)$. Calculate the least squares (16-118) and errors-in-variables (16-119) estimates for increasing values of N. Compare the sample variance of the estimates with the values predicted by (16-123b) and (16-123c).

16.28. Let $E(k)$ be the DFT spectrum of iid noise $e(t)$ with existing moments of any order. Consider the sum $S(N) = \sum_{k=0}^{N-1} E(k)$. Show that $N^{-1/2}S(N)$ is not asymptotically normally distributed. Explain. (Hint: use expression (16-182) to show that the cumulants of $N^{-1/2}S(N)$ do not tend to those of a normal random variable; note that $S(N) = e(0)$.)

16.18 APPENDIXES

Appendix 16.A Indecomposable Sets

Consider a table with I rows and, possibly, a different number of columns per row

$$\begin{matrix} v_{11} & v_{12} & \cdots & v_{1J_1} \\ v_{21} & v_{22} & \cdots & v_{2J_2} \\ \cdots & \cdots & \cdots & \cdots \\ v_{I1} & v_{I2} & \cdots & v_{IJ_I} \end{matrix} \qquad (16\text{-}137)$$

Although some of the entries of this table may take the same numerical value, all elements are considered as distinguishable. Let $\mathbb{P} = \mathbb{P}_1 \cup \mathbb{P}_2 \cup \cdots \cup \mathbb{P}_K$ be a partition of table (16-137). The sets $\mathbb{P}_i$ and $\mathbb{P}_j$ of the partition *hook* if there exist a $v_{i_1 j_1} \in \mathbb{P}_i$ and a $v_{i_2 j_2} \in \mathbb{P}_j$ such that $i_1 = i_2$. Two sets $\mathbb{P}_i$ and $\mathbb{P}_j$ *communicate* if there exists a sequence of sets $\mathbb{P}_i = \mathbb{P}_{m_1}, \mathbb{P}_{m_2}, \ldots, \mathbb{P}_{m_R} = \mathbb{P}_j$ such that $\mathbb{P}_{m_r}$ and $\mathbb{P}_{m_{r+1}}$ hook for $r = 1, 2, \ldots, R-1$. The partition $\mathbb{P}$ is *indecomposable* if all sets of the partition communicate.

Example 16.34: Consider a 2 by 2 table

$$\begin{matrix} v_{11} & v_{12} \\ v_{21} & v_{22} \end{matrix} \qquad (16\text{-}138)$$

All the indecomposable partitions of (16-138) are given by

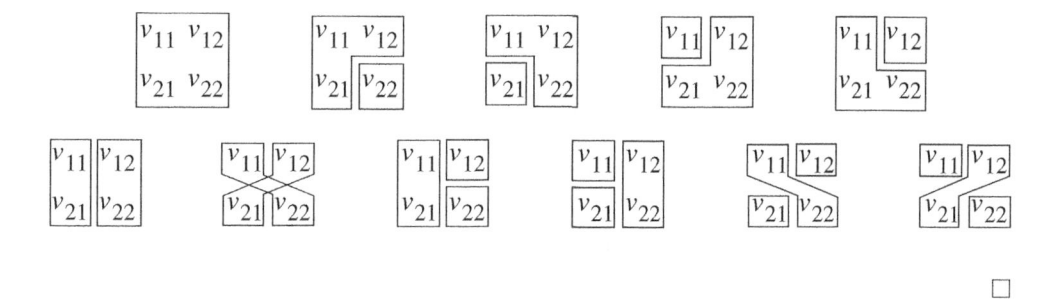

Lemma 16.35: Let $y_i = \prod_{j=1}^{J_i} v_{ij}$, $i = 1, 2, \ldots, I$, then

$$\text{cum}(y_1, y_2, \ldots, y_I) = \sum_{\mathbb{P}} \text{cum}(v_{ij} \in \mathbb{P}_1) \text{cum}(v_{ij} \in \mathbb{P}_2) \cdots \text{cum}(v_{ij} \in \mathbb{P}_K) \qquad (16\text{-}139)$$

where the summation is taken over all indecomposable partitions $\mathbb{P} = \mathbb{P}_1 \cup \mathbb{P}_2 \cup \cdots \cup \mathbb{P}_K$ of table (16-137), and with $\text{cum}(v_{ij} \in \mathbb{P}_r)$ the joint cumulant of all the elements of $\mathbb{P}_r$.

Proof. See Leonov and Shiryaev (1959).

Example 16.36: Suppose we want to calculate $\text{cum}(x_1 x_2, x_3 x_4)$. Note that the definition $y_i = \prod_{j=1}^{J_i} v_{ij}$ in Lemma 16.35 defines the structure of table (16-137): the index i defines the row and j defines the column. Hence, in this case we have on the first row x_1 and x_2 (the order is not important) and on the second row x_3 and x_4 (the order is not important). Using Lemma 16.35 and the result of Example 16.34 with $v_{11} = x_1$, $v_{12} = x_2$, $v_{21} = x_3$, and $v_{22} = x_4$, we find

$$\begin{aligned}
\text{cum}(x_1x_2, x_3x_4) = {} & \text{cum}(x_1, x_2, x_3, x_4) + \\
& \text{cum}(x_1, x_2, x_3)\text{cum}(x_4) + \text{cum}(x_1, x_2, x_4)\text{cum}(x_3) + \\
& \text{cum}(x_2, x_3, x_4)\text{cum}(x_1) + \text{cum}(x_1, x_3, x_4)\text{cum}(x_2) + \\
& \text{cum}(x_1, x_3)\text{cum}(x_2, x_4) + \text{cum}(x_1, x_4)\text{cum}(x_2, x_3) + \\
& \text{cum}(x_1, x_3)\text{cum}(x_2)\text{cum}(x_4) + \text{cum}(x_2, x_4)\text{cum}(x_1)\text{cum}(x_3) + \\
& \text{cum}(x_1, x_4)\text{cum}(x_2)\text{cum}(x_3) + \text{cum}(x_2, x_3)\text{cum}(x_1)\text{cum}(x_4)
\end{aligned} \qquad (16\text{-}140)$$

□

Appendix 16.B Proof of Lemma 16.5

For $k = 1, 2, \ldots, P$ we have

$$\max_{t_k} \sum_{t_1, t_2, \ldots, t_{k-1} = 0}^{\infty} |\text{cum}(y(t_1), y(t_2), \ldots, y(t_k))| \qquad (16\text{-}141)$$

$$\leq \max_{t_k} \sum_{u_1, \ldots, u_k = 0}^{\infty} \sum_{t_1, \ldots, t_{k-1} = 0}^{\infty} |h(t_1, u_1)| \ldots |h(t_k, u_k)| |\text{cum}(x(u_1), x(u_2), \ldots, x(u_k))| \qquad (16\text{-}142)$$

$$\leq C^{k-1} \max_{t_k} \sum_{u_k = 0}^{\infty} |h(t_k, u_k)| \sum_{u_1, \ldots, u_{k-1} = 0}^{\infty} |\text{cum}(x(u_1), x(u_2), \ldots, x(u_k))| \qquad (16\text{-}143)$$

$$\leq C^{k-1} \left(\max_{u_k} \sum_{u_1, \ldots, u_{k-1} = 0}^{\infty} |\text{cum}(x(u_1), x(u_2), \ldots, x(u_k))| \right) \left(\max_{t_k} \sum_{u_k = 0}^{\infty} |h(t_k, u_k)| \right) \qquad (16\text{-}144)$$

$$\leq C^k \max_{u_k} \sum_{u_1, \ldots, u_{k-1} = 0}^{\infty} |\text{cum}(x(u_1), x(u_2), \ldots, x(u_k))| \qquad (16\text{-}145)$$

$$< \infty \qquad (16\text{-}146)$$

Inequality (16-143) is obtained by applying $k-1$ times (16-38a), inequality (16-145) uses (16-38b), and inequality (16-146) uses the mixing of order P property of x. □

Appendix 16.C Proof of Lemma 16.8

The proof follows exactly the same lines of the proof of Theorem 2.9.1 of Brillinger (1981), modified for the more general mixing definition (16-33) as in the proof of Lemma 16.5. Instead of repeating the proof of Brillinger (1981), we will illustrate the differences on the second-order cumulant of a second-degree Volterra system. Background information concerning the new concepts required for this proof can be found in Appendix 16.A. Assuming that the system is causal and that the input is zero for negative times, we have

Section 16.18 ■ Appendixes

$$y(t) = \sum_{u_1, u_2 = 0}^{t} h_2(t - u_1, t - u_2) x(u_1) x(u_2) = \sum_{u_1, u_2 = 0}^{t} h_2(u_1, u_2) x(t - u_1) x(t - u_2) \quad (16\text{-}147)$$

(for noncausal systems and noncausal inputs the sums in (16-147) are extended from $-\infty$ to $+\infty$). The second-order cumulant of $y(t)$ is

$$\text{cum}(y(t_1), y(t_2)) = \sum_{u_{11}, u_{12} = 0}^{t_1} \sum_{u_{21}, u_{22} = 0}^{t_2} h_2(u_{11}, u_{12}) h_2(u_{21}, u_{22}) \cdot \\ \text{cum}(x(t_1 - u_{11}) x(t_1 - u_{12}), x(t_2 - u_{21}) x(t_2 - u_{22})) \quad (16\text{-}148)$$

The mixing condition of order 2 becomes

$$\max_{t_2 \geq 0} \sum_{t_1 = 0}^{\infty} |\text{cum}(y(t_1), y(t_2))| \leq \left(\sum_{u_{11}, u_{12} = 0}^{\infty} |h_2(u_{11}, u_{12})| \right) \left(\sum_{u_{21}, u_{22} = 0}^{\infty} |h_2(u_{21}, u_{22})| \right) \cdot \\ \max_{\substack{u_{11}, u_{12}, u_{12}, u_{22} \\ t_2 \geq 0}} \left(\sum_{t_1 = 0}^{\infty} |\text{cum}(x(t_1 - u_{11}) x(t_1 - u_{12}), x(t_2 - u_{21}) x(t_2 - u_{22}))| \right) \quad (16\text{-}149) \\ \leq C_2^2 \max_{u_1, u_2, v_2} \sum_{v_1 = 0}^{\infty} |\text{cum}(x(v_1) x(v_1 + u_1), x(v_2) x(v_2 + u_2))|$$

where $v_1 = t_1 - u_{11}$, $u_1 = u_{11} - u_{12}$, $v_2 = t_2 - u_{21}$, and $u_2 = u_{21} - u_{22}$. We will now prove that

$$\max_{u_1, u_2, v_2} \sum_{v_1 = 0}^{\infty} |\text{cum}(x(v_1) x(v_1 + u_1), x(v_2) x(v_2 + u_2))| \leq C < \infty \quad (16\text{-}150)$$

which shows that $y(t)$ is mixing of order 2. From Example 16.36 it follows that the cumulant in (16-150) contains four kinds of contributions. For each type of contribution we will show that (16-150) is valid. We find for the type I contribution (fourth-order cumulant);

$$\max_{u_1, u_2, v_2} \sum_{v_1 = 0}^{\infty} |\text{cum}(x(v_1), x(v_1 + u_1), x(v_2), x(v_2 + u_2))| \leq \\ \max_{v_2} \sum_{v_1, u_1, u_2 = 0}^{\infty} |\text{cum}(x(v_1), x(u_1), x(v_2), x(u_2))| \leq C \quad (16\text{-}151)$$

for the type II contributions (product of third- and first-order cumulants)

$$\max_{u_1, u_2, v_2} \sum_{v_1 = 0}^{\infty} |\text{cum}(x(v_1), x(v_2), x(v_2 + u_2)) \text{cum}(x(v_1 + u_1))| \leq \\ \left(\max_{v_2} \sum_{v_1, u_2 = 0}^{\infty} |\text{cum}(x(v_1), x(v_2), x(u_2))| \right) \left(\max_{u_1} |\text{cum}(x(u_1))| \right) \leq C \quad (16\text{-}152)$$

and similarly for the three other third-order terms; for the type III contributions (product of two second-order cumulants)

$$\max_{u_1, u_2, v_2} \sum_{v_1=0}^{\infty} |\text{cum}(x(v_1), x(v_2+u_2))\text{cum}(x(v_1+u_1), x(v_2))| \le$$
$$\left(\max_{u_2} \sum_{v_1=0}^{\infty} |\text{cum}(x(v_1), x(u_2))|\right)\left(\max_{v_2} \sum_{u_1=0}^{\infty} |\text{cum}(x(u_1), x(v_2))|\right) \le C \quad (16\text{-}153)$$

and similarly for the other term; and finally for the type IV contributions (product of second-order cumulant and two first-order cumulants)

$$\max_{u_1, u_2, v_2} \sum_{v_1=0}^{\infty} |\text{cum}(x(v_1+u_1), x(v_2+u_2))\text{cum}(x(v_1))\text{cum}(x(v_2))| \le$$
$$\left(\max_{u_2} \sum_{v_1=0}^{\infty} |\text{cum}(x(v_1), x(u_2))|\right)\left(\max_{v_1} |\text{cum}(x(v_1))|\right)\left(\max_{v_2} |\text{cum}(x(v_2))|\right) \le C \quad (16\text{-}154)$$

and similarly for the three other terms. The last inequalities in (16-151), (16-152), (16-153), and (16-154) are due to the fourth-order mixing property of $x(t)$. The basic reason why all the terms in (16-140) have a finite contribution to (16-150) is that they all stem from indecomposable partitions.

Extending the summations in all the equations from $-\infty$ to $+\infty$ shows that the proof is also valid for noncausal systems and noncausal inputs. $\square$

Appendix 16.D Almost Sure Convergence Implies Convergence in Probability

$x(N)$ converges w.p. 1 to x, hence, for any $\varepsilon, \delta > 0$ there exists an N such that

$$1 - \delta \le \text{Prob}(\sup_{k \ge N}|x(k) - x| \le \varepsilon) = \text{Prob}(\bigcap_{k=N}^{\infty} |x(k) - x| \le \varepsilon) \le \text{Prob}(|x(N) - x| \le \varepsilon)$$

The last inequality shows that $x(N)$ converges in probability to x. $\square$

Appendix 16.E Convergence in Mean Square Implies Convergence in Probability

$x(N)$ converges in mean square to x, hence, for any $\varepsilon, \delta > 0$ there exists a N_0 such that for every $N > N_0$, $\mathbb{E}\{|x(N) - x|^2\} \le \delta\varepsilon^2$. Using Chebyshev's inequality (16-20) we find

$$\text{Prob}(|x(N) - x| \le \varepsilon) = 1 - \text{Prob}(|x(N) - x| > \varepsilon) \ge 1 - \frac{1}{\varepsilon^2}\mathbb{E}\{|x(N) - x|^2\} \ge 1 - \delta$$

which shows that $x(N)$ converges in probability to x. $\square$

Appendix 16.F The Borel-Cantelli Lemma

If $\text{Prob}(A \cup B) = 1$, then $\text{Prob}(A \cap B) = 1 - \text{Prob}(\overline{A \cap B}) = 1 - \text{Prob}(\overline{A} \cup \overline{B})$. Hence,

$$\text{Prob}(\sup_{k \geq N}|x(k)-x| \leq \varepsilon) = \text{Prob}(\bigcap_{k=N}^{\infty} |x(k)-x| \leq \varepsilon) = 1 - \text{Prob}(\bigcup_{k=N}^{\infty} |x(k)-x| > \varepsilon)$$

Using $\text{Prob}(A \cup B) \leq \text{Prob}(A) + \text{Prob}(B)$ and Chebyshev's inequality (16-20) we find

$$\text{Prob}(\sup_{k \geq N}|x(k)-x| \leq \varepsilon) \geq 1 - \sum_{k=N}^{\infty} \text{Prob}(|x(k)-x| > \varepsilon) \geq 1 - \frac{1}{\varepsilon^2}\sum_{k=N}^{\infty} \mathbb{E}\{|x(k)-x|^2\} \quad (16\text{-}155)$$

From (16-56) it follows that for any $\varepsilon, \delta_1, \delta_2 > 0$ there exists an N such that for every $k \geq N$

$$\sum_{k=N}^{\infty} \text{Prob}(|x(k)-x| > \varepsilon) < \delta_1 \quad \text{and} \quad \sum_{k=N}^{\infty} \mathbb{E}\{|x(k)-x|^2\} < \delta_2 \varepsilon^2 \quad (16\text{-}156)$$

Putting (16-156) in (16-155) proves the theorem. $\square$

Appendix 16.G Proof of the (Strong) Law of Large Numbers for Mixing Sequences

We will give the proof for the general case where the sequence in the partial sum depends on N. To simplify the notations, we introduce the zero mean variables $y(N) = (S(N) - \mathbb{E}\{S(N)\})/N$ and $z_N(k) = x_N(k) - \mathbb{E}\{x_N(k)\}$. The law of large numbers then becomes $\underset{N \to \infty}{\text{l.i.m.}}\, y(N) = 0$. The calculation of this limit in mean square requires an expression for $\text{cum}(y(N), y(N))$ (see (16-51)). Using the multilinearity of the cumulant (see Section 16.1, property 2) and the second-order mixing condition of $x_N(k)$, we find

$$\text{cum}(y(N), y(N)) = \frac{1}{N^2}\sum_{k,l=1}^{N} \text{cum}(x_N(k), x_N(l)) = O(N^{-1}) \quad (16\text{-}157)$$

where the last equality is due to (16-36). From (16-157), it follows that $\lim_{N \to \infty}\mathbb{E}\{y^2(N)\} = 0$, which already proves the law of large numbers for mixing sequences, and that $\text{var}(y(N)) = O(N^{-1})$. Using properties 6 and 7 of Section 16.8, it follows that the convergence rate of the limit in mean square is an $O_{\text{m.s.}}(N^{-1/2})$.

Result (16-157) is not sufficient to prove the strong convergence of $y(N)$ via the Borel-Cantelli Lemma 16.10 ($\sum_{N=1}^{\infty} 1/N = \infty$). Therefore, we will first prove that the subsequence $y(N^2)$ converges w.p. 1 to zero. Next, we will show that the deviation of any element in the main sequence $y(N)$ with a nearby element in the subsequence $y(N^2)$ converges to zero w.p. 1. It is easy to see that $\text{cum}(y(N^2), y(N^2)) = O(N^{-2})$ and therefore

$$\sum_{N=1}^{\infty} \mathbb{E}\{y^2(N^2)\} = \sum_{N=1}^{\infty} O(N^{-2}) < \infty \quad (16\text{-}158)$$

Applying the Borel-Cantelli Lemma 16.10 to (16-158) shows that $\underset{N \to \infty}{\text{a.s.lim}}\, y(N^2) = 0$. It remains to be proved that this implies the strong convergence of the whole sequence $y(N)$. Therefore, it is sufficient to show that the maximal difference between the subsequence and the complete sequence

$$\sup_{N^2 < k \leq (N+1)^2} |y(k) - y(N^2)| \quad (16\text{-}159)$$

converges to zero w.p. 1. The difference $y(k) - y(N^2)$ can be rewritten as the sum of three contributions

$$y(k) - y(N^2) = \Delta_1(k) + \Delta_2(k) + \Delta_3(k) \tag{16-160}$$

with

$$\Delta_1(k) = -\frac{(k-N^2)}{k} y(N^2), \quad \Delta_2(k) = \frac{1}{k} \sum_{r=N^2+1}^{k} z_k(r) \text{ and } \Delta_3(k) = \frac{1}{k} \sum_{r=1}^{N^2} (z_k(r) - z_{N^2}(r))$$

Using $\sup_k |a(k) + b(k)| \leq \sup_k |a(k)| + \sup_k |b(k)|$, (16-159) is bounded above by

$$\sup_{N^2 < k \leq (N+1)^2} |\Delta_1(k)| + \sup_{N^2 < k \leq (N+1)^2} |\Delta_2(k)| + \sup_{N^2 < k \leq (N+1)^2} |\Delta_3(k)| \tag{16-161}$$

The strong convergence to zero of the first term in (16-161) is first established. Using $\sup_{r < k \leq s} |a(k)| \leq \sum_{k=r+1}^{s} |a(k)|$, it is bounded above by

$$\sum_{k=N^2+1}^{(N+1)^2} |\Delta_1(k)| \leq |y(N^2)| \left(\max_{N^2 < k \leq (N+1)^2} \left| \frac{k-N^2}{k} \right| \right) ((N+1)^2 - N^2) \tag{16-162}$$

Using $\max_{N^2 < k \leq (N+1)^2} \left| \frac{k-N^2}{k} \right| = \frac{(N+1)^2 - N^2}{(N+1)^2}$ for $N \geq 1$, (16-162) becomes

$$\sum_{k=N^2+1}^{(N+1)^2} |\Delta_1(k)| \leq |y(N^2)| \left(\frac{2N+1}{N+1} \right)^2 = |y(N^2)| O(N^0) \tag{16-163}$$

Because the subsequence $y(N^2)$ converges strongly to zero, so does (16-163).

The strong convergence to zero of the second term in (16-161) will be established by application of the Borel-Cantelli Lemma 16.10. This requires that its variance decreases sufficiently rapidly to zero as $N \to \infty$. Using $\text{var}(a(k)) \leq \mathbb{E}\{a^2(k)\}$, $(\sup_k |a(k)|)^2 = \sup_k |a(k)|^2$, and $\sup_{r < k \leq s} |b(k)| \leq \sum_{k=r+1}^{s} |b(k)|$, we find

$$\text{var}\left(\sup_{N^2 < k \leq (N+1)^2} |\Delta_2(k)| \right) \leq \mathbb{E}\left\{ \sup_{N^2 < k \leq (N+1)^2} |\Delta_2(k)|^2 \right\}$$

$$\leq \sum_{k=N^2+1}^{(N+1)^2} \mathbb{E}\{|\Delta_2(k)|^2\} \tag{16-164}$$

$$\leq \sum_{k=N^2+1}^{(N+1)^2} \frac{1}{k^2} \sum_{r,s=N^2+1}^{k} \text{cum}(z_k(r), z_k(s))$$

As $z_k(r)$ is mixing of order 2, (16-164) is bounded above by (see (16-36))

$$\text{var}(\sup_{N^2 < k \le (N+1)^2}|\Delta_2(k)|) \le \sum_{k=N^2+1}^{(N+1)^2} \frac{1}{k^2}O(k-N^2)$$

$$\le C\sum_{k=N^2+1}^{(N+1)^2} \frac{k-N^2}{k^2}$$

(16-165)

Using $\max_{N^2 < k \le (N+1)^2}\left|\frac{k-N^2}{k^2}\right| = \frac{(N+1)^2 - N^2}{(N+1)^4}$ for $N \ge 3$, (16-165) becomes

$$\text{var}(\sup_{N^2 < k \le (N+1)^2}|\Delta_2(k)|) \le C\left(\max_{N^2 < k \le (N+1)^2}\left|\frac{k-N^2}{k^2}\right|\right)((N+1)^2 - N^2)$$

$$\le O(N^{-2})$$

(16-166)

Applying the Borel-Cantelli lemma to (16-166) shows that the second term in (16-161) converges to zero w.p. 1.

The strong convergence to zero of the third term in (16-161) is shown following exactly the same lines as that of the second term. Similar to (16-164), we find

$$\text{var}(\sup_{N^2 < k \le (N+1)^2}|\Delta_3(k)|) \le \sum_{k=N^2+1}^{(N+1)^2} \frac{1}{k^2}\text{var}(\sum_{r=1}^{N^2}(z_k(r) - z_{N^2}(r)))$$

(16-167)

Using the assumption $\text{var}(\sum_{r=1}^{N^2}(z_k(r) - z_{N^2}(r))) = O(k - N^2)$ and following the lines of (16-165) and (16-166), (16-167) becomes

$$\text{var}(\sup_{N^2 < k \le (N+1)^2}|\Delta_3(k)|) \le O(N^{-2})$$

(16-168)

Applying the Borel-Cantelli lemma to (16-168) shows that the third term in (16-161) converges to zero w.p. 1. Finally, it follows that $\text{a.s.}\lim_{N \to \infty} y(N) = 0$, which proves the strong law of large numbers for mixing sequences. $\square$

Appendix 16.H Proof of the Central Limit Theorem for Mixing Sequences

We will show that the cumulants of $S(N)/\sqrt{N}$ converge for $N \to \infty$ to those of a normal distribution. This concludes the proof because the normal distribution is uniquely determined by its moments (see the Fréchet-Shohat Lemma 16.11).

The proof will be given for the general case where the sequence in the partial sum depends on N. Using the multilinearity of the cumulant (see Section 16.1, property 2), we find for the Jth order cumulant $c_J(N)$ of $S(N)/\sqrt{N}$

$$c_J(N) = \frac{1}{N^{J/2}}\sum_{k_1, k_2, \ldots, k_J = 1}^{N} \text{cum}(x_N(k_1), x_N(k_2), \ldots, x_N(k_J))$$

(16-169)

Because $x_N(k)$ is mixing of order J, $J = 1, 2, \ldots$, the summation in (16-169) is an $O(N)$ (see (16-36)), so that

$$\lim_{N \to \infty} c_J(N) = \lim_{N \to \infty} O(N^{1-J/2}) = 0 \text{ for } J > 2 \qquad (16\text{-}170)$$

By assumption var$(S(N)) = O(N)$ so that

$$\lim_{N \to \infty} c_2(N) = C \text{ with } 0 < C < \infty \qquad (16\text{-}171)$$

We conclude from (16-170) and (16-171) that the $c_J(N)$ converge to the cumulants of a normal distribution (see Example 16.2). The mixing assumption guarantees that the second-order cumulants of $x_N(k)$ are uniformly bounded. This ensures that the number of random variables $x_N(k)$ that have an $O(N^0) \geq C > 0$ contribution to var$(S(N)) = O(N)$ increases as $O(N)$.

The cumulants are the coefficients in the Taylor series expansion of the logarithm of the characteristic function $\phi(t)$ (Brillinger, 1981; Stuart and Ord, 1987). From (16-170), it follows that $\ln(\phi(t))$ corresponding to $S(N)$ equals that of a normal random variable within an $O(N^{-1/2})$, uniformly in t. Because the characteristic function is related to the probability density function by the Fourier integral, it follows that the distribution function $F_N(y)$ of $S(N)$ equals that of a normal random variable within an $O(N^{-1/2})$, uniformly in y.

Appendix 16.I Generalized Cauchy-Schwarz Inequality for Random Vectors

Let $U \in \mathbb{C}^n$ and $V \in \mathbb{C}^k$ be complex random vectors. Define the n by n matrix $M = \mathbb{E}\{(U - \Gamma V)(U - \Gamma V)^H\}$ with $\Gamma \in \mathbb{C}^{n \times k}$. By construction, M is positive semidefinite

$$\mathbb{E}\{(U - \Gamma V)(U - \Gamma V)^H\} \geq 0 \qquad (16\text{-}172)$$

Elaborating inequality (16-172) with $\Gamma = \mathbb{E}\{UV^H\}[\mathbb{E}\{VV^H\}]^+$ using property 2 of the pseudoinverse + (see Section 15.5) gives the *generalized Cauchy-Schwarz inequality* for random vectors

$$\mathbb{E}\{UU^H\} - \mathbb{E}\{UV^H\}[\mathbb{E}\{VV^H\}]^+ \mathbb{E}\{VU^H\} \geq 0 \qquad (16\text{-}173)$$

We will show that (16-173) reaches the lower bound if and only if there exists a nonrandom matrix Γ such that $U = \Gamma V$. Equality holds in (16-172), and, hence, also in (16-173), if and only if all the eigenvalues of the matrix $M = \mathbb{E}\{(U - \Gamma V)(U - \Gamma V)^H\}$ are zero. The positive semidefiniteness of M implies that all its eigenvalues are zero if and only if their sum equals zero. Hence, tr$(M) = 0$ (see Exercise 15.12.), so that

$$\text{tr}(\mathbb{E}\{(U - \Gamma V)(U - \Gamma V)^H\}) = \mathbb{E}\{(U - \Gamma V)^H(U - \Gamma V)\} = \mathbb{E}\{\|U - \Gamma V\|_2^2\} = 0 \quad (16\text{-}174)$$

The last equality in (16-174) is true if and only if $U = \Gamma V$. $\square$

Appendix 16.J Proof of the Generalized Cramér-Rao Inequality (Theorem 16.18)

The generalized Cramér-Rao inequality (16-84) follows as a special case of the generalized Cauchy-Schwarz inequality for random vectors (see Appendix 16.I). In what follows, all the expectations are taken w.r.t. the measurements z. Choosing $U = \hat{G}(\hat{\theta}(z)) - \mathbb{E}\{\hat{G}(\hat{\theta}(z))\}$ and $V^H = \partial \ln f_z(z, \theta_0)/\partial \theta_0$ in (16-173), taking into account that

$$\mathbb{E}\left\{\frac{\partial \ln f_z(z, \theta_0)}{\partial \theta_0}\right\} = \frac{\partial}{\partial \theta_0}\int_Z f_z(z, \theta_0)dz = \frac{\partial 1}{\partial \theta_0} = 0 \tag{16-175}$$

$$\mathbb{E}\left\{\hat{G}(\hat{\theta}(z))\frac{\partial \ln f_z(z, \theta_0)}{\partial \theta_0}\right\} = \frac{\partial}{\partial \theta_0}\int_Z \hat{G}(\hat{\theta}(z))f_z(z, \theta_0)dz = \frac{\partial G(\theta_0)}{\partial \theta_0} + \frac{\partial b_G}{\partial \theta_0} \tag{16-176}$$

gives

$$\mathrm{Cov}(\hat{G}(\hat{\theta}(z))) \geq \left(\frac{\partial G(\theta_0)}{\partial \theta_0} + \frac{\partial b_G}{\partial \theta_0}\right)Fi^+\left(\frac{\partial G(\theta_0)}{\partial \theta_0} + \frac{\partial b_G}{\partial \theta_0}\right)^H \tag{16-177}$$

Adding $b_G b_G^T$ to both sides of (16-177) gives (16-84). The second equality in (16-85) is obtained by differentiating (16-175) w.r.t. θ_0

$$\frac{\partial}{\partial \theta_0}\int_Z \frac{\partial \ln f_z(z, \theta_0)}{\partial \theta_0}f_z(z, \theta_0)dz = 0$$

$$\Downarrow \tag{16-178}$$

$$\mathbb{E}\left\{\frac{\partial^2 \ln f_z(z, \theta_0)}{\partial \theta_0^2}\right\} + \mathbb{E}\left\{\left(\frac{\partial \ln f_z(z, \theta_0)}{\partial \theta_0}\right)^T\left(\frac{\partial \ln f_z(z, \theta_0)}{\partial \theta_0}\right)\right\} = 0$$

Equations (16-175), (16-176), and (16-178) assume that the necessary regularity conditions to allow for the reversal of the order of differentiation and integration are satisfied (Z is the θ_0-independent range of integration). The suitable regularity conditions for the existence of the expected values and the derivatives can be found in Caines (1988). The necessary and sufficient condition (16-86) to attain the lower bound is a direct consequence of the generalized Cauchy-Schwarz inequality (see Appendix 16.I).

Appendix 16.K Proof of Lemma 16.23

Applying (6-110) on page 165 to the polynomial $J(z^{-1}, \theta) = \sum_{m=0}^{n_j} j_m z^{-m}$ in the noise model (16-128) gives

$$J(z^{-1}, \theta) = N^{-1/2}\left(\sum_{m=1}^{n_c}\sum_{t=1}^{m}c_m\Delta_N e(t)z^{t-m} - \sum_{n=1}^{n_d}\sum_{t=1}^{n}d_n\Delta_N v(t)z^{t-n}\right) \tag{16-179}$$

where $\Delta_N x(t) = x(-t) - x(N-t)$ with $x = e, v$. It shows that the coefficients j_m, $m = 0, 1, \ldots, n_j$, of $J(z^{-1}, \theta)$ depend linearly on $2(n_c + n_d)$ (finite number independent of N) random variables. Because $e(t)$ has uniformly bounded absolute moments of order $2 + \delta$, we also have that $\mathbb{E}\{|v(t)|^{2+\delta}\} \leq c < \infty$ (bounded input–bounded output property of a stable system $H(z^{-1})$). Hence, the coefficients j_m of $J(z^{-1}, \theta)$ can be written as

$$j_m = N^{-1/2}x(N) \text{ with } \mathbb{E}\{x(N)\} = 0 \text{ and } \mathbb{E}\{|x(N)|^{2+\delta}\} \le c < \infty \qquad (16\text{-}180)$$

(c is independent of t). Because $\text{var}(j_m) = O(N^{-1})$ it follows that $j_m = O_{\text{m.s.}}(N^{-1/2})$, which implies that $j_m = O_p(N^{-1/2})$ (see Section 16.7, interrelation 2). Applying Markov's inequality (16-21) with $p = 2 + \delta$ to $j_m(N)$ (16-180), we find

$$\sum_{N=1}^{\infty} \text{Prob}(|j_m(N)| > \varepsilon) \le \sum_{N=1}^{\infty} \frac{1}{\varepsilon^{2+\delta}} \mathbb{E}\{|x(N)|^{2+\delta}\} N^{-(1+\delta/2)} < \infty \qquad (16\text{-}181)$$

which shows that $j_m(N)$ converges w.p. 1 to zero (see the Borel-Cantelli Lemma 16.10). Using properties 1 and 3 of the stochastic limits (see Section 16.8), it follows that the results for j_m are also valid for $J(z^{-1}, \theta)$. □

Appendix 16.L Proof of Lemma 16.24

We will show that the joint cumulants of $E(k)$ tend for $N \to \infty$ to those of an independent, circular complex normally distributed random variable (the joint cumulants of order 3 and larger of a multivariate complex normal random variable are zero, see Example 16.2). This concludes the proof because the normal distribution is uniquely determined by its moments (see the Fréchet-Shohat Lemma 16.11).

The Jth order joint cumulant of $E(k)$ is

$$\text{cum}(E(k_1), E(k_2), \ldots, E(k_J)) = \frac{1}{N^{J/2}} \sum_{t_1, t_2, \ldots, t_J = 0}^{N-1} e^{-\frac{2\pi}{N} j \sum_{i=1}^{J} k_i t_i} \text{cum}(e(t_1), e(t_2), \ldots, e(t_J))$$

with $k_i = 0, 1, \ldots, N-1$, $i = 1, 2, \ldots, J$, and $J = 1, 2, \ldots$. Because the noise $e(t)$ is iid, $\text{cum}(e(t_1), e(t_2), \ldots, e(t_J))$ is different from zero only when $t_1 = t_2 = \cdots = t_J$ (see Section 16.1, property 3 of the cumulants). Putting $C_J = \text{cum}(e(t), e(t), \ldots, e(t))$ and using $\sum_{k=0}^{N-1} x^k = (1-x^N)/(1-x)$, we find

$$\text{cum}(E(k_1), E(k_2), \ldots, E(k_J)) = \frac{C_J}{N^{J/2-1}} \delta\left(\left(\sum_{i=1}^{J} k_i\right) \bmod N\right) \qquad (16\text{-}182)$$

with $\delta(k)$ the Kronecker delta. From (16-182) it follows that for any N

$$\begin{array}{lll}
\text{cum}(E(k)) = 0 & k \ne 0 & \text{(a)} \\
\text{cum}(E(k_1), E(k_2)) = 0 & (k_1 + k_2) \bmod N \ne 0 & \text{(b)} \\
\text{cum}(E(k_1), E(N-k_1)) = \text{cum}(E(k_1), \bar{E}(k_1)) = C_2 & & \text{(c)}
\end{array} \qquad (16\text{-}183)$$

All the cumulants of order larger than 2 tend to zero as $N \to \infty$ (16-182) and $\text{var}(E(k)) = O(N^0)$ (16-183c). We conclude from (16-182) and (16-183) that the cumulants equal, asymptotically, those of a zero mean (DC not included) independent circular complex normally distributed random variable.

The proof of the convergence rate of the distribution function is similar to that given in Appendix 16.H. □

Appendix 16.M Proof of Lemma 16.26

Because $|E(k)|^2$ is given by

$$|E(k)|^2 = \frac{1}{N} \sum_{t,u=0}^{N-1} e(t)e(u) e^{-\frac{2\pi}{N}jk(t-u)} \qquad (16\text{-}184)$$

we find for the Jth order joint cumulant of $|E(k)|^2$

$$c_J(k_1, k_2, \ldots, k_J) = \text{cum}(|E(k_1)|^2, |E(k_2)|^2, \ldots, |E(k_J)|^2)$$

$$= \frac{1}{N^J} \sum_{\substack{t_1, t_2, \ldots, t_J = 0 \\ u_1, u_2, \ldots, u_J = 0}}^{N-1} e^{-\frac{2\pi}{N}j \sum_{i=1}^{J} k_i(t_i - u_i)} c(t_1, \ldots, t_J, u_1, \ldots, u_J) \qquad (16\text{-}185)$$

with $c(t_1, \ldots, t_J, u_1, \ldots, u_J) = \text{cum}(e(t_1)e(u_1), e(t_2)e(u_2), \ldots, e(t_J)e(u_J))$. Application of Lemma 16.35 with $v_{i1} = e(t_i)$ and $v_{i2} = e(u_i)$ gives

$$c(t_1, \ldots, t_J, u_1, \ldots, u_J) = \sum_{\mathbb{P}} \text{cum}(e(t_{ij}) \in \mathbb{P}_1)\text{cum}(e(t_{ij}) \in \mathbb{P}_2)\ldots\text{cum}(e(t_{ij}) \in \mathbb{P}_K) \qquad (16\text{-}186)$$

with $e(t_{ij})$ an element of table (16-137) with $J_1 = J_2 = \cdots = J_J = 2$ and $I = J$, and where the summation extends over all indecomposable partitions $\mathbb{P} = \mathbb{P}_1 \cup \mathbb{P}_2 \cup \ldots \cup \mathbb{P}_K$ of this table. Since $e(t)$ is iid, $\text{cum}(e(t_{ij}) \in \mathbb{P}_k)$ is different from zero if and only if for all $e(t_{ij}) \in \mathbb{P}_k$, $e(t_{ij}) = e(u_k)$ (see Section 16.1, property 3 of the cumulants), and $\text{cum}(e(t_{ij} = u_k) \in \mathbb{P}_k) = c(\mathbb{P}_k)$ is independent of u_k. Putting these results in (16-185) gives

$$c_J(k_1, \ldots, k_J) = \frac{1}{N^J} \sum_{\mathbb{P}} c(\mathbb{P}_1)\ldots c(\mathbb{P}_K) \sum_{u_1, \ldots, u_K = 0}^{N-1} \exp(-\frac{2\pi}{N}j \sum_{i=1}^{J} k_i(u_{r_i} - u_{s_i})) \qquad (16\text{-}187)$$

with $r_i, s_i \in \{1, 2, \ldots, K\}$. Because the partition $\mathbb{P} = \mathbb{P}_1 \cup \mathbb{P}_2 \cup \ldots \cup \mathbb{P}_K$ is indecomposable, all the differences $u_{r_i} - u_{s_i}$, $i = 1, 2, \ldots, J$, are obtained by addition and subtraction of the $K-1$ independent differences $u_K - u_k$, $k = 1, 2, \ldots, K-1$ (see Lemma 2.3.1, p. 20, Brillinger, 1981)

$$u_{r_i} - u_{s_i} = \sum_{k=1}^{K-1} A_{[i,k]}(u_K - u_k) \text{ for } i = 1, 2, \ldots, J \qquad (16\text{-}188)$$

with $A \in \{-1, 0, 1\}^{J \times (K-1)}$ and $\text{rank}(A) = K - 1$. Using (16-188), the second summation in the right-hand side of (16-187) becomes

$$\sum_{u_1, \ldots, u_K = 0}^{N-1} \exp(-\frac{2\pi}{N}j \sum_{i=1}^{J} k_i(u_{r_i} - u_{s_i})) = \sum_{u_K = 0}^{N-1} \exp(-\frac{2\pi}{N}j u_K \sum_{k=1}^{K-1} \sum_{i=1}^{J} k_i A_{[i,k]}) \cdot$$

$$\prod_{k=1}^{K-1} \sum_{u_k=0}^{N-1} \exp(\frac{2\pi}{N}j u_k \sum_{i=1}^{J} k_i A_{[i,k]}) \qquad (16\text{-}189)$$

Applying K times $\sum_{k=0}^{N-1} x^k = (1-x^N)/(1-x)$ to (16-189) shows that (16-189) is zero unless $K-1$ linear independent constraints are satisfied (rank$(A) = K-1$)

$$\sum_{u_1,\ldots,u_K=0}^{N-1} \exp(-\frac{2\pi}{N} j \sum_{i=1}^{J} k_i(u_{r_i} - u_{s_i})) = N^K \Leftrightarrow (k^T A) \bmod N = 0 \qquad (16\text{-}190)$$

with $k^T = [k_1, k_2, \ldots, k_J]$. Using (16-190), (16-187), we find

$$c_J(k_1, \ldots, k_J) = \begin{cases} \dfrac{1}{N^{J-K}} \sum_{\mathbb{P}} c(\mathbb{P}_1) \ldots c(\mathbb{P}_K) \Leftrightarrow (k^T A) \bmod N = 0 \\ 0 \qquad\qquad\qquad\qquad\qquad\qquad\qquad \text{elsewhere} \end{cases}$$

Hence, the mixing condition of the Jth order joint cumulant becomes

$$\max_{k_J} \sum_{k_1,\ldots,k_{J-1}=1}^{N-1} |c_J(k_1,\ldots,k_J)| = \left|\sum_{\mathbb{P}} (N-1)^{J-K} N^{K-J} c(\mathbb{P}_1)\ldots c(\mathbb{P}_K)\right| = O(N^0)$$

where the last equality is due to the fact that the number of indecomposable sets is independent of N. □

Appendix 16.N Proof of Lemma 16.27

Using (16-183b) and (16-183c) we find $\max_{k_2} \sum_{k_1=1}^{N-1} |\text{cum}(E(k_1), E(k_2))| = C_2$, so that $E(k)$ is mixing of order 2.

The proof that $|E(k)|^2$ is mixing of order 2 is similar to the proof of Lemma 16.26, except that the third- and fourth-order cumulants of $e(t)$ are now not necessarily stationary. Hence, it is sufficient to study only the contribution of these nonstationary cumulants to the mixing condition. The second-order cumulant of $|E(k)|^2$ is given by (16-185) with $J = 2$. Equation (16-140) with $x_1 = e(t_1)$, $x_2 = e(u_1)$, $x_3 = e(t_2)$, and $x_4 = e(u_2)$ gives an explicit expression for (16-186) with $J = 2$. The eight terms in (16-140) containing a first-order cumulant have a zero contribution to (16-185) because

$$\sum_{t=0}^{N-1} \text{cum}(e(t)) e^{-\frac{2\pi}{N} jkt} = \mu \sum_{t=0}^{N-1} e^{-\frac{2\pi}{N} jkt} = 0 \Leftrightarrow k \neq 0$$

This eliminates all the (nonstationary) third-order cumulants. Because $e(t)$ is independent over t, the contribution of the (nonstationary) fourth-order cumulant to (16-185) becomes

$$\frac{1}{N^2} \sum_{\substack{t_1,t_2=0 \\ u_1,u_2=0}}^{N-1} \text{cum}(e(t_1), e(u_1), e(t_2), e(u_2)) e^{-\frac{2\pi}{N} j \sum_{i=1}^{2} k_i(t_i - u_i)} = \frac{1}{N^2} \sum_{t=0}^{N-1} C_4(t) \qquad (16\text{-}191)$$

Taking into account that $|C_4(t)|$ is uniformly bounded, it can be seen that (16-191) has a $O(N^0)$ contribution to the mixing condition. □

Appendix 16.O Proof of Theorem 16.28

Applying Lemma 16.23 to the partial sums (16-129) gives

$$\frac{1}{F}\sum_{k \in \mathbb{F}} W_k V(k) \rightarrow \frac{1}{F}\sum_{k \in \mathbb{F}} W_k H(z_k^{-1}) E(k) \qquad \text{w.p. 1}$$

$$\frac{1}{F}\sum_{k \in \mathbb{F}} |W_k V(k)|^2 \rightarrow \frac{1}{F}\sum_{k \in \mathbb{F}} |W_k H(z_k^{-1}) E(k)|^2 \qquad \text{w.p. 1}$$

(16-192)

at the rate $O_p(F^{-1/2})$. Hence, it is sufficient to study the following partial sums:

$$S(F) = \sum_{k \in \mathbb{F}} w_k E(k) \quad \text{and} \quad S(F) = \sum_{k \in \mathbb{F}} |w_k E(k)|^2 \qquad (16\text{-}193)$$

where $w_k = W_k H(z_k^{-1})$ is uniformly bounded. From Lemmas 16.3 and 16.27, it follows that $w_k E(k)$ and $|w_k E(k)|^2$ are mixing of order 2. Hence, the partial sums $S(F)/F$ in (16-193) converge in mean square sense at the rate $O_{\text{m.s.}}(F^{-1/2})$ to their expected value (see Section 16.9, version 3 of the law of large numbers). Note that the noise $E(k)$ in (16-193) depends on the number of time domain samples N and, hence, also on the number of frequencies $F = O(N)$. Therefore, it should be denoted more precisely as $E_N(k)$, and to prove the strong convergence of $S(F)/F$, we must also verify that

$$\text{var}(\sum_{k=1}^{s} x_r(k) - x_s(k)) = O(r-s) \text{ with } r \geq s \qquad (16\text{-}194)$$

is satisfied for $x_r(k) = w_k E_r(k)$ and $x_r(k) = |w_k E_r(k)|^2$ (see Section 16.9, version 3 of the law of large numbers). To verify this condition we rearrange the order of the frequencies in (16-193) such that the new added frequencies appear at the end of the sum. In (16-194) we compare terms of the sums (16-193), at the same physical frequencies kf_s/s and not at the same DFT line numbers k; otherwise, the comparison makes no sense. This imposes a condition on the number of time domain samples $r \geq s$: r must be chosen such that the physical frequencies kf_s/s, $k = 0, 1, ..., s-1$, form a subset of the physical frequencies kf_s/r, $k = 0, 1, ..., r-1$. This condition is satisfied for the choice $r = ms$ with $m = 1, 2, 3, ...$. It means that we compare time domain experiments where the number of samples N is increased linearly as mN, $m = 1, 2, 3, ...$. Note that in a classical time domain analysis the number of samples is increased linearly as $N+m$, $m = 1, 2, 3, ...$.

The first partial sum in (16-193) converges strongly to its expected value if

$$\text{var}(\sum_{k \in \mathbb{F}_s} w_k (E_{ms}(mk) - E_s(k))) = O((m-1)s) \qquad (16\text{-}195)$$

with $m \in \mathbb{N}_0$, $\mathbb{F}_s$ a set of $F_s = O(s)$ DFT frequencies, and

$$E_{ms}(mk) = \frac{1}{\sqrt{ms}} \sum_{t=0}^{ms-1} e(t) e^{-2\pi jkt/s}$$

$$E_s(k) = \frac{1}{\sqrt{s}} \sum_{t=0}^{s-1} e(t) e^{-2\pi jkt/s}$$

(16-196)

Because $\sum_{t=0}^{s-1} e^{-2\pi jkt/s} = 0$ for $k \neq 0$, we can replace $e(t)$ by $e(t) - \mu$ in (16-196) without changing $E_{ms}(mk)$ and $E_s(k)$ for $k \neq 0$. Hence, we may assume in the sequel of the analysis that $e(t)$ has zero mean. If DC ($k = 0$) belongs to the set $\mathbb{F}_s$, then $\mu = \mathbb{E}\{e(t)\}$ should be zero, otherwise the expected values of $E_{ms}(0)$ and $E_s(0)$ are not zero. Elaborating the variance expression in (16-195) gives

$$\operatorname{var}(\sum_{k \in \mathbb{F}_s} w_k(E_{ms}(mk) - E_s(k))) = \sum_{k_1, k_2 \in \mathbb{F}_s} w_{k_1} \bar{w}_{k_2} \mathbb{E}\{(E_{ms}(mk_1) - E_s(k_1))\overline{(E_{ms}(mk_2) - E_s(k_2))}\} \quad (16\text{-}197)$$

Because $e(t)$ is an independent random variable with zero mean and variance σ^2, we have

$$\mathbb{E}\{E_{ms}(mk_1)\bar{E}_{ms}(mk_2)\} = \mathbb{E}\{E_s(k_1)\bar{E}_s(k_2)\} = \sigma^2 \delta(k_1 - k_2)$$
$$\mathbb{E}\{E_{ms}(mk_1)\bar{E}_s(k_2)\} = \mathbb{E}\{E_s(k_1)\bar{E}_{ms}(mk_2)\} = \frac{\sigma^2}{\sqrt{m}} \delta(k_1 - k_2) \quad (16\text{-}198)$$

with $\delta(k)$ the Kronecker delta. Using (16-197) and (16-198), we find

$$\operatorname{var}(\sum_{k \in \mathbb{F}} w_k(E_{ms}(mk) - E_s(k))) = 2\sigma^2 \frac{(m-1)s}{\sqrt{m}(\sqrt{m}+1)s} \frac{1}{s} \sum_{k \in \mathbb{F}_s} |w_k|^2 \leq O((m-1)s) \quad (16\text{-}199)$$

where the last inequality is due to $|w_k| \leq c < \infty$ for any k.

The variance expression for the second partial sum in (16-193) equals

$$\operatorname{var}(\sum_{k \in \mathbb{F}} |w_k|^2 (|E_{ms}(mk)|^2 - |E_s(k)|^2)) = \sum_{k_1, k_2 \in \mathbb{F}_s} |w_{k_1}|^2 |w_{k_2}|^2 \mathbb{E}\{(|E_{ms}(mk_1)|^2 - |E_s(k_1)|^2)(|E_{ms}(mk_2)|^2 - |E_s(k_2)|^2)\} \quad (16\text{-}200)$$

Because $e(t)$ is an independent random variable with zero mean, variance σ^2, and uniformly bounded fourth-order moment $\mu_4(t)$, we have

$$\mathbb{E}\{|E_{ms}(mk_1)|^2 |E_{ms}(mk_2)|^2\} = \sigma^4 + \sigma^4 \delta(k_1 - k_2) + \frac{\kappa_4(ms)}{ms}$$
$$\mathbb{E}\{|E_s(k_1)|^2 |E_s(k_2)|^2\} = \sigma^4 + \sigma^4 \delta(k_1 - k_2) + \frac{\kappa_4(s)}{s} \quad (16\text{-}201)$$
$$\mathbb{E}\{|E_{ms}(mk_1)|^2 |E_s(k_2)|^2\} = \mathbb{E}\{|E_{ms}(mk_1)|^2 |E_s(k_2)|^2\} = \sigma^4 + \frac{\sigma^4}{m} \delta(k_1 - k_2) + \frac{\kappa_4(s)}{ms}$$

where $\kappa_4(r) = \sum_{t=0}^{r-1} \mu_4(t)/r - 3\sigma^4$ is an $O(s^0)$. Using (16-200) and (16-201), we find

$$\mathrm{var}(\sum_{k \in \mathbb{F}} |w_k|^2 (|E_{ms}(mk)|^2 - |E_s(k)|^2)) = \frac{(m-1)s}{m} \frac{2\sigma^2}{s} \sum_{k \in \mathbb{F}_s} |w_k|^4$$
$$+ \left(\frac{(m-1)s}{m} \kappa_4(s) + \frac{1}{m} \sum_{t=s}^{ms-1} \mu_4(t)\right) \frac{1}{s^2} \sum_{k_1, k_2 \in \mathbb{F}_s} |w_{k_1}|^2 |w_{k_2}|^2 \qquad (16\text{-}202)$$

Because $\kappa_4(s) = O(s^0)$, $|w_k| \leq c < \infty$ is uniformly bounded, and

$$\left|\sum_{t=s}^{ms-1} \mu_4(t)\right| \leq (\max_t |\mu_4(t)|)((m-1)s)$$

(16-202) is bounded above by

$$\mathrm{var}(\sum_{k \in \mathbb{F}} |w_k|^2 (|E_{ms}(mk)|^2 - |E_s(k)|^2)) \leq O((m-1)s)$$

which concludes the proof for the second partial sum. □

Appendix 16.P Proof of Theorem 16.29

To prove the theorem it is sufficient to replace $V(k)$ by $H(z_k^{-1})E(k)$ (Lemma 16.23). The rest of the proof follows the lines of Appendix 16.L. Using (16-182) and property 2 of the cumulants (see Section 16.1), the Jth order cumulant of $N^{-1/2}S(N)$ becomes

$$C_J(N) = \frac{C_J}{N^{J-1}} \sum_{k_1, k_2, \ldots, k_{J-1} = 0}^{N-1} Y(k_1) Y(k_2) \ldots Y(k_{J-1}) Y(N - \sum_{i=1}^{J-1} k_i) \qquad (16\text{-}203)$$

with $Y(k) = \bar{H}(z_k^{-1})X(k)$. The right-hand side of (16-203) can be written as $J-1$ consecutive circular DFT convolutions of $Y(k)$ with itself

$$C_J(N) = \frac{C_J}{N^{(J-1)/2}} Y_J(N) \qquad (16\text{-}204)$$

with $Y_J(k) = Y(k)*(Y(k)*(\ldots *Y(k)))$ and $Y(k)*Z(k) = N^{-1/2} \sum_{r=0}^{N-1} Y(r)Z(k-r)$. Using the property that the inverse discrete Fourier transform (IDFT) of a circular convolution of DFT spectra equals the product of the corresponding time signals, we can write $Y_J(k)$ as

$$Y_J(k) = \mathrm{DFT}(\mathrm{IDFT}(Y_J(k))) = \mathrm{DFT}(y^J(t)) = \frac{1}{\sqrt{N}} \sum_{t=0}^{N-1} y^J(t) e^{-\frac{2\pi}{N}jkt} \qquad (16\text{-}205)$$

Hence, $Y_J(N) = N^{-1/2} \sum_{t=0}^{N-1} y^J(t)$, and

$$C_J(N) = \frac{C_J}{N^{J/2 - 1}} \frac{1}{N} \sum_{t=0}^{N-1} y^J(t) \qquad (16\text{-}206)$$

Applying the bounded-input, bounded-output property of stable linear systems (Kailath, 1980) to $\bar{Y}(k) = H(z_k^{-1})\bar{X}(k)$, with $\mathrm{IDFT}(\bar{Z}(k)) = z(-t)$ ($Z = X, Y$, $z = x, y$), shows that $\max_t |y(-t)| \leq c_1 < \infty$ if $\max_t |x(-t)| \leq c < \infty$ where c and c_1 are independent of N

($\max_t|z(-t)| = \max_t|z(t)|$, $z = x, y$). Similarly, $N^{-1}\sum_{t=0}^{N-1} y^2(t) = O(N^0)$ since $N^{-1}\sum_{t=0}^{N-1} x^2(t) = O(N^0)$. Therefore, (16-206) becomes for $J = 2$

$$C_2(N) = \text{var}(N^{-1/2}S(N)) = \frac{C_2}{N^0}O(N^0) = O(N^0) \tag{16-207}$$

while for $J > 2$ (16-206) can be bounded above by $(N^{-1}|\sum_{t=0}^{N-1} y^J(t)| \leq \max_t |y(t)|^J)$

$$|C_J(N)| \leq \frac{C_J}{N^{J/2-1}} c_1^J \leq O(N^{-J/2+1}) \tag{16-208}$$

It follows that the cumulants of order $J = 3, 4, \ldots$ are asymptotically zero, while $C_2(N) = O(N^0)$. According to the Fréchet-Shohat Lemma 16.11, $N^{-1/2}S(N)$ is asymptotically normally distributed. The convergence rate to the normal distribution function is established as in Appendix 16.H. □

Appendix 16.Q Proof of Corollary 16.30

The proof follows the same lines as for Theorem 16.29. The only difference lies in the upper bound (16-208). Because (16-207) remains valid (we consider only signals with finite power), the signal can reach its peak value $O(\sqrt{\ln N})$ at most $O(N/\ln N)$ times, while the remaining $N - O(N/\ln N)$ samples have the value $O(N^0)$. Therefore, (16-206) with $J > 2$, is bounded above by

$$|C_J(N)| \leq \frac{C_J}{N^{J/2-1}} O((\ln N)^{J/2-1}) \leq O\left(\left(\frac{\ln N}{N}\right)^{J/2-1}\right) \tag{16-209}$$

which concludes the proof. □

Appendix 16.R Proof of Lemma 16.31

First we prove the theorem for $X(k) = E_2(k)$ and $H_1(z_k^{-1}) = 1$. The generalization to the colored case follows directly from the uniformly boundedness of $H_1(z_k^{-1})$ and $H_2(z_k^{-1})$. Applying Lemma 16.35 with $v_{i1} = E_2(k)$ and $v_{i2} = \bar{E}_1(k)$ to the Jth order joint cumulant of $E_2(k)\bar{E}_1(k)$ gives

$$\text{cum}(E_2(k_1)\bar{E}_1(k_1), E_2(k_2)\bar{E}_1(k_2), \ldots, E_2(k_J)\bar{E}_1(k_J)) =$$
$$\sum_{\mathbb{P}} \text{cum}(E(k_{ij}) \in \mathbb{P}_1)\text{cum}(E(k_{ij}) \in \mathbb{P}_2)\cdots\text{cum}(E(k_{ij}) \in \mathbb{P}_K) \tag{16-210}$$

where $E(k)$ equals $E_2(k)$ and/or $\bar{E}_1(k)$ and where the summation extends over all indecomposable partitions $\mathbb{P} = \mathbb{P}_1 \cup \mathbb{P}_2 \cup \cdots \cup \mathbb{P}_K$ of the table (16-137) with $J_1 = J_2 = \cdots = J_J = 2$. We study the mixing condition

$$\max_{k_J} \sum_{k_1, \ldots, k_{J-1}=0}^{N-1} |\text{cum}(E(k_{ij}) \in \mathbb{P}_1)\text{cum}(E(k_{ij}) \in \mathbb{P}_2)\cdots\text{cum}(E(k_{ij}) \in \mathbb{P}_K)| \tag{16-211}$$

for each term in the summation (16-210). As $\bar{E}_1(k)$ and $E_2(k)$ are mutually independent random variables, the partitions in (16-210) are limited to those where all $E(k_{ij}) \in \mathbb{P}_r$ are equal to $E_2(k)$ or to $\bar{E}_1(k)$. Therefore, we can apply formula (16-182) to each $\text{cum}(E(k_{ij}) \in \mathbb{P}_r)$ in (16-211), which gives

$$\max_{k_J} \sum_{k_1, \ldots, k_{J-1} = 0}^{N-1} \left| \prod_{r=1}^{K} \frac{C_{J_r}}{N^{J_r/2 - 1}} \right| \qquad (16\text{-}212)$$

with $\sum_{r=1}^{K} J_r = 2J$ (each partition contains all elements of the set) and where K constraints of the form

$$(\sum k_{ij}) \bmod N = 0, \quad E(k_{ij}) \in \mathbb{P}_r, \quad r = 1, 2, \ldots, K \qquad (16\text{-}213)$$

should be satisfied. Because the partition $\mathbb{P} = \mathbb{P}_1 \cup \mathbb{P}_2 \cup \cdots \cup \mathbb{P}_K$ is indecomposable, there are exactly $K - 1$ independent constraints in (16-213) (see Lemma 2.3.1, p. 20, Brillinger, 1981), so that (16-212) can be bounded above by

$$N^{(J-1)-(K-1)} \frac{\prod_{r=1}^{K} |C_{J_r}|}{N^{J-K}} = O(N^0) \qquad (16\text{-}214)$$

This concludes the proof for the (colored) white noise case.

The result (16-214) is also valid when $X(k)$ is the DFT spectrum of an integer number of periods of normalized periodic noise or a normalized random multisine. To prove this statement, it is sufficient to note that

$$\text{cum}(X(k_1), X(k_2), \ldots, X(k_J)) = C_J \delta(k_2 - k_1) \delta(k_3 - k_1) \cdots \delta(k_J - k_1)$$

by construction of these periodic signals (see Definitions 3.2 and 3.4). □

Appendix 16.S Proof of Theorem 16.32

The proof follows the lines of Appendix 16.O.

16.S.1 First Partial Sum of (16-135). We distinguish two cases: (i) $X(k)$ is the DFT spectrum of filtered white noise, $x(t) = H_1(q)e_1(t)$, where $H_1(z^{-1})$ is stable and $e_1(t)$ is independently distributed noise with mean $\mu_1 < \infty$ and variance $\sigma_1^2 < \infty$, and (ii) $X(k)$ is the DFT spectrum of an integer number of periods of a normalized random multisine or normalized periodic noise.

Applying Lemma 16.23 to the first partial sum of (16-135) gives, for case (i),

$$\frac{1}{F} \sum_{k \in \mathbb{F}} W_k X(k) \bar{V}(k) \to \frac{1}{F} \sum_{k \in \mathbb{F}} w_k E_1(k) \bar{E}(k) \qquad \text{w.p. 1}$$

at the rate $O_p(F^{-1/2})$, where $w_k = W_k H_1(z_k^{-1})\bar{H}(z_k^{-1})$ is uniformly bounded and where, at the same physical frequencies kf_s/N, w_k is independent of the number of time domain samples N. Hence, it is sufficient to study

$$S(F) = \sum_{k \in \mathbb{F}} w_k E_1(k)\bar{E}(k)$$

Using formulas (16-198) and the fact that $E_1(k)$ and $E(k)$ are independent, we find

$$\text{var}(\sum_{k \in \mathbb{F}} w_k(E_{1ms}(mk)\bar{E}_{ms}(mk) - E_{1s}(k)\bar{E}_s(k))) = 2\sigma_1^2 \sigma^2 \frac{(m-1)s}{m} \frac{1}{s} \sum_{k \in \mathbb{F}_s} |w_k|^2$$

$$\leq O((m-1)s)$$

where the last inequality is due to $|w_k| \leq c < \infty$ for any k.

Applying Lemma 16.23 to the first partial sum of (16-135) gives, for case (ii),

$$\frac{1}{F}\sum_{k \in \mathbb{F}} W_k X(k)\bar{V}(k) \to \frac{1}{F}\sum_{k \in \mathbb{F}} w_k X(k)\bar{E}(k) \qquad \text{w.p. 1}$$

at the rate $O_p(F^{-1/2})$, where $w_k = W_k \bar{H}(z_k^{-1})$ is uniformly bounded and where, at the same physical frequencies kf_s/N, w_k and $X(k)$ are independent of the number of time domain samples N. Hence, we must study

$$\text{var}(\sum_{k \in \mathbb{F}_s} w_k X(k)(\bar{E}_{ms}(mk) - \bar{E}_s(k))) \qquad (16\text{-}215)$$

Because $X(k)$ is independent of $E(k)$, formula (16-199) of Appendix 16.O remains valid for (16-215), if $|w_k|^2$ is replaced by $|w_k|^2 \mathbb{E}\{|X(k)|^2\}$.

16.S.2 Second Partial Sum of (16-135). As the case where $X(k)$ is the DFT spectrum of filtered white noise is already covered by Theorem 16.28, it is sufficient to study the case where $X(k)$ is the DFT spectrum of an integer number of periods of a periodic signal. For normalized random multisines, $|X(k)|^2$ is a uniformly bounded nonrandom number, while for normalized periodic noise, $|X(k)|^2$ is a random variable with uniformly bounded fourth-order moments. In both cases, at the same physical frequencies kf_s/N, $X(k)$ is independent of the number of time domain samples N. Hence, $S(F)/F$ obeys the strong law of large numbers (see Section 16.9, version 3 of the law of large numbers). □

Appendix 16.T Proof of Theorem 16.33

Applying Lemma 16.23 to the first two partial sums of (16-136) gives

$$S(F)/F \to \frac{1}{F}\sum_{k \in \mathbb{F}} W_k Y(k)\overline{H(z_k^{-1})E(k)} \qquad \text{w.p. 1}$$

$$S(F)/F \to \frac{1}{F}\sum_{k \in \mathbb{F}} |W_k H(z_k^{-1})E(k)|^2 \qquad \text{w.p. 1}$$

(16-216)

at the rate $O_p(F^{-1/2})$. $Y(k) = H_1(z_k^{-1})E_1(k)$ for filtered iid noise, and $Y(k) = X(k)$ for the periodic signals. Because W_k, $H_1(z_k^{-1})$, and $H(z_k^{-1})$ are uniformly bounded, $W_k Y(k)\overline{H(z_k^{-1})E(k)}$ and $|W_k H(z_k^{-1})E(k)|^2$ are mixing of order infinity (proof: apply Lemmas 16.3, 16.26, and 16.31). Hence, according to version 4 of the central limit theorem (see Section 16.10), the sums in (16-216) are asymptotically normally distributed (convergence in law at the rate $O(F^{-1/2})$).

For the third partial sum of (16-136), we need only handle the case where $X(k)$ is the DFT spectrum of the periodic signals (the filtered iid case is already covered by the second partial sum of (16-136)). Because $|X(k)|^2$ is, by construction, independent over k (see Definitions 3.2 and 3.4), and W_k, $|X(k)|^2$ have, by assumption, existing moments of any order, $W_k|X(k)|^2$ is mixing of order infinity. Therefore, $W_k|X(k)|^2$ is asymptotically normally distributed at the rate $O(F^{-1/2})$ (see Section 16.10, version 4 of the central limit theorem). For a normalized random multisine, $|X(k)|^2$ is a nonrandom number and the normal distribution is degenerate. □

17

Properties of Least Squares Estimators with Deterministic Weighting

Abstract: This chapter studies the asymptotic stochastic properties (strong convergence, strong consistency, convergence rate, asymptotic bias, and asymptotic normality) of nonlinear least squares estimators with a deterministic weighting. The presented theory is applicable to a large class of estimators such as the quadratic prediction error methods, the Gaussian maximum likelihood estimators, and the total least squares–based methods. Readers who are unfamiliar with the analysis of the stochastic properties of estimators should first read Sections 16.11 to 16.13.

17.1 INTRODUCTION

In this chapter we consider the identification of a parametric plant and/or noise model $M(\theta, z_0, n_z)$ through the minimization of a weighted nonlinear least squares cost function

$$V_N(\theta, z) = \frac{1}{N} z^T W_N(\theta) z \qquad (17\text{-}1)$$

with $W_N(\theta) \in \mathbb{R}^{N \times N}$ a deterministic positive semidefinite weighting matrix, $z \in \mathbb{R}^N$ the noisy measurements, $z = z_0 + n_z$, and $\theta \in \mathbb{R}^{n_\theta}$ the plant and/or noise model parameters with n_θ independent of N. Because the nonsymmetric part of $W_N(\theta)$ does not contribute to the quadratic form (17-1) (see Exercise 15.7), we can assume without any loss of generality that the weighting matrix $W_N(\theta)$ is symmetric. Note that all the elements of $W_N(\theta)$ may change as N increases. Because the true (unknown) observations z_0 can be a random variable, the expected values are taken everywhere w.r.t. the disturbing noise n_z and the true observations z_0.

The analysis of the stochastic properties of the minimizer(s) of (17-1) requires a closed and bounded (= compact) set of parameters, where the cost function (17-1) and/or its higher order derivatives exist and are finite. Such a regular compact set is constructed as follows. Let $\mathbb{P} \subset \mathbb{R}^{n_\theta}$, with $\dim(\mathbb{P}) = n_\theta$, be a compact parameter set. Define $\mathbb{P}_s \subset \mathbb{P}$ as the singular set of parameter values for which the cost function (17-1) does not exist or is infinite.

Usually, the topological dimension of this singular set is smaller than n_θ. The regular set $\mathbb{P}_r$ are the parameters in $\mathbb{P}$ that are not within an ε-distance of the singular set $\mathbb{P}_s$

$$\mathbb{P}_r = \mathbb{P}\setminus\{\theta \in \mathbb{P} | \|\theta - \theta_s\| < \varepsilon, \theta_s \in \mathbb{P}_s\} \qquad (17\text{-}2)$$

$\mathbb{P}_r$ is compact (closed and bounded) by construction. Using the same reasoning, a regular compact set is constructed where the (higher order) derivatives of the cost function exist and are finite. Note that for the maximum likelihood estimation of ARMAX models the compactness assumption of the parameter space can be avoided (Hannan and Deistler, 1988).

Following the same lines as in Section 16.13, the asymptotic ($N \to \infty$) properties (strong convergence, strong consistency, convergence rate, asymptotic bias, and asymptotic normality) of the (set of) minimizer(s)

$$\hat{\theta}(z) = \arg\min_{\theta \in \mathbb{P}_r} V_N(\theta, z) \qquad (17\text{-}3)$$

will be analyzed. Replacing $z \in \mathbb{C}^N$ by $z_{re} \in \mathbb{R}^{2N}$ in (17-1) and/or $\theta \in \mathbb{C}^{n_\theta}$ by $\theta_{re} \in \mathbb{R}^{2n_\theta}$ (see Section 15.8 for the definition of $(\ )_{re}$), it follows directly that the results also apply to complex measurements and/or complex parameters $\theta \in \mathbb{C}^{n_\theta}$. The chapter ends with an overview of the asymptotic properties of $\hat{\theta}(z)$.

17.2 STRONG CONVERGENCE

The first step in the analysis consists of detecting the stochastic sum(s) w in the cost function (17-1) that averages the noise. It can easily be seen that there is only one such sum, namely the cost function itself. Following the notations of Section 16.13, we have

$$w(\theta, z_0, n_z) = V_N(\theta, z) \text{ and } \mu_w(\theta, z_0) = \mathbb{E}\{V_N(\theta, z)\} \qquad (17\text{-}4)$$

so that

$$\tilde{\theta}(z_0) = \arg\min_{\theta \in \mathbb{P}_r} V_N(\theta) \qquad (17\text{-}5)$$

with $V_N(\theta) = \mathbb{E}\{V_N(\theta, z)\}$.

In the second step (see Section 17.2.1), the uniform convergence (w.r.t. θ) of the stochastic sum w toward its expected value μ_w is established

$$\text{a.s.}\lim_{N \to \infty}(V_N(\theta, z) - V_N(\theta)) = 0 \text{ or a.s.}\lim_{N \to \infty} V_N(\theta, z) = \lim_{N \to \infty} V_N(\theta) = V_*(\theta) \qquad (17\text{-}6)$$

This requires some assumptions concerning the true observations z_0, the disturbing noise n_z, the weighting matrix $W_N(\theta)$, and the strategy of adding the measurements. The convergence should be uniform w.r.t. the model parameters $\theta \in \mathbb{P}_r$ to ensure that the convergence of the cost functions implies the convergence of the minimizers. Figure 17-1 shows a counterexample where the cost function $V_N(\theta)$ converges nonuniformly to its limit value $V_*(\theta)$. It can be seen that the global minimum of the sequence $V_N(\theta)$ does not converge to the global minimum of $V_*(\theta)$.

Section 17.2 ■ Strong Convergence

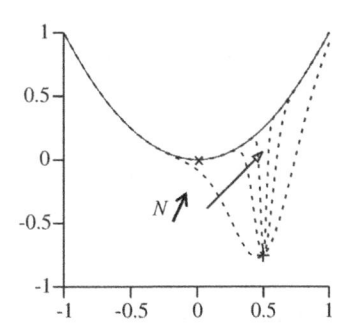

Figure 17-1. Although $V_N(\theta, z)$ (dashed lines) converges nonuniformly to $V_*(\theta)$ (solid line), their global minimizers (×) and (+) differ.

In a third step (see Section 17.2.2) the strong convergence of the minimizer(s) is established from the strong uniform convergence of the cost function

$$\text{a.s.}\lim_{N \to \infty} (\hat{\theta}(z) - \tilde{\theta}(z_0)) = 0 \text{ or a.s.}\lim_{N \to \infty} \hat{\theta}(z) = \lim_{N \to \infty} \tilde{\theta}(z_0) = \theta_* \tag{17-7}$$

It requires that adding measurements to z ($N \to \infty$) increases the knowledge about the model parameters θ such that θ is uniquely identifiable. If this is the case, then the data are said to be *persistently exciting*. The weakest assumption that satisfies this condition is that the asymptotic cost function $V_*(\theta)$ has a unique global minimum θ_*. If this assumption is not fulfilled, then the uniform convergence of the cost functions does not imply the convergence of their minimizer(s). The following counterexample shows this. Consider, for example, the cost functions $V_{1N}(\theta) = 1 - N^{-1}\sin(\theta)$ and $V_{2N}(\theta) = 1 - N^{-1}\cos(\theta)$ with respective minimizers $\hat{\theta}_1(N) = \pi/2 + 2k\pi$ and $\hat{\theta}_2(N) = 2k\pi$, $k \in \mathbb{Z}$. Although $V_{1N}(\theta)$ converges uniformly in θ to $V_{2N}(\theta)$, $\hat{\theta}_1(N)$ does not converge to $\hat{\theta}_2(N)$: $\hat{\theta}_1(\infty) \neq \hat{\theta}_2(\infty)$. The problem with this counterexample is that all θ-values minimize the limit cost function $V_*(\theta) = 1$.

17.2.1 Strong Convergence of the Cost Function

Assumption 17.1 (Mixing Condition of Order P): The true observations $z_0 \in \mathbb{R}^N$ are disturbed by zero mean additive noise $z = z_0 + n_z$. The noise n_z is stochastically independent of z_0. Both n_z and z are mixing of order P.

Assumption 17.2 (Constraints on the Cost Function): (a) The weighting matrix $W_N(\theta) \in \mathbb{R}^{N \times N}$ in (17-1) is a symmetric positive semidefinite matrix, satisfying $\|W_N(\theta)\|_1 \leq c < \infty$, with c an N-independent constant, for all N, ∞ included, and all $\theta \in \mathbb{P}_r$. $W_N(\theta)$ is a continuous matrix function of θ in the compact set $\mathbb{P}_r$. (b) There is an N_0 such that for any $r \geq s \geq N_0$, $\|W_{r[1:s, 1:s]}(\theta) - W_s(\theta)\|_1^2 = O((r-s)/r)$ in $\mathbb{P}_r$.

Note that Assumption 17.1 makes it possible to handle, simultaneously, the cases z_0 random and/or z_0 deterministic. Condition (b) in Assumption 17.2 limits the variation of the elements of $W_N(\theta)$ as N increases to infinity. This is necessary to ensure the strong convergence of the cost function (see proof of Lemma 17.3). If (b) is not satisfied, then only mean square convergence of the cost function can be shown. All lemmas and theorems of this chapter remain valid except that the strong convergence (w.p. 1) must be replaced by weak convergence (in prob.).

Lemma 17.3 (Strong Convergence of the Cost Function): Under Assumptions 17.1 ($P = 4$) and 17.2 the cost function $V_N(\theta, z)$ converges uniformly w.p. 1 to its expected value

$V_N(\theta)$ in the compact set $\mathbb{P}_r$. The uniform mean square convergence rate in $\mathbb{P}_r$ is $O_{\text{m.s.}}(N^{-1/2})$: $V_N(\theta, z) = V_N(\theta) + O_{\text{m.s.}}(N^{-1/2})$.

Proof. See Appendix 17.A. □

Lemma 17.3 does not guarantee that the limit cost function $V_*(\theta) = \lim_{N \to \infty} V_N(\theta)$ exists. Because $V_N(\theta)$ depends on $W_N(\theta)$ and z, the existence of $V_*(\theta)$ imposes some conditions on $W_N(\theta)$ and z that should be verified for each particular choice of the weighting $W_N(\theta)$ and for every experiment (strategy of adding measurements to z). Therefore, we make the following assumption.

Assumption 17.4 (Constraint on the Experiment): The expected value of the cost function $V_N(\theta)$ converges uniformly to the limit cost function $V_*(\theta)$ in the compact set $\mathbb{P}_r$.

17.2.2 Strong Convergence of the Minimizer

Assumption 17.5 (Persistence of Excitation): There exists an N_0 such that for any $N \geq N_0$, ∞ included, the expected value of the cost function $V_N(\theta)$ has a unique global minimum $\tilde{\theta}(z_0)$, which is an interior point of $\mathbb{P}_r$.

Theorem 17.6 (Strong Convergence of the Minimizer): Under Assumptions 17.1 ($P = 4$), 17.2, and 17.5, the minimizer(s) $\hat{\theta}(z)$ converge(s) strongly to $\tilde{\theta}(z_0)$: $\text{a.s.} \lim_{N \to \infty} (\hat{\theta}(z) - \tilde{\theta}(z_0)) = 0$.

Proof. See Appendix 17.B. □

Theorem 17.6 does not guarantee that $\tilde{\theta}(z_0)$ converges to some limit value θ_*. Assumptions 17.4 and 17.7 ensure the existence of this limit value.

Assumption 17.7 (Persistence of Excitation): The asymptotic cost function $V_*(\theta)$ has a unique global minimum θ_*, which is an interior point of $\mathbb{P}_r$.

If $V_N(\theta)$ and/or $V_*(\theta)$ are not convex, then in the presence of model errors it may happen that $V_N(\theta)$ and/or $V_*(\theta)$ have more than one global minimum. An example of this is given in Kabaila (1983) for the identification of particular parametric noise models (MA processes). To handle these cases we restrict the compact set $\mathbb{P}_r$ in Assumptions 17.5 and 17.7 such that $V_N(\theta)$ and/or $V_*(\theta)$ contain a unique global minimum in $\mathbb{P}_r$.

Theorem 17.8 (Strong Convergence of the Minimizer): Under Assumptions 17.1 ($P = 4$), 17.2, 17.4, and 17.7, $\tilde{\theta}(z_0)$ converges to θ_* and $\hat{\theta}(z)$ converges strongly to θ_*: $\lim_{N \to \infty} \tilde{\theta}(z_0) = \theta_*$ and $\text{a.s.} \lim_{N \to \infty} \hat{\theta}(z) = \theta_*$.

Proof. Note that $V_*(\theta)$ is a continuous function in $\mathbb{P}_r$ because it is the deterministic limit of a uniformly convergent sequence of continuous functions $V_N(\theta)$ in $\mathbb{P}_r$ (Theorem 2.1 of Henrici, 1974). The proof of the two limits then follows the same lines as for Theorem 17.6. □

Note that Theorems 17.6 and 17.8 do not require the existence of the derivative(s) of the cost function and are valid in the presence of model errors (the true model cannot be represented by $M(\theta, z_0, n_z)$).

17.3 STRONG CONSISTENCY

Consistency can be proved only if the true model belongs to the considered model set. Therefore, the following assumption is made.

Assumption 17.9 (True Model Belongs to Model Set): There is a $\theta_0 \in \mathbb{P}_r$ such that $M(\theta_0, z_0, n_z)$ represents the true model.

Using the results of Section 17.2, it follows directly that the estimates $\hat{\theta}(z)$ are strongly consistent if either $\theta_* = \theta_0$ (weakest assumption) or $\hat{\theta}(z_0) = \theta_0$ for any $N \geq N_0$ (stronger assumption). This imposes some conditions on the expected value of the cost function, which can be written as

$$V_N(\theta) = \frac{1}{N}\mathbb{E}\{z_0^T W_N(\theta) z_0\} + \frac{1}{N}\text{trace}(W_N(\theta) C_{n_z}) \qquad (17\text{-}8)$$

with C_{n_z} the covariance matrix of the disturbing noise n_z (see Exercise 17.2). The following theorems are in order of reduced conditions on $V_N(\theta)$.

Assumption 17.10 (Consistency Condition on the Cost Function): There exists an N_0 such that for any $N \geq N_0$, ∞ included, $\mathbb{E}\{z_0^T W_N(\theta) z_0\}$ is minimal in the true parameter values $\theta_0 \in \mathbb{P}_r$ and $\text{trace}(W_N(\theta) C_{n_z})$ is a θ-independent constant for any $\theta \in \mathbb{P}_r$.

Theorem 17.11 (Strong Consistency): Under the assumptions of Theorem 17.6 and Assumptions 17.9 and 17.10, the estimate $\hat{\theta}(z)$ is strongly consistent: $\text{a.s.}\lim_{N \to \infty} \hat{\theta}(z) = \theta_0$.

Proof. It follows directly from Theorem 17.6 and Assumptions 17.9 and 17.10. □

Assumption 17.12 (Consistency Condition on the Cost Function): There exists an N_0 such that for any $N \geq N_0$, ∞ included, the expected value of the cost function $V_N(\theta)$ is minimal in the true parameter values $\theta_0 \in \mathbb{P}_r$.

Theorem 17.13 (Strong Consistency): Under the assumptions of Theorem 17.6 and Assumptions 17.9 and 17.12, the estimate $\hat{\theta}(z)$ is strongly consistent: $\text{a.s.}\lim_{N \to \infty} \hat{\theta}(z) = \theta_0$.

Proof. It follows directly from Theorem 17.6 and Assumptions 17.9 and 17.12. □

Assumption 17.14 (Consistency Condition on the Cost Function): The asymptotic cost function $V_*(\theta)$ is minimal in the true parameter values $\theta_0 \in \mathbb{P}_r$.

Theorem 17.15 (Strong Consistency): Under the assumptions of Theorem 17.8 and Assumptions 17.9 and 17.14, the estimate $\hat{\theta}(z)$ is strongly consistent: $\text{a.s.}\lim_{N \to \infty} \hat{\theta}(z) = \theta_0$.

Proof. It follows directly from Theorem 17.8 and Assumptions 17.9 and 17.14. □

Although Assumptions 17.10 and 17.12 are stronger than Assumption 17.14, they are satisfied very often in practice. Assumption 17.10 often applies when a nonparametric noise model is identified (see, for example, Exercise 17.3 and Chapter 9).

17.4 CONVERGENCE RATE

Sections 17.2 and 17.3 study the conditions under which the estimate $\hat{\theta}(z)$ converges. This section studies how fast the estimate $\hat{\theta}(z)$ converges toward its limit value.

In a first step, the convergence rate of $\hat{\theta}(z)$ to $\tilde{\theta}(z_0)$ is analyzed. This is already sufficient for the strongly consistent estimators ($\tilde{\theta}(z_0) = \theta_0$) of Theorems 17.11 and 17.13 and the strongly converging estimator ($\tilde{\theta}(z_0) \neq \theta_0$) of Theorem 17.6. The key idea of the analysis consists of applying the mean value theorem (Kaplan, 1993) to the derivative of the cost function $V_N'(\theta, z)$ at the points $\hat{\theta}(z)$ and $\tilde{\theta}(z_0)$

$$V_N'(\hat{\theta}(z), z) = V_N'(\tilde{\theta}(z_0), z) + (\hat{\theta}(z) - \tilde{\theta}(z_0))^T V_N''(\widehat{\theta}, z) \qquad (17\text{-}9)$$

where $\widehat{\theta}$ is a point on the straight line connecting $\hat{\theta}(z)$ to $\tilde{\theta}(z_0)$

$$\widehat{\theta} = t\hat{\theta}(z) + (1-t)\tilde{\theta}(z_0) \text{ with } t \in [0, 1] \qquad (17\text{-}10)$$

Taking into account that $V_N'(\hat{\theta}(z), z) = 0$ ($\hat{\theta}(z)$ is the minimizer of $V_N(\theta, z)$), an expression for $\hat{\theta}(z) - \tilde{\theta}(z_0)$ is found ($V_N''(\widehat{\theta}, z)$ is symmetric)

$$\hat{\theta}(z) - \tilde{\theta}(z_0) = -V_N''^{-1}(\widehat{\theta}, z) V_N'^T(\tilde{\theta}(z_0), z) \qquad (17\text{-}11)$$

From (17-11) it follows that the convergence rate of $\hat{\theta}(z)$ to $\tilde{\theta}(z_0)$ is determined by the convergence rates of the first- and second-order derivatives of the cost function. Therefore, in Section 17.4.1 we will first analyze under which conditions these derivatives converge to their expected values. Next, in Section 17.4.2 the convergence rate of the minimizer will be established from the convergence rate of the derivatives of the cost function.

In a second step, the convergence of $\tilde{\theta}(z_0)$ to θ_* is analyzed (see Section 17.4.3). This second step is necessary for the strongly consistent estimator of Theorem 17.15 ($\theta_* = \theta_0$) and the strongly converging estimator of Theorem 17.8 ($\theta_* \neq \theta_0$). Following the same lines as in the first step, we find

$$\tilde{\theta}(z_0) - \theta_* = -V_N''^{-1}(\widehat{\theta_*}) V_N'^T(\theta_*) \qquad (17\text{-}12)$$

with $\widehat{\theta_*}$ a point on the straight line connecting $\tilde{\theta}(z_0)$ to θ_*

$$\widehat{\theta_*} = t\tilde{\theta}(z_0) + (1-t)\theta_* \text{ with } t \in [0, 1] \qquad (17\text{-}13)$$

From (17-12), it follows that the convergence rate of $\tilde{\theta}(z_0)$ to θ_* is determined by the deterministic convergence rates of the first- and second-order derivatives of the expected value of the cost function.

Section 17.4 ■ Convergence Rate

17.4.1 Convergence of the Derivatives of the Cost Function

The proof of the mean square converge of the derivatives of the cost function follows the same lines as in Section 17.2.1. Therefore, suitable assumptions concerning the derivatives of $W_N(\theta)$ w.r.t. θ should be made.

Assumption 17.16 (Constraints on First- and Second-Order Derivatives Cost Function): The weighting $W_N(\theta)$ has continuous first- and second-order derivatives w.r.t. θ with bounded 1-norm

$$\left\|\frac{\partial W_N(\theta)}{\partial \theta_{[i]}}\right\|_1 \leq c_1 < \infty, \quad \left\|\frac{\partial^2 W_N(\theta)}{\partial \theta_{[i]} \partial \theta_{[j]}}\right\|_1 \leq c_2 < \infty, \quad i, j = 1, 2, \ldots, n_\theta$$

for $N = 1, 2, \ldots, \infty$ and for any $\theta \in \mathbb{P}_r$. c_1, c_2 are N-independent constants.

Lemma 17.17 (Convergence of the Derivatives of the Cost Function): Under Assumptions 17.1 ($P = 4$) and 17.16, the derivatives of the cost function $V_N'(\theta, z)$ and $V_N''(\theta, z)$ converge uniformly in mean square to their expected values $V_N'(\theta)$ and $V_N''(\theta)$ in the compact set $\mathbb{P}_r$. The uniform mean square convergence rate in $\mathbb{P}_r$ is $O_{m.s.}(N^{-1/2})$: $V_N'(\theta, z) = V_N'(\theta) + O_{m.s.}(N^{-1/2})$ and $V_N''(\theta, z) = V_N''(\theta) + O_{m.s.}(N^{-1/2})$.

Proof. Similar to Lemma 17.3. □

Lemma 17.17 does not guarantee that the Hessian (second-order derivative) of the expected value of the cost function is regular. This is, however, necessary to ensure the existence of the matrix inverse in (17-11) and (17-12). From (17-10) and the convergence of $\hat{\theta}(z)$ to $\tilde{\theta}(z_0)$, it follows that $\widehat{\overline{\theta}}$ converges to $\tilde{\theta}(z_0)$. Hence, it is sufficient to assume that the Hessian of the expected value of the cost function is regular at $\tilde{\theta}(z_0)$. This assumption imposes some conditions on the data set z, and, therefore, it is also a persistence-of-excitation condition that is stronger than Assumption 17.5.

Assumption 17.18 (Persistence of Excitation): There exists an N_0 such that for any $N \geq N_0$, ∞ included, the Hessian of the expected value of the cost function is regular at the unique global minimizer $\tilde{\theta}(z_0)$, which is an interior point of $\mathbb{P}_r$: $c_1 I_{n_\theta} \leq V_N''(\tilde{\theta}(z_0)) \leq c_2 I_{n_\theta}$, where $0 < c_1 \leq c_2 < \infty$ and c_1, c_2 are N-independent constants.

17.4.2 Convergence Rate of $\hat{\theta}(z)$ to $\tilde{\theta}(z_0)$

Theorem 17.19 (Convergence Rate of $\hat{\theta}(z)$ to $\tilde{\theta}(z_0)$): Under Assumptions 17.1 ($P = 4$), 17.2(a), 17.16, and 17.18 the convergence rate in probability of $\hat{\theta}(z)$ equals $O_p(N^{-1/2})$: $\hat{\theta}(z) - \tilde{\theta}(z_0) = O_p(N^{-1/2})$.

Proof. See Appendix 17.D. □

Note that Assumption 17.18 is essential for the convergence rate $O_p(N^{-1/2})$. If the Hessian $V_N''(\tilde{\theta}(z_0))$ is not of full rank, then the convergence rate will decrease (see Exercise 17.4). Using the convergence rate of the minimizer $\hat{\theta}(z)$ and Assumption 17.20, we can strengthen Theorem 17.19.

Assumption 17.20 (Constraint on Third-Order Derivative Cost Function): The weighting $W_N(\theta)$ has continuous third-order derivatives w.r.t. θ with bounded 2-norm

$$\left\|\frac{\partial^3 W_N(\theta)}{\partial \theta_{[i]} \partial \theta_{[j]} \partial \theta_{[k]}}\right\|_2 \leq c_3 < \infty, \; i,j,k = 1, 2, \ldots, n_\theta$$

for $N = 1, 2, \ldots, \infty$ and for any $\theta \in \mathbb{P}_r$.

Theorem 17.21 (Improved Convergence Rate of $\hat{\theta}(z)$ to $\tilde{\theta}(z_0)$): Under Assumptions 17.1 ($P = 4$), 17.2 (a), 17.16, 17.18, and 17.20 the minimizer $\hat{\theta}(z)$ can be written as

$$\begin{aligned}\hat{\theta}(z) &= \tilde{\theta}(z_0) + \delta_\theta(z) + b_\theta(z) \\ \delta_\theta(z) &= -V_N''^{-1}(\tilde{\theta}(z_0)) V_N'^T(\tilde{\theta}(z_0), z)\end{aligned} \qquad (17\text{-}14)$$

where $\mathbb{E}\{\delta_\theta(z)\} = 0$, $\delta_\theta(z) = O_p(N^{-1/2})$, and $b_\theta(z) = O_p(N^{-1})$.

Proof. See Appendix 17.E. □

17.4.3 Convergence Rate of $\tilde{\theta}(z_0)$ to θ_*

Under Assumption 17.18, it follows from (17-12) that the convergence rate of $\tilde{\theta}(z_0)$ to θ_* is entirely determined by the deterministic convergence rate of $V_N'(\theta)$ to $V_*'(\theta)$. The latter depends on the way new data are added to the cost function and should be calculated for each particular weighting $W_N(\theta)$ and for every strategy of adding measurements to z. This is summarized in the following assumptions.

Assumption 17.22 (Constraint on the Experiment): The first- and second-order derivatives of the expected value of the cost function, $V_N'(\theta)$ and $V_N''(\theta)$, converge uniformly to their limit values, $V_*'(\theta)$ and $V_*''(\theta)$, in the compact set $\mathbb{P}_r$.

Assumption 17.23 (Convergence Rate $V_N'(\theta)$): The convergence rate of the derivative of the expected value of the cost function is an $O(N^{-K})$ in $\mathbb{P}_r$: $V_N'(\theta) = V_*'(\theta) + O(N^{-K})$.

Theorem 17.24 (Convergence Rate $\tilde{\theta}(z_0)$ to θ_*): Under Assumptions 17.2(a), 17.4, 17.18, 17.22, and 17.23 the deterministic convergence rate of $\tilde{\theta}(z_0)$ equals $O(N^{-K})$: $\tilde{\theta}(z_0) - \theta_* = O(N^{-K})$.

Proof. Similar to Theorem 17.19. □

17.5 ASYMPTOTIC BIAS

It makes sense to speak about the bias on the estimates $\hat{\theta}(z)$ only if a true model exists and if the true model belongs to the considered model set. Under Assumptions 17.9 and 17.10 or 17.12 or 17.14, Theorem 17.21, eventually combined with Theorem 17.24, gives information about the systematic errors on the estimates. This leads to the following two corollaries.

Section 17.5 ■ Asymptotic Bias

Corollary 17.25 (Improved Convergence Rate of $\hat{\theta}(z)$ to θ_0): Under the assumptions of Theorem 17.21 and Assumptions 17.9 and 17.10 or 17.12, the minimizer $\hat{\theta}(z)$ can be written as

$$\hat{\theta}(z) = \theta_0 + \delta_\theta(z) + b_\theta(z)$$
$$\delta_\theta(z) = -V_N''^{-1}(\theta_0) V_N'^T(\theta_0, z) \tag{17-15}$$

where $\mathbb{E}\{\delta_\theta(z)\} = 0$, $\delta_\theta(z) = O_p(N^{-1/2})$, and $b_\theta(z) = O_p(N^{-1})$.

Proof. It follows directly from Theorem 17.21 and Assumptions 17.9 and 17.10 or 17.12. □

Corollary 17.26 (Improved Convergence Rate of $\hat{\theta}(z)$ to θ_0 – relaxed condition): Under the assumptions of Theorems 17.21 and 17.24 and Assumptions 17.9 and 17.14, the minimizer $\hat{\theta}(z)$ can be written as

$$\hat{\theta}(z) = \theta_0 + \delta_\theta(z) + b_\theta(z)$$
$$\delta_\theta(z) = -V_N''^{-1}(\theta_0) V_N'^T(\theta_0, z) \tag{17-16}$$

where $\mathbb{E}\{\delta_\theta(z)\} = 0$, $\delta_\theta(z) = O_p(N^{-1/2})$, and $b_\theta(z) = O_p(N^{-1}) + O(N^{-K})$.

Proof. It follows directly from Theorems 17.21 and 17.24 and Assumptions 17.9 and 17.14. □

Comparing Corollary 17.25 with Corollary 17.26 shows that the deterministic convergence rate of $\tilde{\theta}(z_0)$ to θ_0 influences only $b_\theta(z)$. Because $\delta_\theta(z)$ is a zero mean random variable, it can be concluded from Corollaries 17.25 and 17.26 that the $b_\theta(z)$ term is responsible for the systematic error on the estimate $\hat{\theta}(z)$ and that, in general (Corollary 17.25, Corollary 17.26 with $K > 1/2$), $b_\theta(z)$ tends faster to zero than $\delta_\theta(z)$ as N increases to infinity. Hence, in probability, no systematic errors should be expected when N is sufficiently large.

Although the previous analysis of the systematic errors is already sufficient for our purposes, we will also, briefly, discuss the bias error on $\hat{\theta}(z)$. It is very tempting to conclude from both corollaries that the bias $b_\theta = \mathbb{E}\{b_\theta(z)\}$ decreases to zero as an $O(N^{-1})$ ($O(N^{-1}) + O(N^{-K})$). However, the expected value of $\hat{\theta}(z)$, and, hence, of $b_\theta(z)$, may, in general, not exist. This is due to the fact that convergence in probability does not exclude realizations of z for which $\hat{\theta}(z)$ tends to infinity. Additional assumptions on the measurements z are required to ensure the existence of $\mathbb{E}\{\hat{\theta}(z)\}$. For example, for quadratic prediction error methods, the eighth-order moments of the disturbing errors should be bounded (see Appendix 9.B of Ljung, 1999). The following pragmatic approach also ensures the existence of $\mathbb{E}\{\hat{\theta}(z)\}$. Define the truncated estimator $\underline{\hat{\theta}}(z)$ as

$$\underline{\hat{\theta}}(z) = \begin{cases} \hat{\theta}(z) & \|\hat{\theta}(z) - \tilde{\theta}(z_0)\|_2 \leq L \\ 0 & \|\hat{\theta}(z) - \tilde{\theta}(z_0)\|_2 > L \end{cases} \tag{17-17}$$

where L is a(n) (arbitrarily) large number ($0 < L < \infty$) independent of N. Note that this is exactly what we do in practice: if the estimate is unacceptably large, then we reject it.

Lemma 17.27 (Equivalence between $\hat{\underline{\theta}}(z)$ and $\hat{\theta}(z)$): Under the assumptions of Theorem 17.6 or Theorem 17.8, there exists an N_0 such that for any $N \geq N_0$, $\hat{\underline{\theta}}(z) = \hat{\theta}(z)$ w.p. 1. Moreover, the results of Theorem 17.21 are still valid.

Proof. See Appendix 17.F. □

The estimate $\hat{\underline{\theta}}(z)$ is uniformly bounded and, hence, its expected value exists. This leads to the following theorem.

Theorem 17.28 (Asymptotic Bias on $\hat{\underline{\theta}}(z)$): Under the assumptions of Corollary 17.25 (Corollary 17.26) the asymptotic bias $b_\theta = \mathbb{E}\{b_\theta(z)\}$ of $\hat{\underline{\theta}}(z)$, and its derivative w.r.t. θ_0, $\partial b_\theta / \partial \theta_0$, are an $O(N^{-1})$ ($O(N^{-1}) + O(N^{-K})$) for all $\theta_0 \in \mathbb{P}_r$.

Proof. See Appendix 17.G. □

17.6 ASYMPTOTIC NORMALITY

It makes no sense to estimate parameters if no quality stamp on the result can be given. Otherwise, any random guess is a valuable estimate. For example, the quality stamp can be a region around the estimated value where the limit (true) value lies within some confidence level. Therefore, we would like to know the distribution function of $\hat{\theta}(z)$ for finite N. In most cases it is impossible to calculate, and we can only make statements about the asymptotic distribution function of $\hat{\theta}(z)$.

Theorem 17.29 (Asymptotic Normality of $\sqrt{N}(\hat{\theta}(z) - \tilde{\theta}(z_0))$): Under the assumptions of Theorem 17.21 and Assumption 17.1 ($P = \infty$), $\sqrt{N}(\hat{\theta}(z) - \tilde{\theta}(z_0))$ converges in law at the rate $O(N^{-1/2})$ to a Gaussian random variable with zero mean and covariance matrix $\text{Cov}(\sqrt{N}\delta_\theta(z))$

$$\text{Cov}(\sqrt{N}\delta_\theta(z)) = V_N''^{-1}(\tilde{\theta}(z_0))Q_N(\tilde{\theta}(z_0))V_N''^{-1}(\tilde{\theta}(z_0))$$
$$Q_N(\tilde{\theta}(z_0)) = N\mathbb{E}\{V_N'^T(\tilde{\theta}(z_0), z)V_N'(\tilde{\theta}(z_0), z)\}$$
(17-18)

Proof. See Appendix 17.I. □

It follows that the uncertainty on the estimated parameters is small if the eigenvalues of the Hessian matrix $V_N''(\tilde{\theta}(z_0))$ are large. Although Theorem 17.29 guarantees neither the convergence of $\text{Cov}(\sqrt{N}\hat{\theta}(z))$ to $\text{Cov}(\sqrt{N}\delta_\theta(z))$ nor the existence of $\text{Cov}(\sqrt{N}\hat{\theta}(z))$, it makes it possible to construct uncertainty regions on $\hat{\theta}(z)$ with a given confidence level. This is sufficient for our purposes.

Additional assumptions on the measurements z are required to ensure the existence and the convergence of $\text{Cov}(\sqrt{N}\hat{\theta}(z))$ (for example, for quadratic prediction error methods the eighth-order moments of the disturbing errors should be bounded: see Appendix 9.B of Ljung, 1999). Another way to ensure its existence is to truncate the estimate $\hat{\theta}(z)$, see (17-17). It makes it possible to strengthen Theorem 17.29 as follows.

Theorem 17.30 (Asymptotic Covariance Matrix of $\hat{\underline{\theta}}(z)$): Under the assumptions of Theorem 17.21 and Assumption 17.1 ($P = \infty$), the covariance matrix $\text{Cov}(\sqrt{N}\hat{\underline{\theta}}(z))$ exists and converges to $\text{Cov}(\sqrt{N}\delta_\theta(z))$ at the rate $O(N^{-1/2})$.

Proof. See Appendix 17.J. □

Section 17.8 ■ Overview of the Asymptotic Properties

TABLE 17-1 Overview of the Notations Used

Cost function	$V_N(\theta, z)$	$V_N(\theta) = \mathbb{E}\{V_N(\theta, z)\}$	$V_*(\theta) = \lim_{N \to \infty} V_N(\theta)$
Minimizer	$\hat{\theta}(z)$	$\tilde{\theta}(z_0)$	θ_*

Note that Theorems 17.29 and 17.30 are also valid for the strongly consistent estimators of Section 17.3 if in addition Assumption 17.9 and Assumption 17.10 or 17.12 or 17.14 (consistency conditions) are satisfied.

17.7 ASYMPTOTIC EFFICIENCY

Analyzing the (asymptotic) efficiency is possible only if a true model exists and if the probability density function of the disturbing noise is known and satisfies some regularity conditions (see Theorem 16.18). In the ideal case, the covariance matrix of the estimate $\hat{\theta}(z)$ should be compared with the generalized Cramér-Rao lower bound (16-84). Because an explicit expression of the bias and its derivative w.r.t. the model parameters is mostly not available, the analysis is simplified to comparing the covariance matrix of $\delta_\theta(z)$ to the unbiased Cramér-Rao lower bound (16-87). Although the classical definition of (asymptotic) efficiency applies only to estimators with finite second-order moments, the concept of efficiency is often extended to (weakly or strongly) consistent estimators. The (weakly or strongly) consistent estimate $\hat{\theta}(z)$ is then said to be asymptotically efficient if

$$\lim_{N \to \infty} (\text{Cov}(\sqrt{N}\delta_\theta(z)) - NFi^{-1}(\theta_0)) = 0 \qquad (17\text{-}19)$$

where $\text{Cov}(\sqrt{N}\delta_\theta(z))$ is given by (17-18) with $\tilde{\theta}(z_0) = \theta_0$, and with $Fi(\theta_0)$ the Fisher information matrix of the model parameters. If (17-19) is valid for finite N and $b_\theta(z) = 0$, then the estimate is efficient. Note that (17-19) can be valid while $\text{Cov}(\hat{\theta}(z))$ may not exist.

This classical definition of asymptotic efficiency can be applied to the truncated estimator $\underline{\hat{\theta}}(z)$ (17-17). Because the bias of $\underline{\hat{\theta}}(z)$ and its derivative w.r.t. θ_0 are asymptotically zero (Theorem 17.28), we can compare the covariance matrix of $\underline{\hat{\theta}}(z)$ to the unbiased Cramér-Rao lower bound (16-87). The asymptotically unbiased estimate $\underline{\hat{\theta}}(z)$ is asymptotically efficient if

$$\lim_{N \to \infty} (\text{Cov}(\sqrt{N}\underline{\hat{\theta}}(z)) - NFi^{-1}(\theta_0)) = 0 \qquad (17\text{-}20)$$

Note that Theorem 17.30 can be used to verify (17-20).

17.8 OVERVIEW OF THE ASYMPTOTIC PROPERTIES

In this section we give an overview of the asymptotic properties of the minimizer $\hat{\theta}(z)$ (17-3) of a cost function $V_N(\theta, z)$ (17-1) that is quadratic-in-the-measurements. In the analysis of the stochastic properties of $\hat{\theta}(z)$, it turned out that the expected value of the cost function $V_N(\theta)$, its limit value $V_*(\theta)$, and the corresponding minimizers $\tilde{\theta}(z_0)$ and θ_*, play an important role. Therefore, we summarize the notations in Table 17-1.

The minimizer $\hat{\theta}(z)$ (17-3) of the cost function $V_N(\theta, z)$ (17-1) has the following asymptotic ($N \to \infty$) properties:

1. *Stochastic convergence:* $\hat{\theta}(z)$ converges strongly to $\tilde{\theta}(z_0)$ (Theorem 17.6).
2. *Stochastic convergence rate:* $\hat{\theta}(z)$ converges in probability at the rate $O_p(N^{-1/2})$ to $\tilde{\theta}(z_0)$ (Theorem 17.19).
3. *Systematic and stochastic errors:* $\hat{\theta}(z)$ converges in probability to $\tilde{\theta}(z_0)$ with

$$\hat{\theta}(z) = \tilde{\theta}(z_0) + \delta_\theta(z) + b_\theta(z)$$
$$\delta_\theta(z) = -V_N''^{-1}(\tilde{\theta}(z_0)) V_N'^T(\tilde{\theta}(z_0), z) \quad (17\text{-}21)$$

where $\delta_\theta(z) = O_p(N^{-1/2})$, with $\mathbb{E}\{\delta_\theta(z)\} = 0$, is the dominating stochastic error and where $b_\theta(z) = O_p(N^{-1})$ contains the contribution of the systematic errors (Theorem 17.21).

4. *Asymptotic normality:* $\sqrt{N}(\hat{\theta}(z) - \tilde{\theta}(z_0))$ converges in law at the rate $O(N^{-1/2})$ to a Gaussian random variable with zero mean and covariance matrix $\text{Cov}(\sqrt{N}\delta_\theta(z))$

$$\text{Cov}(\sqrt{N}\delta_\theta(z)) = V_N''^{-1}(\tilde{\theta}(z_0)) Q_N(\tilde{\theta}(z_0)) V_N''^{-1}(\tilde{\theta}(z_0))$$
$$Q_N(\tilde{\theta}(z_0)) = N \, \mathbb{E}\{V_N'^T(\tilde{\theta}(z_0), z) V_N'(\tilde{\theta}(z_0), z)\} \quad (17\text{-}22)$$

(Theorem 17.29).

5. *Deterministic convergence:* $\tilde{\theta}(z_0)$ converges to θ_* (Theorem 17.8) at the rate $O(N^{-K})$ (Theorem 17.24).

If in addition $V_N(\theta, z)$ satisfies the consistency conditions then,

6. *Consistency:* $\hat{\theta}(z)$ is strongly consistent; in properties 1 to 4 replace $\tilde{\theta}(z_0)$ or $\lim_{F \to \infty} \tilde{\theta}(Z_0) = \theta_*$ by θ_0 (Theorems 17.11, 17.13, and 17.15).
7. *Asymptotic bias:* the asymptotic bias $b_\theta = \mathbb{E}\{b_\theta(z)\}$ and its derivative w.r.t. θ_0, $\partial b_\theta / \partial \theta_0$, of $\hat{\theta}(z)$ are an $O(N^{-1})$ or an $O(N^{-1}) + O(N^{-K})$ (Theorem 17.28).

Properties 1 to 7 make it possible to predict the stochastic behavior (uncertainty, bias, ...) of the estimate $\hat{\theta}(z)$ if, for example, nine times more data are collected. Property 1 ensures that $\hat{\theta}(z)$ will be closer to the minimizer $\tilde{\theta}(z_0)$ of the expected value of the cost function. Property 2 tells us that $\hat{\theta}(z)$ will be (in probability) three times closer to $\tilde{\theta}(z_0)$. From property 3 it follows that the systematic and stochastic errors in the residual $\hat{\theta}(z) - \tilde{\theta}(z_0)$ decrease with a factor of 9 and 3, respectively. Finally, property 4 ensures that the distribution function of $\hat{\theta}(z)$ is three times closer to a normal distribution. Similar results are obtained when no model errors are present $\tilde{\theta}(z_0) = \theta_0$.

17.9 EXERCISES

17.1. Prove the strong convergence of the nonlinear least squares cost function $\sum_{k=1}^{N}(y(k) - f(\theta, u_0(k)))^2 / (N\sigma_k^2)$ with $y(k) = y_0(k) + n_y(k)$, $n_y(k)$ mixing of order 4, and $\sigma_k^2 = \text{var}(n_y(k))$. Under which condition(s) is the convergence uniform w.r.t. θ (hint: follow the lines of Lemma 17.3)?

17.2. Show that the expected value of the cost function is given by (17-8) (hint: use $x^T A x = \text{trace}(Axx^T)$ for any $x \in \mathbb{R}^N$ and $A \in \mathbb{R}^{N \times N}$).

17.3. Show that the term trace($W_N(\theta)C_{n_z}$) in the nonlinear least squares cost function of Exercise 17.1 is θ independent.

17.4. Let $\hat{\theta}(z) \in \mathbb{R}$ be the minimizer of $V_N(\theta, z)$ and θ_0 the unique global minimizer of $V_N(\theta)$ for any N, ∞ included. Assume that the cost function has continuous third-order derivative w.r.t. θ for all $\theta \in \mathbb{P}_r$ with $\|\partial^3 W_N(\theta)/\partial\theta^3\|_1 \le c < \infty$. Assume, furthermore, that $V_N''(\theta_0) = 0$, $V_N'''(\theta_0) \ne 0$, and that Assumptions 17.1 ($P = 4$), 17.2, and 17.16 are satisfied. Show that $\hat{\theta}(z) = \theta_0 + O_p(N^{-1/4})$ (hint: follow the lines of the proof of Theorem 17.19 using $V_N'(\hat{\theta}(z), z) = V_N'(\theta_0, z) + V_N''(\theta_0, z)(\hat{\theta}(z) - \theta_0) + 0.5V_N'''(\widehat{\theta}, z)(\hat{\theta}(z) - \theta_0)^2$).

17.5. Prove Theorem 17.24 (hint: follow the lines of the proof of Theorem 17.19).

17.6. Consider the nonlinear least square estimator of Exercise 17.1 and assume that $n_y(k)$ is independent over k. Assume, furthermore, that the true model is included in the model set ($y_0(k) = f(\theta_0, u_0(k))$). Show that the covariance matrix of $\delta_{\hat{\theta}}(z)$ is given by $(\sum_{k=1}^{N} \sigma_k^{-2} f_{0k}'^T f_{0k}')^{-1}$, with $f_{0k}' = \partial f(\theta, u_0(k))/\partial\theta|_{\theta = \theta_0}$ (hint: use (17-18)).

17.7. Show the weak convergence and the weak consistency of the estimates $\hat{\theta}(z)$ (convergence in prob.) when Assumption 17.2(b) is not fulfilled (hint: use the mean square convergence of the cost function and interrelation 5 of Section 16.7).

17.10 APPENDIXES

Appendix 17.A Proof of the Strong Convergence of the Cost Function (Lemma 17.3)

The cost function (17-1) can be written as

$$V_N(\theta, z) = \frac{1}{N} z^T y_N = \frac{1}{N} \sum_{t=1}^{N} x_N(t) \qquad (17\text{-}23)$$

with $y_N = W_N(\theta)z$ and $x_N(t) = z_{[t]} y_{N[t]}$. Because $W_N(\theta)$ has a bounded 1-norm for $N = 1, 2, \ldots, \infty$ (Assumption 17.2) and $z_{[t]}$ is mixing over t of order 4, $N = 1, 2, \ldots, \infty$ (Assumption 17.1), the conditions of Corollary 16.7 are satisfied with $P = 4$. Hence, $y_{N[t]}$ is mixing over t of order 4, $N = 1, 2, \ldots, \infty$, so that $x_N(t) = z_{[t]} y_{N[t]}$ is mixing over t of order 2, $N = 1, 2, \ldots, \infty$ (Lemma 16.9). This proves that the law of large numbers for mixing sequences and the corresponding convergence rate is valid for (17-23) (see Section 16.9, version 3 of (16-69) and (16-72))

$$V_N(\theta, z) = V_N(\theta) + O_{\text{m.s.}}(N^{-1/2}) \qquad (17\text{-}24)$$

If in addition it can be shown that $\text{var}(\sum_{t=1}^{s} x_r(t) - x_s(t)) = O(r-s)$, $r \ge s$, then also the strong law of large numbers for mixing sequences applies to (17-23) (see Section 16.9, version 3 of (16-69)). Because $\sum_{t=1}^{s} x_r(t) - x_s(t) = \Delta_1 + \Delta_2$ with

$$\Delta_1 = \sum_{t=1}^{s} z_{[t]} \sum_{l=s+1}^{r} W_{r[t,l]}(\theta) z_{[l]}$$

$$\Delta_2 = \sum_{t=1}^{s} z_{[t]} \sum_{l=1}^{s} (W_{r[t,l]}(\theta) - W_{s[t,l]}(\theta)) z_{[l]}$$

and $\text{var}(\Delta_1 + \Delta_2) \le 2\text{var}(\Delta_1) + 2\text{var}(\Delta_2)$ (see Exercise 16.1), it follows that it is sufficient to show that the variance of Δ_1 and Δ_2 are both an $O(r-s)$.

1. Study of Δ_1
   Rewriting Δ_1 as $\Delta_1 = \sum_{l=s+1}^{r} z_{[l]} y_s(l)$, where $y_s(l) = \sum_{t=1}^{s} z_{[t]} W_{r[t,l]}(\theta)$ is mixing over l of order 4 (Corollary 16.7), and $z_{[l]} y_s(l)$ is mixing over l of order 2 (Lemma 16.9), gives

$$\text{var}(\Delta_1) = \text{cum}(\Delta_1, \Delta_1) = \sum_{l_1, l_2 = s+1}^{r} \text{cum}(z_{[l_1]} y_s(l_1), z_{[l_2]} y_s(l_2)) \leq O(r-s)$$

where the last inequality is due to property (16-36) of the mixing condition.

2. Study of Δ_2
   If $\|W_{r[1:s, 1:s]}(\theta) - W_s(\theta)\|_1^2 = O((r-s)/r)$, then there exists a matrix $T(\theta)$ such that

$$W_{r[1:s, 1:s]}(\theta) - W_s(\theta) = T(\theta)\sqrt{(r-s)/r}$$

with $\|T(\theta)\|_1 \leq c < \infty$ for any $r \geq s$, infinity included. It facilitates rewriting Δ_2 as

$$\Delta_2 = \sqrt{(r-s)/r} \sum_{t=1}^{s} z_{[t]} y_s(t)$$

where $y_s(t) = \sum_{l=1}^{s} T_{[t, l]}(\theta) z_{[l]}$ is mixing of order 4 (Corollary 16.7), and $z_{[t]} y_s(t)$ is mixing of order 2 (Lemma 16.9). Hence, we find for any $r \geq s$

$$\text{var}(\Delta_2) = \text{cum}(\Delta_2, \Delta_2) = \frac{r-s}{r} \sum_{t_1, t_2 = 1}^{s} \text{cum}(z_{[t_1]} y_s(t_1), z_{[t_2]} y_s(t_2)) \leq O(r-s)$$

where the last inequality is due to property (16-36) of the mixing condition.

We conclude that the cost function (17-23) converges w.p. 1 to its expected value, making (17-6) valid. It remains to be proved that the mean square convergence (17-24) and the almost sure convergence (17-6) are uniform in $\mathbb{P}_r$. Because $V_N(\theta, z)$ is continuous in the compact set $\mathbb{P}_r$ (Assumption 17.2), it is uniformly bounded in $\mathbb{P}_r$. Hence, the maximum over θ can be taken, where necessary, in the inequalities above and in those of the proof of the strong law of large numbers for mixing sequences (see Appendix 16.G). □

Appendix 17.B Proof of the Strong Convergence of the Minimizer (Theorem 17.6)

The proof follows the lines of the proof of Theorem 2 in Söderström (1974). Because the assumptions of Lemma 17.3 are satisfied, we can consider only those realizations for which $V_N(\theta, z)$ converges to $V_N(\theta)$. These realizations have probability measure 1. Choose an arbitrary $\varepsilon > 0$ and construct the set $\mathbb{P}_\varepsilon = \{\theta \mid \|\theta - \tilde{\theta}(z_0)\|_2 < \varepsilon\} \subset \mathbb{P}_r$. We will show that the global minimizer(s) of $V_N(\theta, z)$ is (are) located in $\mathbb{P}_\varepsilon$ for N sufficiently large. This proves the theorem because ε can be made arbitrarily small.

Because $V_N(\theta)$ is a continuous function in $\mathbb{P}_r$ (Assumption 17.2), we can choose a $\delta > 0$ such that

$$\min_{\theta \in \mathbb{P}_r \backslash \mathbb{P}_\varepsilon} V_N(\theta) \geq V_N(\tilde{\theta}(z_0)) + \delta \qquad (17\text{-}25)$$

As $V_N(\theta, z)$ converges uniformly to $V_N(\theta)$ in $\mathbb{P}_r$, there exists an N_0 such that for any $N \geq N_0$ and any $\theta \in \mathbb{P}_r$

$$-\delta/3 \leq V_N(\theta, z) - V_N(\theta) \leq \delta/3 \qquad (17\text{-}26)$$

Using the upper inequality in (17-26), evaluated in $\tilde{\theta}(z_0)$, we get

$$\min_{\theta \in \mathbb{P}_r} V_N(\theta, z) \leq V_N(\tilde{\theta}(z_0), z) \leq V_N(\tilde{\theta}(z_0)) + \delta/3 \qquad (17\text{-}27)$$

Using the lower inequality in (17-26) and result (17-25), we find

$$\min_{\theta \in \mathbb{P}_r \backslash \mathbb{P}_\varepsilon} V_N(\theta, z) \geq \min_{\theta \in \mathbb{P}_r \backslash \mathbb{P}_\varepsilon} V_N(\theta) - \delta/3 \geq V_N(\tilde{\theta}(z_0)) + 2\delta/3 \qquad (17\text{-}28)$$

From (17-27) and (17-28) it follows that

$$\min_{\theta \in \mathbb{P}_r} V_N(\theta, z) < \min_{\theta \in \mathbb{P}_r \backslash \mathbb{P}_\varepsilon} V_N(\theta, z)$$

which shows that the minimizer(s) of $V_N(\theta, z)$ is (are) located in $\mathbb{P}_\varepsilon$. $\square$

Appendix 17.C Lemmas

In this appendix we study the asymptotic ($N \to \infty$) properties of the function $f_N(\hat{\theta}, z)$ where $\hat{\theta} \in \mathbb{R}^{n_\theta}$ is a stochastic vector of finite dimension (n_θ is an N-independent integer) and $z \in \mathbb{R}^N$ are the noisy observations. The convergence, the convergence rate, the asymptotic bias, and the asymptotic distribution function are analyzed. For the bias analysis, the concept of the truncated estimate of Section 17.5 is used. Although all the theorems are proved assuming convergence w.p. 1, they are also valid for convergence in probability (see Section 16.7, interrelation 5).

Lemma 17.31 (Strong or Weak Convergence): Let $f_N(\theta, z) \in \mathbb{R}$ be a continuous function of θ in $\mathbb{P}_r$, a compact subset of $\mathbb{R}^{n_\theta}$, and $z \in \mathbb{R}^N$ a stochastic variable. If

1. $f_N(\theta, z)$ converges uniformly w.p. 1 (in prob.) to $f(\theta)$ in $\mathbb{P}_r$,
2. $\hat{\theta}$ converges w.p. 1 (in prob.) to θ_*, an interior point of $\mathbb{P}_r$,

then $f_N(\hat{\theta}, z)$ converges w.p. 1 (in prob.) to $f(\theta_*)$.

Proof (Strong Convergence). Consider the stochastic realizations of z for which $f_N(\theta, z)$ converges uniformly to $f(\theta)$ in $\mathbb{P}_r$ and $\hat{\theta}$ converges to θ_*. Due to the almost sure convergence, these realizations have probability measure one. Choose an arbitrary $\varepsilon > 0$ and construct the set $\mathbb{P}_\varepsilon = \{\theta \mid \|\theta - \theta_*\|_2 < \varepsilon\} \subset \mathbb{P}_r$. Because for any of the considered realizations z, $f_N(\theta, z)$ converges uniformly to $f(\theta)$ in $\mathbb{P}_r$ and $\hat{\theta}$ converges to θ_*, there exists, for any $\delta > 0$, an N_0 independent of θ such that for any $N \geq N_0$

$$\hat{\theta} \in \mathbb{P}_\varepsilon \text{ and } |f_N(\theta, z) - f(\theta)| \leq \delta/2 \text{ for any } \theta \in \mathbb{P}_r \qquad (17\text{-}29)$$

The function $f(\theta)$ is continuous in $\mathbb{P}_r$ because it is the limit of a uniformly convergent sequence of continuous functions (see Kaplan, 1993, Theorem 31, Remark 2). Hence, there exists an ε such that

$$|f(\theta) - f(\theta_*)| \leq \delta/2 \text{ for any } \theta \in \mathbb{P}_\varepsilon \qquad (17\text{-}30)$$

Combining (17-29) and (17-30) shows that for any $\delta > 0$ there exist an ε and an N_0 such that for any $N \geq N_0$

$$|f_N(\hat{\theta}, z) - f(\theta_*)| \leq |f_N(\hat{\theta}, z) - f(\hat{\theta})| + |f(\hat{\theta}) - f(\theta_*)| \leq \delta \qquad (17\text{-}31)$$

Making δ arbitrarily small and noting that the considered realizations z occur w.p. 1 reveals directly that $\text{a.s.}\lim_{N \to \infty} (f_N(\hat{\theta}, z) - f(\theta_*)) = 0$ or $f_N(\hat{\theta}, z) = f(\theta_*) + o_{\text{a.s.}}(N^0)$. □

Corollary 17.32 (Strong or Weak Convergence): Let $z \in \mathbb{R}^N$ be a stochastic variable. Let $f_N(\theta, \psi, \eta, z) \in \mathbb{R}$ be a jointly continuous function of $\theta \in \mathbb{P}_r$, $\psi \in \mathbb{S}_r$, and $\eta \in \mathbb{E}_r$. $\mathbb{P}_r$, $\mathbb{S}_r$, and $\mathbb{E}_r$ are compact subsets of, respectively, $\mathbb{R}^{n_\theta}$, $\mathbb{R}^{n_\psi}$, and $\mathbb{R}^{n_\eta}$. If

1. $f_N(\theta, \psi, \eta, z)$ converges uniformly w.p. 1 (in prob.) to $f(\theta, \psi, \eta)$ in $\mathbb{P}_r$, $\mathbb{S}_r$, and $\mathbb{E}_r$.
2. $\hat{\theta}$, $\hat{\psi}$ converge w.p. 1 (in prob.) to θ_*, ψ_*, interior points of $\mathbb{P}_r$, $\mathbb{S}_r$,

then $f_N(\hat{\theta}, \hat{\psi}, \eta, z)$ converges uniformly w.p. 1 (in prob.) to $f(\theta_*, \psi_*, \eta)$ in $\mathbb{E}_r$.

Lemma 17.33 (Strong or Weak Convergence): Let $f_N(\theta, z) \in \mathbb{R}$ be a continuous function of θ in $\mathbb{P}_r$, a compact subset of $\mathbb{R}^{n_\theta}$, and $z \in \mathbb{R}^N$ a stochastic variable. If

1. $f_N(\theta, z) = O_{\text{a.s.}}(N^k)$ $(O_p(N^k))$ uniformly in $\mathbb{P}_r$,
2. $\hat{\theta}$ converges w.p. 1 (in prob.) to θ_*, an interior point of $\mathbb{P}_r$,

then $f_N(\hat{\theta}, z) = O_{\text{a.s.}}(N^k)$ $(O_p(N^k))$.

Proof. Similar to that of Lemma 17.31. □

Lemma 17.34 (Convergence Rate): Let $z \in \mathbb{R}^N$ be a stochastic variable. Let $f_N(\theta, z) \in \mathbb{R}$ and $f_N'(\theta, z)$, its derivative w.r.t. θ, be continuous functions of θ in $\mathbb{P}_r$, a compact subset of $\mathbb{R}^{n_\theta}$. If

1. $f_N(\theta, z)$ converges uniformly w.p. 1 (in prob.) to $f(\theta)$ in $\mathbb{P}_r$ at the rate $O_p(N^{-1/2})$,
2. $\|f_N'(\theta, z)\|_2 \leq O_{\text{a.s.}}(N^0)$ $(O_p(N^0))$ uniformly in $\mathbb{P}_r$,
3. $\hat{\theta}$ converges w.p. 1 (in prob.) to θ_* at the rate $O_p(N^{-1/2})$, with θ_* an interior point of $\mathbb{P}_r$,

then $f_N(\hat{\theta}, z)$ converges w.p. 1 (in prob.) to $f(\theta_*)$ at the rate $O_p(N^{-1/2})$.

Section 17.10 ■ Appendixes

Proof (w.p. 1). Note that the conditions of Lemma 17.31 are satisfied so that only the convergence rate must be proved. Applying the mean value theorem (Kaplan, 1993) to $f_N(\theta, z)$ at the points $\hat{\theta}$, θ_* gives

$$f_N(\hat{\theta}, z) = f_N(\theta_*, z) + f_N'(\widetilde{\theta}, z)(\hat{\theta} - \theta_*) \tag{17-32}$$

with $\widetilde{\theta}$ a point on the straight line connecting $\hat{\theta}$ to θ_* ($\widetilde{\theta} = t\hat{\theta} + (1-t)\theta_*$ with $t \in [0, 1]$). $\widetilde{\theta}$ converges w.p. 1 to θ_* because

$$\text{a.s.}\lim_{N \to \infty} (\widetilde{\theta} - \theta_*) = \lim_{N \to \infty} t \text{ a.s.}\lim_{N \to \infty} (\hat{\theta} - \theta_*) = 0 \tag{17-33}$$

Consider the realizations z for which $\widetilde{\theta}$ converges to θ_* and $\|f_N'(\theta, z)\|_2 \leq O(N^0)$ uniformly in $\mathbb{P}_r$. For these realizations, there is an N_0 such that for any $N \geq N_0$, $\widetilde{\theta} \in \mathbb{P}_r$ and, hence, $\|f_N'(\widetilde{\theta}, z)\|_2 \leq O(N^0)$. Because these realizations occur w.p. 1, we have $\|f_N'(\widetilde{\theta}, z)\|_2 \leq O_{\text{a.s.}}(N^0)$ and, hence, $\|f_N'(\widetilde{\theta}, z)\|_2 \leq O_p(N^0)$ (see Section 16.7). Putting this result in (17-32), taking into account that $f_N(\theta_*, z) = f(\theta_*) + O_p(N^{-1/2})$ (condition 1), $\hat{\theta} = \theta_* + O_p(N^{-1/2})$ (condition 3), and that n_θ is an N-independent integer proves the lemma. □

Corollary 17.35 (Convergence Rate): Let $z \in \mathbb{R}^N$ be a stochastic variable. Let $f_N(\theta, \psi, \eta, z) \in \mathbb{R}$ and its derivatives w.r.t. θ and ψ be jointly continuous functions of $\theta \in \mathbb{P}_r$, $\psi \in \mathbb{S}_r$, and $\eta \in \mathbb{E}_r$, $\mathbb{P}_r$, $\mathbb{S}_r$, and $\mathbb{E}_r$ are compact subsets of, respectively, $\mathbb{R}^{n_\theta}$, $\mathbb{R}^{n_\psi}$, and $\mathbb{R}^{n_\eta}$. If

1. $f_N(\theta, \psi, \eta, z)$ converges uniformly w.p. 1 (in prob.) to $f(\theta, \psi, \eta)$ in $\mathbb{P}_r$, $\mathbb{S}_r$, and $\mathbb{E}_r$ at the rate $O_p(N^{-1/2})$,
2. $\|\partial f_N(\theta, \psi, \eta, z)/\partial\theta\|_2 \leq O_{\text{a.s.}}(N^0)$ $(O_p(N^0))$, $\|\partial f_N(\theta, \psi, \eta, z)/\partial\psi\|_2 \leq O_{\text{a.s.}}(N^0)$ $(O_p(N^0))$ uniformly in $\mathbb{P}_r$, $\mathbb{S}_r$, and $\mathbb{E}_r$,
3. $\hat{\theta}$, $\hat{\psi}$ converge w.p. 1 (in prob.) to θ_*, ψ_* at the rate $O_p(N^{-1/2})$, where θ_*, ψ_* are interior points of $\mathbb{P}_r$, $\mathbb{S}_r$,

then $f_N(\hat{\theta}, \hat{\psi}, \eta, z)$ converges uniformly w.p. 1 (in prob.) to $f(\theta_*, \psi_*, \eta)$ in $\mathbb{E}_r$ at the rate $O_p(N^{-1/2})$.

Lemma 17.36 (Asymptotic Bias): Let $z \in \mathbb{R}^N$ be a stochastic variable. Let $f_N(\theta, z) \in \mathbb{R}$ and $f_N^{(k)}(\theta, z)$, $k = 1, 2$, its derivatives w.r.t. θ, be continuous functions of θ in $\mathbb{P}_r$, a compact subset of $\mathbb{R}^{n_\theta}$. If

1. $f_N(\theta, z)$, $f_N'(\theta, z)$ converge uniformly w.p. 1 (in prob.) to $\mathbb{E}\{f_N(\theta, z)\} = f_N(\theta)$, $\mathbb{E}\{f_N'(\theta, z)\} = f_N'(\theta)$ in $\mathbb{P}_r$ at the rate $O_p(N^{-1/2})$,
2. $\|f_N''(\theta, z)\|_2 \leq O_{\text{a.s.}}(N^0)$ $(O_p(N^0))$ uniformly in $\mathbb{P}_r$,
3. $\hat{\theta}$ converges w.p. 1 (in prob.) to θ_0 at the rate $O_p(N^{-1/2})$, with θ_0 an interior point of $\mathbb{P}_r$,
4. The bias of the truncated estimate $\hat{\underline{\theta}}$ is an $O(N^{-1})$: $\mathbb{E}\{\hat{\underline{\theta}}\} = \theta_0 + O(N^{-1})$,

then $f_N(\hat{\theta}, z)$ converges w.p. 1 (in prob.) to $f(\theta_0)$ at the rate $O_p(N^{-1/2})$ and the bias of the truncated estimate $f_N(\hat{\underline{\theta}}, z)$ is an $O(N^{-1})$: $\mathbb{E}\{f_N(\hat{\underline{\theta}}, z)\} = f_N(\theta_0) + O(N^{-1})$.

Proof (w.p. 1). The conditions of Lemma 17.34 are fulfilled for $f_N(\hat{\theta}, z)$ so that only the claim about the asymptotic bias must be proved. Applying the mean value theorem to $f_N(\theta, z)$ at the points $\hat{\theta}$, θ_0 gives

$$f_N(\hat{\theta}, z) = f_N(\theta_0, z) + f_N'(\overline{\theta}, z)(\hat{\theta} - \theta_0) \tag{17-34}$$

where $\overline{\theta} = t\hat{\theta} + (1-t)\theta_0$ with $t \in [0, 1]$. Note that $f_N'(\overline{\theta}, z)$ satisfies the conditions of Lemma 17.34 because $\overline{\theta}$ converges w.p. 1 to θ_0 at the rate $O_p(N^{-1/2})$ (proof: similar to (17-33)). Referring to the equivalence between the truncated and the original estimate (Lemma 17.27), there is an N_0 such that for any $N \geq N_0$ w.p. 1, $\underline{\hat{\theta}} = \hat{\theta}$, $\underline{f_N}(\hat{\theta}, z) = f_N(\hat{\theta}, z)$, and $\underline{f_N'}(\overline{\theta}, z) = f_N'(\overline{\theta}, z)$. Assume now that $N \geq N_0$ and define $\mathbb{Z}_L$ as the set of realizations z for which $\hat{\theta}$ converges to θ_0, $f_N(\theta_0, z)$ converges to $f_N(\theta_0)$, $f_N'(\overline{\theta}, z)$ converges to $f_N'(\theta_0)$, $\underline{\hat{\theta}} = \hat{\theta}$, $\underline{f_N}(\hat{\theta}, z) = f_N(\hat{\theta}, z)$, and $\underline{f_N'}(\overline{\theta}, z) = f_N'(\overline{\theta}, z)$. The set $\mathbb{Z}_L$ has probability measure one. For $z \in \mathbb{Z}_L$, (17-34) can be written as

$$\underline{f_N}(\hat{\theta}, z) = \underline{f_N}(\theta_0, z) + \underline{f_N'}(\overline{\theta}, z)(\underline{\hat{\theta}} - \theta_0) \tag{17-35}$$

Using the convergence rates of $\hat{\theta}$ (condition 3) and $\underline{f_N'}(\overline{\theta}, z)$ (Lemma 17.34), (17-35) becomes

$$\underline{f_N}(\hat{\theta}, z) = \underline{f_N}(\theta_0, z) + f_N'(\theta_0)(\underline{\hat{\theta}} - \theta_0) + O_p(N^{-1}) \tag{17-36}$$

where $O_p(N^{-1})$ is a uniformly bounded random variable. Calculating the expected value of (17-36) over all realizations $z \in \mathbb{Z}_L$, taking into account that $\mathbb{E}\{O_p(N^{-1}) | z \in \mathbb{Z}_L\} = O(N^{-1})$ (see Section 16.8, (16-63)), and that by definition of the truncated estimate, $\mathbb{E}\{\underline{f_N}(\hat{\theta}, z) | z \in \mathbb{Z}_L\} = \mathbb{E}\{\underline{f_N}(\hat{\theta}, z)\}$, gives

$$\mathbb{E}\{\underline{f_N}(\hat{\theta}, z)\} = \mathbb{E}\{f_N(\theta_0, z) | z \in \mathbb{Z}_L\} + O(N^{-1})$$

Because $\text{Prob}(z \notin \mathbb{Z}_L) = 0$ for $N \geq N_0$ and $\mathbb{E}\{f_N(\theta_0, z)\}$ exists and is finite, we have

$$\mathbb{E}\{f_N(\theta_0, z) | z \in \mathbb{Z}_L\} = \mathbb{E}\{f_N(\theta_0, z)\} = f_N(\theta_0)$$

which concludes the proof. □

Corollary 17.37 (Asymptotic Bias): Let $z \in \mathbb{R}^N$ be a stochastic variable. Let $f_N(\theta, \psi, \eta, z) \in \mathbb{R}$, and its first- and second-order derivatives w.r.t. θ and ψ, be jointly continuous functions of $\theta \in \mathbb{P}_r$, $\psi \in \mathbb{S}_r$, and $\eta \in \mathbb{E}_r$. $\mathbb{P}_r$, $\mathbb{S}_r$, and $\mathbb{E}_r$ are compact subsets of, respectively, $\mathbb{R}^{n_\theta}$, $\mathbb{R}^{n_\psi}$, and $\mathbb{R}^{n_\eta}$. If

1. $f_N(\theta, \psi, \eta, z)$, $\partial f_N(\theta, \psi, \eta, z)/\partial \theta$, $\partial f_N(\theta, \psi, \eta, z)/\partial \psi$ converge uniformly w.p. 1 (in prob.) to their expected values $f_N(\theta, \psi, \eta)$, $\partial f_N(\theta, \psi, \eta)/\partial \theta$, $\partial f_N(\theta, \psi, \eta)/\partial \psi$ in $\mathbb{P}_r$, $\mathbb{S}_r$, and $\mathbb{E}_r$ at the rate $O_p(N^{-1/2})$,

2. $\left\|\dfrac{\partial^2 f_N(\theta, \psi, \eta, z)}{\partial \theta^2}\right\|_2 \leq O_{a.s.}(N^0)\ (O_p(N^0)),\ \left\|\dfrac{\partial^2 f_N(\theta, \psi, \eta, z)}{\partial \psi^2}\right\|_2 \leq O_{a.s.}(N^0)\ (O_p(N^0))$

uniformly in $\mathbb{P}_r$, $\mathbb{S}_r$, and $\mathbb{E}_r$,

3. $\hat{\theta}$, $\hat{\psi}$ converge w.p. 1 (in prob.) to θ_0, ψ_0 at the rate $O_p(N^{-1/2})$, where θ_0, ψ_0 are interior points of $\mathbb{P}_r$, $\mathbb{S}_r$,

4. The bias of the truncated estimates $\hat{\theta}$, $\hat{\psi}$ is an $O(N^{-1})$,

then $f_N(\hat{\theta}, \hat{\psi}, \eta, z)$ converges uniformly w.p. 1 (in prob.) to $f_N(\theta_0, \psi_0, \eta)$ in $\mathbb{E}_r$ at the rate $O_p(N^{-1/2})$, and the bias of the truncated estimate $f_N(\hat{\theta}, \hat{\psi}, \eta, z)$ is an $O(N^{-1})$.

Lemma 17.38 (Asymptotic Distribution Function): Let $z \in \mathbb{R}^N$ be a stochastic variable. Let $f_N(\theta, z) \in \mathbb{R}$, and $f_N^{(k)}(\theta, z)$, $k = 1, 2$, its derivatives w.r.t. θ, be continuous functions of θ in $\mathbb{P}_r$, a compact subset of $\mathbb{R}^{n_\theta}$. If

1. $f_N(\theta, z)$ converges uniformly w.p. 1 (in prob.) to $f(\theta)$ in $\mathbb{P}_r$ at the rate $O_p(N^{-1/2})$ and is asymptotically normally distributed at the rate $O(N^{-1/2})$,

2. $f_N'(\theta, z)$ converges uniformly w.p. 1 (in prob.) to $f'(\theta)$ in $\mathbb{P}_r$ at the rate $O_p(N^{-1/2})$,

3. $\|f_N''(\theta, z)\|_2 \leq O_{a.s.}(N^0)\ (O_p(N^0))$ uniformly in $\mathbb{P}_r$,

4. $\hat{\theta}$ converges w.p. 1 (in prob.) to θ_* at the rate $O_p(N^{-1/2})$, with θ_* an interior point of $\mathbb{P}_r$, and is asymptotically normally distributed at the rate $O(N^{-1/2})$,

then $f_N(\hat{\theta}, z)$ converges w.p. 1 (in prob.) to $f(\theta_*)$ at the rate $O_p(N^{-1/2})$ and is asymptotically normally distributed at the rate $O(N^{-1/2})$. Moreover, we have

$$f_N(\hat{\theta}, z) = f_N(\theta_*, z) + f'(\theta_*)(\hat{\theta} - \theta_*) + O_p(N^{-1})$$

Proof (w.p. 1). Condition 2 implies that $\|f_N'(\theta, z)\|_2 \leq O_{a.s.}(N^0)$ and, hence, all the assumptions of Lemma 17.34 are fulfilled. Therefore, only the asymptotic normality must be proved. Applying the mean value theorem to $f_N(\theta, z)$ at the points $\hat{\theta}$, θ_* gives

$$f_N(\hat{\theta}, z) = f_N(\theta_*, z) + f_N'(\widehat{\theta}, z)(\hat{\theta} - \theta_*) \qquad (17\text{-}37)$$

where $\widehat{\theta} = t\hat{\theta} + (1-t)\theta_*$ with $t \in [0, 1]$. Because $\widehat{\theta}$ converges w.p. 1 to θ_* at the same rate as $\hat{\theta}$ (proof: see (17-33)) it follows from conditions 2 and 3 that $f_N'(\widehat{\theta}, z) = f'(\theta_*) + O_p(N^{-1/2})$ (Lemma 17.34). Putting this result in (17-37) taking into account conditions 1 ($f_N(\theta_*, z) - f(\theta_*) = O_p(N^{-1/2})$) and 4 ($\hat{\theta} - \theta_* = O_p(N^{-1/2})$) gives

$$\begin{aligned}f_N(\hat{\theta}, z) - f(\theta_*) &= \delta_N(z) + O_p(N^{-1}) \\ \delta_N(z) &= f_N(\theta_*, z) - f(\theta_*) + f'(\theta_*)(\hat{\theta} - \theta_*)\end{aligned} \qquad (17\text{-}38)$$

where $\delta_N(z) = O_p(N^{-1/2})$ is asymptotically normally distributed at the rate $O(N^{-1/2})$ (a finite linear combination of asymptotically normally distributed random variables is asymptotically normally distributed and the convergence rate is preserved). Multiplying (17-38) by $\sqrt{N}$ and taking the limit gives

$$\plim_{N \to \infty} \sqrt{N}(f_N(\hat{\theta}, z) - f(\theta_*) - \delta_N(z)) = 0 \qquad (17\text{-}39)$$

Because convergence in probability implies convergence in law (see Section 16.7, interrelation 3), it follows from (17-39) that $\sqrt{N}(f_N(\hat{\theta}, z) - f(\theta_*))$ is asymptotically normally distributed at the rate $O(N^{-1/2})$. □

Corollary 17.39 (Asymptotic Distribution Function): Let $z \in \mathbb{R}^N$ be a stochastic variable. Let $f_N(\theta, \psi, \eta, z) \in \mathbb{R}$, and its first- and second-order derivatives w.r.t. θ and ψ, be jointly continuous functions of $\theta \in \mathbb{P}_r$, $\psi \in \mathbb{S}_r$, and $\eta \in \mathbb{E}_r$. $\mathbb{P}_r$, $\mathbb{S}_r$, and $\mathbb{E}_r$ are compact subsets of, respectively, $\mathbb{R}^{n_\theta}$, $\mathbb{R}^{n_\psi}$, and $\mathbb{R}^{n_\eta}$. If

1. $f_N(\theta, \psi, \eta, z)$ converges uniformly w.p. 1 (in prob.) to $f(\theta, \psi, \eta)$ in $\mathbb{P}_r$, $\mathbb{S}_r$, and $\mathbb{E}_r$ at the rate $O_p(N^{-1/2})$ and is asymptotically normally distributed at the rate $O(N^{-1/2})$,

2. $\partial f_N(\theta, \psi, \eta, z)/\partial \theta$, $\partial f_N(\theta, \psi, \eta, z)/\partial \psi$ converge uniformly w.p. 1 (in prob.) to $\partial f(\theta, \psi, \eta)/\partial \theta$, $\partial f(\theta, \psi, \eta)/\partial \psi$ in $\mathbb{P}_r$, $\mathbb{S}_r$, and $\mathbb{E}_r$ at the rate $O_p(N^{-1/2})$,

3. $\left\| \dfrac{\partial^2 f_N(\theta, \psi, \eta, z)}{\partial \theta^2} \right\|_2 \leq O_{a.s.}(N^0)$ $(O_p(N^0))$, $\left\| \dfrac{\partial^2 f_N(\theta, \psi, \eta, z)}{\partial \psi^2} \right\|_2 \leq O_{a.s.}(N^0)$ $(O_p(N^0))$
uniformly in $\mathbb{P}_r$, $\mathbb{S}_r$, and $\mathbb{E}_r$,

4. $\hat{\theta}$, $\hat{\psi}$ converge w.p. 1 (in prob.) to θ_*, ψ_* at the rate $O_p(N^{-1/2})$, where θ_*, ψ_* are interior points of $\mathbb{P}_r$, $\mathbb{S}_r$, and are asymptotically normally distributed at the rate $O(N^{-1/2})$,

then $f_N(\hat{\theta}, \hat{\psi}, \eta, z)$ converges uniformly w.p. 1 (in prob.) to $f(\theta_*, \psi_*, \eta)$ in $\mathbb{E}_r$ at the rate $O_p(N^{-1/2})$ and is asymptotically normally distributed at the rate $O(N^{-1/2})$. Moreover, we have

$$f_N(\hat{\theta}, \hat{\psi}, \eta, z) = f_N(\theta_*, \psi_*, \eta, z) + \frac{\partial f_N(\theta, \psi_*, \eta)}{\partial \theta_*}(\hat{\theta} - \theta_*) + \frac{\partial f_N(\theta_*, \psi, \eta)}{\partial \psi_*}(\hat{\psi} - \psi_*) + O_p(N^{-1})$$

Appendix 17.D Proof of the Convergence Rate of the Minimizer (Theorem 17.19)

The proof consists of the following three basic steps: first, the convergence rate of $V_N'(\tilde{\theta}(z_0), z)$ is studied; next, the convergence rate of $V_N''(\overline{\theta}, z)$; and finally, both results are combined to establish the convergence rate of $\hat{\theta}(z)$ to $\tilde{\theta}(z_0)$.

1. Convergence rate of $V_N'(\tilde{\theta}(z_0), z)$

    Under Assumptions 17.1 ($P = 4$) and 17.16, $V_N'(\theta, z)$ and $V_N''(\theta, z)$ converge uniformly in mean square at the rate $O_{m.s.}(N^{-1/2})$ to their expected values $V_N'(\theta)$ and $V_N''(\theta)$, respectively (Lemma 17.17)

    $$V_N'(\theta, z) = V_N'(\theta) + O_{m.s.}(N^{-1/2}) \qquad (17\text{-}40)$$

    Because $\tilde{\theta}(z_0)$ is deterministic we may evaluate (17-40) at $\theta = \tilde{\theta}(z_0)$. Taking into account that $\tilde{\theta}(z_0)$ minimizes $V_N(\theta)$, this gives

Section 17.10 ■ Appendixes 647

$$V_N'(\widehat{\theta}(z_0), z) = V_N'(\tilde{\theta}(z_0)) + O_{\text{m.s.}}(N^{-1/2}) = O_{\text{m.s.}}(N^{-1/2}) \qquad (17\text{-}41)$$

2. Convergence rate of $V_N''(\widehat{\theta}, z)$

   Here, $\widehat{\theta}$ is a random variable (see (17-10)) so that the reasoning applied to $V_N'(\widehat{\theta}(z_0), z)$ does not hold for $V_N''(\widehat{\theta}, z)$. Indeed, the convergence rate of $V_N''(\widehat{\theta}, z)$ depends on the stochastic properties of $\widehat{\theta}$. Under Assumptions 17.1 ($P = 4$), 17.2(a), and 17.18, $\hat{\theta}(z)$ converges in prob. to $\tilde{\theta}(z_0)$ (Theorem 17.6 without Assumption 17.2(b) shows convergence in prob., see Exercise 17.7). From (17-10) it follows also that $\widehat{\theta}$ converges in probability to $\tilde{\theta}(z_0)$

$$\underset{N \to \infty}{\text{a.s.lim}}\, (\widehat{\theta} - \tilde{\theta}(z_0)) = \underset{N \to \infty}{\text{a.s.lim}}\, t \; \underset{N \to \infty}{\text{a.s.lim}}\, (\hat{\theta}(z) - \tilde{\theta}(z_0)) = 0 \qquad (17\text{-}42)$$

   Hence, $V_N''(\widehat{\theta}, z)$ converges in probability to $V_N''(\tilde{\theta}(z_0))$ (see Appendix 17.C, Lemma 17.31)

$$V_N''(\widehat{\theta}, z) = V_N''(\tilde{\theta}(z_0)) + o_p(N^0) \qquad (17\text{-}43)$$

3. Convergence rate of $\hat{\theta}(z)$ to $\tilde{\theta}(z_0)$

   Because convergence in mean square and almost sure convergence imply convergence in probability (see Section 16.7, interrelations 1 and 2) and a continuous function and the limit in probability may be interchanged (see Section 16.8, property 3), it follows from (17-11), (17-41), and (17-43) that

$$\hat{\theta}(z) - \tilde{\theta}(z_0) = (V_N''(\tilde{\theta}(z_0)) + o_p(N^0))^{-1} O_p(N^{-1/2})$$
$$= (V_N''^{-1}(\tilde{\theta}(z_0)) + o_p(N^0)) O_p(N^{-1/2})$$

   Under Assumption 17.18, $V_N''^{-1}(\tilde{\theta}(z_0))$ is an $O(N^0)$, so that $\hat{\theta}(z) - \tilde{\theta}(z_0) = O_p(N^{-1/2})$. □

Appendix 17.E Proof of the Improved Convergence Rate of the Minimizer (Theorem 17.21)

Using Assumption 17.20 and the result of Theorem 17.19, the convergence rate of $V_N''(\widehat{\theta}, z)$ will be established. Combined with the convergence rate of $V_N'(\tilde{\theta}(z_0), z)$ (see Appendix 17.D), this will lead to a refined expression for $\hat{\theta}(z) - \tilde{\theta}(z_0)$.

To establish the convergence rate of $V_N''(\widehat{\theta}, z)$, we verify that all the conditions of Lemma 17.34 (see Appendix 17.C) are satisfied. Consistent with the assumptions of Lemma 17.17, $V_N''(\theta, z)$ converges uniformly in mean square to $V_N''(\theta)$ in $\mathbb{P}_r$ at the rate $O_{\text{m.s.}}(N^{-1/2})$, which implies $O_p(N^{-1/2})$ (condition 1, Lemma 17.34). From (17-10) it follows that $\widehat{\theta} - \tilde{\theta}(z_0) = t(\hat{\theta}(z) - \tilde{\theta}(z_0))$ with $t \in [0, 1]$. Hence, $\widehat{\theta} - \tilde{\theta}(z_0)$ converges w.p. 1 to zero at the rate of $\hat{\theta}(z) - \tilde{\theta}(z_0)$, which is given by Theorem 17.19, and, hence, $\widehat{\theta} - \tilde{\theta}(z_0) = O_p(N^{-1/2})$ (condition 3, Lemma 17.34). We will now show under Assumptions 17.1 ($P = 4$), 17.18, and 17.20 that $V_N'''(\theta, z)$ is an $O_{\text{a.s.}}(N^0)$ uniformly in $\mathbb{P}_r$ (condition 2, Lemma 17.34). The absolute value of the third-order derivative of the cost function $V_N(\theta, z)$ is bounded by

$$\left|\frac{\partial^3 V_N(\theta, z)}{\partial \theta_{[i]} \partial \theta_{[j]} \partial \theta_{[k]}}\right| \leq \frac{\|z\|_2^2}{N} \left\|\frac{\partial^3 W_N(\theta)}{\partial \theta_{[i]} \partial \theta_{[j]} \partial \theta_{[k]}}\right\|_2 \quad \text{for } i, j, k = 1, 2, \ldots, n_\theta \qquad (17\text{-}44)$$

Assumption 17.20 guarantees that the second factor in the right-hand side of (17-44) is an $O(N^0)$ uniformly in $\mathbb{P}_r$. Under Assumption 17.1 ($P = 4$), $\|z\|_2^2/N$ obeys the strong law of large numbers for mixing sequences (see Section 16.9, version 3 of (16-69)). Under Assumptions 17.1 ($P = 2$) the expected value of $\|z\|_2^2/N$ is an $O(N^0)$ so that $\|z\|_2^2/N = O_{\text{a.s.}}(N^0)$. This shows that the right-hand side of (17-44) is an $O_{\text{a.s.}}(N^0)$. The three conditions of Lemma 17.34 (see Appendix 17.C) are satisfied and, hence,

$$V_N''(\hat{\theta}, z) = V_N''(\tilde{\theta}(z_0)) + O_p(N^{-1/2})$$

Following the same lines as in the third step in the proof of Theorem 17.19 (see Appendix 17.D) we conclude from (17-11), (17-41), and that

$$\hat{\theta}(z) - \tilde{\theta}(z_0) = -(V_N''^{-1}(\tilde{\theta}(z_0)) + O_p(N^{-1/2}))V_N'^T(\tilde{\theta}(z_0), z) = \delta_\theta(z) + O_p(N^{-1})$$

with $\delta_\theta(z) = -V_N''^{-1}(\tilde{\theta}(z_0))V_N'^T(\tilde{\theta}(z_0), z) = O_p(N^{-1/2})$. As $\mathbb{E}\{V_N'(\theta, z)\} = V_N'(\theta)$ in $\mathbb{P}_r$ and $V_N'(\tilde{\theta}(z_0)) = 0$, we have that $\mathbb{E}\{\delta_\theta(z)\} = 0$, which concludes the proof. □

Appendix 17.F Equivalence between the Truncated and the Original Minimizer (Lemma 17.27)

Define $\mathbb{Z}_L$ as the set of realizations z for which the estimate $\hat{\theta}(z)$ lies within the hyperball with center $\tilde{\theta}(z_0)$ and radius L

$$\mathbb{Z}_L = \{z \mid \|\hat{\theta}(z) - \tilde{\theta}(z_0)\|_2 \leq L\}$$

Under the assumptions of Theorem 17.6 or Theorem 17.8, $\hat{\theta}(z)$ converges strongly to $\tilde{\theta}(z_0)$: there exists an N_0 such that for any $N \geq N_0$, $\|\hat{\theta}(z) - \tilde{\theta}(z_0)\|_2 \leq L$ w.p. 1. Hence, for $N \geq N_0$, the realizations $z \in \mathbb{Z}_L$ happen with probability measure 1. From the definition of $\hat{\underline{\theta}}(z)$ (17-17) we conclude that $\hat{\underline{\theta}}(z) = \hat{\theta}(z)$ w.p. 1 for any $N \geq N_0$.

To prove that the results of Corollaries 17.25 and 17.26 are valid, we still need to show that the expected value of $\delta_\theta(z)$ over the set $\mathbb{Z}_L$ is zero. Using $\mathbb{E}\{\delta_\theta(z)\} = 0$, we get

$$\mathbb{E}\{\delta_\theta(z) \mid z \in \mathbb{Z}_L\} = -\mathbb{E}\{\delta_\theta(z) \mid z \notin \mathbb{Z}_L\} = 0$$

where the last equality is due to the fact that $\text{Prob}(z \notin \mathbb{Z}_L) = 0$ for $N \geq N_0$ and that the second-order moments of z are uniformly bounded. □

Appendix 17.G Proof of the Asymptotic Bias on the Truncated Minimizer (Theorem 17.28)

Define $\mathbb{Z}_L$ as in the proof of Lemma 17.27 with $\tilde{\theta}(z_0) = \theta_0$ (see Appendix 17.F). Under the assumptions of Corollary 17.25 (Corollary 17.26) and (17-17), it follows for any $z \in \mathbb{Z}_L$, that $\hat{\underline{\theta}}(z) = \hat{\theta}(z) = \theta_0 + \delta_\theta(z) + b_\theta(z)$ and that $b_\theta(z)$ is uniformly bounded. The interchangeability property of the expected value and the limit in probability for uniformly bounded random variables (see Section 16.8, (16-63)) guarantees that the expected value of $b_\theta(z)$ taken over $\mathbb{Z}_L$, $\mathbb{E}\{b_\theta(z)|z \in \mathbb{Z}_L\}$, is an $O(N^{-1})$ ($O(N^{-1}) + O(N^{-K})$). From Lemma 17.27 it follows that there exists an N_0 such that for any $N \geq N_0$, $\mathbb{E}\{\delta_\theta(z)|z \in \mathbb{Z}_L\} = 0$. This shows that $\mathbb{E}\{b_\theta(z)|z \in \mathbb{Z}_L\}$ is the bias error of $\hat{\theta}(z)$ for $N \geq N_0$.

If we can show that the derivative of the bias w.r.t. θ_0 is continuous for all $\theta_0 \in \mathbb{P}_r$, then it is also uniformly bounded in the compact set $\mathbb{P}_r$, and, hence, it behaves as an $O(N^{-1})$ ($O(N^{-1}) + O(N^{-K})$). Therefore, it is sufficient to show that the derivative of $\hat{\theta}(z)$ w.r.t. θ_0 is continuous in $\mathbb{P}_r$. Consider (17-9) with $\tilde{\theta}(z_0)$ replaced by θ_0, giving

$$V_N'(\hat{\theta}(z), z) = V_N'(\theta_0, z) + (\hat{\theta}(z) - \theta_0)^T V_N''(\widehat{\theta}, z) \tag{17-45}$$

where $\widehat{\theta} = t\hat{\theta}(z) + (1-t)\theta_0$ with $t \in [0, 1]$. Because $V_N(\theta, z)$ has continuous first-, second-, and third-order derivatives for all $\theta \in \mathbb{P}_r$ (Assumptions 17.16 and 17.20), it follows from the implicit function theorem (Kaplan, 1993) that $\widehat{\theta}$ is a continuous function of $\hat{\theta}(z)$ and θ_0 with continuous partial derivatives. Putting $\widehat{\theta} = g(\hat{\theta}(z), \theta_0)$ in (17-45), taking into account that $V_N'(\hat{\theta}(z), z) = 0$, gives

$$0 = V_N'(\theta_0, z) + (\hat{\theta}(z) - \theta_0)^T V_N''(g(\hat{\theta}(z), \theta_0), z) \tag{17-46}$$

(17-46) defines $\hat{\theta}(z)$ implicitly as a function of θ_0. Applying, again, the implicit function theorem shows that $\partial \hat{\theta}(z)/\partial \theta$ is continuous in $\mathbb{P}_r$. By definition of $\hat{\theta}(z)$ (17-17), this is also true for $\partial \underline{\hat{\theta}}(z)/\partial \theta$. □

Appendix 17.H Cumulants of the Partial Sum of a Mixing Sequence

Let $S(N) = \sum_{t=1}^{N} x_N(t)$ with $x_N(t)$ mixing over t of order P for $N = 1, 2, \ldots, \infty$. The kth order cumulant of $S(N)$ is an $O(N)$, $k = 1, 2, \ldots, P$.

Proof. The kth order cumulant C_k of $S(N)$ is given by

$$C_k = \sum_{t_1, t_2, \ldots, t_k = 1}^{N} \text{cum}(x_N(t_1), x_N(t_2), \ldots, x_N(t_k)) \tag{17-47}$$

Applying (16-36) to (17-47) gives $|C_k| = O(N)$. □

Appendix 17.I Proof of the Asymptotic Distribution of the Minimizer (Theorem 17.29)

From Theorem 17.21, it follows that $\sqrt{N}(\hat{\theta}(z) - \tilde{\theta}(z_0))$ converges in probability and, hence, also in law (see Section 16.7, interrelation 3) to $\sqrt{N}\delta_\theta(z)$. According to (17-14) the stochastic part of $\delta_\theta(z)$ is given by $V_N'^T(\tilde{\theta}(z_0), z)$. Since the matrix dimensions of the Hessian of the cost function are independent of N, we can study the stochastic behavior of $V_N'^T(\tilde{\theta}(z_0), z)$ separately from $V_N''^{-1}(\tilde{\theta}(z_0))$. We will show that the cumulants of $\sqrt{N} V_N'^T(\tilde{\theta}(z_0), z)$ tend to those of a Gaussian random variable. Because a normal distribution is uniquely determined by its moments, it follows from the Fréchet-Shohat Lemma 16.11 that $\sqrt{N} V_N'^T(\tilde{\theta}(z_0), z)$ converges in law to a Gaussian random variable. A linear combination of a finite number (independent of N) of Gaussian random variables is also a Gaussian random variable, which shows that $\sqrt{N}\delta_\theta(z)$ is asymptotically normally distributed. Expression (17-18) for the covariance matrix follows directly from the definition of $\delta_\theta(z)$ (17-14) and the fact that $\mathbb{E}\{\delta_\theta(z)\} = 0$.

Under Assumptions 17.1 ($P = 2K$), 17.2, and 17.16, the derivative of the cost function can be written for any $\theta \in \mathbb{P}_r$ as

$$\frac{\partial V_N(\theta, z)}{\partial \theta_{[i]}} = \frac{1}{N} \sum_{t=1}^{N} x_N(t) \tag{17-48}$$

$i = 1, 2, \ldots, n_\theta$, with $x_N(t)$ mixing over t of order K, $N = 1, 2, \ldots, \infty$ (proof: similar to that of $V_N(\theta, z)$ in Appendix 17.A). Hence, the kth order joint cumulant of $\sqrt{N} V_N'^T(\tilde{\theta}(z_0), z)$ is an $O(N^{1-k/2})$, $k = 1, 2, \ldots, K$ (see Appendix 17.H). Under Assumption 17.1 ($P = \infty$) this is valid for $K = 1, 2, \ldots, \infty$. It shows that the covariance matrix (second-order cumulant) is an $O(N^0)$ and that all the joint cumulants of order $k = 3, 4, \ldots, \infty$ are asymptotically ($N \to \infty$) zero. This concludes the proof because the joint cumulants, of order 3 and larger, of a Gaussian random variable are zero (see Example 16.2).

The proof of the convergence rate follows the lines of Appendix 16.H. □

Appendix 17.J Proof of the Existence and the Convergence of the Covariance Matrix of the Truncated Minimizer (Theorem 17.30)

Lemma 17.27 is valid under the assumptions of Theorem 17.21. It states that there exists an N_0 such that for any $N \geq N_0$, $\hat{\theta}(z) = \tilde{\theta}(z_0) + \delta_\theta(z) + b_\theta(z)$ w.p. 1. Because $\hat{\theta}(z)$, defined by (17-17), is uniformly bounded, its expected value and covariance matrix exist. The same is true for $b_\theta(z)$ for all realizations $z \in \mathbb{Z}_L$, where $\mathbb{Z}_L$ is defined in Appendix 17.F. Taking into account that for all $N \geq N_0$, $\mathbb{E}\{\delta_\theta(z)|z \in \mathbb{Z}_L\} = 0$, and $\mathbb{E}\{b_\theta(z)|z \in \mathbb{Z}_L\} = O(N^{-1})$ (proof: similar to Appendix 17.G), we find

$$\sqrt{N}\hat{\theta}(z) - \mathbb{E}\{\sqrt{N}\hat{\theta}(z)|z \in \mathbb{Z}_L\} = \sqrt{N}\delta_\theta(z) + O_p(N^{-1/2}) \tag{17-49}$$

where $O_p(N^{-1/2})$ is a uniformly bounded random variable. Calculating the covariance matrix of (17-49), taking into account that by definition of $\hat{\theta}(z)$

$$\text{Cov}(\sqrt{N}\hat{\theta}(z)) = \text{Cov}(\sqrt{N}\hat{\theta}(z)|z \in \mathbb{Z}_L) \tag{17-50}$$

gives

$$\text{Cov}(\sqrt{N}\hat{\theta}(z)) = \text{Cov}(\sqrt{N}\delta_\theta(z)|z \in \mathbb{Z}_L) + O(N^{-1/2}) \tag{17-51}$$

Because $\text{Prob}(z \notin \mathbb{Z}_L) = 0$ for $N \geq N_0$ (see Appendix 17.F) and the fourth-order moments of z are uniformly bounded, we have

$$\text{Cov}(\sqrt{N}\delta_\theta(z)|z \in \mathbb{Z}_L) = \text{Cov}(\sqrt{N}\delta_\theta(z)) \tag{17-52}$$

Combining (17-51) and (17-52) proves that $\text{Cov}(\sqrt{N}\hat{\theta}(z))$ converges to $\text{Cov}(\sqrt{N}\delta_\theta(z))$ at the rate $O(N^{-1/2})$. □

18

Properties of Least Squares Estimators with Stochastic Weighting

Abstract: This chapter studies the asymptotic stochastic properties (strong convergence, strong consistency, convergence rate, asymptotic bias, and asymptotic normality) of nonlinear least squares estimators with a stochastic weighting. The presented theory is applicable to, for example, (total) least squares estimators using nonparametric noise models. Because this chapter relies strongly on the results of Chapter 17, it cannot be read independently of that chapter. Readers who are unfamiliar with the analysis of the stochastic properties of estimators should, in addition, first read Sections 16.11 to 16.13.

18.1 INTRODUCTION—NOTATIONAL CONVENTIONS

In this chapter we consider the identification of a parametric plant and/or noise model $M(\theta, z_0, n_z)$ through the minimization of a weighted nonlinear least square cost function

$$V_N(\theta, z) = \frac{1}{N} z^T W_N(\theta, \eta(z), w(\theta, \eta(z), z)) z \tag{18-1}$$

with $z \in \mathbb{R}^N$ the noisy measurements, $z = z_0 + n_z$, $\theta \in \mathbb{R}^{n_\theta}$ the plant model parameters, and $\eta(z) \in \mathbb{R}^q$ a stochastic vector. $w(\theta, \eta(z), z) \in \mathbb{R}^p$ is the vector of the stochastic sums that average the noisy measurements z. $W_N \in \mathbb{R}^{N \times N}$ is a stochastic positive semidefinite weighting matrix depending on θ, $\eta(z)$, and $w(\theta, \eta(z), z)$. n_θ, p and q are N-independent integers. Just as in Chapter 17, we can assume without any loss of generality that the weighting matrix W_N is symmetric. We will often rewrite (18-1) as

$$V_N(\theta, z) = f_N(\theta, \eta(z), w(\theta, \eta(z), z), z) \tag{18-2}$$

and denote the function by

$$f_N(\theta, \eta, w, z) = \frac{1}{N} z^T W_N(\theta, \eta, w) z \tag{18-3}$$

where the stochastic vectors $\eta(z)$, $w(\theta, \eta(z), z)$ have been replaced by the deterministic vectors η, w. Note that (18-3) is a nonlinear least squares cost function with deterministic weighting. Similarly to $\mathbb{P}_r$ (see Section 17.1), we define $\mathbb{W} \subset \mathbb{R}^p$ as a compact (closed and bounded) set of w values for which the cost function (18-3) and/or its higher order derivatives w.r.t. w exist and are finite. From (18-2) and (18-3) it follows that it is not possible to put (18-1) within the framework (16-95) of Section 16.13, even without $\eta(z)$. The reason for this is that (18-1) is a stochastic sum which depends, itself, on other stochastic sums $w(\theta, \eta(z), z)$ and a stochastic vector $\eta(z)$.

Following the same lines as in Chapter 17, the asymptotic ($N \to \infty$) properties (strong convergence, strong consistency, convergence rate, asymptotic bias, and asymptotic normality) of the (set of) minimizer(s)

$$\hat{\theta}(z) = \arg \min_{\theta \in \mathbb{P}_r} V_N(\theta, z) \qquad (18\text{-}4)$$

will be analyzed ($\mathbb{P}_r$, see Section 17.1, is a compact set of parameters where the cost function and/or its higher order derivatives exist and are finite). It is clear that the asymptotic properties of the minimizer (18-4) strongly depend on the stochastic behavior of $\eta(z)$ and $w(\theta, \eta(z), z)$. Assumptions similar to those of Chapter 17 guarantee the stochastic properties of (18-3). The main difference from Chapter 17 is that additional assumptions concerning $\eta(z)$ and $w(\theta, \eta(z), z)$ have to be made to ensure that (18-4) has asymptotic properties similar to those of the nonlinear least squares estimator with deterministic weighting (17-3). The analysis of the nonlinear least squares estimator with stochastic weighting (18-4) relies heavily on the stochastic properties of a converging sequence of functions that also depends on some converging random vector(s). Therefore, it is recommended to read Appendix 17.C first, before going through the proofs of this chapter. The chapter ends with an overview of the asymptotic properties of $\hat{\theta}(z)$.

18.2 STRONG CONVERGENCE

The strong convergence of the minimizer is a direct consequence of the strong convergence of the cost function (see Section 17.2). The cost function (18-1) can converge only if $\eta(z)$ tends to some nonrandom number and if $w(\theta, \eta(z), z)$ tends to some deterministic function of θ. Therefore, it is natural to make the following assumptions (see Lemma 17.31, Appendix 17.C).

Assumption 18.1 (Strong Convergence $\eta(z)$): The stochastic vector $\eta(z)$ converges w.p. 1 to a nonrandom value η_*.

Define $\mathbb{E}_\varepsilon$ as a compact set of η values in the neighborhood of η_*

$$\mathbb{E}_\varepsilon = \{\eta \in \mathbb{R}^q | \|\eta - \eta_*\|_2 \leq \varepsilon\} \qquad (18\text{-}5)$$

and let $w(\theta, \eta, z)$ denote the stochastic sums, where the stochastic vector $\eta(z)$ is replaced by the deterministic vector η. Although the strong convergence of the minimizer puts conditions only on the convergence of $w(\theta, \eta(z), z)$, the assumption will be stated more generally for the kth order derivative w.r.t. θ.

Assumption 18.2 (Strong or Weak Convergence $w^{(k)}(\theta, \eta, z)$): $w^{(k)}(\theta, \eta, z)$, the kth order derivative of $w(\theta, \eta, z)$ w.r.t. θ, converges uniformly w.p. 1 (in prob.) to $\mathbb{E}\{w^{(k)}(\theta, \eta, z)\} = w^{(k)}(\theta, \eta)$, in $\mathbb{P}_r$, $\mathbb{E}_\varepsilon$. For any N, ∞ included, $w^{(k)}(\theta, \eta, z)$ is a jointly continuous function of θ, η in $\mathbb{P}_r$, $\mathbb{E}_\varepsilon$.

18.2.1 Strong Convergence of the Cost Function

Assumption 18.3 (Constraints on the Cost Function): (a) The weighting matrix $W_N(\theta, \eta, w)$ is a symmetric positive semidefinite matrix, satisfying $\|W_N(\theta, \eta, w)\|_1 \leq c < \infty$, with c an N-independent constant, for all N (∞ included) and all $\theta \in \mathbb{P}_r$, $\eta \in \mathbb{E}_\varepsilon$, $w \in \mathbb{W}$. $W_N(\theta, \eta, w)$ is a jointly continuous matrix function of θ, η, w in the compact sets $\mathbb{P}_r$, $\mathbb{E}_\varepsilon$, $\mathbb{W}$. (b) There exists an N_0 such that for any $r \geq s \geq N_0$, $\|W_{r[1:s,1:s]}(\theta, \eta, w) - W_s(\theta, \eta, w)\|_1^2 = O((r-s)/r)$ in $\mathbb{P}_r$, $\mathbb{E}_\varepsilon$, $\mathbb{W}$.

Condition (b) is necessary to ensure the strong convergence of the cost function. If it is not fulfilled, then all the lemmas and theorems of this chapter remain valid except that the strong convergence (convergence w.p. 1) must be replaced everywhere by weak convergence (convergence in prob.).

Lemma 18.4 (Strong Convergence of the Cost Function): Under Assumptions 17.1 ($P = 4$), 18.1, 18.2 (w.p. 1, $k = 0$), and 18.3 the cost function $V_N(\theta, z)$ converges uniformly w.p. 1 to

$$V_N(\theta) = \mathbb{E}\{f_N(\theta, \eta_*, w(\theta, \eta_*), z)\} = f_N(\theta, \eta_*, w(\theta, \eta_*)) \tag{18-6}$$

in the compact set $\mathbb{P}_r$. $V_N(\theta, z)$ and $V_N(\theta)$ are continuous functions of θ in $\mathbb{P}_r$.

Proof. See Appendix 18.A. □

Note that $V_N(\theta) = f_N(\theta, \eta_*, w(\theta, \eta_*))$ is obtained as follows: first replace the stochastic weighting $W_N(\theta, \eta(z), w(\theta, \eta(z), z))$ in (18-1) by the deterministic weighting $W_N(\theta, \eta_*, w(\theta, \eta_*))$, and next, take the expected value. This shows that under Assumptions 18.1 and 18.2 (w.p. 1), the stochastic behavior of the cost function with stochastic weighting is similar to that with deterministic weighting.

18.2.2 Strong Convergence of the Minimizer

Using definition (18-6) of $V_N(\theta)$, the theorems of Section 17.2.2 (strong convergence of the minimizer of nonlinear least squares cost functions with deterministic weighting) remain valid under Assumptions 18.1, 18.2 (w.p. 1, $k = 0$).

Theorem 18.5 (Strong Convergence of the Minimizer): Under Assumptions 17.1 ($P = 4$), 18.1, 18.2 (w.p. 1, $k = 0$), 18.3, and 17.5 the minimizer(s) $\hat{\theta}(z)$ converge(s) w.p. 1 to

$$\tilde{\theta}(z_0) = \arg\min_{\theta \in \mathbb{P}_r} f_N(\theta, \eta_*, w(\theta, \eta_*)) \tag{18-7}$$

Proof. Similar to Theorem 17.6 (see Appendix 17.B). □

Theorem 18.6 (Strong Convergence of the Minimizer): Under Assumptions 17.1 ($P = 4$), 18.1, 18.2 (w.p. 1, $k = 0$), 18.3, 17.4, and 17.7, $\tilde{\theta}(z_0)$ converges to θ_* and $\hat{\theta}(z)$ converges strongly θ_*, with

$$\theta_* = \arg\min_{\theta \in \mathbb{P}_r} V_*(\theta) \text{ and } V_*(\theta) = \lim_{N \to \infty} f_N(\theta, \eta_*, w(\theta, \eta_*)) \tag{18-8}$$

Proof. Similar to Theorem 17.8. □

18.3 STRONG CONSISTENCY

The cost function (18-6) can be written as

$$V_N(\theta) = \frac{1}{N} \mathbb{E}\{z_0^T W_N(\theta, \eta_*, w(\theta, \eta_*))z_0\} + \frac{1}{N} \text{trace}(W_N(\theta, \eta_*, w(\theta, \eta_*)) C_{n_z}) \quad (18\text{-}9)$$

and equals $V_N(\theta)$ in (17-8) where $W_N(\theta)$ is replaced by $W_N(\theta, \eta_*, w(\theta, \eta_*))$. Hence, replacing $W_N(\theta)$ by $W_N(\theta, \eta_*, w(\theta, \eta_*))$ in the assumptions of Section 17.3 shows that the theorems of Section 17.3 (strong consistency of the minimizer of nonlinear least squares cost functions with deterministic weighting) remain valid under Assumptions 18.1, 18.2 (w.p. 1, $k = 0$). We cite them in order of reduced conditions on $V_N(\theta)$.

Theorem 18.7 (Strong Consistency): Under the assumptions of Theorem 18.5 and Assumptions 17.9 and 17.10, the estimate $\hat{\theta}(z)$ converges w.p. 1 to θ_0.

Proof. It follows directly from Theorem 18.5 and Assumptions 17.9 and 17.10. □

Theorem 18.8 (Strong Consistency): Under the assumptions of Theorem 18.5 and Assumptions 17.9 and 17.12, the estimate $\hat{\theta}(z)$ converges w.p. 1 to θ_0.

Proof. It follows directly from Theorem 18.5 and Assumptions 17.9 and 17.12. □

Theorem 18.9 (Strong Consistency): Under the assumptions of Theorem 18.6 and Assumptions 17.9 and 17.14, the estimate $\hat{\theta}(z)$ converges w.p. 1 to θ_0.

Proof. It follows directly from Theorem 18.6 and Assumptions 17.9 and 17.14. □

18.4 CONVERGENCE RATE

The convergence rate of the minimizer is a direct consequence of the convergence rate of the first- and second-order derivatives of the cost function w.r.t. θ (see Section 17.4). These derivatives can be written as

$$\begin{aligned} V_N'^T(\theta, z) &= g_N(\theta, \eta(z), w(\theta, \eta(z), z), w'(\theta, \eta(z), z), z) \\ V_N''(\theta, z) &= h_N(\theta, \eta(z), w(\theta, \eta(z), z), w'(\theta, \eta(z), z), w''(\theta, \eta(z), z), z) \end{aligned} \quad (18\text{-}10)$$

From (18-10) it follows that the convergence rate of the derivatives of the cost function is influenced by the convergence rates of $\eta(z)$ and $w^{(k)}(\theta, \eta(z), z)$, $k = 0, 1, 2$. Therefore, it is natural to make the following assumptions (see Lemma 17.34, Appendix 17.C).

Assumption 18.10 (Convergence Rate $\eta(z)$): The convergence rate in probability of $\eta(z)$ equals $O_p(N^{-1/2})$: $\eta(z) - \eta_* = O_p(N^{-1/2})$.

Assumption 18.11 (Convergence Rate $w^{(k)}(\theta, \eta, z)$): The convergence rate in probability of the kth order derivative $w^{(k)}(\theta, \eta, z)$ equals $O_p(N^{-1/2})$ uniformly in $\mathbb{P}_r$, $\mathbb{E}_\varepsilon$: $w^{(k)}(\theta, \eta, z) - w^{(k)}(\theta, \eta) = O_p(N^{-1/2})$.

Section 18.4 ■ Convergence Rate   655

Assumption 18.12 (Constraint on the Derivative of $w^{(k)}(\theta, \eta, z)$): The derivative of $w^{(k)}(\theta, \eta, z)$ w.r.t. η satisfies

$$\left\| \frac{\partial}{\partial \eta} \frac{\partial^k w(\theta, \eta, z)}{\partial \theta_{[i_1]} \partial \theta_{[i_1]} \ldots \partial \theta_{[i_k]}} \right\|_2 \leq O_p(N^0)$$

uniformly in $\mathbb{P}_r$, $\mathbb{E}_\varepsilon$, for $i_1, i_2, \ldots, i_k = 1, 2, \ldots, n_\theta$ and for any N (∞ included).

We discuss only the convergence rate of $\hat{\theta}(z)$ to $\tilde{\theta}(z_0)$. The reader is referred to Section 17.4.3 for a discussion of the convergence rate of $\tilde{\theta}(z_0)$ to θ_*. The first step in the analysis of the convergence rate of $\hat{\theta}(z)$ is the weak convergence of the derivatives (18-10) of the cost function.

18.4.1 Convergence of the Derivatives of the Cost Function

The derivatives of the cost function (18-10) can be written more explicitly as

$$g_{N[i]}(\theta, \eta(z), w(\theta, \eta(z), z), w'(\theta, \eta(z), z), z) = \frac{1}{N} z^T \frac{dW_N(\theta, \eta(z), w(\theta, \eta(z), z))}{d\theta_{[i]}} z$$

$$h_{N[i,j]}(\theta, \eta(z), w(\theta, \eta(z), z), w'(\theta, \eta(z), z), w''(\theta, \eta(z), z), z) = \frac{1}{N} z^T \frac{d^2 W_N(\theta, \eta(z), w(\theta, \eta(z), z))}{d\theta_{[i]} d\theta_{[j]}} z$$

(18-11)

for $i, j = 1, 2, \ldots, n_\theta$, with

$$\frac{dW_N(\theta, \eta(z), w(\theta, \eta(z), z))}{d\theta_{[i]}} = W_{1i}(\theta, \eta(z), w(\theta, \eta(z), z), w'(\theta, \eta(z), z))$$

$$\frac{d^2 W_N(\theta, \eta(z), w(\theta, \eta(z), z))}{d\theta_{[i]} d\theta_{[j]}} = W_{2ij}(\theta, \eta(z), w(\theta, \eta(z), z), w'(\theta, \eta(z), z), w''(\theta, \eta(z), z))$$

(18-12)

Similarly to (18-3), we will denote the functions (18-10) by

$$g_N(\theta, \eta, w, w_1, z) \text{ and } h_N(\theta, \eta, w, w_1, w_2, z) \quad (18\text{-}13)$$

where the random variables $\eta(z)$ and $w^{(k)}(\theta, \eta(z), z)$, $k = 0, 1, 2$, have been replaced by the deterministic variables η, w, w_1, w_2. Define $\mathbb{W}_1 \subset \mathbb{R}^{p \times n_\theta}$ and $\mathbb{W}_2 \subset \mathbb{R}^{p \times n_\theta \times n_\theta}$ as the compact sets of, respectively, w_1 and w_2 values for which the functions (18-13) and/or their higher order derivatives w.r.t. w_1 and w_2 exist and are finite.

The proof of the convergence in probability follows the same lines as for Lemma 18.4. Therefore, the following assumptions concerning W_{1i}, W_{2ij} must be made.

Assumption 18.13 (Constraints on the Derivatives of the Cost Function): The matrices $W_{1i}(\theta, \eta, w, w_1)$ and $W_{2ij}(\theta, \eta, w, w_1, w_2)$ have bounded 1-norm

$$\|W_{1i}(\theta, \eta, w, w_1)\|_1 \leq c_1 < \infty, \quad \|W_{2ij}(\theta, \eta, w, w_1, w_2)\|_1 \leq c_2 < \infty$$

for $i, j = 1, 2, \ldots, n_\theta$, and are jointly continuous functions of θ, η, w, w_1, w_2 in the compact sets $\mathbb{P}_r$, $\mathbb{E}_\varepsilon$, $\mathbb{W}$, $\mathbb{W}_1$, $\mathbb{W}_2$ for any N (∞ included). c_1, c_2 are N-independent constants.

Lemma 18.14 (Convergence of the Derivatives of the Cost Function): Under Assumptions 17.1 ($P = 4$), 18.1, 18.2 (in prob., $k = 0, 1, 2$), and 18.13, the derivatives of the cost function $V_N'(\theta, z)$ and $V_N''(\theta, z)$ converge uniformly in probability, respectively, to

$$V_N'^T(\theta) = \mathbb{E}\{g_N(\theta, \eta_*, w(\theta, \eta_*), w'(\theta, \eta_*), z)\} = g_N(\theta, \eta_*, w(\theta, \eta_*), w'(\theta, \eta_*))$$
$$V_N''(\theta) = \mathbb{E}\{h_N(\theta, \eta_*, w(\theta, \eta_*), w'(\theta, \eta_*), w''(\theta, \eta_*), z)\} = h_N(\theta, \eta_*, w(\theta, \eta_*), w'(\theta, \eta_*), w''(\theta, \eta_*))$$

in the compact set $\mathbb{P}_r$. $V_N'(\theta, z)$ and $V_N''(\theta, z)$ are continuous functions of θ in $\mathbb{P}_r$.

Proof. Similar to Lemma 18.4. □

18.4.2 Convergence Rate of $\hat{\theta}(z)$ to $\tilde{\theta}(z_0)$

Assumption 18.15 (Constraint on Derivative $g_N(\theta, \eta, w, w_1)$ w.r.t. η, w, w_1): The weighting matrices in $g_N(\theta, \eta, w, w_1)$ have continuous first-order derivatives satisfying

$$\left\| \frac{\partial W_{1i}(\theta, \eta, w, w_1)}{\partial x_{[j]}} \right\|_2 \leq c < \infty \qquad (18\text{-}14)$$

uniformly in $\mathbb{P}_r$, $\mathbb{E}_\varepsilon$, $\mathbb{W}$, and $\mathbb{W}_1$, for $i = 1, 2, \ldots, n_\theta$, $j = 1, 2, \ldots, \dim(x)$, and for any N (∞ included). x is a vector that contains all the elements of η, w, w_1 ($\dim(x) = q + p + p n_\theta$), and c is an N-independent constant.

Theorem 18.16 (Convergence Rate of $\hat{\theta}(z)$ to $\tilde{\theta}(z_0)$): Under Assumptions 17.1 ($P = 4$), 17.18, 18.1, 18.2 (w.p. 1, $k = 0$; in prob., $k = 1, 2$), 18.3(a), 18.10, 18.11 ($k = 0, 1$), 18.12 ($k = 0, 1$), 18.13, and 18.15 the convergence rate of $\hat{\theta}(z)$ to $\tilde{\theta}(z_0)$ equals $O_p(N^{-1/2})$: $\hat{\theta}(z) - \tilde{\theta}(z_0) = O_p(N^{-1/2})$.

Proof. See Appendix 18.B. □

An additional assumption on the third-order derivatives of the cost function allows the refinement of the expression for the convergence rate.

Assumption 18.17 (Constraint on Derivative $h_N(\theta, \eta, w, w_1, w_2)$ w.r.t. $\theta, \eta, w, w_1, w_2$): The weighting matrices in $h_N(\theta, \eta, w, w_1, w_2)$ have jointly continuous first-order derivatives satisfying

$$\left\| \frac{\partial W_{2ij}(\theta, \eta, w, w_1, w_2)}{\partial x_{[k]}} \right\|_2 \leq c < \infty \qquad (18\text{-}15)$$

uniformly in $\mathbb{P}_r$, $\mathbb{E}_\varepsilon$, $\mathbb{W}$, $\mathbb{W}_1$, and $\mathbb{W}_2$, for $i, j = 1, 2, \ldots, n_\theta$, $k = 1, 2, \ldots, \dim(x)$ and for any N (∞ included). x is a vector that contains all the elements of θ, η, w, w_1, w_2 ($\dim(x) = n_\theta + q + p + p n_\theta + p n_\theta^2$), and c is an N-independent constant.

Section 18.5 ■ Asymptotic Bias

Theorem 18.18 (Improved Convergence Rate of $\hat{\theta}(z)$ to $\tilde{\theta}(z_0)$): Under Assumptions 17.1 ($P = 4$), 17.18, 18.1, 18.2 (w.p. 1, $k = 0$; in prob., $k = 1, 2$), 18.3(a), 18.10, 18.11 ($k = 0, 1, 2$), 18.12 ($k = 0, 1, 2$), 18.13, 18.15, and 18.17, the minimizer $\hat{\theta}(z)$ can be written as

$$\hat{\theta}(z) = \tilde{\theta}(z_0) + \delta_\theta(z) + b_\theta(z)$$
$$\delta_\theta(z) = -V_N''^{-1}(\tilde{\theta}(z_0)) V_N'^T(\tilde{\theta}(z_0), z)$$
(18-16)

with $\delta_\theta(z) = O_p(N^{-1/2})$ and $b_\theta(z) = O_p(N^{-1})$.

Proof. Follow the lines of the proof of Theorem 17.21, generalized as in Theorem 18.16 for the stochastic weighting. □

Note that compared with the deterministic weighting (Theorem 17.21), the expected value of $\delta_\theta(z)$ is, in general, not zero and may not even exist. Indeed, $\tilde{\theta}(z_0)$ is not the minimizer of $\mathbb{E}\{V_N(\tilde{\theta}(z_0), z)\}$ (see (18-2) with (18-6) and (18-7)), and the expected value of the weighting $\mathbb{E}\{W_N(\theta, \eta(z), w(\theta, \eta(z), z))\}$ may not exist.

18.5 ASYMPTOTIC BIAS

We assume in this section that a true model exists and that it belongs to the considered model set. Compared with the case with deterministic weighting (see Section 17.5), we need additional assumptions to guarantee that the asymptotic bias behaves as an $O(N^{-1})$. This is due to the fact that the expected value of $\delta_\theta(z)$ in (18-16) is, in general, not zero or equivalently $\mathbb{E}\{V_N'(\theta_0, z)\} \neq 0$. Therefore, using the concept of the truncated estimate (see Section 17.5), the asymptotic bias of $V_N'(\theta_0, z)$ will be analyzed in more details. This requires the following assumptions (see Lemma 17.36).

Assumption 18.19 (Asymptotic Bias $\underline{\eta}(z)$): The bias of the truncated estimate $\underline{\eta}(z)$ is an $O(N^{-1})$: $\mathbb{E}\{\underline{\eta}(z)\} = \eta_* + O(N^{-1})$.

Assumption 18.20 (Derivative $w^{(k)}(\theta, \eta, z)$ w.r.t. η): The derivative of $w^{(k)}(\theta, \eta, z)$ w.r.t. η, $\partial w^{(k)}(\theta, \eta, z)/\partial \eta$, converges uniformly in probability to its expected value $\partial w^{(k)}(\theta, \eta)/\partial \eta$ in $\mathbb{P}_r$, $\mathbb{E}_\varepsilon$ at the rate $O_p(N^{-1/2})$. The second-order derivative of $w^{(k)}(\theta, \eta, z)$ w.r.t. η is uniformly bounded in $\mathbb{P}_r$, $\mathbb{E}_\varepsilon$ for any N (∞ included):

$$\left\| \frac{\partial^2 w^{(k)}(\theta, \eta)}{\partial \eta_{[i]} \partial \eta_{[j]}} \right\|_2 \leq O_p(N^0), \quad i, j = 1, 2, \ldots, q.$$

Assumption 18.21 (Constraint on Derivatives $g_N(\theta, \eta, w, w_1)$ w.r.t. η, w, w_1): The weighting matrices in $g_N(\theta, \eta, w, w_1)$ have jointly continuous first- and second-order derivatives satisfying

$$\left\| \frac{\partial W_{1i}(\theta, \eta, w, w_1)}{\partial x_{[j]}} \right\|_1 \leq c_1 < \infty, \quad \left\| \frac{\partial^2 W_{1i}(\theta, \eta, w, w_1)}{\partial x_{[j]} \partial x_{[k]}} \right\|_2 \leq c_2 < \infty$$

uniformly in $\mathbb{P}_r$, $\mathbb{E}_\varepsilon$, $\mathbb{W}$, and $\mathbb{W}_1$, for $i = 1, 2, ..., n_\theta$, $j, k = 1, 2, ..., \dim(x)$ and for any N, ∞ included. x is a vector that contains all the elements of η, w, w_1 ($\dim(x) = q + p + pn_\theta$) and c_1, c_2 are N-independent constants.

Theorem 18.22 (Asymptotic Bias of $\hat{\underline{\theta}}(z)$): Under the assumptions of Theorem 18.18 and Assumptions 17.9, 17.10 or 17.12, 18.19, 18.20 ($k = 0, 1$), and 18.21, the bias $b_\theta = \mathbb{E}\{b_\theta(z)\}$ of the truncated estimate $\hat{\underline{\theta}}(z)$ is an $O(N^{-1})$. If, in addition, $w^{(3)}(\theta, \eta(z), z)$ is continuous in $\mathbb{P}_r$ then the derivative of the bias w.r.t. θ_0, $\partial b_\theta / \partial \theta_0$, is an $O(N^{-1})$.

Proof. See Appendix 18.C. □

18.6 ASYMPTOTIC NORMALITY

The asymptotic distribution function of the minimizer is determined by the asymptotic distribution function of the first derivative of the cost function w.r.t. θ (see Section 17.6). The asymptotic distribution function of the derivative of the cost function (18-10) is influenced by the asymptotic distribution functions of $\eta(z)$ and $w^{(k)}(\theta, \eta(z), z)$, $k = 0, 1$. Therefore, it is natural to make the following assumptions (see Lemma 17.38, Appendix 17.C).

Assumption 18.23 (Asymptotic Distribution $\eta(z)$): $\eta(z)$ can be written as $\eta(z) = \tilde{\eta}(z) + O_p(N^{-1})$ with $\tilde{\eta}(z) = O_p(N^{-1/2})$, and where $\tilde{\eta}(z)$ has finite second-order moments and converges in law at the rate $O(N^{-1/2})$ to a Gaussian random variable with mean value η_*.

Assumption 18.24 (Asymptotic Distribution $w^{(k)}(\theta, \eta, z)$): $w^{(k)}(\theta, \eta, z)$, $k = 0, 1$, can be written as $w^{(k)}(\theta, \eta, z) = \tilde{w}^{(k)}(\theta, \eta, z) + O_p(N^{-1})$ with $\tilde{w}^{(k)}(\theta, \eta, z) = O_p(N^{-1/2})$, and where $\tilde{w}^{(k)}(\theta, \eta, z)$ has finite second-order moments and converges in law at the rate $O(N^{-1/2})$ to a Gaussian random variable with mean value $w^{(k)}(\theta, \eta)$. The convergence is uniform in $\mathbb{P}_r$, $\mathbb{E}_\varepsilon$.

Theorem 18.25 (Asymptotic Normality of $\sqrt{N}(\hat{\theta}(z) - \tilde{\theta}(z_0)))$: Under the assumptions of Theorem 18.18 and Assumptions 17.1 ($P = \infty$), 18.20 ($k = 0, 1$), 18.21, 18.23, and 18.24, $\sqrt{N}(\hat{\theta}(z) - \tilde{\theta}(z_0))$ converges in law at the rate $O(N^{-1/2})$ to a Gaussian random variable. The expression for the covariance matrix (17-18) is still valid if $\delta_\theta(z)$ is replaced by

$$d_\theta(z) = -V_N'''^{-1}(\tilde{\theta}(z_0))d_N(z)$$

$$d_N(z) = g_N(\tilde{\theta}(z_0), \eta_*, w(\tilde{\theta}(z_0), \eta_*), w'(\tilde{\theta}(z_0), \eta_*), z)$$

$$+ \left.\frac{dg_N(\tilde{\theta}(z_0), \eta, w(\tilde{\theta}(z_0), \eta), w'(\tilde{\theta}(z_0), \eta))}{dx}\right|_{x = x_*} (\tilde{x}(z) - x_*)$$

(18-17)

with

$$x^T = [\eta^T \; w^T(\tilde{\theta}(z_0), \eta) \; \text{vec}^T(w'(\tilde{\theta}(z_0), \eta))]$$

$$x_*^T = [\eta_*^T \; w^T(\tilde{\theta}(z_0), \eta_*) \; \text{vec}^T(w'(\tilde{\theta}(z_0), \eta_*))]$$

$$\tilde{x}^T(z) = [\tilde{\eta}^T(z) \; \tilde{w}^T(\tilde{\theta}(z_0), \eta_*, z) \; \text{vec}^T(\tilde{w}'(\tilde{\theta}(z_0), \eta_*, z))]$$

$d_\theta(z) = O_p(N^{-1/2})$, and $\mathbb{E}\{d_\theta(z)\} = 0$. The functions $g_N(\ )$ are defined in (18-10) and Lemma 18.14, and $\tilde{\eta}(z)$, $\tilde{w}^{(k)}(\tilde{\theta}(z_0), \eta, z)$ are defined in Assumptions 18.23 and 18.24. The derivative w.r.t. η in (18-17) is calculated using the chain rule

$$\frac{dg_N}{d\eta} = \frac{\partial g_N}{\partial \eta} + \frac{\partial g_N}{\partial w}\frac{\partial w}{\partial \eta} + \frac{\partial g_N}{\partial \text{vec}(w')}\frac{\partial \text{vec}(w')}{\partial \eta} \tag{18-18}$$

Proof. See Appendix 18.D. □

18.7 OVERVIEW OF THE ASYMPTOTIC PROPERTIES

In this section we give an overview of the asymptotic properties of the minimizer $\hat{\theta}(z)$ (18-4) of a cost function $V_N(\theta, z)$ (18-1), which is quadratic-in-the-measurements when the stochastic vectors $\eta(z)$, $w(\theta, \eta(z), z)$ in the weighting matrix W_N are replaced by deterministic vectors η, w. In the analysis of the stochastic properties of $\hat{\theta}(z)$, the cost functions and minimizers of Table 18-1 play an important role.

The minimizer $\hat{\theta}(z)$ (18-4) of the cost function $V_N(\theta, z)$ (18-1) has the following asymptotic ($N \to \infty$) properties:

1. *Stochastic convergence:* $\hat{\theta}(z)$ converges strongly to $\tilde{\theta}(z_0)$ (Theorem 18.5).
2. *Stochastic convergence rate:* $\hat{\theta}(z)$ converges in probability at the rate $O_p(N^{-1/2})$ to $\tilde{\theta}(z_0)$ (Theorem 18.16).
3. *Systematic and stochastic errors:* $\hat{\theta}(z)$ converges in probability to $\tilde{\theta}(z_0)$ with

$$\begin{aligned}\hat{\theta}(z) &= \tilde{\theta}(z_0) + \delta_\theta(z) + b_\theta(z) \\ \delta_\theta(z) &= -V_N''^{-1}(\tilde{\theta}(z_0))V_N'^T(\tilde{\theta}(z_0), z)\end{aligned} \tag{18-19}$$

where $\delta_\theta(z) = O_p(N^{-1/2})$ is the dominating stochastic error and where $b_\theta(z) = O_p(N^{-1})$ contains the contribution of the systematic errors (Theorem 18.18).

4. *Asymptotic normality:* $\sqrt{N}(\hat{\theta}(z) - \tilde{\theta}(z_0))$ converges in law at the rate $O(N^{-1/2})$ to a Gaussian random variable with zero mean and covariance matrix $\text{Cov}(\sqrt{N}\,d_\theta(z))$

$$\begin{aligned}\text{Cov}(\sqrt{N}\,d_\theta(z)) &= V_N''^{-1}(\tilde{\theta}(z_0))Q_N(\tilde{\theta}(z_0))V_N''^{-1}(\tilde{\theta}(z_0)) \\ Q_N(\tilde{\theta}(z_0)) &= N\,\mathbb{E}\{d_N(z)d_N^T(z)\}\end{aligned} \tag{18-20}$$

where $d_N(z)$ is defined in (18-17) (Theorem 18.25).

5. *Deterministic convergence:* $\tilde{\theta}(z_0)$ converges to θ_* (Theorem 18.6).

TABLE 18-1 Overview of the Notations Used: $\eta(z)$ and $w(\theta, \eta(z), z)$ Are Stochastic Vectors, and η_*, $w(\theta, \eta_*) = \mathbb{E}\{w(\theta, \eta_*, z)\}$ Are the Corresponding Limit Values

Cost function	$V_N(\theta, z) =$ $f_N(\theta, \eta(z), w(\theta, \eta(z), z), z)$	$V_N(\theta) = \mathbb{E}\{f_N(\theta, \eta_*, w(\theta, \eta_*), z)\}$ $= f_N(\theta, \eta_*, w(\theta, \eta_*))$	$V_*(\theta) = \lim_{N \to \infty} V_N(\theta)$
Minimizer	$\hat{\theta}(z)$	$\tilde{\theta}(z_0)$	θ_*

If in addition $V_N(\theta, z)$ satisfies the consistency conditions, then

6. *Consistency:* $\hat{\theta}(z)$ is strongly consistent; replace in properties 1 to 4 $\hat{\theta}(z_0)$ or $\lim_{F \to \infty} \hat{\theta}(Z_0) = \theta_*$ by θ_0 (Theorems 18.7, 18.8, and 18.9).

7. *Asymptotic bias:* the asymptotic bias $b_\theta = \mathbb{E}\{b_\theta(z)\}$, and its derivative w.r.t. θ_0, $\partial b_\theta / \partial \theta_0$, of $\hat{\theta}(z)$ are an $O(N^{-1})$ (Theorem 18.22).

Note the similarity to the properties of Section 17.8. Compared with the deterministic weighting, the uncertainty (18-20) is increased. This is due to the stochastic vectors $\eta(z)$ and $w(\theta, \eta(z), z)$ in the weighting W_N (compare $d_N(z)$ in (18-17) with $V_N'^T(\tilde{\theta}(z_0), z)$ in (17-18)).

18.8 EXERCISES

18.1. Consider the model equation $y_0(k) = f(\theta, u_0(k))$. Assume that $M \geq 2$ independent repeated experiments of $N \geq n_\theta$ measurements each are available: $y^{[r]}(k) = y_0(k) + n_y^{[r]}(k)$ for $r = 1, 2, ..., M$ and $k = 1, 2, ..., N$, where the disturbing noise $n_y^{[r]}(k)$ is independent and identically distributed (over r, k) with finite fourth-order moment and $\mathrm{var}(n_y^{[r]}(k)) = \sigma_y^2$. Consider the nonlinear least squares cost function with stochastic weighting

$$\left[\sum_{k=1}^N \hat{y}(k) - f(\theta, u_0(k))^2 \right] / \left[\sum_{k=1}^N \hat{\sigma}_y^2(k) \right]$$

where $\hat{y}(k)$ and $\hat{\sigma}_y^2(k)$ are, respectively, the sample mean and sample variance of the kth measurement over the M experiments. Show that $\hat{\theta}(z)$ is a strongly consistent estimate if $f(\theta, u_0(k))$ is a continuous function of θ. Under what conditions on $f(\theta, u_0(k))$ are the other results of this chapter valid? What additional assumption on $n_y(k)$ is required for the asymptotic normality property? (Hint: first write the cost function as $z^T W_N(\theta, w(z)) z / N$ with $z^T = [\hat{y}(1)\ 1\ \hat{y}(2)\ 1\ ...\ \hat{y}(N)\ 1]$, $w(z) = \sum_{k=1}^N \hat{\sigma}_y^2(k)/N$, $W_N(\theta, w(z)) = \mathrm{diag}(C_1, C_2, ..., C_N)$, and

$$C_k = \frac{1}{w(z)} \begin{bmatrix} 1 & -f(\theta, u_0(k)) \\ -f(\theta, u_0(k)) & f^2(\theta, u_0(k)) \end{bmatrix}).$$

18.2. Consider the linear model $y_0(k) = a_0 u_0(k) + b_0$ with $\theta^T = [a\ b]$. Assume that $M \geq 2$ independent repeated experiments of $N \geq 2$ measurements each are available: $y^{[r]}(k) = y_0(k) + n_y^{[r]}(k)$, $u^{[r]}(k) = u_0(k) + n_u^{[r]}(k)$ for $r = 1, 2, ..., M$ and $k = 1, 2, ..., N$. $n_y^{[r]}(k)$, $n_u^{[r]}(k)$ are independent (over r, k) uniformly bounded random variables. Consider the nonlinear least squares cost function with stochastic weighting

$$\left[\sum_{k=1}^N (\hat{y}(k) - a\hat{u}(k) - b)^2 \right] / \left[\sum_{k=1}^N (\hat{\sigma}_y^2(k) + a^2 \hat{\sigma}_u^2(k)) \right]$$

where $\hat{y}(k)$, $\hat{u}(k)$ and $\hat{\sigma}_y^2(k)$, $\hat{\sigma}_u^2(k)$ are, respectively, the sample means and sample variances of the kth measurement over the M experiments. Show that $\hat{\theta}(z)$ is a strongly consistent estimate. Show that all the other results of this chapter are also valid (hint: follow the lines of Exercise 18.1).

18.3. Consider the linear model $y_0(k) = a_0 u_0(k) + b_0$ with $\theta^T = [a\ b]$. Assume that N noisy observations of the input and output are available: $y(k) = y_0(k) + n_y(k)$ and $u(k) = u_0(k) + n_u(k)$, $k = 1, 2, ..., N$. $n_y(k)$, $n_u(k)$ are independent Gaussian random variables. The maximum likelihood solution of this problem minimizes

$$\sum_{k=1}^N (y(k) - au(k) - b)^2 / (\sigma_y^2(k) + a^2 \sigma_u^2(k))$$

w.r.t. θ. This requires a nonlinear minimization and, therefore, the following weighted linear least squares approximation, also called iterative quadratic maximum likelihood (IQML), is often calculated:

$$\hat{\theta}_{\text{IQML}} = \arg\min_{\theta} \sum_{k=1}^{N} (y(k) - au(k) - b)^2 / (\sigma_y^2(k) + \hat{a}_{\text{LS}}^2 \sigma_u^2(k))$$

where $\hat{\theta}_{\text{LS}}$ minimizes the linear least squares cost function $\sum_{k=1}^{N}(y(k) - au(k) - b)^2$. Show that $\hat{\theta}_{\text{IQML}}$ is an inconsistent estimate (hint: apply Theorem 18.5 with $\eta(z) = \hat{a}_{\text{LS}}$ and $w(\theta, \eta(z), z) = 1$, and show that $\tilde{\theta}(z_0) \neq \theta_0$ for any N, ∞ included).

18.9 APPENDIXES

Appendix 18.A Proof of the Strong Convergence of the Cost Function (Lemma 18.4)

We will show that $V_N(\theta, z) = f_N(\theta, \eta(z), w(\theta, \eta(z), z), z)$ satisfies the two conditions of Corollary 17.32 (see Appendix 17.C). Under Assumptions 17.1 ($P = 4$) and 18.3, $f_N(\theta, \eta, w, z)$ converges uniformly w.p. 1 to $\mathbb{E}\{f_N(\theta, \eta, w, z)\} = f_N(\theta, \eta, w)$ in $\mathbb{P}_r$, $\mathbb{E}_\varepsilon$, and $\mathbb{W}$ (Lemma 17.3), so that condition 1 of Corollary 17.32 is satisfied. Under Assumptions 18.1 and 18.2 (w.p. 1, $k = 0$), $w(\theta, \eta(z), z)$ converges uniformly w.p. 1 to $w(\theta, \eta_*)$, an interior point of $\mathbb{W}$, in $\mathbb{P}_r$ (Lemma 17.31, Appendix 17.C). Combining this result with Assumption 18.1 shows that condition 2 of Corollary 17.32 is satisfied. We conclude that $V_N(\theta, z)$ converges uniformly w.p. 1 to $f_N(\theta, \eta_*, w(\theta, \eta_*))$ in $\mathbb{P}_r$.

$V_N(\theta, z)$, $V_N(\theta)$ are continuous in $\mathbb{P}_r$ because $f_N(\theta, \eta, w, z)$, $f_N(\theta, \eta, w)$ are jointly continuous functions of θ, w in $\mathbb{P}_r$, $\mathbb{W}$ and $w(\theta, \eta, z)$, $w(\theta, \eta_*)$ are continuous functions of θ in $\mathbb{P}_r$. □

Appendix 18.B Proof of the Convergence Rate of the Minimizer (Theorem 18.16)

The proof follows the same lines as for Theorem 17.19 (see Appendix 17.D). Applying the mean value theorem to the derivative of the cost function $V_N'(\theta, z)$ at the points $\hat{\theta}(z)$ and $\tilde{\theta}(z_0)$ gives

$$V_N'(\hat{\theta}(z), z) = V_N'(\tilde{\theta}(z_0), z) + (\hat{\theta}(z) - \tilde{\theta}(z_0))^T V_N''(\widehat{\theta}, z) \tag{18-21}$$

where $V_N'(\hat{\theta}(z), z) = 0$ by definition of $\hat{\theta}(z)$, and $\widehat{\theta} = t\hat{\theta}(z) + (1-t)\tilde{\theta}(z_0)$ with $t \in [0, 1]$. The proof consists of three main steps. In a first step, the convergence rate of $V_N'(\tilde{\theta}(z_0), z)$ is shown using Corollary 17.35 of Appendix 17.C. Because $V_N'(\tilde{\theta}(z_0)) = 0$, we find

$$V_N'(\tilde{\theta}(z_0), z) = V_N'(\tilde{\theta}(z_0)) + O_p(N^{-1/2}) = O_p(N^{-1/2}) \tag{18-22}$$

uniformly in $\mathbb{P}_r$. In a second step, the convergence of $V_N''(\widehat{\theta}, z)$ is shown using Corollary 17.32 of Appendix 17.C. Because $V_N''(\tilde{\theta}(z_0)) = O(N^0)$ (Assumption 17.18), we get

$$V_N''(\widehat{\theta}, z) = V_N''(\tilde{\theta}(z_0)) + o_{\text{a.s.}}(N^0) = O_{\text{a.s.}}(N^0) \tag{18-23}$$

uniformly in $\mathbb{P}_r$. In the third and last step (18-21), (18-22), and (18-23) are combined, giving

$$\hat{\theta}(z) - \tilde{\theta}(z_0) = V_N''^{-1}(\widehat{\theta}, z)V_N'^{T}(\tilde{\theta}(z_0), z) = O_p(N^{-1/2}) \qquad (18\text{-}24)$$

In the first step we verify that $V_N'(\tilde{\theta}(z_0), z)$ fulfills all the conditions of Corollary 17.35. Under Assumptions 17.1 ($P = 4$) and 18.13 $g_N(\theta, \eta, w, w_1, z)$ converges uniformly in prob. to $g_N(\theta, \eta, w, w_1)$ in $\mathbb{P}_r$, $\mathbb{E}_\varepsilon$, $\mathbb{W}$, and $\mathbb{W}_1$ at the rate $O_p(N^{-1/2})$ (Lemma 17.17), so that condition 1 of Corollary 17.35 is satisfied. The conditions of Lemma 17.34 are satisfied so that $w^{(k)}(\theta, \eta(z), z)$, $k = 0, 1$, converges uniformly in prob. to $w^{(k)}(\theta, \eta_*)$ in $\mathbb{P}_r$ at the rate $O_p(N^{-1/2})$. This, together with Assumption 18.10, guarantees that condition 3 of Corollary 17.35 is satisfied. Following the same lines as in Appendix 17.E, (17-44), we conclude from Assumptions 17.1 ($P = 4$) and 18.15 that

$$\left\| \frac{\partial g_{N[i]}(\theta, \eta, w, w_1, z)}{\partial x_{[j]}} \right\|_2 \leq \frac{\|z\|_2^2}{N} \left\| \frac{\partial W_{1i}(\theta, \eta, w, w_1)}{\partial x_{[j]}} \right\|_2 = O_{\text{a.s.}}(N^0) \qquad (18\text{-}25)$$

$i = 1, 2, \ldots, n_\theta$. Hence, condition 2 of Corollary 17.35 is satisfied, thus concluding the first step of the proof.

In the second step we verify that $V_N''(\widehat{\theta}, z)$ fulfills all the conditions of Corollary 17.32. The assumptions of Lemma 18.14 are satisfied and, hence, $V_N''(\theta, z)$ converges uniformly in prob. to $V_N''(\theta)$ in $\mathbb{P}_r$ (condition 1 of Corollary 17.32). The assumptions of Theorem 18.5 (Assumption 17.18 is stronger than Assumption 17.5) are satisfied so that $\hat{\theta}(z)$, and, hence, also $\widehat{\theta}$, converges in prob. to $\tilde{\theta}(z_0)$ (Theorem 18.5 without Assumption 18.3(b) shows convergence in prob.). The conditions of Lemma 17.31 are satisfied so that $w^{(k)}(\theta, \eta(z), z)$, $k = 0, 1, 2$, converges uniformly in prob. to $w^{(k)}(\theta, \eta_*)$ in $\mathbb{P}_r$. Together with Assumption 18.1, it shows that condition 2 of Corollary 17.32 is satisfied, which concludes the second step of the proof. $\square$

Appendix 18.C Proof of the Asymptotic Bias of the Truncated Minimizer (Theorem 18.22)

The results of Theorem 18.18 are valid so that only $V_N'(\theta, z)$ must be studied. We will show that $V_N'(\theta, z) = g_N(\theta, \eta(z), w(\theta, \eta(z), z), w'(\theta, \eta(z), z), z)$ satisfies the conditions of Corollary 17.37 (see Appendix 17.C), which proves the theorem.

Under Assumptions 18.1, 18.2 ($k = 0, 1$), 18.10, 18.11 ($k = 0, 1$), 18.19, and 18.20 ($k = 0, 1$), $w^{(k)}(\theta, \eta(z), z)$ satisfies the conditions of Lemma 17.36. Hence, it converges uniformly in prob. to $w^{(k)}(\theta, \eta_*)$ at the rate $O_p(N^{-1/2})$ with bias

$$\mathbb{E}\{\underline{w}^{(k)}(\theta, \eta(z), z)\} = w^{(k)}(\theta, \eta_*) + O(N^{-1})$$

It follows that conditions 3 and 4 of Corollary 17.37 are satisfied for $\eta(z)$ and $w^{(k)}(\theta, \eta(z), z)$.

Under Assumptions 17.1 ($P = 4$) and 18.21, $\partial g_N(\theta, \eta, w, w_1, z)/\partial x_{[j]}$ converges uniformly in prob. to $\partial g_N(\theta, \eta, w, w_1)/\partial x_{[j]}$ at the rate $O_p(N^{-1/2})$ (proof: similar to Lemma 17.17). Under the same assumptions, we also have

$$\left\| \frac{\partial^2 g_{N[i]}(\theta, \eta, w, w_1, z)}{\partial x_{[j]} \partial x_{[k]}} \right\|_2 \leq O_{\text{a.s.}}(N^0)$$

$i = 1, 2, \ldots, n_\theta$ (proof: similar to (18-25)). From the proof of Theorem 18.18, it follows that $g_N(\theta, \eta, w, w_1, z)$ satisfies condition 1 of Corollary 17.35. Hence, conditions 1 and 2 of Corollary 17.37 are satisfied, thus concluding the proof for the bias.

If $w^{(3)}(\theta, \eta(z), z)$ is continuous in $\mathbb{P}_r$, then under Assumptions 18.2 ($k = 0, 1, 2$) and 18.17, $V_N(\theta, z)$ has continuous first-, second-, and third-order derivatives w.r.t. θ in $\mathbb{P}_r$. This is sufficient to show that $\partial b_\theta / \partial \theta_0$ is an $O(N^{-1})$ (see the proof of Theorem 17.28 in Appendix 17.G).

Appendix 18.D Proof of the Asymptotic Normality of the Minimizer (Theorem 18.25)

Multiplying (18-16) by $\sqrt{N}$ and taking the limit for $N \to \infty$ gives

$$\plim_{N \to \infty} \sqrt{N}(\hat{\theta}(z) - \tilde{\theta}(z_0) - \delta_\theta(z)) = 0 \tag{18-26}$$

Because convergence in probability implies convergence in law (see Section 16.7, interrelation 3), it follows directly from (18-26) that $\sqrt{N}(\hat{\theta}(z) - \tilde{\theta}(z_0))$ has the same asymptotic distribution function as $\delta_\theta(z)$. We will show that $\delta_\theta(z)$ converges in law at the rate $O(N^{-1/2})$ to a Gaussian random variable. To prove this it is sufficient to show that the stochastic part of $\delta_\theta(z)$, namely,

$$V_N'^T(\tilde{\theta}(z_0), z) = g_N(\tilde{\theta}(z_0), \eta(z), w(\tilde{\theta}(z_0), \eta(z), z), w'(\tilde{\theta}(z_0), \eta(z), z), z) \tag{18-27}$$

satisfies the conditions of Corollary 17.39.

Condition 1 of Corollary 17.39 is satisfied under Assumptions 17.1 ($P = \infty$), 18.3(a), and 18.13 (proof: similar to Theorem 17.29). Conditions 2 and 3 of Corollary 17.39 are satisfied under Assumptions 17.1 ($P = 4$), and 18.21 (proof: see Theorem 18.22, Appendix 18.C). Under Assumptions 18.1, 18.2 (in prob., $k = 0, 1$), 18.10, 18.11 ($k = 0, 1$), 18.20 ($k = 0, 1$), 18.23, and 18.24, $w^{(k)}(\theta, \eta(z), z)$ satisfies the conditions of Lemma 17.38. Hence, $w^{(k)}(\theta, \eta(z), z)$ converges uniformly in prob. to $w^{(k)}(\theta, \eta_*)$ at the rate $O_p(N^{-1/2})$, is asymptotically normally distributed at the rate $O(N^{-1/2})$, and can be written as

$$w^{(k)}(\theta, \eta(z), z) = w^{(k)}(\theta, \eta_*, z) + \frac{\partial w^{(k)}(\theta, \eta)}{\partial \eta_*}(\eta(z) - \eta_*) + O_p(N^{-1}) \tag{18-28}$$

Combined with Assumptions 18.1, 18.10, and 18.23, it shows that condition 4 of Corollary 17.39 is also fulfilled. We conclude from Corollary 17.39 and (18-28) that (18-27) can be written as

$$V_N'^T(\tilde{\theta}(z_0), z) = g_N(\tilde{\theta}(z_0), \eta_*, w(\tilde{\theta}(z_0), \eta_*), w'(\tilde{\theta}(z_0), \eta_*), z)$$
$$+ \left.\frac{dg_N(\tilde{\theta}(z_0), \eta, w(\tilde{\theta}(z_0), \eta), w'(\tilde{\theta}(z_0), \eta))}{dx}\right|_{x = x_*} (x(z) - x_*) + O(N^{-1}) \tag{18-29}$$

with $x^T(z) = \tilde{x}^T(z) = [\eta^T(z)\ w^T(\tilde{\theta}(z_0), \eta_*, z)\ \text{vec}^T(w'(\tilde{\theta}(z_0), \eta_*, z))]$ and where the derivative w.r.t. η is calculated using the chain rule (18-18). The first two terms in the right-hand side of (18-29) are asymptotically normally distributed. Because $x(z) = \tilde{x}(z) + O_p(N^{-1})$, we can replace $x(z)$ by $\tilde{x}(z)$ in (18-29), which concludes the proof.

19

Identification of Semilinear Models

Abstract: Many signal and system modeling problems lead to parametric models that are linear-in-the-measurements. This chapter treats the identification (parameter estimation and model selection) of such models using the Markov estimator. The asymptotic properties (consistency, convergence rate, asymptotic bias, asymptotic normality, and asymptotic efficiency) of the Markov estimates are analyzed. The different aspects of model selection, such as model validation, and detection of undermodeling and overmodeling are discussed. Explicit expressions for the Cramér-Rao lower bound are derived and conditions for the asymptotic efficiency of the Gaussian maximum likelihood estimator are given. The presented theory is applicable to general signal modeling and system identification problems. Readers who are unfamiliar with the analysis of the stochastic properties of estimators should first read Sections 16.11 to 16.13 and Chapter 17.

19.1 THE SEMILINEAR MODEL

Consider the following general model based on N observations:

$$M_0(\theta) + M_1(\theta)z = 0 \qquad (19\text{-}1)$$

which is linear-in-the-measurements $z \in \mathbb{R}^{sN}$ and (non)linear-in-the-model-parameters $\theta \in \mathbb{R}^{n_\theta}$. $M_0(\theta) \in \mathbb{R}^{rN}$, $M_1(\theta) \in \mathbb{R}^{rN \times sN}$ with $s \geq r$ has rank rN and n_θ, s and r fixed integers, independent of the number of observations N. Each time a new observation is added, the number of model equations and the number of measurements increase with, respectively, r and s. In frequency domain applications, (19-1) often has a block diagonal structure

$$M_{0k}(\theta) + M_{1k}(\theta)z_k = 0 \text{ for } k = 1, 2, ..., N \qquad (19\text{-}2)$$

with $M_{0k}(\theta) \in \mathbb{R}^r$, $M_{1k}(\theta) \in \mathbb{R}^{r \times s}$, and $z_k \in \mathbb{R}^s$. The relationship with (19-1) is given by

$$M_0^T(\theta) = [M_{01}^T(\theta)...M_{0N}^T(\theta)], \ M_1(\theta) = \text{diag}(M_{10}(\theta), ..., M_{1N}(\theta)), \text{ and } z^T = [z_1^T...z_N^T]$$

Two special cases of model (19-1) are worth mentioning.

19.1.1 Signal Model

Putting $s = r$ and $M_1(\theta) = -I_{sN}$ in (19-1) gives the signal model

$$z = M_0(\theta) \tag{19-3}$$

This is typically the form encountered when estimating a linear combination of basis signals such as sine waves (Pintelon and Schoukens, 1996), cisoids (Cadzow, 1990), and exponential functions (Van den Bos and Swarte, 1993).

19.1.2 Transfer Function Model

Putting $M_0(\theta) = 0$, $M_1(\theta) = [A(\theta) \; -B(\theta)]$, and $z^T = [y^T \; u^T]$ gives the transfer function model

$$A(\theta)y = B(\theta)u \tag{19-4}$$

where $y \in \mathbb{R}^{rN}$ and $u \in \mathbb{R}^{qN}$ with $r + q = s$. $A(\theta) \in \mathbb{R}^{rN \times rN}$ is a regular matrix. The model equations of a linear time-invariant discrete-time multivariable system can be written in the time domain under this form (Exercise 19.1). u and y are, respectively, the stacked input and output signals of the system, while r and q are, respectively, the numbers of outputs n_y and inputs n_u. The left matrix fraction description (6-54) can be written under the block diagonal form (19-2) with

$$M_{0k}(\theta) = 0, \quad M_{1k}(\theta) = \left[A_{\mathrm{Re}}(\Omega_k, \theta) \; -B_{\mathrm{Re}}(\Omega_k, \theta)\right], \quad \text{and} \quad z_k^T = \left[Y_{\mathrm{re}}^T(k) \; U_{\mathrm{re}}^T(k)\right]$$

and where the operators $(\;)_{\mathrm{Re}}$ and $(\;)_{\mathrm{re}}$ are defined in Section 15.8 (proof: apply Lemma 15.4 to (6-54)). $Y(k)$ and $U(k)$ are, respectively, the n_y by 1 output and the n_u by 1 input DFT spectra at frequency k ($s = 2(n_y + n_u)$). If the initial conditions are included in the model, then $M_{0k}(\theta) = I_{\mathrm{re}}(\Omega_k, \theta)$, with $I(\Omega_k, \theta)$ the n_y by 1 vector of the equivalent initial conditions (see Section 6.6), and (19-2) becomes (Exercise 19.3)

$$A_{\mathrm{Re}}(\Omega_k, \theta)Y_{\mathrm{re}}(k) = B_{\mathrm{Re}}(\Omega_k, \theta)U_{\mathrm{re}}(k) + I_{\mathrm{re}}(\Omega_k, \theta) \tag{19-5}$$

19.2 THE MARKOV ESTIMATOR

First we construct the Markov estimates for real observations and real model parameters. Afterward, the results are generalized to complex observations and complex model parameters.

19.2.1 Real Case

An estimate $\hat{\theta}$ of the model parameters θ of the semilinear model (19-1) is calculated using noisy observations $z = z_0 + n_z$ of the true (unknown) values z_0. Because z_0 is unknown, it should also be estimated and is parameterized as z_p. Under Assumption 17.1(2), the Markov estimator minimizes the squared residuals $(z - z_p)$ weighted with the noise co-

Section 19.2 ■ The Markov Estimator

variance matrix $C_{n_z} = \text{Cov}(n_z)$, taking into account the model equations (19-1). This constrained minimization problem, with parameters θ and z_p, can be solved using Lagrange multipliers $\lambda \in \mathbb{R}^{rN}$ (Kaplan, 1993)

$$\frac{1}{2}(z - z_p)^T C_{n_z}^+(z - z_p) + \lambda^T(M_0(\theta) + M_1(\theta)z_p) \tag{19-6}$$

with + the Moore-Penrose pseudoinverse (see Section 15.5). Singular noise covariance matrices are allowed to cover the case where parts of the measurements may be known exactly. This is, for example, the case in transfer function modeling with known inputs. Because (19-6) is quadratic in z_p and linear in λ, z_p and λ can be explicitly eliminated. This gives the following expression for z_p (see Appendix 19.A):

$$C_{n_z} C_{n_z}^+ z_p = C_{n_z} C_{n_z}^+ z - C_{n_z} M_1^T(\theta)(M_1(\theta) C_{n_z} M_1^T(\theta))^{-1}(M_0(\theta) + M_1(\theta)z) \tag{19-7}$$

It makes it possible to eliminate the parameters z_p in (19-6), which results in a significant reduction of the size of the minimization problem. The following Markov cost function is found:

$$V_{\text{Markov}}(\theta, z) = \frac{1}{2} e^T(\theta, z) C_e^{-1}(\theta) e(\theta, z) = \frac{1}{2} \varepsilon^T(\theta, z) \varepsilon(\theta, z) \tag{19-8}$$

with

$$e(\theta, z) = M_0(\theta) + M_1(\theta)z, \quad C_e(\theta) = M_1(\theta) C_{n_z} M_1^T(\theta), \quad \varepsilon(\theta, z) = \Lambda(\theta) e(\theta, z) \tag{19-9}$$

and where $\Lambda(\theta) \in \mathbb{R}^{rN \times rN}$ satisfies $\Lambda^T(\theta)\Lambda(\theta) = C_e^{-1}(\theta)$ (see Appendix 19.A). Note that $\text{Cov}(e(\theta, n_z)) = C_e(\theta)$ and $\text{Cov}(\varepsilon(\theta, n_z)) = I_{rN}$. Minimizing the cost function (19-8) w.r.t. θ gives the Markov estimates of the model parameters

$$\hat{\theta}(z) = \arg \min_{\theta \in \mathbb{P}_r} V_{\text{Markov}}(\theta, z) \tag{19-10}$$

The Markov estimates $\hat{z}$ of the true observations z_0 are found by evaluating (19-7) at $\theta = \hat{\theta}(z)$

$$C_{n_z} C_{n_z}^+ \hat{z} = C_{n_z} C_{n_z}^+ z - C_{n_z} M_1^T(\hat{\theta}(z)) C_e^{-1}(\hat{\theta}(z)) e(\hat{\theta}(z), z) \tag{19-11}$$

Note that this formula estimates the observations lying in the regular space of C_{n_z}. Those lying in the null space of C_{n_z} are known exactly.

The Markov estimates require knowledge of the noise covariance matrix C_{n_z}. It can be estimated from independent repeated experiments (see Chapter 10 for transfer function modeling of SISO systems). This consumes a great deal of computer time and memory space if N is large. In frequency domain identification only a (block) diagonal version of C_{n_z} is required. Using the block diagonal structure (19-2) of the model equations and replacing C_{n_z} in (19-8) by $\text{diag}(C_{n_{z1}}, \ldots, C_{n_{zN}})$, with $C_{n_{zk}} = \text{Cov}(n_{zk})$, gives the simplified Markov cost function

$$V_{\text{Markov}}(\theta, z) = \frac{1}{2}\sum_{k=1}^{N} e_k^T(\theta, z_k) C_{e_k}^{-1}(\theta) e_k(\theta, z_k) \qquad (19\text{-}12)$$

with $e_k(\theta, z_k) = M_{0k}(\theta) + M_{1k}(\theta)z_k$, $C_{e_k}(\theta) = \text{Cov}(e_k(\theta, n_{zk})) = M_{1k}(\theta)C_{n_{zk}}M_{1k}^T(\theta)$, and $z_k = z_{0k} + n_{zk}$. Neglecting the nondiagonal terms of $C_{e_k}(\theta)$ in (19-12), the Markov cost function can even be simplified further to

$$V_{\text{Markov}}(\theta, z) = \frac{1}{2}\sum_{k=1}^{N}\sum_{i=1}^{r} \frac{e_{k[i]}^2(\theta, z_k)}{\text{var}(e_{k[i]}(\theta, n_{zk}))} \qquad (19\text{-}13)$$

The stochastic properties of the minimizers of (19-8), (19-12), and (19-13) are analyzed in Section 19.4.

19.2.2 Complex Case

Expressions (19-8), (19-12), and (19-13) are still valid for *complex-valued observations* $z \in \mathbb{C}^{sN}$ and *complex-valued model parameters* $\theta \in \mathbb{C}^{n_\theta}$, if applied to z_{re} and θ_{re}. If in addition, *the errors n_z are circular complex distributed* (see (16-12) and (16-13)), then (19-8) can be written as (see Example 15.5)

$$\begin{aligned} V_{\text{Markov}}(\theta, z) &= e^H(\theta, z) C_e^{-1}(\theta) e(\theta, z) \\ &= \varepsilon^H(\theta, z) \varepsilon(\theta, z) \end{aligned} \qquad (19\text{-}14)$$

with $\varepsilon(\theta, z) = \Lambda(\theta) e(\theta, z)$ and where $\Lambda(\theta) \in \mathbb{C}^{rN \times rN}$ satisfies $\Lambda^H(\theta)\Lambda(\theta) = C_e^{-1}(\theta)$.

19.3 CRAMÉR-RAO LOWER BOUND

We first derive the Carmér-Rao lower bound for real observations and real model parameters. Afterward, the results are generalized to complex observations and complex parameters.

19.3.1 Real Case

The concept of Cramér-Rao lower bound requires the existence of a true model

$$M_0(\theta_0) + M_1(\theta_0)z_0 = 0 \qquad (19\text{-}15)$$

with z_0 the true observations and θ_0 the true model parameters, and knowledge of the probability density function of the measurements $z = z_0 + n_z$. The Cramér-Rao lower bound for unbiased estimators (16-87) is constructed under the following assumption.

Assumption 19.1 (Gaussian Errors): The observations z_0 are deterministic and the errors n_z are normally distributed with known covariance matrix C_{n_z}.

The log-likelihood function becomes

$$\ln f_z(z, z_p, \theta) = -\frac{1}{2}(z - z_p)^T C_{n_z}^+(z - z_p) + c \qquad (19\text{-}16)$$

with c a constant independent of z_p and θ, and where the parameters z_p and θ satisfy the constraint (19-1)

$$M_0(\theta) + M_1(\theta)z_p = 0 \qquad (19\text{-}17)$$

Straightforward calculation of the Fisher information matrix $Fi(z_0, \theta_0)$ (see (16-85)) from (19-16) is impossible because z_p contains too many unknowns. Indeed, the parameters of z_p lying in the singular space of C_{n_z} are known exactly and should not appear in the CR bound. Next, (19-17) puts rN linear constraints on the entries of z_p. The resulting rN linear dependent variables should also not appear in the CR bound.

Hence, the first step in calculating the CR bound consists of reducing the sN parameters z_p to the $\text{rank}(C_{n_z}) - rN$ parameters x: $z_p = z_p(x, \theta)$ (see Appendix 19.B). It facilitates writing the log-likelihood function as

$$\ln f_z(z, x, \theta) = -\frac{1}{2}(z - z_p(x, \theta))^T C_{n_z}^+ (z - z_p(x, \theta)) + c \qquad (19\text{-}18)$$

The corresponding Fisher information matrix $Fi(x_0, \theta_0)$ is

$$Fi(x_0, \theta_0) = \begin{bmatrix} F_{xx} & F_{x\theta} \\ F_{x\theta}^T & F_{\theta\theta} \end{bmatrix} \qquad (19\text{-}19)$$

Applying the inverse of block matrices (15-8) to $CR(x_0, \theta_0) = Fi^{-1}(x_0, \theta_0)$ gives the Cramér-Rao lower bound on the model parameters

$$CR(\theta_0) = Fi^{-1}(\theta_0) = (F_{\theta\theta} - F_{x\theta}^T F_{xx}^{-1} F_{x\theta})^{-1} \qquad (19\text{-}20)$$

Filling out the explicit expressions of $F_{\theta\theta}$, F_{xx}, and $F_{x\theta}$ in (19-20) gives, after some calculations (see Appendix 19.B),

$$\begin{aligned} Fi(\theta_0) &= V_{\text{Markov}}''(\theta_0, z_0) \\ &= \left(\frac{\partial e(\theta, z_0)}{\partial \theta_0}\right)^T C_e^{-1}(\theta_0)\left(\frac{\partial e(\theta, z_0)}{\partial \theta_0}\right) \\ &= \left(\frac{\partial \varepsilon(\theta, z_0)}{\partial \theta_0}\right)^T \left(\frac{\partial \varepsilon(\theta, z_0)}{\partial \theta_0}\right) \end{aligned} \qquad (19\text{-}21)$$

This shows that the Fisher information matrix of the model parameters $Fi(\theta_0)$ equals the Hessian of the Markov cost function (19-8), evaluated at the true observations and the true model parameters. Hence, the larger the eigenvalues of the Hessian matrix, the smaller the Cramér-Rao bound of the model parameters.

19.3.2 Complex Case

We first consider the case where *the observations $z \in \mathbb{C}^{sN}$ are complex, the errors $n_z \in \mathbb{C}^{sN}$ are circular complex distributed* (see (16-12) and (16-13)), and *the model parameters θ are real*. This is, for example, true in frequency domain system identification. Formula

(19-21) of the real case still applies to z_{re} and $e_{re}(\theta_0, z_0)$. Using the isomorphism between complex and real matrices (see Section 15.8), (19-21) becomes

$$\begin{aligned} Fi(\theta_0) &= V_{Markov}''(\theta_0, z_0) \\ &= 2\text{Re}\left(\left(\frac{\partial e(\theta, z_0)}{\partial \theta_0}\right)^H C_e^{-1}(\theta_0)\left(\frac{\partial e(\theta, z_0)}{\partial \theta_0}\right)\right) \\ &= 2\text{Re}\left(\left(\frac{\partial \varepsilon(\theta, z_0)}{\partial \theta_0}\right)^H \left(\frac{\partial \varepsilon(\theta, z_0)}{\partial \theta_0}\right)\right) \end{aligned} \quad (19\text{-}22)$$

with $V_{Markov}(\theta, z)$ is given by (19-14) (see Exercise 19.6).

Next, we consider the case where, also, the *model parameters* θ *are complex*. Applications where θ is complex can be found in nuclear magnetic resonance spectroscopy (Kumaresan et al., 1990) and the diagnosis of asymmetry of rotating machinery (Lee and Joh, 1994; Peeters et al., 2000). Formula (19-22) is still valid for θ_{re}. If $e(\theta, z)$ is an analytic function of $\theta \in \mathbb{C}^{n_\theta}$, then

$$\frac{\partial e(\theta, z)}{\partial \theta_{re}} = \left[\frac{\partial e(\theta, z)}{\partial \text{Re}(\theta)} \quad \frac{\partial e(\theta, z)}{\partial \text{Im}(\theta)}\right] = \left[\frac{\partial e(\theta, z)}{\partial \theta} \quad j\frac{\partial e(\theta, z)}{\partial \theta}\right]$$

and (19-22) can be rewritten as

$$\begin{aligned} Fi(\theta_0) &= 2\left(\frac{\partial e(\theta, z_0)}{\partial \theta_0}\right)^H C_e^{-1}(\theta_0)\left(\frac{\partial e(\theta, z_0)}{\partial \theta_0}\right) \\ &= 2\left(\frac{\partial \varepsilon(\theta, z_0)}{\partial \theta_0}\right)^H \left(\frac{\partial \varepsilon(\theta, z_0)}{\partial \theta_0}\right) \end{aligned} \quad (19\text{-}23)$$

(see Exercise 19.7).

19.4 PROPERTIES OF THE MARKOV ESTIMATOR

The properties of the Markov estimator will be studied for real observations and real model parameters. Following the lines of Sections 19.2.2 and 19.3.2, it is easy to see that the results are also valid for complex observations and complex model parameters. Note that the Markov cost function (19-8) fits within the framework of Chapter 17. Indeed, (19-8) can be written as

$$V_N(\theta, z) = \frac{1}{N} V_{Markov}(\theta, z) = \frac{1}{N}\begin{bmatrix}1\\z\end{bmatrix}^T W_N(\theta) \begin{bmatrix}1\\z\end{bmatrix} \quad (19\text{-}24)$$

with $W_N(\theta)$ an $(sN+1)$ by $(sN+1)$ weighting matrix

$$W_N(\theta) = \frac{1}{2}[M_0(\theta) \quad M_1(\theta)]^T C_e^{-1}(\theta)[M_0(\theta) \quad M_1(\theta)] \quad (19\text{-}25)$$

Cost function (19-24) has exactly the same form as (17-1) and, hence, all the results of Chapter 17 apply to the Markov estimator (19-10). The same is true for the Markov estimates based on the simplified cost functions (19-12) and (19-13). Only the assumptions and the properties that can be worked out more specifically for the Markov estimator are discussed here.

Approximate expressions for "large" signal-to-noise ratios ($\|z_0\|_2/\|n_z\|_2 \gg 1$) and "small" model errors are obtained by replacing n_z by υn_z (and, hence, C_{n_z} by $\upsilon^2 C_{n_z}$) with $\upsilon \to 0$ and $e(\tilde{\theta}(z), z_0)$ by $\mu e(\tilde{\theta}(z), z_0)$ with $\mu \to 0$.

19.4.1 Consistency

Besides the model parameters, (a part of) the measurements are also estimated. First, the consistency of the estimates $\hat{\theta}(z)$ (19-10) of the model parameters is analyzed. Next, the consistency of the estimates $\hat{z}$ (19-11) of the measurements is discussed.

19.4.1.1 Model Parameters. Assumption 17.2 requires that $C_e(\theta)$ is positive definite in the compact set $\mathbb{P}_r$. The condition $\|W_N(\theta)\|_1 \leq c < \infty$ imposes restrictions on the one and infinity norm of $M_1(\theta)$ and C_{n_z}, while the condition $\|W_{m[1:n, 1:n]}(\theta) - W_n(\theta)\|_1^2 = O((m-n)/m)$, with $m \geq n$, limits the variation of $C_e^{-1}(\theta)$ as N increases. Note that both conditions are automatically satisfied for the simplified Markov estimators (19-12) and (19-13). Under Assumption 17.1, using

$$e(\theta, z) = e(\theta, z_0) + M_1(\theta)n_z \quad \text{and} \quad \text{trace}(e^T C_e^{-1} e) = \text{trace}(C_e^{-1} e e^T)$$

we find the expected value of the cost function (19-8)

$$V_{\text{Markov}}(\theta) = \mathbb{E}\{V_{\text{Markov}}(\theta, z)\} = \mathbb{E}\{V_{\text{Markov}}(\theta, z_0)\} + rN/2 \qquad (19\text{-}26)$$

Assumption 17.10 is satisfied because $\mathbb{E}\{V_{\text{Markov}}(\theta_0, z_0)\} = 0$ and $rN/2$ is θ-independent. Hence, it follows from Theorem 17.11 that the Markov estimate (19-10) is strongly consistent.

Note that the expected values of the simplified cost functions (19-12) and (19-13) are also given by (19-26). We conclude that the Markov estimates of the (block) diagonal model (19-2) based on the simplified cost functions (19-12), (19-13) are still strongly consistent. Removing some parts of the nondiagonal elements of C_{n_z} does not influence the consistency property: the minimal requirement is that each residual is weighted with its variance. For frequency domain system identification, this means that the correlation of the errors n_z over the frequencies and between the different outputs can be neglected. However, the correlation of the errors n_z between an output and all the inputs may not be removed because it influences $\text{var}(e_{k[i]}(\theta, z_k))$; otherwise consistency is lost.

19.4.1.2 Observations. In general, the estimates $\hat{z}$ (19-11) of the observations are inconsistent. This is due to the fact that the uncertainty of z in (19-11) is not decreased by making more observations (no averaging effect occurs in $C_{n_z} C_{n_z}^+ z$). z cancels in (19-11) for signal models (19-3) and transfer function models (19-4) with known input (excitation) signals u_0. In these cases, strongly consistent estimates of the observations are obtained through, respectively, $\hat{z} = M_0(\hat{\theta}(z))$ and $\hat{y} = A^{-1}(\hat{\theta}(z))B(\hat{\theta}(z))u_0$ (see Appendix 19.C).

19.4.2 Strong Convergence

If model errors are present ($e(\tilde{\theta}(z_0), z_0) \neq 0$), one can wonder why the Markov estimator should be preferred over, for example, the nonlinear least squares estimator

$$V_{\text{NLS}}(\theta, z) = \frac{1}{2}e^T(\theta, z)e(\theta, z) \qquad (19\text{-}27)$$

The reason for this is that the Markov estimate $\hat{\theta}(z)$ converges to a value $\tilde{\theta}(z_0)$ that is independent of the signal-to-noise ratio. Indeed, replacing n_z by υn_z (and, hence, C_{n_z} by $\upsilon^2 C_{n_z}$) in the expected value of the cost function (19-26) gives

$$V_{\text{Markov}}(\theta) = \upsilon^2 \, \mathbb{E}\{V_{\text{Markov}}(\theta, z_0)\} + rN/2 \qquad (19\text{-}28)$$

It shows that $\tilde{\theta}(z_0)$, the minimizing argument of (19-28), is independent of υ. The same is true for the simplified Markov estimators (19-12) and (19-13). This is not the case for the nonlinear least squares estimator (19-27). Indeed, the expected value of (19-27) equals

$$V_{\text{NLS}}(\theta) = \mathbb{E}\{V_{\text{NLS}}(\theta, z)\} = \mathbb{E}\{V_{\text{NLS}}(\theta, z_0)\} + \text{trace}(C_e(\theta)) \qquad (19\text{-}29)$$

Replacing n_z by υn_z (and, hence, C_{n_z} by $\upsilon^2 C_{n_z}$) in (19-29) gives

$$V_{\text{NLS}}(\theta) = \mathbb{E}\{V_{\text{NLS}}(\theta, z)\} = \mathbb{E}\{V_{\text{NLS}}(\theta, z_0)\} + \upsilon^2 \text{trace}(C_e(\theta)) \qquad (19\text{-}30)$$

Clearly, the minimizer of (19-30) depends on υ.

19.4.3 Convergence Rate

Expression (17-14) for the difference $\hat{\theta}(z) - \tilde{\theta}(z_0)$ can be elaborated for the Markov estimator. It will be used to study the statistical properties of the residuals $\varepsilon(\hat{\theta}(z), z)$ and the global minimum of the cost function $V_{\text{Markov}}(\hat{\theta}(z), z)$.

Theorem 19.2 (Convergence Rate $\hat{\theta}(z)$ to $\tilde{\theta}(z_0)$): Under the assumptions of Theorem 17.21, large signal-to-noise ratios ($\upsilon \to 0$) and small model errors ($\mu \to 0$), the minimizer $\hat{\theta}(z)$ can be written as

$$\hat{\theta}(z) - \tilde{\theta}(z_0) = \Delta_\theta(z) + \partial_\theta(z) + b_\theta(z)$$

$$\Delta_\theta(z) = -\left[\left(\frac{\partial \varepsilon(\theta, z_0)}{\partial \tilde{\theta}(z_0)}\right)^T \left(\frac{\partial \varepsilon(\theta, z_0)}{\partial \tilde{\theta}(z_0)}\right)\right]^{-1} \left(\frac{\partial \varepsilon(\theta, z_0)}{\partial \tilde{\theta}(z_0)}\right)^T \left(\frac{\partial \varepsilon(\tilde{\theta}(z_0), z)}{\partial z}\right) n_z \qquad (19\text{-}31)$$

where $\mathbb{E}\{\Delta_\theta(z)\} = 0$, $\mathbb{E}\{\partial_\theta(z)\} = 0$ and

$$\Delta_\theta(z) = \upsilon O_p(N^{-1/2})$$
$$\partial_\theta(z) = (\upsilon^2 + \upsilon\mu + \mu\lambda(z_0))O_p(N^{-1/2})$$
$$b_\theta(z) = (\upsilon^2 + (\upsilon + \mu)\lambda(z_0))O_p(N^{-1})$$

with $\lambda(z_0) = 1$ for random z_0 and $\lambda(z_0) = 0$ for deterministic z_0.

Proof. See Appendix 19.D. □

From (19-31), it follows that in the presence of model errors ($\mu \neq 0$), $\partial_\theta(z)$ and $b_\theta(z)$ do not decrease to zero for random z_0 as the noise level v tends to zero. In the absence of model errors ($\mu = 0$), $\Delta_\theta(z)$ and $b_\theta(z)$ are, for deterministic z_0, an $vO_p(N^{-1/2})$ and $v^2 O_p(N^{-1})$, respectively. It shows that the bias error decreases as v^2 while the stochastic error decreases as v. Although $b_\theta(z) = (v^2 + v\lambda(z_0))O_p(N^{-1})$ for random z_0, the conclusion remains valid because the expected value of $v\lambda(z_0)O_p(N^{-1})$ in $b_\theta(z)$ is zero (see Appendix 19.D).

19.4.4 Asymptotic Normality

If the true model belongs to the model set, then expression (17-18) of the covariance matrix in Theorem 17.29 (asymptotic normality of $\sqrt{N}(\hat{\theta}(z) - \tilde{\theta}(z_0))$) can be elaborated.

Theorem 19.3 (Asymptotic Normality of $\sqrt{N}(\hat{\theta}(z) - \theta_0)$): Under the assumptions of Theorem 17.21 and Assumptions 17.1 ($P = \infty$) and 17.9, $\sqrt{N}(\hat{\theta}(z) - \theta_0)$ converges in law, at the rate $O(N^{-1/2})$, to a Gaussian random variable with zero mean and covariance matrix $\text{Cov}(\sqrt{N}\delta_\theta(z))$

$$\text{Cov}(\sqrt{N}\delta_\theta(z)) = V_N''^{-1}(\theta_0) + V_N''^{-1}(\theta_0)q_N(\theta_0)V_N''^{-1}(\theta_0)$$
$$q_N(\theta_0) = N\, \mathbb{E}\{v_N'^T(\theta_0, n_z)v_N'(\theta_0, n_z)\} \quad (19\text{-}32)$$
$$+ 2\text{herm}(\mathbb{E}\{\left(\frac{\partial \varepsilon(\theta, z_0)}{\partial \theta_0}\right)^T\}\mathbb{E}\{\Delta(\theta_0, n_z)v_N'(\theta_0, n_z)\})$$

with $v_N(\theta, n_z) = \frac{1}{2N}\Delta^T(\theta, n_z)\Delta(\theta, n_z)$ and $\Delta(\theta, n_z) = \Lambda(\theta)M_1(\theta)n_z$.

Proof. See Appendix 19.E. □

The expression (19-32) for $\text{Cov}(\delta_\theta(z))$ is not tractable and will be approximated. Replacing n_z by vn_z (and, hence, C_{n_z} by $v^2 C_{n_z}$), it can be seen that the second term in the expression of $\text{Cov}(\sqrt{N}\delta_\theta(z))$ decreases to zero faster than $V_N''^{-1}(\theta_0)$ as the signal-to-noise ratio increases to infinity ($v \to 0$). It makes it possible to approximate (19-32) for "sufficiently large" signal-to-noise ratios as

$$\text{Cov}(\delta_\theta(z)) = V_{\text{Markov}}''^{-1}(\theta_0)(I_{n_\theta} + O(v)) \quad (19\text{-}33)$$

with $V_{\text{Markov}}''^{-1}(\theta_0) = v^2 O(N^{-1})$ (Exercise 19.9).

If modeling errors are present, $e(\tilde{\theta}(z), z_0) \neq 0$, then the full expression (17-18) of the covariance matrix should be used. Replacing $e(\tilde{\theta}(z), z_0)$ by $\mu e(\tilde{\theta}(z), z_0)$ and n_z by vn_z, an approximation for "small" model errors ($\mu \to 0$) and "large" signal-to-noise ratios ($v \to 0$) is given by (Exercise 19.10)

$$\text{Cov}(\delta_\theta(z)) = C_\theta(I_{n_\theta} + O(v) + O(\mu) + O(\mu^2 v^{-2})\lambda(z_0))$$
$$C_\theta = \left[\mathbb{E}\{\left(\frac{\partial \varepsilon(\theta, z_0)}{\partial \tilde{\theta}(z_0)}\right)^T \left(\frac{\partial \varepsilon(\theta, z_0)}{\partial \tilde{\theta}(z_0)}\right)\}\right]^{-1} = v^2 O(N^{-1}) \quad (19\text{-}34)$$

where $\lambda(z_0) = 1$ for random z_0 and $\lambda(z_0) = 0$ for deterministic z_0. Formula (19-34) shows that in the presence of model errors ($\mu \neq 0$), the uncertainty of the estimated model parameters does not decrease to zero for random z_0 as the noise level υ tends to zero. Intuitively, this can be understood as follows: in the absence of observation noise, $n_z = 0$, the model errors still depend on the particular realization of z_0. Hence, $\hat{\theta}(z)$ depends on z_0, and $\text{Cov}(\delta_{\hat{\theta}}(z)) = \mu^2 O(N^{-1})$.

19.4.5 Asymptotic Efficiency

Comparing the Cramér-Rao lower bound (19-21) for normally distributed errors n_z and deterministic z_0 with the asymptotic covariance matrix (19-32) shows that the Markov estimates are, in general, asymptotically inefficient ($V_N''(\theta_0) = V_N''(\theta_0, z_0)$ for deterministic z_0). The inefficiency term $V_N''^{-1}(\theta_0) q_N(\theta_0) V_N''^{-1}(\theta_0)$ is, however, small w.r.t. $V_N''^{-1}(\theta_0)$ for sufficiently large signal-to-noise ratios (Exercise 19.9). For some noise covariance matrices the inefficiency term is zero.

Theorem 19.4 (Asymptotic Efficiency of $\hat{\theta}(z)$): Under the assumptions of Theorem 19.3 and Assumption 19.1, the Markov estimates (19-10) are asymptotically efficient if $\text{rank}(C_{n_z}) = rN$ for any $N \geq N_0$.

Proof. See Appendix 19.F. □

The condition $\text{rank}(C_{n_z}) = rN$ is automatically satisfied for signal models (Exercise 19.11). In frequency domain identification of multivariable systems, it implies that the number of noncoherent noise sources must be equal to the number of outputs. Note that Theorem 19.4 is not valid for the estimates based on the simplified Markov cost functions (19-12) and (19-13).

19.4.6 Robustness

The consistency, asymptotic normality, convergence rate, and asymptotic bias properties of the Markov estimator (19-10) for (block) diagonal models (19-2) are robust w.r.t. to the knowledge of some nondiagonal parts of C_{n_z} (compare the simplified cost functions (19-12) and (19-13) with (19-8)). This is not the case for the asymptotic efficiency: removing the nondiagonal elements of C_{n_z} increases the uncertainty of the estimates.

19.4.7 Practical Calculation of Uncertainty Bounds

Theorems 17.29 and 19.4 and formulas (19-33), (19-34) require knowledge of the true observations z_0 and the (true) model parameters θ_0 or $\tilde{\theta}(z)$, which are not available. Approximations of the asymptotic covariance matrix are obtained by replacing z_0 by z and θ_0 or $\tilde{\theta}(z_0)$ by $\hat{\theta}(z)$. Mostly the following approximation is used:

$$\text{Cov}(\hat{\theta}(z)) \approx \left[\left(\frac{\partial \varepsilon(\theta, z)}{\partial \hat{\theta}(z)}\right)^T \left(\frac{\partial \varepsilon(\theta, z)}{\partial \hat{\theta}(z)}\right)\right]^{-1} \quad (19\text{-}35)$$

Note that the right-hand side of (19-35) is calculated in Newton-based minimization methods of the cost function (19-8).

Together with the results of Section 16.2, (19-35) allows the calculation of uncertainty bounds of any model-related quantity. For example, the uncertainty of $f(z, \hat{\theta}(z)) \in \mathbb{R}^m$ is found by linearizing $f(z, \hat{\theta}(z))$ at z_0, $\tilde{\theta}(z_0)$

Section 19.5 ■ Residuals of the Model Equation

$$f(z, \hat{\theta}(z)) \approx \frac{\partial f(z, \tilde{\theta}(z_0))}{\partial z_0} n_z + \frac{\partial f(z_0, \theta)}{\partial \tilde{\theta}(z_0)}(\hat{\theta}(z) - \tilde{\theta}(z_0)) \qquad (19\text{-}36)$$

where $\hat{\theta}(z) - \tilde{\theta}(z_0)$ is given by (19-31). Calculating the covariance matrix of $f(z, \hat{\theta}(z))$ and replacing z_0 afterward by z and $\tilde{\theta}(z_0)$ by $\hat{\theta}(z)$ in this expression gives

$$\begin{aligned}\text{Cov}(f(z, \hat{\theta}(z))) &\approx \left(\frac{\partial f(z, \hat{\theta}(z))}{\partial z}\right) C_{n_z} \left(\frac{\partial f(z, \hat{\theta}(z))}{\partial z}\right)^T + \left(\frac{\partial f(z, \theta)}{\partial \hat{\theta}(z)}\right) \text{Cov}(\hat{\theta}(z)) \left(\frac{\partial f(z, \theta)}{\partial \hat{\theta}(z)}\right)^T \\ &\quad + 2\text{herm}\left(\left(\frac{\partial f(z, \hat{\theta}(z))}{\partial z}\right) \text{Cov}(n_z, \hat{\theta}(z) - \tilde{\theta}(z_0)) \left(\frac{\partial f(z, \theta)}{\partial \hat{\theta}(z)}\right)^T\right)\end{aligned} \qquad (19\text{-}37)$$

where, using (19-31) and (19-35), $\text{Cov}(n_z, \hat{\theta}(z) - \tilde{\theta}(z_0))$ can be approximated as

$$\text{Cov}(n_z, \hat{\theta}(z) - \tilde{\theta}(z_0)) \approx -C_{n_z} \left(\frac{\partial \varepsilon(\hat{\theta}(z), z)}{\partial z}\right)^T \left(\frac{\partial \varepsilon(\theta, z)}{\partial \hat{\theta}(z)}\right) \text{Cov}(\hat{\theta}(z))$$

19.5 RESIDUALS OF THE MODEL EQUATION

First we study the residuals for real observations and real model parameters. Afterward, the results are generalized to complex observations and complex model parameters.

19.5.1 Real Case

The weighted residual of the model equation, $\varepsilon(\hat{\theta}(z), z)$, is a random vector that depends directly on the errors n_z through the observations z and indirectly on these errors through the estimate $\hat{\theta}(z)$, which is a nonlinear function of n_z. To analyze its stochastic properties, we need assumptions on the square root $C_e^{-1/2}(\theta)[M_0(\theta)\ M_1(\theta)]$ of the weighting $W_N(\theta)$ (19-25) (convergence analysis) and on the true observations z_0 (existence of some moments).

Assumption 19.5 (Constraint on the Square Root of the Weighting): The $rN+1$ by $sN+1$ matrix $R_N(\theta) = \Lambda(\theta)[M_0(\theta)\ M_1(\theta)]$, with $\Lambda^T(\theta)\Lambda(\theta) = C_e^{-1}(\theta)$, satisfies $\|R_N(\theta)\|_p \leq c < \infty$, with $p = 1, \infty$ and c an N-independent constant, for all N (∞ included) and any $\theta \in \mathbb{P}_r$. $R_N(\theta)$ is a continuous matrix function of θ in the compact set $\mathbb{P}_r$.

Assumption 19.6 (Constraint on the Derivatives of the Square Root of the Weighting): The $rN+1$ by $sN+1$ matrix $R_N(\theta) = \Lambda(\theta)[M_0(\theta)\ M_1(\theta)]$, with $\Lambda^T(\theta)\Lambda(\theta) = C_e^{-1}(\theta)$, has continuous first- and second-order derivatives w.r.t. θ satisfying

(a) $\left\|\dfrac{\partial R_N(\theta)}{\partial \theta_{[i]}}\right\|_p \leq c_1 < \infty, \ i = 1, 2, \ldots, n_\theta$

(b) $\left\|\dfrac{\partial^2 R_N(\theta)}{\partial \theta_{[i]} \partial \theta_{[j]}}\right\|_p \leq c_2 < \infty, \ i,j = 1, 2, \ldots, n_\theta$

with $p = 1, \infty$ and c_1, c_2 N-independent constants, for $N = 1, 2, \ldots, \infty$ and $\theta \in \mathbb{P}_r$.

Assumption 19.7 (True Observations): The true observations z_0 are uniformly bounded.

Lemma 19.8 (Convergence Rate Residuals): Under the assumptions of Theorem 17.21 and Assumptions 19.5 and 19.6(a), the residual $\varepsilon_{[i]}(\hat{\theta}(z), z)$ converges uniformly in prob. to $\varepsilon_{[i]}(\tilde{\theta}(z_0), z)$ at the rate $O_p(N^{-1/2})$ in $\mathbb{P}_r$ as $N \to \infty$, $i = 1, 2, \ldots, rN$.

Proof. See Appendix 19.G. □

For large signal-to-noise ratios and small model errors, the convergence rate of the residuals can be refined.

Lemma 19.9 (Improved Convergence Rate Residual): Under the assumptions of Theorem 17.21 and Assumptions 19.5, 19.6, and 19.7, large signal-to-noise ratios ($\upsilon \to 0$), and small model errors ($\mu \to 0$), the residual $\varepsilon_{[i]}(\hat{\theta}(z), z)$, $i = 1, 2, \ldots, rN$, can be written as

$$\varepsilon_{[i]}(\hat{\theta}(z), z) = \varepsilon_{[i]}(\tilde{\theta}(z_0), z_0) + (Q_\varepsilon(z_0)\delta_\varepsilon(z))_{[i]} + O_p(N^{-1/2})(\upsilon + \mu + \mu\upsilon^{-1}\lambda(z_0))$$

$$Q_\varepsilon(z_0) = I_{rN} - \left(\frac{\partial\varepsilon(\theta, z_0)}{\partial\tilde{\theta}(z_0)}\right)\left[\left(\frac{\partial\varepsilon(\theta, z_0)}{\partial\tilde{\theta}(z_0)}\right)^T\left(\frac{\partial\varepsilon(\theta, z_0)}{\partial\tilde{\theta}(z_0)}\right)\right]^{-1}\left(\frac{\partial\varepsilon(\theta, z_0)}{\partial\tilde{\theta}(z_0)}\right)^T \quad (19\text{-}38)$$

$$\delta_\varepsilon(z) = \Delta(\tilde{\theta}(z_0), n_z) = \Lambda(\tilde{\theta}(z_0))M_1(\tilde{\theta}(z_0))n_z$$

where $\mathbb{E}\{\delta_\varepsilon(z)\} = 0$, $\mathrm{Cov}(\delta_\varepsilon(z)) = I_{rN}$ and

$$\varepsilon_{[i]}(\tilde{\theta}(z_0), z_0) = \mu\upsilon^{-1}O_p(N^0)$$

$$(Q_\varepsilon(z_0)\delta_\varepsilon(z))_{[i]} = O_p(N^0)$$

$$Q_{\varepsilon[i,j]}(z_0) = I_{rN[i,j]} + O_p(N^{-1})$$

for $i = 1, 2, \ldots, rN$, with $\lambda(z_0) = 1$ for random z_0 and $\lambda(z_0) = 0$ for deterministic z_0. $Q_\varepsilon(z_0)$ is a symmetric idempotent matrix of rank $rN - n_\theta$.

Proof. See Appendix 19.I. □

If no model errors are present, $\varepsilon_{[i]}(\tilde{\theta}(z_0), z_0) = 0$, it follows from Lemma 19.8 that the residuals $\varepsilon(\hat{\theta}(z), z)$ are asymptotically white: $\mathrm{Cov}(\varepsilon(\theta_0, z)) = I_{rN}$. Therefore, we could think of verifying the presence of model errors through the sample correlation of the residuals

$$\hat{R}_{\varepsilon\varepsilon}(k) = \frac{1}{rN - |k|}\sum_{i=1}^{rN - |k|}\varepsilon_{[i]}(\hat{\theta}(z), z)\varepsilon_{[i+k]}(\hat{\theta}(z), z) \quad (19\text{-}39)$$

The following theorem shows that this makes sense, indeed.

Section 19.5 ■ Residuals of the Model Equation

Theorem 19.10 (Properties Sample Correlation): Under the assumptions of Theorem 17.21 and Assumptions 19.5, 19.6(a), the sample correlation $\hat{R}_{\varepsilon\varepsilon}(k)$ converges in prob. to

$$\frac{1}{rN-|k|}\sum_{i=1}^{rN-|k|} \mathbb{E}\{\varepsilon_{[i]}(\tilde{\theta}(z_0),z_0)\varepsilon_{[i+k]}(\tilde{\theta}(z_0),z_0)\} + \delta(k) \tag{19-40}$$

at the rate $O_p(N^{-1/2})$ as $N \to \infty$ ($\delta(k)$ is the Kronecker delta and k is fixed independent of N). If, in addition, Assumptions 17.1 ($P = \infty$) and 19.6(b) are valid, then $\hat{R}_{\varepsilon\varepsilon}(k)$ is asymptotically normally distributed. If no model errors are present (Assumptions 17.9, 17.10) and n_z is normally distributed (Assumption 19.1), then the standard deviation of the truncated sample correlation $\underline{\hat{R}}_{\varepsilon\varepsilon}(k)$ equals, asymptotically,

$$\frac{1+\delta(k)}{\sqrt{rN-|k|}} \tag{19-41}$$

Proof. See Appendix 19.J. □

Under the null hypothesis that no model errors are present, Theorem 19.10 makes it possible to verify whether or not the sample correlation is white within its uncertainty. This procedure is known as the *whiteness test on the residuals*.

Lemma 19.9 shows that in the absence of model errors, $\varepsilon_{[i]}(\tilde{\theta}(z_0), z_0) = 0$ and $\mu = 0$, n_θ linear dependences exist among the residuals $\varepsilon_{[i]}(\tilde{\theta}(z), z)$, $i = 1, 2, \ldots, rN$ (rank($Q_\varepsilon(z_0)) = rN - n_\theta$). It explains why the expected value of $\hat{R}_{\varepsilon\varepsilon}(0)$ approximately ($N \to \infty$, $\upsilon, \mu \to 0$) equals $\mathbb{E}\{\sum_{i=1}^{rN}((Q_\varepsilon(z_0)\delta_\varepsilon(z))_{[i]})^2/(rN)\} = (rN - n_\theta)/(rN)$ (see Exercise 19.12) while Theorem 19.10 predicts the value 1. To compensate for this bias at lag zero, $\hat{R}_{\varepsilon\varepsilon}(k)$ and its standard deviation are often multiplied by $rN/(rN - n_\theta)$.

For deterministic z_0, the covariance matrix of the truncated residuals is given approximately by

$$\text{Cov}(\underline{\varepsilon}(\hat{\theta}(z), z)) \approx \text{Cov}(Q_\varepsilon(z_0)\delta_\varepsilon(z)) \approx I_{rN} - \left(\frac{\partial\varepsilon(\theta,z_0)}{\partial\tilde{\theta}(z_0)}\right)\text{Cov}(\underline{\hat{\theta}}(z))\left(\frac{\partial\varepsilon(\theta,z_0)}{\partial\tilde{\theta}(z_0)}\right)^T \tag{19-42}$$

(see Exercise 19.13). It follows that the total uncertainty equals the uncertainty due to the observation noise n_z minus the uncertainty due to the estimated model parameters $\hat{\theta}$.

19.5.2 Complex Case

For *complex observations* $z \in \mathbb{C}^{sN}$ and *real or complex model parameters* the sample correlation of the residuals is defined as

$$\hat{R}_{\varepsilon\varepsilon}(k) = \frac{1}{rN-|k|}\sum_{i=1}^{rN-|k|}\varepsilon_{[i]}(\hat{\theta}(z),z)\bar{\varepsilon}_{[i+k]}(\hat{\theta}(z),z) \tag{19-43}$$

Theorem 19.10 is still valid for *circular complex distributed errors* $n_z \in \mathbb{C}^{sN}$ (see (16-12) and (16-13)) with the following modifications (proof: all formulas of Section 19.5.1 are valid for z_{re}, $n_{z\text{re}}$, and ε_{re}). $\hat{R}_{\varepsilon\varepsilon}(k)$ is asymptotically circular complex normally distributed except at lag zero, where it is asymptotically normally distributed. If no model errors are present, then $\hat{R}_{\varepsilon\varepsilon}(k)$ ($k \neq 0$) is asymptotically circular complex normally distributed and the variance of the truncated sample correlation $\hat{R}_{\varepsilon\varepsilon}(k)$ equals, asymptotically, $1/(rN-|k|)$ (see Exercise 19.14). The multiplicative bias correcting factor for $\hat{R}_{\varepsilon\varepsilon}(k)$ (and its standard deviation) equals $rN/(rN-n_\theta/2)$ for real model parameters and $rN/(rN-n_\theta)$ for complex model parameters (see Exercise 19.15).

19.6 MEAN AND VARIANCE OF THE COST FUNCTION

First, the case of real observations and real model parameters is handled. Afterward, these results are generalized to complex observations and complex model parameters.

19.6.1 Real Case

This section studies the stochastic properties of the minimum of the cost function $V_{\text{Markov}}(\hat{\theta}(z), z)$. In particular, the contribution of the model errors ($e(\tilde{\theta}(z_0), z_0) \ne 0$) and the noise n_z to $V_{\text{Markov}}(\hat{\theta}(z), z)$ is analyzed. The following lemma gives the properties of $V_{\text{Markov}}(\hat{\theta}(z), z)$ for a large number of observations ($N \to \infty$), large signal-to-noise ratios ($\upsilon \to 0$), and small model errors ($\mu \to 0$).

Lemma 19.11 (Convergence Rate Cost Function): Under the assumptions of Theorem 17.21 and Assumptions 19.5, 19.6, and 19.7, large signal-to-noise ratios ($\upsilon \to 0$), and small model errors ($\mu \to 0$), the minimum of the cost function $V_{\text{Markov}}(\hat{\theta}(z), z)$ can be written as

$$
\begin{aligned}
V_{\text{Markov}}(\hat{\theta}(z), z) &= L(\tilde{\theta}(z_0), z) + (\upsilon + \mu + \mu\upsilon^{-2}(\upsilon+\mu)\lambda(z_0))O_p(N^0) \\
L(\tilde{\theta}(z_0), z) &= V_{\text{Markov}}(\tilde{\theta}(z_0), z_0) + \varepsilon^T(\tilde{\theta}(z_0), z_0)Q_\varepsilon(z_0)\delta_\varepsilon(z) + \frac{1}{2}\delta_\varepsilon^T(z)Q_\varepsilon(z_0)\delta_\varepsilon(z)
\end{aligned}
\tag{19-44}
$$

where

$$
\begin{aligned}
V_{\text{Markov}}(\tilde{\theta}(z_0), z_0) &= \mu^2 \upsilon^{-2} O_p(N) \\
\delta_\varepsilon^T(z)Q_\varepsilon(z_0)\delta_\varepsilon(z) &= O_p(N) \\
\varepsilon^T(\tilde{\theta}(z_0), z_0)\delta_\varepsilon(z) &= \mu\upsilon^{-1}O_p(N^{1/2})
\end{aligned}
$$

with $\lambda(z_0) = 1$ for random z_0 and $\lambda(z_0) = 0$ for deterministic z_0.

Proof. See Appendix 19.K. □

As is the case for the estimates $\hat{\theta}(z)$, in general it is very difficult or impossible to show the existence of the expected value and the variance of the cost function $V_{\text{Markov}}(\hat{\theta}(z), z)$. However, the first- and second-order moments of $L(\tilde{\theta}(z_0), z)$ exist.

Theorem 19.12 (Properties Cost Function): Under Assumption 17.1 ($P = \infty$) and the assumptions of Lemma 19.11, $V_{\text{Markov}}(\hat{\theta}(z), z)$ is asymptotically normally distributed. Under the assumptions of Lemma 19.11, we have

$$
\mathbb{E}\{L(\tilde{\theta}(z_0), z)\} = \mathbb{E}\{V_{\text{Markov}}(\tilde{\theta}(z_0), z_0)\} + (rN - n_\theta)/2
\tag{19-45}
$$

with n_θ the number of identifiable model parameters. If, in addition, the errors n_z are normally distributed (Assumption 19.1), then

$$\text{var}(L(\tilde{\theta}(z_0), z)) = \mathbb{E}\{\varepsilon^T(\tilde{\theta}(z_0), z_0) Q_\varepsilon(z_0) \varepsilon(\tilde{\theta}(z_0), z_0)\} + (rN - n_\theta)/2 \quad (19\text{-}46)$$
$$+ \text{var}(V_{\text{Markov}}(\tilde{\theta}(z_0), z_0))$$

For deterministic z_0, (19-46) reduces to

$$\text{var}(L(\tilde{\theta}(z_0), z)) = \varepsilon^T(\tilde{\theta}(z_0), z_0) \varepsilon(\tilde{\theta}(z_0), z_0) + (rN - n_\theta)/2 \quad (19\text{-}47)$$

Proof. See Appendix 19.L. □

Theorem 19.12 shows that the model errors ($\varepsilon(\tilde{\theta}(z_0), z_0) \neq 0$) increase not only the expected value of the cost function but also its uncertainty. This increase in uncertainty is larger for random than for deterministic observations z_0 ($\text{var}(V_{\text{Markov}}(\tilde{\theta}(z_0), z_0)) = 0$).

Under the null hypothesis that no model errors are present ($\mu = 0$), Theorem 19.12 shows that $V_{\text{Markov}}(\hat{\theta}(z), z)$ is asymptotically $N((rN - n_\theta)/2, (rN - n_\theta)/2)$ distributed. It makes it possible to verify whether or not the cost function $V_{\text{Markov}}(\hat{\theta}(z), z)$ equals $(rN - n_\theta)/2$ within a given confidence level.

In the case of deterministic observations z_0, Theorem 19.12 allows estimation of the uncertainty of the cost function in the presence of model errors. Indeed, from (19-45) and Lemma 19.11 it follows that the contribution of the model errors to the cost function $V_{\text{Markov}}(\tilde{\theta}(z_0), z_0)$ can be estimated as

$$V_{\text{Markov}}(\tilde{\theta}(z_0), z_0) \approx \begin{cases} V_{\text{Markov}}(\hat{\theta}(z), z) - \dfrac{(rN - n_\theta)}{2} & V_{\text{Markov}}(\hat{\theta}(z), z) \geq \dfrac{(rN - n_\theta)}{2} \\ 0 & \text{elsewhere} \end{cases} \quad (19\text{-}48)$$

Substituting this expression in (19-47) gives an estimate of the variance of the cost function

$$\text{var}(L_{\text{Markov}}(\tilde{\theta}(z_0), z_0)) \approx 2 V_{\text{Markov}}(\hat{\theta}(z), z) - (rN - n_\theta)/2 \quad (19\text{-}49)$$

As a null hypothesis test already makes it possible to verify the presence of model errors, one could wonder why it is useful to know the uncertainty of the cost function in the presence of model errors. Formula (19-49) is useful for comparing the cost functions of two independent experiments, for example, to decide whether or not the model errors are significantly different in both experiments.

The variance expression (19-46) becomes intractable for non-Gaussian observation errors n_z. However, for deterministic observations z_0 and non-Gaussian n_z, it is still possible to give upper and lower bounds on the variance (Pintelon et al., 1997a).

19.6.2 Complex Case

For *complex observations* $z \in \mathbb{C}^{sN}$ and *circular complex errors* n_z, Theorem 19.12 and formulas (19-48), (19-49) are still valid with the following modifications. Replace $rN - n_\theta$ by $2rN - n_\theta$ for *real model parameters* and $rN - n_\theta$ by $2(rN - n_\theta)$ for *complex model parameters* (see Exercise 19.16).

19.7 MODEL SELECTION AND MODEL VALIDATION

An identification procedure typically consists of applying iteratively model selection and parameter estimation. The model selection (order estimation) is still the most critical step in the identification process and consists of detecting overmodeling as well as undermodeling. Overmodeling occurs if the considered model set includes the true model and if it is described by too many parameters. Undermodeling is, for example, due to unmodeled dynamics and/or nonlinear distortions in system identification or, for example, due to too small a number of sine waves and/or nonperiodic deterministic disturbances in signal modeling. This section describes the properties of several (classical) model selection methods.

19.7.1 Real Case

19.7.1.1 Detection of Overmodeling. The Akaike information criterion (AIC) and minimum description length (MDL) method select the model $M(\hat{\theta}(z), \hat{z})$ out of the model set $\mathbb{M}$ that minimizes the sum of the negative log-likelihood function of the parameters and a function that penalizes the use of a large number of parameters (Akaike, 1974; Rissanen, 1978; Liang et al., 1993). For model (19-1) and Gaussian-distributed errors n_z, they take the form

$$\text{AIC:} \quad V_{\text{Markov}}(\hat{\theta}(z), z) + n_\theta \qquad (19\text{-}50)$$

$$\text{MDL:} \quad V_{\text{Markov}}(\hat{\theta}(z), z) + \frac{n_\theta}{2}\ln(\text{rank}(C_{n_z})) \qquad (19\text{-}51)$$

with n_θ the number of identifiable model parameters (Appendix 19.M). Minimizing (19-50) and (19-51) over the set of models $\mathbb{M}$ ($V_{\text{Markov}}(\hat{\theta}(z), z)$ and n_θ vary over $\mathbb{M}$) gives the optimal model according to the AIC and MDL criteria, respectively. The AIC criterion is inconsistent because it selects too complex models (Kashyap, 1980), while the MDL criterion gives strongly consistent estimates of the order of ARMA models (Hannan, 1980).

Example 19.13: Consider the identification of the amplitudes A_k, phases ϕ_k, and frequency f_0 of the sum of h harmonically related sine waves (signal model (19-3) with $s = 1$): $M_{0[n]}(\theta) = \sum_{k=1}^{h} A_k \sin(k\omega_0 n T_s + \phi_k)$ with $n = 0, 1, ..., N-1$ and $\theta^T = [A_1...A_h\, \phi_1...\phi_h\, f_0]$. Hence, (19-50) and (19-51) apply with $n_\theta = 2h+1$ and $\text{rank}(C_{n_z}) = N$. According to the AIC or MDL principle, the optimal value of h is found by minimizing (19-50) or (19-51) w.r.t. $h \in \mathbb{N}$ ($V_{\text{Markov}}(\hat{\theta}(z), z)$ is a function of h). □

The AIC and MDL criteria have been derived by assuming implicitly, or explicitly, that the true model belongs to the model set (see Ljung, 1999 for AIC) and, therefore, are unable to detect undermodeling. In the presence of model errors (for example, nonlinear distortions) the additive penalty terms in (19-50) and (19-51) are no longer valid. In the sequel of this section we show that under some suitable assumptions on the model error, a multiplicative penalty term should be used.

Assumption 19.14 (Behavior of the Model Error): The model error contribution $m_{[i]} = \varepsilon_{[i]}(\tilde{\theta}(z_0), z_0)$ and the noise contribution $v_{[i]} = (Q_\varepsilon(z_0)\delta_\varepsilon(z))_{[i]}$ in the normalized residual $\varepsilon_{[i]} = \varepsilon_{[i]}(\hat{\theta}(z), z)$ (19-38) have the following properties:

1. The model error $m_{[i]}$ and the noise error $v_{[i]}$ are uncorrelated: $\mathbb{E}\{m_{[i]}v_{[j]}\} = 0$.
2. The model error $m_{[i]}$ is "uncorrelated": $\mathbb{E}\{m_{[i]}m_{[j]}\} = m^2\delta(i-j)$, with $\delta(k)$ the Kronecker delta.
3. The model error $m_{[i]}$ and the derivative of the normalized residual w.r.t. the model parameters $\varepsilon'_{[i]}$ are asymptotically ($N \to \infty$) uncorrelated:

$$\mathbb{E}\{\varepsilon'^T_{[i]}\varepsilon'_{[j]}m_{[i]}m_{[j]}\} \approx \mathbb{E}\{\varepsilon'^T_{[i]}\varepsilon'_{[j]}\}\mathbb{E}\{m_{[i]}m_{[j]}\}.$$

4. The noise error $v_{[i]}$ and the derivative of the normalized residual w.r.t. the model parameters $\varepsilon'_{[i]}$ are uncorrelated for large signal-to-noise ratios ($v \to \infty$):

$$\mathbb{E}\{\varepsilon'^T_{[i]}\varepsilon'_{[j]}v_{[i]}v_{[j]}\} \approx \mathbb{E}\{\varepsilon'^T_{[i]}\varepsilon'_{[j]}\}\mathbb{E}\{v_{[i]}v_{[j]}\}$$

where $\mathbb{E}\{v_{[i]}v_{[j]}\} = \delta(i-j) + O_p(N^{-1})$ (see Lemma 19.9).

At first glance it might seem hard to assume uncorrelated model errors. However, correlated model errors would be detected in the whitness test of the residuals (see Section 19.5), and it can be argued that one should not bother about AIC or MDL under these conditions.

Theorem 19.15 (AIC and MDL in the Presence of Model Errors): Under the assumptions of Lemma 19.9 and Assumption 19.14, the modified AIC and MDL rules equal

$$\text{modified AIC:} \quad V_{\text{Markov}}(\hat{\theta}(z), z)\left(1 + \frac{2n_\theta}{rN}\right) \quad (19\text{-}52)$$

$$\text{modified MDL:} \quad V_{\text{Markov}}(\hat{\theta}(z), z)\left(1 + \frac{n_\theta}{rN}\ln(\text{rank}(C_{n_z}))\right) \quad (19\text{-}53)$$

Proof. See Appendix 19.N. □

In the presence of "uncorrelated" model errors the modified AIC and MDL rules will help to select the best model within the considered model class. An example of "uncorrelated" model errors are the stochastic nonlinear distortions when approximating a nonlinear dynamic system by a linear transfer function model (see Chapter 3).

19.7.1.2 Detection of Undermodeling. Undermodeling can be detected by a null hypothesis test on the cost function (see Section 19.6): if $V_{\text{Markov}}(\hat{\theta}(z), z) > (rN - n_\theta)/2 + 2\sqrt{(rN - n_\theta)/2}$ then, with 95% confidence, model errors are present.

19.7.1.3 Model Validation. The whiteness test of the residuals (see Section 19.5) can be used as a model validation tool. If the sample correlation is not a delta function within its uncertainty, then model errors are present. If it is white within its uncertainty, then the model passes the validation test. This does not, however, mean that no model errors are present. Indeed, the whiteness test is insensitive to errors that behave as white noise in the residuals. Nonlinear distortions in system identification are an example of such errors (see Chapter 11). The presence of model errors in a validated model can be detected by a null hypothesis test on the cost function (see paragraph 19.7.1.2 of this section).

19.7.1.4 Model Selection Procedure. The following iterative model selection procedure results.

1. Choose an initial model set (model order).
2. Estimate the model parameters.
3. Validate the model using a whiteness test on the residuals (see Section 19.5). If the residuals are white or a user-defined criterion is satisfied, then go to 4, else increase the model complexity and go to 2.
4. Detect undermodeling using a null hypothesis test on the cost function (see Section 19.6). If the cost function lies in the interval $(rN - n_\theta)/2 \pm 2\sqrt{(rN - n_\theta)/2}$, then go to 5, else stop.
5. Detect overmodeling using the MDL criterion (19-51).

Possible user-defined criteria in step 3 are, for example, that the estimated contribution of the model errors $V_{\text{Markov}}(\tilde{\theta}(z_0), z_0)$ to the cost function (see (19-48)) is below a given level C, or, in system identification, that the maximal (relative) transfer function error is less than a given value ε. The proposed procedure starts with simple models and gradually increases the model complexity. Practice has shown that in most identification problems the iterative procedure stops at step 4 (the validated model still contains some model errors). This is quite natural because the proposed model set reflects our belief in what reality is. This belief is mostly (always?) an approximation of the true behavior.

19.7.2 Complex Case

The results of the real case are still valid for *complex observations* $z \in \mathbb{C}^{sN}$ and *circular complex errors* n_z with the following modifications: replace rN by $2rN$, and $\text{rank}(C_{n_z})$ by $2\text{rank}(C_{n_z})$. For *complex model parameters* n_θ is replaced by $2n_\theta$. For example, (19-51) becomes

$$\text{MDL:} \quad V_{\text{Markov}}(\hat{\theta}(z), z) + \frac{n_\theta}{2} \ln(2\text{rank}(C_{n_z})) \qquad (19\text{-}54)$$

for *real model parameters* $\theta \in \mathbb{R}^{n_\theta}$, and (19-53) becomes

$$\text{modified MDL:} \quad V_{\text{Markov}}(\hat{\theta}(z), z)\left(1 + \frac{n_\theta}{rN} \ln(2\text{rank}(C_{n_z}))\right) \qquad (19\text{-}55)$$

for *complex model parameters* $\theta \in \mathbb{C}^{n_\theta}$.

Example 19.16: Consider the frequency domain identification of a linear time-invariant MIMO system from periodic steady-state measurements, and assume that the input and output spectra are observed at F frequencies (model (19-5) with $r = n_y$, $q = n_u$, and $I_k(\theta) = 0$; see also Chapter 6, left matrix fraction description (6-54). In this case (19-54) applies with $n_\theta = n_a n_y^2 + (n_b + 1)n_y n_u$ and $\text{rank}(C_{n_z}) = n_{nc} F$ with n_{nc} the number of noncoherent noise sources. For example, $n_{nc} = n_u + n_y$ for errors-in-variables problems (all observations are disturbed by noise) and $n_{nc} = n_y$ for output error problems (the inputs are exactly known). According to the MDL principle, the optimal values of n_a and n_b are found by minimizing (19-54) w.r.t. $n_a, n_b \in \mathbb{N}$ ($V_{\text{Markov}}(\hat{\theta}(z), z)$ is a function of n_a, n_b). □

19.8 EXERCISES

19.1. Consider a scalar (SISO) discrete-time system and assume that N samples of the input and output signals are available. Show that the time domain model can be written in the form (19-4) with $r = q = 1$.

19.2. Consider a multivariable system with n_u inputs and n_y outputs. Assume that the input and output signals are periodic and that the DFT spectra are available at F frequencies. Show that the frequency domain model equations can be written in the form (19-2) with $r = 2n_y$, $s = 2(n_u + n_y)$, and $N = F$ (hint: use the left matrix fraction description (6-54) and apply Lemma 15.4).

19.3. Repeat Exercise 19.2 for arbitrary excitations (hint: use the left matrix fraction description (6-54) generalized for arbitrary excitations).

19.4. Solve the constraint minimization problem (19-6) assuming that C_{n_z} is a regular matrix.

19.5. Assume that $M_0(\theta)$ in (19-1) is linear in some of the model parameters, say $\psi \in \mathbb{R}^{n_\psi}$, and that $M_1(\theta)$ is independent of ψ: $M_0(\theta) = M(\xi)\psi$ and $M_1(\theta) = M_1(\xi)$ where $\xi \in \mathbb{R}^{n_\theta - n_\psi}$, $M(\xi) \in \mathbb{R}^{rN \times n_\psi}$ and $\theta^T = [\psi^T \xi^T]$. This form, with $M_1(\theta) = I_{rN}$, is typically encountered in signal models (Cadzow, 1990; Van den Bos and Swarte, 1993). First, show that the cost function (19-8), after elimination of ψ, becomes

$$V_{Markov}(\xi, z) = 0.5(R(\xi)\Lambda(\xi)M_1(\xi)z)^T(R(\xi)\Lambda(\xi)M_1(\xi)z)$$

where $\Lambda(\xi) \in \mathbb{R}^{rN \times rN}$ satisfies $\Lambda^T(\xi)\Lambda(\xi) = C_e(\xi)$ and $R(\xi) \in \mathbb{R}^{rN \times rN}$ is an idempotent matrix of rank $rN - n_\psi$, $R(\xi) = I_{rN} - P(\xi)(P^T(\xi)P(\xi))^{-1}P^T(\xi)$ with $P(\xi) = \Lambda(\xi)M(\xi)$. Show that the Markov estimate of ψ is given by

$$\hat{\psi} = -(M^T(\hat{\xi})C_e^{-1}(\hat{\xi})M(\hat{\xi}))^{-1}M^T(\hat{\xi})C_e^{-1}(\hat{\xi})M_1(\hat{\xi})z$$

Next, show that the cost function $V_{Markov}(\xi, z)$ can be written as

$$V_{Markov}(\xi, z) = 0.5\varepsilon^T(\xi, z)\varepsilon(\xi, z)$$

where $\varepsilon(\xi, z) \in \mathbb{R}^{rN - n_\psi}$ equals $\varepsilon(\xi, z) = [I_{rN - n_\psi} \ 0]V^T(\xi)\Lambda(\xi)M_1(\xi)z$, and $V(\xi)$ is the orthogonal matrix of the eigenvectors of $R(\xi)$. Finally, show that $Cov(\varepsilon(\xi, n_z)) = I_{rN - n_\psi}$.

19.6. Prove the Cramér-Rao lower bound (19-22) for complex observations z, circular complex distributed errors n_z, and real model parameters θ (hint: first show that $V_{Markov}(\theta, z) = 0.5 e_{re}^T(\theta, z)C_{e_{re}}^{-1}(\theta)e_{re}(\theta, z) = e^H(\theta, z)C_e^{-1}(\theta)e(\theta, z)$ using Lemmas 15.3 and 15.4).

19.7. Prove the Cramér-Rao lower bound (19-23) for complex observations z, circular complex distributed errors n_z, and complex model parameters θ, assuming that $e(\theta, z)$ is an analytic function of θ (hint: rewrite (19-22) using $\partial f(\theta)/\partial \text{Re}(\theta) = \partial f(\theta)/\partial \theta$, $\partial f(\theta)/\partial \text{Im}(\theta) = j\partial f(\theta)/\partial \theta$, $\text{Re}(jX) = \text{Im}(X)$ and definition (15-40)).

19.8. Consider the model of Exercise 19.5 and show that the CR bound of the parameters ξ is given by

$$CR^{-1}(\xi_0) = Fi(\xi_0) = V_{Markov}''(\xi_0, z_0) = \left(\frac{\partial \varepsilon(\xi, z_0)}{\partial \xi_0}\right)^T\left(\frac{\partial \varepsilon(\xi, z_0)}{\partial \xi_0}\right)$$

(hint: start from the Fisher information matrix $Fi(\psi_0, \xi_0)$, apply the inverse of block matrices (15-8), and use the results of Exercise 19.5).

19.9. Show that the asymptotic covariance matrix is given by (19-33) as the signal-to-noise ratio increases to infinity (hint: show that $V_N''^{-1}(\theta_0) = O(\upsilon^2)$, $q_N(\theta_0) = O(\upsilon^{-1})$).

19.10. Show that the asymptotic covariance matrix of the model parameters is given by (19-34) for "small" model errors and "large" signal-to-noise ratios (hint: use $\varepsilon(\theta, z) = \varepsilon(\theta, z_0) + \Delta(\theta, n_z)$, $\mathbb{E}\{\Delta(\theta, n_z)\} = 0$, $\mathbb{E}\{\Delta(\theta, n_z)\Delta^T(\theta, n_z)\} = I_{n_\theta}$, and $V'(\hat{\theta}(z_0), z_0) = 0$ for deterministic z_0, $\varepsilon(\hat{\theta}(z_0), z_0) = O(\mu v^{-1})$, and $\Delta(\hat{\theta}(z_0), n_z) = O(\mu^0 v^0)$).

19.11. Show that the rank condition $\text{rank}(C_{n_z}) = rN$ in Theorem 19.4 is automatically satisfied for signal models.

19.12. Show that $\mathbb{E}\{\sum_{i=1}^{rN}((Q_\varepsilon(z_0)\delta_\varepsilon(z))_{[i]})^2/(rN)\} = (rN - n_\theta)/(rN)$ (hint: use the properties of $Q_\varepsilon(z_0)$ and $\delta_\varepsilon(z)$ given in Lemma 19.9).

19.13. Show that the covariance matrix of $Q_\varepsilon(z_0)\delta_\varepsilon(z)$ is given by (19-42) (hint: use the properties of $Q_\varepsilon(z_0)$ and $\delta_\varepsilon(z)$ given in Lemma 19.9 and approximation (19-34)).

19.14. Consider the case of complex observations z and circular complex distributed errors n_z. Prove that in the absence of model errors $\hat{R}_{\varepsilon\varepsilon}(k)$, $k \neq 0$, is asymptotically circular complex normally distributed $N^c(0, 1/(rN-k))$ (hint: follow the lines of part 3 of Appendix 19.J and show that $\text{cum}(\tilde{R}_{\varepsilon\varepsilon}(k), \overline{\tilde{R}_{\varepsilon\varepsilon}(k)}) = 1/(rN - k)$, $\text{cum}(\tilde{R}_{\varepsilon\varepsilon}(k), \tilde{R}_{\varepsilon\varepsilon}(k)) = 0$ with $\tilde{R}_{\varepsilon\varepsilon}(k) = (rN-k)^{-1}\sum_{i=1}^{rN-k}\delta_{\varepsilon[i]}(z)\bar{\delta}_{\varepsilon[i+k]}(z)$, and where $\delta_\varepsilon(z)$ defined in Lemma 19.9 is circular complex normally distributed).

19.15. Consider the case of complex observations z and circular complex distributed errors n_z. Show, using Lemma 19.9, that the expected value of $\hat{R}_{\varepsilon\varepsilon}(0)$ is approximately ($N \to \infty$, $v, \mu \to 0$) $(rN - n_\theta/2)/rN$ and $(rN - n_\theta)/rN$ for, respectively, real and complex model parameters (hint: replace z by z_{re}, and θ by θ_{re} for complex model parameters, and show that $\varepsilon(\theta, z_{\text{re}}) = \sqrt{2}\varepsilon_{\text{re}}(\theta, z)$ ($\varepsilon(\theta_{\text{re}}, z_{\text{re}}) = \sqrt{2}\varepsilon_{\text{re}}(\theta, z)$), next follow the lines of Section 19.5).

19.16. Consider the case of complex observations z and circular complex distributed errors n_z. Show, using Lemma 19.11, that Theorem 19.12 and formulas (19-48), (19-49) are still valid, where $rN - n_\theta$ is replaced by $2rN - n_\theta$ for real model parameters and $rN - n_\theta$ by $2(rN - n_\theta)$ for complex model parameters (hint: use the hint of Exercise 19.15).

19.9 APPENDIXES

Appendix 19.A Constrained Minimization (19-6)

Expressing the stationarity of the cost function (19-6) w.r.t. z_p and λ gives

$$-C_{n_z}^+(z - z_p) + M_1^T(\theta)\lambda = 0 \qquad (19\text{-}56)$$

$$M_0(\theta) + M_1(\theta)z_p = 0 \qquad (19\text{-}57)$$

λ is solved from (19-56) by left multiplication with $M_1(\theta)C_{n_z}$

$$\lambda = (M_1(\theta)C_{n_z}M_1^T(\theta))^{-1}M_1(\theta)C_{n_z}C_{n_z}^+(z - z_p) \qquad (19\text{-}58)$$

Because C_{n_z} is a symmetric positive semidefinite matrix, it can be decomposed into singular values as $C_{n_z} = U\Sigma U^T$, where U is an orthogonal matrix ($U^T U = UU^T = I_{sN}$) and Σ a diagonal matrix ($\text{rank}(\Sigma) = \text{rank}(C_{n_z})$) containing the sorted singular values (see Section 15.4). Defining

$$U^T z_p = \begin{bmatrix} W_{1p} \\ W_{20} \end{bmatrix} \qquad (19\text{-}59)$$

where $\dim(W_{1p}) = \text{rank}(C_{n_z})$, it follows that

$$z_p = U\begin{bmatrix}W_{1p}\\W_{20}\end{bmatrix} \text{ and } C_{n_z}C_{n_z}^+ z_p = U\begin{bmatrix}I_{r_c} & 0\\0 & 0\end{bmatrix}U^T z_p = U\begin{bmatrix}W_{1p}\\0\end{bmatrix} \quad (19\text{-}60)$$

with $r_c = \text{rank}(C_{n_z})$. W_{1p} stands for the (linear combination of) measurements lying in the regular space of C_{n_z} and is estimated, while W_{20} represents the (linear combination of) measurements lying in the null space of C_{n_z} and is known exactly. Hence, we have

$$z = U\begin{bmatrix}W_1\\W_{20}\end{bmatrix} \text{ and } C_{n_z}C_{n_z}^+ z = U\begin{bmatrix}W_1\\0\end{bmatrix} \quad (19\text{-}61)$$

Rewriting (19-57) and (19-58) using (19-60) and (19-61) makes it possible to eliminate W_{1p} in (19-58)

$$\lambda = (M_1(\theta)C_{n_z}M_1^T(\theta))^{-1}(M_0(\theta) + M_1(\theta)z) \quad (19\text{-}62)$$

Substituting (19-62) into (19-56) and left multiplication of (19-56) by C_{n_z} gives (19-7). Using (19-60), it can be seen that (19-7) is independent of W_{20}, which is known exactly, and, hence, makes it possible to estimate the measurements W_{1p} lying in the regular space of C_{n_z} only. Because $C_{n_z}^+ = (C_{n_z}^+ C_{n_z})C_{n_z}^+(C_{n_z}C_{n_z}^+)$ (see Section 15.5, properties 1 and 2) and $C_{n_z}^{+T} = C_{n_z}^+$, the cost function (19-6), where z_p satisfies (19-57), can be written as

$$0.5[C_{n_z}C_{n_z}^+(z - z_p)]^T C_{n_z}^+[C_{n_z}C_{n_z}^+(z - z_p)] \quad (19\text{-}63)$$

Substituting (19-7) into (19-63) using $C_{n_z}C_{n_z}^+ C_{n_z} = C_{n_z}$ gives (19-8) directly. □

Appendix 19.B Proof of the Cramér-Rao Lower Bound for Semilinear Models

The proof consists of two parts. In the first part we reduce the sN parameters z_p to the $r_c - rN$ parameters x_0, with $r_c = \text{rank}(C_{n_z})$. In the second part the Fisher information matrix $Fi(x_0, \theta_0)$ of the observations x and the model parameters θ is reduced to the Fisher information matrix $Fi(\theta_0)$.

(i) The known observations lie in the null space of C_{n_z} ($sN - r_c$ parameters) and are separated from the unknown observations by decomposing the parameter vector z_p as in (19-60). Using (19-60), with $U = [U_1 \; U_2]$ and $U_1 \in \mathbb{R}^{sN \times r_c}$, the constraint (19-57) can be written as

$$M_1(\theta)U_1 W_{1p} = -M_0(\theta) - M_1(\theta)U_2 W_{20} \quad (19\text{-}64)$$

$C_e(\theta) = M_1(\theta)C_{n_z}M_1^T(\theta)$ has, by assumption, full rank rN and, therefore, $r_c \geq rN$ and $\text{rank}(M_1(\theta)U_1) = rN$. Because $r_c \geq rN$, (19-64) has, in general, infinitely many solutions for W_{1p}. They can be found by adding one particular solution of (19-64) to the solution of the homogenous part of (19-64). The solution

W_{1h} of the homogenous part lies in the null space of $M_1(\theta)U_1 \in \mathbb{R}^{rN \times r_c}$. Because $M_1(\theta)U_1$ has full rank rN, it can be written as

$$W_{1h} = Q(\theta)x \tag{19-65}$$

with $x \in \mathbb{R}^{r_c - rN}$ and where $Q(\theta) \in \mathbb{R}^{r_c \times (r_c - rN)}$ satisfies $M_1(\theta)U_1Q(\theta) = 0$ (see Section 15.4.1). It can easily be verified that

$$\tilde{W}_{1p}(\theta) = -M^T(\theta)(M(\theta)M^T(\theta))^{-1}(M_0(\theta) + M_1(\theta)U_2W_{20}) \tag{19-66}$$

where $M(\theta) = M_1(\theta)U_1$, is a particular solution of (19-64). The complete solution of (19-64) equals $W_{1p} = W_{1h} + \tilde{W}_{1p}(\theta)$ so that in (19-60) z_p can be written as

$$z_p = z_p(x, \theta) = U_1 Q(\theta)x + U_1 \tilde{W}_{1p}(\theta) + U_2 W_{20} \tag{19-67}$$

with $x \in \mathbb{R}^{r_c - rN}$, which concludes the first part of the proof. The following property of the function $z_p(x, \theta)$ will be used in the second part of the proof:

$$M_1(\theta)\frac{\partial z_p(x, \theta)}{\partial x} = M_1(\theta)U_1 Q(\theta) = 0 \tag{19-68}$$

(ii) The second part of the proof starts with the calculation of the Fisher information matrix $Fi(x_0, \theta_0)$. Applying (16-85) to (19-16), using $z_p(x_0, \theta_0) = z_0$, gives

$$\begin{aligned}
F_{xx} &= \left(\frac{\partial z_p(x, \theta)}{\partial x}\right)^T C_{n_z}^+ \left(\frac{\partial z_p(x, \theta)}{\partial x}\right)\bigg|_{x = x_0, \theta = \theta_0} \\
F_{x\theta} &= \left(\frac{\partial z_p(x, \theta)}{\partial x}\right)^T C_{n_z}^+ \left(\frac{\partial z_p(x, \theta)}{\partial \theta}\right)\bigg|_{x = x_0, \theta = \theta_0} \\
F_{\theta\theta} &= \left(\frac{\partial z_p(x, \theta)}{\partial \theta}\right)^T C_{n_z}^+ \left(\frac{\partial z_p(x, \theta)}{\partial \theta}\right)\bigg|_{x = x_0, \theta = \theta_0}
\end{aligned} \tag{19-69}$$

C_{n_z} and $C_{n_z}^+$ can be decomposed into singular values as

$$\begin{aligned}
C_{n_z} &= U\begin{bmatrix}\Sigma_1 & 0 \\ 0 & 0\end{bmatrix}U^T = U_1 \Sigma_1 U_1^T \\
C_{n_z}^+ &= U\begin{bmatrix}\Sigma_1^{-1} & 0 \\ 0 & 0\end{bmatrix}U^T = U_1 \Sigma_1^{-1} U_1^T
\end{aligned} \tag{19-70}$$

where Σ_1 contains the nonzero singular values of C_{n_z}. Putting (19-69) into (19-20) by using (19-70) gives, after some calculations,

$$Fi(\theta_0) = G^T[I_{r_c} - F(F^T F)^{-1}F^T]G \tag{19-71}$$

with $G = \Sigma_1^{-1/2} U_1^T \partial z_p(x_0, \theta)/\partial \theta_0$ and $F = \Sigma_1^{-1/2} Q(\theta_0)$. Defining $E = M_1(\theta_0) U_1 \Sigma_1^{1/2}$, it follows from (19-68) that $EF = 0$. Because the matrices $E \in \mathbb{R}^{rN \times r_c}$ and $F \in \mathbb{R}^{r_c \times (r_c - rN)}$ have, respectively, full rank rN and $r_c - rN$ and $EF = 0$, it follows that the conditions of Theorem 15.2 are fulfilled for the symmetric idempotent matrices $F(F^T F)^{-1} F^T$ and $E^T(EE^T)^{-1}E$. Hence,

$$I_{r_c} - F(F^T F)^{-1} F^T = E^T(EE^T)^{-1}E \tag{19-72}$$

so that (19-71) can be simplified as

$$Fi(\theta_0) = (EG)^T (EE^T)^{-1} (EG) \tag{19-73}$$

Working out EG gives

$$EG = M_1(\theta_0) \frac{\partial U_1 U_1^T z_p(x_0, \theta)}{\partial \theta_0} \qquad (U_1 \text{ is independent of } \theta)$$

$$= M_1(\theta_0) \frac{\partial(z_p(x_0, \theta) - U_2 W_{20})}{\partial \theta_0} \quad ((19\text{-}67) \text{ with } U_1^T U_1 = I_{r_c} \text{ and } U_1^T U_2 = 0)$$

$$= -M_1(\theta_0) \frac{\partial(z_0 - z_p(x_0, \theta))}{\partial \theta_0} \qquad (U_2, W_{20}, z_0 \text{ are independent of } \theta)$$

$$= -\frac{\partial M_1(\theta)(z_0 - z_p(x_0, \theta))}{\partial \theta_0} \qquad (z_0 = z_p(x_0, \theta_0))$$

$$= -\frac{\partial (M_0(\theta) + M_1(\theta) z_0)}{\partial \theta_0} \qquad ((19\text{-}57): M_1(\theta) z_p(x_0, \theta) = -M_0(\theta))$$

Substituting this result into (19-73), taking into account that $EE^T = C_e(\theta_0)$ (see (19-9)), gives

$$Fi(\theta_0) = \left(\frac{\partial e(\theta, z_0)}{\partial \theta_0}\right)^T C_e^{-1}(\theta_0) \left(\frac{\partial e(\theta, z_0)}{\partial \theta_0}\right) \tag{19-74}$$

Using $e(\theta_0, z_0) = 0$, the two other expressions in (19-21) follow directly. □

Appendix 19.C Markov Estimates of the Observations for Signal Models and Transfer Function Models with Known Input

For signal models, we have $M_1(\theta) = -I_{sN}$, $C_e(\theta) = C_{n_z}$ (see (19-9)), and $C_{n_z}^+ = C_{n_z}^{-1}$ (Assumption 17.2 implies that $C_e(\theta)$ is regular), and (19-11) becomes

$$\hat{z} = M_0(\hat{\theta}(z)) \tag{19-75}$$

For transfer function models with known input, we have

$$C_{n_z} = \begin{bmatrix} C_{n_y} & 0 \\ 0 & 0 \end{bmatrix} \Rightarrow C_{n_z} C_{n_z}^+ = \begin{bmatrix} I_{rN} & 0 \\ 0 & 0 \end{bmatrix} \tag{19-76}$$

Taking into account (19-4), (19-11) becomes

$$\begin{bmatrix} \hat{y} \\ 0 \end{bmatrix} = \begin{bmatrix} A^{-1}(\hat{\theta}(z))B(\hat{\theta}(z))u_0 \\ 0 \end{bmatrix} \tag{19-77}$$

which concludes the proof. □

Appendix 19.D Proof of the Convergence Rate of the Markov Estimates for Large Signal-to-Noise Ratios and Small Model Errors (Theorem 19.2)

$\delta_\theta(z)$ defined in expression (17-14) will be elaborated for the Markov estimator, assuming large signal-to-noise ratios and small model errors. Therefore, n_z is replaced by υn_z (and, hence, C_{n_z} by $\upsilon^2 C_{n_z}$) with $\upsilon \to 0$, and $e(\tilde{\theta}(z), z_0)$ by $\mu e(\tilde{\theta}(z), z_0)$ with $\mu \to 0$. The proof consists of two parts: part one studies $\delta_\theta(z) = \Delta_\theta(z) + \partial_\theta(z)$ and part two $b_\theta(z)$.

(i) The Hessian of the expected value of the cost function can be written as

$$V_N''(\tilde{\theta}(z_0)) = \frac{1}{N}\left(\frac{\partial \varepsilon(\theta, z_0)}{\partial \tilde{\theta}(z_0)}\right)^T \left(\frac{\partial \varepsilon(\theta, z_0)}{\partial \tilde{\theta}(z_0)}\right) + \frac{1}{N}\sum_{k=1}^{rN} \varepsilon_{[k]}(\tilde{\theta}(z_0), z_0)\frac{\partial^2 \varepsilon_{[k]}(\theta, z_0)}{\partial \tilde{\theta}(z_0)^2} \tag{19-78}$$

where the first and second terms in the right-hand side are, respectively, an $O(\upsilon^{-2})$ and $O(\upsilon^{-2}\mu)$. Hence,

$$V_N''^{-1}(\tilde{\theta}(z_0)) = \left(\frac{1}{N}\left(\frac{\partial \varepsilon(\theta, z_0)}{\partial \tilde{\theta}(z_0)}\right)^T \left(\frac{\partial \varepsilon(\theta, z_0)}{\partial \tilde{\theta}(z_0)}\right)\right)^{-1} + \upsilon^2 O(\mu) \tag{19-79}$$

where the first term in the right-hand side is an $O(\upsilon^2)$. Using $\varepsilon(\theta, z) = \varepsilon(\theta, z_0) + \Delta(\theta, n_z)$, with $\Delta(\theta, n_z) = \Lambda(\theta)M_1(\theta)n_z$ (see (19-9)), gives

$$\begin{aligned} V_N'^T(\tilde{\theta}(z_0), z) = &\frac{1}{N}\left(\frac{\partial \varepsilon(\theta, z_0)}{\partial \tilde{\theta}(z_0)}\right)^T \varepsilon(\tilde{\theta}(z_0), z_0) + \frac{1}{N}\left(\frac{\partial \varepsilon(\theta, z_0)}{\partial \tilde{\theta}(z_0)}\right)^T \Delta(\tilde{\theta}(z_0), n_z) \\ &+ \frac{1}{N}\left(\frac{\partial \Delta(\theta, n_z)}{\partial \tilde{\theta}(z_0)}\right)^T \varepsilon(\tilde{\theta}(z_0), z_0) + \frac{1}{N}\left(\frac{\partial \Delta(\theta, n_z)}{\partial \tilde{\theta}(z_0)}\right)^T \Delta(\tilde{\theta}(z_0), n_z) \end{aligned} \tag{19-80}$$

Note that the expected value of each of the terms in the right-hand side of (19-80) is zero. This is evident for the first three terms. For the transpose of the fourth term, we find

$$\mathbb{E}\{\Delta^T(\tilde{\theta}(z_0), n_z)\frac{\partial \Delta(\theta, n_z)}{\partial \tilde{\theta}(z_0)}\} = \frac{\partial \mathbb{E}\{\Delta^T(\theta, n_z)\Delta(\theta, n_z)\}}{\partial \tilde{\theta}(z_0)} = \frac{\partial rN/2}{\partial \tilde{\theta}(z_0)} = 0$$

Applying the law of large numbers for mixing sequences (see Section 16.9, version 3) to each of the terms in the right-hand side of (19-80) shows that they are, respectively, an $O_p(\mu v^{-2}N^{-1/2})$, $O_p(v^{-1}N^{-1/2})$, $O_p(v^{-1}\mu N^{-1/2})$, and $O_p(N^{-1/2})$. Because $V_N'(\tilde{\theta}(z_0), z_0) = 0$ for deterministic z_0, (19-80) becomes

$$V_N'^T(\tilde{\theta}(z_0), z) = v^{-2}(v + \mu\lambda(z_0))O_p(N^{-1/2}) \quad (19\text{-}81)$$

with $\lambda(z_0) = 1$ and 0 for, respectively, random and deterministic z_0. Combining (19-79) and (19-80) with $\Delta(\theta, n_z) = (\partial \varepsilon(\tilde{\theta}(z_0), z)/\partial z)n_z$, gives $\delta_\theta(z) = \Delta_\theta(z) + \partial_\theta(z)$ with $\mathbb{E}\{\Delta_\theta(z)\} = 0$, $\mathbb{E}\{\partial_\theta(z)\} = 0$, $\Delta_\theta(z) = vO_p(N^{-1/2})$, and $\partial_\theta(z) = (v^2 + v\mu + \mu\lambda(z_0))O_p(N^{-1/2})$.

(ii) From the proof of Theorem 17.21 (see Appendix 17.E), it follows that the $b_\theta(z)$ term stems from the difference $V_N''(\widehat{\theta}, z) - V_N''(\tilde{\theta}(z_0)) = O_p(N^{-1/2})$,

$$b_\theta(z) = [V_N''^{-1}(\widehat{\theta}, z) - V_N''^{-1}(\tilde{\theta}(z_0))]V_N'^T(\tilde{\theta}(z_0), z) \quad (19\text{-}82)$$

This expression will be refined. Applying the mean value theorem to $V_N''(\widehat{\theta}, z)$ at the points $\widehat{\theta}$, $\tilde{\theta}(z_0)$ gives

$$V_N''(\widehat{\theta}, z) = V_N''(\tilde{\theta}(z_0), z) + \sum_{k=1}^{n_\theta} \frac{\partial V_N''(\theta, z)}{\partial \theta_{1[k]}}(\widehat{\theta}_{[k]} - \tilde{\theta}_{[k]}(z_0)) \quad (19\text{-}83)$$

with $\theta_1 = t_1\widehat{\theta} + (1 - t_1)\tilde{\theta}(z_0)$ and $t_1 \in [0, 1]$. Because $\widehat{\theta} - \tilde{\theta}(z_0) = t(\hat{\theta}(z) - \tilde{\theta}(z_0))$ (see (17-10)), and $\hat{\theta}(z) - \tilde{\theta}(z_0) = \Delta_\theta(z) + \partial_\theta(z) + b_\theta(z)$ with $b_\theta(z) = O_p(N^{-1})$ (see previous paragraph), we have $\widehat{\theta} - \tilde{\theta}(z_0) = (v + \mu\lambda(z_0))O_p(N^{-1/2})$. Combined with $\partial V_N''(\theta, z)/\partial \theta_{1[k]} = v^{-2}O_p(N^0)$ (see Appendix 17.E), the second term in (19-83) becomes

$$\sum_{k=1}^{n_\theta} \frac{\partial V_N''(\theta, z)}{\partial \theta_{1[k]}}(\widehat{\theta}_{[k]} - \tilde{\theta}_{[k]}(z_0)) = v^{-2}(v + \mu\lambda(z_0))O_p(N^{-1/2}) \quad (19\text{-}84)$$

The first term in (19-83) can be written as

$$V_N''(\tilde{\theta}(z_0), z) = V_N''(\tilde{\theta}(z_0), z_0) + \frac{1}{N}\frac{\partial^2 \varepsilon^T(\theta, z_0)\Delta(\theta, n_z)}{\partial \tilde{\theta}(z_0)^2} + v_N''(\tilde{\theta}(z_0), n_z) \quad (19\text{-}85)$$

with $v_N(\theta, n_z) = (2N)^{-1}\Delta^T(\theta, z_0)\Delta(\theta, n_z)$. $V_N''(\tilde{\theta}(z_0), z_0)$ converges w.p. 1 to

$V_N''(\tilde{\theta}(z_0))$ for random z_0 and $V_N''(\tilde{\theta}(z_0), z_0) = V_N''(\tilde{\theta}(z_0))$ for deterministic z_0 (Lemma 17.17):

$$V_N''(\tilde{\theta}(z_0), z_0) = V_N''(\tilde{\theta}(z_0)) + \upsilon^{-2}\lambda(z_0)O_p(N^{-1/2}) \tag{19-86}$$

Using Lemma 17.17, it follows that the second and third terms in the right-hand side of (19-85) are, respectively, an $O_p(\upsilon^{-1}N^{-1/2})$ and $O_p(\upsilon^0 N^{-1/2})$. Hence,

$$V_N''(\tilde{\theta}(z_0), z) = V_N''(\tilde{\theta}(z_0)) + (\upsilon^{-1} + \upsilon^{-2}\lambda(z_0))O_p(N^{-1/2}) \tag{19-87}$$

Combining (19-83), (19-84), and (19-87), using $V_N''(\tilde{\theta}(z_0)) = \upsilon^{-2}O_p(N^0)$, gives

$$\begin{aligned} V_N''(\widehat{\theta}, z) &= V_N''(\tilde{\theta}(z_0)) + (\upsilon^{-1} + \upsilon^{-2}\lambda(z_0))O_p(N^{-1/2}) \\ \Rightarrow V_N''^{-1}(\widehat{\theta}, z) &= V_N''^{-1}(\tilde{\theta}(z_0)) + \upsilon(\upsilon^2 + \upsilon\lambda(z_0))O_p(N^{-1/2}) \end{aligned} \tag{19-88}$$

Collecting (19-81), (19-82), and (19-88) finally gives

$$b_\theta(z) = (\upsilon^2 + (\upsilon + \mu)\lambda(z_0))O_p(N^{-1}) \tag{19-89}$$

The term $\upsilon\lambda(z_0)O_p(N^{-1})$ in $b_\theta(z)$ stems from the product of the second term in (19-80) with the term $\upsilon^2\lambda(z_0)O_p(N^{-1/2})$ in (19-88). As the latter depends only on z_0 (see (19-86)), the expected value of this product is zero (by assumptions, z_0 and n_z are independent). □

Appendix 19.E Proof of the Asymptotic Distribution of the Markov Estimates without Model Errors

Expression (17-18) will be elaborated for the Markov estimator assuming that no model errors are present ($e(\theta_0, z_0) = 0$). Using $\varepsilon(\theta, z) = \varepsilon(\theta, z_0) + \Delta(\theta, n_z)$, with $\Delta(\theta, n_z) = \Lambda(\theta)M_1(\theta)n_z$, and $\varepsilon(\theta_0, z_0) = 0$ (see (19-8)), we find

$$V_N'^T(\theta_0, z) = \frac{1}{N}\left(\frac{\partial \varepsilon(\theta, z_0)}{\partial \theta_0}\right)^T \Delta(\theta_0, n_z) + v_N'^T(\theta_0, n_z) \tag{19-90}$$

where $v_N(\theta, n_z) = (2N)^{-1}\Delta^T(\theta, n_z)\Delta(\theta, n_z)$. Because $\text{Cov}(\Delta(\theta, n_z)) = I_{rN}$ (see (19-8)), $\varepsilon(\theta_0, z_0) = 0$ and z_0, n_z are mutually independent (Assumption 17.1), $Q_N(\theta_0)$ (17-18) becomes

$$\begin{aligned} Q_N(\theta_0) &= N\, \mathbb{E}\{V_N'^T(\theta_0, z)V_N'(\theta_0, z)\} \\ &= V_N''(\theta_0) + N\, \mathbb{E}\{v_N'^T(\theta_0, n_z)v_N'(\theta_0, n_z)\} \\ &\quad + 2\,\text{herm}(\mathbb{E}\{\left(\frac{\partial \varepsilon(\theta, z_0)}{\partial \theta_0}\right)^T\} \mathbb{E}\{\Delta(\theta_0, n_z)v_N'(\theta_0, n_z)\}) \end{aligned} \tag{19-91}$$

Putting (19-91) into (17-18) gives (19-32). □

Appendix 19.F Proof of the Asymptotic Efficiency of the Markov Estimates (Theorem 19.4)

The second term in the expression of $q_N(\theta_0)$ (see (19-32)) is a function of the third-order moments of n_z and, hence, is zero for Gaussian errors n_z. The first term is positive semidefinite and is zero if and only if $v_N(\theta, n_z)$ is independent of θ for any n_z. From (19-70) and Assumption 19.1, it follows that we can write n_z as $n_z = U_1 \Sigma_1^{1/2} \varepsilon_z$ with $\varepsilon_z \in N_{r_c}(0, I_{r_c})$ (see Exercise 16.9). Hence, $M_1(\theta) n_z = E(\theta) \varepsilon_z$, with $E(\theta) = M_1(\theta) U_1 \Sigma_1^{1/2}$, and $C_e(\theta) = E(\theta) E^T(\theta)$, so that

$$v_N(\theta, n_z) = \frac{1}{2N} \varepsilon_z^T E^T(\theta)(E(\theta) E^T(\theta))^{-1} E(\theta) \varepsilon_z \tag{19-92}$$

The matrix $E(\theta) \in \mathbb{R}^{rN \times r_c}$ has full rank rN ($rN \leq r_c \leq sN$) because $C_e(\theta)$ has full rank rN. If $r_c = rN$, then $(E(\theta) E^T(\theta))^{-1} = E^{-T}(\theta) E^{-1}(\theta)$ and $v_N(\theta, n_z) = \varepsilon_z^T \varepsilon_z/(2N)$. This concludes the proof because ε_z is independent of θ. □

Appendix 19.G Proof of the Convergence Rate of the Residuals (Lemma 19.8)

Applying the mean value theorem to $\varepsilon_{[i]}(\hat{\theta}(z), z)$ at the points $\hat{\theta}(z)$, $\tilde{\theta}(z_0)$ gives

$$\varepsilon_{[i]}(\hat{\theta}(z), z) = \varepsilon_{[i]}(\tilde{\theta}(z_0), z) + \frac{\partial \varepsilon_{[i]}(\theta, z)}{\partial \theta_1}(\hat{\theta}(z) - \tilde{\theta}(z_0)) \tag{19-93}$$

where $\theta_1 = t_1 \hat{\theta}(z) + (1 - t_1)\tilde{\theta}(z_0)$ with $t_1 \in [0, 1]$. Under Assumptions 19.5, 19.6, and 17.1 ($P = 2$) $\varepsilon(\theta, z)$ and $\partial \varepsilon(\theta, z)/\partial \theta_{[j]}$ are both mixing of order 2 in $\mathbb{P}_r$ (Corollary 16.7), so that $\text{var}(\varepsilon_{[i]}(\theta, z)) = O(N^0)$ and $\text{Covar}(\partial \varepsilon_{[i]}(\theta, z)/\partial \theta) = O(N^0)$ uniformly in $\mathbb{P}_r$. Hence, $\varepsilon_{[i]}(\tilde{\theta}(z_0), z)$ is an $O_{\text{m.s.}}(N^0)$ ($O_p(N^0)$). From Theorem 17.21, it follows that $\hat{\theta}(z)$ converges in prob. to $\tilde{\theta}(z_0)$ at the rate $O_p(N^{-1/2})$. The same is true for θ_1, so that $\partial \varepsilon_{[i]}(\theta, z)/\partial \theta_1 = O_p(N^0)$ (Lemma 17.33). We conclude that $\varepsilon_{[i]}(\hat{\theta}(z), z) - \varepsilon_{[i]}(\tilde{\theta}(z_0), z) = O_p(N^{-1/2})$, which proves the lemma. □

Appendix 19.H Properties of the Projection Matrix in Lemma 19.9

Part one of this appendix studies the stochastic properties of the projection matrix $Q_\varepsilon(z_0)$ in (19-38) for random z_0. Part two calculates the covariance matrix of $Q_\varepsilon(z_0) \delta_\varepsilon(z)$.

(i) Under Assumptions 17.1 ($P = 4$) and 19.6(a), $\partial \varepsilon(\theta, z_0)/\partial \theta_{[j]}$ is mixing of order 4. Therefore, $\partial \varepsilon_{[i]}(\theta, z_0)/\partial \theta_{[j]} = O_p(N^0)$ and

$$\frac{1}{N}\left(\frac{\partial \varepsilon(\theta, z_0)}{\partial \theta}\right)^T \left(\frac{\partial \varepsilon(\theta, z_0)}{\partial \theta}\right) = \frac{1}{N} \mathbb{E}\left\{\left(\frac{\partial \varepsilon(\theta, z_0)}{\partial \theta}\right)^T \left(\frac{\partial \varepsilon(\theta, z_0)}{\partial \theta}\right)\right\} + O_p(N^{-1/2}) \tag{19-94}$$

uniformly in $\mathbb{P}_r$ as $N \to \infty$ (proof of (19-94) is similar to Lemma 17.3). Because Assumption 17.18 is valid for any $\mu \to 0$, it guarantees that $\mathbb{E}\{\varepsilon'^T(\tilde{\theta}(z_0), z_0) \varepsilon'(\tilde{\theta}(z_0), z_0)\} = O(N)$. Putting these results in (19-38) finally gives $Q_{\varepsilon[i,j]}(z_0) = I_{rN[i,j]} + O_p(N^{-1})$. As z_0 is uniformly bounded, there exists an N_0 s.t. for any $N \geq N_0$: $\mathbb{E}\{Q_{\varepsilon[i,j]}(z_0)\} = I_{rN[i,j]} + O(N^{-1})$.

(ii) Using $\mathbb{E}\{\delta_\varepsilon(z)\} = 0$, $\text{Cov}(\delta_\varepsilon(z)) = I_{rN}$, and the fact that n_z and z_0 are stochastically independent gives $\text{Cov}(Q_\varepsilon(z_0)\delta_\varepsilon(z)) = \mathbb{E}\{Q_\varepsilon(z_0)\} = O(N^0)$, where the last equality follows from part one. □

Appendix 19.I Proof of the Improved Convergence Rate of the Residuals (Lemma 19.9)

Taylor series expansion of $\varepsilon_{[i]}(\hat{\theta}(z), z)$ at $\tilde{\theta}(z_0), z_0$ gives

$$\varepsilon_{[i]}(\hat{\theta}(z), z) = \varepsilon_{[i]}(\tilde{\theta}(z_0), z_0) + \frac{\partial \varepsilon_{[i]}(\theta, z_0)}{\partial \tilde{\theta}(z_0)} \Delta\theta + \frac{\partial \varepsilon_{[i]}(\tilde{\theta}(z_0), z)}{\partial z_0} n_z$$
$$+ \frac{1}{2} \Delta\theta^T \frac{\partial^2 \varepsilon_{[i]}(\theta, z_1)}{\partial \theta_1^2} \Delta\theta + \Delta\theta^T \frac{\partial^2 \varepsilon_{[i]}(\theta, z_1)}{\partial \theta_1 \partial z_1} n_z \qquad (19\text{-}95)$$

with $\Delta\theta = \hat{\theta}(z) - \tilde{\theta}(z_0)$, $\theta_1 = t_1\hat{\theta}(z) + (1-t_1)\tilde{\theta}(z_0)$, $z_1 = t_1 z + (1-t_1)z_0$, and $t_1 \in [0, 1]$. We analyze each term in the right-hand side of (19-95). Under Assumptions 17.1 ($P = 2$) and 19.5, the first term of (19-95) is mixing of order two (Corollary 16.7) and, hence, it behaves as $v^{-1}\mu O_p(N^0)$. Using (19-31), the sum of the second and third terms of (19-95) becomes

$$(Q_\varepsilon(z_0)\delta_\varepsilon(z))_{[i]} + (v + \mu + \mu\lambda(z_0))O_p(N^{-1/2}) \qquad (19\text{-}96)$$

with $\lambda(z_0) = 1$ for random z_0 and $\lambda(z_0) = 0$ for deterministic z_0. Because $\mathbb{E}\{(Q_\varepsilon(z_0)\delta_\varepsilon(z))_{[i]}\} = 0$ and $\text{var}((Q_\varepsilon(z_0)\delta_\varepsilon(z))_{[i]}) = O(N^0)$ (see Appendix 19.H), we have $(Q_\varepsilon(z_0)\delta_\varepsilon(z))_{[i]} = O_p(N^0)$. Using $\Delta\theta = (v + \mu\lambda(z_0))O_p(N^{-1/2})$ (Theorem 19.2) and Assumption 19.6, it can be seen that the last two terms of (19-95) are, respectively, a $v^{-1}(v + \mu\lambda(z_0))^2 O_p(N^{-1})$ and $(v + \mu\lambda(z_0))O_p(N^{-1/2})$. Putting all these results in (19-95) proves the lemma. □

Appendix 19.J Proof of the Properties of the Sample Correlation of the Residuals (Theorem 19.10)

The proof consists of three parts: part one shows the weak convergence and the convergence rate of $\hat{R}_{\varepsilon\varepsilon}(k)$, part two proves the asymptotic normality of $\hat{R}_{\varepsilon\varepsilon}(k)$, and part three gives an asymptotic expression ($N \to \infty$) for the variance of the truncated sample correlation $\hat{R}_{\varepsilon\varepsilon}(k)$. In the proof, it is essential that $N - |k| = O(N)$. Hence, the results are valid for constant values of k or constant fractions $k = \alpha N$, with α independent of N.

(i) Define the following functions:

$$f_N(\theta, z, k) = \frac{1}{rN - |k|} \sum_{i=1}^{rN-|k|} \varepsilon_{[i]}(\theta, z) \varepsilon_{[i+k]}(\theta, z)$$
$$f_N(\theta, k) = \mathbb{E}\{f_N(\theta, z, k)\}$$

$$f_N'(\theta, z, k) = \frac{1}{rN-|k|}\sum_{i=1}^{rN-|k|}\left(\varepsilon_{[i+k]}(\theta, z)\frac{\partial \varepsilon_{[i]}(\theta, z)}{\partial \theta} + \varepsilon_{[i]}(\theta, z)\frac{\partial \varepsilon_{[i+k]}(\theta, z)}{\partial \theta}\right)$$

with $f_N(\hat{\theta}(z), z, k) = \hat{R}_{\varepsilon\varepsilon}(k)$. Under Assumptions 19.5, 19.6, and 17.1 ($P = 4$) $\varepsilon(\theta, z)$ and $\partial \varepsilon(\theta, z)/\partial \theta_{[j]}$ are jointly mixing of order 4 in $\mathbb{P}_r$. Hence, according to the weak law of large numbers (16-68), $f_N(\theta, z, k)$ and $f_N'(\theta, z, k)$ converge in prob. to their expected values at the rate $O_p(N^{-1/2})$, and $\|f_N'(\theta, z, k)\|_2 = O_p(N^0)$. From Theorem 17.21, it follows that $\hat{\theta}(z)$ converges in prob. to $\tilde{\theta}(z_0)$ at the rate $O_p(N^{-1/2})$. All the conditions of Lemma 17.34 are fulfilled so that $f_N(\hat{\theta}(z), z, k)$ converges in prob. to $f_N(\tilde{\theta}(z_0), k)$ at the rate $O_p(N^{-1/2})$. Using $\varepsilon_{[i]}(\hat{\theta}(z_0), z) = \varepsilon_{[i]}(\tilde{\theta}(z_0), z_0) + \delta_{\varepsilon[i]}(z)$ in $f_N(\tilde{\theta}(z_0), k)$ gives (19-40).

(ii) Now we show that $\hat{R}_{\varepsilon\varepsilon}(k)$ is asymptotically normally distributed. Therefore, it is sufficient to verify that all the conditions of Lemma 17.38 are fulfilled. Under Assumptions 17.1 ($P = \infty$) and 19.5, $\varepsilon(\theta, z)$ is mixing of order infinity. Hence, according the central limit theorem ((16-74), version 4), $f_N(\theta, z, k)$ is asymptotically normally distributed (condition 1 of Lemma 17.38). Condition 2 of Lemma 17.38 is already satisfied (see part one of the proof). Under Assumption 19.6(b) we have $\|f_N''(\theta, z, k)\|_2 = O_p(N^0)$ uniformly in $\mathbb{P}_r$ (condition 3 of Lemma 17.38). The proof is similar to that of $f_N'(\theta, z, k)$ in part one. The assumptions of Theorem 17.29 are satisfied so that $\hat{\theta}(z)$ is asymptotically normally distributed (condition 4 of Lemma 17.38).

(iii) In the absence of model errors (Assumptions 17.9 and 17.10), $\varepsilon(\tilde{\theta}(z_0), z_0) = 0$ and

$$\varepsilon(\tilde{\theta}(z_0), z) = \Delta(\tilde{\theta}(z_0), n_z) = \delta_\varepsilon(z) \qquad (19\text{-}97)$$

where $\delta_\varepsilon(z)$ is defined in Lemma 19.9. Using (19-97), (19-93) becomes

$$\varepsilon_{[i]}(\hat{\theta}(z), z) = \delta_{\varepsilon[i]}(z) + O_p(N^{-1/2}) \qquad (19\text{-}98)$$

where $\delta_{\varepsilon[i]}(z)$ is an $O_p(N^0)$. The asymptotic variance ($N \to \infty$) of the truncated sample correlation $\hat{R}_{\varepsilon\varepsilon}(k)$ is, hence, given by

$$\text{var}(\frac{1}{rN-|k|}\sum_{i=1}^{rN-|k|}\delta_{\varepsilon[i]}(z)\delta_{\varepsilon[i+k]}(z))$$

Under Assumption 19.1, $\delta_{\varepsilon[i]}(z)$ is normally distributed $N_{rN}(0, I_{rN})$, so that the cumulants of $\delta_{\varepsilon[i]}(z)$ of order 3 and higher are zero (see Example 16.2). Using this result together with that of Example 16.36 and $\text{var}(x) = \text{cum}(x, x)$ (see Section 16.1), we find

$$\text{var}(\frac{1}{rN-|k|}\sum_{i=1}^{rN-|k|}\delta_{\varepsilon[i]}(z)\delta_{\varepsilon[i+k]}(z)) = \frac{1}{rN-|k|}(1 + \delta(k)) \qquad \square$$

Appendix 19.K Proof of the Convergence Rate of the Minimum of the Cost Function (Lemma 19.11)

Taylor series expansion with remainder of $V_N(\hat{\theta}(z), z) = N^{-1} V_{\text{Markov}}(\hat{\theta}(z), z)$ at $\tilde{\theta}(z_0)$, z_0 gives

$$V_N(\hat{\theta}(z), z) = V_N(\tilde{\theta}(z_0), z_0) + \frac{\partial V_N(\theta, z_0)}{\partial \tilde{\theta}(z_0)} \Delta\theta + \frac{\partial V_N(\tilde{\theta}(z_0), z)}{\partial z_0} n_z$$

$$+ \frac{1}{2}\Delta\theta^T \frac{\partial^2 V_N(\theta, z_0)}{\partial \tilde{\theta}(z_0)^2} \Delta\theta + \frac{1}{2} n_z^T \frac{\partial^2 V_N(\tilde{\theta}(z_0), z)}{\partial z_0^2} n_z + \Delta\theta^T \frac{\partial^2 V_N(\theta, z)}{\partial \tilde{\theta}(z_0) \partial z_0} n_z \quad (19\text{-}99)$$

$$+ R_1 + R_2 + R_3 + R_4$$

where

$$R_1 = \frac{1}{6}\left(\sum_{i=1}^{n_\theta} \Delta\theta_{[i]} \frac{\partial}{\partial \theta_{1[i]}}\right)^3 V_N(\theta, z_1)$$

$$R_2 = \frac{1}{6}\left(\sum_{i=1}^{sN} n_{z[i]} \frac{\partial}{\partial z_{1[i]}}\right)^3 V_N(\theta_1, z)$$

$$R_3 = \frac{1}{2}\left(\sum_{i=1}^{n_\theta} \Delta\theta_{[i]} \frac{\partial}{\partial \theta_{1[i]}}\right)\left(\sum_{i=1}^{sN} n_{z[i]} \frac{\partial}{\partial z_{1[i]}}\right)^2 V_N(\theta, z)$$

$$R_4 = \frac{1}{2}\left(\sum_{i=1}^{n_\theta} \Delta\theta_{[i]} \frac{\partial}{\partial \theta_{1[i]}}\right)^2\left(\sum_{i=1}^{sN} n_{z[i]} \frac{\partial}{\partial z_{1[i]}}\right) V_N(\theta, z)$$

and $\theta_1 = t_1\hat{\theta}(z) + (1 - t_1)\tilde{\theta}(z_0)$, $z_1 = t_1 z + (1 - t_1)z_0$ with $t_1 \in [0, 1]$. Each term in the right-hand side of (19-99) will be studied.

(i) Under Assumptions 17.1 ($P = 4$) and 17.2(a), the first term $V_N(\tilde{\theta}(z_0), z_0)$ is mixing of order two so that $\text{var}(V_N(\tilde{\theta}(z_0), z_0)) = O(N^{-1})$ and, hence,

$$V_N(\tilde{\theta}(z_0), z_0) = \mu^2 \upsilon^{-2} O_p(N^0) \quad (19\text{-}100)$$

(ii) Using Theorem 19.2, the sum of the second and the third term can be written as

$$\frac{1}{N} \varepsilon^T(\tilde{\theta}(z_0), z_0) Q_\varepsilon(z_0) \delta_\varepsilon(z) + V_N'(\tilde{\theta}(z_0), z_0)(\mu\upsilon + \upsilon^2 + \mu\lambda(z_0))O_p(N^{-1/2}) \quad (19\text{-}101)$$

For deterministic z_0 we have $V_N'(\tilde{\theta}(z_0), z_0) = V_N'(\tilde{\theta}(z_0)) = 0$ (see (19-26)). For random z_0, $V_N'(\tilde{\theta}(z_0), z_0)$ converges in prob. to $V_N'(\tilde{\theta}(z_0))$ at the rate $O_p(N^{-1/2})$ (Lemma 17.17). $V_N'(\tilde{\theta}(z_0)) = 0$ by definition of $\tilde{\theta}(z_0)$, so that $V_N'(\tilde{\theta}(z_0), z_0) = \mu\upsilon^{-2} O_p(N^{-1/2})$. Putting these results in (19-101) gives

Section 19.9 ■ Appendixes

$$\frac{1}{N}\varepsilon^T(\tilde{\theta}(z_0), z_0)Q_\varepsilon(z_0)\delta_\varepsilon(z) + \frac{1}{N}\mu^2 v^{-2}\lambda(z_0)O_p(N^0) \quad (19\text{-}102)$$

with $\lambda(z_0) = 1$ for random z_0 and $\lambda(z_0) = 0$ for deterministic z_0. Note also that $\varepsilon^T(\tilde{\theta}(z_0), z_0)Q_\varepsilon(z_0) = \varepsilon^T(\tilde{\theta}(z_0), z_0)$ for deterministic z_0.

(iii) The second-order derivative in the fourth term of (19-99) can be written as

$$\frac{\partial^2 V_N(\theta, z_0)}{\partial \tilde{\theta}(z_0)^2} = \frac{1}{N}\left(\frac{\partial \varepsilon(\theta, z_0)}{\partial \tilde{\theta}(z_0)}\right)^T\left(\frac{\partial \varepsilon(\theta, z_0)}{\partial \tilde{\theta}(z_0)}\right) + \frac{1}{N}\sum_{k=1}^{rN}\varepsilon_{[k]}(\tilde{\theta}(z_0), z_0)\frac{\partial^2 \varepsilon_{[k]}(\theta, z_0)}{\partial \tilde{\theta}(z_0)^2} \quad (19\text{-}103)$$

Under Assumptions 19.5, 19.6, and 17.1 ($P = 4$), both terms in the right-hand side of (19-103) converge in prob. to their expected value (proof: similar to Lemma 17.17). Assumption 17.18, which is by assumption valid for any $\mu \to 0$, guarantees that both terms behave as an $O_p(N^0)$. We conclude that the first and second terms are, respectively, an $v^{-2}O_p(N^0)$ and $\mu v^{-2}O_p(N^0)$.

(iv) The sixth term of (19-99) can be written as

$$\Delta\theta^T(g(z) + f_N(\tilde{\theta}(z_0), z))$$

$$g(z) = \frac{1}{N}\varepsilon'^T(\tilde{\theta}(z_0), z_0)\delta_\varepsilon(z) \quad (19\text{-}104)$$

$$f_N(\tilde{\theta}(z_0), z) = \frac{1}{N}\sum_{k=1}^{rN}\varepsilon_{[k]}(\tilde{\theta}(z_0), z_0)\left(\frac{\partial \delta_{\varepsilon[k]}(z)}{\partial \tilde{\theta}(z_0)}\right)^T$$

where $\delta_\varepsilon(z)$ is defined in Lemma 19.9. Because $\mathbb{E}\{g(z)\} = 0$ and

$$\text{Cov}(g(z)) = \frac{1}{N^2}\mathbb{E}\left\{\left(\frac{\partial \varepsilon(\theta, z_0)}{\partial \tilde{\theta}(z_0)}\right)^T\left(\frac{\partial \varepsilon(\theta, z_0)}{\partial \tilde{\theta}(z_0)}\right)\right\} = O(N^{-1})$$

we conclude that $g(z) = v^{-1}O_p(N^{-1/2})$. Under Assumptions 19.5, 19.6, and 17.1 ($P = 4$), $\varepsilon(\theta, z_0)$ and $\partial \delta_\varepsilon(z)/\partial \theta_{[i]}$ are jointly mixing of order 4, so that $f_N(\theta, z)$ converges in prob. to its expected value $f_N(\theta)$ at the rate $O_p(N^{-1/2})$, uniformly in $\mathbb{P}_r$ (proof: similar to Lemma 17.17). Because $f_N(\tilde{\theta}(z_0)) = 0$, it follows that $f_N(\tilde{\theta}(z_0), z)$ is a $\mu v^{-1}O_p(N^{-1/2})$. Combining these results with Theorem 19.2 gives the following expression for the sum of the fourth, fifth, and sixth terms in (19-99):

$$\frac{1}{2N}\delta_\varepsilon^T(z)Q_\varepsilon(z_0)\delta_\varepsilon(z) + \frac{1}{N}(\mu + v + \mu v^{-2}(\mu + v)\lambda(z_0))O_p(N^0) \quad (19\text{-}105)$$

(v) The term R_1 is bounded above by

$$|R_1| \leq \sum_{i,j,k=1}^{n_\theta}|\Delta\theta_{[i]}\Delta\theta_{[j]}\Delta\theta_{[k]}|\frac{(\|z_1\|_2^2 + 1)}{N}\left\|\frac{\partial^3 W_N(\theta)}{\partial \theta_{1[i]}\partial \theta_{1[j]}\partial \theta_{1[k]}}\right\|_2 \quad (19\text{-}106)$$

where, by Assumption 17.20 ($\theta_1 \in \mathbb{P}_r$), the 2-norm of the third-order derivative of $W_N(\theta)$ w.r.t. θ is an $\upsilon^{-2}O_p(N^0)$. Under Assumption 17.1 ($P = 4$), $\|z\|_2^2/N$ and $\|z_0\|_2^2/N$ converge both in prob. to their expected values, which are both an $O(N^0)$ (weak law of large numbers, see Section 16.9). Hence, $\|z\|_2^2/N$ and $\|z_0\|_2^2/N$ are both an $O_p(N^0)$. This is also true for $\|z_1\|_2^2/N$, because $z_1 = t_1 z + (1 - t_1)z_0$ with $t_1 \in [0, 1]$. Combining these results with Theorem 19.2 gives

$$|R_1| \leq \frac{1}{N} \upsilon^{-2}(\upsilon + \mu\lambda(z_0))^3 O_p(N^{-1/2}) \qquad (19\text{-}107)$$

(vi) Because $V_N(\theta, z)$ is a quadratic function of z, it follows directly that

$$R_2 = 0 \qquad (19\text{-}108)$$

(vii) The term R_3 can be written as

$$R_3 = v_N'(\theta_1, n_z)\Delta\theta \qquad (19\text{-}109)$$

with $v_N(\theta, n_z) = (2N)^{-1}\Delta^T(\theta, n_z)\Delta(\theta, n_z)$ and $\Delta(\theta, n_z) = \Lambda(\theta)M_1(\theta)n_z$. We will verify that all the conditions of Lemma 17.34 are satisfied for $v_N'(\theta_1, n_z)$. Under Assumptions 17.1 ($P = 4$) and 17.16, $v_N'(\theta, n_z)$ converges in prob. to its expected value $\mathbb{E}\{v_N'(\theta, n_z)\} = \partial(rN/2)/\partial\theta = 0$ at the rate $O_p(N^{-1/2})$ uniformly in $\mathbb{P}_r$ (Lemma 17.17). $\hat{\theta}(z)$ converges in prob. to $\tilde{\theta}(z_0)$ at the rate $O_p(N^{-1/2})$. This is also true for θ_1 because $\theta_1 = t_1\hat{\theta}(z) + (1 - t_1)\tilde{\theta}(z_0)$ with $t_1 \in [0, 1]$. Under Assumptions 19.1 ($P = 4$) and 17.16, we have

$$\|v_N''(\theta, n_z)\|_2 \leq \frac{\|n_z\|_2^2}{N} \left\| \frac{\partial^2 W_N(\theta)}{\partial\theta_{[i]}\partial\theta_{[j]}\partial\theta_{[k]}} \right\|_2 = O_p(N^0)$$

uniformly in $\mathbb{P}_r$. We conclude from Lemma 17.34 that $v_N'(\theta_1, n_z) = \upsilon^0 O_p(N^{-1/2})$. Combining this result with Theorem 19.2 gives

$$R_3 = \frac{1}{N}(\upsilon + \mu\lambda(z_0))O_p(N^0) \qquad (19\text{-}110)$$

(viii) The term R_4 can be written as

$$R_4 = \sum_{i,j=1}^{n_\theta} \Delta\theta_{[i]}\Delta\theta_{[j]}\frac{\partial^2 F(\theta_1, z)}{\partial\theta_{1[i]}\partial\theta_{1[j]}} \qquad (19\text{-}111)$$

with $F(\theta, z) = (2N)^{-1}\Delta^T(\theta, n_z)\varepsilon(\theta, z_1)$. z_1 is mixing of order four because this is the case for z and z_0. Therefore, under Assumptions 17.1 ($P = 4$) and 17.16, $F''(\theta, z)$ converges uniformly in prob. to its expected value $F''(\theta)$, which is an $O(N^0)$ (proof similar to Lemma 17.17). Because θ_1 converges in prob. to $\tilde{\theta}(z_0)$ it follows that $F''(\theta_1, z)$ converges in prob. to $F''(\tilde{\theta}(z_0)) = O(N^0)$ (Lemma 17.31), so that $F''(\theta_1, z) = \upsilon^{-1}O_p(N^0)$. Combining this result with Theorem 19.2 gives

$$R_4 = N^{-1} \upsilon^{-1}(\upsilon + \mu\lambda(z_0))^2 O_p(N^0) \qquad (19\text{-}112)$$

(ix) Putting (19-100), (19-102), (19-105), (19-107), (19-108), (19-110), and (19-112) into (19-99), taking into account that $V_N(\hat{\theta}(z), z) = V_{\text{Markov}}(\hat{\theta}(z), z)/N$, proves (19-44).

Appendix 19.L Proof of the Properties of the Cost Function (Theorem 19.12)

To prove the asymptotic normality of $V_N(\hat{\theta}(z), z)$, we show that all conditions of Lemma 17.38 are satisfied. In fact, it is sufficient to show the asymptotic normality of $V_N(\theta, z)$ because all the other conditions are satisfied under the assumptions of Lemma 19.11. $\varepsilon_{[k]}^2(\theta, z)$ is mixing of order infinity under Assumptions 17.1 and 19.5 and, therefore, the asymptotic normality of $V_N(\theta, z)$ follows directly from version 4 of the central limit theorem (see Section 16.10).

Using $\delta_\varepsilon^T(z) Q_\varepsilon(z_0) \delta_\varepsilon(z) = \text{trace}(Q_\varepsilon(z_0) \delta_\varepsilon(z) \delta_\varepsilon^T(z))$ and Assumption 17.1 ($P = 2$), we find

$$\mathbb{E}\{L_{\text{Markov}}(\tilde{\theta}(z_0), z_0)\} = \mathbb{E}\{V_{\text{Markov}}(\tilde{\theta}(z_0), z_0)\} + \mathbb{E}\{\text{trace}(Q_\varepsilon(z_0))\}$$

$Q_\varepsilon(z_0)$ is an idempotent matrix of rank $rN - n_\theta$ so that $\text{trace}(Q_\varepsilon(z_0)) = rN - n_\theta$. Because $\delta_\varepsilon(z) \in N^{rN}(0, I_{rN})$ under Assumption 19.1, we have $\delta_\varepsilon^T(z) Q_\varepsilon(z_0) \delta_\varepsilon(z) \in \chi^2(rN - n_\theta)$ (see Exercise 16.10), so that $\text{var}(\delta_\varepsilon^T(z) Q_\varepsilon(z_0) \delta_\varepsilon(z)) = 2(rN - n_\theta)$ (Stuart and Ord, 1987). Formula (19-46) follows directly from this result. For deterministic z_0, we have $\text{var}(V_{\text{Markov}}(\tilde{\theta}(z_0), z_0)) = 0$ and $\varepsilon^T(\tilde{\theta}(z_0), z_0) Q_\varepsilon(z_0) = \varepsilon^T(\tilde{\theta}(z_0), z_0)$, so that (19-46) reduces to (19-47). □

Appendix 19.M Model Selection Criteria

For model (19-1), the AIC and MDL criteria have the form

$$-\ln f_z(z, x, \theta) + g(k, m) \qquad (19\text{-}113)$$

with $f_z(z, x, \theta)$ the likelihood function, x the number of unknown, independent variables in z_p (see Section 19.3), k the number of free parameters in the model to get the estimates $\hat{\theta}(z)$ and $\hat{x}$, m the number of independent noisy measurements, $g(k, m) = k$ for AIC, and $g(k, m) = 0.5 k \ln m$ for MDL. The number of free parameters equals the total number of identifiable parameters $\dim(\hat{\theta}(z)) + \dim(\hat{x})$, so that $k = n_\theta + \dim(\hat{x})$. The number of independent noisy measurements equals the number of measurements $r_c = \text{rank}(C_{n_z})$ lying in the regular space of C_{n_z}. Taking into account that $\dim(\hat{x}) = r_c - rN$ is constant over the model set $\mathbb{M}$, (19-113) reduces for Gaussian-distributed errors n_z to (19-50) and (19-51). □

Appendix 19.N Proof of the Modified AIC and MDL Criteria (Theorem 19.15)

The proof follows the same lines of Schoukens et al. (2002). The AIC-rule (19-113) is based on the Kullback-Leibler distance between two distributions. When the cost function is chosen to be the negative log likelihood function, it reduces to a simple correction term on the weighted least squares cost function in case of normally distributed noise n_z (see (19-50)). An alternative interpretation of the AIC rule is to consider it as a prediction of the value of the cost function based on an infinite amount of data starting from the actual value of the cost function based on the finite data set. For $N \gg n_\theta$ the AIC-rule takes then the form (see Ljung, 1999, Theorem 16.1),

$$\mathbb{E}\{V_{\text{Markov}}(\hat{\theta}(z), z)\} + \text{tr}(V_{\text{Markov}}''(\tilde{\theta}(z_0))\text{Cov}(\delta_\theta(z))) \qquad (19\text{-}114)$$

In the absence of model errors $\text{Cov}(\delta_\theta(z)) \approx V_{\text{Markov}}''^{-1}(\tilde{\theta}(z_0))$ for large signal-to-noise ratios (see (19-33)), and (19-114), where the expected value of the cost function is replaced by its actual value, reduces to the AIC-rule (19-50). In the presence of model errors $\text{Cov}(\delta_\theta(z))$ is given by (17-18)

$$\begin{aligned}\text{Cov}(\delta_\theta(z)) &= V_{\text{Markov}}''^{-1}(\tilde{\theta}(z_0)) \, Q(\tilde{\theta}(z_0)) V_{\text{Markov}}''^{-1}(\tilde{\theta}(z_0)) \\ Q(\tilde{\theta}(z_0)) &= \mathbb{E}\{V_{\text{Markov}}'^T(\tilde{\theta}(z_0), z) V_{\text{Markov}}'(\tilde{\theta}(z_0), z)\}\end{aligned} \qquad (19\text{-}115)$$

Using Assumption 19.14 we will approximate now $V_{\text{Markov}}''(\tilde{\theta}(z_0))$ and $Q(\tilde{\theta}(z_0))$ for large signal-to-noise ratios ($\upsilon \to 0$) and small model errors ($\mu \to 0$).

The second order derivative of the cost function can be approximated as

$$V_{\text{Markov}}''(\tilde{\theta}(z_0)) = \mathbb{E}\{\sum_{i=1}^{rN} \varepsilon_{[i]}'^T \varepsilon_{[i]}' + \sum_{i=1}^{rN} \varepsilon_{[i]} \varepsilon_{[i]}''\} \approx \sum_{i=1}^{rN} \mathbb{E}\{\varepsilon_{[i]}'^T \varepsilon_{[i]}'\} \qquad (19\text{-}116)$$

for $\upsilon, \mu \to 0$. Using $\varepsilon_{[i]} \approx v_{[i]} + m_{[i]}$ for $\upsilon, \mu \to 0$ (see (19-38)), $Q(\tilde{\theta}(z_0))$ in (19-115) becomes

$$\begin{aligned}Q(\tilde{\theta}(z_0)) &= \sum_{i=1}^{rN} \sum_{j=1}^{rN} \mathbb{E}\{\varepsilon_{[i]}'^T \varepsilon_{[j]}' \varepsilon_{[i]} \varepsilon_{[j]}\} \\ &\approx \sum_{i=1}^{rN} \sum_{j=1}^{rN} \mathbb{E}\{\varepsilon_{[i]}'^T \varepsilon_{[j]}'(v_{[i]} + m_{[i]})(v_{[j]} + m_{[j]})\}\end{aligned} \qquad (19\text{-}117)$$

Under Assumption 19.14 (19-117) is further simplified as

$$Q(\tilde{\theta}(z_0)) \approx (m^2 + 1) \sum_{i=1}^{rN} \mathbb{E}\{\varepsilon_{[i]}'^T \varepsilon_{[i]}'\} \qquad (19\text{-}118)$$

Combining (19-115), (19-116), and (19-118) gives

$$\text{tr}(V_{\text{Markov}}''(\tilde{\theta}(z_0))\text{Cov}(\delta_\theta(z))) \approx \text{tr}((m^2 + 1)I_{n_\theta}) = (m^2 + 1)n_\theta \qquad (19\text{-}119)$$

For $\upsilon, \mu \to 0$ and $N \gg n_\theta$, the following expression is found for the expected value of the cost function

$$\mathbb{E}\{V_{\text{Markov}}(\hat{\theta}(z), z)\} \approx 0.5rNm^2 + rN - 0.5n_\theta \approx 0.5rN(m^2 + 1) \qquad (19\text{-}120)$$

(see Lemma 19.11 and Theorem 19.12). Using (19-119) and (19-120), (19-114) becomes

$$\mathbb{E}\{V_{\text{Markov}}(\hat{\theta}(z), z)\}(1 + 2n_\theta/(rN)) \approx V_{\text{Markov}}(\hat{\theta}(z), z)(1 + 2n_\theta/(rN)) \qquad (19\text{-}121)$$

which proves (19-52).

Without formal derivation, the MDL-rule (19-53) is proposed solely using the analogy with the AIC-rule.

20

Identification of Invariants of (Over)Parameterized Models

Abstract: This chapter deals with the identification of invariants of (over)parameterized models. First, it is shown that the generalized Cramér-Rao lower bound on the estimate of invariants of (over)parameterized models is independent of the particular (over)parameterization chosen and equals that of the identifiable form. This result is useful for (asymptotically) efficient estimators. Next, it is shown that a certain class of estimates of invariants of (over)parameterized models are with probability one, independent of the particular (over)parameterization chosen. The result is nonasymptotic and the estimators considered minimize a cost function that is invariant with respect to the same parameter transformation as the overparameterized model.

20.1 INTRODUCTION

In many identification problems, one is faced with the choice of the parameterization of a model out of a large number of possibilities (see, for example, Guidorzi, 1975; Van Overbeek and Ljung, 1982). Some of these representations contain a redundant number of parameters and lead to the so-called overparameterized models. Such models result in singular Fisher information matrices (Shapiro, 1986). This situation is often encountered in practical parameter estimation problems.

Example 20.1 (Rational Transfer Function Model for SISO Systems): Consider the identification of the numerator and denominator coefficients of a rational transfer function model of a single input, single output (SISO) continuous-time system

$$G(s, \theta) = \frac{B(s, \theta)}{A(s, \theta)} = \frac{\sum_{r=0}^{n_b} b_r s^r}{\sum_{r=0}^{n_a} a_r s^r} \tag{20-1}$$

where $\theta^T = (a_0, a_1, ..., a_{n_a}, b_0, b_1, ..., b_{n_b})$. Transfer function model (20-1) is overparameterized because θ is unidentifiable: $G(s, \lambda\theta) = G(s, \theta)$ for any $\lambda \in \mathbb{R}_0$. Assuming that the true model order is (n_a, n_b), the dimension of the null space of the corresponding Fisher information matrix (16-85) equals 1. Identifiable parameterizations ψ are obtained by fixing one coefficient of the numerator or denominator, for example, $a_{n_a} = 1$ for a monic denominator polynomial. Invariants of model (20-1) are, for example, the poles and zeros of the transfer function or the value of the transfer function itself. □

Example 20.2 (State Space Models for Multivariable Systems): Consider the identification of a proper multivariable discrete-time system parameterized by its state space representation (A, B, C, D)

$$G(z^{-1}, \theta) = z^{-1}C(I_n - z^{-1}A)^{-1}B + D \qquad (20\text{-}2)$$

with n the order of the state space model, $A \in \mathbb{R}^{n \times n}$, $B \in \mathbb{R}^{n \times n_u}$, $C \in \mathbb{R}^{n_y \times n}$, and $D \in \mathbb{R}^{n_y \times n_u}$. The overparameterized model parameter θ is related to the (A, B, C, D) matrices as

$$\theta^T = [\text{vec}^T(A) \ \text{vec}^T(B) \ \text{vec}^T(C) \ \text{vec}^T(D)] \qquad (20\text{-}3)$$

where vec(.) transforms a matrix into a column vector by stacking the columns of the matrix on top of each other. Transfer function (20-2) is invariant w.r.t. a regular transformation $T \in \mathbb{R}^{n \times n}$: replacing (A, B, C, D) by $(TAT^{-1}, TB, CT^{-1}, D)$ in (20-2), with $\det(T) \neq 0$, leaves $G(z^{-1}, \theta)$ unchanged. Hence, n^2 dependences exist between the entries of θ. Assuming that the true model order is n, the dimension of the null space of the Fisher information matrix equals n^2. Identifiable parameterizations ψ are obtained by constraining the matrices (A, B, C) (Van Overbeek and Ljung, 1982). Invariants of model (20-2) are, for example, the eigenvalues of A. □

First, we define the considered (over)parameterized models and their invariants. Next, the Cramér-Rao lower bound on the estimates of the invariants is analyzed. Further, the finite sample behavior of a certain class of estimates of invariants is studied. Finally, the chapter concludes with a numerical example of identification methods whose estimates are, respectively, dependent on and independent of the particular (over)parameterization chosen. Although throughout this chapter the theory is mainly illustrated on the identification of continuous-time single input, single output systems, it is also valid for discrete-time, multivariable, and nonlinear systems.

20.2 (OVER)PARAMETERIZED MODELS AND THEIR INVARIANTS

The system model is described as a general vector function $M(\theta, z_0)$ of the true signal $z_0 \in \mathbb{R}^N$ and the overparameterized model parameters $\theta \in \mathbb{R}^{n_\theta}$ with $n_\theta < N$. The model $M(\theta, z_0)$ is defined for every $\theta \in \mathbb{D}_\theta$, a subset of $\mathbb{R}^{n_\theta}$. The complement of $\mathbb{D}_\theta$ in $\mathbb{R}^{n_\theta}$ is $\mathbb{S}_\theta$, the set of singular points. $\mathbb{S}_\theta = \mathbb{R}^{n_\theta} \setminus \mathbb{D}_\theta$ has topological dimensions less than n_θ ($\dim(\mathbb{S}_\theta) < n_\theta$). If no system model errors are present, then $M(\theta_0, z_0) = 0$ with θ_0 the true model parameters. Note that for overparameterized models, θ_0 is not a single point but a subspace of $\mathbb{R}^{n_\theta}$.

In practice, only noisy observations $z = z_0 + n_z$ of the true signal z_0 are available. If a parametric noise model for the observation noise n_z is identified, then the system model $M(\theta, z_0)$ must be extended with the noise model. To simplify the notations, the discussion is, without any loss of generality, limited to the case without a parametric noise model.

Section 20.2 ■ (Over)Parameterized Models and Their Invariants

Assumption 20.3 (Invariance Model): The overparameterized model $M(\theta, z_0)$ is invariant with respect to a parameter transformation $g(\theta, \lambda) \in \mathbb{R}^{n_\theta}$ with $\lambda \in \mathbb{R}^{n_\lambda}$ and $0 < n_\lambda < n_\theta$: for any $\theta \in \mathbb{D}_\theta$ and $\lambda \in \mathbb{D}_\lambda \subset \mathbb{R}^{n_\lambda}$, with $\dim(\mathbb{R}^{n_\lambda} \setminus \mathbb{D}_\lambda) < n_\lambda$, we have $\mathrm{rank}(\partial g(\theta, \lambda)/\partial \lambda) = n_\lambda$ and $M(g(\theta, \lambda), z_0) = M(\theta, z_0)$.

Assumption 20.4 (Identifiable Models): The model $M(\theta, z_0)$ can be parameterized in β overlapping identifiable parameter sets $\psi_k \in \mathbb{R}^{n_\psi}$, satisfying

1. $\forall \theta \in \mathbb{D}_k, \exists \lambda_k(\theta) \in \mathbb{D}_\lambda : h_k(\psi_k) = g(\theta, \lambda_k(\theta))$
2. $\mathbb{D}_\theta = \cup_{k=1}^{\beta} \mathbb{D}_k$
3. $\dim(\mathbb{D}_k) = n_\theta$ and $\dim(\mathbb{D}_k \setminus \mathbb{D}_l) < n_\theta$

with $n_\psi = n_\theta - n_\lambda$ and $k, l = 1, 2, \ldots, \beta$. The function $\psi_k(\theta)$ and its derivative w.r.t. θ, $\psi_k'(\theta)$, are continuous in $\mathbb{D}_k$. $\psi_k'(\theta)$ has full rank n_ψ in $\mathbb{D}_k$.

Note that Assumptions 20.3 and 20.4 define a manifold with boundary given by an equivalence relationship. Assumptions 20.4 (1) and (2) guarantee that $M(\theta, z_0)$ can be parameterized in at least one identifiable parameter set ψ_k, $k = 1, 2, \ldots, \beta$, for any value of $\theta \in \mathbb{D}_\theta$. Assumption 20.4 (3) implies that the parameterizations are overlapping: each identifiable parameter set ψ_k can represent any model $M(\theta, z_0)$, except those corresponding to θ-values lying in some lower dimensional ($< n_\theta$) subspaces of $\mathbb{D}_\theta$.

Definition 20.5: An invariant of the model $M(\theta, z_0)$ is each model-related quantity $I(\theta)$ that is invariant w.r.t. the same parameter transformation $g(\theta, \lambda)$ as the model itself: $I(g(\theta, \lambda)) = I(\theta)$ (see Assumption 20.3).

The following two examples show that Assumptions 20.3 and 20.4 are satisfied in many practical identification problems.

Example 20.6 (Rational Transfer Function Model): Consider transfer function model (20-1) of Example 20.1 at angular frequencies ω_f, $f = 1, 2, \ldots, N$. Clearly, $G(j\omega_f, \lambda\theta) = G(j\omega_f, \theta)$. If, for example, coefficient a_0 is fixed to one, then the resulting identifiable parameter set ψ_0 can describe all models of the form (20-1), except those for which $a_0 = 0$. If a_1 is fixed to one, then the resulting identifiable set ψ_1 covers the subspace $\{a_0 = 0\}$ but cannot describe models with $a_1 = 0$. These observations are now stated more formally in the framework of Assumptions 20.3 and 20.4.

Assumption 20.3 is valid with

$$M(\theta, z_0) = \left[G(j\omega_1, \theta) - G_0(j\omega_1) \; G(j\omega_2, \theta) - G_0(j\omega_2) \; \ldots \; G(j\omega_N, \theta) - G_0(j\omega_N)\right]^T$$

$z_0 = [G_0(j\omega_1) \; G_0(j\omega_2) \; \ldots \; G_0(j\omega_N)]^T$, $G_0(j\omega)$ the true frequency response function, $n_\theta = n_a + n_b + 2$, $g(\theta, \lambda) = \lambda\theta$, $n_\lambda = 1$, $\mathbb{D}_\lambda = \mathbb{R}_0$, and

$$\mathbb{S}_\theta = \{\theta | \exists f \in \{1, 2, \ldots, N\} \text{ s.t. } G(j\omega_f, \theta) = \infty \text{ or } 0/0\}$$

Assumption 20.4 is valid with $\beta = n_\theta$, $n_\psi = n_\theta - 1$, $\lambda_k(\theta) = 1/\theta_k$,

$$h_k^T(\psi_k) = [\psi_{k[1]} \; \ldots \; \psi_{k[k-1]} \; 1 \; \psi_{k[k+1]} \; \ldots \; \psi_{k[n_\psi]}]$$

and $\mathbb{D}_k = \mathbb{D}_\theta \setminus \{\theta_{[k]} = 0\}$. The relationship between the identifiable parameters ψ_k and the overparameterized set θ is given by

$$\psi_k^T(\theta) = \frac{1}{\theta_{[k]}}\left[\theta_{[1]} \; \cdots \; \theta_{[k-1]} \; \theta_{[k+1]} \; \cdots \; \theta_{[n_\theta]}\right]$$

□

Example 20.7 (State Space Model): Consider transfer function model (20-2) at the frequencies $z_f = e^{2\pi j f/N}$, $f = 0, 1, \ldots, N-1$. The model $M(\theta, z_0)$ can be defined in exactly the same way as in Example 20.6. Assumption 20.3 is valid with $\lambda = \text{vec}(T)$, $n_\lambda = n^2$, and $\mathbb{D}_\lambda = \mathbb{R}^{n_\lambda} \setminus \{\lambda \,|\, \det(T) = 0\}$. Applying property (15-38) of the Kronecker product to (20-3), where (A, B, C, D) is replaced by $(TAT^{-1}, TB, CT^{-1}, D)$, gives an explicit expression for $g(\theta, \lambda)$

$$g(\theta, \lambda) = \text{diag}(T^{-T} \otimes T, I_{n_u} \otimes T, T^{-T} \otimes I_{n_y}, I_{n_y n_u})\theta$$

with diag(.) a block diagonal matrix. The existence and the properties of overlapping identifiable parameterizations, as required in Assumption 20.4, have been shown for linear multivariable systems in Hazewinkel (1977), Delchamps and Byrnes (1982), and Delchamps (1985), while Van Overbeek and Ljung (1982) discuss the numerical aspects when switching from one identifiable form to another.

□

20.3 CRAMÉR-RAO LOWER BOUND FOR INVARIANTS OF (OVER)PARAMETERIZED MODELS

Consider the identification of a particular model $M(\theta, z_0)$ using noisy observations $z \in \mathbb{R}^N$, and assume that the true model belongs to the considered model set. The Cramér-Rao lower bounds on the covariance matrix of an estimator $\hat{I}(z)$ using (over)parameterization θ and ψ_k are given, respectively, by (see Theorem 16.18)

$$\text{Cov}(\hat{I}(\hat{\theta}(z))) \geq \left(\frac{\partial I_0}{\partial \theta_0} + \frac{\partial b_I}{\partial \theta_0}\right) Fi^+(\theta_0) \left(\frac{\partial I_0}{\partial \theta_0} + \frac{\partial b_I}{\partial \theta_0}\right)^H \quad (a)$$

$$\text{Cov}(\hat{I}(h_k(\hat{\psi}_k(z)))) \geq \left(\frac{\partial I_0}{\partial \psi_{k0}} + \frac{\partial b_I}{\partial \psi_{k0}}\right) Fi^{-1}(\psi_{k0}) \left(\frac{\partial I_0}{\partial \psi_{k0}} + \frac{\partial b_I}{\partial \psi_{k0}}\right)^H \quad (b)$$

(20-4)

with $I_0 = I(\theta_0)$, $b_I = \mathbb{E}\{\hat{I}(z)\} - I_0$ the bias that might be present in the estimate, $Fi(\theta_0)$ the singular Fisher information matrix of the overparameterized parameter vector θ_0, and $Fi(\psi_{k0})$ the regular Fisher information matrix of the identifiable parameter vector ψ_{k0}. Because we will compare (20-4a) with (20-4b), the Fisher information matrix $Fi(\theta_0)$ should be constrained to exclude the $\hat{\theta}(z)$-values lying in the lower dimensional subspace $\mathbb{R}^{n_\theta} \setminus \mathbb{D}_k$. In Gorman and Hero (1990) it has been proved under some suitable regularity conditions that the constrained Cramér-Rao lower bound equals the unconstrained case if the constraints are not active in θ_0. We conclude that we can compare (20-4a) with (20-4b) if $\theta_0 \notin \mathbb{R}^{n_\theta} \setminus \mathbb{D}_k$.

Theorem 20.8 (Cramér-Rao Bound of Overparameterized Models): Under Assumptions 20.3 and 20.4, the Cramér-Rao lower bounds (20-4a) and (20-4b) of any estimator $\hat{I}(z)$ of an invariant $I(\theta)$ of the model $M(\theta, z_0)$ are independent of the particular (over)parameterization $(\theta) \; \psi_k$ chosen.

Proof. See Appendix 20.A.

□

The theorem has been proved by comparing an overparameterized parameter set with an identifiable parameter set. By applying the same reasoning twice, the conclusions are also valid when comparing two different overparameterized parameter sets.

Intuitively, the theorem can be understood as follows. The null space of the Fisher information matrix spanned by the redundant model parameters is not affected by the noise. Hence, the redundant parameters will not increase the variance of identifiable model parameters.

20.4 ESTIMATES OF INVARIANTS OF (OVER)PARAMETERIZED MODELS – FINITE SAMPLE RESULTS

Consider the identification of a particular model $M(\theta, z_0)$ using noisy observations $z \in \mathbb{R}^N$ of the true signal z_0. The basic question now arises whether estimates of invariants $\hat{I}(\hat{\theta}(z))$, $\hat{I}(h_k(\hat{\psi}_k(z)))$, of model $M(\theta, z_0)$, calculated from estimates $\hat{\theta}(z)$, $\hat{\psi}_k(z)$, depend on the particular (over)parameterization θ, ψ_k. In general, the answer is yes: not only each realization but also the statistical properties (bias, uncertainty, ...) of the estimated invariants strongly depend on this choice (De Moor et al., 1994). In this section we show that estimators whose corresponding (equivalent) cost functions are invariant with respect to the same parameter transformation as the overparameterized model lead to estimates of invariants that *do not* depend on this choice. While in Section 20.3 it has been proved that the uncertainty of (asymptotically) efficient estimators is independent of the (over)parameterization θ, ψ_k, the result presented in this section is valid for any finite sample property (the distribution function, the sample mean, the sample variance, ..., and if the moments exist, the bias, variance, ...) of the estimated invariants and is not restricted to the class of (asymptotically) efficient estimators.

20.4.1 The Estimators

The estimators considered in this section minimize a cost function $V(\theta, z)$ that is defined for any $\theta \in \mathbb{D}_\theta$, a subspace of $\mathbb{R}^{n_\theta}$, and any $z \in \mathbb{D}_z$, a subspace of $\mathbb{R}^N$. The complements of $\mathbb{D}_\theta$ in $\mathbb{R}^{n_\theta}$ and $\mathbb{D}_z$ in $\mathbb{R}^N$ are, respectively, $\mathbb{S}_\theta$ and $\mathbb{S}_z$, the sets of singular points. $\mathbb{S}_\theta = \mathbb{R}^{n_\theta} \setminus \mathbb{D}_\theta$ and $\mathbb{S}_z = \mathbb{R}^N \setminus \mathbb{D}_z$ have topological dimensions less than, respectively, n_θ and N.

Assumption 20.9 (Invariance Cost Function): The (equivalent) cost function $V(\theta, z)$ minimized by the estimator is invariant with respect to the same parameter transformation $g(\theta, \lambda)$ as the model itself (see Assumption 20.3): $V(g(\theta, \lambda), z) = V(\theta, z)$, for any $\lambda \in \mathbb{D}_\lambda$, $\theta \in \mathbb{D}_\theta$, and $z \in \mathbb{D}_z$.

Very often the cost function is a function of some transformed form of the system model $M(\theta, z_0)$ and, therefore, Assumption 20.9 is, in general, not true. This is illustrated in the following example.

Example 20.10 (Frequency Domain Identification): Consider the identification of model (20-1) starting from frequency response function measurements $G(j\omega_f)$, $f = 1, 2, ..., N$. The linear least squares estimate minimizes (see Section 9.8)

$$V_{\text{LS}}(\theta, z) = \sum_{f=1}^{N} |A(j\omega_f, \theta) G(j\omega_f) - B(j\omega_f, \theta)|^2 \qquad (20\text{-}5)$$

w.r.t. θ. Because $V_{LS}(\lambda\theta, z) = \lambda^2 V_{LS}(\theta, z)$, it follows directly that Assumption 20.9 is not fulfilled. Often, the frequency response function $G(j\omega_f)$ is obtained as the ratio of the output to the input DFT spectra $G(j\omega_f) = Y(f)/U(f)$. The cost function is infinitely large for $U(f) = 0$ and, hence, $\mathbb{S}_z = \{z | U(f) = 0\}$. □

For each particular observation $z \in \mathbb{D}_z$, θ-values in $\mathbb{S}_\theta$ may exist for which the cost function $V(\theta, z)$ is not defined or infinitely large. At these singular points, the first- and second-order derivative of the cost function w.r.t. the model parameters either do not exist or are rank deficient. These θ-values should, therefore, be excluded during the minimization of $V(\theta, z)$. This can be done by introducing a regularization parameter μ in the cost function.

Assumption 20.11 (Regularized Cost Function): The cost $V(\theta, z)$ can be regularized as $V(\theta, z, \mu)$, with $\mu \in \mathbb{R}$, such that $V(\theta, z, 0) = V(\theta, z)$ and

1. $V(\theta, z, \mu)$ is a continuous function of μ, and $V(g(\theta, \lambda), z, \mu) = V(\theta, z, \mu)$
2. $H_{\psi_k\psi_k} = \partial^2 V(h_k(\psi_k), z, \mu)/\partial \psi_k^2$ has rank n_ψ and is a jointly continuous function of ψ_k, z
3. $H_{\psi_k z} = \partial^2 V(h_k(\psi_k), z, \mu)/\partial \psi_k \partial z$ has rank n_ψ and is a jointly continuous function of ψ_k, z

for any $\psi_k \in \mathbb{R}^{n_\psi}$, $k = 1, 2, \ldots, \beta$, for any $z \in \mathbb{R}_z \subset \mathbb{R}^N$ ($\dim(\mathbb{R}^N \setminus \mathbb{R}_z) < N$), and for any $\mu \in \mathbb{R}_0$.

Due to Assumption 20.11(1), the regularization parameter μ can always be chosen such that the difference $V(\theta, z, \mu) - V(\theta, z)$ is arbitrarily small in the regular space $\mathbb{D}_\theta$. Proceeding in this way, μ is active only in the singular subspace $\mathbb{S}_\theta$ and this is exactly how the regularization is applied in practice. This motivates the following definition.

Definition 20.12: The estimates $\hat{\theta} = \hat{\theta}(z)$, $\hat{\psi}_k = \hat{\psi}_k(z)$ are the minimizing arguments of $V(\theta, z, \mu)$ and $V(h_k(\psi_k), z, \mu)$, respectively,

$$\hat{\theta}(z) = \arg\min_\theta V(\theta, z, \mu) \quad \text{and} \quad \hat{\psi}_k(z) = \arg\min_{\psi_k} V(h_k(\psi_k), z, \mu)$$

Estimators satisfying Assumptions 20.9 and 20.11 are, for example, the nonlinear least squares (see Section 9.9), the generalized total least squares (see Section 9.10 and Van Huffel and Vandewalle, 1991), the one-step bootstrapped total least squares (see Section 9.10), and the maximum likelihood (see Section 9.11; Vandersteen et al., 1996a). Counterexamples in system identification are the linear least squares (see Section 9.8; Kalman, 1958; De Moor et al., 1994), the (iteratively) weighted linear least squares (see Section 9.8; Steigliz and McBride, 1965), and parametric time series analysis using, for example, the conditional maximum likelihood method (Box and Jenkins, 1976).

The z-values for which the cost function is singular ($\forall z \in \mathbb{S}_z$) and/or its second-order partial derivatives are not of full rank ($\forall z \in \mathbb{R}^N \setminus \mathbb{R}_z$, see Assumption 20.11) lie in lower dimensional subspaces of $\mathbb{R}^N$. To ensure that these values occur with probability zero, the following assumption is made.

Assumption 20.13 (pdf Noise): The probability density function of the disturbing errors $n_z = z - z_0$ is continuous.

Under Assumption 20.13, it is clear that $\text{Prob}(z \in \mathbb{S}_z) = \text{Prob}(z \in \mathbb{R}^N \setminus \mathbb{R}_z) = 0$. Assumptions 20.9 and 20.11 are illustrated in the following example.

Example 20.14 (Frequency Domain Identification): Consider the identification of model (20-1) starting from frequency response function measurements $G(j\omega_f)$, $f = 1, 2, \ldots, N$. The nonlinear least squares estimate minimizes (see Section 9.9)

$$V_{\text{NLS}}(\theta, z) = \sum_{f=1}^{N} |G(j\omega_f) - G(j\omega_f, \theta)|^2 \qquad (20\text{-}6)$$

Because $G(j\omega_f, \lambda\theta) = G(j\omega_f, \theta)$, it follows directly that $V_{\text{NLS}}(\lambda\theta, z) = V_{\text{NLS}}(\theta, z)$. The cost function (20-6) is infinitely large for θ-values satisfying $a_k = 0$, $k = 0, 1, \ldots, n_a$. It is undefined for $\theta = 0$ and θ-values such that $G(j\omega_f, \theta)$ has a common pole-zero pair at $j\omega_f$. Hence,

$$\mathbb{D}_\theta = \mathbb{R}^{n_\theta} \setminus \{\theta \mid f \in \{1, 2, \ldots, N\} \text{ s.t. } G(j\omega_f, \theta) = \infty \text{ or } 0/0\}$$

Using $G(j\omega_f, \theta) = B(j\omega_f, \theta)/A(j\omega_f, \theta)$, the regularized cost function becomes

$$V_{\text{NLS}}(\theta, z, \mu) = \sum_{f=1}^{N} \frac{|A(j\omega_f, \theta)G(j\omega_f) - B(j\omega_f, \theta)|^2}{|A(j\omega_f, \theta)|^2 + \mu^2 \theta^T \theta} \qquad \square$$

Quadratic cost functions satisfying Assumption 20.9 can always be written as $V(\theta, z) = \varepsilon^T(\theta, z)\varepsilon(\theta, z)/2$, where the residual $\varepsilon(\theta, z)$ is also invariant w.r.t. the parameter transformation $g(\theta, \lambda)$: $\varepsilon(g(\theta, \lambda), z) = \varepsilon(\theta, z)$. For such cost functions, it is possible to make statements about the null space of the Jacobian matrix $\partial\varepsilon(\theta, z)/\partial\theta$ w.r.t. the overparameterized θ.

Theorem 20.15 (Jacobian Matrix of Overparameterized Models): Under Assumptions 20.9 and 20.11, the Jacobian matrix $\partial\varepsilon(\theta, z)/\partial\theta$ of the quadratic cost function $V(\theta, z) = \varepsilon^T(\theta, z)\varepsilon(\theta, z)/2$ has a null space of dimension n_λ for any $\theta \in \mathbb{D}_k$ and $z \in \mathbb{D}_z$.

Proof. See Appendix 20.B. $\square$

The dimension of the null space of the Jacobian matrix w.r.t. θ is independent of the noise on the observations z and of the model errors. For quadratic cost functions that violate Assumption 20.9, the dimension of the null space of the Jacobian matrix w.r.t. θ is affected by the noise and the model errors (see Exercises 20.2 and 20.4).

20.4.2 Main Result

Theorem 20.16: Under Assumptions 20.3, 20.4, 20.9, 20.11, and 20.13, the estimate of an invariant $I(\theta)$ of model $M(\theta, z_0)$ is with probability one independent of the particular (over)parameterization chosen: $\hat{I}(\hat{\theta}(z)) = \hat{I}(h_k(\hat{\psi}_k(z)))$ for $k = 1, 2, \ldots, \beta$.

Proof. See Appendix 20.C. $\square$

Because $N = \dim(z)$ is fixed in the analysis, the theorem is valid for finite values of N. As it has not been assumed that the true model is within the model set, the theorem is also valid when there are model errors. As it has not been assumed that model $M(\theta, z_0)$ is linear in z_0, the theorem is also valid for nonlinear systems. An application of the nonlinear case can be found in Vandersteen et al. (1996a). The theorem also applies to estimators minimizing a series of cost functions where each cost function satisfies Assumptions 20.9 and 20.11. An example of such an estimator is the multistep bootstrapped total least squares method (see Section 9.12.3). Because the estimates of the invariants are independent of the (over)parameterization chosen, one should use the particular (over)parameterization that leads to the numerically best conditioned set of equations (see, for example, Van Overbeek and Ljung, 1982 and also Practical remark at the end of Section 20.5).

20.5 A SIMPLE NUMERICAL EXAMPLE

Consider again Example 20.14 with $N = 100$ data points, ω_f, $f = 1, 2, \ldots, N$ equally distributed in the band $[1, 1500]2\pi$ rad/s, and a true plant transfer function

$$G_0(s) = \frac{1 + 2 \times 10^{-4} s}{1 + 1 \times 10^{-3} s} \tag{20-7}$$

Independent, circular complex distributed noise $n_G(j\omega)$ with zero mean and variance 2×10^{-2} is added to the true transfer function $G(j\omega_f) = G_0(j\omega_f) + n_G(j\omega_f)$, $f = 1, 2, \ldots, N$. A hundred independent sets of 100 noisy transfer function values are generated. For each noisy data set the nonlinear least squares estimate (20-6), which satisfies Assumption 20.9 ($V_{\mathrm{NLS}}(\lambda\theta, z) = V_{\mathrm{NLS}}(\theta, z)$), and the linear least squares estimate (20-5), which violates Assumption 20.9 ($V_{\mathrm{LS}}(\lambda\theta, z) \neq V_{\mathrm{LS}}(\theta, z)$), are calculated for two models; one without model errors (20-8) and the other one with model errors (20-9).

$$G(s, \theta) = \frac{b_0 + b_1 s}{a_0 + a_1 s} \tag{20-8}$$

$$G(s, \theta) = \frac{b_0}{a_0 + a_1 s} \tag{20-9}$$

20.5.1.1 Model without Model Errors (20-8). The four identifiable forms of the overparameterized model (20-8) that satisfy Assumption 20.4 are

$$G(s, \psi_1) = \frac{\psi_{1[2]} + \psi_{1[3]} s}{1 + \psi_{1[1]} s}, \quad G(s, \psi_2) = \frac{\psi_{2[2]} + \psi_{2[3]} s}{\psi_{2[1]} + s}, \quad G(s, \psi_3) = \frac{1 + \psi_{3[3]} s}{\psi_{3[1]} + \psi_{3[2]} s}, \text{ and}$$

$$G(s, \psi_4) = \frac{\psi_{4[3]} + s}{\psi_{4[1]} + \psi_{4[2]} s}$$

For each of the 100 noisy data sets, the linear and nonlinear least squares estimates of the identifiable model parameters ψ_k, $k = 1, \ldots, 4$, and the overparameterized model parameter θ are calculated. The latter are obtained by solving the normal equations using the

Section 20.5 ■ A Simple Numerical Example   707

pseudoinverse. Three invariants of model (20-8) are the gain $K = b_0/a_0$ at $s = 0$ and the time constants $\tau_1 = b_1/b_0$ and $\tau_2 = a_1/a_0$. These invariants are calculated for each of the 100 least squares and nonlinear least squares estimates of $\hat{\psi}_k$, $k = 1, \ldots, 4$, and $\hat{\theta}$. The results are shown in Table 20-1. It follows that the linear least squares estimates of the invariants are strongly dependent on the particular (over)parameterization chosen. This is explained by the fact that the linear least squares cost function $V_{LS}(\theta, z)$ is not invariant w.r.t. the transformation $g(\theta, \lambda) = \lambda\theta$: $V_{LS}(\lambda\theta, z) \neq V_{LS}(\theta, z)$. Note that the estimates based on parameterizations ψ_2 (monic denominator polynomial) and θ perform much better than the other ones. From Table 20-1, it also follows that the nonlinear least squares estimates and their sample deviations are equal within the numerical precision of the calculations (12 digits). This is also true (but not shown in Table 20-1) for the estimates of each of the 100 data sets separately.

20.5.1.2 Model with Model Errors (20-9).
The three identifiable forms of the overparameterized model (20-9) that satisfy Assumption 20.4 are

$$G(s, \psi_1) = \frac{\psi_{1[2]}}{1 + \psi_{1[1]}s}, \quad G(s, \psi_2) = \frac{\psi_{2[2]}}{\psi_{2[1]} + s}, \quad \text{and} \quad G(s, \psi_3) = \frac{1}{\psi_{3[1]} + \psi_{3[2]}s}$$

For each of the 100 noisy data sets, the linear and nonlinear least squares estimates of the identifiable model parameters ψ_k, $k = 1, 2, 3$, and the overparameterized model parameter θ are calculated. Two invariants of model (20-9) are the gain $K = b_0/a_0$ at $s = 0$ and the time constant $\tau = a_1/a_0$. These invariants are calculated for each of the 100 least squares

TABLE 20-1 Estimated Invariants K, τ_1, and τ_2 of Model (20-8)

Invariant	Least Squares		Nonlinear Least Squares	
	Sample Mean	Sample Sandard Deviation	Sample Mean	Sample Standard Deviation
$\tau_1(\hat{\psi}_1)$	-8.59e-07	7.0e-06	1.9956e-04	1.9e-05
$\tau_1(\hat{\psi}_2)$	2.236e-04	2.9e-05	1.9956e-04	1.9e-05
$\tau_1(\hat{\psi}_3)$	6.394e-05	8.5e-06	1.9956e-04	1.9e-05
$\tau_1(\hat{\psi}_4)$	4.805e-04	5.3e-05	1.9956e-04	1.9e-05
$\tau_1(\hat{\theta})$	1.879e-04	2.7e-05	1.9956e-04	1.9e-05
$\tau_2(\hat{\psi}_1)$	1.011e-04	1.4e-05	1.0055e-03	6.6e-05
$\tau_2(\hat{\psi}_2)$	1.327e-03	1.5e-04	1.0055e-03	6.6e-05
$\tau_2(\hat{\psi}_3)$	3.156e-04	3.2e-05	1.0055e-03	6.6e-05
$\tau_2(\hat{\psi}_4)$	7.7e-03	2.3e-01	1.0055e-03	6.6e-05
$\tau_2(\hat{\theta})$	8.84e-04	1.1e-04	1.0055e-03	6.6e-05
$K(\hat{\psi}_1)$	3.886e-01	1.6e-02	1.0024	3.8e-02
$K(\hat{\psi}_2)$	1.2198	7.8e-02	1.0024	3.8e-02
$K(\hat{\psi}_3)$	7.410e-01	3.2e-02	1.0024	3.8e-02
$K(\hat{\psi}_4)$	4	140	1.0024	3.8e-02
$K(\hat{\theta})$	9.417e-01	6.4e-02	1.0024	3.8e-02

TABLE 20-2 Estimated Invariants K and τ of Model (20-9)

Invariant	Least Squares		Nonlinear Least Squares	
	Sample Mean	Sample Standard Deviation	Sample Mean	Sample Standard Deviation
$\tau(\hat{\psi}_1)$	1.017e-04	1.1e-05	5.3026e-04	5.9e-05
$\tau(\hat{\psi}_2)$	6.068e-04	4.4e-05	5.3026e-04	5.9e-05
$\tau(\hat{\psi}_3)$	1.673e-04	1.9e-05	5.3026e-04	5.9e-05
$\tau(\hat{\theta})$	2.184e-04	3.0e-05	5.3026e-04	5.9e-05
$K(\hat{\psi}_1)$	3.892e-01	1.4e-02	7.6796e-01	4.4e-02
$K(\hat{\psi}_2)$	8.399e-01	3.2e-02	7.6796e-01	4.4e-02
$K(\hat{\psi}_3)$	6.395e-01	2.2e-02	7.6796e-01	4.4e-02
$K(\hat{\theta})$	4.920e-01	2.2e-02	7.6796e-01	4.4e-02

and nonlinear least squares estimates of $\hat{\psi}_k$, $k = 1, 2, 3$, and $\hat{\theta}$. From Table 20-2 it follows that the conclusions of the preceding section are also valid for model errors.

20.5.1.3 Practical Remark. From the many simulations that have been conducted, it can be concluded that the overparameterized models lead to one of the best, but not always the best, condition numbers of the normal equations. This is in agreement with the results of McKelvey and Helmersson (1997). It also followed that the overparameterized form often gives the best linear least squares estimate (see, for example, Table 20-1), which is in agreement with the results of De Moor et al. (1994). Hence, even for estimators that violate Assumption 20.9, the overparameterized models are strongly recommended.

20.6 EXERCISES

20.1. Consider Examples 20.1, 20.6, and 20.14. Show that the null space of the Jacobian matrix of the nonlinear least squares estimate (20-6) has dimension 1 (hint: show that rank$(\partial g(\theta, \lambda)/\partial \lambda) = 1$).

20.2. Consider Examples 20.1, 20.6, and 20.14. Show that the null space of the Jacobian matrix of the linear least squares estimate (20-5) has dimension 0 unless it is evaluated in the noiseless data z_0 and the true model parameters θ_0 (hint: use $\varepsilon(\theta, z) = (\partial \varepsilon(\theta, z)/\partial \theta)\theta$).

20.3. Consider Examples 20.2 and 20.7. Show that the null space of the Jacobian matrix of the nonlinear least squares estimate (20-6) has dimension n^2 (hint: show that rank$(\partial g(\theta, \lambda)/\partial \lambda) = n^2$).

20.4. Redo the simulations of Section 20.5. Verify that the Jacobian matrix $\partial e(\theta, z)/\partial \theta$ of the nonlinear least squares estimate has a null space of dimension 1 for each noise realization. Verify that the Jacobian matrix $\partial e(\theta, z)/\partial \theta$ of the linear least squares estimate is not rank deficient for noisy observations and/or in the presence of model errors.

20.7 APPENDIXES

Appendix 20.A Proof of Theorem 20.8 (Cramér-Rao Bound of (Over)Parameterized Models)

Applying the chain rule for the partial derivatives on $Fi(\theta_0)$, (16-85) and $\partial I_0/\partial \theta_0$, makes it possible to rewrite (20-4a) in $\mathbb{D}_k$ as

Section 20.7 ■ Appendixes

$$\mathrm{Cov}(\hat{I}(\hat{\theta}(z))) \geq \left(\frac{\partial I}{\partial \psi_k} + \frac{\partial b_I}{\partial \psi_k}\right)\left(\frac{\partial \psi_k}{\partial \theta}\right)\left[\left(\frac{\partial \psi_k}{\partial \theta}\right)^T Fi(\psi_k)\left(\frac{\partial \psi_k}{\partial \theta}\right)\right]^+ \left(\frac{\partial \psi_k}{\partial \theta}\right)^T \left(\frac{\partial I}{\partial \psi_k} + \frac{\partial b_I}{\partial \psi_k}\right)^H \quad (20\text{-}10)$$

If we can show that

$$\left(\frac{\partial \psi_k}{\partial \theta}\right)\left[\left(\frac{\partial \psi_k}{\partial \theta}\right)^T Fi(\psi_k)\left(\frac{\partial \psi_k}{\partial \theta}\right)\right]^+ \left(\frac{\partial \psi_k}{\partial \theta}\right)^T = Fi^{-1}(\psi_k) \quad (20\text{-}11)$$

in $\mathbb{D}_k$, then the right-hand sides of (20-4a) and (20-4b) are equal in $\mathbb{D}_k$, which proves the theorem. Using the first Moore-Penrose condition of the pseudoinverse $AA^+A = A$ (see-Section 15.5) with $A = (\partial \psi/\partial \theta)^T Fi(\psi_k)(\partial \psi/\partial \theta)$ gives

$$\left(\frac{\partial \psi_k}{\partial \theta}\right)^T Fi(\psi_k)\left(\frac{\partial \psi_k}{\partial \theta}\right)\left[\left(\frac{\partial \psi_k}{\partial \theta}\right)^T Fi(\psi_k)\left(\frac{\partial \psi_k}{\partial \theta}\right)\right]^+ \left(\frac{\partial \psi_k}{\partial \theta}\right)^T Fi(\psi_k)\left(\frac{\partial \psi_k}{\partial \theta}\right) = \left(\frac{\partial \psi_k}{\partial \theta}\right)^T Fi(\psi_k)\left(\frac{\partial \psi_k}{\partial \theta}\right) \quad (20\text{-}12)$$

Left multiplication with $\partial \psi_k/\partial \theta$ and right multiplication with $(\partial \psi_k/\partial \theta)^T$ of (20-12) give

$$Q Fi(\psi_k)\left(\frac{\partial \psi_k}{\partial \theta}\right)\left[\left(\frac{\partial \psi_k}{\partial \theta}\right)^T Fi(\psi_k)\left(\frac{\partial \psi_k}{\partial \theta}\right)\right]^+ \left(\frac{\partial \psi_k}{\partial \theta}\right)^T Fi(\psi_k)Q = Q Fi(\psi_k)Q \quad (20\text{-}13)$$

where $Q = (\partial \psi_k/\partial \theta)(\partial \psi_k/\partial \theta)^T$ is a regular n_ψ by n_ψ matrix in $\mathbb{D}_k$ (Assumption 20.4). Left division by $QFi(\psi_k)$ and right division by $Fi(\psi_k)Q$ of (20-13) give (20-11). □

Appendix 20.B Proof of Theorem 20.15 (Jacobian Matrix of (Over)Parameterized Models)

Because the residual $\varepsilon(\theta, z)$ is independent of λ, we have $\partial\varepsilon(\theta, z)/\partial\lambda = 0$. Using $\varepsilon(g(\theta, \lambda), z) = \varepsilon(\theta, z)$, this equality can be written as

$$\frac{\partial\varepsilon(g, z)}{\partial g}\frac{\partial g(\theta, \lambda)}{\partial \lambda} = 0 \quad (20\text{-}14)$$

for any $\theta \in \mathbb{D}_\theta$ and $\lambda \in \mathbb{D}_\lambda$, and with $\partial\varepsilon(g, z)/\partial g \in \mathbb{R}^{N \times n_\theta}$ ($N \geq n_\theta$) the Jacobian matrix. Because $\mathrm{rank}(\partial g(\theta, \lambda)/\partial\lambda) = n_\lambda$ with $n_\lambda < n_\theta$ (Assumption 20.3), it follows from (20-14) that $\dim(\mathrm{null}(\partial\varepsilon(g, z)/\partial g)) = n_\lambda$. □

Appendix 20.C Proof of Theorem 20.16

The regularized cost function $V(\theta, z, \mu)$ is an invariant of model $M(\theta, z_0)$ (Assumption 20.11). This implies that for any $\hat{\theta} \in \mathbb{D}_k \cap \mathbb{D}_l$ ($k \neq l$) $h_k(\hat{\psi}_k) = g(\hat{\theta}, \lambda_k(\hat{\theta}))$ and $g(\hat{\theta}, \lambda_l(\hat{\theta})) = h_l(\hat{\psi}_l)$; otherwise $\hat{\psi}_k$ and $\hat{\psi}_l$ would not be the minimizers of $V(h_k(\psi_k), z, \mu)$ and $V(h_l(\psi_l), z, \mu)$, respectively. Using $I(g(\theta, \lambda(\theta))) = I(\theta)$ (Definition 20.5), it follows that for any $\hat{\theta} \in \mathbb{D}_k \cap \mathbb{D}_l$,

$$\hat{I}(h_k(\hat{\psi}_k)) = \hat{I}(g(\hat{\theta}, \lambda_k(\hat{\theta}))) = \hat{I}(\hat{\theta}) \quad \text{and} \quad \hat{I}(h_l(\hat{\psi}_l)) = \hat{I}(g(\hat{\theta}, \lambda_l(\hat{\theta}))) = \hat{I}(\hat{\theta}) \quad (20\text{-}15)$$

and, hence, $\hat{I}(h_k(\hat{\psi}_k)) = \hat{I}(h_l(\hat{\psi}_l))$. The theorem is proved if it can be shown that the event '$\hat{\theta} \in \mathbb{D}_k \cap \mathbb{D}_l$' occurs with probability one. Therefore, it is sufficient to prove that the probability density functions (pdf's) of the estimates $\hat{\psi}_k$, $k = 1, 2, ..., \beta$, are continuous. Indeed, accumulation of probability mass in a lower dimensional space can occur only if the distribution function is degenerate (Papoulis, 1981). If the pdf of $\hat{\psi}_k$ is continuous, then the probability that $\hat{\psi}_k$ lies in a lower dimensional subspace of $\mathbb{R}^{n_\psi}$ equals zero. Under Assumption 20.4, the θ-values that cannot be represented by, for example, ψ_k, represent values of, for example, ψ_l, lying in a lower dimensional subspace of $\mathbb{R}^{n_\psi}$, so that Prob($\hat{\theta} \in \mathbb{D}_k \setminus \mathbb{D}_l$) = 0. Similarly, Prob($\hat{\theta} \in \mathbb{D}_l \setminus \mathbb{D}_k$) = 0 and, thus, Prob($\hat{\theta} \in \mathbb{D}_k \cap \mathbb{D}_l$) = 1.

By Assumption 20.13, the pdf of the noisy data z is continuous. To prove that the probability density functions of the estimates $\hat{\psi}_k$, $k = 1, 2, ..., \beta$, are continuous it is, therefore, sufficient to show that $\hat{\psi}_k = \hat{\psi}_k(z)$ is a continuous function of z with continuous derivatives satisfying

$$\dim(\{z | \text{rank}(\frac{\partial \hat{\psi}_k(z)}{\partial z}) < n_\psi\}) < N \tag{20-16}$$

(Papoulis, 1981). If (20-16) is not satisfied, then the distribution function of $\hat{\psi}_k$ may be degenerate in the subspace $\mathbb{D}_k \setminus \mathbb{D}_l$, so that accumulation of probability mass may occur in $\mathbb{D}_k \setminus \mathbb{D}_l$. In this case, Prob($\hat{\theta} \in \mathbb{D}_k \setminus \mathbb{D}_l$) $\neq 0$ and the theorem is no longer valid. The function $\hat{\psi}_k(z)$ is implicitly known by the definition of the minimizer $\hat{\psi}_k$

$$\left(\frac{\partial V(h_k(\hat{\psi}_k), z, \mu)}{\partial \hat{\psi}_k}\right)^T = 0 \text{ or } F(\hat{\psi}_k, z, \mu) = 0 \tag{20-17}$$

Under Assumption 20.11, $F(\hat{\psi}_k, z, \mu) \in \mathbb{R}^{n_\psi}$ has continuous, full rank, first-order partial derivatives w.r.t. $\hat{\psi}_k$ and z for any $\hat{\psi}_k \in \mathbb{R}^{n_\psi}$ and $z \in \mathbb{R}_z$ with $\dim(\mathbb{R}^N \setminus \mathbb{R}_z) < N$. Applying the implicit function theorem (Kaplan, 1993) to $F(\hat{\psi}_k, z, \mu) = 0$ shows that $\hat{\psi}_k = \hat{\psi}_k(z)$ is a continuous function of z with full rank, continuous derivative

$$\begin{aligned}\frac{\partial \hat{\psi}_k(z)}{\partial z} &= -\left(\frac{\partial F(\hat{\psi}_k, z, \mu)}{\partial \hat{\psi}_k}\right)^{-1}\left(\frac{\partial F(\hat{\psi}_k, z, \mu)}{\partial z}\right) \\ &= -\left(\frac{\partial^2 V(h_k(\hat{\psi}_k), z, \mu)}{\partial \hat{\psi}_k^2}\right)^{-1}\left(\frac{\partial^2 V(h_k(\hat{\psi}_k), z, \mu)}{\partial \hat{\psi}_k \partial z}\right)\end{aligned} \tag{20-18}$$

for any $\hat{\psi}_k \in \mathbb{R}^{n_\psi}$ and $z \in \mathbb{R}_z$ with $\dim(\mathbb{R}^N \setminus \mathbb{R}_z) < N$, which concludes the proof. $\square$

References

Abel, J. P. (1993). A bound on mean-square-estimate error. *IEEE Trans. Information Theory,* vol. 39, no. 5, pp. 1675–1680.

Abramowitz, M., I. A. Stegun (1970). *Handbook of Mathematical functions.* Dover Publications, New York.

Agüero, J. C., and G. C. Goodwin (2008). Identifiability of errors in variables dynamic systems. *Automatica,* vol. 44, no. 2, pp. 371–382.

Agüero, J. C., J. I. Yuz, G. C. Goodwin, and R. A. Delgado (2010). On the equivalence of time and frequency domain maximum likelihood estimation. *Automatica,* vol. 46, no. 2, pp. 260–270.

Akaike, H. (1974). A new look at the statistical model identification. *IEEE Trans. Autom. Contr.,* vol. 19, pp. 716–723.

Albertos, P., R. Sanchis, and A. Sala (1999). Output prediction under scarce data operation: control applications. *Automatica,* vol. 35, no. 10, pp. 1671–1681.

Anderson, B. D. O. and M. Deistler (1984). Identifiability in dynamic errors-in-variables models. *J. Time Series Analysis,* vol. 5, pp. 1–13.

Anderson, E., Z. Bai, C. Bishof, J. Demmel, J. Dongarra, J. Du Croz, A. Greenbaum, S. Hammastry, A. McKenney, S. Ostrouchov, and D. Sorensen (1992). *LAPACK User's Guide.* SIAM Press, Philadelphia.

Anderson, T. W. (1958). *An Introduction to Multivariate Statistical Analysis.* Wiley, New York.

Antoni, J., and J. Schoukens (2007). A comprehensive study of the bias and variance of frequency-response-function measurements: optimal window selection and overlapping strategies. *Automatica,* vol. 43, no. 10, pp. 1723–1736.

Antoni, J., and J. Schoukens (2009). Optimal settings for measuring frequency response functions with weighted overlapped segment averaging. *IEEE Trans. Instrum. and Meas.,* vol. 58, no. 9, pp. 3276–3287.

Antoulas, A. C., and B. D. O. Anderson (1999). On the choice of inputs in identification for robust control. *Automatica,* vol. 35, no. 6, pp. 1009–1031.

Åström, K. J. (1970). *Introduction to Stochastic Control Theory.* Academic Press, New York.

Åström, K. J., P. Hagander, and J. Sternby (1984). Zeros of sampled systems. *Automatica,* vol. 20, pp. 31–38.

Bai, Z., and J. Demmel (1993). Computing the generalized singular value decomposition. *SIAM J. Sci. Stat. Comput.*, vol. 14, no. 6, pp. 1464–1486.

Balabanian, N., and T. A. Bickart (1969). *Electrical Network Theory.* Wiley, New York.

Baratchart, L., J. Leblond, J. R. Partington, and N. Torkhani (1997). Robust identification from band-limited data. *IEEE Trans. Autom. Contr.*, vol. 42, no. 9, pp. 1318–1325.

Barenthin M., and H. Hjalmarsson (2008). Identification and control: Joint input design and H-infinity state feedback with ellipsoidal parametric uncertainty via LMIs. *Automatica*, vol. 44, no. 2, pp. 543–551.

Barenthin M., X. Bombois, H. Hjalmarsson, G. Scorletti (2008). Identification for control of multivariable systems: Controller validation and experiment design via LMIs. *Automatica*, vol. 44, no. 12, pp. 3070–3078.

Barker, H. A., and M. Zhuang (1997). Design of pseudo-random perturbation signals for frequency-domain identification of nonlinear systems. *Preprints of SYSID'97, 11th IFAC Symposium on System Identification,* Kitakyushu, Japan, pp. 1635–1640.

Barker, W. P., L. B. Eldred, and A. N. Palazotto (1996). Viscoelastic material response with a fractional-derivative constitutive model. *AIAA Journal,* vol. 34, no. 3, pp. 596–600.

Battaglia, J. L., O. Cois, L. Puigsegur, and A. Oustaloup (2001). Solving an inverse heat conduction problem using a non-integer identified model. *International Journal of Heat and Mass Transfer,* vol. 44, no. 14, pp. 2671–2680.

Bayard, D. S. (1994a). High-order multivariable transfer-function curve-fitting-algorithms, sparse-matrix methods and experimental results, *Automatica*, vol. 30, no. 3, pp. 1439–1444.

Bayard, D. S. (1994b). An algorithm for state-space frequency domain identification without windowing distortions, *IEEE Trans. Autom. Contr.*, vol., 39, no. 9, pp. 1880–1885.

Beck, J. V., and K. J. Arnold (1977). *Parameter Estimation in Engineering and Science.* Wiley, New York.

Bendat, J. S. (1998). *Nonlinear Systems Techniques and Applications.* Wiley, New York.

Bendat, J. S., and A. G. Piersol (1980). *Engineering Applications of Correlations and Spectral Analysis.* Wiley, New York.

Ben-Israel, A., and T. N. E. Greville (1974). *Generalized Inverses: Theory and Applications.* Wiley, London.

Bergström, A. R. (1990). *Continuous Time Econometric Modeling.* Oxford Univ. Press, Oxford.

Beya, K., R. Pintelon, J. Schoukens, P. Lataire, B. Mpanda-Mabwe, and M. Delhaye (1994). Identification of synchronous machines parameters using broadband excitation. *IEEE Trans. Energy Conv.,* vol. 9, no. 2, pp. 270–280.

Billings, S. A. (1980). Identification of nonlinear systems: A survey. *Proc. Inst. Elec. Eng.*, vol. 127, pt. D, pp. 272–285.

Billings, S. A., and S. Y. Fakhour (1982). Identification of systems containing linear dynamic and static nonlinear elements. *Automatica*, vol. 18, no. 1, pp. 15–26.

Billingsley, P. (1995). *Probability and Measure.* Wiley, New York.

Bohlin, T. (1971). On the problem of ambiguities in maximum likelihood identification. *Automatica,* vol. 7, pp. 199–210.

Bombois, X., G. Scorletti, M. Gevers, P. M. J. Van den Hof, and R. Hildebrand (2006). Least costly identification experiment for control. *Automatica*, vol. 42, no. 10, pp. 1651–1662.

Box, G. E. P., and G. M. Jenkins (1976). *Time Series Analysis: Forecasting and Control.* Holden-Day, Oakland, CA.

Boyd, S., Y. S. Tang, and L. O. Chua (1983). Measuring Volterra kernels. *IEEE Trans. Circuits Systems,* vol. CAS-30, no. 9, pp. 648–651.

Brewer, J. W. (1978). Kronecker products and matrix calculus in system theory. *IEEE Trans. Circuits Systems,* vol. 25, pp. 772–781.

Brigham, E. O. (1974). *The Fast Fourier Transform.* Prentice-Hall, Englewood Cliffs, NJ.

References

Brillinger, D. R. (1981). *Time Series: Data Analysis and Theory.* McGraw-Hill, New York.

Broersen, P. M. T. (1995). A comparison of transfer function estimators. *IEEE Trans. Instrumentation Measurement*, vol. 44, pp. 657–661.

Brown, D., G. Carbon, and K. Ramsey (1977). Survey of excitation techniques applicable to the testing of automotive structures. *Int. Automotive Eng. Congress and Exposition*, Cobo Hall, Detroit, MI.

Bultheel, A. M., and M. Van Barel (1995). Vector orthogonal polynomials and least squares approximation, *SIAM J. Matrix Anal. Appl.*, vol. 16, no. 3, pp. 863–885.

Bultheel, A., M., Van Barel, Y. Rolain, and R. Pintelon (2005). Numerically robust transfer function modeling from noisy frequency response data, *IEEE Trans. Autom. Contr.*, vol. 50, no. 11, pp. 1835–1839.

Cadzow, J. A. (1990). Signal processing via least squares error modeling. *IEEE ASSP Magazine*, pp. 12–31.

Cadzow, J. A., and O. M. Solomon (1987). Linear modeling and the coherence function. *IEEE Trans. Acoustics, Speech and Signal Processing*, vol. 35, no. 1, pp. 19–28.

Caines, P. E. (1988). *Linear Stochastic Systems.* Wiley, New York.

Carter, G. C., and A. H. Nuttall (1980). On the weighted overlapped segment averaging method for power spectral estimation, *Proceedings of the IEEE*, vol. 68, no. 10, pp. 1352–1353.

Cauberghe, B., P. Guillaume, P. Verboven, and E. Parloo (2003). Identification of modal parameters including unmeasured forces and transient effects. *Journ. of Sound and Vibration*, vol. 265, no. 3, pp. 609–625.

Chow, Y. S., and H. Teicher (1988). *Probability Theory: Independence, Interchangeability, Martingales* (2nd ed.). Springer-Verlag, New York.

Chua, L. O., and C.-Y. Ng (1979). Frequency domain analysis of nonlinear systems: general theory. *Electron. Circuits Syst.*, vol. 3, no. 4, pp. 165–185.

Crochiere, R. E., and L. R. Rabiner (1983). *Multirate Digital Signal Processing.* Prentice-Hall, Englewood Cliffs, NJ.

de Callafon, R. A., D. de Roover, P. M. J. Van den Hof (1996). Multivariable least squares frequency domain identification using polynomial matrix fraction descriptions, *Proceedings of the 35th IEEE Conference on Decision and Control*, Kobe (Japan), pp. 2030–2035.

Delbaen, F. (1990). Optimizing the determinant of a positive definite matrix. *Bull. Soc. Math. Belg. Tijdschr. Belg. Wisk. Gen.*, vol. 42, no. 3, ser. B, pp. 333–346.

Delchamps, D. F. (1985). Global structure of families of multivariable linear systems with an application to identification. *Mathematical Systems Theory*, vol. 18, pp. 329–380.

Delchamps, D. F., and C. I. Byrnes (1982). Critical point behavior of objective functions defined on spaces of multivariable systems. *Proceedings of the 21st IEEE Conference on Decision and Control*, Orlando (Florida), vol. 2, pp. 937–943.

De Moor, B., M. Gevers, and G. Goodwin (1994). L_2-underbiased, and L_2-unbiased estimation of transfer functions. *Automatica*, vol. 30, no. 5, pp. 893–898.

D'haene, T., R. Pintelon, and G. Vandersteen (2006). An iterative method to stabilize a transfer function in the s- and z-domains. *IEEE Trans. Instrum. and Meas.*, vol. 55, no. 4, pp. 1192–1196.

D'haene, T., and R. Pintelon (2008). Passivity enforcement of transfer functions. *IEEE Trans. Instrum. and Meas.*, vol. 57, no. 10, pp. 2181–2187.

Dobrowiecki, T. P., and J. Schoukens (2006). Robustness of the related linear dynamic system estimates in cascaded nonlinear MIMO systems. *Instrumentation and Measurement Technology Conference Proceedings*, Apr. 24–27, Sorrento (Italy), vol. 1, pp. 117–122.

Dobrowiecki, T. P., and J. Schoukens (2007a). Linear approximation of weakly nonlinear MIMO systems. *IEEE Trans. Instrum. and Meas.*, vol. 56, no. 3, pp. 887–894.

Dobrowiecki, T. P., and J. Schoukens (2007b). Measuring a linear approximation to weakly nonlinear MIMO systems. *Automatica*, vol. 43, no. 10, pp. 1737–1751.

Dobrowiecki, T. P., J. Schoukens, and P. Guillaume (2006). Optimized excitation signals for MIMO frequency response function measurements. *IEEE Trans. Instrum. and Meas.,* vol. 55, no. 6, pp. 2072–2079.

Durbha, M., M. E. Orazem, and B. Tribollet (1999). A mathematical model for the radially dependent impedance of a rotating disk electrode. *Journal of the Electrochemical Society,* vol. 146, no. 6, pp. 2199–2208.

Enqvist, M. (2005). *Linear Models of Nonlinear Systems.* Ph.d. dissertation, Dept. of Electrical Engineering, Linköping University, Sweden.

Enqvist, M., and L. Ljung (2005). Linear approximations of nonlinear FIR systems for separable input processes. *Automatica,* vol. 41, no. 3, pp. 459–473.

Evans, C. (1998). *Identification of linear and nonlinear systems using multisine test signals.* Ph.D. dissertation, Dept. of Electronics and IT, University of Glamorgan, Wales.

Evans, C., and D. Rees (2000). Nonlinear distortions and multisine signals—Part I: measuring the best linear approximation. *IEEE Trans. Instrum. and Meas.,* vol. 49, no. 3, pp. 602–609.

Evans, C., D. Rees, and D. L. Jones (1994). Identifying linear models of systems suffering nonlinear distortions. *Proceedings IEE Control'94,* Coventry, UK, pp. 288–296.

Ewins, D. J. (1991). *Modal Testing: Theory and Practice.* Wiley, New York.

Eykhoff, P. (1974). *System Identification, Parameter and State Estimation.* Wiley, New York.

Fan, H., T. Söderström, M. Mossberg, B. Carlsson B, and Y. J. Zou (1999). Estimation of continuous-time AR process parameters from discrete-time data. *IEEE Trans. Sign. Proc.,* vol. 47, no. 5, pp. 1232–1244.

Fan, J. Q., and I. Gijbels (1995). Data-driven bandwidth selection in local polynomial fitting – variable bandwidth and spatial adaptation. *Journal of the Royal Statistical Society Series B – Methodological,* vol. 57, no. 2, pp. 371–394.

Fedorov, V. V. (1972). *Theory of Optimal Experiments.* Academic Press, New York.

Feller, W. (1968). *An Introduction to Probability Theory and Its Applications.* Wiley, London.

Figwer, J., and A. Niederlinski (1995). Using the DFT to synthesize multivariate orthogonal white noise series. *Transactions of the Society for Computer Simulations,* vol. 12, no. 3, pp. 261–285.

Fletcher, R. (1991). *Practical Methods of Optimization* (2nd ed.). Wiley, New York.

Forssell, U., and L. Ljung (1999). Closed-loop identification revisited. *Automatica,* vol. 35, no. 7, pp. 1215–1241.

Forssell, U., and L. Ljung (2000a). Identification of unstable systems using output error and Box-Jenkins model structures. *IEEE Trans. Autom. Contr.,* vol. 45, no. 1, pp. 137–141.

Forssell, U., and L. Ljung (2000b). Some results on optimal experiment design. *Automatica,* vol. 36, no. 5, pp. 749–756.

Forssell, U., and L. Ljung (2000c). A projection method for closed-loop identification. *IEEE Trans. Autom. Contr.,* vol. 45, no. 11, pp. 2101–2105.

Forsythe, E. G. (1957). Generation and use of orthogonal polynomials for data-fitting with a digital computer. *J. Soc. Indust. Appl. Math.,* vol. 5, no. 3, pp. 74–88.

Forsythe, E. G., and E. G. Straus (1955). On best conditioned matrices. *Proc. Am. Math. Soc.,* vol. 6, pp. 340–345.

Fuller, W. A. (1987). *Measurement Error Models.* Wiley, New York.

Gaikwad, S. V., and D. E. Rivera (1997). Multivariable frequency-reponse curve fitting with application to control-relevant parameter estimation, *Automatica,* vol. 33, no. 6, pp. 1169–1174.

Gantmacher, F. R. (1990). *The Theory of Matrices.* Chelsea Publishing Company, New York.

Gevers, M. (2005). Identification for control: From the early achievements to the revival of experiment design. *European Journ. of Contr.,* vol. 11, no. 4–5, pp. 335–352.

Gevers, M., and L. Ljung (1986). Optimal experiment designs with respect to the intended model application. *Automatica,* vol. 22, no. 5, pp. 543–554.

Gevers, M., R. Pintelon, and J. Schoukens (2011). The local polynomial method for nonparametric system identification: improvements and experimentation. *Proceedings 50th IEEE Conference on Decision and Control and European Control Conference,* Orlando, Florida (USA), Dec. 12–15, pp. 4302–4307.

Gevers, M., and V. Wertz (1984). Uniquely identifiable state-space and ARMA parametrizations for multivariable linear systems. *Automatica,* vol. 20, no. 3, pp. 333–347.

Giri, N. (1965). On the complex analogues of T^2 and R^2 tests. *Annals of Mathematical Statistics,* vol. 36, pp. 664–670.

Godfrey, K. R. (1969). The theory of the correlation method of dynamic analysis and its application to industrial processes and nuclear power plant. *Measurement and Control,* vol. 2, pp. T65–T72.

Godfrey, K. R. (1980). Correlation methods. *Automatica,* vol. 16, pp. 527–534.

Godfrey, K. R., editor (1993a). *Perturbation Signals for System Identification.* Prentice-Hall, London.

Godfrey, K. R. (1993b). Introduction to perturbation signals for time-domain system identification. In K. R. Godfrey, editor, *Perturbation Signals for System Identification.* Prentice-Hall, Hemel Hempstead, England, pp. 1–59.

Goethals, I., T. Van Gestel, J. Suykens, P. Van Dooren, and B. De Moor (2003). Identification of positive real models in subspace identification by using regularization. *IEEE Trans. Autom. Contr.,* vol. 48, no. 10, pp. 1843–1847.

Golub, G. H., and C. F. Van Loan (1996). *Matrix Computations* (3rd ed.). John Hopkins University Press, Baltimore.

Goodman, N. R. (1963). Statistical analysis based upon a certain multivariate complex Gaussian distribution (an introduction). *Annals of the Mathematical Statistics,* vol. 34, no. 1, pp. 152–177.

Goodwin, G. C., and G. J. Adams (1994). Multi-rate techniques in non-zero-order hold identification. *Preprints of the 10th IFAC Symposium on System Identification,* Copenhagen, Denmark, vol. 3, pp. 125–130.

Goodwin, G. C., and R. L. Payne (1977). *Dynamic System Identification. Experimental Design and Data Analysis.* Academic Press, New York.

Gorman, J. D., and A. O. Hero (1990). Lower bounds for parametric estimation with constraints. *IEEE Trans. Inform. Theory,* vol. IT-26, no. 6, pp. 1285–1301.

Gradshteyn, L. S., and I. M. Ryzhik (1980). *Table of Integrals, Series, and Products* (corrected and enlarged edition). Academic Press, New York.

Grivet-Talocia, S., and M. Bandinu (2006). Improving the convergence of vector fitting for equivalent circuit extraction from noisy frequency responses. *IEEE Trans. Electromagn. Compt.,* vol. 48, no. 1, pp. 104–120.

Grivet-Talocia, S., and A. Ubolli (2007). Passivity enforcement with relative error control. *IEEE Trans. Microw. Theory Techn.,* vol. 55, no. 11, pp. 2374–2383.

Guidorzi, R. (1975). Canonical structures in the identification of multivariable systems. *Automatica,* vol. 11, pp. 361–374.

Guillaume, P. (1998). Frequency response measurements of multivariable systems using nonlinear averaging techniques. *IEEE Trans. Instrum. and Meas.,* vol. 47, no. 3, pp. 796–800.

Guillaume, P., I. Kollár, and R. Pintelon (1996a). Statistical analysis of nonparametric transfer function estimates. *IEEE Trans. Instrum. and Meas.Instrum. and Meas.,* vol. 45, no. 2, pp. 594–600.

Guillaume, P., and R. Pintelon (1996). A Gauss-Newton-like optimization algorithm for "weighted" nonlinear least-squares problems. *IEEE Trans. Sign. Proc.,* vol. 44, no. 9, pp. 2222–2228.

Guillaume, P., R. Pintelon, and J. Schoukens (1992a). Parametric identification of two-port models in the frequency domain. *IEEE Trans. Instrum. and Meas.,* vol. 41, no. 2, pp. 233–239.

Guillaume, P., R. Pintelon, and J. Schoukens (1992b). Non-parametric frequency response function estimators based on nonlinear averaging techniques. *IEEE Trans. Instrum. and Meas.,* vol. 41, no. 6, pp. 739–746.

Guillaume, P., R. Pintelon, and J. Schoukens (1995). Robust parametric transfer function estimation using complex logarithmic frequency response data. *IEEE Trans. Autom. Contr.,* vol. 40, no. 7, pp. 1180–1190.

Guillaume, P., R. Pintelon, and J. Schoukens (1996b). Accurate estimation of multivariable frequency response functions. *Proceedings of the 13th IFAC Triennial World Conference,* San Francisco, pp. 423–428.

Guillaume, P., R. Pintelon, and J. Schoukens (1996c). Parametric identification of multivariable systems in the frequency domain—a survey, *Proceedings ISMA21—Noise and Vibration Engineering,* Leuven (Belgium), vol. II, pp. 1069–1082.

Guillaume, P., J. Schoukens, and R. Pintelon (1989). Sensitivity of roots to errors in the coefficients of polynomials obtained by frequency domain estimation methods. *IEEE Trans. Instrum. and Meas.,* vol. 38, no. 6, pp. 1050–1056.

Guillaume, P., J. Schoukens, R. Pintelon, and I. Kollár (1991). Crest-factor minimization using nonlinear chebyshev approximation methods. *IEEE Trans. Instrum. and Meas.,* vol. 40, no. 6, pp. 982–989.

Gustafsson, F., and J. Schoukens (1998). Utilizing periodic excitation in prediction error based system identification. *Proceedings of the 37th IEEE Conference on Decision and Control,* Tampa, FL, pp. 3926–3931.

Gustavsen, B. (2008). Passivity enforcement of rational models via modal perturbation. *IEEE Trans. Power Delivery,* vol. 23, no. 2, pp. 768–775.

Haber, R. (1985). Nonlinearity tests for dynamic processes. *7th IFAC/IFORS Symposium on Identification and System Parameter Estimation,* York, UK, pp. 409–414.

Halvorsen, W. G., and D. L. Brown (1977). Impulse technique for structural frequency response testing. *Sound and Vibration,* vol. 11, pp. 8–21.

Hannan, E. J. (1980). The estimation of the order of an ARMA process. *Annals of Statistics,* vol. 8, no. 5, pp. 1071–1081.

Hannan, E. J., and M. Deistler (1988). *Linear Systems.* Wiley, New York.

Harris, F. (1978). On the use of windows for harmonic analysis with discrete Fourier transform. *Proceedings of the IEEE,* vol. 66, pp. 51–83.

Hazewinkel, M. (1977). Moduli and canonical forms for linear dynamical systems II: The topological case. *Mathematical Systems Theory,* vol. 10, pp. 363–385.

Heath, W. P. (2001). Bias of indirect non-parametric transfer function estimates for plants in closed loop. *Automatica,* vol. 37, no. 10, pp. 1529–1540.

Henrici, P. (1974). *Applied and Computational Complex Analysis.* Wiley, New York, vol. 1.

Herlufsen, H. (1984). Dual channel FFT analysis (part I). *Tech. Rev.,* Brüel & Kjær, no. 1, Nærum, Denmark.

Heylen, W., S. Lammens, and P. Sas (1997). *Modal Analysis Theory and Testing.* Society for Experimental Mechanics, Bethel (USA).

Hotelling, H. (1933). The generalization of student's ratio. *Annals of Mathematical Statistics,* vol. 2, no. 3, pp. 360–378.

Huber, P. J. (1981). *Robust Statistics.* Wiley, New York.

Isaksson, A. J. (1993). Identification of ARX-models subject to missing data. *IEEE Trans. Autom. Contr.,* vol. 38, no. 5, pp. 813–819.

Jazwinski, A. H. (1970). *Stochastic Processes and Filtering Theory.* Academic Press, London.

Kabaila, P. (1983). Parameter values of ARMA models minimizing the one-step-ahead prediction error when the true system is not in the model set. *J. Appl. Prob.,* vol. 20, no. 2, pp. 405–408.

Kahane, J. P. (1980). Sur les polynomes à coefficients unimodulaires. *Bull. London Math. Soc.,* no. 12, pp. 321–342.

Kahane, J. P. (1985). *Some Random Series of Functions.* University Press, Cambridge (UK).

Kailath, T. (1980). *Linear Systems.* Prentice-Hall, Englewood Cliffs, NJ.

Kalman, R. E. (1958). Design of a self-optimizing control system. *ASME Trans.*, vol. 80, no. 2, pp. 468–478.

Kaplan, W. (1993). *Advanced Calculus.* Addison-Wesley, Reading, MA.

Kashyap, R. L. (1980). Inconsistency of the AIC rule for estimating the order of autoregressive models. *IEEE Trans. Autom. Contr.*, vol. 25, no. 5, pp. 996–998.

Kay, S. M. (1988). *Modern Spectral Estimation: Theory and Application.* Prentice-Hall, Englewood Cliffs, NJ.

Kendall, M., and A. Stuart (1979). *Inference and Relationship,* vol. 2 of *The Advanced Theory of Statistics* (4th ed.). Charles Griffin, London.

Kollár, I. (1994). *Frequency-Domain System Identification Toolbox for Use with Matlab.* The Mathworks, Natick, MA.

Kollár, I., R. Pintelon, Y. Rolain, and J. Schoukens (1991). Correspondence: Another step towards an ideal data acquisition channel. *IEEE Trans. Instrum. and Meas.*, vol. 40, no. 3, pp. 659–660.

Kumaresan, R., C.S. Ramalingam, and D. Van Ormondt (1990). Estimating the parameters of NMR signals by transforming to the frequency domain. *Journal of Magnetic Resonance,* vol. 89, pp. 562–567.

Kwakernaak, H., and R. Sivan (1991). *Modern Signals and Systems.* Prentice-Hall, London.

Lancaster, P., and M. Tismenetsky (1985). *The Theory of Matrices.* Academic Press, Orlando, FL.

Lataire, J., and R. Pintelon (2009). Estimating a non-parametric, colored noise model for linear, slowly time-varying systems, *IEEE Trans. Instrum. and Meas.*, vol. 58, no. 5, pp. 1535–1545.

Lee, C. W. (1993). *Vibration Analysis of Rotors.* Kluwer Academic, Dordrecht.

Lee, C. W., and C.Y. Joh (1994). Development of the use of directional frequency response functions for the diagnosis of anisotropy and asymmetry in rotating machinery: Theory. *Mechanical Systems and Signal Processing,* vol. 8, no. 6, pp. 665–678.

Leonov, V. P., and A. N. Shiryaev (1959). On a method of calculation of semi-invariants. *Theory of Probability and its Applications,* vol. IV, no. 3, pp. 319–329.

Levy, E. C. (1959). Complex curve fitting. *IEEE Trans. Autom. Contr.*, vol. 4, pp. 37–43.

Liang, G., D. M. Wilkes, and J. A. Cadzow (1993). ARMA model order estimation based on the eigenvalues of the covariance matrix. *IEEE Trans. Sign. Proc.*, vol. 41, pp. 3003–3009.

Little, R. J. A., and D. B. Rubin (1987). *Statistical Analysis with Missing Data.* Wiley, New York.

Ljung, L. (1985). Asymptotic variance expressions for identified black-box transfer function models. *IEEE Trans. Autom. Contr.*, vol. 30, no. 9, pp. 834–844.

Ljung, L. (1993). Some results on identifying linear systems using frequency domain data. *Proc. 32nd Conf. Decis. Contr.,* San Antonio, TX, pp. 3534–3538.

Ljung, L. (1995). *System Identification Toolbox User's Guide.* The MathWorks, Natick, MA.

Ljung, L. (1999). *System Identification: Theory for the User* (2nd ed.). Prentice-Hall, Upper Saddle River, NJ.

Ljung, L., and T. Söderström (1983). *Theory and Practice of Recursive Identification.* MIT Press, Cambridge, MA

Lowen, S. B., and M. C. Teich (1990). Power-law shot noise. *IEEE Trans. Inform. Theory,* vol. 36, no. 6, pp. 1302–1318.

Lukacs, E. (1975). *Stochastic Convergence.* Academic Press, New York.

Mannetje: see 't Mannetje.

Mahata, K., R. Pintelon and J. Schoukens (2006). On parameter estimation using non-parametric noise models. *IEEE Trans. Autom. Contr.*, vol. 51, no. 10, pp. 1602–1612.

Maiwald, D., and D. Kraus (2000). Calculation of moments of complex Wishart and complex inverse Wishart distributed matrices. *IEE Proc. Radar, Sonar and Navigation,* vol. 147, no. 4, pp. 162–168.

Mathai, A. M., and S. B. Provost (1992). *Quadratic Forms in Random Variables.* Marcel Dekker, New York.

Mehra, R. (1974). Optimal input signals for parameter estimation in dynamic systems–survey and new results. *Trans. Autom. Contr.*, vol. 19, no. 6, pp. 753–768.

McCormack, A. S., J. O. Flower, and K. R. Godfrey (1994a). The suppression of drift and transient effects for frequecy-domain identification. *IEEE Trans. Instrum. and Meas.*, vol. 43, no. 2, pp. 232–237.

McCormack, A. S., K. R. Godfrey, and J. O. Flower (1994b). The detection of and compensation for nonlinear effects using periodic input signals. *Proceedings IEE Control '94*, Coventry, UK, pp. 297–302.

McCormack, A. S., K. R. Godfrey, and J. O. Flower, (1995). Design of multilevel multiharmonic signals for system identification. *IEE Proceedings, Control Theory and Applications,* vol. 142, no. 3, pp. 247–252.

McKelvey, T. (2000). Frequency domain identification. *Preprints 12th IFAC Symposium on System Identification,* Santa Barbara, California (USA), June 21–23.

McKelvey, T., H. Akçay, and L. Ljung (1996). Subspace-based multivariable system identification from frequency response data, *IEEE Trans. Autom. Contr.,* vol. 41, no. 7, pp. 960–979.

McKelvey, T., and A. Helmersson (1997). System identification using an over-parametrized model class—Improving the optimization algorithm. *Proceedings of the 36th Conference on Decision and Control,* San Diego, CA, pp. 2984–2989.

McKelvey, T., A. Helmersson, and T. Ribarits (2004). Data driven local coordinates for multivariable linear systems and their application to system identification. *Automatica,* vol. 40, no. 9, pp. 1629–1635.

Mendel, J. M. (1991). Tutorial on higher-order statistics (spectra) in signal processing and system theory: Theoretical results and some applications. *Proc. IEEE*, vol. 79, pp. 278–305.

Middleton, R. H., and G. C. Goodwin (1990). *Digital Control and Estimation.* Prentice-Hall, London.

Moreau, X., C. Ramus-Serment, and A. Oustaloup (2002). Fractional differentiation in passive vibration control. *Nonlinear Dynamics,* vol. 29, no. 1–4, pp. 343–362.

Natke, H. G., J.-N. Juang, and W. Gawronski (1988). A brief review on the identification of nonlinear mechanical systems. *Proc. of the 6th International Modal Analysis Conference,* Kissimee, FL, pp. 1569–1574.

Nikias, C. L., and J.M. Mendel (1993). Signal processing with higher-order spectra. *IEEE Signal Processing Magazine,* July 1993, pp. 10–36.

Nikias, C. L., and A. P. Petropulu (1993). *Higher-Order Spectra Analysis.* Prentice-Hall, Englewood Cliffs, NJ.

Ninness, B. M., and G. C. Goodwin (1991). The relationship between discrete time and continuous time linear estimation. In N. K. Sinha and G. P. Rao, editors, *Identification of Continuous-Time Systems.* Kluwer Academic Publishers, Dordrecht, pp. 79–122.

Norton, J. P. (1986). *An Introduction to Identification.* Academic Press, London.

Nuttall, A. H., and G. C. Carter (1982). Spectral estimation combined time and lag weighting. *Proceedings of the IEEE,* vol. 70, no. 9, pp. 1111–1125.

Oldham, K. B., and J. Spanier (1974). *The Fractional Calculus.* Academic Press, New York.

Oppenheim, A. V., and R. W. Schafer (1975). *Digital Signal Processing.* Prentice-Hall, New York.

Oppenheim, A. V., A. S Willsky, and S. H. Nawab (1997). *Signals and Systems.* Prentice-Hall, London.

Orey, S. (1958). A central limit theorem for m-dependent random variables. *Duke Math. J.,* vol. 25, pp. 543–546.

Paehlike, K. D., and H. Rake (1979). Binary multifrequency signals-synthesis and application. *Proc. 5th IFAC Symp.,* Darmstadt, FRG, pp. 589–597.

Paige, C. C. (1986). Computing the generalized singular value decomposition. *SIAM J. Sci. Stat. Comput.,* vol. 7, no. 4, pp. 1126–1146.

Papoulis, A. (1981). *Probability, Random Variables, and Stochastic Processes.* McGraw-Hill, New York.

Peeters, F., R. Pintelon, J. Schoukens, and Y. Rolain (2000). Parametric identification of rotor-bearing systems in the frequency domain. *Proceedings of the 18th International Modal Analysis Conference,* San Antonio, TX, pp. 1355–1361.

Peirlinckx, L., P. Guillaume, and R. Pintelon (1996). Accurate and fast estimation of the Fourier coefficients of periodic signals disturbed by a trend. *IEEE Trans. Instrum. and Meas.,* vol. 45, no. 1, pp. 5–11.

Phadke, M. S., and S. M. Wu (1974). Modeling of continuous-time stochastic processes from discrete observations with application to sunspots data. *Journal of the American Statistical Association,* vol. 69, no. 346, pp. 325–329.

Picinbono, B. (1993). *Random Signals and Systems.* Prentice-Hall, Englewood Cliffs, NJ.

Pintelon, R. (1990). Phase correction of linear time invariant systems with digital all-pas filters, *IEEE Trans. Instrum. and Meas.,* vol. 39, no. 2, pp. 324–330.

Pintelon, R. (1991). Comments on "Design of IIR filters in the complex domain." *IEEE Trans. Sign. Proc.,* vol. 39, no. 6, pp. 1454–1455.

Pintelon, R., K. Barbé, G. Vandersteen, and J. Schoukens (2011a). Improved (non-)parametric identification of dynamic systems excited by periodic signals. *Mechanical Systems and Signal Processing,* vol. 25, no. 7, pp. 2683–2704.

Pintelon, R., P. Guillaume, Y. Rolain, J. Schoukens, and H. Van hamme (1994). Parametric identification of transfer functions in the frequency domain—A survey. *IEEE Trans. Autom. Contr.,* vol. 39, no. 11, pp. 2245–2260.

Pintelon, R., P. Guillaume, Y. Rolain, and F. Verbeyst (1992). Identification of linear systems captured in a feedback loop. *IEEE Trans. Instrum. and Meas.,* vol. 41, no. 6, pp. 747–754.

Pintelon, R., P. Guillaume, and J. Schoukens (1996a). Measurement of noise (cross-) power spectra for frequency-domain system identification purposes: Large-sample results. *IEEE Trans. Instrum. and Meas.,* vol. 45, no. 1, pp. 12–21.

Pintelon, R., P. Guillaume, and J. Schoukens (2007b). Uncertainty calculation in (operational) modal analysis. *Mechanical Systems and Signal Processing,* vol. 21, no. 6, pp. 2359–2373.

Pintelon, R., P. Guillaume, G. Vandersteen, and Y. Rolain (1998). Analyses, development and applications of TLS algorithms in frequency-domain system identification. *SIAM J. Matrix Anal. Appl.,* vol. 19, no. 4, pp. 983–1004.

Pintelon, R., P. Guillaume, S. Vanlanduit, K. De Belder and Y. Rolain (2004). Identification of Young's modulus from broadband modal analysis experiments. *Mechanical Systems and Signal Processing,* vol. 18, no. 4, pp. 699–726.

Pintelon, R. and M. Hong (2007). Asymptotic uncertainty of transfer function estimates using nonparametric noise models, *IEEE Trans. Instrum. Meas.,* vol. 56, no. 6, pp. 2599–2605.

Pintelon, R., and I. Kollár (2005). On the frequency scaling in continuous-time modeling, *IEEE Trans. Instrum. and Meas.,* vol. 54, no. 1, pp. 318–321.

Pintelon, R., Y. Rolain, and J. Schoukens (2006a). Box-Jenkins identification revisited—Part II: applications, *Automatica,* vol. 42, no. 1, pp. 77–84.

Pintelon, R., Y. Rolain, M. Vanden Bossche, and J. Schoukens (1990). Towards an ideal data acquisition channel. *IEEE Trans. Instrum. and Meas.,* vol. 39, no. 1, pp. 116–120.

Pintelon, R., Y. Rolain, G. Vandersteen, and J. Schoukens (2004b). Experimental characterization of operational amplifiers: a system identification approach—Part II: Calibration and measurements. *IEEE Trans. Instrum. and Meas.,* vol. 53, no. 3, pp. 863–876.

Pintelon, R., Y. Rolain and W. Van Moer (2003). Probability density function for frequency response function measurements using periodic signals. *IEEE Trans. Instrum. and Meas.,* vol. 52, no. 1, pp. 61–68.

Pintelon, R., and J. Schoukens (1990a). Robust identification of transfer functions in the s- and z-domains. *IEEE Trans. Instrum. and Meas.,* vol. 39, no. 4, pp. 565–573.

Pintelon, R., and J. Schoukens (1990b). Real-time integration and differentiation of analog signals by means of digital filtering. *IEEE Trans. Instrum. and Meas.*, vol. 39, no. 6, pp. 923–927.

Pintelon, R., and J. Schoukens (1996). An improved sine-wave fitting procedure for characterizing data acquisition channels. *IEEE Trans. Instrum. and Meas.*, vol. 45, no. 2, pp. 588–593.

Pintelon, R., and J. Schoukens (1997a). Frequency-domain identification of linear time-invariant systems under nonstandard conditions. *IEEE Trans. Instrum. and Meas.*, vol. 46, no. 1, pp. 65–71.

Pintelon, R., and J. Schoukens (1997b). Identification of continuous-time systems using arbitrary signals. *Automatica*, vol. 33, no. 5, pp. 991–994.

Pintelon, R., and J. Schoukens (1999a). Time series analysis in the frequency domain. *IEEE Trans. Sign. Proc.*, vol. 47, no. 1, pp. 206–210.

Pintelon, R., and J. Schoukens (1999b). Identification of continuous-time systems with missing data. *IEEE Trans. Instrum. and Meas.*, vol. 48, no. 3, pp. 736–740.

Pintelon, R., and J. Schoukens (2000). Frequency domain system identification with missing data. *IEEE Trans. Autom. Contr.*, vol. 45, no. 2, pp. 364–369.

Pintelon, R., and J. Schoukens (2001). Measurement of frequency response functions using periodic excitations, corrupted by correlated input/output errors. *IEEE Trans. Instrum. and Meas.*, vol. 50, no. 6, pp. 1753–1760.

Pintelon, R., and J. Schoukens (2006). Box-Jenkins identification revisited—Part I: theory, *Automatica*, vol. 42, no. 1, pp. 63–75.

Pintelon, R., and J. Schoukens (2007). Frequency domain maximum likelihood estimation of linear dynamic errors-in-variables models, *Automatica*, vol. 43, no. 4, pp. 621–630.

Pintelon, R., J. Schoukens, and P. Guillaume (1989). Parametric frequency domain modeling in modal analysis. *Mechanical Systems and Signal Processing*, vol. 3, no. 4, pp. 389–403.

Pintelon, R., J. Schoukens, and P. Guillaume (2006b). Continuous-time noise modelling from sampled data, *IEEE Trans. Instrum. and Meas.*, vol. 55, no. 6, pp. 2253–2258.

Pintelon, R., J. Schoukens, and P. Guillaume (2007a). Box-Jenkins identification revisited—Part III: multivariable systems, *Automatica*, vol. 43, no. 5, pp. 868–875.

Pintelon, R., J. Schoukens, T. McKelvey, and Y. Rolain (1996b). Minimum variance bounds for overparametrized models. *IEEE Trans. Autom. Contr.*, vol. 41, no. 5, pp. 719–720.

Pintelon, R., J. Schoukens, L. Pauwels, and E. Van Gheem (2005). Diffusion systems: stability, modeling, and identification. *IEEE Trans. Instrum. and Meas.*, vol. 54, no. 5, pp. 2061–2067.

Pintelon, R., J. Schoukens, and J. Renneboog (1988). The geometric mean of power (amplitude) spectra has a much smaller bias than the classical arithmetic (rms) averaging. *IEEE Trans. Instrum. and Meas.*, vol. 37, no. 2, pp. 213–218.

Pintelon, R., J. Schoukens, and Y. Rolain (2000). Box-Jenkins continuous-time modeling, *Automatica*, vol. 36, no. 7, pp. 983–991.

Pintelon, R., J. Schoukens, and G. Vandersteen (1997a). Model selection through a statistical analysis of the global minimum of a weighted nonlinear least squares cost function. *IEEE Trans. Sign. Proc.*, vol. 45, no. 3, pp. 686–693.

Pintelon, R., J. Schoukens, and G. Vandersteen (1997b). Frequency domain system identification using arbitrary signals. *IEEE Trans. Autom. Contr.*, vol. 42, no. 12, pp. 1717–1720.

Pintelon, R., J. Schoukens, G. Vandersteen, and K. Barbé (2010a). Estimation of nonparametric noise and FRF models for multivariable systems—Part I: theory. *Mechanical Systems and Signal Processing*, vol. 24, no. 3, pp. 573–595.

Pintelon, R., J. Schoukens, G. Vandersteen, and K. Barbé (2010b). Estimation of nonparametric noise and FRF models for multivariable systems—Part II: extensions, applications. *Mechanical Systems and Signal Processing*, vol. 24, no. 3, pp. 596–616.

Pintelon, R., J. Schoukens, G. Vandersteen, and Y. Rolain (1999). Identification of invariants of (over)parametrized models: finite sample results. *IEEE Trans. Autom. Contr.*, vol. 44, no. 5, pp. 1073–1077.

Pintelon, R., and L. Van Biesen (1990). Identification of transfer functions with time delay and its application to cable fault location. *IEEE Trans. Instrum. and Meas.*, vol. 39, no. 3, pp. 479–484.

Pintelon, R., G. Vandersteen, L. De Locht, Y. Rolain, and J. Schoukens (2004a). Experimental characterization of operational amplifiers: a system identification approach—Part I: Theory and simulations. *IEEE Trans. Instrum. and Meas.*, vol. 53, no. 3, pp. 854–862.

Pintelon, R., G. Vandersteen, J. Schoukens, and Y. Rolain (2011b). Improved (non-)parametric identification of dynamic systems excited by periodic signals – The multivariate case. *Mechanical Systems and Signal Processing*, vol. 25, no. 8, pp. 2892–2922.

Pyati, V. P. (1992). An exact expression for the noise voltage across a resistor shunted by a capacitor. *IEEE Trans. Circuits Systems-I*, vol. 39, no. 12, pp. 1027–1029.

Rabiner, L. R., and B. Gold (1975). *Theory and Application of Digital Signal Processing*. Prentice-Hall, New York.

Ralston, A., and P. Rabinowitz (1984). *A First Course in Numerical Analysis*. McGraw-Hill, New York.

Richardson, M. H., and D. L. Formenti (1982). Parameter estimation from frequency response measurements using rational fraction polynomials. *Proc. First Int. Modal Analysis Conf.*, Orlando, FL, vol. 1, pp. 167–181.

Rissanen, J. (1978). Modeling by shortest data description. *Automatica*, vol. 14, pp. 465–471.

Rivera, D. E., H. Lee, H. D. Mittelmann, and M. W. Braun (2007). High-purity distillation – Using plant-friendly multisine signals to identify a strongly interactive process. *IEEE Control Systems Magazine*, vol. 27, no. 5, pp. 72–89.

Rivera, D. E., H. Lee, H. D. Mittelmann, and M. W. Braun (2009). Constrained multisine input signals for plant-friendly identification of chemical process systems. *Journal of Process Control*, vol. 19, no. 4, pp. 623–635.

Rizzi, P. A. (1988). *Microwave Engineering*. Prentice-Hall, London.

Rolain, Y., and R. Pintelon (1999). Generating robust starting values for frequency-domain transfer function estimation. *Automatica*, vol. 35, pp. 965–973.

Rolain, Y., R. Pintelon, K. Q. Xu, and H. Vold (1995). Best conditioned parametric identification of transfer function models in the frequency domain. *IEEE Trans. Autom. Contr.*, vol. 40, no. 11, pp. 1954–1960.

Rolain, Y., and J. Schoukens (1990). Design and implementation of a fast logarithmic stepped sine for a fixed sample rate digital network analyzer. *IEEE Trans. Instrum. and Meas.*, vol. 39, no. 1, pp. 151–156.

Rolain Y., J. Schoukens, and R. Pintelon (1997). Order estimation for linear time-invariant systems using frequency domain identification methods. *IEEE Trans. Autom. Contr.*, vol. 42, no. 10, pp. 1408–1417.

Rolain, Y., J. Schoukens, and G. Vandersteen (1998). Signal reconstruction for non-equidistant finite length sample sets: A "KIS" approach. *IEEE Trans. Instrum. and Meas.*, vol. 47, no. 5, pp. 1046–1052.

Rosén, B. (1967). On the central limit theorem for sums of dependent random variables, *Z. Wahrsch. verw. Geb.*, vol. 7, pp. 48–82.

Sakakibara, S. (1997). Properties of vibration with fractional derivative damping of order 1/2. *JSME Int. Journal*, Series C, vol. 40, no. 3, pp. 393–399.

Sanathanan, C. K., and J. Koerner (1963). Transfer function synthesis as a ratio of two complex polynomials. *IEEE Trans. Autom. Contr.*, vol. 8, pp. 56–58.

Schetzen, M. (1980). *The Volterra and Wiener Theories of Nonlinear Systems*. Wiley, New York.

Schoukens, J. (1990). Modeling of continuous time systems using a discrete time representation. *Automatica*, vol. 26, no. 3, pp. 579–583.

Schoukens, J., K. Barbé, L. Vanbeylen, and R. Pintelon (2010). Nonlinear induced variance of the frequency response function measurements, *IEEE Trans. Instrum. and Meas.*, vol. 59, no. 9, pp. 2468–2474.

Schoukens, J., T. Dobrowiecki, and R. Pintelon (1998a). Identification of linear systems in the presence of nonlinear distortions. A frequency domain approach. *IEEE Trans. Autom. Contr.*, vol. 43, no. 2, pp. 176–190.

Schoukens J., P. Guillaume, and R. Pintelon (1993). Design of broadband excitation signals. In K. R. Godfrey, editor, *Perturbation Signals for System Identification*. Prentice-Hall, Hemel Hempstead, pp. 126–160.

Schoukens, J., P. Guillaume, and R. Pintelon (1995). Generating piecewise-constant excitations with an arbitrary power spectrum. *IEE Proc. Control Theory Appl.*, vol. 142, no. 3, pp. 241–246.

Schoukens, J., J. Lataire, R. Pintelon, G. Vandersteen, and T. Dobrowiecki (2009). Robustness issues of the equivalent linear representation of a nonlinear system. *IEEE Trans. Instrum. and Meas.*, vol. 58, no. 5, pp. 1737–1745.

Schoukens, J., F. Louage, and Y. Rolain (1996a). Study of the influence of clock instabilities in synchronized data acquisition systems. *IEEE Trans. Instrum. and Meas.*, vol. 45, no. 2, pp. 601–604.

Schoukens, J., and R. Pintelon (1990). Measurement of frequency response functions in noisy environments. *IEEE Trans. Instrum. and Meas.*, vol. 39, no. 6, pp. 905–909.

Schoukens, J., and R. Pintelon (1991). *Identification of Linear Systems: A Practical Guideline to Accurate Modeling*. Pergamon, Oxford.

Schoukens, J., and R. Pintelon (2010a). High quality frequency response function measurement without user interaction. *Proceedings UKACC International Conference on Control*, Coventry (UK), Sept. 7–10, pp. 932–936.

Schoukens, J., and R. Pintelon (2010b). Study of the variance of parametric estimates of the best linear approximation of nonlinear systems. *IEEE Trans. Instrum. and Meas.*, vol. 59, no. 12, pp. 3159–3167.

Schoukens, J., R. Pintelon, T. Dobrowiecki, and Y. Rolain (2005). Identification of linear systems with nonlinear distortions, *Automatica*, vol. 41, no. 2, pp. 491–504.

Schoukens, J., R. Pintelon, and J. Renneboog (1988a). A maximum likelihood estimator for linear and nonlinear systems—A practical application of estimation techniques in measurement problems. *IEEE Trans. Instrum. and Meas.*, vol. 37, no. 1, pp. 10–17.

Schoukens, J., R. Pintelon, and Y. Rolain (1997a). Maximum likelihood estimation of errors-in-variables models using a sample covariance matrix obtained from small data sets. In S. Van Huffel, editor, *Recent Advances in Total Least Squares Techniques and Errors-in-Variables Modeling*. SIAM, Philadelphia, pp. 59–68.

Schoukens, J., R. Pintelon, and Y. Rolain (1999a). Study of conditional ML estimators in time and frequency domain system identification. *Automatica*, vol. 35, no. 1, pp. 91–100.

Schoukens, J., R. Pintelon, and Y. Rolain (2000). Broadband versus stepped sine FRF measurements. *IEEE Trans. on Instrum. and Meas.*, vol. 49, no. 2, pp. 275–278.

Schoukens, J., R. Pintelon, E. Van der Ouderaa, and J. Renneboog (1988b). Survey of excitation signals for F.F.T. based signal analyzers. *IEEE Trans. Instrum. and Meas.*, vol. 37, no. 3, pp. 342–351.

Schoukens, J., R. Pintelon, and H. Van hamme (1994). Identification of linear dynamic systems using piecewise constant excitations: Use, misuse and alternatives. *Automatica*, vol. 30, no. 7, pp. 1153–1169.

Schoukens, J., R. Pintelon, G. Vandersteen, and P. Guillaume (1997b). Frequency-domain system identification using non-parametric noise models estimated from a small number of data sets. *Automatica*, vol. 33, no. 6, pp. 1073–1086.

Schoukens, J., and J. Renneboog (1986). Modeling the noise influence on the Fourier coefficients after a discrete Fourier transform. *IEEE Trans. Instrum. and Meas.*, vol. 35, no. 3, pp. 278–286.

Schoukens, J., Y. Rolain, and P. Guillaume (1996b). Design of narrow band, high resolution multisines. *IEEE Trans, Instrum. and Meas.*, vol. 45, no. 3, pp. 750–753.

References

Schoukens, J., Y. Rolain, F. Gustafsson, and R. Pintelon (1998b). Fast calculation of least-squares estimates for system identification. *Proceedings of the 37th IEEE Conference on Decision and Control,* Tampa, FL, pp. 3408–3410.

Schoukens, J., Y. Rolain, and R. Pintelon (1998c). Improved frequency response function measurements using random noise excitations. *IEEE Trans. Instrum. and Meas.,* vol. 47, no. 1, pp. 322–326.

Schoukens, J., Y. Rolain and R. Pintelon (2002). Modified AIC rule for model selection in combination with prior estimated noise models, *Automatica,* vol. 38, no. 5, pp. 903–906.

Schoukens, J., Y. Rolain, and R. Pintelon (2006a). Analysis of windowing/leakage effects in frequency response function measurements. *Automatica,* vol. 42, no. 1, pp. 27–38.

Schoukens, J., Y. Rolain, and R. Pintelon (2006b). Leakage reduction in frequency response function measurements. *IEEE Trans. Instrum. and Meas.,* vol. 55, no. 6, pp. 2286–2291.

Schoukens, J., G. Vandersteen, R. Pintelon, and P. Guillaume (1999b). Frequency domain identification of linear systems using arbitrary excitations and a nonparametric noise model. *IEEE Trans. Autom. Contr.,* vol. 44, no. 2, pp. 343–347.

Schoukens, J., Y. Rolain, G. Vandersteen, and R. Pintelon (2011). User friendly Box-Jenkins identification using nonparametric noise models, *Proceedings 50th IEEE Conference on Decision and Control and European Control Conference,* Orlando, Florida (USA), Dec. 12–15, pp. 2148–2153.

Schroeder, M. R. (1970). Synthesis of low peak factor signals and binary sequences with low autocorrelation. *IEEE Trans. Inform. Theory,* vol. IT-16, pp. 85–89.

Selby, S. M. (1973). *Standard Mathematical Tables.* The Chemical Rubber Company, Cleveland, OH.

Shapiro, A. (1986). Asymptotic theory of overparametrized structural models. *J. Am. Statistical Assoc.,* vol. 81, no. 393, pp. 142–149.

Shibata, R. (1980). Asymptotically efficient selection of the order of the model for estimating parameters of a linear process. *The Annals of Statistics,* vol. 8, no. 1, pp. 147–164.

Sidman, M. D., F. E. DeAngelis, and G. C. Verghese (1991). Parametric system identification on logarithmic frequency response data. *IEEE Trans. Autom. Contr.,* vol. 36, no. 9, pp. 1065–1070.

Sinha, N. K., and G. P. Rao, editors (1991). *Identification of Continuous-Time Systems: Methodology and Computer Implementation.* Kluwer, Dordrecht.

Söderström, T. (1974). Convergence properties of the generalized least squares identification method. *Automatica,* vol. 10, pp. 617–626.

Söderström, T. (2007). Errors-in-variables identification in system identification. *Automatica,* vol. 43, no. 6, pp. 939–958.

Söderström, T., and M. Mossberg (2000). Performance evaluation of methods for identifying continuous-time autoregressive processes. *Automatica,* vol. 36, pp. 53–59.

Söderström, T., H. Fan, B. Carlsson, and S. Bigi (1997a). Least squares parameter estimation of continuous-time ARX models from discrete-time data. *IEEE Trans. Autom. Contr.,* vol. 42, pp. 659–673.

Söderström, T., H. Fan, M. Mossberg, and B. Carlsson (1997b). Bias-compensation schemes for estimating continuous-time AR process parameters. *Proceedings of the 11th IFAC Symposium System Identification,* Kitakyushu, Japan, vol. 3, pp. 1337–1342.

Söderström, T., and P. Stoica (1981). Comparison of some instrumental variable methods—consistency and accuracy aspects. *Automatica,* vol. 17, pp. 101–115.

Söderström, T., and P. Stoica (1989). *System Identification.* Prentice-Hall, Englewood Cliffs, NJ, p. 256.

Sorenson, H. W. (1980). *Parameter Estimation: Principles and Problems.* Marcel Dekker, New York.

Souders, T. M., D. R. Flach, C. Hagwood, and G. L. Yang (1990). The effect of time jitter in sampling systems. *IEEE Trans. Instrum. and Meas.,* vol. 39, pp. 80–85.

Spiegel, M. R. (1965). *Theory and Problems of Laplace Transforms.* McGraw-Hill, New York.

Steigliz, K. and L. E. McBride (1965). A technique for the identification of linear systems. *IEEE Trans. Autom. Contr.,* vol. 10, pp. 461–464.

Stoica, P., P. Eykhoff, P. Janssen, and T. Söderström (1986). Model-structure selection by cross-validation. *Int. J. Control,* vol. 43, no. 6, pp. 1841–1878.

Stoica, P., and R. L. Moses (1990). On biased estimators and the unbiased Cramér-Rao lower bound. *Signal Processing,* vol. 21, pp. 349–350.

Stout, W. F. (1974). *Almost Sure Convergence.* Academic Press, New York.

Stuart, A., and J. K. Ord (1987). *Distribution Theory,* vol. 1 of *Kendall's Advanced Theory of Statistics.* Charles Griffin, London.

Swevers, J., B. De Moor, and H. Van Brussel (1992). Stepped sine system identification, errors-in-variables and the quotient singular value decomposition. *Mechanical Systems and Signal Processing,* vol. 6, no. 2, pp. 121–134.

Tan, A. H., K. R. Godfrey, and H. A. Barker (2005). Design of computer-optimized pseudorandom maximum length signals for linear identification in the presence of nonlinear distortions. *IEEE Trans. Instrum. and Meas.,* vol. 54, no. 6, pp. 2513–2519.

Tan, A. H., K. R. Godfrey, and H. A. Barker (2009). Design of ternary signals for MIMO identification in the presence of noise and nonlinear distortion. *IEEE Trans. Contr. Systems Techn.,* vol. 17, no. 4, pp. 926–933.

Temes, G. C., and J. W. LaPatra (1977). *Circuit Synthesis and Design.* McGraw-Hill, New York.

't Mannetje, J. J. (1973). Transfer-function identification using a complex curve-fitting technique. *J. Mech. Eng. Sci.,* vol. 15, no. 5, pp. 339–345.

Tomlinson, G. R. (1987). Developments in the use of the Hilbert transform for detecting and quantifying non-linearity associated with frequency response functions. *Mechanical Systems and Signal Processing,* vol. 1, no. 2, pp. 151–171.

Torfs, D., R. Vuerinckx, J. Swevers, and J. Schoukens (1998). Comparison of two feedforward design methods aiming at accurate trajectory tracking of the end point of a flexible robot arm. *IEEE Trans. Control Systems Technology,* vol. 6, no. 1, pp. 2–14.

Van Barel, M., and A. Bultheel (1992). A parallel algorithm for discrete least squares rational approximation. *Numer. Math.,* vol. 63, pp. 99–121.

Van Barel, M., and A. Bultheel (1994). Discrete linearized least-squares rational approximation on the unit circle. *Journal of Computational and Applied Mathematics,* vol. 50, pp. 545–563.

Van Brussel, H. (1975). Comparative assessment of harmonic, random, swept sine and shock excitation methods for the identification of machine tool structures with rotating spindles. *Ann. CIRP,* pp. 291–296.

Van den Bos, A. (1974). *Estimation of Parameters of Linear Systems using Periodic Test Signals.* Doctoral thesis, T.U. Delft, The Netherlands, Delftse Universitaire Pers.

Van den Bos, A. (1985). Nonlinear least-absolute-values and minimax model fitting. *7th IFAC Symposium on Identification and System Parameter Estimation,* York, UK, pp. 173–177A.

Van den Bos, A. (1987). A new method for synthesis of low-peak-factor signals. *IEEE Trans. Acoustics, Speech and Signal Processing,* vol. 35, pp. 120–122.

Van den Bos, A. (1991). Identification of continuous-time systems using multiharmonic test signals. In N.K. Sinha and G.P. Rao, editors, *Identification of Continuous-Time Systems.* Kluwer Academic Publishers, Dordrecht, pp. 489–508.

Van den Bos, A., and R. G. Krol (1979). Synthesis of discrete-interval binary signals with specified Fourier amplitude spectra. *Int. J. Contr.,* vol. 30, no. 5, pp. 871–884.

Van den Bos, A., and J. H. Swarte (1993). Resolvability of the parameters of multiexponentials and other sum models, *IEEE Trans. Sign. Proc.,* vol. 41, pp. 313–322.

Vanden Bossche, M., J. Schoukens, and J. Renneboog (1986). Dynamic testing and diagnostics of A/D converters. *IEEE Trans. Circuits and Systems,* vol. CAS-33, no. 8, pp. 775–785.

References

Van den Hof, P. M. J., and R. J. P. Schrama (1995). Identification and control—Closed loop issues, *Automatica,* vol. 31, no. 12, pp. 1751–1770.

Van den Eijnde, E., and J. Schoukens (1991). On the design of optimal excitation signals. *Preprints of the 9th IFAC/IFORS Symposium on Identification and System Parameter Estimation,* Budapest, Hungary, pp. 827–832.

Van den Enden, A. W. M., G. C. Groendael, and E. Van de Zee (1977). An improved complex-curve fitting method. *Proc. Conf. Computer Aided Design of Electronic, Microwave Circuits and Syst.,* Hull, UK, pp. 53–58.

Van den Enden, A. W. M., and G. A. L. Leenknegt (1986). Design of optimal filters with arbitrary amplitude and phase requirements. In P. Young et al., editors, *Signal Processing III: Theories and Applications,* Elsevier Science, North Holland, pp. 183–186.

Van der Ouderaa, E., and J. Renneboog (1988). Logtone crest factors. *IEEE Trans. Instrum. and Meas.,* vol. 37, pp. 656–657.

Van der Ouderaa, E., J. Schoukens, and J. Renneboog (1988a). Peak factor minimization using a time-frequency domain swapping algorithm. *IEEE Trans. Instrum. and Meas.,* vol. 37, no. 1, pp. 145–147.

Van der Ouderaa, E., J. Schoukens, and J. Renneboog (1988b). Peak factor minimization of input and output signals of linear systems. *IEEE Trans. Instrum. and Meas.,* vol. 37, no. 2, pp. 207–212.

Van Gestel, T., J. A. K. Suykens, P. Van Dooren, and B. De Moor (2001). Identification of stable models in subspace identification by using regularization. *IEEE Trans. Autom. Contr.,* vol. 46, no. 9, pp. 1416–1420.

Van hamme, H., R. Pintelon, and J. Schoukens (1991). Discrete-time modeling and identification of continuous time systems: a general framework. In M. K. Sinha and G. P. Rao, editors, *Identification of Continuous-Time Systems.* Kluwer Academic Publishers, Dordrecht, pp. 17–77.

Vanhoenacker, K., J. Schoukens, J. Swevers, and D. Vaes (2002). Summary and comparing overview of techniques for the detection of non-linear distortions. *Proceedings of ISMA 2002: International Conference on Noise and Vibration Engineering,* Leuven (Belgium), Sept. 16–18, vols. 1–5, pp. 1241–1255.

Van Huffel, S., and J. Vandewalle (1991). *The Total Least Squares Problem: Computational Aspects and Analysis.* Frontiers in Applied Mathematics. SIAM, Philadelphia.

Van Overbeek, A. J. M., and L. Ljung (1982). On-line structure selection for multivariable state-space models. *Automatica,* vol. 18, no. 5, pp. 529–543.

Van Overschee, P., and B. De Moor (1994). N4SID: subspace algorithms for the identification of combined deterministic-stochastic systems. *Automatica,* vol. 30, no. 1, pp. 75–93.

Van Overschee, P., and B. De Moor (1996a). Continuous-time frequency domain subspace system identification, *Signal Processing,* vol. 52, no. 2, pp. 179–194.

Van Overschee, P., and B. De Moor (1996b). *Subspace Identification of Linear Systems: Theory, Implementation, Applications.* Kluwer Academic Publishers, Dordrecht.

Vandersteen, G., Y. Rolain, J. Schoukens, and R. Pintelon (1996a). On the use of system identification for accurate parametric modelling of non-linear systems using noisy measurements. *IEEE Trans. Instrum. and Meas.,* vol. 45, no. 2, pp. 605–609.

Vandersteen, G., H. Van hamme, and R. Pintelon (1996b). General framework for asymptotic properties of generalized weighted nonlinear least-squares estimators with deterministic and stochastic weighting. *IEEE Trans. Autom. Contr.,* vol. 41, no. 10, pp. 1501–1507.

Vanhoenacker, K., and J. Schoukens (1999). Frequency response function measurements in the presence of nonlinear distortions. *WISP '99, IEEE International Workshop on Intelligent Signal Processing,* Budapest, Hungary, pp. 87–92.

Verbeeck, J., R. Pintelon, and P. Guillaume (1999a). Determination of synchronous machine parameters using network synthesis techniques. *IEEE Trans. Energy Conv.,* vol. 14, no. 3, pp. 310–314.

Verbeeck, J., R. Pintelon, and P. Lataire (1999b). Identification of synchronous machine parameters using a multiple input multiple output approach. *IEEE Trans. Energy Conv.*, vol. 14, no. 4, pp. 909–917.

Verboven, P., P. Guillaume, and B. Cauberghe (2005). Multivariable frequency-response curve fitting with application to modal parameter estimation, *Automatica*, vol. 41, no. 10, pp. 1773–1782.

Verboven, P., P. Guillaume, B. Cauberghe, E. Parloo, and S. Vanlanduit (2004). Frequency-domain generalized total least-squares identification for modal analysis, *Journal of Sound and Vibration*, vol. 278, no. 1–2, pp. 21–38.

Verhaegen, M. (1994). Identification of the deterministic part of MIMO state space models given in innovations form from input-output data. *Automatica*, vol. 30, no. 1, pp. 61–74.

Verschueren, A., Y. Rolain, R. Vuerinckx, and G. Vandersteen (1998). Identifying S-parameter models in the Laplace domain for high frequency multiple port linear networks. *IEEE-MTT-S International Microwave Symposium Digest*, Baltimore, pp. 25–28.

Viberg, M., B. Wahlberg, and B. Ottersten (1997). Analysis of state space system identification methods based on instrumental variables and subspace fitting. *Automatica*, vol. 33, no. 9, pp. 1603–1616.

Vuerinckx, R., Y. Rolain, J. Schoukens, and R. Pintelon (1996). Design of stable IIR filters in the complex domain by automatic delay selection. *IEEE Trans. Sign. Proc.*, vol. 44, no. 9, pp. 2339–2344.

Vuerinckx, R., R. Pintelon, J. Schoukens, and Y. Rolain (1998). Obtaining accurate confidence regions for the estimated zeros and poles in system identification problems. *Proceedings of the 37th IEEE Conference on Decision and Control*, Tampa, FL, pp. 4464–4469.

Walter, E., and L. Pronzato (1997). *Identification of Parametric Models from Experimental Data*. Springer, Paris.

Wang, J. C. (1987). Realizations of generalized Warburg impedance with RC ladder networks and transmission lines. *J. Electrochem. Soc.*, vol. 134, no. 8, pp. 1915–1920.

Wellstead, P. E. (1977). Reference signals for closed-loop identification. *Int. J. Control*, vol. 26, no. 6, pp. 945–962.

Wellstead, P. E. (1981). Non-parametric methods of system identification. *Automatica*, vol. 17, no. 1, pp. 55–69.

Welsh, P. (1967). The use of the fast fourier transform for the estimation of power spectra: a method based on time averaging over short, modified periodograms. *IEEE Trans. Audio and Electroacoustics*, vol. 15, pp. 70–73.

Welsh, J. S., and G. C. Goodwin (2002). Finite sample properties of indirect nonparametric closed-loop identification. *IEEE Trans. Autom. Contr.*, vol. 47, no. 8, pp. 1277–1292.

Wernholt, E., and S. Gunnarsson (2008). Estimation of nonlinear effects in frequency domain identification of industrial robots. *IEEE Trans. Instrum. and Meas.*, vol. 57, no. 4, pp. 856–863.

Wills, A., and B. Ninness (2008). On gradient-based search for multivariable system estimates. *IEEE Trans. Autom. Contr.*, vol. 53, no. 1, pp. 298–306.

Wills, A., B. Ninness, and S. Gibson (2009). Maximum likelihood estimation of state space models from frequency domain data, *IEEE Trans. Autom. Contr.*, vol. 54, no. 1, pp. 19–33.

Wilkinson, J. H. (1988). *The Algebraic Eigenvalue Problem*. Oxford University Press, Oxford.

Young, P. C., H. Garnier, and M. Gilson (2006). An optimal instrumental variable approach for identifying hybrid continuous-time Box-Jenkins models. *14th IFAC Symposium on System Identification*, Newcastle, Australia, pp. 225–230.

Young, P. C., H. Garnier, M. Gilson (2008). Refined instrumental variable identification of continuous-time hybrid Box-Jenkins models. In H. Garnier and L. Wang, editors, *Identification of Continuous-Time Models from Sampled Data*. Springer, London, England, pp. 91–131.

Zarrop, M. B. (1979). Optimal experiment design for dynamic system identification. *Series of Lecture Notes in Control and Information Sciences*, vol. 21. Springer-Verlag, Berlin.

Zhang, E., J. Antoni, R. Pintelon, and J. Schoukens (2010). Fast detection of system nonlinearity using non-stationary signals. *Mechanical Systems and Signal Processing,* vol. 24, no. 7, pp. 2065–2075.

Zhu, Y., and P. Stec (2006). Simple control-relevant identification test methods for a class of ill-conditioned processes. *Journal of Process Control,* vol. 16, no. 10, pp. 1113–1120.

Subject Index

A

AIC criterion, 439, 479, 680, 681
   measurement example, 480–481
   simulation examples, 439–441
Akaike's information theoretic criterion
   See *AIC criterion*
alias error
   See *discrete Fourier transform*
almost surely
   See *limits—with probability 1*
aluminum plate
   See *measurement examples*
anti-alias filter
   See *band-limited assumption*
approximate maximum likelihood
   See *maximum likelihood*
ARMA
   continuous-time, 201
   definition, 200
   identification, 401–410
ARMAX
   continuous-time, 201
   definition, 200
   identification, 401–410
ARX
   continuous-time, 201
   definition, 200
   identification, 401–410
asymptotic bias
   frequency domain estimators, 297, 299, 387, 407, 472
   how to calculate?, 594
   semilinear models, 673

   simple example, 600
   SML estimator, 387, 472
asymptotic covariance
   See *covariance matrix*
asymptotic efficiency
   consistent estimators, 590
   definition, 587
   frequency domain estimators, 297, 299, 318, 408
   how to prove?, 595
   maximum likelihood, 25, 300, 408
   semilinear models, 674, 691
   simple example, 599–600
   See also *Cramér-Rao lower bound* and *Fisher information matrix*
asymptotic normality
   definition, 587
   frequency domain estimators, 295, 298, 387, 391, 393, 407, 472
   how to prove?, 595
   maximum likelihood, 25, 318, 407
   semilinear models, 673–674
   simple example, 598
   SML estimator, 387, 472
asymptotic properties
   frequency domain estimators, 298–301, 329, 339–341, 387–396, 399–401, 472–474
   general introduction, 12–16, 586–588
   how to prove?, 591–595
   maximum likelihood, 25, 317–318, 406–410
   pitfalls, 595–596
   semilinear models, 670–675

simple example, 596–601
SML estimator, 387–390, 472–474
weighted nonlinear least squares with deterministic weighting, 637–638
weighted nonlinear least squares with stochastic weighting, 659–660
asymptotically unbiased, 587–588
definition, 587
versus consistency, 595–596
autocorrelation, 568
auto-power spectrum, 568

B

band-limited assumption, 44, 499
measurement example, 504–508
band-limited white noise
definition, 197
Bayes estimator, 25–27
definition, 25
simple examples, 26–27
Berry and Esseen theorem, 585
best linear approximation, 78–92
asymptotic variance, 87
closed loop, 93–94, 133, 138–139, 242, 254
definiton, 79
direct measuring method, 83
discrete-time systems, 88–89
identification, 346, 396–398, 479–480
indirect measuring method, 93, 133–134, 254–260
measurement examples, 90–92, 126–130, 246–248, 261–263, 264–266, 410–411, 480–482
multivariable systems, 92–93, 241–243, 254–260
properties, 83–86
See also *Riemann equivalence class*
bias
definition, 568
on FRF, See *frequency response function measurement*
on FRM, See *frequency response matrix measurement*
test, 572–573
See also *asymptotic bias* and *asymptotically unbiased*
BJ
See *Box-Jenkins*
BLA
See *best linear approximation*
black box models, 17
BL-assumption
See *band-limited assumption*
bootstrapped total least squares, 320–321
definition, 320
properties, 320–321, 339–341

Borel-Cantelli lemma, 580
bounded real
See *transfer function models*
Box-Jenkins
continuous-time, 201
definition, 201
hybrid, 201
identification, 401–410
BTLS See *bootstrapped total least squares*

C

C_ARMAX
See *ARMAX, continuous-time*
C-ARMA
See *ARMA, continuous-time*
C-ARX
See *ARX, continuous-time*
Cauchy-Schwarz inequality, 614
C-BJ
See *Box-Jenkins, continuous-time*
central limit theorems, 585–586
convergence rate, 585
example, 586
Chebyshev's inequality, 570
chirp
See *periodic signals—swept sine*
circular complex
definition, 569
normally distributed, 569
closed loop identification
See *feedback*
C-OE
See *output error model, continuous-time*
coherence, 53–54
combining experiments, 541–542
compact parameter space, 293, 628
concatenated data sets
identifiability, 193
identification, 289–330, 472
models, 189–190
See also *frequency response matrix measurement*
condition number, 549
conditional maximum likelihood
See *maximum likelihood*
consistency, 13–14
definitions, 587
frequency domain estimators, 297, 299, 329, 387, 391, 393, 396, 407, 472
how to prove?, 591
maximum likelihood, 25, 318, 407
simple example, 597
SML estimator, 387, 472
versus asymptotic unbiasedness, 595–596
continuous-time systems
identification, 301–330, 386–396

Subject Index

measurement examples, 334–339, 398–399
models, 182–194
noise models, 197–199
convergence
    See *limits*
convergence in mean square
    See *limits—in mean square*
convergence rate
    central limit theorem, 585
    how to prove?, 592
    law of large numbers, 583
    simple example, 598
correlation residuals
    See *sample correlation residuals*
cost function, 19, 279, 289
    errors-in-variables, 281
    interpretation of estimators, 11–12
covariance, 567
covariance matrix
    calculation of, 571
    definition, 568
    frequency domain estimators, 298, 299, 388, 389, 391, 394, 408, 473
    FRM, See *frequency response matrix measurement*
    linear least squares, 21
    lower bound, see also *Cramér-Rao lower bound*
    maximum likelihood, 318–319, 408
    nonlinear least squares, 21
    semilinear models, 673–674
    SML estimator, 389, 475–476
    weighted nonlinear least squares, 23
Cramér-Rao lower bound, 14–16, 588–590
    biased estimators, 590
    frequency domain estimators, 318, 408, 409
    generalized bound, 589, 614
    overparameterized models, 702–703
    semilinear models, 668–670, 685
    unbiased estimators, 589
crest factor, 153
cross correlation, 568
cross-covariance matrix, 568
cross-power spectrum, 568
cumulant
    definition, 569
    Gaussian random variables, 570
    properties, 570

D

data acquisition
    See *experimental setup* and *intersample behavior*
delay
    identification, 344
    model, 184
derivatives
    fractional, 208

w.r.t. a matrix, 555
w.r.t. a vector, 554
detrending, 520
DFT
    See *discrete Fourier transform*
difference equation, 182
differential equation, 181
    fractional, 208
diffusion phenomena
    identification, 301–330, 386–396
    model, 179
discrete Fourier transform, 34–43
    alias error, 35
    basic theory, 35–40
    leakage error, 36–37, 40–42
    of burst signals, 42–43
    of noise, 601–605
    of periodic signals, 40–42
    windowing, 36–37, 40–42
discrete interval binary sequence
    See *periodic signals*
discrete-time systems
    identification, 301–330, 386–396
    models, 182–194
    noise models, 196–197
dispersion function, 169
distribution
    see *asymptotic normality*
D-optimal
    See *optimal experiments*
drift, 536

E

efficiency, 14–16
    definition, 16
    See also *asymptotic efficiency*
eigenvalues, 546
eigenvectors, 546
equation error, 289
errors-in-variables, 280–281, 527
estimator
    properties, See *asymptotic properties*
    See *frequency domain identification—estimators* and *Markov estimator*
exact maximum likelihood
    See *maximum likelihood*
excitation signals
    general purpose, 155–162
    nonparametric measurements, 152–167
    optimized, 162–167, 167–172
    parametric measurements, 167–172
    periodic versus random, 520–522
    quality indicators, 153–154
    quasi-stationary, 294
    single sine versus broadband, 154–155

See also *periodic signals, random signals* and *pulse signals*
expected value, 567
experiment design
 See *optimal experiments*
experimental setup
 band-limited, 44, 198, 501
 zero-order-hold, 196, 500
exponentials (complex, damped), 190

F

factorization in poles and zeros
 See *transfer function models*
fast nonparametric method
 comparison standard procedure – LPM, 259
 local polynomial method (LPM), 253, 258–260
 measurement examples, 142–144, 261–262
 standard procedure, 135–139
 use in parametric modeling, 468–482
feedback, 202, 240, 242, 254
 best linear approximation, 93–94, 202
 identification, 345–346, 403–410, 411–413, 474
 indirect method, 61–62, 133, 138–139
 measurement example, 141–145
Fisher information matrix
 definition, 589
 frequency domain estimators, 318, 408
 overparameterized models, 699–703
 semilinear models, 669–670
 simple example, 15–16
flight flutter analysis
 See *measurement examples*
Fréchet-Shohat lemma, 580
frequency domain experiment, 287
frequency domain identification
 assumptions, 293–298, 385–386
 asymptotic properties, 298–301, 391, 393, 396, 406–410, 472–474
 combining experiments, 243–248, 541–542
 data, 286–288
 estimators, 301–330, 386–396, 401–410, 470–474, 476
 guidelines, 531–543
 intuitive introduction, 279–284
 model selection, 437–441, 449–452, 479
 model validation, 432–449, 477–479
 multivariate systems, 348–349, 410, 470–482
 numerical algorithms, 289–291, 474–475
 overview properties, 339–341, 399–401
 plant models, 288–289
 quick analysis tools, 291–292
 versus time domain identification, 522–527
frequency response function measurement
 bias, 46, 55–56

confidence region, 48–49, 51
 guidelines, 68–69
 in the presence of nonlinear distortions, 78–92
 indirect method, 61–62
 instrumental variables, 62
 leakage, 59–60
 multivariable systems, See *frequency response matrix measurement*
 uncertainty bound, See *confidence region*
 using overlapping segments, 62–64
 using periodic excitations, 44–54
 using random excitations, 54–60
 variance, 46–47, 56–58
 windowing, 60
frequency response matrix measurement
 bias, 231, 236, 260
 concatenated data sets, 243–248
 confidence regions, 236–237
 covariance, 66, 232, 236, 241
 ill-conditioned, 67
 in the presence of nonlinear distortions, 92–93, 241–243, 254–260
 indirect method, 240–241
 local polynomial method, 228–233
 spectral analysis method, 233–236
 using periodic excitations, 64–67, 92–93, 248–263
 using random excitations, 227–248
FRF measurement
 See *frequency response function measurement*
FRM measurement
 See *frequency response matrix measurement*
Frobenius norm, 547
F-test, 51, 477, 573

G

generalized right singular vectors, 549
generalized singular value decomposition, 549
generalized singular values, 549
generalized total least squares, 313–314
 definition, 313
 properties, 313, 339–341
Gram-Schmidt orthogonalization
 definition, 558
 examples, 558–560
GSVD
 See *generalized singular value decomposition*
GTLS
 See *generalized total least squares*
guidelines
 advanced FRM measurements, 268–269
 FRF measurements, 68–69
 identification, 531–543
 model selection, 452

H

half-sine window, 63
Hammerstein system, 85–86
Hanning window, 40
Hermitian part, 547
Hessian, 295
high-order systems
    identification, 341–344
    models, 183, 186
    See also *orthogonal polynomials*
Hotelling's T^2-test, 237, 477, 572
hybrid Box-Jenkins model
    definition, 201
    identification, 410
    measurement example, 410–411

I

idempotent matrix, 551
identifiability, 191–193
    definition, 191
    global, 191
    local, 191
identification
    basic choices, 497–529
    basic steps, 17–19
    guidelines, 531–543
    imposing constraints, 528–529
    introduction, 1–28
    simple example, 2–12, 27, 576–578, 596–601
    time versus frequency domain, 522–527
ill-conditioned, 549
Illustration, 126
indecomposable partitions
    See *indecomposable sets*
indecomposable sets, 606–607
infinity-norm, 548
initial conditions
    See *equivalent initial conditions*
initial parameters
    See *starting values*
inner product
    definition, 556
    examples, 556–558
instrumental variables, 27–28, 323–324
    definition, 28, 323
    frequency response function measurement, 62
    properties, 323–324, 339–341
    simple example, 27
intersample behavior, 498–512
    band-limited assumption, 499
    impact on the identification methods, 512
    impact on the measurement setup, 511–512
    impact on the model, 500–504, 509–511
    measurement example, 504–508
    violation of the assumption, 502–504, 530
    zero-order-hold assumption, 498
IQML
    See *iterative quadratic maximum likelihood*
iterative quadratic maximum likelihood, 319–320
    definition, 319
    properties, 320, 339–341
IV
    See *instrumental variables*
IWLS
    See *weighted linear least squares*

J

Jacobian
    (frequency) scaling, 290–291, 475
    definition, 290
    maximum likelihood estimator, 365
    multivariate systems, 474
    pseudo-, 474
    SML estimator, 474
joint input-output method, 410

K

Kronecker product, 552

L

law of the iterated logarithm, 583
laws of large numbers, 583–585
    convergence rate, 583
    example, 584
    law, 583
    strong law, 583
    weak law, 583
leakage error
    See *discrete Fourier transform* and *frequency response function measurement*
left singular vectors, 546, 548
Levenberg-Marquardt, 291, 366
likelihood function, 14
limits, 578–583
    definitions, 578
    examples, 580–581
    in distribution, 578
    in law, 578
    in mean square, 578
    in probability (in prob.), 578
    interrelations, 579–581
    preliminary example, 576–578
    properties, 582–583
    with probability 1 (w.p. 1), 578
linear least squares, 21–22, 301–305
    definition, 21, 301
    properties, 301–303, 339–341
    simple example, 22
linear time invariant second order equivalent, 89
local polynomial method

bias-variance trade-off, 239–240
comparison with spectral analysis method, 237–239
concatenated data sets, 243–248
errors-in-variables, 240–241
fast method, 253, 258–260
for periodic excitations, 248–263
for random excitations, 228–233
robust method, 252–253, 254–258
LOG
See *logarithmic least squares*
logarithmic least squares, 308–309
definition, 309
properties, 309, 339–341
LS
See *linear least squares*
LTI-SOE
See *linear time invariant second order equivalent*

M

Markov estimator
frequency domain, 317, 350
model selection, 680–682
properties, 670–675
semilinear models, 666–668
Markov's inequality, 571
matrix norms, 547
maximum length binary sequence
See *periodic signals*
maximum likelihood, 23–25, 314–318, 406–410
approximate, 409
conditional, 409
covariance matrix, 318–319
definition, 23, 315
exact, 409
mean and variance cost function, 443
numerical implementation, 365–366
properties, 25, 316–318, 339–341, 406–410
MDL criterion, 439, 479, 680, 681
measurement example, 480–481
simulation examples, 439–441
mean, 567
See also *sample variables*
mean square error
definition, 568
minimum MSE estimators, 588
measurement examples
3.4 MW synchronous motor, 334–339
aluminum plate, 264–266, 480–482
flight flutter analysis, 334–339, 398–399
nonlinear electrical circuit, 90–92, 410–411
octave bandpass filter, 171–172
operational amplifier, 141–145
parallel Wiener system, 126–130
plexiglass beam, 261–263

RC-circuit, 504–508
robot arm, 448–449
steel beam, 246–248, 515
measurement setup
See *experimental setup*
MIMO
See *multivariable systems*
minimum phase region, 181
missing data
identifiability, 192
identification, 346–348
models, 188–189
mixing random variables
(strong) law of large numbers, 584, 611
central limit theorem, 586, 613
definition, 573–574
properties, 574–576
model
See *models*
model errors
classification, 444–448
detection, 442–444
model selection
frequency domain, 437–441, 449–452, 479
guidelines, 452
measurement example, 448–449, 480–482
semilinear models, 680–682
model validation
frequency domain, 432–436, 477–479
measurement example, 448–449, 480–482
semilinear models, 681
models
damped (complex) exponentials, 190–191
equivalent initial conditions, 185
identifiability, 191–193
linear-in-the-parameters, 18
nonparametric, 17, 33, 195
over-parameterized, 699–708
parametric, 17, 182–191, 193–194, 195–201
semilinear, 665–666
transfer function, 177–202
white box versus black box, 17
Moore-Penrose pseudoinverse, 550–551
MSE
See *mean square error*
multiple-input, multiple-output
See *multivariable systems*
multisines
crest factor optimized, 162–164
Fisher optimized, 168–170
full random orthogonal, 93
full random phase, 121
Hadamard, 66
input/output optimized, 165–166

Subject Index

log-tones, 141, 164
nonlinearity detection, 130–140
odd random phase, 121
orthogonal, 67
random orthogonal, 92
random phase, 76
random phase with random harmonic grid, 121
Riemann equivalent, 123
Schroeder, 156–157, 504
snow, 163
zippered, 64
multivariable systems
FRF measurement, 64–67
identification, 348–349, 410, 470–479
measurement example, 480–482
models, 193–194
rank residue matrix, 194

N

Newton-Gauss algorithm, 290–291
multivariate systems, 474–475
NLS, NLS-FRF, NLS-IO
See *nonlinear least squares*
noise after a DFT, 601–605
asymptotic normality, 601
central limit theorem, 602, 605
mixing, 602, 604
strong law of large numbers, 602, 604
noise model, 385
continuous-time, 197–199
discrete-time, 196–197
model structures, 200–201
parametric versus nonparametric, 526
See also *nonparametric noise model*
See also *parametric noise model*
nonlinear distortions
See *nonlinear systems*
nonlinear least squares, 20–21, 305–308
definition, 20, 305, 306
properties, 306, 307–308, 339–341
simple example, 310
See also *weighted nonlinear least squares*
nonlinear systems
continuous-time, 74–78
discrete-time, 180–181
intuitive introduction, 74–75
model, 202
parallel Wiener system, 126
Volterra, 75, 180–181
Wiener-Hammerstein, 85–86
See also *best linear approximation*
nonlinearity detection, 130–140
fast method, 135–139, 258–260
literature overview, 119–120
measurement example, 141–145

robust method, 130–134, 254–258
nonparametric noise model
estimation, 130–138, 227–231, 234, 252–260
use in parametric modeling, 383–401, 463–484
See also *measurement examples*
normal equations
definition, 290
sensitivity of the solution, 562
normalized model parameter
See *scaling*
null space, 545, 549

O

OE
See *output error model*
offset, 536
1-norm, 547
operational amplifier
See *measurement examples*
optimal experiments
D-optimal, 66, 170
for controller design, 173
measurement example, 171–172
nonparametric measurements, 162–167
parametric measurements, 167–172
orthogonal matrix, 546
orthogonal polynomials
calculation of the roots, 560–561
frequency domain estimators, 341–344
scalar, 558, 559
vector, 559
outliers, 537
output error, 288
output error method
identification, 401–410
output error model
continuous-time, 201
definition, 201
overmodeling
frequency domain, 437–441
semilinear models, 680
simulation examples, 439–441
See also *model selection*

P

parallel Wiener system
See *measurement examples*
parametric noise model
identification, 401–410
maximum likelihood solution, 403–410, 524
measurement example, 410–411
model, 195–201
partial fraction expansion
See *transfer function models*
passivity

See *transfer function models*
periodic noise
   See *periodic signals*
periodic signals
   comparison, 161–162
   DFT properties, 40–42
   discrete interval binary sequence, 164–165
   maximum length binary sequence, 158
   optimized multisines, 162–164, 165–166, 167–172
   periodic noise, 77, 122
   pseudorandom binary sequence, 157–158
   random phase multisine, 77, 121–122
   reducing FRF measurement errors, 49–53
   Schroeder multisine, 156–157
   single sine, 156
   spectral representation, 43–44
   swept sine, 156
   ternary sequence, 65, 166
   versus random, 520–522
   why should they be used?, 516–520
   See also *multisines*
persistence of excitation
   frequency domain, 295
   weighted nonlinear least squares, 630, 633
perturbation signals
   See *excitation signals*
plexiglass beam
   See *measurement examples*
positive (semi-)definite, 546
positive real
   See *transfer function models*
prediction error
   definition, 402
prediction error method
   cost function, 402
   frequency domain interpretation, 402
   maximum likelihood solution, 524
   relation with frequency domain ML solution, 405–406
prefiltering, 517–518
preprocessing, 536–539
pseudorandom binary sequence
   See *periodic signals*
pulse signals, 160–161

Q
QR Factorization, 550
quasi-stationary signals, 294

R
random phase multisine
   See *periodic signals*
random signals
   random burst, 160
   random noise, 159
   versus periodic, 520–522
range space, 546, 549
rank, 545, 549
rational form
   See *transfer function models*
RC-circuit
   See *measurement examples*
reciprocity
   See *transfer function models*
rectangular window, 36
residuals
   uncertainty bounds, 433–434
   white-colored, 452
resistance measurement problem
   See *simulation examples*
Riemann equivalence class, 120–130
   definition, 122–124
   invariance BLA, 124–126
   measurement example, 126–130
right singular vectors, 546, 548
robot arm
   See *measurement examples*
robust nonparametric method
   comparison standard procedure – LPM, 256, 263
   local polynomial method (LPM), 252–253, 254–258
   measurement examples, 144–145, 261–262
   standard procedure, 130–134
   use in parametric modeling, 396–398, 468–480
robustness
   definition, 587
   maximum likelihood estimator, 318
   minimum mean square estimators, 588
   semilinear models, 674
   simple example, 600

S
sample bootstrapped total least squares, 392–395
   definition, 393
   multivariate, 476
   properties, 393–395, 399–401
sample correlation residuals
   frequency domain, 445–448, 478–479
   semilinear models, 676–678
   standard deviation (no modeling errors), 448, 479, 677
   variance (modeling errors), 459
sample generalized total least squares, 390–392
   definition, 390
   multivariate, 476
   properties, 391–392, 399–401
sample maximum likelihood, 386–390, 470–480
   covariance matrix, 389, 475–476
   definition, 386, 470–472

Subject Index

mean and variance cost function, 389, 390, 478
multivariate, 470–479
properties, 387–390, 399–401, 472–474
sample subspace algorithms, 395–396
definition, 395
properties, 396, 399–401
sample variables
frequency domain estimators, 50, 384, 397, 464–470
generalized sample covariance, 464–470
generalized sample mean, 464–470
sample covariance matrix, 572
sample mean, 572
sample variance, 572
stochastic properties, 572–573
See also *noise model*
sampling, 34–35
SBTLS
See *sample bootstrapped total least squares*
scaling
frequency, 290
Jacobian matrix, 290, 475
model parameter, 290
SGTLS
See *sample generalized total least squares*
similarity transformation, 546
simulation examples
comparison frequency domain estimators, 330–332
comparison starting value algorithms, 332–333
detection of overmodeling, 439–441
influence parameter constraint on transfer function estimates, 706–708
leakage errors on FRF measurement, 59
noise removal in periodic signals, 518
on-line simulation example for frequency domain estimators, 286, 302, 303, 308, 351
resistance measurement problem, 2–12, 27, 576–578, 596–601
uncertainty poles and zeros, 435–436
uncertainty transfer function residuals, 434
undermodeling, 442
singular value decomposition, 548–549
singular values, 546, 548
SML
See *sample maximum likelihood*
spectral analysis method
See *frequency response matrix measurement*
square root of a matrix, 550
SSUB
See *sample subspace algorithms*
stability
See *transfer function models*
stability region, 181

starting values, 282–283, 332–333
state space equations, 182
state space representation
See *transfer function models*
stationarity, 568
steel beam
See *measurement examples*
step-invariant transformation, 180, 500
stochastic limit
See *limits*
strong consistency
See *consistency*
strong convergence
See *limit with probability 1*
SUB
See *subspace algorithms*
subspace algorithms, 324–330
definition, 325–329
properties, 329–330, 339–341
SVD
See *singular value decomposition*
systems
continuous-time, 74–78, 177–179
discrete-time, 179–181
distributed continuous-time, 178
lumped continuous-time, 177
nonlinear, 74–78, 180–181

T

ternary sequence
See *periodic signals*
time delay
See *delay*
time domain experiment, 286
time domain identification
versus frequency domain identification, 522–527
time factor, 154
total least squares, 310–312
definition, 312
introduction, 310–312
properties, 312, 339–341
trace, 547
transfer function models, 182–190, 195–201
bounded real, 528
factorization in poles and zeros, 183
identifiability, 191–193
imposing constraints, 528–529
minimum phase region, 181
multivariable systems, 193–194
noise models, 195–201
orthogonal polynomials, 183, 186
partial fraction expansion, 182, 186
passivity (positive real), 528
plant models, 182–184

rational form, 182, 186
reciprocity, 528
relation with DFT spectra, 184–190
stability, 528
stability region, 181
state space representation, 184, 186
time delay, 184
trends, 520, 536
truncated estimator, 297, 635
2-norm, 548

U

unbiased, 13–14
definition, 587
See also *asymptotically unbiased*
undermodeling
frequency domain, 441–449
measurement example, 448–449
semilinear models, 681
simulation example, 442
uniform convergence, 628
unitary matrix, 546

V

validation, 542–543
See also *model validation*
variance, 567
on FRF measurements, 46–47, 56–58
See also *sample variables—sample variance*
Volterra systems
continuous-time, 75
discrete-time, 180–181

W

weak consistency
See *consistency*
weak convergence
See *limit in probability*
weighted generalized total least squares, 313–314, 319–323
definition, 314, 321–322
weighted linear least squares, 303–304, 319–323
definition, 303, 304, 321–322
properties, 303–304, 323, 339–341
simple example, 304–305
weighted nonlinear least squares
deterministic weighting, 627–638
intuitive introduction, 22–23
stochastic weighting, 651–660
well-conditioned, 549
WGTLS
See *weighted generalized total least squares*
white box models, 17
whiteness test residuals, 445–448, 478–479, 523, 677, 681
Wiener-Hammerstein systems, 85–86
Wiener-Hopf equation, 55, 523
window
half-sine, 63
Hanning, 40
rectangular, 36
windowing
See *discrete Fourier transform* and *frequency response function measurement*
Wishart distribution
complex, 569
WLS
See *weighted linear least squares*

Z

zero-order-hold assumption, 498
measurement example, 504–508
ZOH-assumption
See *zero-order-hold assumption*

Author Index

A
Abel (1993) 16, 589, 590
Abramowitz and Stegun (1970) 206
Aguero and Goodwin (2008) 515, 527
Agüero *et al.* (2010) 409, 524
Akaike (1974) 680
Albertos *et al.* (1999) 188, 348
Anderson (1958) 567, 569, 572, 573
Anderson and Deistler (1984) 527
Anderson *et al.* (1992) 550
Antoni and Schoukens (2007) 235, 238, 239
Antoni and Schoukens (2009) 63
Antoni *et al.* (2007) 63
Antoulas and Anderson (1999) 173
Åström (1970) 195, 197
Åström *et al.* (1984) 500

B
Bai and Demmel (1993) 451, 550
Baker *et al.* (1996) 179
Balabanian and Bickart (1969) 193
Baratchard *et al.* (1997) 528
Barenthin and Hjalmarsson (2008) 173
Barenthin *et al.* (2008) 173
Barker and Zhuang (1997) 167
Battaglia *et al.* (2001) 179
Bayard (1994a) 348, 476
Bayard (1994b) 348
Beck and Arnold (1977) 1
Bendat (1998) 74, 82, 83
Bendat and Piersol (1980) 53, 54, 55, 60, 83, 234, 523
Ben-Israel and Greville (1974) 550, 551

Bergström (1990) 196
Billings (1980) 83
Billings and Fakhour (1982) 86
Billingsley (1995) 567, 570, 582, 585
Bohlin (1971) 527
Bombois *et al.* (2006) 173
Box and Jenkins (1976) 704
Boyd *et al.* (1983) 167
Brewer (1978) 552
Brigham (1974) 34, 39, 185, 208
Brillinger (1981) 83, 122, 127, 181, 234, 488, 567, 569, 570, 573, 575, 586, 601, 602, 608, 614, 617, 623
Broersen (1995) 47, 55, 280
Brown (1977) 161
Brown *et al.* (1977) 156, 159
Bultheel and Van Barel (1995) 343
Bultheel *et al.* (2005) 343

C
Cadzow (1990) 666, 683
Cadzow and Solomon (1987) 53
Caines (1988) 24, 407, 409, 427, 615
Carter and Nuttall (1980) 63
Cauberghe *et al.* (2003) 196
Chow and Teicher (1988) 567, 580, 584, 585
Chua and Ng (1979) 75
Crochiere and Rabiner (1983) 527

D
D'haene and Pintelon (2008) 528, 529
D'haene *et al.* (2006) 528, 529
de Callafon *et al.* (1996) 348, 476

De Moor *et al.* (1994) 303, 703, 704, 708
Delbaen (1990) 170
Delchamps (1985) 702
Delchamps and Byrnes (1982) 702
Dobrowiecki and Schoukens (2007a) 92
Dobrowiecki and Schoukens (2007b) 92, 118
Dobrowiecki *et al.* (2006) 66, 67
Durbha *et al.* (1999) 179

E
Enqvist (2005) 83, 89
Enqvist and Ljung (2005) 89
Evans (1998) 167
Evans and Rees (2000) 81
Evans *et al.* (1994) 120
Ewins (1991) 182, 190, 193
Eykhoff (1974) 25, 31, 55, 83, 158, 497

F
Fan and Gijbels (1995) 239
Fan *et al.* (1999) 196
Federov (1972) 168, 169
Feller (1968) 585
Figwer and Niederlinski (1995) 64
Fletcher (1991) 20, 291, 366
Forssell and Ljung (2000a) 526
Forssell and Ljung (2000b) 173
Forssell and Ljung (2000c) 82, 527
Forsythe (1957) 558
Forsythe and Strauss (1955) 342

G
Gaikwad and Rivera (1997) 348, 476
Gantmacher (1990) 221, 545
Gevers (2005) 173
Gevers and Ljung (1986) 173
Gevers and Wertz (1984) 475
Gevers *et al.* (2011) 240
Giri (1965) 237, 572, 573
Godfrey (1969) 158
Godfrey (1980) 158
Godfrey (1993a) 157
Godfrey (1993b) 157, 158
Goethals *et al.* (2003) 528
Golub and Van Loan (1996) 219, 233, 545, 546, 548, 549, 550, 551, 558, 562
Goodman (1963) 488, 569
Goodwin and Adams (1994) 188, 348
Goodwin and Payne (1977) 1, 24, 168, 169, 173
Gorman and Hero (1990) 702
Gradshteyn and Ryzhik (1980) 52, 212, 309, 605
Grivet-Talocia and Bandinu (2006) 528
Grivet-Talocia and Ubolli (2007) 528
Guidorzi (1975) 699
Guillaume (1998) 66

Guillaume and Pintelon (1996) 348, 475
Guillaume *et al.* (1989) 436
Guillaume *et al.* (1991) 163, 166, 174
Guillaume *et al.* (1992a) 348
Guillaume *et al.* (1992b) 46, 47, 52, 359
Guillaume *et al.* (1995) 309
Guillaume *et al.* (1996a) 280, 307, 308, 309
Guillaume *et al.* (1996b) 66
Gustafsson and Schoukens (1998) 518, 522
Gustavsen (2008) 528

H
Haber (1985) 119
Halvorsen and Brown (1977) 160
Hannan (1980) 680
Hannan and Deistler (1988) 195, 293, 628
Harris (1978) 39, 40, 41, 235
Hazewinkel (1977) 702
Heath (2001) 62
Henrici (1974) 178, 182, 365, 381, 630
Herlufsen (1984) 160
Heylen *et al.* (1997) 194
Hotelling (1933) 573
Huber (1981) 12

I
Isaksson (1993) 348

J
Jazwinski (1970) 197, 567, 582

K
Kabaila (1983) 295, 630
Kahane (1980) 603
Kahane (1985) 604
Kailath (1980) 177, 182, 194, 210, 367, 368, 547, 621
Kalman (1958) 704
Kaplan (1993) 632, 642, 643, 649, 667, 710
Kashyap (1980) 680
Kay (1988) 596
Kendall and Stuart (1979) 1, 415, 491, 588
Kollár *et al.* (1991) 512
Kumaresan *et al.* (1990) 190, 349, 670
Kwakernaak and Sivan (1991) 177, 222

L
Lancaster and Tismenetsky (1985) 545, 551, 552, 556
Lataire 126
Lataire and Pintelon (2009) 538
Lee (1993) 349
Lee and Joh (1994) 670
Leonov and Shiryaev (1959) 607
Levy (1959) 301
Liang *et al.* (1993) 680

Little and Rubin (1987) 348
Ljung (1985) 438
Ljung (1993) 522, 524
Ljung (1995) 12
Ljung (1999) 59, 195, 201, 283, 294, 348, 384, 401, 402, 409, 410, 411, 413, 442, 497, 500, 509, 512, 514, 516, 527, 532, 543, 635, 636, 680
Ljung and Söderström (1983) 529
Lowen and Teich (1990) 195
Lukacs (1975) 567, 578, 579, 581, 582, 584

M
Mahata *et al.* (2006) 231, 489, 490
Maiwald and Kraus (2000) 488, 489
Mathai and Provost (1992) 567, 569, 570
McCormack *et al.* (1994a) 69, 537
McCormack *et al.* (1994b) 54, 120
McCormack *et al.* (1995) 167
McKelvey (1996) 497
McKelvey (2000) 409
McKelvey and Helmersson (1997) 708
McKelvey *et al.* (1996) 326, 349, 372, 476
McKelvey *et al.* (2004) 475
Mehra (1974) 173
Mendel (1991) 83
Middleton and Goodwin (1990) 195, 500
Moreau *et al.* (2002) 179

N
Natke *et al.* (1988) 119
Nikias and Mendel (1993) 83
Nikias and Petropulu (1993) 83, 86
Ninness and Goodwin (1991) 510
Norton (1986) 1, 26, 28, 62, 158, 497, 588
Nutall and Carter (1982) 63

O
Oldham and Spanier (1974) 208
Oppenheim and Schafer (1975) 158
Oppenheim *et al.* (1997) 88, 177, 185
Orey (1958) 586

P
Paehlike and Rake (1979) 164
Paige (1986) 451, 549
Papoulis (1981) 54, 223, 710
Peeters *et al.* (2000) 348, 349, 670
Peirlinckx *et al.* (1996) 69, 520, 537
Phadke and Wu (1974) 196
Picinbono (1993) 108, 271, 569
Pintelon and Hong (2007) 318

Pintelon and Kollár (2005) 290
Pintelon and Schoukens (1996) 666
Pintelon and Schoukens (1997b) 188
Pintelon and Schoukens (1999) 193
Pintelon and Schoukens (2000) 537
Pintelon and Schoukens (2001) 46
Pintelon and Schoukens (2006) 410
Pintelon and Schoukens (2007) 401, 515, 527
Pintelon and Van Biesen (1990) 344
Pintelon *et al.* (1990) 512
Pintelon *et al.* (1994) 304
Pintelon *et al.* (1997a) 679
Pintelon *et al.* (1998) 348, 476
Pintelon *et al.* (1999) 475
Pintelon *et al.* (2000) 201
Pintelon *et al.* (2003) 48
Pintelon *et al.* (2004) 261
Pintelon *et al.* (2004b) 141
Pintelon *et al.* (2005) 179
Pintelon *et al.* (2006a) 405, 406, 410
Pintelon *et al.* (2006b) 196
Pintelon *et al.* (2007a) 410
Pintelon *et al.* (2007b) 482
Pintelon *et al.* (2010a) 231, 235, 236, 238, 513
Pintelon *et al.* (2010b) 513, 527
Pintelon *et al.* (2011a) 513
Pintelon *et al.* (2011b) 264, 513
Pintelon, Schoukens, and Guillaume (2007) 410
Pyati (1992) 195

R
Rabiner and Gold (1975) 40, 158
Ralston and Rabinowitz (1984) 30, 31, 113, 356
Richardson and Formenti (1982) 342
Rissanen (1978) 680
Rivera *et al.* (2007) 65
Rivera *et al.* (2009) 65, 67, 268
Rizzi (1988) 179
Rolain *et al.* (1998) 537
Rolain and Pintelon (1999) 322
Rolain *et al.* (1995) 327, 342, 373
Rolain *et al.* (1997) 449, 451
Rosén (1967) 586

S
Sakakibara (1997) 179
Sanathanan and Koerner (1963) 303, 304
Schetzen (1980) 75, 77, 78, 85, 110, 180
Schoukens and Pintelon (1990) 569
Schoukens and Pintelon (1991) 169, 446
Schoukens and Pintelon (2010a) 239
Schoukens and Pintelon (2010b) 398, 480, 495

Schoukens *et al.* (1994) 517
Schoukens *et al.* (1995) 159
Schoukens *et al.* (1996a) 534
Schoukens *et al.* (1996b) 603
Schoukens *et al.* (1997b) 513, 527
Schoukens *et al.* (1998b) 522
Schoukens *et al.* (1999a) 386
Schoukens *et al.* (2000) 155
Schoukens *et al.* (2002) 439, 697
Schoukens *et al.* (2005) 407
Schoukens *et al.* (2006) 235
Schoukens *et al.* (2006a) 238
Schoukens *et al.* (2006b) 238
Schoukens *et al.* (2009) 126, 264
Schoukens *et al. (2011)* 514
Schroeder (1970) 157
Selby (1973) 179, 205
Shapiro (1986) 699
Shibata (1980) 439
Sidman *et al.* (1991) 309
Sinha and Rao (1991) 510, 525
Söderström (1974) 640
Söderström (2007) 401, 515, 527
Söderström and Mossberg (2000) 525
Söderström and Stoica (1981) 401, 402
Söderström and Stoica (1989) 195, 409, 410, 411, 497, 523
Söderström *et al.* (1997a) 525
Söderström et al. (1997b) 525
Sörenson (1980) 1, 27
Souders *et al.* (1990) 52
Spiegel (1965) 205
Steigliz and McBride (1965) 704
Stoica and Moses (1990) 588, 590
Stoica *et al.* (1986) 449
Stout (1974) 567, 580, 584
Stuart and Ord (1987) 13, 49, 51, 109, 136, 236, 271, 388, 417, 485, 567, 570, 571, 575, 605, 614, 697
Swevers 448
Swevers *et al.* (1992) 313

T

Tan *et al.* (2005) 167
Tan *et al.* (2009) 65, 167
't Mannetje (1973) 304
Tomlinson (1987) 120
Torfs 448
Torfs *et al.* (1998) 448

V

Van Barel and Bultheel (1992) 559
Van Barel and Bultheel (1994) 343, 559, 560
Van Brussel (1975) 159
Van den Bos (1974) 164
Van den Bos (1985) 12
Van den Bos (1987) 174
Van den Bos (1991) 323
Van den Bos and Krol (1979) 164
Van den Bos and Swarte (1993) 666, 683
Van den Eijnde and Schoukens (1991) 169, 170
Van den Enden and Leenknegt (1986) 306
Van den Enden *et al.* (1977) 306
Van den Hof and Schrama (1995) 413
Van der Ouderaa and Renneboog (1988) 164
Van der Ouderaa *et al.* (1988a) 174
Van der Ouderaa *et al.* (1988b) 174
Van Gestel *et al.* (2001) 528
Van hamme *et al.* (1991) 510
Van Huffel and Vandewalle (1991) 290, 311, 704
Van Overbeek and Ljung (1982) 191, 699, 700, 702, 706
Van Overschee and De Moor (1994) 497
Van Overschee and De Moor (1996a) 326, 329, 349
Van Overschee and De Moor (1996b) 476, 497
Vandersteen *et al.* (1996a) 704, 706
Vanhounacker *et al.* (2002) 119
Verbeeck *et al.* (1999b) 64, 65, 66
Verboven *et al.* (2004) 348, 476
Verboven *et al.* (2005) 348, 476
Verhaegen (1994) 326, 497
Viberg *et al.* (1997) 497
Vuerinckx *et al.* (1998) 436

W

Walter and Pronzato (1997) 168, 170, 436
Wang (1987) 179
Wellstead (1977) 61
Wellstead (1981) 61
Welsh (1967) 62
Welsh and Goodwin (2002) 62
Wernholt and Gunnarsson (2008) 92
Wilkinson (1988) 545, 546
Wills and Ninness (2008) 475
Wills *et al.* (2009) 348

Y

Young *et al.* (2006) 201
Young *et al.* (2008) 201

Z

Zarrop (1979) 168, 169
Zhang *et al.* (2010) 243, 514, 519
Zhu and Stec (2006) 67, 268

About the Authors

Rik Pintelon received a master's degree in electrical engineering in 1982 and a doctorate (PhD) in engineering in 1988 from the Vrije Universiteit Brussel (VUB), Brussels, Belgium.

From 1982 to 2000, Dr. Pintelon was a researcher with the Belgian National Fund for Scientific Research (FWO-Vlaanderen) at the Electrical Engineering (ELEC) Department of the VUB. From 1991 to 2000 he was a part-time lecturer at the same department, where he is currently a full-time professor in electrical engineering. His main research interests include system identification, signal processing, and measurement techniques. Dr. Pintelon is the coauthor of three books and the coauthor of about 200 articles in refereed international journals. He has been a Fellow of IEEE since 1998. Dr. Pintelon is the recipient of the 2012 IEEE Joseph F. Keithley Award in Instrumentation and Measurement (IEEE Technical Field Award).

Johan Schoukens received a master's degree in electrical engineering in 1980 and a doctorate (PhD) in engineering in 1985 from the Vrije Universiteit Brussel (VUB), Brussels, Belgium.

From 1981 to 2000, Dr. Schoukens was a researcher with the Belgian National Fund for Scientific Research (FWO-Vlaanderen) at the Electrical Engineering (ELEC) Department of the VUB. From 1986 to 2000 he was a part-time lecturer in the same department, where he is currently a full-time professor in electrical engineering. His main research interests include system identification, signal processing, and measurement techniques. Dr. Schoukens is the coauthor of three books and the coauthor of about 200 articles in refereed international journals. He has been a Fellow of IEEE since 1997. Dr. Schoukens was the recipient of the 2002 Andrew R. Chi Best Paper Award of the IEEE Transactions on Instrumentation and Measurement, the 2002 Society Distinguished Service Award from the IEEE Instrumentation and Measurement Society, and the 2007 Belgian Francqui Chair at the Université Libre de Bruxelles (Belgium). Since 2010, he is a member of Royal Flemish Academy of Belgium for Sciences and Arts. In 2011 he received a doctorate degree honoris causa from the Budapest University of Technology and Economics (Hungary).